Thermophysical properties of fluids. I. Argon, ethylene, parahydrogen, nitrogen, nitrogen trifluoride, and oxygen

Journal of

Physical and Chemical Reference Data

David R. Lide, Jr., Editor

The Journal of Physical and Chemical Reference Data (ISSN 0047-2689) is published quarterly by the American Chemical Society (1155 16th St., N. W., Washington, DC 20036) and the American Institute of Physics (335 E. 45th St., New York, NY 10017) for the National Bureau of Standards. Second-class postage paid at Washington, DC and additional mailing offices. POSTMASTER: Send address changes to Membership and Subscription Services, P. O. Box 3337, Columbus, Ohio 43210.

The objective of the Journal is to provide critically evaluated physical and chemical property data, fully documented as to the original sources and the criteria used for evaluation. Critical reviews of measurement techniques, whose aim is to assess the accuracy of available data in a given technical area, are also included. The Journal is not intended as a publication outlet for original experimental measurements such as are normally reported in the primary research literature, nor for review articles of a descriptive or primarily theoretical nature.

Supplements to the Journal are published at irregular intervals and are not included in subscriptions to the Journal. They contain compilations which are are too lengthly for a journal format.

The Editor welcomes appropriate manuscripts for consideration by the Editorial Board. Potential contributors who are interested in preparing a compilation are invited to submit an outline of the nature and scope of the proposed compilation, with criteria for evaluation of the data and other pertinent factors, to:

David R. Lide, Jr., Editor
J. Phys. Chem. Ref. Data
National Bureau of Standards
Washington, D.C. 20234

One source of contributions to the Journal is The National Standard Reference Data System (NSRDS), which was established in 1963 as a means of coordinating on a national scale the production and dissemination of critically evaluated reference data in the physical sciences. Under the Standard Reference Data Act (Public Law 90-396) the National Bureau of Standards of the U.S. Department of Commerce has the primary responsibility in the Federal Government for providing reliable scientific and technical reference data. The Office of Standard Reference Data of NBS coordinates a complex of data evaluation centers, located in university, industrial, and other Government laboratories as well as within the National Bureau of Standards, which are engaged in the compilation and critical properties retrieved from the world scientific literature. The participants in this NBS-sponsored program, together with similar groups under private or other Government support which are pursuing the same ends, comprise the National Standard Reference Data System.

The primary focus of the NSRDS is on well-defined physical and chemical properties of well-characterized materials or systems. An effort is made to assess the accuracy of data reported in the primary research literature and to prepare compilations of critically evaluated data which will serve as reliable and convenient reference sources for the scientific and technical community.

Information for Contributors

Manuscripts submitted for publication must be prepared in accordance with *Instructions for Preparation of Manuscripts for the Journal of Physical and Chemical Reference Data*, available on request from the Editor.

New and renewal subscriptions should be sent with payment to the Office of the Controller at the American Chemical Society, 1155 Sixteenth Street, N.W., Washington, DC 20036. **Address changes**, with at least six weeks advance notice, should sent to Manager, Membership and Subscription Services, American Chemical Society, P.O. Box 3337, Columbus, OH 43210. Changes of address must include both old and new addresses and ZIP codes and, if possible, the address label from the mailing wrapper of a recent issue. Claims for missing numbers will not be allowed: if loss was due to failure of the change-of-address notice to be received in the time specified; if claim is dated (a) North America: more than 90 days beyond issue date, (b) all other foreign: more than one year beyond issue date.

Members of AIP member and affiliate societies requesting member subscription rates should direct subscriptions, renewals, and address changes to American Institute of Physics, Dept. S/F, 335 E. 45th St., NY 10017.

Subscription Prices (1982) *(not including supplements)*	U.S.A. Canada Mexico	Foreign (surface mail)	Optional air freight Europe Mideast N. Africa	Asia and Oceania
Members (of ACS or of AIP member or affiliated society)	$38.00	$44.00	$52.00	$61.00
Regular rate	$150.00	$156.00	$164.00	$173.00

Rates above do not apply to nonmember subscribers in Japan, who must enter subscription orders with Maruzen Company Ltd., 3-10 Nihonbashi 2-chome, Chuo-ku, Tokyo 103, Japan. Tel: (03) 272-7211.

Back numbers are available at a cost of $44 per single copy.

Orders for reprints, supplements, and back numbers should be addressed to the American Chemical Society, 1155 Sixteenth Street, N. W., Washington, DC 20036. Prices for reprints and supplements are listed at the end of this issue.

Copying Fees: The code that appears on the first page of articles in this journal gives the fee for each copy of the article made beyond the free copying permitted by AIP. (See statement under "Copyright" elsewhere in this journal.) If no code appears, no fee applies. The fee for pre-1978 articles is $0.25 per copy. With the exception of copying for advertising and promotional purposes, the express permission of AIP is not required provided the fee is paid through the *Copyright Clearance Center, Inc. (CCC), 21 Congress Street, Salem, MA 01970.* Contact the CCC for information on how to report copying and remit payment.

Microfilm subscriptions of the *Journal of Physical and Chemical Reference Data* are available on 16 mm and 35 mm. This journal also appears in Sec. I of *Current Physics Microform* (CPM) along with 26 other journals published by the American Institute of Physics and its member societies. *A Microfilm Catalog* is available on request.

Journal of
**Physical and
Chemical
Reference Data**

Volume 11, 1982
Supplement No. 1

Thermophysical properties of fluids. I. Argon, ethylene, parahydrogen, nitrogen, nitrogen trifluoride, and oxygen

B. A. Younglove

*Thermophysical Properties Division,
National Engineering Laboratory
National Bureau of Standards, Boulder, CO 80303*

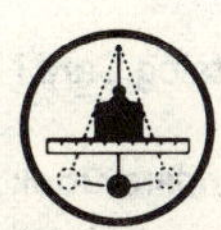

Published by the **American Chemical Society**
and the **American Institute of Physics** for
the **National Bureau of Standards**

Library of Congress Catalog Card Number 82-074456

International Standard Book Numbers
0-88318-415-X (Hard cover)

American Institute of Physics, Inc.
335 East 45th Street
New York, New York 10017

Printed in the United States of America

Foreword

The *Journal of Physical and Chemical Reference Data* is published jointly by the American Institute of Physics and the American Chemical Society for the National Bureau of Standards. Its objective is to provide critically evaluated physical and chemical property data, fully documented as to the original sources and the criteria used for evaluation. One of the principal sources of material for the journal is the National Standard Reference Data System (NSRDS), a program coordinated by NBS for the purpose of promoting the compilation and critical evaluation of property data.

The regular issues of the *Journal of Physical and Chemical Reference Data* are published quarterly and contain compilations and critical data reviews of moderate length. Longer monographs, volumes of collected tables, and other material unsuited to a periodical format are published separately as *Supplements to the Journal*. This monograph, "Thermophysical Properties of Fluids. I. Argon, Ethylene, Parahydrogen, Nitrogen, Nitrogen Trifluoride, and Oxygen", by B. A. Younglove, is presented as Supplement No. 1 to Volume 11 of the *Journal of Physical and Chemical Reference Data*. Its objective is to provide tables of evaluated data on properties of important fluids in a uniform format, designed for convenient application to any problem requiring thermodynamic or transport properties. Tables on additional fluids of industrial importance, presented in the same format, will be published in subsequent volumes. It is anticipated that this series will become a standard reference source of fluid property data.

David R. Lide, Jr., Editor
Journal of Physical and Chemical Reference Data

Thermophysical properties of fluids. I. Argon, ethylene, parahydrogen, nitrogen, nitrogen trifluoride, and oxygen

B. A. Younglove

Thermophysical Properties Division, National Engineering Laboratory
National Bureau of Standards, Boulder, CO 80303

The thermophysical properties of argon, ethylene, parahydrogen, nitrogen, nitrogen trifluoride and oxygen are presented. Properties are given in tables and a standard set of equations is described. The tables list pressure, density, temperature, internal energy, enthalpy, entropy, heat capacity at constant volume, heat capacity at constant pressure, and sound velocity. Also included are viscosity, thermal conductivity, and dielectric constant, for some of the fluids. The equation and related properties of this report represent a compilation from the cooperative efforts of two research groups: the NBS Thermophysical Properties Division and the Center for Applied Thermodynamics Studies of the University of Idaho.

Key words: argon; critically evaluated data; density; ethylene; heat capacity; parahydrogen; nitrogen; nitrogen trifluoride; oxygen; thermodynamic properties; thermophysical properties.

Contents

List of Figures.

List of Tables

1. Introduction

This work presents the thermophysical properties of several fluids in a useful and convenient form. The properties are presented in tabular form and can be computed by a standard set of equations, which is also provided. Special care was given in providing all of the property tables and equations in a uniform manner, so that one who is familiar with the form for one of the fluids can readily use the tables and equations for any fluid.

The equation used in the single-phase region is a 32-term modified Benedict-Webb-Rubin (MBWR) equation of state from a published source [5–10][1]. The MBWR, which has the advantages of high computational speed and accuracy, is widely used for the correlation of thermodynamic property data and has efficient computer programs for the calculation of thermophysical properties. This relation computes pressure from density and temperature and is also used with appropriate ancillary equations for the calculation of derived properties internal energy, enthalpy, entropy, heat capacity, and sound velocity as shown below. Other relations used include the equations for vapor pressure, saturated liquid and saturated vapor densities, and melting pressure. In addition, viscosity and thermal conductivity are computed for argon, nitrogen, and oxygen. Dielectric constant was calculated for the latter fluids and for hydrogen. Computation of the thermophysical properties of these fluids can be accomplished using a computer program which is an interactive system with prompting to aid in its use, described in a companion publication [1]. This program is a modification of program "Fluids Pack" by R. D. McCarty [2]. An early version of a program to represent thermophysical properties of several fluids is program "GASP" (gas properties) by R. C. Hendricks et al. [3]. The considerable improvements in the properties data that have occurred since that time are reflected in this report. Also, by retaining the parametric form of the equations for all fluids, the computer programming is much simplified. Only the numerical constants are changed from one fluid to the next.

This work includes three sets of tables; the properties for the liquid-vapor boundary, the properties of the liquid on the melting line, and properties for isobars on the pressure-density-temperature surface. The isobar tables comprise the largest portion of the tables. The temperature intervals on each isobar were selected by computer calculation of statistical significance to allow accurate linear interpolation of the properties. The temperature intervals were selected on the basis of relative rates of change and basic accuracies of density and heat capacity.

The MBWR equations and related properties found in this work originate from the cooperative efforts of two research groups: the NBS Thermophysical Properties Division and the Center for Applied Thermodynamics Studies of the University of Idaho. The group from this laboratory includes R. D. McCarty, L. A. Weber, H. M. Roder, and R. D. Goodwin, and the University of Idaho group includes R. B. Stewart and R. T. Jacobsen.

The authors and their work are summarized as follows:

Fluid	A	C_2H_4	H_2	N_2	NF_3	O_2
R. D. Goodwin					X	
R. T. Jacobsen	X	X		X		
R. D. McCarty		X	X	X		
H. M. Roder			X	X		
R. B. Stewart	X			X		
L. A. Weber					X	X
Reference	[10]	[5]	[6]	[7]	[9]	[8]

2. Thermodynamic and Related Properties

2.1. The Equation of State

The relation for computing pressure as a function of temperature and density is a 32-term modified Benedict-Webb-Rubin (MBWR) equation of state. Its versatility and adaptability to efficient computer technique make it appropriate for use with multiproperty fitting techniques, where deviations from the *PVT* surface and heat capacity or sound velocity data are minimized simultaneously, resulting in a more accurate model than would be the case if the *PVT* data alone were used. This technique has been described by R. D. McCarty [2,4]. Other advantages of the MBWR are: (1) accurate representation of the thermodynamic surface over wide ranges of temperature and pressure, (2) adaptability to least squares fitting methods using many different kinds of experimental data, and (3) convenience in correlating data from different sources. An analytic function cannot present the proper behavior near the critical point, so that the MBWR is not valid in the critical region (see sec. 5).

The mathematical form of the MBWR is,

$$\begin{aligned}P = {} & \rho RT \\ & + \rho^2(G(1)T+G(2)T^{1/2}+G(3)+G(4)/T+G(5)/T^2) \\ & + \rho^3(G(6)T+G(7)+G(8)/T+G(9)/T^2) \\ & + \rho^4(G(10)T+G(11)+G(12)/T) + \rho^5(G(13)) \\ & + \rho^6(G(14)/T+G(15)/T^2) + \rho^7(G(16)/T) \\ & + \rho^8(G(17)/T+G(18)T^2) + \rho^9(G(19)/T^2) \\ & + \rho^3(G(20)/T^2+G(21)/T^3)\exp(\gamma\rho^2) \\ & + \rho^5(G(22)/T^2+G(23)/T^4)\exp(\gamma\rho^2) \\ & + \rho^7(G(24)/T^2+G(25)/T^3)\exp(\gamma\rho^2) \\ & + \rho^9(G(26)/T^2+G(27)/T^4)\exp(\gamma\rho^2) \\ & + \rho^{11}(G(28)/T^2+G(29)/T^3)\exp(\gamma\rho^2) \\ & + \rho^{13}(G(30)/T^2+G(31)/T^3+G(32)/T^4)\exp(\gamma\rho^2)\,. \end{aligned} \quad (1)$$

[1]Figures in brackets indicate literature references.

The nonlinear coefficient gamma, is defined generally as $\gamma=-1/\rho_c^2$, and is held constant during the fitting process to determine the linear coefficients $G(i)$.

Coefficients for each of the fluids are given in appendix L. Maximum temperatures and pressures, beyond which extrapolation is not advised, for these surfaces are given in appendix C.

2.2. Two-Phase Boundaries

2.2.1. Vapor Pressure

The temperature of the liquid–vapor boundary for each isobar in the tables is computed from a vapor pressure equation,

$$\ln P = \ln P_t + V_p(1)x + V_p(2)x^2 + V_p(3)x^3 + V_p(4)x^4 + V_p(5)x(1-x)^{V_p(6)} \quad (2)$$

where

$$x = (1-T_t/T)/(1-T_t/T_c) \ . \quad (3)$$

2.2.2. Vapor Densities at Coexistence

Densities of the vapor at coexistence with liquid, which are used to generate all the saturated densities given in the tables, are given by

$$\rho = \rho_c + (\rho_{tv}-\rho_c)\exp\{f(T)\} \quad (4)$$

and

$$f(T) = A(1)\ln x + \sum_{i=2}^{4} A(i)(1-x^{(i-5)/3}) + \sum_{i=5}^{13} A(i)(1-x)^{(i-4)/3} \quad (5)$$

where

$$x = (T-T_c)/(T_t-T_c) \ . \quad (6)$$

2.2.3. Liquid Densities at Coexistence

The liquid density at coexistence is calculated from,

$$\rho = \rho_c + (\rho_{tl}-\rho_c)\exp\{f(T)\} \quad (7)$$

and

$$f(T) = A(14)\ln x + \sum_{i=15}^{17} A(i)(1-x^{(i-18)/3}) + \sum_{i=18}^{20} A(i)(1-x^{(i-17)/3}) \quad (8)$$

where x is defined in equation 6.

2.2.4. The Melting Line

The pressures at melting are given by,

$$P = A + BT^C \ . \quad (9)$$

2.3. Derived Thermodynamic Properties

The properties derived from the equation of state and ideal gas heat capacity are entropy, enthalpy, internal energy, specific heat at constant volume and at constant pressure, and sound velocity.

2.3.1. Entropy

The entropy is computed from

$$S(T,\rho) = S°(298.15) + \int_{298.15}^{T} \{C_p^\circ/T\} - R\ln(RT\rho/P_0) + \int_0^{\rho} \left\{R/\rho(1/\rho^2)\left(\frac{\partial P}{\partial T}\right)_\rho\right\}_T d\rho \quad (10)$$

2.3.2. Ideal Gas Specific Heat

The ideal gas specific heat, C_p° is computed from the following,

$$C_p^\circ/R = G_i(1)/T^3 + G_i(2)/T^2 + G_i(3)/T + G + G_i(5)T + G_i(6)T^2 + G_i(7)T^3 + G_i(8)u^2e^u/(e^u-1)^2 \quad (11)$$

where,

$$u = G_i(9)/T \ . \quad (12)$$

2.3.3. Reference State

The reference states $S°$ and $H°$ at $T=298.15$ K (given below in table 1) are from Wagman et al. [11], except for parahydrogen which is taken from Woolley, Scott, and Brickwedde [12], since reference [12], since reference [11] gives only the normal hydrogen values.

TABLE 1. Reference values for temperature, entropy, and enthalpy. The reference value for temperature is 298.15 K for all fluids. Values of S° and H° are from NBS TN 270–3 [11], except hydrogen [12].

	S° J/mol·K	H° J/mol
Argon	154.7335	6169.5
Ethylene	219.451	10564.6
Parahydrogen	130.407	8409.8
Nitrogen	191.502	8669.0
Nitrogen trifluoride	260.621	11828.0
Oxygen	205.029	8680.1

2.3.4. Enthalpy

The enthalpy is computed from

$$H(T,\rho) = H^\circ(T^\circ) + (P - \rho RT)/\rho + \int_0^\rho \left\{\frac{P}{\rho^2} - \frac{T}{\rho^2}\left(\frac{\partial P}{\partial T}\right)_\rho\right\}_T d\rho + \int_{298.15}^T C_p^\circ \, dT . \quad (13)$$

2.3.5. Internal Energy

The internal energy is,

$$E(T,\rho) = H(T,\rho) - P/\rho . \quad (14)$$

2.3.6. Specific Heat at Constant Volume

The specific heat at constant volume is,

$$C_v(T,\rho) = C_p^\circ - R \int_0^\rho \left\{\frac{T}{\rho^2}\left(\frac{\partial^2 P}{\partial T^2}\right)_\rho\right\}_T d\rho . \quad (15)$$

2.3.7. Specific Heat at Constant Pressure

The specific heat as constant pressure is,

$$C_p(T,\rho) = C_v(T,\rho) + \left\{\left(\frac{T}{\rho^2}\right)\left(\frac{\partial P}{\partial T}\right)_\rho^2 \Big/ \left(\frac{\partial P}{\partial \rho}\right)_T\right\} . \quad (16)$$

2.3.8. Sound Velocity

From this the sound velocity can be computed as,

$$W(T,\rho) = \left\{\frac{C_p}{C_v}\left(\frac{\partial P}{\partial \rho}\right)_T\right\}^{1/2} . \quad (17)$$

3. Transport Properties: Viscosity and Thermal Conductivity

Viscosity and thermal conductivity are given for argon, nitrogen, and oxygen. The theory and discussion of the relations used here are from the work of Hanley, McCarty, and Haynes [13]. The functional forms for viscosity and thermal conductivity are,

$$\eta = \eta_0(T) + \eta_1(T)\rho + \eta_2(\rho,T) \quad (18)$$

and

$$\lambda = \lambda_0(T) + \lambda_1(T)\rho + \lambda_2(\rho,T) + \lambda_c(\rho,T) . \quad (19)$$

The first terms of eqs (18) and (19) are the contributions of the dilute gas,

$$\eta_0(T) = \sum_{i=1}^{9} G_v(i)T^{(4-i)/3}$$

and (20)

$$\lambda_0(T) = \sum_{i=1}^{9} G_t(i)T^{(4-i)/3} .$$

The second terms in eqs (18) and (19) represent the contribution to the transport coefficients of the moderately dense gas.

$$\eta_1(T) = F_v(1) + F_v(2)\,\{F_v(3) - \ln\,(T/F_v(4))\}^2$$

and (21)

$$\lambda_1(T) = F_t(1) + F_t(2)\,\{F_t(3) - \ln\,(T/F_t(4))\}^2 .$$

The third terms in these equations are the contribution of the dense gas,

$$\eta_2(\rho,T) = \exp\{F(\rho,T)\} - \exp\{G(T)\} \quad (22)$$

$$F(\rho,T) = E_v(1) + E_v(2)H(\rho) + E_v(3)\rho^{0.1} + E_v(4)H(\rho)/T^2 + E_v(5)\rho^{0.1}/T^{1.5} + E_v(6)/T + E_v(7)H(\rho)/T \quad (23)$$

$$G(T) = E_v(1) + E_v(2)/T \quad (24)$$

where

$$H(\rho) = \rho^{0.5}\,(\rho - E_v(8))/E_v(8) . \quad (25)$$

The functional form of eqs (22), (23), (24), and (25) is used for the corresponding thermal conductivity equations with the coefficients E_t substituted for the E_v's.

The last term of eq. 19 is the critical enhancement term and is described in appendix D. Critical effects can be very large, especially in the range $(\rho-\rho_c)/\rho < 0.25$ and $(T-T_c)/T < 0.025$ and in fact, as the critical point is approached, this term may be larger than the sum of the other terms by several orders of magnitude. We include λ_c to give the proper behavior in the critical region and to produce a smooth transition from outside the critical to within it.

4. Dielectric Constant

The equation used to represent the dielectric constant is,

$$C_m = A + B\rho + C\rho^2 + D\rho^3 + ET + FP \quad (26)$$

where,

$$C_m = \left\{\frac{\epsilon - 1}{\epsilon + 2}\right\} / \rho . \quad (27)$$

Equation 27 is the well-known Clausius-Mossotti relation. Coefficients are given for parahydrogen, nitrogen, and oxygen. The coefficients of Stewart [14] for parahydrogen were modified to conform to eq. (27). The data of Ely and Straty for nitrogen [15] were used in a fit of eq. (27). The oxygen coefficients are from Younglove [16].

Although the range of the Clausius-Mossotti function for most liquids is small, the very high precision of the data requires the terms of eq. (27). The temperature and pressure terms are useful in fitting the compressed liquid states. The pressure term allows the flexibility which is useful in fluids with low compressibilities, usually found at low temperatures and high pressures [17].

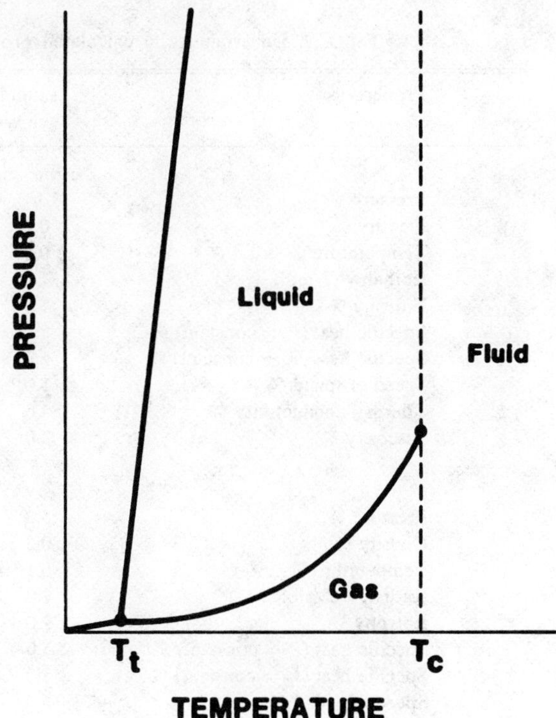

FIG. 1. Regions for associated uncertainties. (See table 2).

5. Summary of Uncertainties

The following uncertainties are taken primarily from the source documents and from [2]. The basic references for each of the fluids should be referred to for comparisons of the mathematical representations to the experimental data. These uncertainties are based on the authors claims for accuracies of the experimental data, and the inaccuracies introduced in fitting the data. They are maximum deviations and in most cases deviations are substantially less. Figure 1 shows the regions indicated for fluid, gas, and liquid.

We stress that the uncertainties in table 2, for the various properties, are not intended to represent the critical region, which we assess to be $\rho_c \pm 0.3\ \rho_c$ and $T_c \pm 0.05\ T_c$. However, for practical calculation the MBWR equation may be used in this region to compute properties with the understanding that the uncertainties will be larger since this equation has not been adapted to produce the proper behavior near the critical point.

Uncertainties for nitrogen trifluoride were not indicated in reference [9], because of the sole source nature of the data. However, based on our experience, the maximum deviations arising from the fitting uncertainties and experimental data uncertainties are probably no greater than those shown for argon.

6. Units

Except as noted, all numerical constants listed in appendices F through L are consistent with the units for the various properties given in appendix A. As stated in the appendices, the densities for viscosity and thermal conductivity calculations must be supplied in units of grams per cubic centimeter.

TABLE 2. Uncertainties in calculated properties. Maximum deviations of properties in percent[a]

Property	Liquid below T_c	Gas below T_c	Fluid above T_c
	Argon and Nitrogen Trifluoride		
Pressure %	10.	0.3	0.3
Density %	0.25	0.3	0.3
Temperature %	0.25	0.3	0.3
Enthalpy (J/mol)	2.0	1.0	1.5
Entropy %	1.0	1.0	1.0
Specific heat (P = constant) %	5.0	5.0	5.0
Specific heat (V = constant) %	5.0	5.0	5.0
Speed of sound %	5.0	5.0	5.0
Thermal conductivity %	4.0	4.0	6.0
Viscosity %	2.0	2.0	2.0
	Ethylene		
Pressure %	5.0	0.25	0.2
Density %	0.3	0.3	0.3
Temperature %	0.1	0.2	0.2
Enthalpy (J/mol)	2.0	2.0	2.0
Entropy %	2.0	1.0	1.0
Specific heat (P = constant) %	5.0	5.0	5.0
Specific heat (V = constant) %	5.0	5.0	5.0
Speed of sound %	5.0	5.0	5.0
	para Hydrogen		
Pressure %	5.0	0.25	0.2
Density %	0.1	0.25	0.2
Temperature %	0.1	0.25	0.2
Enthalpy (J/mol)	1	3	5.1
Entropy %	1.0	1.0	1.0
Specific heat (P = constant) %	3.0	2.0	3.0
Specific heat (V = constant) %	3.0	2.0	3.0
Speed of sound %	2.0	1.0	1.0
	Nitrogen		
Pressure %	5.0	0.3	0.3
Density %	0.5	0.3	0.2
Temperature %	0.5	0.3	0.2
Enthalpy (J/mol)	3	1	1
Entropy %	2.0	1.0	1.0
Specific heat (P = constant) %	5.0	5.0	5.0
Specific heat (V = constant) %	5.0	5.0	5.0
Speed of sound %	2.0	0.25	1.0
Thermal conductivity %	4.0	4.0	6.0
Viscosity %	2.0	2.0	6.0
Dielectric constant %	0.01	0.01	0.01
	Oxygen		
Pressure %	5.0	0.25	0.15
Density %	0.1	0.25	0.15
Temperature %	0.1	0.2	0.1
Enthalpy (J/mol)	0.5	0.25	0.5
Entropy %	0.5	0.25	0.5
Specific heat (P = constant) %	3.0	5.0	3.0
Specific heat (V = constant) %	3.0	5.0	3.0
Speed of sound %	2.0	0.5	0.5
Thermal conductivity %	4.0	4.0	6.0
Viscosity %	2.0	2.0	2.0
Dielectric constant %	0.01	0.01	0.01

[a]All uncertainties are in percent except those for enthalpy, which are given in J/mol.

7. Acknowledgements

We are indebted to the office of Standard Reference Data of the National Bureau of Standards for their generous support of this work, and in particular to Dr. H. J. White, Jr., who originally conceived of this volume.

References

[1] Younglove, B. A., "Interactive Program to Calculate Thermophysical Properties of Six Fluids," Natl. Bur. Stand. (U.S.), Tech. Note 1048 (1982).

[2] McCarty, R. D., "Interactive Fortran IV Computer Programs for the Thermodynamic and Transport Properties of Selected Cryogens (Fluids Pack)," Natl. Bur. Stand. (U.S.), Tech. Note 1025 (1980).

[3] Hendricks, R. D., Baron, A. K., and Paller, I. C., "GASP - A Computer Code for Calculating the Thermodynamic and Transport Properties for Ten Fluids: Parahydrogen, Helium, Neon, Methane, Nitrogen, Carbon Monoxide, Oxygen, Fluorine, Argon, and Carbon Dioxide," NASA Tech. Note D-7808 (1975).

[4] McCarty, R. D., "Determination of Thermodynamic Properties from the Experimental Thermodynamics," Vol. II, Experimental Thermodynamics of Non-Reacting Fluids. B. Le Neindre and B. Vodar, eds., Butterworth and Co. Ltd., London, England (1975).

[5] McCarty, R. D., and Jacobsen, R. T., "An Equation of State for Fluid Ethylene," Natl. Bur. Stand. (U.S.), Tech. Note 1045 (1981).

[6] Roder, H. M., and McCarty, R. D., "A Modified Benedict-Webb-Rubin Equation of State for Parahydrogen - II," Natl. Bur. Stand. (U.S.), Interagency Report NBSIR 75-814 (1975).

[7] Jacobsen, R. T., Stewart, R. B., McCarty, R. D., and Hanley, H. J. M., "Thermophysical Properties of Nitrogen from the Fusion Line to 3500 R [1944 K] for Pressures to 150000 PSIA," Natl. Bur. Stand. (U.S.), Tech. Note 648 (1973).

[8] Weber, L. A., "A Modified Benedict-Webb-Rubin Equation of State for Gaseous and Liquid Oxygen," Natl. Bur. Stand. (U.S.), Interagency Report NBSIR 78-882 (1978).

[9] Goodwin, R. D., and Weber, L. A., "Thermophysical Properties of Nitrogen Trifluoride from 66 to 500 K at Pressures to 500 Bar," Natl. Bur. Stand. (U.S.), Interagency Report NBSIR 80-1632 (1980).

[10] Stewart, R. B., Jacobsen, R. T., and Becker, J. H., "A Thermodynamic Property Formulation for Argon Using a 32 Term Pressure Explicit Equation of State," Center for Applied Thermodynamic Studies, Report No. 81-3, University of Idaho, Moscow, ID (July 1981).

[11] Wagman, D. D., Evans, W. H., Parker, V. B., Halow, I., Baily, S. M., and Schumm, R. H., "Selected Values of Chemical Thermodynamic Properties," Natl. Bur. Stand. (U.S.), Tech. Note 270-3 (1968).

[12] Woolley, H. W., Scott., R. B., and Brickwedde, F. G., "Compilation of Thermal Properties of Hydrogen in Its Various Isotopic and Ortho-Para Modifications," Natl. Bur. Stand. (U.S.), Research Paper RP1932 (1948).

[13] Hanley, H. J. M., McCarty, R. D., and Haynes, W. M., "The Viscosity and Thermal Conductivity Coefficients for Dense Gaseous and Liquid Argon, Krypton, Xenon, Nitrogen, and Oxygen." J. Phys. Chem. Ref. Data **3**, 979-1017 (1979).

[14] Stewart, J. W., "Dielectric Polarizability of Fluid Para-Hydrogen," J. Chem. Phys. **40**, 3297-3306 (1964).

[15] Ely, J. F., and Straty, G. C., "Dielectric constants and molar polarizabilities of saturated and compressed fluid nitrogen," J. Chem. Phys. **61**, 1480-1485 (1974).

[16] Younglove, B. A., "Dielectric Constant of Compressed Gaseous and Liquid Oxygen," J. Res. Natl. Bur. Stand. (U.S.), **76A**, 37-40 (1972).

[17] Sengers, J. V., and Levelt Sengers, J. M. H., "Concepts and methods for describing critical phenomena in fluids," NASA Contractor Report 149665 (NASA-Lewis Research Center, Cleveland, OH). With minor modifications this report was published in *Progress in Liquid Physics*, C. A. Croxton, ed. (Wiley, NY 1978), p. 103.

[18] Remy, H., "Internationale Atomgewichte 1968", Chem. Berichte **101**, I-VI (1968). The atomic weights are based on the carbon 12 scale.

Appendix A. List of Symbols and Units

Primary Thermophysical Quantities

P, P_c, P_t	pressure, P at critical point, P at triple point, MPa
ρ, ρ_c	density, ρ at critical point, mol/dm^3
ρ_{tl}, ρ_{tv}	ρ of liquid at triple point, ρ of gas at triple point
T, T_c, T_t	temperature, T at critical point, T at triple point, K
C_p, C_P°	specific heat at constant pressure, C_p of ideal gas, J/mol·K
C_v, C_v°	specific heat at constant volume, C_v for ideal gas, J/mol·K
E	internal energy, J/mol
H, H°	enthalpy, reference value of H, J/mol
S, S°	entropy, reference value of S, J/mol·K
W	sound velocity, m/s
η	viscosity, μPa·s
λ	thermal conductivity, W/m·K
ϵ	dielectric constant

Other Variables and Constants

R	the gas constant, 8.31434 J/mol·K
K_t	isothermal bulk modulus, (MPa)$^{-1}$
M_r	molecular weight, g/mol
N_A	Avogadro's number, mol^{-1}
C_m	Clausius-Mossotti function
T_0	reference temperature, 298.15 K

Appendix B. Conversion Factors

	From	To	Multiplied by
Pressure	MPa	psia	145.03774
Temperature	K	R	1.8
Density	mol/dm^3	lb/ft^3	$0.0624280 \cdot M_r$
Internal energy	J/mol	BTU/lb	$0.43021035 \cdot M_r$
Enthalpy	J/mol	BTU/lb	$0.43021035 \cdot M_r$
Entropy	J/mol·K	BTU/lb·R	$0.23900575 \cdot M_r$
Specific heat	J/mol·K	BTU/lb·R	$0.23900575 \cdot M_r$
Thermal conductivity	W/m·K	BTU/ft·h·R	0.57817602
Viscosity	Pa·s	lb/ft·s	0.6719690
Velocity of sound	m/s	ft/s	3.28084

M_r for		
	Argon	39.948
	Ethylene	28.054
	Hydrogen	2.01594
	Nitrogen	28.013
	Nitrogen trifluoride	71.019
	Oxygen	31.9988

M_r is molecular weight [18].
lb is pound mass.

Appendix C. Triple Point, Critical Point, Maximum Pressures, and Temperatures

Triple Point

fluid	P_t (MPa)	ρ_{tl} (mol/dm^3)	ρ_{tv} (mol/dm^3)	T_t (K)
Ar	0.68906×10^{-1}	35.400	0.10292	83.80
C_2H_4	0.121295×10^{-3}	23.343	0.142546×10^{-3}	103.986
H_2	0.7042×10^{-2}	38.2143	0.632230×10^{-1}	13.8
N_2	0.12463×10^{-1}	30.977	0.24282×10^{-1}	63.15
NF_3	0.185425×10^{-6}	26.320	0.33612×10^{-8}	66.36
O_2	0.14800×10^{-3}	40.820	0.33189×10^{-3}	54.359

Critical Point

fluid	P_c (MPa)	ρ_c (mol/dm^3)	T_c (K)
Ar	4.9058	13.41	150.86
C_2H_4	5.0404	7.650	282.3428
H_2	1.28377	15.556	32.938
N_2	3.39908	11.21	126.26
NF_3	4.4607	7.92	234.0
O_2	5.043	13.630	154.581

Maximum Pressures and Temperatures[a]

fluid	P_{max} (MPa)	T_{max} (K)
Ar	101	400
C_2H_4	40	400
H_2	121	400
N_2	1013	1900
NF_3	50	500
O_2	121	400

[a]The extrapolation of the MBWR and associated equations beyond these values is not advised.

Appendix D. Critical Point Enhancement to Thermal Conductivity

The last term of eq. (19) is the contribution from critical point effects [13].

$$\Delta\lambda_c = \Delta\lambda' \exp\left\{-18.66\left(\frac{\rho-\rho_c}{\rho_c}\right)^4 - 4.25\left(\frac{T-T_c}{T_c}\right)^2\right\} \tag{D1}$$

where

$$\Delta\lambda' = \left\{KT^2 \left(\frac{\partial P}{\partial T}\right)_\rho K_T^{\ 1/2} Y\right\} 10^{-5} \tag{D2}$$

and

$$Y = \{6\pi\eta l(kT)^{1/2} \rho^{1/2} (N_A/M_r)^{1/2}\}^{-1} \tag{D3}$$

Here K_T is isothermal compressibility, and l represents a length approximating the hard sphere diameter of the molecule,

$$l = \{F\gamma_m^5 \rho(N_A/M_r(\epsilon/k)/T^{1/2}\} \tag{D4}$$

where ϵ is the intermolecular potential at minimum, at which the molecular separation is r_m, and F is a parameter depending on the potential function.

The compressibility is computed from the MBWR surface unless the density is within 25 percent of the critical density and the temperature is within 2.5 percent of the critical temperature, in which case the compressibility is computed from a special scaled equation of state. Also note that the critical point of the scaled equation may be slightly different from that of the corresponding MBWR.

The reduced parameters used are,

$$\begin{aligned} \Delta\rho &= |\rho - \rho_c|/\rho_c \\ \Delta T &= |T - T_c|/T_c \\ \rho^* &= \rho/\rho_c \\ x &= \Delta T/(\Delta\rho)^{1/\beta} \\ K_T^* &= P_c K_t \end{aligned} \tag{D5}$$

The compressibility is computed from,

$$[\rho^{*2} K_T^*]^{-1} = \Delta\rho^{\delta-1} [\delta h(x) - (x/\beta)h'(x)] \tag{D6}$$

with

$$h(x) = E_1 \left(\frac{x+x_0}{x_0}\right) \left[1 + E_1 \left(\frac{x+x_0}{x_c}\right)^{2\beta}\right]^{(\gamma-1)/2\beta} \tag{D7}$$

and

$$h'(x) = \frac{E_1}{x_0} \left[1 + E_2 \left(\frac{x+x_0}{x_0}\right)^{2\beta}\right]^{(\gamma-1)/2\beta} \tag{D8}$$

$$+ \frac{\gamma-1}{x_0} E_1 E_2 \left(\frac{x+x_0}{x_0}\right)^{2\beta} \left[1 + E_2 \left(\frac{x+x_0}{x_0}\right)^{2\beta}\right]^{(\gamma-1-2\beta)/2\beta}$$

Values of the parameters used in the equations for critical enhancement are given below in table D1. Units for density when computing viscosity and thermal conductivity are g/cm^3.

TABLE D1. Critical enhancement parameters [13,17]. When these numerical values are used in equations D1 through D4, the calculated values of $\Delta\lambda_c$ has units of W/(m·K).

	Argon	Nitrogen	Oxygen
ϵ/k (K)	52.8	118.	113.0
r_m (cm)	3.669×10^{-8}	3.933×10^{-8}	3.8896×10^{-8}
T_c (K)	150.725	126.24	154.575
ρ_c (g/cm^3)	0.533	0.3139	0.4362
F	1.7124	1.67108	2.210636
P_c (MPa)	4.8619	3.443	5.0429
x_0	0.183	0.164	0.183
β	0.355	0.355	0.355
δ	4.352	4.352	4.352
E_1	2.27	2.17	2.21
E_2	0.287	0.287	0.287
γ	1.190	1.190	1.190

Appendix E. Summary of Basic Equations

1. The modified Benedict-Webb-Ruben equation of state:

$$\begin{aligned}P = {} & \rho RT \\ & + \rho^2(G(1)T+G(2)T^{1/2}+G(3)+G(4)/T+G(5)/T^2) \\ & + \rho^3(G(6)T + G(7) + G(8)/T + G(9)/T^2) \\ & + \rho^4(G(10)T + G(11) + G(12)/T) + \rho^5(G(13)) \\ & + \rho^6(G(14)/T + G(15)/T^2) + \rho^7(G(16)/T) \\ & + \rho^8(G(17)/T + G(18)T^2) + \rho^9(G(19)/T^2) \\ & + \rho^3(G(20)/T^2 + G(21)/T^3)\exp(\gamma\rho^2) \\ & + \rho^5(G(22)/T^2 + G(23)/T^4)\exp(\gamma\rho^2) \\ & + \rho^7(G(24)/T^2 + G(25)/T^3)\exp(\gamma\rho^2) \\ & + \rho^9(G(26)/T^2 + G(27)/T^4)\exp(\gamma\rho^2) \\ & + \rho^{11}(G(28)/T^2 + G(29)/T^3)\exp(\gamma\rho^2) \\ & + \rho^{13}(G(30)/T^2 + G(31)/T^3 + G(32)/T^4)\exp(\gamma\rho^2)\,. \end{aligned} \tag{1}$$

2. The vapor pressure:

$$\ln P = \ln P_t + V_p(1)x + V_p(2)x^2 + V_p(3)x^3 + V_p(4)x^4 + V_p(5)x(1-x)^{V_P(6)} \tag{2}$$

$$x = (1-T_t/T)/(1-T_t/T_c)\,. \tag{3}$$

3. The vapor densities at coexistence:

$$\rho = \rho_c + (\rho_{tv}-\rho_c)\exp\{f(T)\} \tag{4}$$

$$f(T) = A(1)\ln x + \sum_{i=2}^{4} A(i)(1 - x^{(i-5)/3}) + \sum_{i=5}^{13} A(i)(1 - x)^{(i-4)/3} \tag{5}$$

$$x = (T-T_c)/(T_t-T_c)\,. \tag{6}$$

4. The liquid densities at coexistence:

$$\rho = \rho_c + (\rho_{tl}-\rho_c)\exp\{f(T)\} \tag{7}$$

$$f(T) = A(14)\ln x + \sum_{i=15}^{17} A(i)(1 - x^{(i-18)/3}) + \sum_{i=18}^{20} A(i)(1 - x^{(i-17)/3})\,. \tag{8}$$

5. The melting line:

$$P = A + BT^C\,. \tag{9}$$

6. The viscosity:

$$\eta = \eta_0(T) + \eta_1(T)\rho + \eta_2(\rho,T)\,. \tag{18}$$

7. The thermal conductivity:

$$\lambda = \lambda_0(T) + \lambda_1(T)\rho + \lambda_2(\rho,T) + \lambda_c(\rho,T)\,. \tag{19}$$

8. The dielectric constant:

$$C_m = \left\{\frac{\epsilon - 1}{\epsilon + 2}\right\}\Big/\rho \tag{27}$$

$$C_m = A + B\rho + C\rho^2 + D\rho^3 + ET + FP\,. \tag{26}$$

Appendix F. Thermophysical Properties of Argon

Thermophysical properties of coexisting gaseous and liquid argon

T K	Pres. MPa	Density kg/m³	Density mol/dm³	E J/mol	H J/mol	S J/(mol·K)	C_v J/(mol·K)	C_p J/(mol·K)	Sound m/s	Visc. μPa·s	Therm. W/(m·K)
83.800[a]	.06891	1414.0	35.40	−1373.0	−1371.0	54.85	23.60	44.87	820.6	288.0	.135
83.800[a]	.06891	.01325	.0003328	4533.0	212200.0	180.7	12.47	20.79	170.5	6.92	.00540
84.0	.07052	1413.0	35.37	−1364.0	−1362.0	54.96	23.50	44.85	820.4	287.0	.134
84.0	.07052	4.194	.1050	4498.0	5170.0	132.7	13.08	22.35	168.1	6.98	.00560
86.0	.08826	1401.0	35.07	−1274.0	−1272.0	56.01	22.69	44.77	816.8	270.0	.130
86.0	.08826	5.109	.1279	4516.0	5206.0	131.3	13.16	22.59	169.7	7.15	.00575
88.0	.1093	1389.0	34.77	−1185.0	−1182.0	57.03	22.04	44.81	811.1	254.0	.127
88.0	.1093	6.190	.1549	4533.0	5239.0	130.0	13.25	22.86	171.2	7.31	.00592
90.0	.1339	1376.0	34.45	−1096.0	−1092.0	58.04	21.51	44.96	803.7	240.0	.124
90.0	.1339	7.448	.1864	4549.0	5268.0	128.7	13.34	23.16	172.6	7.48	.00609
92.0	.1626	1364.0	34.14	−1006.0	−1002.0	59.02	21.06	45.17	794.8	227.0	.120
92.0	.1626	8.897	.2227	4564.0	5294.0	127.4	13.44	23.50	174.0	7.65	.00627
94.0	.1956	1351.0	33.82	−916.4	−910.7	59.99	20.69	45.44	784.7	215.0	.117
94.0	.1956	10.55	.2641	4578.0	5319.0	126.2	13.55	23.86	175.2	7.82	.00645
96.0	.2335	1338.0	33.49	−826.0	−819.0	60.94	20.37	45.76	773.5	203.0	.114
96.0	.2335	12.42	.3110	4591.0	5342.0	125.1	13.66	24.27	176.4	8.00	.00664
98.0	.2765	1325.0	33.16	−734.9	−726.5	61.88	20.08	46.13	761.3	193.0	.111
98.0	.2765	14.53	.3637	4602.0	5363.0	124.0	13.77	24.71	177.5	8.18	.00685
100.0	.3252	1311.0	32.82	−643.2	−633.3	62.81	19.83	46.52	748.4	183.0	.109
100.0	.3252	16.89	.4228	4613.0	5382.0	122.9	13.89	25.19	178.6	8.36	.00706
102.0	.3799	1297.0	32.48	−550.8	−539.1	63.72	19.60	46.96	734.9	174.0	.106
102.0	.3799	19.52	.4887	4621.0	5399.0	121.9	14.02	25.72	179.6	8.54	.00727
104.0	.4411	1283.0	32.13	−457.6	−443.9	64.63	19.39	47.44	720.7	165.0	.103
104.0	.4411	22.45	.5620	4629.0	5414.0	120.9	14.15	26.30	180.5	8.73	.00750
106.0	.5092	1269.0	31.77	−363.7	−347.6	65.52	19.20	47.96	706.1	157.0	.101
106.0	.5092	25.70	.6432	4634.0	5426.0	120.0	14.29	26.94	181.3	8.93	.00774
108.0	.5846	1255.0	31.40	−268.9	−250.3	66.41	19.02	48.53	691.0	150.0	.0982
108.0	.5846	29.28	.7331	4639.0	5436.0	119.0	14.43	27.65	182.1	9.13	.00799
110.0	.6678	1240.0	31.03	−173.2	−151.7	67.29	18.84	49.17	675.4	142.0	.0957
110.0	.6678	33.25	.8322	4641.0	5443.0	118.1	14.58	28.43	182.7	9.33	.00826
112.0	.7591	1224.0	30.65	−76.6	−51.8	68.16	18.68	49.87	659.5	136.0	.0932
112.0	.7591	37.61	.9415	4641.0	5448.0	117.2	14.74	29.30	183.3	9.54	.00854
114.0	.8592	1209.0	30.26	21.1	49.5	69.03	18.53	50.66	643.2	129.0	.0908
114.0	.8592	42.41	1.062	4640.0	5449.0	116.4	14.90	30.28	183.9	9.76	.00883
116.0	.9683	1193.0	29.86	119.9	152.3	69.89	18.38	51.55	626.5	123.0	.0884
116.0	.9683	47.70	1.194	4636.0	5447.0	115.5	15.07	31.37	184.3	9.99	.00914
118.0	1.087	1176.0	29.45	219.9	256.9	70.75	18.24	52.56	609.4	117.0	.0861
118.0	1.087	53.51	1.340	4629.0	5441.0	114.7	15.25	32.60	184.7	10.2	.00947
120.0	1.216	1159.0	29.02	321.4	363.3	71.61	18.11	53.70	592.0	112.0	.0837
120.0	1.216	59.91	1.500	4620.0	5431.0	113.8	15.45	34.01	185.0	10.5	.00981
122.0	1.355	1142.0	28.59	424.4	471.8	72.46	17.99	55.02	574.1	106.0	.0813
122.0	1.355	66.94	1.676	4608.0	5417.0	113.0	15.65	35.62	185.2	10.7	.0102
124.0	1.505	1124.0	28.13	529.1	582.6	73.32	17.88	56.54	555.9	101.0	.0790
124.0	1.505	74.70	1.870	4593.0	5398.0	112.1	15.87	37.49	185.4	11.0	.0106
126.0	1.667	1105.0	27.67	635.9	696.1	74.18	17.77	58.32	537.1	96.0	.0766
126.0	1.667	83.27	2.085	4574.0	5374.0	111.3	16.10	39.68	185.5	11.3	.0110
128.0	1.841	1086.0	27.18	744.9	812.7	75.05	17.68	60.41	517.9	91.2	.0742
128.0	1.841	92.77	2.322	4551.0	5344.0	110.4	16.35	42.29	185.4	11.6	.0115
130.0	2.027	1065.0	26.66	856.7	932.8	75.93	17.61	62.91	498.1	86.5	.0718
130.0	2.027	103.3	2.586	4524.0	5308.0	109.6	16.62	45.42	185.3	11.9	.0121

Thermophysical properties of coexisting gaseous and liquid argon—Continued

T K	Pres. MPa	Density kg/m³	Density mol/dm³	E J/mol	H J/mol	S J/(mol·K)	C_v J/(mol·K)	C_p J/(mol·K)	Sound m/s	Visc. μPa·s	Therm. W/(m·K)
132.0	2.227	1044.0	26.13	971.7	1057.0	76.82	17.55	65.94	477.7	81.9	.0694
132.0	2.227	115.1	2.881	4491.0	5264.0	108.7	16.92	49.28	185.2	12.3	.0127
134.0	2.441	1021.0	25.55	1091.0	1186.0	77.72	17.50	69.69	456.5	77.4	.0670
134.0	2.441	128.3	3.213	4451.0	5211.0	107.8	17.24	54.12	184.9	12.7	.0134
136.0	2.668	996.5	24.94	1214.0	1321.0	78.66	17.49	74.45	434.4	73.0	.0645
136.0	2.668	143.3	3.588	4405.0	5148.0	106.8	17.59	60.38	184.6	13.1	.0143
138.0	2.911	970.2	24.29	1343.0	1463.0	79.62	17.50	80.68	411.4	68.6	.0621
138.0	2.911	160.5	4.017	4349.0	5073.0	105.8	17.99	68.75	184.1	13.6	.0154
140.0	3.170	941.4	23.57	1480.0	1615.0	80.63	17.55	89.22	387.0	64.1	.0597
140.0	3.170	180.3	4.514	4281.0	4983.0	104.7	18.43	80.50	183.6	14.2	.0167
142.0	3.446	909.4	22.76	1627.0	1778.0	81.71	17.66	101.7	361.1	59.6	.0575
142.0	3.446	203.7	5.099	4199.0	4875.0	103.5	18.93	98.07	183.1	14.8	.0185
144.0	3.740	872.9	21.85	1787.0	1958.0	82.87	17.84	121.6	333.3	55.0	.0554
144.0	3.740	232.1	5.809	4097.0	4741.0	102.2	19.50	126.9	182.5	15.6	.0210
146.0	4.053	829.6	20.77	1967.0	2163.0	84.18	18.14	158.6	302.9	50.0	.0537
146.0	4.053	268.0	6.708	3966.0	4570.0	100.7	20.15	182.3	181.9	16.7	.0247
148.0	4.386	774.3	19.38	2184.0	2411.0	85.75	18.66	250.7	268.8	44.4	.0525
148.0	4.386	317.8	7.956	3780.0	4331.0	98.73	20.91	326.3	181.5	18.3	.0311
150.86[b]	4.905	535.7	13.41	3004.0	3370.0	91.95					

[a]Triple point.
[b]Critical point.

Thermophysical properties of argon on the melting line

T K	Pres. MPa	Density kg/m^3	Density mol/dm^3	E J/mol	H J/mol	S J/(mol·K)	C_v J/(mol·K)	C_p J/(mol·K)	Sound m/s	Visc. μPa·s	Therm. W/(m·K)
83.80[a]	.06891	1414.0	35.40	−1373.0	−1371.0	54.85	23.60	44.87	820.6	288.0	.135
84.0	.8731	1416.0	35.45	−1375.0	−1350.0	54.83	23.53	44.67	824.5	290.0	.135
84.5	2.888	1419.0	35.51	−1372.0	−1291.0	54.86	23.36	44.30	831.5	291.0	.135
85.0	4.910	1421.0	35.58	−1370.0	−1232.0	54.89	23.19	43.96	838.5	292.0	.135
85.5	6.939	1424.0	35.64	−1367.0	−1172.0	54.91	23.04	43.64	845.6	293.0	.135
86.0	8.976	1426.0	35.71	−1364.0	−1113.0	54.94	22.90	43.34	852.7	294.0	.136
86.5	11.02	1429.0	35.77	−1361.0	−1053.0	54.97	22.77	43.06	859.7	295.0	.136
87.0	13.07	1431.0	35.83	−1358.0	−993.7	54.99	22.65	42.80	866.8	296.0	.136
87.5	15.13	1434.0	35.89	−1355.0	−933.9	55.02	22.54	42.56	873.8	297.0	.136
88.0	17.19	1436.0	35.95	−1352.0	−874.1	55.05	22.44	42.33	880.8	298.0	.137
88.5	19.26	1438.0	36.01	−1349.0	−814.1	55.08	22.34	42.12	887.7	299.0	.137
89.0	21.34	1441.0	36.07	−1346.0	−754.0	55.10	22.26	41.93	894.6	300.0	.137
89.5	23.43	1443.0	36.12	−1342.0	−693.8	55.13	22.18	41.74	901.4	301.0	.137
90.0	25.52	1445.0	36.18	−1339.0	−633.4	55.16	22.11	41.57	908.2	302.0	.137
90.5	27.62	1448.0	36.24	−1335.0	−573.0	55.18	22.04	41.41	914.8	303.0	.138
91.0	29.73	1450.0	36.29	−1332.0	−512.4	55.21	21.99	41.26	921.5	304.0	.138
91.5	31.84	1452.0	36.34	−1328.0	−451.7	55.24	21.94	41.12	928.0	304.0	.138
92.0	33.96	1454.0	36.40	−1324.0	−390.8	55.27	21.89	40.99	934.5	305.0	.138
92.5	36.09	1456.0	36.45	−1320.0	−329.8	55.30	21.85	40.87	940.9	306.0	.139
93.0	38.22	1458.0	36.50	−1316.0	−268.7	55.32	21.82	40.75	947.2	307.0	.139
93.5	40.36	1460.0	36.55	−1312.0	−207.4	55.35	21.79	40.65	953.4	307.0	.139
94.0	42.51	1462.0	36.61	−1307.0	−146.0	55.38	21.77	40.55	959.6	308.0	.139
94.5	44.67	1464.0	36.66	−1303.0	−84.5	55.41	21.75	40.45	965.6	309.0	.140
95.0	46.83	1466.0	36.71	−1299.0	−22.8	55.44	21.73	40.37	971.6	309.0	.140
95.5	49.00	1468.0	36.76	−1294.0	39.0	55.47	21.72	40.28	977.5	310.0	.140
96.0	51.17	1470.0	36.80	−1289.0	101.0	55.50	21.71	40.21	983.3	310.0	.140
96.5	53.35	1472.0	36.85	−1285.0	163.1	55.53	21.71	40.13	989.1	311.0	.141
97.0	55.54	1474.0	36.90	−1280.0	225.3	55.56	21.71	40.07	994.7	312.0	.141
97.5	57.74	1476.0	36.95	−1275.0	287.7	55.59	21.71	40.00	1000.0	312.0	.141
98.0	59.94	1478.0	36.99	−1270.0	350.2	55.62	21.72	39.94	1006.0	313.0	.141
98.5	62.15	1480.0	37.04	−1265.0	412.9	55.65	21.73	39.88	1011.0	313.0	.141
99.0	64.36	1482.0	37.09	−1260.0	475.7	55.68	21.74	39.83	1017.0	314.0	.142
99.5	66.59	1483.0	37.13	−1255.0	538.7	55.71	21.76	39.78	1022.0	314.0	.142
100.0	68.82	1485.0	37.18	−1249.0	601.8	55.74	21.77	39.73	1027.0	315.0	.142
100.5	71.05	1487.0	37.22	−1244.0	665.0	55.77	21.79	39.69	1032.0	315.0	.142
101.0	73.29	1489.0	37.27	−1238.0	728.4	55.80	21.81	39.65	1037.0	316.0	.143
101.5	75.54	1490.0	37.31	−1233.0	791.9	55.84	21.83	39.61	1042.0	316.0	.143
102.0	77.80	1492.0	37.35	−1227.0	855.6	55.87	21.86	39.57	1047.0	317.0	.143
102.5	80.06	1494.0	37.40	−1221.0	919.4	55.90	21.88	39.53	1052.0	317.0	.144
103.0	82.33	1496.0	37.44	−1216.0	983.3	55.93	21.91	39.50	1057.0	317.0	.144
103.5	84.61	1497.0	37.48	−1210.0	1047.0	55.96	21.94	39.47	1062.0	318.0	.144
104.0	86.89	1499.0	37.52	−1204.0	1112.0	56.00	21.96	39.44	1066.0	318.0	.144
104.5	89.18	1501.0	37.57	−1198.0	1176.0	56.03	21.99	39.41	1071.0	319.0	.145
105.0	91.47	1502.0	37.61	−1192.0	1241.0	56.06	22.03	39.38	1076.0	319.0	.145
105.5	93.77	1504.0	37.65	−1186.0	1305.0	56.10	22.06	39.35	1080.0	320.0	.145
106.0	96.08	1506.0	37.69	−1179.0	1370.0	56.13	22.09	39.33	1085.0	320.0	.145
106.5	98.40	1507.0	37.73	−1173.0	1435.0	56.16	22.12	39.30	1089.0	320.0	.146

[a]Triple point.

Thermophysical properties of argon—Continued

T K	Density kg/m³	Density mol/dm³	E J/mol	H J/mol	S J/(mol·K)	C_v J/(mol·K)	C_p J/(mol·K)	Sound m/s	Visc. μPa·s	Therm. W/(m·K)
				.08 MPa isobar						
83.80[a]	1415.0	35.42	−1375.0	−1373.0	54.82	23.60	44.82	821.8	289.0	.135
85.112[b]	1406.0	35.20	−1314.0	−1312.0	55.55	23.03	44.79	818.6	277.0	.132
85.112[b]	4.683	.1172	4508.0	5191.0	131.9	13.13	22.48	169.0	7.07	.00568
90.0	4.378	.1096	4575.0	5305.0	133.2	12.98	22.13	174.3	7.45	.00596
95.0	4.132	.1034	4641.0	5415.0	134.4	12.87	21.88	179.5	7.84	.00625
100.0	3.914	.09798	4707.0	5523.0	135.5	12.79	21.69	184.5	8.23	.00654
105.0	3.719	.09309	4772.0	5631.0	136.6	12.72	21.54	189.3	8.62	.00684
110.0	3.542	.08868	4837.0	5739.0	137.6	12.68	21.42	194.0	9.01	.00714
115.0	3.383	.08468	4901.0	5846.0	138.5	12.64	21.33	198.6	9.41	.00744
120.0	3.237	.08104	4965.0	5952.0	139.4	12.61	21.26	203.0	9.80	.00774
130.0	2.982	.07464	5092.0	6164.0	141.1	12.57	21.15	211.6	10.6	.00835
140.0	2.764	.06920	5219.0	6375.0	142.7	12.54	21.07	219.7	11.4	.00895
150.0	2.577	.06451	5345.0	6586.0	144.2	12.52	21.02	227.6	12.2	.00956
160.0	2.413	.06042	5471.0	6796.0	145.5	12.51	20.98	235.2	12.9	.0102
170.0	2.270	.05682	5597.0	7005.0	146.8	12.50	20.95	242.5	13.7	.0108
180.0	2.142	.05363	5723.0	7215.0	148.0	12.50	20.92	249.6	14.5	.0114
190.0	2.029	.05078	5848.0	7424.0	149.1	12.49	20.90	256.5	15.2	.0119
200.0	1.926	.04822	5974.0	7633.0	150.2	12.49	20.89	263.3	16.0	.0125
210.0	1.834	.04591	6099.0	7841.0	151.2	12.49	20.88	269.8	16.7	.0131
220.0	1.750	.04381	6224.0	8050.0	152.2	12.48	20.87	276.2	17.5	.0137
230.0	1.674	.04190	6349.0	8259.0	153.1	12.48	20.86	282.4	18.2	.0142
240.0	1.604	.04014	6474.0	8467.0	154.0	12.48	20.85	288.5	18.9	.0148
250.0	1.539	.03853	6599.0	8676.0	154.8	12.48	20.85	294.5	19.6	.0153
260.0	1.480	.03704	6724.0	8884.0	155.6	12.48	20.84	300.3	20.3	.0158
270.0	1.425	.03567	6849.0	9093.0	156.4	12.48	20.84	306.1	20.9	.0164
280.0	1.374	.03439	6974.0	9301.0	157.2	12.48	20.83	311.7	21.6	.0169
300.0	1.282	.03209	7224.0	9717.0	158.6	12.48	20.82	322.7	22.9	.0179
320.0	1.202	.03008	7474.0	10130.0	160.0	12.48	20.82	333.3	24.2	.0189
340.0	1.131	.02831	7724.0	10550.0	161.2	12.48	20.81	343.5	25.4	.0199
360.0	1.068	.02673	7974.0	10970.0	162.4	12.48	20.81	353.5	26.6	.0208
380.0	1.012	.02532	8223.0	11380.0	163.6	12.48	20.81	363.2	27.8	.0217
400.0	.9610	.02406	8473.0	11800.0	164.6	12.48	20.81	372.6	28.9	.0226
				.10 MPa isobar						
83.81[a]	1415.0	35.42	−1375.0	−1373.0	54.82	23.60	44.81	821.8	289.0	.135
87.158[b]	1394.0	34.89	−1223.0	−1220.0	56.61	22.30	44.78	813.7	261.0	.128
87.158[b]	5.713	.1430	4526.0	5225.0	130.5	13.21	22.74	170.6	7.24	.00585
90.0	5.509	.1379	4565.0	5291.0	131.3	13.11	22.50	173.7	7.46	.00601
95.0	5.194	.1300	4633.0	5402.0	132.5	12.97	22.18	179.0	7.85	.00629
100.0	4.915	.1230	4700.0	5512.0	133.6	12.87	21.93	184.0	8.24	.00658
105.0	4.667	.1168	4766.0	5622.0	134.7	12.79	21.74	188.9	8.63	.00687
110.0	4.444	.1112	4831.0	5730.0	135.7	12.73	21.59	193.7	9.02	.00717
115.0	4.242	.1062	4896.0	5837.0	136.6	12.68	21.47	198.3	9.42	.00747
120.0	4.058	.1016	4960.0	5945.0	137.5	12.65	21.38	202.7	9.81	.00777
130.0	3.735	.09350	5088.0	6158.0	139.2	12.59	21.24	211.4	10.6	.00837
140.0	3.461	.08665	5215.0	6370.0	140.8	12.56	21.14	219.6	11.4	.00898
150.0	3.226	.08075	5342.0	6581.0	142.3	12.54	21.07	227.5	12.2	.00958
160.0	3.020	.07561	5468.0	6791.0	143.6	12.52	21.02	235.1	13.0	.0102
170.0	2.840	.07109	5594.0	7001.0	144.9	12.51	20.99	242.4	13.7	.0108
180.0	2.680	.06709	5720.0	7211.0	146.1	12.50	20.96	249.6	14.5	.0114
190.0	2.537	.06352	5846.0	7420.0	147.2	12.50	20.93	256.5	15.2	.0120
200.0	2.409	.06031	5971.0	7629.0	148.3	12.49	20.91	263.2	16.0	.0125
210.0	2.294	.05742	6097.0	7839.0	149.3	12.49	20.90	269.8	16.7	.0131
220.0	2.189	.05479	6222.0	8047.0	150.3	12.49	20.89	276.2	17.5	.0137
230.0	2.093	.05239	6347.0	8256.0	151.2	12.49	20.88	282.4	18.2	.0142
240.0	2.005	.05019	6473.0	8465.0	152.1	12.48	20.87	288.5	18.9	.0148
250.0	1.924	.04817	6598.0	8674.0	153.0	12.48	20.86	294.5	19.6	.0153
260.0	1.850	.04631	6723.0	8882.0	153.8	12.48	20.85	300.3	20.3	.0158
270.0	1.781	.04459	6848.0	9091.0	154.6	12.48	20.85	306.1	20.9	.0164

Thermophysical properties of argon—Continued

T K	Density kg/m³	Density mol/dm³	E J/mol	H J/mol	S J/(mol·K)	C_v J/(mol·K)	C_p J/(mol·K)	Sound m/s	Visc. μPa·s	Therm. W/(m·K)
280.0	1.717	.04299	6973.0	9299.0	155.3	12.48	20.84	311.7	21.6	.0169
300.0	1.603	.04012	7223.0	9716.0	156.8	12.48	20.83	322.7	22.9	.0179
320.0	1.502	.03760	7473.0	10130.0	158.1	12.48	20.83	333.3	24.2	.0189
340.0	1.414	.03539	7723.0	10550.0	159.4	12.48	20.82	343.5	25.4	.0199
360.0	1.335	.03342	7973.0	10970.0	160.6	12.48	20.82	353.5	26.6	.0208
380.0	1.265	.03165	8223.0	11380.0	161.7	12.48	20.81	363.2	27.8	.0217
400.0	1.201	.03007	8472.0	11800.0	162.8	12.48	20.81	372.6	28.9	.0226
				.101325 MPa isobar						
83.81[a]	1415.0	35.42	−1375.0	−1373.0	54.82	23.60	44.81	821.8	289.0	.135
87.282[b]	1393.0	34.88	−1217.0	−1214.0	56.67	22.26	44.78	813.4	260.0	.128
87.282[b]	5.781	.1447	4527.0	5227.0	130.4	13.22	22.76	170.7	7.25	.00586
90.0	5.584	.1398	4565.0	5290.0	131.1	13.12	22.53	173.6	7.46	.00601
95.0	5.265	.1318	4633.0	5401.0	132.4	12.98	22.20	178.9	7.85	.00630
100.0	4.982	.1247	4699.0	5512.0	133.5	12.87	21.94	184.0	8.24	.00658
105.0	4.730	.1184	4765.0	5621.0	134.6	12.79	21.75	188.9	8.63	.00688
110.0	4.504	.1127	4830.0	5729.0	135.6	12.73	21.60	193.6	9.02	.00717
115.0	4.299	.1076	4895.0	5837.0	136.5	12.69	21.48	198.2	9.42	.00747
120.0	4.112	.1029	4960.0	5944.0	137.4	12.65	21.39	202.7	9.81	.00777
130.0	3.785	.09476	5088.0	6157.0	139.1	12.60	21.24	211.3	10.6	.00837
140.0	3.508	.08781	5215.0	6369.0	140.7	12.56	21.15	219.6	11.4	.00898
150.0	3.269	.08182	5342.0	6580.0	142.2	12.54	21.08	227.5	12.2	.00958
160.0	3.061	.07662	5468.0	6791.0	143.5	12.52	21.03	235.1	13.0	.0102
170.0	2.878	.07204	5594.0	7001.0	144.8	12.51	20.99	242.4	13.7	.0108
180.0	2.716	.06798	5720.0	7211.0	146.0	12.50	20.96	249.6	14.5	.0114
190.0	2.571	.06436	5846.0	7420.0	147.1	12.50	20.94	256.5	15.2	.0120
200.0	2.441	.06111	5971.0	7629.0	148.2	12.49	20.92	263.2	16.0	.0125
210.0	2.324	.05818	6097.0	7838.0	149.2	12.49	20.90	269.8	16.7	.0131
220.0	2.218	.05551	6222.0	8047.0	150.2	12.49	20.89	276.2	17.5	.0137
230.0	2.121	.05308	6347.0	8256.0	151.1	12.49	20.88	282.4	18.2	.0142
240.0	2.032	.05086	6473.0	8465.0	152.0	12.48	20.87	288.5	18.9	.0148
250.0	1.950	.04881	6598.0	8674.0	152.9	12.48	20.86	294.5	19.6	.0153
260.0	1.875	.04693	6723.0	8882.0	153.7	12.48	20.85	300.3	20.3	.0158
270.0	1.805	.04518	6848.0	9091.0	154.5	12.48	20.85	306.1	20.9	.0164
280.0	1.740	.04356	6973.0	9299.0	155.2	12.48	20.84	311.7	21.6	.0169
300.0	1.624	.04065	7223.0	9716.0	156.7	12.48	20.83	322.7	22.9	.0179
320.0	1.522	.03810	7473.0	10130.0	158.0	12.48	20.83	333.3	24.2	.0189
340.0	1.432	.03585	7723.0	10550.0	159.3	12.48	20.82	343.5	25.4	.0199
360.0	1.353	.03386	7973.0	10970.0	160.5	12.48	20.82	353.5	26.6	.0208
380.0	1.281	.03207	8223.0	11380.0	161.6	12.48	20.81	363.2	27.8	.0217
400.0	1.217	.03047	8472.0	11800.0	162.7	12.48	20.81	372.7	28.9	.0226
				.20 MPa isobar						
83.83[a]	1415.0	35.43	−1375.0	−1370.0	54.82	23.59	44.80	822.2	289.0	.135
90.0	1377.0	34.48	−1099.0	−1094.0	58.00	21.52	44.89	805.0	241.0	.124
94.246[b]	1349.0	33.78	−905.3	−899.4	60.11	20.65	45.48	783.4	213.0	.117
94.246[b]	10.77	.2696	4580.0	5322.0	126.1	13.56	23.91	175.4	7.84	.00648
95.0	10.69	.2677	4591.0	5338.0	126.3	13.52	23.82	176.2	7.90	.00652
100.0	10.07	.2522	4662.0	5455.0	127.5	13.30	23.24	181.7	8.29	.00679
105.0	9.531	.2386	4732.0	5571.0	128.6	13.13	22.81	186.9	8.68	.00706
110.0	9.049	.2265	4801.0	5684.0	129.6	13.00	22.48	191.9	9.07	.00734
115.0	8.618	.2157	4868.0	5795.0	130.6	12.91	22.22	196.8	9.46	.00763
120.0	8.228	.2060	4935.0	5906.0	131.6	12.83	22.02	201.4	9.86	.00792
125.0	7.875	.1971	5001.0	6016.0	132.5	12.77	21.85	205.9	10.2	.00821
130.0	7.552	.1891	5067.0	6125.0	133.3	12.72	21.72	210.3	10.6	.00850
140.0	6.983	.1748	5197.0	6341.0	134.9	12.65	21.52	218.8	11.4	.00909
150.0	6.497	.1626	5325.0	6555.0	136.4	12.60	21.37	226.9	12.2	.00969
160.0	6.076	.1521	5453.0	6768.0	137.8	12.57	21.27	234.6	13.0	.0103
170.0	5.708	.1429	5581.0	6981.0	139.1	12.55	21.19	242.1	13.8	.0109
180.0	5.382	.1347	5708.0	7192.0	140.3	12.53	21.13	249.3	14.5	.0114

Thermophysical properties of argon—Continued

T K	Density kg/m^3	Density mol/dm^3	E J/mol	H J/mol	S J/(mol·K)	C_v J/(mol·K)	C_p J/(mol·K)	Sound m/s	Visc. μPa·s	Therm. W/(m·K)
190.0	5.092	.1275	5834.0	7403.0	141.4	12.52	21.08	256.3	15.3	.0120
200.0	4.833	.1210	5961.0	7614.0	142.5	12.51	21.05	263.0	16.0	.0126
210.0	4.599	.1151	6087.0	7824.0	143.5	12.51	21.01	269.6	16.8	.0132
220.0	4.386	.1098	6213.0	8034.0	144.5	12.50	20.99	276.1	17.5	.0137
230.0	4.193	.1050	6339.0	8244.0	145.4	12.50	20.97	282.4	18.2	.0143
240.0	4.016	.1005	6464.0	8454.0	146.3	12.50	20.95	288.5	18.9	.0148
250.0	3.854	.09648	6590.0	8663.0	147.2	12.49	20.93	294.5	19.6	.0154
260.0	3.704	.09273	6715.0	8872.0	148.0	12.49	20.92	300.4	20.3	.0159
270.0	3.566	.08927	6841.0	9081.0	148.8	12.49	20.91	306.1	21.0	.0164
280.0	3.438	.08605	6966.0	9290.0	149.5	12.49	20.90	311.8	21.6	.0169
300.0	3.207	.08028	7217.0	9708.0	151.0	12.49	20.88	322.8	22.9	.0180
320.0	3.005	.07524	7467.0	10130.0	152.3	12.48	20.87	333.4	24.2	.0189
340.0	2.828	.07079	7718.0	10540.0	153.6	12.48	20.86	343.7	25.4	.0199
360.0	2.670	.06684	7968.0	10960.0	154.8	12.48	20.85	353.7	26.6	.0208
380.0	2.529	.06331	8218.0	11380.0	155.9	12.48	20.84	363.4	27.8	.0218
400.0	2.402	.06014	8468.0	11790.0	157.0	12.48	20.84	372.8	28.9	.0227

.30 MPa isobar

T K	Density kg/m^3	Density mol/dm^3	E J/mol	H J/mol	S J/(mol·K)	C_v J/(mol·K)	C_p J/(mol·K)	Sound m/s	Visc. μPa·s	Therm. W/(m·K)
83.86[a]	1415.0	35.43	−1375.0	−1367.0	54.82	23.58	44.78	822.5	289.0	.135
90.0	1378.0	34.49	−1101.0	−1092.0	57.98	21.52	44.87	805.4	241.0	.124
98.995[b]	1318.0	32.99	−689.3	−680.2	62.34	19.95	46.32	755.0	188.0	.110
98.995[b]	15.67	.3923	4608.0	5372.0	123.4	13.83	24.94	178.1	8.27	.00695
100.0	15.52	.3884	4623.0	5395.0	123.7	13.77	24.78	179.2	8.34	.00700
105.0	14.62	.3660	4697.0	5517.0	124.9	13.50	24.03	184.8	8.73	.00726
110.0	13.84	.3463	4769.0	5636.0	126.0	13.30	23.48	190.1	9.12	.00752
115.0	13.14	.3290	4840.0	5752.0	127.0	13.14	23.05	195.2	9.51	.00780
120.0	12.52	.3135	4909.0	5866.0	128.0	13.02	22.71	200.1	9.90	.00807
125.0	11.96	.2995	4977.0	5979.0	128.9	12.93	22.45	204.8	10.3	.00836
130.0	11.46	.2868	5045.0	6091.0	129.8	12.85	22.23	209.3	10.7	.00864
140.0	10.57	.2646	5178.0	6311.0	131.4	12.74	21.91	218.0	11.5	.00922
150.0	9.816	.2457	5309.0	6529.0	132.9	12.67	21.69	226.2	12.2	.00979
160.0	9.168	.2295	5438.0	6745.0	134.3	12.62	21.52	234.1	13.0	.0104
170.0	8.604	.2154	5567.0	6960.0	135.6	12.59	21.40	241.7	13.8	.0110
180.0	8.107	.2029	5695.0	7174.0	136.8	12.57	21.31	249.0	14.6	.0115
190.0	7.665	.1919	5823.0	7386.0	138.0	12.55	21.24	256.0	15.3	.0121
200.0	7.270	.1820	5950.0	7598.0	139.1	12.54	21.18	262.9	16.0	.0127
210.0	6.915	.1731	6077.0	7810.0	140.1	12.53	21.13	269.5	16.8	.0132
220.0	6.594	.1651	6203.0	8021.0	141.1	12.52	21.09	276.0	17.5	.0138
230.0	6.301	.1577	6330.0	8232.0	142.0	12.51	21.06	282.3	18.2	.0143
240.0	6.034	.1510	6456.0	8442.0	142.9	12.51	21.03	288.5	18.9	.0149
250.0	5.789	.1449	6582.0	8652.0	143.8	12.50	21.01	294.5	19.6	.0154
260.0	5.563	.1393	6708.0	8862.0	144.6	12.50	20.99	300.4	20.3	.0159
270.0	5.354	.1340	6834.0	9072.0	145.4	12.50	20.97	306.2	21.0	.0165
280.0	5.161	.1292	6960.0	9282.0	146.2	12.50	20.96	311.8	21.6	.0170
300.0	4.813	.1205	7211.0	9701.0	147.6	12.49	20.93	322.9	22.9	.0180
320.0	4.510	.1129	7462.0	10120.0	148.9	12.49	20.91	333.5	24.2	.0190
340.0	4.243	.1062	7713.0	10540.0	150.2	12.49	20.90	343.8	25.4	.0199
360.0	4.006	.1003	7963.0	10950.0	151.4	12.49	20.88	353.8	26.6	.0209
380.0	3.794	.09498	8214.0	11370.0	152.5	12.49	20.87	363.5	27.8	.0218
400.0	3.604	.09021	8464.0	11790.0	153.6	12.49	20.86	373.0	28.9	.0227

.40 MPa isobar

T K	Density kg/m^3	Density mol/dm^3	E J/mol	H J/mol	S J/(mol·K)	C_v J/(mol·K)	C_p J/(mol·K)	Sound m/s	Visc. μPa·s	Therm. W/(m·K)
83.88[a]	1415.0	35.43	−1375.0	−1364.0	54.82	23.57	44.76	822.9	289.0	.135
90.0	1378.0	34.49	−1102.0	−1090.0	57.97	21.53	44.85	805.8	241.0	.124
100.0	1312.0	32.85	−647.0	−634.8	62.77	19.85	46.44	749.7	184.0	.109
102.681[b]	1293.0	32.36	−519.1	−506.8	64.03	19.53	47.12	730.1	171.0	.105
102.681[b]	20.49	.5128	4624.0	5404.0	121.6	14.06	25.91	179.9	8.61	.00735
105.0	19.97	.4998	4660.0	5460.0	122.1	13.90	25.45	182.6	8.79	.00747
110.0	18.83	.4712	4737.0	5585.0	123.3	13.61	24.60	188.2	9.17	.00771
115.0	17.83	.4463	4811.0	5707.0	124.3	13.39	23.96	193.6	9.56	.00797

Thermophysical properties of argon—Continued

T K	Density kg/m³	Density mol/dm³	E J/mol	H J/mol	S J/(mol·K)	C_v J/(mol·K)	C_p J/(mol·K)	Sound m/s	Visc. μPa·s	Therm. W/(m·K)
120.0	16.95	.4243	4883.0	5825.0	125.4	13.23	23.47	198.7	9.95	.00823
125.0	16.16	.4046	4953.0	5942.0	126.3	13.09	23.09	203.5	10.3	.00850
130.0	15.45	.3868	5022.0	6056.0	127.2	12.99	22.78	208.2	10.7	.00878
140.0	14.22	.3560	5158.0	6282.0	128.9	12.84	22.32	217.1	11.5	.00934
150.0	13.19	.3301	5291.0	6503.0	130.4	12.74	22.01	225.6	12.3	.00990
160.0	12.30	.3079	5423.0	6722.0	131.8	12.68	21.78	233.6	13.1	.0105
170.0	11.53	.2886	5553.0	6939.0	133.1	12.63	21.62	241.3	13.8	.0110
180.0	10.85	.2717	5682.0	7155.0	134.4	12.60	21.49	248.7	14.6	.0116
190.0	10.26	.2567	5811.0	7369.0	135.5	12.57	21.39	255.8	15.3	.0122
200.0	9.723	.2434	5939.0	7583.0	136.6	12.56	21.31	262.7	16.1	.0127
210.0	9.243	.2314	6067.0	7795.0	137.7	12.54	21.25	269.4	16.8	.0133
220.0	8.810	.2205	6194.0	8008.0	138.6	12.53	21.20	275.9	17.5	.0138
230.0	8.417	.2107	6321.0	8219.0	139.6	12.53	21.15	282.3	18.2	.0144
240.0	8.058	.2017	6448.0	8431.0	140.5	12.52	21.12	288.5	18.9	.0149
250.0	7.729	.1935	6574.0	8642.0	141.3	12.51	21.08	294.5	19.6	.0155
260.0	7.426	.1859	6700.0	8852.0	142.2	12.51	21.06	300.4	20.3	.0160
270.0	7.146	.1789	6827.0	9063.0	143.0	12.51	21.03	306.2	21.0	.0165
280.0	6.887	.1724	6953.0	9273.0	143.7	12.50	21.01	311.9	21.7	.0170
300.0	6.422	.1608	7205.0	9693.0	145.2	12.50	20.98	322.9	23.0	.0180
320.0	6.016	.1506	7456.0	10110.0	146.5	12.50	20.95	333.6	24.2	.0190
340.0	5.659	.1417	7707.0	10530.0	147.8	12.49	20.93	343.9	25.5	.0200
360.0	5.342	.1337	7958.0	10950.0	149.0	12.49	20.91	353.9	26.7	.0209
380.0	5.059	.1266	8209.0	11370.0	150.1	12.49	20.90	363.7	27.8	.0218
400.0	4.805	.1203	8460.0	11790.0	151.2	12.49	20.89	373.1	29.0	.0227
					.60 MPa isobar					
83.93[a]	1416.0	35.44	−1375.0	−1358.0	54.83	23.55	44.72	823.6	290.0	.135
90.0	1379.0	34.51	−1104.0	−1087.0	57.94	21.54	44.81	806.7	242.0	.124
100.0	1313.0	32.87	−650.1	−631.9	62.74	19.86	46.38	750.8	184.0	.109
105.0	1278.0	31.98	−415.9	−397.1	65.03	19.32	47.57	715.2	162.0	.102
108.385[b]	1252.0	31.33	−250.5	−231.4	66.58	18.98	48.65	688.0	148.0	.0977
108.385[b]	30.02	.7514	4639.0	5438.0	118.9	14.46	27.80	182.2	9.17	.00804
110.0	29.53	.7392	4666.0	5477.0	119.2	14.32	27.35	184.2	9.29	.00812
115.0	27.77	.6951	4748.0	5611.0	120.4	13.95	26.13	190.1	9.67	.00834
120.0	26.25	.6572	4826.0	5739.0	121.5	13.67	25.23	195.7	10.1	.00857
125.0	24.93	.6239	4902.0	5864.0	122.5	13.45	24.54	201.0	10.4	.00882
130.0	23.75	.5945	4975.0	5985.0	123.5	13.29	24.00	206.0	10.8	.00907
135.0	22.69	.5680	5047.0	6104.0	124.4	13.15	23.57	210.8	11.2	.00933
140.0	21.74	.5442	5118.0	6221.0	125.2	13.05	23.22	215.5	11.6	.00959
150.0	20.08	.5026	5256.0	6450.0	126.8	12.89	22.70	224.3	12.4	.0101
160.0	18.67	.4675	5392.0	6675.0	128.3	12.79	22.33	232.6	13.1	.0107
170.0	17.47	.4373	5525.0	6897.0	129.6	12.72	22.07	240.5	13.9	.0112
180.0	16.42	.4110	5657.0	7117.0	130.9	12.66	21.87	248.1	14.6	.0118
190.0	15.49	.3878	5787.0	7335.0	132.0	12.63	21.71	255.3	15.4	.0123
200.0	14.67	.3672	5917.0	7551.0	133.1	12.60	21.59	262.4	16.1	.0129
210.0	13.94	.3488	6046.0	7766.0	134.2	12.58	21.49	269.2	16.9	.0134
220.0	13.27	.3322	6175.0	7981.0	135.2	12.57	21.41	275.8	17.6	.0140
230.0	12.67	.3172	6303.0	8195.0	136.1	12.55	21.34	282.2	18.3	.0145
240.0	12.12	.3035	6431.0	8408.0	137.0	12.54	21.28	288.4	19.0	.0150
250.0	11.62	.2910	6558.0	8620.0	137.9	12.54	21.24	294.5	19.7	.0156
260.0	11.16	.2795	6685.0	8832.0	138.7	12.53	21.20	300.5	20.4	.0161
270.0	10.74	.2688	6812.0	9044.0	139.5	12.53	21.16	306.3	21.0	.0166
280.0	10.35	.2590	6939.0	9256.0	140.3	12.52	21.13	312.0	21.7	.0171
290.0	9.983	.2499	7066.0	9467.0	141.1	12.52	21.10	317.6	22.3	.0176
300.0	9.644	.2414	7192.0	9678.0	141.8	12.51	21.08	323.1	23.0	.0181
320.0	9.031	.2261	7445.0	10100.0	143.1	12.51	21.04	333.8	24.3	.0191
340.0	8.493	.2126	7697.0	10520.0	144.4	12.51	21.00	344.2	25.5	.0200
360.0	8.016	.2007	7949.0	10940.0	145.6	12.50	20.98	354.2	26.7	.0210
380.0	7.590	.1900	8200.0	11360.0	146.7	12.50	20.96	364.0	27.8	.0219
400.0	7.207	.1804	8452.0	11780.0	147.8	12.50	20.94	373.4	29.0	.0228

Thermophysical properties of argon—Continued

T K	Density kg/m³	Density mol/dm³	E J/mol	H J/mol	S J/(mol·K)	C_v J/(mol·K)	C_p J/(mol·K)	Sound m/s	Visc. μPa·s	Therm. W/(m·K)
				.80 MPa isobar						
83.98[a]	1416.0	35.45	−1375.0	−1352.0	54.83	23.54	44.68	824.2	290.0	.135
90.0	1379.0	34.53	−1107.0	−1083.0	57.92	21.55	44.76	807.6	242.0	.124
100.0	1314.0	32.89	−653.3	−629.0	62.70	19.88	46.31	751.9	185.0	.109
105.0	1279.0	32.01	−419.6	−394.6	64.99	19.33	47.48	716.5	162.0	.102
110.0	1241.0	31.07	−179.3	−153.6	67.23	18.87	49.00	677.5	143.0	.0959
112.839[b]	1218.0	30.49	−35.8	−9.5	68.53	18.62	50.19	652.7	133.0	.0922
112.839[b]	39.57	.9905	4641.0	5449.0	116.9	14.80	29.70	183.6	9.63	.00866
115.0	38.63	.9671	4679.0	5506.0	117.4	14.59	28.92	186.3	9.80	.00874
120.0	36.27	.9079	4765.0	5646.0	118.6	14.17	27.38	192.5	10.2	.00894
125.0	34.25	.8574	4847.0	5780.0	119.7	13.85	26.26	198.3	10.5	.00915
130.0	32.50	.8135	4926.0	5909.0	120.7	13.61	25.41	203.7	10.9	.00938
135.0	30.95	.7748	5002.0	6035.0	121.6	13.42	24.75	208.8	11.3	.00962
140.0	29.57	.7402	5076.0	6157.0	122.5	13.27	24.23	213.7	11.7	.00986
145.0	28.32	.7090	5149.0	6277.0	123.4	13.14	23.80	218.4	12.1	.0101
150.0	27.19	.6808	5220.0	6395.0	124.2	13.05	23.45	222.9	12.4	.0104
160.0	25.22	.6312	5360.0	6627.0	125.7	12.90	22.92	231.6	13.2	.0109
170.0	23.53	.5891	5496.0	6854.0	127.0	12.80	22.54	239.7	14.0	.0114
180.0	22.08	.5527	5631.0	7078.0	128.3	12.73	22.26	247.5	14.7	.0119
190.0	20.80	.5208	5764.0	7300.0	129.5	12.68	22.04	254.9	15.5	.0125
200.0	19.68	.4926	5895.0	7519.0	130.6	12.65	21.87	262.0	16.2	.0130
210.0	18.67	.4675	6026.0	7737.0	131.7	12.62	21.73	268.9	16.9	.0136
220.0	17.77	.4449	6156.0	7954.0	132.7	12.60	21.62	275.6	17.6	.0141
230.0	16.96	.4245	6285.0	8170.0	133.7	12.58	21.53	282.1	18.3	.0146
240.0	16.22	.4059	6414.0	8385.0	134.6	12.57	21.45	288.4	19.0	.0151
250.0	15.54	.3890	6542.0	8599.0	135.5	12.56	21.39	294.6	19.7	.0157
260.0	14.92	.3735	6670.0	8812.0	136.3	12.55	21.33	300.6	20.4	.0162
270.0	14.35	.3592	6798.0	9026.0	137.1	12.54	21.29	306.4	21.1	.0167
280.0	13.82	.3459	6926.0	9238.0	137.9	12.54	21.24	312.2	21.7	.0172
290.0	13.33	.3337	7053.0	9451.0	138.6	12.53	21.21	317.8	22.4	.0177
300.0	12.87	.3223	7180.0	9662.0	139.3	12.53	21.18	323.3	23.0	.0182
320.0	12.05	.3017	7434.0	10090.0	140.7	12.52	21.12	334.1	24.3	.0192
340.0	11.33	.2836	7687.0	10510.0	142.0	12.52	21.08	344.4	25.5	.0201
360.0	10.69	.2676	7939.0	10930.0	143.2	12.51	21.04	354.5	26.7	.0210
380.0	10.12	.2534	8192.0	11350.0	144.3	12.51	21.01	364.3	27.9	.0219
400.0	9.610	.2406	8443.0	11770.0	145.4	12.51	20.99	373.7	29.0	.0228
				1.00 MPa isobar						
84.03[a]	1416.0	35.45	−1374.0	−1346.0	54.83	23.52	44.64	824.9	290.0	.135
90.0	1380.0	34.54	−1109.0	−1080.0	57.89	21.56	44.72	808.4	243.0	.124
100.0	1315.0	32.91	−656.5	−626.1	62.67	19.89	46.25	753.0	185.0	.109
105.0	1280.0	32.03	−423.3	−392.1	64.96	19.35	47.40	717.8	163.0	.103
110.0	1242.0	31.10	−183.7	−151.5	67.19	18.89	48.89	679.0	143.0	.0961
115.0	1202.0	30.10	64.4	97.7	69.41	18.47	50.90	636.9	127.0	.0898
116.550[b]	1188.0	29.75	147.3	180.9	70.13	18.34	51.81	621.8	121.0	.0878
116.550[b]	49.24	1.233	4634.0	5445.0	115.3	15.12	31.69	184.4	10.1	.00923
120.0	47.18	1.181	4699.0	5545.0	116.1	14.74	30.13	189.1	10.3	.00934
125.0	44.26	1.108	4789.0	5691.0	117.3	14.30	28.36	195.4	10.7	.00951
130.0	41.78	1.046	4874.0	5830.0	118.4	13.96	27.08	201.2	11.0	.00971
135.0	39.63	.9921	4955.0	5963.0	119.4	13.70	26.11	206.7	11.4	.00992
140.0	37.75	.9449	5033.0	6091.0	120.3	13.50	25.36	211.9	11.8	.0101
145.0	36.06	.9028	5109.0	6216.0	121.2	13.34	24.76	216.9	12.2	.0104
150.0	34.55	.8649	5183.0	6339.0	122.1	13.21	24.28	221.6	12.5	.0106
160.0	31.93	.7994	5327.0	6578.0	123.6	13.02	23.56	230.5	13.3	.0111
170.0	29.73	.7442	5467.0	6811.0	125.0	12.89	23.04	238.9	14.0	.0116
180.0	27.84	.6969	5604.0	7039.0	126.3	12.80	22.67	246.9	14.8	.0121
190.0	26.19	.6557	5739.0	7265.0	127.5	12.74	22.38	254.4	15.5	.0126
200.0	24.75	.6195	5873.0	7487.0	128.7	12.69	22.16	261.7	16.3	.0132
210.0	23.46	.5873	6005.0	7708.0	129.7	12.66	21.99	268.7	17.0	.0137
220.0	22.31	.5585	6137.0	7927.0	130.8	12.63	21.84	275.5	17.7	.0142
230.0	21.27	.5326	6267.0	8145.0	131.7	12.61	21.73	282.0	18.4	.0147

Thermophysical properties of argon—Continued

T K	Density kg/m^3	Density mol/dm^3	E J/mol	H J/mol	S J/(mol·K)	C_v J/(mol·K)	C_p J/(mol·K)	Sound m/s	Visc. μPa·s	Therm. W/(m·K)
240.0	20.33	.5090	6397.0	8362.0	132.7	12.59	21.63	288.4	19.1	.0153
250.0	19.48	.4875	6526.0	8577.0	133.5	12.58	21.55	294.6	19.8	.0158
260.0	18.69	.4679	6655.0	8793.0	134.4	12.57	21.47	300.6	20.4	.0163
270.0	17.97	.4498	6784.0	9007.0	135.2	12.56	21.41	306.5	21.1	.0168
280.0	17.30	.4331	6912.0	9221.0	136.0	12.55	21.36	312.3	21.8	.0173
290.0	16.68	.4177	7040.0	9434.0	136.7	12.55	21.31	318.0	22.4	.0178
300.0	16.11	.4033	7168.0	9647.0	137.4	12.54	21.27	323.5	23.1	.0183
320.0	15.08	.3774	7422.0	10070.0	138.8	12.54	21.21	334.3	24.3	.0192
340.0	14.17	.3547	7676.0	10500.0	140.1	12.53	21.15	344.7	25.5	.0202
360.0	13.37	.3346	7930.0	10920.0	141.3	12.52	21.11	354.8	26.7	.0211
380.0	12.65	.3167	8183.0	11340.0	142.4	12.52	21.07	364.6	27.9	.0220
400.0	12.01	.3007	8435.0	11760.0	143.5	12.52	21.04	374.1	29.0	.0229
				1.50 MPa isobar						
84.16[a]	1417.0	35.47	−1374.0	−1331.0	54.84	23.47	44.55	826.7	290.0	.135
90.0	1381.0	34.58	−1115.0	−1071.0	57.83	21.58	44.62	810.5	244.0	.125
100.0	1317.0	32.96	−664.3	−618.7	62.59	19.93	46.09	755.7	186.0	.110
110.0	1245.0	31.17	−194.5	−146.3	67.10	18.93	48.60	682.7	145.0	.0965
115.0	1206.0	30.18	51.5	101.2	69.30	18.52	50.49	641.3	128.0	.0903
120.0	1163.0	29.11	308.4	360.0	71.50	18.15	53.20	596.4	113.0	.0841
123.932[b]	1125.0	28.15	525.5	578.8	73.29	17.88	56.48	556.5	101.0	.0790
123.932[b]	74.43	1.863	4594.0	5399.0	112.2	15.86	37.42	185.4	11.0	.0106
125.0	73.47	1.839	4616.0	5432.0	112.4	15.71	36.56	187.0	11.1	.0106
126.0	72.28	1.809	4639.0	5468.0	112.7	15.55	35.71	188.5	11.1	.0106
128.0	70.07	1.754	4683.0	5538.0	113.3	15.28	34.24	191.5	11.3	.0106
130.0	68.05	1.703	4725.0	5605.0	113.8	15.04	33.00	194.4	11.4	.0107
135.0	63.68	1.594	4823.0	5764.0	115.0	14.55	30.64	201.0	11.7	.0108
140.0	60.03	1.503	4914.0	5913.0	116.1	14.18	28.96	207.1	12.1	.0109
145.0	56.89	1.424	5001.0	6054.0	117.1	13.89	27.70	212.7	12.4	.0111
150.0	54.16	1.356	5084.0	6190.0	118.0	13.67	26.74	218.1	12.8	.0113
155.0	51.74	1.295	5164.0	6322.0	118.9	13.49	25.97	223.1	13.2	.0115
160.0	49.57	1.241	5241.0	6450.0	119.7	13.34	25.36	227.9	13.5	.0117
165.0	47.62	1.192	5317.0	6576.0	120.4	13.23	24.85	232.5	13.9	.0119
170.0	45.83	1.147	5391.0	6699.0	121.2	13.13	24.44	236.9	14.3	.0121
180.0	42.70	1.069	5536.0	6940.0	122.6	12.98	23.78	245.4	15.0	.0126
190.0	40.02	1.002	5678.0	7175.0	123.8	12.88	23.30	253.4	15.7	.0130
200.0	37.70	.9436	5816.0	7406.0	125.0	12.81	22.93	260.9	16.4	.0135
210.0	35.65	.8924	5953.0	7634.0	126.1	12.75	22.64	268.2	17.1	.0140
220.0	33.83	.8469	6088.0	7859.0	127.2	12.71	22.41	275.1	17.8	.0145
230.0	32.21	.8062	6222.0	8082.0	128.2	12.68	22.23	281.9	18.5	.0150
240.0	30.74	.7695	6354.0	8304.0	129.1	12.65	22.07	288.4	19.2	.0155
250.0	29.41	.7362	6486.0	8524.0	130.0	12.63	21.94	294.7	19.9	.0160
260.0	28.20	.7058	6617.0	8743.0	130.9	12.62	21.83	300.9	20.6	.0165
270.0	27.08	.6779	6748.0	8960.0	131.7	12.61	21.74	306.9	21.2	.0170
280.0	26.06	.6523	6878.0	9177.0	132.5	12.59	21.66	312.7	21.9	.0175
290.0	25.11	.6286	7007.0	9394.0	133.2	12.59	21.58	318.4	22.5	.0180
300.0	24.24	.6067	7137.0	9609.0	134.0	12.58	21.52	324.0	23.2	.0185
320.0	22.66	.5672	7394.0	10040.0	135.4	12.57	21.42	334.9	24.4	.0194
340.0	21.28	.5327	7650.0	10470.0	136.6	12.56	21.33	345.4	25.6	.0204
360.0	20.07	.5023	7906.0	10890.0	137.9	12.55	21.27	355.5	26.8	.0213
380.0	18.99	.4753	8161.0	11320.0	139.0	12.55	21.21	365.3	28.0	.0222
400.0	18.02	.4510	8415.0	11740.0	140.1	12.54	21.16	374.9	29.1	.0230
				2.00 MPa isobar						
84.28[a]	1418.0	35.48	−1373.0	−1317.0	54.85	23.43	44.46	828.4	290.0	.135
90.0	1383.0	34.62	−1121.0	−1063.0	57.76	21.60	44.52	812.7	245.0	.125
100.0	1319.0	33.01	−671.9	−611.4	62.52	19.96	45.93	758.4	187.0	.110
110.0	1248.0	31.24	−205.0	−141.0	67.00	18.97	48.33	686.3	146.0	.0969
115.0	1209.0	30.26	38.8	104.9	69.18	18.57	50.11	645.6	129.0	.0908
120.0	1167.0	29.21	292.9	361.3	71.37	18.20	52.61	601.7	114.0	.0847

Thermophysical properties of argon—Continued

T K	Density kg/m³	Density mol/dm³	E J/mol	H J/mol	S J/(mol·K)	C_v J/(mol·K)	C_p J/(mol·K)	Sound m/s	Visc. μPa·s	Therm. W/(m·K)
125.0	1120.0	28.04	561.8	633.1	73.58	17.87	56.40	553.4	99.9	.0785
126.0	1110.0	27.78	618.0	690.0	74.04	17.81	57.40	543.1	97.2	.0772
128.0	1089.0	27.25	733.7	807.1	74.96	17.70	59.76	521.7	91.9	.0746
129.714[b]	1068.0	26.74	840.6	915.4	75.80	17.62	62.53	501.0	87.2	.0721
129.714[b]	101.7	2.547	4528.0	5313.0	109.7	16.58	44.94	185.4	11.9	.0120
130.0	101.4	2.539	4535.0	5323.0	109.8	16.53	44.56	185.8	11.9	.0120
132.0	97.53	2.441	4589.0	5408.0	110.4	16.14	41.52	189.4	12.0	.0119
134.0	94.12	2.356	4640.0	5489.0	111.0	15.80	39.15	192.7	12.1	.0119
136.0	91.08	2.280	4688.0	5565.0	111.6	15.50	37.24	195.8	12.2	.0119
138.0	88.33	2.211	4734.0	5638.0	112.1	15.24	35.66	198.8	12.3	.0119
140.0	85.83	2.149	4777.0	5708.0	112.6	15.02	34.33	201.6	12.5	.0119
145.0	80.41	2.013	4880.0	5873.0	113.8	14.55	31.79	208.2	12.8	.0120
150.0	75.88	1.900	4974.0	6027.0	114.8	14.20	29.97	214.3	13.1	.0121
155.0	72.00	1.802	5064.0	6174.0	115.8	13.92	28.62	219.9	13.5	.0122
160.0	68.61	1.718	5149.0	6314.0	116.7	13.71	27.57	225.2	13.8	.0123
165.0	65.61	1.642	5232.0	6450.0	117.5	13.53	26.73	230.2	14.1	.0125
170.0	62.92	1.575	5312.0	6582.0	118.3	13.39	26.06	235.0	14.5	.0127
175.0	60.49	1.514	5390.0	6710.0	119.1	13.27	25.50	239.5	14.8	.0129
180.0	58.28	1.459	5466.0	6837.0	119.8	13.18	25.03	243.9	15.2	.0131
190.0	54.39	1.361	5614.0	7083.0	121.1	13.03	24.30	252.3	15.9	.0135
200.0	51.06	1.278	5758.0	7323.0	122.3	12.92	23.76	260.2	16.6	.0139
210.0	48.16	1.206	5900.0	7559.0	123.5	12.85	23.34	267.7	17.3	.0144
220.0	45.61	1.142	6039.0	7790.0	124.6	12.79	23.01	274.9	18.0	.0149
230.0	43.34	1.085	6176.0	8019.0	125.6	12.75	22.75	281.8	18.7	.0153
240.0	41.31	1.034	6311.0	8246.0	126.5	12.71	22.53	288.4	19.4	.0158
250.0	39.47	.9881	6446.0	8470.0	127.5	12.69	22.35	294.9	20.0	.0163
260.0	37.80	.9463	6579.0	8693.0	128.3	12.67	22.20	301.1	20.7	.0168
270.0	36.28	.9082	6712.0	8914.0	129.2	12.65	22.07	307.2	21.4	.0173
280.0	34.88	.8732	6844.0	9134.0	130.0	12.64	21.95	313.2	22.0	.0177
290.0	33.60	.8410	6975.0	9353.0	130.7	12.62	21.86	318.9	22.6	.0182
300.0	32.40	.8112	7106.0	9571.0	131.5	12.61	21.77	324.6	23.3	.0187
320.0	30.27	.7577	7366.0	10010.0	132.9	12.60	21.63	335.6	24.5	.0196
340.0	28.41	.7112	7624.0	10440.0	134.2	12.59	21.52	346.1	25.7	.0205
360.0	26.77	.6702	7882.0	10870.0	135.4	12.58	21.43	356.3	26.9	.0214
380.0	25.32	.6339	8138.0	11290.0	136.6	12.57	21.35	366.1	28.1	.0223
400.0	24.02	.6013	8394.0	11720.0	137.7	12.56	21.29	375.7	29.2	.0232
				2.50 MPa isobar						
84.40[a]	1418.0	35.50	−1373.0	−1302.0	54.85	23.39	44.37	830.1	291.0	.135
90.0	1385.0	34.66	−1126.0	−1054.0	57.70	21.62	44.42	814.8	247.0	.125
100.0	1321.0	33.06	−679.5	−603.9	62.44	19.99	45.79	761.1	188.0	.110
110.0	1251.0	31.30	−215.3	−135.5	66.90	19.02	48.07	689.9	147.0	.0974
115.0	1212.0	30.34	26.5	108.9	69.07	18.61	49.75	649.9	130.0	.0913
120.0	1171.0	29.31	277.8	363.1	71.24	18.24	52.07	606.8	115.0	.0852
125.0	1125.0	28.16	542.7	631.4	73.43	17.92	55.52	559.9	101.0	.0791
126.0	1115.0	27.92	597.8	687.4	73.87	17.86	56.41	550.0	98.6	.0778
128.0	1095.0	27.40	711.0	802.2	74.78	17.75	58.50	529.3	93.3	.0753
130.0	1073.0	26.85	828.7	921.8	75.71	17.65	61.12	507.5	88.2	.0727
132.0	1049.0	26.26	952.0	1047.0	76.66	17.57	64.48	484.2	83.0	.0700
134.0	1023.0	25.61	1083.0	1180.0	77.66	17.51	69.01	459.0	77.8	.0672
134.535[b]	1014.0	25.40	1123.0	1222.0	77.97	17.50	70.85	450.7	76.2	.0663
134.535[b]	132.2	3.309	4440.0	5195.0	107.5	17.33	55.63	184.8	12.8	.0136
135.0	130.4	3.265	4457.0	5223.0	107.7	17.19	54.01	185.9	12.8	.0136
136.0	127.3	3.187	4491.0	5275.0	108.1	16.94	51.18	187.9	12.8	.0135
138.0	121.9	3.051	4554.0	5373.0	108.8	16.49	46.77	191.8	12.9	.0133
140.0	117.2	2.934	4611.0	5463.0	109.5	16.11	43.47	195.3	13.0	.0132
142.0	113.1	2.832	4665.0	5548.0	110.1	15.78	40.90	198.6	13.1	.0131
144.0	109.5	2.740	4715.0	5627.0	110.6	15.49	38.82	201.7	13.2	.0131
146.0	106.2	2.658	4762.0	5703.0	111.1	15.25	37.11	204.7	13.3	.0130
150.0	100.4	2.514	4852.0	5846.0	112.1	14.83	34.46	210.2	13.5	.0130
155.0	94.42	2.364	4954.0	6012.0	113.2	14.43	32.07	216.6	13.8	.0130

Thermophysical properties of argon—Continued

T K	Density kg/m^3	Density mol/dm^3	E J/mol	H J/mol	S J/(mol·K)	C_v J/(mol·K)	C_p J/(mol·K)	Sound m/s	Visc. μPa·s	Therm. W/(m·K)
160.0	89.35	2.237	5050.0	6168.0	114.2	14.12	30.33	222.4	14.1	.0131
165.0	84.98	2.127	5141.0	6316.0	115.1	13.87	29.01	227.9	14.4	.0132
170.0	81.14	2.031	5227.0	6458.0	115.9	13.67	27.97	233.0	14.8	.0133
175.0	77.72	1.946	5311.0	6596.0	116.7	13.51	27.14	237.9	15.1	.0134
180.0	74.65	1.869	5392.0	6730.0	117.5	13.38	26.45	242.6	15.4	.0136
185.0	71.87	1.799	5471.0	6861.0	118.2	13.28	25.89	247.0	15.8	.0137
190.0	69.33	1.736	5548.0	6989.0	118.9	13.19	25.41	251.3	16.1	.0139
200.0	64.85	1.623	5699.0	7239.0	120.2	13.05	24.65	259.6	16.8	.0143
210.0	61.00	1.527	5845.0	7483.0	121.4	12.95	24.08	267.3	17.5	.0147
220.0	57.64	1.443	5989.0	7721.0	122.5	12.87	23.64	274.7	18.2	.0152
230.0	54.68	1.369	6129.0	7956.0	123.5	12.82	23.29	281.8	18.8	.0156
240.0	52.04	1.303	6268.0	8187.0	124.5	12.77	23.00	288.6	19.5	.0161
250.0	49.66	1.243	6405.0	8416.0	125.4	12.74	22.77	295.1	20.2	.0166
260.0	47.52	1.189	6541.0	8642.0	126.3	12.71	22.57	301.5	20.8	.0170
270.0	45.56	1.141	6675.0	8867.0	127.2	12.69	22.40	307.6	21.5	.0175
280.0	43.78	1.096	6809.0	9091.0	128.0	12.68	22.26	313.6	22.1	.0180
290.0	42.13	1.055	6942.0	9313.0	128.8	12.66	22.13	319.5	22.8	.0184
300.0	40.62	1.017	7074.0	9533.0	129.5	12.65	22.02	325.2	23.4	.0189
310.0	39.21	.9816	7206.0	9753.0	130.2	12.64	21.93	330.8	24.0	.0194
320.0	37.91	.9489	7337.0	9972.0	130.9	12.63	21.84	336.2	24.6	.0198
340.0	35.55	.8900	7598.0	10410.0	132.3	12.61	21.70	346.8	25.8	.0207
360.0	33.49	.8383	7858.0	10840.0	133.5	12.60	21.59	357.1	27.0	.0216
380.0	31.66	.7925	8116.0	11270.0	134.7	12.59	21.49	366.9	28.1	.0225
400.0	30.02	.7516	8374.0	11700.0	135.8	12.58	21.41	376.5	29.3	.0233

3.00 MPa isobar

T K	Density kg/m^3	Density mol/dm^3	E J/mol	H J/mol	S J/(mol·K)	C_v J/(mol·K)	C_p J/(mol·K)	Sound m/s	Visc. μPa·s	Therm. W/(m·K)
84.53[a]	1419.0	35.52	−1372.0	−1287.0	54.86	23.35	44.28	831.9	291.0	.135
90.0	1386.0	34.70	−1132.0	−1045.0	57.63	21.64	44.33	816.9	248.0	.125
100.0	1323.0	33.11	−687.0	−596.4	62.36	20.03	45.64	763.8	190.0	.111
110.0	1253.0	31.37	−225.5	−129.9	66.81	19.06	47.83	693.4	148.0	.0978
115.0	1215.0	30.42	14.4	113.0	68.97	18.65	49.40	654.1	131.0	.0917
120.0	1175.0	29.40	263.1	365.2	71.11	18.29	51.57	611.9	116.0	.0857
125.0	1130.0	28.28	524.3	630.3	73.28	17.96	54.72	566.2	102.0	.0797
130.0	1079.0	27.02	804.3	915.3	75.51	17.69	59.68	515.6	89.6	.0734
132.0	1057.0	26.45	924.0	1037.0	76.45	17.60	62.57	493.4	84.6	.0708
134.0	1032.0	25.83	1050.0	1166.0	77.41	17.53	66.34	469.7	79.6	.0681
136.0	1005.0	25.15	1184.0	1304.0	78.43	17.49	71.49	444.0	74.5	.0653
138.0	973.8	24.38	1331.0	1454.0	79.53	17.50	79.08	415.4	69.2	.0624
138.698[b]	960.4	24.04	1390.0	1515.0	79.97	17.51	83.34	403.0	67.0	.0613
138.698[b]	167.1	4.182	4326.0	5044.0	105.4	18.14	72.38	184.0	13.8	.0158
140.0	159.3	3.989	4390.0	5142.0	106.1	17.65	63.82	187.3	13.7	.0154
142.0	151.0	3.781	4468.0	5261.0	107.0	17.08	55.91	191.7	13.7	.0150
144.0	144.2	3.610	4537.0	5368.0	107.7	16.62	50.51	195.6	13.8	.0147
146.0	138.4	3.465	4599.0	5464.0	108.4	16.22	46.57	199.3	13.8	.0145
148.0	133.4	3.340	4656.0	5554.0	109.0	15.89	43.53	202.6	13.9	.0144
150.0	129.0	3.228	4710.0	5639.0	109.6	15.60	41.12	205.8	14.0	.0142
152.0	125.0	3.128	4760.0	5719.0	110.1	15.34	39.15	208.8	14.1	.0142
154.0	121.4	3.038	4808.0	5796.0	110.6	15.12	37.51	211.6	14.2	.0141
156.0	118.1	2.955	4854.0	5869.0	111.1	14.92	36.11	214.4	14.3	.0140
160.0	112.2	2.809	4941.0	6009.0	111.9	14.58	33.89	219.5	14.5	.0140
165.0	106.0	2.653	5043.0	6173.0	113.0	14.25	31.81	225.5	14.8	.0139
170.0	100.7	2.520	5138.0	6328.0	113.9	13.99	30.25	231.0	15.1	.0140
175.0	96.02	2.404	5228.0	6476.0	114.7	13.78	29.04	236.3	15.4	.0140
180.0	91.91	2.301	5315.0	6619.0	115.5	13.61	28.07	241.2	15.7	.0141
185.0	88.23	2.209	5399.0	6757.0	116.3	13.47	27.29	245.9	16.0	.0143
190.0	84.90	2.125	5481.0	6892.0	117.0	13.35	26.63	250.4	16.4	.0144
195.0	81.88	2.050	5560.0	7024.0	117.7	13.26	26.09	254.8	16.7	.0146
200.0	79.11	1.980	5638.0	7153.0	118.4	13.18	25.62	259.0	17.0	.0147
210.0	74.19	1.857	5790.0	7405.0	119.6	13.05	24.87	267.0	17.7	.0151
220.0	69.94	1.751	5938.0	7651.0	120.7	12.96	24.30	274.6	18.4	.0155
230.0	66.22	1.658	6082.0	7892.0	121.8	12.89	23.85	281.8	19.0	.0160
240.0	62.93	1.575	6224.0	8128.0	122.8	12.84	23.49	288.7	19.7	.0164

Thermophysical properties of argon—Continued

T K	Density kg/m^3	Density mol/dm^3	E J/mol	H J/mol	S J/(mol·K)	C_v J/(mol·K)	C_p J/(mol·K)	Sound m/s	Visc. μPa·s	Therm. W/(m·K)
250.0	59.99	1.502	6364.0	8362.0	123.8	12.79	23.19	295.4	20.3	.0168
260.0	57.33	1.435	6502.0	8592.0	124.7	12.76	22.95	301.8	21.0	.0173
270.0	54.93	1.375	6639.0	8821.0	125.5	12.74	22.74	308.1	21.6	.0177
280.0	52.73	1.320	6775.0	9047.0	126.4	12.71	22.56	314.1	22.3	.0182
290.0	50.72	1.270	6909.0	9272.0	127.1	12.70	22.41	320.0	22.9	.0187
300.0	48.87	1.223	7043.0	9496.0	127.9	12.68	22.28	325.8	23.5	.0191
310.0	47.16	1.181	7177.0	9718.0	128.6	12.67	22.16	331.4	24.1	.0196
320.0	45.57	1.141	7309.0	9939.0	129.3	12.66	22.06	336.9	24.7	.0200
340.0	42.71	1.069	7572.0	10380.0	130.7	12.64	21.89	347.6	25.9	.0209
360.0	40.21	1.007	7834.0	10810.0	131.9	12.63	21.75	357.9	27.1	.0218
380.0	38.00	.9512	8094.0	11250.0	133.1	12.62	21.63	367.8	28.2	.0226
400.0	36.02	.9018	8353.0	11680.0	134.2	12.61	21.54	377.4	29.3	.0235
				3.50 MPa isobar						
84.65[a]	1419.0	35.53	−1371.0	−1273.0	54.87	23.31	44.19	833.6	291.0	.135
90.0	1388.0	34.74	−1137.0	−1037.0	57.57	21.66	44.23	819.0	249.0	.126
100.0	1325.0	33.16	−694.4	−588.8	62.29	20.06	45.50	766.4	191.0	.111
110.0	1256.0	31.43	−235.5	−124.1	66.72	19.09	47.59	696.9	149.0	.0982
115.0	1218.0	30.50	2.6	117.4	68.86	18.69	49.08	658.2	132.0	.0922
120.0	1178.0	29.50	248.9	367.6	70.99	18.33	51.10	616.8	117.0	.0863
125.0	1135.0	28.40	506.6	629.8	73.13	18.00	53.99	572.2	104.0	.0803
130.0	1085.0	27.17	781.1	910.0	75.33	17.72	58.42	523.4	91.1	.0742
132.0	1064.0	26.63	897.8	1029.0	76.24	17.63	60.93	502.1	86.1	.0716
134.0	1040.0	26.04	1020.0	1154.0	77.18	17.56	64.12	479.7	81.2	.0690
136.0	1015.0	25.40	1149.0	1286.0	78.16	17.51	68.33	455.6	76.3	.0663
138.0	986.0	24.68	1287.0	1429.0	79.20	17.49	74.17	429.3	71.2	.0635
140.0	953.3	23.86	1438.0	1585.0	80.32	17.52	83.01	399.9	66.0	.0607
141.0	934.6	23.40	1522.0	1671.0	80.94	17.56	89.49	383.5	63.2	.0593
142.0	913.7	22.87	1612.0	1765.0	81.60	17.64	98.55	365.6	60.2	.0578
142.2	909.2	22.76	1631.0	1785.0	81.74	17.65	100.8	361.7	59.6	.0575
142.376[b]	902.9	22.60	1656.0	1811.0	81.92	17.69	104.7	356.1	58.8	.0571
142.376[b]	208.6	5.222	4182.0	4852.0	103.3	19.03	102.4	183.0	15.0	.0189
145.0	186.3	4.665	4341.0	5092.0	104.9	17.86	71.64	190.4	14.7	.0173
146.0	180.8	4.525	4387.0	5160.0	105.4	17.55	65.94	192.8	14.7	.0169
148.0	171.4	4.290	4468.0	5283.0	106.3	17.01	57.81	197.0	14.6	.0164
150.0	163.7	4.097	4539.0	5393.0	107.0	16.56	52.22	200.8	14.7	.0160
152.0	157.1	3.934	4604.0	5493.0	107.7	16.19	48.11	204.4	14.7	.0157
154.0	151.4	3.791	4663.0	5586.0	108.3	15.87	44.94	207.7	14.7	.0155
156.0	146.4	3.665	4719.0	5674.0	108.8	15.58	42.42	210.8	14.8	.0153
158.0	141.9	3.552	4771.0	5756.0	109.4	15.34	40.35	213.8	14.9	.0152
160.0	137.8	3.450	4821.0	5835.0	109.9	15.12	38.62	216.6	15.0	.0151
162.0	134.1	3.356	4868.0	5911.0	110.3	14.93	37.16	219.3	15.1	.0150
165.0	129.0	3.230	4936.0	6020.0	111.0	14.67	35.34	223.1	15.2	.0149
170.0	121.7	3.048	5042.0	6190.0	112.0	14.33	33.01	229.1	15.5	.0148
175.0	115.5	2.892	5141.0	6351.0	112.9	14.06	31.28	234.7	15.7	.0147
180.0	110.2	2.757	5234.0	6504.0	113.8	13.85	29.93	240.0	16.0	.0147
185.0	105.4	2.638	5324.0	6650.0	114.6	13.67	28.86	244.9	16.3	.0148
190.0	101.2	2.532	5410.0	6792.0	115.4	13.53	27.99	249.6	16.6	.0149
195.0	97.32	2.436	5494.0	6931.0	116.1	13.41	27.27	254.2	17.0	.0150
200.0	93.85	2.349	5576.0	7065.0	116.8	13.31	26.67	258.5	17.3	.0152
210.0	87.74	2.196	5733.0	7327.0	118.0	13.16	25.71	266.8	17.9	.0155
220.0	82.52	2.066	5886.0	7580.0	119.2	13.04	24.99	274.5	18.6	.0159
230.0	77.98	1.952	6034.0	7827.0	120.3	12.96	24.44	281.9	19.2	.0163
240.0	73.99	1.852	6180.0	8069.0	121.3	12.90	23.99	289.0	19.9	.0167
250.0	70.44	1.763	6323.0	8308.0	122.3	12.85	23.63	295.7	20.5	.0171
260.0	67.25	1.684	6463.0	8542.0	123.2	12.81	23.33	302.3	21.1	.0176
270.0	64.37	1.611	6602.0	8774.0	124.1	12.78	23.09	308.6	21.8	.0180
280.0	61.76	1.546	6740.0	9004.0	124.9	12.75	22.87	314.7	22.4	.0184
290.0	59.36	1.486	6877.0	9232.0	125.7	12.73	22.69	320.7	23.0	.0189
300.0	57.17	1.431	7012.0	9458.0	126.5	12.72	22.53	326.5	23.6	.0193
310.0	55.14	1.380	7147.0	9683.0	127.3	12.70	22.40	332.1	24.2	.0198
320.0	53.26	1.333	7281.0	9906.0	128.0	12.69	22.27	337.7	24.8	.0202

Thermophysical properties of argon—Continued

T K	Density kg/m^3	Density mol/dm^3	E J/mol	H J/mol	S J/(mol·K)	C_v J/(mol·K)	C_p J/(mol·K)	Sound m/s	Visc. μPa·s	Therm. W/(m·K)
340.0	49.89	1.249	7547.0	10350.0	129.3	12.67	22.07	348.4	26.0	.0211
360.0	46.94	1.175	7810.0	10790.0	130.6	12.65	21.91	358.7	27.2	.0219
380.0	44.34	1.110	8072.0	11230.0	131.7	12.64	21.77	368.6	28.3	.0228
400.0	42.02	1.052	8333.0	11660.0	132.9	12.63	21.66	378.2	29.4	.0236
				4.00 MPa isobar						
84.78[a]	1420.0	35.55	-1371.0	-1258.0	54.87	23.27	44.11	835.4	292.0	.135
90.0	1389.0	34.77	-1143.0	-1028.0	57.51	21.68	44.14	821.2	250.0	.126
100.0	1326.0	33.21	-701.7	-581.2	62.21	20.09	45.36	769.1	192.0	.111
110.0	1258.0	31.50	-245.2	-118.2	66.63	19.13	47.36	700.4	150.0	.0986
115.0	1221.0	30.57	-8.9	121.9	68.76	18.73	48.77	662.2	133.0	.0926
120.0	1182.0	29.59	235.1	370.3	70.87	18.37	50.66	621.6	119.0	.0868
125.0	1139.0	28.51	489.5	629.8	72.99	18.04	53.32	578.1	105.0	.0809
130.0	1091.0	27.32	759.2	905.6	75.16	17.76	57.31	530.7	92.4	.0749
132.0	1070.0	26.79	873.0	1022.0	76.05	17.67	59.51	510.4	87.6	.0724
134.0	1048.0	26.23	991.5	1144.0	76.96	17.59	62.25	488.9	82.8	.0699
136.0	1024.0	25.62	1116.0	1272.0	77.91	17.53	65.76	466.2	78.0	.0673
138.0	996.9	24.96	1248.0	1408.0	78.90	17.49	70.43	441.7	73.1	.0646
140.0	967.0	24.21	1390.0	1555.0	79.96	17.49	77.03	415.0	68.2	.0619
141.0	950.5	23.79	1466.0	1634.0	80.52	17.51	81.52	400.4	65.6	.0605
142.0	932.4	23.34	1547.0	1718.0	81.12	17.55	87.28	384.9	62.9	.0591
143.0	912.4	22.84	1634.0	1809.0	81.76	17.62	95.02	368.0	60.1	.0577
144.0	889.7	22.27	1730.0	1909.0	82.45	17.72	106.2	349.3	57.1	.0563
145.0	862.9	21.60	1839.0	2024.0	83.25	17.88	124.2	328.0	53.8	.0549
145.5	847.0	21.20	1901.0	2089.0	83.70	18.00	138.5	315.8	51.9	.0543
145.672[b]	837.3	20.96	1936.0	2127.0	83.95	18.08	150.5	308.1	50.8	.0539
145.672[b]	261.4	6.543	3990.0	4601.0	100.9	20.04	170.0	182.0	16.5	.0240
150.0	209.7	5.250	4317.0	5078.0	104.2	17.85	75.22	195.1	15.6	.0190
152.0	197.7	4.949	4409.0	5217.0	105.1	17.26	64.33	199.5	15.5	.0181
154.0	188.1	4.708	4489.0	5338.0	105.9	16.78	57.19	203.4	15.5	.0175
156.0	180.0	4.506	4560.0	5447.0	106.6	16.38	52.10	207.0	15.5	.0171
158.0	173.1	4.333	4624.0	5547.0	107.2	16.04	48.26	210.4	15.5	.0168
160.0	167.1	4.182	4684.0	5641.0	107.8	15.75	45.24	213.6	15.5	.0165
162.0	161.7	4.047	4740.0	5729.0	108.4	15.49	42.80	216.6	15.6	.0162
164.0	156.8	3.926	4793.0	5812.0	108.9	15.26	40.79	219.4	15.7	.0161
166.0	152.4	3.815	4844.0	5892.0	109.4	15.06	39.10	222.1	15.7	.0159
170.0	144.7	3.621	4938.0	6043.0	110.3	14.71	36.41	227.3	15.9	.0157
175.0	136.5	3.416	5047.0	6218.0	111.3	14.37	33.92	233.3	16.1	.0155
180.0	129.5	3.242	5149.0	6383.0	112.2	14.10	32.07	238.8	16.4	.0154
185.0	123.5	3.090	5245.0	6540.0	113.1	13.89	30.64	244.0	16.7	.0154
190.0	118.1	2.957	5337.0	6690.0	113.9	13.71	29.49	248.9	16.9	.0155
195.0	113.4	2.838	5426.0	6835.0	114.6	13.57	28.57	253.6	17.2	.0155
200.0	109.1	2.731	5511.0	6976.0	115.3	13.45	27.80	258.1	17.5	.0156
205.0	105.2	2.634	5595.0	7113.0	116.0	13.35	27.15	262.5	17.8	.0158
210.0	101.7	2.545	5676.0	7248.0	116.7	13.26	26.60	266.6	18.2	.0159
220.0	95.36	2.387	5833.0	7509.0	117.9	13.13	25.72	274.6	18.8	.0162
230.0	89.95	2.252	5986.0	7763.0	119.0	13.03	25.04	282.1	19.4	.0166
240.0	85.21	2.133	6135.0	8010.0	120.1	12.96	24.51	289.3	20.0	.0170
250.0	81.01	2.028	6281.0	8253.0	121.0	12.90	24.08	296.2	20.7	.0174
260.0	77.27	1.934	6424.0	8492.0	122.0	12.86	23.73	302.8	21.3	.0178
270.0	73.90	1.850	6566.0	8728.0	122.9	12.82	23.43	309.1	21.9	.0182
280.0	70.84	1.773	6705.0	8961.0	123.7	12.79	23.19	315.3	22.5	.0187
290.0	68.05	1.704	6844.0	9192.0	124.5	12.77	22.97	321.3	23.2	.0191
300.0	65.50	1.640	6981.0	9421.0	125.3	12.75	22.79	327.2	23.8	.0195
310.0	63.14	1.581	7117.0	9648.0	126.0	12.73	22.63	332.9	24.4	.0200
320.0	60.97	1.526	7252.0	9873.0	126.8	12.72	22.49	338.4	25.0	.0204
340.0	57.07	1.429	7521.0	10320.0	128.1	12.70	22.26	349.2	26.1	.0212
360.0	53.67	1.343	7787.0	10760.0	129.4	12.68	22.07	359.5	27.3	.0221
380.0	50.68	1.269	8050.0	11200.0	130.6	12.66	21.91	369.5	28.4	.0229
400.0	48.01	1.202	8313.0	11640.0	131.7	12.65	21.79	379.1	29.5	.0238

Thermophysical properties of argon—Continued

T K	Density kg/m³	Density mol/dm³	E J/mol	H J/mol	S J/(mol·K)	C_v J/(mol·K)	C_p J/(mol·K)	Sound m/s	Visc. μPa·s	Therm. W/(m·K)
				4.50 MPa isobar						
84.90[a]	1421.0	35.57	−1370.0	−1244.0	54.88	23.23	44.02	837.1	292.0	.135
90.0	1391.0	34.81	−1148.0	−1019.0	57.45	21.70	44.05	823.3	252.0	.126
100.0	1328.0	33.25	−708.8	−573.5	62.14	20.12	45.23	771.7	193.0	.112
110.0	1261.0	31.56	−254.8	−112.2	66.54	19.17	47.15	703.8	151.0	.0989
115.0	1224.0	30.65	−20.2	126.7	68.66	18.77	48.48	666.2	135.0	.0930
120.0	1185.0	29.68	221.6	373.2	70.76	18.41	50.24	626.3	120.0	.0872
125.0	1143.0	28.62	473.0	630.2	72.86	18.08	52.70	583.7	106.0	.0815
130.0	1097.0	27.46	738.2	902.1	74.99	17.80	56.30	537.8	93.8	.0756
132.0	1077.0	26.95	849.6	1017.0	75.86	17.70	58.26	518.2	89.0	.0731
134.0	1055.0	26.41	965.0	1135.0	76.76	17.61	60.64	497.7	84.3	.0707
136.0	1032.0	25.83	1085.0	1260.0	77.67	17.55	63.62	476.1	79.6	.0681
138.0	1007.0	25.20	1212.0	1390.0	78.63	17.50	67.47	453.1	74.9	.0656
140.0	979.1	24.51	1347.0	1530.0	79.64	17.48	72.63	428.3	70.1	.0629
142.0	947.8	23.73	1493.0	1682.0	80.72	17.51	80.04	401.2	65.2	.0603
143.0	930.3	23.29	1572.0	1765.0	81.29	17.54	85.14	386.4	62.7	.0590
144.0	911.2	22.81	1656.0	1853.0	81.91	17.60	91.79	370.6	60.0	.0576
145.0	889.8	22.27	1747.0	1949.0	82.57	17.69	100.9	353.3	57.2	.0563
146.0	865.1	21.66	1849.0	2057.0	83.31	17.83	114.5	334.1	54.1	.0550
146.5	851.1	21.31	1905.0	2116.0	83.72	17.92	124.2	323.5	52.5	.0544
147.0	835.5	20.91	1966.0	2181.0	84.16	18.03	137.3	312.0	50.7	.0537
147.5	817.5	20.46	2035.0	2254.0	84.66	18.19	156.5	299.2	48.8	.0531
148.0	796.1	19.93	2114.0	2340.0	85.24	18.39	187.9	284.5	46.6	.0526
148.2	786.0	19.67	2150.0	2379.0	85.50	18.50	207.3	277.9	45.6	.0524
148.4	774.5	19.39	2191.0	2423.0	85.80	18.62	234.3	270.6	44.5	.0523
148.6	761.2	19.05	2238.0	2474.0	86.14	18.78	275.1	262.4	43.2	.0524
148.652[b]	751.5	18.81	2270.0	2509.0	86.38	18.91	319.1	256.4	42.3	.0526
148.652[b]	340.0	8.511	3698.0	4226.0	97.93	21.19	438.6	181.5	19.1	.0345
150.0	286.8	7.179	3959.0	4586.0	100.3	19.81	163.0	187.6	17.6	.0259
150.5	277.1	6.936	4013.0	4662.0	100.8	19.49	140.6	189.3	17.4	.0247
151.0	268.9	6.732	4059.0	4728.0	101.3	19.21	125.0	190.9	17.2	.0239
151.5	261.9	6.556	4101.0	4787.0	101.7	18.96	113.3	192.3	17.0	.0232
152.0	255.7	6.401	4138.0	4841.0	102.0	18.73	104.3	193.7	16.9	.0226
153.0	245.2	6.137	4205.0	4939.0	102.7	18.32	90.92	196.3	16.7	.0216
154.0	236.3	5.916	4264.0	5025.0	103.2	17.96	81.51	198.7	16.6	.0209
155.0	228.7	5.726	4317.0	5102.0	103.7	17.65	74.46	200.9	16.5	.0203
156.0	222.1	5.559	4365.0	5174.0	104.2	17.37	68.94	203.0	16.4	.0198
158.0	210.8	5.277	4450.0	5303.0	105.0	16.89	60.82	206.9	16.3	.0190
160.0	201.4	5.043	4526.0	5419.0	105.7	16.48	55.08	210.5	16.3	.0184
162.0	193.5	4.843	4595.0	5525.0	106.4	16.13	50.78	213.9	16.3	.0179
164.0	186.5	4.669	4659.0	5623.0	107.0	15.83	47.42	217.0	16.3	.0176
166.0	180.3	4.514	4718.0	5715.0	107.6	15.57	44.73	220.0	16.3	.0172
168.0	174.8	4.376	4773.0	5802.0	108.1	15.34	42.51	222.8	16.3	.0170
170.0	169.8	4.250	4826.0	5885.0	108.6	15.13	40.65	225.6	16.4	.0168
172.0	165.2	4.136	4876.0	5965.0	109.0	14.95	39.07	228.2	16.5	.0166
175.0	159.0	3.980	4948.0	6079.0	109.7	14.71	37.09	231.9	16.6	.0164
180.0	150.1	3.757	5060.0	6258.0	110.7	14.37	34.55	237.8	16.8	.0162
185.0	142.5	3.567	5164.0	6425.0	111.6	14.11	32.65	243.2	17.0	.0161
190.0	135.9	3.402	5262.0	6585.0	112.5	13.90	31.16	248.4	17.3	.0160
195.0	130.1	3.256	5355.0	6737.0	113.3	13.73	29.98	253.2	17.6	.0161
200.0	124.9	3.126	5445.0	6885.0	114.0	13.59	29.02	257.9	17.8	.0161
205.0	120.2	3.009	5532.0	7028.0	114.7	13.47	28.22	262.3	18.1	.0162
210.0	116.0	2.903	5617.0	7167.0	115.4	13.37	27.55	266.6	18.4	.0163
220.0	108.5	2.716	5780.0	7437.0	116.7	13.22	26.48	274.7	19.0	.0166
230.0	102.1	2.556	5937.0	7698.0	117.8	13.10	25.67	282.4	19.6	.0170
240.0	96.59	2.418	6090.0	7951.0	118.9	13.02	25.04	289.7	20.2	.0173
250.0	91.72	2.296	6239.0	8199.0	119.9	12.95	24.54	296.6	20.8	.0177
260.0	87.39	2.187	6385.0	8442.0	120.9	12.90	24.13	303.3	21.5	.0181
270.0	83.50	2.090	6529.0	8682.0	121.8	12.86	23.79	309.7	22.1	.0185
280.0	79.99	2.002	6671.0	8918.0	122.6	12.83	23.50	316.0	22.7	.0189
290.0	76.79	1.922	6811.0	9152.0	123.4	12.80	23.26	322.0	23.3	.0193
300.0	73.87	1.849	6950.0	9384.0	124.2	12.78	23.05	327.9	23.9	.0197

Thermophysical properties of argon—Continued

T K	Density kg/m³	Density mol/dm³	E J/mol	H J/mol	S J/(mol·K)	C_v J/(mol·K)	C_p J/(mol·K)	Sound m/s	Visc. μPa·s	Therm. W/(m·K)
310.0	71.18	1.782	7088.0	9613.0	125.0	12.76	22.87	333.6	24.5	.0202
320.0	68.70	1.720	7224.0	9841.0	125.7	12.75	22.71	339.2	25.1	.0206
330.0	66.40	1.662	7360.0	10070.0	126.4	12.73	22.57	344.7	25.7	.0210
340.0	64.26	1.609	7495.0	10290.0	127.1	12.72	22.44	350.0	26.2	.0214
360.0	60.41	1.512	7763.0	10740.0	128.3	12.70	22.23	360.4	27.4	.0223
380.0	57.01	1.427	8029.0	11180.0	129.5	12.68	22.05	370.4	28.5	.0231
400.0	54.00	1.352	8292.0	11620.0	130.7	12.67	21.91	380.0	29.6	.0239
				4.80 MPa isobar						
84.97[a]	1421.0	35.57	−1370.0	−1235.0	54.88	23.20	43.97	838.2	292.0	.135
90.0	1392.0	34.84	−1152.0	−1014.0	57.41	21.71	44.00	824.5	252.0	.126
100.0	1330.0	33.28	−713.1	−568.9	62.10	20.14	45.15	773.2	194.0	.112
110.0	1262.0	31.60	−260.5	−108.6	66.48	19.19	47.02	705.8	152.0	.0992
115.0	1226.0	30.69	−26.8	129.6	68.60	18.80	48.31	668.5	135.0	.0933
120.0	1188.0	29.73	213.7	375.1	70.69	18.43	50.00	629.0	120.0	.0875
125.0	1146.0	28.69	463.4	630.7	72.78	18.10	52.35	587.1	107.0	.0818
130.0	1100.0	27.54	726.1	900.4	74.89	17.82	55.75	541.9	94.5	.0760
132.0	1080.0	27.04	836.1	1014.0	75.76	17.72	57.57	522.7	89.8	.0736
134.0	1059.0	26.51	949.9	1131.0	76.64	17.63	59.78	502.7	85.1	.0711
136.0	1037.0	25.95	1068.0	1253.0	77.54	17.56	62.50	481.7	80.5	.0687
138.0	1012.0	25.34	1192.0	1381.0	78.48	17.51	65.96	459.4	75.9	.0661
140.0	985.6	24.67	1323.0	1518.0	79.46	17.48	70.49	435.7	71.2	.0636
142.0	956.0	23.93	1464.0	1665.0	80.50	17.49	76.76	410.0	66.5	.0610
143.0	939.6	23.52	1539.0	1743.0	81.05	17.51	80.93	396.1	64.0	.0597
144.0	921.9	23.08	1619.0	1827.0	81.63	17.55	86.15	381.4	61.5	.0584
145.0	902.5	22.59	1704.0	1916.0	82.25	17.62	92.93	365.7	58.9	.0571
146.0	880.8	22.05	1796.0	2013.0	82.92	17.71	102.2	348.6	56.1	.0558
147.0	855.9	21.42	1898.0	2122.0	83.66	17.86	115.9	329.7	53.1	.0545
147.5	841.7	21.07	1954.0	2182.0	84.07	17.95	125.6	319.3	51.5	.0539
148.0	825.9	20.67	2016.0	2248.0	84.52	18.07	138.6	308.0	49.7	.0533
148.5	807.9	20.22	2084.0	2322.0	85.01	18.23	157.2	295.7	47.8	.0527
149.0	786.6	19.69	2163.0	2407.0	85.59	18.43	186.5	281.7	45.7	.0522
149.2	776.7	19.44	2199.0	2446.0	85.85	18.53	203.9	275.4	44.7	.0520
149.4	765.6	19.17	2238.0	2489.0	86.14	18.65	226.9	268.7	43.7	.0518
149.6	753.0	18.85	2283.0	2537.0	86.46	18.80	259.5	261.3	42.6	.0518
149.8	738.1	18.48	2334.0	2594.0	86.84	18.98	309.9	253.0	41.2	.0519
149.9	729.4	18.26	2364.0	2627.0	87.06	19.09	347.6	248.4	40.5	.0521
150.0	719.4	18.01	2397.0	2664.0	87.31	19.22	401.0	243.3	39.6	.0524
150.1	707.6	17.71	2437.0	2708.0	87.60	19.38	484.2	237.5	38.7	.0530
150.2	692.6	17.34	2486.0	2763.0	87.97	19.59	637.2	230.7	37.5	.0543
150.3	670.6	16.79	2557.0	2843.0	88.50	19.90	1058.0	221.6	35.8	.0578
150.304[b]	641.3	16.05	2650.0	2949.0	89.20	20.33	3726.0	211.0	33.7	.0569
150.304[b]	450.8	11.28	3291.0	3717.0	94.31	21.88	4306.0	183.5	23.3	.0596
155.0	264.2	6.613	4156.0	4882.0	102.0	18.45	97.41	198.2	17.4	.0230
156.0	254.1	6.362	4219.0	4974.0	102.6	18.08	86.70	200.5	17.2	.0221
157.0	245.6	6.148	4276.0	5056.0	103.1	17.76	78.76	202.7	17.1	.0214
158.0	238.1	5.961	4327.0	5132.0	103.6	17.48	72.60	204.8	17.0	.0208
160.0	225.6	5.647	4417.0	5268.0	104.4	16.98	63.61	208.7	16.8	.0199
162.0	215.2	5.388	4497.0	5388.0	105.2	16.57	57.31	212.3	16.8	.0192
164.0	206.5	5.169	4569.0	5498.0	105.9	16.21	52.64	215.6	16.7	.0186
166.0	198.9	4.978	4635.0	5599.0	106.5	15.91	49.01	218.8	16.7	.0182
168.0	192.1	4.810	4696.0	5694.0	107.1	15.64	46.11	221.8	16.7	.0179
170.0	186.1	4.660	4754.0	5784.0	107.6	15.40	43.73	224.6	16.8	.0176
172.0	180.7	4.524	4808.0	5870.0	108.1	15.19	41.74	227.3	16.8	.0173
174.0	175.8	4.400	4860.0	5951.0	108.6	15.01	40.05	229.9	16.8	.0171
180.0	163.1	4.083	5004.0	6179.0	109.8	14.55	36.23	237.2	17.1	.0167
185.0	154.4	3.865	5113.0	6355.0	110.8	14.26	33.98	242.8	17.3	.0165
190.0	146.9	3.678	5215.0	6520.0	111.7	14.02	32.26	248.1	17.5	.0164
195.0	140.4	3.515	5312.0	6678.0	112.5	13.83	30.90	253.0	17.8	.0164
200.0	134.6	3.370	5405.0	6829.0	113.3	13.68	29.80	257.8	18.0	.0164
205.0	129.4	3.240	5494.0	6976.0	114.0	13.55	28.90	262.3	18.3	.0165
210.0	124.7	3.122	5581.0	7119.0	114.7	13.44	28.14	266.6	18.6	.0166

Thermophysical properties of argon—Continued

T K	Density kg/m³	Density mol/dm³	E J/mol	H J/mol	S J/(mol·K)	C_v J/(mol·K)	C_p J/(mol·K)	Sound m/s	Visc. μPa·s	Therm. W/(m·K)
220.0	116.5	2.916	5748.0	7394.0	116.0	13.27	26.95	274.9	19.2	.0168
230.0	109.5	2.742	5908.0	7659.0	117.1	13.15	26.06	282.6	19.8	.0172
240.0	103.5	2.591	6063.0	7916.0	118.2	13.06	25.37	289.9	20.4	.0175
250.0	98.20	2.458	6214.0	8166.0	119.3	12.99	24.82	296.9	21.0	.0179
260.0	93.50	2.341	6361.0	8412.0	120.2	12.93	24.37	303.7	21.6	.0183
270.0	89.29	2.235	6507.0	8654.0	121.1	12.89	24.00	310.1	22.2	.0187
280.0	85.50	2.140	6650.0	8893.0	122.0	12.85	23.69	316.4	22.8	.0191
290.0	82.05	2.054	6791.0	9128.0	122.8	12.83	23.43	322.5	23.4	.0195
300.0	78.90	1.975	6931.0	9361.0	123.6	12.80	23.20	328.4	24.0	.0199
310.0	76.01	1.903	7070.0	9592.0	124.4	12.78	23.01	334.1	24.6	.0203
320.0	73.35	1.836	7207.0	9822.0	125.1	12.76	22.84	339.7	25.2	.0207
330.0	70.88	1.774	7344.0	10050.0	125.8	12.75	22.68	345.2	25.7	.0211
340.0	68.58	1.717	7479.0	10280.0	126.5	12.74	22.55	350.5	26.3	.0215
360.0	64.45	1.613	7749.0	10720.0	127.8	12.72	22.32	360.9	27.5	.0224
380.0	60.81	1.522	8015.0	11170.0	129.0	12.70	22.14	370.9	28.6	.0232
400.0	57.59	1.442	8280.0	11610.0	130.1	12.68	21.98	380.6	29.7	.0240

5.00 MPa isobar

T K	Density kg/m³	Density mol/dm³	E J/mol	H J/mol	S J/(mol·K)	C_v J/(mol·K)	C_p J/(mol·K)	Sound m/s	Visc. μPa·s	Therm. W/(m·K)
85.02[a]	1421.0	35.58	−1369.0	−1229.0	54.89	23.19	43.94	838.8	292.0	.135
90.0	1392.0	34.85	−1154.0	−1010.0	57.38	21.71	43.96	825.4	253.0	.127
100.0	1330.0	33.30	−715.9	−565.8	62.07	20.15	45.10	774.3	194.0	.112
110.0	1263.0	31.62	−264.3	−106.1	66.45	19.21	46.94	707.1	152.0	.0993
115.0	1227.0	30.72	−31.2	131.6	68.56	18.81	48.20	670.1	136.0	.0935
120.0	1189.0	29.76	208.5	376.5	70.64	18.45	49.85	630.9	121.0	.0877
125.0	1148.0	28.73	457.0	631.1	72.72	18.12	52.12	589.3	107.0	.0820
130.0	1102.0	27.59	718.1	899.3	74.83	17.83	55.40	544.6	95.0	.0762
135.0	1051.0	26.31	997.9	1188.0	77.01	17.60	60.46	495.7	83.4	.0702
136.0	1040.0	26.02	1057.0	1249.0	77.46	17.57	61.81	485.3	81.1	.0690
138.0	1016.0	25.43	1179.0	1376.0	78.38	17.51	65.04	463.5	76.5	.0665
140.0	989.8	24.78	1308.0	1510.0	79.35	17.48	69.22	440.4	72.0	.0639
142.0	961.1	24.06	1446.0	1654.0	80.37	17.48	74.89	415.5	67.3	.0614
144.0	928.4	23.24	1596.0	1811.0	81.47	17.53	83.10	388.1	62.4	.0588
145.0	910.0	22.78	1677.0	1897.0	82.06	17.58	88.83	373.2	59.9	.0576
146.0	889.8	22.27	1765.0	1989.0	82.70	17.66	96.37	357.2	57.3	.0563
147.0	867.0	21.70	1860.0	2091.0	83.39	17.77	106.8	339.7	54.4	.0551
148.0	840.6	21.04	1967.0	2205.0	84.16	17.94	122.6	320.3	51.4	.0538
148.5	825.3	20.66	2027.0	2269.0	84.59	18.05	134.2	309.6	49.7	.0532
149.0	808.2	20.23	2093.0	2340.0	85.07	18.19	150.0	297.9	47.9	.0526
149.5	788.3	19.73	2167.0	2420.0	85.61	18.37	173.4	285.0	45.9	.0520
150.0	764.2	19.13	2254.0	2516.0	86.25	18.62	212.5	270.2	43.6	.0516
150.2	752.7	18.84	2295.0	2560.0	86.54	18.75	236.9	263.6	42.6	.0514
150.4	739.5	18.51	2341.0	2611.0	86.88	18.90	270.9	256.4	41.4	.0514
150.6	724.1	18.12	2394.0	2670.0	87.27	19.09	322.4	248.3	40.1	.0515
150.7	715.1	17.90	2425.0	2704.0	87.50	19.21	359.7	243.8	39.4	.0517
150.8	704.8	17.64	2459.0	2742.0	87.75	19.34	410.6	239.0	38.5	.0519
150.9	692.9	17.35	2499.0	2787.0	88.05	19.49	484.8	233.7	37.6	.0524
151.0	678.5	16.98	2546.0	2841.0	88.40	19.69	604.1	227.6	36.5	.0533
151.1	659.5	16.51	2608.0	2911.0	88.87	19.95	831.6	220.4	35.1	.0546
151.2	630.5	15.78	2702.0	3019.0	89.58	20.34	1454.0	211.0	33.1	.0589
151.3	558.9	13.99	2936.0	3293.0	91.40	21.18	4808.0	195.4	28.7	.0764
151.4	481.6	12.06	3203.0	3618.0	93.54	21.63	1932.0	187.7	24.8	.0654
151.5	450.0	11.27	3320.0	3764.0	94.51	21.60	1142.0	186.6	23.4	.0542
151.6	430.3	10.77	3397.0	3861.0	95.15	21.51	829.6	186.4	22.6	.0483
151.7	415.8	10.41	3454.0	3935.0	95.63	21.40	660.5	186.5	22.0	.0445
151.8	404.4	10.12	3501.0	3995.0	96.03	21.29	553.5	186.7	21.6	.0419
151.9	394.9	9.884	3541.0	4046.0	96.37	21.19	479.4	186.9	21.2	.0386
152.0	386.7	9.681	3575.0	4092.0	96.67	21.08	424.8	187.2	21.0	.0372
152.2	373.3	9.346	3633.0	4168.0	97.17	20.88	349.3	187.9	20.5	.0352
152.4	362.5	9.074	3682.0	4233.0	97.59	20.70	299.4	188.6	20.1	.0336
152.6	353.4	8.845	3724.0	4289.0	97.96	20.53	263.7	189.2	19.9	.0324
152.8	345.5	8.648	3761.0	4339.0	98.29	20.37	236.8	189.9	19.6	.0313
153.0	338.5	8.473	3794.0	4384.0	98.59	20.22	215.8	190.5	19.4	.0304

Thermophysical properties of argon—Continued

T K	Density kg/m^3	Density mol/dm^3	E J/mol	H J/mol	S J/(mol·K)	C_v J/(mol·K)	C_p J/(mol·K)	Sound m/s	Visc. μPa·s	Therm. W/(m·K)
153.5	324.0	8.111	3866.0	4482.0	99.22	19.88	178.7	192.1	19.0	.0287
154.0	312.3	7.819	3925.0	4565.0	99.76	19.58	154.3	193.6	18.7	.0274
154.5	302.6	7.574	3977.0	4637.0	100.2	19.31	136.9	195.0	18.4	.0264
155.0	294.2	7.364	4024.0	4703.0	100.7	19.06	123.8	196.3	18.2	.0255
155.5	286.8	7.179	4065.0	4762.0	101.0	18.84	113.6	197.6	18.1	.0248
156.0	280.2	7.014	4104.0	4816.0	101.4	18.63	105.3	198.9	17.9	.0242
157.0	268.8	6.729	4172.0	4915.0	102.0	18.25	92.74	201.2	17.7	.0231
158.0	259.2	6.489	4233.0	5003.0	102.6	17.91	83.58	203.4	17.5	.0223
159.0	250.9	6.282	4287.0	5083.0	103.1	17.62	76.58	205.5	17.4	.0216
160.0	243.6	6.099	4337.0	5157.0	103.5	17.35	71.02	207.5	17.3	.0211
161.0	237.1	5.936	4383.0	5225.0	104.0	17.10	66.50	209.4	17.2	.0206
162.0	231.2	5.788	4426.0	5290.0	104.4	16.88	62.74	211.3	17.2	.0202
164.0	220.9	5.530	4505.0	5409.0	105.1	16.49	56.82	214.7	17.1	.0195
166.0	212.1	5.310	4577.0	5518.0	105.8	16.15	52.36	218.0	17.0	.0189
168.0	204.4	5.118	4642.0	5619.0	106.4	15.85	48.86	221.1	17.0	.0185
170.0	197.6	4.948	4703.0	5714.0	106.9	15.59	46.05	224.0	17.0	.0181
172.0	191.6	4.795	4761.0	5804.0	107.5	15.37	43.73	226.8	17.0	.0178
174.0	186.0	4.657	4816.0	5889.0	107.9	15.16	41.78	229.4	17.1	.0176
176.0	181.0	4.531	4867.0	5971.0	108.4	14.98	40.12	232.0	17.1	.0174
178.0	176.4	4.415	4917.0	6050.0	108.9	14.81	38.68	234.5	17.2	.0172
180.0	172.1	4.307	4965.0	6126.0	109.3	14.67	37.44	236.9	17.2	.0171
185.0	162.6	4.070	5078.0	6307.0	110.3	14.35	34.92	242.6	17.4	.0168
190.0	154.5	3.867	5183.0	6476.0	111.2	14.10	33.02	247.9	17.7	.0167
195.0	147.4	3.691	5283.0	6637.0	112.0	13.90	31.53	252.9	17.9	.0166
200.0	141.2	3.535	5378.0	6792.0	112.8	13.74	30.34	257.7	18.2	.0166
205.0	135.7	3.396	5469.0	6941.0	113.5	13.60	29.36	262.3	18.4	.0167
210.0	130.6	3.270	5557.0	7086.0	114.2	13.48	28.55	266.7	18.7	.0168
215.0	126.1	3.156	5643.0	7227.0	114.9	13.39	27.86	270.9	19.0	.0169
220.0	121.9	3.052	5726.0	7365.0	115.5	13.31	27.27	275.0	19.3	.0170
230.0	114.5	2.866	5888.0	7632.0	116.7	13.18	26.32	282.8	19.8	.0173
240.0	108.1	2.707	6045.0	7892.0	117.8	13.08	25.59	290.1	20.4	.0176
250.0	102.5	2.567	6197.0	8145.0	118.9	13.01	25.01	297.2	21.0	.0180
260.0	97.60	2.443	6346.0	8392.0	119.8	12.95	24.54	303.9	21.6	.0184
270.0	93.17	2.332	6492.0	8636.0	120.7	12.91	24.15	310.4	22.2	.0188
280.0	89.19	2.233	6636.0	8875.0	121.6	12.87	23.82	316.7	22.8	.0192
290.0	85.57	2.142	6778.0	9112.0	122.5	12.84	23.55	322.8	23.4	.0196
300.0	82.27	2.059	6919.0	9347.0	123.2	12.81	23.31	328.7	24.0	.0200
310.0	79.24	1.984	7058.0	9579.0	124.0	12.79	23.10	334.4	24.6	.0204
320.0	76.45	1.914	7196.0	9809.0	124.7	12.78	22.92	340.0	25.2	.0208
340.0	71.47	1.789	7469.0	10260.0	126.1	12.75	22.62	350.9	26.4	.0216
360.0	67.14	1.681	7739.0	10710.0	127.4	12.73	22.39	361.3	27.5	.0224
380.0	63.35	1.586	8007.0	11160.0	128.6	12.71	22.19	371.3	28.6	.0233
400.0	59.99	1.502	8272.0	11600.0	129.7	12.69	22.03	381.0	29.7	.0241
				5.10 MPa isobar						
85.05[a]	1422.0	35.58	−1369.0	−1226.0	54.89	23.18	43.93	839.2	292.0	.135
90.0	1392.0	34.86	−1155.0	−1009.0	57.37	21.72	43.95	825.8	253.0	.127
100.0	1331.0	33.31	−717.3	−564.2	62.05	20.15	45.08	774.8	194.0	.112
110.0	1264.0	31.63	−266.1	−104.9	66.43	19.21	46.90	707.8	153.0	.0994
115.0	1228.0	30.73	−33.4	132.6	68.54	18.82	48.14	670.8	136.0	.0935
120.0	1190.0	29.78	205.9	377.1	70.62	18.46	49.77	631.8	121.0	.0878
125.0	1148.0	28.75	453.9	631.3	72.70	18.13	52.01	590.3	108.0	.0821
130.0	1103.0	27.62	714.2	898.9	74.80	17.84	55.23	545.9	95.3	.0763
135.0	1052.0	26.34	992.8	1186.0	76.97	17.61	60.17	497.4	83.7	.0704
136.0	1041.0	26.06	1052.0	1247.0	77.41	17.57	61.48	487.1	81.4	.0691
138.0	1017.0	25.47	1173.0	1373.0	78.33	17.51	64.60	465.5	76.9	.0667
140.0	991.8	24.83	1301.0	1506.0	79.29	17.48	68.63	442.7	72.3	.0641
142.0	963.5	24.12	1437.0	1649.0	80.30	17.48	74.03	418.1	67.7	.0616
144.0	931.5	23.32	1585.0	1804.0	81.39	17.52	81.74	391.3	62.9	.0591
145.0	913.6	22.87	1665.0	1888.0	81.97	17.57	87.04	376.8	60.4	.0578
146.0	894.0	22.38	1751.0	1978.0	82.59	17.63	93.90	361.2	57.8	.0566
147.0	872.1	21.83	1843.0	2077.0	83.26	17.74	103.2	344.4	55.1	.0553

Thermophysical properties of argon—Continued

T K	Density kg/m^3	Density mol/dm^3	E J/mol	H J/mol	S J/(mol·K)	C_v J/(mol·K)	C_p J/(mol·K)	Sound m/s	Visc. μPa·s	Therm. W/(m·K)
148.0	847.0	21.20	1946.0	2186.0	84.00	17.88	116.8	325.8	52.1	.0541
148.5	832.7	20.85	2002.0	2247.0	84.41	17.98	126.3	315.7	50.5	.0535
149.0	816.9	20.45	2064.0	2313.0	84.86	18.10	138.8	304.8	48.8	.0529
149.5	799.1	20.00	2132.0	2387.0	85.35	18.25	156.2	293.0	47.0	.0523
150.0	778.2	19.48	2209.0	2471.0	85.91	18.45	182.5	279.8	45.0	.0517
150.2	768.6	19.24	2244.0	2509.0	86.16	18.55	197.2	274.0	44.1	.0515
150.4	758.2	18.98	2281.0	2550.0	86.44	18.66	215.8	267.9	43.1	.0514
150.6	746.5	18.69	2322.0	2595.0	86.74	18.79	240.3	261.4	42.1	.0512
150.8	733.2	18.35	2369.0	2647.0	87.08	18.95	274.1	254.3	40.9	.0512
151.0	717.6	17.96	2422.0	2706.0	87.47	19.14	324.0	246.4	39.6	.0512
151.2	698.6	17.49	2487.0	2778.0	87.95	19.38	405.8	237.5	38.1	.0516
151.3	687.0	17.20	2525.0	2822.0	88.24	19.53	470.3	232.5	37.2	.0520
151.4	673.2	16.85	2571.0	2873.0	88.58	19.71	565.5	227.0	36.1	.0526
151.5	656.3	16.43	2627.0	2937.0	89.00	19.94	719.1	220.8	34.9	.0533
151.6	633.8	15.87	2700.0	3021.0	89.56	20.23	998.1	213.5	33.3	.0553
151.7	601.1	15.05	2807.0	3146.0	90.38	20.64	1545.0	204.9	31.2	.0592
151.8	553.9	13.86	2963.0	3331.0	91.60	21.13	2031.0	196.2	28.5	.0637
151.9	509.3	12.75	3116.0	3516.0	92.82	21.43	1602.0	191.2	26.2	.0610
152.0	478.5	11.98	3227.0	3653.0	93.72	21.50	1166.0	189.2	24.7	.0551
152.1	456.6	11.43	3309.0	3755.0	94.39	21.48	899.2	188.4	23.7	.0503
152.2	439.9	11.01	3373.0	3836.0	94.92	21.41	729.8	188.1	23.0	.0467
152.3	426.6	10.68	3426.0	3903.0	95.36	21.33	614.9	188.1	22.5	.0440
152.4	415.6	10.40	3470.0	3960.0	95.74	21.24	532.4	188.1	22.1	.0419
152.5	406.2	10.17	3509.0	4010.0	96.07	21.15	470.6	188.3	21.7	.0402
152.6	398.0	9.963	3543.0	4055.0	96.36	21.06	422.6	188.5	21.4	.0378
152.8	384.3	9.619	3602.0	4132.0	96.86	20.88	353.0	189.1	20.9	.0358
153.0	373.0	9.338	3651.0	4197.0	97.29	20.71	304.9	189.6	20.6	.0343
153.2	363.5	9.099	3694.0	4255.0	97.67	20.55	269.7	190.2	20.3	.0331
153.4	355.2	8.893	3732.0	4306.0	98.00	20.39	242.8	190.8	20.0	.0320
153.6	347.9	8.710	3767.0	4352.0	98.30	20.25	221.4	191.4	19.8	.0311
154.0	335.5	8.399	3827.0	4434.0	98.83	19.98	189.8	192.6	19.4	.0297
154.5	322.8	8.080	3891.0	4522.0	99.40	19.67	162.7	194.1	19.0	.0282
155.0	312.2	7.816	3945.0	4598.0	99.89	19.40	143.5	195.5	18.8	.0271
155.5	303.2	7.589	3994.0	4666.0	100.3	19.15	129.2	196.8	18.5	.0262
156.0	295.3	7.391	4038.0	4728.0	100.7	18.92	118.1	198.1	18.4	.0254
157.0	281.9	7.057	4115.0	4837.0	101.4	18.51	101.8	200.5	18.1	.0242
158.0	270.9	6.782	4181.0	4933.0	102.0	18.15	90.46	202.8	17.8	.0232
159.0	261.6	6.548	4240.0	5019.0	102.6	17.83	82.00	204.9	17.7	.0224
160.0	253.4	6.344	4294.0	5098.0	103.1	17.54	75.43	207.0	17.5	.0218
161.0	246.2	6.163	4343.0	5170.0	103.5	17.28	70.18	208.9	17.4	.0212
162.0	239.7	6.001	4389.0	5238.0	103.9	17.04	65.86	210.8	17.4	.0207
164.0	228.5	5.721	4472.0	5363.0	104.7	16.63	59.17	214.3	17.2	.0199
166.0	219.0	5.483	4546.0	5476.0	105.4	16.27	54.21	217.6	17.2	.0193
168.0	210.8	5.278	4614.0	5581.0	106.0	15.96	50.37	220.7	17.2	.0188
170.0	203.6	5.097	4678.0	5678.0	106.6	15.69	47.30	223.7	17.1	.0184
172.0	197.1	4.935	4737.0	5770.0	107.1	15.45	44.79	226.5	17.2	.0181
174.0	191.3	4.789	4793.0	5858.0	107.6	15.24	42.69	229.2	17.2	.0178
176.0	186.0	4.656	4846.0	5941.0	108.1	15.05	40.92	231.8	17.2	.0176
178.0	181.1	4.534	4897.0	6021.0	108.6	14.88	39.39	234.3	17.3	.0174
180.0	176.6	4.422	4946.0	6099.0	109.0	14.73	38.07	236.8	17.3	.0173
185.0	166.7	4.174	5060.0	6282.0	110.0	14.40	35.41	242.5	17.5	.0170
190.0	158.3	3.963	5167.0	6454.0	110.9	14.14	33.42	247.9	17.7	.0168
195.0	151.0	3.780	5268.0	6617.0	111.8	13.94	31.86	252.9	18.0	.0167
200.0	144.6	3.619	5364.0	6773.0	112.6	13.77	30.62	257.7	18.2	.0167
205.0	138.8	3.475	5456.0	6924.0	113.3	13.62	29.60	262.3	18.5	.0168
210.0	133.6	3.345	5545.0	7070.0	114.0	13.51	28.76	266.7	18.8	.0169
215.0	128.9	3.227	5631.0	7211.0	114.7	13.41	28.04	270.9	19.0	.0170
220.0	124.6	3.120	5715.0	7350.0	115.3	13.33	27.44	275.0	19.3	.0171
230.0	117.0	2.929	5878.0	7619.0	116.5	13.19	26.45	282.8	19.9	.0174
240.0	110.5	2.765	6035.0	7880.0	117.6	13.09	25.70	290.2	20.5	.0177
250.0	104.7	2.621	6188.0	8134.0	118.7	13.02	25.10	297.3	21.1	.0181
260.0	99.65	2.494	6338.0	8382.0	119.6	12.96	24.62	304.0	21.7	.0184
270.0	95.12	2.381	6485.0	8626.0	120.6	12.91	24.22	310.6	22.3	.0188

Thermophysical properties of argon—Continued

T K	Density kg/m^3	Density mol/dm^3	E J/mol	H J/mol	S J/(mol·K)	C_v J/(mol·K)	C_p J/(mol·K)	Sound m/s	Visc. μPa·s	Therm. W/(m·K)
280.0	91.03	2.279	6629.0	8867.0	121.4	12.88	23.89	316.8	22.9	.0192
290.0	87.33	2.186	6771.0	9104.0	122.3	12.85	23.60	322.9	23.5	.0196
300.0	83.95	2.102	6912.0	9339.0	123.1	12.82	23.36	328.8	24.1	.0200
310.0	80.85	2.024	7052.0	9572.0	123.8	12.80	23.15	334.6	24.7	.0204
320.0	78.00	1.953	7190.0	9802.0	124.6	12.78	22.97	340.2	25.2	.0208
340.0	72.91	1.825	7464.0	10260.0	125.9	12.75	22.66	351.1	26.4	.0216
360.0	68.49	1.714	7734.0	10710.0	127.2	12.73	22.42	361.5	27.5	.0225
380.0	64.62	1.617	8002.0	11160.0	128.4	12.71	22.22	371.5	28.6	.0233
400.0	61.18	1.532	8268.0	11600.0	129.6	12.70	22.06	381.1	29.7	.0241

5.20 MPa isobar

T K	Density kg/m^3	Density mol/dm^3	E J/mol	H J/mol	S J/(mol·K)	C_v J/(mol·K)	C_p J/(mol·K)	Sound m/s	Visc. μPa·s	Therm. W/(m·K)
85.07[a]	1422.0	35.59	−1369.0	−1223.0	54.89	23.17	43.91	839.5	292.0	.135
90.0	1393.0	34.86	−1156.0	−1007.0	57.36	21.72	43.93	826.2	253.0	.127
100.0	1331.0	33.32	−718.7	−562.7	62.04	20.16	45.05	775.3	195.0	.112
110.0	1264.0	31.65	−268.0	−103.7	66.41	19.22	46.86	708.5	153.0	.0995
115.0	1228.0	30.75	−35.6	133.6	68.52	18.83	48.09	671.6	136.0	.0936
120.0	1190.0	29.80	203.3	377.8	70.60	18.46	49.70	632.7	121.0	.0879
125.0	1149.0	28.77	450.8	631.5	72.67	18.13	51.90	591.4	108.0	.0822
130.0	1104.0	27.64	710.3	898.4	74.76	17.84	55.06	547.2	95.5	.0765
135.0	1054.0	26.37	987.7	1185.0	76.93	17.61	59.88	499.1	84.0	.0705
136.0	1043.0	26.10	1046.0	1245.0	77.37	17.58	61.16	488.8	81.7	.0693
138.0	1019.0	25.51	1167.0	1371.0	78.29	17.52	64.18	467.5	77.2	.0668
140.0	993.8	24.88	1294.0	1503.0	79.24	17.48	68.05	444.9	72.6	.0643
142.0	965.9	24.18	1429.0	1644.0	80.24	17.47	73.21	420.7	68.1	.0618
144.0	934.5	23.39	1575.0	1797.0	81.31	17.51	80.47	394.4	63.3	.0593
145.0	917.0	22.96	1653.0	1880.0	81.88	17.55	85.39	380.2	60.9	.0581
146.0	898.0	22.48	1737.0	1968.0	82.49	17.61	91.67	365.1	58.3	.0568
147.0	876.8	21.95	1827.0	2064.0	83.14	17.70	100.0	348.8	55.7	.0556
148.0	852.9	21.35	1926.0	2169.0	83.86	17.83	111.8	331.0	52.8	.0544
149.0	824.8	20.65	2038.0	2290.0	84.67	18.03	130.1	311.1	49.7	.0531
149.5	808.4	20.24	2101.0	2358.0	85.12	18.16	143.7	300.1	48.0	.0525
150.0	789.8	19.77	2171.0	2434.0	85.63	18.32	162.8	288.1	46.1	.0520
150.5	767.8	19.22	2252.0	2522.0	86.22	18.54	191.9	274.7	44.0	.0514
150.6	762.9	19.10	2270.0	2542.0	86.35	18.59	199.8	271.9	43.6	.0513
150.8	752.3	18.83	2307.0	2584.0	86.63	18.70	218.4	265.8	42.6	.0511
151.0	740.5	18.54	2349.0	2630.0	86.93	18.83	242.6	259.4	41.6	.0510
151.2	727.2	18.20	2396.0	2681.0	87.27	18.99	275.5	252.5	40.4	.0509
151.4	711.7	17.82	2449.0	2741.0	87.67	19.18	322.6	244.9	39.1	.0509
151.6	693.0	17.35	2512.0	2812.0	88.14	19.41	395.8	236.4	37.7	.0512
151.8	669.0	16.75	2592.0	2903.0	88.73	19.72	523.1	226.8	35.9	.0519
151.9	653.9	16.37	2642.0	2960.0	89.11	19.91	626.7	221.4	34.8	.0521
152.0	635.5	15.91	2703.0	3029.0	89.57	20.15	777.1	215.5	33.5	.0532
152.1	612.5	15.33	2778.0	3117.0	90.15	20.43	987.8	209.2	32.0	.0548
152.2	584.1	14.62	2872.0	3228.0	90.87	20.75	1214.0	202.9	30.3	.0566
152.3	552.3	13.83	2979.0	3355.0	91.71	21.05	1292.0	197.6	28.4	.0574
152.4	522.5	13.08	3082.0	3479.0	92.53	21.25	1177.0	194.0	26.9	.0560
152.5	497.4	12.45	3171.0	3588.0	93.24	21.35	1003.0	191.9	25.6	.0533
152.6	477.1	11.94	3245.0	3681.0	93.85	21.37	847.9	190.7	24.7	.0502
152.7	460.4	11.52	3308.0	3759.0	94.36	21.35	725.0	190.1	24.0	.0475
152.8	446.5	11.18	3361.0	3827.0	94.80	21.30	629.4	189.8	23.4	.0451
152.9	434.7	10.88	3408.0	3886.0	95.19	21.24	554.7	189.7	22.9	.0431
153.0	424.5	10.63	3449.0	3938.0	95.53	21.16	495.4	189.7	22.5	.0415
153.2	407.6	10.20	3518.0	4028.0	96.12	21.01	408.3	190.0	21.8	.0388
153.4	393.9	9.861	3576.0	4103.0	96.61	20.85	348.2	190.4	21.4	.0362
153.6	382.6	9.577	3625.0	4168.0	97.03	20.69	304.5	190.8	21.0	.0348
153.8	372.9	9.333	3668.0	4226.0	97.41	20.54	271.3	191.4	20.6	.0336
154.0	364.4	9.121	3707.0	4277.0	97.74	20.39	245.4	191.9	20.4	.0326
154.2	356.8	8.933	3742.0	4324.0	98.04	20.25	224.6	192.5	20.1	.0317
154.5	347.0	8.685	3789.0	4388.0	98.46	20.06	200.0	193.3	19.8	.0306
155.0	333.2	8.341	3856.0	4480.0	99.05	19.75	170.5	194.7	19.4	.0290
155.5	321.8	8.057	3914.0	4560.0	99.57	19.48	149.8	196.0	19.1	.0278

Thermophysical properties of argon—Continued

T K	Density kg/m³	Density mol/dm³	E J/mol	H J/mol	S J/(mol·K)	C_v J/(mol·K)	C_p J/(mol·K)	Sound m/s	Visc. μPa·s	Therm. W/(m·K)
156.0	312.2	7.814	3965.0	4630.0	100.0	19.22	134.4	197.3	18.8	.0269
156.5	303.7	7.603	4011.0	4695.0	100.4	18.99	122.5	198.6	18.6	.0260
157.0	296.3	7.417	4052.0	4753.0	100.8	18.77	112.9	199.8	18.5	.0253
158.0	283.5	7.098	4126.0	4859.0	101.5	18.39	98.57	202.1	18.2	.0242
159.0	272.9	6.831	4191.0	4952.0	102.1	18.04	88.25	204.3	18.0	.0232
160.0	263.7	6.602	4248.0	5036.0	102.6	17.74	80.42	206.4	17.8	.0225
161.0	255.7	6.402	4301.0	5113.0	103.1	17.46	74.28	208.4	17.7	.0218
162.0	248.6	6.224	4349.0	5185.0	103.5	17.21	69.30	210.3	17.6	.0213
164.0	236.4	5.918	4437.0	5316.0	104.3	16.77	61.72	213.9	17.4	.0204
166.0	226.2	5.662	4515.0	5433.0	105.0	16.40	56.19	217.3	17.4	.0197
168.0	217.4	5.442	4586.0	5541.0	105.7	16.07	51.97	220.4	17.3	.0192
170.0	209.7	5.249	4651.0	5642.0	106.3	15.79	48.62	223.4	17.3	.0188
172.0	202.9	5.078	4712.0	5736.0	106.8	15.54	45.91	226.3	17.3	.0184
174.0	196.7	4.924	4769.0	5826.0	107.3	15.32	43.65	229.0	17.3	.0181
176.0	191.1	4.784	4824.0	5911.0	107.8	15.12	41.75	231.6	17.3	.0178
178.0	186.0	4.656	4876.0	5993.0	108.3	14.95	40.13	234.2	17.4	.0176
180.0	181.3	4.538	4926.0	6072.0	108.7	14.79	38.72	236.6	17.4	.0174
185.0	171.0	4.279	5043.0	6258.0	109.8	14.45	35.92	242.4	17.6	.0171
190.0	162.2	4.060	5151.0	6432.0	110.7	14.19	33.82	247.8	17.8	.0170
195.0	154.6	3.870	5253.0	6597.0	111.5	13.97	32.19	252.9	18.0	.0169
200.0	147.9	3.703	5350.0	6754.0	112.3	13.80	30.90	257.7	18.3	.0168
205.0	142.0	3.554	5443.0	6906.0	113.1	13.65	29.84	262.3	18.5	.0169
210.0	136.6	3.420	5533.0	7053.0	113.8	13.53	28.96	266.7	18.8	.0169
215.0	131.8	3.299	5620.0	7196.0	114.5	13.43	28.23	271.0	19.1	.0170
220.0	127.4	3.188	5704.0	7336.0	115.1	13.34	27.60	275.1	19.4	.0172
230.0	119.5	2.992	5868.0	7606.0	116.3	13.21	26.59	282.9	19.9	.0174
240.0	112.8	2.823	6026.0	7868.0	117.4	13.11	25.81	290.3	20.5	.0178
250.0	106.9	2.676	6180.0	8123.0	118.5	13.03	25.20	297.4	21.1	.0181
260.0	101.7	2.546	6330.0	8372.0	119.4	12.97	24.70	304.2	21.7	.0185
270.0	97.06	2.430	6477.0	8617.0	120.4	12.92	24.29	310.7	22.3	.0189
280.0	92.88	2.325	6622.0	8858.0	121.2	12.88	23.95	317.0	22.9	.0192
290.0	89.09	2.230	6765.0	9096.0	122.1	12.85	23.66	323.1	23.5	.0196
300.0	85.64	2.144	6906.0	9332.0	122.9	12.83	23.41	329.0	24.1	.0200
310.0	82.47	2.064	7046.0	9565.0	123.6	12.81	23.20	334.8	24.7	.0205
320.0	79.55	1.991	7185.0	9796.0	124.4	12.79	23.01	340.4	25.3	.0209
340.0	74.35	1.861	7459.0	10250.0	125.8	12.76	22.70	351.2	26.4	.0217
360.0	69.84	1.748	7730.0	10700.0	127.1	12.73	22.45	361.6	27.5	.0225
380.0	65.88	1.649	7998.0	11150.0	128.3	12.72	22.25	371.7	28.6	.0233
400.0	62.38	1.561	8264.0	11590.0	129.4	12.70	22.08	381.3	29.7	.0241
				5.30 MPa isobar						
85.10[a]	1422.0	35.59	−1369.0	−1220.0	54.89	23.16	43.89	839.9	292.0	.135
90.0	1393.0	34.87	−1157.0	−1005.0	57.35	21.72	43.91	826.7	254.0	.127
100.0	1331.0	33.33	−720.1	−561.1	62.02	20.17	45.03	775.8	195.0	.112
110.0	1265.0	31.66	−269.8	−102.4	66.39	19.23	46.82	709.1	153.0	.0996
115.0	1229.0	30.76	−37.7	134.6	68.50	18.83	48.03	672.4	136.0	.0937
120.0	1191.0	29.81	200.7	378.5	70.58	18.47	49.63	633.6	121.0	.0880
125.0	1150.0	28.79	447.7	631.8	72.65	18.14	51.80	592.5	108.0	.0823
130.0	1105.0	27.67	706.5	898.0	74.73	17.85	54.90	548.5	95.8	.0766
135.0	1055.0	26.41	982.7	1183.0	76.89	17.62	59.61	500.7	84.2	.0707
136.0	1044.0	26.13	1041.0	1244.0	77.33	17.58	60.85	490.6	82.0	.0695
138.0	1021.0	25.55	1161.0	1368.0	78.24	17.52	63.78	469.5	77.5	.0670
140.0	995.8	24.93	1287.0	1499.0	79.18	17.48	67.51	447.2	73.0	.0645
142.0	968.2	24.24	1420.0	1639.0	80.17	17.47	72.43	423.3	68.4	.0620
144.0	937.4	23.47	1564.0	1790.0	81.23	17.50	79.29	397.4	63.8	.0595
145.0	920.3	23.04	1642.0	1872.0	81.80	17.54	83.87	383.5	61.3	.0583
146.0	901.8	22.57	1724.0	1958.0	82.39	17.59	89.65	368.8	58.8	.0571
147.0	881.4	22.06	1811.0	2052.0	83.03	17.67	97.21	353.0	56.2	.0558
148.0	858.4	21.49	1907.0	2154.0	83.72	17.79	107.6	335.9	53.5	.0546
149.0	831.9	20.82	2014.0	2269.0	84.49	17.96	123.0	317.0	50.5	.0534
149.5	816.7	20.44	2073.0	2333.0	84.92	18.07	134.0	306.7	48.9	.0528

Thermophysical properties of argon—Continued

T K	Density kg/m³	Density mol/dm³	E J/mol	H J/mol	S J/(mol·K)	C_v J/(mol·K)	C_p J/(mol·K)	Sound m/s	Visc. μPa·s	Therm. W/(m·K)
150.0	799.8	20.02	2138.0	2403.0	85.39	18.21	148.7	295.5	47.1	.0522
150.5	780.4	19.53	2211.0	2482.0	85.92	18.39	169.6	283.4	45.2	.0516
151.0	757.3	18.96	2295.0	2575.0	86.53	18.62	201.7	269.8	43.1	.0511
151.5	728.4	18.23	2397.0	2688.0	87.28	18.94	258.2	254.4	40.6	.0507
151.6	721.5	18.06	2421.0	2715.0	87.46	19.02	275.0	250.9	40.0	.0506
151.8	706.2	17.68	2474.0	2774.0	87.85	19.20	318.4	243.7	38.7	.0506
152.0	688.0	17.22	2536.0	2843.0	88.31	19.43	382.2	235.7	37.3	.0508
152.2	665.6	16.66	2611.0	2929.0	88.87	19.71	481.6	227.0	35.6	.0508
152.3	652.1	16.32	2656.0	2981.0	89.21	19.88	552.7	222.3	34.7	.0512
152.4	636.5	15.93	2707.0	3040.0	89.60	20.07	642.7	217.3	33.6	.0517
152.5	618.4	15.48	2767.0	3110.0	90.05	20.29	750.5	212.3	32.4	.0524
152.6	597.7	14.96	2836.0	3190.0	90.58	20.52	861.0	207.3	31.1	.0532
152.7	574.9	14.39	2913.0	3281.0	91.18	20.76	938.8	202.7	29.8	.0538
152.8	551.5	13.81	2992.0	3376.0	91.80	20.96	951.5	199.0	28.4	.0536
152.9	529.3	13.25	3069.0	3469.0	92.41	21.11	904.9	196.3	27.3	.0526
153.0	509.2	12.75	3140.0	3556.0	92.98	21.20	828.8	194.4	26.3	.0510
153.1	491.6	12.31	3204.0	3635.0	93.49	21.24	746.8	193.1	25.4	.0491
153.2	476.3	11.92	3261.0	3705.0	93.95	21.24	670.3	192.3	24.7	.0472
153.3	462.9	11.59	3312.0	3769.0	94.37	21.22	602.9	191.8	24.1	.0454
153.4	451.1	11.29	3357.0	3826.0	94.74	21.18	545.0	191.5	23.6	.0437
153.6	431.4	10.80	3435.0	3926.0	95.39	21.07	453.6	191.3	22.8	.0409
153.8	415.5	10.40	3500.0	4009.0	95.93	20.94	387.0	191.4	22.2	.0387
154.0	402.3	10.07	3555.0	4082.0	96.40	20.80	337.2	191.7	21.7	.0370
154.2	391.1	9.790	3604.0	4145.0	96.81	20.65	299.1	192.1	21.3	.0351
154.4	381.3	9.546	3647.0	4202.0	97.18	20.51	269.1	192.6	21.0	.0340
154.6	372.8	9.331	3685.0	4253.0	97.51	20.38	245.0	193.1	20.7	.0330
154.8	365.1	9.140	3720.0	4300.0	97.82	20.24	225.3	193.6	20.4	.0322
155.0	358.2	8.967	3752.0	4343.0	98.10	20.12	208.8	194.1	20.2	.0314
155.5	343.5	8.599	3823.0	4440.0	98.72	19.82	177.6	195.4	19.8	.0298
156.0	331.4	8.295	3884.0	4523.0	99.25	19.54	155.6	196.7	19.4	.0285
156.5	321.1	8.038	3937.0	4596.0	99.72	19.29	139.2	198.0	19.2	.0275
157.0	312.2	7.814	3984.0	4662.0	100.1	19.05	126.6	199.2	18.9	.0266
157.5	304.3	7.617	4027.0	4723.0	100.5	18.84	116.5	200.4	18.7	.0259
158.0	297.2	7.440	4067.0	4779.0	100.9	18.64	108.2	201.5	18.6	.0252
159.0	285.0	7.135	4138.0	4881.0	101.5	18.27	95.48	203.8	18.3	.0241
160.0	274.7	6.877	4201.0	4971.0	102.1	17.94	86.10	205.9	18.1	.0233
161.0	265.8	6.654	4257.0	5054.0	102.6	17.65	78.87	207.9	18.0	.0225
162.0	258.0	6.457	4309.0	5130.0	103.1	17.38	73.11	209.9	17.8	.0219
163.0	250.9	6.282	4357.0	5200.0	103.5	17.14	68.41	211.7	17.7	.0214
164.0	244.6	6.124	4401.0	5267.0	103.9	16.92	64.49	213.5	17.7	.0209
166.0	233.6	5.847	4483.0	5389.0	104.7	16.53	58.32	216.9	17.5	.0202
168.0	224.1	5.611	4556.0	5501.0	105.3	16.19	53.66	220.1	17.5	.0196
170.0	215.9	5.406	4624.0	5604.0	105.9	15.89	50.01	223.2	17.4	.0191
172.0	208.7	5.224	4687.0	5701.0	106.5	15.63	47.07	226.1	17.4	.0187
174.0	202.2	5.061	4746.0	5793.0	107.0	15.40	44.65	228.8	17.4	.0183
176.0	196.3	4.914	4802.0	5880.0	107.5	15.20	42.62	231.5	17.5	.0181
178.0	190.9	4.780	4855.0	5964.0	108.0	15.02	40.89	234.0	17.5	.0178
180.0	186.0	4.656	4906.0	6044.0	108.5	14.85	39.40	236.5	17.5	.0176
185.0	175.2	4.386	5025.0	6233.0	109.5	14.50	36.43	242.3	17.7	.0173
190.0	166.1	4.158	5135.0	6410.0	110.4	14.23	34.23	247.7	17.9	.0171
195.0	158.2	3.960	5238.0	6576.0	111.3	14.01	32.53	252.9	18.1	.0170
200.0	151.3	3.787	5336.0	6735.0	112.1	13.83	31.18	257.7	18.4	.0170
205.0	145.2	3.634	5430.0	6889.0	112.9	13.68	30.08	262.3	18.6	.0170
210.0	139.6	3.496	5520.0	7037.0	113.6	13.55	29.18	266.8	18.9	.0170
215.0	134.6	3.371	5608.0	7181.0	114.3	13.45	28.41	271.0	19.1	.0171
220.0	130.1	3.256	5693.0	7321.0	114.9	13.36	27.77	275.2	19.4	.0172
230.0	122.0	3.055	5858.0	7593.0	116.1	13.22	26.72	283.0	20.0	.0175
240.0	115.1	2.882	6017.0	7856.0	117.2	13.12	25.92	290.4	20.6	.0178
250.0	109.1	2.731	6171.0	8112.0	118.3	13.04	25.29	297.5	21.2	.0182
260.0	103.8	2.598	6322.0	8362.0	119.3	12.98	24.78	304.3	21.8	.0185
270.0	99.01	2.478	6470.0	8608.0	120.2	12.93	24.36	310.8	22.3	.0189
280.0	94.73	2.371	6615.0	8850.0	121.1	12.89	24.01	317.1	22.9	.0193

Thermophysical properties of argon—Continued

T K	Density kg/m³	Density mol/dm³	E J/mol	H J/mol	S J/(mol·K)	C_v J/(mol·K)	C_p J/(mol·K)	Sound m/s	Visc. μPa·s	Therm. W/(m·K)
290.0	90.86	2.274	6758.0	9089.0	121.9	12.86	23.72	323.2	23.5	.0197
300.0	87.32	2.186	6900.0	9324.0	122.7	12.83	23.46	329.2	24.1	.0201
310.0	84.09	2.105	7040.0	9558.0	123.5	12.81	23.24	334.9	24.7	.0205
320.0	81.11	2.030	7179.0	9789.0	124.2	12.79	23.05	340.6	25.3	.0209
340.0	75.79	1.897	7454.0	10250.0	125.6	12.76	22.73	351.4	26.4	.0217
360.0	71.19	1.782	7725.0	10700.0	126.9	12.74	22.48	361.8	27.6	.0225
380.0	67.15	1.681	7994.0	11150.0	128.1	12.72	22.27	371.8	28.7	.0234
400.0	63.57	1.591	8260.0	11590.0	129.2	12.70	22.10	381.5	29.8	.0242
				5.40 MPa isobar						
85.12[a]	1422.0	35.59	−1369.0	−1217.0	54.89	23.16	43.88	840.2	292.0	.135
90.0	1393.0	34.88	−1158.0	−1003.0	57.34	21.73	43.90	827.1	254.0	.127
100.0	1332.0	33.34	−721.5	−559.5	62.01	20.17	45.00	776.3	195.0	.112
110.0	1265.0	31.67	−271.7	−101.2	66.38	19.23	46.78	709.8	153.0	.0996
115.0	1229.0	30.78	−39.9	135.6	68.48	18.84	47.98	673.1	136.0	.0938
120.0	1192.0	29.83	198.2	379.2	70.56	18.48	49.55	634.5	122.0	.0881
125.0	1151.0	28.81	444.6	632.0	72.62	18.15	51.69	593.5	108.0	.0824
130.0	1106.0	27.70	702.7	897.6	74.70	17.86	54.74	549.8	96.0	.0767
135.0	1056.0	26.44	977.8	1182.0	76.85	17.62	59.34	502.4	84.5	.0708
136.0	1045.0	26.17	1036.0	1242.0	77.29	17.59	60.54	492.3	82.3	.0696
138.0	1022.0	25.59	1155.0	1366.0	78.19	17.52	63.39	471.4	77.8	.0672
140.0	997.7	24.97	1280.0	1496.0	79.13	17.48	66.98	449.3	73.3	.0647
142.0	970.5	24.29	1412.0	1634.0	80.11	17.47	71.69	425.8	68.8	.0622
144.0	940.2	23.54	1554.0	1784.0	81.16	17.49	78.18	400.4	64.2	.0597
146.0	905.5	22.67	1711.0	1949.0	82.30	17.57	87.80	372.4	59.3	.0573
147.0	885.7	22.17	1797.0	2040.0	82.92	17.65	94.67	357.1	56.8	.0561
148.0	863.6	21.62	1890.0	2139.0	83.59	17.75	103.9	340.6	54.1	.0549
149.0	838.5	20.99	1992.0	2249.0	84.33	17.90	117.2	322.5	51.2	.0537
149.5	824.3	20.63	2048.0	2310.0	84.74	18.00	126.3	312.7	49.7	.0531
150.0	808.6	20.24	2109.0	2376.0	85.18	18.12	138.0	302.3	48.0	.0525
150.5	791.0	19.80	2176.0	2449.0	85.66	18.27	153.9	291.0	46.3	.0519
151.0	770.8	19.30	2251.0	2531.0	86.21	18.46	176.5	278.8	44.3	.0513
151.5	746.7	18.69	2339.0	2627.0	86.85	18.71	211.5	265.2	42.2	.0508
152.0	716.2	17.93	2446.0	2747.0	87.63	19.05	272.8	249.7	39.6	.0504
152.2	701.2	17.55	2498.0	2805.0	88.02	19.23	312.0	242.8	38.4	.0503
152.4	683.7	17.11	2557.0	2873.0	88.46	19.44	366.3	235.4	37.0	.0503
152.6	662.7	16.59	2628.0	2953.0	88.99	19.69	443.5	227.4	35.5	.0501
152.8	637.1	15.95	2714.0	3052.0	89.64	20.00	551.2	219.0	33.7	.0506
153.0	605.6	15.16	2819.0	3175.0	90.44	20.36	677.7	210.6	31.6	.0513
153.2	569.4	14.25	2941.0	3320.0	91.39	20.72	756.2	203.3	29.5	.0515
153.4	533.3	13.35	3066.0	3470.0	92.37	20.98	731.5	198.3	27.5	.0502
153.6	501.8	12.56	3179.0	3609.0	93.27	21.10	647.4	195.4	26.0	.0478
153.8	475.7	11.91	3276.0	3729.0	94.05	21.12	556.2	193.9	24.7	.0451
154.0	454.4	11.38	3357.0	3832.0	94.72	21.06	477.2	193.2	23.8	.0425
154.2	436.8	10.94	3427.0	3921.0	95.30	20.97	413.4	192.9	23.1	.0403
154.4	422.1	10.57	3487.0	3998.0	95.80	20.85	362.8	193.0	22.5	.0385
154.6	409.5	10.25	3540.0	4067.0	96.24	20.73	322.5	193.2	22.0	.0365
154.8	398.6	9.977	3587.0	4128.0	96.64	20.60	290.2	193.5	21.7	.0353
155.0	389.0	9.737	3629.0	4183.0	96.99	20.47	263.8	193.9	21.3	.0343
155.2	380.4	9.523	3667.0	4234.0	97.32	20.34	242.0	194.3	21.0	.0333
155.4	372.8	9.331	3701.0	4280.0	97.62	20.22	223.8	194.8	20.8	.0325
156.0	353.6	8.851	3791.0	4401.0	98.40	19.86	183.6	196.2	20.2	.0305
156.5	340.8	8.530	3854.0	4487.0	98.95	19.59	160.7	197.4	19.8	.0292
157.0	329.9	8.258	3909.0	4563.0	99.43	19.34	143.6	198.7	19.5	.0281
157.5	320.5	8.023	3958.0	4632.0	99.87	19.11	130.4	199.8	19.2	.0272
158.0	312.2	7.815	4003.0	4694.0	100.3	18.89	119.8	201.0	19.0	.0264
159.0	298.1	7.462	4082.0	4805.0	101.0	18.50	103.9	203.3	18.7	.0251
160.0	286.4	7.170	4150.0	4903.0	101.6	18.15	92.58	205.4	18.4	.0241
161.0	276.4	6.920	4211.0	4991.0	102.1	17.84	84.02	207.5	18.2	.0233
162.0	267.7	6.702	4266.0	5072.0	102.6	17.56	77.33	209.4	18.1	.0226
163.0	260.1	6.510	4317.0	5147.0	103.1	17.30	71.95	211.3	18.0	.0220

Thermophysical properties of argon—Continued

T K	Density kg/m³	Density mol/dm³	E J/mol	H J/mol	S J/(mol·K)	C_v J/(mol·K)	C_p J/(mol·K)	Sound m/s	Visc. μPa·s	Therm. W/(m·K)
164.0	253.2	6.337	4364.0	5216.0	103.5	17.07	67.51	213.1	17.9	.0215
166.0	241.2	6.038	4450.0	5344.0	104.3	16.66	60.60	216.6	17.7	.0206
168.0	231.1	5.785	4526.0	5460.0	105.0	16.30	55.46	219.8	17.6	.0200
170.0	222.3	5.566	4596.0	5567.0	105.6	16.00	51.48	222.9	17.6	.0194
172.0	214.6	5.373	4661.0	5666.0	106.2	15.73	48.30	225.8	17.6	.0190
174.0	207.8	5.201	4722.0	5760.0	106.7	15.49	45.69	228.6	17.6	.0186
176.0	201.6	5.046	4779.0	5849.0	107.2	15.27	43.52	231.3	17.6	.0183
178.0	195.9	4.905	4833.0	5934.0	107.7	15.08	41.67	233.9	17.6	.0181
180.0	190.8	4.776	4885.0	6016.0	108.2	14.91	40.09	236.4	17.7	.0178
185.0	179.5	4.494	5007.0	6208.0	109.2	14.55	36.96	242.2	17.8	.0175
190.0	170.0	4.256	5118.0	6387.0	110.2	14.27	34.65	247.7	18.0	.0172
195.0	161.9	4.052	5223.0	6556.0	111.1	14.04	32.88	252.8	18.2	.0171
200.0	154.7	3.873	5322.0	6716.0	111.9	13.86	31.47	257.7	18.4	.0171
205.0	148.4	3.714	5417.0	6871.0	112.6	13.70	30.33	262.4	18.7	.0171
210.0	142.7	3.572	5508.0	7020.0	113.4	13.58	29.39	266.8	18.9	.0171
215.0	137.5	3.443	5597.0	7165.0	114.0	13.47	28.60	271.1	19.2	.0172
220.0	132.8	3.325	5682.0	7306.0	114.7	13.38	27.93	275.2	19.5	.0173
230.0	124.6	3.118	5848.0	7580.0	115.9	13.24	26.86	283.1	20.0	.0176
240.0	117.5	2.941	6008.0	7844.0	117.0	13.13	26.04	290.6	20.6	.0179
250.0	111.3	2.786	6163.0	8101.0	118.1	13.05	25.39	297.7	21.2	.0182
260.0	105.8	2.649	6314.0	8352.0	119.1	12.99	24.87	304.4	21.8	.0186
270.0	101.0	2.527	6462.0	8599.0	120.0	12.94	24.44	311.0	22.4	.0190
280.0	96.59	2.418	6608.0	8841.0	120.9	12.90	24.08	317.3	23.0	.0193
290.0	92.62	2.319	6752.0	9081.0	121.7	12.87	23.78	323.4	23.6	.0197
300.0	89.01	2.228	6894.0	9317.0	122.5	12.84	23.52	329.3	24.1	.0201
310.0	85.70	2.145	7034.0	9551.0	123.3	12.82	23.29	335.1	24.7	.0205
320.0	82.66	2.069	7173.0	9783.0	124.0	12.80	23.10	340.7	25.3	.0209
340.0	77.23	1.933	7449.0	10240.0	125.4	12.77	22.77	351.6	26.5	.0218
360.0	72.53	1.816	7720.0	10690.0	126.7	12.74	22.51	362.0	27.6	.0226
380.0	68.41	1.713	7989.0	11140.0	127.9	12.73	22.30	372.0	28.7	.0234
400.0	64.77	1.621	8256.0	11590.0	129.1	12.71	22.13	381.7	29.8	.0242

5.50 MPa isobar

T K	Density kg/m³	Density mol/dm³	E J/mol	H J/mol	S J/(mol·K)	C_v J/(mol·K)	C_p J/(mol·K)	Sound m/s	Visc. μPa·s	Therm. W/(m·K)
85.15[a]	1422.0	35.60	−1369.0	−1214.0	54.89	23.15	43.86	840.6	292.0	.135
90.0	1394.0	34.89	−1159.0	−1002.0	57.32	21.73	43.88	827.5	254.0	.127
100.0	1332.0	33.35	−722.9	−558.0	62.00	20.18	44.98	776.9	195.0	.112
110.0	1266.0	31.68	−273.5	−99.9	66.36	19.24	46.74	710.4	153.0	.0997
115.0	1230.0	30.79	−42.0	136.6	68.46	18.85	47.93	673.9	137.0	.0939
120.0	1192.0	29.85	195.7	379.9	70.53	18.49	49.48	635.3	122.0	.0882
125.0	1152.0	28.83	441.6	632.3	72.59	18.16	51.59	594.6	109.0	.0826
130.0	1107.0	27.72	698.9	897.3	74.67	17.87	54.58	551.1	96.3	.0768
135.0	1058.0	26.48	972.9	1181.0	76.81	17.63	59.07	504.0	84.8	.0710
136.0	1047.0	26.20	1030.0	1240.0	77.25	17.59	60.25	494.0	82.6	.0698
138.0	1024.0	25.64	1149.0	1363.0	78.15	17.53	63.01	473.3	78.1	.0674
140.0	999.6	25.02	1273.0	1493.0	79.08	17.48	66.47	451.5	73.7	.0649
142.0	972.8	24.35	1404.0	1630.0	80.05	17.47	70.99	428.3	69.2	.0624
144.0	943.0	23.61	1545.0	1778.0	81.09	17.48	77.14	403.3	64.6	.0600
146.0	909.0	22.75	1699.0	1940.0	82.21	17.56	86.10	375.9	59.8	.0575
147.0	889.8	22.27	1783.0	2029.0	82.82	17.62	92.39	361.0	57.3	.0563
148.0	868.5	21.74	1873.0	2126.0	83.47	17.72	100.7	345.0	54.7	.0551
149.0	844.5	21.14	1972.0	2232.0	84.18	17.85	112.2	327.7	51.9	.0539
150.0	816.5	20.44	2083.0	2352.0	84.99	18.05	129.6	308.5	48.9	.0527
150.5	800.4	20.04	2146.0	2420.0	85.44	18.17	142.2	298.0	47.2	.0521
151.0	782.2	19.58	2214.0	2495.0	85.94	18.33	159.1	286.7	45.4	.0515
151.5	761.2	19.06	2292.0	2580.0	86.50	18.53	183.4	274.4	43.5	.0510
152.0	736.1	18.43	2382.0	2681.0	87.16	18.79	220.9	260.8	41.3	.0504
152.5	704.2	17.63	2494.0	2806.0	87.98	19.15	285.4	245.4	38.6	.0500
152.6	696.7	17.44	2520.0	2835.0	88.18	19.24	304.0	242.1	38.0	.0499
152.8	679.9	17.02	2577.0	2900.0	88.60	19.44	349.6	235.2	36.8	.0498
153.0	660.4	16.53	2643.0	2976.0	89.10	19.67	409.4	228.0	35.3	.0496
153.2	637.5	15.96	2720.0	3065.0	89.68	19.94	484.4	220.6	33.7	.0497
153.4	610.7	15.29	2810.0	3170.0	90.36	20.24	564.9	213.3	32.0	.0500

Thermophysical properties of argon—Continued

T K	Density kg/m^3	Density mol/dm^3	E J/mol	H J/mol	S J/(mol·K)	C_v J/(mol·K)	C_p J/(mol·K)	Sound m/s	Visc. μPa·s	Therm. W/(m·K)
153.6	581.0	14.54	2911.0	3289.0	91.14	20.53	621.9	206.9	30.2	.0499
153.8	550.5	13.78	3016.0	3415.0	91.96	20.78	628.4	202.0	28.5	.0491
154.0	522.1	13.07	3117.0	3538.0	92.76	20.93	591.3	198.7	27.0	.0475
154.2	496.9	12.44	3208.0	3651.0	93.49	20.99	535.0	196.6	25.8	.0456
154.4	475.4	11.90	3289.0	3752.0	94.14	20.99	476.5	195.4	24.8	.0435
154.6	456.9	11.44	3361.0	3842.0	94.73	20.94	423.2	194.8	24.0	.0405
154.8	441.2	11.04	3423.0	3921.0	95.24	20.86	377.2	194.6	23.3	.0390
155.0	427.5	10.70	3479.0	3993.0	95.70	20.76	338.5	194.6	22.8	.0377
155.2	415.6	10.40	3529.0	4057.0	96.12	20.64	306.2	194.7	22.3	.0364
155.4	405.1	10.14	3573.0	4116.0	96.50	20.53	279.1	195.0	22.0	.0354
155.6	395.7	9.907	3614.0	4169.0	96.84	20.41	256.3	195.3	21.6	.0344
155.8	387.3	9.696	3651.0	4219.0	97.16	20.29	237.0	195.6	21.3	.0336
156.0	379.7	9.506	3686.0	4264.0	97.45	20.18	220.5	196.0	21.1	.0328
156.5	363.4	9.097	3761.0	4366.0	98.10	19.90	188.3	197.1	20.5	.0311
157.0	350.0	8.760	3826.0	4454.0	98.66	19.63	165.0	198.3	20.1	.0298
157.5	338.6	8.475	3883.0	4532.0	99.16	19.38	147.4	199.4	19.8	.0287
158.0	328.7	8.228	3934.0	4602.0	99.60	19.15	133.7	200.6	19.5	.0277
158.5	320.0	8.011	3979.0	4666.0	100.0	18.94	122.8	201.7	19.3	.0269
159.0	312.3	7.817	4022.0	4725.0	100.4	18.73	113.8	202.8	19.1	.0262
160.0	298.9	7.483	4097.0	4832.0	101.0	18.37	100.00	205.0	18.8	.0250
161.0	287.7	7.202	4163.0	4926.0	101.6	18.04	89.84	207.1	18.6	.0241
162.0	278.1	6.960	4222.0	5012.0	102.2	17.74	82.03	209.1	18.4	.0233
163.0	269.6	6.748	4276.0	5091.0	102.7	17.47	75.84	211.0	18.2	.0226
164.0	262.0	6.560	4326.0	5164.0	103.1	17.22	70.79	212.8	18.1	.0220
166.0	249.1	6.235	4415.0	5298.0	103.9	16.79	63.05	216.3	17.9	.0211
168.0	238.2	5.963	4495.0	5418.0	104.6	16.42	57.38	219.6	17.8	.0204
170.0	228.9	5.730	4568.0	5528.0	105.3	16.10	53.03	222.7	17.8	.0198
172.0	220.7	5.525	4635.0	5630.0	105.9	15.82	49.58	225.6	17.7	.0193
174.0	213.5	5.344	4697.0	5727.0	106.4	15.57	46.78	228.4	17.7	.0189
176.0	207.0	5.180	4756.0	5818.0	107.0	15.35	44.45	231.1	17.7	.0186
178.0	201.0	5.032	4812.0	5905.0	107.4	15.15	42.49	233.7	17.7	.0183
180.0	195.6	4.897	4865.0	5988.0	107.9	14.98	40.81	236.2	17.8	.0181
182.0	190.7	4.772	4916.0	6068.0	108.4	14.82	39.35	238.7	17.8	.0179
185.0	183.9	4.603	4988.0	6183.0	109.0	14.61	37.51	242.2	17.9	.0176
190.0	174.0	4.356	5102.0	6365.0	110.0	14.31	35.08	247.7	18.1	.0174
195.0	165.5	4.144	5208.0	6535.0	110.8	14.08	33.23	252.8	18.3	.0172
200.0	158.1	3.959	5308.0	6697.0	111.7	13.89	31.76	257.7	18.5	.0172
205.0	151.6	3.795	5404.0	6853.0	112.4	13.73	30.58	262.4	18.7	.0172
210.0	145.7	3.648	5496.0	7004.0	113.2	13.60	29.61	266.9	19.0	.0172
215.0	140.4	3.515	5585.0	7150.0	113.8	13.49	28.79	271.2	19.3	.0173
220.0	135.6	3.394	5671.0	7292.0	114.5	13.40	28.10	275.3	19.5	.0174
225.0	131.2	3.284	5756.0	7431.0	115.1	13.32	27.51	279.3	19.8	.0175
230.0	127.1	3.182	5838.0	7567.0	115.7	13.25	26.99	283.2	20.1	.0177
240.0	119.8	3.000	5999.0	7832.0	116.8	13.14	26.15	290.7	20.7	.0180
250.0	113.5	2.841	6154.0	8090.0	117.9	13.06	25.48	297.8	21.2	.0183
260.0	107.9	2.701	6306.0	8343.0	118.9	13.00	24.95	304.6	21.8	.0186
270.0	102.9	2.576	6455.0	8590.0	119.8	12.95	24.51	311.1	22.4	.0190
280.0	98.44	2.464	6601.0	8833.0	120.7	12.91	24.14	317.4	23.0	.0194
290.0	94.39	2.363	6745.0	9073.0	121.5	12.87	23.83	323.6	23.6	.0198
300.0	90.70	2.270	6887.0	9310.0	122.4	12.85	23.57	329.5	24.2	.0202
310.0	87.32	2.186	7028.0	9544.0	123.1	12.82	23.34	335.3	24.8	.0206
320.0	84.21	2.108	7168.0	9777.0	123.9	12.80	23.14	340.9	25.3	.0210
340.0	78.68	1.969	7443.0	10240.0	125.3	12.77	22.81	351.8	26.5	.0218
360.0	73.88	1.849	7716.0	10690.0	126.5	12.75	22.54	362.2	27.6	.0226
380.0	69.68	1.744	7985.0	11140.0	127.8	12.73	22.33	372.2	28.7	.0234
400.0	65.96	1.651	8252.0	11580.0	128.9	12.71	22.15	381.9	29.8	.0242

5.60 MPa isobar

T K	Density kg/m^3	Density mol/dm^3	E J/mol	H J/mol	S J/(mol·K)	C_v J/(mol·K)	C_p J/(mol·K)	Sound m/s	Visc. μPa·s	Therm. W/(m·K)
85.17[a]	1422.0	35.60	−1369.0	−1211.0	54.89	23.14	43.85	840.9	292.0	.135
90.0	1394.0	34.89	−1160.0	−999.8	57.31	21.73	43.86	827.9	254.0	.127
100.0	1333.0	33.36	−724.3	−556.4	61.98	20.18	44.95	777.4	195.0	.112

Thermophysical properties of argon—Continued

T K	Density kg/m³	Density mol/dm³	*E* J/mol	*H* J/mol	*S* J/(mol·K)	C_v J/(mol·K)	C_p J/(mol·K)	Sound m/s	Visc. μPa·s	Therm. W/(m·K)
110.0	1266.0	31.69	−275.4	−98.7	66.34	19.25	46.70	711.1	154.0	.0998
115.0	1231.0	30.80	−44.2	137.6	68.44	18.86	47.88	674.6	137.0	.0940
120.0	1193.0	29.86	193.1	380.7	70.51	18.50	49.41	636.2	122.0	.0883
125.0	1153.0	28.85	438.5	632.6	72.57	18.16	51.49	595.7	109.0	.0827
130.0	1108.0	27.75	695.1	897.0	74.64	17.87	54.43	552.4	96.5	.0770
135.0	1059.0	26.51	968.0	1179.0	76.77	17.63	58.82	505.6	85.1	.0711
136.0	1048.0	26.24	1025.0	1239.0	77.21	17.60	59.96	495.7	82.8	.0699
138.0	1026.0	25.68	1143.0	1361.0	78.10	17.53	62.64	475.2	78.4	.0675
140.0	1001.0	25.07	1266.0	1490.0	79.03	17.48	65.99	453.6	74.0	.0651
142.0	975.0	24.41	1396.0	1626.0	79.99	17.46	70.31	430.7	69.5	.0626
144.0	945.7	23.67	1535.0	1772.0	81.02	17.48	76.15	406.1	65.0	.0602
146.0	912.4	22.84	1687.0	1932.0	82.12	17.54	84.54	379.3	60.3	.0578
147.0	893.7	22.37	1769.0	2019.0	82.72	17.60	90.31	364.7	57.8	.0566
148.0	873.2	21.86	1857.0	2113.0	83.35	17.69	97.81	349.2	55.3	.0554
149.0	850.2	21.28	1953.0	2216.0	84.04	17.81	108.0	332.5	52.6	.0542
150.0	823.7	20.62	2059.0	2331.0	84.81	17.98	122.8	314.3	49.7	.0530
150.5	808.7	20.24	2118.0	2395.0	85.24	18.09	133.0	304.4	48.1	.0524
151.0	792.1	19.83	2182.0	2464.0	85.70	18.22	146.3	293.8	46.4	.0518
151.5	773.3	19.36	2252.0	2542.0	86.21	18.39	164.3	282.5	44.6	.0512
152.0	751.6	18.81	2332.0	2630.0	86.79	18.60	190.1	270.2	42.6	.0506
152.5	725.5	18.16	2426.0	2734.0	87.48	18.88	229.5	256.7	40.4	.0501
153.0	692.6	17.34	2541.0	2864.0	88.32	19.25	294.8	241.7	37.8	.0496
153.2	676.6	16.94	2595.0	2926.0	88.73	19.43	332.7	235.3	36.5	.0494
153.4	658.4	16.48	2657.0	2997.0	89.20	19.64	379.3	228.7	35.2	.0490
153.6	637.7	15.96	2728.0	3078.0	89.73	19.88	433.4	222.1	33.8	.0490
153.8	614.4	15.38	2806.0	3171.0	90.33	20.13	487.9	215.7	32.3	.0490
154.0	589.1	14.75	2893.0	3273.0	90.99	20.38	528.9	210.0	30.7	.0488
154.2	563.0	14.09	2983.0	3380.0	91.69	20.60	542.9	205.4	29.2	.0481
154.4	537.7	13.46	3072.0	3488.0	92.38	20.76	529.2	201.9	27.9	.0470
154.6	514.4	12.88	3156.0	3591.0	93.05	20.85	497.4	199.5	26.7	.0442
154.8	493.5	12.35	3233.0	3686.0	93.67	20.88	458.0	197.9	25.7	.0427
155.0	475.1	11.89	3303.0	3774.0	94.23	20.87	417.6	197.0	24.8	.0412
155.2	458.9	11.49	3366.0	3854.0	94.75	20.82	379.7	196.4	24.1	.0398
155.4	444.6	11.13	3423.0	3926.0	95.21	20.74	345.5	196.2	23.5	.0385
155.6	432.0	10.81	3474.0	3992.0	95.64	20.65	315.6	196.1	23.0	.0374
155.8	420.8	10.53	3521.0	4053.0	96.03	20.56	289.5	196.2	22.6	.0363
156.0	410.8	10.28	3564.0	4108.0	96.38	20.45	267.0	196.4	22.2	.0354
156.2	401.8	10.06	3603.0	4160.0	96.71	20.34	247.5	196.7	21.9	.0345
156.5	389.7	9.756	3656.0	4230.0	97.16	20.18	223.0	197.2	21.5	.0333
157.0	372.8	9.333	3733.0	4333.0	97.82	19.91	191.5	198.1	20.9	.0317
157.5	358.9	8.984	3800.0	4423.0	98.39	19.66	168.3	199.2	20.5	.0303
158.0	347.0	8.687	3858.0	4502.0	98.90	19.41	150.5	200.3	20.1	.0292
158.5	336.7	8.430	3910.0	4574.0	99.35	19.19	136.6	201.4	19.8	.0283
159.0	327.7	8.203	3957.0	4639.0	99.76	18.97	125.4	202.5	19.6	.0274
160.0	312.4	7.819	4040.0	4756.0	100.5	18.58	108.5	204.6	19.2	.0260
161.0	299.7	7.502	4112.0	4858.0	101.1	18.24	96.40	206.7	18.9	.0249
162.0	288.9	7.233	4175.0	4950.0	101.7	17.92	87.26	208.7	18.7	.0240
163.0	279.6	6.998	4233.0	5033.0	102.2	17.64	80.12	210.6	18.5	.0233
164.0	271.3	6.792	4286.0	5110.0	102.7	17.38	74.38	212.5	18.4	.0226
165.0	263.9	6.607	4335.0	5182.0	103.1	17.14	69.65	214.3	18.2	.0221
166.0	257.3	6.440	4380.0	5250.0	103.5	16.93	65.69	216.0	18.1	.0216
168.0	245.6	6.148	4464.0	5375.0	104.3	16.54	59.42	219.3	18.0	.0208
170.0	235.6	5.898	4539.0	5489.0	104.9	16.21	54.66	222.5	17.9	.0201
172.0	227.0	5.681	4608.0	5594.0	105.6	15.91	50.93	225.4	17.9	.0196
174.0	219.3	5.489	4673.0	5693.0	106.1	15.66	47.91	228.3	17.8	.0192
176.0	212.4	5.317	4733.0	5786.0	106.7	15.43	45.42	231.0	17.8	.0188
178.0	206.2	5.162	4790.0	5875.0	107.2	15.22	43.33	233.6	17.9	.0185
180.0	200.5	5.020	4844.0	5960.0	107.6	15.04	41.55	236.1	17.9	.0183
182.0	195.3	4.890	4896.0	6041.0	108.1	14.88	40.01	238.6	17.9	.0181
185.0	188.3	4.713	4970.0	6158.0	108.7	14.66	38.06	242.1	18.0	.0178
190.0	178.0	4.456	5085.0	6342.0	109.7	14.36	35.52	247.6	18.2	.0175
195.0	169.2	4.237	5193.0	6514.0	110.6	14.11	33.58	252.8	18.4	.0174
200.0	161.6	4.045	5294.0	6678.0	111.4	13.92	32.06	257.8	18.6	.0173

Thermophysical properties of argon—Continued

T K	Density kg/m³	Density mol/dm³	E J/mol	H J/mol	S J/(mol·K)	C_v J/(mol·K)	C_p J/(mol·K)	Sound m/s	Visc. μPa·s	Therm. W/(m·K)
205.0	154.8	3.876	5391.0	6835.0	112.2	13.76	30.83	262.4	18.8	.0173
210.0	148.8	3.725	5483.0	6987.0	112.9	13.62	29.83	266.9	19.1	.0173
215.0	143.3	3.588	5573.0	7134.0	113.6	13.51	28.98	271.2	19.3	.0174
220.0	138.4	3.464	5660.0	7277.0	114.3	13.42	28.27	275.4	19.6	.0175
225.0	133.8	3.350	5745.0	7417.0	114.9	13.33	27.66	279.4	19.9	.0176
230.0	129.6	3.245	5828.0	7554.0	115.5	13.27	27.13	283.3	20.1	.0177
240.0	122.2	3.059	5990.0	7821.0	116.7	13.16	26.26	290.8	20.7	.0180
250.0	115.7	2.896	6146.0	8080.0	117.7	13.07	25.58	297.9	21.3	.0184
260.0	110.0	2.753	6298.0	8333.0	118.7	13.01	25.03	304.7	21.9	.0187
270.0	104.9	2.625	6447.0	8581.0	119.6	12.96	24.58	311.3	22.5	.0191
280.0	100.3	2.511	6594.0	8824.0	120.5	12.91	24.21	317.6	23.0	.0194
290.0	96.16	2.407	6739.0	9065.0	121.4	12.88	23.89	323.7	23.6	.0198
300.0	92.39	2.313	6881.0	9302.0	122.2	12.85	23.62	329.7	24.2	.0202
310.0	88.94	2.226	7022.0	9537.0	123.0	12.83	23.39	335.4	24.8	.0206
320.0	85.77	2.147	7162.0	9770.0	123.7	12.81	23.18	341.1	25.4	.0210
340.0	80.12	2.006	7438.0	10230.0	125.1	12.78	22.84	352.0	26.5	.0218
360.0	75.23	1.883	7711.0	10680.0	126.4	12.75	22.57	362.4	27.6	.0226
380.0	70.95	1.776	7981.0	11130.0	127.6	12.73	22.36	372.4	28.7	.0234
400.0	67.16	1.681	8248.0	11580.0	128.7	12.72	22.18	382.1	29.8	.0242
				5.70 MPa isobar						
85.19[a]	1422.0	35.60	−1369.0	−1208.0	54.90	23.13	43.83	841.3	293.0	.135
90.0	1394.0	34.90	−1161.0	−998.0	57.30	21.74	43.85	828.3	255.0	.127
100.0	1333.0	33.37	−725.7	−554.8	61.97	20.19	44.93	777.9	196.0	.112
110.0	1267.0	31.71	−277.2	−97.4	66.33	19.26	46.66	711.8	154.0	.0999
115.0	1231.0	30.82	−46.3	138.7	68.42	18.86	47.83	675.4	137.0	.0940
120.0	1194.0	29.88	190.6	381.4	70.49	18.50	49.34	637.1	122.0	.0884
125.0	1153.0	28.87	435.5	632.9	72.54	18.17	51.39	596.7	109.0	.0828
130.0	1109.0	27.77	691.4	896.6	74.61	17.88	54.27	553.7	96.8	.0771
135.0	1060.0	26.54	963.2	1178.0	76.73	17.64	58.57	507.2	85.3	.0713
136.0	1050.0	26.27	1020.0	1237.0	77.17	17.60	59.68	497.3	83.1	.0701
138.0	1027.0	25.71	1137.0	1359.0	78.06	17.53	62.28	477.0	78.7	.0677
140.0	1003.0	25.11	1260.0	1487.0	78.98	17.49	65.52	455.7	74.3	.0653
142.0	977.1	24.46	1389.0	1622.0	79.94	17.46	69.67	433.1	69.9	.0628
144.0	948.3	23.74	1526.0	1766.0	80.95	17.47	75.22	408.9	65.4	.0604
146.0	915.7	22.92	1675.0	1924.0	82.04	17.53	83.08	382.6	60.8	.0580
147.0	897.5	22.47	1756.0	2010.0	82.62	17.58	88.42	368.4	58.3	.0568
148.0	877.6	21.97	1842.0	2101.0	83.24	17.66	95.24	353.3	55.8	.0556
149.0	855.5	21.42	1935.0	2201.0	83.91	17.77	104.3	337.2	53.2	.0545
150.0	830.4	20.79	2037.0	2311.0	84.65	17.92	117.0	319.7	50.4	.0533
150.5	816.3	20.43	2093.0	2372.0	85.05	18.02	125.6	310.3	48.9	.0527
151.0	800.9	20.05	2153.0	2437.0	85.49	18.13	136.4	300.4	47.3	.0521
152.0	764.4	19.14	2291.0	2588.0	86.48	18.45	169.4	278.5	43.8	.0509
152.5	742.0	18.57	2372.0	2679.0	87.08	18.67	196.3	266.3	41.8	.0503
153.0	715.1	17.90	2468.0	2787.0	87.79	18.96	236.7	253.0	39.6	.0497
153.5	681.4	17.06	2586.0	2920.0	88.65	19.33	300.0	238.6	36.9	.0491
153.6	673.6	16.86	2613.0	2951.0	88.85	19.42	316.4	235.6	36.4	.0490
153.8	656.7	16.44	2671.0	3018.0	89.29	19.61	352.9	229.5	35.1	.0485
154.0	637.8	15.97	2735.0	3092.0	89.77	19.82	392.9	223.6	33.8	.0484
154.2	617.2	15.45	2806.0	3175.0	90.31	20.04	431.7	217.9	32.5	.0482
154.4	595.0	14.90	2881.0	3264.0	90.89	20.25	461.7	212.7	31.1	.0479
154.6	572.3	14.33	2960.0	3358.0	91.50	20.44	476.0	208.4	29.8	.0465
154.8	549.8	13.76	3039.0	3453.0	92.11	20.59	472.8	205.0	28.5	.0454
155.0	528.5	13.23	3116.0	3546.0	92.71	20.70	455.8	202.4	27.4	.0442
155.2	508.8	12.74	3188.0	3635.0	93.29	20.76	430.6	200.6	26.4	.0429
155.4	491.0	12.29	3255.0	3718.0	93.82	20.77	401.7	199.3	25.6	.0416
155.6	474.9	11.89	3316.0	3796.0	94.32	20.75	372.3	198.5	24.9	.0404
155.8	460.4	11.53	3373.0	3867.0	94.78	20.70	344.1	198.0	24.3	.0392
156.0	447.5	11.20	3425.0	3934.0	95.20	20.63	318.1	197.8	23.7	.0381
156.2	435.8	10.91	3472.0	3995.0	95.60	20.55	294.5	197.7	23.2	.0371
156.4	425.3	10.65	3516.0	4052.0	95.96	20.46	273.4	197.8	22.8	.0362
156.6	415.8	10.41	3557.0	4104.0	96.30	20.37	254.7	197.9	22.5	.0353

Thermophysical properties of argon—Continued

T K	Density kg/m³	Density mol/dm³	E J/mol	H J/mol	S J/(mol·K)	C_v J/(mol·K)	C_p J/(mol·K)	Sound m/s	Visc. μPa·s	Therm. W/(m·K)
157.0	399.1	9.991	3629.0	4200.0	96.91	20.17	223.3	198.4	21.9	.0338
157.5	381.9	9.559	3707.0	4303.0	97.57	19.91	193.3	199.2	21.3	.0322
158.0	367.5	9.199	3775.0	4394.0	98.14	19.67	170.6	200.2	20.8	.0308
158.5	355.2	8.892	3834.0	4475.0	98.65	19.43	153.0	201.2	20.4	.0297
159.0	344.6	8.626	3887.0	4548.0	99.11	19.21	139.0	202.2	20.1	.0287
159.5	335.2	8.391	3935.0	4614.0	99.53	19.00	127.7	203.3	19.9	.0279
160.0	326.9	8.182	3979.0	4676.0	99.91	18.80	118.3	204.3	19.6	.0271
161.0	312.5	7.822	4058.0	4786.0	100.6	18.44	103.8	206.4	19.3	.0259
162.0	300.4	7.521	4127.0	4885.0	101.2	18.11	93.09	208.4	19.0	.0249
163.0	290.1	7.261	4188.0	4973.0	101.8	17.81	84.84	210.3	18.8	.0240
164.0	281.0	7.034	4245.0	5055.0	102.3	17.54	78.29	212.2	18.6	.0233
165.0	273.0	6.833	4296.0	5130.0	102.7	17.29	72.96	214.0	18.5	.0227
166.0	265.7	6.652	4344.0	5201.0	103.1	17.07	68.53	215.8	18.4	.0221
168.0	253.2	6.337	4431.0	5331.0	103.9	16.66	61.59	219.1	18.2	.0212
170.0	242.5	6.071	4510.0	5449.0	104.6	16.31	56.39	222.3	18.1	.0205
172.0	233.3	5.841	4581.0	5557.0	105.2	16.01	52.34	225.3	18.0	.0199
174.0	225.2	5.638	4647.0	5658.0	105.8	15.74	49.09	228.1	18.0	.0195
176.0	218.0	5.457	4709.0	5754.0	106.4	15.50	46.42	230.9	18.0	.0191
178.0	211.5	5.294	4768.0	5844.0	106.9	15.29	44.20	233.5	18.0	.0188
180.0	205.5	5.145	4823.0	5931.0	107.4	15.10	42.31	236.0	18.0	.0185
182.0	200.1	5.009	4876.0	6014.0	107.8	14.93	40.68	238.5	18.0	.0183
185.0	192.7	4.824	4951.0	6133.0	108.5	14.71	38.63	242.0	18.1	.0180
190.0	182.1	4.558	5068.0	6319.0	109.5	14.40	35.97	247.6	18.2	.0177
195.0	173.0	4.330	5177.0	6493.0	110.4	14.15	33.95	252.8	18.4	.0175
200.0	165.1	4.132	5280.0	6659.0	111.2	13.95	32.36	257.8	18.6	.0174
205.0	158.1	3.958	5377.0	6818.0	112.0	13.78	31.09	262.5	18.9	.0174
210.0	151.9	3.802	5471.0	6970.0	112.7	13.65	30.05	267.0	19.1	.0174
215.0	146.3	3.661	5562.0	7118.0	113.4	13.53	29.18	271.3	19.4	.0175
220.0	141.2	3.534	5649.0	7262.0	114.1	13.43	28.44	275.5	19.6	.0176
225.0	136.5	3.417	5735.0	7403.0	114.7	13.35	27.81	279.5	19.9	.0177
230.0	132.2	3.309	5818.0	7541.0	115.3	13.28	27.27	283.4	20.2	.0178
240.0	124.5	3.118	5980.0	7809.0	116.5	13.17	26.38	290.9	20.8	.0181
250.0	117.9	2.951	6137.0	8069.0	117.5	13.08	25.68	298.0	21.3	.0184
260.0	112.0	2.805	6290.0	8323.0	118.5	13.02	25.11	304.9	21.9	.0188
270.0	106.8	2.674	6440.0	8571.0	119.5	12.96	24.66	311.4	22.5	.0191
280.0	102.2	2.557	6587.0	8816.0	120.4	12.92	24.27	317.8	23.1	.0195
290.0	97.93	2.451	6732.0	9057.0	121.2	12.89	23.95	323.9	23.7	.0199
300.0	94.08	2.355	6875.0	9295.0	122.0	12.86	23.67	329.8	24.2	.0203
310.0	90.56	2.267	7016.0	9531.0	122.8	12.84	23.43	335.6	24.8	.0207
320.0	87.33	2.186	7156.0	9764.0	123.5	12.82	23.22	341.3	25.4	.0211
340.0	81.56	2.042	7433.0	10220.0	124.9	12.78	22.88	352.1	26.5	.0219
360.0	76.58	1.917	7706.0	10680.0	126.2	12.76	22.61	362.6	27.7	.0227
380.0	72.21	1.808	7976.0	11130.0	127.4	12.74	22.38	372.6	28.8	.0235
400.0	68.35	1.711	8244.0	11580.0	128.6	12.72	22.20	382.3	29.8	.0243

5.80 MPa isobar

T K	Density kg/m³	Density mol/dm³	E J/mol	H J/mol	S J/(mol·K)	C_v J/(mol·K)	C_p J/(mol·K)	Sound m/s	Visc. μPa·s	Therm. W/(m·K)
85.22[a]	1422.0	35.61	−1368.0	−1206.0	54.90	23.13	43.81	841.6	293.0	.135
90.0	1395.0	34.91	−1162.0	−996.2	57.29	21.74	43.83	828.8	255.0	.127
100.0	1333.0	33.37	−727.1	−553.3	61.95	20.19	44.91	778.4	196.0	.112
110.0	1267.0	31.72	−279.0	−96.2	66.31	19.26	46.62	712.4	154.0	.0999
115.0	1232.0	30.83	−48.4	139.7	68.41	18.87	47.77	676.2	137.0	.0941
120.0	1194.0	29.89	188.1	382.1	70.47	18.51	49.27	638.0	123.0	.0885
125.0	1154.0	28.89	432.5	633.3	72.52	18.18	51.29	597.7	109.0	.0829
130.0	1110.0	27.80	687.7	896.4	74.58	17.89	54.12	554.9	97.0	.0772
135.0	1062.0	26.57	958.5	1177.0	76.70	17.64	58.32	508.7	85.6	.0714
136.0	1051.0	26.31	1015.0	1236.0	77.13	17.60	59.41	499.0	83.4	.0702
138.0	1029.0	25.75	1132.0	1357.0	78.02	17.54	61.93	478.8	79.0	.0679
140.0	1005.0	25.16	1253.0	1484.0	78.93	17.49	65.06	457.7	74.6	.0654
142.0	979.2	24.51	1381.0	1618.0	79.88	17.46	69.06	435.4	70.2	.0630
144.0	950.8	23.80	1517.0	1761.0	80.88	17.47	74.35	411.5	65.8	.0606
146.0	918.9	23.00	1664.0	1916.0	81.95	17.52	81.73	385.7	61.2	.0582
147.0	901.1	22.56	1743.0	2001.0	82.53	17.56	86.68	371.9	58.8	.0570

Thermophysical properties of argon—Continued

T K	Density kg/m^3	Density mol/dm^3	E J/mol	H J/mol	S J/(mol·K)	C_v J/(mol·K)	C_p J/(mol·K)	Sound m/s	Visc. μPa·s	Therm. W/(m·K)
148.0	881.8	22.07	1827.0	2090.0	83.14	17.63	92.91	357.3	56.4	.0559
149.0	860.5	21.54	1918.0	2187.0	83.79	17.73	101.1	341.6	53.8	.0547
150.0	836.6	20.94	2016.0	2293.0	84.50	17.87	112.2	324.8	51.1	.0535
151.0	808.9	20.25	2127.0	2413.0	85.29	18.06	128.4	306.4	48.2	.0524
152.0	775.5	19.41	2255.0	2553.0	86.22	18.33	154.4	285.9	44.9	.0511
152.5	755.6	18.91	2329.0	2635.0	86.76	18.51	174.1	274.7	43.0	.0505
153.0	732.5	18.34	2413.0	2729.0	87.37	18.74	201.8	262.6	41.0	.0499
153.5	704.9	17.65	2510.0	2839.0	88.09	19.03	242.1	249.7	38.8	.0493
154.0	671.0	16.80	2629.0	2974.0	88.97	19.40	300.9	236.0	36.2	.0486
154.5	628.7	15.74	2774.0	3143.0	90.06	19.85	374.8	222.4	33.3	.0477
155.0	579.4	14.50	2945.0	3345.0	91.37	20.31	423.7	211.2	30.2	.0460
155.5	530.6	13.28	3119.0	3555.0	92.72	20.59	409.2	204.1	27.6	.0435
155.6	521.6	13.06	3152.0	3596.0	92.98	20.62	400.6	203.2	27.1	.0429
155.8	504.6	12.63	3215.0	3674.0	93.49	20.65	380.6	201.8	26.3	.0418
156.0	489.0	12.24	3274.0	3748.0	93.96	20.65	358.6	200.8	25.6	.0407
156.2	474.7	11.88	3329.0	3817.0	94.41	20.63	336.4	200.1	24.9	.0397
156.4	461.7	11.56	3381.0	3883.0	94.82	20.58	314.7	199.6	24.4	.0387
156.6	449.8	11.26	3428.0	3943.0	95.21	20.52	294.3	199.4	23.9	.0377
156.8	439.0	10.99	3473.0	4000.0	95.57	20.45	275.4	199.3	23.4	.0368
157.0	429.2	10.74	3514.0	4054.0	95.91	20.36	258.2	199.3	23.0	.0360
157.5	407.9	10.21	3605.0	4173.0	96.67	20.14	221.9	199.7	22.2	.0341
158.0	390.4	9.773	3683.0	4277.0	97.33	19.90	193.8	200.4	21.6	.0326
158.5	375.7	9.405	3751.0	4368.0	97.91	19.67	171.9	201.3	21.2	.0313
159.0	363.1	9.090	3812.0	4450.0	98.42	19.44	154.7	202.2	20.8	.0301
159.5	352.2	8.817	3865.0	4523.0	98.88	19.23	140.8	203.2	20.4	.0291
160.0	342.6	8.575	3914.0	4591.0	99.31	19.02	129.5	204.2	20.2	.0283
161.0	326.2	8.165	4001.0	4711.0	100.1	18.64	112.2	206.2	19.7	.0269
162.0	312.6	7.826	4076.0	4817.0	100.7	18.30	99.57	208.2	19.4	.0257
163.0	301.1	7.538	4142.0	4911.0	101.3	17.98	90.03	210.1	19.1	.0248
164.0	291.2	7.288	4202.0	4997.0	101.8	17.70	82.56	212.0	18.9	.0240
165.0	282.4	7.068	4256.0	5077.0	102.3	17.44	76.54	213.8	18.8	.0233
166.0	274.5	6.872	4307.0	5151.0	102.7	17.21	71.58	215.5	18.6	.0227
168.0	261.0	6.533	4398.0	5286.0	103.6	16.78	63.90	218.9	18.4	.0217
170.0	249.6	6.249	4480.0	5408.0	104.3	16.42	58.21	222.1	18.3	.0209
172.0	239.8	6.004	4554.0	5520.0	104.9	16.11	53.82	225.1	18.2	.0203
174.0	231.3	5.789	4622.0	5624.0	105.5	15.83	50.32	228.0	18.1	.0198
176.0	223.7	5.599	4685.0	5721.0	106.1	15.58	47.47	230.8	18.1	.0193
178.0	216.8	5.427	4745.0	5814.0	106.6	15.37	45.10	233.4	18.1	.0190
180.0	210.6	5.272	4802.0	5902.0	107.1	15.17	43.09	236.0	18.1	.0187
182.0	204.9	5.130	4856.0	5986.0	107.6	14.99	41.38	238.4	18.1	.0185
184.0	199.7	4.999	4907.0	6068.0	108.0	14.84	39.89	240.8	18.2	.0182
190.0	186.2	4.660	5051.0	6296.0	109.2	14.44	36.42	247.6	18.3	.0178
195.0	176.7	4.424	5162.0	6473.0	110.2	14.19	34.31	252.8	18.5	.0176
200.0	168.6	4.220	5265.0	6640.0	111.0	13.98	32.67	257.8	18.7	.0175
205.0	161.4	4.040	5364.0	6800.0	111.8	13.81	31.35	262.5	18.9	.0175
210.0	155.0	3.879	5459.0	6954.0	112.5	13.67	30.27	267.1	19.2	.0175
215.0	149.2	3.735	5550.0	7103.0	113.2	13.55	29.37	271.4	19.4	.0176
220.0	144.0	3.604	5638.0	7248.0	113.9	13.45	28.61	275.6	19.7	.0176
225.0	139.2	3.484	5724.0	7389.0	114.5	13.37	27.97	279.6	20.0	.0178
230.0	134.8	3.373	5808.0	7528.0	115.1	13.30	27.41	283.5	20.2	.0179
240.0	126.9	3.177	5971.0	7797.0	116.3	13.18	26.49	291.0	20.8	.0182
250.0	120.1	3.007	6129.0	8058.0	117.4	13.09	25.77	298.2	21.4	.0185
260.0	114.1	2.857	6282.0	8313.0	118.4	13.02	25.20	305.0	21.9	.0188
270.0	108.8	2.723	6433.0	8562.0	119.3	12.97	24.73	311.6	22.5	.0192
280.0	104.0	2.604	6580.0	8808.0	120.2	12.93	24.34	317.9	23.1	.0195
290.0	99.70	2.496	6725.0	9049.0	121.0	12.89	24.01	324.1	23.7	.0199
300.0	95.78	2.398	6869.0	9288.0	121.9	12.87	23.72	330.0	24.3	.0203
310.0	92.18	2.308	7010.0	9524.0	122.6	12.84	23.48	335.8	24.8	.0207
320.0	88.88	2.225	7151.0	9758.0	123.4	12.82	23.27	341.4	25.4	.0211
340.0	83.01	2.078	7428.0	10220.0	124.8	12.79	22.92	352.3	26.6	.0219
360.0	77.93	1.951	7701.0	10670.0	126.1	12.76	22.64	362.8	27.7	.0227
380.0	73.48	1.839	7972.0	11130.0	127.3	12.74	22.41	372.8	28.8	.0235
400.0	69.55	1.741	8240.0	11570.0	128.4	12.73	22.23	382.5	29.9	.0243

Thermophysical properties of argon—Continued

T K	Density kg/m³	Density mol/dm³	E J/mol	H J/mol	S J/(mol·K)	C_v J/(mol·K)	C_p J/(mol·K)	Sound m/s	Visc. μPa·s	Therm. W/(m·K)
				6.00 MPa isobar						
85.27[a]	1423.0	35.61	−1368.0	−1200.0	54.90	23.11	43.78	842.3	293.0	.135
90.0	1395.0	34.92	−1165.0	−992.7	57.26	21.75	43.80	829.6	255.0	.127
100.0	1334.0	33.39	−729.8	−550.1	61.92	20.21	44.86	779.4	196.0	.113
110.0	1268.0	31.74	−282.7	−93.6	66.27	19.28	46.54	713.7	155.0	.100
115.0	1233.0	30.86	−52.6	141.8	68.37	18.89	47.67	677.7	138.0	.0943
120.0	1196.0	29.93	183.2	383.6	70.43	18.53	49.13	639.7	123.0	.0887
125.0	1156.0	28.93	426.5	633.9	72.47	18.19	51.09	599.8	110.0	.0831
130.0	1112.0	27.84	680.4	895.9	74.52	17.90	53.83	557.4	97.5	.0775
135.0	1064.0	26.64	949.1	1174.0	76.62	17.66	57.85	511.8	86.2	.0717
136.0	1054.0	26.38	1005.0	1233.0	77.06	17.61	58.88	502.2	84.0	.0705
138.0	1032.0	25.83	1121.0	1353.0	77.93	17.54	61.27	482.4	79.6	.0682
140.0	1009.0	25.25	1241.0	1478.0	78.83	17.49	64.20	461.7	75.3	.0658
142.0	983.3	24.61	1366.0	1610.0	79.77	17.46	67.90	439.9	70.9	.0634
144.0	955.7	23.92	1500.0	1751.0	80.75	17.46	72.72	416.8	66.5	.0610
146.0	925.0	23.15	1643.0	1902.0	81.80	17.49	79.30	391.9	62.1	.0586
148.0	889.7	22.27	1800.0	2070.0	82.94	17.59	88.88	364.7	57.4	.0563
149.0	869.8	21.77	1886.0	2162.0	83.56	17.67	95.58	350.0	54.9	.0552
150.0	847.7	21.22	1979.0	2262.0	84.22	17.78	104.4	334.3	52.4	.0540
151.0	822.8	20.60	2080.0	2372.0	84.95	17.93	116.4	317.4	49.6	.0529
152.0	793.9	19.87	2194.0	2496.0	85.78	18.14	134.0	299.0	46.7	.0517
153.0	758.9	19.00	2327.0	2643.0	86.74	18.43	161.8	278.8	43.4	.0505
153.5	738.0	18.48	2404.0	2729.0	87.30	18.63	182.3	267.8	41.6	.0498
154.0	714.1	17.87	2491.0	2826.0	87.93	18.86	209.8	256.3	39.6	.0491
154.5	685.9	17.17	2591.0	2940.0	88.67	19.15	246.5	244.4	37.4	.0484
155.0	652.7	16.34	2707.0	3074.0	89.54	19.50	291.4	232.4	35.0	.0474
155.5	614.3	15.38	2841.0	3231.0	90.55	19.87	333.7	221.5	32.4	.0462
156.0	573.2	14.35	2986.0	3404.0	91.66	20.19	352.3	213.0	30.0	.0445
156.5	533.6	13.36	3129.0	3578.0	92.77	20.39	340.4	207.5	27.8	.0424
157.0	498.6	12.48	3261.0	3741.0	93.81	20.44	310.6	204.4	26.1	.0402
157.5	468.9	11.74	3377.0	3888.0	94.75	20.38	275.9	202.9	24.8	.0381
158.0	444.2	11.12	3478.0	4018.0	95.57	20.23	243.1	202.4	23.8	.0363
158.5	423.4	10.60	3566.0	4132.0	96.29	20.05	214.8	202.5	23.0	.0346
159.0	406.0	10.16	3643.0	4233.0	96.93	19.85	191.2	203.0	22.3	.0332
159.5	391.0	9.788	3711.0	4324.0	97.50	19.64	171.8	203.6	21.8	.0320
160.0	378.0	9.463	3772.0	4406.0	98.01	19.43	156.0	204.4	21.4	.0309
160.5	366.7	9.178	3826.0	4480.0	98.48	19.23	142.8	205.2	21.0	.0299
161.0	356.6	8.926	3877.0	4549.0	98.90	19.03	131.8	206.1	20.7	.0290
162.0	339.4	8.495	3965.0	4672.0	99.66	18.67	114.7	207.9	20.2	.0276
163.0	325.1	8.137	4042.0	4780.0	100.3	18.33	102.0	209.8	19.9	.0264
164.0	312.9	7.834	4111.0	4877.0	100.9	18.03	92.27	211.6	19.6	.0254
165.0	302.4	7.570	4172.0	4965.0	101.5	17.75	84.59	213.4	19.3	.0246
166.0	293.1	7.338	4229.0	5046.0	101.9	17.49	78.39	215.2	19.1	.0238
167.0	284.9	7.131	4281.0	5122.0	102.4	17.25	73.26	216.9	19.0	.0232
168.0	277.4	6.944	4329.0	5193.0	102.8	17.03	68.97	218.6	18.9	.0227
170.0	264.4	6.618	4417.0	5324.0	103.6	16.64	62.15	221.8	18.7	.0217
172.0	253.3	6.342	4497.0	5443.0	104.3	16.30	56.99	224.9	18.5	.0210
174.0	243.8	6.102	4569.0	5552.0	104.9	16.01	52.94	227.8	18.5	.0204
176.0	235.3	5.890	4636.0	5655.0	105.5	15.74	49.68	230.6	18.4	.0199
178.0	227.8	5.701	4699.0	5752.0	106.1	15.51	46.99	233.3	18.4	.0195
180.0	221.0	5.531	4758.0	5843.0	106.6	15.30	44.74	235.8	18.4	.0192
182.0	214.8	5.376	4815.0	5931.0	107.1	15.11	42.82	238.3	18.4	.0189
184.0	209.1	5.234	4868.0	6015.0	107.5	14.95	41.17	240.8	18.4	.0186
190.0	194.5	4.868	5017.0	6250.0	108.8	14.53	37.36	247.6	18.5	.0181
195.0	184.4	4.615	5130.0	6430.0	109.7	14.26	35.07	252.9	18.7	.0179
200.0	175.7	4.397	5237.0	6601.0	110.6	14.04	33.29	257.9	18.9	.0177
205.0	168.0	4.206	5337.0	6764.0	111.4	13.86	31.88	262.7	19.1	.0177
210.0	161.2	4.036	5434.0	6920.0	112.1	13.71	30.72	267.2	19.3	.0177
215.0	155.1	3.883	5526.0	7072.0	112.8	13.59	29.77	271.6	19.6	.0177
220.0	149.6	3.744	5616.0	7218.0	113.5	13.49	28.96	275.8	19.8	.0178
225.0	144.5	3.618	5703.0	7361.0	114.2	13.40	28.28	279.8	20.1	.0179

Thermophysical properties of argon—Continued

T K	Density kg/m³	Density mol/dm³	E J/mol	H J/mol	S J/(mol·K)	C_v J/(mol·K)	C_p J/(mol·K)	Sound m/s	Visc. μPa·s	Therm. W/(m·K)
230.0	139.9	3.502	5788.0	7501.0	114.8	13.32	27.69	283.8	20.3	.0180
240.0	131.7	3.296	5953.0	7773.0	115.9	13.20	26.72	291.3	20.9	.0183
250.0	124.5	3.118	6112.0	8036.0	117.0	13.11	25.97	298.5	21.5	.0186
260.0	118.3	2.961	6267.0	8293.0	118.0	13.04	25.37	305.3	22.0	.0189
270.0	112.7	2.822	6418.0	8544.0	119.0	12.99	24.87	311.9	22.6	.0193
280.0	107.7	2.697	6566.0	8791.0	119.9	12.94	24.47	318.3	23.2	.0196
290.0	103.3	2.585	6712.0	9034.0	120.7	12.91	24.12	324.4	23.8	.0200
300.0	99.16	2.482	6856.0	9273.0	121.5	12.88	23.83	330.4	24.3	.0204
310.0	95.43	2.389	6999.0	9510.0	122.3	12.85	23.57	336.2	24.9	.0208
320.0	92.00	2.303	7139.0	9745.0	123.1	12.83	23.35	341.8	25.5	.0212
330.0	88.83	2.224	7279.0	9977.0	123.8	12.81	23.16	347.3	26.0	.0216
340.0	85.89	2.150	7418.0	10210.0	124.5	12.80	22.99	352.7	26.6	.0220
360.0	80.62	2.018	7692.0	10660.0	125.8	12.77	22.70	363.1	27.7	.0228
380.0	76.01	1.903	7963.0	11120.0	127.0	12.75	22.47	373.2	28.8	.0236
400.0	71.93	1.801	8232.0	11560.0	128.1	12.73	22.27	382.9	29.9	.0244
				6.20 MPa isobar						
85.32[a]	1423.0	35.62	−1368.0	−1194.0	54.90	23.10	43.75	843.0	293.0	.135
90.0	1396.0	34.94	−1167.0	−989.2	57.24	21.75	43.76	830.4	256.0	.127
100.0	1335.0	33.41	−732.6	−547.0	61.90	20.22	44.81	780.5	197.0	.113
110.0	1269.0	31.77	−286.3	−91.1	66.24	19.29	46.47	715.0	155.0	.100
115.0	1234.0	30.89	−56.8	143.9	68.33	18.90	47.57	679.2	138.0	.0945
120.0	1197.0	29.96	178.2	385.2	70.38	18.54	49.00	641.5	123.0	.0888
125.0	1157.0	28.97	420.7	634.7	72.42	18.21	50.90	601.9	110.0	.0833
130.0	1114.0	27.89	673.2	895.4	74.46	17.91	53.55	559.9	98.0	.0777
135.0	1067.0	26.70	939.9	1172.0	76.55	17.67	57.40	514.9	86.7	.0720
136.0	1056.0	26.44	995.6	1230.0	76.98	17.62	58.38	505.4	84.5	.0708
138.0	1035.0	25.90	1110.0	1349.0	77.85	17.55	60.64	485.9	80.2	.0685
140.0	1012.0	25.33	1228.0	1473.0	78.74	17.49	63.40	465.6	75.9	.0661
142.0	987.3	24.71	1352.0	1603.0	79.66	17.46	66.84	444.3	71.6	.0638
144.0	960.4	24.04	1483.0	1741.0	80.63	17.45	71.25	421.8	67.3	.0614
146.0	930.7	23.30	1623.0	1889.0	81.65	17.47	77.15	397.7	62.9	.0591
148.0	897.1	22.46	1775.0	2051.0	82.75	17.55	85.49	371.7	58.4	.0568
149.0	878.3	21.98	1857.0	2139.0	83.34	17.62	91.12	357.7	56.0	.0557
150.0	857.7	21.47	1945.0	2234.0	83.98	17.71	98.26	342.9	53.5	.0545
151.0	834.8	20.90	2040.0	2337.0	84.66	17.83	107.6	327.2	51.0	.0534
152.0	809.0	20.25	2144.0	2450.0	85.41	17.99	120.5	310.4	48.3	.0523
153.0	778.9	19.50	2262.0	2580.0	86.26	18.22	139.1	292.2	45.3	.0511
154.0	742.6	18.59	2398.0	2732.0	87.25	18.53	167.6	272.5	42.0	.0497
154.5	721.2	18.05	2477.0	2820.0	87.82	18.73	187.7	262.1	40.2	.0490
155.0	696.8	17.44	2565.0	2920.0	88.47	18.96	212.8	251.4	38.3	.0483
155.5	669.1	16.75	2664.0	3034.0	89.20	19.24	242.6	240.7	36.2	.0474
156.0	637.7	15.96	2775.0	3163.0	90.03	19.54	273.0	230.6	34.0	.0463
156.5	603.6	15.11	2896.0	3306.0	90.94	19.83	295.5	221.9	31.9	.0449
157.0	568.8	14.24	3020.0	3456.0	91.90	20.06	301.9	215.2	29.8	.0433
157.5	535.6	13.41	3143.0	3605.0	92.85	20.20	292.4	210.7	28.0	.0415
158.0	505.5	12.65	3257.0	3747.0	93.75	20.24	273.0	207.9	26.5	.0397
158.5	479.1	11.99	3360.0	3877.0	94.57	20.19	249.6	206.3	25.3	.0380
159.0	456.3	11.42	3453.0	3996.0	95.32	20.08	226.2	205.6	24.4	.0364
159.5	436.5	10.93	3536.0	4104.0	96.00	19.93	204.5	205.5	23.6	.0349
160.0	419.5	10.50	3611.0	4201.0	96.61	19.76	185.4	205.7	22.9	.0336
160.5	404.6	10.13	3677.0	4290.0	97.16	19.57	168.9	206.2	22.4	.0325
161.0	391.5	9.801	3738.0	4370.0	97.66	19.39	154.8	206.8	21.9	.0314
161.5	380.0	9.511	3793.0	4445.0	98.12	19.20	142.8	207.5	21.6	.0305
162.0	369.6	9.252	3843.0	4514.0	98.55	19.02	132.5	208.2	21.2	.0296
163.0	351.8	8.807	3933.0	4637.0	99.31	18.67	116.1	209.9	20.7	.0282
164.0	337.0	8.435	4012.0	4747.0	99.98	18.35	103.6	211.6	20.3	.0270
165.0	324.3	8.118	4082.0	4846.0	100.6	18.05	93.92	213.3	20.0	.0260
166.0	313.3	7.842	4145.0	4936.0	101.1	17.77	86.19	215.0	19.7	.0251
167.0	303.6	7.599	4203.0	5018.0	101.6	17.52	79.89	216.7	19.5	.0244
168.0	294.9	7.383	4256.0	5096.0	102.1	17.29	74.68	218.4	19.4	.0237

Thermophysical properties of argon—Continued

T K	Density kg/m^3	Density mol/dm^3	E J/mol	H J/mol	S J/(mol·K)	C_v J/(mol·K)	C_p J/(mol·K)	Sound m/s	Visc. μPa·s	Therm. W/(m·K)
170.0	280.0	7.009	4352.0	5236.0	102.9	16.87	66.53	221.6	19.1	.0226
172.0	267.5	6.697	4437.0	5363.0	103.7	16.50	60.47	224.7	18.9	.0218
174.0	256.8	6.428	4515.0	5479.0	104.3	16.19	55.79	227.6	18.8	.0211
176.0	247.4	6.193	4586.0	5587.0	104.9	15.91	52.06	230.5	18.7	.0205
178.0	239.1	5.985	4652.0	5688.0	105.5	15.66	49.01	233.2	18.7	.0200
180.0	231.6	5.798	4714.0	5783.0	106.0	15.44	46.49	235.8	18.6	.0196
182.0	224.9	5.629	4773.0	5874.0	106.5	15.24	44.35	238.3	18.6	.0193
184.0	218.7	5.475	4828.0	5961.0	107.0	15.06	42.52	240.7	18.6	.0190
186.0	213.0	5.333	4882.0	6044.0	107.5	14.90	40.94	243.1	18.6	.0188
190.0	202.9	5.080	4982.0	6203.0	108.3	14.62	38.34	247.6	18.7	.0184
195.0	192.1	4.810	5099.0	6388.0	109.3	14.33	35.85	253.0	18.9	.0182
200.0	182.8	4.577	5208.0	6562.0	110.2	14.10	33.94	258.0	19.0	.0180
205.0	174.7	4.374	5310.0	6728.0	111.0	13.92	32.42	262.8	19.2	.0179
210.0	167.5	4.193	5408.0	6887.0	111.7	13.76	31.19	267.4	19.4	.0179
215.0	161.1	4.032	5503.0	7040.0	112.5	13.63	30.17	271.8	19.7	.0179
220.0	155.2	3.886	5593.0	7189.0	113.2	13.52	29.32	276.0	19.9	.0180
225.0	149.9	3.753	5682.0	7334.0	113.8	13.43	28.59	280.1	20.2	.0181
230.0	145.1	3.632	5768.0	7475.0	114.4	13.35	27.97	284.0	20.4	.0182
240.0	136.5	3.416	5934.0	7749.0	115.6	13.23	26.95	291.6	21.0	.0184
250.0	129.0	3.229	6095.0	8015.0	116.7	13.13	26.16	298.8	21.5	.0187
260.0	122.5	3.066	6251.0	8273.0	117.7	13.06	25.53	305.6	22.1	.0190
270.0	116.7	2.921	6403.0	8526.0	118.6	13.00	25.02	312.2	22.7	.0194
280.0	111.5	2.791	6552.0	8774.0	119.5	12.96	24.60	318.6	23.2	.0197
290.0	106.8	2.674	6699.0	9018.0	120.4	12.92	24.24	324.7	23.8	.0201
300.0	102.6	2.567	6844.0	9259.0	121.2	12.89	23.93	330.7	24.4	.0205
310.0	98.68	2.470	6987.0	9497.0	122.0	12.87	23.67	336.5	25.0	.0209
320.0	95.11	2.381	7128.0	9732.0	122.7	12.84	23.44	342.2	25.5	.0213
330.0	91.83	2.299	7268.0	9966.0	123.5	12.83	23.24	347.7	26.1	.0217
340.0	88.78	2.222	7407.0	10200.0	124.2	12.81	23.06	353.1	26.7	.0221
360.0	83.32	2.086	7683.0	10660.0	125.5	12.78	22.76	363.5	27.8	.0228
380.0	78.54	1.966	7954.0	11110.0	126.7	12.76	22.52	373.6	28.9	.0236
400.0	74.32	1.860	8224.0	11560.0	127.8	12.74	22.32	383.2	29.9	.0244
				6.40 MPa isobar						
85.37[a]	1423.0	35.63	−1368.0	−1188.0	54.91	23.08	43.72	843.7	293.0	.135
90.0	1396.0	34.95	−1169.0	−985.6	57.21	21.76	43.73	831.3	256.0	.127
100.0	1335.0	33.43	−735.3	−543.8	61.87	20.23	44.76	781.5	197.0	.113
110.0	1270.0	31.79	−289.8	−88.5	66.21	19.30	46.39	716.3	155.0	.100
120.0	1198.0	29.99	173.4	386.8	70.34	18.55	48.87	643.2	124.0	.0890
125.0	1159.0	29.01	414.8	635.5	72.37	18.22	50.72	603.9	111.0	.0835
130.0	1116.0	27.94	666.0	895.1	74.41	17.93	53.28	562.3	98.5	.0779
135.0	1069.0	26.76	931.0	1170.0	76.48	17.68	56.97	517.9	87.2	.0723
136.0	1059.0	26.51	986.1	1228.0	76.91	17.63	57.91	508.5	85.0	.0711
138.0	1038.0	25.98	1099.0	1345.0	77.77	17.56	60.05	489.4	80.7	.0688
140.0	1015.0	25.41	1216.0	1468.0	78.65	17.50	62.64	469.4	76.5	.0665
142.0	991.1	24.81	1338.0	1596.0	79.56	17.46	65.85	448.6	72.2	.0641
144.0	964.9	24.15	1467.0	1732.0	80.51	17.44	69.91	426.6	68.0	.0618
146.0	936.1	23.43	1604.0	1877.0	81.51	17.46	75.25	403.3	63.7	.0595
148.0	903.9	22.63	1751.0	2034.0	82.58	17.52	82.58	378.2	59.3	.0572
149.0	886.0	22.18	1831.0	2119.0	83.15	17.57	87.40	364.9	57.0	.0561
150.0	866.7	21.70	1915.0	2210.0	83.75	17.65	93.36	350.9	54.6	.0550
151.0	845.5	21.16	2004.0	2306.0	84.40	17.74	100.9	336.2	52.2	.0539
152.0	821.9	20.57	2101.0	2412.0	85.09	17.88	110.8	320.6	49.6	.0528
153.0	795.2	19.91	2208.0	2529.0	85.86	18.05	124.3	303.9	46.9	.0516
154.0	764.2	19.13	2328.0	2663.0	86.73	18.29	143.5	286.1	44.0	.0504
155.0	727.0	18.20	2468.0	2819.0	87.74	18.61	171.4	267.2	40.8	.0490
155.5	705.4	17.66	2547.0	2909.0	88.33	18.81	189.8	257.5	39.0	.0482
156.0	681.3	17.05	2634.0	3009.0	88.97	19.03	210.8	247.8	37.2	.0474
156.5	654.6	16.39	2730.0	3120.0	89.68	19.28	232.8	238.5	35.3	.0464
157.0	625.7	15.66	2833.0	3242.0	90.45	19.53	251.9	230.1	33.3	.0452
158.0	565.5	14.16	3051.0	3504.0	92.11	19.92	264.7	217.6	29.7	.0423
158.5	536.9	13.44	3158.0	3634.0	92.94	20.02	257.0	213.7	28.2	.0408

Thermophysical properties of argon—Continued

T K	Density kg/m^3	Density mol/dm^3	E J/mol	H J/mol	S J/(mol·K)	C_v J/(mol·K)	C_p J/(mol·K)	Sound m/s	Visc. μPa·s	Therm. W/(m·K)
159.0	510.7	12.78	3259.0	3759.0	93.73	20.05	243.3	211.2	26.9	.0392
159.5	487.0	12.19	3352.0	3877.0	94.47	20.01	226.9	209.7	25.8	.0377
160.0	466.1	11.67	3438.0	3986.0	95.15	19.92	209.7	208.8	24.9	.0363
160.5	447.5	11.20	3516.0	4087.0	95.78	19.79	193.0	208.5	24.1	.0351
161.0	431.1	10.79	3586.0	4179.0	96.35	19.64	177.7	208.6	23.5	.0339
161.5	416.6	10.43	3651.0	4265.0	96.88	19.48	164.0	208.9	22.9	.0328
162.0	403.6	10.10	3710.0	4344.0	97.37	19.32	151.8	209.3	22.5	.0318
162.5	392.0	9.814	3765.0	4417.0	97.82	19.15	141.2	209.9	22.1	.0309
163.0	381.6	9.552	3815.0	4485.0	98.24	18.98	131.8	210.5	21.8	.0301
164.0	363.5	9.098	3905.0	4609.0	99.00	18.65	116.4	212.0	21.2	.0287
165.0	348.2	8.717	3985.0	4719.0	99.67	18.34	104.5	213.5	20.8	.0275
166.0	335.1	8.389	4056.0	4819.0	100.3	18.05	94.99	215.2	20.4	.0265
167.0	323.7	8.104	4120.0	4910.0	100.8	17.79	87.34	216.8	20.2	.0256
168.0	313.7	7.852	4179.0	4994.0	101.3	17.54	81.05	218.4	19.9	.0249
169.0	304.7	7.626	4233.0	5072.0	101.8	17.31	75.80	220.0	19.7	.0242
170.0	296.5	7.423	4284.0	5146.0	102.2	17.09	71.36	221.6	19.6	.0236
172.0	282.4	7.069	4376.0	5281.0	103.0	16.70	64.27	224.6	19.3	.0226
174.0	270.4	6.768	4458.0	5404.0	103.7	16.37	58.87	227.6	19.2	.0218
176.0	260.0	6.508	4534.0	5517.0	104.4	16.07	54.61	230.4	19.0	.0211
178.0	250.8	6.278	4603.0	5623.0	105.0	15.81	51.17	233.2	19.0	.0206
180.0	242.6	6.074	4669.0	5722.0	105.5	15.57	48.34	235.8	18.9	.0201
182.0	235.3	5.889	4730.0	5817.0	106.0	15.36	45.96	238.3	18.9	.0198
184.0	228.6	5.722	4788.0	5906.0	106.5	15.17	43.94	240.8	18.9	.0195
186.0	222.5	5.569	4843.0	5992.0	107.0	15.00	42.20	243.2	18.9	.0192
188.0	216.8	5.427	4896.0	6075.0	107.4	14.85	40.69	245.5	18.9	.0190
190.0	211.6	5.296	4947.0	6155.0	107.9	14.71	39.36	247.7	18.9	.0188
195.0	200.0	5.007	5067.0	6345.0	108.8	14.41	36.66	253.1	19.0	.0184
200.0	190.1	4.759	5178.0	6523.0	109.8	14.17	34.60	258.2	19.2	.0182
205.0	181.5	4.543	5283.0	6692.0	110.6	13.97	32.97	263.0	19.4	.0181
210.0	173.9	4.353	5383.0	6853.0	111.4	13.81	31.66	267.6	19.6	.0181
215.0	167.1	4.183	5479.0	7009.0	112.1	13.67	30.58	272.0	19.8	.0181
220.0	161.0	4.029	5571.0	7159.0	112.8	13.56	29.67	276.2	20.1	.0182
225.0	155.4	3.890	5660.0	7306.0	113.4	13.46	28.91	280.3	20.3	.0182
230.0	150.3	3.762	5747.0	7449.0	114.1	13.38	28.25	284.3	20.6	.0183
240.0	141.3	3.536	5916.0	7726.0	115.3	13.25	27.19	291.9	21.1	.0186
250.0	133.5	3.341	6078.0	7993.0	116.3	13.15	26.36	299.1	21.6	.0188
260.0	126.7	3.171	6235.0	8253.0	117.4	13.08	25.70	306.0	22.2	.0192
270.0	120.6	3.019	6388.0	8508.0	118.3	13.02	25.17	312.6	22.8	.0195
280.0	115.2	2.884	6538.0	8757.0	119.2	12.97	24.73	318.9	23.3	.0198
290.0	110.4	2.763	6686.0	9002.0	120.1	12.94	24.35	325.1	23.9	.0202
300.0	106.0	2.652	6831.0	9244.0	120.9	12.90	24.04	331.1	24.5	.0206
310.0	101.9	2.551	6975.0	9483.0	121.7	12.88	23.76	336.9	25.0	.0210
320.0	98.23	2.459	7117.0	9720.0	122.4	12.85	23.53	342.5	25.6	.0213
330.0	94.82	2.374	7258.0	9954.0	123.2	12.84	23.32	348.1	26.1	.0217
340.0	91.67	2.295	7397.0	10190.0	123.9	12.82	23.14	353.5	26.7	.0221
360.0	86.01	2.153	7673.0	10650.0	125.2	12.79	22.83	363.9	27.8	.0229
380.0	81.07	2.029	7946.0	11100.0	126.4	12.77	22.58	373.9	28.9	.0237
400.0	76.71	1.920	8216.0	11550.0	127.6	12.75	22.37	383.6	30.0	.0245

6.60 MPa isobar

T K	Density kg/m^3	Density mol/dm^3	E J/mol	H J/mol	S J/(mol·K)	C_v J/(mol·K)	C_p J/(mol·K)	Sound m/s	Visc. μPa·s	Therm. W/(m·K)
85.42[a]	1423.0	35.63	−1367.0	−1182.0	54.91	23.07	43.69	844.4	293.0	.135
90.0	1397.0	34.97	−1171.0	−982.1	57.19	21.77	43.70	832.1	257.0	.128
100.0	1336.0	33.45	−738.0	−540.7	61.84	20.24	44.72	782.5	198.0	.113
110.0	1271.0	31.81	−293.4	−85.9	66.17	19.32	46.32	717.6	156.0	.101
120.0	1199.0	30.02	168.5	388.4	70.30	18.57	48.74	644.9	124.0	.0892
125.0	1160.0	29.04	409.1	636.3	72.32	18.24	50.54	605.9	111.0	.0837
130.0	1118.0	27.99	659.0	894.9	74.35	17.94	53.01	564.7	98.9	.0782
135.0	1071.0	26.82	922.1	1168.0	76.41	17.69	56.56	520.8	87.7	.0726
136.0	1061.0	26.57	976.8	1225.0	76.83	17.64	57.45	511.6	85.6	.0714
138.0	1041.0	26.05	1089.0	1342.0	77.69	17.57	59.49	492.7	81.3	.0691
140.0	1018.0	25.49	1205.0	1463.0	78.56	17.50	61.94	473.2	77.1	.0668

Thermophysical properties of argon—Continued

T K	Density kg/m^3	Density mol/dm^3	*E* J/mol	*H* J/mol	*S* J/(mol·K)	C_v J/(mol·K)	C_p J/(mol·K)	Sound m/s	Visc. μPa·s	Therm. W/(m·K)
142.0	994.8	24.90	1325.0	1590.0	79.46	17.46	64.93	452.7	72.9	.0645
144.0	969.2	24.26	1452.0	1724.0	80.39	17.44	68.69	431.3	68.7	.0622
146.0	941.3	23.56	1586.0	1866.0	81.37	17.44	73.54	408.6	64.4	.0599
148.0	910.3	22.79	1729.0	2019.0	82.41	17.49	80.05	384.5	60.1	.0576
150.0	874.9	21.90	1886.0	2188.0	83.55	17.60	89.31	358.4	55.6	.0555
151.0	855.1	21.40	1972.0	2280.0	84.16	17.68	95.57	344.4	53.3	.0544
152.0	833.3	20.86	2063.0	2379.0	84.82	17.78	103.5	329.8	50.9	.0533
153.0	809.0	20.25	2162.0	2488.0	85.53	17.93	113.8	314.3	48.3	.0522
154.0	781.6	19.57	2271.0	2608.0	86.31	18.11	127.7	297.9	45.6	.0510
155.0	749.8	18.77	2393.0	2745.0	87.20	18.36	146.7	280.7	42.8	.0497
156.0	712.3	17.83	2534.0	2904.0	88.22	18.68	172.8	262.9	39.6	.0482
157.0	667.5	16.71	2698.0	3093.0	89.42	19.07	204.8	245.4	36.2	.0465
158.0	616.1	15.42	2884.0	3312.0	90.82	19.48	231.6	230.4	32.8	.0442
159.0	563.0	14.09	3080.0	3548.0	92.31	19.78	236.2	220.0	29.7	.0416
159.5	537.9	13.47	3175.0	3665.0	93.04	19.85	229.8	216.7	28.3	.0402
160.0	514.6	12.88	3265.0	3777.0	93.74	19.87	219.7	214.4	27.2	.0388
160.5	493.3	12.35	3350.0	3884.0	94.41	19.83	207.5	212.9	26.2	.0375
161.0	474.0	11.87	3429.0	3985.0	95.04	19.75	194.6	212.0	25.3	.0363
161.5	456.7	11.43	3501.0	4079.0	95.62	19.65	181.7	211.6	24.6	.0351
162.0	441.1	11.04	3569.0	4167.0	96.16	19.52	169.4	211.5	24.0	.0340
162.5	427.1	10.69	3631.0	4248.0	96.67	19.37	158.0	211.6	23.5	.0330
163.0	414.4	10.37	3688.0	4325.0	97.13	19.22	147.7	212.0	23.0	.0321
164.0	392.5	9.826	3791.0	4463.0	97.98	18.92	130.0	212.9	22.2	.0305
165.0	374.3	9.369	3881.0	4586.0	98.73	18.61	115.9	214.2	21.7	.0291
166.0	358.8	8.981	3961.0	4696.0	99.39	18.32	104.6	215.6	21.2	.0280
167.0	345.4	8.646	4033.0	4796.0	99.99	18.04	95.51	217.1	20.9	.0269
168.0	333.7	8.353	4097.0	4887.0	100.5	17.78	88.05	218.6	20.6	.0261
169.0	323.3	8.094	4157.0	4972.0	101.0	17.54	81.85	220.2	20.3	.0253
170.0	314.0	7.861	4212.0	5052.0	101.5	17.31	76.64	221.7	20.1	.0246
172.0	298.0	7.460	4311.0	5196.0	102.4	16.91	68.40	224.7	19.8	.0235
174.0	284.6	7.123	4400.0	5326.0	103.1	16.55	62.18	227.7	19.6	.0226
176.0	273.0	6.834	4480.0	5446.0	103.8	16.23	57.34	230.5	19.4	.0218
178.0	262.9	6.581	4554.0	5556.0	104.4	15.96	53.46	233.2	19.3	.0212
180.0	254.0	6.357	4622.0	5660.0	105.0	15.71	50.30	235.8	19.2	.0207
182.0	245.9	6.157	4686.0	5758.0	105.5	15.49	47.66	238.4	19.2	.0202
184.0	238.7	5.975	4746.0	5851.0	106.0	15.29	45.43	240.9	19.1	.0199
186.0	232.1	5.809	4804.0	5940.0	106.5	15.11	43.51	243.3	19.1	.0196
188.0	226.0	5.657	4859.0	6025.0	107.0	14.94	41.86	245.6	19.1	.0193
190.0	220.4	5.517	4911.0	6107.0	107.4	14.80	40.41	247.8	19.1	.0191
195.0	208.0	5.207	5034.0	6302.0	108.4	14.48	37.49	253.2	19.2	.0187
200.0	197.5	4.944	5149.0	6484.0	109.3	14.23	35.27	258.3	19.4	.0185
205.0	188.4	4.715	5256.0	6655.0	110.2	14.03	33.53	263.2	19.5	.0183
210.0	180.3	4.514	5357.0	6819.0	111.0	13.86	32.14	267.8	19.7	.0183
215.0	173.2	4.335	5455.0	6977.0	111.7	13.72	30.99	272.2	19.9	.0183
220.0	166.7	4.173	5548.0	7130.0	112.4	13.60	30.04	276.5	20.2	.0183
225.0	160.9	4.027	5639.0	7278.0	113.1	13.50	29.23	280.6	20.4	.0184
230.0	155.5	3.893	5727.0	7422.0	113.7	13.41	28.54	284.6	20.7	.0185
240.0	146.1	3.657	5897.0	7702.0	114.9	13.28	27.42	292.2	21.2	.0187
250.0	138.0	3.453	6060.0	7972.0	116.0	13.18	26.56	299.4	21.7	.0190
260.0	130.9	3.276	6219.0	8234.0	117.1	13.10	25.87	306.3	22.3	.0193
270.0	124.6	3.119	6373.0	8489.0	118.0	13.04	25.32	312.9	22.8	.0196
280.0	119.0	2.978	6524.0	8740.0	118.9	12.99	24.86	319.3	23.4	.0199
290.0	113.9	2.852	6673.0	8987.0	119.8	12.95	24.47	325.5	24.0	.0203
300.0	109.4	2.737	6819.0	9230.0	120.6	12.92	24.14	331.5	24.5	.0207
310.0	105.2	2.633	6963.0	9470.0	121.4	12.89	23.86	337.3	25.1	.0210
320.0	101.4	2.537	7106.0	9707.0	122.2	12.87	23.61	342.9	25.6	.0214
330.0	97.82	2.449	7247.0	9942.0	122.9	12.85	23.40	348.4	26.2	.0218
340.0	94.56	2.367	7387.0	10180.0	123.6	12.83	23.21	353.8	26.8	.0222
360.0	88.71	2.221	7664.0	10640.0	124.9	12.80	22.89	364.3	27.9	.0230
380.0	83.60	2.093	7937.0	11090.0	126.1	12.78	22.63	374.3	28.9	.0238
400.0	79.09	1.980	8208.0	11540.0	127.3	12.76	22.42	384.0	30.0	.0245

Thermophysical properties of argon—Continued

T K	Density kg/m³	Density mol/dm³	E J/mol	H J/mol	S J/(mol·K)	C_v J/(mol·K)	C_p J/(mol·K)	Sound m/s	Visc. μPa·s	Therm. W/(m·K)
				6.80 MPa isobar						
85.47[a]	1424.0	35.64	−1367.0	−1176.0	54.91	23.05	43.66	845.1	293.0	.135
90.0	1397.0	34.98	−1173.0	−978.5	57.17	21.77	43.67	833.0	257.0	.128
100.0	1337.0	33.47	−740.7	−537.5	61.81	20.25	44.67	783.5	198.0	.113
110.0	1272.0	31.84	−296.9	−83.3	66.14	19.33	46.25	718.9	156.0	.101
120.0	1201.0	30.05	163.8	390.0	70.26	18.58	48.61	646.6	125.0	.0894
125.0	1162.0	29.08	403.4	637.2	72.27	18.25	50.36	607.9	111.0	.0839
130.0	1120.0	28.03	652.1	894.7	74.29	17.95	52.76	567.1	99.4	.0784
135.0	1074.0	26.88	913.5	1166.0	76.34	17.70	56.16	523.7	88.2	.0728
140.0	1022.0	25.57	1193.0	1459.0	78.47	17.51	61.27	476.8	77.6	.0671
142.0	998.3	24.99	1312.0	1584.0	79.36	17.46	64.08	456.8	73.5	.0648
144.0	973.4	24.37	1437.0	1716.0	80.28	17.43	67.56	435.8	69.4	.0625
146.0	946.2	23.69	1568.0	1855.0	81.24	17.43	72.00	413.8	65.2	.0603
148.0	916.3	22.94	1708.0	2005.0	82.26	17.47	77.83	390.4	61.0	.0581
150.0	882.5	22.09	1860.0	2168.0	83.35	17.55	85.89	365.3	56.6	.0559
151.0	863.8	21.62	1942.0	2257.0	83.94	17.62	91.19	352.1	54.4	.0548
152.0	843.4	21.11	2029.0	2351.0	84.56	17.71	97.72	338.2	52.0	.0538
154.0	796.3	19.93	2222.0	2564.0	85.95	17.98	116.5	308.5	47.1	.0515
155.0	768.2	19.23	2333.0	2687.0	86.75	18.17	130.4	292.6	44.4	.0503
156.0	736.0	18.42	2457.0	2826.0	87.65	18.42	148.7	276.1	41.6	.0490
157.0	698.6	17.49	2597.0	2986.0	88.67	18.72	171.8	259.6	38.6	.0475
158.0	655.5	16.41	2756.0	3170.0	89.84	19.08	196.5	244.1	35.5	.0456
159.0	608.3	15.23	2929.0	3376.0	91.14	19.41	213.2	231.3	32.4	.0434
160.0	560.9	14.04	3107.0	3591.0	92.48	19.64	213.7	222.5	29.6	.0409
161.0	517.7	12.96	3274.0	3799.0	93.78	19.69	200.5	217.5	27.4	.0384
161.5	498.4	12.48	3352.0	3897.0	94.39	19.66	191.1	216.1	26.5	.0372
162.0	480.6	12.03	3425.0	3990.0	94.96	19.59	181.1	215.2	25.7	.0361
162.5	464.4	11.63	3493.0	4078.0	95.50	19.49	170.9	214.7	25.0	.0351
163.0	449.7	11.26	3557.0	4161.0	96.01	19.38	161.0	214.5	24.4	.0341
163.5	436.2	10.92	3616.0	4239.0	96.49	19.25	151.6	214.5	23.9	.0332
164.0	423.9	10.61	3672.0	4312.0	96.94	19.12	142.8	214.7	23.5	.0323
165.0	402.4	10.07	3772.0	4447.0	97.76	18.84	127.4	215.5	22.7	.0308
166.0	384.2	9.619	3861.0	4568.0	98.49	18.55	114.6	216.6	22.1	.0295
167.0	368.7	9.228	3940.0	4677.0	99.15	18.28	104.2	217.8	21.6	.0283
168.0	355.1	8.889	4012.0	4777.0	99.74	18.01	95.53	219.2	21.3	.0273
169.0	343.2	8.591	4077.0	4869.0	100.3	17.77	88.35	220.6	20.9	.0265
170.0	332.6	8.325	4137.0	4954.0	100.8	17.53	82.32	222.1	20.7	.0257
171.0	323.1	8.087	4193.0	5034.0	101.3	17.31	77.21	223.5	20.5	.0250
172.0	314.4	7.871	4245.0	5109.0	101.7	17.10	72.82	225.0	20.3	.0244
174.0	299.4	7.494	4340.0	5247.0	102.5	16.73	65.73	227.8	20.0	.0233
176.0	286.6	7.173	4425.0	5373.0	103.2	16.40	60.24	230.6	19.8	.0225
178.0	275.4	6.895	4502.0	5489.0	103.9	16.11	55.89	233.4	19.6	.0218
180.0	265.7	6.650	4574.0	5597.0	104.5	15.84	52.36	236.0	19.5	.0212
182.0	256.9	6.432	4641.0	5698.0	105.0	15.61	49.44	238.5	19.4	.0207
184.0	249.1	6.235	4704.0	5795.0	105.6	15.40	46.98	241.0	19.4	.0203
186.0	241.9	6.056	4764.0	5887.0	106.1	15.21	44.88	243.4	19.4	.0200
188.0	235.4	5.892	4821.0	5975.0	106.5	15.04	43.08	245.7	19.4	.0197
190.0	229.4	5.742	4875.0	6059.0	107.0	14.89	41.51	248.0	19.4	.0195
195.0	216.2	5.411	5002.0	6258.0	108.0	14.56	38.35	253.4	19.4	.0190
200.0	205.0	5.131	5119.0	6444.0	109.0	14.29	35.97	258.6	19.5	.0187
205.0	195.3	4.889	5228.0	6619.0	109.8	14.08	34.11	263.4	19.7	.0186
210.0	186.8	4.677	5332.0	6786.0	110.6	13.90	32.63	268.0	19.9	.0185
215.0	179.3	4.488	5431.0	6946.0	111.4	13.76	31.41	272.5	20.1	.0185
220.0	172.5	4.319	5526.0	7100.0	112.1	13.63	30.41	276.8	20.3	.0185
225.0	166.4	4.165	5617.0	7250.0	112.8	13.53	29.56	280.9	20.5	.0186
230.0	160.8	4.025	5707.0	7396.0	113.4	13.44	28.83	284.9	20.8	.0186
240.0	150.9	3.778	5878.0	7678.0	114.6	13.30	27.66	292.5	21.3	.0188
250.0	142.5	3.566	6043.0	7950.0	115.7	13.20	26.76	299.7	21.8	.0191
260.0	135.1	3.381	6203.0	8214.0	116.7	13.12	26.04	306.6	22.4	.0194
270.0	128.6	3.218	6358.0	8471.0	117.7	13.05	25.46	313.3	22.9	.0197
280.0	122.7	3.072	6510.0	8723.0	118.6	13.00	24.99	319.7	23.5	.0200
290.0	117.5	2.941	6659.0	8971.0	119.5	12.96	24.59	325.8	24.0	.0204

Thermophysical properties of argon—Continued

T K	Density kg/m³	Density mol/dm³	E J/mol	H J/mol	S J/(mol·K)	C_v J/(mol·K)	C_p J/(mol·K)	Sound m/s	Visc. μPa·s	Therm. W/(m·K)
300.0	112.8	2.823	6806.0	9215.0	120.3	12.93	24.24	331.8	24.6	.0208
310.0	108.4	2.714	6951.0	9456.0	121.1	12.90	23.95	337.6	25.1	.0211
320.0	104.5	2.615	7094.0	9695.0	121.9	12.88	23.70	343.3	25.7	.0215
330.0	100.8	2.524	7236.0	9930.0	122.6	12.86	23.48	348.8	26.3	.0219
340.0	97.45	2.439	7377.0	10160.0	123.3	12.84	23.28	354.2	26.8	.0223
360.0	91.40	2.288	7654.0	10630.0	124.6	12.81	22.95	364.7	27.9	.0231
380.0	86.12	2.156	7929.0	11080.0	125.9	12.79	22.68	374.7	29.0	.0238
400.0	81.47	2.039	8200.0	11530.0	127.0	12.77	22.47	384.4	30.1	.0246
				7.00 MPa isobar						
85.51[a]	1424.0	35.64	−1367.0	−1170.0	54.91	23.04	43.63	845.8	293.0	.135
90.0	1398.0	35.00	−1175.0	−974.9	57.14	21.78	43.63	833.8	258.0	.128
100.0	1338.0	33.48	−743.4	−534.3	61.78	20.26	44.62	784.5	199.0	.113
110.0	1273.0	31.86	−300.5	−80.7	66.11	19.34	46.18	720.2	157.0	.101
120.0	1202.0	30.09	159.0	391.7	70.21	18.60	48.49	648.3	125.0	.0896
125.0	1163.0	29.12	397.8	638.2	72.23	18.27	50.19	609.8	112.0	.0841
130.0	1122.0	28.08	645.3	894.6	74.24	17.97	52.51	569.4	99.9	.0786
135.0	1076.0	26.93	905.0	1165.0	76.28	17.71	55.78	526.5	88.7	.0731
140.0	1025.0	25.65	1182.0	1455.0	78.39	17.52	60.64	480.3	78.2	.0674
142.0	1002.0	25.08	1300.0	1579.0	79.27	17.46	63.28	460.7	74.1	.0652
144.0	977.4	24.47	1422.0	1709.0	80.17	17.43	66.52	440.2	70.0	.0629
146.0	951.0	23.81	1551.0	1846.0	81.12	17.42	70.59	418.7	65.9	.0607
148.0	922.0	23.08	1688.0	1992.0	82.11	17.45	75.86	396.0	61.8	.0585
150.0	889.6	22.27	1836.0	2150.0	83.18	17.52	82.96	371.9	57.5	.0563
152.0	852.6	21.34	1998.0	2326.0	84.34	17.64	93.00	346.0	53.1	.0542
154.0	809.0	20.25	2180.0	2526.0	85.64	17.86	108.1	318.1	48.4	.0521
155.0	783.7	19.62	2282.0	2639.0	86.38	18.02	118.8	303.2	45.9	.0509
156.0	755.3	18.91	2394.0	2764.0	87.18	18.22	132.3	287.9	43.3	.0497
157.0	723.0	18.10	2518.0	2905.0	88.08	18.46	149.3	272.4	40.6	.0483
158.0	686.2	17.18	2656.0	3063.0	89.09	18.75	168.8	257.2	37.7	.0467
159.0	645.2	16.15	2808.0	3242.0	90.21	19.06	187.0	243.5	34.8	.0448
160.0	601.8	15.06	2970.0	3435.0	91.42	19.33	197.1	232.6	32.1	.0426
161.0	559.2	14.00	3132.0	3632.0	92.65	19.50	195.4	225.1	29.6	.0403
162.0	520.2	13.02	3285.0	3822.0	93.83	19.52	184.6	220.6	27.6	.0381
163.0	486.1	12.17	3424.0	4000.0	94.92	19.43	169.2	218.3	26.1	.0360
164.0	457.0	11.44	3549.0	4161.0	95.91	19.24	152.9	217.4	24.9	.0341
165.0	432.4	10.82	3659.0	4306.0	96.79	19.01	137.7	217.5	23.9	.0325
166.0	411.4	10.30	3757.0	4437.0	97.58	18.75	124.3	218.1	23.1	.0310
167.0	393.4	9.848	3844.0	4555.0	98.29	18.49	112.8	219.0	22.5	.0298
168.0	377.8	9.458	3923.0	4663.0	98.93	18.23	103.2	220.1	22.1	.0286
169.0	364.2	9.117	3994.0	4762.0	99.52	17.98	95.12	221.3	21.7	.0277
170.0	352.1	8.815	4059.0	4854.0	100.1	17.74	88.29	222.7	21.3	.0268
171.0	341.4	8.545	4120.0	4939.0	100.6	17.51	82.48	224.0	21.1	.0260
172.0	331.7	8.303	4176.0	5019.0	101.0	17.30	77.51	225.4	20.8	.0253
174.0	314.8	7.881	4277.0	5165.0	101.9	16.91	69.48	228.2	20.5	.0242
176.0	300.6	7.526	4368.0	5298.0	102.6	16.56	63.31	230.9	20.2	.0232
178.0	288.4	7.220	4450.0	5419.0	103.3	16.25	58.45	233.6	20.0	.0224
180.0	277.7	6.952	4525.0	5532.0	104.0	15.98	54.53	236.2	19.9	.0218
182.0	268.2	6.714	4595.0	5638.0	104.5	15.74	51.30	238.8	19.8	.0213
184.0	259.7	6.501	4661.0	5738.0	105.1	15.52	48.60	241.2	19.7	.0208
186.0	252.0	6.308	4723.0	5833.0	105.6	15.32	46.31	243.6	19.6	.0204
188.0	245.0	6.132	4782.0	5923.0	106.1	15.14	44.34	246.0	19.6	.0201
190.0	238.5	5.971	4838.0	6010.0	106.5	14.98	42.64	248.2	19.6	.0198
195.0	224.4	5.618	4968.0	6214.0	107.6	14.63	39.23	253.7	19.6	.0193
200.0	212.6	5.321	5088.0	6404.0	108.6	14.36	36.68	258.8	19.7	.0190
205.0	202.3	5.065	5200.0	6582.0	109.4	14.13	34.70	263.7	19.9	.0188
210.0	193.4	4.841	5306.0	6752.0	110.3	13.95	33.12	268.3	20.0	.0187
215.0	185.5	4.643	5406.0	6914.0	111.0	13.80	31.84	272.8	20.2	.0187
220.0	178.4	4.465	5503.0	7070.0	111.7	13.67	30.78	277.1	20.4	.0187
225.0	171.9	4.304	5596.0	7222.0	112.4	13.56	29.89	281.2	20.7	.0187
230.0	166.1	4.158	5686.0	7370.0	113.1	13.47	29.12	285.2	20.9	.0188

Thermophysical properties of argon—Continued

T K	Density kg/m^3	Density mol/dm^3	E J/mol	H J/mol	S J/(mol·K)	C_v J/(mol·K)	C_p J/(mol·K)	Sound m/s	Visc. μPa·s	Therm. W/(m·K)
235.0	160.7	4.024	5774.0	7513.0	113.7	13.39	28.47	289.1	21.1	.0189
240.0	155.8	3.900	5860.0	7654.0	114.3	13.33	27.90	292.8	21.4	.0190
250.0	147.0	3.679	6026.0	7928.0	115.4	13.22	26.96	300.1	21.9	.0192
260.0	139.3	3,487	6187.0	8194.0	116.4	13.13	26.21	307.0	22.4	.0195
270.0	132.5	3.318	6343.0	8453.0	117.4	13.07	25.61	313.6	23.0	.0198
280.0	126.5	3.167	6496.0	8707.0	118.3	13.02	25.12	320.0	23.5	.0202
290.0	121.1	3.031	6646.0	8956.0	119.2	12.98	24.70	326.2	24.1	.0205
300.0	116.2	2.908	6794.0	9201.0	120.0	12.94	24.35	332.2	24.6	.0209
310.0	111.7	2.796	6939.0	9443.0	120.8	12.91	24.05	338.0	25.2	.0212
320.0	107.6	2.693	7083.0	9682.0	121.6	12.89	23.78	343.7	25.8	.0216
330.0	103.8	2.599	7225.0	9919.0	122.3	12.87	23.55	349.2	26.3	.0220
340.0	100.3	2.512	7366.0	10150.0	123.0	12.85	23.35	354.6	26.9	.0224
360.0	94.10	2.355	7645.0	10620.0	124.4	12.82	23.01	365.1	28.0	.0231
380.0	88.65	2.219	7920.0	11070.0	125.6	12.79	22.74	375.1	29.0	.0239
400.0	83.85	2.099	8192.0	11530.0	126.8	12.77	22.51	384.8	30.1	.0247

7.50 MPa isobar

T K	Density kg/m^3	Density mol/dm^3	E J/mol	H J/mol	S J/(mol·K)	C_v J/(mol·K)	C_p J/(mol·K)	Sound m/s	Visc. μPa·s	Therm. W/(m·K)
85.64[a]	1425.0	35.66	−1366.0	−1156.0	54.92	23.00	43.55	847.5	293.0	.136
90.0	1399.0	35.03	−1180.0	−966.0	57.08	21.79	43.55	835.9	259.0	.128
100.0	1339.0	33.53	−750.0	−526.3	61.71	20.28	44.51	787.1	200.0	.114
110.0	1275.0	31.92	−309.1	−74.2	66.02	19.37	46.00	723.4	158.0	.101
120.0	1205.0	30.16	147.3	396.0	70.11	18.63	48.19	652.5	126.0	.0900
125.0	1167.0	29.21	384.0	640.7	72.11	18.30	49.78	614.7	113.0	.0846
130.0	1126.0	28.19	628.7	894.7	74.10	18.00	51.93	575.1	101.0	.0792
135.0	1082.0	27.07	884.3	1161.0	76.11	17.74	54.90	533.4	90.0	.0738
140.0	1032.0	25.83	1156.0	1446.0	78.18	17.53	59.20	488.9	79.6	.0682
142.0	1010.0	25.28	1270.0	1567.0	79.04	17.47	61.49	470.1	75.6	.0660
144.0	986.8	24.70	1389.0	1692.0	79.92	17.43	64.23	450.6	71.6	.0638
146.0	962.0	24.08	1512.0	1824.0	80.83	17.41	67.58	430.3	67.6	.0616
148.0	935.0	23.41	1643.0	1963.0	81.77	17.41	71.77	409.2	63.6	.0594
150.0	905.5	22.67	1781.0	2112.0	82.77	17.44	77.14	387.0	59.6	.0574
152.0	872.6	21.84	1929.0	2273.0	83.84	17.52	84.24	363.6	55.5	.0553
154.0	835.3	20.91	2092.0	2450.0	85.00	17.66	94.00	338.8	51.3	.0533
156.0	791.8	19.82	2273.0	2651.0	86.29	17.88	107.8	312.7	46.8	.0511
157.0	767.1	19.20	2373.0	2764.0	87.01	18.03	116.9	299.2	44.5	.0500
158.0	739.9	18.52	2481.0	2886.0	87.79	18.20	127.4	285.7	42.1	.0487
159.0	710.0	17.77	2597.0	3019.0	88.63	18.40	139.1	272.5	39.7	.0473
160.0	677.5	16.96	2722.0	3164.0	89.54	18.62	150.9	260.1	37.2	.0457
162.0	607.4	15.20	2990.0	3483.0	91.52	19.01	165.2	240.4	32.6	.0420
163.0	572.6	14.33	3125.0	3648.0	92.54	19.12	164.4	233.9	30.6	.0401
164.0	539.9	13.52	3255.0	3810.0	93.53	19.15	158.9	229.4	28.9	.0382
165.0	510.2	12.77	3378.0	3965.0	94.47	19.09	150.3	226.7	27.4	.0364
166.0	483.6	12.11	3491.0	4111.0	95.35	18.97	140.3	225.2	26.2	.0348
167.0	460.1	11.52	3595.0	4246.0	96.16	18.80	130.0	224.6	25.3	.0333
168.0	439.5	11.00	3689.0	4371.0	96.90	18.60	120.1	224.6	24.5	.0320
169.0	421.3	10.55	3775.0	4486.0	97.59	18.39	111.1	225.0	23.8	.0308
170.0	405.3	10.14	3854.0	4593.0	98.22	18.16	103.0	225.7	23.3	.0297
171.0	391.0	9.787	3926.0	4693.0	98.80	17.94	95.91	226.5	22.8	.0287
172.0	378.2	9.468	3993.0	4785.0	99.34	17.73	89.68	227.6	22.4	.0279
173.0	366.7	9.180	4055.0	4872.0	99.85	17.52	84.23	228.7	22.1	.0271
174.0	356.3	8.919	4113.0	4954.0	100.3	17.32	79.45	229.8	21.8	.0264
176.0	338.1	8.464	4219.0	5105.0	101.2	16.95	71.54	232.3	21.4	.0252
178.0	322.7	8.079	4313.0	5241.0	102.0	16.62	65.32	234.7	21.0	.0242
180.0	309.4	7.746	4398.0	5367.0	102.7	16.32	60.34	237.2	20.8	.0233
182.0	297.8	7.454	4477.0	5483.0	103.3	16.05	56.28	239.7	20.6	.0226
184.0	287.4	7.195	4550.0	5592.0	103.9	15.80	52.92	242.1	20.5	.0220
186.0	278.2	6.963	4618.0	5695.0	104.4	15.58	50.09	244.5	20.4	.0215
188.0	269.8	6.754	4682.0	5793.0	105.0	15.39	47.69	246.8	20.3	.0211
190.0	262.2	6.563	4743.0	5886.0	105.5	15.21	45.62	249.0	20.2	.0208
192.0	255.2	6.387	4801.0	5976.0	105.9	15.04	43.82	251.2	20.2	.0205
195.0	245.7	6.149	4884.0	6104.0	106.6	14.82	41.54	254.5	20.2	.0201
200.0	231.9	5.806	5012.0	6303.0	107.6	14.52	38.53	259.6	20.2	.0197

Thermophysical properties of argon—Continued

T K	Density kg/m^3	Density mol/dm^3	E J/mol	H J/mol	S J/(mol·K)	C_v J/(mol·K)	C_p J/(mol·K)	Sound m/s	Visc. μPa·s	Therm. W/(m·K)
205.0	220.2	5.513	5130.0	6490.0	108.5	14.27	36.22	264.5	20.3	.0194
210.0	210.1	5.259	5240.0	6666.0	109.4	14.07	34.40	269.1	20.4	.0192
215.0	201.1	5.035	5345.0	6835.0	110.2	13.90	32.94	273.6	20.6	.0192
220.0	193.2	4.835	5445.0	6996.0	110.9	13.76	31.73	277.9	20.8	.0191
225.0	186.0	4.656	5541.0	7152.0	111.6	13.64	30.72	282.0	21.0	.0191
230.0	179.5	4.493	5634.0	7304.0	112.3	13.54	29.87	286.1	21.2	.0192
235.0	173.5	4.344	5725.0	7451.0	112.9	13.46	29.14	290.0	21.4	.0193
240.0	168.1	4.208	5813.0	7595.0	113.5	13.39	28.50	293.7	21.7	.0193
250.0	158.4	3.964	5983.0	7875.0	114.7	13.27	27.46	301.0	22.1	.0196
260.0	149.9	3.753	6147.0	8145.0	115.7	13.18	26.64	307.9	22.7	.0198
270.0	142.5	3.568	6306.0	8408.0	116.7	13.11	25.98	314.6	23.2	.0201
280.0	135.9	3.403	6461.0	8665.0	117.7	13.05	25.44	321.0	23.7	.0204
290.0	130.0	3.255	6613.0	8917.0	118.5	13.01	24.99	327.2	24.3	.0207
300.0	124.7	3.122	6763.0	9165.0	119.4	12.97	24.61	333.2	24.8	.0211
310.0	119.9	3.000	6910.0	9410.0	120.2	12.94	24.28	339.0	25.4	.0214
320.0	115.4	2.889	7055.0	9651.0	120.9	12.91	24.00	344.7	25.9	.0218
330.0	111.3	2.787	7199.0	9890.0	121.7	12.89	23.75	350.2	26.5	.0222
340.0	107.6	2.693	7341.0	10130.0	122.4	12.87	23.53	355.6	27.0	.0225
360.0	100.8	2.524	7622.0	10590.0	123.7	12.84	23.17	366.1	28.1	.0233
380.0	94.96	2.377	7898.0	11050.0	125.0	12.82	22.87	376.1	29.1	.0241
400.0	89.80	2.248	8172.0	11510.0	126.1	12.79	22.63	385.8	30.2	.0248
				8.00 MPa isobar						
85.76[a]	1425.0	35.68	−1366.0	−1141.0	54.93	22.97	43.48	849.3	294.0	.136
90.0	1401.0	35.07	−1185.0	−957.1	57.02	21.81	43.48	838.0	260.0	.128
100.0	1341.0	33.57	−756.6	−518.3	61.65	20.31	44.40	789.6	201.0	.114
110.0	1277.0	31.97	−317.7	−67.5	65.94	19.41	45.84	726.5	159.0	.102
120.0	1208.0	30.24	135.9	400.5	70.01	18.67	47.91	656.5	127.0	.0904
125.0	1170.0	29.30	370.5	643.6	72.00	18.33	49.40	619.4	114.0	.0851
130.0	1130.0	28.30	612.5	895.3	73.97	18.03	51.38	580.7	102.0	.0798
135.0	1087.0	27.21	864.6	1159.0	75.96	17.76	54.10	540.0	91.2	.0744
140.0	1039.0	26.00	1130.0	1438.0	77.99	17.55	57.94	497.0	80.9	.0690
142.0	1018.0	25.48	1242.0	1556.0	78.83	17.48	59.93	478.9	77.0	.0668
144.0	995.6	24.92	1357.0	1678.0	79.68	17.43	62.29	460.3	73.1	.0646
146.0	972.0	24.33	1477.0	1805.0	80.56	17.40	65.11	441.1	69.2	.0625
148.0	946.7	23.70	1601.0	1939.0	81.47	17.38	68.54	421.2	65.4	.0604
150.0	919.3	23.01	1732.0	2080.0	82.41	17.39	72.78	400.5	61.5	.0583
152.0	889.4	22.26	1871.0	2231.0	83.41	17.44	78.13	378.9	57.6	.0563
154.0	856.3	21.43	2020.0	2394.0	84.48	17.52	85.05	356.4	53.7	.0544
156.0	818.9	20.50	2182.0	2572.0	85.63	17.66	94.17	333.0	49.6	.0524
158.0	776.2	19.43	2361.0	2772.0	86.90	17.87	106.2	309.0	45.5	.0502
160.0	726.9	18.20	2560.0	2999.0	88.33	18.15	121.2	285.3	41.2	.0477
162.0	670.8	16.79	2781.0	3257.0	89.93	18.47	136.1	263.9	36.9	.0448
164.0	611.0	15.29	3016.0	3539.0	91.66	18.73	143.6	247.6	33.0	.0415
166.0	553.6	13.86	3247.0	3824.0	93.39	18.82	139.6	237.7	29.8	.0382
167.0	527.5	13.21	3355.0	3961.0	94.21	18.77	134.3	234.8	28.5	.0366
168.0	503.6	12.61	3458.0	4092.0	94.99	18.68	127.8	233.0	27.4	.0352
169.0	481.9	12.06	3553.0	4216.0	95.73	18.55	120.8	231.9	26.4	.0338
170.0	462.3	11.57	3642.0	4334.0	96.42	18.40	113.7	231.5	25.6	.0326
171.0	444.7	11.13	3725.0	4444.0	97.07	18.22	106.8	231.4	24.9	.0315
172.0	428.8	10.73	3802.0	4547.0	97.67	18.04	100.4	231.8	24.4	.0305
173.0	414.4	10.37	3874.0	4645.0	98.24	17.85	94.46	232.3	23.9	.0295
174.0	401.4	10.05	3940.0	4736.0	98.77	17.66	89.09	233.0	23.5	.0287
175.0	389.5	9.751	4003.0	4823.0	99.26	17.47	84.25	233.8	23.1	.0279
176.0	378.7	9.480	4061.0	4905.0	99.73	17.29	79.90	234.8	22.8	.0272
178.0	359.7	9.003	4169.0	5057.0	100.6	16.95	72.49	236.8	22.3	.0260
180.0	343.4	8.595	4265.0	5196.0	101.4	16.63	66.49	239.0	21.9	.0250
182.0	329.2	8.242	4353.0	5324.0	102.1	16.34	61.58	241.3	21.6	.0241
184.0	316.8	7.931	4434.0	5443.0	102.7	16.08	57.52	243.5	21.4	.0234
186.0	305.8	7.654	4509.0	5554.0	103.3	15.84	54.13	245.8	21.2	.0228
188.0	295.9	7.406	4580.0	5660.0	103.9	15.63	51.25	248.0	21.0	.0222
190.0	286.9	7.182	4646.0	5760.0	104.4	15.43	48.79	250.2	20.9	.0218

Thermophysical properties of argon—Continued

T K	Density kg/m³	Density mol/dm³	E J/mol	H J/mol	S J/(mol·K)	C_v J/(mol·K)	C_p J/(mol·K)	Sound m/s	Visc. μPa·s	Therm. W/(m·K)
192.0	278.7	6.977	4709.0	5855.0	104.9	15.25	46.66	252.4	20.9	.0214
194.0	271.2	6.789	4768.0	5947.0	105.4	15.09	44.81	254.5	20.8	.0211
196.0	264.3	6.616	4825.0	6035.0	105.8	14.94	43.18	256.6	20.8	.0208
200.0	251.9	6.306	4933.0	6202.0	106.7	14.68	40.46	260.6	20.7	.0204
205.0	238.6	5.973	5058.0	6397.0	107.6	14.41	37.81	265.5	20.8	.0200
210.0	227.1	5.686	5174.0	6581.0	108.5	14.19	35.73	270.1	20.9	.0198
215.0	217.1	5.435	5283.0	6755.0	109.4	14.01	34.07	274.6	21.0	.0197
220.0	208.2	5.212	5387.0	6922.0	110.1	13.85	32.71	278.9	21.1	.0196
225.0	200.3	5.013	5487.0	7082.0	110.8	13.73	31.58	283.0	21.3	.0196
230.0	193.0	4.833	5582.0	7238.0	111.5	13.62	30.63	287.1	21.5	.0196
235.0	186.5	4.668	5675.0	7389.0	112.2	13.53	29.82	291.0	21.7	.0196
240.0	180.5	4.518	5766.0	7536.0	112.8	13.45	29.11	294.7	21.9	.0197
250.0	169.8	4.251	5940.0	7821.0	114.0	13.32	27.97	302.0	22.4	.0199
260.0	160.6	4.021	6107.0	8096.0	115.0	13.22	27.07	309.0	22.9	.0201
270.0	152.6	3.819	6269.0	8363.0	116.0	13.15	26.36	315.6	23.4	.0204
280.0	145.4	3.640	6426.0	8624.0	117.0	13.09	25.77	322.0	23.9	.0207
290.0	139.0	3.480	6580.0	8879.0	117.9	13.04	25.28	328.2	24.4	.0210
300.0	133.3	3.336	6732.0	9130.0	118.7	13.00	24.87	334.2	25.0	.0213
310.0	128.0	3.205	6880.0	9377.0	119.6	12.97	24.52	340.1	25.5	.0217
320.0	123.2	3.085	7027.0	9620.0	120.3	12.94	24.21	345.7	26.1	.0220
330.0	118.8	2.975	7172.0	9861.0	121.1	12.92	23.95	351.3	26.6	.0224
340.0	114.8	2.874	7315.0	10100.0	121.8	12.90	23.71	356.7	27.1	.0227
360.0	107.6	2.692	7598.0	10570.0	123.1	12.86	23.32	367.1	28.2	.0235
380.0	101.3	2.535	7877.0	11030.0	124.4	12.84	23.01	377.2	29.3	.0242
400.0	95.74	2.397	8152.0	11490.0	125.5	12.81	22.75	386.9	30.3	.0250
				8.50 MPa isobar						
85.88[a]	1426.0	35.69	−1365.0	−1127.0	54.93	22.93	43.41	851.0	294.0	.136
90.0	1402.0	35.10	−1190.0	−948.1	56.96	21.82	43.40	840.1	262.0	.129
100.0	1343.0	33.62	−763.1	−510.3	61.58	20.33	44.30	792.1	202.0	.114
110.0	1279.0	32.03	−326.1	−60.7	65.86	19.44	45.67	729.6	160.0	.102
120.0	1211.0	30.31	124.7	405.1	69.91	18.70	47.64	660.6	128.0	.0909
125.0	1174.0	29.39	357.4	646.7	71.88	18.37	49.03	624.0	115.0	.0855
130.0	1135.0	28.40	596.9	896.2	73.84	18.06	50.88	586.1	103.0	.0803
135.0	1092.0	27.34	845.5	1156.0	75.81	17.79	53.36	546.4	92.4	.0750
140.0	1045.0	26.16	1107.0	1431.0	77.81	17.57	56.81	504.7	82.2	.0697
145.0	992.7	24.85	1385.0	1727.0	79.88	17.41	61.78	460.4	72.6	.0644
146.0	981.3	24.56	1443.0	1790.0	80.31	17.39	63.04	451.1	70.7	.0633
148.0	957.3	23.96	1564.0	1918.0	81.19	17.37	65.91	432.3	67.0	.0612
150.0	931.7	23.32	1689.0	2054.0	82.09	17.36	69.36	412.8	63.3	.0592
152.0	904.0	22.63	1821.0	2196.0	83.04	17.38	73.57	392.7	59.5	.0573
154.0	873.9	21.88	1960.0	2348.0	84.03	17.43	78.79	371.9	55.8	.0554
156.0	840.7	21.05	2108.0	2512.0	85.09	17.52	85.33	350.5	52.0	.0535
158.0	803.8	20.12	2268.0	2691.0	86.23	17.65	93.55	328.7	48.2	.0515
160.0	762.4	19.08	2442.0	2888.0	87.47	17.84	103.6	306.9	44.3	.0493
162.0	716.1	17.92	2632.0	3106.0	88.82	18.07	114.6	286.1	40.5	.0469
164.0	665.5	16.66	2835.0	3345.0	90.29	18.30	124.1	268.0	36.7	.0440
166.0	613.3	15.35	3045.0	3599.0	91.82	18.48	128.0	254.3	33.3	.0410
168.0	563.5	14.11	3250.0	3853.0	93.34	18.52	125.0	245.4	30.5	.0381
170.0	519.0	12.99	3441.0	4095.0	94.78	18.40	116.9	240.6	28.3	.0354
171.0	499.1	12.49	3529.0	4210.0	95.45	18.30	111.9	239.2	27.4	.0342
172.0	480.8	12.03	3613.0	4319.0	96.09	18.17	106.7	238.5	26.6	.0330
173.0	464.0	11.61	3691.0	4423.0	96.69	18.03	101.5	238.1	26.0	.0320
174.0	448.6	11.23	3765.0	4522.0	97.26	17.87	96.44	238.1	25.4	.0310
175.0	434.5	10.88	3834.0	4616.0	97.80	17.71	91.65	238.4	24.9	.0301
176.0	421.6	10.55	3900.0	4705.0	98.31	17.54	87.17	238.8	24.4	.0293
178.0	398.8	9.982	4020.0	4871.0	99.25	17.22	79.19	240.1	23.7	.0279
180.0	379.3	9.494	4128.0	5023.0	100.1	16.90	72.49	241.8	23.1	.0267
182.0	362.4	9.072	4225.0	5162.0	100.9	16.61	66.90	243.6	22.7	.0256
184.0	347.7	8.704	4314.0	5291.0	101.6	16.33	62.22	245.6	22.4	.0248
186.0	334.7	8.378	4397.0	5411.0	102.2	16.08	58.28	247.7	22.1	.0240

Thermophysical properties of argon—Continued

T K	Density kg/m³	Density mol/dm³	E J/mol	H J/mol	S J/(mol·K)	C_v J/(mol·K)	C_p J/(mol·K)	Sound m/s	Visc. μPa·s	Therm. W/(m·K)
188.0	323.1	8.088	4474.0	5525.0	102.8	15.85	54.94	249.8	21.9	.0234
190.0	312.6	7.826	4545.0	5632.0	103.4	15.65	52.08	251.8	21.7	.0229
192.0	303.2	7.589	4613.0	5733.0	103.9	15.45	49.61	253.9	21.6	.0224
194.0	294.5	7.373	4677.0	5830.0	104.4	15.28	47.47	255.9	21.5	.0220
196.0	286.6	7.175	4738.0	5923.0	104.9	15.12	45.59	258.0	21.4	.0217
200.0	272.5	6.821	4853.0	6099.0	105.8	14.84	42.47	261.9	21.3	.0212
205.0	257.4	6.445	4985.0	6304.0	106.8	14.54	39.44	266.7	21.3	.0207
210.0	244.6	6.123	5106.0	6495.0	107.7	14.31	37.09	271.3	21.3	.0204
215.0	233.4	5.842	5220.0	6675.0	108.6	14.11	35.22	275.8	21.4	.0202
220.0	223.5	5.595	5328.0	6847.0	109.4	13.94	33.71	280.0	21.5	.0201
225.0	214.7	5.375	5431.0	7013.0	110.1	13.81	32.45	284.2	21.7	.0200
230.0	206.8	5.176	5530.0	7172.0	110.8	13.69	31.40	288.2	21.8	.0200
235.0	199.6	4.996	5626.0	7327.0	111.5	13.59	30.50	292.1	22.0	.0200
240.0	193.0	4.831	5718.0	7478.0	112.1	13.51	29.73	295.8	22.2	.0201
250.0	181.4	4.540	5896.0	7768.0	113.3	13.37	28.48	303.1	22.7	.0202
260.0	171.4	4.290	6067.0	8048.0	114.4	13.27	27.51	310.1	23.1	.0204
270.0	162.6	4.071	6231.0	8319.0	115.4	13.19	26.73	316.7	23.6	.0207
280.0	154.9	3.878	6391.0	8583.0	116.4	13.12	26.09	323.1	24.1	.0209
290.0	148.0	3.706	6547.0	8841.0	117.3	13.07	25.57	329.3	24.6	.0212
300.0	141.8	3.550	6701.0	9095.0	118.1	13.03	25.13	335.3	25.2	.0215
310.0	136.2	3.409	6851.0	9344.0	119.0	13.00	24.75	341.1	25.7	.0219
320.0	131.1	3.281	6999.0	9590.0	119.7	12.97	24.42	346.8	26.2	.0222
330.0	126.4	3.163	7145.0	9833.0	120.5	12.94	24.14	352.3	26.7	.0226
340.0	122.0	3.054	7290.0	10070.0	121.2	12.92	23.89	357.7	27.3	.0229
360.0	114.3	2.861	7575.0	10550.0	122.6	12.89	23.47	368.2	28.3	.0236
380.0	107.6	2.693	7855.0	11010.0	123.8	12.86	23.14	378.2	29.4	.0244
400.0	101.7	2.545	8132.0	11470.0	125.0	12.83	22.87	387.9	30.4	.0251
				9.00 MPa isobar						
86.01[a]	1426.0	35.71	−1364.0	−1112.0	54.94	22.90	43.34	852.7	294.0	.136
90.0	1404.0	35.14	−1195.0	−939.1	56.91	21.84	43.33	842.1	263.0	.129
100.0	1345.0	33.66	−769.5	−502.1	61.51	20.36	44.19	794.6	203.0	.114
110.0	1282.0	32.08	−334.4	−53.9	65.78	19.47	45.52	732.7	161.0	.102
120.0	1214.0	30.38	113.8	410.0	69.82	18.73	47.38	664.5	129.0	.0913
125.0	1177.0	29.47	344.6	650.0	71.78	18.40	48.69	628.6	116.0	.0860
130.0	1139.0	28.50	581.7	897.5	73.72	18.09	50.40	591.4	104.0	.0808
135.0	1097.0	27.46	827.2	1155.0	75.66	17.82	52.69	552.6	93.5	.0756
140.0	1051.0	26.32	1084.0	1426.0	77.63	17.59	55.81	512.1	83.5	.0704
145.0	1001.0	25.05	1356.0	1715.0	79.66	17.42	60.18	469.4	74.0	.0651
146.0	989.9	24.78	1413.0	1776.0	80.08	17.39	61.27	460.5	72.2	.0641
148.0	967.1	24.21	1529.0	1901.0	80.93	17.36	63.71	442.5	68.5	.0621
150.0	942.9	23.60	1650.0	2031.0	81.80	17.33	66.58	424.1	64.9	.0601
152.0	917.0	22.96	1775.0	2167.0	82.70	17.33	70.00	405.2	61.3	.0582
154.0	889.2	22.26	1907.0	2311.0	83.64	17.36	74.11	385.7	57.7	.0563
156.0	859.0	21.50	2046.0	2464.0	84.63	17.41	79.08	365.9	54.2	.0545
158.0	826.1	20.68	2193.0	2628.0	85.68	17.50	85.09	345.7	50.6	.0526
160.0	789.9	19.77	2350.0	2806.0	86.79	17.62	92.23	325.6	47.0	.0506
162.0	750.2	18.78	2519.0	2998.0	87.99	17.78	100.3	306.0	43.4	.0485
164.0	706.9	17.70	2698.0	3207.0	89.27	17.96	108.2	287.8	39.9	.0460
166.0	661.2	16.55	2886.0	3430.0	90.62	18.13	114.2	272.4	36.6	.0434
168.0	615.0	15.40	3076.0	3661.0	92.00	18.24	116.1	260.6	33.6	.0406
170.0	571.0	14.29	3261.0	3891.0	93.36	18.24	113.6	252.6	31.1	.0379
172.0	530.9	13.29	3436.0	4113.0	94.66	18.14	107.7	247.9	29.1	.0355
174.0	495.7	12.41	3595.0	4321.0	95.86	17.94	100.1	245.4	27.5	.0333
176.0	465.3	11.65	3740.0	4513.0	96.96	17.69	92.07	244.6	26.3	.0314
178.0	439.1	10.99	3870.0	4689.0	97.96	17.40	84.50	244.8	25.3	.0298
180.0	416.5	10.43	3988.0	4851.0	98.86	17.11	77.71	245.7	24.6	.0284
182.0	396.9	9.936	4095.0	5001.0	99.69	16.83	71.78	247.0	24.0	.0272
184.0	379.8	9.508	4192.0	5139.0	100.4	16.55	66.69	248.5	23.5	.0262
186.0	364.7	9.131	4282.0	5268.0	101.1	16.30	62.33	250.3	23.1	.0254
188.0	351.3	8.795	4365.0	5389.0	101.8	16.06	58.59	252.1	22.8	.0246

Thermophysical properties of argon—Continued

T K	Density kg/m^3	Density mol/dm^3	E J/mol	H J/mol	S J/(mol·K)	C_v J/(mol·K)	C_p J/(mol·K)	Sound m/s	Visc. μPa·s	Therm. W/(m·K)
190.0	339.3	8.494	4443.0	5503.0	102.4	15.85	55.37	254.0	22.6	.0240
192.0	328.5	8.223	4516.0	5610.0	103.0	15.64	52.59	255.9	22.4	.0235
194.0	318.6	7.976	4585.0	5713.0	103.5	15.46	50.17	257.8	22.2	.0230
196.0	309.6	7.750	4650.0	5811.0	104.0	15.29	48.04	259.7	22.1	.0226
198.0	301.3	7.542	4712.0	5905.0	104.5	15.13	46.17	261.7	22.0	.0222
200.0	293.6	7.350	4772.0	5996.0	104.9	14.99	44.52	263.6	21.9	.0219
205.0	276.7	6.927	4911.0	6210.0	106.0	14.68	41.11	268.2	21.8	.0214
210.0	262.4	6.567	5038.0	6409.0	106.9	14.42	38.48	272.7	21.8	.0210
215.0	249.9	6.257	5157.0	6596.0	107.8	14.21	36.40	277.1	21.8	.0208
220.0	239.0	5.984	5269.0	6773.0	108.6	14.03	34.72	281.4	21.9	.0206
225.0	229.4	5.741	5376.0	6943.0	109.4	13.89	33.33	285.5	22.0	.0205
230.0	220.7	5.524	5478.0	7107.0	110.1	13.76	32.18	289.4	22.2	.0205
235.0	212.8	5.327	5576.0	7265.0	110.8	13.66	31.20	293.3	22.3	.0204
240.0	205.6	5.148	5671.0	7419.0	111.4	13.56	30.35	297.1	22.5	.0205
250.0	193.0	4.832	5853.0	7716.0	112.7	13.42	28.99	304.3	22.9	.0206
260.0	182.2	4.561	6027.0	8000.0	113.8	13.31	27.94	311.2	23.4	.0207
270.0	172.8	4.325	6194.0	8275.0	114.8	13.22	27.10	317.9	23.8	.0210
280.0	164.5	4.117	6356.0	8543.0	115.8	13.16	26.42	324.3	24.3	.0212
290.0	157.1	3.932	6515.0	8804.0	116.7	13.10	25.86	330.4	24.8	.0215
300.0	150.4	3.765	6670.0	9060.0	117.6	13.06	25.38	336.4	25.3	.0218
310.0	144.4	3.614	6822.0	9312.0	118.4	13.02	24.98	342.3	25.9	.0221
320.0	138.9	3.477	6971.0	9560.0	119.2	12.99	24.63	347.9	26.4	.0224
330.0	133.9	3.351	7119.0	9805.0	119.9	12.97	24.33	353.4	26.9	.0228
340.0	129.2	3.235	7265.0	10050.0	120.7	12.94	24.07	358.8	27.4	.0231
360.0	121.0	3.029	7552.0	10520.0	122.0	12.91	23.62	369.3	28.5	.0238
380.0	113.8	2.850	7834.0	10990.0	123.3	12.88	23.27	379.3	29.5	.0245
400.0	107.6	2.693	8112.0	11450.0	124.5	12.85	22.98	389.0	30.5	.0253

10.00 MPa isobar

T K	Density kg/m^3	Density mol/dm^3	E J/mol	H J/mol	S J/(mol·K)	C_v J/(mol·K)	C_p J/(mol·K)	Sound m/s	Visc. μPa·s	Therm. W/(m·K)
86.25[a]	1428.0	35.74	−1363.0	−1083.0	54.95	22.84	43.20	856.2	295.0	.136
90.0	1406.0	35.21	−1205.0	−921.1	56.79	21.86	43.19	846.3	265.0	.129
100.0	1348.0	33.74	−782.2	−485.8	61.38	20.41	43.99	799.5	205.0	.115
110.0	1286.0	32.19	−350.6	−40.0	65.62	19.53	45.22	738.8	163.0	.103
120.0	1219.0	30.53	92.5	420.1	69.63	18.79	46.90	672.2	131.0	.0921
125.0	1184.0	29.64	319.9	657.3	71.56	18.46	48.05	637.4	118.0	.0869
130.0	1146.0	28.70	552.7	901.2	73.48	18.15	49.54	601.5	106.0	.0818
135.0	1106.0	27.69	792.4	1154.0	75.38	17.87	51.49	564.4	95.7	.0768
140.0	1063.0	26.61	1041.0	1417.0	77.30	17.63	54.07	526.0	85.9	.0717
145.0	1016.0	25.42	1302.0	1696.0	79.25	17.43	57.54	486.1	76.6	.0666
150.0	962.6	24.10	1580.0	1995.0	81.28	17.31	62.32	444.5	67.9	.0618
152.0	939.4	23.52	1696.0	2122.0	82.12	17.28	64.73	427.3	64.5	.0599
155.0	902.2	22.58	1879.0	2322.0	83.43	17.27	69.05	401.2	59.5	.0572
160.0	831.6	20.82	2209.0	2690.0	85.76	17.36	78.56	357.0	51.3	.0528
162.0	799.8	20.02	2352.0	2852.0	86.76	17.43	83.22	339.5	48.1	.0510
164.0	765.9	19.17	2501.0	3023.0	87.82	17.52	88.15	322.7	45.0	.0490
166.0	730.0	18.27	2657.0	3204.0	88.91	17.62	92.92	307.0	42.0	.0469
170.0	654.5	16.38	2980.0	3590.0	91.21	17.79	99.05	281.5	36.5	.0422
172.0	617.1	15.45	3141.0	3789.0	92.37	17.81	99.12	272.4	34.1	.0398
174.0	581.3	14.55	3298.0	3985.0	93.51	17.77	97.06	265.9	32.1	.0376
176.0	548.2	13.72	3447.0	4176.0	94.60	17.67	93.38	261.5	30.4	.0355
178.0	518.2	12.97	3587.0	4358.0	95.63	17.51	88.70	258.9	29.0	.0336
180.0	491.2	12.30	3717.0	4530.0	96.59	17.31	83.58	257.5	27.8	.0320
182.0	467.2	11.69	3837.0	4692.0	97.49	17.09	78.45	257.0	26.9	.0305
184.0	445.8	11.16	3948.0	4844.0	98.32	16.86	73.57	257.2	26.1	.0293
186.0	426.8	10.68	4051.0	4987.0	99.09	16.62	69.08	257.9	25.5	.0282
188.0	409.9	10.26	4146.0	5121.0	99.80	16.40	65.02	258.9	25.0	.0272
190.0	394.7	9.879	4235.0	5247.0	100.5	16.18	61.40	260.1	24.5	.0264
192.0	380.9	9.536	4318.0	5367.0	101.1	15.97	58.19	261.5	24.2	.0257
194.0	368.5	9.224	4396.0	5480.0	101.7	15.78	55.34	263.0	23.9	.0250
196.0	357.2	8.941	4470.0	5588.0	102.2	15.60	52.82	264.6	23.7	.0245
198.0	346.8	8.681	4540.0	5692.0	102.8	15.43	50.58	266.2	23.5	.0240
200.0	337.2	8.442	4606.0	5791.0	103.3	15.27	48.58	267.9	23.3	.0236

Thermophysical properties of argon—Continued

T K	Density kg/m³	Density mol/dm³	E J/mol	H J/mol	S J/(mol·K)	C_v J/(mol·K)	C_p J/(mol·K)	Sound m/s	Visc. μPa·s	Therm. W/(m·K)
202.0	328.4	8.221	4670.0	5886.0	103.7	15.13	46.80	269.5	23.2	.0233
205.0	316.3	7.919	4760.0	6023.0	104.4	14.93	44.46	272.1	23.0	.0228
210.0	298.8	7.480	4900.0	6237.0	105.4	14.64	41.29	276.3	22.9	.0223
215.0	283.8	7.103	5029.0	6437.0	106.4	14.40	38.78	280.4	22.8	.0219
220.0	270.7	6.776	5150.0	6626.0	107.2	14.21	36.76	284.5	22.8	.0216
225.0	259.1	6.487	5264.0	6805.0	108.1	14.04	35.11	288.5	22.8	.0215
230.0	248.8	6.229	5372.0	6977.0	108.8	13.90	33.74	292.3	22.9	.0214
235.0	239.6	5.997	5476.0	7143.0	109.5	13.78	32.59	296.1	23.0	.0213
240.0	231.2	5.787	5575.0	7304.0	110.2	13.68	31.60	299.8	23.2	.0213
245.0	223.5	5.595	5672.0	7459.0	110.8	13.59	30.76	303.4	23.3	.0213
250.0	216.5	5.419	5766.0	7611.0	111.5	13.52	30.02	307.0	23.5	.0213
260.0	203.9	5.105	5946.0	7905.0	112.6	13.39	28.80	313.8	23.9	.0214
270.0	193.1	4.834	6119.0	8188.0	113.7	13.30	27.84	320.4	24.3	.0215
280.0	183.6	4.596	6287.0	8463.0	114.7	13.22	27.07	326.7	24.8	.0218
290.0	175.2	4.385	6449.0	8730.0	115.6	13.16	26.43	332.9	25.2	.0220
300.0	167.6	4.195	6608.0	8991.0	116.5	13.12	25.89	338.8	25.7	.0223
310.0	160.8	4.024	6763.0	9248.0	117.3	13.08	25.44	344.6	26.2	.0226
320.0	154.6	3.869	6916.0	9500.0	118.1	13.04	25.05	350.2	26.7	.0229
330.0	148.9	3.727	7066.0	9749.0	118.9	13.01	24.71	355.7	27.2	.0232
340.0	143.7	3.596	7214.0	9995.0	119.6	12.99	24.42	361.1	27.7	.0235
360.0	134.4	3.364	7506.0	10480.0	121.0	12.95	23.92	371.5	28.7	.0242
380.0	126.4	3.164	7792.0	10950.0	122.3	12.92	23.53	381.5	29.7	.0249
400.0	119.4	2.988	8073.0	11420.0	123.5	12.89	23.21	391.1	30.8	.0256
				12.00 MPa isobar						
86.74[a]	1430.0	35.80	−1360.0	−1025.0	54.98	22.71	42.94	863.1	296.0	.136
90.0	1412.0	35.34	−1224.0	−884.8	56.56	21.91	42.92	854.5	270.0	.131
100.0	1355.0	33.91	−806.6	−452.7	61.12	20.49	43.62	809.2	209.0	.116
110.0	1294.0	32.40	−381.7	−11.4	65.32	19.64	44.68	750.6	167.0	.104
120.0	1230.0	30.79	52.2	441.9	69.26	18.91	46.05	687.0	135.0	.0937
130.0	1161.0	29.05	498.9	911.9	73.03	18.27	48.10	620.5	110.0	.0837
135.0	1123.0	28.12	729.1	1156.0	74.87	17.97	49.54	586.1	99.9	.0789
140.0	1084.0	27.12	965.6	1408.0	76.70	17.71	51.39	551.1	90.3	.0741
145.0	1041.0	26.06	1210.0	1671.0	78.54	17.49	53.73	515.3	81.4	.0694
150.0	994.7	24.90	1464.0	1946.0	80.41	17.31	56.71	478.9	73.1	.0647
155.0	944.2	23.64	1731.0	2239.0	82.33	17.18	60.49	442.0	65.4	.0603
160.0	888.4	22.24	2013.0	2553.0	84.32	17.12	65.19	405.1	58.0	.0563
165.0	826.6	20.69	2312.0	2892.0	86.41	17.12	70.74	369.5	51.2	.0523
170.0	759.0	19.00	2629.0	3260.0	88.61	17.16	76.28	337.2	44.9	.0480
175.0	687.9	17.22	2955.0	3652.0	90.88	17.18	79.70	311.2	39.4	.0432
180.0	618.7	15.49	3275.0	4050.0	93.12	17.09	78.89	293.8	34.9	.0386
182.0	592.8	14.84	3398.0	4206.0	93.99	17.01	77.36	289.2	33.5	.0369
184.0	568.4	14.23	3516.0	4359.0	94.82	16.91	75.32	285.8	32.2	.0353
186.0	545.5	13.66	3629.0	4507.0	95.62	16.78	72.92	283.3	31.1	.0339
188.0	524.3	13.12	3736.0	4651.0	96.39	16.64	70.31	281.6	30.1	.0326
190.0	504.6	12.63	3839.0	4789.0	97.12	16.48	67.60	280.5	29.3	.0314
192.0	486.4	12.18	3936.0	4921.0	97.81	16.32	64.90	280.0	28.6	.0304
194.0	469.7	11.76	4028.0	5048.0	98.47	16.16	62.27	279.9	28.0	.0295
196.0	454.2	11.37	4115.0	5170.0	99.10	16.00	59.76	280.1	27.5	.0287
198.0	440.0	11.01	4198.0	5287.0	99.69	15.84	57.39	280.6	27.0	.0280
200.0	426.8	10.68	4277.0	5400.0	100.3	15.69	55.19	281.2	26.6	.0273
202.0	414.6	10.38	4352.0	5508.0	100.8	15.54	53.14	282.1	26.3	.0268
204.0	403.3	10.10	4424.0	5613.0	101.3	15.40	51.26	283.0	26.0	.0263
210.0	373.8	9.357	4623.0	5905.0	102.7	15.02	46.46	286.5	25.4	.0251
215.0	353.2	8.842	4772.0	6129.0	103.8	14.75	43.29	289.7	25.0	.0244
220.0	335.5	8.398	4910.0	6339.0	104.7	14.52	40.70	293.0	24.8	.0239
225.0	320.0	8.010	5039.0	6537.0	105.6	14.32	38.58	296.4	24.7	.0236
230.0	306.3	7.667	5160.0	6725.0	106.5	14.16	36.81	299.9	24.6	.0233
235.0	294.0	7.360	5275.0	6906.0	107.2	14.02	35.32	303.3	24.6	.0231
240.0	283.0	7.085	5385.0	7079.0	108.0	13.90	34.05	306.7	24.6	.0229
245.0	273.0	6.834	5491.0	7246.0	108.7	13.79	32.97	310.1	24.7	.0228
250.0	263.9	6.606	5592.0	7409.0	109.3	13.70	32.03	313.4	24.8	.0228

Thermophysical properties of argon—Continued

T K	Density kg/m^3	Density mol/dm^3	*E* J/mol	*H* J/mol	*S* J/(mol·K)	C_v J/(mol·K)	C_p J/(mol·K)	Sound m/s	Visc. μPa·s	Therm. W/(m·K)
260.0	247.8	6.204	5787.0	7721.0	110.5	13.55	30.49	319.9	25.0	.0227
270.0	234.0	5.858	5971.0	8020.0	111.7	13.44	29.29	326.2	25.3	.0228
280.0	222.0	5.557	6148.0	8308.0	112.7	13.35	28.33	332.3	25.7	.0229
290.0	211.4	5.292	6319.0	8587.0	113.7	13.28	27.54	338.3	26.1	.0231
300.0	202.0	5.056	6485.0	8859.0	114.6	13.23	26.88	344.1	26.5	.0233
310.0	193.5	4.843	6647.0	9125.0	115.5	13.18	26.33	349.8	27.0	.0235
320.0	185.8	4.651	6806.0	9386.0	116.3	13.14	25.86	355.3	27.4	.0237
330.0	178.8	4.476	6961.0	9642.0	117.1	13.11	25.45	360.7	27.9	.0240
340.0	172.4	4.316	7115.0	9895.0	117.9	13.08	25.09	366.0	28.3	.0243
350.0	166.5	4.168	7265.0	10140.0	118.6	13.05	24.78	371.2	28.8	.0246
360.0	161.1	4.032	7415.0	10390.0	119.3	13.03	24.50	376.3	29.3	.0249
380.0	151.3	3.788	7708.0	10880.0	120.6	12.99	24.03	386.2	30.3	.0255
400.0	142.8	3.575	7996.0	11350.0	121.8	12.96	23.65	395.7	31.2	.0262
				14.00 MPa isobar						
87.23[a]	1432.0	35.86	−1357.0	−966.6	55.01	22.60	42.69	869.9	297.0	.136
100.0	1361.0	34.07	−830.0	−419.0	60.86	20.57	43.28	818.8	213.0	.117
110.0	1302.0	32.60	−411.2	18.3	65.03	19.74	44.20	762.0	170.0	.106
120.0	1240.0	31.04	14.6	465.6	68.92	19.02	45.32	701.0	139.0	.0952
130.0	1174.0	29.38	449.9	926.3	72.61	18.37	46.93	638.0	114.0	.0855
140.0	1101.0	27.57	899.3	1407.0	76.17	17.80	49.39	573.4	94.3	.0763
145.0	1062.0	26.59	1132.0	1658.0	77.93	17.55	51.07	540.6	85.7	.0718
150.0	1021.0	25.55	1370.0	1918.0	79.70	17.34	53.11	507.7	77.7	.0674
155.0	976.3	24.44	1617.0	2190.0	81.48	17.17	55.55	474.9	70.3	.0631
160.0	928.6	23.24	1872.0	2475.0	83.29	17.03	58.39	442.5	63.4	.0592
165.0	877.4	21.96	2137.0	2774.0	85.13	16.94	61.57	411.2	57.0	.0555
170.0	822.8	20.60	2411.0	3090.0	87.02	16.88	64.82	381.9	51.1	.0518
180.0	707.5	17.71	2973.0	3764.0	90.86	16.76	68.97	335.1	41.3	.0437
185.0	651.1	16.30	3249.0	4108.0	92.75	16.63	68.40	319.6	37.5	.0398
190.0	598.9	14.99	3511.0	4444.0	94.55	16.43	65.91	309.6	34.5	.0363
195.0	552.5	13.83	3753.0	4765.0	96.21	16.16	62.16	303.8	32.2	.0335
200.0	512.2	12.82	3973.0	5065.0	97.73	15.86	57.93	301.1	30.5	.0313
205.0	477.6	11.96	4174.0	5344.0	99.11	15.56	53.80	300.4	29.2	.0295
210.0	448.0	11.22	4356.0	5604.0	100.4	15.27	50.04	301.1	28.3	.0282
215.0	422.5	10.58	4522.0	5846.0	101.5	15.00	46.75	302.7	27.6	.0272
220.0	400.4	10.02	4675.0	6072.0	102.5	14.77	43.94	304.7	27.1	.0264
225.0	381.0	9.539	4818.0	6286.0	103.5	14.56	41.55	307.2	26.8	.0259
230.0	363.9	9.111	4952.0	6488.0	104.4	14.38	39.52	309.8	26.5	.0254
235.0	348.7	8.729	5078.0	6681.0	105.2	14.22	37.78	312.6	26.4	.0250
240.0	335.0	8.387	5197.0	6867.0	106.0	14.09	36.30	315.5	26.3	.0248
245.0	322.7	8.078	5312.0	7045.0	106.7	13.97	35.02	318.4	26.2	.0245
250.0	311.5	7.797	5421.0	7217.0	107.4	13.87	33.91	321.4	26.2	.0244
255.0	301.2	7.539	5527.0	7384.0	108.1	13.78	32.94	324.3	26.2	.0243
260.0	291.7	7.303	5629.0	7547.0	108.7	13.70	32.08	327.3	26.3	.0242
270.0	274.9	6.881	5826.0	7860.0	109.9	13.57	30.66	333.1	26.5	.0241
280.0	260.3	6.516	6012.0	8161.0	111.0	13.47	29.52	338.9	26.7	.0241
290.0	247.5	6.196	6192.0	8451.0	112.0	13.39	28.59	344.6	27.1	.0242
300.0	236.2	5.912	6365.0	8733.0	113.0	13.33	27.82	350.2	27.4	.0243
310.0	226.0	5.658	6534.0	9008.0	113.9	13.27	27.18	355.6	27.8	.0244
320.0	216.8	5.428	6698.0	9277.0	114.7	13.23	26.62	361.0	28.2	.0246
330.0	208.5	5.220	6859.0	9541.0	115.5	13.19	26.15	366.2	28.6	.0249
340.0	200.9	5.029	7017.0	9800.0	116.3	13.16	25.74	371.4	29.0	.0251
350.0	193.9	4.855	7172.0	10060.0	117.1	13.13	25.37	376.5	29.5	.0254
360.0	187.5	4.693	7325.0	10310.0	117.8	13.11	25.05	381.4	29.9	.0256
380.0	176.0	4.405	7625.0	10800.0	119.1	13.07	24.51	391.1	30.8	.0262
400.0	166.0	4.155	7919.0	11290.0	120.4	13.03	24.07	400.5	31.7	.0268
				16.00 MPa isobar						
87.71[a]	1435.0	35.91	−1354.0	−908.6	55.03	22.50	42.46	876.7	298.0	.136
100.0	1367.0	34.22	−852.4	−384.9	60.62	20.65	42.98	828.1	218.0	.119

Thermophysical properties of argon—Continued

T K	Density kg/m³	Density mol/dm³	E J/mol	H J/mol	S J/(mol·K)	C_v J/(mol·K)	C_p J/(mol·K)	Sound m/s	Visc. μPa·s	Therm. W/(m·K)
110.0	1310.0	32.78	–439.3	48.8	64.75	19.83	43.77	773.1	174.0	.107
120.0	1250.0	31.28	–20.6	490.9	68.60	19.13	44.70	714.3	142.0	.0966
130.0	1186.0	29.68	404.8	943.8	72.22	18.47	45.96	654.3	118.0	.0872
140.0	1117.0	27.97	840.3	1412.0	75.69	17.89	47.84	593.6	98.1	.0783
145.0	1081.0	27.06	1063.0	1654.0	77.39	17.63	49.08	563.2	89.6	.0740
150.0	1043.0	26.10	1290.0	1903.0	79.08	17.39	50.55	532.8	81.8	.0697
155.0	1002.0	25.09	1523.0	2160.0	80.77	17.19	52.25	502.9	74.6	.0657
160.0	960.0	24.03	1760.0	2426.0	82.46	17.01	54.16	473.5	68.0	.0619
170.0	868.4	21.74	2252.0	2989.0	85.86	16.75	58.30	418.5	56.3	.0548
180.0	770.0	19.28	2759.0	3589.0	89.30	16.55	61.54	372.8	46.7	.0477
190.0	672.6	16.84	3258.0	4208.0	92.64	16.28	61.40	341.7	39.5	.0407
195.0	627.7	15.71	3493.0	4511.0	94.22	16.09	59.74	332.0	36.8	.0377
200.0	586.6	14.68	3714.0	4804.0	95.70	15.87	57.30	325.6	34.6	.0351
205.0	549.8	13.76	3921.0	5083.0	97.08	15.64	54.45	321.7	32.9	.0331
210.0	517.1	12.94	4112.0	5348.0	98.36	15.39	51.50	319.8	31.6	.0314
215.0	488.3	12.22	4290.0	5599.0	99.53	15.15	48.65	319.2	30.5	.0301
220.0	462.8	11.59	4454.0	5835.0	100.6	14.93	46.03	319.5	29.7	.0291
225.0	440.3	11.02	4608.0	6059.0	101.6	14.73	43.68	320.6	29.1	.0283
230.0	420.3	10.52	4752.0	6272.0	102.6	14.55	41.60	322.2	28.7	.0276
235.0	402.3	10.07	4887.0	6476.0	103.4	14.39	39.77	324.0	28.3	.0271
240.0	386.2	9.668	5016.0	6670.0	104.3	14.25	38.17	326.2	28.1	.0267
245.0	371.6	9.303	5138.0	6858.0	105.0	14.12	36.77	328.5	27.9	.0264
250.0	358.4	8.971	5255.0	7039.0	105.8	14.01	35.55	330.9	27.8	.0261
255.0	346.2	8.667	5368.0	7213.0	106.5	13.92	34.47	333.4	27.7	.0259
260.0	335.1	8.389	5476.0	7383.0	107.1	13.83	33.51	335.9	27.7	.0257
270.0	315.3	7.893	5683.0	7710.0	108.3	13.69	31.91	341.1	27.7	.0255
280.0	298.2	7.465	5879.0	8023.0	109.5	13.58	30.62	346.4	27.9	.0253
290.0	283.2	7.090	6067.0	8323.0	110.5	13.49	29.57	351.7	28.1	.0253
300.0	270.0	6.759	6247.0	8615.0	111.5	13.42	28.70	356.9	28.4	.0254
310.0	258.2	6.463	6422.0	8898.0	112.5	13.36	27.97	362.1	28.7	.0255
320.0	247.5	6.196	6592.0	9174.0	113.3	13.32	27.34	367.2	29.0	.0256
330.0	237.9	5.955	6758.0	9445.0	114.2	13.27	26.81	372.2	29.4	.0258
340.0	229.1	5.735	6921.0	9711.0	115.0	13.24	26.34	377.2	29.7	.0260
350.0	221.0	5.533	7080.0	9972.0	115.7	13.21	25.94	382.1	30.1	.0262
360.0	213.6	5.347	7237.0	10230.0	116.4	13.18	25.58	387.0	30.6	.0264
380.0	200.3	5.015	7544.0	10730.0	117.8	13.14	24.97	396.4	31.4	.0269
400.0	188.9	4.727	7845.0	11230.0	119.1	13.10	24.48	405.6	32.3	.0275

18.00 MPa isobar

T K	Density kg/m³	Density mol/dm³	E J/mol	H J/mol	S J/(mol·K)	C_v J/(mol·K)	C_p J/(mol·K)	Sound m/s	Visc. μPa·s	Therm. W/(m·K)
88.20[a]	1437.0	35.97	–1351.0	–850.7	55.06	22.40	42.25	883.5	299.0	.137
100.0	1373.0	34.37	–874.0	–350.3	60.38	20.72	42.69	837.3	222.0	.120
110.0	1317.0	32.97	–466.0	80.0	64.48	19.92	43.39	783.8	178.0	.108
120.0	1258.0	31.50	–53.8	517.6	68.29	19.23	44.15	727.0	146.0	.0980
130.0	1197.0	29.97	363.1	963.8	71.86	18.57	45.15	669.6	121.0	.0888
140.0	1132.0	28.34	786.8	1422.0	75.26	17.97	46.60	612.1	102.0	.0802
150.0	1062.0	26.58	1221.0	1898.0	78.54	17.45	48.62	555.3	85.6	.0719
160.0	986.1	24.68	1667.0	2396.0	81.76	17.03	51.22	500.4	72.0	.0643
170.0	904.0	22.63	2128.0	2923.0	84.95	16.70	54.09	449.5	60.7	.0574
180.0	817.2	20.46	2597.0	3477.0	88.11	16.43	56.45	405.9	51.3	.0509
190.0	729.9	18.27	3061.0	4046.0	91.19	16.16	57.06	373.0	43.9	.0444
195.0	688.1	17.23	3285.0	4330.0	92.66	16.01	56.40	361.0	41.0	.0414
200.0	648.6	16.24	3501.0	4609.0	94.08	15.83	55.12	352.0	38.6	.0387
205.0	612.0	15.32	3706.0	4880.0	95.42	15.64	53.36	345.5	36.6	.0365
210.0	578.4	14.48	3899.0	5142.0	96.68	15.43	51.30	341.1	34.9	.0346
215.0	548.0	13.72	4081.0	5393.0	97.86	15.23	49.11	338.3	33.6	.0331
220.0	520.6	13.03	4252.0	5633.0	98.97	15.03	46.93	336.9	32.5	.0318
225.0	495.9	12.41	4413.0	5863.0	100.00	14.85	44.85	336.4	31.7	.0308
230.0	473.7	11.86	4564.0	6082.0	101.0	14.67	42.91	336.6	31.0	.0300
235.0	453.6	11.36	4707.0	6292.0	101.9	14.52	41.15	337.4	30.5	.0293
240.0	435.5	10.90	4843.0	6494.0	102.7	14.38	39.57	338.6	30.1	.0287
245.0	419.0	10.49	4972.0	6688.0	103.5	14.25	38.14	340.1	29.7	.0283
250.0	403.9	10.11	5095.0	6876.0	104.3	14.14	36.87	341.8	29.5	.0279

Thermophysical properties of argon—Continued

T K	Density kg/m³	Density mol/dm³	E J/mol	H J/mol	S J/(mol·K)	C_v J/(mol·K)	C_p J/(mol·K)	Sound m/s	Visc. µPa·s	Therm. W/(m·K)
255.0	390.1	9.766	5214.0	7057.0	105.0	14.04	35.73	343.7	29.3	.0275
260.0	377.4	9.448	5328.0	7233.0	105.7	13.95	34.72	345.8	29.2	.0273
270.0	354.9	8.884	5545.0	7571.0	107.0	13.80	33.00	350.2	29.1	.0269
280.0	335.4	8.396	5750.0	7894.0	108.1	13.68	31.60	354.8	29.1	.0267
290.0	318.3	7.969	5945.0	8204.0	109.2	13.59	30.45	359.5	29.2	.0265
300.0	303.3	7.591	6133.0	8504.0	110.2	13.51	29.50	364.3	29.4	.0265
310.0	289.8	7.255	6313.0	8795.0	111.2	13.45	28.69	369.1	29.6	.0265
320.0	277.7	6.952	6489.0	9078.0	112.1	13.39	28.01	373.9	29.9	.0266
330.0	266.8	6.678	6660.0	9355.0	112.9	13.35	27.42	378.7	30.2	.0267
340.0	256.8	6.429	6827.0	9627.0	113.7	13.31	26.91	383.5	30.5	.0268
350.0	247.7	6.200	6990.0	9893.0	114.5	13.28	26.46	388.2	30.9	.0270
360.0	239.3	5.990	7151.0	10160.0	115.3	13.25	26.06	392.8	31.2	.0272
380.0	224.3	5.616	7465.0	10670.0	116.7	13.20	25.40	402.0	32.0	.0277
400.0	211.4	5.292	7771.0	11170.0	117.9	13.16	24.86	410.9	32.8	.0282
				20.00 MPa isobar						
88.68[a]	1439.0	36.03	−1348.0	−792.8	55.09	22.31	42.05	890.1	300.0	.137
100.0	1379.0	34.52	−894.7	−315.3	60.15	20.78	42.43	846.3	226.0	.121
120.0	1267.0	31.71	−85.2	545.4	68.00	19.32	43.66	739.3	149.0	.0993
130.0	1208.0	30.23	324.2	985.8	71.52	18.66	44.45	684.0	125.0	.0903
140.0	1145.0	28.67	737.9	1436.0	74.85	18.06	45.57	629.3	105.0	.0819
150.0	1079.0	27.01	1158.0	1899.0	78.05	17.52	47.11	575.7	89.1	.0740
160.0	1008.0	25.24	1587.0	2379.0	81.15	17.07	49.04	524.2	75.8	.0665
170.0	933.3	23.36	2024.0	2880.0	84.18	16.69	51.11	476.6	64.6	.0598
180.0	854.8	21.40	2466.0	3400.0	87.16	16.38	52.85	435.1	55.4	.0536
190.0	775.6	19.42	2903.0	3933.0	90.04	16.09	53.55	401.9	48.0	.0475
200.0	699.9	17.52	3324.0	4466.0	92.77	15.78	52.63	378.2	42.3	.0420
205.0	664.6	16.64	3524.0	4726.0	94.06	15.62	51.57	369.9	40.1	.0397
210.0	631.5	15.81	3716.0	4981.0	95.28	15.44	50.19	363.6	38.2	.0377
215.0	600.9	15.04	3898.0	5228.0	96.45	15.27	48.60	359.0	36.6	.0360
220.0	572.7	14.34	4072.0	5467.0	97.54	15.09	46.91	355.8	35.4	.0346
225.0	546.9	13.69	4236.0	5697.0	98.58	14.92	45.19	353.8	34.3	.0334
230.0	523.3	13.10	4392.0	5919.0	99.55	14.76	43.52	352.7	33.4	.0324
235.0	501.7	12.56	4540.0	6132.0	100.5	14.61	41.94	352.3	32.7	.0315
240.0	482.0	12.07	4681.0	6338.0	101.3	14.48	40.46	352.4	32.1	.0308
245.0	464.0	11.62	4815.0	6537.0	102.2	14.35	39.09	353.1	31.7	.0302
250.0	447.5	11.20	4944.0	6729.0	102.9	14.24	37.84	354.0	31.3	.0297
255.0	432.3	10.82	5068.0	6916.0	103.7	14.14	36.71	355.3	31.0	.0293
260.0	418.3	10.47	5186.0	7097.0	104.4	14.05	35.68	356.7	30.8	.0289
270.0	393.2	9.843	5413.0	7444.0	105.7	13.89	33.90	360.1	30.5	.0284
280.0	371.5	9.301	5625.0	7776.0	106.9	13.77	32.44	363.9	30.4	.0280
290.0	352.6	8.825	5828.0	8094.0	108.0	13.67	31.23	368.0	30.4	.0278
300.0	335.8	8.405	6021.0	8401.0	109.1	13.59	30.21	372.3	30.4	.0276
310.0	320.8	8.030	6208.0	8699.0	110.0	13.53	29.35	376.7	30.6	.0276
320.0	307.3	7.693	6389.0	8988.0	110.9	13.47	28.61	381.2	30.8	.0276
330.0	295.1	7.388	6564.0	9271.0	111.8	13.42	27.98	385.6	31.0	.0276
340.0	284.0	7.110	6735.0	9548.0	112.6	13.38	27.43	390.1	31.3	.0277
350.0	273.9	6.856	6903.0	9820.0	113.4	13.35	26.94	394.6	31.6	.0278
360.0	264.6	6.622	7067.0	10090.0	114.2	13.32	26.52	399.0	31.9	.0280
380.0	247.9	6.207	7388.0	10610.0	115.6	13.27	25.79	407.8	32.7	.0284
400.0	233.6	5.847	7699.0	11120.0	116.9	13.22	25.21	416.5	33.4	.0288
				40.00 MPa isobar						
93.42[a]	1460.0	36.55	−1312.0	−217.8	55.35	21.79	40.66	952.4	307.0	.139
200.0	956.7	23.95	2430.0	4101.0	86.22	16.00	40.14	571.6	69.4	.0634
220.0	867.5	21.72	3054.0	4896.0	90.01	15.43	39.34	532.8	59.4	.0558
240.0	786.2	19.68	3638.0	5670.0	93.38	14.99	38.00	505.7	52.5	.0499
250.0	749.1	18.75	3913.0	6046.0	94.91	14.81	37.19	495.8	49.9	.0475
260.0	714.5	17.89	4178.0	6414.0	96.36	14.65	36.33	488.0	47.8	.0455
270.0	682.4	17.08	4431.0	6773.0	97.71	14.51	35.45	482.1	46.1	.0438

Thermophysical properties of argon—Continued

T K	Density kg/m³	Density mol/dm³	E J/mol	H J/mol	S J/(mol·K)	C_v J/(mol·K)	C_p J/(mol·K)	Sound m/s	Visc. μPa·s	Therm. W/(m·K)
280.0	652.8	16.34	4675.0	7123.0	98.98	14.39	34.58	477.6	44.7	.0423
290.0	625.5	15.66	4910.0	7464.0	100.2	14.29	33.73	474.4	43.5	.0411
300.0	600.3	15.03	5136.0	7798.0	101.3	14.20	32.93	472.3	42.6	.0401
310.0	577.0	14.44	5354.0	8123.0	102.4	14.12	32.18	471.1	41.9	.0393
320.0	555.6	13.91	5565.0	8442.0	103.4	14.05	31.48	470.6	41.3	.0386
330.0	535.7	13.41	5770.0	8753.0	104.3	14.00	30.84	470.7	40.9	.0380
340.0	517.3	12.95	5970.0	9059.0	105.3	13.94	30.25	471.3	40.6	.0375
360.0	484.4	12.12	6354.0	9653.0	107.0	13.86	29.20	473.7	40.2	.0368
380.0	455.7	11.41	6721.0	10230.0	108.5	13.79	28.31	477.3	40.1	.0364
400.0	430.6	10.78	7075.0	10790.0	109.9	13.73	27.56	481.6	40.1	.0362
				60.00 MPa isobar						
98.01[a]	1478.0	37.00	−1270.0	352.0	55.62	21.72	39.94	1006.0	313.0	.141
120.0	1386.0	34.69	−504.3	1225.0	63.66	20.64	39.46	922.2	208.0	.120
140.0	1305.0	32.66	169.3	2007.0	69.68	19.42	38.63	851.6	157.0	.106
160.0	1226.0	30.68	814.8	2771.0	74.79	18.26	37.79	790.4	125.0	.0947
180.0	1149.0	28.75	1433.0	3519.0	79.20	17.29	37.11	737.6	104.0	.0852
200.0	1074.0	26.89	2024.0	4256.0	83.07	16.53	36.51	692.5	88.3	.0771
220.0	1003.0	25.11	2590.0	4979.0	86.52	15.94	35.86	655.3	77.0	.0700
240.0	936.2	23.44	3129.0	5689.0	89.61	15.48	35.08	625.6	68.7	.0641
260.0	874.4	21.89	3641.0	6382.0	92.39	15.14	34.20	602.7	62.6	.0592
280.0	818.1	20.48	4127.0	7057.0	94.89	14.87	33.26	585.4	58.0	.0553
300.0	767.2	19.21	4588.0	7712.0	97.15	14.67	32.31	572.9	54.6	.0522
320.0	721.5	18.06	5027.0	8349.0	99.20	14.51	31.38	564.2	52.2	.0497
340.0	680.6	17.04	5447.0	8968.0	101.1	14.39	30.52	558.5	50.4	.0478
360.0	644.0	16.12	5849.0	9570.0	102.8	14.29	29.73	555.2	49.1	.0463
380.0	611.1	15.30	6236.0	10160.0	104.4	14.21	29.01	553.7	48.2	.0452
400.0	581.5	14.56	6610.0	10730.0	105.9	14.14	28.36	553.6	47.6	.0443
				80.00 MPa isobar						
102.50[a]	1494.0	37.40	−1222.0	917.7	55.90	21.88	39.53	1052.0	317.0	.143
120.0	1427.0	35.71	−636.6	1604.0	62.08	21.16	38.78	989.7	233.0	.128
140.0	1354.0	33.88	7.5	2369.0	67.98	19.99	37.69	925.5	178.0	.114
160.0	1284.0	32.13	621.3	3111.0	72.94	18.82	36.58	870.8	144.0	.104
180.0	1216.0	30.45	1206.0	3833.0	77.19	17.83	35.64	823.7	120.0	.0954
200.0	1152.0	28.84	1764.0	4538.0	80.90	17.04	34.86	783.0	104.0	.0878
220.0	1091.0	27.31	2298.0	5228.0	84.19	16.44	34.18	748.1	91.6	.0812
240.0	1033.0	25.85	2810.0	5905.0	87.14	15.96	33.52	718.7	82.4	.0755
260.0	978.2	24.49	3302.0	6569.0	89.79	15.60	32.86	694.5	75.2	.0706
280.0	927.4	23.21	3773.0	7219.0	92.20	15.32	32.18	674.9	69.7	.0664
300.0	880.4	22.04	4226.0	7856.0	94.40	15.10	31.51	659.4	65.5	.0629
320.0	837.1	20.95	4662.0	8480.0	96.41	14.93	30.85	647.3	62.2	.0599
340.0	797.3	19.96	5082.0	9090.0	98.27	14.79	30.21	638.0	59.6	.0575
360.0	760.8	19.05	5488.0	9688.0	99.98	14.67	29.61	631.2	57.6	.0555
380.0	727.4	18.21	5881.0	10270.0	101.6	14.58	29.05	626.3	56.1	.0539
400.0	696.7	17.44	6264.0	10850.0	103.0	14.50	28.52	623.1	54.9	.0525
				100.00 MPa isobar						
106.80[a]	1508.0	37.76	−1169.0	1480.0	56.18	22.15	39.28	1092.0	321.0	.146
120.0	1461.0	36.58	−742.6	1991.0	60.70	21.64	38.45	1048.0	257.0	.135
140.0	1394.0	34.89	−118.5	2747.0	66.53	20.51	37.18	988.2	198.0	.122
160.0	1330.0	33.29	474.6	3478.0	71.41	19.34	35.92	937.5	161.0	.112
180.0	1269.0	31.77	1038.0	4186.0	75.58	18.34	34.84	894.1	136.0	.104
200.0	1211.0	30.32	1575.0	4873.0	79.20	17.53	33.96	856.3	118.0	.0970
220.0	1156.0	28.94	2090.0	5545.0	82.40	16.90	33.24	823.5	105.0	.0907
240.0	1104.0	27.64	2585.0	6203.0	85.27	16.42	32.60	795.1	94.4	.0853
260.0	1055.0	26.40	3061.0	6849.0	87.85	16.03	32.01	770.9	86.5	.0804
280.0	1008.0	25.23	3521.0	7484.0	90.21	15.73	31.45	750.5	80.3	.0762
300.0	964.5	24.14	3966.0	8108.0	92.36	15.50	30.91	733.6	75.3	.0725

Thermophysical properties of argon—Continued

T K	Density kg/m^3	Density mol/dm^3	E J/mol	H J/mol	S J/(mol·K)	C_v J/(mol·K)	C_p J/(mol·K)	Sound m/s	Visc. μPa·s	Therm. W/(m·K)
320.0	923.8	23.13	4396.0	8721.0	94.34	15.31	30.38	719.7	71.4	.0693
340.0	885.9	22.18	4814.0	9323.0	96.16	15.15	29.87	708.5	68.2	.0666
360.0	850.7	21.29	5219.0	9915.0	97.85	15.02	29.38	699.6	65.7	.0642
380.0	817.9	20.47	5614.0	10500.0	99.43	14.91	28.92	692.6	63.7	.0622
400.0	787.4	19.71	5999.0	11070.0	100.9	14.82	28.48	687.3	62.1	.0606

[a]At melting line.
[b]At liquid–vapor boundary.

Appendix G. Thermophysical Properties of Ethylene

Thermophysical properties of coexisting gaseous and liquid ethylene

T K	Pres. MPa	Density kg/m^3	Density mol/dm^3	E J/mol	H J/mol	S J/(mol·K)	C_v J/(mol·K)	C_p J/(mol·K)	Sound m/s
103.986[a]	.00012	654.9	23.34	−12400.0	−12400.0	84.86	39.01	73.26	1770.0
103.986[a]	.00012	.00401	.0001440	2641.0	3496.0	237.7	24.99	33.32	202.7
104.0	.00012	654.8	23.34	−12400.0	−12400.0	84.87	39.02	73.25	1770.0
104.0	.00012	.00401	.0001440	2641.0	3498.0	237.7	24.99	33.32	202.7
108.0	.00024	649.8	23.16	−12140.0	−12140.0	87.32	41.55	70.75	1733.0
108.0	.00024	.00819	.0002927	2740.0	3566.0	232.8	25.01	33.35	206.5
112.0	.00045	644.7	22.98	−11880.0	−11880.0	89.74	42.91	69.32	1706.0
112.0	.00045	.01454	.0005192	2839.0	3707.0	228.9	25.04	33.39	210.2
116.0	.00080	639.6	22.80	−11610.0	−11610.0	92.10	43.51	68.40	1683.0
116.0	.00080	.02423	.0008647	2938.0	3866.0	225.5	25.07	33.44	213.9
120.0	.00137	634.5	22.62	−11340.0	−11340.0	94.40	43.63	67.75	1659.0
120.0	.00137	.03897	.001390	3037.0	4020.0	222.4	25.12	33.50	217.5
124.0	.00224	629.4	22.43	−11070.0	−11070.0	96.63	43.44	67.29	1634.0
124.0	.00224	.06101	.002176	3135.0	4165.0	219.5	25.17	33.59	221.0
128.0	.00355	624.2	22.25	−10800.0	−10800.0	98.79	43.06	66.97	1608.0
128.0	.00355	.09312	.003320	3233.0	4302.0	216.7	25.24	33.70	224.4
132.0	.00545	618.9	22.06	−10530.0	−10530.0	100.9	42.57	66.76	1581.0
132.0	.00545	.1386	.004941	3329.0	4433.0	214.2	25.33	33.84	227.6
136.0	.00813	613.7	21.88	−10260.0	−10250.0	102.9	42.03	66.66	1554.0
136.0	.00813	.2012	.007174	3425.0	4559.0	211.8	25.43	34.00	230.8
140.0	.01182	608.4	21.69	−9985.0	−9984.0	104.9	41.47	66.63	1526.0
140.0	.01182	.2853	.01017	3521.0	4683.0	209.6	25.56	34.20	233.9
144.0	.01679	603.0	21.50	−9714.0	−9714.0	106.8	40.93	66.68	1497.0
144.0	.01679	.3956	.01410	3615.0	4805.0	207.6	25.71	34.44	236.8
148.0	.02334	597.6	21.30	−9444.0	−9443.0	108.6	40.40	66.79	1468.0
148.0	.02334	.5373	.01915	3707.0	4926.0	205.7	25.87	34.71	239.7
152.0	.03182	592.2	21.11	−9175.0	−9173.0	110.4	39.91	66.94	1438.0
152.0	.03182	.7161	.02553	3799.0	5046.0	204.0	26.06	35.03	242.3
156.0	.04260	586.7	20.91	−8905.0	−8903.0	112.2	39.45	67.13	1409.0
156.0	.04260	.9383	.03345	3889.0	5163.0	202.4	26.27	35.38	244.9
160.0	.05612	581.1	20.71	−8635.0	−8632.0	113.9	39.04	67.34	1379.0
160.0	.05612	1.210	.04315	3978.0	5279.0	200.8	26.50	35.78	247.3
164.0	.07281	575.5	20.51	−8365.0	−8361.0	115.6	38.66	67.59	1349.0
164.0	.07281	1.540	.05488	4065.0	5392.0	199.4	26.75	36.22	249.5
168.0	.09318	569.8	20.31	−8094.0	−8089.0	117.2	38.32	67.85	1319.0
168.0	.09318	1.933	.06892	4151.0	5503.0	198.1	27.02	36.71	251.6
172.0	.1177	564.0	20.11	−7822.0	−7817.0	118.8	38.02	68.14	1289.0
172.0	.1177	2.400	.08555	4235.0	5610.0	196.9	27.32	37.24	253.5
176.0	.1470	558.2	19.90	−7550.0	−7543.0	120.3	37.76	68.46	1259.0
176.0	.1470	2.948	.1051	4316.0	5715.0	195.7	27.63	37.83	255.3
180.0	.1816	552.2	19.68	−7277.0	−7268.0	121.9	37.54	68.80	1229.0
180.0	.1816	3.586	.1278	4396.0	5817.0	194.6	27.97	38.48	256.8
184.0	.2221	546.2	19.47	−7003.0	−6992.0	123.4	37.35	69.18	1198.0
184.0	.2221	4.324	.1541	4473.0	5914.0	193.5	28.32	39.19	258.2
188.0	.2691	540.1	19.25	−6727.0	−6713.0	124.9	37.19	69.59	1167.0
188.0	.2691	5.172	.1844	4548.0	6008.0	192.5	28.70	39.96	259.4
192.0	.3232	533.8	19.03	−6450.0	−6433.0	126.3	37.06	70.07	1136.0
192.0	.3232	6.142	.2189	4620.0	6097.0	191.6	29.09	40.81	260.4
196.0	.3850	527.5	18.80	−6171.0	−6151.0	127.8	36.97	70.60	1105.0
196.0	.3850	7.245	.2582	4690.0	6181.0	190.7	29.51	41.74	261.1

Thermophysical properties of coexisting gaseous and liquid ethylene—Continued

T K	Pres. MPa	Density kg/m^3	Density mol/dm^3	E J/mol	H J/mol	S J/(mol·K)	C_v J/(mol·K)	C_p J/(mol·K)	Sound m/s
200.0	.4553	521.0	18.57	−5890.0	−5866.0	129.2	36.90	71.20	1073.0
200.0	.4553	8.493	.3027	4756.0	6260.0	189.8	29.95	42.76	261.7
204.0	.5347	514.4	18.34	−5607.0	−5578.0	130.6	36.86	71.90	1040.0
204.0	.5347	9.902	.3530	4819.0	6334.0	189.0	30.41	43.88	262.1
208.0	.6238	507.6	18.09	−5321.0	−5287.0	132.0	36.85	72.70	1007.0
208.0	.6238	11.49	.4094	4879.0	6403.0	188.2	30.89	45.12	262.2
212.0	.7234	500.7	17.85	−5032.0	−4992.0	133.4	36.87	73.62	973.8
212.0	.7234	13.26	.4727	4934.0	6465.0	187.4	31.39	46.49	262.1
216.0	.8342	493.5	17.59	−4741.0	−4693.0	134.7	36.92	74.68	939.6
216.0	.8342	15.25	.5435	4986.0	6521.0	186.6	31.92	48.03	261.7
220.0	.9568	486.2	17.33	−4446.0	−4390.0	136.1	37.00	75.91	904.9
220.0	.9568	17.47	.6226	5032.0	6569.0	185.9	32.48	49.74	261.1
224.0	1.092	478.6	17.06	−4147.0	−4083.0	137.4	37.10	77.34	869.5
224.0	1.092	19.94	.7108	5074.0	6610.0	185.2	33.06	51.68	260.3
228.0	1.240	470.8	16.78	−3843.0	−3769.0	138.8	37.24	79.00	833.5
228.0	1.240	22.70	.8092	5110.0	6642.0	184.4	33.68	53.88	259.2
232.0	1.403	462.8	16.49	−3535.0	−3450.0	140.1	37.41	80.93	796.8
232.0	1.403	25.78	.9190	5139.0	6666.0	183.7	34.33	56.42	257.8
236.0	1.580	454.3	16.20	−3222.0	−3124.0	141.5	37.61	83.20	759.4
236.0	1.580	29.22	1.042	5161.0	6678.0	183.0	35.01	59.37	256.2
240.0	1.773	445.6	15.88	−2902.0	−2790.0	142.8	37.86	85.87	721.4
240.0	1.773	33.08	1.179	5175.0	6679.0	182.3	35.74	62.84	254.2
244.0	1.983	436.3	15.55	−2575.0	−2448.0	144.2	38.14	89.05	682.7
244.0	1.983	37.40	1.333	5179.0	6667.0	181.5	36.52	67.01	252.0
248.0	2.210	426.6	15.21	−2240.0	−2095.0	145.6	38.48	92.88	643.2
248.0	2.210	42.28	1.507	5172.0	6639.0	180.8	37.35	72.10	249.5
252.0	2.455	416.3	14.84	−1895.0	−1730.0	146.9	38.87	97.60	602.9
252.0	2.455	47.82	1.704	5152.0	6592.0	180.0	38.24	78.48	246.7
256.0	2.719	405.2	14.44	−1538.0	−1350.0	148.4	39.33	103.6	561.6
256.0	2.719	54.16	1.931	5115.0	6523.0	179.1	39.21	86.73	243.5
260.0	3.004	393.1	14.01	−1166.0	−951.9	149.8	39.88	111.4	519.0
260.0	3.004	61.51	2.193	5057.0	6427.0	178.2	40.28	97.85	240.1
264.0	3.310	379.8	13.54	−774.9	−530.3	151.4	40.55	122.3	474.8
264.0	3.310	70.18	2.502	4973.0	6296.0	177.2	41.47	113.7	236.3
268.0	3.639	364.6	13.00	−356.4	−76.4	153.0	41.37	138.9	428.3
268.0	3.639	80.67	2.875	4851.0	6116.0	176.1	42.83	138.4	232.1
272.0	3.993	346.8	12.36	102.5	425.5	154.7	42.43	168.3	378.5
272.0	3.993	93.85	3.345	4674.0	5868.0	174.7	44.40	182.0	227.4
276.0	4.373	324.2	11.56	630.5	1009.0	156.7	43.91	236.9	324.0
276.0	4.373	111.5	3.976	4407.0	5507.0	173.0	46.29	280.0	222.2
282.34[b]	5.040	214.6	7.650	2603.0	3262.0	164.5			

[a]Triple point.
[b]Critical point.

Thermophysical properties of ethylene on the melting line

T K	Pres. MPa	Density kg/m³	Density mol/dm³	E J/mol	H J/mol	S J/(mol·K)	C_v J/(mol·K)	C_p J/(mol·K)	Sound m/s
103.99[a]	.00012	654.9	23.34	−12400.0	−12400.0	84.86	39.01	73.26	1770.0
104.0	.1064	656.2	23.39	−12430.0	−12430.0	84.61	39.51	74.45	1758.0
104.2	1.529	656.7	23.41	−12430.0	−12370.0	84.59	39.93	74.97	1749.0
104.4	2.955	657.2	23.43	−12440.0	−12310.0	84.57	40.33	75.49	1741.0
104.6	4.384	657.8	23.45	−12440.0	−12250.0	84.55	40.72	76.03	1734.0
104.8	5.816	658.3	23.47	−12440.0	−12190.0	84.53	41.09	76.59	1727.0
105.0	7.250	658.8	23.48	−12440.0	−12130.0	84.51	41.45	77.16	1721.0
105.2	8.688	659.4	23.50	−12440.0	−12070.0	84.49	41.79	77.75	1715.0
105.4	10.13	659.9	23.52	−12450.0	−12010.0	84.46	42.12	78.36	1709.0
105.6	11.57	660.5	23.54	−12450.0	−11960.0	84.44	42.44	78.99	1704.0
105.8	13.02	661.0	23.56	−12450.0	−11900.0	84.42	42.74	79.65	1700.0
106.0	14.47	661.6	23.58	−12450.0	−11840.0	84.39	43.03	80.32	1695.0
106.2	15.92	662.2	23.60	−12450.0	−11780.0	84.37	43.31	81.03	1691.0
106.4	17.37	662.7	23.62	−12460.0	−11720.0	84.34	43.58	81.76	1688.0
106.6	18.83	663.3	23.64	−12460.0	−11660.0	84.32	43.83	82.52	1684.0
106.8	20.29	663.9	23.66	−12460.0	−11600.0	84.29	44.08	83.31	1681.0
107.0	21.76	664.5	23.69	−12460.0	−11540.0	84.27	44.31	84.14	1679.0
107.2	23.22	665.1	23.71	−12460.0	−11480.0	84.24	44.54	85.01	1676.0
107.4	24.69	665.7	23.73	−12470.0	−11430.0	84.21	44.75	85.92	1674.0
107.6	26.16	666.3	23.75	−12470.0	−11370.0	84.18	44.95	86.87	1673.0
107.8	27.64	666.9	23.77	−12470.0	−11310.0	84.15	45.15	87.87	1671.0
108.0	29.12	667.5	23.79	−12470.0	−11250.0	84.11	45.34	88.92	1670.0
108.2	30.60	668.1	23.82	−12480.0	−11190.0	84.08	45.52	90.03	1669.0
108.4	32.08	668.7	23.84	−12480.0	−11130.0	84.05	45.69	91.20	1668.0
108.6	33.57	669.4	23.86	−12480.0	−11070.0	84.01	45.86	92.43	1668.0
108.8	35.06	670.0	23.88	−12480.0	−11020.0	83.97	46.01	93.74	1668.0
109.0	36.55	670.6	23.91	−12490.0	−10960.0	83.93	46.17	95.13	1668.0
109.2	38.05	671.3	23.93	−12490.0	−10900.0	83.89	46.31	96.61	1668.0
109.4	39.55	671.9	23.95	−12490.0	−10840.0	83.85	46.45	98.18	1669.0

[a]Triple point.

Thermophysical properties of ethylene

T K	Density kg/m³	Density mol/dm³	E J/mol	H J/mol	S J/(mol·K)	C_v J/(mol·K)	C_p J/(mol·K)	Sound m/s
			.02 MPa isobar					
104.00[a]	656.2	23.39	−12430.0	−12430.0	84.62	39.49	74.42	1758.0
105.0	654.4	23.33	−12360.0	−12360.0	85.33	40.08	73.09	1752.0
110.0	646.7	23.05	−12000.0	−12000.0	88.64	42.19	69.75	1721.0
120.0	633.2	22.57	−11320.0	−11310.0	94.60	43.41	67.71	1656.0
130.0	620.6	22.12	−10640.0	−10640.0	99.99	42.70	66.94	1590.0
140.0	607.9	21.67	−9975.0	−9974.0	104.9	41.43	66.71	1522.0
146.094[b]	600.2	21.40	−9573.0	−9572.0	107.7	40.65	66.73	1482.0
146.094[b]	.4655	.01659	3663.0	4869.0	206.6	25.79	34.58	238.3
150.0	.4543	.01620	3765.0	5000.0	207.5	25.80	34.53	241.5
160.0	.4251	.01515	4025.0	5345.0	209.7	25.89	34.52	249.5
170.0	.3995	.01424	4286.0	5691.0	211.8	26.07	34.64	257.1
180.0	.3769	.01344	4549.0	6038.0	213.8	26.35	34.86	264.3
190.0	.3568	.01272	4816.0	6388.0	215.7	26.71	35.19	271.2
200.0	.3387	.01207	5085.0	6742.0	217.5	27.15	35.60	277.7
210.0	.3224	.01149	5360.0	7100.0	219.3	27.66	36.09	284.0
220.0	.3076	.01097	5640.0	7464.0	220.9	28.24	36.65	290.1
230.0	.2941	.01049	5926.0	7834.0	222.6	28.88	37.28	295.9
240.0	.2818	.01005	6219.0	8210.0	224.2	29.59	37.98	301.5
260.0	.2600	.009268	6827.0	8985.0	227.3	31.15	39.52	312.1
280.0	.2413	.008604	7467.0	9792.0	230.3	32.89	41.25	322.2
300.0	.2252	.008028	8144.0	10640.0	233.2	34.76	43.11	331.7
320.0	.2111	.007525	8859.0	11520.0	236.0	36.72	45.06	340.9
340.0	.1986	.007081	9614.0	12440.0	238.8	38.72	47.06	349.7
360.0	.1876	.006687	10410.0	13400.0	241.6	40.74	49.08	358.3
380.0	.1777	.006334	11240.0	14400.0	244.3	42.75	51.08	366.7
400.0	.1688	.006017	12120.0	15440.0	246.9	44.73	53.06	374.9
			.04 MPa isobar					
104.00[a]	656.2	23.39	−12430.0	−12430.0	84.62	39.49	74.43	1758.0
105.0	654.4	23.33	−12360.0	−12350.0	85.33	40.08	73.09	1752.0
110.0	646.7	23.05	−12000.0	−12000.0	88.64	42.19	69.76	1721.0
120.0	633.2	22.57	−11320.0	−11310.0	94.60	43.41	67.72	1656.0
130.0	620.6	22.12	−10640.0	−10640.0	99.99	42.70	66.94	1590.0
140.0	607.9	21.67	−9975.0	−9973.0	104.9	41.43	66.70	1522.0
150.0	594.8	21.20	−9308.0	−9306.0	109.5	40.14	66.87	1452.0
155.117[b]	587.9	20.96	−8964.0	−8963.0	111.8	39.55	67.08	1415.0
155.117[b]	.8852	.03155	3870.0	5137.0	202.7	26.22	35.30	244.3
160.0	.8571	.03055	3999.0	5309.0	203.8	26.22	35.20	248.3
170.0	.8042	.02867	4265.0	5660.0	205.9	26.32	35.15	256.0
180.0	.7578	.02701	4531.0	6012.0	207.9	26.53	35.26	263.4
190.0	.7167	.02555	4800.0	6366.0	209.8	26.85	35.49	270.4
200.0	.6800	.02424	5072.0	6722.0	211.7	27.26	35.84	277.1
210.0	.6469	.02306	5348.0	7083.0	213.4	27.74	36.29	283.5
220.0	.6169	.02199	5629.0	7448.0	215.1	28.30	36.81	289.6
230.0	.5897	.02102	5917.0	7819.0	216.8	28.93	37.42	295.4
240.0	.5648	.02013	6210.0	8197.0	218.4	29.63	38.09	301.1
250.0	.5419	.01932	6511.0	8581.0	220.0	30.38	38.82	306.5
260.0	.5209	.01857	6819.0	8974.0	221.5	31.18	39.61	311.8
270.0	.5014	.01787	7136.0	9374.0	223.0	32.02	40.44	316.9
280.0	.4833	.01723	7461.0	9783.0	224.5	32.91	41.32	321.9
290.0	.4665	.01663	7795.0	10200.0	226.0	33.83	42.23	326.8
300.0	.4509	.01607	8138.0	10630.0	227.4	34.77	43.16	331.5
310.0	.4362	.01555	8491.0	11060.0	228.8	35.74	44.13	336.1
320.0	.4225	.01506	8854.0	11510.0	230.3	36.72	45.10	340.7
330.0	.4096	.01460	9226.0	11970.0	231.7	37.72	46.10	345.2
340.0	.3975	.01417	9609.0	12430.0	233.0	38.72	47.10	349.6
350.0	.3861	.01376	10000.0	12910.0	234.4	39.73	48.10	353.9
360.0	.3754	.01338	10400.0	13390.0	235.8	40.74	49.10	358.2
370.0	.3652	.01302	10820.0	13890.0	237.2	41.75	50.11	362.4

Thermophysical properties of ethylene—Continued

T K	Density kg/m^3	Density mol/dm^3	*E* J/mol	*H* J/mol	*S* J/(mol·K)	C_v J/(mol·K)	C_p J/(mol·K)	Sound m/s
380.0	.3555	.01267	11240.0	14400.0	238.5	42.75	51.11	366.6
390.0	.3464	.01235	11670.0	14910.0	239.8	43.75	52.10	370.7
400.0	.3377	.01204	12110.0	15440.0	241.2	44.73	53.09	374.8
				.06 MPa isobar				
104.00[a]	656.2	23.39	−12430.0	−12430.0	84.62	39.50	74.44	1758.0
105.0	654.5	23.33	−12360.0	−12350.0	85.33	40.09	73.10	1752.0
110.0	646.7	23.05	−12000.0	−12000.0	88.63	42.19	69.76	1721.0
120.0	633.3	22.57	−11320.0	−11310.0	94.60	43.41	67.72	1656.0
130.0	620.6	22.12	−10640.0	−10640.0	99.98	42.71	66.94	1590.0
140.0	607.9	21.67	−9975.0	−9973.0	104.9	41.43	66.70	1522.0
150.0	594.8	21.20	−9308.0	−9305.0	109.5	40.14	66.87	1452.0
160.0	581.2	20.72	−8637.0	−8634.0	113.9	39.04	67.32	1380.0
161.007[b]	579.7	20.66	−8567.0	−8564.0	114.3	38.94	67.40	1372.0
161.007[b]	1.288	.04590	4000.0	5307.0	200.5	26.56	35.88	247.9
165.0	1.254	.04470	4108.0	5451.0	201.4	26.55	35.77	251.1
170.0	1.214	.04329	4243.0	5629.0	202.4	26.57	35.68	255.0
180.0	1.143	.04074	4513.0	5986.0	204.5	26.72	35.66	262.5
190.0	1.080	.03850	4784.0	6343.0	206.4	26.99	35.81	269.6
200.0	1.024	.03650	5059.0	6702.0	208.2	27.36	36.09	276.4
210.0	.9735	.03470	5336.0	7065.0	210.0	27.83	36.49	282.9
220.0	.9281	.03308	5619.0	7433.0	211.7	28.37	36.98	289.1
230.0	.8867	.03161	5907.0	7805.0	213.4	28.99	37.55	295.0
240.0	.8490	.03026	6201.0	8184.0	215.0	29.67	38.20	300.7
250.0	.8144	.02903	6503.0	8569.0	216.6	30.41	38.92	306.2
260.0	.7826	.02790	6812.0	8962.0	218.1	31.20	39.69	311.5
270.0	.7532	.02685	7129.0	9363.0	219.6	32.05	40.52	316.6
280.0	.7260	.02588	7454.0	9773.0	221.1	32.93	41.38	321.6
290.0	.7006	.02498	7789.0	10190.0	222.6	33.84	42.28	326.5
300.0	.6770	.02413	8132.0	10620.0	224.0	34.78	43.22	331.3
310.0	.6550	.02335	8486.0	11060.0	225.4	35.75	44.17	335.9
320.0	.6343	.02261	8849.0	11500.0	226.9	36.73	45.15	340.5
330.0	.6150	.02192	9221.0	11960.0	228.3	37.73	46.13	345.0
340.0	.5967	.02127	9604.0	12420.0	229.7	38.73	47.13	349.4
350.0	.5796	.02066	9997.0	12900.0	231.0	39.74	48.13	353.8
360.0	.5634	.02008	10400.0	13390.0	232.4	40.75	49.13	358.0
370.0	.5481	.01954	10810.0	13880.0	233.8	41.75	50.14	362.3
380.0	.5336	.01902	11240.0	14390.0	235.1	42.75	51.13	366.5
390.0	.5198	.01853	11670.0	14910.0	236.5	43.75	52.12	370.6
400.0	.5067	.01806	12110.0	15430.0	237.8	44.74	53.11	374.7
				.08 MPa isobar				
104.00[a]	656.2	23.39	−12430.0	−12430.0	84.61	39.50	74.44	1758.0
105.0	654.5	23.33	−12360.0	−12350.0	85.32	40.09	73.11	1752.0
110.0	646.7	23.05	−12000.0	−12000.0	88.63	42.20	69.76	1721.0
120.0	633.3	22.57	−11320.0	−11310.0	94.60	43.42	67.72	1656.0
140.0	607.9	21.67	−9976.0	−9972.0	104.9	41.43	66.70	1523.0
160.0	581.3	20.72	−8637.0	−8634.0	113.9	39.04	67.32	1380.0
165.502[b]	573.4	20.44	−8263.0	−8259.0	116.2	38.53	67.69	1338.0
165.502[b]	1.679	.05987	4098.0	5434.0	198.9	26.85	36.40	250.3
170.0	1.630	.05812	4221.0	5598.0	199.9	26.83	36.24	253.9
180.0	1.533	.05463	4495.0	5959.0	202.0	26.92	36.09	261.6
190.0	1.447	.05157	4769.0	6320.0	203.9	27.14	36.14	268.9
200.0	1.371	.04885	5045.0	6682.0	205.8	27.47	36.35	275.7
210.0	1.302	.04643	5324.0	7047.0	207.6	27.91	36.69	282.3
220.0	1.241	.04424	5608.0	7417.0	209.3	28.43	37.15	288.5
230.0	1.185	.04225	5897.0	7791.0	210.9	29.04	37.69	294.5
240.0	1.135	.04044	6192.0	8171.0	212.6	29.71	38.32	300.3
260.0	1.045	.03726	6804.0	8951.0	215.7	31.23	39.78	311.2
280.0	.9693	.03455	7448.0	9763.0	218.7	32.94	41.45	321.4

Thermophysical properties of ethylene—Continued

T K	Density kg/m³	Density mol/dm³	E J/mol	H J/mol	S J/(mol·K)	C_v J/(mol·K)	C_p J/(mol·K)	Sound m/s
300.0	.9037	.03221	8127.0	10610.0	221.6	34.80	43.27	331.1
320.0	.8465	.03018	8843.0	11490.0	224.5	36.74	45.19	340.3
340.0	.7962	.02838	9599.0	12420.0	227.3	38.74	47.17	349.3
360.0	.7516	.02679	10400.0	13380.0	230.0	40.75	49.16	357.9
380.0	.7118	.02537	11230.0	14380.0	232.7	42.76	51.16	366.4
400.0	.6759	.02409	12110.0	15430.0	235.4	44.74	53.13	374.6

.10 MPa isobar

T K	Density kg/m³	Density mol/dm³	E J/mol	H J/mol	S J/(mol·K)	C_v J/(mol·K)	C_p J/(mol·K)	Sound m/s
104.00[a]	656.2	23.39	−12430.0	−12430.0	84.61	39.51	74.45	1758.0
105.0	654.5	23.33	−12360.0	−12350.0	85.32	40.09	73.12	1752.0
110.0	646.7	23.05	−12000.0	−12000.0	88.63	42.20	69.77	1721.0
120.0	633.3	22.57	−11320.0	−11310.0	94.60	43.42	67.72	1656.0
140.0	607.9	21.67	−9976.0	−9971.0	104.9	41.43	66.70	1523.0
160.0	581.3	20.72	−8638.0	−8633.0	113.9	39.04	67.32	1380.0
169.186[b]	568.1	20.25	−8013.0	−8008.0	117.7	38.23	67.94	1310.0
169.186[b]	2.064	.07357	4176.0	5535.0	197.7	27.11	36.86	252.2
170.0	2.052	.07316	4199.0	5566.0	197.9	27.10	36.82	252.9
180.0	1.927	.06868	4476.0	5932.0	200.0	27.11	36.52	260.7
190.0	1.817	.06476	4753.0	6297.0	202.0	27.28	36.47	268.1
200.0	1.720	.06131	5031.0	6662.0	203.8	27.58	36.61	275.1
210.0	1.633	.05823	5312.0	7030.0	205.6	28.00	36.90	281.7
220.0	1.556	.05545	5597.0	7401.0	207.4	28.50	37.32	288.0
230.0	1.485	.05295	5887.0	7776.0	209.0	29.09	37.83	294.1
240.0	1.421	.05066	6184.0	8158.0	210.7	29.75	38.44	299.9
260.0	1.309	.04665	6797.0	8940.0	213.8	31.26	39.86	310.8
280.0	1.213	.04325	7441.0	9754.0	216.8	32.96	41.51	321.1
300.0	1.131	.04031	8121.0	10600.0	219.7	34.81	43.32	330.8
320.0	1.059	.03775	8838.0	11490.0	222.6	36.75	45.23	340.1
340.0	.9960	.03550	9595.0	12410.0	225.4	38.74	47.20	349.1
360.0	.9401	.03351	10390.0	13380.0	228.1	40.76	49.19	357.8
380.0	.8901	.03173	11230.0	14380.0	230.9	42.76	51.18	366.2
400.0	.8453	.03013	12100.0	15420.0	233.5	44.74	53.15	374.5

.101325 MPa isobar

T K	Density kg/m³	Density mol/dm³	E J/mol	H J/mol	S J/(mol·K)	C_v J/(mol·K)	C_p J/(mol·K)	Sound m/s
104.00[a]	656.2	23.39	−12430.0	−12430.0	84.61	39.51	74.45	1758.0
105.0	654.5	23.33	−12360.0	−12350.0	85.32	40.09	73.12	1752.0
110.0	646.7	23.05	−12000.0	−12000.0	88.63	42.20	69.77	1721.0
120.0	633.3	22.57	−11320.0	−11310.0	94.60	43.42	67.72	1656.0
140.0	607.9	21.67	−9976.0	−9971.0	104.9	41.43	66.70	1523.0
160.0	581.3	20.72	−8638.0	−8633.0	113.9	39.05	67.32	1380.0
169.409[b]	567.8	20.24	−7998.0	−7993.0	117.7	38.21	67.95	1309.0
169.409[b]	2.089	.07447	4181.0	5541.0	197.7	27.13	36.89	252.3
170.0	2.080	.07416	4197.0	5563.0	197.8	27.12	36.86	252.8
180.0	1.953	.06961	4475.0	5930.0	199.9	27.13	36.55	260.6
190.0	1.842	.06564	4752.0	6295.0	201.9	27.29	36.49	268.0
200.0	1.743	.06214	5030.0	6661.0	203.7	27.59	36.63	275.0
210.0	1.655	.05901	5311.0	7028.0	205.5	28.00	36.92	281.7
220.0	1.577	.05620	5597.0	7399.0	207.3	28.51	37.33	288.0
230.0	1.505	.05366	5887.0	7775.0	208.9	29.09	37.84	294.0
240.0	1.440	.05134	6183.0	8157.0	210.5	29.75	38.45	299.8
260.0	1.326	.04728	6796.0	8939.0	213.7	31.26	39.87	310.8
280.0	1.229	.04382	7441.0	9753.0	216.7	32.96	41.52	321.1
300.0	1.146	.04085	8120.0	10600.0	219.6	34.81	43.32	330.8
320.0	1.073	.03826	8838.0	11490.0	222.5	36.75	45.23	340.1
340.0	1.009	.03598	9594.0	12410.0	225.3	38.74	47.20	349.1
360.0	.9526	.03396	10390.0	13370.0	228.0	40.76	49.20	357.8
380.0	.9020	.03215	11230.0	14380.0	230.7	42.76	51.19	366.2
400.0	.8565	.03053	12100.0	15420.0	233.4	44.75	53.16	374.5

Thermophysical properties of ethylene—Continued

T K	Density kg/m^3	Density mol/dm^3	E J/mol	H J/mol	S J/(mol·K)	C_v J/(mol·K)	C_p J/(mol·K)	Sound m/s
			.20 MPa isobar					
104.00[a]	656.2	23.39	−12430.0	−12420.0	84.61	39.54	74.49	1757.0
110.0	646.8	23.05	−12000.0	−11990.0	88.62	42.21	69.79	1720.0
120.0	633.3	22.58	−11320.0	−11310.0	94.59	43.43	67.72	1656.0
140.0	608.0	21.67	−9977.0	−9968.0	104.9	41.44	66.69	1523.0
160.0	581.4	20.72	−8640.0	−8630.0	113.9	39.05	67.30	1381.0
180.0	552.4	19.69	−7281.0	−7271.0	121.9	37.54	68.76	1230.0
181.894[b]	549.4	19.58	−7147.0	−7137.0	122.6	37.44	68.97	1214.0
181.894[b]	3.922	.1398	4433.0	5863.0	194.1	28.13	38.81	257.5
185.0	3.843	.1370	4524.0	5984.0	194.7	28.09	38.58	260.0
190.0	3.725	.1328	4669.0	6176.0	195.8	28.07	38.31	264.0
200.0	3.511	.1252	4959.0	6557.0	197.7	28.17	38.04	271.6
210.0	3.324	.1185	5250.0	6938.0	199.6	28.44	38.03	278.7
220.0	3.158	.1126	5542.0	7319.0	201.4	28.85	38.22	285.4
230.0	3.009	.1073	5838.0	7702.0	203.1	29.36	38.58	291.7
240.0	2.875	.1025	6139.0	8091.0	204.7	29.96	39.06	297.8
260.0	2.641	.09412	6759.0	8884.0	207.9	31.39	40.31	309.2
280.0	2.443	.08709	7408.0	9705.0	210.9	33.05	41.85	319.7
300.0	2.274	.08107	8092.0	10560.0	213.9	34.87	43.58	329.7
320.0	2.128	.07585	8812.0	11450.0	216.7	36.79	45.44	339.2
340.0	1.999	.07127	9571.0	12380.0	219.6	38.78	47.38	348.3
360.0	1.886	.06722	10370.0	13340.0	222.3	40.78	49.34	357.1
380.0	1.785	.06362	11210.0	14350.0	225.0	42.78	51.32	365.7
400.0	1.694	.06038	12080.0	15400.0	227.7	44.76	53.27	374.1
			.40 MPa isobar					
104.00[a]	656.3	23.39	−12430.0	−12410.0	84.61	39.60	74.56	1756.0
110.0	646.9	23.06	−12000.0	−11990.0	88.60	42.24	69.82	1720.0
120.0	633.5	22.58	−11320.0	−11300.0	94.57	43.45	67.72	1656.0
140.0	608.1	21.68	−9980.0	−9962.0	104.9	41.45	66.67	1524.0
160.0	581.5	20.73	−8643.0	−8624.0	113.8	39.06	67.27	1382.0
180.0	552.7	19.70	−7286.0	−7265.0	121.8	37.56	68.71	1232.0
190.0	537.2	19.15	−6594.0	−6573.0	125.6	37.13	69.77	1154.0
196.895[b]	526.0	18.75	−6108.0	−6087.0	128.1	36.95	70.73	1098.0
196.895[b]	7.511	.2677	4705.0	6199.0	190.5	29.60	41.96	261.3
200.0	7.352	.2621	4803.0	6330.0	191.2	29.52	41.56	264.0
210.0	6.906	.2462	5116.0	6741.0	193.2	29.45	40.71	272.2
220.0	6.521	.2325	5425.0	7145.0	195.0	29.61	40.33	279.8
230.0	6.185	.2205	5734.0	7548.0	196.8	29.95	40.26	286.8
240.0	5.886	.2098	6045.0	7951.0	198.5	30.42	40.43	293.4
250.0	5.619	.2003	6360.0	8357.0	200.2	31.00	40.78	299.7
260.0	5.377	.1917	6681.0	8768.0	201.8	31.68	41.27	305.7
280.0	4.956	.1767	7341.0	9605.0	204.9	33.24	42.56	316.9
300.0	4.600	.1640	8032.0	10470.0	207.9	34.99	44.13	327.4
320.0	4.294	.1531	8759.0	11370.0	210.8	36.88	45.88	337.3
340.0	4.028	.1436	9522.0	12310.0	213.6	38.84	47.74	346.8
360.0	3.795	.1353	10320.0	13280.0	216.4	40.83	49.65	355.8
380.0	3.587	.1279	11170.0	14290.0	219.2	42.83	51.58	364.6
400.0	3.402	.1213	12050.0	15340.0	221.9	44.80	53.50	373.2
			.60 MPa isobar					
104.10[a]	656.4	23.40	−12430.0	−12410.0	84.61	39.66	74.63	1755.0
110.0	647.0	23.06	−12010.0	−11980.0	88.58	42.27	69.86	1720.0
120.0	633.6	22.58	−11320.0	−11290.0	94.55	43.47	67.72	1657.0
140.0	608.3	21.68	−9983.0	−9955.0	104.9	41.47	66.65	1525.0
160.0	581.7	20.74	−8647.0	−8618.0	113.8	39.07	67.23	1383.0
180.0	552.9	19.71	−7291.0	−7260.0	121.8	37.57	68.67	1234.0
200.0	521.4	18.58	−5897.0	−5865.0	129.1	36.91	71.11	1075.0
206.974[b]	509.4	18.16	−5395.0	−5362.0	131.6	36.85	72.48	1016.0

Thermophysical properties of ethylene—Continued

T K	Density kg/m³	Density mol/dm³	E J/mol	H J/mol	S J/(mol·K)	C_v J/(mol·K)	C_p J/(mol·K)	Sound m/s
206.974[b]	11.06	.3943	4864.0	6386.0	188.4	30.76	44.79	262.2
210.0	10.82	.3856	4966.0	6522.0	189.0	30.64	44.22	265.0
220.0	10.14	.3613	5297.0	6958.0	191.1	30.49	42.94	273.7
230.0	9.558	.3407	5622.0	7383.0	193.0	30.61	42.28	281.6
240.0	9.056	.3228	5946.0	7805.0	194.8	30.93	42.03	288.9
250.0	8.615	.3071	6271.0	8225.0	196.5	31.40	42.08	295.7
260.0	8.221	.2930	6599.0	8647.0	198.1	31.99	42.35	302.1
270.0	7.866	.2804	6933.0	9073.0	199.7	32.67	42.78	308.2
280.0	7.544	.2689	7272.0	9503.0	201.3	33.43	43.33	314.1
300.0	6.981	.2488	7972.0	10380.0	204.3	35.13	44.72	325.1
320.0	6.502	.2318	8704.0	11290.0	207.3	36.97	46.35	335.4
340.0	6.089	.2170	9473.0	12240.0	210.1	38.91	48.12	345.2
360.0	5.727	.2042	10280.0	13220.0	212.9	40.88	49.97	354.5
380.0	5.408	.1928	11120.0	14240.0	215.7	42.87	51.85	363.5
400.0	5.124	.1827	12010.0	15290.0	218.4	44.84	53.74	372.3
				.80 MPa isobar				
104.10[a]	656.5	23.40	-12430.0	-12400.0	84.60	39.72	74.70	1753.0
110.0	647.1	23.07	-12010.0	-11970.0	88.56	42.30	69.90	1719.0
120.0	633.7	22.59	-11320.0	-11290.0	94.53	43.49	67.73	1657.0
140.0	608.4	21.69	-9985.0	-9949.0	104.9	41.48	66.63	1526.0
160.0	581.9	20.74	-8651.0	-8612.0	113.8	39.08	67.20	1385.0
180.0	553.1	19.72	-7296.0	-7255.0	121.8	37.58	68.62	1235.0
200.0	521.7	18.60	-5904.0	-5861.0	129.1	36.92	71.03	1078.0
210.0	504.6	17.99	-5186.0	-5141.0	132.6	36.87	73.01	993.6
214.809[b]	495.7	17.67	-4828.0	-4783.0	134.3	36.90	74.35	949.9
214.809[b]	14.63	.5216	4971.0	6505.0	186.9	31.76	47.55	261.9
215.0	14.59	.5200	4979.0	6517.0	186.9	31.75	47.48	262.1
220.0	14.07	.5015	5156.0	6751.0	188.0	31.53	46.30	267.0
230.0	13.17	.4694	5502.0	7206.0	190.0	31.37	44.75	275.9
240.0	12.41	.4424	5840.0	7649.0	191.9	31.50	43.93	284.0
250.0	11.76	.4191	6177.0	8086.0	193.7	31.84	43.58	291.5
260.0	11.18	.3986	6515.0	8522.0	195.4	32.33	43.56	298.4
270.0	10.67	.3804	6855.0	8958.0	197.0	32.94	43.78	304.9
280.0	10.21	.3641	7201.0	9398.0	198.6	33.65	44.18	311.1
300.0	9.420	.3358	7910.0	10290.0	201.7	35.26	45.35	322.7
320.0	8.753	.3120	8650.0	11210.0	204.7	37.07	46.83	333.4
340.0	8.181	.2916	9424.0	12170.0	207.6	38.98	48.51	343.6
360.0	7.684	.2739	10230.0	13150.0	210.4	40.94	50.30	353.2
380.0	7.248	.2584	11080.0	14180.0	213.2	42.91	52.13	362.4
400.0	6.861	.2446	11970.0	15240.0	215.9	44.87	53.98	371.4
				1.00 MPa isobar				
104.10[a]	656.5	23.40	-12430.0	-12390.0	84.60	39.78	74.77	1752.0
110.0	647.2	23.07	-12010.0	-11970.0	88.54	42.33	69.94	1719.0
120.0	633.8	22.59	-11330.0	-11280.0	94.52	43.51	67.73	1657.0
140.0	608.6	21.69	-9988.0	-9942.0	104.8	41.49	66.61	1527.0
160.0	582.1	20.75	-8654.0	-8606.0	113.8	39.10	67.17	1386.0
180.0	553.4	19.73	-7301.0	-7250.0	121.7	37.59	68.57	1237.0
200.0	522.0	18.61	-5911.0	-5857.0	129.1	36.93	70.95	1080.0
210.0	505.0	18.00	-5194.0	-5139.0	132.6	36.87	72.89	996.4
220.0	486.6	17.35	-4454.0	-4396.0	136.0	37.00	75.76	907.7
221.321[b]	483.7	17.24	-4347.0	-4289.0	136.5	37.03	76.36	893.3
221.321[b]	18.25	.6507	5047.0	6584.0	185.7	32.67	50.36	260.9
225.0	17.71	.6312	5186.0	6770.0	186.5	32.44	49.13	264.8
230.0	17.08	.6087	5370.0	7013.0	187.6	32.25	47.87	269.8
240.0	15.99	.5698	5727.0	7482.0	189.6	32.15	46.21	278.8
250.0	15.07	.5371	6078.0	7939.0	191.4	32.32	45.33	287.0
260.0	14.28	.5090	6426.0	8390.0	193.2	32.70	44.95	294.5
270.0	13.59	.4843	6775.0	8840.0	194.9	33.23	44.90	301.5

Thermophysical properties of ethylene—Continued

T K	Density kg/m^3	Density mol/dm^3	E J/mol	H J/mol	S J/(mol·K)	C_v J/(mol·K)	C_p J/(mol·K)	Sound m/s
280.0	12.97	.4624	7127.0	9289.0	196.5	33.87	45.11	308.1
290.0	12.42	.4428	7484.0	9742.0	198.1	34.61	45.50	314.3
300.0	11.92	.4250	7847.0	10200.0	199.7	35.41	46.02	320.3
310.0	11.47	.4087	8216.0	10660.0	201.2	36.27	46.64	326.0
320.0	11.05	.3938	8594.0	11130.0	202.7	37.17	47.35	331.5
340.0	10.31	.3674	9373.0	12100.0	205.6	39.05	48.92	341.9
360.0	9.667	.3446	10190.0	13090.0	208.4	40.99	50.63	351.9
380.0	9.108	.3246	11040.0	14120.0	211.2	42.95	52.42	361.3
400.0	8.613	.3070	11930.0	15190.0	214.0	44.91	54.23	370.5
				1.50 MPa isobar				
104.20[a]	656.7	23.41	-12430.0	-12370.0	84.59	39.92	74.96	1749.0
110.0	647.5	23.08	-12020.0	-11950.0	88.50	42.41	70.03	1718.0
120.0	634.1	22.60	-11330.0	-11260.0	94.47	43.56	67.74	1658.0
140.0	608.9	21.70	-9995.0	-9926.0	104.8	41.52	66.56	1529.0
160.0	582.5	20.76	-8663.0	-8591.0	113.7	39.12	67.09	1389.0
180.0	554.0	19.75	-7313.0	-7237.0	121.7	37.61	68.45	1241.0
200.0	522.8	18.64	-5928.0	-5847.0	129.0	36.95	70.75	1086.0
210.0	506.0	18.04	-5214.0	-5131.0	132.5	36.89	72.61	1003.0
220.0	487.8	17.39	-4478.0	-4392.0	135.9	37.01	75.32	916.0
230.0	467.8	16.68	-3710.0	-3620.0	139.4	37.31	79.47	821.8
234.229[b]	458.1	16.33	-3361.0	-3269.0	140.9	37.52	82.15	776.0
234.229[b]	27.65	.9857	5152.0	6674.0	183.3	34.71	58.01	256.9
235.0	27.42	.9773	5187.0	6722.0	183.5	34.62	57.45	258.0
240.0	26.23	.9351	5398.0	7002.0	184.7	34.21	54.75	264.0
245.0	25.21	.8985	5601.0	7271.0	185.8	33.95	52.78	269.6
250.0	24.29	.8660	5799.0	7531.0	186.9	33.81	51.32	274.7
260.0	22.74	.8104	6183.0	8034.0	188.8	33.80	49.40	284.0
270.0	21.43	.7640	6559.0	8522.0	190.7	34.06	48.37	292.4
280.0	20.32	.7242	6932.0	9003.0	192.4	34.52	47.88	300.1
290.0	19.34	.6894	7305.0	9481.0	194.1	35.11	47.78	307.3
300.0	18.48	.6586	7682.0	9959.0	195.7	35.81	47.93	314.0
310.0	17.70	.6310	8063.0	10440.0	197.3	36.59	48.28	320.3
320.0	17.00	.6060	8450.0	10930.0	198.8	37.43	48.76	326.4
330.0	16.36	.5833	8844.0	11420.0	200.4	38.32	49.35	332.2
340.0	15.78	.5624	9245.0	11910.0	201.8	39.23	50.02	337.8
360.0	14.74	.5254	10070.0	12930.0	204.7	41.13	51.52	348.5
380.0	13.84	.4935	10930.0	13970.0	207.6	43.06	53.16	358.6
400.0	13.06	.4656	11830.0	15050.0	210.3	45.00	54.86	368.2
				2.00 MPa isobar				
104.30[a]	656.9	23.42	-12430.0	-12350.0	84.59	40.06	75.14	1747.0
110.0	647.8	23.09	-12020.0	-11940.0	88.45	42.48	70.13	1717.0
120.0	634.4	22.61	-11340.0	-11250.0	94.43	43.61	67.75	1658.0
140.0	609.2	21.72	-10000.0	-9909.0	104.7	41.55	66.51	1531.0
160.0	583.0	20.78	-8672.0	-8576.0	113.6	39.15	67.01	1393.0
180.0	554.6	19.77	-7325.0	-7224.0	121.6	37.64	68.33	1246.0
200.0	523.6	18.67	-5944.0	-5837.0	128.9	36.97	70.55	1091.0
210.0	506.9	18.07	-5234.0	-5123.0	132.4	36.91	72.33	1010.0
220.0	489.0	17.43	-4503.0	-4388.0	135.8	37.02	74.91	924.1
230.0	469.4	16.73	-3740.0	-3621.0	139.2	37.30	78.78	832.0
235.0	458.6	16.35	-3343.0	-3220.0	140.9	37.53	81.49	782.7
240.0	447.0	15.93	-2930.0	-2805.0	142.7	37.83	85.00	730.5
244.311[b]	435.6	15.53	-2549.0	-2421.0	144.3	38.17	89.32	679.6
244.311[b]	37.76	1.346	5179.0	6665.0	181.5	36.58	67.37	251.8
245.0	37.45	1.335	5214.0	6712.0	181.7	36.47	66.53	252.9
250.0	35.55	1.267	5454.0	7033.0	183.0	35.87	61.82	260.0
255.0	33.96	1.211	5681.0	7333.0	184.1	35.48	58.57	266.3
260.0	32.59	1.162	5898.0	7620.0	185.3	35.24	56.24	272.1
265.0	31.38	1.119	6108.0	7896.0	186.3	35.13	54.51	277.5
270.0	30.30	1.080	6314.0	8166.0	187.3	35.11	53.23	282.4
280.0	28.44	1.014	6716.0	8689.0	189.2	35.30	51.54	291.5

Thermophysical properties of ethylene—Continued

T K	Density kg/m³	Density mol/dm³	E J/mol	H J/mol	S J/(mol·K)	C_v J/(mol·K)	C_p J/(mol·K)	Sound m/s
290.0	26.88	.9581	7112.0	9199.0	191.0	35.71	50.65	299.8
300.0	25.53	.9100	7505.0	9703.0	192.7	36.27	50.25	307.4
310.0	24.34	.8678	7901.0	10210.0	194.4	36.96	50.20	314.5
320.0	23.29	.8302	8299.0	10710.0	196.0	37.72	50.39	321.2
330.0	22.35	.7965	8703.0	11210.0	197.5	38.55	50.76	327.5
340.0	21.49	.7660	9113.0	11720.0	199.0	39.43	51.25	333.6
350.0	20.71	.7381	9529.0	12240.0	200.5	40.34	51.83	339.4
360.0	19.99	.7125	9954.0	12760.0	202.0	41.27	52.49	345.0
380.0	18.71	.6670	10830.0	13820.0	204.9	43.17	53.95	355.8
400.0	17.61	.6277	11730.0	14920.0	207.7	45.08	55.53	365.9
				2.50 MPa isobar				
104.30[a]	657.1	23.42	−12430.0	−12330.0	84.58	40.20	75.32	1744.0
110.0	648.1	23.10	−12030.0	−11920.0	88.40	42.55	70.23	1717.0
120.0	634.6	22.62	−11340.0	−11230.0	94.38	43.65	67.76	1659.0
140.0	609.6	21.73	−10010.0	−9893.0	104.7	41.58	66.47	1534.0
160.0	583.4	20.80	−8681.0	−8561.0	113.6	39.18	66.93	1396.0
180.0	555.1	19.79	−7337.0	−7211.0	121.5	37.67	68.22	1250.0
200.0	524.4	18.69	−5961.0	−5827.0	128.8	37.00	70.36	1097.0
210.0	507.9	18.10	−5253.0	−5115.0	132.3	36.93	72.07	1016.0
220.0	490.2	17.47	−4526.0	−4383.0	135.7	37.02	74.51	931.9
230.0	470.9	16.79	−3770.0	−3621.0	139.1	37.30	78.15	841.7
235.0	460.4	16.41	−3377.0	−3224.0	140.8	37.51	80.65	793.8
240.0	449.1	16.01	−2969.0	−2813.0	142.5	37.79	83.84	743.3
245.0	436.7	15.57	−2545.0	−2384.0	144.3	38.15	88.06	689.5
250.0	423.0	15.08	−2096.0	−1930.0	146.1	38.62	93.93	631.5
252.704[b]	414.4	14.77	−1833.0	−1664.0	147.2	38.95	98.54	595.7
252.704[b]	48.87	1.742	5147.0	6582.0	179.8	38.41	79.78	246.2
255.0	47.41	1.690	5279.0	6758.0	180.5	37.95	75.25	250.2
260.0	44.70	1.593	5546.0	7116.0	181.9	37.22	68.30	258.1
265.0	42.49	1.515	5794.0	7445.0	183.2	36.74	63.76	265.1
270.0	40.63	1.448	6029.0	7755.0	184.3	36.45	60.59	271.3
275.0	39.02	1.391	6255.0	8052.0	185.4	36.30	58.30	277.0
280.0	37.60	1.340	6474.0	8339.0	186.4	36.26	56.60	282.3
290.0	35.19	1.254	6900.0	8893.0	188.4	36.41	54.37	291.9
300.0	33.18	1.183	7316.0	9430.0	190.2	36.81	53.13	300.6
310.0	31.46	1.121	7729.0	9958.0	191.9	37.37	52.51	308.5
320.0	29.97	1.068	8141.0	10480.0	193.6	38.05	52.29	315.9
330.0	28.65	1.021	8556.0	11000.0	195.2	38.81	52.35	322.8
340.0	27.47	.9790	8976.0	11530.0	196.8	39.64	52.61	329.4
350.0	26.40	.9410	9400.0	12060.0	198.3	40.51	53.02	335.6
360.0	25.43	.9064	9832.0	12590.0	199.8	41.41	53.54	341.6
380.0	23.72	.8456	10720.0	13670.0	202.7	43.28	54.79	353.0
400.0	22.26	.7936	11630.0	14780.0	205.6	45.17	56.23	363.6
				3.00 MPa isobar				
104.40[a]	657.3	23.43	−12440.0	−12310.0	84.57	40.34	75.51	1741.0
110.0	648.4	23.11	−12030.0	−11900.0	88.35	42.62	70.33	1716.0
120.0	634.9	22.63	−11350.0	−11210.0	94.34	43.70	67.77	1660.0
140.0	609.9	21.74	−10010.0	−9877.0	104.7	41.61	66.42	1536.0
160.0	583.8	20.81	−8690.0	−8546.0	113.5	39.20	66.86	1399.0
180.0	555.7	19.81	−7349.0	−7197.0	121.5	37.69	68.11	1254.0
200.0	525.2	18.72	−5977.0	−5816.0	128.7	37.02	70.18	1102.0
210.0	508.8	18.14	−5272.0	−5107.0	132.2	36.95	71.81	1023.0
220.0	491.4	17.51	−4549.0	−4378.0	135.6	37.04	74.14	939.6
230.0	472.4	16.84	−3799.0	−3620.0	139.0	37.30	77.56	851.2
240.0	451.0	16.08	−3007.0	−2821.0	142.4	37.76	82.80	755.4
245.0	439.1	15.65	−2589.0	−2398.0	144.1	38.09	86.59	703.6
250.0	425.9	15.18	−2150.0	−1953.0	145.9	38.52	91.70	648.2
255.0	411.0	14.65	−1682.0	−1477.0	147.8	39.08	99.10	587.8

Thermophysical properties of ethylene—Continued

T K	Density kg/m³	Density mol/dm³	E J/mol	H J/mol	S J/(mol·K)	C_v J/(mol·K)	C_p J/(mol·K)	Sound m/s
256.0	407.7	14.53	–1583.0	–1377.0	148.2	39.22	101.0	574.9
258.0	400.7	14.28	–1381.0	–1171.0	149.0	39.52	105.5	548.1
259.947[b]	393.3	14.02	–1171.0	–957.3	149.8	39.88	111.3	519.6
259.947[b]	61.41	2.189	5058.0	6429.0	178.2	40.27	97.68	240.1
260.0	61.41	2.189	5060.0	6431.0	178.2	40.26	97.61	240.2
262.0	59.35	2.116	5200.0	6618.0	179.0	39.70	89.97	244.5
264.0	57.56	2.052	5330.0	6792.0	179.6	39.24	84.25	248.4
266.0	55.98	1.996	5452.0	6956.0	180.2	38.85	79.79	252.0
270.0	53.28	1.899	5681.0	7261.0	181.4	38.26	73.25	258.5
275.0	50.52	1.801	5946.0	7612.0	182.7	37.76	67.78	265.8
280.0	48.21	1.718	6196.0	7941.0	183.8	37.46	64.03	272.2
285.0	46.22	1.648	6434.0	8254.0	185.0	37.31	61.35	278.2
290.0	44.49	1.586	6664.0	8556.0	186.0	37.27	59.37	283.6
295.0	42.95	1.531	6889.0	8849.0	187.0	37.31	57.89	288.7
300.0	41.56	1.481	7110.0	9135.0	188.0	37.43	56.77	293.5
310.0	39.14	1.395	7545.0	9695.0	189.8	37.84	55.29	302.4
320.0	37.09	1.322	7974.0	10240.0	191.5	38.41	54.51	310.5
330.0	35.30	1.258	8403.0	10790.0	193.2	39.10	54.17	318.1
340.0	33.73	1.202	8833.0	11330.0	194.8	39.87	54.15	325.1
350.0	32.33	1.153	9267.0	11870.0	196.4	40.70	54.33	331.8
360.0	31.07	1.108	9707.0	12420.0	197.9	41.57	54.68	338.2
370.0	29.93	1.067	10150.0	12960.0	199.4	42.47	55.14	344.3
380.0	28.88	1.029	10600.0	13520.0	200.9	43.39	55.69	350.2
400.0	27.02	.9633	11530.0	14640.0	203.8	45.26	56.96	361.4
				3.50 MPa isobar				
104.50[a]	657.4	23.43	–12440.0	–12290.0	84.56	40.48	75.70	1738.0
110.0	648.6	23.12	–12040.0	–11890.0	88.30	42.69	70.44	1715.0
120.0	635.2	22.64	–11350.0	–11200.0	94.29	43.74	67.78	1661.0
140.0	610.3	21.75	–10020.0	–9860.0	104.6	41.64	66.38	1538.0
160.0	584.3	20.83	–8699.0	–8531.0	113.5	39.23	66.78	1402.0
180.0	556.3	19.83	–7360.0	–7184.0	121.4	37.72	68.00	1258.0
200.0	526.0	18.75	–5992.0	–5806.0	128.7	37.04	70.00	1108.0
210.0	509.8	18.17	–5291.0	–5098.0	132.1	36.96	71.57	1029.0
220.0	492.5	17.55	–4572.0	–4372.0	135.5	37.05	73.79	947.1
230.0	473.8	16.89	–3827.0	–3619.0	138.8	37.29	77.00	860.4
240.0	452.9	16.14	–3044.0	–2827.0	142.2	37.73	81.86	767.0
245.0	441.3	15.73	–2632.0	–2409.0	143.9	38.04	85.28	717.0
250.0	428.7	15.28	–2201.0	–1972.0	145.7	38.44	89.79	663.9
255.0	414.5	14.77	–1745.0	–1509.0	147.5	38.94	96.06	606.8
256.0	411.4	14.66	–1650.0	–1412.0	147.9	39.07	97.63	594.7
258.0	404.9	14.43	–1455.0	–1213.0	148.7	39.33	101.2	569.8
260.0	398.0	14.19	–1253.0	–1006.0	149.5	39.63	105.6	543.7
262.0	390.5	13.92	–1041.0	–789.8	150.3	39.98	111.1	516.0
264.0	382.2	13.63	–817.4	–560.6	151.2	40.40	118.4	486.4
266.0	372.9	13.29	–577.4	–314.1	152.1	40.91	128.7	454.0
266.2	371.9	13.26	–552.2	–288.2	152.2	40.96	130.0	450.5
266.3	371.4	13.24	–539.5	–275.2	152.3	40.99	130.7	448.8
266.345[b]	371.1	13.23	–533.5	–268.9	152.3	41.01	131.0	447.9
266.345[b]	76.07	2.711	4907.0	6197.0	176.6	42.25	126.7	233.9
270.0	70.42	2.510	5213.0	6607.0	178.1	40.92	101.7	242.9
272.0	68.03	2.425	5359.0	6803.0	178.8	40.39	93.72	247.2
274.0	65.95	2.351	5495.0	6984.0	179.5	39.95	87.71	251.0
276.0	64.12	2.286	5623.0	7154.0	180.1	39.58	83.02	254.6
280.0	61.00	2.174	5862.0	7472.0	181.2	39.03	76.14	261.1
285.0	57.80	2.060	6138.0	7837.0	182.5	38.57	70.37	268.4
290.0	55.13	1.965	6397.0	8178.0	183.7	38.31	66.42	274.8
295.0	52.85	1.884	6645.0	8503.0	184.8	38.18	63.58	280.8
300.0	50.85	1.813	6884.0	8815.0	185.9	38.17	61.49	286.2
305.0	49.08	1.749	7118.0	9119.0	186.9	38.24	59.91	291.4
310.0	47.48	1.693	7347.0	9415.0	187.9	38.38	58.72	296.2
320.0	44.71	1.594	7798.0	9994.0	189.7	38.81	57.14	305.1

Thermophysical properties of ethylene—Continued

T K	Density kg/m³	Density mol/dm³	E J/mol	H J/mol	S J/(mol·K)	C_v J/(mol·K)	C_p J/(mol·K)	Sound m/s
330.0	42.36	1.510	8242.0	10560.0	191.4	39.41	56.27	313.3
340.0	40.32	1.437	8686.0	11120.0	193.1	40.11	55.86	321.0
350.0	38.53	1.373	9130.0	11680.0	194.7	40.89	55.78	328.1
360.0	36.93	1.316	9578.0	12240.0	196.3	41.73	55.92	334.9
370.0	35.49	1.265	10030.0	12800.0	197.8	42.60	56.22	341.4
380.0	34.19	1.219	10490.0	13360.0	199.3	43.50	56.64	347.5
400.0	31.90	1.137	11430.0	14500.0	202.3	45.35	57.72	359.2
				4.00 MPa isobar				
104.50[a]	657.6	23.44	−12440.0	−12270.0	84.56	40.62	75.89	1736.0
110.0	648.9	23.13	−12040.0	−11870.0	88.25	42.76	70.55	1714.0
120.0	635.5	22.65	−11360.0	−11180.0	94.25	43.79	67.79	1661.0
140.0	610.6	21.77	−10030.0	−9844.0	104.6	41.67	66.33	1540.0
160.0	584.7	20.84	−8708.0	−8516.0	113.4	39.25	66.71	1406.0
180.0	556.9	19.85	−7372.0	−7171.0	121.3	37.74	67.89	1262.0
200.0	526.8	18.78	−6008.0	−5795.0	128.6	37.06	69.83	1113.0
210.0	510.7	18.20	−5309.0	−5090.0	132.0	36.98	71.33	1035.0
220.0	493.6	17.59	−4594.0	−4366.0	135.4	37.06	73.45	954.4
230.0	475.1	16.94	−3854.0	−3618.0	138.7	37.30	76.49	869.3
240.0	454.7	16.21	−3079.0	−2832.0	142.1	37.71	80.99	778.2
245.0	443.5	15.81	−2672.0	−2419.0	143.8	38.00	84.11	729.7
250.0	431.2	15.37	−2249.0	−1989.0	145.5	38.37	88.14	678.7
255.0	417.7	14.89	−1804.0	−1536.0	147.3	38.83	93.55	624.3
260.0	402.3	14.34	−1329.0	−1050.0	149.2	39.44	101.3	565.3
262.0	395.4	14.09	−1127.0	−843.1	150.0	39.74	105.6	539.9
264.0	387.9	13.83	−916.1	−626.8	150.8	40.08	110.9	513.2
266.0	379.8	13.54	−694.0	−398.5	151.7	40.48	117.7	484.7
268.0	370.6	13.21	−456.9	−154.2	152.6	40.97	127.2	453.9
270.0	360.1	12.84	−198.2	113.4	153.6	41.58	141.5	419.7
271.0	354.0	12.62	−57.1	259.9	154.1	41.96	152.0	400.8
272.081[b]	346.4	12.35	112.4	436.4	154.8	42.46	169.1	377.5
272.081[b]	94.16	3.356	4670.0	5861.0	174.7	44.44	183.2	227.3
275.0	86.02	3.066	5008.0	6313.0	176.4	42.82	131.8	236.7
276.0	83.98	2.994	5104.0	6440.0	176.8	42.41	122.6	239.3
278.0	80.52	2.870	5277.0	6671.0	177.6	41.74	109.2	244.1
280.0	77.63	2.767	5434.0	6879.0	178.4	41.19	99.94	248.4
282.0	75.15	2.679	5579.0	7072.0	179.1	40.75	93.05	252.3
284.0	72.98	2.601	5715.0	7253.0	179.7	40.38	87.72	255.9
286.0	71.04	2.532	5844.0	7424.0	180.3	40.07	83.46	259.3
288.0	69.28	2.470	5967.0	7587.0	180.9	39.82	79.99	262.5
290.0	67.69	2.413	6086.0	7744.0	181.4	39.61	77.09	265.5
295.0	64.23	2.289	6368.0	8115.0	182.7	39.24	71.63	272.5
300.0	61.32	2.186	6633.0	8463.0	183.9	39.04	67.82	278.8
305.0	58.83	2.097	6887.0	8795.0	185.0	38.97	65.06	284.6
310.0	56.63	2.019	7133.0	9115.0	186.0	38.99	63.01	290.0
315.0	54.68	1.949	7374.0	9426.0	187.0	39.10	61.46	295.1
320.0	52.92	1.887	7610.0	9730.0	188.0	39.27	60.27	299.8
330.0	49.86	1.777	8074.0	10320.0	189.8	39.75	58.68	308.7
340.0	47.26	1.685	8532.0	10910.0	191.5	40.38	57.80	316.9
350.0	45.01	1.604	8988.0	11480.0	193.2	41.10	57.37	324.5
360.0	43.02	1.534	9446.0	12050.0	194.8	41.90	57.26	331.7
370.0	41.25	1.471	9907.0	12630.0	196.4	42.74	57.38	338.5
380.0	39.66	1.414	10370.0	13200.0	197.9	43.62	57.65	345.0
390.0	38.21	1.362	10840.0	13780.0	199.4	44.52	58.04	351.2
400.0	36.89	1.315	11320.0	14360.0	200.9	45.44	58.53	357.1
				4.50 MPa isobar				
104.60[a]	657.8	23.45	−12440.0	−12250.0	84.55	40.75	76.08	1733.0
110.0	649.2	23.14	−12050.0	−11850.0	88.20	42.83	70.66	1714.0
120.0	635.8	22.66	−11360.0	−11160.0	94.20	43.83	67.81	1662.0

Thermophysical properties of ethylene—Continued

T K	Density kg/m^3	Density mol/dm^3	E J/mol	H J/mol	S J/(mol·K)	C_v J/(mol·K)	C_p J/(mol·K)	Sound m/s
140.0	610.9	21.78	-10030.0	-9828.0	104.5	41.69	66.29	1543.0
160.0	585.1	20.86	-8716.0	-8500.0	113.4	39.28	66.63	1409.0
180.0	557.4	19.87	-7384.0	-7157.0	121.3	37.77	67.79	1266.0
200.0	527.5	18.80	-6024.0	-5784.0	128.5	37.08	69.66	1118.0
210.0	511.6	18.23	-5328.0	-5081.0	131.9	37.00	71.11	1041.0
220.0	494.7	17.63	-4615.0	-4360.0	135.3	37.07	73.13	961.6
230.0	476.5	16.98	-3880.0	-3615.0	138.6	37.30	76.00	877.9
240.0	456.5	16.27	-3112.0	-2836.0	141.9	37.69	80.20	788.9
245.0	445.5	15.88	-2711.0	-2428.0	143.6	37.97	83.06	741.9
250.0	433.6	15.46	-2295.0	-2004.0	145.3	38.31	86.68	692.6
255.0	420.7	14.99	-1859.0	-1559.0	147.1	38.73	91.42	640.5
260.0	406.1	14.48	-1398.0	-1087.0	148.9	39.28	97.96	584.8
262.0	399.7	14.25	-1204.0	-887.7	149.7	39.54	101.4	561.2
264.0	392.8	14.00	-1002.0	-681.0	150.5	39.83	105.4	536.7
266.0	385.5	13.74	-792.8	-465.3	151.3	40.17	110.5	510.9
268.0	377.4	13.45	-572.7	-238.2	152.1	40.56	116.9	483.6
270.0	368.5	13.14	-338.8	3.8	153.0	41.03	125.5	454.4
272.0	358.3	12.77	-85.9	266.4	154.0	41.60	138.0	422.5
273.0	352.6	12.57	50.5	408.5	154.5	41.94	146.7	405.1
274.0	346.2	12.34	196.1	560.8	155.1	42.34	158.4	386.5
275.0	339.0	12.08	354.5	726.9	155.7	42.81	174.9	366.1
276.0	330.5	11.78	531.7	913.7	156.4	43.39	201.0	343.3
276.5	325.6	11.61	631.2	1019.0	156.7	43.74	221.1	330.5
277.0	319.8	11.40	741.7	1136.0	157.2	44.17	250.7	316.4
277.274[b]	315.3	11.24	823.2	1224.0	157.5	44.52	282.7	305.4
277.274[b]	118.7	4.232	4292.0	5356.0	172.4	46.98	342.1	220.3
280.0	103.7	3.696	4782.0	5999.0	174.7	44.62	176.9	231.9
281.0	100.4	3.577	4908.0	6166.0	175.3	44.06	157.1	235.1
282.0	97.53	3.476	5021.0	6315.0	175.8	43.59	142.9	238.0
284.0	92.88	3.311	5221.0	6580.0	176.8	42.81	123.7	243.1
286.0	89.12	3.177	5398.0	6814.0	177.6	42.19	111.0	247.6
288.0	85.96	3.064	5558.0	7027.0	178.3	41.69	102.0	251.7
290.0	83.24	2.967	5707.0	7224.0	179.0	41.28	95.23	255.5
292.0	80.84	2.882	5847.0	7409.0	179.6	40.94	89.92	259.0
294.0	78.70	2.805	5980.0	7584.0	180.2	40.66	85.65	262.3
296.0	76.76	2.736	6107.0	7752.0	180.8	40.43	82.14	265.4
300.0	73.36	2.615	6348.0	8069.0	181.9	40.08	76.72	271.2
305.0	69.80	2.488	6631.0	8440.0	183.1	39.82	71.91	277.8
310.0	66.77	2.380	6900.0	8790.0	184.2	39.71	68.48	283.8
315.0	64.14	2.286	7158.0	9126.0	185.3	39.70	65.96	289.4
320.0	61.82	2.204	7409.0	9451.0	186.3	39.78	64.05	294.7
325.0	59.74	2.130	7654.0	9767.0	187.3	39.92	62.60	299.6
330.0	57.87	2.063	7896.0	10080.0	188.3	40.13	61.49	304.3
340.0	54.59	1.946	8372.0	10680.0	190.1	40.66	59.98	313.0
350.0	51.79	1.846	8842.0	11280.0	191.8	41.32	59.13	321.1
360.0	49.36	1.759	9311.0	11870.0	193.4	42.07	58.72	328.6
370.0	47.21	1.683	9781.0	12450.0	195.1	42.88	58.62	335.8
380.0	45.30	1.615	10250.0	13040.0	196.6	43.74	58.72	342.5
390.0	43.57	1.553	10730.0	13630.0	198.1	44.62	58.98	349.0
400.0	42.00	1.497	11220.0	14220.0	199.6	45.52	59.36	355.2

5.00 MPa isobar

T K	Density kg/m^3	Density mol/dm^3	E J/mol	H J/mol	S J/(mol·K)	C_v J/(mol·K)	C_p J/(mol·K)	Sound m/s
104.70[a]	658.0	23.45	-12440.0	-12230.0	84.54	40.88	76.27	1731.0
110.0	649.5	23.15	-12050.0	-11840.0	88.16	42.90	70.77	1713.0
120.0	636.0	22.67	-11370.0	-11150.0	94.16	43.88	67.82	1663.0
140.0	611.3	21.79	-10040.0	-9811.0	104.5	41.72	66.25	1545.0
160.0	585.6	20.87	-8725.0	-8485.0	113.3	39.30	66.56	1412.0
180.0	558.0	19.89	-7395.0	-7144.0	121.2	37.79	67.69	1270.0
200.0	528.3	18.83	-6039.0	-5773.0	128.4	37.11	69.50	1123.0
210.0	512.4	18.27	-5346.0	-5072.0	131.9	37.02	70.89	1047.0
220.0	495.7	17.67	-4637.0	-4354.0	135.2	37.09	72.82	968.6
230.0	477.8	17.03	-3906.0	-3613.0	138.5	37.30	75.54	886.4

Thermophysical properties of ethylene—Continued

T K	Density kg/m³	Density mol/dm³	E J/mol	H J/mol	S J/(mol·K)	C_v J/(mol·K)	C_p J/(mol·K)	Sound m/s
240.0	458.2	16.33	−3145.0	−2839.0	141.8	37.68	79.47	799.3
245.0	447.4	15.95	−2749.0	−2435.0	143.4	37.94	82.10	753.5
250.0	435.9	15.54	−2339.0	−2017.0	145.1	38.26	85.38	705.8
255.0	423.4	15.09	−1911.0	−1580.0	146.9	38.65	89.58	655.8
260.0	409.6	14.60	−1461.0	−1119.0	148.7	39.15	95.18	602.9
265.0	393.8	14.04	−980.3	−624.2	150.5	39.78	103.2	546.0
266.0	390.4	13.92	−879.3	−520.0	150.9	39.93	105.2	534.0
268.0	383.1	13.66	−671.2	−305.0	151.7	40.26	110.0	509.2
270.0	375.2	13.37	−453.3	−79.4	152.6	40.64	115.9	483.1
272.0	366.5	13.06	−222.9	159.8	153.5	41.08	123.7	455.4
274.0	356.7	12.71	24.2	417.4	154.4	41.61	134.5	425.6
276.0	345.3	12.31	295.5	701.7	155.4	42.27	151.1	392.8
277.0	338.7	12.07	444.5	858.6	156.0	42.68	163.3	374.8
278.0	331.3	11.81	606.6	1030.0	156.6	43.16	180.6	355.3
279.0	322.5	11.50	788.0	1223.0	157.3	43.75	207.3	333.7
279.5	317.4	11.32	889.4	1331.0	157.7	44.10	227.3	321.8
280.0	311.6	11.11	1001.0	1452.0	158.1	44.52	255.6	308.8
280.5	304.7	10.86	1129.0	1589.0	158.6	45.02	299.6	294.5
280.6	303.2	10.81	1157.0	1620.0	158.7	45.14	311.6	291.5
280.8	299.8	10.69	1217.0	1685.0	159.0	45.39	340.6	285.0
281.0	296.1	10.55	1283.0	1757.0	159.2	45.67	379.4	278.1
281.2	291.8	10.40	1357.0	1838.0	159.5	46.00	434.4	270.7
281.4	286.7	10.22	1443.0	1933.0	159.8	46.39	519.3	262.4
281.6	280.2	9.989	1549.0	2050.0	160.3	46.88	671.8	253.0
281.7	276.1	9.843	1615.0	2123.0	160.5	47.20	808.2	247.6
281.8	271.0	9.659	1697.0	2215.0	160.8	47.59	1048.0	241.5
281.9	263.7	9.400	1811.0	2343.0	161.3	48.12	1616.0	233.9
281.985[b]	266.4	9.498	1774.0	2301.0	161.2	47.90	1290.0	237.1
281.985[b]	161.9	5.771	3561.0	4428.0	168.7	49.94	1579.0	213.6
285.0	122.7	4.372	4560.0	5703.0	173.2	46.07	232.8	229.8
286.0	117.5	4.190	4723.0	5916.0	173.9	45.39	196.4	233.3
287.0	113.4	4.043	4863.0	6100.0	174.6	44.82	172.7	236.4
288.0	109.9	3.919	4988.0	6264.0	175.2	44.34	155.9	239.3
289.0	107.0	3.813	5102.0	6414.0	175.7	43.91	143.3	241.9
290.0	104.3	3.719	5207.0	6552.0	176.2	43.54	133.4	244.3
292.0	99.84	3.559	5398.0	6803.0	177.0	42.91	118.8	248.8
294.0	96.11	3.426	5570.0	7030.0	177.8	42.40	108.5	252.9
296.0	92.91	3.312	5729.0	7238.0	178.5	41.98	100.8	256.7
298.0	90.10	3.212	5877.0	7434.0	179.2	41.63	94.78	260.2
300.0	87.61	3.123	6017.0	7618.0	179.8	41.35	89.99	263.5
302.0	85.37	3.043	6151.0	7794.0	180.4	41.11	86.07	266.7
305.0	82.38	2.936	6342.0	8045.0	181.2	40.83	81.37	271.1
310.0	78.14	2.785	6642.0	8437.0	182.5	40.52	75.64	277.8
315.0	74.58	2.658	6924.0	8804.0	183.6	40.38	71.61	284.0
320.0	71.51	2.549	7193.0	9155.0	184.7	40.35	68.66	289.7
325.0	68.83	2.453	7454.0	9492.0	185.8	40.41	66.44	295.0
330.0	66.44	2.368	7709.0	9820.0	186.8	40.54	64.75	300.0
335.0	64.29	2.292	7958.0	10140.0	187.7	40.73	63.45	304.8
340.0	62.33	2.222	8205.0	10450.0	188.7	40.97	62.44	309.3
350.0	58.90	2.099	8690.0	11070.0	190.5	41.56	61.08	317.9
360.0	55.95	1.994	9171.0	11680.0	192.2	42.26	60.31	325.8
370.0	53.38	1.903	9652.0	12280.0	193.8	43.03	59.94	333.2
380.0	51.10	1.822	10130.0	12880.0	195.4	43.86	59.85	340.3
390.0	49.07	1.749	10620.0	13480.0	197.0	44.72	59.97	347.0
400.0	47.23	1.683	11110.0	14080.0	198.5	45.61	60.23	353.4

Thermophysical properties of ethylene—Continued

T K	Density kg/m^3	Density mol/dm^3	E J/mol	H J/mol	S J/(mol·K)	C_v J/(mol·K)	C_p J/(mol·K)	Sound m/s
			5.03 MPa isobar					
104.70[a]	658.0	23.46	−12440.0	−12220.0	84.54	40.89	76.28	1731.0
110.0	649.5	23.15	−12050.0	−11840.0	88.15	42.91	70.78	1713.0
120.0	636.1	22.67	−11370.0	−11150.0	94.16	43.88	67.82	1663.0
140.0	611.3	21.79	−10040.0	−9810.0	104.5	41.72	66.25	1545.0
160.0	585.6	20.87	−8725.0	−8484.0	113.3	39.30	66.56	1412.0
180.0	558.0	19.89	−7396.0	−7143.0	121.2	37.79	67.68	1271.0
200.0	528.3	18.83	−6040.0	−5773.0	128.4	37.11	69.49	1124.0
210.0	512.5	18.27	−5347.0	−5071.0	131.8	37.02	70.88	1048.0
220.0	495.8	17.67	−4638.0	−4353.0	135.2	37.09	72.80	969.1
230.0	477.9	17.03	−3908.0	−3613.0	138.5	37.30	75.51	886.9
240.0	458.3	16.33	−3147.0	−2839.0	141.8	37.68	79.43	799.9
245.0	447.6	15.95	−2751.0	−2436.0	143.4	37.94	82.05	754.2
250.0	436.1	15.54	−2341.0	−2018.0	145.1	38.26	85.31	706.6
255.0	423.6	15.10	−1914.0	−1581.0	146.9	38.65	89.48	656.7
260.0	409.8	14.61	−1465.0	−1120.0	148.6	39.14	95.03	603.9
265.0	394.1	14.05	−985.0	−626.9	150.5	39.77	102.9	547.2
266.0	390.7	13.93	−884.2	−523.0	150.9	39.92	104.9	535.3
268.0	383.4	13.67	−676.7	−308.6	151.7	40.24	109.6	510.6
270.0	375.6	13.39	−459.6	−83.8	152.6	40.62	115.4	484.7
272.0	366.9	13.08	−230.3	154.3	153.4	41.05	123.1	457.2
274.0	357.2	12.73	15.3	410.3	154.4	41.58	133.6	427.7
276.0	346.0	12.33	284.3	692.1	155.4	42.23	149.4	395.3
277.0	339.5	12.10	431.5	847.1	156.0	42.62	161.1	377.5
278.0	332.3	11.84	591.0	1016.0	156.6	43.08	177.2	358.4
279.0	323.8	11.54	768.4	1204.0	157.2	43.65	201.7	337.3
279.5	318.9	11.37	866.8	1309.0	157.6	43.99	219.5	325.7
280.0	313.4	11.17	974.5	1425.0	158.0	44.38	244.1	313.3
280.5	306.9	10.94	1096.0	1555.0	158.5	44.84	280.7	299.7
281.0	299.1	10.66	1238.0	1710.0	159.0	45.42	342.1	284.3
281.2	295.3	10.53	1304.0	1782.0	159.3	45.70	380.8	277.5
281.4	291.0	10.37	1378.0	1863.0	159.6	46.03	435.3	270.1
281.6	285.9	10.19	1464.0	1958.0	159.9	46.42	518.7	262.0
281.8	279.5	9.963	1570.0	2075.0	160.3	46.91	665.5	252.8
281.9	275.5	9.819	1635.0	2147.0	160.6	47.21	792.9	247.6
282.0	270.5	9.642	1714.0	2236.0	160.9	47.59	1007.0	241.7
282.1	263.7	9.400	1821.0	2356.0	161.3	48.08	1460.0	234.7
282.2	251.9	8.979	2004.0	2565.0	162.1	48.90	3280.0	225.2
282.251[b]	253.0	9.019	1990.0	2547.0	162.0	48.81	2844.0	226.2
282.251[b]	175.4	6.251	3306.0	4111.0	167.6	50.51	3494.0	212.1
285.0	125.6	4.479	4492.0	5615.0	172.9	46.37	251.5	228.7
285.5	122.6	4.371	4583.0	5734.0	173.3	45.99	226.9	230.6
286.0	120.0	4.277	4667.0	5843.0	173.7	45.64	208.0	232.3
287.0	115.5	4.118	4815.0	6036.0	174.3	45.04	180.7	235.5
288.0	111.8	3.986	4945.0	6207.0	174.9	44.53	161.8	238.4
289.0	108.7	3.873	5063.0	6361.0	175.5	44.09	147.9	241.1
290.0	105.9	3.774	5171.0	6503.0	176.0	43.70	137.1	243.6
292.0	101.2	3.607	5366.0	6761.0	176.8	43.05	121.3	248.2
294.0	97.32	3.469	5542.0	6992.0	177.6	42.52	110.4	252.4
296.0	94.01	3.351	5703.0	7204.0	178.4	42.08	102.3	256.2
298.0	91.12	3.248	5854.0	7402.0	179.0	41.73	95.99	259.7
300.0	88.56	3.157	5996.0	7589.0	179.6	41.43	91.00	263.1
302.0	86.26	3.075	6131.0	7767.0	180.2	41.18	86.92	266.2
305.0	83.20	2.966	6324.0	8020.0	181.1	40.89	82.05	270.7
310.0	78.87	2.811	6625.0	8415.0	182.4	40.58	76.14	277.5
315.0	75.24	2.682	6909.0	8784.0	183.5	40.42	71.99	283.7
320.0	72.12	2.571	7180.0	9136.0	184.6	40.38	68.96	289.4
325.0	69.40	2.474	7442.0	9475.0	185.7	40.44	66.69	294.7
330.0	66.97	2.387	7697.0	9804.0	186.7	40.56	64.96	299.8
335.0	64.79	2.309	7947.0	10130.0	187.7	40.75	63.63	304.6
340.0	62.81	2.239	8194.0	10440.0	188.6	40.99	62.60	309.1

Thermophysical properties of ethylene—Continued

T K	Density kg/m^3	Density mol/dm^3	E J/mol	H J/mol	S J/(mol·K)	C_v J/(mol·K)	C_p J/(mol·K)	Sound m/s
350.0	59.33	2.115	8681.0	11060.0	190.4	41.57	61.20	317.7
360.0	56.36	2.009	9163.0	11670.0	192.1	42.27	60.41	325.6
370.0	53.76	1.916	9644.0	12270.0	193.8	43.04	60.02	333.1
380.0	51.46	1.834	10130.0	12870.0	195.4	43.87	59.92	340.1
390.0	49.40	1.761	10610.0	13470.0	196.9	44.73	60.03	346.8
400.0	47.54	1.695	11100.0	14070.0	198.4	45.62	60.28	353.3
				5.10 MPa isobar				
104.70[a]	658.0	23.46	−12440.0	−12220.0	84.54	40.91	76.31	1731.0
110.0	649.5	23.15	−12050.0	−11830.0	88.15	42.91	70.79	1713.0
120.0	636.1	22.67	−11370.0	−11140.0	94.15	43.89	67.82	1663.0
140.0	611.3	21.79	−10040.0	−9808.0	104.5	41.73	66.24	1545.0
160.0	585.7	20.88	−8726.0	−8482.0	113.3	39.31	66.55	1413.0
180.0	558.1	19.89	−7397.0	−7141.0	121.2	37.80	67.67	1271.0
200.0	528.4	18.84	−6042.0	−5771.0	128.4	37.11	69.47	1124.0
210.0	512.6	18.27	−5349.0	−5070.0	131.8	37.03	70.85	1049.0
220.0	495.9	17.68	−4641.0	−4352.0	135.2	37.09	72.76	970.0
230.0	478.0	17.04	−3911.0	−3612.0	138.5	37.30	75.45	888.0
240.0	458.5	16.34	−3152.0	−2839.0	141.8	37.68	79.33	801.3
245.0	447.8	15.96	−2756.0	−2437.0	143.4	37.94	81.92	755.8
250.0	436.4	15.56	−2347.0	−2019.0	145.1	38.25	85.14	708.4
255.0	424.0	15.11	−1921.0	−1584.0	146.8	38.64	89.25	658.7
260.0	410.2	14.62	−1473.0	−1125.0	148.6	39.12	94.69	606.3
265.0	394.7	14.07	−995.7	−633.2	150.5	39.74	102.4	550.1
266.0	391.3	13.95	−895.5	−529.9	150.9	39.89	104.3	538.3
268.0	384.1	13.69	−689.3	−316.9	151.7	40.21	108.8	513.9
270.0	376.4	13.42	−474.0	−93.9	152.5	40.57	114.4	488.3
272.0	367.9	13.11	−247.1	141.8	153.4	40.99	121.6	461.3
274.0	358.4	12.78	−4.8	394.3	154.3	41.50	131.5	432.4
276.0	347.5	12.39	259.0	670.7	155.3	42.12	145.9	400.8
277.0	341.3	12.17	402.4	821.5	155.8	42.49	156.3	383.6
278.0	334.4	11.92	556.6	984.4	156.4	42.92	170.2	365.3
279.0	326.5	11.64	725.8	1164.0	157.1	43.44	190.5	345.2
280.0	317.0	11.30	918.0	1369.0	157.8	44.08	223.0	322.9
280.5	311.4	11.10	1027.0	1487.0	158.2	44.48	248.6	310.5
281.0	304.8	10.87	1151.0	1620.0	158.7	44.95	286.6	297.0
281.5	296.8	10.58	1296.0	1778.0	159.3	45.54	350.5	281.8
281.6	294.9	10.51	1329.0	1814.0	159.4	45.68	369.0	278.4
281.8	290.8	10.37	1400.0	1892.0	159.7	45.99	416.2	271.5
282.0	286.0	10.19	1481.0	1982.0	160.0	46.35	484.5	263.9
282.2	280.2	9.988	1578.0	2088.0	160.4	46.79	593.7	255.5
282.3	276.7	9.864	1635.0	2152.0	160.6	47.05	677.8	250.9
282.4	272.7	9.719	1700.0	2225.0	160.8	47.35	799.2	245.9
282.5	267.7	9.543	1779.0	2314.0	161.2	47.71	991.3	240.5
282.6	261.3	9.314	1881.0	2429.0	161.6	48.16	1342.0	234.3
282.7	251.9	8.978	2029.0	2597.0	162.2	48.80	2174.0	226.9
282.8	233.3	8.316	2322.0	2936.0	163.4	49.85	5410.0	217.7
282.9	198.0	7.058	2918.0	3640.0	165.9	50.74	5932.0	212.3
283.0	180.5	6.433	3245.0	4038.0	167.3	50.48	2705.0	213.1
283.1	171.8	6.123	3419.0	4252.0	168.0	50.16	1720.0	214.1
283.2	166.0	5.918	3538.0	4399.0	168.5	49.87	1282.0	215.1
283.3	161.7	5.765	3630.0	4514.0	168.9	49.61	1035.0	216.0
283.4	158.3	5.642	3705.0	4609.0	169.3	49.39	876.5	216.8
283.6	152.9	5.450	3828.0	4763.0	169.8	48.99	683.0	218.3
283.8	148.7	5.302	3926.0	4888.0	170.3	48.64	568.3	219.6
284.0	145.3	5.180	4009.0	4993.0	170.6	48.33	491.8	220.8
284.2	142.4	5.076	4081.0	5086.0	171.0	48.06	436.8	221.9
284.5	138.7	4.945	4176.0	5207.0	171.4	47.69	377.7	223.5
285.0	133.8	4.768	4309.0	5379.0	172.0	47.15	313.7	225.8
285.5	129.8	4.625	4422.0	5525.0	172.5	46.69	272.1	227.9
286.0	126.4	4.505	4521.0	5653.0	172.9	46.28	242.7	229.8

Thermophysical properties of ethylene—Continued

T K	Density kg/m^3	Density mol/dm^3	E J/mol	H J/mol	S J/(mol·K)	C_v J/(mol·K)	C_p J/(mol·K)	Sound m/s
287.0	120.9	4.310	4691.0	5874.0	173.7	45.59	203.4	233.3
288.0	116.5	4.154	4837.0	6064.0	174.4	45.02	178.0	236.4
289.0	112.9	4.023	4965.0	6233.0	175.0	44.52	160.1	239.3
290.0	109.7	3.911	5082.0	6386.0	175.5	44.09	146.7	241.9
292.0	104.5	3.725	5290.0	6659.0	176.4	43.38	127.9	246.7
294.0	100.2	3.573	5474.0	6902.0	177.3	42.80	115.2	251.0
296.0	96.65	3.445	5642.0	7123.0	178.0	42.33	106.0	254.9
298.0	93.55	3.335	5798.0	7327.0	178.7	41.95	98.98	258.6
300.0	90.81	3.237	5944.0	7519.0	179.3	41.63	93.46	262.0
302.0	88.37	3.150	6083.0	7702.0	179.9	41.36	88.99	265.2
304.0	86.16	3.071	6215.0	7876.0	180.5	41.14	85.31	268.3
310.0	80.59	2.873	6587.0	8362.0	182.1	40.70	77.34	276.6
315.0	76.80	2.738	6874.0	8737.0	183.3	40.52	72.91	282.9
320.0	73.56	2.622	7148.0	9093.0	184.4	40.47	69.70	288.7
325.0	70.73	2.521	7413.0	9435.0	185.5	40.51	67.29	294.1
330.0	68.22	2.432	7670.0	9767.0	186.5	40.63	65.47	299.2
335.0	65.97	2.352	7922.0	10090.0	187.5	40.80	64.06	304.0
340.0	63.93	2.279	8170.0	10410.0	188.4	41.03	62.97	308.6
350.0	60.36	2.152	8659.0	11030.0	190.2	41.61	61.49	317.2
360.0	57.30	2.043	9143.0	11640.0	191.9	42.30	60.64	325.2
370.0	54.64	1.948	9625.0	12240.0	193.6	43.06	60.22	332.7
380.0	52.29	1.864	10110.0	12850.0	195.2	43.89	60.09	339.8
390.0	50.18	1.789	10600.0	13450.0	196.8	44.74	60.17	346.6
400.0	48.29	1.721	11090.0	14050.0	198.3	45.63	60.40	353.0
				5.20 MPa isobar				
104.70[a]	658.1	23.46	−12440.0	−12220.0	84.54	40.93	76.35	1730.0
110.0	649.6	23.15	−12060.0	−11830.0	88.14	42.93	70.82	1713.0
120.0	636.2	22.68	−11370.0	−11140.0	94.14	43.90	67.83	1663.0
140.0	611.4	21.79	−10040.0	−9805.0	104.4	41.73	66.23	1546.0
160.0	585.7	20.88	−8728.0	−8479.0	113.3	39.31	66.53	1413.0
180.0	558.2	19.90	−7400.0	−7138.0	121.2	37.80	67.65	1272.0
200.0	528.6	18.84	−6045.0	−5769.0	128.4	37.12	69.44	1125.0
210.0	512.8	18.28	−5353.0	−5068.0	131.8	37.03	70.80	1050.0
220.0	496.1	17.68	−4645.0	−4351.0	135.2	37.09	72.70	971.4
230.0	478.3	17.05	−3917.0	−3612.0	138.4	37.30	75.37	889.7
240.0	458.8	16.35	−3158.0	−2840.0	141.7	37.68	79.19	803.3
245.0	448.2	15.98	−2763.0	−2438.0	143.4	37.93	81.74	758.0
250.0	436.8	15.57	−2355.0	−2022.0	145.1	38.24	84.90	710.9
255.0	424.5	15.13	−1931.0	−1587.0	146.8	38.62	88.92	661.6
260.0	410.9	14.65	−1485.0	−1130.0	148.6	39.10	94.21	609.7
265.0	395.5	14.10	−1011.0	−641.9	150.4	39.70	101.6	554.2
266.0	392.2	13.98	−911.3	−539.4	150.8	39.85	103.5	542.5
268.0	385.2	13.73	−707.1	−328.3	151.6	40.16	107.7	518.5
270.0	377.6	13.46	−494.1	−107.7	152.4	40.51	113.0	493.4
272.0	369.3	13.16	−270.3	124.7	153.3	40.92	119.7	467.0
274.0	360.1	12.84	−32.5	372.6	154.2	41.39	128.7	438.9
276.0	349.6	12.46	224.7	642.0	155.2	41.98	141.5	408.4
277.0	343.7	12.25	363.3	787.7	155.7	42.32	150.4	392.0
278.0	337.2	12.02	511.1	943.6	156.3	42.72	162.0	374.5
279.0	329.9	11.76	671.0	1113.0	156.9	43.18	178.0	355.8
280.0	321.5	11.46	848.4	1302.0	157.5	43.74	201.9	335.3
280.5	316.6	11.29	946.5	1407.0	157.9	44.07	219.0	324.1
281.0	311.1	11.09	1053.0	1522.0	158.3	44.44	242.0	312.2
281.5	304.9	10.87	1172.0	1651.0	158.8	44.89	275.2	299.3
282.0	297.3	10.60	1310.0	1801.0	159.3	45.43	327.5	285.1
282.2	293.8	10.47	1372.0	1869.0	159.6	45.69	358.2	278.9
282.4	289.9	10.33	1441.0	1944.0	159.8	45.98	398.4	272.4
282.6	285.4	10.17	1518.0	2029.0	160.1	46.31	453.7	265.4
282.8	280.1	9.986	1607.0	2128.0	160.5	46.69	534.5	257.9
283.0	273.7	9.756	1713.0	2246.0	160.9	47.16	663.7	249.7
283.1	269.8	9.618	1777.0	2317.0	161.1	47.44	762.0	245.2

Thermophysical properties of ethylene—Continued

T K	Density kg/m^3	Density mol/dm^3	*E* J/mol	*H* J/mol	*S* J/(mol·K)	C_v J/(mol·K)	C_p J/(mol·K)	Sound m/s
283.2	265.3	9.455	1850.0	2400.0	161.4	47.76	900.4	240.5
283.3	259.8	9.260	1938.0	2500.0	161.8	48.14	1106.0	235.4
283.4	252.9	9.013	2048.0	2625.0	162.2	48.59	1427.0	230.1
283.5	243.8	8.690	2193.0	2791.0	162.8	49.13	1931.0	224.5
283.6	231.7	8.261	2387.0	3016.0	163.6	49.74	2573.0	219.5
283.7	217.4	7.750	2625.0	3296.0	164.6	50.25	2933.0	216.0
283.8	203.5	7.253	2867.0	3584.0	165.6	50.49	2745.0	214.6
283.9	192.1	6.846	3075.0	3835.0	166.5	50.47	2254.0	214.4
284.0	183.4	6.538	3240.0	4036.0	167.2	50.31	1778.0	214.9
284.1	176.9	6.304	3370.0	4195.0	167.8	50.10	1426.0	215.5
284.2	171.7	6.121	3475.0	4324.0	168.2	49.89	1179.0	216.2
284.3	167.5	5.972	3562.0	4433.0	168.6	49.68	1004.0	216.9
284.4	164.0	5.848	3637.0	4527.0	168.9	49.49	875.2	217.6
284.6	158.4	5.647	3762.0	4683.0	169.5	49.13	701.1	218.9
284.8	154.0	5.489	3864.0	4811.0	169.9	48.80	589.7	220.1
285.0	150.3	5.358	3950.0	4921.0	170.3	48.51	512.5	221.2
285.2	147.2	5.247	4026.0	5017.0	170.6	48.24	455.7	222.3
285.5	143.2	5.105	4126.0	5144.0	171.1	47.87	394.0	223.8
286.0	137.9	4.915	4265.0	5323.0	171.7	47.34	326.4	226.0
286.5	133.6	4.763	4383.0	5475.0	172.2	46.88	282.3	228.1
287.0	130.0	4.635	4486.0	5608.0	172.7	46.47	251.1	230.0
287.5	126.9	4.525	4578.0	5727.0	173.1	46.10	227.8	231.8
288.0	124.2	4.428	4662.0	5836.0	173.5	45.77	209.5	233.5
289.0	119.6	4.263	4812.0	6031.0	174.2	45.19	182.8	236.6
290.0	115.8	4.126	4944.0	6204.0	174.8	44.69	164.0	239.4
291.0	112.5	4.008	5064.0	6361.0	175.3	44.26	150.0	242.0
292.0	109.6	3.905	5174.0	6505.0	175.8	43.88	139.1	244.5
294.0	104.7	3.731	5373.0	6766.0	176.7	43.23	123.1	249.0
296.0	100.6	3.587	5551.0	7001.0	177.5	42.71	112.0	253.1
298.0	97.17	3.464	5715.0	7216.0	178.2	42.28	103.7	256.9
300.0	94.16	3.356	5868.0	7417.0	178.9	41.92	97.29	260.4
302.0	91.49	3.261	6012.0	7606.0	179.5	41.63	92.19	263.8
304.0	89.10	3.176	6149.0	7786.0	180.1	41.38	88.02	266.9
310.0	83.10	2.962	6530.0	8286.0	181.7	40.89	79.14	275.5
315.0	79.08	2.819	6824.0	8669.0	183.0	40.67	74.28	281.9
320.0	75.65	2.697	7102.0	9031.0	184.1	40.59	70.78	287.8
325.0	72.67	2.590	7370.0	9378.0	185.2	40.61	68.18	293.3
330.0	70.04	2.497	7631.0	9714.0	186.2	40.71	66.21	298.4
335.0	67.68	2.413	7885.0	10040.0	187.2	40.88	64.69	303.3
340.0	65.55	2.337	8136.0	10360.0	188.1	41.10	63.52	307.9
350.0	61.83	2.204	8628.0	10990.0	190.0	41.66	61.91	316.6
360.0	58.66	2.091	9115.0	11600.0	191.7	42.33	60.98	324.7
370.0	55.91	1.993	9599.0	12210.0	193.4	43.09	60.50	332.3
380.0	53.47	1.906	10080.0	12810.0	195.0	43.91	60.32	339.4
390.0	51.30	1.829	10570.0	13420.0	196.5	44.77	60.37	346.2
400.0	49.35	1.759	11060.0	14020.0	198.1	45.65	60.58	352.7
				5.30 MPa isobar				
104.70[a]	658.1	23.46	−12440.0	−12210.0	84.54	40.96	76.39	1730.0
110.0	649.6	23.16	−12060.0	−11830.0	88.13	42.94	70.84	1712.0
120.0	636.2	22.68	−11370.0	−11140.0	94.13	43.90	67.83	1664.0
140.0	611.5	21.80	−10040.0	−9801.0	104.4	41.74	66.22	1546.0
160.0	585.8	20.88	−8730.0	−8476.0	113.3	39.32	66.52	1414.0
180.0	558.3	19.90	−7402.0	−7135.0	121.2	37.81	67.63	1273.0
200.0	528.7	18.85	−6048.0	−5767.0	128.4	37.12	69.41	1126.0
210.0	513.0	18.28	−5356.0	−5066.0	131.8	37.03	70.76	1051.0
220.0	496.3	17.69	−4649.0	−4350.0	135.1	37.10	72.64	972.8
230.0	478.6	17.06	−3922.0	−3611.0	138.4	37.31	75.28	891.3
240.0	459.1	16.37	−3164.0	−2840.0	141.7	37.67	79.06	805.3
245.0	448.6	15.99	−2771.0	−2439.0	143.4	37.93	81.57	760.2
250.0	437.3	15.59	−2364.0	−2024.0	145.0	38.23	84.67	713.4

Thermophysical properties of ethylene—Continued

T K	Density kg/m^3	Density mol/dm^3	E J/mol	H J/mol	S J/(mol·K)	C_v J/(mol·K)	C_p J/(mol·K)	Sound m/s
255.0	425.0	15.15	–1941.0	–1591.0	146.7	38.61	88.60	664.5
260.0	411.5	14.67	–1497.0	–1136.0	148.5	39.08	93.74	613.0
265.0	396.4	14.13	–1025.0	–650.3	150.4	39.67	100.9	558.2
266.0	393.1	14.01	–926.8	–548.6	150.7	39.81	102.6	546.7
268.0	386.1	13.76	–724.4	–339.3	151.5	40.11	106.7	523.0
270.0	378.7	13.50	–513.7	–121.1	152.3	40.45	111.7	498.4
272.0	370.6	13.21	–292.9	108.3	153.2	40.84	118.0	472.5
274.0	361.7	12.89	–59.0	352.1	154.1	41.30	126.2	445.1
276.0	351.6	12.53	192.3	615.2	155.0	41.85	137.7	415.6
278.0	339.8	12.11	469.2	906.7	156.1	42.53	155.2	383.2
279.0	333.0	11.87	621.7	1068.0	156.7	42.95	168.3	365.5
280.0	325.3	11.60	788.1	1245.0	157.3	43.44	186.7	346.4
281.0	316.2	11.27	974.7	1445.0	158.0	44.05	215.0	325.4
281.5	310.9	11.08	1079.0	1557.0	158.4	44.41	235.9	313.9
282.0	304.9	10.87	1194.0	1682.0	158.9	44.83	265.0	301.6
282.5	297.8	10.62	1325.0	1825.0	159.4	45.33	308.6	288.3
283.0	289.1	10.30	1481.0	1996.0	160.0	45.96	381.4	273.5
283.2	284.9	10.15	1554.0	2076.0	160.2	46.26	426.4	267.0
283.4	280.0	9.983	1636.0	2167.0	160.6	46.61	487.8	260.2
283.6	274.4	9.781	1731.0	2273.0	160.9	47.01	575.9	253.0
283.8	267.5	9.535	1845.0	2401.0	161.4	47.49	710.5	245.2
284.0	258.7	9.221	1988.0	2563.0	162.0	48.08	928.9	237.0
284.1	253.3	9.027	2076.0	2663.0	162.3	48.42	1086.0	232.8
284.2	246.9	8.800	2179.0	2781.0	162.7	48.80	1280.0	228.7
284.3	239.5	8.537	2299.0	2920.0	163.2	49.20	1494.0	224.8
284.4	231.2	8.240	2436.0	3079.0	163.8	49.58	1683.0	221.6
284.5	222.3	7.923	2585.0	3254.0	164.4	49.91	1796.0	219.2
284.6	213.4	7.605	2738.0	3435.0	165.0	50.14	1811.0	217.6
284.7	204.9	7.304	2888.0	3613.0	165.7	50.26	1741.0	216.8
284.8	197.2	7.031	3027.0	3781.0	166.2	50.27	1610.0	216.5
284.9	190.5	6.791	3154.0	3934.0	166.8	50.21	1448.0	216.5
285.0	184.7	6.584	3266.0	4071.0	167.3	50.09	1284.0	216.8
285.1	179.8	6.407	3364.0	4192.0	167.7	49.95	1136.0	217.2
285.2	175.5	6.255	3451.0	4299.0	168.1	49.80	1009.0	217.7
285.4	168.5	6.007	3598.0	4480.0	168.7	49.48	815.5	218.8
285.6	163.0	5.811	3717.0	4629.0	169.2	49.17	681.9	219.8
285.8	158.6	5.652	3818.0	4755.0	169.7	48.88	586.8	220.9
286.0	154.8	5.518	3905.0	4865.0	170.0	48.60	516.6	221.9
286.2	151.6	5.403	3982.0	4963.0	170.4	48.35	462.9	222.9
286.5	147.4	5.255	4084.0	5092.0	170.8	48.00	402.8	224.3
287.0	141.8	5.056	4227.0	5276.0	171.5	47.48	335.0	226.5
287.5	137.3	4.895	4348.0	5431.0	172.0	47.03	290.0	228.5
288.0	133.6	4.761	4454.0	5568.0	172.5	46.62	257.8	230.4
288.5	130.3	4.645	4549.0	5690.0	172.9	46.26	233.6	232.1
289.0	127.5	4.543	4636.0	5802.0	173.3	45.93	214.7	233.8
290.0	122.6	4.370	4789.0	6002.0	174.0	45.35	186.9	236.8
291.0	118.6	4.227	4925.0	6179.0	174.6	44.85	167.4	239.6
292.0	115.1	4.104	5047.0	6339.0	175.2	44.41	152.9	242.3
293.0	112.1	3.997	5159.0	6486.0	175.7	44.03	141.6	244.7
294.0	109.4	3.901	5264.0	6622.0	176.1	43.68	132.5	247.0
296.0	104.8	3.737	5455.0	6873.0	177.0	43.10	118.9	251.3
298.0	101.0	3.600	5628.0	7100.0	177.7	42.62	109.0	255.3
300.0	97.66	3.481	5788.0	7311.0	178.4	42.23	101.6	258.9
302.0	94.74	3.377	5938.0	7508.0	179.1	41.90	95.69	262.3
304.0	92.14	3.284	6080.0	7694.0	179.7	41.63	90.97	265.5
306.0	89.79	3.201	6216.0	7872.0	180.3	41.41	87.08	268.6
310.0	85.69	3.054	6473.0	8208.0	181.4	41.07	81.06	274.3
315.0	81.40	2.902	6772.0	8599.0	182.6	40.82	75.73	280.8
320.0	77.77	2.772	7056.0	8968.0	183.8	40.72	71.91	286.8
325.0	74.64	2.660	7328.0	9320.0	184.9	40.72	69.10	292.4
330.0	71.88	2.562	7591.0	9660.0	185.9	40.80	66.97	297.6
335.0	69.41	2.474	7848.0	9990.0	186.9	40.96	65.34	302.6
340.0	67.19	2.395	8101.0	10310.0	187.9	41.16	64.07	307.3

Thermophysical properties of ethylene—Continued

T K	Density kg/m³	Density mol/dm³	E J/mol	H J/mol	S J/(mol·K)	C_v J/(mol·K)	C_p J/(mol·K)	Sound m/s
350.0	63.32	2.257	8597.0	10940.0	189.7	41.71	62.34	316.0
360.0	60.03	2.140	9086.0	11560.0	191.5	42.37	61.32	324.2
370.0	57.18	2.038	9573.0	12170.0	193.1	43.12	60.78	331.8
380.0	54.67	1.949	10060.0	12780.0	194.7	43.93	60.56	339.0
390.0	52.43	1.869	10550.0	13380.0	196.3	44.79	60.58	345.8
400.0	50.42	1.797	11040.0	13990.0	197.9	45.67	60.76	352.4
				5.40 MPa isobar				
104.70[a]	658.1	23.46	–12440.0	–12210.0	84.54	40.98	76.43	1729.0
110.0	649.7	23.16	–12060.0	–11820.0	88.12	42.96	70.86	1712.0
120.0	636.3	22.68	–11370.0	–11130.0	94.13	43.91	67.83	1664.0
140.0	611.5	21.80	–10050.0	–9798.0	104.4	41.74	66.21	1547.0
160.0	585.9	20.89	–8732.0	–8473.0	113.3	39.32	66.51	1415.0
180.0	558.4	19.91	–7404.0	–7133.0	121.2	37.81	67.61	1274.0
200.0	528.9	18.85	–6051.0	–5765.0	128.4	37.12	69.37	1128.0
210.0	513.1	18.29	–5360.0	–5064.0	131.8	37.04	70.72	1052.0
220.0	496.5	17.70	–4653.0	–4348.0	135.1	37.10	72.58	974.2
230.0	478.8	17.07	–3927.0	–3610.0	138.4	37.31	75.19	892.9
240.0	459.5	16.38	–3171.0	–2841.0	141.7	37.67	78.92	807.3
245.0	448.9	16.00	–2778.0	–2440.0	143.3	37.92	81.39	762.4
250.0	437.7	15.60	–2372.0	–2026.0	145.0	38.23	84.44	715.9
255.0	425.5	15.17	–1950.0	–1594.0	146.7	38.60	88.29	667.4
260.0	412.2	14.69	–1509.0	–1141.0	148.5	39.06	93.29	616.3
265.0	397.2	14.16	–1040.0	–658.5	150.3	39.64	100.2	562.1
266.0	393.9	14.04	–942.0	–557.5	150.7	39.77	101.9	550.7
268.0	387.1	13.80	–741.3	–349.9	151.5	40.06	105.8	527.4
270.0	379.8	13.54	–532.7	–133.8	152.3	40.39	110.5	503.2
272.0	371.9	13.26	–314.6	92.7	153.1	40.77	116.3	477.9
274.0	363.2	12.95	–84.5	332.6	154.0	41.21	123.9	451.1
276.0	353.4	12.60	161.6	590.2	154.9	41.73	134.3	422.5
278.0	342.2	12.20	430.3	873.0	155.9	42.37	149.6	391.3
279.0	335.8	11.97	576.7	1028.0	156.5	42.75	160.5	374.5
280.0	328.7	11.72	734.4	1195.0	157.1	43.19	175.3	356.5
281.0	320.5	11.42	907.8	1381.0	157.8	43.72	196.6	337.0
282.0	310.7	11.07	1105.0	1592.0	158.5	44.37	230.2	315.6
282.5	304.9	10.87	1217.0	1714.0	158.9	44.77	255.9	303.9
283.0	298.2	10.63	1342.0	1850.0	159.4	45.24	292.7	291.3
283.5	290.2	10.34	1488.0	2010.0	160.0	45.80	349.9	277.6
284.0	279.9	9.978	1666.0	2207.0	160.7	46.52	449.9	262.5
284.2	274.9	9.798	1752.0	2303.0	161.0	46.87	513.3	256.0
284.4	269.0	9.588	1851.0	2414.0	161.4	47.28	600.1	249.2
284.6	262.0	9.337	1967.0	2545.0	161.9	47.74	721.2	242.3
284.8	253.4	9.033	2108.0	2705.0	162.4	48.27	886.7	235.3
285.0	243.0	8.661	2279.0	2902.0	163.1	48.86	1086.0	228.9
285.2	230.8	8.227	2481.0	3138.0	163.9	49.42	1256.0	223.7
285.4	217.9	7.766	2702.0	3398.0	164.9	49.84	1322.0	220.5
285.6	205.5	7.326	2922.0	3659.0	165.8	50.04	1277.0	218.9
285.8	194.6	6.937	3125.0	3903.0	166.6	50.03	1155.0	218.5
286.0	185.5	6.611	3302.0	4119.0	167.4	49.87	1001.0	218.8
286.2	178.0	6.344	3453.0	4304.0	168.0	49.65	855.7	219.4
286.4	171.8	6.126	3581.0	4463.0	168.6	49.39	735.2	220.2
286.6	166.8	5.944	3692.0	4600.0	169.1	49.13	639.9	221.1
286.8	162.4	5.790	3788.0	4720.0	169.5	48.88	565.3	222.0
287.0	158.7	5.658	3873.0	4827.0	169.9	48.64	506.3	222.9
287.2	155.5	5.542	3949.0	4924.0	170.2	48.40	459.0	223.8
287.5	151.3	5.392	4051.0	5053.0	170.6	48.07	403.8	225.1
288.0	145.5	5.187	4196.0	5237.0	171.3	47.58	339.0	227.1
288.5	140.8	5.021	4319.0	5395.0	171.8	47.14	294.7	229.1
289.0	136.9	4.881	4427.0	5534.0	172.3	46.75	262.5	230.9
289.5	133.5	4.760	4524.0	5659.0	172.7	46.39	238.1	232.6
290.0	130.6	4.654	4613.0	5773.0	173.1	46.06	218.8	234.2

Thermophysical properties of ethylene—Continued

T K	Density kg/m³	Density mol/dm³	E J/mol	H J/mol	S J/(mol·K)	C_v J/(mol·K)	C_p J/(mol·K)	Sound m/s
291.0	125.5	4.474	4770.0	5977.0	173.8	45.48	190.4	237.2
292.0	121.3	4.325	4908.0	6156.0	174.5	44.98	170.4	240.0
293.0	117.8	4.198	5033.0	6319.0	175.0	44.55	155.4	242.6
294.0	114.6	4.086	5147.0	6468.0	175.5	44.16	143.8	245.0
296.0	109.4	3.899	5353.0	6738.0	176.4	43.51	126.9	249.5
298.0	105.0	3.744	5537.0	6979.0	177.2	42.98	115.1	253.6
300.0	101.3	3.612	5705.0	7200.0	178.0	42.55	106.3	257.4
302.0	98.13	3.498	5862.0	7406.0	178.7	42.19	99.55	260.9
304.0	95.30	3.397	6010.0	7599.0	179.3	41.89	94.17	264.2
306.0	92.75	3.306	6150.0	7783.0	179.9	41.64	89.80	267.3
308.0	90.45	3.224	6284.0	7959.0	180.5	41.43	86.16	270.3
310.0	88.35	3.149	6414.0	8128.0	181.0	41.27	83.11	273.2
315.0	83.78	2.986	6720.0	8528.0	182.3	40.98	77.24	279.8
320.0	79.94	2.850	7008.0	8903.0	183.5	40.85	73.10	285.9
325.0	76.64	2.732	7284.0	9261.0	184.6	40.83	70.05	291.6
330.0	73.74	2.629	7551.0	9605.0	185.6	40.90	67.76	296.9
335.0	71.17	2.537	7811.0	9939.0	186.7	41.03	66.00	301.8
340.0	68.85	2.454	8066.0	10270.0	187.6	41.23	64.64	306.6
350.0	64.83	2.311	8565.0	10900.0	189.5	41.76	62.77	315.5
360.0	61.42	2.189	9057.0	11520.0	191.2	42.41	61.67	323.7
370.0	58.47	2.084	9546.0	12140.0	192.9	43.16	61.07	331.3
380.0	55.87	1.992	10040.0	12750.0	194.5	43.96	60.80	338.6
390.0	53.57	1.909	10530.0	13350.0	196.1	44.81	60.79	345.5
400.0	51.49	1.836	11020.0	13960.0	197.6	45.68	60.94	352.0
				5.50 MPa isobar				
104.80[a]	658.2	23.46	−12440.0	−12200.0	84.53	41.01	76.47	1729.0
110.0	649.8	23.16	−12060.0	−11820.0	88.11	42.97	70.89	1712.0
120.0	636.3	22.68	−11370.0	−11130.0	94.12	43.92	67.83	1664.0
140.0	611.6	21.80	−10050.0	−9795.0	104.4	41.75	66.21	1547.0
160.0	586.0	20.89	−8733.0	−8470.0	113.3	39.33	66.49	1415.0
180.0	558.5	19.91	−7406.0	−7130.0	121.2	37.82	67.59	1274.0
200.0	529.0	18.86	−6054.0	−5762.0	128.4	37.13	69.34	1129.0
210.0	513.3	18.30	−5363.0	−5063.0	131.8	37.04	70.68	1053.0
220.0	496.7	17.71	−4658.0	−4347.0	135.1	37.10	72.53	975.5
230.0	479.1	17.08	−3932.0	−3610.0	138.4	37.31	75.11	894.5
240.0	459.8	16.39	−3177.0	−2841.0	141.6	37.67	78.79	809.3
250.0	438.1	15.62	−2380.0	−2028.0	145.0	38.22	84.22	718.4
255.0	426.1	15.19	−1960.0	−1598.0	146.7	38.59	87.98	670.2
260.0	412.8	14.71	−1520.0	−1146.0	148.4	39.04	92.86	619.6
265.0	398.0	14.19	−1054.0	−666.4	150.2	39.60	99.50	565.9
266.0	394.8	14.07	−956.9	−566.1	150.6	39.73	101.1	554.7
268.0	388.0	13.83	−757.8	−360.2	151.4	40.02	104.9	531.8
270.0	380.9	13.58	−551.3	−146.2	152.2	40.34	109.3	508.0
272.0	373.1	13.30	−335.8	77.8	153.0	40.70	114.8	483.1
274.0	364.6	13.00	−109.0	314.1	153.9	41.12	121.8	456.9
276.0	355.2	12.66	132.3	566.7	154.8	41.62	131.3	429.1
278.0	344.5	12.28	393.9	841.8	155.8	42.21	144.8	399.1
279.0	338.4	12.06	535.1	991.1	156.3	42.57	154.1	382.9
280.0	331.7	11.82	685.8	1151.0	156.9	42.97	166.3	365.9
281.0	324.2	11.56	849.2	1325.0	157.5	43.44	183.0	347.6
282.0	315.5	11.24	1030.0	1520.0	158.2	44.01	207.6	327.9
282.5	310.5	11.07	1130.0	1627.0	158.6	44.34	224.9	317.3
283.0	304.9	10.87	1239.0	1745.0	159.0	44.72	247.7	306.1
283.5	298.5	10.64	1360.0	1877.0	159.5	45.15	279.2	294.1
284.0	291.1	10.37	1497.0	2027.0	160.0	45.67	325.3	281.4
284.5	281.9	10.05	1659.0	2206.0	160.6	46.30	398.6	267.6
285.0	270.0	9.626	1863.0	2434.0	161.4	47.10	526.3	252.8
285.2	264.1	9.414	1963.0	2547.0	161.8	47.49	603.3	246.7
285.4	257.3	9.170	2077.0	2677.0	162.3	47.91	699.4	240.6
285.6	249.3	8.887	2209.0	2828.0	162.8	48.37	811.2	234.9
286.0	230.5	8.216	2526.0	3195.0	164.1	49.26	1006.0	225.9

Thermophysical properties of ethylene—Continued

T K	Density kg/m^3	Density mol/dm^3	E J/mol	H J/mol	S J/(mol·K)	C_v J/(mol·K)	C_p J/(mol·K)	Sound m/s
286.2	220.4	7.856	2700.0	3401.0	164.8	49.59	1042.0	223.1
286.4	210.5	7.505	2876.0	3609.0	165.5	49.78	1032.0	221.5
286.6	201.4	7.178	3044.0	3811.0	166.3	49.84	982.4	220.7
286.8	193.1	6.885	3201.0	4000.0	166.9	49.79	907.1	220.6
287.0	185.9	6.628	3343.0	4173.0	167.5	49.66	820.3	220.8
287.2	179.7	6.406	3470.0	4328.0	168.1	49.48	734.4	221.3
287.4	174.3	6.215	3582.0	4467.0	168.5	49.27	656.8	221.9
287.6	169.7	6.049	3682.0	4592.0	169.0	49.05	589.9	222.6
287.8	165.7	5.905	3772.0	4704.0	169.4	48.83	533.4	223.4
288.0	162.1	5.777	3854.0	4806.0	169.7	48.62	486.2	224.1
288.5	154.7	5.515	4028.0	5025.0	170.5	48.10	398.1	226.1
289.0	148.9	5.308	4172.0	5208.0	171.1	47.64	338.7	228.0
289.5	144.1	5.138	4296.0	5367.0	171.7	47.22	296.4	229.8
290.0	140.1	4.995	4405.0	5507.0	172.1	46.84	265.1	231.5
290.5	136.6	4.871	4504.0	5633.0	172.6	46.49	240.9	233.2
291.0	133.6	4.761	4593.0	5748.0	173.0	46.17	221.7	234.7
292.0	128.4	4.575	4753.0	5955.0	173.7	45.60	193.1	237.7
293.0	124.0	4.421	4893.0	6137.0	174.3	45.10	172.8	240.4
294.0	120.3	4.289	5020.0	6302.0	174.9	44.67	157.6	243.0
295.0	117.1	4.174	5136.0	6454.0	175.4	44.28	145.8	245.4
296.0	114.2	4.072	5244.0	6595.0	175.9	43.94	136.3	247.7
298.0	109.3	3.897	5441.0	6852.0	176.7	43.36	122.0	251.9
300.0	105.2	3.750	5619.0	7085.0	177.5	42.88	111.6	255.8
302.0	101.7	3.624	5783.0	7300.0	178.2	42.48	103.8	259.5
304.0	98.58	3.514	5936.0	7502.0	178.9	42.15	97.68	262.9
306.0	95.83	3.416	6082.0	7692.0	179.5	41.88	92.74	266.1
308.0	93.35	3.327	6220.0	7873.0	180.1	41.65	88.68	269.1
310.0	91.09	3.247	6353.0	8047.0	180.7	41.46	85.29	272.0
315.0	86.22	3.073	6667.0	8456.0	182.0	41.14	78.84	278.8
320.0	82.15	2.928	6960.0	8839.0	183.2	40.98	74.33	285.0
325.0	78.67	2.804	7240.0	9202.0	184.3	40.94	71.04	290.7
330.0	75.64	2.696	7510.0	9550.0	185.4	40.99	68.57	296.1
335.0	72.94	2.600	7773.0	9888.0	186.4	41.11	66.68	301.1
340.0	70.53	2.514	8030.0	10220.0	187.4	41.30	65.23	305.9
350.0	66.34	2.365	8534.0	10860.0	189.2	41.81	63.22	314.9
360.0	62.81	2.239	9028.0	11480.0	191.0	42.45	62.02	323.2
370.0	59.76	2.130	9519.0	12100.0	192.7	43.19	61.36	330.9
380.0	57.08	2.035	10010.0	12710.0	194.3	43.99	61.05	338.2
390.0	54.71	1.950	10500.0	13320.0	195.9	44.83	61.00	345.1
400.0	52.57	1.874	11000.0	13930.0	197.4	45.70	61.13	351.7
			5.60 MPa isobar					
104.80[a]	658.2	23.46	−12440.0	−12200.0	84.53	41.04	76.50	1728.0
110.0	649.8	23.16	−12060.0	−11820.0	88.10	42.98	70.91	1712.0
120.0	636.4	22.68	−11370.0	−11130.0	94.11	43.93	67.84	1664.0
140.0	611.7	21.80	−10050.0	−9792.0	104.4	41.75	66.20	1548.0
160.0	586.1	20.89	−8735.0	−8467.0	113.3	39.33	66.48	1416.0
180.0	558.6	19.91	−7409.0	−7127.0	121.1	37.82	67.57	1275.0
200.0	529.1	18.86	−6057.0	−5760.0	128.3	37.13	69.31	1130.0
210.0	513.5	18.30	−5367.0	−5061.0	131.8	37.04	70.64	1054.0
220.0	497.0	17.71	−4662.0	−4346.0	135.1	37.11	72.47	976.9
230.0	479.3	17.09	−3937.0	−3609.0	138.4	37.31	75.03	896.2
240.0	460.1	16.40	−3183.0	−2842.0	141.6	37.67	78.66	811.2
250.0	438.6	15.63	−2388.0	−2030.0	144.9	38.21	84.00	720.8
255.0	426.6	15.20	−1969.0	−1601.0	146.6	38.57	87.69	672.9
260.0	413.4	14.74	−1531.0	−1151.0	148.4	39.02	92.44	622.8
265.0	398.7	14.21	−1068.0	−674.1	150.2	39.57	98.86	569.7
270.0	381.9	13.61	−569.4	−158.0	152.1	40.29	108.2	512.6
272.0	374.3	13.34	−356.3	63.4	152.9	40.64	113.4	488.2
274.0	366.0	13.05	−132.7	296.5	153.8	41.04	119.9	462.6
276.0	356.9	12.72	104.3	544.6	154.7	41.51	128.5	435.5
278.0	346.5	12.35	359.6	812.9	155.7	42.08	140.6	406.4

Thermophysical properties of ethylene—Continued

T K	Density kg/m^3	Density mol/dm^3	E J/mol	H J/mol	S J/(mol·K)	C_v J/(mol·K)	C_p J/(mol·K)	Sound m/s
280.0	334.5	11.92	641.3	1111.0	156.7	42.78	159.0	374.6
281.0	327.5	11.67	796.7	1276.0	157.3	43.20	172.5	357.4
282.0	319.5	11.39	966.2	1458.0	158.0	43.70	191.5	338.9
283.0	310.2	11.06	1156.0	1662.0	158.7	44.30	219.9	318.9
284.0	298.8	10.65	1378.0	1904.0	159.5	45.07	267.4	296.9
284.5	291.8	10.40	1508.0	2046.0	160.0	45.54	305.4	284.9
285.0	283.5	10.11	1658.0	2212.0	160.6	46.10	361.7	272.2
285.5	273.2	9.738	1838.0	2413.0	161.3	46.79	449.6	258.8
285.6	270.8	9.653	1879.0	2459.0	161.5	46.95	472.7	256.0
285.8	265.6	9.469	1968.0	2559.0	161.8	47.28	525.9	250.6
286.0	259.9	9.263	2066.0	2670.0	162.2	47.63	588.5	245.2
286.2	253.4	9.032	2175.0	2795.0	162.7	48.01	658.9	240.0
286.4	246.2	8.777	2296.0	2934.0	163.2	48.40	731.4	235.3
287.0	221.9	7.911	2714.0	3422.0	164.9	49.36	863.4	225.7
287.5	202.3	7.213	3072.0	3848.0	166.3	49.63	821.5	222.8
287.6	198.8	7.087	3139.0	3929.0	166.6	49.63	800.4	222.7
287.8	192.2	6.852	3267.0	4085.0	167.2	49.56	750.9	222.6
288.0	186.3	6.640	3386.0	4229.0	167.7	49.44	695.9	222.8
288.2	181.0	6.451	3495.0	4363.0	168.1	49.29	640.5	223.2
288.4	176.3	6.283	3595.0	4486.0	168.6	49.12	588.1	223.7
288.6	172.1	6.133	3686.0	4599.0	168.9	48.93	540.4	224.2
288.8	168.3	5.999	3769.0	4702.0	169.3	48.74	498.1	224.9
289.0	164.9	5.878	3846.0	4798.0	169.6	48.55	461.0	225.5
289.5	157.8	5.623	4013.0	5009.0	170.4	48.09	387.4	227.3
290.0	152.0	5.418	4155.0	5189.0	171.0	47.66	334.6	229.0
290.5	147.2	5.247	4279.0	5346.0	171.5	47.26	295.5	230.7
291.0	143.1	5.102	4388.0	5486.0	172.0	46.90	265.6	232.3
291.5	139.6	4.975	4487.0	5613.0	172.4	46.56	242.2	233.9
292.0	136.4	4.863	4578.0	5729.0	172.8	46.25	223.4	235.4
293.0	131.1	4.673	4739.0	5937.0	173.6	45.69	195.1	238.3
294.0	126.6	4.514	4881.0	6122.0	174.2	45.20	174.8	241.0
295.0	122.8	4.378	5010.0	6289.0	174.7	44.78	159.4	243.5
296.0	119.5	4.260	5127.0	6442.0	175.3	44.39	147.5	245.9
298.0	113.9	4.060	5339.0	6718.0	176.2	43.75	129.9	250.3
300.0	109.3	3.896	5528.0	6965.0	177.0	43.22	117.6	254.3
302.0	105.4	3.757	5700.0	7191.0	177.8	42.78	108.5	258.0
304.0	102.0	3.636	5860.0	7401.0	178.5	42.42	101.5	261.5
306.0	99.01	3.529	6011.0	7598.0	179.1	42.12	95.93	264.8
308.0	96.33	3.434	6154.0	7785.0	179.7	41.87	91.39	268.0
310.0	93.91	3.348	6291.0	7964.0	180.3	41.67	87.62	270.9
315.0	88.71	3.162	6612.0	8383.0	181.6	41.31	80.53	277.8
320.0	84.41	3.009	6911.0	8773.0	182.9	41.11	75.62	284.1
325.0	80.75	2.878	7196.0	9141.0	184.0	41.05	72.06	289.9
330.0	77.56	2.765	7469.0	9495.0	185.1	41.08	69.41	295.3
335.0	74.75	2.664	7735.0	9837.0	186.1	41.19	67.39	300.5
340.0	72.23	2.575	7994.0	10170.0	187.1	41.37	65.83	305.3
350.0	67.88	2.419	8502.0	10820.0	189.0	41.86	63.67	314.3
360.0	64.21	2.289	8999.0	11450.0	190.8	42.49	62.38	322.7
370.0	61.06	2.177	9493.0	12070.0	192.5	43.22	61.65	330.5
380.0	58.30	2.078	9985.0	12680.0	194.1	44.01	61.29	337.8
390.0	55.85	1.991	10480.0	13290.0	195.7	44.85	61.21	344.8
400.0	53.66	1.913	10980.0	13900.0	197.2	45.72	61.31	351.4
				5.70 MPa isobar				
104.80[a]	658.3	23.46	−12440.0	−12200.0	84.53	41.06	76.54	1728.0
110.0	649.9	23.16	−12060.0	−11810.0	88.09	43.00	70.93	1712.0
120.0	636.4	22.69	−11380.0	−11120.0	94.10	43.94	67.84	1664.0
140.0	611.7	21.81	−10050.0	−9788.0	104.4	41.76	66.19	1548.0
160.0	586.2	20.89	−8737.0	−8464.0	113.2	39.34	66.46	1416.0
180.0	558.7	19.92	−7411.0	−7125.0	121.1	37.82	67.55	1276.0
200.0	529.3	18.87	−6060.0	−5758.0	128.3	37.14	69.28	1131.0

Thermophysical properties of ethylene—Continued

T K	Density kg/m³	Density mol/dm³	E J/mol	H J/mol	S J/(mol·K)	C_v J/(mol·K)	C_p J/(mol·K)	Sound m/s
210.0	513.6	18.31	−5370.0	−5059.0	131.7	37.05	70.60	1056.0
220.0	497.2	17.72	−4666.0	−4344.0	135.1	37.11	72.41	978.2
230.0	479.6	17.09	−3942.0	−3608.0	138.3	37.31	74.94	897.8
240.0	460.4	16.41	−3189.0	−2842.0	141.6	37.67	78.53	813.2
250.0	439.0	15.65	−2396.0	−2032.0	144.9	38.20	83.79	723.2
255.0	427.1	15.22	−1979.0	−1604.0	146.6	38.56	87.40	675.7
260.0	414.0	14.76	−1543.0	−1156.0	148.3	39.00	92.03	626.0
265.0	399.5	14.24	−1082.0	−681.5	150.1	39.54	98.24	573.4
270.0	382.9	13.65	−587.1	−169.5	152.1	40.24	107.2	517.1
272.0	375.4	13.38	−376.3	49.7	152.9	40.58	112.1	493.1
274.0	367.4	13.09	−155.6	279.7	153.7	40.97	118.2	468.0
276.0	358.5	12.78	77.5	523.6	154.6	41.42	126.1	441.6
278.0	348.5	12.42	327.2	786.0	155.5	41.95	136.9	413.4
280.0	337.0	12.01	600.1	1075.0	156.6	42.60	152.8	382.9
281.0	330.5	11.78	748.9	1233.0	157.1	42.99	164.2	366.5
282.0	323.1	11.52	909.3	1404.0	157.7	43.44	179.3	349.1
283.0	314.7	11.22	1085.0	1593.0	158.4	43.97	200.7	330.4
284.0	304.8	10.87	1285.0	1809.0	159.2	44.61	233.4	310.3
285.0	292.5	10.42	1521.0	2068.0	160.1	45.43	289.0	288.2
285.5	284.8	10.15	1661.0	2223.0	160.6	45.94	333.6	276.4
286.0	275.6	9.825	1824.0	2405.0	161.3	46.54	398.1	264.1
286.5	264.3	9.423	2021.0	2626.0	162.0	47.25	491.8	251.7
287.0	250.3	8.921	2262.0	2901.0	163.0	48.05	612.1	240.2
287.5	233.6	8.325	2550.0	3235.0	164.2	48.82	713.6	231.5
288.0	216.1	7.702	2862.0	3602.0	165.4	49.31	739.7	226.6
288.5	200.0	7.129	3163.0	3962.0	166.7	49.43	693.0	224.8
288.6	197.1	7.025	3219.0	4031.0	166.9	49.41	677.5	224.7
288.8	191.6	6.829	3328.0	4163.0	167.4	49.34	642.8	224.6
289.0	186.5	6.650	3431.0	4288.0	167.8	49.23	605.2	224.8
289.2	181.9	6.485	3526.0	4405.0	168.2	49.10	567.1	225.1
289.4	177.8	6.336	3615.0	4515.0	168.6	48.96	530.1	225.5
290.0	167.3	5.963	3847.0	4803.0	169.6	48.47	434.2	227.1
290.5	160.4	5.719	4007.0	5004.0	170.3	48.05	373.7	228.6
291.0	154.8	5.518	4146.0	5179.0	170.9	47.65	327.6	230.2
291.5	150.0	5.348	4267.0	5333.0	171.4	47.28	292.1	231.7
292.0	145.9	5.202	4376.0	5472.0	171.9	46.93	264.4	233.3
293.0	139.2	4.960	4566.0	5715.0	172.7	46.31	224.0	236.2
294.0	133.7	4.766	4728.0	5924.0	173.4	45.77	196.3	239.0
295.0	129.2	4.604	4872.0	6110.0	174.1	45.29	176.1	241.6
296.0	125.3	4.465	5001.0	6278.0	174.6	44.87	160.8	244.1
297.0	121.8	4.343	5120.0	6433.0	175.2	44.49	148.8	246.4
298.0	118.8	4.236	5231.0	6577.0	175.6	44.15	139.1	248.6
300.0	113.6	4.051	5432.0	6839.0	176.5	43.57	124.4	252.8
302.0	109.3	3.896	5614.0	7077.0	177.3	43.10	113.8	256.6
304.0	105.6	3.764	5782.0	7296.0	178.0	42.70	105.7	260.2
306.0	102.3	3.647	5938.0	7501.0	178.7	42.37	99.40	263.6
308.0	99.43	3.544	6086.0	7695.0	179.3	42.10	94.30	266.8
310.0	96.82	3.451	6227.0	7879.0	179.9	41.88	90.11	269.8
315.0	91.27	3.253	6556.0	8308.0	181.3	41.47	82.31	276.9
320.0	86.71	3.091	6862.0	8706.0	182.6	41.25	76.97	283.2
325.0	82.85	2.953	7151.0	9081.0	183.7	41.16	73.13	289.1
330.0	79.51	2.834	7428.0	9439.0	184.8	41.18	70.28	294.6
335.0	76.57	2.729	7696.0	9785.0	185.9	41.28	68.11	299.8
340.0	73.94	2.636	7958.0	10120.0	186.8	41.44	66.44	304.7
345.0	71.58	2.551	8216.0	10450.0	187.8	41.65	65.15	309.3
350.0	69.42	2.475	8469.0	10770.0	188.7	41.91	64.14	313.8
360.0	65.63	2.339	8970.0	11410.0	190.5	42.53	62.75	322.2
370.0	62.37	2.223	9466.0	12030.0	192.2	43.25	61.95	330.0
380.0	59.52	2.122	9960.0	12650.0	193.9	44.04	61.54	337.4
390.0	57.00	2.032	10460.0	13260.0	195.5	44.87	61.42	344.4
400.0	54.74	1.951	10950.0	13880.0	197.0	45.74	61.50	351.1

Thermophysical properties of ethylene—Continued

T K	Density kg/m^3	Density mol/dm^3	E J/mol	H J/mol	S J/(mol·K)	C_v J/(mol·K)	C_p J/(mol·K)	Sound m/s
				5.80 MPa isobar				
104.80[a]	658.3	23.47	−12440.0	−12190.0	84.53	41.09	76.58	1727.0
110.0	649.9	23.17	−12060.0	−11810.0	88.08	43.01	70.96	1712.0
120.0	636.5	22.69	−11380.0	−11120.0	94.09	43.95	67.84	1664.0
140.0	611.8	21.81	−10050.0	−9785.0	104.4	41.76	66.18	1549.0
160.0	586.2	20.90	−8738.0	−8461.0	113.2	39.34	66.45	1417.0
180.0	558.9	19.92	−7413.0	−7122.0	121.1	37.83	67.53	1277.0
200.0	529.4	18.87	−6063.0	−5756.0	128.3	37.14	69.25	1132.0
210.0	513.8	18.32	−5374.0	−5057.0	131.7	37.05	70.56	1057.0
220.0	497.4	17.73	−4670.0	−4343.0	135.0	37.11	72.36	979.6
230.0	479.8	17.10	−3947.0	−3607.0	138.3	37.31	74.86	899.4
240.0	460.7	16.42	−3195.0	−2842.0	141.6	37.67	78.41	815.1
250.0	439.4	15.66	−2404.0	−2034.0	144.9	38.20	83.58	725.6
255.0	427.6	15.24	−1988.0	−1608.0	146.6	38.55	87.11	678.4
260.0	414.6	14.78	−1554.0	−1161.0	148.3	38.98	91.63	629.1
265.0	400.3	14.27	−1095.0	−688.8	150.1	39.51	97.65	577.1
270.0	383.9	13.68	−604.4	−180.5	152.0	40.19	106.2	521.5
272.0	376.6	13.42	−395.7	36.4	152.8	40.52	110.9	497.9
274.0	368.7	13.14	−177.7	263.6	153.6	40.90	116.6	473.3
276.0	360.0	12.83	51.8	503.7	154.5	41.33	123.9	447.5
278.0	350.4	12.49	296.3	760.7	155.4	41.83	133.6	420.1
280.0	339.4	12.10	561.6	1041.0	156.4	42.44	147.6	390.7
281.0	333.2	11.88	705.0	1193.0	157.0	42.80	157.3	375.0
282.0	326.3	11.63	858.0	1357.0	157.5	43.21	169.7	358.5
283.0	318.7	11.36	1024.0	1534.0	158.2	43.68	186.6	340.9
284.0	309.8	11.04	1207.0	1732.0	158.9	44.24	210.7	322.2
285.0	299.2	10.67	1416.0	1960.0	159.7	44.92	247.9	302.0
286.0	285.9	10.19	1667.0	2237.0	160.6	45.79	311.3	280.2
286.5	277.6	9.895	1818.0	2404.0	161.2	46.32	360.9	268.8
287.0	267.7	9.543	1993.0	2600.0	161.9	46.94	428.4	257.4
287.5	255.8	9.120	2199.0	2835.0	162.7	47.63	513.4	246.5
288.0	241.9	8.623	2441.0	3114.0	163.7	48.32	597.0	237.5
288.5	226.8	8.084	2709.0	3426.0	164.8	48.89	644.9	231.3
289.0	211.9	7.552	2982.0	3750.0	165.9	49.19	642.7	228.0
289.5	198.3	7.069	3242.0	4062.0	167.0	49.22	600.7	226.7
290.0	186.8	6.657	3476.0	4347.0	168.0	49.03	536.4	226.8
290.5	177.2	6.316	3680.0	4598.0	168.8	48.72	468.3	227.5
291.0	169.3	6.035	3856.0	4817.0	169.6	48.36	407.9	228.7
291.5	162.8	5.803	4008.0	5008.0	170.2	47.98	358.3	230.0
292.0	157.3	5.607	4142.0	5177.0	170.8	47.62	318.7	231.4
292.5	152.6	5.440	4262.0	5328.0	171.3	47.27	287.0	232.9
293.0	148.5	5.294	4369.0	5465.0	171.8	46.95	261.5	234.3
294.0	141.7	5.052	4558.0	5706.0	172.6	46.35	223.5	237.1
295.0	136.2	4.855	4721.0	5915.0	173.3	45.82	196.7	239.8
296.0	131.6	4.690	4865.0	6102.0	174.0	45.36	177.0	242.3
297.0	127.6	4.548	4996.0	6271.0	174.5	44.95	161.8	244.7
298.0	124.1	4.424	5115.0	6426.0	175.1	44.58	149.9	247.0
300.0	118.3	4.215	5331.0	6707.0	176.0	43.94	132.2	251.3
302.0	113.4	4.043	5524.0	6958.0	176.8	43.42	119.7	255.3
304.0	109.3	3.897	5700.0	7188.0	177.6	42.99	110.4	258.9
306.0	105.8	3.770	5863.0	7401.0	178.3	42.63	103.2	262.4
308.0	102.6	3.658	6016.0	7602.0	178.9	42.34	97.44	265.7
310.0	99.82	3.558	6162.0	7792.0	179.6	42.09	92.77	268.8
312.0	97.29	3.468	6301.0	7973.0	180.1	41.88	88.90	271.7
315.0	93.89	3.347	6500.0	8233.0	181.0	41.64	84.19	275.9
320.0	89.06	3.175	6811.0	8638.0	182.2	41.39	78.38	282.4
325.0	85.00	3.030	7105.0	9019.0	183.4	41.28	74.24	288.3
330.0	81.49	2.905	7386.0	9382.0	184.5	41.28	71.17	293.9
335.0	78.42	2.795	7657.0	9732.0	185.6	41.36	68.85	299.1
340.0	75.68	2.698	7922.0	10070.0	186.6	41.51	67.07	304.1
345.0	73.22	2.610	8181.0	10400.0	187.6	41.71	65.69	308.8
350.0	70.98	2.530	8437.0	10730.0	188.5	41.97	64.61	313.3

Thermophysical properties of ethylene—Continued

T K	Density kg/m^3	Density mol/dm^3	E J/mol	H J/mol	S J/(mol·K)	C_v J/(mol·K)	C_p J/(mol·K)	Sound m/s
360.0	67.06	2.390	8941.0	11370.0	190.3	42.57	63.12	321.7
370.0	63.69	2.270	9439.0	11990.0	192.0	43.28	62.25	329.6
380.0	60.76	2.166	9935.0	12610.0	193.7	44.06	61.80	337.1
390.0	58.16	2.073	10430.0	13230.0	195.3	44.89	61.63	344.1
400.0	55.84	1.990	10930.0	13850.0	196.8	45.75	61.68	350.8
				5.90 MPa isobar				
104.80[a]	658.3	23.47	−12440.0	−12190.0	84.53	41.11	76.62	1727.0
110.0	650.0	23.17	−12060.0	−11810.0	88.07	43.02	70.98	1712.0
120.0	636.5	22.69	−11380.0	−11120.0	94.08	43.96	67.85	1664.0
140.0	611.9	21.81	−10050.0	−9782.0	104.4	41.77	66.17	1549.0
160.0	586.3	20.90	−8740.0	−8458.0	113.2	39.35	66.44	1418.0
180.0	559.0	19.92	−7415.0	−7119.0	121.1	37.83	67.51	1278.0
200.0	529.6	18.88	−6066.0	−5753.0	128.3	37.15	69.22	1133.0
220.0	497.6	17.74	−4674.0	−4341.0	135.0	37.11	72.30	980.9
230.0	480.1	17.11	−3951.0	−3607.0	138.3	37.31	74.78	900.9
240.0	461.0	16.43	−3201.0	−2842.0	141.5	37.66	78.28	817.0
250.0	439.8	15.68	−2412.0	−2036.0	144.8	38.19	83.37	728.0
255.0	428.0	15.26	−1997.0	−1611.0	146.5	38.54	86.84	681.1
260.0	415.2	14.80	−1565.0	−1166.0	148.2	38.96	91.24	632.2
265.0	401.0	14.29	−1109.0	−695.8	150.0	39.49	97.08	580.7
270.0	384.8	13.72	−621.3	−191.2	151.9	40.15	105.3	525.8
272.0	377.7	13.46	−414.6	23.7	152.7	40.47	109.7	502.6
274.0	369.9	13.19	−199.2	248.2	153.5	40.83	115.1	478.5
276.0	361.5	12.88	26.9	484.8	154.4	41.24	121.8	453.2
278.0	352.1	12.55	266.9	737.0	155.3	41.72	130.7	426.6
280.0	341.6	12.18	525.5	1010.0	156.3	42.29	143.2	398.1
281.0	335.7	11.97	664.2	1157.0	156.8	42.63	151.5	383.0
282.0	329.3	11.74	811.1	1314.0	157.4	43.00	162.0	367.2
283.0	322.1	11.48	968.5	1482.0	158.0	43.43	175.7	350.6
284.0	314.1	11.20	1140.0	1667.0	158.6	43.92	194.4	333.1
285.0	304.7	10.86	1331.0	1874.0	159.3	44.51	221.3	314.3
286.0	293.5	10.46	1550.0	2114.0	160.2	45.23	263.0	294.3
286.5	286.8	10.22	1676.0	2253.0	160.7	45.66	293.1	283.9
287.0	279.2	9.952	1816.0	2409.0	161.2	46.14	332.5	273.2
287.5	270.4	9.637	1975.0	2587.0	161.8	46.68	383.5	262.5
288.0	260.0	9.270	2158.0	2794.0	162.5	47.28	446.1	252.3
288.5	248.2	8.845	2367.0	3034.0	163.4	47.90	511.6	243.4
289.0	235.0	8.378	2599.0	3303.0	164.3	48.46	561.2	236.5
289.5	221.6	7.899	2843.0	3590.0	165.3	48.86	579.7	232.0
290.0	208.7	7.440	3085.0	3878.0	166.3	49.04	567.1	229.6
290.5	197.1	7.025	3313.0	4153.0	167.2	49.01	531.2	228.7
291.0	186.9	6.663	3521.0	4407.0	168.1	48.83	482.3	228.7
291.5	178.3	6.356	3707.0	4635.0	168.9	48.56	430.7	229.4
292.0	171.0	6.097	3870.0	4838.0	169.6	48.24	383.2	230.4
292.5	164.9	5.876	4015.0	5019.0	170.2	47.90	342.5	231.6
293.0	159.6	5.688	4144.0	5182.0	170.8	47.57	308.6	232.8
293.5	155.0	5.524	4261.0	5329.0	171.3	47.25	280.6	234.2
294.0	150.9	5.380	4367.0	5463.0	171.7	46.94	257.5	235.5
295.0	144.1	5.138	4554.0	5702.0	172.5	46.37	222.0	238.1
296.0	138.6	4.940	4716.0	5911.0	173.2	45.87	196.4	240.7
297.0	133.9	4.773	4861.0	6097.0	173.9	45.41	177.3	243.1
298.0	129.9	4.629	4992.0	6266.0	174.4	45.01	162.4	245.5
299.0	126.3	4.503	5113.0	6423.0	175.0	44.65	150.6	247.7
300.0	123.2	4.391	5225.0	6568.0	175.5	44.32	141.0	249.9
302.0	117.8	4.198	5429.0	6835.0	176.3	43.76	126.2	253.9
304.0	113.2	4.037	5614.0	7076.0	177.1	43.29	115.5	257.7
306.0	109.4	3.898	5785.0	7298.0	177.9	42.90	107.3	261.2
308.0	106.0	3.777	5944.0	7506.0	178.5	42.58	100.8	264.5
310.0	102.9	3.669	6094.0	7703.0	179.2	42.31	95.63	267.7
312.0	100.2	3.572	6238.0	7889.0	179.8	42.08	91.34	270.7
315.0	96.58	3.443	6442.0	8155.0	180.6	41.82	86.18	275.0

Thermophysical properties of ethylene—Continued

T K	Density kg/m³	Density mol/dm³	E J/mol	H J/mol	S J/(mol·K)	C_v J/(mol·K)	C_p J/(mol·K)	Sound m/s
320.0	91.46	3.260	6760.0	8569.0	181.9	41.53	79.86	281.5
325.0	87.18	3.108	7058.0	8957.0	183.1	41.40	75.38	287.6
330.0	83.50	2.977	7343.0	9325.0	184.3	41.37	72.10	293.2
335.0	80.29	2.862	7618.0	9679.0	185.3	41.44	69.62	298.5
340.0	77.44	2.760	7885.0	10020.0	186.3	41.58	67.71	303.5
345.0	74.88	2.669	8147.0	10360.0	187.3	41.78	66.24	308.2
350.0	72.56	2.586	8404.0	10690.0	188.3	42.02	65.09	312.7
360.0	68.49	2.442	8911.0	11330.0	190.1	42.61	63.49	321.3
370.0	65.02	2.318	9412.0	11960.0	191.8	43.31	62.55	329.2
380.0	61.99	2.210	9910.0	12580.0	193.5	44.09	62.05	336.7
390.0	59.32	2.115	10410.0	13200.0	195.1	44.91	61.85	343.8
400.0	56.94	2.029	10910.0	13820.0	196.6	45.77	61.87	350.5
				6.00 MPa isobar				
104.80[a]	658.4	23.47	−12440.0	−12180.0	84.53	41.14	76.66	1726.0
110.0	650.0	23.17	−12060.0	−11800.0	88.06	43.04	71.00	1711.0
120.0	636.6	22.69	−11380.0	−11110.0	94.07	43.96	67.85	1665.0
140.0	611.9	21.81	−10050.0	−9778.0	104.4	41.77	66.16	1550.0
160.0	586.4	20.90	−8742.0	−8455.0	113.2	39.35	66.42	1418.0
180.0	559.1	19.93	−7418.0	−7116.0	121.1	37.84	67.49	1278.0
200.0	529.7	18.88	−6069.0	−5751.0	128.3	37.15	69.19	1134.0
220.0	497.8	17.74	−4678.0	−4340.0	135.0	37.12	72.24	982.3
230.0	480.3	17.12	−3956.0	−3606.0	138.3	37.32	74.70	902.5
240.0	461.3	16.44	−3207.0	−2843.0	141.5	37.66	78.16	818.9
250.0	440.2	15.69	−2420.0	−2038.0	144.8	38.18	83.17	730.4
255.0	428.5	15.27	−2006.0	−1614.0	146.5	38.53	86.57	683.8
260.0	415.8	14.82	−1575.0	−1170.0	148.2	38.95	90.87	635.2
265.0	401.7	14.32	−1122.0	−702.7	150.0	39.46	96.53	584.2
270.0	385.8	13.75	−637.8	−201.5	151.9	40.11	104.4	530.1
272.0	378.7	13.50	−433.1	11.4	152.6	40.42	108.6	507.2
274.0	371.1	13.23	−220.1	233.5	153.5	40.76	113.6	483.5
276.0	362.9	12.93	3.0	466.8	154.3	41.16	120.0	458.8
278.0	353.8	12.61	238.8	714.5	155.2	41.62	128.1	432.8
280.0	343.7	12.25	491.4	981.2	156.1	42.16	139.2	405.2
282.0	331.9	11.83	767.8	1275.0	157.2	42.82	155.5	375.5
283.0	325.3	11.59	918.4	1436.0	157.8	43.21	166.9	359.7
284.0	317.8	11.33	1080.0	1610.0	158.4	43.65	181.9	343.0
285.0	309.4	11.03	1258.0	1802.0	159.1	44.17	202.5	325.5
286.0	299.6	10.68	1456.0	2018.0	159.8	44.79	232.3	306.9
287.0	287.6	10.25	1686.0	2271.0	160.7	45.53	277.9	287.3
287.5	280.6	10.00	1818.0	2418.0	161.2	45.97	309.9	277.2
288.0	272.5	9.715	1965.0	2583.0	161.8	46.46	349.9	267.2
288.5	263.4	9.387	2130.0	2769.0	162.4	46.99	397.8	257.6
289.0	253.0	9.017	2316.0	2981.0	163.2	47.54	449.0	248.9
290.0	229.5	8.179	2738.0	3471.0	164.9	48.50	519.5	236.5
291.0	206.3	7.354	3175.0	3991.0	166.6	48.87	507.3	231.3
291.5	196.1	6.990	3380.0	4238.0	167.5	48.81	477.1	230.6
292.0	187.1	6.669	3567.0	4467.0	168.3	48.64	438.8	230.7
292.5	179.3	6.390	3737.0	4676.0	169.0	48.39	398.5	231.2
293.0	172.5	6.149	3890.0	4866.0	169.6	48.11	360.5	232.1
293.5	166.7	5.941	4028.0	5038.0	170.2	47.80	326.8	233.1
294.0	161.6	5.760	4152.0	5194.0	170.8	47.50	297.9	234.3
294.5	157.1	5.600	4265.0	5336.0	171.2	47.20	273.3	235.5
295.0	153.1	5.459	4368.0	5467.0	171.7	46.91	252.5	236.8
296.0	146.4	5.219	4553.0	5703.0	172.5	46.38	219.8	239.2
297.0	140.8	5.020	4715.0	5910.0	173.2	45.89	195.6	241.7
298.0	136.1	4.852	4859.0	6096.0	173.8	45.46	177.1	244.0
299.0	132.0	4.707	4991.0	6265.0	174.4	45.06	162.7	246.3
300.0	128.5	4.579	5112.0	6422.0	174.9	44.71	151.0	248.5
302.0	122.4	4.362	5330.0	6706.0	175.8	44.10	133.6	252.6
304.0	117.4	4.184	5526.0	6960.0	176.7	43.59	121.1	256.5
306.0	113.1	4.032	5704.0	7192.0	177.4	43.17	111.8	260.1

Thermophysical properties of ethylene—Continued

T K	Density kg/m^3	Density mol/dm^3	E J/mol	H J/mol	S J/(mol·K)	C_v J/(mol·K)	C_p J/(mol·K)	Sound m/s
308.0	109.4	3.900	5870.0	7408.0	178.1	42.82	104.5	263.5
310.0	106.1	3.783	6025.0	7611.0	178.8	42.53	98.69	266.7
312.0	103.2	3.679	6173.0	7804.0	179.4	42.28	93.95	269.7
314.0	100.6	3.585	6314.0	7988.0	180.0	42.08	90.01	272.6
320.0	93.91	3.348	6707.0	8500.0	181.6	41.68	81.40	280.7
325.0	89.40	3.187	7011.0	8894.0	182.8	41.52	76.58	286.8
330.0	85.55	3.049	7300.0	9268.0	184.0	41.47	73.05	292.5
335.0	82.19	2.930	7578.0	9626.0	185.1	41.53	70.40	297.9
340.0	79.22	2.824	7848.0	9973.0	186.1	41.65	68.38	302.9
345.0	76.56	2.729	8112.0	10310.0	187.1	41.84	66.80	307.7
350.0	74.15	2.643	8371.0	10640.0	188.0	42.07	65.58	312.2
360.0	69.94	2.493	8881.0	11290.0	189.8	42.65	63.88	320.8
370.0	66.36	2.365	9384.0	11920.0	191.6	43.34	62.86	328.8
380.0	63.24	2.254	9885.0	12550.0	193.3	44.11	62.31	336.3
390.0	60.49	2.156	10390.0	13170.0	194.9	44.93	62.07	343.5
400.0	58.04	2.069	10890.0	13790.0	196.4	45.79	62.06	350.3
				6.20 MPa isobar				
104.90[a]	658.4	23.47	−12440.0	−12180.0	84.52	41.19	76.74	1725.0
110.0	650.1	23.17	−12070.0	−11800.0	88.04	43.06	71.05	1711.0
120.0	636.7	22.70	−11380.0	−11110.0	94.06	43.98	67.86	1665.0
140.0	612.1	21.82	−10060.0	−9772.0	104.4	41.78	66.15	1550.0
160.0	586.6	20.91	−8745.0	−8449.0	113.2	39.36	66.39	1420.0
180.0	559.3	19.94	−7422.0	−7111.0	121.1	37.85	67.45	1280.0
200.0	530.0	18.89	−6075.0	−5747.0	128.2	37.16	69.13	1136.0
220.0	498.2	17.76	−4686.0	−4337.0	135.0	37.12	72.14	984.9
230.0	480.8	17.14	−3966.0	−3604.0	138.2	37.32	74.54	905.7
240.0	462.0	16.47	−3220.0	−2843.0	141.5	37.66	77.92	822.7
250.0	441.0	15.72	−2435.0	−2041.0	144.7	38.17	82.78	735.1
255.0	429.5	15.31	−2024.0	−1619.0	146.4	38.51	86.04	689.0
260.0	416.9	14.86	−1596.0	−1179.0	148.1	38.91	90.15	641.2
265.0	403.1	14.37	−1147.0	−715.8	149.9	39.41	95.48	591.2
270.0	387.6	13.82	−669.9	−221.1	151.7	40.03	102.8	538.3
272.0	380.7	13.57	−468.7	−11.9	152.5	40.32	106.6	516.1
274.0	373.4	13.31	−260.1	205.7	153.3	40.65	111.1	493.2
276.0	365.5	13.03	−42.7	433.2	154.1	41.01	116.6	469.4
278.0	357.0	12.72	185.8	673.1	155.0	41.43	123.6	444.6
280.0	347.5	12.39	428.3	928.9	155.9	41.92	132.7	418.5
282.0	336.7	12.00	689.7	1206.0	156.9	42.50	145.3	390.8
283.0	330.8	11.79	829.7	1356.0	157.4	42.83	153.7	376.2
284.0	324.3	11.56	977.9	1514.0	158.0	43.21	164.2	361.0
286.0	309.0	11.01	1308.0	1871.0	159.2	44.11	195.2	328.7
287.0	299.8	10.69	1497.0	2078.0	160.0	44.66	219.5	311.5
288.0	288.9	10.30	1711.0	2313.0	160.8	45.32	253.8	293.6
289.0	275.9	9.834	1960.0	2590.0	161.7	46.09	302.7	275.7
290.0	259.9	9.265	2254.0	2924.0	162.9	46.97	365.9	259.0
291.0	241.1	8.594	2599.0	3320.0	164.2	47.82	423.0	245.9
292.0	221.2	7.886	2970.0	3757.0	165.7	48.39	442.2	238.1
293.0	202.8	7.231	3333.0	4190.0	167.2	48.52	419.9	235.0
293.5	194.7	6.940	3502.0	4395.0	167.9	48.43	398.2	234.5
294.0	187.3	6.678	3660.0	4588.0	168.6	48.27	373.0	234.6
294.5	180.8	6.443	3806.0	4768.0	169.2	48.07	346.7	235.0
295.0	174.9	6.234	3940.0	4935.0	169.8	47.83	321.3	235.6
295.5	169.7	6.048	4064.0	5090.0	170.3	47.58	297.7	236.5
296.0	165.0	5.882	4179.0	5233.0	170.8	47.33	276.4	237.4
297.0	157.1	5.598	4383.0	5491.0	171.6	46.82	240.9	239.5
298.0	150.5	5.365	4562.0	5718.0	172.4	46.35	213.6	241.7
299.0	145.0	5.168	4721.0	5920.0	173.1	45.91	192.3	243.9
300.0	140.3	5.000	4864.0	6104.0	173.7	45.50	175.6	246.1
301.0	136.1	4.853	4995.0	6272.0	174.3	45.14	162.1	248.2
302.0	132.5	4.723	5116.0	6429.0	174.8	44.80	151.1	250.3

Thermophysical properties of ethylene—Continued

T K	Density kg/m^3	Density mol/dm^3	E J/mol	H J/mol	S J/(mol·K)	C_v J/(mol·K)	C_p J/(mol·K)	Sound m/s
304.0	126.3	4.501	5336.0	6713.0	175.7	44.22	134.3	254.2
306.0	121.1	4.317	5533.0	6969.0	176.6	43.73	122.0	257.9
308.0	116.7	4.161	5713.0	7203.0	177.3	43.33	112.8	261.4
310.0	112.9	4.024	5881.0	7421.0	178.0	42.99	105.5	264.7
312.0	109.5	3.904	6038.0	7626.0	178.7	42.70	99.69	267.9
314.0	106.5	3.796	6188.0	7821.0	179.3	42.46	94.92	270.9
316.0	103.8	3.699	6330.0	8007.0	179.9	42.27	90.94	273.7
320.0	98.98	3.528	6600.0	8357.0	181.0	41.97	84.71	279.2
325.0	93.97	3.349	6915.0	8766.0	182.3	41.76	79.10	285.4
330.0	89.73	3.198	7212.0	9151.0	183.4	41.68	75.06	291.2
335.0	86.06	3.068	7497.0	9518.0	184.5	41.70	72.04	296.7
340.0	82.84	2.953	7773.0	9872.0	185.6	41.80	69.75	301.8
345.0	79.97	2.850	8041.0	10220.0	186.6	41.96	67.97	306.6
350.0	77.38	2.758	8305.0	10550.0	187.6	42.18	66.59	311.3
360.0	72.88	2.598	8822.0	11210.0	189.4	42.74	64.66	320.0
370.0	69.06	2.462	9330.0	11850.0	191.2	43.41	63.49	328.1
380.0	65.75	2.344	9834.0	12480.0	192.8	44.16	62.83	335.7
390.0	62.84	2.240	10340.0	13110.0	194.5	44.97	62.51	342.9
400.0	60.26	2.148	10840.0	13730.0	196.1	45.82	62.44	349.7
				6.40 MPa isobar				
104.90[a]	658.5	23.47	−12440.0	−12170.0	84.52	41.24	76.82	1724.0
110.0	650.3	23.18	−12070.0	−11790.0	88.02	43.09	71.10	1711.0
120.0	636.8	22.70	−11380.0	−11100.0	94.04	44.00	67.86	1665.0
140.0	612.2	21.82	−10060.0	−9765.0	104.3	41.79	66.13	1551.0
160.0	586.7	20.91	−8748.0	−8442.0	113.2	39.37	66.37	1421.0
180.0	559.5	19.94	−7426.0	−7105.0	121.0	37.86	67.41	1282.0
200.0	530.3	18.90	−6081.0	−5742.0	128.2	37.17	69.07	1138.0
220.0	498.6	17.77	−4694.0	−4334.0	134.9	37.13	72.03	987.6
230.0	481.3	17.16	−3976.0	−3603.0	138.2	37.32	74.39	908.8
240.0	462.6	16.49	−3231.0	−2843.0	141.4	37.66	77.69	826.4
250.0	441.8	15.75	−2451.0	−2044.0	144.7	38.16	82.40	739.6
255.0	430.4	15.34	−2042.0	−1625.0	146.3	38.49	85.55	694.2
260.0	418.0	14.90	−1617.0	−1188.0	148.0	38.88	89.47	647.1
265.0	404.5	14.42	−1172.0	−728.2	149.8	39.36	94.51	597.9
270.0	389.3	13.88	−700.7	−239.6	151.6	39.95	101.3	546.2
272.0	382.7	13.64	−502.8	−33.6	152.4	40.23	104.8	524.6
274.0	375.6	13.39	−298.1	179.9	153.1	40.54	108.8	502.4
276.0	368.0	13.12	−85.6	402.3	154.0	40.88	113.7	479.5
278.0	359.9	12.83	136.6	635.5	154.8	41.27	119.8	455.7
280.0	350.9	12.51	370.7	882.4	155.7	41.71	127.4	430.8
282.0	341.0	12.15	620.4	1147.0	156.6	42.23	137.6	404.7
284.0	329.6	11.75	890.9	1436.0	157.6	42.84	151.9	377.0
286.0	316.3	11.28	1192.0	1759.0	158.8	43.60	173.4	347.4
288.0	299.9	10.69	1539.0	2138.0	160.1	44.55	208.7	315.8
289.0	290.0	10.34	1740.0	2359.0	160.9	45.13	235.4	299.4
290.0	278.4	9.923	1967.0	2612.0	161.7	45.80	270.7	283.1
291.0	264.7	9.437	2226.0	2904.0	162.7	46.53	314.1	267.8
292.0	249.1	8.879	2519.0	3240.0	163.9	47.27	356.5	255.0
293.0	232.2	8.277	2838.0	3611.0	165.2	47.86	381.7	245.9
294.0	215.6	7.685	3162.0	3994.0	166.5	48.17	381.2	240.8
295.0	200.5	7.147	3471.0	4366.0	167.7	48.17	359.5	238.7
296.0	187.5	6.685	3752.0	4709.0	168.9	47.93	325.7	238.4
297.0	176.7	6.300	4001.0	5017.0	169.9	47.56	289.4	239.3
298.0	167.8	5.981	4219.0	5289.0	170.8	47.13	256.3	240.7
299.0	160.3	5.716	4412.0	5531.0	171.7	46.69	228.5	242.5
300.0	154.1	5.492	4583.0	5748.0	172.4	46.27	205.9	244.4
301.0	148.7	5.300	4737.0	5945.0	173.0	45.88	187.6	246.4
302.0	144.0	5.133	4878.0	6125.0	173.6	45.51	172.7	248.3
303.0	139.9	4.986	5007.0	6291.0	174.2	45.17	160.4	250.3
304.0	136.2	4.856	5128.0	6446.0	174.7	44.86	150.2	252.2
306.0	129.9	4.631	5348.0	6730.0	175.6	44.31	134.3	256.0

Thermophysical properties of ethylene—Continued

T K	Density kg/m³	Density mol/dm³	E J/mol	H J/mol	S J/(mol·K)	C_v J/(mol·K)	C_p J/(mol·K)	Sound m/s
308.0	124.7	4.444	5546.0	6986.0	176.5	43.85	122.4	259.5
310.0	120.2	4.283	5727.0	7221.0	177.2	43.46	113.4	262.9
312.0	116.2	4.144	5896.0	7441.0	177.9	43.13	106.2	266.1
314.0	112.8	4.020	6055.0	7647.0	178.6	42.86	100.4	269.2
316.0	109.7	3.909	6206.0	7843.0	179.2	42.63	95.68	272.2
318.0	106.8	3.809	6350.0	8030.0	179.8	42.43	91.70	275.0
320.0	104.3	3.717	6488.0	8210.0	180.4	42.28	88.33	277.7
325.0	98.70	3.518	6816.0	8635.0	181.7	42.01	81.83	284.1
330.0	94.04	3.352	7122.0	9032.0	182.9	41.89	77.20	290.0
335.0	90.04	3.210	7415.0	9409.0	184.0	41.88	73.77	295.5
340.0	86.55	3.085	7696.0	9771.0	185.1	41.95	71.18	300.7
345.0	83.45	2.975	7970.0	10120.0	186.1	42.09	69.19	305.7
350.0	80.67	2.876	8238.0	10460.0	187.1	42.29	67.64	310.4
360.0	75.85	2.704	8761.0	11130.0	189.0	42.82	65.46	319.2
370.0	71.79	2.559	9274.0	11780.0	190.8	43.47	64.13	327.4
380.0	68.29	2.434	9783.0	12410.0	192.5	44.22	63.36	335.0
390.0	65.22	2.325	10290.0	13040.0	194.1	45.02	62.96	342.3
400.0	62.49	2.228	10800.0	13670.0	195.7	45.86	62.82	349.2

6.60 MPa isobar

T K	Density kg/m³	Density mol/dm³	E J/mol	H J/mol	S J/(mol·K)	C_v J/(mol·K)	C_p J/(mol·K)	Sound m/s
104.90[a]	658.6	23.48	−12440.0	−12160.0	84.52	41.29	76.90	1724.0
110.0	650.4	23.18	−12070.0	−11790.0	88.00	43.11	71.15	1711.0
120.0	636.9	22.70	−11380.0	−11090.0	94.02	44.01	67.87	1666.0
140.0	612.3	21.83	−10060.0	−9759.0	104.3	41.80	66.12	1552.0
160.0	586.9	20.92	−8752.0	−8436.0	113.1	39.38	66.34	1422.0
180.0	559.7	19.95	−7431.0	−7100.0	121.0	37.87	67.37	1283.0
200.0	530.6	18.91	−6086.0	−5738.0	128.2	37.18	69.01	1140.0
220.0	499.0	17.79	−4702.0	−4331.0	134.9	37.14	71.92	990.2
230.0	481.8	17.17	−3985.0	−3601.0	138.1	37.33	74.24	911.8
240.0	463.2	16.51	−3243.0	−2843.0	141.4	37.66	77.46	830.1
250.0	442.6	15.78	−2466.0	−2047.0	144.6	38.15	82.03	744.1
255.0	431.3	15.37	−2059.0	−1630.0	146.3	38.47	85.07	699.2
260.0	419.1	14.94	−1637.0	−1195.0	147.9	38.86	88.82	652.8
265.0	405.8	14.47	−1196.0	−739.9	149.7	39.32	93.60	604.5
270.0	391.0	13.94	−730.5	−256.9	151.5	39.88	99.93	553.9
275.0	374.1	13.34	−231.2	263.7	153.4	40.60	108.9	500.3
276.0	370.4	13.20	−126.1	373.8	153.8	40.76	111.2	489.1
278.0	362.6	12.92	90.6	601.3	154.6	41.12	116.5	466.1
280.0	354.1	12.62	317.6	840.6	155.5	41.53	123.1	442.3
282.0	344.7	12.29	557.7	1095.0	156.4	42.00	131.5	417.5
284.0	334.3	11.92	814.6	1368.0	157.3	42.55	142.8	391.5
286.0	322.4	11.49	1095.0	1669.0	158.4	43.19	158.7	364.0
288.0	308.3	10.99	1408.0	2009.0	159.6	43.98	182.7	335.0
289.0	300.0	10.70	1582.0	2199.0	160.2	44.45	199.4	320.0
290.0	290.8	10.37	1772.0	2409.0	161.0	44.97	220.8	304.9
291.0	280.3	9.992	1982.0	2643.0	161.8	45.55	247.4	289.9
292.0	268.3	9.565	2216.0	2906.0	162.7	46.18	278.8	275.8
293.0	254.9	9.085	2474.0	3201.0	163.7	46.81	310.4	263.4
294.0	240.3	8.567	2753.0	3524.0	164.8	47.37	333.6	253.9
295.0	225.6	8.042	3042.0	3863.0	165.9	47.75	341.8	247.6
296.0	211.6	7.541	3327.0	4202.0	167.1	47.90	334.4	244.0
297.0	198.8	7.088	3596.0	4528.0	168.2	47.83	315.5	242.4
298.0	187.7	6.691	3844.0	4831.0	169.2	47.61	290.2	242.2
299.0	178.2	6.352	4068.0	5107.0	170.1	47.29	263.4	242.9
300.0	170.1	6.062	4269.0	5358.0	171.0	46.92	238.3	244.1
301.0	163.1	5.815	4450.0	5585.0	171.7	46.54	216.3	245.6
302.0	157.2	5.602	4614.0	5792.0	172.4	46.17	197.7	247.3
303.0	152.0	5.417	4763.0	5981.0	173.0	45.82	182.1	249.0
304.0	147.4	5.253	4900.0	6157.0	173.6	45.49	169.0	250.8
306.0	139.7	4.978	5147.0	6473.0	174.6	44.89	148.6	254.4
308.0	133.3	4.753	5366.0	6755.0	175.6	44.37	133.7	258.0

Thermophysical properties of ethylene—Continued

T K	Density kg/m^3	Density mol/dm^3	E J/mol	H J/mol	S J/(mol·K)	C_v J/(mol·K)	C_p J/(mol·K)	Sound m/s
310.0	128.0	4.563	5564.0	7010.0	176.4	43.94	122.4	261.4
312.0	123.4	4.400	5746.0	7246.0	177.2	43.57	113.6	264.6
314.0	119.4	4.258	5916.0	7466.0	177.9	43.26	106.6	267.7
316.0	115.9	4.131	6076.0	7673.0	178.5	42.99	100.9	270.7
318.0	112.7	4.018	6228.0	7870.0	179.1	42.77	96.24	273.6
320.0	109.8	3.915	6373.0	8059.0	179.7	42.59	92.30	276.4
325.0	103.6	3.694	6713.0	8500.0	181.1	42.27	84.77	282.9
330.0	98.49	3.511	7030.0	8910.0	182.3	42.10	79.48	288.9
335.0	94.13	3.355	7330.0	9297.0	183.5	42.06	75.60	294.5
340.0	90.35	3.220	7618.0	9668.0	184.6	42.10	72.69	299.8
345.0	87.01	3.101	7897.0	10030.0	185.7	42.22	70.46	304.8
350.0	84.02	2.995	8169.0	10370.0	186.7	42.41	68.72	309.5
355.0	81.33	2.899	8437.0	10710.0	187.6	42.64	67.36	314.1
360.0	78.88	2.812	8700.0	11050.0	188.6	42.91	66.29	318.4
370.0	74.56	2.658	9219.0	11700.0	190.3	43.54	64.79	326.7
380.0	70.86	2.526	9732.0	12350.0	192.1	44.27	63.90	334.5
390.0	67.62	2.410	10240.0	12980.0	193.7	45.06	63.41	341.8
400.0	64.75	2.308	10750.0	13610.0	195.3	45.90	63.21	348.8

6.80 MPa isobar

T K	Density kg/m^3	Density mol/dm^3	E J/mol	H J/mol	S J/(mol·K)	C_v J/(mol·K)	C_p J/(mol·K)	Sound m/s
104.90[a]	658.7	23.48	−12440.0	−12150.0	84.51	41.34	76.98	1723.0
110.0	650.5	23.19	−12070.0	−11780.0	87.98	43.14	71.20	1710.0
120.0	637.0	22.71	−11390.0	−11090.0	94.00	44.03	67.88	1666.0
140.0	612.5	21.83	−10060.0	−9752.0	104.3	41.81	66.10	1553.0
160.0	587.1	20.93	−8755.0	−8430.0	113.1	39.39	66.31	1423.0
180.0	559.9	19.96	−7435.0	−7095.0	121.0	37.88	67.33	1285.0
200.0	530.9	18.92	−6092.0	−5733.0	128.2	37.19	68.95	1142.0
220.0	499.4	17.80	−4710.0	−4328.0	134.8	37.14	71.82	992.8
230.0	482.2	17.19	−3995.0	−3599.0	138.1	37.33	74.09	914.9
240.0	463.8	16.53	−3255.0	−2843.0	141.3	37.66	77.24	833.8
250.0	443.4	15.80	−2480.0	−2050.0	144.5	38.14	81.68	748.6
255.0	432.2	15.41	−2076.0	−1635.0	146.2	38.46	84.61	704.2
260.0	420.2	14.98	−1657.0	−1203.0	147.9	38.83	88.21	658.4
265.0	407.1	14.51	−1220.0	−751.1	149.6	39.28	92.74	610.8
270.0	392.6	13.99	−759.2	−273.3	151.4	39.82	98.67	561.3
275.0	376.2	13.41	−267.7	239.4	153.3	40.50	106.9	509.1
276.0	372.6	13.28	−164.7	347.3	153.6	40.66	108.9	498.2
278.0	365.1	13.01	47.3	569.8	154.4	40.99	113.7	476.1
280.0	357.0	12.72	268.2	802.6	155.3	41.37	119.4	453.2
282.0	348.2	12.41	500.3	1048.0	156.2	41.80	126.5	429.5
284.0	338.4	12.06	746.4	1310.0	157.1	42.29	135.8	404.8
286.0	327.5	11.67	1011.0	1593.0	158.1	42.86	148.2	379.0
288.0	315.0	11.23	1300.0	1906.0	159.2	43.54	165.7	352.0
290.0	300.1	10.70	1626.0	2261.0	160.4	44.35	191.4	324.0
291.0	291.5	10.39	1807.0	2461.0	161.1	44.82	208.8	309.9
292.0	281.9	10.05	2003.0	2680.0	161.8	45.34	229.6	296.1
293.0	271.1	9.665	2218.0	2921.0	162.7	45.88	253.2	283.0
294.0	259.3	9.242	2451.0	3187.0	163.6	46.43	277.2	271.3
296.0	233.4	8.321	2960.0	3777.0	165.6	47.33	307.4	254.6
297.0	220.6	7.864	3221.0	4086.0	166.6	47.56	307.4	250.0
298.0	208.5	7.433	3474.0	4389.0	167.6	47.62	298.1	247.4
299.0	197.6	7.043	3714.0	4680.0	168.6	47.52	282.1	246.2
300.0	187.9	6.697	3937.0	4952.0	169.5	47.31	262.4	246.0
301.0	179.4	6.394	4141.0	5204.0	170.3	47.03	241.9	246.5
302.0	172.0	6.130	4327.0	5436.0	171.1	46.71	222.4	247.5
303.0	165.5	5.900	4497.0	5650.0	171.8	46.38	204.8	248.8
304.0	159.9	5.698	4653.0	5846.0	172.5	46.05	189.4	250.3
305.0	154.9	5.520	4797.0	6029.0	173.1	45.74	176.1	251.9
306.0	150.4	5.361	4931.0	6199.0	173.6	45.44	164.7	253.5
308.0	142.8	5.090	5173.0	6509.0	174.6	44.89	146.4	256.8
310.0	136.5	4.866	5390.0	6788.0	175.5	44.42	132.6	260.1

Thermophysical properties of ethylene—Continued

T K	Density kg/m^3	Density mol/dm^3	E J/mol	H J/mol	S J/(mol·K)	C_v J/(mol·K)	C_p J/(mol·K)	Sound m/s
312.0	131.2	4.675	5587.0	7042.0	176.4	44.01	121.9	263.3
314.0	126.5	4.511	5769.0	7277.0	177.1	43.66	113.5	266.5
316.0	122.5	4.366	5940.0	7497.0	177.8	43.36	106.8	269.5
318.0	118.9	4.237	6100.0	7705.0	178.5	43.12	101.2	272.4
320.0	115.6	4.122	6253.0	7903.0	179.1	42.91	96.62	275.2
322.0	112.7	4.017	6399.0	8092.0	179.7	42.73	92.73	277.9
325.0	108.7	3.876	6608.0	8363.0	180.5	42.53	87.93	281.8
330.0	103.1	3.675	6936.0	8786.0	181.8	42.32	81.90	287.8
335.0	98.34	3.505	7244.0	9184.0	183.0	42.24	77.54	293.5
340.0	94.23	3.359	7539.0	9563.0	184.1	42.26	74.27	298.9
345.0	90.64	3.231	7824.0	9928.0	185.2	42.36	71.78	303.9
350.0	87.44	3.117	8100.0	10280.0	186.2	42.52	69.85	308.7
355.0	84.56	3.014	8371.0	10630.0	187.2	42.73	68.33	313.3
360.0	81.95	2.921	8638.0	10970.0	188.1	42.99	67.14	317.8
370.0	77.37	2.758	9163.0	11630.0	189.9	43.61	65.46	326.1
380.0	73.45	2.618	9680.0	12280.0	191.7	44.32	64.45	333.9
390.0	70.04	2.497	10200.0	12920.0	193.3	45.10	63.87	341.3
400.0	67.02	2.389	10710.0	13560.0	195.0	45.93	63.61	348.3
			7.00 MPa isobar					
105.00[a]	658.7	23.48	−12440.0	−12140.0	84.51	41.39	77.06	1722.0
110.0	650.6	23.19	−12070.0	−11770.0	87.96	43.17	71.25	1710.0
120.0	637.2	22.71	−11390.0	−11080.0	93.99	44.05	67.88	1666.0
140.0	612.6	21.84	−10070.0	−9746.0	104.3	41.82	66.08	1554.0
160.0	587.2	20.93	−8758.0	−8424.0	113.1	39.40	66.29	1425.0
180.0	560.2	19.97	−7440.0	−7089.0	121.0	37.89	67.30	1286.0
200.0	531.2	18.93	−6098.0	−5728.0	128.1	37.20	68.90	1144.0
220.0	499.7	17.81	−4718.0	−4325.0	134.8	37.15	71.72	995.3
230.0	482.7	17.21	−4004.0	−3597.0	138.0	37.33	73.94	917.9
240.0	464.4	16.55	−3266.0	−2843.0	141.3	37.66	77.02	837.3
250.0	444.1	15.83	−2495.0	−2053.0	144.5	38.13	81.34	753.0
255.0	433.1	15.44	−2093.0	−1639.0	146.1	38.44	84.17	709.1
260.0	421.2	15.01	−1676.0	−1210.0	147.8	38.81	87.62	663.8
265.0	408.3	14.56	−1243.0	−761.7	149.5	39.24	91.94	617.1
270.0	394.1	14.05	−786.9	−288.7	151.3	39.76	97.51	568.4
275.0	378.2	13.48	−302.6	216.7	153.1	40.41	105.1	517.5
276.0	374.7	13.36	−201.4	322.7	153.5	40.56	106.9	507.0
278.0	367.5	13.10	6.3	540.7	154.3	40.88	111.2	485.5
280.0	359.7	12.82	221.9	767.9	155.1	41.23	116.2	463.4
282.0	351.3	12.52	447.2	1006.0	156.0	41.62	122.4	440.7
284.0	342.2	12.20	684.4	1258.0	156.8	42.07	130.1	417.1
286.0	332.1	11.84	936.7	1528.0	157.8	42.58	140.1	392.7
288.0	320.7	11.43	1209.0	1821.0	158.8	43.17	153.6	367.4
290.0	307.5	10.96	1507.0	2146.0	159.9	43.87	172.3	341.2
292.0	292.0	10.41	1843.0	2515.0	161.2	44.69	198.7	314.7
294.0	273.4	9.744	2228.0	2947.0	162.7	45.63	233.8	289.6
296.0	251.4	8.962	2669.0	3451.0	164.4	46.57	268.5	269.1
298.0	228.0	8.127	3145.0	4006.0	166.3	47.22	282.5	256.4
299.0	216.7	7.725	3381.0	4287.0	167.2	47.34	278.8	252.9
300.0	206.2	7.351	3610.0	4562.0	168.1	47.34	269.3	250.8
301.0	196.6	7.009	3826.0	4825.0	169.0	47.23	255.9	249.9
302.0	188.0	6.702	4028.0	5073.0	169.8	47.03	240.2	249.7
303.0	180.4	6.429	4216.0	5305.0	170.6	46.78	224.1	250.2
304.0	173.6	6.187	4390.0	5521.0	171.3	46.51	208.5	251.0
305.0	167.6	5.972	4550.0	5722.0	172.0	46.22	194.2	252.1
306.0	162.2	5.782	4699.0	5910.0	172.6	45.93	181.4	253.4
307.0	157.4	5.611	4838.0	6086.0	173.1	45.65	170.0	254.8
308.0	153.1	5.458	4968.0	6251.0	173.7	45.38	160.1	256.3
310.0	145.7	5.192	5206.0	6554.0	174.7	44.88	143.7	259.3
312.0	139.4	4.970	5420.0	6828.0	175.5	44.44	131.1	262.4
314.0	134.1	4.780	5616.0	7080.0	176.3	44.06	121.1	265.5

Thermophysical properties of ethylene—Continued

T K	Density kg/m^3	Density mol/dm^3	E J/mol	H J/mol	S J/(mol·K)	C_v J/(mol·K)	C_p J/(mol·K)	Sound m/s
316.0	129.5	4.615	5797.0	7314.0	177.1	43.74	113.2	268.5
318.0	125.4	4.469	5967.0	7534.0	177.8	43.46	106.7	271.3
320.0	121.7	4.339	6128.0	7742.0	178.4	43.22	101.3	274.1
322.0	118.4	4.222	6281.0	7940.0	179.0	43.03	96.83	276.9
325.0	114.0	4.065	6500.0	8221.0	179.9	42.79	91.33	280.8
330.0	107.8	3.844	6839.0	8660.0	181.3	42.54	84.49	286.9
335.0	102.7	3.659	7157.0	9070.0	182.5	42.42	79.57	292.7
340.0	98.22	3.501	7458.0	9458.0	183.6	42.42	75.93	298.1
345.0	94.34	3.363	7749.0	9830.0	184.7	42.49	73.16	303.2
350.0	90.92	3.241	8030.0	10190.0	185.8	42.64	71.01	308.0
355.0	87.84	3.131	8306.0	10540.0	186.8	42.83	69.34	312.7
360.0	85.07	3.032	8576.0	10880.0	187.7	43.08	68.02	317.1
370.0	80.21	2.859	9106.0	11550.0	189.6	43.67	66.15	325.6
380.0	76.07	2.712	9628.0	12210.0	191.3	44.37	65.01	333.4
390.0	72.48	2.584	10150.0	12860.0	193.0	45.15	64.34	340.9
400.0	69.31	2.471	10660.0	13500.0	194.6	45.97	64.01	347.9
				7.50 MPa isobar				
105.00[a]	658.9	23.49	−12440.0	−12120.0	84.50	41.51	77.26	1720.0
110.0	650.9	23.20	−12080.0	−11760.0	87.91	43.23	71.37	1709.0
120.0	637.4	22.72	−11390.0	−11060.0	93.94	44.09	67.90	1667.0
140.0	612.9	21.85	−10070.0	−9729.0	104.2	41.85	66.04	1556.0
160.0	587.7	20.95	−8767.0	−8409.0	113.0	39.42	66.22	1428.0
180.0	560.7	19.99	−7450.0	−7075.0	120.9	37.91	67.20	1290.0
200.0	531.9	18.96	−6112.0	−5717.0	128.1	37.22	68.76	1148.0
220.0	500.7	17.85	−4738.0	−4317.0	134.7	37.17	71.47	1002.0
230.0	483.9	17.25	−4027.0	−3593.0	137.9	37.35	73.59	925.3
240.0	465.8	16.60	−3295.0	−2843.0	141.1	37.66	76.50	846.1
250.0	446.0	15.90	−2531.0	−2059.0	144.3	38.11	80.54	763.6
255.0	435.2	15.51	−2133.0	−1650.0	146.0	38.41	83.14	720.9
260.0	423.7	15.10	−1723.0	−1226.0	147.6	38.75	86.27	677.0
265.0	411.3	14.66	−1297.0	−785.9	149.3	39.16	90.11	632.0
270.0	397.8	14.18	−852.7	−323.7	151.0	39.64	94.93	585.5
275.0	382.8	13.64	−383.8	165.9	152.8	40.22	101.2	537.3
280.0	365.8	13.04	117.1	692.3	154.7	40.93	109.9	486.9
282.0	358.3	12.77	329.1	916.5	155.5	41.26	114.4	466.1
284.0	350.2	12.48	549.6	1150.0	156.3	41.63	119.7	444.8
286.0	341.5	12.17	780.0	1396.0	157.2	42.03	126.3	422.9
288.0	332.0	11.83	1023.0	1656.0	158.1	42.49	134.3	400.5
290.0	321.5	11.46	1281.0	1935.0	159.1	43.00	144.6	377.7
292.0	309.8	11.04	1558.0	2237.0	160.1	43.58	157.7	354.5
294.0	296.6	10.57	1859.0	2568.0	161.2	44.23	174.5	331.4
296.0	281.5	10.03	2190.0	2937.0	162.5	44.95	195.0	309.4
298.0	264.4	9.426	2553.0	3349.0	163.9	45.68	216.3	290.0
300.0	246.0	8.767	2943.0	3799.0	165.4	46.31	231.9	275.2
302.0	227.4	8.105	3343.0	4268.0	166.9	46.69	235.3	265.7
304.0	210.1	7.489	3730.0	4732.0	168.5	46.77	226.5	260.8
306.0	195.0	6.950	4090.0	5169.0	169.9	46.58	209.9	259.0
308.0	182.2	6.494	4415.0	5569.0	171.2	46.23	190.5	259.1
310.0	171.5	6.115	4705.0	5931.0	172.4	45.82	171.8	260.5
312.0	162.6	5.798	4965.0	6258.0	173.4	45.39	155.7	262.5
314.0	155.1	5.530	5200.0	6556.0	174.4	44.99	142.2	264.9
316.0	148.7	5.301	5414.0	6829.0	175.2	44.62	131.1	267.4
318.0	143.2	5.103	5612.0	7082.0	176.0	44.29	122.1	270.0
320.0	138.3	4.929	5796.0	7318.0	176.8	44.01	114.6	272.6
322.0	133.9	4.774	5970.0	7541.0	177.5	43.76	108.4	275.2
324.0	130.1	4.636	6134.0	7752.0	178.1	43.55	103.1	277.8
326.0	126.5	4.511	6291.0	7954.0	178.7	43.37	98.70	280.3
330.0	120.4	4.292	6586.0	8334.0	179.9	43.10	91.61	285.2
335.0	114.0	4.063	6929.0	8775.0	181.2	42.89	85.13	291.0
340.0	108.6	3.871	7251.0	9188.0	182.4	42.82	80.39	296.5
345.0	104.0	3.705	7557.0	9581.0	183.6	42.83	76.84	301.7

Thermophysical properties of ethylene—Continued

T K	Density kg/m³	Density mol/dm³	*E* J/mol	*H* J/mol	*S* J/(mol·K)	C_v J/(mol·K)	C_p J/(mol·K)	Sound m/s
350.0	99.89	3.561	7852.0	9958.0	184.7	42.93	74.11	306.6
355.0	96.29	3.432	8138.0	10320.0	185.7	43.09	71.98	311.3
360.0	93.06	3.317	8417.0	10680.0	186.7	43.30	70.31	315.9
370.0	87.46	3.118	8963.0	11370.0	188.6	43.84	67.94	324.4
380.0	82.74	2.949	9497.0	12040.0	190.4	44.51	66.44	332.4
390.0	78.68	2.805	10030.0	12700.0	192.1	45.25	65.53	340.0
400.0	75.12	2.678	10550.0	13350.0	193.8	46.06	65.02	347.1
				8.00 MPa isobar				
105.10[a]	659.1	23.49	−12440.0	−12100.0	84.50	41.63	77.47	1718.0
110.0	651.2	23.21	−12080.0	−11740.0	87.86	43.30	71.50	1709.0
120.0	637.7	22.73	−11400.0	−11050.0	93.90	44.13	67.92	1668.0
140.0	613.2	21.86	−10080.0	−9713.0	104.2	41.87	66.01	1559.0
160.0	588.1	20.96	−8775.0	−8393.0	113.0	39.44	66.15	1431.0
180.0	561.2	20.01	−7461.0	−7061.0	120.8	37.93	67.11	1294.0
200.0	532.6	18.98	−6127.0	−5705.0	128.0	37.24	68.62	1153.0
220.0	501.7	17.88	−4757.0	−4309.0	134.6	37.18	71.23	1008.0
230.0	485.0	17.29	−4050.0	−3587.0	137.8	37.36	73.26	932.6
240.0	467.2	16.65	−3322.0	−2842.0	141.0	37.66	76.02	854.7
250.0	447.8	15.96	−2565.0	−2064.0	144.2	38.10	79.79	773.9
260.0	426.1	15.19	−1768.0	−1241.0	147.4	38.71	85.06	689.6
265.0	414.1	14.76	−1349.0	−807.2	149.1	39.09	88.51	646.0
270.0	401.1	14.30	−913.9	−354.4	150.8	39.53	92.75	601.4
275.0	386.9	13.79	−457.9	122.1	152.5	40.06	98.09	555.4
280.0	371.1	13.23	24.4	629.2	154.3	40.69	105.1	508.1
282.0	364.2	12.98	226.6	842.8	155.1	40.98	108.6	488.6
284.0	356.9	12.72	435.0	1064.0	155.9	41.29	112.6	468.9
290.0	331.9	11.83	1109.0	1786.0	158.4	42.41	129.4	407.9
292.0	322.2	11.48	1356.0	2052.0	159.3	42.86	137.4	387.1
294.0	311.6	11.11	1616.0	2336.0	160.3	43.36	147.1	366.2
296.0	299.9	10.69	1894.0	2642.0	161.3	43.90	158.7	345.7
298.0	287.0	10.23	2191.0	2972.0	162.4	44.47	172.1	326.1
300.0	272.9	9.727	2508.0	3331.0	163.6	45.05	185.9	308.3
305.0	234.4	8.355	3365.0	4322.0	166.9	46.15	204.9	278.0
306.0	226.8	8.086	3538.0	4527.0	167.6	46.24	203.8	274.7
308.0	212.6	7.578	3873.0	4929.0	168.9	46.28	197.3	270.3
310.0	199.8	7.121	4190.0	5313.0	170.1	46.16	186.5	268.2
312.0	188.5	6.721	4483.0	5673.0	171.3	45.92	173.7	267.7
314.0	178.8	6.374	4753.0	6008.0	172.3	45.62	160.7	268.3
316.0	170.4	6.074	5000.0	6317.0	173.3	45.30	148.7	269.6
318.0	163.1	5.813	5227.0	6603.0	174.2	44.99	138.1	271.3
320.0	156.7	5.586	5438.0	6870.0	175.1	44.69	128.9	273.3
322.0	151.1	5.386	5635.0	7120.0	175.8	44.42	121.1	275.5
324.0	146.1	5.209	5819.0	7355.0	176.6	44.18	114.4	277.7
326.0	141.7	5.050	5994.0	7578.0	177.3	43.97	108.7	280.0
328.0	137.6	4.906	6160.0	7791.0	177.9	43.79	103.8	282.3
330.0	134.0	4.776	6319.0	7994.0	178.5	43.64	99.59	284.6
335.0	126.1	4.495	6690.0	8470.0	180.0	43.36	91.30	290.2
340.0	119.6	4.263	7034.0	8911.0	181.3	43.22	85.31	295.6
345.0	114.1	4.066	7358.0	9326.0	182.5	43.18	80.85	300.8
350.0	109.3	3.895	7667.0	9721.0	183.6	43.23	77.45	305.7
355.0	105.1	3.746	7965.0	10100.0	184.7	43.35	74.81	310.5
360.0	101.3	3.612	8255.0	10470.0	185.7	43.52	72.75	315.0
365.0	97.98	3.493	8539.0	10830.0	186.7	43.75	71.11	319.4
370.0	94.93	3.384	8817.0	11180.0	187.7	44.02	69.81	323.6
380.0	89.58	3.193	9364.0	11870.0	189.5	44.64	67.94	331.7
390.0	85.01	3.030	9903.0	12540.0	191.3	45.36	66.76	339.4
400.0	81.03	2.888	10440.0	13210.0	192.9	46.14	66.06	346.6
				8.50 MPa isobar				
105.20[a]	659.3	23.50	−12440.0	−12080.0	84.49	41.75	77.67	1716.0

Thermophysical properties of ethylene—Continued

T K	Density kg/m³	Density mol/dm³	*E* J/mol	*H* J/mol	*S* J/(mol·K)	C_v J/(mol·K)	C_p J/(mol·K)	Sound m/s
110.0	651.4	23.22	−12090.0	−11720.0	87.81	43.36	71.64	1708.0
120.0	638.0	22.74	−11400.0	−11030.0	93.86	44.17	67.94	1669.0
140.0	613.6	21.87	−10080.0	−9696.0	104.1	41.90	65.97	1561.0
160.0	588.5	20.98	−8783.0	−8378.0	112.9	39.46	66.09	1434.0
180.0	561.7	20.02	−7472.0	−7047.0	120.8	37.95	67.02	1298.0
200.0	533.3	19.01	−6141.0	−5694.0	127.9	37.26	68.49	1158.0
220.0	502.6	17.92	−4776.0	−4301.0	134.5	37.20	71.00	1014.0
230.0	486.1	17.33	−4072.0	−3582.0	137.7	37.37	72.94	939.7
240.0	468.6	16.70	−3349.0	−2840.0	140.9	37.66	75.56	863.0
250.0	449.5	16.02	−2598.0	−2068.0	144.0	38.09	79.11	783.8
260.0	428.4	15.27	−1810.0	−1253.0	147.2	38.67	83.97	701.6
265.0	416.8	14.86	−1398.0	−826.1	148.9	39.03	87.09	659.4
270.0	404.3	14.41	−971.3	−381.5	150.5	39.44	90.86	616.3
275.0	390.7	13.93	−526.3	83.9	152.2	39.93	95.48	572.3
280.0	375.8	13.40	−59.1	575.4	154.0	40.50	101.3	527.3
285.0	359.2	12.80	436.5	1100.0	155.9	41.17	109.0	481.2
290.0	340.2	12.13	969.7	1671.0	157.8	41.98	119.7	434.0
295.0	317.9	11.33	1555.0	2305.0	160.0	42.96	135.1	386.3
296.0	312.9	11.16	1680.0	2442.0	160.5	43.18	139.0	376.8
298.0	302.4	10.78	1940.0	2728.0	161.4	43.64	147.4	358.3
300.0	291.1	10.38	2213.0	3033.0	162.5	44.11	156.8	340.6
302.0	278.9	9.940	2501.0	3356.0	163.5	44.59	166.3	324.3
305.0	259.3	9.243	2954.0	3874.0	165.2	45.25	178.3	303.8
310.0	226.4	8.069	3730.0	4783.0	168.2	45.86	181.4	283.3
312.0	214.3	7.638	4028.0	5141.0	169.4	45.88	176.1	279.4
314.0	203.2	7.244	4313.0	5486.0	170.5	45.79	168.4	277.2
316.0	193.3	6.890	4580.0	5814.0	171.5	45.63	159.4	276.3
318.0	184.5	6.575	4831.0	6124.0	172.5	45.41	150.1	276.4
320.0	176.6	6.296	5065.0	6415.0	173.4	45.18	141.1	277.1
322.0	169.7	6.049	5283.0	6689.0	174.2	44.94	132.8	278.3
324.0	163.5	5.829	5488.0	6947.0	175.0	44.71	125.4	279.8
326.0	158.0	5.632	5681.0	7191.0	175.8	44.49	118.8	281.6
328.0	153.0	5.455	5864.0	7422.0	176.5	44.30	113.0	283.4
330.0	148.5	5.295	6038.0	7643.0	177.2	44.13	108.0	285.4
332.0	144.5	5.149	6204.0	7855.0	177.8	43.98	103.5	287.4
335.0	139.0	4.954	6441.0	8156.0	178.7	43.80	97.90	290.5
340.0	131.2	4.676	6809.0	8627.0	180.1	43.60	90.58	295.6
345.0	124.6	4.443	7152.0	9065.0	181.4	43.51	85.14	300.6
350.0	119.1	4.244	7477.0	9480.0	182.6	43.52	81.01	305.4
355.0	114.2	4.071	7789.0	9877.0	183.7	43.60	77.82	310.1
360.0	109.9	3.917	8090.0	10260.0	184.8	43.75	75.32	314.7
365.0	106.1	3.781	8383.0	10630.0	185.8	43.95	73.34	319.0
370.0	102.6	3.657	8670.0	10990.0	186.8	44.19	71.76	323.3
380.0	96.57	3.442	9230.0	11700.0	188.7	44.78	69.48	331.4
390.0	91.46	3.260	9779.0	12390.0	190.5	45.47	68.03	339.0
400.0	87.03	3.102	10320.0	13060.0	192.2	46.23	67.12	346.3

Thermophysical properties of ethylene—Continued

T K	Density kg/m^3	Density mol/dm^3	E J/mol	H J/mol	S J/(mol·K)	C_v J/(mol·K)	C_p J/(mol·K)	Sound m/s
			9.00 MPa isobar					
105.20[a]	659.5	23.51	−12440.0	−12060.0	84.48	41.86	77.88	1714.0
110.0	651.7	23.23	−12100.0	−11710.0	87.76	43.42	71.77	1707.0
120.0	638.3	22.75	−11410.0	−11010.0	93.81	44.20	67.96	1670.0
140.0	613.9	21.88	−10090.0	−9680.0	104.1	41.92	65.93	1563.0
160.0	588.9	20.99	−8791.0	−8362.0	112.9	39.48	66.02	1437.0
180.0	562.3	20.04	−7483.0	−7033.0	120.7	37.98	66.93	1302.0
200.0	533.9	19.03	−6155.0	−5682.0	127.8	37.29	68.36	1163.0
220.0	503.5	17.95	−4794.0	−4293.0	134.5	37.22	70.78	1020.0
230.0	487.2	17.37	−4094.0	−3576.0	137.6	37.38	72.64	946.6
240.0	469.9	16.75	−3375.0	−2838.0	140.8	37.67	75.13	871.1
250.0	451.1	16.08	−2630.0	−2071.0	143.9	38.08	78.47	793.4
260.0	430.5	15.35	−1851.0	−1265.0	147.1	38.64	82.98	713.1
265.0	419.3	14.94	−1445.0	−842.7	148.7	38.98	85.83	672.1
270.0	407.2	14.52	−1025.0	−405.4	150.3	39.37	89.20	630.4
275.0	394.3	14.05	−590.0	50.4	152.0	39.82	93.27	588.1
280.0	380.1	13.55	−135.4	528.8	153.7	40.34	98.26	545.1
285.0	364.6	13.00	342.7	1035.0	155.5	40.94	104.6	501.4
290.0	347.2	12.38	850.5	1578.0	157.4	41.65	112.8	457.2
295.0	327.4	11.67	1397.0	2168.0	159.4	42.48	124.0	412.8
300.0	304.4	10.85	1994.0	2824.0	161.6	43.43	138.9	369.6
305.0	277.8	9.902	2652.0	3561.0	164.0	44.44	155.9	331.5
310.0	248.6	8.863	3357.0	4372.0	166.7	45.25	166.6	304.1
315.0	220.6	7.864	4057.0	5201.0	169.3	45.56	162.6	289.7
316.0	215.5	7.681	4191.0	5362.0	169.8	45.55	160.2	288.1
318.0	205.8	7.336	4450.0	5677.0	170.8	45.49	154.3	285.9
320.0	197.0	7.021	4697.0	5979.0	171.8	45.37	147.7	284.8
322.0	188.9	6.735	4931.0	6268.0	172.7	45.22	140.7	284.5
324.0	181.7	6.477	5153.0	6542.0	173.5	45.05	133.8	284.9
326.0	175.2	6.244	5362.0	6803.0	174.3	44.87	127.3	285.6
328.0	169.3	6.034	5560.0	7052.0	175.1	44.69	121.3	286.8
330.0	163.9	5.843	5748.0	7289.0	175.8	44.53	115.8	288.1
332.0	159.1	5.670	5928.0	7515.0	176.5	44.38	110.9	289.6
334.0	154.6	5.512	6100.0	7733.0	177.2	44.25	106.5	291.3
340.0	143.3	5.109	6577.0	8339.0	179.0	43.96	96.00	296.6
345.0	135.7	4.837	6941.0	8802.0	180.3	43.84	89.61	301.2
350.0	129.2	4.606	7283.0	9237.0	181.6	43.80	84.72	305.9
355.0	123.6	4.407	7609.0	9651.0	182.7	43.85	80.95	310.4
360.0	118.7	4.232	7921.0	10050.0	183.8	43.97	77.99	314.8
365.0	114.4	4.077	8224.0	10430.0	184.9	44.14	75.65	319.1
370.0	110.5	3.938	8520.0	10810.0	185.9	44.36	73.78	323.3
380.0	103.7	3.697	9094.0	11530.0	187.8	44.91	71.07	331.4
390.0	98.02	3.494	9654.0	12230.0	189.7	45.58	69.31	339.0
400.0	93.13	3.320	10210.0	12920.0	191.4	46.32	68.19	346.3
			9.50 MPa isobar					
105.30[a]	659.7	23.51	−12440.0	−12040.0	84.47	41.98	78.09	1712.0
110.0	652.0	23.24	−12100.0	−11690.0	87.72	43.48	71.91	1707.0
120.0	638.5	22.76	−11410.0	−11000.0	93.77	44.24	67.99	1670.0
140.0	614.2	21.89	−10100.0	−9663.0	104.0	41.94	65.89	1566.0
160.0	589.3	21.01	−8799.0	−8347.0	112.8	39.50	65.96	1440.0
180.0	562.8	20.06	−7493.0	−7019.0	120.7	38.00	66.85	1306.0
200.0	534.6	19.06	−6168.0	−5670.0	127.8	37.31	68.23	1167.0
220.0	504.4	17.98	−4812.0	−4284.0	134.4	37.24	70.57	1026.0
230.0	488.3	17.41	−4116.0	−3570.0	137.5	37.40	72.35	953.4
240.0	471.2	16.80	−3401.0	−2835.0	140.7	37.68	74.72	879.0
250.0	452.8	16.14	−2662.0	−2073.0	143.8	38.08	77.87	802.6
260.0	432.6	15.42	−1890.0	−1274.0	146.9	38.61	82.07	724.2
265.0	421.6	15.03	−1490.0	−857.5	148.5	38.94	84.68	684.2

Thermophysical properties of ethylene—Continued

T K	Density kg/m^3	Density mol/dm^3	E J/mol	H J/mol	S J/(mol·K)	C_v J/(mol·K)	C_p J/(mol·K)	Sound m/s
270.0	410.0	14.61	−1077.0	−426.6	150.1	39.31	87.74	643.8
275.0	397.5	14.17	−649.6	20.8	151.7	39.73	91.36	602.9
280.0	384.1	13.69	−205.8	488.1	153.4	40.21	95.69	561.6
285.0	369.4	13.17	257.9	979.3	155.2	40.76	101.0	520.0
290.0	353.3	12.59	745.8	1500.0	157.0	41.39	107.6	478.1
295.0	335.3	11.95	1264.0	2059.0	158.9	42.11	116.2	436.3
300.0	314.9	11.23	1820.0	2666.0	160.9	42.92	127.1	395.6
305.0	291.9	10.40	2420.0	3333.0	163.1	43.80	139.9	358.0
310.0	266.4	9.496	3062.0	4062.0	165.5	44.61	151.1	327.3
315.0	240.3	8.567	3721.0	4830.0	168.0	45.15	154.4	307.0
320.0	216.3	7.712	4357.0	5589.0	170.4	45.29	147.7	296.5
322.0	207.8	7.406	4597.0	5880.0	171.3	45.24	143.1	294.4
324.0	199.8	7.123	4828.0	6161.0	172.1	45.16	138.0	293.2
326.0	192.5	6.863	5048.0	6432.0	173.0	45.05	132.6	292.6
328.0	185.8	6.624	5258.0	6692.0	173.8	44.93	127.2	292.7
330.0	179.7	6.406	5458.0	6941.0	174.5	44.80	122.0	293.1
332.0	174.1	6.207	5649.0	7180.0	175.2	44.68	117.1	293.9
334.0	169.0	6.024	5833.0	7410.0	175.9	44.56	112.6	294.9
336.0	164.3	5.856	6008.0	7631.0	176.6	44.45	108.4	296.1
340.0	156.0	5.559	6340.0	8049.0	177.8	44.27	101.2	299.0
345.0	147.1	5.245	6726.0	8537.0	179.3	44.13	94.05	302.9
350.0	139.7	4.980	7085.0	8993.0	180.6	44.07	88.49	307.1
355.0	133.3	4.753	7425.0	9424.0	181.8	44.09	84.15	311.4
360.0	127.8	4.555	7750.0	9836.0	182.9	44.18	80.73	315.6
365.0	122.9	4.380	8063.0	10230.0	184.0	44.33	78.02	319.8
370.0	118.5	4.224	8368.0	10620.0	185.1	44.53	75.85	323.9
375.0	114.6	4.084	8665.0	10990.0	186.1	44.77	74.11	327.9
380.0	111.0	3.956	8957.0	11360.0	187.1	45.04	72.69	331.8
390.0	104.7	3.732	9528.0	12070.0	188.9	45.68	70.62	339.3
400.0	99.31	3.540	10090.0	12770.0	190.7	46.41	69.28	346.6

10.00 MPa isobar

T K	Density kg/m^3	Density mol/dm^3	E J/mol	H J/mol	S J/(mol·K)	C_v J/(mol·K)	C_p J/(mol·K)	Sound m/s
105.40[a]	659.9	23.52	−12450.0	−12020.0	84.47	42.09	78.31	1710.0
110.0	652.3	23.25	−12110.0	−11680.0	87.67	43.54	72.05	1706.0
120.0	638.8	22.77	−11420.0	−10980.0	93.73	44.28	68.01	1671.0
140.0	614.5	21.91	−10100.0	−9647.0	104.0	41.97	65.86	1568.0
160.0	589.7	21.02	−8807.0	−8331.0	112.8	39.53	65.90	1443.0
180.0	563.3	20.08	−7503.0	−7005.0	120.6	38.02	66.76	1309.0
200.0	535.3	19.08	−6182.0	−5658.0	127.7	37.33	68.11	1172.0
220.0	505.3	18.01	−4830.0	−4275.0	134.3	37.26	70.36	1032.0
230.0	489.4	17.44	−4137.0	−3563.0	137.4	37.41	72.07	960.1
240.0	472.5	16.84	−3426.0	−2832.0	140.6	37.69	74.33	886.7
250.0	454.3	16.19	−2692.0	−2074.0	143.6	38.08	77.31	811.6
260.0	434.6	15.49	−1928.0	−1283.0	146.8	38.59	81.23	734.8
270.0	412.6	14.71	−1125.0	−445.5	149.9	39.25	86.44	656.5
275.0	400.6	14.28	−705.7	−5.4	151.5	39.65	89.68	616.9
280.0	387.7	13.82	−271.4	452.3	153.2	40.10	93.50	577.1
285.0	373.8	13.32	180.2	930.8	154.9	40.60	98.05	537.2
290.0	358.6	12.78	652.1	1434.0	156.6	41.17	103.6	497.2
295.0	342.0	12.19	1148.0	1969.0	158.4	41.82	110.4	457.6
300.0	323.6	11.53	1674.0	2541.0	160.4	42.53	118.8	418.9
305.0	303.1	10.80	2233.0	3159.0	162.4	43.30	128.6	382.5
310.0	280.6	10.00	2827.0	3826.0	164.6	44.06	138.2	350.8
315.0	256.9	9.158	3443.0	4535.0	166.8	44.68	144.2	326.7
320.0	233.9	8.336	4057.0	5256.0	169.1	45.02	143.2	311.3
325.0	213.1	7.595	4640.0	5956.0	171.3	45.07	136.0	303.4
326.0	209.3	7.460	4751.0	6091.0	171.7	45.06	134.0	302.5
328.0	202.1	7.204	4967.0	6355.0	172.5	45.00	129.9	301.2
330.0	195.4	6.966	5176.0	6611.0	173.3	44.93	125.6	300.6
332.0	189.3	6.746	5376.0	6858.0	174.0	44.85	121.3	300.4
334.0	183.5	6.543	5568.0	7096.0	174.8	44.76	117.1	300.6
336.0	178.3	6.354	5753.0	7327.0	175.4	44.67	113.1	301.1

Thermophysical properties of ethylene—Continued

T K	Density kg/m^3	Density mol/dm^3	E J/mol	H J/mol	S J/(mol·K)	C_v J/(mol·K)	C_p J/(mol·K)	Sound m/s
340.0	168.9	6.019	6103.0	7764.0	176.7	44.51	105.8	302.8
345.0	158.9	5.663	6508.0	8274.0	178.2	44.37	98.23	305.8
350.0	150.5	5.364	6885.0	8749.0	179.6	44.30	92.15	309.4
355.0	143.3	5.108	7239.0	9197.0	180.9	44.31	87.32	313.1
360.0	137.1	4.885	7577.0	9624.0	182.1	44.38	83.48	317.1
365.0	131.6	4.690	7901.0	10030.0	183.2	44.51	80.41	321.0
370.0	126.7	4.516	8215.0	10430.0	184.3	44.69	77.95	324.9
375.0	122.3	4.360	8520.0	10810.0	185.3	44.91	75.95	328.8
380.0	118.4	4.219	8819.0	11190.0	186.3	45.17	74.34	332.6
390.0	111.4	3.972	9402.0	11920.0	188.2	45.79	71.95	340.0
400.0	105.6	3.763	9973.0	12630.0	190.0	46.49	70.37	347.2
			15.00 MPa isobar					
106.10[a]	661.8	23.59	−12450.0	−11820.0	84.38	43.13	80.58	1694.0
110.0	655.2	23.35	−12160.0	−11520.0	87.17	44.12	73.67	1700.0
120.0	641.5	22.87	−11470.0	−10810.0	93.30	44.62	68.29	1680.0
140.0	617.7	22.02	−10160.0	−9481.0	103.6	42.17	65.53	1591.0
160.0	593.6	21.16	−8884.0	−8175.0	112.3	39.72	65.32	1474.0
180.0	568.2	20.26	−7603.0	−6862.0	120.0	38.23	65.98	1346.0
200.0	541.6	19.31	−6310.0	−5533.0	127.0	37.55	67.02	1216.0
220.0	513.6	18.31	−4997.0	−4178.0	133.5	37.47	68.64	1086.0
240.0	483.8	17.25	−3650.0	−2780.0	139.5	37.83	71.31	955.2
250.0	467.9	16.68	−2958.0	−2058.0	142.5	38.16	73.18	889.8
260.0	451.2	16.08	−2248.0	−1315.0	145.4	38.57	75.46	824.5
270.0	433.4	15.45	−1518.0	−547.6	148.3	39.07	78.18	759.8
280.0	414.5	14.77	−765.5	249.8	151.2	39.65	81.37	696.4
290.0	394.1	14.05	13.5	1081.0	154.1	40.30	85.00	635.4
300.0	372.2	13.27	820.5	1951.0	157.1	41.04	89.03	577.5
310.0	348.8	12.43	1656.0	2863.0	160.1	41.83	93.34	524.0
320.0	323.9	11.55	2518.0	3818.0	163.1	42.64	97.56	476.4
340.0	272.5	9.713	4283.0	5827.0	169.2	44.11	101.9	406.2
350.0	248.4	8.854	5145.0	6839.0	172.1	44.67	100.1	385.9
355.0	237.3	8.459	5562.0	7336.0	173.5	44.91	98.35	379.1
360.0	227.0	8.090	5968.0	7822.0	174.9	45.13	96.21	374.1
365.0	217.4	7.749	6362.0	8297.0	176.2	45.35	93.87	370.7
370.0	208.6	7.435	6743.0	8761.0	177.5	45.57	91.47	368.5
375.0	200.5	7.147	7113.0	9212.0	178.7	45.80	89.14	367.4
380.0	193.1	6.884	7473.0	9652.0	179.8	46.04	86.95	367.0
390.0	180.1	6.420	8165.0	10500.0	182.0	46.58	83.11	368.1
400.0	169.1	6.027	8828.0	11320.0	184.1	47.19	80.08	370.8
			20.00 MPa isobar					
106.80[a]	663.8	23.66	−12460.0	−11610.0	84.30	44.03	83.15	1682.0
110.0	658.1	23.46	−12210.0	−11360.0	86.66	44.65	75.73	1695.0
120.0	644.2	22.96	−11520.0	−10640.0	92.87	44.92	68.67	1690.0
140.0	620.7	22.12	−10220.0	−9314.0	103.1	42.33	65.26	1614.0
160.0	597.3	21.29	−8955.0	−8016.0	111.8	39.88	64.82	1504.0
180.0	572.8	20.42	−7695.0	−6715.0	119.5	38.42	65.32	1381.0
200.0	547.4	19.51	−6426.0	−5401.0	126.4	37.77	66.14	1256.0
220.0	521.0	18.57	−5144.0	−4067.0	132.7	37.69	67.37	1133.0
240.0	493.4	17.59	−3839.0	−2702.0	138.7	38.05	69.29	1012.0
260.0	464.1	16.54	−2498.0	−1289.0	144.3	38.74	72.12	894.2
280.0	432.5	15.42	−1108.0	189.3	149.8	39.69	75.88	781.2
290.0	415.8	14.82	−390.9	958.7	152.5	40.25	78.02	728.0
300.0	398.3	14.20	341.2	1750.0	155.2	40.85	80.24	677.8
320.0	361.5	12.89	1847.0	3399.0	160.5	42.18	84.56	588.8
340.0	323.3	11.52	3390.0	5125.0	165.7	43.56	87.84	517.8
360.0	286.0	10.20	4934.0	6896.0	170.8	44.91	88.69	467.6
370.0	268.7	9.578	5692.0	7780.0	173.2	45.57	87.97	450.3

Thermophysical properties of ethylene—Continued

T K	Density kg/m³	Density mol/dm³	E J/mol	H J/mol	S J/(mol·K)	C_v J/(mol·K)	C_p J/(mol·K)	Sound m/s
380.0	252.6	9.004	6432.0	8653.0	175.6	46.22	86.64	437.8
390.0	238.0	8.482	7153.0	9511.0	177.8	46.88	84.96	429.1
400.0	224.7	8.011	7856.0	10350.0	179.9	47.57	83.18	423.7
				30.00 MPa isobar				
108.10[a]	667.9	23.81	−12470.0	−11210.0	84.09	45.45	89.57	1669.0
110.0	664.2	23.67	−12320.0	−11050.0	85.57	45.58	82.01	1686.0
115.0	656.2	23.39	−11950.0	−10670.0	88.99	45.69	73.40	1707.0
120.0	649.5	23.15	−11610.0	−10310.0	92.03	45.39	69.78	1711.0
140.0	626.4	22.33	−10320.0	−8978.0	102.3	42.59	64.88	1659.0
160.0	604.1	21.53	−9086.0	−7693.0	110.9	40.16	63.99	1561.0
180.0	581.2	20.72	−7860.0	−6412.0	118.4	38.77	64.23	1446.0
200.0	557.7	19.88	−6631.0	−5122.0	125.2	38.19	64.79	1328.0
250.0	496.7	17.70	−3520.0	−1825.0	139.9	38.89	67.47	1053.0
260.0	484.1	17.25	−2885.0	−1146.0	142.6	39.27	68.35	1002.0
280.0	458.2	16.33	−1596.0	240.8	147.7	40.21	70.43	904.6
300.0	431.4	15.38	−277.9	1673.0	152.7	41.31	72.79	815.8
350.0	362.5	12.92	3128.0	5450.0	164.3	44.59	77.90	644.1
360.0	348.9	12.44	3820.0	6232.0	166.5	45.30	78.55	619.1
380.0	322.6	11.50	5204.0	7812.0	170.8	46.76	79.35	577.8
400.0	298.2	10.63	6580.0	9402.0	174.9	48.28	79.52	547.3
				40.00 MPa isobar				
109.50[a]	672.1	23.96	−12490.0	−10820.0	83.83	46.49	98.67	1669.0
110.0	670.9	23.91	−12450.0	−10770.0	84.31	46.46	94.37	1676.0
112.0	666.8	23.77	−12280.0	−10590.0	85.91	46.36	84.47	1697.0
115.0	661.8	23.59	−12050.0	−10350.0	88.03	46.20	77.18	1717.0
120.0	654.8	23.34	−11700.0	−9984.0	91.18	45.74	71.46	1732.0
130.0	642.8	22.91	−11040.0	−9297.0	96.68	44.31	66.72	1729.0
140.0	631.8	22.52	−10420.0	−8641.0	101.5	42.76	64.68	1703.0
200.0	566.7	20.20	−6807.0	−4827.0	124.2	38.61	63.81	1393.0
250.0	510.9	18.21	−3793.0	−1596.0	138.6	39.50	65.67	1132.0
300.0	454.1	16.19	−698.4	1772.0	150.9	42.07	69.33	914.8
350.0	396.9	14.15	2516.0	5344.0	161.9	45.47	73.40	752.2
360.0	385.6	13.75	3171.0	6081.0	164.0	46.20	74.07	727.1
380.0	363.7	12.96	4489.0	7575.0	168.0	47.71	75.21	683.9
400.0	342.8	12.22	5815.0	9088.0	171.9	49.27	76.09	649.3

[a]At melting line.
[b]At liquid–vapor boundary.

Appendix H. Thermophysical Properties of Hydrogen

Thermophysical properties of coexisting gaseous and liquid hydrogen

T K	Pres. MPa	Density kg/m^3	Density mol/dm^3	E J/mol	H J/mol	S J/(mol·K)	C_v J/(mol·K)	C_p J/(mol·K)	Sound m/s	Diel. const.
13.800[a]	.00704	77.04	38.21	–622.6	–622.4	10.00	10.97	15.56	1376.0	1.25164
13.800[a]	.00704	.1275	.06322	169.6	281.0	75.45	12.74	21.53	305.0	1.00039
14.0	.00790	76.87	38.13	–619.5	–619.3	10.22	10.58	15.17	1367.0	1.25105
14.0	.00790	.1404	.06967	171.8	285.1	74.82	12.75	21.57	307.0	1.00042
14.5	.01038	76.44	37.92	–612.1	–611.8	10.74	9.95	14.57	1343.0	1.24955
14.5	.01038	.1778	.08820	177.3	295.0	73.28	12.79	21.70	311.8	1.00054
15.0	.01343	76.01	37.70	–604.9	–604.6	11.23	9.65	14.37	1319.0	1.24803
15.0	.01343	.2226	.1104	182.7	304.3	71.82	12.83	21.85	316.4	1.00067
15.5	.01712	75.57	37.49	–597.7	–597.3	11.70	9.58	14.45	1295.0	1.24648
15.5	.01712	.2756	.1367	187.9	313.1	70.44	12.86	22.02	320.8	1.00083
16.0	.02153	75.12	37.26	–590.5	–589.9	12.16	9.65	14.71	1272.0	1.24491
16.0	.02153	.3374	.1674	193.0	321.7	69.14	12.91	22.20	325.1	1.00102
16.5	.02674	74.66	37.04	–583.0	–582.3	12.62	9.80	15.09	1249.0	1.24330
16.5	.02674	.4087	.2028	198.0	329.8	67.90	12.95	22.40	329.1	1.00123
17.0	.03284	74.19	36.80	–575.4	–574.5	13.08	10.00	15.56	1227.0	1.24165
17.0	.03284	.4904	.2433	202.8	337.8	66.74	12.99	22.62	333.0	1.00148
17.5	.03992	73.71	36.56	–567.6	–566.5	13.53	10.23	16.08	1206.0	1.23997
17.5	.03992	.5831	.2892	207.4	345.4	65.64	13.04	22.86	336.7	1.00176
18.0	.04808	73.22	36.32	–559.4	–558.1	13.99	10.46	16.64	1185.0	1.23825
18.0	.04808	.6877	.3411	211.8	352.8	64.59	13.08	23.12	340.2	1.00207
18.5	.05739	72.71	36.07	–551.0	–549.4	14.45	10.69	17.23	1166.0	1.23649
18.5	.05739	.8051	.3994	216.1	359.8	63.60	13.13	23.41	343.6	1.00243
19.0	.06796	72.19	35.81	–542.4	–540.5	14.91	10.92	17.84	1147.0	1.23468
19.0	.06796	.9361	.4644	220.1	366.5	62.65	13.18	23.72	346.7	1.00282
19.5	.07989	71.66	35.55	–533.4	–531.2	15.38	11.13	18.47	1129.0	1.23281
19.5	.07989	1.082	.5366	224.0	372.9	61.74	13.23	24.07	349.7	1.00326
20.0	.09326	71.11	35.27	–524.1	–521.5	15.85	11.32	19.12	1111.0	1.23090
20.0	.09326	1.243	.6166	227.6	378.8	60.87	13.28	24.45	352.6	1.00375
20.5	.1082	70.54	34.99	–514.6	–511.5	16.32	11.51	19.79	1093.0	1.22893
20.5	.1082	1.421	.7049	230.9	384.4	60.02	13.34	24.87	355.2	1.00429
21.0	.1247	69.96	34.70	–504.7	–501.1	16.80	11.68	20.48	1075.0	1.22690
21.0	.1247	1.617	.8020	234.0	389.5	59.21	13.39	25.33	357.7	1.00488
21.5	.1431	69.35	34.40	–494.6	–490.4	17.28	11.83	21.19	1057.0	1.22480
21.5	.1431	1.832	.9087	236.9	394.4	58.43	13.45	25.84	360.1	1.00553
22.0	.1632	68.73	34.09	–484.1	–479.3	17.76	11.98	21.93	1038.0	1.22264
22.0	.1632	2.067	1.025	239.4	398.5	57.66	13.52	26.40	362.2	1.00624
22.5	.1853	68.08	33.77	–473.3	–467.8	18.25	12.11	22.71	1020.0	1.22041
22.5	.1853	2.325	1.153	241.6	402.3	56.92	13.58	27.03	364.3	1.00702
23.0	.2094	67.42	33.44	–462.1	–455.8	18.74	12.23	23.53	1001.0	1.21809
23.0	.2094	2.606	1.293	243.5	405.5	56.19	13.65	27.73	366.1	1.00787
23.5	.2357	66.72	33.10	–450.6	–443.5	19.24	12.35	24.41	981.2	1.21570
23.5	.2357	2.912	1.445	245.0	408.2	55.48	13.72	28.51	367.8	1.00880
24.0	.2642	66.00	32.74	–438.7	–430.7	19.75	12.46	25.34	961.2	1.21321
24.0	.2642	3.246	1.610	246.1	410.2	54.78	13.80	29.38	369.3	1.00982
24.5	.2951	65.25	32.37	–426.5	–417.3	20.26	12.56	26.36	940.6	1.21063
24.5	.2951	3.610	1.791	246.8	411.7	54.09	13.88	30.37	370.7	1.01092
25.0	.3285	64.47	31.98	–413.8	–403.5	20.77	12.65	27.46	919.3	1.20794
25.0	.3285	4.006	1.987	247.1	412.4	53.40	13.97	31.50	371.9	1.01212
25.5	.3643	63.66	31.58	–400.7	–389.2	21.29	12.75	28.67	897.2	1.20515
25.5	.3643	4.437	2.201	246.9	412.4	52.73	14.07	32.79	373.0	1.01343

Thermophysical properties of coexisting gaseous and liquid hydrogen—Continued

T K	Pres. MPa	Density kg/m^3	Density mol/dm^3	E J/mol	H J/mol	S J/(mol·K)	C_v J/(mol·K)	C_p J/(mol·K)	Sound m/s	Diel. const.
26.0	.4029	62.80	31.15	−387.2	−374.3	21.83	12.84	30.02	874.4	1.20222
26.0	.4029	4.907	2.434	246.1	411.6	52.05	14.17	34.27	373.9	1.01486
26.5	.4443	61.91	30.71	−373.2	−358.7	22.37	12.93	31.53	850.8	1.19916
26.5	.4443	5.421	2.689	244.8	410.0	51.37	14.28	35.99	374.7	1.01643
27.0	.4885	60.97	30.24	−358.6	−342.5	22.92	13.02	33.25	826.2	1.19595
27.0	.4885	5.982	2.967	242.8	407.4	50.69	14.40	38.03	375.3	1.01814
27.5	.5357	59.98	29.75	−343.5	−325.4	23.49	13.12	35.24	800.6	1.19257
27.5	.5357	6.598	3.273	240.0	403.7	50.00	14.52	40.45	375.7	1.02002
28.0	.5861	58.93	29.23	−327.7	−307.6	24.07	13.22	37.58	774.0	1.18899
28.0	.5861	7.276	3.609	236.5	398.9	49.29	14.66	43.38	376.0	1.02210
28.5	.6397	57.80	28.67	−311.1	−288.8	24.67	13.33	40.39	746.1	1.18517
28.5	.6397	8.026	3.981	232.0	392.6	48.57	14.82	47.00	376.2	1.02439
29.0	.6967	56.60	28.08	−293.7	−268.9	25.29	13.46	43.84	716.8	1.18109
29.0	.6967	8.860	4.395	226.3	384.8	47.83	14.98	51.58	376.2	1.02695
29.5	.7573	55.29	27.43	−275.3	−247.7	25.94	13.60	48.24	686.0	1.17667
29.5	.7573	9.797	4.860	219.4	375.2	47.05	15.17	57.55	376.0	1.02983
30.0	.8214	53.86	26.72	−255.6	−224.8	26.63	13.76	54.05	653.5	1.17185
30.0	.8214	10.86	5.387	210.8	363.2	46.23	15.38	65.64	375.7	1.03311
30.5	.8895	52.27	25.93	−234.4	−200.1	27.36	13.96	62.18	619.0	1.16651
30.5	.8895	12.09	5.996	200.1	348.4	45.34	15.61	77.23	375.2	1.03690
31.0	.9615	50.48	25.04	−211.2	−172.8	28.16	14.19	74.37	582.4	1.16049
31.0	.9615	13.54	6.715	186.6	329.8	44.36	15.88	95.15	374.5	1.04139
31.5	1.038	48.39	24.00	−185.2	−141.9	29.05	14.49	94.71	543.4	1.15352
31.5	1.038	15.31	7.597	169.0	305.7	43.25	16.20	126.4	373.7	1.04692
32.0	1.119	45.74	22.69	−153.9	−104.6	30.11	14.91	138.5	499.4	1.14472
32.0	1.119	17.65	8.754	144.8	272.6	41.90	16.57	194.1	372.8	1.05420
32.5	1.204	38.55	19.12	−80.7	−17.7	32.67	16.29	2151.0	407.9	1.12105
32.5	1.204	24.50	12.16	65.3	164.4	38.28	17.48	7867.0	367.6	1.07581
32.94[b]	1.283	31.36	15.56	−3.3	79.2	35.49				1.09773

[a]Triple point.
[b]Critical point.

Thermophysical properties of hydrogen on the melting line

T K	Pres. MPa	Density kg/m^3	Density mol/dm^3	E J/mol	H J/mol	S J/(mol·K)	C_v J/(mol·K)	C_p J/(mol·K)	Sound m/s	Diel. const.
13.80[a]	.00704	77.04	38.21	−622.6	−622.4	10.00	10.97	15.56	1376.0	1.25164
15.0	3.737	78.91	39.14	−617.3	−521.8	10.29	10.60	13.49	1332.0	1.25823
15.5	5.396	79.76	39.56	−614.5	−478.1	10.39	10.31	13.03	1356.0	1.26124
16.0	7.112	80.61	39.99	−611.6	−433.7	10.47	10.04	12.75	1393.0	1.26424
16.5	8.883	81.44	40.40	−608.4	−388.5	10.54	9.84	12.61	1439.0	1.26718
17.0	10.71	82.25	40.80	−605.1	−342.6	10.60	9.73	12.60	1490.0	1.27004
17.5	12.58	83.02	41.18	−601.5	−296.0	10.65	9.70	12.68	1541.0	1.27281
18.0	14.51	83.77	41.56	−597.8	−248.6	10.69	9.75	12.83	1591.0	1.27548
18.5	16.49	84.50	41.91	−593.8	−200.4	10.74	9.87	13.03	1640.0	1.27805
19.0	18.51	85.20	42.26	−589.5	−151.5	10.78	10.03	13.24	1685.0	1.28055
19.5	20.59	85.87	42.60	−585.0	−101.7	10.83	10.22	13.47	1728.0	1.28296
20.0	22.71	86.53	42.92	−580.2	−51.2	10.88	10.44	13.70	1767.0	1.28531
20.5	24.87	87.17	43.24	−575.0	.2	10.93	10.67	13.92	1805.0	1.28760
21.0	27.08	87.79	43.55	−569.6	52.3	10.99	10.90	14.13	1839.0	1.28984
21.5	29.34	88.41	43.85	−563.8	105.2	11.05	11.14	14.33	1872.0	1.29204
22.0	31.64	89.01	44.15	−557.7	158.9	11.11	11.37	14.51	1903.0	1.29419
22.5	33.98	89.60	44.44	−551.2	213.3	11.18	11.60	14.68	1932.0	1.29632
23.0	36.36	90.18	44.73	−544.5	268.4	11.26	11.81	14.84	1960.0	1.29841
23.5	38.79	90.75	45.02	−537.4	324.3	11.33	12.02	14.97	1987.0	1.30048
24.0	41.26	91.32	45.30	−530.0	380.8	11.41	12.21	15.10	2013.0	1.30253
24.5	43.77	91.89	45.58	−522.3	438.1	11.49	12.39	15.21	2038.0	1.30456
25.0	46.33	92.44	45.86	−514.2	496.0	11.58	12.56	15.31	2062.0	1.30657
25.5	48.92	92.99	46.13	−505.9	554.5	11.66	12.72	15.39	2085.0	1.30857
26.0	51.55	93.54	46.40	−497.3	613.6	11.75	12.86	15.47	2108.0	1.31055
26.5	54.22	94.09	46.67	−488.4	673.3	11.84	13.00	15.54	2129.0	1.31252
27.0	56.93	94.63	46.94	−479.2	733.5	11.93	13.12	15.60	2150.0	1.31448
27.5	59.67	95.16	47.21	−469.7	794.3	12.02	13.23	15.66	2171.0	1.31643
28.0	62.46	95.70	47.47	−460.0	855.7	12.11	13.33	15.71	2191.0	1.31838
28.5	65.28	96.23	47.74	−450.0	917.6	12.20	13.42	15.75	2210.0	1.32032
29.0	68.14	96.77	48.00	−439.7	979.9	12.29	13.51	15.79	2229.0	1.32226
29.5	71.04	97.30	48.26	−429.1	1043.0	12.38	13.59	15.83	2247.0	1.32419
30.0	73.98	97.83	48.53	−418.3	1106.0	12.47	13.66	15.87	2265.0	1.32612
30.5	76.95	98.35	48.79	−407.3	1170.0	12.56	13.72	15.90	2282.0	1.32805
31.0	79.97	98.88	49.05	−396.0	1234.0	12.64	13.78	15.93	2298.0	1.32998
31.5	83.02	99.41	49.31	−384.5	1299.0	12.73	13.83	15.97	2314.0	1.33190
32.0	86.10	99.94	49.57	−372.7	1364.0	12.82	13.88	16.00	2330.0	1.33383
32.5	89.23	100.5	49.84	−360.7	1430.0	12.90	13.93	16.04	2344.0	1.33577
33.0	92.39	101.0	50.10	−348.5	1496.0	12.98	13.97	16.08	2358.0	1.33770
33.5	95.59	101.5	50.36	−336.1	1562.0	13.06	14.01	16.13	2372.0	1.33964
34.0	98.82	102.1	50.62	−323.4	1629.0	13.14	14.05	16.17	2385.0	1.34159
34.5	102.1	102.6	50.89	−310.6	1696.0	13.22	14.09	16.23	2397.0	1.34354
35.0	105.4	103.1	51.15	−297.5	1763.0	13.29	14.12	16.29	2408.0	1.34550
35.5	108.7	103.7	51.42	−284.2	1831.0	13.36	14.16	16.36	2419.0	1.34747
36.0	112.1	104.2	51.68	−270.7	1899.0	13.43	14.20	16.44	2429.0	1.34945
36.5	115.5	104.7	51.95	−257.0	1967.0	13.49	14.23	16.53	2439.0	1.35145
37.0	119.0	105.3	52.22	−243.1	2036.0	13.55	14.27	16.64	2447.0	1.35345

[a]Triple point.

Thermophysical properties of hydrogen

T K	Density kg/m³	Density mol/dm³	*E* J/mol	*H* J/mol	*S* J/(mol·K)	C_v J/(mol·K)	C_p J/(mol·K)	Sound m/s	Diel. const.
				.005 MPa isobar					
14.0	.08755	.04343	173.1	288.2	78.79	12.64	21.27	308.2	1.00026
15.0	.08157	.04046	185.8	309.4	80.25	12.60	21.16	319.4	1.00025
16.0	.07636	.03788	198.5	330.5	81.61	12.56	21.08	330.2	1.00023
17.0	.07179	.03561	211.2	351.6	82.89	12.54	21.02	340.6	1.00022
18.0	.06774	.03360	223.8	372.6	84.09	12.52	20.98	350.6	1.00021
19.0	.06413	.03181	236.4	393.5	85.22	12.51	20.95	360.4	1.00019
20.0	.06089	.03020	248.9	414.5	86.30	12.50	20.93	369.9	1.00018
21.0	.05796	.02875	261.5	435.4	87.32	12.50	20.91	379.2	1.00018
22.0	.05530	.02743	274.0	456.3	88.29	12.49	20.89	388.2	1.00017
23.0	.05288	.02623	286.5	477.2	89.22	12.49	20.88	397.0	1.00016
24.0	.05066	.02513	299.1	498.0	90.11	12.48	20.87	405.6	1.00015
26.0	.04674	.02318	324.1	539.8	91.78	12.48	20.85	422.3	1.00014
28.0	.04338	.02152	349.1	581.4	93.32	12.48	20.84	438.3	1.00013
30.0	.04048	.02008	374.1	623.1	94.76	12.48	20.83	453.8	1.00012
32.0	.03794	.01882	399.1	664.8	96.10	12.47	20.82	468.7	1.00012
34.0	.03570	.01771	424.1	706.4	97.36	12.47	20.82	483.2	1.00011
36.0	.03371	.01672	449.0	748.0	98.55	12.47	20.82	497.3	1.00010
38.0	.03193	.01584	474.0	789.7	99.68	12.47	20.81	510.9	1.00010
40.0	.03033	.01505	499.0	831.3	100.7	12.48	20.82	524.1	1.00009
42.0	.02889	.01433	524.0	873.0	101.8	12.50	20.84	536.9	1.00009
44.0	.02757	.01368	549.0	914.7	102.7	12.53	20.86	549.3	1.00008
46.0	.02637	.01308	574.1	956.4	103.7	12.56	20.89	561.4	1.00008
48.0	.02527	.01254	599.3	998.2	104.6	12.59	20.92	573.2	1.00008
50.0	.02426	.01203	624.6	1040.0	105.4	12.64	20.97	584.7	1.00007
52.0	.02332	.01157	649.9	1082.0	106.2	12.69	21.02	595.7	1.00007
54.0	.02246	.01114	675.4	1124.0	107.0	12.76	21.09	606.4	1.00007
56.0	.02166	.01074	701.0	1166.0	107.8	12.85	21.17	616.8	1.00007
58.0	.02091	.01037	726.8	1209.0	108.5	12.95	21.28	626.7	1.00006
60.0	.02021	.01003	752.8	1252.0	109.3	13.07	21.40	636.2	1.00006
65.0	.01866	.009254	819.1	1359.0	111.0	13.47	21.79	658.5	1.00006
70.0	.01732	.008593	887.7	1470.0	112.6	13.98	22.30	678.6	1.00005
75.0	.01617	.008020	959.1	1583.0	114.2	14.60	22.92	696.8	1.00005
80.0	.01516	.007519	1034.0	1699.0	115.7	15.32	23.64	713.5	1.00005
85.0	.01426	.007076	1112.0	1819.0	117.1	16.10	24.42	729.1	1.00004
90.0	.01347	.006683	1195.0	1943.0	118.6	16.94	25.26	743.9	1.00004
95.0	.01276	.006331	1282.0	2072.0	119.9	17.81	26.13	758.1	1.00004
100.0	.01212	.006015	1373.0	2205.0	121.3	18.68	27.00	772.0	1.00004
105.0	.01155	.005728	1469.0	2342.0	122.6	19.54	27.86	785.7	1.00004
110.0	.01102	.005468	1569.0	2483.0	124.0	20.36	28.68	799.4	1.00003
115.0	.01054	.005230	1672.0	2628.0	125.3	21.13	29.45	813.0	1.00003
120.0	.01010	.005012	1780.0	2777.0	126.5	21.83	30.15	826.7	1.00003
125.0	.00970	.004812	1890.0	2930.0	127.8	22.45	30.77	840.6	1.00003
130.0	.00933	.004627	2004.0	3085.0	129.0	22.98	31.30	854.5	1.00003
140.0	.00866	.004296	2238.0	3402.0	131.3	23.82	32.14	882.6	1.00003
150.0	.00808	.004010	2479.0	3727.0	133.6	24.35	32.67	911.0	1.00003
160.0	.00758	.003759	2724.0	4055.0	135.7	24.62	32.93	939.6	1.00002
170.0	.00713	.003538	2971.0	4385.0	137.7	24.67	32.98	968.3	1.00002
180.0	.00674	.003342	3217.0	4714.0	139.6	24.56	32.88	996.9	1.00002
190.0	.00638	.003166	3462.0	5042.0	141.3	24.35	32.67	1025.0	1.00002
200.0	.00606	.003008	3704.0	5367.0	143.0	24.08	32.39	1053.0	1.00002
210.0	.00577	.002865	3943.0	5689.0	144.6	23.77	32.08	1081.0	1.00002
220.0	.00551	.002734	4179.0	6009.0	146.1	23.45	31.76	1109.0	1.00002
230.0	.00527	.002616	4412.0	6325.0	147.5	23.13	31.45	1136.0	1.00002
240.0	.00505	.002507	4642.0	6638.0	148.8	22.84	31.15	1162.0	1.00002
250.0	.00485	.002406	4869.0	6948.0	150.1	22.57	30.88	1188.0	1.00002
260.0	.00466	.002314	5094.0	7255.0	151.3	22.32	30.64	1213.0	1.00002
270.0	.00449	.002228	5316.0	7561.0	152.4	22.10	30.42	1238.0	1.00001
280.0	.00433	.002149	5536.0	7864.0	153.5	21.91	30.23	1262.0	1.00001
300.0	.00404	.002006	5971.0	8465.0	155.6	21.61	29.92	1309.0	1.00001
320.0	.00379	.001880	6401.0	9061.0	157.5	21.39	29.71	1354.0	1.00001
340.0	.00357	.001770	6827.0	9654.0	159.3	21.24	29.56	1397.0	1.00001

Thermophysical properties of hydrogen—Continued

T K	Density kg/m^3	Density mol/dm^3	E J/mol	H J/mol	S J/(mol·K)	C_v J/(mol·K)	C_p J/(mol·K)	Sound m/s	Diel. const.
360.0	.00337	.001671	7251.0	10240.0	161.0	21.14	29.45	1438.0	1.00001
380.0	.00319	.001584	7673.0	10830.0	162.6	21.07	29.38	1478.0	1.00001
400.0	.00303	.001504	8094.0	11420.0	164.1	21.01	29.33	1517.0	1.00001
				.01 MPa isobar					
13.80[a]	77.04	38.21	−622.5	−622.2	10.01	10.96	15.55	1375.0	1.25163
14.430[b]	76.50	37.95	−613.1	−612.9	10.67	10.01	14.62	1347.0	1.24976
14.430[b]	.1721	.08537	176.5	293.7	73.50	12.78	21.68	311.1	1.00052
15.0	.1647	.08169	184.0	306.4	74.36	12.73	21.56	317.6	1.00050
16.0	.1539	.07636	196.9	327.8	75.75	12.66	21.39	328.7	1.00046
17.0	.1446	.07171	209.7	349.2	77.04	12.61	21.27	339.3	1.00044
18.0	.1363	.06760	222.5	370.4	78.25	12.58	21.19	349.5	1.00041
19.0	.1289	.06395	235.2	391.6	79.40	12.55	21.12	359.4	1.00039
20.0	.1223	.06068	247.8	412.7	80.48	12.54	21.07	369.0	1.00037
21.0	.1164	.05773	260.5	433.7	81.51	12.52	21.03	378.4	1.00035
22.0	.1110	.05506	273.1	454.7	82.48	12.51	21.00	387.5	1.00034
23.0	.1061	.05262	285.7	475.7	83.42	12.50	20.97	396.3	1.00032
24.0	.1016	.05040	298.2	496.7	84.31	12.50	20.95	405.0	1.00031
26.0	.09369	.04648	323.4	538.5	85.98	12.49	20.92	421.8	1.00028
28.0	.08693	.04312	348.4	580.3	87.53	12.48	20.89	437.9	1.00026
30.0	.08109	.04022	373.5	622.1	88.97	12.48	20.88	453.5	1.00025
32.0	.07598	.03769	398.5	663.8	90.32	12.48	20.86	468.4	1.00023
34.0	.07148	.03546	423.5	705.6	91.59	12.48	20.85	483.0	1.00022
36.0	.06749	.03348	448.5	747.2	92.78	12.48	20.84	497.1	1.00020
38.0	.06392	.03171	473.5	788.9	93.90	12.47	20.84	510.7	1.00019
40.0	.06071	.03012	498.5	830.6	94.97	12.48	20.84	524.0	1.00018
42.0	.05781	.02868	523.6	872.3	95.99	12.51	20.86	536.8	1.00018
44.0	.05517	.02737	548.6	914.0	96.96	12.53	20.88	549.2	1.00017
46.0	.05277	.02618	573.8	955.8	97.89	12.56	20.91	561.4	1.00016
48.0	.05056	.02508	599.0	997.7	98.78	12.60	20.94	573.1	1.00015
50.0	.04853	.02408	624.2	1040.0	99.64	12.64	20.98	584.6	1.00015
52.0	.04666	.02315	649.6	1082.0	100.5	12.69	21.03	595.7	1.00014
54.0	.04493	.02229	675.1	1124.0	101.3	12.76	21.10	606.4	1.00014
56.0	.04332	.02149	700.7	1166.0	102.0	12.85	21.18	616.7	1.00013
58.0	.04183	.02075	726.5	1208.0	102.8	12.95	21.29	626.7	1.00013
60.0	.04043	.02006	752.5	1251.0	103.5	13.07	21.41	636.2	1.00012
65.0	.03732	.01851	818.9	1359.0	105.2	13.47	21.80	658.5	1.00011
70.0	.03465	.01719	887.5	1469.0	106.9	13.98	22.31	678.6	1.00011
75.0	.03234	.01604	958.9	1582.0	108.4	14.60	22.93	696.8	1.00010
80.0	.03031	.01504	1034.0	1699.0	109.9	15.32	23.64	713.5	1.00009
85.0	.02853	.01415	1112.0	1819.0	111.4	16.10	24.43	729.1	1.00009
90.0	.02694	.01337	1195.0	1943.0	112.8	16.94	25.27	743.9	1.00008
95.0	.02552	.01266	1282.0	2072.0	114.2	17.81	26.13	758.2	1.00008
100.0	.02425	.01203	1373.0	2204.0	115.5	18.68	27.00	772.1	1.00007
105.0	.02309	.01146	1469.0	2342.0	116.9	19.54	27.86	785.8	1.00007
110.0	.02204	.01093	1568.0	2483.0	118.2	20.36	28.68	799.4	1.00007
115.0	.02108	.01046	1672.0	2628.0	119.5	21.13	29.45	813.0	1.00006
120.0	.02021	.01002	1780.0	2777.0	120.8	21.83	30.15	826.8	1.00006
125.0	.01940	.009623	1890.0	2930.0	122.0	22.45	30.77	840.6	1.00006
130.0	.01865	.009252	2004.0	3085.0	123.2	22.98	31.30	854.6	1.00006
140.0	.01732	.008592	2238.0	3402.0	125.6	23.82	32.14	882.6	1.00005
150.0	.01616	.008019	2479.0	3727.0	127.8	24.35	32.67	911.1	1.00005
160.0	.01515	.007518	2724.0	4055.0	129.9	24.62	32.93	939.6	1.00005
170.0	.01426	.007075	2971.0	4385.0	131.9	24.67	32.99	968.3	1.00004
180.0	.01347	.006682	3217.0	4714.0	133.8	24.56	32.88	996.9	1.00004
190.0	.01276	.006331	3462.0	5042.0	135.6	24.35	32.67	1025.0	1.00004
200.0	.01212	.006014	3704.0	5367.0	137.2	24.08	32.39	1054.0	1.00004
210.0	.01155	.005728	3943.0	5689.0	138.8	23.77	32.08	1081.0	1.00004
220.0	.01102	.005468	4179.0	6009.0	140.3	23.45	31.76	1109.0	1.00003
230.0	.01054	.005230	4412.0	6325.0	141.7	23.13	31.45	1136.0	1.00003
240.0	.01010	.005012	4642.0	6638.0	143.0	22.84	31.15	1162.0	1.00003
250.0	.00970	.004812	4869.0	6948.0	144.3	22.57	30.88	1188.0	1.00003

Thermophysical properties of hydrogen—Continued

T K	Density kg/m^3	Density mol/dm^3	E J/mol	H J/mol	S J/(mol·K)	C_v J/(mol·K)	C_p J/(mol·K)	Sound m/s	Diel. const.
260.0	.00933	.004627	5094.0	7255.0	145.5	22.32	30.64	1213.0	1.00003
270.0	.00898	.004455	5316.0	7561.0	146.7	22.10	30.42	1238.0	1.00003
280.0	.00866	.004296	5536.0	7864.0	147.8	21.91	30.23	1262.0	1.00003
300.0	.00808	.004010	5971.0	8465.0	149.8	21.61	29.92	1309.0	1.00003
320.0	.00758	.003759	6401.0	9061.0	151.8	21.39	29.71	1354.0	1.00002
340.0	.00713	.003538	6827.0	9654.0	153.6	21.24	29.56	1397.0	1.00002
360.0	.00674	.003342	7251.0	10240.0	155.3	21.14	29.45	1438.0	1.00002
380.0	.00638	.003166	7673.0	10830.0	156.8	21.07	29.38	1479.0	1.00002
400.0	.00606	.003008	8093.0	11420.0	158.3	21.02	29.33	1517.0	1.00002
				.02 MPa isobar					
13.81[a]	77.04	38.22	−622.5	−622.0	10.01	10.96	15.54	1375.0	1.25165
14.0	76.88	38.13	−619.5	−619.0	10.22	10.59	15.17	1366.0	1.25107
15.0	76.01	37.70	−604.9	−604.4	11.23	9.65	14.37	1319.0	1.24803
15.836[b]	75.27	37.34	−592.9	−592.3	12.01	9.62	14.61	1279.0	1.24543
15.836[b]	.3162	.1568	191.4	318.9	69.55	12.89	22.14	323.7	1.00095
20.0	.2469	.1225	245.6	409.0	74.61	12.60	21.37	367.2	1.00075
21.0	.2346	.1164	258.5	430.3	75.65	12.58	21.29	376.8	1.00071
22.0	.2236	.1109	271.2	451.5	76.64	12.55	21.22	386.0	1.00067
23.0	.2135	.1059	283.9	472.7	77.58	12.54	21.16	395.1	1.00064
24.0	.2044	.1014	296.6	493.9	78.48	12.52	21.12	403.9	1.00062
26.0	.1883	.09339	321.9	536.0	80.16	12.51	21.05	420.9	1.00057
28.0	.1745	.08658	347.1	578.1	81.72	12.50	21.00	437.1	1.00053
30.0	.1627	.08071	372.3	620.1	83.17	12.49	20.97	452.8	1.00049
32.0	.1524	.07559	397.4	662.0	84.52	12.48	20.94	467.9	1.00046
34.0	.1433	.07109	422.5	703.8	85.79	12.48	20.92	482.5	1.00043
36.0	.1353	.06710	447.6	745.7	86.99	12.48	20.90	496.7	1.00041
38.0	.1281	.06353	472.6	787.4	88.12	12.48	20.89	510.4	1.00039
40.0	.1216	.06033	497.7	829.2	89.19	12.48	20.89	523.7	1.00037
42.0	.1158	.05743	522.7	871.0	90.21	12.51	20.90	536.6	1.00035
44.0	.1105	.05480	547.9	912.8	91.18	12.53	20.92	549.0	1.00033
46.0	.1056	.05240	573.0	954.7	92.11	12.56	20.94	561.2	1.00032
48.0	.1012	.05021	598.2	996.6	93.00	12.60	20.97	573.0	1.00031
50.0	.09715	.04819	623.5	1039.0	93.86	12.64	21.01	584.5	1.00029
52.0	.09339	.04633	648.9	1081.0	94.68	12.70	21.06	595.6	1.00028
54.0	.08992	.04461	674.4	1123.0	95.48	12.76	21.13	606.3	1.00027
56.0	.08670	.04301	700.1	1165.0	96.25	12.85	21.21	616.7	1.00026
58.0	.08370	.04152	725.9	1208.0	97.00	12.95	21.31	626.6	1.00025
60.0	.08090	.04013	752.0	1250.0	97.72	13.08	21.43	636.2	1.00024
65.0	.07466	.03703	818.4	1358.0	99.45	13.47	21.81	658.5	1.00023
70.0	.06931	.03438	887.0	1469.0	101.1	13.98	22.32	678.6	1.00021
75.0	.06468	.03209	958.5	1582.0	102.6	14.60	22.94	696.8	1.00020
80.0	.06063	.03008	1033.0	1698.0	104.1	15.32	23.65	713.6	1.00018
85.0	.05706	.02831	1112.0	1818.0	105.6	16.11	24.44	729.2	1.00017
90.0	.05389	.02673	1195.0	1943.0	107.0	16.94	25.27	744.0	1.00016
95.0	.05105	.02532	1281.0	2071.0	108.4	17.81	26.14	758.2	1.00015
100.0	.04850	.02406	1373.0	2204.0	109.8	18.68	27.01	772.1	1.00015
105.0	.04619	.02291	1468.0	2341.0	111.1	19.54	27.87	785.8	1.00014
110.0	.04408	.02187	1568.0	2483.0	112.4	20.36	28.69	799.5	1.00013
115.0	.04217	.02092	1672.0	2628.0	113.7	21.13	29.46	813.1	1.00013
120.0	.04041	.02005	1779.0	2777.0	115.0	21.83	30.15	826.8	1.00012
125.0	.03879	.01924	1890.0	2930.0	116.2	22.45	30.77	840.7	1.00012
130.0	.03730	.01850	2004.0	3085.0	117.5	22.98	31.30	854.6	1.00011
140.0	.03464	.01718	2238.0	3402.0	119.8	23.82	32.14	882.7	1.00011
150.0	.03233	.01604	2479.0	3726.0	122.0	24.35	32.67	911.2	1.00010
160.0	.03031	.01503	2724.0	4055.0	124.2	24.62	32.94	939.7	1.00009
170.0	.02852	.01415	2971.0	4384.0	126.2	24.67	32.99	968.4	1.00009
180.0	.02694	.01336	3217.0	4714.0	128.0	24.57	32.88	997.0	1.00008
190.0	.02552	.01266	3462.0	5042.0	129.8	24.35	32.67	1025.0	1.00008
200.0	.02424	.01203	3704.0	5367.0	131.5	24.08	32.39	1054.0	1.00007
210.0	.02309	.01145	3943.0	5690.0	133.1	23.77	32.08	1081.0	1.00007
220.0	.02204	.01093	4179.0	6009.0	134.5	23.45	31.76	1109.0	1.00007

Thermophysical properties of hydrogen—Continued

T K	Density kg/m^3	Density mol/dm^3	E J/mol	H J/mol	S J/(mol·K)	C_v J/(mol·K)	C_p J/(mol·K)	Sound m/s	Diel. const.
230.0	.02108	.01046	4412.0	6325.0	135.9	23.13	31.45	1136.0	1.00006
240.0	.02020	.01002	4642.0	6638.0	137.3	22.84	31.15	1162.0	1.00006
250.0	.01940	.009622	4869.0	6948.0	138.5	22.57	30.88	1188.0	1.00006
260.0	.01865	.009252	5094.0	7256.0	139.8	22.32	30.64	1213.0	1.00006
270.0	.01796	.008909	5316.0	7561.0	140.9	22.10	30.42	1238.0	1.00006
280.0	.01732	.008591	5536.0	7864.0	142.0	21.91	30.23	1262.0	1.00005
300.0	.01616	.008018	5971.0	8465.0	144.1	21.61	29.93	1309.0	1.00005
320.0	.01515	.007517	6401.0	9062.0	146.0	21.39	29.71	1354.0	1.00005
340.0	.01426	.007075	6827.0	9654.0	147.8	21.24	29.56	1397.0	1.00004
360.0	.01347	.006682	7251.0	10240.0	149.5	21.14	29.45	1438.0	1.00004
380.0	.01276	.006331	7673.0	10830.0	151.1	21.07	29.38	1479.0	1.00004
400.0	.01212	.006014	8093.0	11420.0	152.6	21.02	29.33	1518.0	1.00004

.04 MPa isobar

T K	Density kg/m^3	Density mol/dm^3	E J/mol	H J/mol	S J/(mol·K)	C_v J/(mol·K)	C_p J/(mol·K)	Sound m/s	Diel. const.
13.81[a]	77.05	38.22	−622.5	−621.4	10.01	10.96	15.53	1374.0	1.25168
14.0	76.89	38.14	−619.6	−618.6	10.21	10.60	15.16	1366.0	1.25112
15.0	76.03	37.71	−605.0	−603.9	11.22	9.66	14.37	1319.0	1.24809
16.0	75.13	37.27	−590.5	−589.5	12.16	9.66	14.70	1272.0	1.24495
17.505[b]	73.71	36.56	−567.5	−566.4	13.54	10.23	16.09	1205.0	1.23996
17.505[b]	.5841	.2897	207.4	345.5	65.63	13.04	22.86	336.7	1.00176
20.0	.5033	.2497	241.1	401.3	68.61	12.76	22.05	363.5	1.00152
21.0	.4772	.2367	254.3	423.2	69.68	12.69	21.85	373.5	1.00144
22.0	.4538	.2251	267.3	445.0	70.69	12.65	21.70	383.1	1.00137
23.0	.4328	.2147	280.3	466.7	71.66	12.61	21.58	392.4	1.00131
24.0	.4136	.2052	293.2	488.2	72.57	12.58	21.48	401.5	1.00125
26.0	.3802	.1886	318.9	531.0	74.29	12.55	21.33	418.9	1.00115
28.0	.3519	.1745	344.4	573.6	75.86	12.52	21.23	435.5	1.00106
30.0	.3276	.1625	369.8	615.9	77.33	12.51	21.16	451.5	1.00099
32.0	.3065	.1520	395.1	658.2	78.69	12.50	21.10	466.8	1.00092
34.0	.2880	.1429	420.4	700.4	79.97	12.49	21.06	481.6	1.00087
36.0	.2716	.1347	445.6	742.4	81.17	12.49	21.02	495.9	1.00082
38.0	.2571	.1275	470.8	784.5	82.31	12.48	21.00	509.7	1.00078
40.0	.2440	.1210	495.9	826.4	83.38	12.49	20.98	523.2	1.00074
42.0	.2322	.1152	521.1	868.4	84.41	12.51	20.99	536.1	1.00070
44.0	.2215	.1099	546.3	910.4	85.38	12.54	21.00	548.6	1.00067
46.0	.2117	.1050	571.5	952.4	86.32	12.57	21.01	560.9	1.00064
48.0	.2028	.1006	596.8	994.4	87.21	12.60	21.03	572.7	1.00061
50.0	.1946	.09654	622.2	1037.0	88.07	12.64	21.07	584.2	1.00059
52.0	.1871	.09279	647.6	1079.0	88.90	12.70	21.11	595.4	1.00056
54.0	.1801	.08933	673.2	1121.0	89.69	12.77	21.17	606.2	1.00054
56.0	.1736	.08611	698.9	1163.0	90.47	12.85	21.25	616.6	1.00052
58.0	.1676	.08312	724.8	1206.0	91.21	12.96	21.35	626.5	1.00051
60.0	.1619	.08033	750.9	1249.0	91.94	13.08	21.47	636.1	1.00049
65.0	.1494	.07412	817.4	1357.0	93.67	13.47	21.85	658.5	1.00045
70.0	.1387	.06880	886.1	1467.0	95.31	13.98	22.35	678.6	1.00042
75.0	.1294	.06420	957.6	1581.0	96.87	14.60	22.96	696.9	1.00039
80.0	.1213	.06017	1033.0	1697.0	98.37	15.32	23.67	713.6	1.00037
85.0	.1141	.05662	1111.0	1818.0	99.83	16.11	24.46	729.3	1.00034
90.0	.1078	.05347	1194.0	1942.0	101.3	16.95	25.29	744.1	1.00033
95.0	.1021	.05065	1281.0	2071.0	102.6	17.81	26.15	758.4	1.00031
100.0	.09700	.04811	1372.0	2203.0	104.0	18.69	27.02	772.3	1.00029
105.0	.09237	.04582	1468.0	2341.0	105.3	19.54	27.88	786.0	1.00028
110.0	.08817	.04374	1568.0	2482.0	106.7	20.37	28.70	799.6	1.00027
115.0	.08433	.04183	1671.0	2628.0	108.0	21.13	29.47	813.3	1.00026
120.0	.08081	.04009	1779.0	2777.0	109.2	21.83	30.16	827.0	1.00024
125.0	.07758	.03848	1890.0	2929.0	110.5	22.45	30.78	840.8	1.00023
130.0	.07459	.03700	2003.0	3084.0	111.7	22.98	31.31	854.8	1.00023
140.0	.06926	.03436	2238.0	3402.0	114.0	23.83	32.15	882.9	1.00021
150.0	.06464	.03207	2479.0	3726.0	116.3	24.35	32.68	911.3	1.00020
160.0	.06060	.03006	2724.0	4055.0	118.4	24.62	32.94	939.9	1.00018
170.0	.05704	.02829	2971.0	4384.0	120.4	24.67	32.99	968.6	1.00017
180.0	.05387	.02672	3217.0	4714.0	122.3	24.57	32.89	997.2	1.00016

Thermophysical properties of hydrogen—Continued

T K	Density kg/m³	Density mol/dm³	E J/mol	H J/mol	S J/(mol·K)	C_v J/(mol·K)	C_p J/(mol·K)	Sound m/s	Diel. const.
190.0	.05103	.02531	3462.0	5042.0	124.1	24.35	32.67	1026.0	1.00015
200.0	.04848	.02405	3704.0	5367.0	125.7	24.08	32.40	1054.0	1.00015
210.0	.04617	.02290	3943.0	5690.0	127.3	23.77	32.08	1082.0	1.00014
220.0	.04407	.02186	4179.0	6009.0	128.8	23.45	31.77	1109.0	1.00013
230.0	.04216	.02091	4412.0	6325.0	130.2	23.13	31.45	1136.0	1.00013
240.0	.04040	.02004	4642.0	6638.0	131.5	22.84	31.16	1162.0	1.00012
250.0	.03879	.01924	4869.0	6948.0	132.8	22.57	30.88	1188.0	1.00012
260.0	.03729	.01850	5093.0	7256.0	134.0	22.32	30.64	1214.0	1.00011
270.0	.03591	.01781	5315.0	7561.0	135.1	22.10	30.42	1238.0	1.00011
280.0	.03463	.01718	5536.0	7864.0	136.2	21.91	30.23	1262.0	1.00011
300.0	.03232	.01603	5971.0	8466.0	138.3	21.61	29.93	1309.0	1.00010
320.0	.03030	.01503	6401.0	9062.0	140.2	21.39	29.71	1354.0	1.00009
340.0	.02852	.01415	6827.0	9654.0	142.0	21.24	29.56	1397.0	1.00009
360.0	.02694	.01336	7251.0	10240.0	143.7	21.14	29.45	1439.0	1.00008
380.0	.02552	.01266	7673.0	10830.0	145.3	21.07	29.38	1479.0	1.00008
400.0	.02424	.01203	8093.0	11420.0	146.8	21.02	29.33	1518.0	1.00007
				.06 MPa isobar					
13.82[a]	77.06	38.23	−622.4	−620.9	10.01	10.96	15.51	1373.0	1.25172
14.0	76.91	38.15	−619.7	−618.1	10.21	10.61	15.16	1365.0	1.25118
15.0	76.04	37.72	−605.1	−603.5	11.22	9.67	14.36	1319.0	1.24815
16.0	75.15	37.28	−590.6	−589.0	12.15	9.66	14.70	1272.0	1.24502
18.0	73.23	36.33	−559.5	−557.9	13.98	10.46	16.63	1186.0	1.23831
18.629[b]	72.58	36.00	−548.8	−547.2	14.57	10.75	17.39	1161.0	1.23602
18.629[b]	.8376	.4155	217.2	361.6	63.35	13.14	23.49	344.4	1.00253
20.0	.7705	.3822	236.3	393.3	64.99	12.93	22.83	359.6	1.00232
21.0	.7286	.3614	249.9	415.9	66.10	12.83	22.49	370.0	1.00220
22.0	.6914	.3430	263.3	438.3	67.14	12.75	22.24	380.1	1.00209
23.0	.6581	.3264	276.6	460.4	68.12	12.69	22.04	389.7	1.00199
24.0	.6281	.3115	289.8	482.4	69.05	12.65	21.87	399.1	1.00189
25.0	.6008	.2980	302.8	504.2	69.94	12.61	21.74	408.2	1.00181
26.0	.5759	.2857	315.8	525.9	70.80	12.59	21.64	417.0	1.00174
28.0	.5321	.2640	341.6	569.0	72.39	12.55	21.47	433.9	1.00161
30.0	.4947	.2454	367.3	611.8	73.87	12.53	21.36	450.1	1.00149
32.0	.4624	.2294	392.8	654.4	75.25	12.51	21.27	465.7	1.00139
34.0	.4341	.2153	418.2	696.9	76.53	12.50	21.20	480.6	1.00131
36.0	.4092	.2030	443.6	739.2	77.74	12.50	21.15	495.1	1.00123
38.0	.3870	.1920	468.9	781.5	78.88	12.49	21.10	509.1	1.00117
40.0	.3671	.1821	494.2	823.6	79.97	12.50	21.08	522.6	1.00111
42.0	.3492	.1732	519.4	865.8	80.99	12.52	21.07	535.6	1.00105
44.0	.3330	.1652	544.7	907.9	81.97	12.54	21.07	548.2	1.00100
46.0	.3183	.1579	570.0	950.1	82.91	12.57	21.08	560.5	1.00096
48.0	.3048	.1512	595.4	992.2	83.81	12.60	21.10	572.5	1.00092
50.0	.2924	.1450	620.8	1034.0	84.67	12.65	21.13	584.0	1.00088
52.0	.2810	.1394	646.3	1077.0	85.50	12.70	21.17	595.2	1.00085
54.0	.2705	.1342	671.9	1119.0	86.30	12.77	21.22	606.0	1.00082
56.0	.2607	.1293	697.7	1162.0	87.07	12.86	21.30	616.4	1.00079
58.0	.2516	.1248	723.6	1204.0	87.82	12.96	21.39	626.5	1.00076
60.0	.2431	.1206	749.8	1247.0	88.55	13.08	21.51	636.1	1.00073
62.0	.2352	.1167	776.2	1290.0	89.26	13.22	21.64	645.3	1.00071
65.0	.2243	.1113	816.4	1356.0	90.28	13.47	21.88	658.5	1.00068
70.0	.2082	.1033	885.2	1466.0	91.92	13.98	22.38	678.6	1.00063
75.0	.1942	.09633	956.8	1580.0	93.49	14.61	22.99	696.9	1.00059
80.0	.1820	.09028	1032.0	1696.0	94.99	15.32	23.70	713.7	1.00055
85.0	.1713	.08495	1110.0	1817.0	96.45	16.11	24.48	729.4	1.00052
90.0	.1617	.08022	1193.0	1941.0	97.88	16.95	25.31	744.2	1.00049
95.0	.1532	.07598	1280.0	2070.0	99.27	17.81	26.17	758.5	1.00046
100.0	.1455	.07217	1372.0	2203.0	100.6	18.69	27.04	772.4	1.00044
105.0	.1386	.06873	1467.0	2340.0	102.0	19.54	27.89	786.1	1.00042
110.0	.1322	.06560	1567.0	2482.0	103.3	20.37	28.71	799.8	1.00040
115.0	.1265	.06274	1671.0	2627.0	104.6	21.14	29.48	813.4	1.00038

Thermophysical properties of hydrogen—Continued

T K	Density kg/m³	Density mol/dm³	E J/mol	H J/mol	S J/(mol·K)	C_v J/(mol·K)	C_p	Sound m/s	Diel. const.
120.0	.1212	.06013	1778.0	2776.0	105.9	21.83	30.17	827.2	1.00037
125.0	.1164	.05772	1889.0	2929.0	107.1	22.45	30.79	841.0	1.00035
130.0	.1119	.05550	2003.0	3084.0	108.3	22.98	31.32	855.0	1.00034
140.0	.1039	.05153	2237.0	3402.0	110.7	23.83	32.16	883.1	1.00031
150.0	.09695	.04809	2478.0	3726.0	112.9	24.35	32.68	911.5	1.00029
160.0	.09089	.04509	2724.0	4054.0	115.0	24.62	32.95	940.1	1.00027
170.0	.08554	.04243	2970.0	4384.0	117.0	24.67	33.00	968.8	1.00026
180.0	.08079	.04008	3217.0	4714.0	118.9	24.57	32.89	997.4	1.00024
190.0	.07654	.03797	3461.0	5042.0	120.7	24.35	32.68	1026.0	1.00023
200.0	.07271	.03607	3704.0	5367.0	122.3	24.08	32.40	1054.0	1.00022
210.0	.06925	.03435	3943.0	5690.0	123.9	23.77	32.09	1082.0	1.00021
220.0	.06610	.03279	4179.0	6009.0	125.4	23.45	31.77	1109.0	1.00020
230.0	.06323	.03136	4412.0	6325.0	126.8	23.14	31.45	1136.0	1.00019
240.0	.06059	.03006	4642.0	6638.0	128.1	22.84	31.16	1163.0	1.00018
250.0	.05817	.02886	4869.0	6948.0	129.4	22.57	30.89	1188.0	1.00018
260.0	.05593	.02775	5093.0	7256.0	130.6	22.32	30.64	1214.0	1.00017
270.0	.05386	.02672	5315.0	7561.0	131.8	22.10	30.42	1238.0	1.00016
280.0	.05194	.02576	5535.0	7864.0	132.9	21.91	30.23	1263.0	1.00016
300.0	.04848	.02405	5971.0	8466.0	134.9	21.61	29.93	1309.0	1.00015
320.0	.04545	.02254	6401.0	9062.0	136.9	21.39	29.71	1354.0	1.00014
340.0	.04277	.02122	6827.0	9655.0	138.7	21.24	29.56	1397.0	1.00013
360.0	.04040	.02004	7251.0	10240.0	140.4	21.14	29.45	1439.0	1.00012
380.0	.03827	.01899	7673.0	10830.0	141.9	21.07	29.38	1479.0	1.00012
400.0	.03636	.01804	8093.0	11420.0	143.5	21.02	29.33	1518.0	1.00011
				.08 MPa isobar					
13.83[a]	77.07	38.23	−622.4	−620.3	10.01	10.96	15.49	1373.0	1.25175
14.0	76.92	38.16	−619.8	−617.7	10.20	10.63	15.16	1365.0	1.25123
15.0	76.06	37.73	−605.2	−603.1	11.21	9.68	14.36	1319.0	1.24821
16.0	75.17	37.29	−590.8	−588.6	12.15	9.67	14.69	1272.0	1.24509
18.0	73.26	36.34	−559.7	−557.5	13.98	10.47	16.62	1187.0	1.23838
19.504[b]	71.65	35.54	−533.3	−531.1	15.38	11.13	18.48	1129.0	1.23280
19.504[b]	1.083	.5373	224.0	372.9	61.73	13.23	24.07	349.8	1.00327
20.0	1.050	.5209	231.2	384.8	62.33	13.13	23.75	355.5	1.00317
21.0	.9898	.4910	245.3	408.2	63.48	12.98	23.23	366.5	1.00299
22.0	.9370	.4648	259.1	431.3	64.55	12.87	22.84	376.9	1.00283
23.0	.8901	.4415	272.8	453.9	65.56	12.78	22.54	387.0	1.00269
24.0	.8481	.4207	286.2	476.4	66.51	12.72	22.30	396.6	1.00256
25.0	.8102	.4019	299.5	498.6	67.42	12.67	22.12	405.9	1.00244
26.0	.7757	.3848	312.7	520.6	68.28	12.63	21.96	415.0	1.00234
28.0	.7154	.3549	338.8	564.3	69.90	12.58	21.73	432.3	1.00216
30.0	.6642	.3295	364.7	607.6	71.39	12.55	21.56	448.8	1.00200
32.0	.6201	.3076	390.5	650.6	72.78	12.53	21.44	464.5	1.00187
34.0	.5817	.2885	416.1	693.3	74.08	12.51	21.35	479.7	1.00175
36.0	.5479	.2718	441.6	736.0	75.29	12.50	21.27	494.3	1.00165
38.0	.5179	.2569	467.0	778.4	76.44	12.50	21.21	508.4	1.00156
40.0	.4910	.2436	492.4	820.8	77.53	12.50	21.17	522.0	1.00148
42.0	.4669	.2316	517.7	863.2	78.56	12.52	21.16	535.1	1.00141
44.0	.4451	.2208	543.1	905.5	79.55	12.54	21.15	547.8	1.00134
46.0	.4253	.2110	568.5	947.8	80.49	12.57	21.15	560.2	1.00128
48.0	.4071	.2020	594.0	990.1	81.39	12.61	21.16	572.2	1.00123
50.0	.3905	.1937	619.4	1032.0	82.25	12.65	21.18	583.8	1.00118
52.0	.3752	.1861	645.0	1075.0	83.08	12.70	21.22	595.0	1.00113
54.0	.3611	.1791	670.7	1117.0	83.88	12.77	21.27	605.9	1.00109
56.0	.3480	.1726	696.5	1160.0	84.66	12.86	21.35	616.3	1.00105
58.0	.3358	.1666	722.5	1203.0	85.41	12.96	21.44	626.4	1.00101
60.0	.3245	.1610	748.7	1246.0	86.14	13.08	21.55	636.0	1.00098
62.0	.3139	.1557	775.1	1289.0	86.85	13.22	21.68	645.3	1.00095
65.0	.2993	.1484	815.3	1354.0	87.88	13.47	21.91	658.5	1.00090
70.0	.2777	.1377	884.2	1465.0	89.52	13.99	22.41	678.7	1.00084
75.0	.2590	.1285	955.9	1579.0	91.08	14.61	23.01	697.0	1.00078
80.0	.2427	.1204	1031.0	1695.0	92.59	15.32	23.72	713.8	1.00073

Thermophysical properties of hydrogen—Continued

T K	Density kg/m^3	Density mol/dm^3	E J/mol	H J/mol	S J/(mol·K)	C_v J/(mol·K)	C_p J/(mol·K)	Sound m/s	Diel. const.
85.0	.2284	.1133	1110.0	1816.0	94.05	16.11	24.50	729.5	1.00069
90.0	.2156	.1070	1192.0	1940.0	95.48	16.95	25.32	744.3	1.00065
95.0	.2042	.1013	1280.0	2069.0	96.87	17.82	26.18	758.6	1.00062
100.0	.1940	.09624	1371.0	2202.0	98.23	18.69	27.05	772.5	1.00059
105.0	.1847	.09164	1467.0	2340.0	99.57	19.55	27.90	786.3	1.00056
110.0	.1763	.08747	1567.0	2481.0	100.9	20.37	28.72	799.9	1.00053
115.0	.1686	.08365	1670.0	2627.0	102.2	21.14	29.49	813.6	1.00051
120.0	.1616	.08016	1778.0	2776.0	103.5	21.84	30.18	827.3	1.00049
125.0	.1551	.07695	1889.0	2928.0	104.7	22.45	30.80	841.2	1.00047
130.0	.1492	.07399	2003.0	3084.0	105.9	22.99	31.33	855.1	1.00045
140.0	.1385	.06870	2237.0	3401.0	108.3	23.83	32.16	883.2	1.00042
150.0	.1293	.06412	2478.0	3726.0	110.5	24.35	32.69	911.7	1.00039
160.0	.1212	.06011	2723.0	4054.0	112.6	24.62	32.95	940.3	1.00037
170.0	.1140	.05657	2970.0	4384.0	114.6	24.67	33.00	968.9	1.00034
180.0	.1077	.05343	3216.0	4714.0	116.5	24.57	32.89	997.5	1.00033
190.0	.1020	.05061	3461.0	5042.0	118.3	24.36	32.68	1026.0	1.00031
200.0	.09693	.04808	3703.0	5367.0	120.0	24.08	32.40	1054.0	1.00029
210.0	.09232	.04579	3943.0	5690.0	121.5	23.77	32.09	1082.0	1.00028
220.0	.08812	.04371	4179.0	6009.0	123.0	23.45	31.77	1109.0	1.00027
230.0	.08429	.04181	4412.0	6325.0	124.4	23.14	31.46	1136.0	1.00025
240.0	.08078	.04007	4642.0	6638.0	125.8	22.84	31.16	1163.0	1.00024
250.0	.07755	.03847	4869.0	6948.0	127.0	22.57	30.89	1189.0	1.00023
260.0	.07457	.03699	5093.0	7256.0	128.2	22.32	30.64	1214.0	1.00023
270.0	.07181	.03562	5315.0	7561.0	129.4	22.11	30.42	1239.0	1.00022
280.0	.06924	.03435	5535.0	7865.0	130.5	21.92	30.23	1263.0	1.00021
300.0	.06463	.03206	5970.0	8466.0	132.6	21.61	29.93	1310.0	1.00020
320.0	.06059	.03006	6400.0	9062.0	134.5	21.39	29.71	1354.0	1.00018
340.0	.05703	.02829	6827.0	9655.0	136.3	21.24	29.56	1397.0	1.00017
360.0	.05386	.02672	7250.0	10240.0	138.0	21.14	29.45	1439.0	1.00016
380.0	.05103	.02531	7673.0	10830.0	139.6	21.07	29.38	1479.0	1.00015
400.0	.04848	.02405	8093.0	11420.0	141.1	21.02	29.33	1518.0	1.00015

.10 MPa isobar

T K	Density kg/m^3	Density mol/dm^3	E J/mol	H J/mol	S J/(mol·K)	C_v J/(mol·K)	C_p J/(mol·K)	Sound m/s	Diel. const.
13.83[a]	77.08	38.23	−622.4	−619.8	10.02	10.96	15.47	1372.0	1.25178
14.0	76.94	38.16	−619.9	−617.2	10.20	10.64	15.15	1365.0	1.25129
15.0	76.08	37.74	−605.3	−602.6	11.21	9.69	14.35	1319.0	1.24827
16.0	75.19	37.30	−590.9	−588.2	12.14	9.68	14.68	1272.0	1.24515
18.0	73.28	36.35	−559.8	−557.0	13.97	10.47	16.61	1187.0	1.23846
20.0	71.12	35.28	−524.2	−521.4	15.85	11.32	19.11	1111.0	1.23093
20.232[b]	70.85	35.14	−519.7	−516.9	16.07	11.41	19.43	1102.0	1.22999
20.232[b]	1.324	.6566	229.2	381.5	60.47	13.31	24.64	353.8	1.00399
25.0	1.025	.5082	296.1	492.8	65.42	12.73	22.52	403.7	1.00309
26.0	.9798	.4860	309.5	515.2	66.30	12.68	22.31	412.9	1.00296
28.0	.9018	.4473	336.0	559.5	67.94	12.61	22.00	430.6	1.00272
30.0	.8360	.4147	362.2	603.3	69.45	12.57	21.78	447.4	1.00252
32.0	.7797	.3867	388.1	646.7	70.85	12.54	21.62	463.4	1.00235
34.0	.7307	.3625	413.9	689.8	72.16	12.52	21.50	478.7	1.00220
36.0	.6877	.3411	439.6	732.7	73.38	12.51	21.40	493.5	1.00207
38.0	.6497	.3223	465.1	775.4	74.54	12.50	21.33	507.7	1.00196
40.0	.6158	.3054	490.6	818.0	75.63	12.51	21.27	521.5	1.00186
42.0	.5853	.2903	516.1	860.5	76.67	12.53	21.25	534.7	1.00177
44.0	.5577	.2767	541.5	903.0	77.66	12.55	21.23	547.4	1.00168
46.0	.5327	.2643	567.0	945.4	78.60	12.58	21.22	559.9	1.00161
48.0	.5099	.2529	592.5	987.9	79.50	12.61	21.23	571.9	1.00154
50.0	.4890	.2425	618.1	1030.0	80.37	12.65	21.24	583.6	1.00147
52.0	.4697	.2330	643.7	1073.0	81.20	12.71	21.28	594.9	1.00142
54.0	.4520	.2242	669.4	1115.0	82.01	12.78	21.32	605.7	1.00136
56.0	.4355	.2160	695.3	1158.0	82.78	12.86	21.39	616.2	1.00131
58.0	.4202	.2084	721.3	1201.0	83.54	12.96	21.48	626.3	1.00127
60.0	.4060	.2014	747.5	1244.0	84.27	13.09	21.59	636.0	1.00122
62.0	.3927	.1948	774.0	1287.0	84.97	13.23	21.72	645.3	1.00118
65.0	.3743	.1857	814.3	1353.0	86.01	13.48	21.95	658.5	1.00113

Thermophysical properties of hydrogen—Continued

T K	Density kg/m³	Density mol/dm³	*E* J/mol	*H* J/mol	*S* J/(mol·K)	C_v J/(mol·K)	C_p J/(mol·K)	Sound m/s	Diel. const.
70.0	.3473	.1723	883.3	1464.0	87.65	13.99	22.44	678.7	1.00105
75.0	.3239	.1607	955.1	1577.0	89.22	14.61	23.04	697.1	1.00098
80.0	.3035	.1506	1030.0	1694.0	90.73	15.33	23.74	713.9	1.00092
85.0	.2855	.1416	1109.0	1815.0	92.19	16.11	24.51	729.6	1.00086
90.0	.2696	.1337	1192.0	1940.0	93.61	16.95	25.34	744.4	1.00081
95.0	.2553	.1267	1279.0	2068.0	95.01	17.82	26.20	758.7	1.00077
100.0	.2425	.1203	1370.0	2202.0	96.37	18.69	27.06	772.7	1.00073
105.0	.2309	.1146	1466.0	2339.0	97.71	19.55	27.92	786.4	1.00070
110.0	.2204	.1093	1566.0	2481.0	99.03	20.37	28.73	800.1	1.00067
115.0	.2108	.1046	1670.0	2626.0	100.3	21.14	29.50	813.7	1.00064
120.0	.2020	.1002	1777.0	2776.0	101.6	21.84	30.19	827.5	1.00061
125.0	.1939	.09618	1888.0	2928.0	102.8	22.46	30.81	841.3	1.00059
130.0	.1864	.09248	2002.0	3083.0	104.1	22.99	31.33	855.3	1.00056
140.0	.1731	.08586	2237.0	3401.0	106.4	23.83	32.17	883.4	1.00052
150.0	.1615	.08013	2478.0	3726.0	108.7	24.35	32.69	911.9	1.00049
160.0	.1514	.07512	2723.0	4054.0	110.8	24.62	32.96	940.4	1.00046
170.0	.1425	.07070	2970.0	4384.0	112.8	24.67	33.01	969.1	1.00043
180.0	.1346	.06677	3216.0	4714.0	114.7	24.57	32.90	997.7	1.00041
190.0	.1275	.06326	3461.0	5042.0	116.4	24.36	32.68	1026.0	1.00039
200.0	.1211	.06010	3703.0	5367.0	118.1	24.08	32.40	1054.0	1.00037
210.0	.1154	.05723	3942.0	5690.0	119.7	23.77	32.09	1082.0	1.00035
220.0	.1101	.05463	4179.0	6009.0	121.2	23.45	31.77	1110.0	1.00033
230.0	.1053	.05226	4412.0	6325.0	122.6	23.14	31.46	1136.0	1.00032
240.0	.1010	.05008	4641.0	6638.0	123.9	22.84	31.16	1163.0	1.00031
250.0	.09692	.04808	4868.0	6948.0	125.2	22.57	30.89	1189.0	1.00029
260.0	.09320	.04623	5093.0	7256.0	126.4	22.32	30.64	1214.0	1.00028
270.0	.08975	.04452	5315.0	7561.0	127.5	22.11	30.42	1239.0	1.00027
280.0	.08654	.04293	5535.0	7865.0	128.6	21.92	30.23	1263.0	1.00026
300.0	.08077	.04007	5970.0	8466.0	130.7	21.61	29.93	1310.0	1.00024
320.0	.07573	.03756	6400.0	9062.0	132.6	21.39	29.71	1355.0	1.00023
340.0	.07127	.03536	6827.0	9655.0	134.4	21.24	29.56	1398.0	1.00022
360.0	.06732	.03339	7250.0	10250.0	136.1	21.14	29.46	1439.0	1.00020
380.0	.06378	.03164	7672.0	10830.0	137.7	21.07	29.38	1479.0	1.00019
400.0	.06059	.03005	8093.0	11420.0	139.2	21.02	29.33	1518.0	1.00018

.101325 MPa isobar

T K	Density kg/m³	Density mol/dm³	*E* J/mol	*H* J/mol	*S* J/(mol·K)	C_v J/(mol·K)	C_p J/(mol·K)	Sound m/s	Diel. const.
13.83[a]	77.08	38.23	−622.4	−619.7	10.02	10.96	15.47	1372.0	1.25178
14.0	76.94	38.16	−619.9	−617.2	10.20	10.64	15.15	1365.0	1.25129
15.0	76.08	37.74	−605.3	−602.6	11.21	9.69	14.35	1319.0	1.24827
16.0	75.19	37.30	−590.9	−588.1	12.14	9.68	14.68	1272.0	1.24516
18.0	73.28	36.35	−559.8	−557.0	13.97	10.47	16.61	1187.0	1.23847
20.0	71.12	35.28	−524.2	−521.4	15.85	11.33	19.11	1111.0	1.23094
20.277[b]	70.80	35.12	−518.9	−516.0	16.11	11.43	19.49	1101.0	1.22982
20.277[b]	1.339	.6644	229.5	382.0	60.40	13.31	24.68	354.1	1.00404
25.0	1.039	.5154	295.8	492.4	65.30	12.74	22.54	403.5	1.00313
26.0	.9935	.4928	309.3	514.9	66.18	12.69	22.33	412.8	1.00300
28.0	.9143	.4535	335.8	559.2	67.82	12.62	22.02	430.5	1.00276
30.0	.8475	.4204	362.0	603.0	69.33	12.57	21.79	447.3	1.00256
32.0	.7903	.3920	388.0	646.4	70.74	12.54	21.63	463.3	1.00238
34.0	.7406	.3674	413.8	689.6	72.04	12.53	21.51	478.7	1.00223
36.0	.6970	.3458	439.4	732.5	73.27	12.51	21.41	493.4	1.00210
38.0	.6585	.3266	465.0	775.2	74.43	12.50	21.33	507.7	1.00199
40.0	.6240	.3096	490.5	817.8	75.52	12.51	21.28	521.4	1.00188
42.0	.5931	.2942	516.0	860.3	76.56	12.53	21.25	534.6	1.00179
44.0	.5652	.2804	541.4	902.8	77.54	12.55	21.23	547.4	1.00170
46.0	.5398	.2678	566.9	945.3	78.49	12.58	21.23	559.8	1.00163
48.0	.5167	.2563	592.4	987.7	79.39	12.61	21.23	571.9	1.00156
50.0	.4955	.2458	618.0	1030.0	80.26	12.65	21.25	583.6	1.00149
52.0	.4760	.2361	643.6	1073.0	81.09	12.71	21.28	594.8	1.00144
54.0	.4580	.2272	669.3	1115.0	81.90	12.78	21.33	605.7	1.00138
56.0	.4413	.2189	695.2	1158.0	82.67	12.86	21.39	616.2	1.00133

Thermophysical properties of hydrogen—Continued

T K	Density kg/m³	Density mol/dm³	E J/mol	H J/mol	S J/(mol·K)	C_v J/(mol·K)	C_p J/(mol·K)	Sound m/s	Diel. const.
58.0	.4258	.2112	721.2	1201.0	83.42	12.96	21.48	626.3	1.00128
60.0	.4114	.2041	747.5	1244.0	84.15	13.09	21.59	636.0	1.00124
62.0	.3979	.1974	774.0	1287.0	84.86	13.23	21.72	645.3	1.00120
65.0	.3793	.1882	814.3	1353.0	85.90	13.48	21.95	658.5	1.00114
70.0	.3519	.1746	883.2	1464.0	87.54	13.99	22.44	678.7	1.00106
75.0	.3282	.1628	955.0	1577.0	89.11	14.61	23.04	697.1	1.00099
80.0	.3075	.1526	1030.0	1694.0	90.62	15.33	23.74	713.9	1.00093
85.0	.2893	.1435	1109.0	1815.0	92.08	16.11	24.52	729.6	1.00087
90.0	.2732	.1355	1192.0	1940.0	93.50	16.95	25.34	744.4	1.00082
95.0	.2587	.1283	1279.0	2068.0	94.90	17.82	26.20	758.7	1.00078
100.0	.2457	.1219	1370.0	2202.0	96.26	18.69	27.07	772.7	1.00074
105.0	.2340	.1161	1466.0	2339.0	97.60	19.55	27.92	786.4	1.00071
110.0	.2233	.1108	1566.0	2481.0	98.92	20.37	28.73	800.1	1.00067
115.0	.2136	.1059	1670.0	2626.0	100.2	21.14	29.50	813.7	1.00064
120.0	.2047	.1015	1777.0	2776.0	101.5	21.84	30.19	827.5	1.00062
125.0	.1965	.09746	1888.0	2928.0	102.7	22.46	30.81	841.3	1.00059
130.0	.1889	.09370	2002.0	3083.0	103.9	22.99	31.33	855.3	1.00057
140.0	.1754	.08700	2236.0	3401.0	106.3	23.83	32.17	883.4	1.00053
150.0	.1637	.08119	2478.0	3726.0	108.5	24.35	32.69	911.9	1.00049
160.0	.1534	.07612	2723.0	4054.0	110.7	24.62	32.96	940.5	1.00046
170.0	.1444	.07164	2970.0	4384.0	112.7	24.67	33.01	969.1	1.00044
180.0	.1364	.06766	3216.0	4714.0	114.5	24.57	32.90	997.7	1.00041
190.0	.1292	.06410	3461.0	5042.0	116.3	24.36	32.68	1026.0	1.00039
200.0	.1228	.06089	3703.0	5367.0	118.0	24.08	32.40	1054.0	1.00037
210.0	.1169	.05799	3942.0	5690.0	119.6	23.77	32.09	1082.0	1.00035
220.0	.1116	.05536	4179.0	6009.0	121.0	23.45	31.77	1110.0	1.00034
230.0	.1067	.05295	4412.0	6325.0	122.5	23.14	31.46	1136.0	1.00032
240.0	.1023	.05074	4641.0	6638.0	123.8	22.84	31.16	1163.0	1.00031
250.0	.09821	.04871	4868.0	6948.0	125.1	22.57	30.89	1189.0	1.00030
260.0	.09443	.04684	5093.0	7256.0	126.3	22.32	30.64	1214.0	1.00029
270.0	.09093	.04511	5315.0	7561.0	127.4	22.11	30.43	1239.0	1.00027
280.0	.08769	.04350	5535.0	7865.0	128.5	21.92	30.23	1263.0	1.00027
300.0	.08184	.04060	5970.0	8466.0	130.6	21.61	29.93	1310.0	1.00025
320.0	.07673	.03806	6400.0	9062.0	132.5	21.39	29.71	1355.0	1.00023
340.0	.07222	.03582	6827.0	9655.0	134.3	21.24	29.56	1398.0	1.00022
360.0	.06821	.03384	7250.0	10250.0	136.0	21.14	29.46	1439.0	1.00021
380.0	.06462	.03206	7672.0	10830.0	137.6	21.07	29.38	1479.0	1.00020
400.0	.06139	.03045	8093.0	11420.0	139.1	21.02	29.33	1518.0	1.00019
				.20 MPa isobar					
13.87[a]	77.12	38.26	–622.3	–617.0	10.02	10.96	15.39	1368.0	1.25195
14.0	77.01	38.20	–620.3	–615.0	10.17	10.71	15.14	1362.0	1.25155
15.0	76.16	37.78	–605.7	–600.4	11.18	9.74	14.33	1318.0	1.24856
16.0	75.28	37.34	–591.4	–586.0	12.11	9.71	14.64	1273.0	1.24548
18.0	73.39	36.41	–560.5	–555.0	13.93	10.49	16.55	1190.0	1.23885
20.0	71.25	35.35	–525.1	–519.5	15.80	11.33	19.02	1116.0	1.23141
22.0	68.79	34.12	–484.5	–478.6	17.74	11.98	21.88	1041.0	1.22285
22.810[b]	67.67	33.57	–466.4	–460.4	18.56	12.19	23.21	1008.0	1.21898
22.810[b]	2.496	1.238	242.8	404.4	56.47	13.62	27.45	365.4	1.00754
25.0	2.187	1.085	277.3	461.6	58.87	13.13	25.15	391.3	1.00661
26.0	2.075	1.029	292.1	486.5	59.84	12.99	24.50	402.0	1.00627
27.0	1.976	.9802	306.7	510.7	60.75	12.89	24.00	412.2	1.00597
28.0	1.887	.9362	320.9	534.5	61.62	12.81	23.61	422.0	1.00570
29.0	1.808	.8966	334.9	557.9	62.44	12.75	23.28	431.3	1.00546
30.0	1.735	.8606	348.7	581.1	63.23	12.70	23.02	440.4	1.00524
32.0	1.608	.7975	375.9	626.7	64.70	12.63	22.61	457.6	1.00485
34.0	1.499	.7438	402.7	671.6	66.06	12.59	22.32	473.9	1.00453
36.0	1.406	.6973	429.2	716.0	67.33	12.56	22.10	489.5	1.00424
38.0	1.324	.6567	455.5	760.0	68.52	12.54	21.92	504.4	1.00400
40.0	1.252	.6208	481.6	803.7	69.64	12.54	21.79	518.7	1.00378
42.0	1.187	.5888	507.6	847.2	70.70	12.55	21.70	532.3	1.00358

Thermophysical properties of hydrogen—Continued

T K	Density kg/m^3	Density mol/dm^3	E J/mol	H J/mol	S J/(mol·K)	C_v J/(mol·K)	C_p J/(mol·K)	Sound m/s	Diel. const.
44.0	1.129	.5601	533.5	890.6	71.71	12.57	21.64	545.5	1.00341
46.0	1.077	.5342	559.4	933.8	72.67	12.59	21.59	558.2	1.00325
48.0	1.029	.5107	585.3	976.9	73.59	12.63	21.56	570.6	1.00311
50.0	.9861	.4892	611.2	1020.0	74.47	12.67	21.55	582.5	1.00298
52.0	.9464	.4695	637.1	1063.0	75.31	12.72	21.55	594.0	1.00286
54.0	.9098	.4513	663.1	1106.0	76.13	12.79	21.58	605.1	1.00274
56.0	.8761	.4346	689.2	1149.0	76.91	12.87	21.63	615.7	1.00264
58.0	.8448	.4191	715.5	1193.0	77.67	12.98	21.70	625.9	1.00255
60.0	.8157	.4046	742.0	1236.0	78.41	13.10	21.79	635.7	1.00246
62.0	.7886	.3912	768.7	1280.0	79.13	13.24	21.90	645.1	1.00238
65.0	.7512	.3726	809.2	1346.0	80.17	13.49	22.12	658.5	1.00227
70.0	.6964	.3454	878.6	1458.0	81.82	14.00	22.58	678.9	1.00210
75.0	.6491	.3220	950.8	1572.0	83.40	14.62	23.16	697.4	1.00196
80.0	.6079	.3015	1026.0	1689.0	84.91	15.33	23.85	714.3	1.00183
85.0	.5716	.2836	1105.0	1811.0	86.38	16.12	24.61	730.1	1.00172
90.0	.5395	.2676	1188.0	1936.0	87.81	16.96	25.43	745.0	1.00163
95.0	.5109	.2534	1276.0	2065.0	89.21	17.83	26.28	759.4	1.00154
100.0	.4851	.2406	1367.0	2198.0	90.58	18.70	27.13	773.3	1.00146
105.0	.4619	.2291	1463.0	2336.0	91.92	19.55	27.98	787.1	1.00139
110.0	.4407	.2186	1563.0	2478.0	93.24	20.38	28.79	800.8	1.00133
115.0	.4215	.2091	1667.0	2624.0	94.54	21.15	29.55	814.5	1.00127
120.0	.4038	.2003	1775.0	2773.0	95.81	21.84	30.24	828.3	1.00122
125.0	.3876	.1923	1886.0	2926.0	97.06	22.46	30.85	842.1	1.00117
130.0	.3727	.1849	2000.0	3082.0	98.28	22.99	31.37	856.1	1.00112
140.0	.3460	.1716	2235.0	3400.0	100.6	23.83	32.20	884.3	1.00104
150.0	.3229	.1602	2476.0	3725.0	102.9	24.36	32.72	912.7	1.00097
160.0	.3027	.1501	2721.0	4053.0	105.0	24.63	32.98	941.3	1.00091
170.0	.2849	.1413	2968.0	4384.0	107.0	24.68	33.03	970.0	1.00086
180.0	.2690	.1334	3215.0	4713.0	108.9	24.57	32.92	998.6	1.00081
190.0	.2549	.1264	3460.0	5042.0	110.7	24.36	32.70	1027.0	1.00077
200.0	.2421	.1201	3702.0	5367.0	112.3	24.08	32.42	1055.0	1.00073
210.0	.2306	.1144	3941.0	5690.0	113.9	23.77	32.11	1083.0	1.00070
220.0	.2201	.1092	4178.0	6009.0	115.4	23.45	31.78	1110.0	1.00066
230.0	.2105	.1044	4411.0	6326.0	116.8	23.14	31.47	1137.0	1.00064
240.0	.2018	.1001	4641.0	6639.0	118.1	22.84	31.17	1164.0	1.00061
250.0	.1937	.09609	4868.0	6949.0	119.4	22.57	30.90	1190.0	1.00058
260.0	.1863	.09240	5092.0	7257.0	120.6	22.33	30.65	1215.0	1.00056
270.0	.1794	.08898	5314.0	7562.0	121.8	22.11	30.43	1240.0	1.00054
280.0	.1730	.08580	5535.0	7866.0	122.9	21.92	30.24	1264.0	1.00052
300.0	.1614	.08009	5970.0	8467.0	124.9	21.61	29.94	1311.0	1.00049
320.0	.1514	.07509	6400.0	9064.0	126.9	21.40	29.72	1355.0	1.00046
340.0	.1425	.07067	6826.0	9656.0	128.7	21.24	29.56	1398.0	1.00043
360.0	.1346	.06675	7250.0	10250.0	130.3	21.14	29.46	1440.0	1.00041
380.0	.1275	.06324	7672.0	10830.0	131.9	21.07	29.39	1480.0	1.00039
400.0	.1211	.06008	8093.0	11420.0	133.4	21.02	29.33	1519.0	1.00037
				.30 MPa isobar					
13.90[a]	77.17	38.28	–622.1	–614.3	10.03	10.96	15.31	1364.0	1.25211
14.0	77.09	38.24	–620.7	–612.8	10.14	10.78	15.13	1360.0	1.25182
15.0	76.25	37.82	–606.2	–598.2	11.15	9.79	14.30	1317.0	1.24886
16.0	75.38	37.39	–591.9	–583.8	12.07	9.75	14.61	1274.0	1.24580
18.0	73.50	36.46	–561.1	–552.9	13.89	10.50	16.49	1194.0	1.23924
20.0	71.39	35.41	–526.0	–517.5	15.76	11.34	18.93	1121.0	1.23187
22.0	68.96	34.21	–485.7	–477.0	17.69	11.98	21.73	1048.0	1.22342
23.0	67.58	33.52	–463.4	–454.4	18.69	12.23	23.35	1008.0	1.21867
24.0	66.08	32.78	–439.3	–430.2	19.72	12.45	25.25	964.4	1.21347
24.576[b]	65.14	32.31	–424.6	–415.3	20.33	12.57	26.52	937.4	1.21023
24.576[b]	3.667	1.819	246.9	411.8	53.98	13.90	30.54	370.9	1.01109
25.0	3.559	1.765	254.6	424.6	54.50	13.74	29.60	376.7	1.01076
26.0	3.334	1.654	271.9	453.3	55.62	13.45	27.92	389.6	1.01008
27.0	3.145	1.560	288.3	480.6	56.65	13.24	26.72	401.4	1.00951
28.0	2.982	1.479	304.0	506.8	57.61	13.09	25.83	412.5	1.00901

Thermophysical properties of hydrogen—Continued

T K	Density kg/m^3	Density mol/dm^3	E J/mol	H J/mol	S J/(mol·K)	C_v J/(mol·K)	C_p J/(mol·K)	Sound m/s	Diel. const.
29.0	2.838	1.408	319.2	532.3	58.50	12.97	25.15	422.9	1.00858
30.0	2.711	1.345	334.1	557.2	59.35	12.88	24.61	432.9	1.00819
31.0	2.596	1.288	348.6	581.6	60.15	12.80	24.18	442.4	1.00784
32.0	2.492	1.236	362.9	605.6	60.91	12.75	23.83	451.6	1.00753
34.0	2.311	1.146	391.0	652.6	62.34	12.67	23.28	469.0	1.00698
36.0	2.157	1.070	418.5	698.8	63.65	12.62	22.88	485.4	1.00652
38.0	2.025	1.004	445.5	744.2	64.88	12.58	22.58	501.0	1.00611
40.0	1.909	.9468	472.3	789.2	66.04	12.57	22.36	515.9	1.00576
42.0	1.806	.8960	498.9	833.7	67.12	12.58	22.20	530.0	1.00545
44.0	1.715	.8507	525.3	878.0	68.15	12.59	22.07	543.6	1.00518
46.0	1.633	.8101	551.7	922.0	69.13	12.61	21.97	556.7	1.00493
48.0	1.559	.7733	577.9	965.9	70.06	12.64	21.90	569.3	1.00471
50.0	1.492	.7399	604.2	1010.0	70.96	12.68	21.86	581.5	1.00450
52.0	1.430	.7094	630.5	1053.0	71.81	12.73	21.84	593.2	1.00432
54.0	1.374	.6814	656.8	1097.0	72.64	12.80	21.84	604.4	1.00415
56.0	1.322	.6557	683.1	1141.0	73.43	12.88	21.86	615.2	1.00399
58.0	1.274	.6318	709.7	1184.0	74.20	12.99	21.92	625.6	1.00384
60.0	1.229	.6097	736.4	1228.0	74.95	13.11	21.99	635.5	1.00371
62.0	1.188	.5892	763.3	1272.0	75.67	13.25	22.09	645.0	1.00358
64.0	1.149	.5700	790.4	1317.0	76.37	13.41	22.22	654.1	1.00347
70.0	1.047	.5195	874.0	1451.0	78.38	14.01	22.73	679.1	1.00316
75.0	.9754	.4839	946.5	1566.0	79.97	14.63	23.29	697.8	1.00294
80.0	.9131	.4529	1022.0	1685.0	81.49	15.34	23.96	714.8	1.00275
85.0	.8583	.4258	1102.0	1806.0	82.97	16.13	24.71	730.6	1.00259
90.0	.8098	.4017	1185.0	1932.0	84.40	16.97	25.51	745.6	1.00244
95.0	.7666	.3803	1272.0	2061.0	85.80	17.83	26.35	760.0	1.00231
100.0	.7278	.3610	1364.0	2195.0	87.18	18.71	27.20	774.0	1.00220
105.0	.6928	.3437	1460.0	2333.0	88.53	19.56	28.04	787.8	1.00209
110.0	.6610	.3279	1561.0	2476.0	89.85	20.38	28.84	801.5	1.00199
115.0	.6321	.3135	1665.0	2622.0	91.15	21.15	29.60	815.3	1.00191
120.0	.6056	.3004	1773.0	2771.0	92.42	21.85	30.28	829.0	1.00183
125.0	.5812	.2883	1884.0	2924.0	93.67	22.47	30.89	842.9	1.00175
130.0	.5587	.2772	1998.0	3080.0	94.89	23.00	31.41	857.0	1.00169
140.0	.5187	.2573	2233.0	3399.0	97.25	23.84	32.24	885.1	1.00156
150.0	.4840	.2401	2474.0	3724.0	99.50	24.37	32.75	913.6	1.00146
160.0	.4537	.2251	2720.0	4053.0	101.6	24.63	33.00	942.2	1.00137
170.0	.4270	.2118	2967.0	4383.0	103.6	24.68	33.05	970.9	1.00129
180.0	.4033	.2000	3213.0	4713.0	105.5	24.58	32.94	999.5	1.00122
190.0	.3820	.1895	3458.0	5042.0	107.3	24.36	32.72	1028.0	1.00115
200.0	.3629	.1800	3701.0	5367.0	109.0	24.09	32.43	1056.0	1.00109
210.0	.3456	.1715	3940.0	5690.0	110.5	23.78	32.12	1084.0	1.00104
220.0	.3299	.1637	4177.0	6010.0	112.0	23.46	31.80	1111.0	1.00100
230.0	.3156	.1566	4410.0	6326.0	113.4	23.14	31.48	1138.0	1.00095
240.0	.3025	.1500	4640.0	6639.0	114.8	22.85	31.18	1165.0	1.00091
250.0	.2904	.1440	4867.0	6950.0	116.0	22.57	30.91	1191.0	1.00088
260.0	.2792	.1385	5092.0	7258.0	117.2	22.33	30.66	1216.0	1.00084
270.0	.2689	.1334	5314.0	7563.0	118.4	22.11	30.44	1241.0	1.00081
280.0	.2593	.1286	5534.0	7866.0	119.5	21.92	30.25	1265.0	1.00078
300.0	.2420	.1201	5969.0	8468.0	121.6	21.62	29.94	1311.0	1.00073
320.0	.2269	.1126	6399.0	9065.0	123.5	21.40	29.72	1356.0	1.00068
340.0	.2136	.1059	6826.0	9657.0	125.3	21.25	29.57	1399.0	1.00064
360.0	.2017	.1001	7250.0	10250.0	127.0	21.14	29.46	1441.0	1.00061
380.0	.1911	.09481	7672.0	10840.0	128.6	21.07	29.39	1481.0	1.00058
400.0	.1816	.09008	8093.0	11420.0	130.1	21.02	29.34	1520.0	1.00055

.40 MPa isobar

T K	Density kg/m^3	Density mol/dm^3	E J/mol	H J/mol	S J/(mol·K)	C_v J/(mol·K)	C_p J/(mol·K)	Sound m/s	Diel. const.
13.93[a]	77.22	38.30	−622.0	−611.6	10.04	10.96	15.23	1361.0	1.25228
14.0	77.17	38.28	−621.0	−610.6	10.11	10.84	15.11	1358.0	1.25209
15.0	76.33	37.86	−606.6	−596.0	11.12	9.84	14.28	1317.0	1.24916
16.0	75.47	37.44	−592.4	−581.7	12.04	9.78	14.57	1275.0	1.24613
18.0	73.61	36.51	−561.8	−550.8	13.86	10.52	16.43	1197.0	1.23962
20.0	71.52	35.48	−526.9	−515.6	15.71	11.34	18.84	1126.0	1.23233

Thermophysical properties of hydrogen—Continued

T K	Density kg/m^3	Density mol/dm^3	E J/mol	H J/mol	S J/(mol·K)	C_v J/(mol·K)	C_p J/(mol·K)	Sound m/s	Diel. const.
22.0	69.12	34.29	–486.9	–475.3	17.63	11.98	21.58	1055.0	1.22399
23.0	67.77	33.62	–464.8	–452.9	18.63	12.23	23.16	1016.0	1.21932
24.0	66.29	32.88	–441.0	–428.9	19.65	12.44	24.98	973.7	1.21421
25.0	64.65	32.07	–415.3	–402.8	20.71	12.64	27.19	927.0	1.20856
25.963[b]	62.87	31.18	–388.2	–375.4	21.79	12.83	29.91	876.1	1.20244
25.963[b]	4.871	2.416	246.2	411.7	52.10	14.16	34.15	373.8	1.01476
30.0	3.784	1.877	317.9	531.1	56.38	13.10	26.73	424.8	1.01145
31.0	3.604	1.788	333.7	557.4	57.25	12.98	25.95	435.3	1.01090
32.0	3.445	1.709	348.9	583.0	58.06	12.89	25.33	445.3	1.01042
33.0	3.302	1.638	363.9	608.1	58.83	12.82	24.83	454.8	1.00999
34.0	3.173	1.574	378.5	632.7	59.57	12.76	24.42	464.0	1.00959
35.0	3.055	1.515	393.0	657.0	60.27	12.72	24.08	472.8	1.00924
36.0	2.947	1.462	407.2	680.9	60.94	12.68	23.79	481.3	1.00891
38.0	2.755	1.366	435.2	728.0	62.22	12.63	23.32	497.7	1.00833
40.0	2.589	1.284	462.8	774.2	63.40	12.60	22.98	513.1	1.00782
42.0	2.444	1.212	490.0	819.9	64.52	12.60	22.73	527.7	1.00738
44.0	2.316	1.149	517.0	865.2	65.57	12.61	22.53	541.7	1.00700
46.0	2.202	1.092	543.8	910.1	66.57	12.63	22.38	555.1	1.00665
48.0	2.099	1.041	570.5	954.7	67.52	12.66	22.27	568.1	1.00634
50.0	2.006	.9950	597.1	999.2	68.42	12.70	22.18	580.5	1.00606
52.0	1.921	.9530	623.7	1043.0	69.29	12.75	22.13	592.4	1.00580
54.0	1.844	.9146	650.4	1088.0	70.13	12.81	22.10	603.9	1.00557
56.0	1.773	.8793	677.0	1132.0	70.93	12.90	22.11	614.8	1.00535
58.0	1.707	.8468	703.8	1176.0	71.71	13.00	22.14	625.3	1.00515
60.0	1.646	.8167	730.7	1220.0	72.46	13.12	22.20	635.4	1.00497
62.0	1.590	.7888	757.9	1265.0	73.19	13.26	22.28	645.0	1.00480
64.0	1.538	.7628	785.2	1310.0	73.90	13.42	22.40	654.2	1.00464
70.0	1.400	.6944	869.3	1445.0	75.92	14.02	22.87	679.4	1.00422
75.0	1.303	.6464	942.2	1561.0	77.52	14.64	23.42	698.2	1.00393
80.0	1.219	.6047	1018.0	1680.0	79.05	15.35	24.07	715.3	1.00368
85.0	1.145	.5682	1098.0	1802.0	80.53	16.14	24.80	731.2	1.00346
90.0	1.080	.5359	1181.0	1928.0	81.97	16.98	25.60	746.2	1.00326
95.0	1.022	.5072	1269.0	2058.0	83.38	17.84	26.43	760.7	1.00309
100.0	.9706	.4814	1361.0	2192.0	84.76	18.71	27.27	774.7	1.00293
105.0	.9237	.4582	1458.0	2331.0	86.11	19.57	28.10	788.6	1.00279
110.0	.8812	.4371	1558.0	2473.0	87.43	20.39	28.90	802.3	1.00266
115.0	.8425	.4179	1662.0	2619.0	88.73	21.16	29.65	816.0	1.00254
120.0	.8071	.4004	1770.0	2769.0	90.01	21.86	30.33	829.8	1.00243
125.0	.7746	.3842	1882.0	2923.0	91.26	22.47	30.93	843.8	1.00234
130.0	.7446	.3694	1996.0	3079.0	92.48	23.01	31.45	857.8	1.00225
140.0	.6912	.3429	2231.0	3397.0	94.85	23.85	32.27	886.0	1.00208
150.0	.6449	.3199	2473.0	3723.0	97.09	24.37	32.78	914.5	1.00195
160.0	.6045	.2999	2718.0	4052.0	99.22	24.64	33.03	943.1	1.00182
170.0	.5689	.2822	2965.0	4383.0	101.2	24.69	33.07	971.8	1.00172
180.0	.5373	.2665	3212.0	4713.0	103.1	24.58	32.95	1000.0	1.00162
190.0	.5090	.2525	3457.0	5041.0	104.9	24.37	32.73	1029.0	1.00154
200.0	.4835	.2399	3700.0	5367.0	106.6	24.09	32.45	1057.0	1.00146
210.0	.4605	.2284	3939.0	5690.0	108.1	23.78	32.13	1085.0	1.00139
220.0	.4396	.2181	4176.0	6010.0	109.6	23.46	31.81	1112.0	1.00133
230.0	.4205	.2086	4409.0	6327.0	111.0	23.15	31.49	1139.0	1.00127
240.0	.4030	.1999	4639.0	6640.0	112.4	22.85	31.19	1166.0	1.00122
250.0	.3869	.1919	4866.0	6950.0	113.6	22.58	30.92	1191.0	1.00117
260.0	.3720	.1846	5091.0	7258.0	114.8	22.33	30.67	1217.0	1.00112
270.0	.3583	.1777	5313.0	7564.0	116.0	22.11	30.45	1241.0	1.00108
280.0	.3455	.1714	5533.0	7867.0	117.1	21.92	30.25	1266.0	1.00104
300.0	.3225	.1600	5969.0	8469.0	119.2	21.62	29.95	1312.0	1.00097
320.0	.3024	.1500	6399.0	9066.0	121.1	21.40	29.73	1357.0	1.00091
340.0	.2846	.1412	6826.0	9659.0	122.9	21.25	29.57	1400.0	1.00086
360.0	.2688	.1334	7249.0	10250.0	124.6	21.14	29.47	1441.0	1.00081
380.0	.2547	.1263	7672.0	10840.0	126.2	21.07	29.39	1482.0	1.00077
400.0	.2420	.1200	8093.0	11420.0	127.7	21.02	29.34	1520.0	1.00073

Thermophysical properties of hydrogen—Continued

T K	Density kg/m³	Density mol/dm³	E J/mol	H J/mol	S J/(mol·K)	C_v J/(mol·K)	C_p J/(mol·K)	Sound m/s	Diel. const.
.50 MPa isobar									
13.97[a]	77.27	38.33	−621.9	−608.9	10.05	10.96	15.15	1357.0	1.25245
14.0	77.24	38.32	−621.4	−608.4	10.08	10.90	15.10	1356.0	1.25235
15.0	76.41	37.91	−607.0	−593.8	11.09	9.88	14.25	1316.0	1.24945
16.0	75.56	37.48	−592.8	−579.5	12.01	9.81	14.54	1276.0	1.24645
18.0	73.72	36.57	−562.4	−548.7	13.82	10.53	16.38	1200.0	1.24000
20.0	71.65	35.54	−527.7	−513.7	15.67	11.35	18.75	1131.0	1.23278
22.0	69.28	34.37	−488.1	−473.5	17.58	11.98	21.44	1061.0	1.22455
23.0	67.95	33.71	−466.2	−451.3	18.56	12.22	22.98	1024.0	1.21994
24.0	66.50	32.99	−442.7	−427.5	19.58	12.44	24.73	982.7	1.21493
25.0	64.90	32.19	−417.3	−401.8	20.63	12.63	26.83	937.6	1.20941
26.0	63.09	31.30	−389.6	−373.6	21.73	12.82	29.50	886.7	1.20321
27.0	61.01	30.26	−359.0	−342.4	22.91	13.02	33.16	827.9	1.19609
27.1	60.78	30.15	−355.7	−339.1	23.03	13.04	33.61	821.4	1.19531
27.125[b]	60.73	30.12	−354.9	−338.3	23.06	13.05	33.72	819.9	1.19513
27.125[b]	6.130	3.041	242.2	406.6	50.52	14.43	38.59	375.4	1.01860
30.0	4.985	2.473	299.8	502.0	53.87	13.40	29.70	416.0	1.01510
31.0	4.714	2.339	317.2	531.0	54.82	13.21	28.30	427.7	1.01428
32.0	4.480	2.222	333.8	558.7	55.70	13.07	27.25	438.6	1.01357
33.0	4.275	2.120	349.8	585.6	56.52	12.96	26.44	448.9	1.01294
34.0	4.092	2.030	365.3	611.7	57.30	12.88	25.79	458.8	1.01238
35.0	3.927	1.948	380.5	637.2	58.04	12.82	25.27	468.2	1.01188
36.0	3.778	1.874	395.4	662.2	58.75	12.76	24.83	477.2	1.01143
37.0	3.642	1.806	410.1	686.9	59.42	12.72	24.46	485.9	1.01102
38.0	3.517	1.744	424.6	711.2	60.07	12.69	24.15	494.3	1.01064
40.0	3.294	1.634	453.0	759.0	61.30	12.65	23.66	510.4	1.00996
42.0	3.102	1.538	480.9	805.9	62.44	12.64	23.30	525.5	1.00938
44.0	2.933	1.455	508.5	852.2	63.52	12.64	23.02	539.9	1.00886
46.0	2.783	1.380	535.9	898.0	64.54	12.65	22.81	553.7	1.00841
48.0	2.649	1.314	563.0	943.5	65.51	12.68	22.64	566.9	1.00801
50.0	2.529	1.254	590.0	988.6	66.43	12.71	22.52	579.6	1.00764
52.0	2.419	1.200	617.0	1034.0	67.31	12.76	22.43	591.7	1.00731
54.0	2.320	1.151	643.9	1078.0	68.15	12.83	22.38	603.3	1.00701
56.0	2.229	1.106	670.9	1123.0	68.97	12.91	22.36	614.5	1.00673
58.0	2.145	1.064	697.9	1168.0	69.75	13.01	22.37	625.1	1.00648
60.0	2.067	1.026	725.1	1213.0	70.51	13.13	22.41	635.3	1.00624
62.0	1.996	.9900	752.4	1258.0	71.25	13.27	22.48	645.0	1.00603
64.0	1.929	.9569	780.0	1303.0	71.96	13.43	22.58	654.3	1.00583
66.0	1.867	.9260	807.9	1348.0	72.66	13.61	22.70	663.1	1.00564
70.0	1.754	.8702	864.6	1439.0	74.00	14.03	23.02	679.7	1.00530
75.0	1.632	.8094	937.9	1556.0	75.61	14.65	23.54	698.6	1.00493
80.0	1.526	.7569	1014.0	1675.0	77.15	15.36	24.18	715.8	1.00461
85.0	1.433	.7109	1094.0	1798.0	78.63	16.15	24.90	731.8	1.00433
90.0	1.351	.6703	1178.0	1924.0	80.08	16.98	25.68	746.9	1.00408
95.0	1.279	.6342	1266.0	2054.0	81.49	17.85	26.50	761.3	1.00386
100.0	1.213	.6019	1358.0	2189.0	82.87	18.72	27.34	775.4	1.00366
105.0	1.155	.5727	1455.0	2328.0	84.22	19.58	28.16	789.3	1.00348
110.0	1.101	.5463	1555.0	2471.0	85.55	20.40	28.95	803.1	1.00332
115.0	1.053	.5223	1660.0	2617.0	86.86	21.17	29.70	816.8	1.00318
120.0	1.009	.5003	1768.0	2767.0	88.14	21.86	30.38	830.7	1.00304
125.0	.9678	.4801	1879.0	2921.0	89.39	22.48	30.97	844.6	1.00292
130.0	.9303	.4615	1994.0	3077.0	90.61	23.01	31.49	858.6	1.00281
140.0	.8635	.4283	2229.0	3396.0	92.98	23.85	32.30	886.9	1.00260
150.0	.8057	.3996	2471.0	3722.0	95.23	24.38	32.81	915.4	1.00243
160.0	.7552	.3746	2717.0	4051.0	97.35	24.64	33.05	944.0	1.00228
170.0	.7107	.3525	2964.0	4382.0	99.36	24.69	33.09	972.7	1.00214
180.0	.6711	.3329	3211.0	4713.0	101.2	24.59	32.97	1001.0	1.00202
190.0	.6358	.3154	3456.0	5041.0	103.0	24.37	32.75	1030.0	1.00192
200.0	.6040	.2996	3699.0	5368.0	104.7	24.09	32.46	1058.0	1.00182
210.0	.5753	.2854	3938.0	5691.0	106.3	23.78	32.15	1086.0	1.00174
220.0	.5491	.2724	4175.0	6010.0	107.8	23.46	31.82	1113.0	1.00166
230.0	.5253	.2606	4408.0	6327.0	109.2	23.15	31.50	1140.0	1.00158

Thermophysical properties of hydrogen—Continued

T K	Density kg/m^3	Density mol/dm^3	E J/mol	H J/mol	S J/(mol·K)	C_v J/(mol·K)	C_p J/(mol·K)	Sound m/s	Diel. const.
240.0	.5034	.2497	4638.0	6640.0	110.5	22.85	31.20	1166.0	1.00152
250.0	.4833	.2397	4866.0	6951.0	111.8	22.58	30.92	1192.0	1.00146
260.0	.4647	.2305	5090.0	7259.0	113.0	22.33	30.67	1218.0	1.00140
270.0	.4476	.2220	5313.0	7565.0	114.1	22.12	30.45	1242.0	1.00135
280.0	.4316	.2141	5533.0	7868.0	115.2	21.93	30.26	1266.0	1.00130
300.0	.4029	.1999	5968.0	8470.0	117.3	21.62	29.95	1313.0	1.00122
320.0	.3778	.1874	6399.0	9067.0	119.2	21.40	29.73	1358.0	1.00114
340.0	.3556	.1764	6825.0	9660.0	121.0	21.25	29.58	1401.0	1.00107
360.0	.3359	.1666	7249.0	10250.0	122.7	21.15	29.47	1442.0	1.00101
380.0	.3182	.1579	7671.0	10840.0	124.3	21.07	29.40	1482.0	1.00096
400.0	.3024	.1500	8092.0	11430.0	125.8	21.02	29.34	1521.0	1.00091
				.60 MPa isobar					
14.00[a]	77.31	38.35	−621.8	−606.1	10.06	10.96	15.08	1354.0	1.25261
15.0	76.50	37.95	−607.4	−591.6	11.06	9.92	14.23	1316.0	1.24974
16.0	75.65	37.53	−593.3	−577.3	11.98	9.84	14.51	1277.0	1.24677
18.0	73.83	36.62	−563.0	−546.6	13.79	10.55	16.32	1203.0	1.24038
20.0	71.78	35.61	−528.6	−511.7	15.62	11.36	18.67	1136.0	1.23323
22.0	69.44	34.44	−489.2	−471.8	17.52	11.98	21.31	1068.0	1.22509
23.0	68.13	33.79	−467.5	−449.7	18.50	12.22	22.80	1031.0	1.22056
24.0	66.70	33.09	−444.3	−426.1	19.51	12.43	24.50	991.4	1.21563
25.0	65.13	32.31	−419.2	−400.6	20.55	12.62	26.51	947.8	1.21022
26.0	63.38	31.44	−392.0	−372.9	21.64	12.80	29.00	898.9	1.20420
27.0	61.37	30.44	−362.1	−342.4	22.79	12.99	32.36	842.9	1.19732
27.5	60.24	29.88	−345.7	−325.6	23.40	13.09	34.58	811.2	1.19346
28.0	58.99	29.26	−328.2	−307.7	24.05	13.22	37.40	776.4	1.18919
28.133[b]	58.64	29.09	−323.3	−302.7	24.23	13.25	38.27	766.7	1.18800
28.133[b]	7.467	3.704	235.4	397.4	49.10	14.70	44.26	376.1	1.02268
30.0	6.368	3.159	278.9	468.9	51.57	13.79	34.15	406.1	1.01932
31.0	5.960	2.956	298.7	501.6	52.64	13.51	31.58	419.4	1.01807
32.0	5.621	2.788	317.1	532.3	53.62	13.30	29.79	431.5	1.01704
33.0	5.332	2.645	334.5	561.4	54.51	13.14	28.48	442.8	1.01616
34.0	5.079	2.520	351.2	589.3	55.35	13.02	27.48	453.4	1.01539
35.0	4.857	2.409	367.4	616.4	56.13	12.93	26.69	463.4	1.01471
36.0	4.657	2.310	383.1	642.8	56.87	12.86	26.05	473.0	1.01410
37.0	4.477	2.221	398.4	668.5	57.58	12.80	25.53	482.2	1.01356
38.0	4.314	2.140	413.5	693.8	58.25	12.75	25.09	491.0	1.01306
40.0	4.026	1.997	442.9	743.3	59.52	12.69	24.41	507.8	1.01218
42.0	3.780	1.875	471.6	791.6	60.70	12.67	23.92	523.4	1.01144
44.0	3.566	1.769	499.9	839.1	61.81	12.67	23.55	538.2	1.01079
46.0	3.378	1.676	527.8	885.9	62.85	12.67	23.26	552.3	1.01022
48.0	3.210	1.593	555.4	932.1	63.83	12.69	23.04	565.8	1.00971
50.0	3.061	1.518	582.8	978.0	64.77	12.73	22.87	578.7	1.00925
52.0	2.925	1.451	610.2	1024.0	65.66	12.78	22.74	591.1	1.00884
54.0	2.802	1.390	637.4	1069.0	66.52	12.84	22.66	602.9	1.00847
56.0	2.690	1.334	664.7	1114.0	67.34	12.92	22.61	614.2	1.00813
58.0	2.587	1.283	692.0	1160.0	68.14	13.02	22.60	624.9	1.00782
60.0	2.492	1.236	719.4	1205.0	68.90	13.14	22.62	635.2	1.00753
62.0	2.405	1.193	747.0	1250.0	69.64	13.28	22.68	645.0	1.00726
64.0	2.323	1.152	774.8	1295.0	70.37	13.44	22.76	654.4	1.00702
66.0	2.247	1.115	802.9	1341.0	71.07	13.62	22.87	663.3	1.00679
70.0	2.110	1.047	860.0	1433.0	72.42	14.04	23.17	680.0	1.00637
75.0	1.962	.9731	933.6	1550.0	74.04	14.66	23.67	699.0	1.00592
80.0	1.833	.9094	1010.0	1670.0	75.58	15.37	24.29	716.3	1.00554
85.0	1.721	.8539	1091.0	1793.0	77.08	16.16	24.99	732.4	1.00520
90.0	1.623	.8049	1175.0	1920.0	78.53	16.99	25.77	747.5	1.00490
95.0	1.535	.7613	1263.0	2051.0	79.94	17.86	26.58	762.0	1.00463
100.0	1.456	.7224	1355.0	2186.0	81.33	18.73	27.40	776.2	1.00440
105.0	1.385	.6873	1452.0	2325.0	82.68	19.58	28.22	790.1	1.00418
110.0	1.321	.6555	1553.0	2468.0	84.01	20.41	29.01	803.8	1.00399
115.0	1.263	.6266	1657.0	2615.0	85.32	21.17	29.75	817.6	1.00381
120.0	1.210	.6001	1766.0	2765.0	86.60	21.87	30.42	831.5	1.00365

Thermophysical properties of hydrogen—Continued

T K	Density kg/m^3	Density mol/dm^3	E J/mol	H J/mol	S J/(mol·K)	C_v J/(mol·K)	C_p J/(mol·K)	Sound m/s	Diel. const.
125.0	1.161	.5759	1877.0	2919.0	87.85	22.49	31.02	845.4	1.00350
130.0	1.116	.5535	1991.0	3075.0	89.08	23.02	31.53	859.5	1.00337
140.0	1.036	.5137	2227.0	3395.0	91.45	23.86	32.33	887.7	1.00312
150.0	.9662	.4793	2469.0	3721.0	93.70	24.38	32.83	916.3	1.00291
160.0	.9056	.4492	2715.0	4051.0	95.83	24.65	33.08	944.9	1.00273
170.0	.8522	.4227	2962.0	4382.0	97.83	24.70	33.11	973.6	1.00257
180.0	.8048	.3992	3210.0	4712.0	99.72	24.59	32.99	1002.0	1.00243
190.0	.7624	.3782	3455.0	5041.0	101.5	24.38	32.77	1031.0	1.00230
200.0	.7243	.3593	3698.0	5368.0	103.2	24.10	32.48	1059.0	1.00218
210.0	.6898	.3422	3937.0	5691.0	104.8	23.79	32.16	1087.0	1.00208
220.0	.6585	.3266	4174.0	6011.0	106.2	23.47	31.83	1114.0	1.00199
230.0	.6299	.3125	4407.0	6327.0	107.6	23.15	31.51	1141.0	1.00190
240.0	.6037	.2995	4637.0	6641.0	109.0	22.86	31.21	1167.0	1.00182
250.0	.5796	.2875	4865.0	6952.0	110.3	22.58	30.93	1193.0	1.00175
260.0	.5573	.2765	5090.0	7260.0	111.5	22.34	30.68	1218.0	1.00168
270.0	.5367	.2662	5312.0	7565.0	112.6	22.12	30.46	1243.0	1.00162
280.0	.5176	.2568	5532.0	7869.0	113.7	21.93	30.27	1267.0	1.00156
300.0	.4832	.2397	5968.0	8471.0	115.8	21.62	29.96	1314.0	1.00146
320.0	.4531	.2247	6398.0	9068.0	117.7	21.40	29.74	1359.0	1.00137
340.0	.4265	.2115	6825.0	9661.0	119.5	21.25	29.58	1402.0	1.00129
360.0	.4028	.1998	7249.0	10250.0	121.2	21.15	29.47	1443.0	1.00122
380.0	.3817	.1893	7671.0	10840.0	122.8	21.08	29.40	1483.0	1.00115
400.0	.3626	.1799	8092.0	11430.0	124.3	21.02	29.34	1522.0	1.00109
				.80 MPa isobar					
14.07[a]	77.41	38.40	−621.5	−600.7	10.07	10.96	14.94	1349.0	1.25296
15.0	76.66	38.03	−608.2	−587.2	11.00	10.00	14.18	1315.0	1.25032
16.0	75.83	37.62	−594.2	−573.0	11.92	9.89	14.44	1279.0	1.24740
18.0	74.04	36.73	−564.2	−542.5	13.72	10.57	16.22	1209.0	1.24112
20.0	72.03	35.73	−530.2	−507.8	15.54	11.36	18.51	1146.0	1.23411
22.0	69.74	34.60	−491.4	−468.3	17.42	11.97	21.05	1081.0	1.22616
23.0	68.47	33.97	−470.1	−446.5	18.39	12.21	22.48	1046.0	1.22175
24.0	67.09	33.28	−447.3	−423.3	19.38	12.42	24.07	1008.0	1.21698
25.0	65.59	32.53	−422.9	−398.3	20.40	12.60	25.91	967.1	1.21178
26.0	63.92	31.71	−396.5	−371.3	21.45	12.76	28.15	921.6	1.20605
27.0	62.04	30.77	−367.8	−341.8	22.57	12.93	31.02	870.4	1.19960
27.5	60.99	30.26	−352.3	−325.8	23.15	13.03	32.83	841.9	1.19603
28.0	59.86	29.69	−335.8	−308.9	23.76	13.13	35.03	811.1	1.19216
28.5	58.61	29.07	−318.2	−290.7	24.41	13.24	37.80	777.3	1.18791
29.0	57.21	28.38	−299.1	−270.9	25.10	13.38	41.48	739.7	1.18316
29.5	55.60	27.58	−278.0	−249.0	25.85	13.55	46.76	696.9	1.17770
29.836[b]	54.35	26.96	−262.2	−232.5	26.40	13.71	51.95	664.3	1.17348
29.836[b]	10.50	5.207	213.8	367.4	46.50	15.31	62.69	375.8	1.03199
30.0	10.25	5.085	220.0	377.4	46.83	15.14	58.62	379.6	1.03123
30.2	9.982	4.951	227.1	388.7	47.21	14.96	54.67	384.0	1.03040
30.5	9.629	4.776	236.9	404.4	47.73	14.73	50.15	390.0	1.02932
31.0	9.136	4.532	251.5	428.0	48.50	14.41	44.88	399.2	1.02780
31.5	8.727	4.329	264.7	449.5	49.18	14.16	41.23	407.6	1.02654
32.0	8.376	4.155	276.9	469.4	49.81	13.95	38.53	415.3	1.02547
32.5	8.068	4.002	288.2	488.1	50.39	13.77	36.46	422.5	1.02452
33.0	7.795	3.867	299.0	505.9	50.93	13.63	34.80	429.3	1.02369
33.5	7.549	3.745	309.3	523.0	51.45	13.50	33.45	435.8	1.02293
34.0	7.325	3.634	319.2	539.4	51.93	13.40	32.32	441.9	1.02225
35.0	6.931	3.438	338.1	570.8	52.84	13.22	30.56	453.6	1.02104
36.0	6.592	3.270	356.0	600.7	53.69	13.09	29.23	464.4	1.02000
37.0	6.295	3.122	373.2	629.4	54.47	12.99	28.21	474.7	1.01910
38.0	6.031	2.992	389.7	657.2	55.21	12.91	27.39	484.4	1.01829
39.0	5.794	2.874	405.9	684.2	55.92	12.84	26.72	493.7	1.01757
40.0	5.580	2.768	421.6	710.6	56.58	12.80	26.17	502.6	1.01691
42.0	5.205	2.582	452.2	762.1	57.84	12.75	25.33	519.3	1.01577
44.0	4.886	2.424	482.0	812.1	59.00	12.73	24.71	535.0	1.01480

Thermophysical properties of hydrogen—Continued

T K	Density kg/m^3	Density mol/dm^3	E J/mol	H J/mol	S J/(mol·K)	C_v J/(mol·K)	C_p J/(mol·K)	Sound m/s	Diel. const.
46.0	4.609	2.287	511.1	861.0	60.09	12.72	24.24	549.8	1.01396
48.0	4.367	2.166	539.8	909.1	61.11	12.73	23.88	563.8	1.01322
50.0	4.152	2.060	568.2	956.6	62.08	12.76	23.61	577.2	1.01257
52.0	3.960	1.964	596.3	1004.0	63.01	12.80	23.40	590.0	1.01198
54.0	3.786	1.878	624.3	1050.0	63.89	12.87	23.25	602.1	1.01146
56.0	3.629	1.800	652.2	1097.0	64.73	12.94	23.14	613.7	1.01098
58.0	3.485	1.729	680.1	1143.0	65.54	13.04	23.08	624.7	1.01054
60.0	3.353	1.663	708.0	1189.0	66.32	13.16	23.06	635.2	1.01014
62.0	3.232	1.603	736.1	1235.0	67.08	13.30	23.08	645.2	1.00977
64.0	3.119	1.547	764.3	1281.0	67.81	13.46	23.13	654.8	1.00943
66.0	3.015	1.496	792.8	1328.0	68.52	13.64	23.21	663.8	1.00911
68.0	2.918	1.447	821.5	1374.0	69.22	13.84	23.33	672.5	1.00882
70.0	2.827	1.403	850.6	1421.0	69.90	14.06	23.47	680.8	1.00855
75.0	2.625	1.302	925.0	1539.0	71.53	14.67	23.92	700.0	1.00793
80.0	2.451	1.216	1002.0	1660.0	73.09	15.39	24.51	717.5	1.00740
85.0	2.299	1.140	1083.0	1785.0	74.60	16.17	25.19	733.6	1.00695
90.0	2.166	1.074	1168.0	1912.0	76.06	17.01	25.94	748.9	1.00654
95.0	2.048	1.016	1256.0	2044.0	77.48	17.87	26.73	763.5	1.00618
100.0	1.942	.9634	1349.0	2180.0	78.87	18.74	27.54	777.6	1.00586
105.0	1.847	.9163	1446.0	2319.0	80.24	19.60	28.34	791.6	1.00558
110.0	1.761	.8737	1547.0	2463.0	81.57	20.42	29.12	805.4	1.00532
115.0	1.683	.8350	1652.0	2611.0	82.88	21.19	29.85	819.2	1.00508
120.0	1.612	.7996	1761.0	2761.0	84.17	21.88	30.51	833.1	1.00487
125.0	1.547	.7671	1873.0	2916.0	85.43	22.50	31.10	847.1	1.00467
130.0	1.486	.7373	1987.0	3072.0	86.66	23.03	31.60	861.2	1.00449
140.0	1.379	.6841	2223.0	3393.0	89.03	23.87	32.40	889.5	1.00416
150.0	1.287	.6382	2466.0	3719.0	91.28	24.39	32.89	918.0	1.00388
160.0	1.206	.5981	2712.0	4049.0	93.41	24.66	33.13	946.7	1.00364
170.0	1.135	.5628	2960.0	4381.0	95.42	24.71	33.15	975.5	1.00342
180.0	1.072	.5315	3207.0	4712.0	97.32	24.60	33.03	1004.0	1.00323
190.0	1.015	.5035	3452.0	5041.0	99.10	24.38	32.80	1033.0	1.00306
200.0	.9643	.4783	3695.0	5368.0	100.8	24.11	32.51	1061.0	1.00291
210.0	.9184	.4556	3935.0	5691.0	102.4	23.79	32.19	1089.0	1.00277
220.0	.8767	.4349	4172.0	6011.0	103.8	23.47	31.85	1116.0	1.00264
230.0	.8387	.4160	4405.0	6328.0	105.2	23.16	31.53	1143.0	1.00253
240.0	.8038	.3987	4636.0	6642.0	106.6	22.86	31.23	1169.0	1.00242
250.0	.7717	.3828	4863.0	6953.0	107.9	22.59	30.95	1195.0	1.00233
260.0	.7421	.3681	5088.0	7261.0	109.1	22.34	30.70	1220.0	1.00224
270.0	.7147	.3545	5311.0	7567.0	110.2	22.12	30.48	1245.0	1.00216
280.0	.6893	.3419	5531.0	7871.0	111.3	21.93	30.28	1269.0	1.00208
300.0	.6435	.3192	5967.0	8473.0	113.4	21.63	29.97	1316.0	1.00194
320.0	.6034	.2993	6397.0	9070.0	115.3	21.41	29.74	1360.0	1.00182
340.0	.5680	.2818	6824.0	9663.0	117.1	21.26	29.59	1403.0	1.00171
360.0	.5365	.2662	7248.0	10250.0	118.8	21.15	29.48	1445.0	1.00162
380.0	.5084	.2522	7671.0	10840.0	120.4	21.08	29.40	1485.0	1.00153
400.0	.4831	.2396	8092.0	11430.0	121.9	21.03	29.35	1523.0	1.00146

1.00 MPa isobar

T K	Density kg/m^3	Density mol/dm^3	E J/mol	H J/mol	S J/(mol·K)	C_v J/(mol·K)	C_p J/(mol·K)	Sound m/s	Diel. const.
14.13[a]	77.51	38.45	–621.3	–595.2	10.09	10.95	14.80	1344.0	1.25330
15.0	76.83	38.11	–609.0	–582.8	10.95	10.08	14.13	1315.0	1.25089
16.0	76.01	37.70	–595.1	–568.6	11.86	9.94	14.37	1281.0	1.24802
18.0	74.25	36.83	–565.4	–538.3	13.65	10.59	16.11	1216.0	1.24185
20.0	72.27	35.85	–531.7	–503.8	15.46	11.37	18.36	1155.0	1.23496
22.0	70.04	34.74	–493.5	–464.7	17.32	11.97	20.82	1093.0	1.22718
23.0	68.80	34.13	–472.5	–443.2	18.28	12.20	22.18	1060.0	1.22289
24.0	67.47	33.47	–450.2	–420.3	19.25	12.40	23.68	1024.0	1.21827
25.0	66.01	32.75	–426.3	–395.8	20.25	12.58	25.39	985.2	1.21326
26.0	64.42	31.96	–400.7	–369.4	21.29	12.74	27.42	942.7	1.20777
27.0	62.64	31.07	–373.0	–340.8	22.37	12.89	29.93	895.3	1.20167
28.0	60.62	30.07	–342.5	–309.3	23.51	13.06	33.26	841.5	1.19476
28.5	59.49	29.51	–326.0	–292.1	24.12	13.15	35.43	811.5	1.19089

Thermophysical properties of hydrogen—Continued

T K	Density kg/m³	Density mol/dm³	E J/mol	H J/mol	S J/(mol·K)	C_v J/(mol·K)	C_p J/(mol·K)	Sound m/s	Diel. const.
29.0	58.24	28.89	−308.4	−273.7	24.76	13.26	38.13	778.8	1.18666
29.5	56.86	28.20	−289.3	−253.8	25.44	13.39	41.66	742.8	1.18196
30.0	55.28	27.42	−268.3	−231.9	26.18	13.55	46.59	702.3	1.17662
30.5	53.40	26.49	−244.6	−206.8	27.00	13.77	54.23	655.5	1.17030
30.6	52.98	26.28	−239.3	−201.3	27.19	13.82	56.34	645.1	1.16887
30.8	52.06	25.82	−228.3	−189.5	27.57	13.94	61.52	622.9	1.16579
31.0	51.02	25.31	−216.1	−176.5	27.99	14.09	68.70	598.4	1.16231
31.2	49.81	24.71	−202.3	−161.8	28.46	14.28	79.53	570.6	1.15825
31.256[b]	49.45	24.53	−198.3	−157.5	28.60	14.33	83.31	562.8	1.15707
31.256[b]	14.40	7.141	178.2	318.3	43.82	16.04	108.7	374.1	1.04406
35.0	9.410	4.668	303.5	517.7	49.90	13.63	36.85	443.0	1.02864
35.5	9.109	4.519	314.5	535.8	50.41	13.51	35.29	449.4	1.02772
36.0	8.837	4.383	324.9	553.1	50.90	13.41	34.00	455.6	1.02688
37.0	8.358	4.146	344.8	586.0	51.80	13.24	31.98	467.1	1.02541
38.0	7.948	3.943	363.6	617.2	52.63	13.11	30.47	477.9	1.02416
39.0	7.590	3.765	381.5	647.1	53.41	13.00	29.30	488.1	1.02306
40.0	7.273	3.608	398.7	675.9	54.14	12.93	28.38	497.6	1.02209
41.0	6.989	3.467	415.4	703.9	54.83	12.88	27.63	506.8	1.02122
42.0	6.732	3.339	431.7	731.2	55.48	12.84	27.01	515.5	1.02043
43.0	6.497	3.223	447.7	757.9	56.11	12.81	26.49	523.9	1.01972
44.0	6.283	3.116	463.3	784.2	56.72	12.79	26.05	532.1	1.01906
46.0	5.901	2.927	493.9	835.6	57.86	12.77	25.34	547.6	1.01790
48.0	5.571	2.764	523.8	885.7	58.93	12.78	24.81	562.2	1.01689
50.0	5.282	2.620	553.2	934.9	59.93	12.80	24.41	576.0	1.01601
52.0	5.025	2.493	582.2	983.4	60.88	12.83	24.10	589.2	1.01523
54.0	4.796	2.379	611.0	1031.0	61.79	12.89	23.86	601.7	1.01453
56.0	4.589	2.276	639.5	1079.0	62.65	12.97	23.69	613.5	1.01390
58.0	4.401	2.183	668.0	1126.0	63.48	13.07	23.58	624.8	1.01332
60.0	4.229	2.098	696.5	1173.0	64.28	13.18	23.51	635.5	1.01280
62.0	4.072	2.020	725.1	1220.0	65.05	13.32	23.49	645.7	1.01232
64.0	3.926	1.948	753.8	1267.0	65.79	13.48	23.51	655.3	1.01188
66.0	3.792	1.881	782.7	1314.0	66.52	13.66	23.56	664.5	1.01147
68.0	3.667	1.819	811.8	1361.0	67.22	13.86	23.65	673.3	1.01109
70.0	3.551	1.762	841.2	1409.0	67.91	14.07	23.77	681.7	1.01074
75.0	3.292	1.633	916.4	1529.0	69.56	14.69	24.18	701.1	1.00996
80.0	3.071	1.523	994.4	1651.0	71.14	15.40	24.73	718.7	1.00928
85.0	2.879	1.428	1076.0	1776.0	72.66	16.19	25.38	735.0	1.00870
90.0	2.710	1.344	1161.0	1905.0	74.13	17.02	26.10	750.3	1.00819
95.0	2.561	1.270	1250.0	2037.0	75.56	17.89	26.88	764.9	1.00774
100.0	2.428	1.204	1343.0	2174.0	76.96	18.76	27.67	779.2	1.00734
105.0	2.309	1.145	1441.0	2314.0	78.33	19.61	28.46	793.2	1.00697
110.0	2.201	1.092	1542.0	2458.0	79.67	20.43	29.23	807.0	1.00665
115.0	2.103	1.043	1647.0	2606.0	80.99	21.20	29.95	820.9	1.00635
120.0	2.013	.9987	1756.0	2758.0	82.28	21.90	30.60	834.8	1.00608
125.0	1.931	.9581	1868.0	2912.0	83.54	22.51	31.18	848.8	1.00583
130.0	1.856	.9207	1983.0	3069.0	84.77	23.04	31.68	862.9	1.00560
140.0	1.722	.8541	2219.0	3390.0	87.15	23.88	32.46	891.2	1.00520
150.0	1.606	.7967	2462.0	3717.0	89.41	24.40	32.94	919.8	1.00485
160.0	1.505	.7466	2709.0	4048.0	91.54	24.67	33.17	948.5	1.00454
170.0	1.416	.7026	2957.0	4380.0	93.55	24.71	33.20	977.3	1.00427
180.0	1.337	.6634	3204.0	4712.0	95.45	24.61	33.07	1006.0	1.00404
190.0	1.267	.6285	3450.0	5041.0	97.23	24.39	32.83	1034.0	1.00382
200.0	1.204	.5971	3693.0	5368.0	98.91	24.11	32.54	1063.0	1.00363
210.0	1.146	.5687	3933.0	5692.0	100.5	23.80	32.21	1090.0	1.00346
220.0	1.094	.5429	4170.0	6012.0	102.0	23.48	31.88	1118.0	1.00330
230.0	1.047	.5193	4404.0	6329.0	103.4	23.16	31.55	1145.0	1.00316
240.0	1.003	.4977	4634.0	6643.0	104.7	22.87	31.25	1171.0	1.00303
250.0	.9634	.4779	4862.0	6954.0	106.0	22.59	30.97	1197.0	1.00291
260.0	.9265	.4596	5087.0	7263.0	107.2	22.35	30.71	1222.0	1.00280
270.0	.8923	.4426	5309.0	7569.0	108.4	22.13	30.49	1247.0	1.00269
280.0	.8605	.4269	5530.0	7873.0	109.5	21.94	30.29	1271.0	1.00260
300.0	.8034	.3985	5966.0	8475.0	111.5	21.63	29.98	1317.0	1.00242
320.0	.7533	.3737	6396.0	9072.0	113.5	21.41	29.75	1362.0	1.00227

Thermophysical properties of hydrogen—Continued

T K	Density kg/m³	Density mol/dm³	E J/mol	H J/mol	S J/(mol·K)	C_v J/(mol·K)	C_p J/(mol·K)	Sound m/s	Diel. const.
340.0	.7092	.3518	6823.0	9666.0	115.3	21.26	29.60	1405.0	1.00214
360.0	.6700	.3323	7248.0	10260.0	117.0	21.15	29.49	1446.0	1.00202
380.0	.6349	.3149	7670.0	10850.0	118.5	21.08	29.41	1486.0	1.00192
400.0	.6032	.2992	8091.0	11430.0	120.1	21.03	29.36	1525.0	1.00182
				1.10 MPa isobar					
14.17[a]	77.56	38.47	−621.1	−592.5	10.10	10.94	14.74	1342.0	1.25347
15.0	76.91	38.15	−609.4	−580.6	10.92	10.11	14.11	1315.0	1.25118
16.0	76.10	37.75	−595.5	−566.4	11.83	9.97	14.34	1282.0	1.24833
18.0	74.35	36.88	−566.0	−536.2	13.61	10.60	16.07	1219.0	1.24221
20.0	72.40	35.91	−532.5	−501.8	15.42	11.37	18.28	1159.0	1.23538
22.0	70.18	34.81	−494.5	−462.9	17.27	11.97	20.71	1099.0	1.22769
24.0	67.65	33.56	−451.5	−418.8	19.19	12.40	23.50	1031.0	1.21889
25.0	66.22	32.85	−427.9	−394.5	20.18	12.57	25.15	993.8	1.21396
26.0	64.66	32.07	−402.7	−368.4	21.21	12.72	27.09	952.6	1.20859
27.0	62.93	31.21	−375.4	−340.1	22.27	12.87	29.46	906.9	1.20264
28.0	60.97	30.24	−345.6	−309.2	23.40	13.03	32.53	855.4	1.19596
29.0	58.70	29.12	−312.4	−274.7	24.61	13.21	36.87	796.2	1.18822
29.5	57.39	28.47	−294.1	−255.5	25.26	13.33	39.88	762.6	1.18379
30.0	55.93	27.74	−274.3	−234.6	25.96	13.46	43.89	725.4	1.17883
30.5	54.24	26.91	−252.2	−211.3	26.73	13.64	49.63	683.5	1.17314
30.6	53.87	26.72	−247.5	−206.3	26.90	13.68	51.12	674.4	1.17188
30.8	53.08	26.33	−237.5	−195.7	27.24	13.78	54.59	655.2	1.16922
31.0	52.21	25.90	−226.9	−184.4	27.61	13.88	58.97	634.6	1.16630
31.2	51.25	25.42	−215.3	−172.1	28.01	14.01	64.74	612.2	1.16307
31.4	50.15	24.88	−202.6	−158.3	28.44	14.17	72.84	587.4	1.15941
31.6	48.86	24.24	−188.0	−142.6	28.94	14.37	85.31	559.3	1.15510
31.7	48.11	23.86	−179.8	−133.7	29.23	14.50	94.66	543.6	1.15259
31.8	47.25	23.44	−170.5	−123.6	29.54	14.64	108.0	526.1	1.14972
31.888[b]	46.42	23.03	−161.8	−114.0	29.85	14.79	124.2	510.3	1.14698
31.888[b]	17.05	8.460	151.0	281.1	42.23	16.48	172.9	373.0	1.05234
35.0	10.88	5.396	283.2	487.1	48.45	13.90	41.71	437.5	1.03316
35.5	10.48	5.197	295.6	507.3	49.03	13.74	39.30	444.4	1.03193
36.0	10.12	5.020	307.3	526.5	49.56	13.61	37.38	451.0	1.03083
36.5	9.801	4.862	318.5	544.7	50.07	13.49	35.81	457.3	1.02985
37.0	9.510	4.718	329.1	562.3	50.55	13.39	34.50	463.3	1.02895
38.0	9.000	4.465	349.3	595.7	51.44	13.23	32.44	474.7	1.02738
39.0	8.563	4.248	368.4	627.3	52.26	13.10	30.89	485.3	1.02604
40.0	8.181	4.058	386.5	657.6	53.03	13.01	29.70	495.3	1.02487
41.0	7.842	3.890	404.0	686.8	53.75	12.94	28.75	504.8	1.02383
42.0	7.538	3.739	421.0	715.2	54.43	12.90	27.97	513.8	1.02290
43.0	7.263	3.603	437.5	742.8	55.08	12.86	27.33	522.4	1.02206
44.0	7.013	3.479	453.7	769.9	55.70	12.83	26.80	530.8	1.02129
46.0	6.571	3.260	485.1	822.6	56.87	12.80	25.95	546.6	1.01994
48.0	6.192	3.072	515.7	873.8	57.96	12.80	25.31	561.5	1.01878
50.0	5.862	2.908	545.6	923.9	58.99	12.81	24.83	575.6	1.01778
52.0	5.570	2.763	575.1	973.2	59.95	12.85	24.46	588.9	1.01689
54.0	5.310	2.634	604.2	1022.0	60.87	12.91	24.18	601.5	1.01609
56.0	5.077	2.518	633.2	1070.0	61.75	12.98	23.98	613.5	1.01538
58.0	4.865	2.413	662.0	1118.0	62.58	13.08	23.83	624.9	1.01474
60.0	4.673	2.318	690.8	1165.0	63.39	13.19	23.74	635.7	1.01415
62.0	4.496	2.230	719.6	1213.0	64.17	13.33	23.70	645.9	1.01361
64.0	4.334	2.150	748.5	1260.0	64.92	13.49	23.70	655.7	1.01312
66.0	4.184	2.075	777.6	1308.0	65.65	13.67	23.74	664.9	1.01266
68.0	4.045	2.006	806.9	1355.0	66.36	13.87	23.81	673.8	1.01224
70.0	3.915	1.942	836.5	1403.0	67.05	14.08	23.92	682.2	1.01185
75.0	3.628	1.800	912.1	1523.0	68.72	14.70	24.31	701.7	1.01097
80.0	3.382	1.678	990.5	1646.0	70.30	15.41	24.84	719.4	1.01023
85.0	3.169	1.572	1072.0	1772.0	71.82	16.20	25.47	735.7	1.00958
90.0	2.983	1.480	1158.0	1901.0	73.30	17.03	26.19	751.0	1.00902
95.0	2.818	1.398	1247.0	2034.0	74.74	17.90	26.95	765.7	1.00852
100.0	2.671	1.325	1340.0	2171.0	76.14	18.77	27.74	780.0	1.00807

Thermophysical properties of hydrogen—Continued

T K	Density kg/m³	Density mol/dm³	E J/mol	H J/mol	S J/(mol·K)	C_v J/(mol·K)	C_p J/(mol·K)	Sound m/s	Diel. const.
105.0	2.539	1.260	1438.0	2311.0	77.51	19.62	28.52	794.0	1.00767
110.0	2.420	1.201	1540.0	2456.0	78.86	20.44	29.28	807.9	1.00731
115.0	2.312	1.147	1645.0	2604.0	80.17	21.21	29.99	821.7	1.00699
120.0	2.214	1.098	1754.0	2756.0	81.46	21.90	30.65	835.7	1.00669
125.0	2.124	1.053	1866.0	2910.0	82.73	22.52	31.22	849.7	1.00641
130.0	2.041	1.012	1981.0	3068.0	83.96	23.05	31.72	863.8	1.00616
140.0	1.893	.9390	2217.0	3389.0	86.34	23.88	32.50	892.1	1.00572
150.0	1.766	.8758	2461.0	3716.0	88.60	24.41	32.97	920.7	1.00533
160.0	1.655	.8207	2707.0	4048.0	90.74	24.67	33.20	949.5	1.00499
170.0	1.557	.7723	2955.0	4380.0	92.75	24.72	33.22	978.2	1.00470
180.0	1.470	.7293	3203.0	4711.0	94.65	24.61	33.08	1007.0	1.00444
190.0	1.393	.6908	3449.0	5041.0	96.43	24.40	32.85	1035.0	1.00420
200.0	1.323	.6563	3692.0	5368.0	98.11	24.12	32.55	1064.0	1.00399
210.0	1.260	.6251	3932.0	5692.0	99.69	23.80	32.22	1091.0	1.00380
220.0	1.203	.5967	4169.0	6013.0	101.2	23.48	31.89	1119.0	1.00363
230.0	1.151	.5709	4403.0	6330.0	102.6	23.17	31.56	1146.0	1.00347
240.0	1.103	.5471	4633.0	6644.0	103.9	22.87	31.26	1172.0	1.00333
250.0	1.059	.5253	4861.0	6955.0	105.2	22.60	30.98	1198.0	1.00320
260.0	1.018	.5052	5086.0	7264.0	106.4	22.35	30.72	1223.0	1.00307
270.0	.9809	.4866	5309.0	7570.0	107.6	22.13	30.50	1248.0	1.00296
280.0	.9460	.4692	5529.0	7874.0	108.7	21.94	30.30	1272.0	1.00285
300.0	.8832	.4381	5965.0	8476.0	110.7	21.63	29.99	1318.0	1.00266
320.0	.8282	.4108	6396.0	9074.0	112.7	21.41	29.76	1363.0	1.00250
340.0	.7797	.3868	6823.0	9667.0	114.5	21.26	29.60	1406.0	1.00235
360.0	.7366	.3654	7247.0	10260.0	116.2	21.16	29.49	1447.0	1.00222
380.0	.6980	.3462	7670.0	10850.0	117.8	21.08	29.41	1487.0	1.00211
400.0	.6633	.3290	8091.0	11430.0	119.3	21.03	29.36	1526.0	1.00200

1.20 MPa isobar

T K	Density kg/m³	Density mol/dm³	E J/mol	H J/mol	S J/(mol·K)	C_v J/(mol·K)	C_p J/(mol·K)	Sound m/s	Diel. const.
14.20[a]	77.61	38.50	−621.0	−589.8	10.11	10.94	14.67	1340.0	1.25365
15.0	76.99	38.19	−609.8	−578.4	10.89	10.14	14.08	1315.0	1.25146
16.0	76.18	37.79	−596.0	−564.2	11.80	9.99	14.31	1283.0	1.24863
18.0	74.45	36.93	−566.5	−534.0	13.58	10.61	16.02	1222.0	1.24256
20.0	72.51	35.97	−533.2	−499.9	15.38	11.37	18.21	1164.0	1.23579
22.0	70.33	34.89	−495.5	−461.1	17.22	11.97	20.60	1105.0	1.22818
24.0	67.82	33.64	−452.9	−417.2	19.13	12.39	23.33	1039.0	1.21950
25.0	66.42	32.95	−429.5	−393.1	20.12	12.56	24.93	1002.0	1.21465
26.0	64.89	32.19	−404.6	−367.3	21.13	12.71	26.78	962.2	1.20938
27.0	63.20	31.35	−377.7	−339.4	22.18	12.85	29.03	918.1	1.20358
28.0	61.30	30.41	−348.5	−309.0	23.28	13.00	31.88	868.7	1.19709
29.0	59.12	29.33	−316.2	−275.3	24.47	13.17	35.79	812.4	1.18966
29.5	57.89	28.72	−298.6	−256.8	25.10	13.27	38.41	780.8	1.18546
30.0	56.52	28.04	−279.6	−236.8	25.77	13.39	41.77	746.4	1.18083
30.5	54.97	27.27	−258.8	−214.8	26.50	13.54	46.33	708.1	1.17560
31.0	53.17	26.37	−235.6	−190.1	27.30	13.74	53.07	664.8	1.16952
31.2	52.35	25.97	−225.3	−179.1	27.66	13.83	56.83	645.5	1.16676
31.4	51.44	25.52	−214.3	−167.3	28.03	13.95	61.62	624.9	1.16373
31.6	50.44	25.02	−202.3	−154.4	28.44	14.08	67.96	602.4	1.16036
31.8	49.29	24.45	−189.0	−139.9	28.90	14.24	76.94	577.6	1.15651
32.0	47.93	23.77	−173.7	−123.3	29.42	14.46	90.90	549.6	1.15197
32.1	47.13	23.38	−165.0	−113.7	29.72	14.59	101.5	533.8	1.14931
32.2	46.20	22.92	−155.2	−102.8	30.06	14.75	116.7	516.5	1.14626
32.3	45.10	22.37	−143.7	−90.0	30.45	14.94	140.9	496.9	1.14261
32.4	43.69	21.67	−129.3	−73.9	30.95	15.21	187.8	473.6	1.13792
32.476[b]	38.89	19.29	−84.2	−22.0	32.55	16.23	1650.0	410.5	1.12216
32.476[b]	24.18	11.99	68.8	168.8	38.42	17.46	4692.0	367.4	1.07478
35.0	12.57	6.236	260.0	452.4	46.97	14.22	48.71	431.7	1.03840
35.5	12.02	5.964	274.5	475.7	47.63	14.01	44.78	439.3	1.03670
36.0	11.55	5.729	287.9	497.3	48.24	13.84	41.81	446.5	1.03524
36.5	11.13	5.523	300.4	517.6	48.80	13.69	39.49	453.2	1.03396
37.0	10.76	5.340	312.2	536.9	49.32	13.56	37.62	459.6	1.03282

Thermophysical properties of hydrogen—Continued

T K	Density kg/m³	Density mol/dm³	E J/mol	H J/mol	S J/(mol·K)	C_v J/(mol·K)	C_p J/(mol·K)	Sound m/s	Diel. const.
38.0	10.13	5.024	334.2	573.0	50.28	13.36	34.78	471.6	1.03086
39.0	9.595	4.759	354.6	606.7	51.16	13.21	32.74	482.7	1.02921
40.0	9.135	4.531	373.8	638.7	51.97	13.10	31.20	493.1	1.02780
41.0	8.733	4.332	392.2	669.2	52.72	13.01	30.00	502.9	1.02656
42.0	8.376	4.155	409.9	698.7	53.43	12.95	29.04	512.2	1.02547
43.0	8.056	3.996	427.1	727.4	54.11	12.91	28.25	521.0	1.02449
44.0	7.766	3.852	443.8	755.3	54.75	12.87	27.60	529.6	1.02360
46.0	7.258	3.600	476.1	809.4	55.95	12.83	26.58	545.8	1.02204
48.0	6.826	3.386	507.4	861.8	57.07	12.82	25.83	560.9	1.02072
50.0	6.452	3.200	537.9	912.9	58.11	12.83	25.27	575.2	1.01958
52.0	6.123	3.037	567.9	962.9	59.09	12.87	24.84	588.7	1.01857
54.0	5.832	2.893	597.5	1012.0	60.02	12.92	24.51	601.5	1.01768
56.0	5.570	2.763	626.8	1061.0	60.91	12.99	24.27	613.6	1.01689
58.0	5.334	2.646	655.9	1109.0	61.76	13.09	24.09	625.1	1.01617
60.0	5.120	2.540	685.0	1157.0	62.57	13.20	23.98	635.9	1.01551
62.0	4.924	2.442	714.0	1205.0	63.36	13.34	23.91	646.3	1.01492
64.0	4.744	2.353	743.2	1253.0	64.12	13.50	23.90	656.1	1.01437
66.0	4.578	2.271	772.5	1301.0	64.85	13.68	23.92	665.4	1.01386
68.0	4.424	2.194	802.0	1349.0	65.57	13.87	23.98	674.3	1.01339
70.0	4.281	2.124	831.8	1397.0	66.26	14.09	24.07	682.7	1.01296
75.0	3.964	1.966	907.8	1518.0	67.94	14.71	24.43	702.3	1.01200
80.0	3.694	1.832	986.5	1641.0	69.53	15.42	24.94	720.0	1.01117
85.0	3.460	1.716	1069.0	1768.0	71.06	16.20	25.57	736.4	1.01046
90.0	3.255	1.615	1154.0	1897.0	72.54	17.04	26.27	751.8	1.00984
95.0	3.075	1.525	1244.0	2031.0	73.98	17.90	27.03	766.5	1.00930
100.0	2.914	1.445	1337.0	2168.0	75.39	18.77	27.80	780.8	1.00881
105.0	2.770	1.374	1435.0	2309.0	76.76	19.63	28.58	794.8	1.00837
110.0	2.640	1.309	1537.0	2453.0	78.11	20.45	29.33	808.7	1.00798
115.0	2.522	1.251	1643.0	2602.0	79.43	21.21	30.04	822.6	1.00762
120.0	2.414	1.197	1752.0	2754.0	80.72	21.91	30.69	836.5	1.00729
125.0	2.315	1.149	1864.0	2909.0	81.99	22.52	31.26	850.5	1.00699
130.0	2.225	1.104	1979.0	3066.0	83.22	23.05	31.75	864.7	1.00672
140.0	2.064	1.024	2216.0	3388.0	85.60	23.89	32.53	893.0	1.00623
150.0	1.925	.9548	2459.0	3716.0	87.87	24.41	33.00	921.6	1.00581
160.0	1.804	.8947	2706.0	4047.0	90.00	24.67	33.22	950.4	1.00545
170.0	1.697	.8419	2954.0	4379.0	92.02	24.72	33.24	979.1	1.00512
180.0	1.603	.7950	3202.0	4711.0	93.92	24.62	33.10	1008.0	1.00484
190.0	1.518	.7531	3448.0	5041.0	95.70	24.40	32.86	1036.0	1.00458
200.0	1.442	.7155	3691.0	5368.0	97.38	24.12	32.57	1064.0	1.00435
210.0	1.374	.6814	3931.0	5692.0	98.96	23.81	32.24	1092.0	1.00415
220.0	1.311	.6505	4168.0	6013.0	100.5	23.49	31.90	1120.0	1.00396
230.0	1.255	.6223	4402.0	6330.0	101.9	23.17	31.57	1146.0	1.00379
240.0	1.202	.5965	4633.0	6644.0	103.2	22.87	31.27	1173.0	1.00363
250.0	1.155	.5727	4860.0	6956.0	104.5	22.60	30.98	1199.0	1.00348
260.0	1.110	.5508	5085.0	7264.0	105.7	22.35	30.73	1224.0	1.00335
270.0	1.069	.5304	5308.0	7570.0	106.8	22.13	30.50	1248.0	1.00323
280.0	1.031	.5116	5529.0	7874.0	107.9	21.94	30.31	1273.0	1.00311
300.0	.9629	.4776	5965.0	8477.0	110.0	21.64	29.99	1319.0	1.00291
320.0	.9030	.4479	6396.0	9075.0	111.9	21.42	29.76	1364.0	1.00272
340.0	.8501	.4217	6823.0	9668.0	113.7	21.26	29.60	1406.0	1.00256
360.0	.8031	.3984	7247.0	10260.0	115.4	21.16	29.49	1448.0	1.00242
380.0	.7611	.3775	7670.0	10850.0	117.0	21.09	29.42	1488.0	1.00230
400.0	.7232	.3587	8091.0	11440.0	118.5	21.03	29.36	1526.0	1.00218

1.29 MPa isobar

T K	Density kg/m³	Density mol/dm³	E J/mol	H J/mol	S J/(mol·K)	C_v J/(mol·K)	C_p J/(mol·K)	Sound m/s	Diel. const.
14.23[a]	77.65	38.52	–620.9	–587.4	10.11	10.93	14.62	1338.0	1.25381
15.0	77.06	38.22	–610.1	–576.4	10.87	10.17	14.06	1315.0	1.25171
16.0	76.26	37.83	–596.3	–562.2	11.78	10.01	14.28	1284.0	1.24891
18.0	74.54	36.98	–567.0	–532.1	13.55	10.62	15.97	1225.0	1.24288
20.0	72.62	36.02	–533.9	–498.1	15.34	11.38	18.15	1168.0	1.23616
22.0	70.45	34.95	–496.3	–459.4	17.18	11.97	20.51	1110.0	1.22862
24.0	67.98	33.72	–454.1	–415.8	19.08	12.39	23.18	1045.0	1.22004

Thermophysical properties of hydrogen—Continued

T K	Density kg/m^3	Density mol/dm^3	E J/mol	H J/mol	S J/(mol·K)	C_v J/(mol·K)	C_p J/(mol·K)	Sound m/s	Diel. const.
25.0	66.60	33.03	−430.9	−391.9	20.06	12.55	24.74	1010.0	1.21526
26.0	65.09	32.29	−406.2	−366.3	21.06	12.70	26.53	970.6	1.21008
27.0	63.44	31.47	−379.7	−338.7	22.10	12.84	28.67	927.8	1.20439
28.0	61.59	30.55	−351.0	−308.7	23.19	12.98	31.35	880.1	1.19807
29.0	59.48	29.51	−319.4	−275.7	24.35	13.14	34.93	826.1	1.19089
29.5	58.30	28.92	−302.3	−257.7	24.96	13.23	37.28	796.2	1.18687
30.0	57.00	28.28	−283.9	−238.3	25.61	13.34	40.21	763.7	1.18246
30.5	55.56	27.56	−264.1	−217.3	26.31	13.47	44.04	728.2	1.17757
31.0	53.90	26.74	−242.3	−194.0	27.07	13.63	49.37	688.5	1.17199
31.2	53.17	26.37	−232.8	−183.9	27.39	13.71	52.17	671.3	1.16951
31.4	52.37	25.98	−222.8	−173.1	27.74	13.80	55.56	653.0	1.16684
31.6	51.51	25.55	−212.1	−161.6	28.10	13.90	59.76	633.5	1.16394
31.8	50.56	25.08	−200.6	−149.2	28.49	14.02	65.16	612.5	1.16075
32.0	49.49	24.55	−188.0	−135.4	28.92	14.17	72.42	589.8	1.15718
32.2	48.26	23.94	−173.9	−120.0	29.41	14.34	82.86	564.6	1.15307
32.3	47.56	23.59	−166.0	−111.3	29.67	14.45	90.08	550.9	1.15074
32.4	46.78	23.21	−157.5	−101.9	29.97	14.57	99.48	536.2	1.14817
32.5	45.90	22.77	−148.0	−91.3	30.29	14.72	112.4	520.3	1.14525
32.6	44.88	22.26	−137.2	−79.2	30.66	14.89	131.3	502.8	1.14187
32.7	43.64	21.65	−124.3	−64.7	31.11	15.12	162.3	483.0	1.13775
32.8	41.99	20.83	−107.7	−45.8	31.69	15.42	224.7	459.6	1.13234
32.9	39.35	19.52	−81.7	−15.6	32.61	15.92	434.1	428.7	1.12365
33.0	25.86	12.83	58.1	158.6	37.89	17.22	1038.0	377.1	1.08011
33.1	22.58	11.20	99.0	214.2	39.57	16.86	353.1	381.1	1.06971
33.2	21.09	10.46	119.4	242.7	40.43	16.58	235.2	384.9	1.06501
33.3	20.09	9.964	133.8	263.3	41.05	16.35	182.9	388.3	1.06185
33.4	19.32	9.585	145.4	280.0	41.55	16.16	152.7	391.4	1.05945
33.5	18.70	9.276	155.1	294.2	41.98	15.99	132.7	394.3	1.05750
33.6	18.17	9.016	163.6	306.7	42.35	15.84	118.4	397.1	1.05585
33.7	17.72	8.790	171.2	318.0	42.68	15.70	107.6	399.7	1.05442
33.8	17.32	8.590	178.1	328.3	42.99	15.58	99.15	402.2	1.05316
34.0	16.63	8.248	190.4	346.8	43.54	15.35	86.61	406.8	1.05101
34.2	16.05	7.962	201.1	363.2	44.02	15.16	77.73	411.2	1.04921
34.4	15.55	7.716	210.8	378.0	44.45	14.98	71.06	415.3	1.04766
34.6	15.12	7.499	219.7	391.7	44.85	14.83	65.85	419.1	1.04630
34.8	14.73	7.306	227.9	404.4	45.21	14.69	61.66	422.9	1.04509
35.0	14.38	7.131	235.5	416.4	45.56	14.57	58.20	426.4	1.04400
35.5	13.62	6.757	252.9	443.8	46.33	14.30	51.71	434.7	1.04166
36.0	13.00	6.448	268.4	468.4	47.02	14.08	47.15	442.4	1.03972
36.5	12.47	6.183	282.5	491.1	47.65	13.89	43.75	449.6	1.03807
37.0	12.00	5.953	295.6	512.3	48.22	13.74	41.12	456.4	1.03663
37.5	11.59	5.749	307.9	532.3	48.76	13.61	39.02	462.8	1.03536
38.0	11.22	5.566	319.6	551.4	49.27	13.49	37.31	469.0	1.03422
38.5	10.89	5.400	330.8	569.7	49.75	13.40	35.87	474.8	1.03319
39.0	10.58	5.248	341.5	587.3	50.20	13.31	34.66	480.5	1.03225
40.0	10.04	4.980	361.9	620.9	51.05	13.18	32.72	491.2	1.03058
41.0	9.571	4.748	381.2	652.9	51.84	13.08	31.25	501.3	1.02914
42.0	9.159	4.543	399.6	683.5	52.58	13.01	30.09	510.8	1.02787
43.0	8.793	4.362	417.4	713.1	53.28	12.95	29.15	519.9	1.02675
44.0	8.463	4.198	434.6	741.9	53.94	12.91	28.38	528.6	1.02574
45.0	8.164	4.050	451.4	769.9	54.57	12.88	27.73	537.0	1.02482
46.0	7.891	3.914	467.8	797.4	55.17	12.86	27.19	545.1	1.02398
48.0	7.408	3.675	499.8	850.9	56.31	12.84	26.33	560.5	1.02250
50.0	6.992	3.468	530.9	902.8	57.37	12.85	25.68	575.0	1.02123
52.0	6.628	3.288	561.3	953.7	58.37	12.88	25.19	588.6	1.02011
54.0	6.306	3.128	591.3	1004.0	59.31	12.93	24.82	601.5	1.01913
56.0	6.019	2.986	620.9	1053.0	60.21	13.00	24.53	613.7	1.01825
58.0	5.760	2.857	650.4	1102.0	61.06	13.10	24.33	625.3	1.01746
60.0	5.525	2.741	679.7	1150.0	61.89	13.21	24.19	636.2	1.01675
62.0	5.311	2.635	709.0	1199.0	62.68	13.35	24.11	646.6	1.01610
64.0	5.115	2.537	738.4	1247.0	63.44	13.51	24.07	656.5	1.01550
66.0	4.934	2.448	767.9	1295.0	64.18	13.69	24.08	665.8	1.01495
68.0	4.767	2.365	797.6	1343.0	64.90	13.88	24.13	674.8	1.01444

Thermophysical properties of hydrogen—Continued

T K	Density kg/m³	Density mol/dm³	E J/mol	H J/mol	S J/(mol·K)	C_v J/(mol·K)	C_p J/(mol·K)	Sound m/s	Diel. const.
70.0	4.612	2.288	827.6	1392.0	65.60	14.10	24.21	683.2	1.01396
72.0	4.467	2.216	857.9	1440.0	66.29	14.33	24.33	691.4	1.01352
74.0	4.332	2.149	888.5	1489.0	66.96	14.59	24.47	699.1	1.01311
76.0	4.205	2.086	919.5	1538.0	67.61	14.85	24.64	706.6	1.01273
80.0	3.975	1.972	983.0	1637.0	68.88	15.43	25.04	720.7	1.01203
85.0	3.722	1.846	1065.0	1764.0	70.42	16.21	25.65	737.1	1.01126
90.0	3.501	1.737	1151.0	1894.0	71.91	17.05	26.35	752.5	1.01059
95.0	3.306	1.640	1241.0	2028.0	73.35	17.91	27.09	767.2	1.01000
100.0	3.133	1.554	1335.0	2165.0	74.76	18.78	27.86	781.5	1.00947
105.0	2.977	1.477	1433.0	2306.0	76.14	19.63	28.63	795.5	1.00900
110.0	2.837	1.407	1535.0	2451.0	77.49	20.45	29.38	809.5	1.00858
115.0	2.710	1.344	1640.0	2600.0	78.81	21.22	30.09	823.3	1.00819
120.0	2.594	1.287	1750.0	2752.0	80.10	21.91	30.73	837.3	1.00784
125.0	2.488	1.234	1862.0	2907.0	81.37	22.53	31.30	851.3	1.00752
130.0	2.391	1.186	1977.0	3065.0	82.61	23.06	31.79	865.5	1.00722
135.0	2.301	1.141	2094.0	3225.0	83.81	23.51	32.20	879.7	1.00695
140.0	2.217	1.100	2214.0	3387.0	84.99	23.89	32.56	893.8	1.00670
150.0	2.068	1.026	2457.0	3715.0	87.26	24.42	33.02	922.5	1.00625
160.0	1.938	.9612	2704.0	4046.0	89.39	24.68	33.24	951.2	1.00585
170.0	1.823	.9044	2953.0	4379.0	91.41	24.73	33.26	980.0	1.00551
180.0	1.722	.8541	3201.0	4711.0	93.31	24.62	33.12	1009.0	1.00520
190.0	1.631	.8091	3447.0	5041.0	95.09	24.40	32.88	1037.0	1.00492
200.0	1.550	.7686	3690.0	5368.0	96.77	24.12	32.58	1065.0	1.00468
210.0	1.476	.7321	3930.0	5692.0	98.35	23.81	32.25	1093.0	1.00445
220.0	1.409	.6989	4167.0	6013.0	99.85	23.49	31.91	1120.0	1.00425
230.0	1.348	.6686	4401.0	6331.0	101.3	23.17	31.58	1147.0	1.00407
240.0	1.292	.6408	4632.0	6645.0	102.6	22.88	31.28	1174.0	1.00390
250.0	1.240	.6153	4860.0	6956.0	103.9	22.60	30.99	1199.0	1.00374
260.0	1.193	.5917	5085.0	7265.0	105.1	22.36	30.74	1225.0	1.00360
270.0	1.149	.5699	5308.0	7571.0	106.2	22.14	30.51	1249.0	1.00347
280.0	1.108	.5496	5528.0	7875.0	107.3	21.94	30.31	1273.0	1.00334
300.0	1.035	.5132	5964.0	8478.0	109.4	21.64	30.00	1320.0	1.00312
320.0	.9702	.4813	6395.0	9076.0	111.3	21.42	29.77	1364.0	1.00293
340.0	.9134	.4531	6822.0	9669.0	113.1	21.27	29.61	1407.0	1.00276
360.0	.8630	.4281	7247.0	10260.0	114.8	21.16	29.50	1448.0	1.00260
380.0	.8178	.4057	7669.0	10850.0	116.4	21.09	29.42	1488.0	1.00247
400.0	.7771	.3855	8091.0	11440.0	117.9	21.03	29.36	1527.0	1.00234

1.30 MPa isobar

T K	Density kg/m³	Density mol/dm³	E J/mol	H J/mol	S J/(mol·K)	C_v J/(mol·K)	C_p J/(mol·K)	Sound m/s	Diel. const.
14.23[a]	77.66	38.52	–620.8	–587.1	10.12	10.93	14.61	1338.0	1.25383
15.0	77.07	38.23	–610.1	–576.1	10.87	10.18	14.06	1315.0	1.25174
16.0	76.27	37.83	–596.4	–562.0	11.78	10.01	14.28	1285.0	1.24894
18.0	74.55	36.98	–567.1	–531.9	13.55	10.62	15.97	1225.0	1.24291
20.0	72.63	36.03	–533.9	–497.9	15.34	11.38	18.14	1168.0	1.23620
22.0	70.47	34.95	–496.4	–459.3	17.18	11.96	20.50	1110.0	1.22867
24.0	68.00	33.73	–454.2	–415.7	19.07	12.39	23.17	1046.0	1.22010
25.0	66.62	33.04	–431.1	–391.7	20.05	12.55	24.72	1010.0	1.21533
26.0	65.11	32.30	–406.4	–366.1	21.05	12.70	26.50	971.5	1.21015
27.0	63.46	31.48	–379.9	–338.6	22.09	12.84	28.63	928.8	1.20448
28.0	61.62	30.57	–351.2	–308.7	23.18	12.98	31.29	881.3	1.19818
29.0	59.52	29.53	–319.8	–275.7	24.34	13.13	34.85	827.6	1.19102
29.5	58.35	28.94	–302.7	–257.8	24.95	13.23	37.16	797.8	1.18702
30.0	57.06	28.30	–284.4	–238.5	25.60	13.33	40.05	765.6	1.18264
30.5	55.62	27.59	–264.7	–217.6	26.29	13.46	43.81	730.3	1.17777
31.0	53.98	26.78	–243.0	–194.4	27.04	13.62	49.02	691.0	1.17225
31.5	52.05	25.82	–218.5	–168.1	27.88	13.83	56.93	646.4	1.16576
31.7	51.16	25.38	–207.5	–156.3	28.26	13.94	61.47	626.5	1.16277
31.8	50.68	25.14	–201.7	–150.0	28.46	14.00	64.22	616.0	1.16116
32.0	49.63	24.62	–189.3	–136.5	28.88	14.14	71.07	593.7	1.15766
32.2	48.43	24.03	–175.5	–121.4	29.35	14.31	80.78	569.1	1.15367

Thermophysical properties of hydrogen—Continued

T K	Density kg/m^3	Density mol/dm^3	E J/mol	H J/mol	S J/(mol·K)	C_v J/(mol·K)	C_p J/(mol·K)	Sound m/s	Diel. const.
32.4	47.01	23.32	–159.6	–103.8	29.89	14.53	95.84	541.5	1.14893
32.5	46.17	22.90	–150.5	–93.7	30.20	14.66	107.2	526.2	1.14616
32.6	45.21	22.43	–140.2	–82.3	30.56	14.83	123.2	509.4	1.14297
32.7	44.07	21.86	–128.3	–68.8	30.97	15.03	148.2	490.8	1.13919
32.8	42.62	21.14	–113.5	–52.0	31.48	15.29	193.3	469.3	1.13441
32.9	40.54	20.11	–92.8	–28.1	32.21	15.68	303.9	442.8	1.12758
33.0	35.84	17.78	–46.9	26.3	33.86	16.52	1251.0	401.9	1.11222
33.1	24.36	12.08	77.4	185.0	38.66	17.05	550.1	379.6	1.07534
33.2	22.16	10.99	105.9	224.2	39.85	16.74	295.1	383.4	1.06838
33.3	20.89	10.36	123.6	249.1	40.59	16.49	213.2	386.9	1.06437
33.4	19.97	9.908	136.9	268.1	41.17	16.28	171.4	390.2	1.06150
33.5	19.26	9.553	147.8	283.9	41.64	16.10	145.6	393.2	1.05925
33.6	18.67	9.260	157.1	297.5	42.04	15.94	128.0	396.0	1.05739
33.7	18.16	9.010	165.3	309.6	42.40	15.79	115.0	398.7	1.05581
33.8	17.72	8.791	172.7	320.6	42.73	15.66	105.1	401.2	1.05443
34.0	16.98	8.421	185.7	340.1	43.30	15.42	90.74	406.0	1.05210
34.2	16.36	8.115	197.0	357.2	43.81	15.22	80.79	410.4	1.05017
34.4	15.83	7.854	207.1	372.6	44.25	15.04	73.44	414.5	1.04853
34.6	15.37	7.626	216.2	386.7	44.66	14.88	67.77	418.5	1.04710
34.8	14.97	7.424	224.7	399.8	45.04	14.74	63.25	422.2	1.04583
35.0	14.60	7.242	232.5	412.0	45.39	14.61	59.54	425.8	1.04469
35.2	14.27	7.076	239.9	423.6	45.72	14.49	56.44	429.3	1.04365
35.5	13.82	6.853	250.3	440.0	46.18	14.33	52.64	434.2	1.04225
36.0	13.17	6.533	266.0	465.0	46.88	14.11	47.84	442.0	1.04025
36.5	12.62	6.261	280.4	488.0	47.52	13.92	44.30	449.2	1.03855
37.0	12.15	6.025	293.7	509.5	48.10	13.76	41.56	456.0	1.03708
37.5	11.72	5.815	306.1	529.7	48.65	13.63	39.39	462.5	1.03578
38.0	11.35	5.628	317.9	548.9	49.15	13.51	37.61	468.7	1.03461
38.5	11.00	5.459	329.2	567.3	49.64	13.41	36.14	474.6	1.03356
39.0	10.69	5.304	340.0	585.1	50.09	13.32	34.89	480.3	1.03260
40.0	10.14	5.031	360.5	618.9	50.95	13.19	32.91	491.0	1.03090
41.0	9.666	4.795	379.9	651.1	51.74	13.09	31.40	501.1	1.02943
42.0	9.248	4.587	398.5	681.8	52.49	13.01	30.21	510.7	1.02815
43.0	8.876	4.403	416.3	711.6	53.19	12.96	29.25	519.8	1.02700
44.0	8.542	4.237	433.6	740.4	53.85	12.92	28.47	528.5	1.02598
45.0	8.239	4.087	450.4	768.5	54.48	12.89	27.81	536.9	1.02505
46.0	7.962	3.950	466.9	796.1	55.09	12.86	27.26	545.0	1.02420
48.0	7.473	3.707	499.0	849.7	56.23	12.85	26.38	560.4	1.02270
50.0	7.052	3.498	530.1	901.7	57.29	12.85	25.73	574.9	1.02141
52.0	6.685	3.316	560.6	952.7	58.29	12.88	25.23	588.6	1.02029
54.0	6.359	3.155	590.6	1003.0	59.23	12.93	24.85	601.5	1.01929
56.0	6.069	3.010	620.3	1052.0	60.13	13.01	24.56	613.7	1.01841
58.0	5.808	2.881	649.8	1101.0	60.99	13.10	24.36	625.3	1.01761
60.0	5.571	2.763	679.1	1150.0	61.81	13.21	24.21	636.3	1.01689
62.0	5.354	2.656	708.5	1198.0	62.61	13.35	24.13	646.7	1.01623
64.0	5.156	2.558	737.9	1246.0	63.37	13.51	24.09	656.5	1.01562
66.0	4.974	2.467	767.4	1294.0	64.11	13.69	24.10	665.9	1.01507
68.0	4.805	2.383	797.1	1343.0	64.83	13.88	24.14	674.8	1.01455
70.0	4.648	2.306	827.1	1391.0	65.53	14.10	24.23	683.3	1.01408
72.0	4.502	2.233	857.4	1439.0	66.22	14.33	24.34	691.4	1.01363
74.0	4.366	2.166	888.1	1488.0	66.89	14.59	24.48	699.2	1.01322
76.0	4.239	2.103	919.1	1537.0	67.54	14.85	24.65	706.6	1.01283
80.0	4.006	1.987	982.6	1637.0	68.82	15.43	25.05	720.7	1.01212
85.0	3.751	1.861	1065.0	1764.0	70.35	16.21	25.66	737.1	1.01135
90.0	3.528	1.750	1151.0	1894.0	71.84	17.05	26.35	752.5	1.01067
95.0	3.332	1.653	1241.0	2027.0	73.28	17.91	27.10	767.3	1.01008
100.0	3.157	1.566	1334.0	2165.0	74.69	18.78	27.87	781.6	1.00955
105.0	3.000	1.488	1432.0	2306.0	76.07	19.63	28.64	795.6	1.00907
110.0	2.859	1.418	1534.0	2451.0	77.42	20.45	29.39	809.5	1.00864
115.0	2.731	1.355	1640.0	2600.0	78.74	21.22	30.09	823.4	1.00825
120.0	2.614	1.297	1749.0	2752.0	80.04	21.91	30.73	837.4	1.00790
125.0	2.507	1.244	1862.0	2907.0	81.30	22.53	31.30	851.4	1.00758
130.0	2.409	1.195	1977.0	3065.0	82.54	23.06	31.79	865.5	1.00728

Thermophysical properties of hydrogen—Continued

T K	Density kg/m^3	Density mol/dm^3	E J/mol	H J/mol	S J/(mol·K)	C_v J/(mol·K)	C_p J/(mol·K)	Sound m/s	Diel. const.
135.0	2.318	1.150	2094.0	3225.0	83.75	23.51	32.20	879.8	1.00700
140.0	2.234	1.108	2214.0	3387.0	84.93	23.90	32.56	893.9	1.00675
150.0	2.084	1.034	2457.0	3715.0	87.19	24.42	33.03	922.6	1.00629
160.0	1.953	.9686	2704.0	4046.0	89.33	24.68	33.24	951.3	1.00590
170.0	1.837	.9114	2953.0	4379.0	91.35	24.73	33.26	980.1	1.00555
180.0	1.735	.8606	3200.0	4711.0	93.24	24.62	33.12	1009.0	1.00524
190.0	1.644	.8153	3446.0	5041.0	95.03	24.40	32.88	1037.0	1.00496
200.0	1.561	.7745	3690.0	5368.0	96.71	24.12	32.58	1065.0	1.00471
210.0	1.487	.7377	3930.0	5693.0	98.29	23.81	32.25	1093.0	1.00449
220.0	1.420	.7042	4167.0	6013.0	99.78	23.49	31.91	1121.0	1.00428
230.0	1.358	.6737	4401.0	6331.0	101.2	23.17	31.58	1147.0	1.00410
240.0	1.302	.6457	4632.0	6645.0	102.5	22.88	31.28	1174.0	1.00393
250.0	1.250	.6200	4860.0	6956.0	103.8	22.60	30.99	1199.0	1.00377
260.0	1.202	.5963	5085.0	7265.0	105.0	22.36	30.74	1225.0	1.00363
270.0	1.158	.5743	5308.0	7571.0	106.2	22.14	30.51	1249.0	1.00349
280.0	1.117	.5539	5528.0	7875.0	107.3	21.95	30.31	1273.0	1.00337
300.0	1.042	.5171	5964.0	8478.0	109.4	21.64	30.00	1320.0	1.00315
320.0	.9777	.4850	6395.0	9076.0	111.3	21.42	29.77	1365.0	1.00295
340.0	.9205	.4566	6822.0	9669.0	113.1	21.27	29.61	1407.0	1.00278
360.0	.8696	.4314	7247.0	10260.0	114.8	21.16	29.50	1449.0	1.00262
380.0	.8241	.4088	7669.0	10850.0	116.4	21.09	29.42	1488.0	1.00249
400.0	.7831	.3884	8091.0	11440.0	117.9	21.03	29.36	1527.0	1.00236

1.32 MPa isobar

T K	Density kg/m^3	Density mol/dm^3	E J/mol	H J/mol	S J/(mol·K)	C_v J/(mol·K)	C_p J/(mol·K)	Sound m/s	Diel. const.
14.24[a]	77.67	38.53	–620.8	–586.6	10.12	10.93	14.60	1338.0	1.25386
15.0	77.08	38.24	–610.2	–575.7	10.86	10.18	14.06	1315.0	1.25180
16.0	76.29	37.84	–596.5	–561.6	11.77	10.01	14.27	1285.0	1.24900
18.0	74.57	36.99	–567.2	–531.5	13.54	10.62	15.96	1226.0	1.24298
20.0	72.65	36.04	–534.1	–497.5	15.33	11.38	18.13	1169.0	1.23629
22.0	70.49	34.97	–496.6	–458.9	17.17	11.96	20.48	1111.0	1.22876
24.0	68.03	33.75	–454.5	–415.3	19.06	12.39	23.14	1048.0	1.22022
25.0	66.65	33.06	–431.4	–391.4	20.04	12.55	24.68	1012.0	1.21546
26.0	65.16	32.32	–406.8	–365.9	21.04	12.70	26.44	973.4	1.21031
27.0	63.51	31.51	–380.3	–338.4	22.07	12.84	28.55	930.9	1.20466
28.0	61.68	30.60	–351.8	–308.6	23.16	12.97	31.18	883.8	1.19839
29.0	59.60	29.56	–320.4	–275.8	24.31	13.13	34.67	830.6	1.19128
29.5	58.43	28.99	–303.5	–257.9	24.92	13.22	36.94	801.1	1.18732
30.0	57.16	28.35	–285.3	–238.8	25.56	13.32	39.75	769.3	1.18298
30.5	55.74	27.65	–265.8	–218.0	26.25	13.44	43.38	734.5	1.17818
31.0	54.13	26.85	–244.3	–195.2	26.99	13.60	48.35	695.9	1.17275
31.5	52.24	25.91	–220.2	–169.3	27.82	13.80	55.78	652.3	1.16641
31.7	51.38	25.49	–209.5	–157.7	28.19	13.91	59.97	633.0	1.16351
31.8	50.91	25.26	–203.9	–151.6	28.38	13.96	62.47	622.8	1.16195
32.0	49.91	24.76	–191.9	–138.5	28.79	14.09	68.63	601.2	1.15859
32.2	48.77	24.19	–178.6	–124.0	29.24	14.25	77.11	577.7	1.15479
32.4	47.44	23.53	–163.5	–107.4	29.76	14.45	89.70	551.6	1.15036
32.5	46.67	23.15	–155.0	–98.0	30.05	14.57	98.72	537.2	1.14781
32.6	45.81	22.72	–145.7	–87.6	30.37	14.71	110.9	521.7	1.14494
32.7	44.81	22.23	–135.1	–75.7	30.73	14.88	128.3	504.8	1.14164
32.8	43.62	21.64	–122.6	–61.6	31.16	15.08	155.6	486.0	1.13769
32.9	42.09	20.88	–107.0	–43.8	31.70	15.36	205.8	464.3	1.13266
33.0	39.86	19.77	–84.9	–18.1	32.48	15.77	332.2	437.7	1.12534
33.1	34.82	17.27	–35.3	41.1	34.27	16.63	1227.0	398.8	1.10892
33.2	25.57	12.68	64.9	168.9	38.13	17.07	630.9	381.2	1.07918
33.3	23.00	11.41	97.3	213.0	39.46	16.79	325.0	384.4	1.07103
33.4	21.57	10.70	116.7	240.1	40.27	16.54	229.6	387.7	1.06651
33.5	20.56	10.20	131.0	260.4	40.88	16.33	182.1	390.9	1.06335
33.6	19.79	9.815	142.6	277.1	41.37	16.14	153.3	393.8	1.06090
33.7	19.15	9.499	152.4	291.4	41.80	15.98	133.9	396.6	1.05891
33.8	18.61	9.232	161.0	304.0	42.17	15.83	119.7	399.3	1.05721
33.9	18.14	8.999	168.8	315.4	42.51	15.69	109.0	401.8	1.05575
34.0	17.73	8.793	175.8	325.9	42.82	15.57	100.5	404.2	1.05445

Thermophysical properties of hydrogen—Continued

T K	Density kg/m³	Density mol/dm³	E J/mol	H J/mol	S J/(mol·K)	C_v J/(mol·K)	C_p J/(mol·K)	Sound m/s	Diel. const.
34.2	17.02	8.441	188.3	344.7	43.37	15.35	87.81	408.8	1.05223
34.4	16.42	8.146	199.2	361.3	43.85	15.16	78.79	413.1	1.05037
34.6	15.91	7.892	209.1	376.3	44.29	14.99	72.01	417.1	1.04877
34.8	15.46	7.669	218.0	390.2	44.69	14.83	66.71	421.0	1.04737
35.0	15.06	7.470	226.4	403.1	45.06	14.69	62.43	424.6	1.04612
35.2	14.70	7.290	234.1	415.2	45.40	14.57	58.91	428.2	1.04499
35.5	14.21	7.050	245.0	432.2	45.88	14.40	54.63	433.2	1.04349
36.0	13.52	6.708	261.3	458.1	46.61	14.16	49.31	441.1	1.04135
36.5	12.94	6.419	276.1	481.7	47.26	13.97	45.44	448.4	1.03954
37.0	12.44	6.170	289.8	503.7	47.86	13.80	42.48	455.3	1.03799
37.5	12.00	5.951	302.5	524.3	48.41	13.66	40.14	461.9	1.03662
38.0	11.60	5.755	314.5	543.9	48.93	13.54	38.25	468.1	1.03540
38.5	11.25	5.578	326.0	562.6	49.42	13.44	36.68	474.1	1.03430
39.0	10.92	5.418	337.0	580.6	49.88	13.35	35.37	479.8	1.03330
40.0	10.35	5.134	357.8	614.9	50.75	13.21	33.28	490.7	1.03154
41.0	9.858	4.890	377.4	647.4	51.55	13.11	31.69	500.8	1.03002
42.0	9.427	4.676	396.1	678.4	52.30	13.03	30.46	510.4	1.02870
43.0	9.044	4.486	414.1	708.3	53.01	12.97	29.46	519.6	1.02752
44.0	8.700	4.316	431.5	737.4	53.67	12.93	28.65	528.3	1.02646
45.0	8.389	4.161	448.5	765.7	54.31	12.89	27.97	536.8	1.02551
46.0	8.105	4.021	465.1	793.4	54.92	12.87	27.40	544.9	1.02464
48.0	7.604	3.772	497.3	847.2	56.06	12.85	26.50	560.4	1.02310
50.0	7.174	3.558	528.5	899.5	57.13	12.86	25.82	574.9	1.02178
52.0	6.798	3.372	559.1	950.6	58.13	12.88	25.31	588.6	1.02063
54.0	6.466	3.207	589.2	1001.0	59.08	12.94	24.92	601.5	1.01962
56.0	6.169	3.060	619.0	1050.0	59.98	13.01	24.62	613.8	1.01871
58.0	5.903	2.928	648.5	1099.0	60.84	13.10	24.41	625.4	1.01790
60.0	5.661	2.808	678.0	1148.0	61.67	13.22	24.26	636.3	1.01716
62.0	5.441	2.699	707.4	1196.0	62.46	13.35	24.17	646.7	1.01649
64.0	5.239	2.599	736.8	1245.0	63.23	13.51	24.13	656.6	1.01588
66.0	5.053	2.507	766.4	1293.0	63.97	13.69	24.13	666.0	1.01531
68.0	4.881	2.421	796.2	1341.0	64.69	13.89	24.18	674.9	1.01479
70.0	4.722	2.342	826.2	1390.0	65.39	14.10	24.26	683.4	1.01430
72.0	4.573	2.269	856.5	1438.0	66.08	14.34	24.37	691.5	1.01385
74.0	4.435	2.200	887.2	1487.0	66.75	14.59	24.51	699.3	1.01343
76.0	4.305	2.135	918.3	1536.0	67.40	14.85	24.67	706.8	1.01303
80.0	4.069	2.018	981.8	1636.0	68.68	15.43	25.08	720.9	1.01231
85.0	3.809	1.890	1064.0	1763.0	70.22	16.21	25.68	737.3	1.01153
90.0	3.583	1.777	1150.0	1893.0	71.70	17.05	26.37	752.7	1.01084
95.0	3.383	1.678	1240.0	2027.0	73.15	17.91	27.11	767.4	1.01023
100.0	3.206	1.590	1334.0	2164.0	74.56	18.78	27.88	781.7	1.00969
105.0	3.046	1.511	1432.0	2305.0	75.94	19.64	28.65	795.8	1.00921
110.0	2.903	1.440	1534.0	2450.0	77.29	20.45	29.40	809.7	1.00877
115.0	2.773	1.375	1640.0	2599.0	78.61	21.22	30.10	823.6	1.00838
120.0	2.654	1.317	1749.0	2751.0	79.91	21.92	30.74	837.5	1.00802
125.0	2.546	1.263	1861.0	2907.0	81.17	22.53	31.31	851.6	1.00769
130.0	2.446	1.213	1976.0	3064.0	82.41	23.06	31.80	865.7	1.00739
135.0	2.354	1.168	2094.0	3224.0	83.62	23.51	32.21	879.9	1.00711
140.0	2.268	1.125	2213.0	3386.0	84.80	23.90	32.57	894.1	1.00685
150.0	2.116	1.049	2457.0	3715.0	87.06	24.42	33.03	922.7	1.00639
160.0	1.982	.9834	2704.0	4046.0	89.20	24.68	33.25	951.5	1.00599
170.0	1.865	.9253	2952.0	4379.0	91.22	24.73	33.26	980.2	1.00563
180.0	1.761	.8737	3200.0	4711.0	93.12	24.62	33.12	1009.0	1.00532
190.0	1.669	.8277	3446.0	5041.0	94.90	24.40	32.88	1037.0	1.00504
200.0	1.585	.7863	3690.0	5368.0	96.58	24.12	32.58	1066.0	1.00479
210.0	1.510	.7489	3930.0	5693.0	98.16	23.81	32.25	1093.0	1.00456
220.0	1.441	.7150	4167.0	6013.0	99.65	23.49	31.91	1121.0	1.00435
230.0	1.379	.6840	4401.0	6331.0	101.1	23.17	31.59	1148.0	1.00416
240.0	1.322	.6556	4632.0	6645.0	102.4	22.88	31.28	1174.0	1.00399
250.0	1.269	.6295	4860.0	6957.0	103.7	22.60	30.99	1200.0	1.00383
260.0	1.220	.6054	5085.0	7265.0	104.9	22.36	30.74	1225.0	1.00368
270.0	1.175	.5830	5307.0	7571.0	106.0	22.14	30.51	1249.0	1.00355
280.0	1.134	.5623	5528.0	7876.0	107.1	21.95	30.31	1274.0	1.00342

Thermophysical properties of hydrogen—Continued

T K	Density kg/m³	Density mol/dm³	E J/mol	H J/mol	S J/(mol·K)	C_v J/(mol·K)	C_p J/(mol·K)	Sound m/s	Diel. const.
300.0	1.058	.5250	5964.0	8478.0	109.2	21.64	30.00	1320.0	1.00319
320.0	.9926	.4924	6395.0	9076.0	111.2	21.42	29.77	1365.0	1.00299
340.0	.9345	.4636	6822.0	9670.0	113.0	21.27	29.61	1407.0	1.00282
360.0	.8829	.4380	7247.0	10260.0	114.6	21.16	29.50	1449.0	1.00266
380.0	.8367	.4150	7669.0	10850.0	116.2	21.09	29.42	1489.0	1.00252
400.0	.7951	.3944	8091.0	11440.0	117.7	21.03	29.36	1527.0	1.00240
				1.34 MPa isobar					
14.24[a]	77.68	38.53	−620.8	−586.0	10.12	10.93	14.59	1337.0	1.25390
15.0	77.10	38.24	−610.3	−575.2	10.86	10.19	14.05	1315.0	1.25185
16.0	76.30	37.85	−596.6	−561.2	11.77	10.02	14.26	1285.0	1.24906
18.0	74.59	37.00	−567.3	−531.1	13.53	10.62	15.95	1226.0	1.24305
20.0	72.68	36.05	−534.2	−497.1	15.32	11.38	18.12	1170.0	1.23637
22.0	70.52	34.98	−496.8	−458.5	17.16	11.96	20.46	1113.0	1.22886
24.0	68.06	33.76	−454.7	−415.0	19.05	12.38	23.10	1049.0	1.22033
25.0	66.69	33.08	−431.7	−391.2	20.02	12.55	24.63	1014.0	1.21559
26.0	65.20	32.34	−407.1	−365.7	21.02	12.70	26.39	975.2	1.21046
27.0	63.57	31.53	−380.8	−338.3	22.06	12.83	28.48	933.0	1.20483
28.0	61.74	30.63	−352.3	−308.6	23.14	12.97	31.07	886.2	1.19860
29.0	59.68	29.60	−321.1	−275.9	24.28	13.12	34.50	833.5	1.19154
29.5	58.52	29.03	−304.2	−258.1	24.89	13.21	36.72	804.3	1.18761
30.0	57.26	28.40	−286.2	−239.1	25.53	13.31	39.45	772.9	1.18332
30.5	55.86	27.71	−266.9	−218.5	26.21	13.43	42.96	738.6	1.17858
31.0	54.27	26.92	−245.7	−195.9	26.95	13.58	47.72	700.7	1.17324
31.5	52.43	26.01	−221.9	−170.4	27.76	13.78	54.71	658.0	1.16703
31.7	51.59	25.59	−211.5	−159.1	28.12	13.87	58.59	639.2	1.16421
31.8	51.14	25.37	−205.9	−153.1	28.31	13.93	60.89	629.3	1.16271
32.0	50.17	24.89	−194.3	−140.4	28.71	14.05	66.46	608.4	1.15946
32.2	49.08	24.35	−181.4	−126.4	29.14	14.20	73.96	585.8	1.15583
32.4	47.83	23.73	−167.1	−110.6	29.63	14.37	84.70	560.9	1.15166
32.5	47.12	23.37	−159.1	−101.8	29.90	14.48	92.10	547.4	1.14929
32.6	46.33	22.98	−150.4	−92.1	30.20	14.61	101.7	533.0	1.14667
32.7	45.44	22.54	−140.8	−81.3	30.53	14.75	114.7	517.4	1.14371
32.8	44.41	22.03	−129.8	−69.0	30.91	14.92	133.5	500.4	1.14030
32.9	43.16	21.41	−116.9	−54.3	31.35	15.14	163.3	481.4	1.13618
33.0	41.55	20.61	−100.5	−35.5	31.92	15.43	218.5	459.7	1.13089
33.1	39.18	19.44	−76.9	−8.0	32.76	15.86	357.9	433.2	1.12312
33.2	34.17	16.95	−27.2	51.8	34.56	16.66	1038.0	398.2	1.10680
33.3	26.68	13.24	53.7	155.0	37.66	17.05	664.6	383.5	1.08274
33.4	23.82	11.82	88.9	202.3	39.08	16.81	350.2	385.7	1.07366
33.5	22.25	11.03	109.8	231.3	39.95	16.57	244.7	388.7	1.06866
33.6	21.15	10.49	125.2	252.9	40.59	16.36	192.3	391.8	1.06521
33.7	20.31	10.08	137.4	270.4	41.11	16.18	160.8	394.6	1.06256
33.8	19.63	9.739	147.8	285.4	41.56	16.01	139.6	397.4	1.06042
33.9	19.06	9.454	156.8	298.5	41.95	15.86	124.3	400.0	1.05862
34.0	18.56	9.208	164.8	310.4	42.30	15.73	112.8	402.5	1.05706
34.2	17.73	8.796	178.8	331.2	42.90	15.49	96.31	407.2	1.05447
34.4	17.05	8.460	190.8	349.2	43.43	15.28	85.09	411.6	1.05234
34.6	16.48	8.175	201.5	365.4	43.90	15.09	76.90	415.8	1.05055
34.8	15.98	7.928	211.1	380.1	44.32	14.93	70.63	419.7	1.04900
35.0	15.54	7.710	219.9	393.7	44.71	14.78	65.67	423.5	1.04762
35.2	15.15	7.514	228.1	406.4	45.08	14.65	61.63	427.1	1.04640
35.4	14.79	7.337	235.8	418.4	45.42	14.53	58.27	430.5	1.04529
36.0	13.89	6.888	256.5	451.0	46.33	14.22	50.89	440.2	1.04247
36.5	13.27	6.582	271.8	475.3	47.00	14.02	46.65	447.6	1.04056
37.0	12.74	6.319	285.8	497.8	47.61	13.85	43.44	454.6	1.03892
37.5	12.27	6.089	298.8	518.9	48.18	13.70	40.93	461.2	1.03748
38.0	11.86	5.884	311.1	538.8	48.71	13.58	38.91	467.5	1.03620
38.5	11.49	5.700	322.8	557.9	49.20	13.47	37.25	473.6	1.03506
39.0	11.15	5.533	333.9	576.1	49.68	13.37	35.86	479.4	1.03402
40.0	10.56	5.239	355.0	610.8	50.55	13.23	33.66	490.3	1.03219
41.0	10.05	4.986	374.9	643.6	51.36	13.12	32.00	500.5	1.03062

Thermophysical properties of hydrogen—Continued

T K	Density kg/m³	Density mol/dm³	E J/mol	H J/mol	S J/(mol·K)	C_v J/(mol·K)	C_p J/(mol·K)	Sound m/s	Diel. const.
42.0	9.607	4.766	393.8	674.9	52.12	13.04	30.71	510.1	1.02925
43.0	9.213	4.570	411.9	705.1	52.83	12.98	29.68	519.3	1.02804
44.0	8.860	4.395	429.5	734.4	53.50	12.93	28.83	528.2	1.02695
45.0	8.540	4.236	446.5	762.8	54.14	12.90	28.13	536.6	1.02597
46.0	8.249	4.092	463.2	790.7	54.75	12.88	27.54	544.8	1.02508
48.0	7.736	3.837	495.6	844.8	55.90	12.85	26.61	560.3	1.02351
50.0	7.295	3.619	527.0	897.3	56.98	12.86	25.91	574.9	1.02216
52.0	6.911	3.428	557.7	948.5	57.98	12.89	25.39	588.6	1.02098
54.0	6.572	3.260	587.9	998.9	58.93	12.94	24.99	601.5	1.01994
56.0	6.270	3.110	617.7	1049.0	59.83	13.01	24.68	613.8	1.01902
58.0	5.998	2.975	647.3	1098.0	60.70	13.10	24.46	625.4	1.01819
60.0	5.752	2.853	676.8	1146.0	61.52	13.22	24.31	636.4	1.01744
62.0	5.528	2.742	706.3	1195.0	62.32	13.35	24.21	646.8	1.01676
64.0	5.322	2.640	735.8	1243.0	63.09	13.51	24.17	656.7	1.01613
66.0	5.133	2.546	765.4	1292.0	63.83	13.69	24.17	666.1	1.01555
68.0	4.958	2.459	795.2	1340.0	64.55	13.89	24.21	675.0	1.01502
70.0	4.796	2.379	825.2	1389.0	65.26	14.10	24.29	683.6	1.01453
72.0	4.645	2.304	855.6	1437.0	65.94	14.34	24.40	691.7	1.01407
74.0	4.504	2.234	886.3	1486.0	66.61	14.59	24.53	699.5	1.01364
76.0	4.372	2.168	917.4	1535.0	67.27	14.86	24.70	706.9	1.01323
80.0	4.131	2.049	981.0	1635.0	68.54	15.43	25.10	721.0	1.01250
85.0	3.868	1.918	1063.0	1762.0	70.08	16.22	25.70	737.4	1.01170
90.0	3.637	1.804	1149.0	1892.0	71.57	17.05	26.39	752.9	1.01100
95.0	3.435	1.704	1239.0	2026.0	73.02	17.91	27.13	767.6	1.01039
100.0	3.254	1.614	1333.0	2163.0	74.43	18.78	27.90	781.9	1.00984
105.0	3.092	1.534	1431.0	2305.0	75.81	19.64	28.66	796.0	1.00935
110.0	2.947	1.462	1533.0	2450.0	77.16	20.46	29.41	809.9	1.00891
115.0	2.815	1.396	1639.0	2599.0	78.48	21.22	30.11	823.8	1.00851
120.0	2.694	1.336	1748.0	2751.0	79.78	21.92	30.75	837.7	1.00814
125.0	2.584	1.282	1861.0	2906.0	81.04	22.53	31.32	851.8	1.00781
130.0	2.483	1.231	1976.0	3064.0	82.28	23.06	31.80	865.9	1.00750
135.0	2.389	1.185	2093.0	3224.0	83.49	23.51	32.22	880.1	1.00722
140.0	2.303	1.142	2213.0	3386.0	84.67	23.90	32.57	894.3	1.00696
150.0	2.147	1.065	2456.0	3714.0	86.93	24.42	33.04	922.9	1.00649
160.0	2.012	.9982	2704.0	4046.0	89.07	24.68	33.25	951.7	1.00608
170.0	1.893	.9392	2952.0	4379.0	91.09	24.73	33.27	980.4	1.00572
180.0	1.788	.8868	3200.0	4711.0	92.99	24.62	33.13	1009.0	1.00540
190.0	1.694	.8401	3446.0	5041.0	94.77	24.41	32.89	1038.0	1.00511
200.0	1.609	.7981	3689.0	5368.0	96.45	24.13	32.59	1066.0	1.00486
210.0	1.532	.7602	3930.0	5693.0	98.04	23.81	32.25	1094.0	1.00463
220.0	1.463	.7257	4167.0	6013.0	99.53	23.49	31.92	1121.0	1.00442
230.0	1.400	.6942	4401.0	6331.0	100.9	23.18	31.59	1148.0	1.00422
240.0	1.341	.6654	4632.0	6645.0	102.3	22.88	31.28	1174.0	1.00405
250.0	1.288	.6389	4859.0	6957.0	103.5	22.60	31.00	1200.0	1.00389
260.0	1.239	.6144	5085.0	7265.0	104.8	22.36	30.74	1225.0	1.00374
270.0	1.193	.5918	5307.0	7572.0	105.9	22.14	30.51	1250.0	1.00360
280.0	1.151	.5708	5528.0	7876.0	107.0	21.95	30.32	1274.0	1.00347
300.0	1.074	.5329	5964.0	8479.0	109.1	21.64	30.00	1320.0	1.00324
320.0	1.008	.4998	6395.0	9076.0	111.0	21.42	29.77	1365.0	1.00304
340.0	.9486	.4705	6822.0	9670.0	112.8	21.27	29.61	1408.0	1.00286
360.0	.8962	.4445	7247.0	10260.0	114.5	21.16	29.50	1449.0	1.00270
380.0	.8493	.4213	7669.0	10850.0	116.1	21.09	29.42	1489.0	1.00256
400.0	.8070	.4003	8091.0	11440.0	117.6	21.03	29.36	1527.0	1.00243
1.36 MPa isobar									
14.25[a]	77.69	38.54	–620.8	–585.5	10.12	10.92	14.58	1337.0	1.25393
15.0	77.11	38.25	–610.4	–574.8	10.85	10.19	14.05	1315.0	1.25191
16.0	76.32	37.86	–596.6	–560.7	11.76	10.02	14.26	1285.0	1.24912
18.0	74.61	37.01	–567.4	–530.7	13.53	10.62	15.94	1227.0	1.24312
20.0	72.70	36.06	–534.4	–496.7	15.32	11.38	18.10	1171.0	1.23645
22.0	70.55	35.00	–497.0	–458.1	17.15	11.96	20.44	1114.0	1.22895
24.0	68.10	33.78	–455.0	–414.7	19.04	12.38	23.07	1050.0	1.22045

Thermophysical properties of hydrogen—Continued

T K	Density kg/m^3	Density mol/dm^3	E J/mol	H J/mol	S J/(mol·K)	C_v J/(mol·K)	C_p J/(mol·K)	Sound m/s	Diel. const.
25.0	66.73	33.10	−432.0	−390.9	20.01	12.55	24.59	1015.0	1.21572
26.0	65.24	32.36	−407.5	−365.4	21.01	12.69	26.34	977.0	1.21061
27.0	63.62	31.56	−381.2	−338.1	22.04	12.83	28.40	935.1	1.20501
28.0	61.80	30.66	−352.8	−308.5	23.12	12.97	30.97	888.6	1.19881
29.0	59.75	29.64	−321.8	−275.9	24.26	13.11	34.34	836.4	1.19180
29.5	58.61	29.07	−305.0	−258.2	24.86	13.20	36.50	807.5	1.18790
30.0	57.36	28.45	−287.1	−239.3	25.50	13.30	39.16	776.4	1.18366
30.5	55.97	27.77	−267.9	−218.9	26.17	13.42	42.56	742.6	1.17898
31.0	54.42	26.99	−247.0	−196.6	26.90	13.56	47.12	705.3	1.17372
31.5	52.61	26.10	−223.6	−171.5	27.70	13.75	53.72	663.6	1.16764
31.7	51.79	25.69	−213.3	−160.4	28.05	13.84	57.33	645.2	1.16489
31.8	51.36	25.47	−207.9	−154.5	28.24	13.89	59.45	635.6	1.16343
32.0	50.42	25.01	−196.5	−142.2	28.63	14.01	64.53	615.4	1.16030
32.2	49.38	24.49	−184.2	−128.6	29.05	14.14	71.22	593.6	1.15681
32.4	48.19	23.90	−170.4	−113.5	29.52	14.31	80.53	569.8	1.15286
32.6	46.79	23.21	−154.7	−96.1	30.05	14.52	94.56	543.3	1.14821
32.7	45.98	22.81	−145.8	−86.2	30.36	14.64	104.7	528.8	1.14552
32.8	45.06	22.35	−135.9	−75.0	30.70	14.79	118.6	513.2	1.14247
32.9	43.99	21.82	−124.6	−62.2	31.09	14.97	138.8	496.1	1.13894
33.0	42.70	21.18	−111.1	−46.9	31.55	15.20	171.0	477.1	1.13466
33.1	41.01	20.34	−94.0	−27.2	32.15	15.50	230.9	455.4	1.12912
33.2	38.52	19.11	−69.2	2.0	33.03	15.94	377.5	429.5	1.12097
33.3	33.76	16.75	−21.5	59.7	34.76	16.66	854.9	398.9	1.10548
33.4	27.62	13.70	44.9	144.1	37.29	17.00	645.5	386.2	1.08572
33.5	24.62	12.21	81.0	192.4	38.74	16.82	367.1	387.3	1.07618
33.6	22.91	11.37	103.3	222.9	39.65	16.60	257.4	390.0	1.07077
33.7	21.74	10.78	119.5	245.6	40.32	16.39	201.5	392.8	1.06705
33.8	20.84	10.34	132.4	263.9	40.86	16.21	167.7	395.6	1.06422
33.9	20.11	9.978	143.2	279.5	41.32	16.04	145.0	398.3	1.06194
34.0	19.51	9.676	152.6	293.1	41.73	15.89	128.7	400.8	1.06003
34.1	18.98	9.416	160.9	305.4	42.09	15.76	116.4	403.3	1.05838
34.2	18.52	9.188	168.5	316.5	42.41	15.63	106.8	405.7	1.05694
34.4	17.74	8.800	181.8	336.4	42.99	15.41	92.57	410.2	1.05449
34.6	17.09	8.478	193.4	353.8	43.50	15.21	82.56	414.5	1.05246
34.8	16.54	8.202	203.8	369.6	43.95	15.03	75.09	418.5	1.05072
35.0	16.05	7.962	213.2	384.0	44.36	14.88	69.29	422.3	1.04921
35.2	15.62	7.748	221.8	397.4	44.74	14.74	64.65	426.0	1.04787
35.4	15.23	7.556	229.9	409.9	45.10	14.61	60.83	429.5	1.04666
35.6	14.88	7.382	237.5	421.7	45.43	14.49	57.64	432.9	1.04557
36.0	14.26	7.074	251.5	443.7	46.05	14.29	52.58	439.3	1.04364
36.5	13.61	6.749	267.3	468.8	46.74	14.07	47.93	446.9	1.04160
37.0	13.05	6.471	281.7	491.8	47.37	13.89	44.46	453.9	1.03987
37.5	12.56	6.230	295.1	513.4	47.94	13.74	41.76	460.6	1.03836
38.0	12.13	6.015	307.6	533.7	48.48	13.61	39.60	467.0	1.03702
38.5	11.74	5.823	319.5	553.0	48.99	13.50	37.84	473.1	1.03582
39.0	11.39	5.649	330.8	571.6	49.47	13.40	36.36	478.9	1.03474
40.0	10.77	5.344	352.2	606.7	50.36	13.25	34.05	489.9	1.03284
41.0	10.25	5.083	372.3	639.9	51.18	13.14	32.31	500.2	1.03122
42.0	9.789	4.856	391.4	671.5	51.94	13.05	30.97	509.9	1.02981
43.0	9.383	4.654	409.7	701.9	52.65	12.99	29.90	519.1	1.02856
44.0	9.020	4.474	427.4	731.3	53.33	12.94	29.02	528.0	1.02745
45.0	8.692	4.312	444.6	760.0	53.97	12.91	28.29	536.5	1.02644
46.0	8.394	4.164	461.3	788.0	54.59	12.88	27.69	544.7	1.02552
47.0	8.120	4.028	477.7	815.4	55.18	12.87	27.17	552.6	1.02468
48.0	7.868	3.903	493.9	842.3	55.75	12.86	26.72	560.2	1.02391
50.0	7.418	3.679	525.4	895.0	56.82	12.86	26.01	574.8	1.02253
52.0	7.025	3.485	556.2	946.5	57.83	12.89	25.47	588.6	1.02133
54.0	6.679	3.313	586.5	997.0	58.78	12.94	25.06	601.6	1.02027
56.0	6.371	3.160	616.4	1047.0	59.69	13.01	24.75	613.9	1.01933
58.0	6.094	3.023	646.1	1096.0	60.55	13.11	24.52	625.5	1.01848
60.0	5.843	2.898	675.6	1145.0	61.38	13.22	24.36	636.5	1.01772
62.0	5.614	2.785	705.1	1193.0	62.18	13.36	24.26	646.9	1.01702
64.0	5.405	2.681	734.7	1242.0	62.95	13.51	24.21	656.8	1.01638

Thermophysical properties of hydrogen—Continued

T K	Density kg/m³	Density mol/dm³	E J/mol	H J/mol	S J/(mol·K)	C_v J/(mol·K)	C_p J/(mol·K)	Sound m/s	Diel. const.
66.0	5.212	2.586	764.4	1290.0	63.69	13.69	24.21	666.2	1.01579
68.0	5.034	2.497	794.2	1339.0	64.42	13.89	24.24	675.2	1.01525
70.0	4.869	2.415	824.3	1387.0	65.12	14.11	24.32	683.7	1.01475
72.0	4.716	2.339	854.7	1436.0	65.81	14.34	24.42	691.8	1.01428
74.0	4.572	2.268	885.4	1485.0	66.48	14.59	24.56	699.6	1.01385
76.0	4.438	2.202	916.6	1534.0	67.13	14.86	24.72	707.1	1.01344
80.0	4.194	2.080	980.2	1634.0	68.41	15.43	25.12	721.2	1.01269
85.0	3.926	1.947	1063.0	1761.0	69.95	16.22	25.72	737.6	1.01188
90.0	3.692	1.831	1149.0	1891.0	71.44	17.05	26.40	753.0	1.01117
95.0	3.486	1.729	1239.0	2025.0	72.89	17.91	27.14	767.8	1.01054
100.0	3.303	1.638	1333.0	2163.0	74.30	18.78	27.91	782.1	1.00999
105.0	3.139	1.557	1431.0	2304.0	75.68	19.64	28.68	796.1	1.00949
110.0	2.991	1.483	1533.0	2450.0	77.03	20.46	29.42	810.0	1.00904
115.0	2.856	1.417	1639.0	2598.0	78.35	21.22	30.12	823.9	1.00863
120.0	2.734	1.356	1748.0	2751.0	79.65	21.92	30.76	837.9	1.00826
125.0	2.622	1.301	1860.0	2906.0	80.92	22.53	31.33	851.9	1.00792
130.0	2.519	1.250	1976.0	3064.0	82.16	23.06	31.81	866.1	1.00761
135.0	2.424	1.203	2093.0	3224.0	83.36	23.51	32.22	880.3	1.00732
140.0	2.337	1.159	2213.0	3386.0	84.54	23.90	32.58	894.5	1.00706
150.0	2.179	1.081	2456.0	3714.0	86.81	24.42	33.04	923.1	1.00658
160.0	2.042	1.013	2703.0	4046.0	88.95	24.68	33.26	951.8	1.00617
170.0	1.921	.9530	2952.0	4379.0	90.97	24.73	33.27	980.6	1.00580
180.0	1.814	.9000	3200.0	4711.0	92.86	24.62	33.13	1009.0	1.00548
190.0	1.719	.8525	3446.0	5041.0	94.65	24.41	32.89	1038.0	1.00519
200.0	1.633	.8099	3689.0	5368.0	96.33	24.13	32.59	1066.0	1.00493
210.0	1.555	.7714	3930.0	5693.0	97.91	23.81	32.26	1094.0	1.00469
220.0	1.485	.7364	4167.0	6014.0	99.40	23.49	31.92	1121.0	1.00448
230.0	1.420	.7045	4401.0	6331.0	100.8	23.18	31.59	1148.0	1.00429
240.0	1.361	.6753	4631.0	6645.0	102.2	22.88	31.28	1174.0	1.00411
250.0	1.307	.6484	4859.0	6957.0	103.4	22.60	31.00	1200.0	1.00394
260.0	1.257	.6235	5084.0	7265.0	104.6	22.36	30.74	1225.0	1.00379
270.0	1.211	.6006	5307.0	7572.0	105.8	22.14	30.52	1250.0	1.00365
280.0	1.168	.5792	5528.0	7876.0	106.9	21.95	30.32	1274.0	1.00352
300.0	1.090	.5408	5964.0	8479.0	109.0	21.64	30.00	1320.0	1.00329
320.0	1.022	.5072	6395.0	9076.0	110.9	21.42	29.77	1365.0	1.00308
340.0	.9626	.4775	6822.0	9670.0	112.7	21.27	29.61	1408.0	1.00290
360.0	.9094	.4511	7247.0	10260.0	114.4	21.16	29.50	1449.0	1.00274
380.0	.8618	.4275	7669.0	10850.0	116.0	21.09	29.42	1489.0	1.00260
400.0	.8190	.4063	8091.0	11440.0	117.5	21.04	29.36	1528.0	1.00247
				1.38 MPa isobar					
14.26[a]	77.70	38.54	−620.7	−584.9	10.12	10.92	14.56	1337.0	1.25397
15.0	77.13	38.26	−610.4	−574.4	10.84	10.20	14.04	1315.0	1.25197
16.0	76.34	37.87	−596.7	−560.3	11.75	10.02	14.25	1286.0	1.24918
18.0	74.63	37.02	−567.5	−530.2	13.52	10.63	15.93	1227.0	1.24319
20.0	72.72	36.07	−534.5	−496.3	15.31	11.38	18.09	1172.0	1.23653
22.0	70.58	35.01	−497.2	−457.8	17.14	11.96	20.42	1115.0	1.22905
24.0	68.13	33.80	−455.2	−414.4	19.03	12.38	23.04	1052.0	1.22057
25.0	66.77	33.12	−432.3	−390.6	20.00	12.55	24.55	1017.0	1.21586
26.0	65.29	32.39	−407.8	−365.2	20.99	12.69	26.28	978.8	1.21076
27.0	63.67	31.58	−381.6	−337.9	22.02	12.83	28.33	937.1	1.20518
28.0	61.87	30.69	−353.4	−308.4	23.10	12.96	30.86	891.0	1.19901
29.0	59.83	29.68	−322.5	−276.0	24.23	13.11	34.18	839.2	1.19205
29.5	58.69	29.11	−305.8	−258.4	24.84	13.19	36.30	810.7	1.18819
30.0	57.45	28.50	−288.0	−239.6	25.47	13.29	38.89	779.9	1.18398
30.5	56.09	27.82	−268.9	−219.3	26.14	13.40	42.17	746.6	1.17936
31.0	54.55	27.06	−248.2	−197.2	26.86	13.54	46.55	709.9	1.17418
31.5	52.78	26.18	−225.2	−172.5	27.65	13.72	52.80	669.0	1.16823
31.7	51.99	25.79	−215.1	−161.6	27.99	13.81	56.18	651.1	1.16555
31.8	51.56	25.58	−209.8	−155.9	28.17	13.86	58.14	641.7	1.16413
32.0	50.66	25.13	−198.7	−143.8	28.55	13.97	62.79	622.1	1.16109

Thermophysical properties of hydrogen—Continued

T K	Density kg/m^3	Density mol/dm^3	*E* J/mol	*H* J/mol	*S* J/(mol·K)	C_v J/(mol·K)	C_p J/(mol·K)	Sound m/s	Diel. const.
32.2	49.66	24.63	−186.7	−130.7	28.96	14.10	68.82	601.0	1.15774
32.4	48.52	24.07	−173.5	−116.1	29.41	14.25	76.99	578.2	1.15397
32.6	47.22	23.42	−158.6	−99.6	29.92	14.44	88.83	553.0	1.14961
32.7	46.47	23.05	−150.2	−90.4	30.20	14.55	97.08	539.4	1.14713
32.8	45.63	22.64	−141.1	−80.1	30.51	14.68	107.9	524.8	1.14436
32.9	44.69	22.17	−130.9	−68.7	30.86	14.84	122.6	509.1	1.14122
33.0	43.58	21.62	−119.2	−55.4	31.26	15.02	144.1	492.0	1.13757
33.1	42.23	20.95	−105.3	−39.4	31.75	15.26	178.6	473.0	1.13313
33.2	40.48	20.08	−87.5	−18.8	32.37	15.56	242.2	451.5	1.12736
33.3	37.91	18.80	−61.8	11.7	33.29	16.00	388.5	426.6	1.11895
33.4	33.49	16.61	−17.2	65.9	34.91	16.63	714.3	400.3	1.10461
33.5	28.35	14.07	38.4	136.5	37.02	16.94	596.0	389.1	1.08808
33.6	25.36	12.58	74.0	183.7	38.43	16.81	373.9	389.2	1.07853
33.7	23.56	11.69	97.1	215.1	39.36	16.61	266.8	391.4	1.07282
33.8	22.31	11.07	114.0	238.7	40.06	16.41	209.2	394.0	1.06886
33.9	21.36	10.59	127.5	257.7	40.62	16.23	173.9	396.7	1.06586
34.0	20.59	10.21	138.7	273.8	41.10	16.07	150.0	399.3	1.06344
34.1	19.95	9.897	148.5	287.9	41.51	15.92	132.9	401.8	1.06142
34.2	19.40	9.623	157.1	300.5	41.88	15.78	119.9	404.2	1.05969
34.3	18.92	9.384	164.9	312.0	42.22	15.66	109.7	406.5	1.05818
34.4	18.49	9.170	172.1	322.6	42.52	15.54	101.6	408.8	1.05683
34.6	17.75	8.803	184.8	341.6	43.08	15.33	89.18	413.2	1.05451
34.8	17.12	8.495	196.0	358.4	43.56	15.14	80.20	417.3	1.05256
35.0	16.59	8.228	206.1	373.8	44.00	14.97	73.38	421.2	1.05089
35.2	16.12	7.994	215.3	387.9	44.40	14.83	68.00	424.9	1.04941
35.4	15.69	7.785	223.8	401.0	44.78	14.69	63.65	428.5	1.04810
35.6	15.31	7.597	231.7	413.4	45.12	14.57	60.04	431.9	1.04691
36.0	14.65	7.267	246.3	436.2	45.76	14.35	54.41	438.5	1.04485
36.5	13.95	6.921	262.7	462.1	46.48	14.13	49.30	446.1	1.04268
37.0	13.36	6.627	277.5	485.8	47.12	13.94	45.53	453.3	1.04084
37.5	12.85	6.373	291.2	507.8	47.71	13.78	42.63	460.0	1.03925
38.0	12.40	6.149	304.1	528.5	48.26	13.64	40.32	466.4	1.03785
38.5	11.99	5.948	316.2	548.2	48.77	13.53	38.44	472.6	1.03660
39.0	11.63	5.767	327.7	567.0	49.26	13.43	36.88	478.5	1.03548
40.0	10.99	5.451	349.4	602.6	50.16	13.27	34.45	489.5	1.03351
41.0	10.44	5.181	369.7	636.1	50.99	13.15	32.64	499.9	1.03183
42.0	9.972	4.947	389.0	668.0	51.76	13.07	31.23	509.6	1.03037
43.0	9.555	4.740	407.5	698.6	52.48	13.00	30.12	518.9	1.02909
44.0	9.182	4.554	425.3	728.3	53.16	12.95	29.21	527.8	1.02794
45.0	8.845	4.388	442.6	757.1	53.81	12.92	28.46	536.3	1.02691
46.0	8.539	4.236	459.4	785.2	54.43	12.89	27.83	544.5	1.02597
47.0	8.259	4.097	475.9	812.8	55.02	12.87	27.30	552.5	1.02511
48.0	8.001	3.969	492.1	839.9	55.59	12.87	26.84	560.2	1.02432
50.0	7.540	3.740	523.8	892.8	56.67	12.87	26.10	574.8	1.02291
52.0	7.139	3.541	554.7	944.4	57.68	12.89	25.55	588.6	1.02168
54.0	6.786	3.366	585.1	995.1	58.64	12.94	25.13	601.6	1.02060
56.0	6.472	3.210	615.1	1045.0	59.54	13.01	24.81	613.9	1.01964
58.0	6.189	3.070	644.9	1094.0	60.41	13.11	24.57	625.5	1.01877
60.0	5.934	2.943	674.5	1143.0	61.24	13.22	24.41	636.6	1.01799
62.0	5.701	2.828	704.0	1192.0	62.04	13.36	24.30	647.0	1.01728
64.0	5.488	2.722	733.6	1241.0	62.81	13.52	24.25	656.9	1.01663
66.0	5.292	2.625	763.3	1289.0	63.56	13.69	24.24	666.3	1.01604
68.0	5.111	2.535	793.2	1338.0	64.28	13.89	24.28	675.3	1.01549
70.0	4.943	2.452	823.4	1386.0	64.98	14.11	24.35	683.8	1.01497
72.0	4.787	2.375	853.8	1435.0	65.67	14.34	24.45	691.9	1.01450
74.0	4.641	2.302	884.6	1484.0	66.34	14.59	24.59	699.7	1.01405
76.0	4.505	2.235	915.8	1533.0	67.00	14.86	24.75	707.2	1.01364
80.0	4.256	2.111	979.5	1633.0	68.28	15.43	25.14	721.3	1.01288
85.0	3.984	1.976	1062.0	1760.0	69.82	16.22	25.74	737.7	1.01206
90.0	3.747	1.859	1148.0	1891.0	71.31	17.05	26.42	753.2	1.01134
95.0	3.537	1.755	1238.0	2025.0	72.76	17.92	27.16	767.9	1.01070
100.0	3.351	1.662	1332.0	2162.0	74.17	18.79	27.92	782.2	1.01013

Thermophysical properties of hydrogen—Continued

T K	Density kg/m^3	Density mol/dm^3	E J/mol	H J/mol	S J/(mol·K)	C_v J/(mol·K)	C_p J/(mol·K)	Sound m/s	Diel. const.
105.0	3.185	1.580	1430.0	2304.0	75.55	19.64	28.69	796.3	1.00963
110.0	3.034	1.505	1532.0	2449.0	76.91	20.46	29.43	810.2	1.00917
115.0	2.898	1.438	1638.0	2598.0	78.23	21.22	30.13	824.1	1.00876
120.0	2.774	1.376	1747.0	2750.0	79.53	21.92	30.77	838.1	1.00838
125.0	2.661	1.320	1860.0	2906.0	80.79	22.53	31.33	852.1	1.00804
130.0	2.556	1.268	1975.0	3063.0	82.03	23.06	31.82	866.3	1.00772
135.0	2.460	1.220	2093.0	3224.0	83.24	23.51	32.23	880.5	1.00743
140.0	2.371	1.176	2212.0	3386.0	84.42	23.90	32.58	894.6	1.00716
150.0	2.211	1.097	2456.0	3714.0	86.68	24.42	33.05	923.3	1.00668
160.0	2.072	1.028	2703.0	4046.0	88.83	24.68	33.26	952.0	1.00626
170.0	1.949	.9669	2951.0	4379.0	90.84	24.73	33.27	980.8	1.00589
180.0	1.841	.9131	3199.0	4711.0	92.74	24.62	33.13	1009.0	1.00556
190.0	1.744	.8650	3446.0	5041.0	94.53	24.41	32.89	1038.0	1.00526
200.0	1.657	.8217	3689.0	5368.0	96.21	24.13	32.59	1066.0	1.00500
210.0	1.578	.7827	3929.0	5693.0	97.79	23.81	32.26	1094.0	1.00476
220.0	1.506	.7472	4167.0	6014.0	99.28	23.49	31.92	1121.0	1.00455
230.0	1.441	.7148	4400.0	6331.0	100.7	23.18	31.59	1148.0	1.00435
240.0	1.381	.6851	4631.0	6646.0	102.0	22.88	31.28	1174.0	1.00417
250.0	1.326	.6578	4859.0	6957.0	103.3	22.60	31.00	1200.0	1.00400
260.0	1.275	.6326	5084.0	7266.0	104.5	22.36	30.74	1225.0	1.00385
270.0	1.228	.6093	5307.0	7572.0	105.7	22.14	30.52	1250.0	1.00371
280.0	1.185	.5877	5528.0	7876.0	106.8	21.95	30.32	1274.0	1.00357
300.0	1.106	.5487	5964.0	8479.0	108.9	21.64	30.00	1321.0	1.00334
320.0	1.037	.5146	6395.0	9077.0	110.8	21.42	29.77	1365.0	1.00313
340.0	.9767	.4845	6822.0	9670.0	112.6	21.27	29.61	1408.0	1.00295
360.0	.9227	.4577	7246.0	10260.0	114.3	21.16	29.50	1449.0	1.00278
380.0	.8744	.4338	7669.0	10850.0	115.9	21.09	29.42	1489.0	1.00264
400.0	.8310	.4122	8090.0	11440.0	117.4	21.04	29.36	1528.0	1.00251
				1.40 MPa isobar					
14.26[a]	77.71	38.55	−620.7	−584.4	10.12	10.92	14.55	1336.0	1.25400
15.0	77.15	38.27	−610.5	−573.9	10.84	10.21	14.04	1315.0	1.25202
16.0	76.36	37.88	−596.8	−559.8	11.75	10.03	14.25	1286.0	1.24924
18.0	74.65	37.03	−567.6	−529.8	13.51	10.63	15.92	1228.0	1.24326
20.0	72.75	36.09	−534.7	−495.9	15.30	11.38	18.08	1173.0	1.23661
22.0	70.60	35.02	−497.4	−457.4	17.13	11.96	20.40	1116.0	1.22915
24.0	68.16	33.81	−455.5	−414.1	19.02	12.38	23.01	1053.0	1.22068
25.0	66.81	33.14	−432.6	−390.3	19.98	12.55	24.51	1018.0	1.21599
26.0	65.33	32.41	−408.2	−365.0	20.98	12.69	26.23	980.5	1.21090
27.0	63.72	31.61	−382.1	−337.8	22.01	12.82	28.26	939.2	1.20535
28.0	61.92	30.72	−353.9	−308.3	23.08	12.96	30.76	893.4	1.19922
29.0	59.90	29.71	−323.1	−276.0	24.21	13.10	34.02	842.0	1.19230
29.5	58.77	29.15	−306.5	−258.5	24.81	13.19	36.09	813.8	1.18847
30.0	57.55	28.55	−288.9	−239.8	25.44	13.28	38.62	783.4	1.18431
30.5	56.20	27.88	−270.0	−219.7	26.10	13.39	41.81	750.5	1.17974
31.0	54.69	27.13	−249.5	−197.8	26.81	13.53	46.01	714.4	1.17464
31.5	52.95	26.27	−226.7	−173.5	27.59	13.70	51.94	674.3	1.16880
31.7	52.18	25.88	−216.8	−162.8	27.93	13.78	55.10	656.8	1.16618
31.8	51.76	25.68	−211.7	−157.2	28.11	13.83	56.93	647.6	1.16480
32.0	50.88	25.24	−200.8	−145.4	28.48	13.93	61.22	628.5	1.16185
32.2	49.92	24.76	−189.1	−132.6	28.87	14.05	66.68	608.1	1.15862
32.4	48.84	24.23	−176.4	−118.6	29.31	14.20	73.93	586.1	1.15501
32.6	47.60	23.61	−162.1	−102.8	29.79	14.37	84.11	562.1	1.15089
32.8	46.14	22.89	−145.7	−84.6	30.35	14.59	99.64	535.5	1.14603
32.9	45.28	22.46	−136.4	−74.1	30.67	14.72	111.0	520.9	1.14319
33.0	44.31	21.98	−125.9	−62.2	31.03	14.88	126.6	505.1	1.13997
33.1	43.16	21.41	−113.9	−48.5	31.45	15.07	149.5	488.1	1.13619
33.2	41.77	20.72	−99.5	−31.9	31.95	15.31	185.9	469.2	1.13160
33.3	39.95	19.82	−81.0	−10.4	32.59	15.62	251.8	448.2	1.12564
33.4	37.34	18.52	−54.8	20.8	33.53	16.05	389.9	424.6	1.11711
33.5	33.30	16.52	−13.6	71.1	35.03	16.60	609.8	402.1	1.10400

Thermophysical properties of hydrogen—Continued

T K	Density kg/m³	Density mol/dm³	E J/mol	H J/mol	S J/(mol·K)	C_v J/(mol·K)	C_p J/(mol·K)	Sound m/s	Diel. const.
33.6	28.92	14.35	33.9	131.4	36.83	16.86	538.9	392.1	1.08990
33.7	26.03	12.91	68.0	176.4	38.17	16.78	371.0	391.4	1.08066
33.8	24.18	11.99	91.3	208.1	39.10	16.60	272.3	393.1	1.07477
33.9	22.87	11.34	108.8	232.2	39.82	16.42	215.1	395.4	1.07063
34.0	21.87	10.85	122.7	251.8	40.39	16.25	179.1	397.9	1.06747
34.1	21.06	10.45	134.4	268.4	40.88	16.09	154.4	400.4	1.06492
34.2	20.39	10.11	144.5	282.9	41.31	15.94	136.6	402.8	1.06281
34.3	19.81	9.828	153.4	295.9	41.69	15.80	123.1	405.2	1.06099
34.4	19.31	9.578	161.4	307.6	42.03	15.68	112.5	407.5	1.05941
34.6	18.46	9.155	175.6	328.5	42.63	15.45	96.99	411.9	1.05673
34.8	17.75	8.807	187.8	346.7	43.16	15.25	86.09	416.1	1.05453
35.0	17.16	8.511	198.6	363.1	43.63	15.07	78.01	420.1	1.05266
35.2	16.64	8.253	208.4	378.0	44.05	14.92	71.75	423.9	1.05104
35.4	16.18	8.025	217.4	391.9	44.45	14.77	66.76	427.5	1.04961
35.6	15.77	7.820	225.8	404.8	44.81	14.64	62.67	431.0	1.04832
35.8	15.39	7.635	233.6	417.0	45.15	14.53	59.27	434.4	1.04716
36.0	15.05	7.466	241.0	428.5	45.47	14.42	56.38	437.6	1.04609
36.5	14.31	7.097	258.0	455.2	46.21	14.18	50.75	445.4	1.04378
37.0	13.68	6.787	273.3	479.6	46.87	13.98	46.66	452.6	1.04184
37.5	13.14	6.520	287.3	502.1	47.48	13.82	43.54	459.4	1.04017
38.0	12.67	6.285	300.4	523.2	48.04	13.68	41.07	465.9	1.03870
38.5	12.25	6.076	312.8	543.2	48.56	13.56	39.08	472.1	1.03740
39.0	11.87	5.887	324.5	562.3	49.05	13.45	37.43	478.0	1.03622
40.0	11.21	5.559	346.6	598.4	49.97	13.29	34.86	489.2	1.03418
41.0	10.64	5.280	367.1	632.3	50.80	13.17	32.96	499.6	1.03245
42.0	10.16	5.038	386.6	664.5	51.58	13.08	31.50	509.4	1.03094
43.0	9.727	4.825	405.2	695.4	52.31	13.01	30.34	518.7	1.02962
44.0	9.344	4.635	423.2	725.2	52.99	12.96	29.40	527.6	1.02844
45.0	8.999	4.464	440.6	754.2	53.64	12.92	28.63	536.2	1.02738
46.0	8.685	4.308	457.6	782.5	54.27	12.90	27.98	544.4	1.02642
47.0	8.398	4.166	474.2	810.2	54.86	12.88	27.43	552.4	1.02554
48.0	8.134	4.035	490.4	837.4	55.43	12.87	26.96	560.1	1.02473
50.0	7.663	3.801	522.2	890.5	56.52	12.87	26.20	574.8	1.02328
52.0	7.254	3.598	553.3	942.3	57.53	12.90	25.63	588.6	1.02203
54.0	6.894	3.420	583.7	993.1	58.49	12.95	25.20	601.6	1.02093
56.0	6.573	3.261	613.8	1043.0	59.40	13.02	24.87	613.9	1.01995
58.0	6.285	3.118	643.6	1093.0	60.27	13.11	24.63	625.6	1.01907
60.0	6.025	2.989	673.3	1142.0	61.10	13.22	24.45	636.6	1.01827
62.0	5.788	2.871	702.9	1191.0	61.90	13.36	24.35	647.1	1.01755
64.0	5.571	2.764	732.6	1239.0	62.67	13.52	24.29	657.0	1.01689
66.0	5.372	2.665	762.3	1288.0	63.42	13.69	24.28	666.4	1.01628
68.0	5.188	2.573	792.3	1336.0	64.15	13.89	24.31	675.4	1.01572
70.0	5.017	2.489	822.4	1385.0	64.85	14.11	24.38	683.9	1.01520
72.0	4.858	2.410	852.9	1434.0	65.54	14.34	24.48	692.1	1.01472
74.0	4.710	2.336	883.7	1483.0	66.21	14.59	24.61	699.9	1.01426
76.0	4.571	2.268	914.9	1532.0	66.87	14.86	24.77	707.3	1.01384
80.0	4.319	2.142	978.7	1632.0	68.15	15.44	25.16	721.5	1.01307
85.0	4.042	2.005	1061.0	1759.0	69.69	16.22	25.76	737.9	1.01223
90.0	3.801	1.886	1147.0	1890.0	71.19	17.05	26.44	753.3	1.01150
95.0	3.589	1.780	1237.0	2024.0	72.63	17.92	27.17	768.1	1.01086
100.0	3.400	1.686	1332.0	2162.0	74.05	18.79	27.93	782.4	1.01028
105.0	3.231	1.603	1430.0	2303.0	75.43	19.64	28.70	796.5	1.00977
110.0	3.078	1.527	1532.0	2449.0	76.78	20.46	29.44	810.4	1.00931
115.0	2.940	1.458	1638.0	2598.0	78.11	21.22	30.14	824.3	1.00889
120.0	2.814	1.396	1747.0	2750.0	79.40	21.92	30.78	838.2	1.00851
125.0	2.699	1.339	1859.0	2905.0	80.67	22.54	31.34	852.3	1.00816
130.0	2.593	1.286	1975.0	3063.0	81.91	23.06	31.83	866.4	1.00784
135.0	2.495	1.238	2092.0	3223.0	83.12	23.51	32.24	880.6	1.00754
140.0	2.405	1.193	2212.0	3385.0	84.30	23.90	32.59	894.8	1.00727
150.0	2.243	1.112	2455.0	3714.0	86.56	24.42	33.05	923.5	1.00677
160.0	2.101	1.042	2703.0	4046.0	88.70	24.68	33.27	952.2	1.00635
170.0	1.977	.9808	2951.0	4379.0	90.72	24.73	33.28	981.0	1.00597
180.0	1.867	.9262	3199.0	4711.0	92.62	24.62	33.14	1010.0	1.00564

Thermophysical properties of hydrogen—Continued

T K	Density kg/m^3	Density mol/dm^3	E J/mol	H J/mol	S J/(mol·K)	C_v J/(mol·K)	C_p J/(mol·K)	Sound m/s	Diel. const.
190.0	1.769	.8774	3445.0	5041.0	94.41	24.41	32.90	1038.0	1.00534
200.0	1.680	.8335	3689.0	5368.0	96.09	24.13	32.59	1066.0	1.00507
210.0	1.600	.7939	3929.0	5693.0	97.67	23.81	32.26	1094.0	1.00483
220.0	1.528	.7579	4166.0	6014.0	99.16	23.49	31.92	1121.0	1.00461
230.0	1.462	.7250	4400.0	6331.0	100.6	23.18	31.60	1148.0	1.00441
240.0	1.401	.6949	4631.0	6646.0	101.9	22.88	31.29	1175.0	1.00423
250.0	1.345	.6673	4859.0	6957.0	103.2	22.61	31.00	1200.0	1.00406
260.0	1.294	.6417	5084.0	7266.0	104.4	22.36	30.75	1226.0	1.00390
270.0	1.246	.6181	5307.0	7572.0	105.5	22.14	30.52	1250.0	1.00376
280.0	1.202	.5961	5528.0	7876.0	106.7	21.95	30.32	1274.0	1.00363
300.0	1.122	.5566	5964.0	8479.0	108.7	21.64	30.00	1321.0	1.00339
320.0	1.052	.5220	6395.0	9077.0	110.7	21.42	29.77	1365.0	1.00317
340.0	.9907	.4914	6822.0	9671.0	112.5	21.27	29.61	1408.0	1.00299
360.0	.9360	.4643	7246.0	10260.0	114.2	21.16	29.50	1449.0	1.00282
380.0	.8870	.4400	7669.0	10850.0	115.7	21.09	29.42	1489.0	1.00268
400.0	.8429	.4181	8090.0	11440.0	117.3	21.04	29.37	1528.0	1.00254
				1.45 MPa isobar					
14.28[a]	77.73	38.56	−620.6	−583.0	10.13	10.92	14.52	1335.0	1.25409
15.0	77.19	38.29	−610.7	−572.8	10.83	10.22	14.02	1315.0	1.25216
16.0	76.40	37.90	−597.0	−558.7	11.73	10.04	14.23	1286.0	1.24939
18.0	74.70	37.05	−567.9	−528.8	13.50	10.63	15.90	1230.0	1.24344
20.0	72.81	36.11	−535.0	−494.9	15.28	11.38	18.04	1175.0	1.23681
22.0	70.67	35.06	−497.9	−456.5	17.11	11.96	20.35	1119.0	1.22938
24.0	68.25	33.85	−456.1	−413.3	18.99	12.38	22.94	1057.0	1.22097
25.0	66.90	33.19	−433.3	−389.6	19.95	12.54	24.42	1022.0	1.21631
26.0	65.44	32.46	−409.0	−364.4	20.94	12.69	26.10	985.0	1.21127
27.0	63.84	31.67	−383.1	−337.3	21.96	12.82	28.08	944.2	1.20578
28.0	62.07	30.79	−355.1	−308.1	23.03	12.95	30.51	899.2	1.19972
29.0	60.08	29.80	−324.7	−276.1	24.15	13.09	33.64	848.9	1.19292
30.0	57.78	28.66	−291.0	−240.4	25.36	13.26	37.99	791.8	1.18510
30.5	56.47	28.01	−272.4	−220.7	26.01	13.36	40.94	760.0	1.18066
31.0	55.01	27.29	−252.4	−199.3	26.71	13.49	44.77	725.2	1.17573
31.5	53.36	26.47	−230.4	−175.7	27.46	13.64	50.02	686.9	1.17015
32.0	51.41	25.50	−205.7	−148.8	28.31	13.85	57.85	643.8	1.16363
32.2	50.52	25.06	−194.7	−136.8	28.68	13.96	62.25	624.8	1.16064
32.4	49.54	24.58	−182.9	−123.9	29.08	14.08	67.83	604.6	1.15737
32.6	48.45	24.03	−169.9	−109.6	29.52	14.22	75.22	582.9	1.15372
32.8	47.20	23.41	−155.5	−93.6	30.01	14.39	85.51	559.2	1.14955
33.0	45.72	22.68	−139.0	−75.0	30.57	14.61	101.0	533.2	1.14465
33.2	43.89	21.77	−119.1	−52.5	31.25	14.90	127.0	504.0	1.13859
33.3	42.76	21.21	−107.2	−38.8	31.67	15.08	148.0	487.9	1.13488
33.4	41.42	20.55	−93.1	−22.6	32.15	15.30	179.2	470.5	1.13045
33.5	39.74	19.71	−75.9	−2.4	32.76	15.58	228.9	451.8	1.12495
33.6	37.55	18.63	−53.6	24.3	33.55	15.92	309.7	432.2	1.11780
33.7	34.66	17.19	−23.9	60.5	34.63	16.31	413.2	414.1	1.10839
33.8	31.37	15.56	11.2	104.4	35.93	16.61	444.0	402.3	1.09775
33.9	28.51	14.14	43.4	145.9	37.15	16.69	378.4	398.0	1.08859
34.0	26.38	13.08	68.9	179.8	38.15	16.61	301.2	397.5	1.08177
34.1	24.80	12.30	88.9	206.8	38.95	16.48	243.2	398.7	1.07674
34.2	23.58	11.70	105.0	229.0	39.60	16.33	202.5	400.6	1.07289
34.3	22.61	11.22	118.4	247.7	40.14	16.18	173.6	402.6	1.06981
34.4	21.81	10.82	129.9	264.0	40.62	16.04	152.4	404.8	1.06728
34.5	21.13	10.48	140.0	278.4	41.03	15.90	136.3	407.0	1.06512
34.6	20.54	10.19	149.0	291.3	41.41	15.77	123.7	409.2	1.06326
34.7	20.01	9.928	157.1	303.2	41.75	15.65	113.6	411.3	1.06163
34.8	19.55	9.698	164.6	314.1	42.07	15.54	105.3	413.4	1.06016
35.0	18.75	9.300	177.9	333.8	42.63	15.34	92.61	417.5	1.05765
35.2	18.08	8.966	189.7	351.4	43.13	15.16	83.26	421.4	1.05554
35.4	17.49	8.678	200.2	367.3	43.58	14.99	76.11	425.2	1.05372
35.6	16.98	8.425	209.8	381.9	43.99	14.84	70.44	428.8	1.05212
35.8	16.53	8.199	218.7	395.5	44.38	14.71	65.84	432.3	1.05070

Thermophysical properties of hydrogen—Continued

T K	Density kg/m³	Density mol/dm³	E J/mol	H J/mol	S J/(mol·K)	C_v J/(mol·K)	C_p J/(mol·K)	Sound m/s	Diel. const.
36.0	16.12	7.996	227.0	408.3	44.73	14.59	62.03	435.6	1.04942
36.2	15.75	7.811	234.8	420.4	45.07	14.48	58.82	438.9	1.04826
36.5	15.24	7.562	245.7	437.4	45.53	14.32	54.84	443.6	1.04670
37.0	14.52	7.204	262.2	463.5	46.24	14.11	49.77	451.0	1.04445
37.5	13.91	6.900	277.3	487.4	46.89	13.92	46.00	458.0	1.04255
38.0	13.38	6.636	291.1	509.6	47.48	13.77	43.09	464.6	1.04090
38.5	12.91	6.403	304.1	530.6	48.02	13.64	40.76	470.9	1.03944
39.0	12.49	6.195	316.4	550.5	48.54	13.52	38.86	477.0	1.03814
39.5	12.11	6.007	328.1	569.5	49.02	13.42	37.28	482.8	1.03697
40.0	11.76	5.835	339.3	587.8	49.48	13.35	35.95	488.3	1.03590
41.0	11.15	5.532	360.5	622.6	50.34	13.21	33.82	498.9	1.03401
42.0	10.63	5.271	380.5	655.6	51.14	13.12	32.20	508.8	1.03239
43.0	10.16	5.042	399.6	687.1	51.88	13.04	30.92	518.2	1.03097
44.0	9.755	4.839	417.9	717.5	52.58	12.99	29.90	527.2	1.02970
45.0	9.387	4.656	435.6	747.0	53.24	12.94	29.06	535.9	1.02857
46.0	9.053	4.491	452.8	775.7	53.87	12.91	28.35	544.2	1.02755
47.0	8.749	4.340	469.6	803.7	54.47	12.89	27.76	552.2	1.02661
48.0	8.470	4.201	486.1	831.2	55.05	12.88	27.25	560.0	1.02575
50.0	7.973	3.955	518.3	884.9	56.15	12.88	26.44	574.7	1.02423
52.0	7.542	3.741	549.6	937.2	57.17	12.91	25.83	588.6	1.02291
54.0	7.163	3.553	580.3	988.3	58.14	12.95	25.37	601.7	1.02175
56.0	6.827	3.387	610.6	1039.0	59.05	13.02	25.02	614.1	1.02072
58.0	6.526	3.237	640.5	1088.0	59.93	13.12	24.76	625.8	1.01980
60.0	6.254	3.102	670.4	1138.0	60.76	13.23	24.58	636.9	1.01897
62.0	6.006	2.979	700.1	1187.0	61.57	13.37	24.45	647.3	1.01822
64.0	5.780	2.867	729.9	1236.0	62.34	13.52	24.39	657.3	1.01752
66.0	5.572	2.764	759.8	1284.0	63.09	13.70	24.37	666.7	1.01689
68.0	5.380	2.669	789.8	1333.0	63.82	13.90	24.39	675.7	1.01630
70.0	5.202	2.580	820.1	1382.0	64.53	14.11	24.46	684.2	1.01576
72.0	5.037	2.498	850.6	1431.0	65.22	14.35	24.55	692.4	1.01526
74.0	4.882	2.422	881.5	1480.0	65.89	14.60	24.68	700.2	1.01479
76.0	4.738	2.350	912.8	1530.0	66.55	14.87	24.84	707.7	1.01435
80.0	4.476	2.220	976.7	1630.0	67.84	15.44	25.22	721.8	1.01355
85.0	4.188	2.078	1059.0	1757.0	69.38	16.22	25.80	738.3	1.01268
90.0	3.938	1.953	1146.0	1888.0	70.88	17.06	26.48	753.7	1.01192
95.0	3.717	1.844	1236.0	2022.0	72.33	17.92	27.21	768.5	1.01125
100.0	3.521	1.747	1330.0	2160.0	73.74	18.79	27.97	782.8	1.01065
105.0	3.346	1.660	1428.0	2302.0	75.12	19.64	28.73	796.9	1.01012
110.0	3.188	1.581	1530.0	2447.0	76.48	20.46	29.47	810.8	1.00964
115.0	3.044	1.510	1636.0	2597.0	77.80	21.23	30.16	824.7	1.00920
120.0	2.914	1.445	1746.0	2749.0	79.10	21.92	30.80	838.7	1.00881
125.0	2.795	1.386	1858.0	2904.0	80.37	22.54	31.36	852.7	1.00845
130.0	2.685	1.332	1974.0	3062.0	81.61	23.07	31.85	866.9	1.00811
135.0	2.584	1.282	2091.0	3223.0	82.82	23.52	32.25	881.1	1.00781
140.0	2.490	1.235	2211.0	3385.0	84.00	23.90	32.61	895.3	1.00752
150.0	2.322	1.152	2455.0	3713.0	86.27	24.42	33.07	923.9	1.00701
160.0	2.176	1.079	2702.0	4045.0	88.41	24.69	33.28	952.7	1.00657
170.0	2.047	1.015	2950.0	4378.0	90.43	24.73	33.29	981.4	1.00618
180.0	1.933	.9589	3198.0	4711.0	92.33	24.63	33.15	1010.0	1.00584
190.0	1.831	.9084	3445.0	5041.0	94.11	24.41	32.90	1039.0	1.00553
200.0	1.740	.8630	3688.0	5369.0	95.79	24.13	32.60	1067.0	1.00525
210.0	1.657	.8219	3929.0	5693.0	97.37	23.82	32.27	1095.0	1.00500
220.0	1.582	.7847	4166.0	6014.0	98.87	23.49	31.93	1122.0	1.00478
230.0	1.513	.7507	4400.0	6332.0	100.3	23.18	31.60	1149.0	1.00457
240.0	1.450	.7195	4631.0	6646.0	101.6	22.88	31.29	1175.0	1.00438
250.0	1.393	.6909	4859.0	6957.0	102.9	22.61	31.01	1201.0	1.00420
260.0	1.339	.6644	5084.0	7266.0	104.1	22.36	30.75	1226.0	1.00404
270.0	1.290	.6399	5307.0	7572.0	105.3	22.14	30.52	1251.0	1.00389
280.0	1.244	.6172	5527.0	7877.0	106.4	21.95	30.32	1275.0	1.00375
300.0	1.162	.5763	5964.0	8480.0	108.4	21.64	30.00	1321.0	1.00351
320.0	1.090	.5405	6395.0	9077.0	110.4	21.42	29.77	1366.0	1.00329
340.0	1.026	.5089	6822.0	9671.0	112.2	21.27	29.61	1409.0	1.00310
360.0	.9692	.4808	7246.0	10260.0	113.9	21.16	29.50	1450.0	1.00292

Thermophysical properties of hydrogen—Continued

T K	Density kg/m^3	Density mol/dm^3	E J/mol	H J/mol	S J/(mol·K)	C_v J/(mol·K)	C_p	Sound m/s	Diel. const.
380.0	.9185	.4556	7669.0	10850.0	115.5	21.09	29.42	1490.0	1.00277
400.0	.8728	.4330	8090.0	11440.0	117.0	21.04	29.37	1528.0	1.00263
				1.50 MPa isobar					
14.30[a]	77.76	38.57	–620.6	–581.7	10.13	10.91	14.49	1335.0	1.25418
15.0	77.23	38.31	–610.9	–571.7	10.81	10.24	14.01	1315.0	1.25230
16.0	76.44	37.92	–597.2	–557.6	11.72	10.05	14.21	1287.0	1.24954
18.0	74.75	37.08	–568.2	–527.7	13.48	10.63	15.88	1231.0	1.24361
20.0	72.86	36.14	–535.4	–493.9	15.26	11.38	18.01	1177.0	1.23701
22.0	70.74	35.09	–498.3	–455.6	17.09	11.96	20.30	1121.0	1.22962
24.0	68.33	33.90	–456.7	–412.5	18.96	12.38	22.86	1060.0	1.22126
25.0	66.99	33.23	–434.0	–388.9	19.92	12.54	24.32	1026.0	1.21663
26.0	65.54	32.51	–409.9	–363.8	20.91	12.68	25.98	989.3	1.21163
27.0	63.96	31.73	–384.1	–336.8	21.92	12.81	27.91	949.2	1.20620
28.0	62.22	30.86	–356.4	–307.8	22.98	12.94	30.27	904.9	1.20021
29.0	60.26	29.89	–326.3	–276.1	24.09	13.07	33.28	855.7	1.19352
30.0	58.01	28.77	–293.0	–240.9	25.28	13.24	37.40	800.0	1.18586
30.5	56.73	28.14	–274.8	–221.5	25.93	13.33	40.16	769.1	1.18154
31.0	55.32	27.44	–255.2	–200.6	26.61	13.45	43.67	735.6	1.17677
31.5	53.73	26.65	–233.9	–177.6	27.34	13.59	48.37	698.9	1.17141
32.0	51.90	25.74	–210.1	–151.9	28.15	13.78	55.11	658.0	1.16524
32.2	51.07	25.33	–199.7	–140.5	28.50	13.87	58.75	640.2	1.16246
32.4	50.16	24.88	–188.6	–128.3	28.88	13.98	63.24	621.4	1.15944
32.6	49.17	24.39	–176.6	–115.1	29.29	14.10	68.92	601.4	1.15613
32.8	48.07	23.84	–163.5	–100.6	29.73	14.24	76.39	579.9	1.15244
33.0	46.80	23.22	–149.0	–84.4	30.22	14.41	86.70	556.7	1.14824
33.2	45.32	22.48	–132.4	–65.6	30.79	14.63	101.9	531.5	1.14332
33.4	43.50	21.58	–112.5	–43.0	31.47	14.91	126.6	503.5	1.13730
33.5	42.40	21.03	–100.8	–29.5	31.88	15.08	145.4	488.4	1.13368
33.6	41.11	20.39	–87.2	–13.7	32.35	15.28	171.7	472.4	1.12944
33.7	39.57	19.63	–71.2	5.3	32.91	15.52	209.1	455.8	1.12438
33.8	37.68	18.69	–51.6	28.6	33.60	15.81	260.7	439.1	1.11821
33.9	35.39	17.55	–27.8	57.6	34.46	16.11	318.3	423.9	1.11076
34.0	32.84	16.29	–.7	91.4	35.45	16.37	350.1	412.6	1.10250
34.1	30.38	15.07	26.5	126.1	36.47	16.51	335.6	406.4	1.09458
34.2	28.29	14.03	50.8	157.7	37.40	16.52	295.1	403.9	1.08788
34.3	26.60	13.20	71.4	185.0	38.19	16.45	252.2	403.7	1.08249
34.4	25.25	12.52	88.6	208.4	38.87	16.34	215.9	404.6	1.07818
34.5	24.14	11.98	103.2	228.4	39.46	16.22	187.3	406.0	1.07467
34.6	23.23	11.52	115.8	246.0	39.97	16.09	165.0	407.7	1.07176
34.7	22.45	11.13	126.9	261.6	40.42	15.96	147.5	409.6	1.06929
34.8	21.77	10.80	136.7	275.6	40.82	15.84	133.6	411.6	1.06716
34.9	21.18	10.51	145.6	288.4	41.19	15.72	122.4	413.5	1.06530
35.0	20.65	10.25	153.8	300.2	41.52	15.61	113.1	415.5	1.06364
35.2	19.75	9.799	168.2	321.3	42.12	15.41	98.81	419.4	1.06080
35.4	19.00	9.426	180.8	340.0	42.65	15.22	88.34	423.2	1.05844
35.6	18.36	9.107	192.1	356.8	43.13	15.06	80.34	426.8	1.05642
35.8	17.80	8.828	202.3	372.2	43.56	14.91	74.04	430.4	1.05466
36.0	17.30	8.581	211.7	386.5	43.96	14.77	68.95	433.8	1.05311
36.2	16.85	8.359	220.4	399.8	44.33	14.64	64.74	437.1	1.05171
36.4	16.45	8.158	228.6	412.4	44.67	14.53	61.21	440.4	1.05044
37.0	15.42	7.649	250.5	446.6	45.61	14.23	53.35	449.6	1.04724
37.5	14.72	7.302	266.7	472.1	46.29	14.03	48.79	456.7	1.04507
38.0	14.12	7.004	281.5	495.6	46.91	13.86	45.33	463.5	1.04320
38.5	13.60	6.745	295.2	517.6	47.49	13.72	42.61	469.9	1.04158
39.0	13.13	6.514	308.0	538.3	48.02	13.59	40.42	476.0	1.04014
39.5	12.71	6.307	320.2	558.0	48.52	13.49	38.61	481.9	1.03884
40.0	12.34	6.120	331.8	577.0	49.00	13.40	37.11	487.5	1.03767
41.0	11.67	5.790	353.8	612.8	49.89	13.26	34.73	498.2	1.03562
42.0	11.11	5.509	374.3	646.6	50.70	13.15	32.93	508.3	1.03387
43.0	10.61	5.263	393.8	678.8	51.46	13.07	31.53	517.8	1.03234
44.0	10.17	5.046	412.5	709.8	52.17	13.01	30.41	526.9	1.03099

Thermophysical properties of hydrogen—Continued

T K	Density kg/m^3	Density mol/dm^3	E J/mol	H J/mol	S J/(mol·K)	C_v J/(mol·K)	C_p J/(mol·K)	Sound m/s	Diel. const.
45.0	9.780	4.851	430.5	739.7	52.84	12.97	29.50	535.6	1.02978
46.0	9.426	4.676	448.0	768.8	53.48	12.93	28.74	544.0	1.02869
47.0	9.104	4.516	465.1	797.2	54.09	12.91	28.10	552.1	1.02770
48.0	8.809	4.370	481.8	825.0	54.68	12.90	27.56	559.9	1.02680
50.0	8.285	4.110	514.3	879.3	55.79	12.89	26.69	574.7	1.02519
52.0	7.832	3.885	545.9	932.0	56.82	12.91	26.04	588.7	1.02380
54.0	7.435	3.688	576.8	983.5	57.79	12.96	25.55	601.8	1.02258
56.0	7.083	3.513	607.3	1034.0	58.72	13.03	25.18	614.2	1.02151
58.0	6.767	3.357	637.5	1084.0	59.59	13.12	24.90	626.0	1.02054
60.0	6.483	3.216	667.4	1134.0	60.43	13.23	24.70	637.1	1.01967
62.0	6.225	3.088	697.3	1183.0	61.24	13.37	24.56	647.6	1.01888
64.0	5.989	2.971	727.2	1232.0	62.02	13.53	24.49	657.6	1.01816
66.0	5.772	2.863	757.2	1281.0	62.77	13.70	24.46	667.0	1.01750
68.0	5.572	2.764	787.4	1330.0	63.50	13.90	24.48	676.0	1.01689
70.0	5.387	2.672	817.7	1379.0	64.21	14.12	24.53	684.6	1.01633
72.0	5.215	2.587	848.4	1428.0	64.91	14.35	24.62	692.8	1.01580
74.0	5.055	2.507	879.3	1478.0	65.58	14.60	24.75	700.6	1.01531
76.0	4.905	2.433	910.7	1527.0	66.24	14.87	24.90	708.1	1.01486
80.0	4.632	2.298	974.7	1628.0	67.53	15.44	25.27	722.2	1.01403
85.0	4.334	2.150	1058.0	1755.0	69.08	16.23	25.85	738.7	1.01312
90.0	4.075	2.021	1144.0	1886.0	70.58	17.06	26.52	754.1	1.01233
95.0	3.846	1.908	1234.0	2021.0	72.03	17.92	27.25	768.9	1.01164
100.0	3.643	1.807	1329.0	2159.0	73.45	18.79	28.00	783.2	1.01102
105.0	3.461	1.717	1427.0	2301.0	74.83	19.65	28.76	797.3	1.01047
110.0	3.297	1.636	1529.0	2446.0	76.18	20.47	29.49	811.2	1.00997
115.0	3.149	1.562	1635.0	2595.0	77.51	21.23	30.19	825.2	1.00952
120.0	3.014	1.495	1745.0	2748.0	78.81	21.93	30.82	839.1	1.00911
125.0	2.890	1.434	1857.0	2904.0	80.08	22.54	31.38	853.2	1.00874
130.0	2.777	1.377	1973.0	3062.0	81.32	23.07	31.86	867.3	1.00839
135.0	2.672	1.325	2090.0	3222.0	82.53	23.52	32.27	881.6	1.00807
140.0	2.575	1.277	2210.0	3384.0	83.71	23.91	32.62	895.7	1.00778
150.0	2.401	1.191	2454.0	3713.0	85.98	24.43	33.08	924.4	1.00725
160.0	2.250	1.116	2701.0	4045.0	88.12	24.69	33.29	953.1	1.00680
170.0	2.117	1.050	2950.0	4378.0	90.14	24.74	33.30	981.9	1.00639
180.0	1.999	.9916	3198.0	4711.0	92.04	24.63	33.16	1011.0	1.00604
190.0	1.894	.9394	3444.0	5041.0	93.83	24.41	32.91	1039.0	1.00572
200.0	1.799	.8924	3688.0	5369.0	95.51	24.13	32.61	1067.0	1.00543
210.0	1.714	.8500	3928.0	5693.0	97.09	23.82	32.27	1095.0	1.00517
220.0	1.636	.8114	4166.0	6014.0	98.58	23.50	31.94	1122.0	1.00494
230.0	1.565	.7763	4399.0	6332.0	100.00	23.18	31.61	1149.0	1.00472
240.0	1.500	.7441	4630.0	6646.0	101.3	22.88	31.29	1175.0	1.00453
250.0	1.440	.7144	4858.0	6958.0	102.6	22.61	31.01	1201.0	1.00435
260.0	1.385	.6871	5083.0	7267.0	103.8	22.36	30.75	1226.0	1.00418
270.0	1.334	.6618	5306.0	7573.0	105.0	22.14	30.53	1251.0	1.00403
280.0	1.287	.6383	5527.0	7877.0	106.1	21.95	30.33	1275.0	1.00388
300.0	1.201	.5960	5963.0	8480.0	108.2	21.64	30.01	1322.0	1.00363
320.0	1.127	.5589	6394.0	9078.0	110.1	21.42	29.78	1366.0	1.00340
340.0	1.061	.5263	6822.0	9672.0	111.9	21.27	29.62	1409.0	1.00320
360.0	1.002	.4972	7246.0	10260.0	113.6	21.16	29.50	1450.0	1.00302
380.0	.9499	.4712	7669.0	10850.0	115.2	21.09	29.42	1490.0	1.00287
400.0	.9027	.4478	8090.0	11440.0	116.7	21.04	29.37	1529.0	1.00272

1.60 MPa isobar

T K	Density kg/m^3	Density mol/dm^3	E J/mol	H J/mol	S J/(mol·K)	C_v J/(mol·K)	C_p J/(mol·K)	Sound m/s	Diel. const.
14.33[a]	77.81	38.60	−620.4	−579.0	10.14	10.90	14.44	1333.0	1.25436
15.0	77.30	38.35	−611.2	−569.5	10.79	10.26	13.99	1316.0	1.25258
16.0	76.53	37.96	−597.6	−555.4	11.69	10.07	14.18	1288.0	1.24984
18.0	74.85	37.13	−568.7	−525.6	13.45	10.64	15.83	1234.0	1.24395
20.0	72.98	36.20	−536.0	−491.8	15.23	11.38	17.95	1181.0	1.23741
22.0	70.87	35.16	−499.2	−453.7	17.04	11.96	20.21	1127.0	1.23008
24.0	68.49	33.98	−457.9	−410.8	18.90	12.37	22.72	1067.0	1.22182
26.0	65.75	32.62	−411.6	−362.5	20.84	12.67	25.74	997.8	1.21235
27.0	64.20	31.85	−386.1	−335.9	21.84	12.80	27.59	958.8	1.20702

Thermophysical properties of hydrogen—Continued

T K	Density kg/m^3	Density mol/dm^3	E J/mol	H J/mol	S J/(mol·K)	C_v J/(mol·K)	C_p J/(mol·K)	Sound m/s	Diel. const.
28.0	62.50	31.00	−358.8	−307.2	22.88	12.92	29.82	916.0	1.20117
29.0	60.60	30.06	−329.3	−276.0	23.98	13.05	32.62	868.7	1.19468
30.0	58.43	28.99	−296.8	−241.6	25.14	13.20	36.35	815.6	1.18731
30.5	57.22	28.38	−279.3	−222.9	25.76	13.28	38.78	786.4	1.18320
31.0	55.89	27.73	−260.5	−202.8	26.42	13.39	41.78	755.0	1.17870
31.5	54.42	26.99	−240.2	−181.0	27.12	13.51	45.64	721.0	1.17373
32.0	52.75	26.17	−218.1	−156.9	27.87	13.66	50.86	683.8	1.16812
32.5	50.81	25.20	−193.2	−129.7	28.72	13.86	58.47	642.5	1.16161
32.7	49.93	24.77	−182.2	−117.6	29.09	13.96	62.62	624.5	1.15865
32.8	49.46	24.53	−176.4	−111.2	29.28	14.01	65.05	615.2	1.15708
33.0	48.45	24.03	−164.2	−97.7	29.69	14.14	70.83	595.7	1.15371
33.2	47.32	23.47	−150.9	−82.8	30.14	14.28	78.31	575.1	1.14997
33.4	46.05	22.84	−136.2	−66.2	30.64	14.45	88.38	553.0	1.14574
33.6	44.57	22.11	−119.5	−47.2	31.21	14.65	102.6	529.4	1.14084
33.8	42.80	21.23	−100.0	−24.7	31.88	14.90	123.6	504.2	1.13501
34.0	40.61	20.14	−76.4	3.0	32.69	15.22	156.0	477.6	1.12779
34.1	39.29	19.49	−62.4	19.7	33.18	15.41	177.8	464.3	1.12349
34.2	37.81	18.76	−46.6	38.7	33.74	15.61	202.4	451.5	1.11864
34.3	36.16	17.94	−29.0	60.2	34.37	15.81	226.3	440.0	1.11327
34.4	34.41	17.07	−10.0	83.7	35.05	15.99	243.2	430.6	1.10757
34.5	32.63	16.19	9.6	108.4	35.77	16.13	248.1	423.7	1.10183
34.6	30.95	15.35	28.7	133.0	36.48	16.21	241.2	419.3	1.09640
34.7	29.41	14.59	46.7	156.4	37.16	16.23	226.6	416.9	1.09147
34.8	28.06	13.92	63.2	178.2	37.78	16.20	208.6	415.9	1.08713
34.9	26.87	13.33	78.1	198.1	38.36	16.14	190.4	415.8	1.08334
35.0	25.84	12.82	91.5	216.3	38.88	16.06	173.4	416.3	1.08006
35.1	24.94	12.37	103.5	232.9	39.35	15.97	158.3	417.3	1.07719
35.2	24.14	11.98	114.4	248.0	39.78	15.87	145.2	418.5	1.07467
35.3	23.44	11.63	124.4	262.0	40.18	15.78	133.9	419.9	1.07244
35.4	22.81	11.32	133.5	274.9	40.54	15.68	124.3	421.4	1.07046
35.6	21.74	10.78	149.7	298.1	41.19	15.49	108.7	424.7	1.06705
35.8	20.84	10.34	163.8	318.6	41.77	15.31	96.90	428.0	1.06422
36.0	20.08	9.959	176.4	337.0	42.28	15.15	87.76	431.3	1.06182
36.2	19.41	9.630	187.7	353.8	42.75	15.00	80.51	434.6	1.05974
36.4	18.83	9.341	198.0	369.3	43.17	14.86	74.62	437.9	1.05791
36.6	18.31	9.083	207.6	383.7	43.57	14.73	69.77	441.0	1.05627
36.8	17.84	8.850	216.5	397.3	43.94	14.61	65.69	444.2	1.05480
37.0	17.41	8.638	224.8	410.1	44.28	14.50	62.23	447.2	1.05347
37.5	16.49	8.180	243.8	439.4	45.07	14.26	55.49	454.5	1.05058
38.0	15.72	7.800	260.7	465.9	45.77	14.06	50.60	461.5	1.04819
38.5	15.07	7.475	276.1	490.2	46.41	13.89	46.88	468.1	1.04615
39.0	14.50	7.191	290.4	512.9	46.99	13.75	43.96	474.4	1.04437
39.5	13.99	6.941	303.7	534.2	47.54	13.62	41.61	480.5	1.04280
40.0	13.54	6.716	316.3	554.6	48.05	13.52	39.69	486.2	1.04140
40.5	13.13	6.513	328.3	574.0	48.53	13.43	38.07	491.8	1.04013
41.0	12.76	6.328	339.8	592.7	48.99	13.35	36.70	497.1	1.03897
42.0	12.10	6.000	361.6	628.2	49.85	13.23	34.51	507.4	1.03693
43.0	11.53	5.718	382.0	661.9	50.64	13.13	32.83	517.1	1.03517
44.0	11.03	5.471	401.5	694.0	51.38	13.06	31.50	526.4	1.03363
45.0	10.58	5.251	420.2	724.9	52.07	13.01	30.43	535.2	1.03226
46.0	10.19	5.053	438.3	754.9	52.73	12.97	29.55	543.7	1.03104
47.0	9.827	4.874	455.8	784.1	53.36	12.94	28.81	551.9	1.02993
48.0	9.498	4.711	473.0	812.6	53.96	12.92	28.19	559.8	1.02891
50.0	8.917	4.423	506.2	867.9	55.09	12.91	27.20	574.8	1.02713
52.0	8.418	4.176	538.4	921.5	56.14	12.93	26.47	588.8	1.02560
54.0	7.982	3.960	569.8	973.9	57.13	12.97	25.91	602.1	1.02426
56.0	7.597	3.769	600.7	1025.0	58.06	13.04	25.49	614.6	1.02308
58.0	7.254	3.598	631.3	1076.0	58.95	13.13	25.17	626.4	1.02203
60.0	6.945	3.445	661.6	1126.0	59.80	13.25	24.94	637.6	1.02108
62.0	6.664	3.306	691.7	1176.0	60.62	13.38	24.78	648.2	1.02023
64.0	6.409	3.179	721.9	1225.0	61.40	13.54	24.69	658.2	1.01944
66.0	6.174	3.063	752.1	1275.0	62.16	13.71	24.64	667.7	1.01873

Thermophysical properties of hydrogen—Continued

T K	Density kg/m³	Density mol/dm³	E J/mol	H J/mol	S J/(mol·K)	C_v J/(mol·K)	C_p J/(mol·K)	Sound m/s	Diel. const.
68.0	5.958	2.956	782.5	1324.0	62.90	13.91	24.64	676.7	1.01807
70.0	5.759	2.857	813.0	1373.0	63.61	14.13	24.69	685.3	1.01746
72.0	5.573	2.765	843.8	1423.0	64.31	14.36	24.77	693.5	1.01689
74.0	5.401	2.679	875.0	1472.0	64.99	14.61	24.88	701.3	1.01637
76.0	5.239	2.599	906.5	1522.0	65.65	14.88	25.02	708.8	1.01588
80.0	4.946	2.454	970.8	1623.0	66.95	15.45	25.38	723.0	1.01498
85.0	4.626	2.295	1054.0	1751.0	68.50	16.24	25.95	739.5	1.01401
90.0	4.348	2.157	1141.0	1883.0	70.00	17.07	26.60	755.0	1.01316
95.0	4.103	2.035	1231.0	2017.0	71.46	17.93	27.32	769.8	1.01242
100.0	3.886	1.927	1326.0	2156.0	72.88	18.80	28.06	784.1	1.01176
105.0	3.691	1.831	1424.0	2298.0	74.27	19.65	28.82	798.2	1.01117
110.0	3.516	1.744	1527.0	2444.0	75.63	20.47	29.55	812.1	1.01064
115.0	3.358	1.666	1633.0	2593.0	76.95	21.24	30.23	826.0	1.01015
120.0	3.213	1.594	1742.0	2746.0	78.25	21.93	30.86	840.0	1.00972
125.0	3.081	1.528	1855.0	2902.0	79.53	22.55	31.42	854.1	1.00932
130.0	2.960	1.468	1971.0	3060.0	80.77	23.08	31.90	868.2	1.00895
135.0	2.848	1.413	2088.0	3221.0	81.98	23.52	32.30	882.5	1.00861
140.0	2.745	1.362	2208.0	3383.0	83.16	23.91	32.65	896.6	1.00830
150.0	2.560	1.270	2452.0	3712.0	85.43	24.43	33.11	925.3	1.00773
160.0	2.398	1.190	2700.0	4044.0	87.57	24.69	33.31	954.1	1.00725
170.0	2.256	1.119	2948.0	4378.0	89.60	24.74	33.32	982.8	1.00682
180.0	2.131	1.057	3197.0	4710.0	91.50	24.63	33.17	1012.0	1.00644
190.0	2.018	1.001	3443.0	5041.0	93.28	24.42	32.93	1040.0	1.00610
200.0	1.918	.9512	3687.0	5369.0	94.97	24.13	32.62	1068.0	1.00579
210.0	1.826	.9060	3927.0	5693.0	96.55	23.82	32.29	1096.0	1.00551
220.0	1.744	.8649	4165.0	6014.0	98.04	23.50	31.95	1123.0	1.00526
230.0	1.668	.8275	4399.0	6332.0	99.46	23.18	31.62	1150.0	1.00504
240.0	1.599	.7931	4630.0	6647.0	100.8	22.89	31.30	1176.0	1.00483
250.0	1.535	.7616	4857.0	6958.0	102.1	22.61	31.02	1202.0	1.00463
260.0	1.477	.7324	5083.0	7267.0	103.3	22.36	30.76	1227.0	1.00446
270.0	1.422	.7054	5306.0	7574.0	104.4	22.14	30.53	1252.0	1.00429
280.0	1.372	.6804	5526.0	7878.0	105.5	21.95	30.33	1276.0	1.00414
300.0	1.281	.6353	5963.0	8481.0	107.6	21.65	30.01	1322.0	1.00387
320.0	1.201	.5959	6394.0	9079.0	109.6	21.42	29.78	1367.0	1.00362
340.0	1.131	.5610	6821.0	9673.0	111.4	21.27	29.62	1410.0	1.00341
360.0	1.069	.5301	7246.0	10260.0	113.0	21.17	29.51	1451.0	1.00322
380.0	1.013	.5024	7669.0	10850.0	114.6	21.09	29.43	1491.0	1.00306
400.0	.9624	.4774	8090.0	11440.0	116.1	21.04	29.37	1529.0	1.00290

1.70 MPa isobar

T K	Density kg/m³	Density mol/dm³	E J/mol	H J/mol	S J/(mol·K)	C_v J/(mol·K)	C_p J/(mol·K)	Sound m/s	Diel. const.
14.36[a]	77.86	38.62	−620.3	−576.3	10.15	10.89	14.38	1332.0	1.25453
15.0	77.38	38.39	−611.5	−567.3	10.76	10.29	13.97	1316.0	1.25286
16.0	76.61	38.00	−598.0	−553.3	11.67	10.08	14.15	1290.0	1.25014
18.0	74.95	37.18	−569.2	−523.5	13.42	10.65	15.79	1237.0	1.24429
20.0	73.09	36.26	−536.7	−489.8	15.19	11.38	17.88	1186.0	1.23780
22.0	71.01	35.22	−500.1	−451.8	17.00	11.96	20.12	1132.0	1.23054
24.0	68.65	34.06	−459.1	−409.2	18.85	12.37	22.58	1073.0	1.22238
26.0	65.95	32.72	−413.2	−361.2	20.77	12.66	25.51	1006.0	1.21304
27.0	64.43	31.96	−388.0	−334.8	21.76	12.79	27.29	968.2	1.20781
28.0	62.77	31.14	−361.1	−306.5	22.79	12.90	29.41	926.7	1.20210
29.0	60.92	30.22	−332.1	−275.9	23.87	13.03	32.02	881.1	1.19578
30.0	58.83	29.18	−300.5	−242.2	25.01	13.16	35.43	830.3	1.18868
30.5	57.68	28.61	−283.4	−224.0	25.61	13.24	37.59	802.6	1.18474
31.0	56.42	27.99	−265.3	−204.5	26.24	13.33	40.21	773.0	1.18048
31.5	55.04	27.30	−245.9	−183.7	26.91	13.44	43.47	741.2	1.17582
32.0	53.50	26.54	−225.0	−160.9	27.63	13.57	47.70	706.9	1.17063
32.5	51.75	25.67	−201.9	−135.7	28.41	13.73	53.47	669.4	1.16476
32.7	50.98	25.29	−191.9	−124.7	28.75	13.81	56.44	653.3	1.16216
32.8	50.57	25.08	−186.8	−119.0	28.92	13.85	58.12	645.1	1.16079
33.0	49.70	24.66	−175.9	−107.0	29.29	13.94	61.95	628.0	1.15791
33.5	47.20	23.41	−145.5	−72.9	30.31	14.23	75.75	581.6	1.14955

Thermophysical properties of hydrogen—Continued

T K	Density kg/m³	Density mol/dm³	E J/mol	H J/mol	S J/(mol·K)	C_v J/(mol·K)	C_p J/(mol·K)	Sound m/s	Diel. const.
33.7	46.00	22.82	-131.4	-57.0	30.79	14.38	84.02	561.4	1.14558
33.8	45.35	22.49	-123.9	-48.3	31.04	14.46	89.08	550.9	1.14341
34.0	43.90	21.77	-107.4	-29.3	31.60	14.65	101.7	529.2	1.13861
34.2	42.21	20.94	-88.5	-7.3	32.25	14.88	118.8	506.7	1.13305
34.4	40.21	19.95	-66.6	18.6	33.01	15.14	141.6	484.2	1.12650
34.6	37.86	18.78	-40.9	49.6	33.90	15.44	168.2	463.0	1.11881
34.8	35.21	17.47	-11.7	85.6	34.94	15.71	189.7	445.9	1.11018
35.0	32.51	16.12	18.9	124.3	36.05	15.90	194.2	434.8	1.10142
35.2	30.03	14.90	48.1	162.2	37.13	15.96	182.6	429.3	1.09346
35.3	28.93	14.35	61.6	180.0	37.63	15.94	173.6	428.1	1.08994
35.4	27.93	13.86	74.2	196.9	38.11	15.91	163.7	427.6	1.08674
35.5	27.02	13.41	85.9	212.8	38.56	15.85	153.9	427.6	1.08383
35.6	26.20	13.00	96.9	227.7	38.98	15.79	144.4	428.0	1.08121
35.7	25.45	12.63	107.0	241.7	39.37	15.72	135.5	428.7	1.07883
35.8	24.78	12.29	116.5	254.8	39.74	15.65	127.4	429.6	1.07667
36.0	23.59	11.70	133.6	278.8	40.41	15.49	113.4	431.9	1.07291
36.2	22.59	11.20	148.6	300.3	41.00	15.33	102.0	434.5	1.06974
36.4	21.73	10.78	162.1	319.8	41.54	15.18	92.79	437.3	1.06703
36.6	20.98	10.41	174.2	337.6	42.03	15.04	85.25	440.2	1.06467
36.8	20.32	10.08	185.3	354.0	42.47	14.91	79.03	443.1	1.06259
37.0	19.74	9.790	195.6	369.2	42.89	14.78	73.82	446.1	1.06075
37.2	19.21	9.527	205.1	383.6	43.27	14.67	69.42	449.0	1.05908
37.4	18.73	9.289	214.0	397.1	43.63	14.56	65.65	451.8	1.05758
37.6	18.29	9.071	222.4	409.8	43.98	14.46	62.40	454.7	1.05620
38.0	17.51	8.686	238.0	433.7	44.61	14.27	57.08	460.2	1.05377
38.5	16.68	8.276	255.5	460.9	45.32	14.08	52.02	466.9	1.05119
39.0	15.98	7.927	271.4	485.9	45.96	13.91	48.15	473.3	1.04899
39.5	15.37	7.622	286.2	509.2	46.56	13.76	45.10	479.4	1.04707
40.0	14.82	7.353	299.9	531.1	47.11	13.65	42.64	485.3	1.04539
40.5	14.34	7.112	312.9	551.9	47.63	13.54	40.62	490.9	1.04388
41.0	13.90	6.895	325.2	571.8	48.11	13.45	38.92	496.4	1.04251
42.0	13.13	6.514	348.3	609.3	49.02	13.31	36.24	506.8	1.04014
43.0	12.48	6.191	369.9	644.5	49.85	13.20	34.23	516.7	1.03812
44.0	11.91	5.910	390.2	677.9	50.61	13.12	32.67	526.0	1.03636
45.0	11.41	5.662	409.7	709.9	51.33	13.05	31.42	535.0	1.03482
46.0	10.97	5.441	428.3	740.8	52.01	13.01	30.40	543.6	1.03344
47.0	10.57	5.241	446.4	770.8	52.66	12.97	29.56	551.8	1.03220
48.0	10.20	5.060	464.0	800.0	53.27	12.95	28.85	559.8	1.03108
49.0	9.867	4.895	481.2	828.5	53.86	12.94	28.25	567.5	1.03005
50.0	9.560	4.742	498.0	856.5	54.43	12.93	27.73	574.9	1.02911
52.0	9.012	4.471	530.8	911.1	55.50	12.95	26.91	589.1	1.02742
54.0	8.536	4.234	562.8	964.3	56.50	12.99	26.28	602.5	1.02596
56.0	8.117	4.027	594.1	1016.0	57.45	13.05	25.81	615.1	1.02467
58.0	7.744	3.842	625.0	1068.0	58.35	13.14	25.45	626.9	1.02353
60.0	7.409	3.675	655.7	1118.0	59.20	13.26	25.19	638.2	1.02251
62.0	7.107	3.525	686.1	1168.0	60.03	13.39	25.01	648.8	1.02158
64.0	6.831	3.388	716.6	1218.0	60.82	13.54	24.89	658.8	1.02073
66.0	6.578	3.263	747.0	1268.0	61.58	13.72	24.83	668.4	1.01996
68.0	6.346	3.148	777.6	1318.0	62.32	13.92	24.81	677.4	1.01925
70.0	6.131	3.041	808.3	1367.0	63.04	14.13	24.84	686.0	1.01860
72.0	5.932	2.943	839.3	1417.0	63.74	14.37	24.91	694.2	1.01799
74.0	5.747	2.851	870.6	1467.0	64.43	14.62	25.01	702.1	1.01743
76.0	5.574	2.765	902.3	1517.0	65.10	14.88	25.15	709.6	1.01690
78.0	5.413	2.685	934.4	1568.0	65.75	15.17	25.31	716.8	1.01640
80.0	5.261	2.610	966.9	1618.0	66.40	15.46	25.49	723.8	1.01594
85.0	4.919	2.440	1050.0	1747.0	67.96	16.24	26.04	740.3	1.01490
90.0	4.621	2.292	1137.0	1879.0	69.46	17.08	26.68	755.8	1.01399
95.0	4.360	2.163	1228.0	2014.0	70.92	17.94	27.39	770.6	1.01320
100.0	4.128	2.048	1323.0	2153.0	72.35	18.81	28.13	785.0	1.01250
105.0	3.921	1.945	1421.0	2295.0	73.74	19.66	28.87	799.0	1.01187
110.0	3.735	1.853	1524.0	2442.0	75.10	20.48	29.60	813.0	1.01130
115.0	3.566	1.769	1630.0	2591.0	76.43	21.24	30.28	826.9	1.01079
120.0	3.413	1.693	1740.0	2744.0	77.73	21.94	30.91	840.9	1.01032

Thermophysical properties of hydrogen—Continued

T K	Density kg/m^3	Density mol/dm^3	E J/mol	H J/mol	S J/(mol·K)	C_v J/(mol·K)	C_p J/(mol·K)	Sound m/s	Diel. const.
125.0	3.272	1.623	1853.0	2900.0	79.00	22.55	31.46	855.0	1.00990
130.0	3.143	1.559	1968.0	3059.0	80.25	23.08	31.94	869.1	1.00950
135.0	3.025	1.500	2086.0	3219.0	81.46	23.53	32.34	883.4	1.00914
140.0	2.915	1.446	2206.0	3382.0	82.64	23.92	32.68	897.6	1.00881
150.0	2.718	1.348	2450.0	3711.0	84.92	24.44	33.13	926.2	1.00821
160.0	2.546	1.263	2698.0	4044.0	87.06	24.70	33.34	955.0	1.00769
170.0	2.396	1.188	2947.0	4377.0	89.08	24.75	33.34	983.8	1.00724
180.0	2.262	1.122	3195.0	4710.0	90.99	24.64	33.19	1012.0	1.00683
190.0	2.143	1.063	3442.0	5041.0	92.77	24.42	32.94	1041.0	1.00647
200.0	2.036	1.010	3686.0	5369.0	94.46	24.14	32.64	1069.0	1.00615
210.0	1.939	.9619	3926.0	5694.0	96.04	23.82	32.30	1097.0	1.00586
220.0	1.851	.9183	4164.0	6015.0	97.53	23.50	31.96	1124.0	1.00559
230.0	1.771	.8786	4398.0	6333.0	98.95	23.19	31.63	1151.0	1.00535
240.0	1.698	.8421	4629.0	6647.0	100.3	22.89	31.31	1177.0	1.00513
250.0	1.630	.8086	4857.0	6959.0	101.6	22.61	31.03	1203.0	1.00492
260.0	1.568	.7777	5082.0	7268.0	102.8	22.37	30.77	1228.0	1.00473
270.0	1.510	.7491	5305.0	7575.0	103.9	22.15	30.54	1253.0	1.00456
280.0	1.456	.7225	5526.0	7879.0	105.0	21.95	30.34	1277.0	1.00440
300.0	1.360	.6746	5962.0	8482.0	107.1	21.65	30.02	1323.0	1.00410
320.0	1.276	.6327	6394.0	9080.0	109.0	21.43	29.79	1368.0	1.00385
340.0	1.201	.5958	6821.0	9674.0	110.8	21.27	29.62	1411.0	1.00362
360.0	1.135	.5629	7245.0	10270.0	112.5	21.17	29.51	1452.0	1.00342
380.0	1.075	.5335	7668.0	10850.0	114.1	21.09	29.43	1491.0	1.00325
400.0	1.022	.5070	8090.0	11440.0	115.6	21.04	29.37	1530.0	1.00308
				1.80 MPa isobar					
14.39[a]	77.91	38.65	−620.2	−573.6	10.15	10.88	14.33	1331.0	1.25471
15.0	77.46	38.42	−611.9	−565.0	10.74	10.32	13.94	1316.0	1.25313
16.0	76.70	38.04	−598.4	−551.1	11.64	10.10	14.12	1291.0	1.25043
18.0	75.04	37.22	−569.7	−521.3	13.39	10.65	15.74	1240.0	1.24463
20.0	73.20	36.31	−537.4	−487.8	15.15	11.38	17.82	1190.0	1.23819
22.0	71.14	35.29	−501.0	−450.0	16.95	11.95	20.03	1138.0	1.23100
24.0	68.81	34.13	−460.3	−407.5	18.80	12.36	22.45	1080.0	1.22292
26.0	66.15	32.81	−414.8	−359.9	20.70	12.65	25.30	1014.0	1.21372
27.0	64.66	32.07	−389.9	−333.8	21.69	12.78	27.01	977.2	1.20858
28.0	63.03	31.26	−363.4	−305.8	22.71	12.89	29.02	937.0	1.20299
29.0	61.23	30.37	−334.8	−275.6	23.77	13.00	31.48	893.0	1.19684
30.0	59.21	29.37	−303.9	−242.6	24.88	13.13	34.62	844.3	1.18997
31.0	56.90	28.23	−269.7	−206.0	26.08	13.28	38.88	789.8	1.18213
31.5	55.60	27.58	−251.1	−185.8	26.73	13.38	41.70	759.9	1.17772
32.0	54.17	26.87	−231.1	−164.2	27.41	13.49	45.23	727.9	1.17288
32.5	52.56	26.07	−209.5	−140.4	28.15	13.62	49.83	693.3	1.16748
33.0	50.73	25.17	−185.6	−114.0	28.95	13.80	56.14	655.8	1.16135
33.5	48.59	24.10	−158.5	−83.8	29.86	14.02	65.41	614.7	1.15417
34.0	45.95	22.79	−126.7	−47.7	30.93	14.32	80.26	569.5	1.14542
34.2	44.70	22.17	−112.0	−30.8	31.43	14.47	88.86	550.3	1.14127
34.4	43.30	21.48	−95.8	−12.0	31.97	14.64	99.69	530.5	1.13663
34.6	41.71	20.69	−77.8	9.2	32.59	14.84	113.2	510.7	1.13140
34.8	39.90	19.79	−57.5	33.4	33.29	15.05	129.1	491.4	1.12547
35.0	37.88	18.79	−34.9	60.9	34.07	15.28	145.5	473.9	1.11886
35.2	35.70	17.71	−10.3	91.3	34.94	15.48	157.9	459.5	1.11175
35.4	33.48	16.61	15.1	123.5	35.85	15.63	162.2	449.2	1.10458
35.6	31.39	15.57	40.0	155.6	36.76	15.71	157.9	443.0	1.09782
35.8	29.50	14.63	63.3	186.3	37.62	15.70	148.1	439.9	1.09176
36.0	27.86	13.82	84.5	214.7	38.41	15.64	136.1	438.9	1.08650
36.2	26.44	13.12	103.5	240.7	39.13	15.54	124.0	439.4	1.08198
36.4	25.23	12.51	120.6	264.4	39.78	15.42	113.0	440.7	1.07810
36.6	24.17	11.99	135.9	286.0	40.37	15.29	103.2	442.5	1.07476
36.8	23.26	11.54	149.8	305.8	40.91	15.16	94.90	444.7	1.07185
37.0	22.45	11.14	162.4	324.1	41.41	15.04	87.79	447.1	1.06930
37.2	21.73	10.78	174.0	341.0	41.86	14.91	81.73	449.6	1.06704
37.4	21.09	10.46	184.8	356.8	42.29	14.80	76.54	452.1	1.06502

Thermophysical properties of hydrogen—Continued

T K	Density kg/m³	Density mol/dm³	*E* J/mol	*H* J/mol	*S* J/(mol·K)	C_v J/(mol·K)	C_p J/(mol·K)	Sound m/s	Diel. const.
37.6	20.52	10.18	194.8	371.7	42.68	14.69	72.08	454.8	1.06320
37.8	19.99	9.916	204.2	385.7	43.06	14.58	68.22	457.4	1.06155
38.0	19.51	9.678	213.0	399.0	43.41	14.48	64.85	460.0	1.06004
38.5	18.47	9.160	233.1	429.6	44.21	14.26	58.08	466.5	1.05676
39.0	17.59	8.727	251.1	457.4	44.92	14.07	53.02	472.8	1.05403
39.5	16.85	8.357	267.5	482.8	45.57	13.91	49.10	479.0	1.05170
40.0	16.20	8.034	282.5	506.6	46.17	13.78	46.00	484.8	1.04966
40.5	15.62	7.749	296.6	528.9	46.73	13.66	43.47	490.5	1.04787
41.0	15.11	7.493	309.9	550.1	47.25	13.56	41.39	496.0	1.04627
41.5	14.64	7.263	322.5	570.4	47.74	13.47	39.64	501.3	1.04482
42.0	14.22	7.053	334.6	589.8	48.20	13.39	38.15	506.5	1.04350
43.0	13.47	6.682	357.3	626.7	49.07	13.27	35.76	516.5	1.04119
44.0	12.83	6.364	378.7	661.5	49.87	13.17	33.92	525.9	1.03920
45.0	12.27	6.085	398.9	694.7	50.62	13.10	32.47	535.0	1.03746
46.0	11.77	5.838	418.2	726.5	51.32	13.05	31.30	543.6	1.03592
47.0	11.32	5.617	436.9	757.3	51.98	13.01	30.34	552.0	1.03454
48.0	10.92	5.416	454.9	787.3	52.61	12.98	29.53	560.0	1.03329
49.0	10.55	5.234	472.5	816.5	53.21	12.96	28.86	567.7	1.03216
50.0	10.21	5.067	489.8	845.0	53.79	12.96	28.28	575.2	1.03112
52.0	9.615	4.770	523.2	900.6	54.88	12.97	27.36	589.5	1.02928
54.0	9.097	4.513	555.7	954.6	55.90	13.00	26.66	603.0	1.02768
56.0	8.643	4.287	587.5	1007.0	56.86	13.07	26.14	615.6	1.02629
58.0	8.239	4.087	618.8	1059.0	57.77	13.15	25.74	627.5	1.02505
60.0	7.878	3.908	649.8	1110.0	58.63	13.27	25.44	638.8	1.02394
62.0	7.552	3.746	680.5	1161.0	59.46	13.40	25.23	649.5	1.02294
64.0	7.255	3.599	711.2	1211.0	60.26	13.55	25.09	659.5	1.02203
66.0	6.984	3.465	741.9	1261.0	61.03	13.73	25.01	669.1	1.02120
68.0	6.735	3.341	772.7	1311.0	61.78	13.93	24.98	678.2	1.02044
70.0	6.506	3.227	803.6	1361.0	62.50	14.14	25.00	686.8	1.01974
72.0	6.293	3.121	834.8	1411.0	63.21	14.38	25.05	695.0	1.01909
74.0	6.095	3.023	866.3	1462.0	63.90	14.63	25.15	702.9	1.01849
76.0	5.910	2.932	898.1	1512.0	64.57	14.89	25.27	710.4	1.01792
78.0	5.738	2.846	930.3	1563.0	65.23	15.17	25.42	717.7	1.01740
80.0	5.576	2.766	963.0	1614.0	65.87	15.47	25.60	724.7	1.01690
85.0	5.211	2.585	1047.0	1743.0	67.44	16.25	26.13	741.2	1.01579
90.0	4.895	2.428	1134.0	1875.0	68.95	17.08	26.77	756.7	1.01483
95.0	4.617	2.290	1225.0	2011.0	70.42	17.95	27.46	771.5	1.01398
100.0	4.371	2.168	1320.0	2150.0	71.84	18.81	28.19	785.8	1.01323
105.0	4.151	2.059	1419.0	2293.0	73.24	19.67	28.93	799.9	1.01256
110.0	3.953	1.961	1521.0	2439.0	74.60	20.49	29.65	813.9	1.01196
115.0	3.775	1.872	1628.0	2589.0	75.93	21.25	30.33	827.8	1.01142
120.0	3.612	1.792	1738.0	2742.0	77.24	21.95	30.95	841.8	1.01093
125.0	3.463	1.718	1851.0	2899.0	78.51	22.56	31.50	855.9	1.01047
130.0	3.326	1.650	1966.0	3057.0	79.76	23.09	31.97	870.0	1.01006
135.0	3.201	1.588	2084.0	3218.0	80.97	23.53	32.37	884.3	1.00968
140.0	3.084	1.530	2204.0	3381.0	82.16	23.92	32.71	898.5	1.00932
150.0	2.876	1.426	2449.0	3710.0	84.43	24.44	33.16	927.2	1.00869
160.0	2.694	1.337	2696.0	4043.0	86.58	24.70	33.36	955.9	1.00814
170.0	2.535	1.257	2946.0	4377.0	88.60	24.75	33.36	984.7	1.00766
180.0	2.394	1.187	3194.0	4710.0	90.50	24.64	33.21	1013.0	1.00723
190.0	2.267	1.125	3441.0	5041.0	92.29	24.42	32.96	1042.0	1.00685
200.0	2.154	1.069	3684.0	5369.0	93.98	24.14	32.65	1070.0	1.00651
210.0	2.052	1.018	3925.0	5694.0	95.56	23.83	32.31	1098.0	1.00620
220.0	1.959	.9717	4163.0	6015.0	97.06	23.51	31.97	1125.0	1.00592
230.0	1.874	.9296	4397.0	6333.0	98.47	23.19	31.64	1152.0	1.00566
240.0	1.796	.8911	4628.0	6648.0	99.81	22.89	31.32	1178.0	1.00542
250.0	1.725	.8556	4856.0	6960.0	101.1	22.62	31.03	1204.0	1.00521
260.0	1.659	.8229	5081.0	7269.0	102.3	22.37	30.78	1229.0	1.00501
270.0	1.598	.7926	5304.0	7575.0	103.5	22.15	30.55	1254.0	1.00482
280.0	1.541	.7645	5525.0	7880.0	104.6	21.96	30.35	1278.0	1.00465
300.0	1.439	.7139	5962.0	8483.0	106.6	21.65	30.02	1324.0	1.00434
320.0	1.350	.6696	6393.0	9081.0	108.6	21.43	29.79	1369.0	1.00407
340.0	1.271	.6305	6820.0	9675.0	110.4	21.27	29.63	1411.0	1.00384

Thermophysical properties of hydrogen—Continued

T K	Density kg/m³	Density mol/dm³	E J/mol	H J/mol	S J/(mol·K)	C_v J/(mol·K)	C_p J/(mol·K)	Sound m/s	Diel. const.
360.0	1.201	.5957	7245.0	10270.0	112.1	21.17	29.51	1452.0	1.00362
380.0	1.138	.5646	7668.0	10860.0	113.7	21.09	29.43	1492.0	1.00343
400.0	1.082	.5366	8090.0	11440.0	115.2	21.04	29.38	1531.0	1.00326
				1.90 MPa isobar					
14.42[a]	77.96	38.67	−620.0	−570.9	10.16	10.87	14.27	1330.0	1.25489
15.0	77.54	38.46	−612.2	−562.8	10.71	10.34	13.92	1317.0	1.25340
16.0	76.78	38.09	−598.7	−548.9	11.61	10.11	14.09	1293.0	1.25073
18.0	75.14	37.27	−570.2	−519.2	13.36	10.66	15.70	1243.0	1.24497
20.0	73.31	36.37	−538.0	−485.8	15.11	11.38	17.76	1194.0	1.23857
22.0	71.27	35.35	−501.8	−448.1	16.91	11.95	19.95	1143.0	1.23145
24.0	68.97	34.21	−461.4	−405.9	18.74	12.36	22.32	1086.0	1.22345
26.0	66.34	32.91	−416.3	−358.6	20.64	12.65	25.09	1022.0	1.21438
27.0	64.87	32.18	−391.7	−332.7	21.61	12.77	26.74	986.0	1.20933
28.0	63.28	31.39	−365.5	−305.0	22.62	12.88	28.67	947.0	1.20386
29.0	61.53	30.52	−337.4	−275.2	23.67	12.99	30.98	904.4	1.19786
30.0	59.58	29.55	−307.1	−242.8	24.76	13.10	33.89	857.5	1.19120
31.0	57.36	28.45	−273.9	−207.1	25.93	13.24	37.73	805.5	1.18367
31.5	56.12	27.84	−255.9	−187.6	26.56	13.33	40.21	777.3	1.17947
32.0	54.77	27.17	−236.7	−166.8	27.21	13.42	43.22	747.2	1.17491
32.5	53.28	26.43	−216.2	−144.3	27.91	13.54	47.02	715.0	1.16990
33.0	51.61	25.60	−193.8	−119.6	28.66	13.68	51.99	680.5	1.16430
33.5	49.71	24.66	−169.0	−92.0	29.49	13.86	58.80	643.3	1.15792
34.0	47.47	23.55	−141.0	−60.3	30.43	14.09	68.69	603.0	1.15045
34.5	44.74	22.19	−108.0	−22.4	31.54	14.39	83.90	559.8	1.14139
34.7	43.46	21.56	−93.0	−4.8	32.05	14.53	92.19	542.1	1.13716
34.8	42.77	21.21	−85.0	4.6	32.32	14.61	96.88	533.2	1.13489
35.0	41.28	20.48	−67.8	25.0	32.90	14.78	107.3	515.7	1.12999
35.2	39.64	19.66	−49.0	47.6	33.55	14.96	118.5	499.1	1.12463
35.4	37.87	18.79	−28.8	72.4	34.25	15.13	129.1	484.2	1.11884
35.6	36.01	17.86	−7.3	99.0	35.00	15.29	136.8	471.9	1.11278
35.8	34.14	16.94	14.6	126.8	35.78	15.41	140.0	462.6	1.10671
36.0	32.34	16.04	36.3	154.7	36.56	15.48	138.4	456.3	1.10089
36.2	30.67	15.21	57.0	181.9	37.31	15.49	133.0	452.5	1.09550
36.4	29.16	14.46	76.4	207.8	38.02	15.46	125.6	450.6	1.09065
36.6	27.81	13.79	94.4	232.1	38.69	15.40	117.4	450.0	1.08634
36.8	26.61	13.20	110.8	254.7	39.30	15.31	109.2	450.5	1.08252
37.0	25.56	12.68	125.9	275.8	39.88	15.21	101.4	451.6	1.07916
37.2	24.62	12.21	139.8	295.4	40.40	15.10	94.45	453.1	1.07618
37.4	23.78	11.80	152.6	313.6	40.89	14.99	88.21	454.9	1.07353
37.6	23.03	11.43	164.4	330.7	41.35	14.89	82.70	457.0	1.07115
37.8	22.36	11.09	175.4	346.7	41.77	14.78	77.84	459.2	1.06902
38.0	21.75	10.79	185.7	361.9	42.17	14.68	73.58	461.4	1.06708
38.2	21.19	10.51	195.4	376.2	42.55	14.58	69.81	463.8	1.06531
38.4	20.67	10.25	204.6	389.8	42.90	14.49	66.48	466.2	1.06370
39.0	19.35	9.600	229.3	427.2	43.87	14.24	58.53	473.3	1.05955
39.5	18.45	9.150	247.5	455.2	44.58	14.06	53.61	479.3	1.05670
40.0	17.67	8.763	264.2	481.0	45.23	13.91	49.75	485.0	1.05426
40.5	16.99	8.426	279.6	505.1	45.83	13.78	46.65	490.6	1.05213
41.0	16.38	8.126	293.9	527.7	46.39	13.66	44.11	496.1	1.05024
41.5	15.84	7.859	307.5	549.2	46.91	13.56	42.00	501.4	1.04856
42.0	15.35	7.616	320.3	569.8	47.40	13.48	40.22	506.6	1.04704
43.0	14.50	7.193	344.4	608.5	48.31	13.34	37.40	516.6	1.04439
44.0	13.78	6.833	366.8	644.8	49.15	13.23	35.26	526.1	1.04213
45.0	13.15	6.521	387.8	679.2	49.92	13.15	33.58	535.2	1.04018
46.0	12.59	6.246	407.9	712.1	50.64	13.09	32.25	543.9	1.03846
47.0	12.10	6.001	427.1	743.8	51.32	13.04	31.15	552.3	1.03693
48.0	11.65	5.780	445.7	774.5	51.97	13.01	30.25	560.4	1.03556
49.0	11.25	5.580	463.8	804.3	52.59	12.99	29.49	568.2	1.03431
50.0	10.88	5.397	481.4	833.5	53.17	12.98	28.84	575.7	1.03317
52.0	10.23	5.073	515.5	890.1	54.28	12.98	27.82	590.1	1.03116

Thermophysical properties of hydrogen—Continued

T K	Density kg/m^3	Density mol/dm^3	E J/mol	H J/mol	S J/(mol·K)	C_v J/(mol·K)	C_p J/(mol·K)	Sound m/s	Diel. const.
54.0	9.664	4.794	548.6	944.9	55.32	13.02	27.05	603.5	1.02943
56.0	9.173	4.550	580.8	998.4	56.29	13.08	26.47	616.2	1.02791
58.0	8.738	4.334	612.5	1051.0	57.21	13.17	26.03	628.2	1.02658
60.0	8.350	4.142	643.8	1103.0	58.09	13.28	25.70	639.5	1.02539
62.0	7.999	3.968	674.9	1154.0	58.93	13.41	25.46	650.2	1.02431
64.0	7.682	3.811	705.9	1204.0	59.73	13.56	25.30	660.3	1.02334
66.0	7.392	3.667	736.8	1255.0	60.51	13.74	25.20	669.9	1.02245
68.0	7.126	3.535	767.8	1305.0	61.26	13.93	25.15	679.0	1.02164
70.0	6.881	3.413	798.9	1356.0	61.99	14.15	25.15	687.6	1.02089
72.0	6.654	3.301	830.3	1406.0	62.70	14.38	25.20	695.9	1.02019
74.0	6.443	3.196	861.9	1456.0	63.39	14.63	25.28	703.7	1.01955
76.0	6.247	3.099	893.9	1507.0	64.07	14.90	25.39	711.3	1.01895
78.0	6.063	3.008	926.3	1558.0	64.73	15.18	25.54	718.5	1.01839
80.0	5.891	2.922	959.1	1609.0	65.38	15.47	25.71	725.5	1.01786
85.0	5.504	2.730	1043.0	1739.0	66.95	16.26	26.23	742.1	1.01668
90.0	5.169	2.564	1131.0	1872.0	68.47	17.09	26.85	757.6	1.01566
95.0	4.874	2.418	1222.0	2008.0	69.94	17.95	27.54	772.4	1.01476
100.0	4.614	2.289	1317.0	2147.0	71.37	18.82	28.26	786.7	1.01397
105.0	4.381	2.173	1416.0	2290.0	72.76	19.67	28.99	800.8	1.01326
110.0	4.172	2.069	1519.0	2437.0	74.13	20.49	29.70	814.8	1.01263
115.0	3.983	1.976	1625.0	2587.0	75.46	21.26	30.38	828.7	1.01205
120.0	3.811	1.890	1735.0	2741.0	76.77	21.95	30.99	842.7	1.01153
125.0	3.654	1.812	1849.0	2897.0	78.05	22.57	31.54	856.8	1.01105
130.0	3.509	1.741	1964.0	3056.0	79.29	23.09	32.01	871.0	1.01061
135.0	3.376	1.675	2082.0	3217.0	80.51	23.54	32.40	885.2	1.01021
140.0	3.253	1.614	2203.0	3380.0	81.69	23.93	32.74	899.4	1.00984
150.0	3.033	1.505	2447.0	3710.0	83.97	24.45	33.19	928.1	1.00917
160.0	2.842	1.410	2695.0	4043.0	86.12	24.71	33.38	956.9	1.00859
170.0	2.674	1.326	2944.0	4377.0	88.14	24.75	33.38	985.6	1.00808
180.0	2.525	1.252	3193.0	4710.0	90.05	24.64	33.23	1014.0	1.00763
190.0	2.392	1.186	3439.0	5041.0	91.84	24.43	32.98	1043.0	1.00723
200.0	2.272	1.127	3683.0	5369.0	93.52	24.15	32.66	1071.0	1.00686
210.0	2.164	1.074	3924.0	5694.0	95.11	23.83	32.33	1099.0	1.00654
220.0	2.066	1.025	4162.0	6016.0	96.60	23.51	31.98	1126.0	1.00624
230.0	1.977	.9806	4396.0	6334.0	98.02	23.19	31.65	1153.0	1.00597
240.0	1.895	.9399	4627.0	6649.0	99.36	22.89	31.33	1179.0	1.00572
250.0	1.819	.9025	4855.0	6960.0	100.6	22.62	31.04	1205.0	1.00549
260.0	1.750	.8680	5081.0	7270.0	101.8	22.37	30.78	1230.0	1.00528
270.0	1.686	.8361	5304.0	7576.0	103.0	22.15	30.55	1255.0	1.00509
280.0	1.626	.8065	5525.0	7881.0	104.1	21.96	30.35	1279.0	1.00491
300.0	1.518	.7531	5961.0	8484.0	106.2	21.65	30.03	1325.0	1.00458
320.0	1.424	.7064	6393.0	9082.0	108.1	21.43	29.80	1369.0	1.00430
340.0	1.341	.6651	6820.0	9677.0	109.9	21.28	29.63	1412.0	1.00405
360.0	1.267	.6285	7245.0	10270.0	111.6	21.17	29.52	1453.0	1.00382
380.0	1.201	.5957	7668.0	10860.0	113.2	21.10	29.44	1493.0	1.00362
400.0	1.141	.5661	8089.0	11450.0	114.7	21.04	29.38	1532.0	1.00344
				2.00 MPa isobar					
14.46[a]	78.01	38.70	−619.9	−568.2	10.17	10.86	14.22	1329.0	1.25507
15.0	77.62	38.50	−612.5	−560.6	10.69	10.36	13.90	1317.0	1.25368
16.0	76.86	38.13	−599.1	−546.7	11.59	10.13	14.06	1294.0	1.25102
18.0	75.23	37.32	−570.7	−517.1	13.32	10.66	15.66	1246.0	1.24530
20.0	73.42	36.42	−538.7	−483.8	15.08	11.38	17.70	1198.0	1.23895
22.0	71.39	35.41	−502.7	−446.2	16.87	11.95	19.86	1148.0	1.23189
24.0	69.12	34.29	−462.5	−404.2	18.69	12.35	22.20	1093.0	1.22398
26.0	66.53	33.00	−417.8	−357.2	20.57	12.64	24.90	1030.0	1.21503
27.0	65.09	32.29	−393.4	−331.5	21.54	12.76	26.49	994.6	1.21006
28.0	63.52	31.51	−367.6	−304.1	22.54	12.86	28.33	956.6	1.20469
29.0	61.81	30.66	−339.9	−274.7	23.57	12.97	30.53	915.4	1.19884
30.0	59.92	29.72	−310.2	−242.9	24.65	13.08	33.23	870.2	1.19237
31.0	57.78	28.66	−277.8	−208.0	25.79	13.21	36.73	820.4	1.18511

Thermophysical properties of hydrogen—Continued

T K	Density kg/m^3	Density mol/dm^3	E J/mol	H J/mol	S J/(mol·K)	C_v J/(mol·K)	C_p J/(mol·K)	Sound m/s	Diel. const.
31.5	56.60	28.08	–260.3	–189.1	26.40	13.28	38.93	793.5	1.18110
32.0	55.32	27.44	–241.9	–169.0	27.03	13.37	41.56	765.1	1.17678
32.5	53.93	26.75	–222.2	–147.4	27.70	13.47	44.78	734.9	1.17207
33.0	52.39	25.99	–201.0	–124.1	28.41	13.59	48.84	702.8	1.16689
33.5	50.66	25.13	–178.0	–98.4	29.18	13.73	54.14	668.6	1.16110
34.0	48.68	24.15	–152.4	–69.6	30.04	13.91	61.34	632.0	1.15450
34.5	46.37	23.00	–123.5	–36.5	31.00	14.14	71.54	593.1	1.14680
35.0	43.58	21.62	–89.8	2.7	32.13	14.44	86.26	552.8	1.13757
35.2	42.30	20.98	–74.7	20.7	32.64	14.57	93.62	536.8	1.13335
35.4	40.91	20.30	–58.4	40.2	33.20	14.71	101.6	521.4	1.12879
35.6	39.43	19.56	–41.0	61.3	33.79	14.86	109.6	507.0	1.12392
36.0	36.23	17.97	–3.4	107.8	35.09	15.12	121.8	483.4	1.11350
36.5	32.28	16.01	44.6	169.5	36.79	15.29	122.0	466.6	1.10069
36.7	30.84	15.30	62.8	193.5	37.45	15.29	118.0	463.4	1.09607
36.8	30.17	14.97	71.6	205.2	37.77	15.28	115.4	462.3	1.09391
37.0	28.91	14.34	88.3	227.7	38.38	15.24	109.8	461.1	1.08987
37.2	27.77	13.78	103.9	249.1	38.95	15.18	103.8	460.9	1.08622
37.4	26.74	13.27	118.5	269.3	39.49	15.10	97.88	461.3	1.08293
37.6	25.81	12.80	132.1	288.3	40.00	15.01	92.24	462.2	1.07997
37.8	24.97	12.39	144.7	306.2	40.47	14.92	87.00	463.6	1.07730
38.0	24.21	12.01	156.6	323.1	40.92	14.83	82.22	465.1	1.07487
38.2	23.51	11.66	167.6	339.1	41.34	14.74	77.89	466.9	1.07267
38.4	22.88	11.35	178.1	354.3	41.74	14.64	73.98	468.8	1.07066
38.6	22.30	11.06	187.9	368.7	42.11	14.56	70.47	470.9	1.06882
39.0	21.26	10.55	206.1	395.7	42.81	14.39	64.47	475.1	1.06556
39.5	20.17	10.00	226.4	426.4	43.59	14.20	58.51	480.6	1.06211
40.0	19.24	9.543	244.8	454.4	44.29	14.04	53.84	486.0	1.05919
40.5	18.44	9.145	261.6	480.3	44.94	13.89	50.10	491.5	1.05667
41.0	17.73	8.795	277.2	504.6	45.53	13.77	47.07	496.8	1.05446
41.5	17.11	8.485	291.8	527.5	46.09	13.66	44.56	502.1	1.05250
42.0	16.54	8.207	305.5	549.2	46.61	13.56	42.46	507.2	1.05075
42.5	16.04	7.955	318.6	570.0	47.10	13.48	40.68	512.2	1.04916
43.0	15.57	7.725	331.1	590.0	47.57	13.41	39.15	517.1	1.04772
44.0	14.75	7.319	354.6	627.8	48.44	13.29	36.67	526.6	1.04517
45.0	14.05	6.970	376.5	663.5	49.24	13.20	34.76	535.7	1.04298
46.0	13.43	6.664	397.4	697.5	49.99	13.13	33.24	544.4	1.04107
47.0	12.89	6.394	417.3	730.1	50.69	13.08	32.01	552.8	1.03938
48.0	12.40	6.151	436.4	761.5	51.35	13.04	30.99	560.9	1.03787
49.0	11.96	5.932	454.9	792.1	51.98	13.02	30.14	568.7	1.03650
50.0	11.56	5.732	473.0	821.9	52.58	13.00	29.43	576.3	1.03526
52.0	10.85	5.380	507.8	879.5	53.71	13.00	28.29	590.7	1.03307
54.0	10.24	5.078	541.4	935.2	54.76	13.03	27.44	604.2	1.03119
56.0	9.708	4.816	574.1	989.4	55.75	13.09	26.80	617.0	1.02956
58.0	9.241	4.584	606.2	1043.0	56.68	13.18	26.32	629.0	1.02812
60.0	8.824	4.377	637.9	1095.0	57.57	13.29	25.95	640.3	1.02684
62.0	8.450	4.192	669.3	1146.0	58.41	13.42	25.69	651.0	1.02569
64.0	8.111	4.023	700.5	1198.0	59.23	13.57	25.50	661.1	1.02465
66.0	7.802	3.870	731.7	1248.0	60.01	13.75	25.38	670.7	1.02371
68.0	7.518	3.729	762.9	1299.0	60.77	13.94	25.32	679.8	1.02284
70.0	7.257	3.600	794.2	1350.0	61.50	14.16	25.31	688.5	1.02204
72.0	7.016	3.480	825.8	1400.0	62.21	14.39	25.34	696.7	1.02130
74.0	6.793	3.369	857.6	1451.0	62.91	14.64	25.41	704.6	1.02062
76.0	6.584	3.266	889.7	1502.0	63.59	14.91	25.52	712.2	1.01998
78.0	6.389	3.169	922.2	1553.0	64.25	15.19	25.65	719.4	1.01939
80.0	6.207	3.079	955.2	1605.0	64.90	15.48	25.82	726.4	1.01883
85.0	5.797	2.876	1040.0	1735.0	66.48	16.27	26.32	742.9	1.01758
90.0	5.442	2.700	1127.0	1868.0	68.00	17.10	26.93	758.5	1.01650
95.0	5.131	2.545	1219.0	2004.0	69.48	17.96	27.61	773.3	1.01555
100.0	4.856	2.409	1314.0	2144.0	70.91	18.83	28.32	787.6	1.01471
105.0	4.611	2.287	1413.0	2288.0	72.31	19.68	29.04	801.7	1.01396
110.0	4.390	2.178	1516.0	2435.0	73.68	20.50	29.75	815.7	1.01329
115.0	4.191	2.079	1623.0	2585.0	75.02	21.26	30.42	829.6	1.01268
120.0	4.009	1.989	1733.0	2739.0	76.32	21.96	31.04	843.6	1.01213

Thermophysical properties of hydrogen—Continued

T K	Density kg/m³	Density mol/dm³	E J/mol	H J/mol	S J/(mol·K)	C_v J/(mol·K)	C_p J/(mol·K)	Sound m/s	Diel. const.
125.0	3.844	1.907	1846.0	2895.0	77.60	22.57	31.58	857.7	1.01163
130.0	3.692	1.831	1962.0	3054.0	78.85	23.10	32.04	871.9	1.01117
135.0	3.552	1.762	2081.0	3216.0	80.07	23.55	32.44	886.1	1.01074
140.0	3.423	1.698	2201.0	3379.0	81.25	23.93	32.77	900.3	1.01035
150.0	3.191	1.583	2445.0	3709.0	83.53	24.45	33.21	929.0	1.00965
160.0	2.989	1.483	2693.0	4042.0	85.68	24.71	33.41	957.8	1.00904
170.0	2.812	1.395	2943.0	4376.0	87.71	24.76	33.40	986.6	1.00850
180.0	2.656	1.317	3191.0	4710.0	89.61	24.65	33.24	1015.0	1.00803
190.0	2.516	1.248	3438.0	5041.0	91.40	24.43	32.99	1044.0	1.00760
200.0	2.390	1.186	3682.0	5369.0	93.09	24.15	32.68	1072.0	1.00722
210.0	2.277	1.129	3923.0	5694.0	94.68	23.84	32.34	1100.0	1.00688
220.0	2.173	1.078	4161.0	6016.0	96.17	23.51	31.99	1127.0	1.00656
230.0	2.079	1.031	4395.0	6334.0	97.59	23.20	31.66	1154.0	1.00628
240.0	1.993	.9887	4626.0	6649.0	98.93	22.90	31.34	1180.0	1.00602
250.0	1.914	.9494	4855.0	6961.0	100.2	22.62	31.05	1206.0	1.00578
260.0	1.841	.9131	5080.0	7270.0	101.4	22.37	30.79	1231.0	1.00556
270.0	1.773	.8796	5303.0	7577.0	102.6	22.15	30.56	1255.0	1.00535
280.0	1.710	.8484	5524.0	7882.0	103.7	21.96	30.36	1279.0	1.00516
300.0	1.597	.7922	5961.0	8485.0	105.8	21.65	30.03	1326.0	1.00482
320.0	1.498	.7431	6392.0	9084.0	107.7	21.43	29.80	1370.0	1.00452
340.0	1.411	.6998	6820.0	9678.0	109.5	21.28	29.64	1413.0	1.00426
360.0	1.333	.6612	7245.0	10270.0	111.2	21.17	29.52	1454.0	1.00402
380.0	1.263	.6267	7668.0	10860.0	112.8	21.10	29.44	1494.0	1.00381
400.0	1.201	.5956	8089.0	11450.0	114.3	21.04	29.38	1532.0	1.00362
				2.20 MPa isobar					
14.52[a]	78.12	38.75	–619.6	–562.8	10.18	10.84	14.12	1328.0	1.25543
15.0	77.77	38.58	–613.1	–556.1	10.64	10.41	13.85	1318.0	1.25422
16.0	77.03	38.21	–599.8	–542.3	11.53	10.15	14.01	1297.0	1.25160
18.0	75.42	37.41	–571.6	–512.8	13.26	10.67	15.57	1252.0	1.24596
20.0	73.63	36.53	–539.9	–479.7	15.01	11.38	17.59	1206.0	1.23970
22.0	71.64	35.54	–504.3	–442.4	16.78	11.94	19.70	1158.0	1.23276
24.0	69.41	34.43	–464.7	–400.8	18.59	12.34	21.96	1105.0	1.22501
26.0	66.89	33.18	–420.7	–354.4	20.45	12.62	24.54	1044.0	1.21628
27.0	65.50	32.49	–396.8	–329.1	21.40	12.74	26.03	1011.0	1.21147
28.0	63.99	31.74	–371.5	–302.2	22.38	12.84	27.73	975.0	1.20630
29.0	62.36	30.93	–344.7	–273.5	23.39	12.94	29.71	936.1	1.20069
30.0	60.56	30.04	–315.9	–242.7	24.43	13.04	32.09	893.9	1.19456
31.0	58.57	29.05	–284.9	–209.2	25.53	13.15	35.06	848.0	1.18777
32.0	56.31	27.93	–251.0	–172.3	26.70	13.28	38.94	797.6	1.18012
32.5	55.06	27.31	–232.7	–152.2	27.32	13.35	41.39	770.6	1.17589
33.0	53.70	26.64	–213.4	–130.8	27.98	13.44	44.32	742.2	1.17132
33.5	52.22	25.90	–192.7	–107.8	28.67	13.55	47.91	712.3	1.16633
34.0	50.58	25.09	–170.4	–82.7	29.41	13.67	52.39	680.9	1.16084
34.5	48.75	24.18	–146.1	–55.2	30.22	13.82	58.13	648.0	1.15472
35.0	46.67	23.15	–119.3	–24.3	31.10	14.01	65.59	613.9	1.14780
35.5	44.28	21.96	–89.4	10.8	32.10	14.22	75.22	579.3	1.13988
36.0	41.53	20.60	–55.6	51.2	33.23	14.47	86.71	546.2	1.13081
36.5	38.44	19.07	–18.0	97.4	34.50	14.71	97.43	517.9	1.12070
37.0	35.22	17.47	21.8	147.7	35.87	14.89	102.5	497.8	1.11022
37.5	32.18	15.96	60.8	198.6	37.24	14.95	99.90	486.6	1.10038
37.7	31.07	15.41	75.6	218.3	37.76	14.94	97.30	484.1	1.09679
37.8	30.54	15.15	82.7	228.0	38.02	14.93	95.79	483.3	1.09509
38.0	29.53	14.65	96.6	246.8	38.52	14.90	92.50	482.1	1.09185
38.2	28.59	14.18	109.8	264.9	38.99	14.85	89.01	481.7	1.08885
38.4	27.73	13.75	122.4	282.4	39.45	14.79	85.47	481.7	1.08607
38.6	26.92	13.35	134.4	299.1	39.88	14.73	81.98	482.1	1.08350
38.8	26.17	12.98	145.8	315.2	40.30	14.66	78.61	482.9	1.08112
39.0	25.48	12.64	156.6	330.6	40.69	14.59	75.41	483.9	1.07892
39.5	23.96	11.89	181.4	366.5	41.61	14.42	68.26	487.4	1.07409
40.0	22.68	11.25	203.5	399.1	42.43	14.26	62.32	491.4	1.07003
40.5	21.59	10.71	223.5	429.0	43.17	14.10	57.43	495.8	1.06659

Thermophysical properties of hydrogen—Continued

T K	Density kg/m^3	Density mol/dm^3	E J/mol	H J/mol	S J/(mol·K)	C_v J/(mol·K)	C_p J/(mol·K)	Sound m/s	Diel. const.
41.0	20.65	10.24	241.8	456.6	43.85	13.97	53.41	500.5	1.06361
41.5	19.82	9.831	258.7	482.5	44.48	13.84	50.08	505.2	1.06101
42.0	19.09	9.468	274.4	506.8	45.06	13.73	47.30	510.0	1.05871
42.5	18.43	9.143	289.2	529.8	45.60	13.64	44.94	514.7	1.05665
43.0	17.84	8.849	303.2	551.8	46.12	13.55	42.94	519.4	1.05480
43.5	17.30	8.583	316.5	572.8	46.60	13.47	41.22	524.1	1.05312
44.0	16.81	8.339	329.2	593.1	47.07	13.41	39.72	528.6	1.05158
45.0	15.94	7.906	353.2	631.5	47.93	13.30	37.27	537.6	1.04886
46.0	15.19	7.533	375.7	667.8	48.73	13.21	35.35	546.2	1.04651
47.0	14.53	7.206	397.0	702.3	49.47	13.15	33.81	554.5	1.04446
48.0	13.94	6.915	417.3	735.5	50.17	13.10	32.55	562.6	1.04264
49.0	13.41	6.654	436.9	767.5	50.83	13.07	31.51	570.4	1.04101
50.0	12.94	6.419	455.8	798.6	51.46	13.05	30.64	578.0	1.03954
51.0	12.51	6.204	474.2	828.8	52.06	13.04	29.90	585.3	1.03820
52.0	12.11	6.007	492.1	858.4	52.63	13.04	29.26	592.4	1.03697
54.0	11.40	5.657	526.9	915.8	53.72	13.06	28.25	605.9	1.03479
56.0	10.79	5.354	560.7	971.6	54.73	13.12	27.49	618.7	1.03290
58.0	10.26	5.089	593.6	1026.0	55.68	13.20	26.91	630.7	1.03125
60.0	9.784	4.853	626.0	1079.0	56.59	13.31	26.47	642.1	1.02979
62.0	9.358	4.642	658.0	1132.0	57.45	13.44	26.15	652.8	1.02848
64.0	8.974	4.452	689.7	1184.0	58.28	13.59	25.91	663.0	1.02731
66.0	8.626	4.279	721.4	1236.0	59.07	13.76	25.75	672.6	1.02623
68.0	8.307	4.121	753.1	1287.0	59.84	13.96	25.66	681.7	1.02526
70.0	8.014	3.975	784.8	1338.0	60.58	14.17	25.62	690.3	1.02436
72.0	7.744	3.841	816.8	1389.0	61.30	14.41	25.63	698.6	1.02353
74.0	7.493	3.717	848.9	1441.0	62.00	14.66	25.68	706.5	1.02276
76.0	7.260	3.601	881.4	1492.0	62.69	14.92	25.76	714.0	1.02205
78.0	7.043	3.494	914.2	1544.0	63.36	15.20	25.88	721.3	1.02138
80.0	6.840	3.393	947.4	1596.0	64.02	15.50	26.03	728.3	1.02076
85.0	6.384	3.167	1032.0	1727.0	65.61	16.28	26.51	744.8	1.01937
90.0	5.990	2.971	1121.0	1861.0	67.14	17.11	27.09	760.3	1.01817
95.0	5.645	2.800	1212.0	1998.0	68.62	17.97	27.75	775.1	1.01711
100.0	5.341	2.649	1308.0	2139.0	70.06	18.84	28.45	789.5	1.01619
105.0	5.070	2.515	1408.0	2283.0	71.47	19.69	29.16	803.6	1.01536
110.0	4.826	2.394	1511.0	2430.0	72.84	20.51	29.85	817.6	1.01462
115.0	4.606	2.285	1618.0	2581.0	74.18	21.27	30.52	831.5	1.01395
120.0	4.406	2.186	1729.0	2735.0	75.50	21.97	31.12	845.5	1.01334
125.0	4.224	2.095	1842.0	2892.0	76.78	22.58	31.66	859.6	1.01279
130.0	4.056	2.012	1958.0	3052.0	78.03	23.11	32.11	873.8	1.01228
135.0	3.902	1.936	2077.0	3213.0	79.25	23.56	32.50	888.0	1.01181
140.0	3.760	1.865	2197.0	3377.0	80.44	23.94	32.84	902.2	1.01138
150.0	3.505	1.739	2442.0	3707.0	82.72	24.46	33.26	930.9	1.01060
160.0	3.284	1.629	2690.0	4041.0	84.87	24.72	33.45	959.7	1.00993
170.0	3.089	1.532	2940.0	4376.0	86.90	24.77	33.44	988.4	1.00934
180.0	2.917	1.447	3189.0	4709.0	88.81	24.66	33.28	1017.0	1.00882
190.0	2.763	1.371	3436.0	5041.0	90.60	24.44	33.02	1046.0	1.00835
200.0	2.625	1.302	3680.0	5370.0	92.29	24.16	32.71	1074.0	1.00793
210.0	2.501	1.240	3921.0	5695.0	93.87	23.84	32.36	1101.0	1.00756
220.0	2.387	1.184	4159.0	6017.0	95.37	23.52	32.01	1129.0	1.00721
230.0	2.284	1.133	4394.0	6335.0	96.79	23.20	31.68	1156.0	1.00690
240.0	2.190	1.086	4625.0	6650.0	98.13	22.90	31.36	1182.0	1.00661
250.0	2.103	1.043	4853.0	6963.0	99.40	22.63	31.07	1207.0	1.00635
260.0	2.022	1.003	5079.0	7272.0	100.6	22.38	30.81	1233.0	1.00611
270.0	1.948	.9663	5302.0	7579.0	101.8	22.16	30.57	1257.0	1.00588
280.0	1.879	.9320	5523.0	7883.0	102.9	21.97	30.37	1281.0	1.00567
300.0	1.755	.8704	5960.0	8487.0	105.0	21.66	30.04	1328.0	1.00530
320.0	1.646	.8165	6391.0	9086.0	106.9	21.44	29.81	1372.0	1.00497
340.0	1.550	.7689	6819.0	9680.0	108.7	21.28	29.64	1415.0	1.00468
360.0	1.465	.7266	7244.0	10270.0	110.4	21.18	29.53	1456.0	1.00442
380.0	1.388	.6887	7667.0	10860.0	112.0	21.10	29.45	1495.0	1.00419
400.0	1.320	.6545	8089.0	11450.0	113.5	21.05	29.39	1534.0	1.00398

Thermophysical properties of hydrogen—Continued

T K	Density kg/m^3	Density mol/dm^3	E J/mol	H J/mol	S J/(mol·K)	C_v J/(mol·K)	C_p J/(mol·K)	Sound m/s	Diel. const.
				2.40 MPa isobar					
14.58[a]	78.22	38.80	−619.3	−557.4	10.20	10.81	14.03	1327.0	1.25580
15.0	77.92	38.65	−613.7	−551.7	10.59	10.44	13.80	1320.0	1.25476
16.0	77.19	38.29	−600.5	−537.8	11.48	10.17	13.95	1300.0	1.25217
18.0	75.60	37.50	−572.6	−508.6	13.20	10.67	15.49	1258.0	1.24661
20.0	73.84	36.63	−541.1	−475.6	14.94	11.37	17.48	1214.0	1.24044
22.0	71.89	35.66	−505.9	−438.6	16.70	11.93	19.55	1168.0	1.23360
24.0	69.70	34.57	−466.7	−397.3	18.49	12.33	21.74	1116.0	1.22600
26.0	67.24	33.35	−423.4	−351.4	20.33	12.61	24.21	1059.0	1.21749
28.0	64.43	31.96	−375.2	−300.2	22.23	12.82	27.19	992.4	1.20782
29.0	62.86	31.18	−349.0	−272.1	23.21	12.91	29.00	955.6	1.20243
30.0	61.15	30.34	−321.2	−242.0	24.23	13.00	31.13	915.9	1.19658
31.0	59.27	29.40	−291.3	−209.7	25.29	13.10	33.71	873.0	1.19017
32.0	57.18	28.36	−259.0	−174.4	26.41	13.21	36.95	826.6	1.18306
33.0	54.81	27.19	−223.7	−135.4	27.61	13.34	41.18	776.3	1.17504
34.0	52.06	25.83	−184.4	−91.5	28.92	13.51	47.02	721.7	1.16581
35.0	48.80	24.21	−139.7	−40.5	30.40	13.74	55.50	662.9	1.15488
35.5	46.90	23.26	−114.6	−11.4	31.23	13.89	61.19	632.5	1.14856
36.0	44.79	22.22	−87.2	20.9	32.13	14.06	68.01	602.2	1.14156
36.5	42.44	21.05	−57.3	56.7	33.12	14.24	75.64	573.3	1.13380
37.0	39.88	19.78	−24.9	96.4	34.20	14.42	82.88	547.7	1.12539
38.0	34.55	17.14	43.4	183.4	36.52	14.65	88.50	513.6	1.10805
38.5	32.10	15.92	76.4	227.1	37.66	14.66	85.64	505.6	1.10012
39.0	29.93	14.85	107.0	268.7	38.73	14.60	80.60	502.0	1.09312
39.5	28.04	13.91	135.0	307.6	39.72	14.50	74.82	501.4	1.08708
40.0	26.42	13.10	160.4	343.5	40.63	14.38	69.15	502.5	1.08190
40.5	25.02	12.41	183.4	376.8	41.45	14.25	63.99	504.9	1.07746
41.0	23.81	11.81	204.5	407.6	42.21	14.12	59.46	508.1	1.07362
41.5	22.76	11.29	223.8	436.4	42.91	13.99	55.56	511.7	1.07028
42.0	21.83	10.83	241.7	463.3	43.55	13.88	52.22	515.6	1.06734
42.5	21.00	10.42	258.3	488.7	44.15	13.77	49.36	519.7	1.06474
43.0	20.27	10.05	274.0	512.7	44.72	13.68	46.91	523.9	1.06241
43.5	19.60	9.722	288.8	535.6	45.25	13.60	44.79	528.2	1.06032
44.0	18.99	9.421	302.8	557.6	45.75	13.52	42.95	532.4	1.05841
45.0	17.93	8.893	329.1	598.9	46.68	13.40	39.93	540.9	1.05508
46.0	17.02	8.443	353.4	637.6	47.53	13.30	37.58	549.2	1.05224
47.0	16.23	8.052	376.2	674.2	48.32	13.22	35.71	557.3	1.04978
48.0	15.54	7.708	397.8	709.2	49.05	13.17	34.19	565.2	1.04761
49.0	14.92	7.401	418.4	742.7	49.74	13.12	32.94	572.9	1.04569
50.0	14.37	7.126	438.3	775.1	50.40	13.10	31.90	580.4	1.04396
51.0	13.86	6.877	457.5	806.6	51.02	13.08	31.02	587.6	1.04240
52.0	13.40	6.649	476.2	837.2	51.61	13.08	30.27	594.7	1.04098
54.0	12.59	6.247	512.3	896.5	52.73	13.09	29.08	608.1	1.03847
56.0	11.90	5.902	547.1	953.7	53.77	13.14	28.19	620.9	1.03631
58.0	11.29	5.601	580.9	1009.0	54.75	13.22	27.51	632.9	1.03444
60.0	10.75	5.335	614.0	1064.0	55.67	13.33	27.00	644.2	1.03278
62.0	10.28	5.097	646.6	1117.0	56.55	13.45	26.61	654.9	1.03131
64.0	9.846	4.884	679.0	1170.0	57.39	13.61	26.33	665.0	1.02998
66.0	9.456	4.691	711.2	1223.0	58.20	13.78	26.13	674.6	1.02879
68.0	9.100	4.514	743.3	1275.0	58.98	13.97	26.00	683.7	1.02769
70.0	8.774	4.353	775.5	1327.0	59.73	14.19	25.93	692.4	1.02669
72.0	8.474	4.204	807.8	1379.0	60.46	14.42	25.91	700.6	1.02577
74.0	8.196	4.066	840.3	1431.0	61.17	14.67	25.94	708.5	1.02492
76.0	7.939	3.938	873.0	1482.0	61.86	14.94	26.01	716.0	1.02413
78.0	7.698	3.819	906.1	1535.0	62.54	15.22	26.11	723.3	1.02339
80.0	7.474	3.707	939.6	1587.0	63.20	15.51	26.25	730.3	1.02270
85.0	6.971	3.458	1025.0	1719.0	64.81	16.29	26.69	746.8	1.02116
90.0	6.538	3.243	1114.0	1854.0	66.35	17.13	27.25	762.3	1.01984
95.0	6.159	3.055	1206.0	1992.0	67.84	17.99	27.89	777.1	1.01868
100.0	5.825	2.890	1302.0	2133.0	69.29	18.85	28.57	791.4	1.01766
105.0	5.528	2.742	1402.0	2278.0	70.70	19.71	29.27	805.5	1.01676
110.0	5.261	2.610	1506.0	2426.0	72.07	20.52	29.96	819.4	1.01594

Thermophysical properties of hydrogen—Continued

T K	Density kg/m³	Density mol/dm³	E J/mol	H J/mol	S J/(mol·K)	C_v J/(mol·K)	C_p	Sound m/s	Diel. const.
115.0	5.020	2.490	1613.0	2577.0	73.42	21.29	30.61	833.4	1.01521
120.0	4.802	2.382	1724.0	2732.0	74.74	21.98	31.20	847.4	1.01454
125.0	4.603	2.283	1838.0	2889.0	76.02	22.59	31.73	861.5	1.01394
130.0	4.420	2.193	1954.0	3049.0	77.27	23.12	32.18	875.6	1.01338
135.0	4.252	2.109	2073.0	3211.0	78.50	23.57	32.57	889.9	1.01287
140.0	4.096	2.032	2193.0	3374.0	79.69	23.95	32.89	904.1	1.01240
150.0	3.819	1.894	2439.0	3706.0	81.97	24.47	33.32	932.8	1.01155
160.0	3.577	1.774	2687.0	4040.0	84.13	24.73	33.50	961.6	1.01082
170.0	3.365	1.669	2937.0	4375.0	86.16	24.78	33.48	990.3	1.01018
180.0	3.178	1.576	3186.0	4709.0	88.07	24.66	33.31	1019.0	1.00961
190.0	3.010	1.493	3434.0	5041.0	89.86	24.45	33.05	1047.0	1.00910
200.0	2.860	1.419	3678.0	5370.0	91.55	24.16	32.73	1076.0	1.00864
210.0	2.724	1.351	3919.0	5695.0	93.14	23.85	32.39	1103.0	1.00823
220.0	2.601	1.290	4157.0	6018.0	94.64	23.52	32.04	1131.0	1.00786
230.0	2.488	1.234	4392.0	6336.0	96.06	23.21	31.70	1157.0	1.00752
240.0	2.385	1.183	4623.0	6652.0	97.40	22.91	31.38	1184.0	1.00721
250.0	2.291	1.136	4852.0	6964.0	98.67	22.63	31.08	1209.0	1.00692
260.0	2.203	1.093	5077.0	7273.0	99.89	22.38	30.82	1234.0	1.00665
270.0	2.122	1.053	5301.0	7580.0	101.0	22.16	30.59	1259.0	1.00641
280.0	2.047	1.016	5522.0	7885.0	102.2	21.97	30.38	1283.0	1.00618
300.0	1.912	.9484	5959.0	8489.0	104.2	21.66	30.06	1329.0	1.00577
320.0	1.794	.8897	6391.0	9088.0	106.2	21.44	29.82	1374.0	1.00542
340.0	1.689	.8379	6818.0	9683.0	108.0	21.29	29.65	1416.0	1.00510
360.0	1.596	.7918	7243.0	10270.0	109.7	21.18	29.53	1457.0	1.00482
380.0	1.513	.7505	7666.0	10860.0	111.3	21.10	29.45	1497.0	1.00457
400.0	1.438	.7134	8088.0	11450.0	112.8	21.05	29.39	1535.0	1.00434

2.60 MPa isobar

T K	Density kg/m³	Density mol/dm³	E J/mol	H J/mol	S J/(mol·K)	C_v J/(mol·K)	C_p	Sound m/s	Diel. const.
14.65[a]	78.32	38.85	−619.0	−552.1	10.21	10.78	13.93	1327.0	1.25616
15.0	78.07	38.73	−614.3	−547.2	10.54	10.48	13.76	1321.0	1.25529
16.0	77.35	38.37	−601.2	−533.4	11.43	10.19	13.89	1303.0	1.25274
18.0	75.79	37.59	−573.4	−504.3	13.15	10.67	15.42	1264.0	1.24725
20.0	74.05	36.73	−542.3	−471.5	14.87	11.37	17.37	1222.0	1.24116
22.0	72.12	35.78	−507.4	−434.7	16.62	11.93	19.41	1177.0	1.23443
24.0	69.98	34.71	−468.7	−393.8	18.40	12.32	21.54	1128.0	1.22697
26.0	67.58	33.52	−426.0	−348.4	20.22	12.60	23.90	1072.0	1.21865
28.0	64.85	32.17	−378.7	−297.9	22.09	12.80	26.71	1009.0	1.20926
29.0	63.34	31.42	−353.1	−270.4	23.05	12.89	28.38	973.8	1.20407
30.0	61.71	30.61	−326.0	−241.1	24.05	12.97	30.31	936.3	1.19847
31.0	59.92	29.72	−297.1	−209.7	25.08	13.06	32.59	896.1	1.19238
32.0	57.96	28.75	−266.2	−175.7	26.15	13.15	35.36	852.9	1.18570
33.0	55.77	27.66	−232.7	−138.7	27.29	13.26	38.84	806.6	1.17828
34.0	53.29	26.44	−196.1	−97.7	28.51	13.40	43.36	756.9	1.16993
35.0	50.44	25.02	−155.4	−51.5	29.86	13.57	49.45	704.2	1.16035
36.0	47.08	23.35	−109.4	1.9	31.36	13.79	57.78	649.6	1.14916
36.5	45.17	22.41	−83.9	32.1	32.19	13.92	62.86	622.6	1.14283
37.0	43.10	21.38	−56.7	64.9	33.08	14.06	68.29	597.0	1.13600
38.0	38.59	19.14	2.3	138.1	35.03	14.32	77.29	554.5	1.12119
39.0	34.08	16.91	62.9	216.7	37.08	14.43	78.26	529.7	1.10652
39.5	32.04	15.89	91.6	255.2	38.06	14.41	75.67	523.7	1.09992
40.0	30.20	14.98	118.6	292.1	38.99	14.36	71.99	520.6	1.09399
40.5	28.56	14.17	143.6	327.1	39.86	14.28	67.90	519.8	1.08875
41.0	27.12	13.45	166.8	360.0	40.66	14.18	63.81	520.5	1.08414
41.5	25.85	12.82	188.2	391.0	41.41	14.08	59.98	522.3	1.08008
42.0	24.72	12.26	208.1	420.1	42.11	13.98	56.50	524.8	1.07650
42.5	23.72	11.77	226.6	447.5	42.76	13.88	53.40	527.7	1.07332
43.0	22.82	11.32	243.9	473.5	43.37	13.79	50.66	531.0	1.07048
43.5	22.02	10.92	260.2	498.2	43.94	13.70	48.25	534.6	1.06793
44.0	21.28	10.56	275.6	521.8	44.48	13.62	46.13	538.2	1.06562
44.5	20.62	10.23	290.2	544.4	44.99	13.55	44.26	542.0	1.06353
45.0	20.01	9.926	304.2	566.1	45.48	13.49	42.61	545.8	1.06161
46.0	18.93	9.390	330.4	607.3	46.38	13.38	39.85	553.5	1.05822

Thermophysical properties of hydrogen—Continued

T K	Density kg/m^3	Density mol/dm^3	E J/mol	H J/mol	S J/(mol·K)	C_v J/(mol·K)	C_p J/(mol·K)	Sound m/s	Diel. const.
47.0	18.00	8.930	354.9	646.0	47.21	13.29	37.65	561.2	1.05531
48.0	17.19	8.527	377.8	682.7	47.99	13.23	35.87	568.8	1.05277
49.0	16.47	8.171	399.7	717.9	48.71	13.18	34.41	576.2	1.05053
50.0	15.83	7.853	420.6	751.6	49.39	13.14	33.19	583.5	1.04852
51.0	15.25	7.566	440.7	784.3	50.04	13.12	32.17	590.6	1.04672
52.0	14.73	7.305	460.1	816.0	50.66	13.11	31.31	597.5	1.04509
53.0	14.25	7.067	479.1	847.0	51.25	13.11	30.57	604.3	1.04359
54.0	13.81	6.848	497.5	877.2	51.81	13.12	29.93	610.8	1.04222
56.0	13.02	6.458	533.4	936.0	52.88	13.17	28.90	623.4	1.03978
58.0	12.34	6.119	568.1	993.0	53.88	13.24	28.12	635.4	1.03767
60.0	11.74	5.822	602.0	1049.0	54.82	13.35	27.53	646.6	1.03581
62.0	11.20	5.557	635.3	1103.0	55.72	13.47	27.08	657.3	1.03417
64.0	10.72	5.320	668.2	1157.0	56.57	13.62	26.74	667.3	1.03269
66.0	10.29	5.105	700.9	1210.0	57.39	13.80	26.50	676.9	1.03136
68.0	9.899	4.910	733.5	1263.0	58.18	13.99	26.34	686.0	1.03015
70.0	9.539	4.732	766.1	1316.0	58.94	14.20	26.24	694.6	1.02904
72.0	9.208	4.567	798.8	1368.0	59.68	14.44	26.20	702.8	1.02802
74.0	8.902	4.416	831.6	1420.0	60.40	14.69	26.20	710.6	1.02708
76.0	8.619	4.275	864.7	1473.0	61.10	14.95	26.25	718.2	1.02621
78.0	8.355	4.144	898.1	1525.0	61.78	15.23	26.34	725.4	1.02540
80.0	8.109	4.022	931.9	1578.0	62.45	15.53	26.46	732.4	1.02465
85.0	7.559	3.750	1018.0	1712.0	64.06	16.31	26.87	748.8	1.02296
90.0	7.085	3.515	1107.0	1847.0	65.61	17.14	27.41	764.3	1.02151
95.0	6.672	3.310	1200.0	1986.0	67.11	18.00	28.03	779.0	1.02025
100.0	6.309	3.129	1297.0	2128.0	68.57	18.87	28.70	793.4	1.01914
105.0	5.985	2.969	1397.0	2273.0	69.98	19.72	29.38	807.4	1.01815
110.0	5.695	2.825	1501.0	2421.0	71.37	20.54	30.06	821.4	1.01727
115.0	5.434	2.696	1609.0	2573.0	72.72	21.30	30.70	835.3	1.01647
120.0	5.197	2.578	1720.0	2728.0	74.03	21.99	31.29	849.3	1.01575
125.0	4.981	2.471	1834.0	2886.0	75.32	22.61	31.81	863.4	1.01509
130.0	4.783	2.372	1950.0	3046.0	76.58	23.13	32.25	877.5	1.01449
135.0	4.600	2.282	2069.0	3208.0	77.80	23.58	32.63	891.8	1.01393
140.0	4.432	2.198	2190.0	3372.0	79.00	23.96	32.95	906.0	1.01342
150.0	4.131	2.049	2435.0	3704.0	81.29	24.48	33.37	934.7	1.01250
160.0	3.870	1.920	2684.0	4039.0	83.45	24.74	33.54	963.5	1.01171
170.0	3.640	1.806	2934.0	4374.0	85.48	24.78	33.52	992.2	1.01101
180.0	3.437	1.705	3184.0	4709.0	87.39	24.67	33.35	1021.0	1.01040
190.0	3.256	1.615	3431.0	5041.0	89.19	24.45	33.08	1049.0	1.00985
200.0	3.094	1.535	3676.0	5370.0	90.88	24.17	32.76	1077.0	1.00935
210.0	2.947	1.462	3917.0	5696.0	92.47	23.86	32.41	1105.0	1.00891
220.0	2.814	1.396	4155.0	6018.0	93.97	23.53	32.06	1132.0	1.00850
230.0	2.692	1.335	4390.0	6337.0	95.38	23.21	31.72	1159.0	1.00814
240.0	2.581	1.280	4622.0	6653.0	96.73	22.91	31.40	1185.0	1.00780
250.0	2.478	1.229	4850.0	6965.0	98.00	22.64	31.10	1211.0	1.00749
260.0	2.384	1.182	5076.0	7275.0	99.22	22.39	30.84	1236.0	1.00720
270.0	2.296	1.139	5299.0	7582.0	100.4	22.17	30.60	1261.0	1.00694
280.0	2.215	1.099	5521.0	7887.0	101.5	21.98	30.40	1285.0	1.00669
300.0	2.069	1.026	5958.0	8491.0	103.6	21.67	30.07	1331.0	1.00625
320.0	1.941	.9628	6390.0	9090.0	105.5	21.44	29.83	1375.0	1.00586
340.0	1.828	.9067	6818.0	9685.0	107.3	21.29	29.66	1418.0	1.00552
360.0	1.727	.8569	7243.0	10280.0	109.0	21.18	29.54	1459.0	1.00522
380.0	1.637	.8123	7666.0	10870.0	110.6	21.11	29.46	1498.0	1.00494
400.0	1.556	.7721	8088.0	11460.0	112.1	21.05	29.40	1537.0	1.00470

2.80 MPa isobar

T K	Density kg/m^3	Density mol/dm^3	E J/mol	H J/mol	S J/(mol·K)	C_v J/(mol·K)	C_p J/(mol·K)	Sound m/s	Diel. const.
14.71[a]	78.42	38.90	–618.7	–546.7	10.23	10.75	13.85	1327.0	1.25652
15.0	78.22	38.80	–614.9	–542.7	10.50	10.51	13.71	1323.0	1.25582
16.0	77.51	38.45	–601.8	–529.0	11.38	10.21	13.83	1307.0	1.25330
18.0	75.97	37.68	–574.3	–500.0	13.09	10.67	15.34	1270.0	1.24788
20.0	74.25	36.83	–543.4	–467.4	14.80	11.36	17.27	1230.0	1.24187
22.0	72.36	35.89	–508.9	–430.9	16.54	11.92	19.27	1186.0	1.23524
24.0	70.25	34.85	–470.6	–390.3	18.31	12.31	21.35	1139.0	1.22791

Thermophysical properties of hydrogen—Continued

T K	Density kg/m^3	Density mol/dm^3	E J/mol	H J/mol	S J/(mol·K)	C_v J/(mol·K)	C_p J/(mol·K)	Sound m/s	Diel. const.
26.0	67.90	33.68	–428.5	–345.4	20.10	12.59	23.62	1085.0	1.21977
28.0	65.25	32.37	–382.1	–295.6	21.95	12.79	26.27	1024.0	1.21064
29.0	63.79	31.64	–357.0	–268.5	22.90	12.87	27.82	991.1	1.20562
30.0	62.22	30.87	–330.5	–239.8	23.87	12.95	29.59	955.5	1.20023
31.0	60.52	30.02	–302.5	–209.2	24.87	13.03	31.64	917.5	1.19442
32.0	58.66	29.10	–272.6	–176.4	25.91	13.11	34.07	877.0	1.18810
33.0	56.62	28.09	–240.6	–140.9	27.01	13.20	37.01	833.9	1.18117
34.0	54.34	26.96	–206.0	–102.2	28.16	13.31	40.69	788.2	1.17348
35.0	51.78	25.68	–168.2	–59.2	29.41	13.44	45.37	740.1	1.16485
36.0	48.85	24.23	–126.5	–11.0	30.77	13.61	51.42	690.3	1.15505
37.0	45.48	22.56	–80.0	44.1	32.28	13.81	58.95	641.1	1.14385
38.0	41.66	20.67	–28.4	107.1	33.95	14.02	66.82	597.0	1.13125
39.0	37.60	18.65	26.6	176.7	35.76	14.19	71.56	564.2	1.11795
40.0	33.73	16.73	80.9	248.2	37.57	14.23	70.54	545.6	1.10538
40.5	31.98	15.87	106.5	283.0	38.44	14.21	68.32	540.8	1.09974
41.0	30.39	15.08	130.7	316.4	39.26	14.16	65.50	538.3	1.09462
41.5	28.96	14.36	153.5	348.4	40.03	14.09	62.44	537.5	1.09001
42.0	27.66	13.72	174.8	378.9	40.76	14.02	59.37	537.9	1.08587
42.5	26.50	13.15	194.8	407.8	41.45	13.94	56.43	539.2	1.08216
43.0	25.46	12.63	213.6	435.3	42.09	13.85	53.70	541.2	1.07883
43.5	24.51	12.16	231.3	461.6	42.70	13.77	51.22	543.6	1.07583
44.0	23.65	11.73	248.0	486.6	43.27	13.70	48.97	546.4	1.07312
44.5	22.88	11.35	263.8	510.6	43.81	13.63	46.95	549.5	1.07065
45.0	22.16	10.99	278.9	533.6	44.33	13.56	45.13	552.7	1.06839
46.0	20.90	10.37	307.1	577.1	45.28	13.45	42.05	559.4	1.06442
47.0	19.83	9.835	333.2	617.9	46.16	13.36	39.57	566.4	1.06103
48.0	18.89	9.370	357.6	656.4	46.97	13.28	37.55	573.5	1.05809
49.0	18.06	8.961	380.6	693.1	47.73	13.23	35.88	580.6	1.05550
50.0	17.33	8.597	402.6	728.3	48.44	13.19	34.50	587.5	1.05321
51.0	16.67	8.270	423.6	762.2	49.11	13.16	33.33	594.4	1.05115
52.0	16.08	7.975	443.9	795.0	49.75	13.15	32.35	601.1	1.04929
53.0	15.53	7.706	463.5	826.9	50.36	13.15	31.51	607.7	1.04760
54.0	15.04	7.459	482.7	858.0	50.94	13.15	30.78	614.1	1.04605
56.0	14.16	7.022	519.6	918.4	52.03	13.19	29.61	626.5	1.04331
58.0	13.39	6.644	555.3	976.7	53.06	13.27	28.73	638.2	1.04095
60.0	12.73	6.313	589.9	1033.0	54.02	13.36	28.06	649.4	1.03888
62.0	12.14	6.020	623.9	1089.0	54.93	13.49	27.54	659.9	1.03705
64.0	11.61	5.758	657.5	1144.0	55.80	13.64	27.16	669.9	1.03542
66.0	11.13	5.522	690.7	1198.0	56.63	13.81	26.87	679.4	1.03395
68.0	10.70	5.308	723.7	1251.0	57.43	14.01	26.67	688.4	1.03262
70.0	10.31	5.112	756.8	1304.0	58.20	14.22	26.55	697.0	1.03140
72.0	9.944	4.933	789.8	1357.0	58.95	14.45	26.48	705.2	1.03029
74.0	9.610	4.767	823.0	1410.0	59.67	14.70	26.46	713.0	1.02926
76.0	9.300	4.613	856.4	1463.0	60.38	14.97	26.49	720.4	1.02831
78.0	9.013	4.471	890.1	1516.0	61.07	15.25	26.56	727.6	1.02742
80.0	8.745	4.338	924.2	1570.0	61.74	15.54	26.67	734.6	1.02660
82.0	8.494	4.213	958.6	1623.0	62.40	15.84	26.80	741.3	1.02583
85.0	8.147	4.041	1011.0	1704.0	63.37	16.32	27.05	751.0	1.02476
90.0	7.633	3.786	1101.0	1840.0	64.93	17.15	27.56	766.4	1.02319
95.0	7.185	3.564	1194.0	1980.0	66.44	18.01	28.16	781.1	1.02182
100.0	6.792	3.369	1291.0	2122.0	67.90	18.88	28.82	795.4	1.02061
105.0	6.442	3.195	1392.0	2268.0	69.32	19.73	29.49	809.4	1.01955
110.0	6.129	3.040	1496.0	2417.0	70.71	20.55	30.15	823.4	1.01859
115.0	5.847	2.900	1604.0	2569.0	72.06	21.31	30.79	837.3	1.01773
120.0	5.591	2.773	1715.0	2725.0	73.38	22.00	31.37	851.2	1.01695
125.0	5.358	2.658	1829.0	2883.0	74.67	22.62	31.88	865.3	1.01624
130.0	5.144	2.552	1946.0	3043.0	75.93	23.14	32.32	879.5	1.01559
135.0	4.948	2.454	2065.0	3206.0	77.16	23.59	32.69	893.7	1.01499
140.0	4.767	2.364	2186.0	3370.0	78.36	23.97	33.01	907.9	1.01444
150.0	4.443	2.204	2432.0	3703.0	80.65	24.49	33.42	936.6	1.01345
160.0	4.161	2.064	2681.0	4038.0	82.81	24.75	33.58	965.4	1.01260
170.0	3.915	1.942	2932.0	4374.0	84.85	24.79	33.56	994.1	1.01185
180.0	3.696	1.834	3181.0	4708.0	86.76	24.68	33.38	1023.0	1.01118

Thermophysical properties of hydrogen—Continued

T K	Density kg/m^3	Density mol/dm^3	E J/mol	H J/mol	S J/(mol·K)	C_v J/(mol·K)	C_p J/(mol·K)	Sound m/s	Diel. const.
190.0	3.502	1.737	3429.0	5041.0	88.56	24.46	33.11	1051.0	1.01059
200.0	3.327	1.650	3674.0	5371.0	90.25	24.18	32.79	1079.0	1.01006
210.0	3.169	1.572	3915.0	5697.0	91.84	23.86	32.44	1107.0	1.00958
220.0	3.026	1.501	4154.0	6019.0	93.34	23.54	32.08	1134.0	1.00915
230.0	2.895	1.436	4389.0	6338.0	94.76	23.22	31.74	1161.0	1.00875
240.0	2.775	1.377	4620.0	6654.0	96.10	22.92	31.41	1187.0	1.00839
250.0	2.665	1.322	4849.0	6967.0	97.38	22.64	31.12	1213.0	1.00805
260.0	2.564	1.272	5075.0	7276.0	98.59	22.39	30.85	1238.0	1.00775
270.0	2.470	1.225	5298.0	7584.0	99.75	22.17	30.61	1262.0	1.00746
280.0	2.382	1.182	5520.0	7889.0	100.9	21.98	30.41	1286.0	1.00720
300.0	2.225	1.104	5957.0	8494.0	103.0	21.67	30.08	1333.0	1.00672
320.0	2.088	1.036	6389.0	9093.0	104.9	21.45	29.84	1377.0	1.00631
340.0	1.966	.9754	6817.0	9687.0	106.7	21.29	29.67	1419.0	1.00594
360.0	1.858	.9218	7242.0	10280.0	108.4	21.19	29.55	1460.0	1.00561
380.0	1.762	.8739	7665.0	10870.0	110.0	21.11	29.46	1500.0	1.00532
400.0	1.675	.8307	8087.0	11460.0	111.5	21.06	29.40	1538.0	1.00506
				3.00 MPa isobar					
14.77[a]	78.53	38.95	–618.4	–541.4	10.24	10.72	13.77	1327.0	1.25689
15.0	78.37	38.88	–615.4	–538.3	10.45	10.53	13.66	1324.0	1.25634
16.0	77.67	38.53	–602.5	–524.6	11.33	10.22	13.78	1310.0	1.25386
18.0	76.14	37.77	–575.1	–495.7	13.03	10.67	15.26	1276.0	1.24850
20.0	74.45	36.93	–544.5	–463.3	14.74	11.36	17.18	1237.0	1.24257
22.0	72.58	36.00	–510.3	–427.0	16.47	11.91	19.14	1196.0	1.23603
24.0	70.51	34.98	–472.5	–386.7	18.22	12.31	21.16	1149.0	1.22883
26.0	68.22	33.84	–430.9	–342.2	20.00	12.58	23.36	1098.0	1.22086
28.0	65.64	32.56	–385.2	–293.1	21.82	12.77	25.88	1039.0	1.21196
30.0	62.71	31.11	–334.8	–238.3	23.70	12.93	28.96	973.5	1.20190
31.0	61.08	30.30	–307.5	–208.5	24.68	13.00	30.82	937.5	1.19632
32.0	59.31	29.42	–278.6	–176.6	25.70	13.07	32.98	899.3	1.19031
33.0	57.39	28.47	–247.8	–142.4	26.75	13.15	35.53	858.9	1.18377
34.0	55.27	27.42	–214.8	–105.4	27.85	13.24	38.62	816.4	1.17661
35.0	52.92	26.25	–179.2	–64.9	29.03	13.35	42.40	771.9	1.16870
36.0	50.30	24.95	–140.5	–20.3	30.28	13.48	47.07	726.1	1.15988
37.0	47.34	23.48	–98.2	29.6	31.65	13.63	52.74	680.3	1.15002
38.0	44.03	21.84	–51.9	85.4	33.14	13.80	59.00	637.2	1.13906
39.0	40.45	20.06	–2.3	147.2	34.74	13.96	64.30	601.0	1.12726
40.0	36.82	18.27	48.7	212.9	36.41	14.06	66.37	575.4	1.11542
41.0	33.46	16.60	97.9	278.7	38.03	14.07	64.57	561.0	1.10450
41.5	31.94	15.84	121.1	310.5	38.80	14.04	62.67	557.2	1.09959
42.0	30.54	15.15	143.2	341.3	39.54	13.99	60.43	555.1	1.09508
42.5	29.26	14.51	164.2	370.9	40.24	13.94	58.05	554.3	1.09098
43.0	28.09	13.93	184.0	399.3	40.90	13.87	55.66	554.6	1.08724
43.5	27.03	13.41	202.8	426.6	41.53	13.81	53.35	555.6	1.08384
44.0	26.06	12.93	220.6	452.7	42.13	13.74	51.18	557.2	1.08075
44.5	25.17	12.49	237.5	477.8	42.70	13.68	49.15	559.2	1.07792
45.0	24.36	12.08	253.6	501.9	43.24	13.62	47.30	561.6	1.07534
46.0	22.92	11.37	283.6	547.5	44.24	13.50	44.05	567.1	1.07079
47.0	21.69	10.76	311.3	590.2	45.16	13.41	41.37	573.1	1.06690
48.0	20.62	10.23	337.2	630.4	46.00	13.34	39.15	579.4	1.06354
49.0	19.69	9.766	361.4	668.6	46.79	13.28	37.31	585.9	1.06060
50.0	18.86	9.355	384.4	705.1	47.53	13.23	35.77	592.4	1.05799
51.0	18.12	8.987	406.4	740.2	48.23	13.20	34.48	598.9	1.05567
52.0	17.45	8.655	427.5	774.1	48.88	13.19	33.38	605.4	1.05358
53.0	16.84	8.354	447.9	807.0	49.51	13.18	32.44	611.7	1.05168
54.0	16.29	8.079	467.7	839.1	50.11	13.18	31.63	617.9	1.04994
56.0	15.31	7.592	505.8	901.0	51.23	13.22	30.33	630.0	1.04689
58.0	14.46	7.174	542.4	960.6	52.28	13.29	29.34	641.5	1.04426
60.0	13.73	6.809	577.9	1018.0	53.26	13.38	28.58	652.5	1.04198
62.0	13.08	6.487	612.6	1075.0	54.19	13.51	28.01	662.9	1.03997
64.0	12.50	6.200	646.7	1131.0	55.07	13.66	27.57	672.8	1.03817

Thermophysical properties of hydrogen—Continued

T K	Density kg/m³	Density mol/dm³	E J/mol	H J/mol	S J/(mol·K)	C_v J/(mol·K)	C_p J/(mol·K)	Sound m/s	Diel. const.
66.0	11.98	5.941	680.5	1185.0	55.91	13.83	27.24	682.2	1.03656
68.0	11.51	5.707	714.0	1240.0	56.72	14.02	27.01	691.1	1.03510
70.0	11.08	5.494	747.4	1293.0	57.50	14.23	26.85	699.6	1.03378
72.0	10.68	5.299	780.9	1347.0	58.26	14.46	26.76	707.7	1.03256
74.0	10.32	5.119	814.4	1401.0	58.99	14.71	26.72	715.4	1.03144
76.0	9.983	4.952	848.2	1454.0	59.71	14.98	26.73	722.9	1.03041
78.0	9.671	4.797	882.2	1508.0	60.40	15.26	26.79	730.0	1.02945
80.0	9.381	4.653	916.5	1561.0	61.08	15.55	26.88	736.9	1.02855
82.0	9.110	4.519	951.1	1615.0	61.74	15.86	27.00	743.6	1.02772
85.0	8.734	4.333	1004.0	1696.0	62.72	16.33	27.23	753.2	1.02657
90.0	8.180	4.057	1094.0	1834.0	64.29	17.17	27.72	768.6	1.02487
95.0	7.697	3.818	1188.0	1974.0	65.80	18.03	28.30	783.2	1.02339
100.0	7.274	3.608	1285.0	2117.0	67.27	18.89	28.94	797.5	1.02209
105.0	6.898	3.422	1386.0	2263.0	68.70	19.74	29.60	811.5	1.02094
110.0	6.561	3.255	1491.0	2413.0	70.09	20.56	30.25	825.4	1.01991
115.0	6.258	3.104	1599.0	2566.0	71.45	21.32	30.87	839.3	1.01899
120.0	5.984	2.968	1711.0	2721.0	72.77	22.02	31.45	853.2	1.01815
125.0	5.734	2.844	1825.0	2880.0	74.07	22.63	31.96	867.3	1.01738
130.0	5.505	2.731	1942.0	3041.0	75.33	23.15	32.39	881.4	1.01669
135.0	5.294	2.626	2061.0	3204.0	76.56	23.60	32.75	895.7	1.01604
140.0	5.100	2.530	2182.0	3368.0	77.76	23.98	33.07	909.9	1.01545
150.0	4.753	2.358	2429.0	3701.0	80.05	24.50	33.47	938.5	1.01440
160.0	4.452	2.208	2678.0	4037.0	82.22	24.76	33.63	967.3	1.01348
170.0	4.188	2.078	2929.0	4373.0	84.26	24.80	33.59	996.0	1.01268
180.0	3.955	1.962	3179.0	4708.0	86.17	24.69	33.42	1025.0	1.01197
190.0	3.746	1.858	3427.0	5041.0	87.97	24.47	33.14	1053.0	1.01133
200.0	3.559	1.766	3672.0	5371.0	89.67	24.18	32.81	1081.0	1.01077
210.0	3.391	1.682	3914.0	5697.0	91.26	23.87	32.46	1109.0	1.01025
220.0	3.237	1.606	4152.0	6020.0	92.76	23.54	32.10	1136.0	1.00979
230.0	3.098	1.537	4387.0	6339.0	94.18	23.23	31.76	1163.0	1.00937
240.0	2.970	1.473	4619.0	6655.0	95.52	22.93	31.43	1189.0	1.00898
250.0	2.852	1.415	4847.0	6968.0	96.80	22.65	31.13	1215.0	1.00862
260.0	2.743	1.361	5073.0	7278.0	98.02	22.40	30.87	1240.0	1.00829
270.0	2.643	1.311	5297.0	7585.0	99.18	22.18	30.63	1264.0	1.00799
280.0	2.549	1.265	5518.0	7891.0	100.3	21.99	30.42	1288.0	1.00770
300.0	2.381	1.181	5956.0	8496.0	102.4	21.68	30.09	1334.0	1.00719
320.0	2.234	1.108	6388.0	9095.0	104.3	21.45	29.85	1379.0	1.00675
340.0	2.105	1.044	6816.0	9690.0	106.1	21.30	29.67	1421.0	1.00636
360.0	1.989	.9867	7242.0	10280.0	107.8	21.19	29.55	1462.0	1.00601
380.0	1.886	.9354	7665.0	10870.0	109.4	21.11	29.47	1501.0	1.00569
400.0	1.793	.8892	8087.0	11460.0	110.9	21.06	29.41	1540.0	1.00541
				3.50 MPa isobar					
14.93[a]	78.79	39.08	−617.6	−528.1	10.28	10.64	13.57	1330.0	1.25780
15.0	78.74	39.06	−616.7	−527.1	10.34	10.58	13.55	1329.0	1.25763
16.0	78.06	38.72	−603.9	−513.6	11.21	10.24	13.64	1319.0	1.25523
18.0	76.58	37.99	−577.1	−485.0	12.90	10.66	15.09	1290.0	1.25002
20.0	74.94	37.17	−547.1	−453.0	14.58	11.34	16.95	1256.0	1.24426
22.0	73.13	36.28	−513.7	−417.2	16.28	11.89	18.84	1217.0	1.23795
24.0	71.15	35.29	−476.8	−377.6	18.01	12.28	20.75	1175.0	1.23102
26.0	68.96	34.21	−436.4	−334.1	19.74	12.56	22.77	1127.0	1.22342
28.0	66.53	33.00	−392.4	−286.4	21.51	12.75	25.02	1074.0	1.21504
30.0	63.82	31.66	−344.3	−233.8	23.33	12.89	27.65	1015.0	1.20570
31.0	62.33	30.92	−318.6	−205.4	24.26	12.95	29.17	982.6	1.20060
32.0	60.74	30.13	−291.5	−175.3	25.21	13.01	30.87	949.0	1.19518
33.0	59.04	29.29	−263.0	−143.5	26.19	13.06	32.79	913.7	1.18939
34.0	57.21	28.38	−233.0	−109.7	27.20	13.13	34.99	877.1	1.18317
36.0	53.08	26.33	−167.4	−34.5	29.35	13.27	40.42	800.4	1.16924
37.0	50.75	25.17	−131.5	7.5	30.50	13.36	43.74	761.3	1.16140
38.0	48.21	23.92	−93.2	53.1	31.71	13.46	47.43	723.1	1.15293
39.0	45.48	22.56	−52.7	102.4	32.99	13.56	51.21	687.2	1.14387
40.0	42.61	21.13	−10.2	155.4	34.33	13.66	54.52	655.6	1.13436

Thermophysical properties of hydrogen—Continued

T K	Density kg/m^3	Density mol/dm^3	E J/mol	H J/mol	S J/(mol·K)	C_v J/(mol·K)	C_p J/(mol·K)	Sound m/s	Diel. const.
42.0	36.84	18.27	76.3	267.9	37.08	13.77	56.80	612.2	1.11547
43.0	34.20	16.97	117.8	324.1	38.40	13.76	55.42	601.0	1.10691
44.0	31.84	15.80	156.8	378.4	39.65	13.72	53.01	595.2	1.09929
45.0	29.77	14.77	193.0	430.0	40.81	13.65	50.20	593.3	1.09263
46.0	27.97	13.87	226.5	478.8	41.88	13.57	47.38	594.0	1.08685
47.0	26.40	13.10	257.5	524.8	42.87	13.49	44.77	596.4	1.08184
48.0	25.03	12.41	286.5	568.4	43.79	13.42	42.43	599.9	1.07747
49.0	23.82	11.82	313.6	609.8	44.64	13.37	40.39	604.2	1.07364
50.0	22.75	11.29	339.2	649.3	45.44	13.32	38.62	609.0	1.07026
51.0	21.80	10.81	363.4	687.1	46.19	13.29	37.10	614.1	1.06724
52.0	20.94	10.39	386.6	723.5	46.90	13.26	35.78	619.4	1.06454
53.0	20.17	10.00	408.9	758.7	47.57	13.25	34.64	624.8	1.06210
54.0	19.46	9.654	430.3	792.9	48.21	13.25	33.66	630.2	1.05988
55.0	18.82	9.333	451.1	826.1	48.82	13.26	32.80	635.7	1.05786
56.0	18.22	9.039	471.3	858.5	49.40	13.28	32.05	641.0	1.05600
58.0	17.17	8.516	510.3	921.3	50.50	13.34	30.82	651.6	1.05269
60.0	16.25	8.062	547.8	982.0	51.53	13.43	29.87	661.8	1.04984
62.0	15.45	7.664	584.3	1041.0	52.50	13.55	29.14	671.7	1.04734
64.0	14.74	7.311	620.0	1099.0	53.41	13.70	28.58	681.1	1.04513
66.0	14.10	6.996	655.1	1155.0	54.29	13.86	28.15	690.1	1.04315
68.0	13.53	6.711	689.8	1211.0	55.12	14.06	27.83	698.7	1.04137
70.0	13.01	6.453	724.3	1267.0	55.92	14.27	27.60	706.9	1.03975
72.0	12.53	6.216	758.7	1322.0	56.70	14.50	27.44	714.7	1.03828
74.0	12.09	6.000	793.2	1377.0	57.45	14.75	27.35	722.3	1.03692
76.0	11.69	5.800	827.7	1431.0	58.18	15.01	27.32	729.5	1.03568
78.0	11.32	5.615	862.5	1486.0	58.89	15.29	27.33	736.5	1.03453
80.0	10.97	5.443	897.5	1541.0	59.58	15.59	27.38	743.2	1.03345
82.0	10.65	5.282	932.8	1595.0	60.26	15.89	27.47	749.8	1.03246
85.0	10.20	5.061	986.5	1678.0	61.25	16.37	27.66	759.3	1.03108
90.0	9.544	4.734	1078.0	1817.0	62.84	17.20	28.09	774.3	1.02906
95.0	8.974	4.452	1173.0	1959.0	64.37	18.06	28.63	788.8	1.02731
100.0	8.475	4.204	1271.0	2104.0	65.86	18.92	29.23	802.9	1.02577
105.0	8.033	3.985	1373.0	2252.0	67.30	19.77	29.86	816.8	1.02441
110.0	7.638	3.789	1479.0	2402.0	68.70	20.59	30.49	830.6	1.02321
115.0	7.283	3.613	1588.0	2556.0	70.07	21.35	31.09	844.4	1.02212
120.0	6.962	3.453	1700.0	2713.0	71.41	22.04	31.64	858.3	1.02113
125.0	6.669	3.308	1815.0	2873.0	72.71	22.65	32.14	872.3	1.02024
130.0	6.402	3.176	1932.0	3034.0	73.98	23.18	32.56	886.4	1.01942
135.0	6.156	3.054	2052.0	3198.0	75.21	23.62	32.91	900.6	1.01867
140.0	5.930	2.941	2174.0	3363.0	76.41	24.01	33.21	914.8	1.01798
150.0	5.526	2.741	2421.0	3698.0	78.72	24.52	33.59	943.4	1.01675
160.0	5.175	2.567	2671.0	4034.0	80.89	24.78	33.73	972.1	1.01568
170.0	4.868	2.415	2922.0	4372.0	82.94	24.82	33.69	1001.0	1.01475
180.0	4.597	2.280	3173.0	4708.0	84.86	24.71	33.50	1029.0	1.01392
190.0	4.355	2.160	3421.0	5041.0	86.66	24.49	33.22	1058.0	1.01318
200.0	4.137	2.052	3667.0	5372.0	88.36	24.20	32.88	1086.0	1.01252
210.0	3.941	1.955	3909.0	5699.0	89.95	23.89	32.52	1114.0	1.01193
220.0	3.764	1.867	4147.0	6022.0	91.46	23.56	32.15	1141.0	1.01139
230.0	3.601	1.786	4383.0	6342.0	92.88	23.24	31.80	1167.0	1.01089
240.0	3.453	1.713	4615.0	6658.0	94.23	22.94	31.47	1194.0	1.01044
250.0	3.316	1.645	4844.0	6972.0	95.50	22.66	31.17	1219.0	1.01003
260.0	3.190	1.582	5070.0	7282.0	96.72	22.41	30.90	1244.0	1.00965
270.0	3.073	1.525	5294.0	7590.0	97.88	22.19	30.66	1269.0	1.00929
280.0	2.965	1.471	5516.0	7895.0	98.99	22.00	30.45	1292.0	1.00896
300.0	2.770	1.374	5954.0	8501.0	101.1	21.69	30.11	1339.0	1.00837
320.0	2.599	1.289	6386.0	9100.0	103.0	21.46	29.87	1383.0	1.00785
340.0	2.449	1.215	6814.0	9696.0	104.8	21.31	29.69	1425.0	1.00740
360.0	2.315	1.148	7240.0	10290.0	106.5	21.20	29.57	1466.0	1.00699
380.0	2.194	1.089	7664.0	10880.0	108.1	21.12	29.48	1505.0	1.00663
400.0	2.086	1.035	8086.0	11470.0	109.6	21.07	29.42	1543.0	1.00630

Thermophysical properties of hydrogen—Continued

T K	Density kg/m^3	Density mol/dm^3	E J/mol	H J/mol	S J/(mol·K)	C_v J/(mol·K)	C_p J/(mol·K)	Sound m/s	Diel. const.
				4.00 MPa isobar					
15.08[a]	79.04	39.21	–616.8	–514.8	10.31	10.55	13.40	1334.0	1.25871
16.0	78.43	38.91	–605.3	–502.5	11.10	10.24	13.51	1328.0	1.25656
18.0	76.99	38.19	–578.9	–474.2	12.76	10.65	14.92	1305.0	1.25149
20.0	75.40	37.40	–549.5	–442.6	14.43	11.32	16.74	1274.0	1.24589
22.0	73.65	36.54	–516.8	–407.3	16.11	11.87	18.56	1238.0	1.23977
24.0	71.74	35.59	–480.7	–368.3	17.80	12.26	20.39	1199.0	1.23310
26.0	69.65	34.55	–441.5	–325.7	19.51	12.54	22.27	1154.0	1.22582
28.0	67.35	33.41	–398.9	–279.1	21.23	12.73	24.32	1105.0	1.21785
30.0	64.80	32.15	–352.7	–228.2	22.99	12.86	26.63	1051.0	1.20909
32.0	61.97	30.74	–302.5	–172.3	24.79	12.97	29.33	991.9	1.19938
33.0	60.43	29.98	–275.7	–142.2	25.72	13.01	30.88	960.5	1.19412
34.0	58.79	29.16	–247.7	–110.5	26.66	13.06	32.59	927.9	1.18854
36.0	55.19	27.38	–187.6	–41.5	28.64	13.16	36.56	860.5	1.17633
38.0	51.09	25.34	–121.5	36.3	30.74	13.27	41.35	792.2	1.16255
40.0	46.49	23.06	–49.3	124.2	32.99	13.41	46.49	728.4	1.14721
42.0	41.58	20.62	27.4	221.4	35.36	13.53	50.28	677.2	1.13097
44.0	36.82	18.26	103.9	322.9	37.72	13.58	50.58	645.4	1.11541
45.0	34.66	17.19	140.3	372.9	38.85	13.56	49.47	636.5	1.10838
46.0	32.69	16.21	174.9	421.6	39.92	13.53	47.85	631.3	1.10201
47.0	30.91	15.33	207.7	468.6	40.93	13.49	46.00	628.8	1.09628
48.0	29.32	14.54	238.6	513.6	41.87	13.45	44.10	628.5	1.09117
49.0	27.90	13.84	267.7	556.8	42.77	13.41	42.28	629.6	1.08662
50.0	26.62	13.21	295.3	598.2	43.60	13.37	40.59	631.9	1.08255
51.0	25.48	12.64	321.5	638.0	44.39	13.34	39.06	634.9	1.07890
52.0	24.44	12.13	346.5	676.4	45.14	13.32	37.68	638.4	1.07562
53.0	23.51	11.66	370.4	713.5	45.84	13.31	36.46	642.3	1.07265
54.0	22.66	11.24	393.5	749.4	46.51	13.31	35.38	646.5	1.06996
55.0	21.88	10.85	415.7	784.3	47.15	13.31	34.42	650.9	1.06749
56.0	21.16	10.50	437.2	818.3	47.77	13.33	33.58	655.4	1.06524
58.0	19.89	9.867	478.6	883.9	48.92	13.39	32.17	664.5	1.06124
60.0	18.80	9.323	518.1	947.1	49.99	13.47	31.07	673.6	1.05779
62.0	17.84	8.848	556.3	1008.0	50.99	13.59	30.20	682.5	1.05479
64.0	16.99	8.428	593.5	1068.0	51.94	13.73	29.53	691.2	1.05214
66.0	16.24	8.054	629.9	1127.0	52.84	13.90	29.01	699.6	1.04979
68.0	15.56	7.717	665.9	1184.0	53.70	14.09	28.61	707.6	1.04767
70.0	14.94	7.413	701.5	1241.0	54.53	14.30	28.31	715.4	1.04576
72.0	14.38	7.135	736.8	1297.0	55.32	14.53	28.09	722.9	1.04402
74.0	13.87	6.881	772.2	1354.0	56.09	14.78	27.95	730.1	1.04243
76.0	13.40	6.647	807.5	1409.0	56.83	15.04	27.87	737.1	1.04096
78.0	12.96	6.431	843.0	1465.0	57.56	15.32	27.85	743.8	1.03961
80.0	12.56	6.230	878.7	1521.0	58.26	15.62	27.87	750.3	1.03836
82.0	12.18	6.044	914.7	1577.0	58.95	15.92	27.92	756.7	1.03720
84.0	11.83	5.869	950.9	1632.0	59.62	16.24	28.02	762.9	1.03611
90.0	10.90	5.408	1062.0	1802.0	61.57	17.23	28.45	780.6	1.03324
95.0	10.24	5.082	1158.0	1945.0	63.12	18.09	28.95	794.8	1.03121
100.0	9.669	4.796	1257.0	2091.0	64.62	18.95	29.51	808.7	1.02944
105.0	9.161	4.544	1360.0	2240.0	66.07	19.80	30.11	822.4	1.02788
110.0	8.707	4.319	1466.0	2392.0	67.49	20.62	30.72	836.1	1.02648
115.0	8.300	4.117	1576.0	2548.0	68.87	21.38	31.30	849.8	1.02524
120.0	7.932	3.935	1689.0	2705.0	70.21	22.07	31.83	863.6	1.02411
125.0	7.598	3.769	1804.0	2866.0	71.52	22.68	32.31	877.5	1.02308
130.0	7.292	3.617	1923.0	3028.0	72.80	23.21	32.72	891.6	1.02215
135.0	7.012	3.478	2043.0	3193.0	74.04	23.65	33.06	905.7	1.02129
140.0	6.753	3.350	2165.0	3359.0	75.24	24.03	33.35	919.8	1.02050
145.0	6.514	3.231	2288.0	3526.0	76.42	24.32	33.56	934.1	1.01977
150.0	6.292	3.121	2413.0	3694.0	77.56	24.55	33.71	948.4	1.01909
160.0	5.893	2.923	2664.0	4032.0	79.74	24.80	33.84	977.1	1.01787
170.0	5.543	2.750	2916.0	4370.0	81.79	24.84	33.78	1006.0	1.01680
180.0	5.234	2.596	3167.0	4707.0	83.72	24.73	33.58	1034.0	1.01586
190.0	4.958	2.460	3415.0	5042.0	85.52	24.51	33.29	1063.0	1.01502
200.0	4.711	2.337	3661.0	5373.0	87.22	24.22	32.94	1091.0	1.01427
210.0	4.488	2.226	3904.0	5701.0	88.82	23.90	32.58	1118.0	1.01359

Thermophysical properties of hydrogen—Continued

T K	Density kg/m^3	Density mol/dm^3	E J/mol	H J/mol	S J/(mol·K)	C_v J/(mol·K)	C_p J/(mol·K)	Sound m/s	Diel. const.
220.0	4.286	2.126	4143.0	6024.0	90.33	23.57	32.21	1145.0	1.01297
230.0	4.102	2.035	4379.0	6345.0	91.75	23.26	31.85	1172.0	1.01241
240.0	3.933	1.951	4611.0	6662.0	93.10	22.95	31.52	1198.0	1.01190
250.0	3.777	1.874	4840.0	6975.0	94.38	22.68	31.21	1224.0	1.01143
260.0	3.634	1.803	5067.0	7286.0	95.60	22.43	30.94	1249.0	1.01099
270.0	3.501	1.737	5291.0	7594.0	96.76	22.20	30.69	1273.0	1.01059
280.0	3.378	1.676	5513.0	7900.0	97.87	22.01	30.48	1297.0	1.01022
300.0	3.156	1.566	5951.0	8506.0	99.97	21.70	30.14	1343.0	1.00954
320.0	2.962	1.469	6384.0	9106.0	101.9	21.47	29.89	1387.0	1.00895
340.0	2.791	1.384	6813.0	9702.0	103.7	21.31	29.71	1429.0	1.00844
360.0	2.638	1.309	7239.0	10290.0	105.4	21.21	29.59	1470.0	1.00797
380.0	2.502	1.241	7662.0	10890.0	107.0	21.13	29.50	1509.0	1.00756
400.0	2.379	1.180	8085.0	11470.0	108.5	21.07	29.43	1547.0	1.00719

5.00 MPa isobar

T K	Density kg/m^3	Density mol/dm^3	E J/mol	H J/mol	S J/(mol·K)	C_v J/(mol·K)	C_p J/(mol·K)	Sound m/s	Diel. const.
15.38[a]	79.56	39.46	−615.2	−488.5	10.37	10.37	13.12	1349.0	1.26052
16.0	79.16	39.27	−607.7	−480.4	10.88	10.22	13.25	1348.0	1.25914
18.0	77.79	38.59	−582.2	−452.6	12.52	10.59	14.61	1333.0	1.25430
20.0	76.28	37.84	−553.8	−421.7	14.15	11.26	16.36	1308.0	1.24898
22.0	74.63	37.02	−522.3	−387.2	15.79	11.82	18.09	1277.0	1.24319
24.0	72.84	36.13	−487.7	−349.4	17.43	12.22	19.77	1242.0	1.23693
26.0	70.90	35.17	−450.3	−308.1	19.08	12.50	21.46	1204.0	1.23018
28.0	68.80	34.13	−410.0	−263.5	20.73	12.69	23.22	1161.0	1.22288
30.0	66.52	33.00	−366.7	−215.2	22.40	12.83	25.11	1115.0	1.21499
32.0	64.03	31.76	−320.4	−162.9	24.09	12.92	27.19	1064.0	1.20644
34.0	61.32	30.42	−270.7	−106.3	25.80	12.99	29.52	1011.0	1.19714
36.0	58.35	28.94	−217.4	−44.7	27.56	13.05	32.13	955.6	1.18703
38.0	55.10	27.33	−160.5	22.4	29.37	13.11	35.01	899.4	1.17604
40.0	51.58	25.59	−99.9	95.4	31.25	13.18	38.05	844.8	1.16420
42.0	47.84	23.73	−36.2	174.5	33.18	13.26	40.93	794.7	1.15169
44.0	43.99	21.82	29.5	258.6	35.13	13.32	43.01	753.1	1.13893
46.0	40.23	19.96	95.1	345.6	37.06	13.36	43.71	722.6	1.12655
48.0	36.76	18.23	158.3	432.5	38.91	13.37	42.98	703.4	1.11521
50.0	33.70	16.72	217.8	516.9	40.64	13.36	41.32	693.4	1.10527
51.0	32.33	16.04	245.9	557.7	41.45	13.36	40.34	691.0	1.10085
52.0	31.06	15.41	273.0	597.6	42.22	13.36	39.33	689.8	1.09677
53.0	29.89	14.83	299.2	636.4	42.96	13.36	38.34	689.7	1.09301
54.0	28.81	14.29	324.4	674.2	43.67	13.37	37.38	690.4	1.08955
55.0	27.82	13.80	348.8	711.2	44.34	13.38	36.49	691.7	1.08636
56.0	26.89	13.34	372.4	747.2	44.99	13.40	35.65	693.6	1.08341
58.0	25.24	12.52	417.7	817.0	46.22	13.46	34.18	698.3	1.07816
60.0	23.81	11.81	460.8	884.1	47.36	13.55	32.95	704.0	1.07361
62.0	22.56	11.19	502.1	949.0	48.42	13.66	31.95	710.2	1.06964
64.0	21.45	10.64	542.1	1012.0	49.42	13.80	31.14	716.6	1.06615
66.0	20.46	10.15	581.1	1074.0	50.37	13.97	30.49	723.2	1.06304
68.0	19.58	9.713	619.3	1134.0	51.27	14.16	29.97	729.7	1.06026
70.0	18.78	9.317	657.0	1194.0	52.13	14.37	29.57	736.2	1.05775
72.0	18.06	8.957	694.2	1252.0	52.96	14.59	29.26	742.6	1.05548
74.0	17.39	8.629	731.3	1311.0	53.76	14.84	29.04	748.8	1.05341
76.0	16.79	8.327	768.2	1369.0	54.53	15.11	28.89	754.9	1.05151
78.0	16.23	8.049	805.1	1426.0	55.28	15.38	28.80	760.9	1.04976
80.0	15.71	7.792	842.2	1484.0	56.01	15.68	28.76	766.8	1.04814
82.0	15.23	7.552	879.3	1541.0	56.72	15.98	28.76	772.5	1.04664
84.0	14.78	7.330	916.8	1599.0	57.41	16.30	28.80	778.2	1.04524
86.0	14.36	7.121	954.5	1657.0	58.09	16.62	28.88	783.8	1.04393
90.0	13.59	6.743	1031.0	1773.0	59.41	17.29	29.12	794.7	1.04157
95.0	12.76	6.329	1129.0	1919.0	61.00	18.15	29.54	808.1	1.03898
100.0	12.03	5.968	1230.0	2068.0	62.52	19.01	30.04	821.4	1.03673
105.0	11.39	5.650	1335.0	2220.0	64.00	19.86	30.59	834.6	1.03475
110.0	10.82	5.368	1443.0	2374.0	65.44	20.67	31.15	847.8	1.03299
115.0	10.31	5.115	1554.0	2531.0	66.83	21.43	31.69	861.2	1.03142
120.0	9.850	4.886	1668.0	2691.0	68.19	22.12	32.19	874.7	1.03000

Thermophysical properties of hydrogen—Continued

T K	Density kg/m^3	Density mol/dm^3	E J/mol	H J/mol	S J/(mol·K)	C_v J/(mol·K)	C_p J/(mol·K)	Sound m/s	Diel. const.
125.0	9.432	4.679	1784.0	2853.0	69.52	22.73	32.64	888.4	1.02871
130.0	9.051	4.490	1903.0	3017.0	70.80	23.26	33.02	902.3	1.02754
135.0	8.701	4.316	2025.0	3183.0	72.06	23.70	33.34	916.3	1.02647
140.0	8.380	4.157	2147.0	3350.0	73.27	24.08	33.61	930.3	1.02548
145.0	8.082	4.009	2272.0	3519.0	74.46	24.37	33.80	944.4	1.02457
150.0	7.806	3.872	2397.0	3688.0	75.61	24.59	33.93	958.6	1.02372
160.0	7.310	3.626	2649.0	4028.0	77.80	24.84	34.04	987.1	1.02220
170.0	6.876	3.411	2903.0	4368.0	79.86	24.88	33.95	1016.0	1.02087
180.0	6.493	3.221	3155.0	4707.0	81.80	24.76	33.73	1044.0	1.01970
190.0	6.152	3.052	3404.0	5043.0	83.61	24.54	33.43	1072.0	1.01866
200.0	5.846	2.900	3651.0	5375.0	85.32	24.25	33.07	1100.0	1.01773
210.0	5.570	2.763	3895.0	5704.0	86.92	23.93	32.69	1128.0	1.01689
220.0	5.320	2.639	4134.0	6029.0	88.43	23.61	32.31	1155.0	1.01612
230.0	5.092	2.526	4371.0	6350.0	89.86	23.28	31.94	1181.0	1.01543
240.0	4.883	2.422	4604.0	6668.0	91.21	22.98	31.60	1207.0	1.01479
250.0	4.691	2.327	4834.0	6982.0	92.50	22.70	31.29	1233.0	1.01421
260.0	4.513	2.239	5061.0	7294.0	93.72	22.45	31.01	1257.0	1.01367
270.0	4.349	2.157	5285.0	7603.0	94.89	22.23	30.76	1282.0	1.01317
280.0	4.197	2.082	5507.0	7909.0	96.00	22.03	30.54	1305.0	1.01270
300.0	3.922	1.946	5946.0	8516.0	98.09	21.72	30.19	1351.0	1.01187
320.0	3.682	1.827	6380.0	9117.0	100.0	21.49	29.93	1395.0	1.01114
340.0	3.470	1.721	6809.0	9714.0	101.8	21.33	29.75	1437.0	1.01049
360.0	3.281	1.628	7236.0	10310.0	103.5	21.22	29.62	1477.0	1.00992
380.0	3.112	1.544	7660.0	10900.0	105.1	21.14	29.52	1517.0	1.00941
400.0	2.959	1.468	8083.0	11490.0	106.7	21.09	29.46	1555.0	1.00895

10.00 MPa isobar

T K	Density kg/m^3	Density mol/dm^3	E J/mol	H J/mol	S J/(mol·K)	C_v J/(mol·K)	C_p J/(mol·K)	Sound m/s	Diel. const.
16.81[a]	81.94	40.65	−606.4	−360.4	10.58	9.76	12.59	1470.0	1.26896
18.0	81.23	40.29	−593.0	−344.8	11.47	10.15	13.47	1471.0	1.26644
20.0	79.97	39.67	−568.4	−316.3	12.97	10.91	15.05	1458.0	1.26198
22.0	78.63	39.00	−541.1	−284.7	14.47	11.56	16.52	1438.0	1.25723
24.0	77.20	38.30	−511.4	−250.3	15.97	12.03	17.86	1415.0	1.25222
26.0	75.71	37.55	−479.6	−213.3	17.45	12.38	19.11	1389.0	1.24696
28.0	74.13	36.77	−445.8	−173.9	18.91	12.63	20.30	1361.0	1.24145
30.0	72.49	35.96	−410.3	−132.2	20.35	12.81	21.45	1332.0	1.23570
32.0	70.77	35.10	−373.0	−88.1	21.77	12.94	22.58	1301.0	1.22971
34.0	68.97	34.21	−334.1	−41.9	23.17	13.03	23.70	1270.0	1.22349
40.0	63.19	31.34	−208.8	110.2	27.28	13.18	26.98	1171.0	1.20354
45.0	58.00	28.77	−96.1	251.5	30.61	13.27	29.45	1091.0	1.18585
50.0	52.70	26.14	21.2	403.7	33.81	13.36	31.31	1022.0	1.16794
55.0	47.57	23.60	139.4	563.2	36.85	13.51	32.31	968.4	1.15079
60.0	42.88	21.27	255.4	725.4	39.68	13.76	32.48	930.9	1.13528
62.0	41.17	20.42	300.7	790.3	40.74	13.90	32.39	920.1	1.12964
64.0	39.56	19.62	345.4	855.0	41.77	14.05	32.24	911.2	1.12436
66.0	38.05	18.87	389.4	919.3	42.76	14.23	32.06	904.0	1.11942
68.0	36.63	18.17	432.9	983.2	43.71	14.42	31.86	898.4	1.11480
70.0	35.31	17.52	475.8	1047.0	44.63	14.63	31.68	894.1	1.11051
72.0	34.08	16.90	518.3	1110.0	45.52	14.86	31.50	890.9	1.10651
74.0	32.92	16.33	560.5	1173.0	46.38	15.11	31.36	888.6	1.10278
76.0	31.85	15.80	602.3	1235.0	47.22	15.37	31.23	887.1	1.09930
78.0	30.84	15.30	644.0	1298.0	48.03	15.65	31.14	886.2	1.09606
80.0	29.90	14.83	685.6	1360.0	48.81	15.94	31.09	886.0	1.09303
82.0	29.01	14.39	727.2	1422.0	49.58	16.25	31.06	886.2	1.09019
85.0	27.79	13.78	789.7	1515.0	50.70	16.72	31.07	887.4	1.08626
90.0	25.97	12.88	894.7	1671.0	52.48	17.55	31.21	891.2	1.08047
95.0	24.40	12.10	1001.0	1828.0	54.17	18.40	31.49	896.7	1.07547
100.0	23.02	11.42	1110.0	1986.0	55.79	19.27	31.85	903.6	1.07109
105.0	21.80	10.81	1221.0	2146.0	57.36	20.11	32.27	911.7	1.06724
110.0	20.71	10.27	1335.0	2309.0	58.87	20.92	32.71	920.8	1.06382
115.0	19.74	9.792	1452.0	2473.0	60.33	21.67	33.14	930.7	1.06076
120.0	18.86	9.357	1571.0	2640.0	61.75	22.36	33.54	941.3	1.05801

Thermophysical properties of hydrogen—Continued

T K	Density kg/m^3	Density mol/dm^3	E J/mol	H J/mol	S J/(mol·K)	C_v J/(mol·K)	C_p J/(mol·K)	Sound m/s	Diel. const.
125.0	18.07	8.962	1693.0	2809.0	63.13	22.96	33.90	952.6	1.05551
130.0	17.34	8.602	1816.0	2979.0	64.46	23.48	34.20	964.5	1.05324
135.0	16.68	8.273	1942.0	3151.0	65.76	23.92	34.44	976.7	1.05116
140.0	16.07	7.969	2068.0	3323.0	67.02	24.30	34.65	989.2	1.04926
145.0	15.50	7.689	2196.0	3497.0	68.23	24.58	34.78	1002.0	1.04749
150.0	14.98	7.429	2325.0	3671.0	69.42	24.80	34.85	1015.0	1.04587
160.0	14.04	6.963	2583.0	4020.0	71.67	25.04	34.85	1042.0	1.04294
170.0	13.21	6.555	2842.0	4367.0	73.77	25.07	34.68	1069.0	1.04039
180.0	12.49	6.195	3099.0	4713.0	75.75	24.94	34.39	1096.0	1.03814
190.0	11.84	5.875	3353.0	5055.0	77.60	24.71	34.02	1123.0	1.03615
200.0	11.26	5.588	3604.0	5393.0	79.33	24.41	33.60	1150.0	1.03436
210.0	10.74	5.329	3850.0	5727.0	80.96	24.08	33.18	1176.0	1.03275
220.0	10.27	5.094	4094.0	6057.0	82.50	23.75	32.75	1203.0	1.03129
230.0	9.837	4.880	4333.0	6382.0	83.94	23.42	32.35	1228.0	1.02996
240.0	9.441	4.683	4569.0	6704.0	85.31	23.11	31.97	1254.0	1.02874
250.0	9.077	4.503	4801.0	7022.0	86.61	22.83	31.63	1278.0	1.02762
260.0	8.741	4.336	5030.0	7336.0	87.84	22.57	31.32	1302.0	1.02659
280.0	8.141	4.038	5481.0	7958.0	90.15	22.14	30.81	1349.0	1.02475
300.0	7.620	3.780	5924.0	8570.0	92.26	21.82	30.42	1394.0	1.02315
320.0	7.163	3.553	6360.0	9175.0	94.21	21.58	30.13	1436.0	1.02175
340.0	6.758	3.352	6793.0	9775.0	96.03	21.42	29.92	1477.0	1.02051
360.0	6.398	3.174	7221.0	10370.0	97.74	21.30	29.77	1516.0	1.01941
380.0	6.075	3.013	7648.0	10970.0	99.34	21.22	29.66	1554.0	1.01842
400.0	5.783	2.869	8073.0	11560.0	100.9	21.16	29.57	1592.0	1.01753

15.00 MPa isobar

T K	Density kg/m^3	Density mol/dm^3	E J/mol	H J/mol	S J/(mol·K)	C_v J/(mol·K)	C_p J/(mol·K)	Sound m/s	Diel. const.
18.12[a]	83.96	41.65	–596.8	–236.6	10.70	9.78	12.88	1604.0	1.27613
20.0	82.90	41.12	–575.9	–211.2	12.04	10.62	14.28	1590.0	1.27236
25.0	79.87	39.62	–510.2	–131.6	15.58	12.12	17.38	1536.0	1.26162
30.0	76.55	37.97	–433.5	–38.4	18.97	12.84	19.84	1478.0	1.24991
35.0	72.96	36.19	–348.2	66.3	22.19	13.20	22.00	1417.0	1.23734
40.0	69.14	34.30	–256.2	181.2	25.26	13.38	23.94	1355.0	1.22407
50.0	61.13	30.32	–57.0	437.7	30.97	13.64	27.17	1236.0	1.19649
60.0	53.25	26.42	152.6	720.4	36.12	14.05	29.17	1139.0	1.16980
65.0	49.61	24.61	258.4	867.9	38.48	14.42	29.77	1102.0	1.15760
70.0	46.26	22.95	364.2	1018.0	40.70	14.91	30.21	1072.0	1.14645
75.0	43.22	21.44	470.2	1170.0	42.80	15.51	30.55	1049.0	1.13639
80.0	40.49	20.09	576.5	1323.0	44.78	16.21	30.87	1032.0	1.12741
85.0	38.05	18.88	683.8	1478.0	46.66	16.98	31.21	1020.0	1.11942
90.0	35.87	17.79	792.5	1635.0	48.45	17.80	31.58	1012.0	1.11233
95.0	33.92	16.83	903.0	1794.0	50.17	18.64	32.01	1008.0	1.10601
100.0	32.18	15.96	1016.0	1956.0	51.83	19.50	32.46	1006.0	1.10037
105.0	30.61	15.18	1131.0	2119.0	53.42	20.34	32.93	1007.0	1.09531
110.0	29.19	14.48	1249.0	2285.0	54.96	21.14	33.39	1010.0	1.09076
115.0	27.90	13.84	1369.0	2453.0	56.46	21.89	33.84	1015.0	1.08664
120.0	26.73	13.26	1492.0	2623.0	57.91	22.57	34.24	1022.0	1.08290
125.0	25.66	12.73	1617.0	2795.0	59.31	23.17	34.59	1029.0	1.07948
130.0	24.68	12.24	1743.0	2969.0	60.67	23.68	34.87	1038.0	1.07636
135.0	23.77	11.79	1872.0	3144.0	61.99	24.11	35.10	1047.0	1.07348
140.0	22.93	11.38	2001.0	3320.0	63.27	24.49	35.28	1057.0	1.07083
150.0	21.44	10.63	2263.0	3674.0	65.72	24.98	35.45	1079.0	1.06610
160.0	20.13	9.987	2526.0	4028.0	68.00	25.21	35.41	1102.0	1.06200
170.0	18.99	9.420	2789.0	4381.0	70.14	25.23	35.20	1127.0	1.05841
180.0	17.98	8.918	3050.0	4732.0	72.15	25.10	34.86	1152.0	1.05523
190.0	17.07	8.470	3307.0	5078.0	74.02	24.86	34.46	1177.0	1.05241
200.0	16.26	8.067	3561.0	5421.0	75.78	24.55	34.01	1202.0	1.04987
210.0	15.53	7.703	3811.0	5758.0	77.43	24.22	33.55	1227.0	1.04758
220.0	14.86	7.372	4057.0	6092.0	78.98	23.88	33.10	1252.0	1.04550
230.0	14.25	7.069	4299.0	6421.0	80.44	23.54	32.67	1277.0	1.04361
240.0	13.69	6.792	4537.0	6745.0	81.82	23.23	32.27	1301.0	1.04187
250.0	13.18	6.537	4771.0	7066.0	83.13	22.94	31.91	1325.0	1.04028
260.0	12.70	6.300	5003.0	7384.0	84.37	22.68	31.58	1348.0	1.03880

Thermophysical properties of hydrogen—Continued

T K	Density kg/m³	Density mol/dm³	E J/mol	H J/mol	S J/(mol·K)	C_v J/(mol·K)	C_p J/(mol·K)	Sound m/s	Diel. const.
280.0	11.85	5.878	5457.0	8009.0	86.69	22.24	31.03	1393.0	1.03616
300.0	11.11	5.510	5903.0	8626.0	88.82	21.91	30.61	1436.0	1.03387
320.0	10.46	5.186	6343.0	9235.0	90.79	21.67	30.30	1477.0	1.03186
340.0	9.878	4.900	6777.0	9838.0	92.62	21.50	30.07	1517.0	1.03008
360.0	9.362	4.644	7208.0	10440.0	94.33	21.38	29.90	1555.0	1.02850
380.0	8.899	4.414	7637.0	11030.0	95.94	21.29	29.77	1592.0	1.02707
400.0	8.480	4.207	8063.0	11630.0	97.47	21.22	29.68	1628.0	1.02579
				20.00 MPa isobar					
19.36[a]	85.68	42.50	−586.3	−115.8	10.81	10.16	13.41	1716.0	1.28229
25.0	82.64	40.99	−518.3	−30.4	14.67	12.09	16.67	1649.0	1.27144
30.0	79.73	39.55	−447.1	58.6	17.91	12.89	18.86	1593.0	1.26113
35.0	76.63	38.01	−368.4	157.7	20.96	13.32	20.74	1540.0	1.25021
40.0	73.38	36.40	−283.9	265.6	23.84	13.57	22.39	1488.0	1.23882
60.0	59.91	29.72	93.1	766.0	33.92	14.35	27.17	1299.0	1.19235
70.0	53.68	26.63	294.5	1046.0	38.22	15.20	28.70	1226.0	1.17124
75.0	50.84	25.22	397.7	1191.0	40.23	15.79	29.36	1196.0	1.16169
80.0	48.19	23.91	502.5	1339.0	42.14	16.47	29.99	1171.0	1.15286
85.0	45.75	22.69	609.4	1491.0	43.98	17.23	30.60	1151.0	1.14475
90.0	43.51	21.58	718.5	1645.0	45.74	18.04	31.21	1135.0	1.13733
95.0	41.45	20.56	830.1	1803.0	47.45	18.88	31.82	1123.0	1.13055
100.0	39.56	19.62	944.3	1963.0	49.09	19.72	32.42	1114.0	1.12436
105.0	37.83	18.77	1061.0	2127.0	50.69	20.55	32.99	1109.0	1.11871
110.0	36.24	17.98	1181.0	2293.0	52.24	21.35	33.54	1106.0	1.11354
115.0	34.78	17.25	1303.0	2462.0	53.74	22.09	34.05	1106.0	1.10879
120.0	33.44	16.59	1428.0	2634.0	55.20	22.77	34.50	1108.0	1.10443
125.0	32.19	15.97	1555.0	2807.0	56.62	23.36	34.88	1112.0	1.10042
130.0	31.04	15.40	1683.0	2982.0	57.99	23.87	35.18	1117.0	1.09670
140.0	28.98	14.37	1945.0	3336.0	60.61	24.66	35.62	1130.0	1.09008
150.0	27.18	13.48	2210.0	3694.0	63.08	25.14	35.79	1147.0	1.08434
160.0	25.61	12.70	2477.0	4052.0	65.39	25.37	35.75	1167.0	1.07932
170.0	24.21	12.01	2743.0	4408.0	67.55	25.38	35.53	1188.0	1.07489
180.0	22.97	11.40	3007.0	4762.0	69.57	25.24	35.19	1210.0	1.07096
190.0	21.86	10.84	3267.0	5112.0	71.46	24.99	34.77	1233.0	1.06744
200.0	20.86	10.35	3524.0	5457.0	73.23	24.68	34.30	1257.0	1.06427
210.0	19.94	9.894	3776.0	5798.0	74.90	24.35	33.83	1280.0	1.06141
220.0	19.11	9.481	4024.0	6134.0	76.46	24.00	33.36	1304.0	1.05879
230.0	18.35	9.104	4268.0	6465.0	77.93	23.66	32.92	1327.0	1.05641
240.0	17.65	8.757	4508.0	6792.0	79.33	23.34	32.51	1350.0	1.05422
250.0	17.01	8.436	4745.0	7115.0	80.64	23.05	32.13	1372.0	1.05220
260.0	16.41	8.140	4978.0	7435.0	81.90	22.78	31.79	1395.0	1.05033
280.0	15.34	7.607	5436.0	8065.0	84.23	22.33	31.22	1438.0	1.04698
300.0	14.40	7.142	5884.0	8684.0	86.37	22.00	30.78	1479.0	1.04406
320.0	13.57	6.733	6326.0	9297.0	88.35	21.75	30.45	1519.0	1.04150
340.0	12.84	6.369	6763.0	9903.0	90.18	21.57	30.20	1557.0	1.03923
360.0	12.18	6.044	7196.0	10510.0	91.90	21.45	30.02	1594.0	1.03720
380.0	11.59	5.751	7626.0	11100.0	93.52	21.36	29.88	1630.0	1.03537
400.0	11.06	5.486	8055.0	11700.0	95.05	21.29	29.77	1665.0	1.03372
				25.00 MPa isobar					
20.53[a]	87.21	43.26	−574.7	3.2	10.93	10.68	13.93	1807.0	1.28773
25.0	84.99	42.16	−522.2	70.7	13.90	12.11	16.18	1747.0	1.27982
30.0	82.39	40.87	−454.9	156.8	17.03	12.95	18.18	1691.0	1.27054
35.0	79.63	39.50	−380.9	252.1	19.97	13.43	19.89	1641.0	1.26078
40.0	76.75	38.07	−301.4	355.3	22.72	13.72	21.39	1595.0	1.25062
60.0	64.83	32.16	54.3	831.6	32.32	14.60	25.88	1427.0	1.20919
70.0	59.19	29.36	247.6	1099.0	36.44	15.46	27.58	1355.0	1.18990
80.0	54.06	26.82	450.6	1383.0	40.23	16.72	29.16	1294.0	1.17251
90.0	49.51	24.56	664.3	1682.0	43.75	18.27	30.68	1249.0	1.15727
100.0	45.55	22.59	889.8	1996.0	47.06	19.94	32.13	1219.0	1.14409

Thermophysical properties of hydrogen—Continued

T K	Density kg/m^3	Density mol/dm^3	E J/mol	H J/mol	S J/(mol·K)	C_v J/(mol·K)	C_p J/(mol·K)	Sound m/s	Diel. const.
110.0	42.11	20.89	1128.0	2324.0	50.18	21.55	33.43	1202.0	1.13274
120.0	39.14	19.41	1376.0	2664.0	53.14	22.95	34.51	1196.0	1.12297
130.0	36.54	18.13	1634.0	3013.0	55.93	24.04	35.28	1197.0	1.11451
140.0	34.27	17.00	1898.0	3369.0	58.57	24.82	35.77	1205.0	1.10714
150.0	32.28	16.01	2166.0	3728.0	61.04	25.30	35.97	1218.0	1.10069
160.0	30.51	15.13	2435.0	4087.0	63.36	25.52	35.95	1233.0	1.09498
170.0	28.93	14.35	2704.0	4446.0	65.54	25.52	35.74	1251.0	1.08992
180.0	27.51	13.65	2970.0	4802.0	67.57	25.37	35.40	1271.0	1.08538
190.0	26.23	13.01	3232.0	5154.0	69.47	25.12	34.98	1291.0	1.08130
200.0	25.07	12.44	3491.0	5501.0	71.26	24.81	34.51	1313.0	1.07761
210.0	24.02	11.91	3745.0	5844.0	72.93	24.46	34.03	1334.0	1.07426
220.0	23.05	11.43	3995.0	6182.0	74.50	24.11	33.56	1356.0	1.07120
230.0	22.16	10.99	4241.0	6515.0	75.98	23.77	33.11	1378.0	1.06839
240.0	21.34	10.59	4483.0	6844.0	77.38	23.44	32.69	1399.0	1.06581
250.0	20.58	10.21	4721.0	7169.0	78.71	23.14	32.30	1421.0	1.06342
260.0	19.88	9.862	4955.0	7490.0	79.97	22.88	31.96	1442.0	1.06120
280.0	18.61	9.234	5416.0	8123.0	82.32	22.42	31.37	1483.0	1.05723
300.0	17.51	8.684	5867.0	8746.0	84.46	22.08	30.92	1523.0	1.05375
320.0	16.53	8.198	6311.0	9361.0	86.45	21.83	30.57	1561.0	1.05069
340.0	15.65	7.765	6750.0	9969.0	88.29	21.65	30.31	1598.0	1.04797
360.0	14.87	7.377	7185.0	10570.0	90.02	21.51	30.12	1633.0	1.04554
380.0	14.17	7.027	7617.0	11170.0	91.64	21.42	29.97	1668.0	1.04334
400.0	13.53	6.710	8047.0	11770.0	93.18	21.35	29.85	1702.0	1.04136

30.00 MPa isobar

T K	Density kg/m^3	Density mol/dm^3	E J/mol	H J/mol	S J/(mol·K)	C_v J/(mol·K)	C_p J/(mol·K)	Sound m/s	Diel. const.
21.65[a]	88.58	43.94	–562.0	120.7	11.07	11.21	14.39	1881.0	1.29267
30.0	84.68	42.01	–458.9	255.3	16.30	13.02	17.66	1777.0	1.27871
40.0	79.58	39.48	–312.5	447.5	21.81	13.85	20.66	1686.0	1.26060
60.0	68.77	34.11	27.4	906.9	31.06	14.82	24.98	1535.0	1.22278
70.0	63.58	31.54	214.4	1166.0	35.04	15.68	26.76	1465.0	1.20489
80.0	58.76	29.15	412.7	1442.0	38.73	16.94	28.49	1403.0	1.18844
90.0	54.40	26.99	623.6	1735.0	42.18	18.48	30.18	1353.0	1.17368
100.0	50.51	25.06	847.9	2045.0	45.45	20.13	31.79	1317.0	1.16061
110.0	47.07	23.35	1085.0	2370.0	48.55	21.73	33.23	1294.0	1.14911
120.0	44.02	21.84	1335.0	2709.0	51.49	23.12	34.41	1281.0	1.13902
130.0	41.32	20.50	1594.0	3057.0	54.28	24.20	35.25	1277.0	1.13014
140.0	38.93	19.31	1859.0	3413.0	56.92	24.98	35.80	1280.0	1.12230
150.0	36.80	18.25	2129.0	3772.0	59.39	25.44	36.04	1288.0	1.11535
160.0	34.89	17.31	2400.0	4133.0	61.72	25.65	36.05	1300.0	1.10915
170.0	33.18	16.46	2670.0	4493.0	63.90	25.65	35.86	1315.0	1.10360
180.0	31.63	15.69	2938.0	4850.0	65.94	25.50	35.53	1332.0	1.09860
190.0	30.22	14.99	3202.0	5203.0	67.85	25.24	35.11	1350.0	1.09408
200.0	28.94	14.36	3462.0	5552.0	69.64	24.92	34.65	1369.0	1.08997
210.0	27.77	13.78	3718.0	5896.0	71.32	24.57	34.17	1389.0	1.08621
220.0	26.69	13.24	3969.0	6235.0	72.90	24.21	33.70	1409.0	1.08277
230.0	25.70	12.75	4217.0	6570.0	74.39	23.87	33.25	1429.0	1.07961
240.0	24.78	12.29	4460.0	6900.0	75.79	23.54	32.83	1449.0	1.07669
260.0	23.13	11.48	4935.0	7549.0	78.39	22.96	32.09	1489.0	1.07147
280.0	21.70	10.76	5398.0	8185.0	80.75	22.51	31.49	1528.0	1.06694
300.0	20.44	10.14	5851.0	8810.0	82.90	22.16	31.03	1566.0	1.06297
320.0	19.33	9.587	6297.0	9427.0	84.89	21.90	30.68	1603.0	1.05946
340.0	18.33	9.092	6738.0	10040.0	86.75	21.71	30.41	1638.0	1.05633
360.0	17.43	8.648	7174.0	10640.0	88.48	21.58	30.21	1672.0	1.05353
380.0	16.62	8.246	7608.0	11250.0	90.11	21.48	30.05	1706.0	1.05100
400.0	15.89	7.881	8039.0	11850.0	91.64	21.40	29.92	1739.0	1.04870

35.00 MPa isobar

T K	Density kg/m^3	Density mol/dm^3	E J/mol	H J/mol	S J/(mol·K)	C_v J/(mol·K)	C_p J/(mol·K)	Sound m/s	Diel. const.
22.72[a]	89.85	44.57	–548.4	236.9	11.21	11.69	14.75	1945.0	1.29722
30.0	86.71	43.01	–460.0	353.8	15.66	13.09	17.24	1854.0	1.28595

Thermophysical properties of hydrogen—Continued

T K	Density kg/m^3	Density mol/dm^3	E J/mol	H J/mol	S J/(mol·K)	C_v J/(mol·K)	C_p J/(mol·K)	Sound m/s	Diel. const.
40.0	82.04	40.69	−319.2	540.9	21.02	13.95	20.08	1766.0	1.26931
60.0	72.08	35.75	8.4	987.4	30.02	15.00	24.30	1627.0	1.23427
70.0	67.24	33.35	190.1	1240.0	33.90	15.88	26.13	1561.0	1.21748
80.0	62.69	31.10	384.4	1510.0	37.51	17.14	27.94	1499.0	1.20185
90.0	58.52	29.03	592.6	1798.0	40.90	18.68	29.75	1448.0	1.18759
100.0	54.73	27.15	815.2	2104.0	44.13	20.32	31.46	1408.0	1.17476
110.0	51.32	25.45	1052.0	2427.0	47.20	21.90	33.00	1380.0	1.16330
120.0	48.26	23.94	1301.0	2764.0	50.13	23.29	34.26	1362.0	1.15308
130.0	45.52	22.58	1561.0	3111.0	52.91	24.35	35.17	1354.0	1.14397
140.0	43.05	21.36	1827.0	3466.0	55.54	25.12	35.77	1353.0	1.13584
150.0	40.84	20.26	2098.0	3825.0	58.02	25.58	36.05	1358.0	1.12855
160.0	38.84	19.27	2370.0	4186.0	60.35	25.78	36.08	1366.0	1.12200
170.0	37.03	18.37	2641.0	4546.0	62.53	25.78	35.92	1378.0	1.11609
180.0	35.38	17.55	2910.0	4904.0	64.57	25.62	35.60	1392.0	1.11074
190.0	33.88	16.81	3175.0	5258.0	66.49	25.35	35.20	1408.0	1.10587
200.0	32.50	16.12	3437.0	5608.0	68.28	25.03	34.75	1425.0	1.10142
210.0	31.24	15.50	3694.0	5953.0	69.97	24.67	34.27	1443.0	1.09734
220.0	30.07	14.92	3947.0	6293.0	71.55	24.31	33.80	1462.0	1.09358
230.0	28.99	14.38	4195.0	6629.0	73.04	23.96	33.35	1481.0	1.09012
240.0	27.99	13.88	4439.0	6960.0	74.45	23.63	32.93	1499.0	1.08691
260.0	26.19	12.99	4917.0	7611.0	77.06	23.05	32.19	1537.0	1.08116
280.0	24.61	12.21	5382.0	8249.0	79.42	22.59	31.59	1574.0	1.07615
300.0	23.22	11.52	5837.0	8876.0	81.58	22.23	31.12	1610.0	1.07174
320.0	21.98	10.90	6285.0	9494.0	83.58	21.97	30.76	1645.0	1.06783
340.0	20.88	10.36	6727.0	10110.0	85.44	21.78	30.49	1679.0	1.06434
360.0	19.88	9.861	7165.0	10710.0	87.17	21.64	30.28	1712.0	1.06120
380.0	18.97	9.412	7600.0	11320.0	88.81	21.54	30.12	1744.0	1.05836
400.0	18.15	9.004	8032.0	11920.0	90.35	21.46	29.99	1776.0	1.05578

40.00 MPa isobar

T K	Density kg/m^3	Density mol/dm^3	E J/mol	H J/mol	S J/(mol·K)	C_v J/(mol·K)	C_p J/(mol·K)	Sound m/s	Diel. const.
23.75[a]	91.03	45.16	−533.8	352.0	11.37	12.12	15.04	2000.0	1.30149
30.0	88.53	43.91	−458.9	452.0	15.10	13.16	16.91	1925.0	1.29248
40.0	84.22	41.78	−322.6	634.8	20.34	14.04	19.60	1839.0	1.27706
60.0	74.94	37.17	−5.1	1071.0	29.12	15.16	23.77	1708.0	1.24428
70.0	70.39	34.92	172.2	1318.0	32.93	16.05	25.62	1646.0	1.22839
80.0	66.07	32.78	362.9	1583.0	36.47	17.32	27.50	1586.0	1.21346
90.0	62.06	30.79	568.4	1868.0	39.82	18.85	29.37	1533.0	1.19968
100.0	58.38	28.96	789.3	2171.0	43.01	20.49	31.17	1491.0	1.18713
110.0	55.03	27.30	1025.0	2491.0	46.05	22.06	32.78	1460.0	1.17578
120.0	51.99	25.79	1274.0	2825.0	48.97	23.44	34.10	1439.0	1.16554
130.0	49.23	24.42	1533.0	3171.0	51.74	24.50	35.07	1428.0	1.15633
140.0	46.74	23.18	1800.0	3525.0	54.36	25.26	35.71	1424.0	1.14802
150.0	44.47	22.06	2071.0	3884.0	56.84	25.71	36.03	1425.0	1.14052
160.0	42.41	21.04	2344.0	4245.0	59.16	25.90	36.09	1431.0	1.13372
170.0	40.53	20.11	2616.0	4605.0	61.35	25.89	35.94	1440.0	1.12755
180.0	38.81	19.25	2886.0	4963.0	63.39	25.73	35.64	1452.0	1.12192
190.0	37.24	18.47	3152.0	5318.0	65.31	25.46	35.25	1466.0	1.11677
200.0	35.79	17.75	3415.0	5668.0	67.11	25.13	34.81	1481.0	1.11205
210.0	34.45	17.09	3673.0	6014.0	68.80	24.77	34.34	1497.0	1.10770
220.0	33.21	16.47	3927.0	6355.0	70.38	24.41	33.88	1514.0	1.10369
240.0	30.99	15.37	4421.0	7024.0	73.29	23.72	33.01	1549.0	1.09653
260.0	29.05	14.41	4901.0	7676.0	75.90	23.13	32.27	1584.0	1.09032
280.0	27.35	13.57	5367.0	8315.0	78.27	22.66	31.67	1619.0	1.08488
300.0	25.85	12.82	5824.0	8944.0	80.44	22.30	31.20	1653.0	1.08008
320.0	24.51	12.16	6274.0	9564.0	82.44	22.04	30.84	1686.0	1.07582
340.0	23.30	11.56	6717.0	10180.0	84.30	21.84	30.56	1719.0	1.07200
360.0	22.21	11.02	7157.0	10790.0	86.04	21.70	30.35	1751.0	1.06855
380.0	21.22	10.53	7593.0	11390.0	87.68	21.59	30.18	1782.0	1.06543
400.0	20.32	10.08	8026.0	11990.0	89.22	21.51	30.04	1812.0	1.06259

Thermophysical properties of hydrogen—Continued

T K	Density kg/m^3	Density mol/dm^3	E J/mol	H J/mol	S J/(mol·K)	C_v J/(mol·K)	C_p J/(mol·K)	Sound m/s	Diel. const.
				50.00 MPa isobar					
25.71[a]	93.22	46.24	−502.4	578.9	11.70	12.78	15.43	2095.0	1.30939
30.0	91.71	45.49	−451.9	647.1	14.15	13.30	16.40	2049.0	1.30394
40.0	87.98	43.64	−322.5	823.3	19.20	14.16	18.81	1967.0	1.29050
100.0	64.50	31.99	752.3	2315.0	41.17	20.78	30.68	1640.0	1.20803
110.0	61.27	30.39	985.7	2631.0	44.18	22.35	32.39	1606.0	1.19699
120.0	58.31	28.92	1233.0	2962.0	47.06	23.71	33.81	1581.0	1.18688
130.0	55.58	27.57	1492.0	3306.0	49.81	24.75	34.85	1565.0	1.17763
140.0	53.07	26.32	1758.0	3658.0	52.42	25.50	35.56	1556.0	1.16918
150.0	50.76	25.18	2030.0	4015.0	54.89	25.94	35.93	1553.0	1.16144
160.0	48.64	24.13	2303.0	4376.0	57.21	26.13	36.03	1554.0	1.15435
170.0	46.68	23.16	2576.0	4735.0	59.39	26.10	35.92	1560.0	1.14784
180.0	44.88	22.26	2847.0	5093.0	61.44	25.93	35.66	1568.0	1.14185
190.0	43.20	21.43	3115.0	5448.0	63.36	25.65	35.29	1578.0	1.13632
200.0	41.65	20.66	3379.0	5799.0	65.16	25.32	34.87	1590.0	1.13121
220.0	38.86	19.28	3894.0	6487.0	68.44	24.58	33.96	1617.0	1.12207
240.0	36.43	18.07	4391.0	7158.0	71.36	23.88	33.11	1647.0	1.11414
260.0	34.29	17.01	4873.0	7813.0	73.98	23.28	32.38	1678.0	1.10720
280.0	32.40	16.07	5343.0	8454.0	76.36	22.80	31.78	1709.0	1.10107
300.0	30.71	15.23	5802.0	9085.0	78.53	22.43	31.31	1739.0	1.09563
320.0	29.19	14.48	6254.0	9707.0	80.54	22.16	30.95	1769.0	1.09076
340.0	27.82	13.80	6700.0	10320.0	82.41	21.96	30.67	1799.0	1.08638
360.0	26.58	13.18	7142.0	10930.0	84.15	21.81	30.45	1828.0	1.08241
380.0	25.45	12.62	7581.0	11540.0	85.80	21.70	30.28	1857.0	1.07880
400.0	24.41	12.11	8016.0	12150.0	87.35	21.61	30.14	1885.0	1.07550
				60.00 MPa isobar					
27.56[a]	95.23	47.24	−468.6	801.6	12.03	13.24	15.66	2173.0	1.31667
40.0	91.16	45.22	−315.7	1011.0	18.27	14.23	18.18	2079.0	1.30194
60.0	83.78	41.56	−26.6	1417.0	26.45	15.57	22.29	1968.0	1.27549
80.0	76.33	37.86	317.8	1902.0	33.40	17.84	26.28	1863.0	1.24914
100.0	69.52	34.48	728.8	2469.0	39.70	21.02	30.30	1769.0	1.22536
110.0	66.42	32.95	959.9	2781.0	42.68	22.58	32.08	1734.0	1.21466
120.0	63.54	31.52	1206.0	3109.0	45.53	23.94	33.55	1707.0	1.20474
130.0	60.86	30.19	1463.0	3451.0	48.27	24.98	34.65	1688.0	1.19558
140.0	58.37	28.96	1729.0	3801.0	50.86	25.72	35.41	1676.0	1.18712
150.0	56.07	27.81	2000.0	4158.0	53.32	26.15	35.82	1670.0	1.17930
160.0	53.93	26.75	2274.0	4517.0	55.64	26.33	35.96	1669.0	1.17207
170.0	51.94	25.76	2547.0	4876.0	57.82	26.30	35.88	1671.0	1.16538
180.0	50.08	24.84	2819.0	5234.0	59.86	26.11	35.64	1677.0	1.15917
200.0	46.74	23.19	3352.0	5939.0	63.58	25.49	34.88	1694.0	1.14805
220.0	43.82	21.74	3868.0	6628.0	66.86	24.74	34.01	1717.0	1.13837
240.0	41.25	20.46	4368.0	7300.0	69.79	24.02	33.17	1742.0	1.12990
260.0	38.97	19.33	4852.0	7956.0	72.41	23.42	32.45	1769.0	1.12242
280.0	36.93	18.32	5324.0	8599.0	74.79	22.93	31.86	1796.0	1.11577
300.0	35.10	17.41	5785.0	9231.0	76.98	22.55	31.39	1824.0	1.10982
320.0	33.45	16.59	6239.0	9855.0	78.99	22.27	31.03	1851.0	1.10447
340.0	31.95	15.85	6687.0	10470.0	80.86	22.07	30.75	1878.0	1.09963
360.0	30.58	15.17	7131.0	11090.0	82.61	21.91	30.53	1905.0	1.09523
380.0	29.33	14.55	7571.0	11690.0	84.26	21.79	30.35	1931.0	1.09122
400.0	28.18	13.98	8008.0	12300.0	85.81	21.70	30.21	1958.0	1.08753
				80.00 MPa isobar					
31.01[a]	98.89	49.05	−395.9	1235.0	12.64	13.78	15.93	2299.0	1.33000
40.0	96.41	47.83	−289.3	1383.0	16.84	14.27	17.23	2262.0	1.32098
60.0	90.26	44.77	−19.0	1768.0	24.58	15.76	21.23	2171.0	1.29869
80.0	83.73	41.53	308.4	2235.0	31.26	18.16	25.48	2075.0	1.27533
100.0	77.53	38.46	707.2	2787.0	37.41	21.39	29.73	1987.0	1.25339
120.0	71.93	35.68	1176.0	3418.0	43.15	24.31	33.16	1923.0	1.23376
140.0	66.96	33.21	1695.0	4104.0	48.43	26.08	35.16	1888.0	1.21651

Thermophysical properties of hydrogen—Continued

T K	Density kg/m^3	Density mol/dm^3	E J/mol	H J/mol	S J/(mol·K)	C_v J/(mol·K)	C_p J/(mol·K)	Sound m/s	Diel. const.
160.0	62.56	31.03	2238.0	4815.0	53.18	26.66	35.82	1874.0	1.20141
180.0	58.69	29.11	2782.0	5530.0	57.40	26.43	35.57	1875.0	1.18817
200.0	55.25	27.41	3316.0	6235.0	61.11	25.78	34.88	1885.0	1.17653
220.0	52.19	25.89	3834.0	6925.0	64.39	25.01	34.04	1902.0	1.16623
240.0	49.45	24.53	4336.0	7597.0	67.32	24.28	33.23	1921.0	1.15707
260.0	46.99	23.31	4823.0	8255.0	69.95	23.66	32.53	1942.0	1.14888
280.0	44.77	22.21	5297.0	8899.0	72.34	23.16	31.95	1964.0	1.14151
300.0	42.76	21.21	5761.0	9533.0	74.53	22.77	31.50	1986.0	1.13485
320.0	40.92	20.30	6218.0	10160.0	76.55	22.48	31.14	2009.0	1.12881
340.0	39.23	19.46	6669.0	10780.0	78.43	22.26	30.86	2031.0	1.12329
360.0	37.69	18.70	7115.0	11390.0	80.19	22.10	30.64	2053.0	1.11824
380.0	36.26	17.99	7558.0	12010.0	81.84	21.97	30.46	2076.0	1.11360
400.0	34.94	17.33	7998.0	12610.0	83.40	21.87	30.32	2098.0	1.10931
100.00 MPa isobar									
34.18[a]	102.2	50.72	−318.8	1653.0	13.17	14.06	16.19	2389.0	1.34230
40.0	100.8	49.99	−252.1	1748.0	15.74	14.23	16.67	2394.0	1.33694
60.0	95.43	47.34	4.3	2117.0	23.16	15.81	20.37	2339.0	1.31740
80.0	89.61	44.45	318.0	2568.0	29.62	18.35	24.81	2253.0	1.29635
100.0	83.89	41.61	706.2	3109.0	35.64	21.65	29.28	2167.0	1.27588
120.0	78.59	38.98	1168.0	3733.0	41.32	24.59	32.87	2105.0	1.25709
140.0	73.79	36.60	1682.0	4414.0	46.56	26.36	34.97	2068.0	1.24026
160.0	69.49	34.47	2222.0	5123.0	51.29	26.94	35.71	2052.0	1.22528
180.0	65.64	32.56	2765.0	5836.0	55.49	26.69	35.52	2049.0	1.21197
200.0	62.18	30.85	3298.0	6540.0	59.20	26.03	34.86	2056.0	1.20010
220.0	59.07	29.30	3817.0	7229.0	62.49	25.24	34.05	2069.0	1.18948
240.0	56.26	27.91	4319.0	7903.0	65.42	24.50	33.27	2084.0	1.17993
260.0	53.70	26.64	4807.0	8561.0	68.05	23.87	32.58	2101.0	1.17131
280.0	51.37	25.48	5283.0	9207.0	70.45	23.35	32.02	2120.0	1.16349
300.0	49.25	24.43	5749.0	9842.0	72.64	22.96	31.57	2138.0	1.15637
320.0	47.29	23.46	6207.0	10470.0	74.66	22.66	31.21	2157.0	1.14986
340.0	45.49	22.56	6660.0	11090.0	76.55	22.43	30.94	2176.0	1.14388
360.0	43.82	21.74	7108.0	11710.0	78.31	22.26	30.72	2195.0	1.13838
380.0	42.28	20.97	7553.0	12320.0	79.97	22.12	30.54	2214.0	1.13329
400.0	40.84	20.26	7994.0	12930.0	81.53	22.02	30.40	2234.0	1.12857
120.00 MPa isobar									
37.14[a]	105.4	52.30	−239.0	2056.0	13.57	14.29	16.67	2449.0	1.35403
60.0	99.79	49.50	37.1	2461.0	22.02	15.76	19.67	2480.0	1.33331
100.0	89.19	44.24	718.0	3430.0	34.19	21.82	28.87	2325.0	1.29485
120.0	84.15	41.74	1173.0	4048.0	39.81	24.81	32.62	2262.0	1.27680
140.0	79.51	39.44	1682.0	4724.0	45.02	26.58	34.82	2226.0	1.26036
160.0	75.30	37.35	2218.0	5431.0	49.74	27.16	35.63	2208.0	1.24555
180.0	71.50	35.47	2760.0	6143.0	53.93	26.91	35.49	2205.0	1.23225
200.0	68.05	33.75	3292.0	6847.0	57.64	26.24	34.86	2210.0	1.22028
220.0	64.92	32.20	3810.0	7537.0	60.93	25.44	34.07	2220.0	1.20947
240.0	62.06	30.79	4313.0	8210.0	63.86	24.69	33.31	2233.0	1.19969
260.0	59.46	29.49	4801.0	8869.0	66.50	24.05	32.63	2248.0	1.19080
280.0	57.07	28.31	5277.0	9516.0	68.90	23.53	32.07	2264.0	1.18268
300.0	54.87	27.22	5744.0	10150.0	71.09	23.12	31.63	2280.0	1.17525
320.0	52.84	26.21	6204.0	10780.0	73.12	22.81	31.28	2296.0	1.16841
340.0	50.96	25.28	6658.0	11400.0	75.01	22.58	31.01	2312.0	1.16210
360.0	49.21	24.41	7107.0	12020.0	76.78	22.40	30.79	2329.0	1.15626
380.0	47.58	23.60	7553.0	12640.0	78.44	22.26	30.61	2345.0	1.15084
400.0	46.07	22.85	7996.0	13250.0	80.00	22.15	30.47	2363.0	1.14579

[a]At melting line.
[b]At liquid–vapor boundary.

Appendix I. Thermophysical Properties of Nitrogen

Thermophysical properties of coexisting gaseous and liquid nitrogen

T K	Pres. MPa	Density kg/m^3	Density mol/dm^3	E J/mol	H J/mol	S J/(mol·K)	C_v J/(mol·K)	C_p J/(mol·K)	Sound m/s	Visc. μPa·s	Therm. W/(m·K)	Diel. const.
63.150[a]	.01246	867.9	30.98	−4212.0	−4212.0	68.01	26.69	54.68	1326.0	283.0	.151	1.46867
63.150[a]	.01246	.6803	.02428	1296.0	1809.0	163.5	21.00	29.64	161.1	4.22	.00568	1.00032
64.0	.01456	864.6	30.86	−4165.0	−4165.0	68.75	27.22	55.25	1288.0	270.0	.150	1.46675
64.0	.01456	.7811	.02788	1313.0	1835.0	162.6	21.02	29.70	162.1	4.28	.00580	1.00037
66.0	.02059	856.9	30.59	−4053.0	−4052.0	70.47	28.02	56.23	1210.0	245.0	.148	1.46215
66.0	.02059	1.068	.03813	1351.0	1891.0	160.6	21.08	29.87	164.3	4.43	.00608	1.00050
68.0	.02847	848.9	30.30	−3940.0	−3939.0	72.16	28.38	56.84	1147.0	223.0	.146	1.45742
68.0	.02847	1.435	.05122	1388.0	1944.0	158.8	21.15	30.06	166.4	4.58	.00634	1.00068
70.0	.03854	840.7	30.01	−3826.0	−3825.0	73.81	28.47	57.21	1092.0	204.0	.143	1.45256
70.0	.03854	1.894	.06761	1424.0	1994.0	157.0	21.22	30.29	168.4	4.73	.00661	1.00089
72.0	.05121	832.3	29.71	−3712.0	−3710.0	75.42	28.39	57.44	1045.0	187.0	.141	1.44757
72.0	.05121	2.459	.08779	1459.0	2043.0	155.4	21.31	30.55	170.3	4.88	.00686	1.00116
74.0	.06689	823.7	29.40	−3597.0	−3595.0	76.99	28.22	57.59	1002.0	172.0	.138	1.44246
74.0	.06689	3.145	.1123	1493.0	2089.0	153.9	21.41	30.85	172.1	5.03	.00712	1.00148
76.0	.08605	814.8	29.08	−3483.0	−3480.0	78.52	27.99	57.71	963.6	160.0	.136	1.43721
76.0	.08605	3.966	.1416	1526.0	2134.0	152.5	21.51	31.19	173.8	5.18	.00739	1.00187
78.0	.1091	805.7	28.76	−3368.0	−3364.0	80.01	27.74	57.84	927.8	148.0	.133	1.43184
78.0	.1091	4.939	.1763	1558.0	2177.0	151.1	21.62	31.58	175.3	5.33	.00766	1.00233
80.0	.1367	796.4	28.43	−3253.0	−3248.0	81.47	27.50	58.00	894.2	138.0	.130	1.42634
80.0	.1367	6.081	.2170	1588.0	2218.0	149.9	21.75	32.02	176.7	5.48	.00793	1.00287
82.0	.1692	786.8	28.08	−3137.0	−3131.0	82.89	27.27	58.22	862.3	128.0	.127	1.42071
82.0	.1692	7.410	.2645	1617.0	2257.0	148.7	21.88	32.51	178.0	5.64	.00822	1.00349
84.0	.2072	777.0	27.73	−3021.0	−3014.0	84.29	27.06	58.50	831.7	120.0	.124	1.41494
84.0	.2072	8.948	.3194	1644.0	2293.0	147.5	22.02	33.06	179.2	5.80	.00851	1.00422
86.0	.2512	767.0	27.38	−2905.0	−2896.0	85.66	26.88	58.87	802.2	112.0	.121	1.40903
86.0	.2512	10.72	.3825	1669.0	2326.0	146.4	22.17	33.69	180.2	5.96	.00882	1.00505
88.0	.3019	756.7	27.01	−2788.0	−2777.0	87.01	26.71	59.34	773.6	105.0	.118	1.40298
88.0	.3019	12.74	.4546	1693.0	2357.0	145.4	22.33	34.39	181.1	6.12	.00914	1.00601
90.0	.3597	746.1	26.63	−2670.0	−2656.0	88.34	26.58	59.91	745.6	98.8	.114	1.39679
90.0	.3597	15.04	.5367	1714.0	2384.0	144.4	22.51	35.19	181.9	6.29	.00948	1.00710
92.0	.4254	735.3	26.24	−2551.0	−2535.0	89.64	26.46	60.59	718.3	92.8	.111	1.39044
92.0	.4254	17.64	.6298	1733.0	2408.0	143.4	22.69	36.10	182.5	6.46	.00984	1.00833
94.0	.4994	724.1	25.85	−2431.0	−2412.0	90.93	26.37	61.40	691.4	87.2	.108	1.38393
94.0	.4994	20.59	.7350	1750.0	2429.0	142.5	22.89	37.13	183.0	6.64	.0102	1.00973
96.0	.5823	712.6	25.44	−2310.0	−2288.0	92.21	26.29	62.35	664.9	82.0	.105	1.37725
96.0	.5823	23.92	.8537	1764.0	2446.0	141.6	23.11	38.31	183.3	6.82	.0106	1.01130
98.0	.6749	700.8	25.02	−2188.0	−2161.0	93.48	26.23	63.46	638.8	77.2	.101	1.37038
98.0	.6749	27.66	.9874	1775.0	2458.0	140.6	23.33	39.67	183.5	7.01	.0111	1.01308
100.0	.7777	688.6	24.58	−2064.0	−2033.0	94.73	26.19	64.75	612.9	72.6	.0981	1.36330
100.0	.7777	31.87	1.138	1783.0	2466.0	139.7	23.58	41.25	183.5	7.21	.0115	1.01508
102.0	.8913	676.0	24.13	−1939.0	−1902.0	95.98	26.17	66.26	587.2	68.3	.0947	1.35601
102.0	.8913	36.61	1.307	1787.0	2469.0	138.8	23.84	43.11	183.4	7.42	.0120	1.01734
104.0	1.016	663.0	23.66	−1811.0	−1768.0	97.23	26.15	68.03	561.6	64.3	.0914	1.34845
104.0	1.016	41.94	1.497	1787.0	2466.0	138.0	24.12	45.32	183.1	7.64	.0126	1.01988
106.0	1.154	649.4	23.18	−1681.0	−1631.0	98.48	26.16	70.15	535.9	60.4	.0880	1.34061
106.0	1.154	47.94	1.711	1783.0	2458.0	137.1	24.43	47.98	182.7	7.87	.0132	1.02275
108.0	1.304	635.1	22.67	−1548.0	−1491.0	99.73	26.17	72.72	510.0	56.7	.0845	1.33244
108.0	1.304	54.73	1.953	1774.0	2442.0	136.1	24.76	51.24	182.1	8.12	.0139	1.02600
110.0	1.467	620.1	22.14	−1412.0	−1346.0	101.0	26.21	75.90	483.8	53.2	.0811	1.32386
110.0	1.467	62.43	2.228	1759.0	2417.0	135.2	25.12	55.32	181.4	8.39	.0146	1.02969

Thermophysical properties of coexisting gaseous and liquid nitrogen—Continued

T K	Pres. MPa	Density kg/m³	Density mol/dm³	E J/mol	H J/mol	S J/(mol·K)	C_v J/(mol·K)	C_p J/(mol·K)	Sound m/s	Visc. μPa·s	Therm. W/(m·K)	Diel. const.
112.0	1.645	604.3	21.57	−1272.0	−1195.0	102.3	26.26	79.95	456.9	49.7	.0776	1.31481
112.0	1.645	71.23	2.542	1737.0	2384.0	134.2	25.52	60.57	180.5	8.69	.0155	1.03393
114.0	1.838	587.3	20.96	−1126.0	−1038.0	103.6	26.34	85.30	429.3	46.4	.0742	1.30515
114.0	1.838	81.36	2.904	1707.0	2340.0	133.2	25.96	67.57	179.5	9.02	.0166	1.03883
116.0	2.046	568.8	20.30	−974.0	−873.2	104.9	26.46	92.71	400.5	43.1	.0708	1.29472
116.0	2.046	93.19	3.326	1666.0	2281.0	132.1	26.45	77.33	178.3	9.40	.0180	1.04456
118.0	2.271	548.4	19.58	−812.9	−696.9	106.3	26.62	103.7	370.1	39.8	.0676	1.28323
118.0	2.271	107.3	3.829	1611.0	2204.0	130.9	27.01	91.87	177.0	9.84	.0198	1.05141
120.0	2.513	525.2	18.75	−638.7	−504.6	107.8	26.86	121.7	337.5	36.5	.0646	1.27022
120.0	2.513	124.5	4.443	1538.0	2103.0	129.6	27.66	115.8	175.6	10.4	.0223	1.05984
122.0	2.774	497.2	17.75	−442.8	−286.4	109.5	27.22	157.0	301.6	32.9	.0620	1.25467
122.0	2.774	146.6	5.234	1435.0	1965.0	128.0	28.43	162.2	174.2	11.1	.0262	1.07076
124.0	3.055	458.9	16.38	−199.3	−12.8	111.6	27.87	260.6	259.1	28.7	.0602	1.23359
124.0	3.055	178.3	6.363	1277.0	1757.0	125.9	29.37	289.2	173.1	12.1	.0331	1.08650
126.26[b]	3.398	314.1	11.21	561.1	864.3	118.4						1.15601

[a]Triple point.
[b]Critical point.

Thermophysical properties of nitrogen on the melting line

T K	Pres. MPa	Density kg/m³	Density mol/dm³	E J/mol	H J/mol	S J/(mol·K)	C_v J/(mol·K)	C_p J/(mol·K)	Sound m/s	Visc. μPa·s	Therm. W/(m·K)	Diel. const.
63.15[a]	.01246	867.9	30.98	-4212.0	-4212.0	68.01	26.69	54.68	1326.0	283.0	.151	1.46867
65.0	8.519	871.2	31.10	-4182.0	-3908.0	68.47	31.20	53.98	1229.0	277.0	.153	1.47034
67.5	20.33	876.9	31.30	-4133.0	-3483.0	69.17	34.98	52.64	1169.0	272.0	.157	1.47326
70.0	32.50	883.4	31.53	-4076.0	-3045.0	69.90	36.78	50.98	1153.0	270.0	.162	1.47666
72.5	45.01	890.4	31.78	-4013.0	-2597.0	70.65	37.24	49.12	1160.0	269.0	.166	1.48031
75.0	57.87	897.6	32.04	-3947.0	-2140.0	71.37	36.88	47.19	1180.0	269.0	.171	1.48406
77.5	71.08	904.8	32.29	-3878.0	-1677.0	72.06	36.05	45.28	1208.0	270.0	.176	1.48781
80.0	84.63	911.9	32.55	-3809.0	-1209.0	72.71	35.00	43.48	1242.0	272.0	.181	1.49149
82.5	98.53	918.9	32.80	-3739.0	-734.6	73.30	33.89	41.84	1278.0	274.0	.186	1.49507
85.0	112.8	925.8	33.04	-3668.0	-255.9	73.86	32.83	40.38	1315.0	276.0	.191	1.49854
87.5	127.3	932.4	33.28	-3598.0	227.6	74.37	31.86	39.11	1353.0	278.0	.196	1.50188
90.0	142.2	938.9	33.51	-3528.0	716.1	74.85	31.02	38.03	1391.0	280.0	.201	1.50510
92.5	157.5	945.3	33.74	-3457.0	1210.0	75.29	30.31	37.11	1429.0	282.0	.206	1.50820
95.0	173.0	951.5	33.96	-3386.0	1709.0	75.71	29.73	36.35	1465.0	284.0	.211	1.51119
97.5	188.9	957.5	34.18	-3314.0	2213.0	76.11	29.28	35.73	1501.0	287.0	.216	1.51407
100.0	205.1	963.5	34.39	-3242.0	2723.0	76.48	28.93	35.22	1535.0	289.0	.221	1.51686
102.5	221.7	969.3	34.60	-3168.0	3240.0	76.84	28.69	34.82	1569.0	292.0	.226	1.51955
105.0	238.6	975.0	34.80	-3093.0	3762.0	77.19	28.53	34.50	1601.0	295.0	.231	1.52215
107.5	255.8	980.6	35.00	-3017.0	4291.0	77.53	28.44	34.25	1632.0	298.0	.236	1.52467
110.0	273.3	986.1	35.20	-2939.0	4825.0	77.86	28.41	34.06	1662.0	301.0	.241	1.52711
112.5	291.1	991.6	35.39	-2859.0	5367.0	78.18	28.43	33.92	1691.0	304.0	.246	1.52949
115.0	309.3	996.9	35.58	-2777.0	5914.0	78.50	28.49	33.82	1719.0	307.0	.251	1.53179
117.5	327.7	1002.0	35.77	-2693.0	6468.0	78.81	28.58	33.74	1746.0	310.0	.256	1.53403
120.0	346.5	1007.0	35.96	-2607.0	7029.0	79.12	28.69	33.69	1773.0	313.0	.261	1.53621
122.5	365.6	1013.0	36.14	-2519.0	7596.0	79.43	28.82	33.66	1799.0	317.0	.266	1.53832
125.0	385.0	1018.0	36.33	-2429.0	8170.0	79.74	28.97	33.63	1824.0	320.0	.271	1.54039
127.5	404.8	1023.0	36.51	-2337.0	8750.0	80.04	29.12	33.62	1849.0	324.0	.276	1.54239
130.0	424.8	1028.0	36.69	-2243.0	9336.0	80.34	29.27	33.61	1874.0	328.0	.281	1.54434
132.5	445.1	1033.0	36.86	-2146.0	9929.0	80.65	29.42	33.60	1898.0	332.0	.286	1.54625
135.0	465.8	1038.0	37.04	-2047.0	10530.0	80.95	29.57	33.59	1922.0	336.0	.291	1.54810
137.5	486.7	1043.0	37.22	-1945.0	11130.0	81.25	29.72	33.58	1945.0	340.0	.297	1.54990
140.0	508.0	1047.0	37.39	-1842.0	11740.0	81.55	29.86	33.56	1968.0	344.0	.302	1.55166
142.5	529.6	1052.0	37.56	-1736.0	12360.0	81.85	29.99	33.54	1991.0	348.0	.307	1.55337
145.0	551.4	1057.0	37.73	-1628.0	12990.0	82.15	30.11	33.51	2014.0	352.0	.312	1.55503
147.5	573.6	1062.0	37.90	-1517.0	13620.0	82.44	30.22	33.47	2036.0	357.0	.318	1.55665
150.0	596.1	1067.0	38.07	-1405.0	14250.0	82.74	30.32	33.43	2058.0	361.0	.323	1.55822
152.5	618.8	1071.0	38.24	-1290.0	14890.0	83.04	30.41	33.38	2080.0	366.0	.328	1.55975
155.0	641.9	1076.0	38.41	-1173.0	15540.0	83.33	30.48	33.32	2102.0	371.0	.334	1.56124
157.5	665.3	1081.0	38.57	-1054.0	16190.0	83.63	30.54	33.25	2124.0	376.0	.339	1.56268
160.0	688.9	1085.0	38.74	-932.6	16850.0	83.92	30.60	33.17	2145.0	380.0	.345	1.56408
162.5	712.9	1090.0	38.90	-809.3	17520.0	84.21	30.63	33.09	2166.0	385.0	.350	1.56545
165.0	737.1	1094.0	39.06	-683.9	18190.0	84.50	30.66	32.99	2188.0	391.0	.356	1.56676
167.5	761.7	1099.0	39.23	-556.4	18860.0	84.79	30.67	32.89	2209.0	396.0	.361	1.56804
170.0	786.5	1103.0	39.39	-426.9	19540.0	85.08	30.68	32.78	2229.0	401.0	.367	1.56928
172.5	811.6	1108.0	39.55	-295.3	20230.0	85.36	30.67	32.66	2250.0	407.0	.373	1.57048
175.0	837.1	1113.0	39.71	-161.8	20920.0	85.65	30.64	32.54	2271.0	412.0	.378	1.57163
177.5	862.8	1117.0	39.87	-26.4	21610.0	85.93	30.61	32.40	2291.0	418.0	.384	1.57275
180.0	888.8	1121.0	40.03	111.0	22310.0	86.21	30.57	32.26	2311.0	423.0	.390	1.57383
182.5	915.1	1126.0	40.19	250.2	23020.0	86.49	30.51	32.11	2332.0	429.0	.396	1.57487
185.0	941.6	1130.0	40.34	391.4	23730.0	86.77	30.45	31.96	2352.0	435.0	.401	1.57587
187.5	968.5	1135.0	40.50	534.4	24450.0	87.04	30.37	31.80	2371.0	441.0	.407	1.57683
190.0	995.7	1139.0	40.66	679.2	25170.0	87.31	30.28	31.63	2391.0	447.0	.413	1.57775

[a]Triple point.

Thermophysical properties of nitrogen

T K	Density kg/m³	Density mol/dm³	*E* J/mol	*H* J/mol	*S* J/(mol·K)	C_v J/(mol·K)	C_p J/(mol·K)	Sound m/s	Visc. μPa·s	Therm. W/(m·K)	Diel. const.
					.01 MPa isobar						
65.0	.5220	.01863	1337.0	1874.0	166.3	20.94	29.48	163.7	4.35	.00599	1.00025
70.0	.4840	.01728	1442.0	2021.0	168.5	20.90	29.39	170.0	4.71	.00658	1.00023
75.0	.4513	.01611	1547.0	2168.0	170.5	20.87	29.33	176.1	5.06	.00711	1.00021
80.0	.4228	.01509	1652.0	2315.0	172.4	20.85	29.28	181.9	5.40	.00761	1.00020
85.0	.3977	.01420	1756.0	2461.0	174.2	20.83	29.25	187.6	5.74	.00810	1.00019
90.0	.3754	.01340	1861.0	2607.0	175.9	20.82	29.22	193.1	6.08	.00857	1.00018
95.0	.3555	.01269	1965.0	2753.0	177.4	20.81	29.20	198.4	6.42	.00904	1.00017
100.0	.3376	.01205	2069.0	2899.0	178.9	20.81	29.18	203.6	6.75	.00951	1.00016
105.0	.3215	.01148	2173.0	3045.0	180.4	20.80	29.17	208.7	7.09	.00997	1.00015
110.0	.3068	.01095	2278.0	3191.0	181.7	20.80	29.16	213.6	7.42	.0104	1.00015
120.0	.2811	.01004	2486.0	3482.0	184.3	20.79	29.15	223.2	8.08	.0114	1.00013
130.0	.2594	.009261	2694.0	3774.0	186.6	20.79	29.13	232.3	8.72	.0123	1.00012
135.0	.2498	.008918	2798.0	3919.0	187.7	20.79	29.13	236.7	9.04	.0128	1.00012
140.0	.2409	.008598	2902.0	4065.0	188.7	20.79	29.13	241.1	9.36	.0132	1.00011
145.0	.2325	.008301	3006.0	4211.0	189.8	20.79	29.12	245.4	9.67	.0137	1.00011
150.0	.2248	.008024	3110.0	4356.0	190.8	20.79	29.12	249.6	9.98	.0141	1.00011
155.0	.2175	.007765	3214.0	4502.0	191.7	20.79	29.12	253.7	10.3	.0146	1.00010
160.0	.2107	.007522	3318.0	4647.0	192.6	20.78	29.12	257.8	10.6	.0150	1.00010
165.0	.2043	.007294	3422.0.	4793.0	193.5	20.78	29.11	261.8	10.9	.0154	1.00010
170.0	.1983	.007079	3526.0	4939.0	194.4	20.78	29.11	265.7	11.2	.0159	1.00009
175.0	.1926	.006876	3630.0	5084.0	195.2	20.78	29.11	269.6	11.5	.0163	1.00009
180.0	.1873	.006685	3734.0	5230.0	196.1	20.78	29.11	273.4	11.8	.0167	1.00009
190.0	.1774	.006333	3942.0	5521.0	197.6	20.78	29.11	280.9	12.3	.0176	1.00008
200.0	.1685	.006016	4149.0	5812.0	199.1	20.78	29.11	288.3	12.9	.0184	1.00008
210.0	.1605	.005730	4357.0	6103.0	200.6	20.78	29.10	295.4	13.4	.0192	1.00008
220.0	.1532	.005469	4565.0	6394.0	201.9	20.78	29.10	302.3	14.0	.0200	1.00007
230.0	.1465	.005231	4773.0	6685.0	203.2	20.78	29.10	309.1	14.5	.0208	1.00007
240.0	.1404	.005013	4981.0	6976.0	204.4	20.78	29.10	315.8	15.0	.0215	1.00007
250.0	.1348	.004812	5189.0	7267.0	205.6	20.78	29.10	322.3	15.5	.0223	1.00006
260.0	.1296	.004627	5396.0	7558.0	206.8	20.78	29.10	328.7	16.0	.0230	1.00006
270.0	.1248	.004456	5604.0	7849.0	207.9	20.78	29.10	334.9	16.5	.0237	1.00006
280.0	.1204	.004297	5812.0	8140.0	208.9	20.78	29.10	341.1	17.0	.0244	1.00006
290.0	.1162	.004149	6020.0	8431.0	209.9	20.79	29.11	347.1	17.5	.0251	1.00006
300.0	.1123	.004010	6228.0	8722.0	210.9	20.79	29.11	353.0	17.9	.0258	1.00005
310.0	.1087	.003881	6436.0	9013.0	211.9	20.80	29.11	358.9	18.4	.0265	1.00005
320.0	.1053	.003760	6644.0	9304.0	212.8	20.80	29.12	364.6	18.8	.0272	1.00005
330.0	.1021	.003646	6852.0	9596.0	213.7	20.81	29.13	370.2	19.3	.0278	1.00005
340.0	.09911	.003539	7060.0	9887.0	214.6	20.82	29.13	375.8	19.7	.0285	1.00005
350.0	.09628	.003437	7268.0	10180.0	215.4	20.83	29.14	381.2	20.2	.0291	1.00005
360.0	.09360	.003342	7477.0	10470.0	216.2	20.84	29.16	386.6	20.6	.0298	1.00004
370.0	.09107	.003252	7685.0	10760.0	217.0	20.85	29.17	391.9	21.0	.0304	1.00004
380.0	.08868	.003166	7894.0	11050.0	217.8	20.87	29.19	397.1	21.4	.0311	1.00004
400.0	.08424	.003008	8312.0	11640.0	219.3	20.91	29.22	407.3	22.2	.0323	1.00004
420.0	.08023	.002865	8730.0	12220.0	220.7	20.95	29.27	417.3	23.0	.0336	1.00004
440.0	.07658	.002734	9150.0	12810.0	222.1	21.01	29.33	426.9	23.8	.0348	1.00004
460.0	.07325	.002616	9571.0	13400.0	223.4	21.08	29.39	436.3	24.6	.0361	1.00004
480.0	.07020	.002507	9993.0	13980.0	224.7	21.15	29.47	445.5	25.3	.0373	1.00003
500.0	.06739	.002406	10420.0	14570.0	225.9	21.23	29.55	454.4	26.1	.0385	1.00003
520.0	.06480	.002314	10840.0	15170.0	227.0	21.32	29.64	463.2	26.8	.0398	1.00003
540.0	.06240	.002228	11270.0	15760.0	228.2	21.42	29.74	471.7	27.5	.0410	1.00003
560.0	.06017	.002149	11700.0	16360.0	229.2	21.53	29.84	480.0	28.2	.0422	1.00003
580.0	.05810	.002075	12130.0	16950.0	230.3	21.64	29.95	488.1	28.9	.0434	1.00003
600.0	.05616	.002006	12560.0	17550.0	231.3	21.76	30.07	496.1	29.5	.0446	1.00003
620.0	.05435	.001941	13000.0	18160.0	232.3	21.88	30.19	503.9	30.2	.0458	1.00003
640.0	.05265	.001880	13440.0	18760.0	233.3	22.00	30.32	511.6	30.9	.0470	1.00003
660.0	.05105	.001823	13880.0	19370.0	234.2	22.13	30.45	519.1	31.5	.0482	1.00002
680.0	.04955	.001770	14330.0	19980.0	235.1	22.26	30.58	526.5	32.1	.0493	1.00002
700.0	.04814	.001719	14770.0	20590.0	236.0	22.39	30.71	533.7	32.8	.0505	1.00002
720.0	.04680	.001671	15220.0	21210.0	236.9	22.53	30.84	540.9	33.4	.0517	1.00002
740.0	.04554	.001626	15670.0	21830.0	237.7	22.66	30.98	547.9	34.0	.0529	1.00002

Thermophysical properties of nitrogen—Continued

T K	Density kg/m³	Density mol/dm³	E J/mol	H J/mol	S J/(mol·K)	C_v J/(mol·K)	C_p J/(mol·K)	Sound m/s	Visc. μPa·s	Therm. W/(m·K)	Diel. const.
760.0	.04434	.001584	16130.0	22450.0	238.5	22.80	31.11	554.8	34.6	.0541	1.00002
780.0	.04320	.001543	16580.0	23070.0	239.3	22.93	31.25	561.6	35.2	.0552	1.00002
800.0	.04212	.001504	17040.0	23700.0	240.1	23.07	31.38	568.3	35.8	.0564	1.00002
820.0	.04109	.001468	17510.0	24330.0	240.9	23.20	31.51	575.0	36.4	.0576	1.00002
840.0	.04011	.001433	17970.0	24960.0	241.7	23.33	31.65	581.5	37.0	.0587	1.00002
860.0	.03918	.001400	18440.0	25590.0	242.4	23.46	31.78	588.0	37.6	.0599	1.00002
880.0	.03829	.001368	18910.0	26230.0	243.1	23.59	31.90	594.3	38.2	.0610	1.00002
900.0	.03744	.001337	19380.0	26870.0	243.9	23.72	32.03	600.6	38.7	.0621	1.00002
920.0	.03663	.001308	19860.0	27510.0	244.6	23.84	32.16	606.8	39.3	.0633	1.00002
940.0	.03585	.001281	20340.0	28150.0	245.3	23.96	32.28	613.0	39.8	.0644	1.00002
1000.0	.03370	.001204	21790.0	30100.0	247.3	24.32	32.63	631.1	41.5	.0678	1.00002
1050.0	.03209	.001146	23010.0	31740.0	248.9	24.59	32.91	645.7	42.8	.0705	1.00002
1100.0	.03063	.001094	24250.0	33390.0	250.4	24.85	33.17	660.1	44.2	.0732	1.00002
1150.0	.02930	.001047	25490.0	35060.0	251.9	25.10	33.41	674.1	45.5	.0759	1.00001
1200.0	.02808	.001003	26760.0	36730.0	253.3	25.33	33.64	687.8	46.7	.0785	1.00001
1250.0	.02696	.0009632	28030.0	38420.0	254.7	25.54	33.86	701.2	48.0	.0811	1.00001
1300.0	.02592	.0009262	29310.0	40120.0	256.0	25.75	34.06	714.4	49.2	.0836	1.00001
1350.0	.02496	.0008919	30600.0	41830.0	257.3	25.93	34.25	727.4	50.4	.0861	1.00001
1400.0	.02407	.0008601	31900.0	43540.0	258.6	26.11	34.42	740.1	51.6	.0886	1.00001
1450.0	.02324	.0008305	33210.0	45270.0	259.8	26.27	34.59	752.7	52.8	.0910	1.00001
1500.0	.02247	.0008028	34530.0	47000.0	260.9	26.43	34.74	765.0	54.0	.0934	1.00001
1550.0	.02174	.0007770	35850.0	48740.0	262.1	26.57	34.89	777.1	55.2	.0958	1.00001
1600.0	.02106	.0007527	37190.0	50490.0	263.2	26.71	35.02	789.1	56.3	.0981	1.00001
1650.0	.02042	.0007299	38530.0	52240.0	264.3	26.83	35.15	800.9	57.4	.100	1.00001
1700.0	.01982	.0007085	39870.0	54000.0	265.3	26.95	35.26	812.5	58.6	.103	1.00001
1750.0	.01926	.0006883	41220.0	55770.0	266.4	27.06	35.37	824.0	59.7	.105	1.00001
1800.0	.01872	.0006692	42580.0	57540.0	267.3	27.16	35.48	835.3	60.8	.107	1.00001
1850.0	.01822	.0006511	43940.0	59320.0	268.3	27.26	35.58	846.4	61.8	.109	1.00001
1900.0	.01774	.0006340	45300.0	61100.0	269.3	27.35	35.67	857.5	62.9	.112	1.00001

.02 MPa isobar

T K	Density kg/m³	Density mol/dm³	E J/mol	H J/mol	S J/(mol·K)	C_v J/(mol·K)	C_p J/(mol·K)	Sound m/s	Visc. μPa·s	Therm. W/(m·K)	Diel. const.
63.15[a]	867.9	30.98	−4212.0	−4211.0	68.01	26.70	54.68	1326.0	283.0	.152	1.46866
65.826[b]	857.6	30.61	−4063.0	−4062.0	70.32	27.97	56.16	1217.0	247.0	.149	1.46255
65.826[b]	1.040	.03713	1347.0	1886.0	160.8	21.07	29.85	164.1	4.42	.00616	1.00049
70.0	.9735	.03475	1436.0	2012.0	162.6	21.01	29.70	169.5	4.72	.00665	1.00046
75.0	.9067	.03237	1542.0	2160.0	164.7	20.95	29.56	175.6	5.07	.00718	1.00043
80.0	.8487	.03029	1647.0	2307.0	166.6	20.91	29.47	181.5	5.41	.00767	1.00040
85.0	.7978	.02848	1752.0	2455.0	168.4	20.88	29.40	187.3	5.75	.00815	1.00038
90.0	.7528	.02687	1857.0	2601.0	170.1	20.86	29.34	192.8	6.09	.00863	1.00036
95.0	.7126	.02544	1962.0	2748.0	171.6	20.85	29.30	198.2	6.43	.00909	1.00034
100.0	.6766	.02415	2066.0	2894.0	173.1	20.83	29.27	203.4	6.76	.00956	1.00032
105.0	.6440	.02299	2171.0	3041.0	174.6	20.83	29.25	208.5	7.09	.0100	1.00030
110.0	.6145	.02193	2275.0	3187.0	175.9	20.82	29.23	213.4	7.42	.0105	1.00029
120.0	.5629	.02009	2484.0	3479.0	178.5	20.81	29.19	223.0	8.08	.0114	1.00027
130.0	.5193	.01854	2692.0	3771.0	180.8	20.80	29.17	232.2	8.73	.0123	1.00025
135.0	.5000	.01785	2796.0	3917.0	181.9	20.80	29.17	236.7	9.04	.0128	1.00024
140.0	.4821	.01721	2900.0	4062.0	183.0	20.80	29.16	241.0	9.36	.0133	1.00023
145.0	.4654	.01661	3004.0	4208.0	184.0	20.79	29.15	245.3	9.67	.0137	1.00022
150.0	.4498	.01606	3108.0	4354.0	185.0	20.79	29.15	249.5	9.98	.0142	1.00021
155.0	.4353	.01554	3212.0	4500.0	185.9	20.79	29.14	253.7	10.3	.0146	1.00021
160.0	.4216	.01505	3316.0	4645.0	186.9	20.79	29.14	257.7	10.6	.0151	1.00020
165.0	.4088	.01459	3420.0	4791.0	187.8	20.79	29.13	261.7	10.9	.0155	1.00019
170.0	.3967	.01416	3524.0	4937.0	188.6	20.79	29.13	265.7	11.2	.0159	1.00019
175.0	.3854	.01376	3628.0	5082.0	189.5	20.79	29.13	269.6	11.5	.0164	1.00018
180.0	.3746	.01337	3732.0	5228.0	190.3	20.79	29.13	273.4	11.8	.0168	1.00018
190.0	.3549	.01267	3940.0	5519.0	191.9	20.78	29.12	280.9	12.3	.0176	1.00017
200.0	.3371	.01203	4148.0	5810.0	193.4	20.78	29.12	288.2	12.9	.0184	1.00016
210.0	.3210	.01146	4356.0	6102.0	194.8	20.78	29.11	295.4	13.5	.0192	1.00015
220.0	.3064	.01094	4564.0	6393.0	196.1	20.78	29.11	302.3	14.0	.0200	1.00015
230.0	.2931	.01046	4772.0	6684.0	197.4	20.78	29.11	309.1	14.5	.0208	1.00014
240.0	.2809	.01003	4980.0	6975.0	198.7	20.78	29.11	315.8	15.0	.0216	1.00013

Thermophysical properties of nitrogen—Continued

T K	Density kg/m³	Density mol/dm³	E J/mol	H J/mol	S J/(mol·K)	C_v J/(mol·K)	C_p J/(mol·K)	Sound m/s	Visc. μPa·s	Therm. W/(m·K)	Diel. const.
250.0	.2696	.009625	5188.0	7266.0	199.9	20.78	29.11	322.3	15.5	.0223	1.00013
260.0	.2592	.009254	5396.0	7557.0	201.0	20.78	29.11	328.7	16.0	.0230	1.00012
270.0	.2496	.008911	5604.0	7848.0	202.1	20.78	29.11	335.0	16.5	.0238	1.00012
280.0	.2407	.008593	5812.0	8139.0	203.2	20.79	29.11	341.1	17.0	.0245	1.00011
290.0	.2324	.008296	6019.0	8430.0	204.2	20.79	29.11	347.1	17.5	.0252	1.00011
300.0	.2247	.008020	6227.0	8722.0	205.2	20.79	29.11	353.1	17.9	.0258	1.00011
310.0	.2174	.007761	6435.0	9013.0	206.1	20.80	29.12	358.9	18.4	.0265	1.00010
320.0	.2106	.007518	6643.0	9304.0	207.0	20.80	29.12	364.6	18.8	.0272	1.00010
330.0	.2042	.007291	6851.0	9595.0	207.9	20.81	29.13	370.3	19.3	.0279	1.00010
340.0	.1982	.007076	7060.0	9886.0	208.8	20.82	29.14	375.8	19.7	.0285	1.00009
350.0	.1926	.006874	7268.0	10180.0	209.7	20.83	29.15	381.3	20.2	.0292	1.00009
360.0	.1872	.006683	7476.0	10470.0	210.5	20.84	29.16	386.6	20.6	.0298	1.00009
370.0	.1821	.006502	7685.0	10760.0	211.3	20.85	29.17	391.9	21.0	.0305	1.00009
380.0	.1773	.006331	7893.0	11050.0	212.1	20.87	29.19	397.2	21.4	.0311	1.00008
400.0	.1685	.006015	8311.0	11640.0	213.6	20.91	29.23	407.4	22.2	.0324	1.00008
420.0	.1605	.005728	8730.0	12220.0	215.0	20.96	29.27	417.3	23.0	.0336	1.00008
440.0	.1532	.005468	9150.0	12810.0	216.3	21.01	29.33	427.0	23.8	.0349	1.00007
460.0	.1465	.005230	9570.0	13400.0	217.7	21.08	29.39	436.4	24.6	.0361	1.00007
480.0	.1404	.005012	9993.0	13980.0	218.9	21.15	29.47	445.5	25.3	.0374	1.00007
500.0	.1348	.004812	10420.0	14570.0	220.1	21.23	29.55	454.5	26.1	.0386	1.00006
520.0	.1296	.004627	10840.0	15170.0	221.3	21.32	29.64	463.2	26.8	.0398	1.00006
540.0	.1248	.004455	11270.0	15760.0	222.4	21.42	29.74	471.7	27.5	.0410	1.00006
560.0	.1203	.004296	11700.0	16360.0	223.5	21.53	29.84	480.0	28.2	.0422	1.00006
580.0	.1162	.004148	12130.0	16950.0	224.5	21.64	29.95	488.2	28.9	.0434	1.00006
600.0	.1123	.004010	12560.0	17550.0	225.5	21.76	30.07	496.1	29.5	.0446	1.00005
620.0	.1087	.003881	13000.0	18160.0	226.5	21.88	30.19	504.0	30.2	.0458	1.00005
640.0	.1053	.003759	13440.0	18760.0	227.5	22.00	30.32	511.6	30.9	.0470	1.00005
660.0	.1021	.003645	13880.0	19370.0	228.4	22.13	30.45	519.1	31.5	.0482	1.00005
680.0	.09910	.003538	14330.0	19980.0	229.3	22.26	30.58	526.5	32.1	.0494	1.00005
700.0	.09627	.003437	14770.0	20590.0	230.2	22.39	30.71	533.8	32.8	.0506	1.00005
720.0	.09360	.003342	15220.0	21210.0	231.1	22.53	30.84	540.9	33.4	.0518	1.00004
740.0	.09107	.003251	15670.0	21830.0	231.9	22.66	30.98	547.9	34.0	.0529	1.00004
760.0	.08867	.003166	16130.0	22450.0	232.8	22.80	31.11	554.8	34.6	.0541	1.00004
780.0	.08640	.003085	16580.0	23070.0	233.6	22.93	31.25	561.7	35.2	.0553	1.00004
800.0	.08424	.003008	17040.0	23700.0	234.4	23.07	31.38	568.4	35.8	.0564	1.00004
820.0	.08218	.002934	17510.0	24330.0	235.1	23.20	31.51	575.0	36.4	.0576	1.00004
840.0	.08023	.002865	17970.0	24960.0	235.9	23.33	31.65	581.5	37.0	.0588	1.00004
860.0	.07836	.002798	18440.0	25590.0	236.6	23.46	31.78	588.0	37.6	.0599	1.00004
880.0	.07658	.002734	18910.0	26230.0	237.4	23.59	31.90	594.3	38.2	.0611	1.00004
900.0	.07488	.002674	19380.0	26870.0	238.1	23.72	32.03	600.6	38.7	.0622	1.00004
920.0	.07325	.002616	19860.0	27510.0	238.8	23.84	32.16	606.9	39.3	.0633	1.00004
940.0	.07169	.002560	20340.0	28150.0	239.5	23.96	32.28	613.0	39.8	.0645	1.00003
1000.0	.06739	.002406	21790.0	30100.0	241.5	24.32	32.63	631.1	41.5	.0678	1.00003
1050.0	.06418	.002292	23010.0	31740.0	243.1	24.59	32.91	645.8	42.8	.0706	1.00003
1100.0	.06126	.002188	24250.0	33390.0	244.6	24.85	33.17	660.1	44.2	.0733	1.00003
1150.0	.05860	.002093	25490.0	35060.0	246.1	25.10	33.41	674.1	45.5	.0759	1.00003
1200.0	.05616	.002006	26760.0	36730.0	247.5	25.33	33.64	687.8	46.7	.0786	1.00003
1250.0	.05391	.001925	28030.0	38420.0	248.9	25.54	33.86	701.2	48.0	.0811	1.00003
1300.0	.05184	.001851	29310.0	40120.0	250.3	25.75	34.06	714.4	49.2	.0837	1.00003
1350.0	.04992	.001783	30600.0	41830.0	251.5	25.93	34.25	727.4	50.4	.0862	1.00002
1400.0	.04814	.001719	31900.0	43540.0	252.8	26.11	34.42	740.2	51.7	.0886	1.00002
1450.0	.04648	.001660	33210.0	45270.0	254.0	26.27	34.59	752.7	52.8	.0911	1.00002
1500.0	.04493	.001605	34530.0	47000.0	255.2	26.43	34.74	765.0	54.0	.0935	1.00002
1550.0	.04348	.001553	35850.0	48740.0	256.3	26.57	34.89	777.2	55.2	.0958	1.00002
1600.0	.04212	.001504	37190.0	50490.0	257.4	26.71	35.02	789.1	56.3	.0981	1.00002
1650.0	.04084	.001459	38530.0	52240.0	258.5	26.83	35.15	800.9	57.4	.100	1.00002
1700.0	.03964	.001416	39870.0	54000.0	259.6	26.95	35.26	812.5	58.6	.103	1.00002
1750.0	.03851	.001376	41220.0	55770.0	260.6	27.06	35.37	824.0	59.7	.105	1.00002
1800.0	.03744	.001337	42580.0	57540.0	261.6	27.16	35.48	835.3	60.8	.107	1.00002
1850.0	.03643	.001301	43940.0	59320.0	262.6	27.26	35.58	846.5	61.8	.109	1.00002
1900.0	.03547	.001267	45300.0	61100.0	263.5	27.35	35.67	857.5	62.9	.112	1.00002

Thermophysical properties of nitrogen—Continued

T K	Density kg/m^3	Density mol/dm^3	E J/mol	H J/mol	S J/(mol·K)	C_v J/(mol·K)	C_p J/(mol·K)	Sound m/s	Visc. μPa·s	Therm. W/(m·K)	Diel. const.
					.05 MPa isobar						
63.16[a]	867.9	30.98	-4212.0	-4210.0	68.01	26.72	54.68	1325.0	283.0	.152	1.46867
71.827[b]	833.0	29.73	-3722.0	-3720.0	75.28	28.40	57.42	1049.0	188.0	.143	1.44801
71.827[b]	2.406	.08589	1456.0	2038.0	155.5	21.30	30.52	170.2	4.87	.00702	1.00113
75.0	2.299	.08205	1525.0	2134.0	156.8	21.22	30.31	174.2	5.09	.00734	1.00108
80.0	2.146	.07661	1633.0	2285.0	158.8	21.11	30.06	180.4	5.43	.00783	1.00101
85.0	2.014	.07188	1739.0	2435.0	160.6	21.04	29.87	186.3	5.77	.00830	1.00095
90.0	1.897	.06772	1846.0	2584.0	162.3	20.99	29.73	191.9	6.11	.00877	1.00089
95.0	1.794	.06403	1951.0	2732.0	163.9	20.94	29.62	197.4	6.44	.00923	1.00085
100.0	1.701	.06073	2057.0	2880.0	165.4	20.91	29.54	202.7	6.78	.00969	1.00080
105.0	1.618	.05776	2162.0	3028.0	166.9	20.89	29.48	207.9	7.11	.0102	1.00076
110.0	1.543	.05508	2267.0	3175.0	168.2	20.87	29.42	212.9	7.44	.0106	1.00073
120.0	1.412	.05040	2477.0	3469.0	170.8	20.85	29.35	222.6	8.10	.0115	1.00067
130.0	1.302	.04647	2686.0	3762.0	173.1	20.83	29.29	231.9	8.74	.0125	1.00061
135.0	1.253	.04473	2791.0	3908.0	174.3	20.82	29.27	236.4	9.06	.0129	1.00059
140.0	1.208	.04311	2895.0	4055.0	175.3	20.82	29.25	240.8	9.37	.0134	1.00057
145.0	1.166	.04161	2999.0	4201.0	176.3	20.81	29.24	245.1	9.68	.0138	1.00055
150.0	1.126	.04021	3104.0	4347.0	177.3	20.81	29.23	249.3	9.99	.0143	1.00053
155.0	1.090	.03890	3208.0	4493.0	178.3	20.80	29.21	253.5	10.3	.0147	1.00051
160.0	1.056	.03768	3312.0	4639.0	179.2	20.80	29.20	257.6	10.6	.0151	1.00050
165.0	1.023	.03653	3416.0	4785.0	180.1	20.80	29.20	261.6	10.9	.0156	1.00048
170.0	.9930	.03544	3521.0	4931.0	181.0	20.80	29.19	265.6	11.2	.0160	1.00047
175.0	.9645	.03443	3625.0	5077.0	181.8	20.80	29.18	269.5	11.5	.0164	1.00045
180.0	.9375	.03346	3729.0	5223.0	182.7	20.79	29.17	273.3	11.8	.0169	1.00044
190.0	.8879	.03169	3937.0	5515.0	184.2	20.79	29.16	280.9	12.4	.0177	1.00042
200.0	.8433	.03010	4145.0	5806.0	185.7	20.79	29.15	288.2	12.9	.0185	1.00040
210.0	.8030	.02866	4353.0	6098.0	187.2	20.79	29.15	295.3	13.5	.0193	1.00038
220.0	.7664	.02736	4562.0	6389.0	188.5	20.79	29.14	302.3	14.0	.0201	1.00036
230.0	.7330	.02616	4770.0	6681.0	189.8	20.79	29.14	309.1	14.5	.0209	1.00035
240.0	.7024	.02507	4978.0	6972.0	191.0	20.79	29.13	315.8	15.0	.0216	1.00033
250.0	.6742	.02407	5186.0	7263.0	192.2	20.79	29.13	322.3	15.5	.0224	1.00032
260.0	.6482	.02314	5394.0	7555.0	193.4	20.79	29.13	328.7	16.0	.0231	1.00031
270.0	.6242	.02228	5602.0	7846.0	194.5	20.79	29.13	335.0	16.5	.0238	1.00029
280.0	.6018	.02148	5810.0	8137.0	195.5	20.79	29.13	341.1	17.0	.0245	1.00028
290.0	.5811	.02074	6018.0	8428.0	196.6	20.79	29.13	347.2	17.5	.0252	1.00027
300.0	.5617	.02005	6226.0	8720.0	197.5	20.79	29.13	353.1	17.9	.0259	1.00026
310.0	.5435	.01940	6434.0	9011.0	198.5	20.80	29.13	358.9	18.4	.0266	1.00026
320.0	.5265	.01879	6642.0	9302.0	199.4	20.80	29.13	364.7	18.9	.0273	1.00025
330.0	.5106	.01822	6850.0	9594.0	200.3	20.81	29.14	370.3	19.3	.0280	1.00024
340.0	.4955	.01769	7058.0	9885.0	201.2	20.82	29.15	375.9	19.7	.0286	1.00023
350.0	.4814	.01718	7266.0	10180.0	202.0	20.83	29.16	381.3	20.2	.0293	1.00023
360.0	.4680	.01670	7475.0	10470.0	202.9	20.84	29.17	386.7	20.6	.0299	1.00022
370.0	.4553	.01625	7683.0	10760.0	203.7	20.86	29.18	392.0	21.0	.0306	1.00021
380.0	.4433	.01583	7892.0	11050.0	204.4	20.87	29.20	397.2	21.4	.0312	1.00021
400.0	.4212	.01503	8310.0	11640.0	205.9	20.91	29.23	407.4	22.2	.0325	1.00020
420.0	.4011	.01432	8729.0	12220.0	207.4	20.96	29.28	417.4	23.0	.0337	1.00019
440.0	.3829	.01367	9149.0	12810.0	208.7	21.01	29.33	427.0	23.8	.0350	1.00018
460.0	.3662	.01307	9570.0	13390.0	210.0	21.08	29.40	436.4	24.6	.0362	1.00017
480.0	.3509	.01253	9992.0	13980.0	211.3	21.15	29.47	445.6	25.3	.0374	1.00017
500.0	.3369	.01203	10420.0	14570.0	212.5	21.23	29.55	454.5	26.1	.0387	1.00016
520.0	.3239	.01156	10840.0	15170.0	213.6	21.32	29.64	463.3	26.8	.0399	1.00015
540.0	.3119	.01114	11270.0	15760.0	214.8	21.42	29.74	471.8	27.5	.0411	1.00015
560.0	.3008	.01074	11700.0	16360.0	215.9	21.53	29.85	480.1	28.2	.0423	1.00014
580.0	.2904	.01037	12130.0	16950.0	216.9	21.64	29.96	488.2	28.9	.0435	1.00014
600.0	.2807	.01002	12560.0	17550.0	217.9	21.76	30.07	496.2	29.5	.0447	1.00013
620.0	.2717	.009699	13000.0	18160.0	218.9	21.88	30.20	504.0	30.2	.0459	1.00013
640.0	.2632	.009396	13440.0	18760.0	219.9	22.00	30.32	511.7	30.9	.0471	1.00012
660.0	.2552	.009111	13880.0	19370.0	220.8	22.13	30.45	519.2	31.5	.0483	1.00012
680.0	.2477	.008843	14320.0	19980.0	221.7	22.26	30.58	526.6	32.1	.0495	1.00012
700.0	.2406	.008590	14770.0	20590.0	222.6	22.39	30.71	533.8	32.8	.0507	1.00011
720.0	.2340	.008352	15220.0	21210.0	223.5	22.53	30.85	541.0	33.4	.0519	1.00011
740.0	.2276	.008126	15670.0	21830.0	224.3	22.66	30.98	548.0	34.0	.0530	1.00011

Thermophysical properties of nitrogen—Continued

T K	Density kg/m³	Density mol/dm³	E J/mol	H J/mol	S J/(mol·K)	C_v J/(mol·K)	C_p J/(mol·K)	Sound m/s	Visc. µPa·s	Therm. W/(m·K)	Diel. const.
760.0	.2216	.007912	16130.0	22450.0	225.1	22.80	31.11	554.9	34.6	.0542	1.00010
780.0	.2160	.007710	16580.0	23070.0	226.0	22.93	31.25	561.7	35.2	.0554	1.00010
800.0	.2106	.007517	17040.0	23700.0	226.7	23.07	31.38	568.4	35.8	.0565	1.00010
820.0	.2054	.007334	17510.0	24330.0	227.5	23.20	31.52	575.1	36.4	.0577	1.00010
840.0	.2005	.007159	17970.0	24960.0	228.3	23.33	31.65	581.6	37.0	.0589	1.00009
860.0	.1959	.006993	18440.0	25590.0	229.0	23.46	31.78	588.0	37.6	.0600	1.00009
880.0	.1914	.006834	18910.0	26230.0	229.8	23.59	31.91	594.4	38.2	.0612	1.00009
900.0	.1872	.006682	19380.0	26870.0	230.5	23.72	32.03	600.7	38.7	.0623	1.00009
920.0	.1831	.006537	19860.0	27510.0	231.2	23.84	32.16	606.9	39.3	.0634	1.00009
940.0	.1792	.006398	20340.0	28150.0	231.9	23.96	32.28	613.1	39.9	.0646	1.00008
1000.0	.1685	.006014	21790.0	30100.0	233.9	24.32	32.63	631.2	41.5	.0679	1.00008
1050.0	.1604	.005728	23010.0	31740.0	235.5	24.59	32.91	645.8	42.8	.0707	1.00008
1100.0	.1531	.005467	24250.0	33390.0	237.0	24.85	33.17	660.1	44.2	.0734	1.00007
1150.0	.1465	.005230	25490.0	35060.0	238.5	25.10	33.41	674.1	45.5	.0760	1.00007
1200.0	.1404	.005012	26750.0	36730.0	239.9	25.33	33.64	687.9	46.7	.0787	1.00007
1250.0	.1348	.004811	28030.0	38420.0	241.3	25.54	33.86	701.3	48.0	.0812	1.00006
1300.0	.1296	.004626	29310.0	40120.0	242.6	25.75	34.06	714.5	49.2	.0838	1.00006
1350.0	.1248	.004455	30600.0	41830.0	243.9	25.93	34.25	727.5	50.4	.0863	1.00006
1400.0	.1203	.004296	31900.0	43540.0	245.2	26.11	34.42	740.2	51.7	.0888	1.00006
1450.0	.1162	.004148	33210.0	45270.0	246.4	26.27	34.59	752.7	52.8	.0912	1.00005
1500.0	.1123	.004010	34530.0	47000.0	247.6	26.43	34.74	765.1	54.0	.0936	1.00005
1550.0	.1087	.003880	35850.0	48740.0	248.7	26.57	34.89	777.2	55.2	.0959	1.00005
1600.0	.1053	.003759	37190.0	50490.0	249.8	26.71	35.02	789.2	56.3	.0983	1.00005
1650.0	.1021	.003645	38530.0	52250.0	250.9	26.83	35.15	801.0	57.4	.101	1.00005
1700.0	.09910	.003538	39870.0	54010.0	251.9	26.95	35.26	812.6	58.6	.103	1.00005
1750.0	.09627	.003437	41220.0	55770.0	253.0	27.06	35.38	824.0	59.7	.105	1.00005
1800.0	.09359	.003342	42580.0	57540.0	254.0	27.16	35.48	835.4	60.8	.107	1.00004
1850.0	.09106	.003251	43940.0	59320.0	254.9	27.26	35.58	846.5	61.8	.110	1.00004
1900.0	.08867	.003166	45300.0	61100.0	255.9	27.35	35.67	857.5	62.9	.112	1.00004
					.10 MPa isobar						
63.17[a]	867.9	30.98	−4212.0	−4208.0	68.01	26.75	54.68	1324.0	283.0	.152	1.46868
70.0	840.9	30.02	−3828.0	−3824.0	73.79	28.53	57.17	1092.0	204.0	.145	1.45269
77.251[b]	809.1	28.88	−3411.0	−3407.0	79.45	27.84	57.79	940.9	152.0	.135	1.43387
77.251[b]	4.556	.1626	1546.0	2161.0	151.6	21.58	31.43	174.8	5.27	.00782	1.00215
80.0	4.380	.1563	1607.0	2247.0	152.7	21.47	31.13	178.3	5.46	.00807	1.00206
85.0	4.095	.1462	1717.0	2402.0	154.6	21.31	30.72	184.5	5.80	.00853	1.00193
90.0	3.847	.1373	1826.0	2554.0	156.3	21.20	30.42	190.5	6.13	.00898	1.00181
95.0	3.630	.1296	1934.0	2706.0	158.0	21.11	30.19	196.1	6.47	.00943	1.00171
100.0	3.437	.1227	2041.0	2856.0	159.5	21.05	30.01	201.6	6.80	.00988	1.00162
105.0	3.265	.1165	2148.0	3006.0	161.0	21.00	29.87	206.9	7.13	.0103	1.00154
110.0	3.109	.1110	2254.0	3155.0	162.4	20.96	29.76	212.1	7.46	.0108	1.00146
120.0	2.841	.1014	2466.0	3452.0	164.9	20.91	29.60	222.0	8.12	.0117	1.00134
130.0	2.616	.09336	2676.0	3747.0	167.3	20.88	29.49	231.4	8.76	.0126	1.00123
135.0	2.516	.08981	2781.0	3895.0	168.4	20.86	29.45	235.9	9.08	.0131	1.00119
140.0	2.424	.08653	2886.0	4042.0	169.5	20.85	29.42	240.4	9.39	.0135	1.00114
145.0	2.339	.08349	2991.0	4189.0	170.5	20.84	29.38	244.7	9.70	.0140	1.00110
150.0	2.260	.08065	3096.0	4336.0	171.5	20.84	29.36	249.0	10.0	.0144	1.00106
155.0	2.185	.07801	3201.0	4483.0	172.5	20.83	29.33	253.2	10.3	.0148	1.00103
160.0	2.116	.07553	3305.0	4629.0	173.4	20.82	29.31	257.3	10.6	.0153	1.00100
165.0	2.051	.07321	3410.0	4776.0	174.3	20.82	29.30	261.4	10.9	.0157	1.00097
170.0	1.990	.07103	3514.0	4922.0	175.2	20.82	29.28	265.4	11.2	.0161	1.00094
175.0	1.932	.06897	3619.0	5069.0	176.0	20.81	29.27	269.3	11.5	.0166	1.00091
180.0	1.878	.06703	3723.0	5215.0	176.9	20.81	29.25	273.2	11.8	.0170	1.00088
190.0	1.778	.06347	3932.0	5507.0	178.4	20.80	29.23	280.7	12.4	.0178	1.00084
200.0	1.688	.06027	4140.0	5800.0	179.9	20.80	29.22	288.1	12.9	.0186	1.00080
210.0	1.607	.05738	4349.0	6092.0	181.4	20.80	29.20	295.3	13.5	.0194	1.00076
220.0	1.534	.05475	4557.0	6384.0	182.7	20.79	29.19	302.3	14.0	.0202	1.00072
230.0	1.467	.05236	4765.0	6675.0	184.0	20.79	29.18	309.1	14.5	.0210	1.00069
240.0	1.405	.05017	4974.0	6967.0	185.3	20.79	29.17	315.8	15.0	.0218	1.00066
250.0	1.349	.04815	5182.0	7259.0	186.5	20.79	29.16	322.3	15.6	.0225	1.00064
260.0	1.297	.04629	5390.0	7550.0	187.6	20.79	29.16	328.8	16.1	.0232	1.00061

Thermophysical properties of nitrogen—Continued

T K	Density kg/m^3	Density mol/dm^3	E J/mol	H J/mol	S J/(mol·K)	C_v J/(mol·K)	C_p J/(mol·K)	Sound m/s	Visc. μPa·s	Therm. W/(m·K)	Diel. const.
270.0	1.249	.04457	5598.0	7842.0	188.7	20.79	29.15	335.0	16.5	.0239	1.00059
280.0	1.204	.04297	5806.0	8134.0	189.8	20.79	29.15	341.2	17.0	.0247	1.00057
290.0	1.162	.04149	6015.0	8425.0	190.8	20.79	29.15	347.2	17.5	.0254	1.00055
300.0	1.123	.04010	6223.0	8717.0	191.8	20.80	29.15	353.2	18.0	.0260	1.00053
310.0	1.087	.03880	6431.0	9008.0	192.7	20.80	29.15	359.0	18.4	.0267	1.00051
320.0	1.053	.03759	6639.0	9300.0	193.7	20.81	29.15	364.8	18.9	.0274	1.00050
330.0	1.021	.03645	6847.0	9591.0	194.6	20.81	29.16	370.4	19.3	.0281	1.00048
340.0	.9910	.03537	7056.0	9883.0	195.4	20.82	29.17	376.0	19.7	.0287	1.00047
350.0	.9627	.03436	7264.0	10170.0	196.3	20.83	29.17	381.4	20.2	.0294	1.00045
360.0	.9359	.03341	7473.0	10470.0	197.1	20.84	29.18	386.8	20.6	.0300	1.00044
370.0	.9105	.03250	7681.0	10760.0	197.9	20.86	29.20	392.1	21.0	.0307	1.00043
380.0	.8866	.03165	7890.0	11050.0	198.7	20.87	29.21	397.3	21.4	.0313	1.00042
400.0	.8422	.03006	8308.0	11630.0	200.2	20.91	29.25	407.6	22.2	.0326	1.00040
420.0	.8021	.02863	8727.0	12220.0	201.6	20.96	29.29	417.5	23.0	.0338	1.00038
440.0	.7656	.02733	9147.0	12810.0	203.0	21.01	29.34	427.2	23.8	.0351	1.00036
460.0	.7323	.02614	9568.0	13390.0	204.3	21.08	29.41	436.6	24.6	.0363	1.00034
480.0	.7017	.02505	9990.0	13980.0	205.5	21.15	29.48	445.7	25.3	.0375	1.00033
500.0	.6737	.02405	10410.0	14570.0	206.7	21.24	29.56	454.7	26.1	.0388	1.00032
520.0	.6478	.02312	10840.0	15170.0	207.9	21.33	29.65	463.4	26.8	.0400	1.00030
540.0	.6238	.02227	11270.0	15760.0	209.0	21.42	29.75	471.9	27.5	.0412	1.00029
560.0	.6015	.02147	11700.0	16360.0	210.1	21.53	29.85	480.2	28.2	.0424	1.00028
580.0	.5807	.02073	12130.0	16950.0	211.1	21.64	29.96	488.4	28.9	.0436	1.00027
600.0	.5614	.02004	12560.0	17550.0	212.2	21.76	30.08	496.4	29.5	.0448	1.00026
620.0	.5433	.01939	13000.0	18160.0	213.1	21.88	30.20	504.2	30.2	.0460	1.00026
640.0	.5263	.01879	13440.0	18760.0	214.1	22.00	30.32	511.8	30.9	.0472	1.00025
660.0	.5103	.01822	13880.0	19370.0	215.0	22.13	30.45	519.3	31.5	.0484	1.00024
680.0	.4953	.01768	14320.0	19980.0	215.9	22.26	30.58	526.7	32.2	.0496	1.00023
700.0	.4812	.01718	14770.0	20590.0	216.8	22.40	30.71	534.0	32.8	.0508	1.00023
720.0	.4678	.01670	15220.0	21210.0	217.7	22.53	30.85	541.1	33.4	.0520	1.00022
740.0	.4552	.01625	15670.0	21830.0	218.6	22.66	30.98	548.1	34.0	.0531	1.00021
760.0	.4432	.01582	16130.0	22450.0	219.4	22.80	31.12	555.0	34.6	.0543	1.00021
780.0	.4318	.01542	16580.0	23070.0	220.2	22.93	31.25	561.9	35.2	.0555	1.00020
800.0	.4210	.01503	17040.0	23700.0	221.0	23.07	31.38	568.6	35.8	.0567	1.00020
820.0	.4108	.01466	17510.0	24330.0	221.8	23.20	31.52	575.2	36.4	.0578	1.00019
840.0	.4010	.01431	17970.0	24960.0	222.5	23.33	31.65	581.7	37.0	.0590	1.00019
860.0	.3917	.01398	18440.0	25590.0	223.3	23.46	31.78	588.2	37.6	.0601	1.00018
880.0	.3828	.01366	18910.0	26230.0	224.0	23.59	31.91	594.5	38.2	.0613	1.00018
900.0	.3743	.01336	19380.0	26870.0	224.7	23.72	32.03	600.8	38.7	.0624	1.00018
920.0	.3661	.01307	19860.0	27510.0	225.4	23.84	32.16	607.1	39.3	.0635	1.00017
940.0	.3584	.01279	20340.0	28160.0	226.1	23.96	32.28	613.2	39.9	.0647	1.00017
1000.0	.3369	.01202	21790.0	30100.0	228.1	24.32	32.63	631.3	41.5	.0680	1.00016
1050.0	.3208	.01145	23010.0	31740.0	229.7	24.59	32.91	645.9	42.8	.0708	1.00015
1100.0	.3062	.01093	24240.0	33390.0	231.3	24.85	33.17	660.3	44.2	.0735	1.00014
1150.0	.2929	.01046	25490.0	35060.0	232.7	25.10	33.41	674.3	45.5	.0762	1.00014
1200.0	.2807	.01002	26750.0	36730.0	234.2	25.33	33.64	688.0	46.7	.0788	1.00013
1250.0	.2695	.009620	28030.0	38420.0	235.5	25.54	33.86	701.4	48.0	.0814	1.00013
1300.0	.2591	.009251	29310.0	40120.0	236.9	25.75	34.06	714.6	49.2	.0839	1.00012
1350.0	.2495	.008908	30600.0	41830.0	238.2	25.93	34.25	727.6	50.5	.0864	1.00012
1400.0	.2406	.008590	31900.0	43550.0	239.4	26.11	34.42	740.3	51.7	.0889	1.00011
1450.0	.2323	.008294	33210.0	45270.0	240.6	26.27	34.59	752.8	52.8	.0913	1.00011
1500.0	.2246	.008018	34530.0	47000.0	241.8	26.43	34.74	765.2	54.0	.0937	1.00011
1550.0	.2173	.007759	35850.0	48740.0	242.9	26.57	34.89	777.3	55.2	.0961	1.00010
1600.0	.2106	.007517	37190.0	50490.0	244.1	26.71	35.02	789.3	56.3	.0984	1.00010
1650.0	.2042	.007289	38530.0	52250.0	245.1	26.83	35.15	801.0	57.4	.101	1.00010
1700.0	.1982	.007075	39870.0	54010.0	246.2	26.95	35.26	812.7	58.6	.103	1.00009
1750.0	.1925	.006873	41220.0	55770.0	247.2	27.06	35.38	824.1	59.7	.105	1.00009
1800.0	.1872	.006682	42580.0	57540.0	248.2	27.16	35.48	835.4	60.8	.107	1.00009
1850.0	.1821	.006501	43940.0	59320.0	249.2	27.26	35.58	846.6	61.8	.110	1.00009
1900.0	.1773	.006330	45300.0	61100.0	250.1	27.35	35.67	857.6	62.9	.112	1.00008
				.101325 MPa isobar							
63.17[a]	867.9	30.98	–4212.0	–4208.0	68.01	26.75	54.68	1324.0	283.0	.152	1.46868

Thermophysical properties of nitrogen—Continued

T K	Density kg/m^3	Density mol/dm^3	E J/mol	H J/mol	S J/(mol·K)	C_v J/(mol·K)	C_p J/(mol·K)	Sound m/s	Visc. μPa·s	Therm. W/(m·K)	Diel. const.
70.0	841.0	30.02	−3828.0	−3824.0	73.79	28.53	57.17	1092.0	204.0	.145	1.45269
77.363[b]	808.6	28.86	−3404.0	−3401.0	79.54	27.82	57.79	938.9	152.0	.135	1.43357
77.363[b]	4.611	.1646	1548.0	2164.0	151.5	21.59	31.45	174.8	5.28	.00783	1.00217
80.0	4.440	.1585	1607.0	2246.0	152.6	21.48	31.16	178.3	5.46	.00808	1.00209
85.0	4.151	.1482	1717.0	2401.0	154.5	21.32	30.74	184.5	5.80	.00853	1.00196
90.0	3.900	.1392	1826.0	2554.0	156.2	21.20	30.44	190.4	6.14	.00898	1.00184
95.0	3.679	.1313	1934.0	2705.0	157.9	21.12	30.20	196.1	6.47	.00943	1.00173
100.0	3.484	.1243	2041.0	2856.0	159.4	21.05	30.02	201.6	6.80	.00988	1.00164
105.0	3.309	.1181	2148.0	3006.0	160.9	21.00	29.88	206.9	7.13	.0103	1.00156
110.0	3.151	.1125	2254.0	3155.0	162.3	20.97	29.77	212.1	7.46	.0108	1.00148
120.0	2.879	.1027	2465.0	3452.0	164.8	20.91	29.61	221.9	8.12	.0117	1.00136
130.0	2.651	.09461	2676.0	3747.0	167.2	20.88	29.50	231.4	8.76	.0126	1.00125
135.0	2.550	.09101	2781.0	3894.0	168.3	20.86	29.46	235.9	9.08	.0131	1.00120
140.0	2.457	.08769	2886.0	4042.0	169.4	20.85	29.42	240.3	9.39	.0135	1.00116
145.0	2.370	.08460	2991.0	4189.0	170.4	20.84	29.39	244.7	9.70	.0140	1.00112
150.0	2.290	.08173	3096.0	4336.0	171.4	20.84	29.36	249.0	10.0	.0144	1.00108
155.0	2.214	.07904	3200.0	4482.0	172.4	20.83	29.34	253.2	10.3	.0148	1.00104
160.0	2.144	.07653	3305.0	4629.0	173.3	20.82	29.32	257.3	10.6	.0153	1.00101
165.0	2.078	.07418	3410.0	4775.0	174.2	20.82	29.30	261.4	10.9	.0157	1.00098
170.0	2.016	.07197	3514.0	4922.0	175.1	20.82	29.28	265.4	11.2	.0162	1.00095
175.0	1.958	.06989	3618.0	5068.0	175.9	20.81	29.27	269.3	11.5	.0166	1.00092
180.0	1.903	.06792	3723.0	5215.0	176.8	20.81	29.26	273.2	11.8	.0170	1.00090
190.0	1.802	.06431	3932.0	5507.0	178.3	20.80	29.23	280.7	12.4	.0178	1.00085
200.0	1.711	.06107	4140.0	5799.0	179.8	20.80	29.22	288.1	12.9	.0187	1.00081
210.0	1.629	.05814	4349.0	6091.0	181.3	20.80	29.20	295.3	13.5	.0195	1.00077
220.0	1.554	.05548	4557.0	6383.0	182.6	20.79	29.19	302.3	14.0	.0202	1.00073
230.0	1.486	.05305	4765.0	6675.0	183.9	20.79	29.18	309.1	14.5	.0210	1.00070
240.0	1.424	.05083	4974.0	6967.0	185.2	20.79	29.17	315.8	15.0	.0218	1.00067
250.0	1.367	.04879	5182.0	7259.0	186.3	20.79	29.17	322.3	15.6	.0225	1.00064
260.0	1.314	.04690	5390.0	7550.0	187.5	20.79	29.16	328.8	16.1	.0232	1.00062
270.0	1.265	.04516	5598.0	7842.0	188.6	20.79	29.16	335.0	16.5	.0240	1.00060
280.0	1.220	.04354	5806.0	8133.0	189.7	20.79	29.15	341.2	17.0	.0247	1.00057
290.0	1.178	.04204	6015.0	8425.0	190.7	20.79	29.15	347.2	17.5	.0254	1.00055
300.0	1.138	.04063	6223.0	8716.0	191.7	20.80	29.15	353.2	18.0	.0260	1.00054
310.0	1.102	.03932	6431.0	9008.0	192.6	20.80	29.15	359.0	18.4	.0267	1.00052
320.0	1.067	.03809	6639.0	9300.0	193.5	20.81	29.16	364.8	18.9	.0274	1.00050
330.0	1.035	.03693	6847.0	9591.0	194.4	20.81	29.16	370.4	19.3	.0281	1.00049
340.0	1.004	.03584	7056.0	9883.0	195.3	20.82	29.17	376.0	19.7	.0287	1.00047
350.0	.9754	.03482	7264.0	10170.0	196.2	20.83	29.17	381.4	20.2	.0294	1.00046
360.0	.9483	.03385	7473.0	10470.0	197.0	20.84	29.18	386.8	20.6	.0300	1.00045
370.0	.9226	.03293	7681.0	10760.0	197.8	20.86	29.20	392.1	21.0	.0307	1.00043
380.0	.8983	.03206	7890.0	11050.0	198.6	20.87	29.21	397.3	21.4	.0313	1.00042
400.0	.8533	.03046	8308.0	11630.0	200.1	20.91	29.25	407.6	22.2	.0326	1.00040
420.0	.8127	.02901	8727.0	12220.0	201.5	20.96	29.29	417.5	23.0	.0338	1.00038
440.0	.7757	.02769	9147.0	12810.0	202.8	21.01	29.34	427.2	23.8	.0351	1.00036
460.0	.7420	.02648	9568.0	13390.0	204.2	21.08	29.41	436.6	24.6	.0363	1.00035
480.0	.7110	.02538	9990.0	13980.0	205.4	21.15	29.48	445.7	25.3	.0375	1.00033
500.0	.6826	.02437	10410.0	14570.0	206.6	21.24	29.56	454.7	26.1	.0388	1.00032
520.0	.6563	.02343	10840.0	15170.0	207.8	21.33	29.65	463.4	26.8	.0400	1.00031
540.0	.6320	.02256	11270.0	15760.0	208.9	21.42	29.75	471.9	27.5	.0412	1.00030
560.0	.6094	.02175	11700.0	16360.0	210.0	21.53	29.85	480.2	28.2	.0424	1.00029
580.0	.5884	.02100	12130.0	16950.0	211.0	21.64	29.96	488.4	28.9	.0436	1.00028
600.0	.5688	.02030	12560.0	17550.0	212.0	21.76	30.08	496.4	29.5	.0448	1.00027
620.0	.5505	.01965	13000.0	18160.0	213.0	21.88	30.20	504.2	30.2	.0460	1.00026
640.0	.5333	.01904	13440.0	18760.0	214.0	22.00	30.32	511.8	30.9	.0472	1.00025
660.0	.5171	.01846	13880.0	19370.0	214.9	22.13	30.45	519.3	31.5	.0484	1.00024
680.0	.5019	.01792	14320.0	19980.0	215.8	22.26	30.58	526.7	32.2	.0496	1.00024
700.0	.4876	.01740	14770.0	20590.0	216.7	22.40	30.71	534.0	32.8	.0508	1.00023
720.0	.4740	.01692	15220.0	21210.0	217.6	22.53	30.85	541.1	33.4	.0520	1.00022
740.0	.4612	.01646	15670.0	21830.0	218.4	22.66	30.98	548.1	34.0	.0531	1.00022
760.0	.4491	.01603	16130.0	22450.0	219.3	22.80	31.12	555.0	34.6	.0543	1.00021
780.0	.4376	.01562	16580.0	23070.0	220.1	22.93	31.25	561.9	35.2	.0555	1.00021
800.0	.4266	.01523	17040.0	23700.0	220.9	23.07	31.38	568.6	35.8	.0567	1.00020

Thermophysical properties of nitrogen—Continued

T K	Density kg/m³	Density mol/dm³	E J/mol	H J/mol	S J/(mol·K)	C_v J/(mol·K)	C_p J/(mol·K)	Sound m/s	Visc. μPa·s	Therm. W/(m·K)	Diel. const.
820.0	.4162	.01486	17510.0	24330.0	221.7	23.20	31.52	575.2	36.4	.0578	1.00020
840.0	.4063	.01450	17970.0	24960.0	222.4	23.33	31.65	581.7	37.0	.0590	1.00019
860.0	.3969	.01417	18440.0	25590.0	223.2	23.46	31.78	588.2	37.6	.0601	1.00019
880.0	.3878	.01384	18910.0	26230.0	223.9	23.59	31.91	594.5	38.2	.0613	1.00018
900.0	.3792	.01354	19380.0	26870.0	224.6	23.72	32.03	600.8	38.7	.0624	1.00018
920.0	.3710	.01324	19860.0	27510.0	225.3	23.84	32.16	607.1	39.3	.0635	1.00017
940.0	.3631	.01296	20340.0	28160.0	226.0	23.96	32.28	613.2	39.9	.0647	1.00017
1000.0	.3413	.01218	21790.0	30100.0	228.0	24.32	32.63	631.3	41.5	.0680	1.00016
1050.0	.3251	.01160	23010.0	31740.0	229.6	24.59	32.91	645.9	42.8	.0708	1.00015
1100.0	.3103	.01108	24240.0	33390.0	231.2	24.85	33.17	660.3	44.2	.0735	1.00015
1150.0	.2968	.01060	25490.0	35060.0	232.6	25.10	33.41	674.3	45.5	.0762	1.00014
1200.0	.2844	.01015	26750.0	36730.0	234.1	25.33	33.64	688.0	46.7	.0788	1.00013
1250.0	.2731	.009748	28030.0	38420.0	235.4	25.54	33.86	701.4	48.0	.0814	1.00013
1300.0	.2626	.009373	29310.0	40120.0	236.8	25.75	34.06	714.6	49.2	.0839	1.00012
1350.0	.2528	.009026	30600.0	41830.0	238.1	25.93	34.25	727.6	50.5	.0864	1.00012
1400.0	.2438	.008704	31900.0	43550.0	239.3	26.11	34.42	740.3	51.7	.0889	1.00011
1450.0	.2354	.008404	33210.0	45270.0	240.5	26.27	34.59	752.8	52.8	.0913	1.00011
1500.0	.2276	.008124	34530.0	47000.0	241.7	26.43	34.74	765.2	54.0	.0937	1.00011
1550.0	.2202	.007862	35850.0	48740.0	242.8	26.57	34.89	777.3	55.2	.0961	1.00010
1600.0	.2133	.007616	37190.0	50490.0	243.9	26.71	35.02	789.3	56.3	.0984	1.00010
1650.0	.2069	.007385	38530.0	52250.0	245.0	26.83	35.15	801.0	57.4	.101	1.00010
1700.0	.2008	.007168	39870.0	54010.0	246.1	26.95	35.26	812.7	58.6	.103	1.00009
1750.0	.1951	.006964	41220.0	55770.0	247.1	27.06	35.38	824.1	59.7	.105	1.00009
1800.0	.1896	.006770	42580.0	57540.0	248.1	27.16	35.48	835.4	60.8	.107	1.00009
1850.0	.1845	.006587	43940.0	59320.0	249.1	27.26	35.58	846.6	61.8	.110	1.00009
1900.0	.1797	.006414	45300.0	61100.0	250.0	27.35	35.67	857.6	62.9	.112	1.00008
					.15 MPa isobar						
63.18[a]	867.9	30.98	−4212.0	−4207.0	68.02	26.78	54.67	1324.0	283.0	.152	1.46869
70.0	841.0	30.02	−3828.0	−3823.0	73.78	28.55	57.15	1092.0	204.0	.145	1.45274
80.859[b]	792.3	28.28	−3203.0	−3198.0	82.08	27.40	58.09	880.3	134.0	.130	1.42394
80.859[b]	6.627	.2366	1601.0	2235.0	149.3	21.80	32.22	177.3	5.55	.00839	1.00312
85.0	6.250	.2231	1695.0	2367.0	150.9	21.60	31.66	182.7	5.83	.00875	1.00295
90.0	5.855	.2090	1806.0	2524.0	152.7	21.42	31.16	188.9	6.16	.00918	1.00276
95.0	5.511	.1967	1916.0	2679.0	154.4	21.29	30.79	194.8	6.50	.00962	1.00260
100.0	5.209	.1859	2025.0	2832.0	156.0	21.19	30.51	200.5	6.83	.0101	1.00245
105.0	4.940	.1763	2133.0	2984.0	157.5	21.12	30.29	206.0	7.16	.0105	1.00233
110.0	4.700	.1678	2241.0	3135.0	158.9	21.06	30.12	211.2	7.49	.0109	1.00221
115.0	4.483	.1600	2348.0	3285.0	160.2	21.01	29.98	216.3	7.81	.0114	1.00211
120.0	4.286	.1530	2454.0	3435.0	161.5	20.98	29.87	221.3	8.14	.0118	1.00202
130.0	3.941	.1407	2667.0	3733.0	163.9	20.92	29.70	230.8	8.78	.0127	1.00186
135.0	3.790	.1353	2772.0	3881.0	165.0	20.90	29.63	235.4	9.10	.0132	1.00179
140.0	3.650	.1303	2878.0	4029.0	166.1	20.89	29.58	239.9	9.41	.0136	1.00172
145.0	3.520	.1256	2983.0	4177.0	167.1	20.87	29.53	244.4	9.72	.0141	1.00166
150.0	3.399	.1213	3088.0	4324.0	168.1	20.86	29.49	248.7	10.0	.0145	1.00160
155.0	3.287	.1173	3193.0	4472.0	169.1	20.85	29.46	252.9	10.3	.0150	1.00155
160.0	3.182	.1136	3298.0	4619.0	170.0	20.85	29.43	257.1	10.6	.0154	1.00150
165.0	3.083	.1100	3403.0	4766.0	170.9	20.84	29.40	261.2	10.9	.0158	1.00145
170.0	2.991	.1067	3508.0	4913.0	171.8	20.83	29.38	265.2	11.2	.0163	1.00141
175.0	2.903	.1036	3612.0	5060.0	172.6	20.83	29.35	269.1	11.5	.0167	1.00137
180.0	2.821	.1007	3717.0	5207.0	173.5	20.82	29.34	273.0	11.8	.0171	1.00133
190.0	2.671	.09533	3926.0	5500.0	175.0	20.82	29.30	280.6	12.4	.0179	1.00126
200.0	2.535	.09050	4135.0	5793.0	176.5	20.81	29.28	288.0	12.9	.0188	1.00119
210.0	2.413	.08614	4344.0	6085.0	178.0	20.81	29.26	295.2	13.5	.0195	1.00114
220.0	2.303	.08219	4553.0	6378.0	179.3	20.80	29.24	302.3	14.0	.0203	1.00108
230.0	2.202	.07858	4761.0	6670.0	180.6	20.80	29.22	309.1	14.5	.0211	1.00104
240.0	2.109	.07528	4970.0	6962.0	181.9	20.80	29.21	315.8	15.1	.0219	1.00099
250.0	2.024	.07225	5178.0	7254.0	183.1	20.80	29.20	322.4	15.6	.0226	1.00095
260.0	1.946	.06946	5387.0	7546.0	184.2	20.80	29.19	328.8	16.1	.0233	1.00092
270.0	1.873	.06687	5595.0	7838.0	185.3	20.80	29.18	335.1	16.5	.0240	1.00088
280.0	1.806	.06447	5803.0	8130.0	186.4	20.80	29.18	341.3	17.0	.0247	1.00085
290.0	1.744	.06224	6012.0	8422.0	187.4	20.80	29.18	347.3	17.5	.0254	1.00082

Thermophysical properties of nitrogen—Continued

T	Density	Density	E	H	S	C_v	C_p	Sound	Visc.	Therm.	Diel.
K	kg/m^3	mol/dm^3	J/mol	J/mol	J/(mol·K)	J/(mol·K)		m/s	μPa·s	W/(m·K)	const.
300.0	1.685	.06016	6220.0	8713.0	188.4	20.80	29.17	353.3	18.0	.0261	1.00079
310.0	1.631	.05821	6428.0	9005.0	189.3	20.81	29.17	359.1	18.4	.0268	1.00077
320.0	1.580	.05638	6636.0	9297.0	190.3	20.81	29.17	364.9	18.9	.0275	1.00074
330.0	1.532	.05467	6845.0	9589.0	191.2	20.82	29.18	370.5	19.3	.0282	1.00072
340.0	1.486	.05306	7053.0	9880.0	192.0	20.83	29.18	376.1	19.7	.0288	1.00070
350.0	1.444	.05154	7262.0	10170.0	192.9	20.84	29.19	381.5	20.2	.0295	1.00068
360.0	1.404	.05010	7470.0	10460.0	193.7	20.85	29.20	386.9	20.6	.0301	1.00066
370.0	1.366	.04875	7679.0	10760.0	194.5	20.86	29.21	392.2	21.0	.0308	1.00064
380.0	1.330	.04746	7888.0	11050.0	195.3	20.88	29.22	397.5	21.4	.0314	1.00063
390.0	1.296	.04624	8097.0	11340.0	196.0	20.89	29.24	402.6	21.8	.0320	1.00061
400.0	1.263	.04509	8306.0	11630.0	196.8	20.91	29.26	407.7	22.3	.0327	1.00059
420.0	1.203	.04294	8725.0	12220.0	198.2	20.96	29.30	417.6	23.0	.0339	1.00057
440.0	1.148	.04098	9145.0	12810.0	199.6	21.02	29.35	427.3	23.8	.0352	1.00054
460.0	1.098	.03920	9566.0	13390.0	200.9	21.08	29.42	436.7	24.6	.0364	1.00052
480.0	1.052	.03757	9989.0	13980.0	202.1	21.15	29.49	445.9	25.3	.0376	1.00049
500.0	1.010	.03606	10410.0	14570.0	203.3	21.24	29.57	454.8	26.1	.0389	1.00047
520.0	.9714	.03467	10840.0	15160.0	204.5	21.33	29.66	463.5	26.8	.0401	1.00046
540.0	.9354	.03339	11270.0	15760.0	205.6	21.43	29.75	472.0	27.5	.0413	1.00044
560.0	.9020	.03220	11700.0	16350.0	206.7	21.53	29.86	480.4	28.2	.0425	1.00042
580.0	.8709	.03109	12130.0	16950.0	207.8	21.64	29.97	488.5	28.9	.0437	1.00041
600.0	.8419	.03005	12560.0	17550.0	208.8	21.76	30.08	496.5	29.6	.0449	1.00040
620.0	.8147	.02908	13000.0	18160.0	209.8	21.88	30.20	504.3	30.2	.0461	1.00038
640.0	.7893	.02817	13440.0	18760.0	210.7	22.00	30.33	512.0	30.9	.0473	1.00037
660.0	.7654	.02732	13880.0	19370.0	211.7	22.13	30.46	519.5	31.5	.0485	1.00036
680.0	.7428	.02652	14320.0	19980.0	212.6	22.26	30.59	526.9	32.2	.0497	1.00035
700.0	.7216	.02576	14770.0	20590.0	213.5	22.40	30.72	534.1	32.8	.0509	1.00034
720.0	.7016	.02504	15220.0	21210.0	214.3	22.53	30.85	541.2	33.4	.0521	1.00033
740.0	.6826	.02437	15670.0	21830.0	215.2	22.66	30.99	548.3	34.0	.0532	1.00032
760.0	.6647	.02373	16130.0	22450.0	216.0	22.80	31.12	555.2	34.6	.0544	1.00031
780.0	.6476	.02312	16580.0	23070.0	216.8	22.93	31.25	562.0	35.2	.0556	1.00030
800.0	.6314	.02254	17040.0	23700.0	217.6	23.07	31.39	568.7	35.8	.0567	1.00030
820.0	.6160	.02199	17510.0	24330.0	218.4	23.20	31.52	575.3	36.4	.0579	1.00029
840.0	.6014	.02147	17970.0	24960.0	219.1	23.33	31.65	581.8	37.0	.0591	1.00028
860.0	.5874	.02097	18440.0	25590.0	219.9	23.46	31.78	588.3	37.6	.0602	1.00028
880.0	.5741	.02049	18910.0	26230.0	220.6	23.59	31.91	594.7	38.2	.0614	1.00027
900.0	.5613	.02004	19380.0	26870.0	221.3	23.72	32.04	601.0	38.7	.0625	1.00026
920.0	.5491	.01960	19860.0	27510.0	222.1	23.84	32.16	607.2	39.3	.0636	1.00026
940.0	.5374	.01918	20340.0	28160.0	222.7	23.96	32.28	613.3	39.9	.0648	1.00025
1000.0	.5052	.01803	21790.0	30100.0	224.8	24.32	32.63	631.4	41.5	.0681	1.00024
1050.0	.4811	.01717	23010.0	31740.0	226.4	24.59	32.91	646.1	42.8	.0709	1.00023
1100.0	.4593	.01639	24240.0	33390.0	227.9	24.85	33.17	660.4	44.2	.0736	1.00021
1150.0	.4393	.01568	25490.0	35060.0	229.4	25.10	33.42	674.4	45.5	.0763	1.00021
1200.0	.4210	.01503	26750.0	36740.0	230.8	25.33	33.65	688.1	46.7	.0789	1.00020
1250.0	.4042	.01443	28030.0	38420.0	232.2	25.55	33.86	701.5	48.0	.0815	1.00019
1300.0	.3886	.01387	29310.0	40120.0	233.5	25.75	34.06	714.7	49.2	.0840	1.00018
1350.0	.3743	.01336	30600.0	41830.0	234.8	25.93	34.25	727.7	50.5	.0865	1.00017
1400.0	.3609	.01288	31900.0	43550.0	236.0	26.11	34.43	740.4	51.7	.0890	1.00017
1450.0	.3485	.01244	33210.0	45270.0	237.3	26.28	34.59	752.9	52.8	.0914	1.00016
1500.0	.3368	.01202	34530.0	47010.0	238.4	26.43	34.74	765.3	54.0	.0938	1.00016
1550.0	.3260	.01164	35850.0	48750.0	239.6	26.57	34.89	777.4	55.2	.0962	1.00015
1600.0	.3158	.01127	37190.0	50490.0	240.7	26.71	35.02	789.4	56.3	.0985	1.00015
1650.0	.3062	.01093	38530.0	52250.0	241.8	26.83	35.15	801.1	57.4	.101	1.00014
1700.0	.2972	.01061	39870.0	54010.0	242.8	26.95	35.27	812.8	58.6	.103	1.00014
1750.0	.2887	.01031	41220.0	55770.0	243.8	27.06	35.38	824.2	59.7	.105	1.00013
1800.0	.2807	.01002	42580.0	57550.0	244.8	27.17	35.48	835.5	60.8	.108	1.00013
1850.0	.2731	.009750	43940.0	59320.0	245.8	27.26	35.58	846.7	61.8	.110	1.00013
1900.0	.2660	.009494	45300.0	61100.0	246.8	27.35	35.67	857.7	62.9	.112	1.00012
				.20 MPa isobar							
63.19[a]	867.9	30.98	−4211.0	−4205.0	68.02	26.81	54.67	1323.0	283.0	.152	1.46869
70.0	841.1	30.02	−3829.0	−3822.0	73.77	28.57	57.14	1092.0	204.0	.145	1.45279

Thermophysical properties of nitrogen—Continued

T K	Density kg/m^3	Density mol/dm^3	E J/mol	H J/mol	S J/(mol·K)	C_v J/(mol·K)	C_p J/(mol·K)	Sound m/s	Visc. μPa·s	Therm. W/(m·K)	Diel. const.
83.644[b]	778.8	27.80	−3042.0	−3035.0	84.04	27.10	58.45	837.1	121.0	.126	1.41597
83.644[b]	8.658	.3090	1639.0	2286.0	147.7	21.99	32.96	179.0	5.77	.00886	1.00408
85.0	8.488	.3030	1671.0	2331.0	148.3	21.91	32.71	180.8	5.86	.00897	1.00400
90.0	7.925	.2829	1786.0	2493.0	150.1	21.66	31.98	187.3	6.19	.00939	1.00374
95.0	7.441	.2656	1898.0	2651.0	151.8	21.47	31.44	193.5	6.52	.00981	1.00351
100.0	7.019	.2506	2009.0	2807.0	153.4	21.34	31.04	199.3	6.85	.0102	1.00331
105.0	6.647	.2373	2119.0	2962.0	154.9	21.23	30.73	204.9	7.18	.0107	1.00313
110.0	6.316	.2254	2227.0	3115.0	156.4	21.15	30.49	210.3	7.51	.0111	1.00298
115.0	6.018	.2148	2336.0	3267.0	157.7	21.09	30.30	215.6	7.84	.0115	1.00284
120.0	5.749	.2052	2443.0	3418.0	159.0	21.04	30.14	220.6	8.16	.0120	1.00271
130.0	5.279	.1884	2657.0	3718.0	161.4	20.97	29.91	230.3	8.80	.0129	1.00249
135.0	5.074	.1811	2763.0	3867.0	162.5	20.95	29.82	235.0	9.11	.0133	1.00239
140.0	4.884	.1743	2869.0	4016.0	163.6	20.92	29.75	239.5	9.43	.0138	1.00230
145.0	4.709	.1681	2975.0	4165.0	164.6	20.91	29.68	244.0	9.74	.0142	1.00222
150.0	4.546	.1622	3080.0	4313.0	165.7	20.89	29.63	248.3	10.0	.0146	1.00214
155.0	4.394	.1568	3186.0	4461.0	166.6	20.88	29.58	252.6	10.3	.0151	1.00207
160.0	4.252	.1518	3291.0	4609.0	167.6	20.87	29.54	256.8	10.7	.0155	1.00200
165.0	4.119	.1470	3396.0	4756.0	168.5	20.86	29.50	260.9	10.9	.0159	1.00194
170.0	3.995	.1426	3501.0	4904.0	169.4	20.85	29.47	265.0	11.2	.0164	1.00188
175.0	3.878	.1384	3606.0	5051.0	170.2	20.84	29.44	269.0	11.5	.0168	1.00183
180.0	3.768	.1345	3711.0	5198.0	171.0	20.84	29.42	272.9	11.8	.0172	1.00177
185.0	3.664	.1308	3816.0	5345.0	171.8	20.83	29.39	276.7	12.1	.0176	1.00172
190.0	3.566	.1273	3921.0	5492.0	172.6	20.83	29.37	280.5	12.4	.0180	1.00168
200.0	3.384	.1208	4130.0	5786.0	174.1	20.82	29.34	288.0	12.9	.0188	1.00159
210.0	3.221	.1150	4339.0	6079.0	175.6	20.81	29.31	295.2	13.5	.0196	1.00152
220.0	3.072	.1097	4548.0	6372.0	176.9	20.81	29.29	302.2	14.0	.0204	1.00145
230.0	2.937	.1048	4757.0	6665.0	178.2	20.81	29.27	309.1	14.6	.0212	1.00138
240.0	2.814	.1004	4966.0	6957.0	179.5	20.80	29.25	315.8	15.1	.0219	1.00132
250.0	2.700	.09637	5174.0	7250.0	180.7	20.80	29.23	322.4	15.6	.0227	1.00127
260.0	2.595	.09264	5383.0	7542.0	181.8	20.80	29.22	328.8	16.1	.0234	1.00122
270.0	2.498	.08918	5592.0	7834.0	182.9	20.80	29.21	335.1	16.6	.0241	1.00118
280.0	2.409	.08598	5800.0	8126.0	184.0	20.80	29.21	341.3	17.0	.0248	1.00113
290.0	2.325	.08299	6009.0	8418.0	185.0	20.80	29.20	347.4	17.5	.0255	1.00109
300.0	2.247	.08021	6217.0	8710.0	186.0	20.80	29.20	353.3	18.0	.0262	1.00106
310.0	2.174	.07761	6425.0	9002.0	186.9	20.81	29.19	359.2	18.4	.0269	1.00102
320.0	2.106	.07518	6634.0	9294.0	187.9	20.81	29.19	364.9	18.9	.0276	1.00099
330.0	2.042	.07289	6842.0	9586.0	188.8	20.82	29.20	370.6	19.3	.0282	1.00096
340.0	1.982	.07074	7051.0	9878.0	189.6	20.83	29.20	376.2	19.8	.0289	1.00093
350.0	1.925	.06871	7259.0	10170.0	190.5	20.84	29.21	381.6	20.2	.0295	1.00090
360.0	1.871	.06680	7468.0	10460.0	191.3	20.85	29.21	387.0	20.6	.0302	1.00088
370.0	1.821	.06499	7677.0	10750.0	192.1	20.86	29.22	392.3	21.0	.0308	1.00086
380.0	1.773	.06328	7886.0	11050.0	192.9	20.88	29.24	397.6	21.4	.0315	1.00083
390.0	1.727	.06165	8095.0	11340.0	193.7	20.90	29.25	402.7	21.9	.0321	1.00081
400.0	1.684	.06011	8304.0	11630.0	194.4	20.92	29.27	407.8	22.3	.0327	1.00079
420.0	1.604	.05724	8723.0	12220.0	195.8	20.96	29.31	417.7	23.1	.0340	1.00075
440.0	1.531	.05463	9143.0	12800.0	197.2	21.02	29.36	427.4	23.8	.0352	1.00072
460.0	1.464	.05226	9565.0	13390.0	198.5	21.08	29.42	436.8	24.6	.0365	1.00069
480.0	1.403	.05008	9987.0	13980.0	199.7	21.16	29.49	446.0	25.3	.0377	1.00066
500.0	1.347	.04807	10410.0	14570.0	201.0	21.24	29.57	454.9	26.1	.0389	1.00063
520.0	1.295	.04622	10840.0	15160.0	202.1	21.33	29.66	463.7	26.8	.0401	1.00061
540.0	1.247	.04451	11270.0	15760.0	203.2	21.43	29.76	472.2	27.5	.0414	1.00059
560.0	1.202	.04292	11690.0	16350.0	204.3	21.53	29.86	480.5	28.2	.0426	1.00056
580.0	1.161	.04144	12130.0	16950.0	205.4	21.64	29.97	488.6	28.9	.0438	1.00054
600.0	1.122	.04006	12560.0	17550.0	206.4	21.76	30.09	496.6	29.6	.0450	1.00053
620.0	1.086	.03877	13000.0	18160.0	207.4	21.88	30.21	504.4	30.2	.0462	1.00051
640.0	1.052	.03756	13440.0	18760.0	208.3	22.01	30.33	512.1	30.9	.0474	1.00049
660.0	1.020	.03642	13880.0	19370.0	209.3	22.13	30.46	519.6	31.5	.0486	1.00048
680.0	.9903	.03535	14320.0	19980.0	210.2	22.26	30.59	527.0	32.2	.0498	1.00046
700.0	.9620	.03434	14770.0	20590.0	211.1	22.40	30.72	534.2	32.8	.0509	1.00045
720.0	.9353	.03338	15220.0	21210.0	211.9	22.53	30.85	541.4	33.4	.0521	1.00044
740.0	.9100	.03248	15670.0	21830.0	212.8	22.67	30.99	548.4	34.0	.0533	1.00043
760.0	.8860	.03163	16120.0	22450.0	213.6	22.80	31.12	555.3	34.6	.0545	1.00042
780.0	.8633	.03082	16580.0	23070.0	214.4	22.93	31.26	562.1	35.2	.0557	1.00040

Thermophysical properties of nitrogen—Continued

T K	Density kg/m^3	Density mol/dm^3	E J/mol	H J/mol	S J/(mol·K)	C_v J/(mol·K)	C_p J/(mol·K)	Sound m/s	Visc. μPa·s	Therm. W/(m·K)	Diel. const.
800.0	.8418	.03005	17040.0	23700.0	215.2	23.07	31.39	568.8	35.8	.0568	1.00039
820.0	.8212	.02931	17510.0	24330.0	216.0	23.20	31.52	575.4	36.4	.0580	1.00038
840.0	.8017	.02862	17970.0	24960.0	216.8	23.33	31.65	582.0	37.0	.0591	1.00038
860.0	.7831	.02795	18440.0	25590.0	217.5	23.46	31.78	588.4	37.6	.0603	1.00037
880.0	.7653	.02732	18910.0	26230.0	218.2	23.59	31.91	594.8	38.2	.0614	1.00036
900.0	.7483	.02671	19380.0	26870.0	219.0	23.72	32.04	601.1	38.7	.0626	1.00035
920.0	.7320	.02613	19860.0	27510.0	219.7	23.84	32.16	607.3	39.3	.0637	1.00034
940.0	.7164	.02557	20340.0	28160.0	220.4	23.97	32.28	613.4	39.9	.0648	1.00034
1000.0	.6735	.02404	21780.0	30100.0	222.4	24.32	32.64	631.5	41.5	.0682	1.00032
1050.0	.6414	.02290	23010.0	31740.0	224.0	24.59	32.91	646.2	42.9	.0710	1.00030
1100.0	.6123	.02186	24240.0	33400.0	225.5	24.86	33.17	660.5	44.2	.0737	1.00029
1150.0	.5857	.02091	25490.0	35060.0	227.0	25.10	33.42	674.5	45.5	.0763	1.00027
1200.0	.5613	.02004	26750.0	36740.0	228.4	25.33	33.65	688.2	46.7	.0790	1.00026
1250.0	.5388	.01923	28030.0	38420.0	229.8	25.55	33.86	701.6	48.0	.0816	1.00025
1300.0	.5181	.01849	29310.0	40120.0	231.1	25.75	34.06	714.8	49.2	.0841	1.00024
1350.0	.4989	.01781	30600.0	41830.0	232.4	25.94	34.25	727.8	50.5	.0866	1.00023
1400.0	.4811	.01717	31900.0	43550.0	233.7	26.11	34.43	740.5	51.7	.0891	1.00022
1450.0	.4646	.01658	33210.0	45270.0	234.9	26.28	34.59	753.0	52.8	.0915	1.00022
1500.0	.4491	.01603	34530.0	47010.0	236.0	26.43	34.74	765.4	54.0	.0939	1.00021
1550.0	.4346	.01551	35850.0	48750.0	237.2	26.57	34.89	777.5	55.2	.0963	1.00020
1600.0	.4210	.01503	37190.0	50500.0	238.3	26.71	35.02	789.4	56.3	.0986	1.00020
1650.0	.4083	.01457	38520.0	52250.0	239.4	26.83	35.15	801.2	57.4	.101	1.00019
1700.0	.3963	.01415	39870.0	54010.0	240.4	26.95	35.27	812.8	58.6	.103	1.00018
1750.0	.3849	.01374	41220.0	55780.0	241.4	27.06	35.38	824.3	59.7	.105	1.00018
1800.0	.3743	.01336	42580.0	57550.0	242.4	27.17	35.48	835.6	60.8	.108	1.00017
1850.0	.3641	.01300	43940.0	59320.0	243.4	27.26	35.58	846.8	61.8	.110	1.00017
1900.0	.3546	.01266	45300.0	61100.0	244.4	27.35	35.67	857.8	62.9	.112	1.00016

.30 MPa isobar

T K	Density kg/m^3	Density mol/dm^3	E J/mol	H J/mol	S J/(mol·K)	C_v J/(mol·K)	C_p J/(mol·K)	Sound m/s	Visc. μPa·s	Therm. W/(m·K)	Diel. const.
63.21[a]	868.0	30.98	−4211.0	−4201.0	68.02	26.87	54.66	1321.0	282.0	.152	1.46871
70.0	841.3	30.03	−3830.0	−3820.0	73.75	28.61	57.11	1092.0	204.0	.145	1.45288
80.0	796.7	28.44	−3255.0	−3244.0	81.44	27.54	57.94	894.9	138.0	.132	1.42655
87.930[b]	757.0	27.02	−2792.0	−2781.0	86.96	26.72	59.32	774.6	106.0	.120	1.40320
87.930[b]	12.66	.4519	1692.0	2356.0	145.4	22.33	34.37	181.1	6.11	.00966	1.00597
90.0	12.28	.4383	1742.0	2426.0	146.2	22.17	33.86	184.0	6.25	.00981	1.00579
95.0	11.46	.4092	1860.0	2593.0	148.0	21.87	32.90	190.7	6.58	.0102	1.00541
100.0	10.77	.3843	1975.0	2756.0	149.7	21.65	32.21	196.9	6.91	.0106	1.00508
105.0	10.16	.3627	2088.0	2915.0	151.3	21.48	31.69	202.9	7.23	.0110	1.00479
110.0	9.628	.3437	2200.0	3073.0	152.7	21.36	31.29	208.6	7.56	.0114	1.00454
115.0	9.154	.3267	2310.0	3228.0	154.1	21.26	30.97	214.0	7.88	.0118	1.00431
120.0	8.728	.3115	2420.0	3383.0	155.4	21.18	30.72	219.3	8.20	.0123	1.00411
130.0	7.994	.2853	2637.0	3688.0	157.9	21.07	30.35	229.3	8.84	.0131	1.00377
135.0	7.674	.2739	2744.0	3839.0	159.0	21.03	30.21	234.0	9.15	.0136	1.00362
140.0	7.381	.2634	2851.0	3990.0	160.1	21.00	30.09	238.7	9.46	.0140	1.00348
145.0	7.110	.2538	2958.0	4140.0	161.2	20.97	29.99	243.3	9.77	.0144	1.00335
150.0	6.859	.2448	3065.0	4290.0	162.2	20.95	29.91	247.7	10.1	.0149	1.00323
155.0	6.626	.2365	3171.0	4439.0	163.2	20.93	29.83	252.0	10.4	.0153	1.00312
160.0	6.409	.2288	3277.0	4588.0	164.1	20.91	29.77	256.3	10.7	.0157	1.00302
165.0	6.206	.2215	3383.0	4737.0	165.0	20.90	29.71	260.5	11.0	.0161	1.00292
170.0	6.016	.2147	3488.0	4885.0	165.9	20.89	29.66	264.6	11.3	.0166	1.00283
175.0	5.838	.2084	3594.0	5034.0	166.8	20.88	29.62	268.6	11.6	.0170	1.00275
180.0	5.670	.2024	3699.0	5182.0	167.6	20.87	29.58	272.6	11.9	.0174	1.00267
185.0	5.512	.1967	3805.0	5330.0	168.4	20.86	29.55	276.5	12.1	.0178	1.00260
190.0	5.362	.1914	3910.0	5477.0	169.2	20.85	29.52	280.3	12.4	.0182	1.00252
200.0	5.087	.1816	4120.0	5772.0	170.7	20.84	29.46	287.8	13.0	.0190	1.00239
210.0	4.840	.1727	4330.0	6066.0	172.1	20.83	29.42	295.1	13.5	.0198	1.00228
220.0	4.615	.1647	4539.0	6360.0	173.5	20.83	29.38	302.2	14.1	.0206	1.00217
230.0	4.411	.1574	4749.0	6654.0	174.8	20.82	29.35	309.1	14.6	.0214	1.00208
240.0	4.224	.1508	4958.0	6948.0	176.1	20.82	29.33	315.9	15.1	.0221	1.00199
250.0	4.053	.1447	5167.0	7241.0	177.3	20.81	29.31	322.5	15.6	.0228	1.00191
260.0	3.895	.1390	5376.0	7534.0	178.4	20.81	29.29	328.9	16.1	.0236	1.00183

Thermophysical properties of nitrogen—Continued

T K	Density kg/m^3	Density mol/dm^3	E J/mol	H J/mol	S J/(mol·K)	C_v J/(mol·K)	C_p J/(mol·K)	Sound m/s	Visc. μPa·s	Therm. W/(m·K)	Diel. const.
270.0	3.750	.1338	5585.0	7826.0	179.5	20.81	29.27	335.2	16.6	.0243	1.00176
280.0	3.614	.1290	5794.0	8119.0	180.6	20.81	29.26	341.4	17.1	.0250	1.00170
290.0	3.489	.1245	6002.0	8412.0	181.6	20.81	29.25	347.5	17.5	.0257	1.00164
300.0	3.371	.1203	6211.0	8704.0	182.6	20.81	29.24	353.5	18.0	.0264	1.00159
310.0	3.262	.1164	6420.0	8997.0	183.6	20.81	29.24	359.4	18.4	.0270	1.00153
320.0	3.159	.1128	6629.0	9289.0	184.5	20.82	29.23	365.1	18.9	.0277	1.00149
330.0	3.063	.1093	6837.0	9581.0	185.4	20.83	29.23	370.8	19.3	.0284	1.00144
340.0	2.973	.1061	7046.0	9873.0	186.3	20.83	29.23	376.4	19.8	.0290	1.00140
350.0	2.887	.1031	7255.0	10170.0	187.1	20.84	29.24	381.9	20.2	.0297	1.00136
360.0	2.807	.1002	7464.0	10460.0	187.9	20.85	29.24	387.3	20.6	.0303	1.00132
370.0	2.731	.09746	7673.0	10750.0	188.7	20.87	29.25	392.6	21.0	.0310	1.00128
380.0	2.658	.09489	7882.0	11040.0	189.5	20.88	29.26	397.8	21.5	.0316	1.00125
390.0	2.590	.09245	8091.0	11340.0	190.3	20.90	29.28	403.0	21.9	.0322	1.00122
400.0	2.525	.09013	8300.0	11630.0	191.0	20.92	29.29	408.0	22.3	.0329	1.00119
420.0	2.405	.08583	8720.0	12220.0	192.4	20.97	29.33	418.0	23.1	.0341	1.00113
440.0	2.295	.08192	9140.0	12800.0	193.8	21.02	29.38	427.7	23.8	.0354	1.00108
460.0	2.195	.07835	9562.0	13390.0	195.1	21.09	29.44	437.1	24.6	.0366	1.00103
480.0	2.104	.07509	9984.0	13980.0	196.4	21.16	29.51	446.2	25.3	.0378	1.00099
500.0	2.019	.07208	10410.0	14570.0	197.6	21.24	29.59	455.2	26.1	.0391	1.00095
520.0	1.942	.06931	10830.0	15160.0	198.7	21.33	29.68	463.9	26.8	.0403	1.00091
540.0	1.870	.06674	11260.0	15760.0	199.9	21.43	29.77	472.4	27.5	.0415	1.00088
560.0	1.803	.06435	11690.0	16350.0	200.9	21.53	29.87	480.8	28.2	.0427	1.00085
580.0	1.741	.06213	12120.0	16950.0	202.0	21.65	29.98	488.9	28.9	.0439	1.00082
600.0	1.683	.06006	12560.0	17550.0	203.0	21.76	30.10	496.9	29.6	.0451	1.00079
620.0	1.628	.05813	13000.0	18160.0	204.0	21.88	30.22	504.7	30.2	.0463	1.00076
640.0	1.578	.05631	13430.0	18760.0	205.0	22.01	30.34	512.3	30.9	.0475	1.00074
660.0	1.530	.05460	13880.0	19370.0	205.9	22.14	30.47	519.9	31.5	.0487	1.00072
680.0	1.485	.05300	14320.0	19980.0	206.8	22.27	30.60	527.2	32.2	.0499	1.00070
700.0	1.442	.05148	14770.0	20590.0	207.7	22.40	30.73	534.5	32.8	.0511	1.00068
720.0	1.402	.05005	15220.0	21210.0	208.6	22.53	30.86	541.6	33.4	.0523	1.00066
740.0	1.364	.04870	15670.0	21830.0	209.4	22.67	30.99	548.6	34.0	.0534	1.00064
760.0	1.329	.04742	16120.0	22450.0	210.2	22.80	31.13	555.5	34.6	.0546	1.00062
780.0	1.294	.04621	16580.0	23070.0	211.1	22.94	31.26	562.4	35.3	.0558	1.00061
800.0	1.262	.04505	17040.0	23700.0	211.8	23.07	31.39	569.1	35.8	.0570	1.00059
820.0	1.231	.04395	17500.0	24330.0	212.6	23.20	31.53	575.7	36.4	.0581	1.00058
840.0	1.202	.04291	17970.0	24960.0	213.4	23.33	31.66	582.2	37.0	.0593	1.00056
860.0	1.174	.04191	18440.0	25600.0	214.1	23.46	31.79	588.7	37.6	.0604	1.00055
880.0	1.147	.04096	18910.0	26230.0	214.9	23.59	31.91	595.0	38.2	.0616	1.00054
900.0	1.122	.04005	19380.0	26870.0	215.6	23.72	32.04	601.3	38.7	.0627	1.00052
920.0	1.098	.03918	19860.0	27510.0	216.3	23.84	32.16	607.5	39.3	.0639	1.00051
940.0	1.074	.03835	20340.0	28160.0	217.0	23.97	32.29	613.7	39.9	.0650	1.00050
1000.0	1.010	.03605	21780.0	30110.0	219.0	24.32	32.64	631.7	41.5	.0683	1.00047
1050.0	.9618	.03433	23010.0	31750.0	220.6	24.60	32.91	646.4	42.9	.0711	1.00045
1100.0	.9181	.03277	24240.0	33400.0	222.1	24.86	33.17	660.7	44.2	.0738	1.00043
1150.0	.8782	.03135	25490.0	35060.0	223.6	25.10	33.42	674.7	45.5	.0765	1.00041
1200.0	.8417	.03004	26750.0	36740.0	225.0	25.33	33.65	688.4	46.7	.0791	1.00039
1250.0	.8080	.02884	28030.0	38430.0	226.4	25.55	33.86	701.8	48.0	.0817	1.00038
1300.0	.7770	.02773	29310.0	40130.0	227.7	25.75	34.06	715.0	49.2	.0843	1.00036
1350.0	.7482	.02671	30600.0	41830.0	229.0	25.94	34.25	728.0	50.5	.0868	1.00035
1400.0	.7215	.02575	31900.0	43550.0	230.3	26.11	34.43	740.7	51.7	.0892	1.00034
1450.0	.6966	.02487	33210.0	45280.0	231.5	26.28	34.59	753.2	52.8	.0917	1.00032
1500.0	.6734	.02404	34530.0	47010.0	232.7	26.43	34.74	765.5	54.0	.0941	1.00031
1550.0	.6517	.02326	35850.0	48750.0	233.8	26.57	34.89	777.7	55.2	.0964	1.00030
1600.0	.6314	.02254	37190.0	50500.0	234.9	26.71	35.02	789.6	56.3	.0987	1.00029
1650.0	.6123	.02185	38520.0	52250.0	236.0	26.83	35.15	801.4	57.5	.101	1.00028
1700.0	.5943	.02121	39870.0	54010.0	237.0	26.95	35.27	813.0	58.6	.103	1.00028
1750.0	.5773	.02061	41220.0	55780.0	238.1	27.06	35.38	824.5	59.7	.106	1.00027
1800.0	.5613	.02003	42580.0	57550.0	239.1	27.17	35.48	835.8	60.8	.108	1.00026
1850.0	.5461	.01949	43940.0	59330.0	240.0	27.26	35.58	846.9	61.8	.110	1.00025
1900.0	.5317	.01898	45300.0	61110.0	241.0	27.35	35.67	858.0	62.9	.112	1.00025

Thermophysical properties of nitrogen—Continued

T K	Density kg/m^3	Density mol/dm^3	E J/mol	H J/mol	S J/(mol·K)	C_v J/(mol·K)	C_p J/(mol·K)	Sound m/s	Visc. μPa·s	Therm. W/(m·K)	Diel. const.
				.40 MPa isobar							
63.24[a]	868.0	30.98	−4211.0	−4198.0	68.03	26.94	54.66	1320.0	282.0	.152	1.46873
70.0	841.5	30.03	−3831.0	−3818.0	73.74	28.65	57.08	1092.0	205.0	.145	1.45298
80.0	797.0	28.45	−3257.0	−3243.0	81.42	27.58	57.90	895.4	138.0	.132	1.42670
91.256[b]	739.3	26.39	−2595.0	−2580.0	89.16	26.50	60.32	728.4	95.0	.115	1.39282
91.256[b]	16.64	.5938	1726.0	2400.0	143.8	22.62	35.74	182.3	6.39	.0104	1.00785
95.0	15.73	.5615	1819.0	2531.0	145.2	22.31	34.64	187.7	6.64	.0106	1.00742
100.0	14.70	.5247	1939.0	2702.0	146.9	21.98	33.55	194.4	6.96	.0110	1.00694
105.0	13.82	.4933	2056.0	2867.0	148.6	21.75	32.76	200.7	7.28	.0113	1.00652
110.0	13.06	.4660	2171.0	3030.0	150.1	21.57	32.16	206.7	7.61	.0117	1.00616
115.0	12.38	.4420	2284.0	3189.0	151.5	21.43	31.70	212.4	7.93	.0121	1.00584
120.0	11.78	.4206	2396.0	3347.0	152.8	21.32	31.34	217.9	8.25	.0125	1.00556
125.0	11.25	.4015	2506.0	3503.0	154.1	21.24	31.05	223.1	8.56	.0130	1.00530
130.0	10.76	.3841	2616.0	3657.0	155.3	21.17	30.81	228.2	8.88	.0134	1.00507
135.0	10.32	.3684	2725.0	3811.0	156.5	21.12	30.62	233.1	9.19	.0138	1.00486
140.0	9.915	.3539	2833.0	3964.0	157.6	21.07	30.45	237.9	9.50	.0142	1.00467
145.0	9.543	.3406	2941.0	4115.0	158.7	21.03	30.31	242.5	9.81	.0147	1.00450
150.0	9.200	.3284	3049.0	4267.0	159.7	21.00	30.19	247.0	10.1	.0151	1.00434
155.0	8.882	.3170	3156.0	4417.0	160.7	20.98	30.09	251.5	10.4	.0155	1.00418
160.0	8.586	.3065	3263.0	4568.0	161.6	20.96	30.01	255.8	10.7	.0159	1.00405
165.0	8.311	.2966	3369.0	4718.0	162.5	20.94	29.93	260.0	11.0	.0163	1.00391
170.0	8.053	.2874	3475.0	4867.0	163.4	20.92	29.86	264.2	11.3	.0168	1.00379
175.0	7.811	.2788	3581.0	5016.0	164.3	20.91	29.80	268.3	11.6	.0172	1.00368
180.0	7.584	.2707	3687.0	5165.0	165.1	20.90	29.75	272.3	11.9	.0176	1.00357
185.0	7.370	.2631	3793.0	5314.0	166.0	20.89	29.70	276.2	12.2	.0180	1.00347
190.0	7.169	.2559	3899.0	5462.0	166.7	20.88	29.66	280.1	12.4	.0184	1.00338
200.0	6.798	.2426	4110.0	5758.0	168.3	20.86	29.59	287.7	13.0	.0192	1.00320
210.0	6.464	.2307	4320.0	6054.0	169.7	20.85	29.53	295.0	13.5	.0200	1.00304
220.0	6.163	.2200	4530.0	6349.0	171.1	20.84	29.48	302.1	14.1	.0207	1.00290
230.0	5.888	.2102	4740.0	6644.0	172.4	20.83	29.44	309.1	14.6	.0215	1.00277
240.0	5.638	.2012	4950.0	6938.0	173.6	20.83	29.41	315.9	15.1	.0222	1.00265
250.0	5.408	.1930	5160.0	7232.0	174.8	20.82	29.38	322.5	15.6	.0230	1.00254
260.0	5.197	.1855	5369.0	7525.0	176.0	20.82	29.35	329.0	16.1	.0237	1.00244
270.0	5.002	.1785	5578.0	7819.0	177.1	20.82	29.33	335.4	16.6	.0244	1.00235
280.0	4.821	.1721	5787.0	8112.0	178.2	20.82	29.31	341.6	17.1	.0251	1.00227
290.0	4.653	.1661	5996.0	8405.0	179.2	20.82	29.30	347.7	17.5	.0258	1.00219
300.0	4.496	.1605	6205.0	8698.0	180.2	20.82	29.29	353.7	18.0	.0265	1.00211
310.0	4.350	.1553	6414.0	8991.0	181.1	20.82	29.28	359.5	18.5	.0272	1.00205
320.0	4.213	.1504	6623.0	9283.0	182.1	20.83	29.27	365.3	18.9	.0278	1.00198
330.0	4.084	.1458	6832.0	9576.0	183.0	20.83	29.27	371.0	19.3	.0285	1.00192
340.0	3.963	.1415	7041.0	9869.0	183.9	20.84	29.27	376.6	19.8	.0292	1.00186
350.0	3.849	.1374	7250.0	10160.0	184.7	20.85	29.27	382.1	20.2	.0298	1.00181
360.0	3.742	.1336	7459.0	10450.0	185.5	20.86	29.27	387.5	20.6	.0305	1.00176
370.0	3.640	.1299	7668.0	10750.0	186.3	20.87	29.28	392.8	21.1	.0311	1.00171
380.0	3.544	.1265	7878.0	11040.0	187.1	20.89	29.29	398.0	21.5	.0317	1.00166
390.0	3.453	.1232	8087.0	11330.0	187.9	20.90	29.30	403.2	21.9	.0324	1.00162
400.0	3.366	.1201	8296.0	11630.0	188.6	20.92	29.32	408.3	22.3	.0330	1.00158
420.0	3.205	.1144	8716.0	12210.0	190.0	20.97	29.35	418.2	23.1	.0343	1.00151
440.0	3.059	.1092	9137.0	12800.0	191.4	21.02	29.40	427.9	23.9	.0355	1.00144
460.0	2.926	.1044	9558.0	13390.0	192.7	21.09	29.46	437.3	24.6	.0367	1.00137
480.0	2.804	.1001	9981.0	13980.0	194.0	21.16	29.53	446.5	25.4	.0380	1.00132
500.0	2.691	.09607	10410.0	14570.0	195.2	21.24	29.60	455.5	26.1	.0392	1.00126
520.0	2.588	.09237	10830.0	15160.0	196.3	21.33	29.69	464.2	26.8	.0404	1.00121
540.0	2.492	.08895	11260.0	15760.0	197.5	21.43	29.78	472.7	27.5	.0416	1.00117
560.0	2.403	.08577	11690.0	16350.0	198.5	21.54	29.88	481.0	28.2	.0428	1.00113
580.0	2.320	.08281	12120.0	16950.0	199.6	21.65	29.99	489.2	28.9	.0440	1.00109
600.0	2.243	.08005	12560.0	17550.0	200.6	21.76	30.11	497.1	29.6	.0452	1.00105
620.0	2.170	.07747	12990.0	18160.0	201.6	21.88	30.22	505.0	30.2	.0464	1.00102
640.0	2.102	.07505	13430.0	18760.0	202.6	22.01	30.35	512.6	30.9	.0476	1.00099
660.0	2.039	.07277	13870.0	19370.0	203.5	22.14	30.47	520.1	31.5	.0488	1.00096
680.0	1.979	.07063	14320.0	19980.0	204.4	22.27	30.60	527.5	32.2	.0500	1.00093
700.0	1.922	.06862	14770.0	20600.0	205.3	22.40	30.73	534.8	32.8	.0512	1.00090
720.0	1.869	.06671	15210.0	21210.0	206.2	22.53	30.87	541.9	33.4	.0524	1.00088

Thermophysical properties of nitrogen—Continued

T K	Density kg/m³	Density mol/dm³	E J/mol	H J/mol	S J/(mol·K)	C_v J/(mol·K)	C_p J/(mol·K)	Sound m/s	Visc. μPa·s	Therm. W/(m·K)	Diel. const.
740.0	1.818	.06491	15670.0	21830.0	207.0	22.67	31.00	548.9	34.0	.0536	1.00085
760.0	1.771	.06320	16120.0	22450.0	207.8	22.80	31.13	555.8	34.7	.0547	1.00083
780.0	1.725	.06158	16580.0	23070.0	208.7	22.94	31.27	562.6	35.3	.0559	1.00081
800.0	1.682	.06005	17040.0	23700.0	209.5	23.07	31.40	569.3	35.9	.0571	1.00079
820.0	1.641	.05858	17500.0	24330.0	210.2	23.20	31.53	575.9	36.4	.0582	1.00077
840.0	1.602	.05719	17970.0	24960.0	211.0	23.34	31.66	582.5	37.0	.0594	1.00075
860.0	1.565	.05586	18440.0	25600.0	211.7	23.47	31.79	588.9	37.6	.0605	1.00073
880.0	1.529	.05459	18910.0	26230.0	212.5	23.59	31.92	595.3	38.2	.0617	1.00072
900.0	1.495	.05338	19380.0	26870.0	213.2	23.72	32.04	601.6	38.7	.0628	1.00070
920.0	1.463	.05222	19860.0	27520.0	213.9	23.85	32.17	607.8	39.3	.0640	1.00068
940.0	1.432	.05111	20330.0	28160.0	214.6	23.97	32.29	613.9	39.9	.0651	1.00067
1000.0	1.346	.04805	21780.0	30110.0	216.6	24.32	32.64	632.0	41.5	.0685	1.00063
1050.0	1.282	.04576	23010.0	31750.0	218.2	24.60	32.92	646.6	42.9	.0712	1.00060
1100.0	1.224	.04368	24240.0	33400.0	219.7	24.86	33.18	660.9	44.2	.0739	1.00057
1150.0	1.171	.04179	25490.0	35070.0	221.2	25.10	33.42	674.9	45.5	.0766	1.00055
1200.0	1.122	.04005	26750.0	36740.0	222.6	25.33	33.65	688.6	46.8	.0792	1.00052
1250.0	1.077	.03845	28030.0	38430.0	224.0	25.55	33.86	702.0	48.0	.0818	1.00050
1300.0	1.036	.03697	29310.0	40130.0	225.4	25.75	34.07	715.2	49.2	.0844	1.00048
1350.0	.9974	.03560	30600.0	41840.0	226.6	25.94	34.25	728.2	50.5	.0869	1.00046
1400.0	.9618	.03433	31900.0	43550.0	227.9	26.11	34.43	740.9	51.7	.0894	1.00045
1450.0	.9286	.03315	33210.0	45280.0	229.1	26.28	34.59	753.4	52.9	.0918	1.00043
1500.0	.8977	.03204	34530.0	47010.0	230.3	26.43	34.75	765.7	54.0	.0942	1.00042
1550.0	.8688	.03101	35850.0	48750.0	231.4	26.57	34.89	777.9	55.2	.0965	1.00040
1600.0	.8416	.03004	37190.0	50500.0	232.5	26.71	35.02	789.8	56.3	.0989	1.00039
1650.0	.8162	.02913	38520.0	52260.0	233.6	26.83	35.15	801.6	57.5	.101	1.00038
1700.0	.7922	.02828	39870.0	54020.0	234.7	26.95	35.27	813.2	58.6	.103	1.00037
1750.0	.7696	.02747	41220.0	55780.0	235.7	27.06	35.38	824.7	59.7	.106	1.00036
1800.0	.7482	.02671	42580.0	57550.0	236.7	27.17	35.48	836.0	60.8	.108	1.00035
1850.0	.7280	.02599	43940.0	59330.0	237.7	27.26	35.58	847.1	61.9	.110	1.00034
1900.0	.7089	.02530	45300.0	61110.0	238.6	27.35	35.67	858.1	62.9	.112	1.00033

.60 MPa isobar

T K	Density kg/m³	Density mol/dm³	E J/mol	H J/mol	S J/(mol·K)	C_v J/(mol·K)	C_p J/(mol·K)	Sound m/s	Visc. μPa·s	Therm. W/(m·K)	Diel. const.
63.28[a]	868.1	30.98	−4210.0	−4191.0	68.04	27.06	54.64	1317.0	282.0	.152	1.46876
70.0	841.8	30.05	−3833.0	−3813.0	73.70	28.73	57.02	1092.0	205.0	.145	1.45317
80.0	797.5	28.47	−3260.0	−3239.0	81.37	27.64	57.82	896.4	139.0	.132	1.42700
90.0	747.0	26.66	−2676.0	−2653.0	88.27	26.63	59.74	748.0	99.3	.117	1.39732
96.399[b]	710.3	25.35	−2286.0	−2262.0	92.46	26.28	62.56	659.7	81.0	.107	1.37589
96.399[b]	24.63	.8791	1766.0	2449.0	141.4	23.15	38.56	183.3	6.86	.0116	1.01164
100.0	23.25	.8301	1861.0	2584.0	142.8	22.75	36.95	189.0	7.08	.0118	1.01099
105.0	21.65	.7728	1988.0	2765.0	144.5	22.34	35.35	196.2	7.40	.0121	1.01023
110.0	20.31	.7249	2111.0	2939.0	146.1	22.04	34.21	202.9	7.71	.0124	1.00959
115.0	19.15	.6837	2230.0	3107.0	147.6	21.81	33.36	209.1	8.03	.0127	1.00904
120.0	18.15	.6478	2346.0	3273.0	149.0	21.63	32.72	215.0	8.34	.0131	1.00857
125.0	17.26	.6161	2461.0	3435.0	150.4	21.49	32.22	220.6	8.65	.0135	1.00815
130.0	16.47	.5878	2574.0	3595.0	151.6	21.38	31.81	226.0	8.96	.0139	1.00777
135.0	15.75	.5623	2686.0	3753.0	152.8	21.30	31.48	231.2	9.27	.0143	1.00743
140.0	15.10	.5391	2797.0	3910.0	154.0	21.23	31.21	236.2	9.58	.0147	1.00712
145.0	14.51	.5180	2907.0	4065.0	155.0	21.17	30.99	241.0	9.88	.0151	1.00684
150.0	13.97	.4986	3016.0	4220.0	156.1	21.12	30.80	245.8	10.2	.0155	1.00659
155.0	13.47	.4808	3125.0	4373.0	157.1	21.08	30.63	250.3	10.5	.0159	1.00635
160.0	13.01	.4642	3234.0	4526.0	158.1	21.05	30.49	254.8	10.8	.0163	1.00613
165.0	12.58	.4489	3342.0	4678.0	159.0	21.02	30.37	259.2	11.1	.0167	1.00593
170.0	12.17	.4346	3449.0	4830.0	159.9	20.99	30.27	263.4	11.4	.0171	1.00574
175.0	11.80	.4212	3556.0	4981.0	160.8	20.97	30.17	267.6	11.7	.0175	1.00556
180.0	11.45	.4087	3663.0	5132.0	161.6	20.95	30.09	271.7	11.9	.0179	1.00540
185.0	11.12	.3969	3770.0	5282.0	162.5	20.94	30.02	275.7	12.2	.0183	1.00524
190.0	10.81	.3858	3877.0	5432.0	163.3	20.92	29.95	279.7	12.5	.0187	1.00509
195.0	10.52	.3754	3983.0	5581.0	164.0	20.91	29.90	283.6	12.8	.0191	1.00495
200.0	10.24	.3655	4089.0	5731.0	164.8	20.90	29.84	287.4	13.0	.0195	1.00482
210.0	9.731	.3473	4301.0	6029.0	166.2	20.88	29.75	294.8	13.6	.0203	1.00458
220.0	9.271	.3309	4513.0	6326.0	167.6	20.87	29.68	302.1	14.1	.0210	1.00437
230.0	8.853	.3160	4724.0	6622.0	168.9	20.86	29.62	309.1	14.6	.0218	1.00417

Thermophysical properties of nitrogen—Continued

T K	Density kg/m^3	Density mol/dm^3	E J/mol	H J/mol	S J/(mol·K)	C_v J/(mol·K)	C_p J/(mol·K)	Sound m/s	Visc. μPa·s	Therm. W/(m·K)	Diel. const.
240.0	8.473	.3024	4934.0	6918.0	170.2	20.85	29.56	316.0	15.1	.0225	1.00399
250.0	8.125	.2900	5145.0	7214.0	171.4	20.84	29.52	322.7	15.7	.0233	1.00382
260.0	7.805	.2786	5355.0	7509.0	172.6	20.84	29.48	329.2	16.1	.0240	1.00367
270.0	7.510	.2680	5565.0	7803.0	173.7	20.84	29.45	335.6	16.6	.0247	1.00353
280.0	7.236	.2583	5775.0	8098.0	174.8	20.83	29.42	341.8	17.1	.0254	1.00340
290.0	6.982	.2492	5984.0	8392.0	175.8	20.83	29.40	348.0	17.6	.0261	1.00328
300.0	6.746	.2408	6194.0	8686.0	176.8	20.83	29.38	354.0	18.0	.0267	1.00317
310.0	6.525	.2329	6403.0	8979.0	177.7	20.83	29.36	359.9	18.5	.0274	1.00307
320.0	6.319	.2256	6613.0	9273.0	178.7	20.84	29.35	365.7	18.9	.0281	1.00297
330.0	6.125	.2186	6822.0	9566.0	179.6	20.84	29.34	371.4	19.4	.0287	1.00288
340.0	5.943	.2121	7031.0	9860.0	180.5	20.85	29.34	377.0	19.8	.0294	1.00279
350.0	5.772	.2060	7241.0	10150.0	181.3	20.86	29.33	382.5	20.2	.0300	1.00271
360.0	5.610	.2003	7450.0	10450.0	182.1	20.87	29.33	387.9	20.7	.0307	1.00264
370.0	5.458	.1948	7660.0	10740.0	182.9	20.88	29.34	393.3	21.1	.0313	1.00256
380.0	5.313	.1896	7869.0	11030.0	183.7	20.90	29.34	398.5	21.5	.0320	1.00250
400.0	5.046	.1801	8289.0	11620.0	185.2	20.93	29.36	408.8	22.3	.0332	1.00237
420.0	4.805	.1715	8709.0	12210.0	186.7	20.98	29.39	418.7	23.1	.0345	1.00226
440.0	4.585	.1637	9130.0	12800.0	188.0	21.03	29.44	428.4	23.9	.0357	1.00215
460.0	4.385	.1565	9552.0	13390.0	189.3	21.09	29.49	437.9	24.6	.0369	1.00206
480.0	4.202	.1500	9975.0	13980.0	190.6	21.17	29.56	447.0	25.4	.0382	1.00197
500.0	4.034	.1440	10400.0	14570.0	191.8	21.25	29.63	456.0	26.1	.0394	1.00189
520.0	3.878	.1384	10830.0	15160.0	193.0	21.34	29.71	464.7	26.8	.0406	1.00182
540.0	3.735	.1333	11260.0	15760.0	194.1	21.44	29.81	473.2	27.5	.0418	1.00175
560.0	3.601	.1285	11690.0	16350.0	195.2	21.54	29.91	481.6	28.2	.0430	1.00169
580.0	3.477	.1241	12120.0	16950.0	196.2	21.65	30.01	489.7	28.9	.0442	1.00163
600.0	3.361	.1200	12550.0	17550.0	197.2	21.77	30.12	497.7	29.6	.0454	1.00158
620.0	3.253	.1161	12990.0	18160.0	198.2	21.89	30.24	505.5	30.2	.0466	1.00152
640.0	3.151	.1125	13430.0	18760.0	199.2	22.01	30.36	513.1	30.9	.0478	1.00148
660.0	3.056	.1091	13870.0	19370.0	200.1	22.14	30.49	520.6	31.6	.0490	1.00143
680.0	2.966	.1059	14320.0	19980.0	201.0	22.27	30.62	528.0	32.2	.0502	1.00139
700.0	2.881	.1028	14760.0	20600.0	201.9	22.40	30.75	535.3	32.8	.0514	1.00135
720.0	2.801	.09999	15210.0	21210.0	202.8	22.54	30.88	542.4	33.4	.0526	1.00131
740.0	2.726	.09729	15660.0	21830.0	203.6	22.67	31.01	549.4	34.1	.0538	1.00128
760.0	2.654	.09473	16120.0	22450.0	204.5	22.81	31.14	556.3	34.7	.0549	1.00124
780.0	2.586	.09230	16580.0	23080.0	205.3	22.94	31.28	563.1	35.3	.0561	1.00121
800.0	2.521	.09000	17040.0	23700.0	206.1	23.07	31.41	569.8	35.9	.0573	1.00118
820.0	2.460	.08780	17500.0	24330.0	206.9	23.21	31.54	576.4	36.5	.0584	1.00115
840.0	2.401	.08572	17970.0	24970.0	207.6	23.34	31.67	582.9	37.0	.0596	1.00112
860.0	2.346	.08373	18430.0	25600.0	208.4	23.47	31.80	589.4	37.6	.0607	1.00110
880.0	2.292	.08183	18900.0	26240.0	209.1	23.60	31.93	595.7	38.2	.0619	1.00107
900.0	2.242	.08001	19380.0	26880.0	209.8	23.72	32.05	602.0	38.8	.0630	1.00105
920.0	2.193	.07827	19850.0	27520.0	210.5	23.85	32.17	608.2	39.3	.0642	1.00102
940.0	2.146	.07661	20330.0	28160.0	211.2	23.97	32.30	614.4	39.9	.0653	1.00100
1000.0	2.018	.07202	21780.0	30110.0	213.2	24.32	32.65	632.4	41.5	.0687	1.00094
1050.0	1.922	.06859	23000.0	31750.0	214.8	24.60	32.92	647.1	42.9	.0714	1.00090
1100.0	1.834	.06548	24240.0	33400.0	216.4	24.86	33.18	661.4	44.2	.0741	1.00086
1150.0	1.755	.06264	25490.0	35070.0	217.8	25.10	33.42	675.3	45.5	.0768	1.00082
1200.0	1.682	.06003	26750.0	36750.0	219.3	25.33	33.65	689.0	46.8	.0794	1.00078
1250.0	1.615	.05763	28020.0	38430.0	220.6	25.55	33.87	702.5	48.0	.0820	1.00075
1300.0	1.553	.05542	29310.0	40130.0	222.0	25.75	34.07	715.6	49.3	.0846	1.00072
1350.0	1.495	.05337	30600.0	41840.0	223.3	25.94	34.26	728.6	50.5	.0871	1.00070
1400.0	1.442	.05147	31900.0	43560.0	224.5	26.11	34.43	741.3	51.7	.0896	1.00067
1450.0	1.392	.04970	33210.0	45280.0	225.7	26.28	34.59	753.8	52.9	.0920	1.00065
1500.0	1.346	.04804	34530.0	47020.0	226.9	26.43	34.75	766.1	54.0	.0944	1.00063
1550.0	1.303	.04649	35850.0	48760.0	228.0	26.57	34.89	778.2	55.2	.0968	1.00061
1600.0	1.262	.04504	37190.0	50510.0	229.2	26.71	35.02	790.2	56.3	.0991	1.00059
1650.0	1.224	.04368	38520.0	52260.0	230.2	26.83	35.15	801.9	57.5	.101	1.00057
1700.0	1.188	.04240	39870.0	54020.0	231.3	26.95	35.27	813.6	58.6	.104	1.00055
1750.0	1.154	.04119	41220.0	55790.0	232.3	27.06	35.38	825.0	59.7	.106	1.00053
1800.0	1.122	.04004	42580.0	57560.0	233.3	27.17	35.48	836.3	60.8	.108	1.00052
1850.0	1.092	.03896	43940.0	59340.0	234.3	27.26	35.58	847.5	61.9	.110	1.00051
1900.0	1.063	.03794	45300.0	61120.0	235.2	27.36	35.67	858.5	62.9	.113	1.00049

Thermophysical properties of nitrogen—Continued

T K	Density kg/m³	Density mol/dm³	*E* J/mol	*H* J/mol	*S* J/(mol·K)	C_v J/(mol·K)	C_p J/(mol·K)	Sound m/s	Visc. μPa·s	Therm. W/(m·K)	Diel. const.
					.80 MPa isobar						
63.32[a]	868.1	30.99	−4209.0	−4184.0	68.05	27.19	54.63	1314.0	282.0	.152	1.46880
70.0	842.1	30.06	−3836.0	−3809.0	73.67	28.81	56.97	1092.0	205.0	.145	1.45336
80.0	798.1	28.49	−3263.0	−3235.0	81.33	27.70	57.74	897.4	139.0	.132	1.42730
90.0	747.8	26.69	−2680.0	−2650.0	88.22	26.68	59.59	750.0	99.7	.117	1.39778
100.409[b]	686.1	24.49	−2039.0	−2006.0	94.99	26.18	65.04	607.6	71.7	.101	1.36183
100.409[b]	32.80	1.171	1784.0	2467.0	139.6	23.63	41.61	183.5	7.25	.0128	1.01552
105.0	30.35	1.083	1913.0	2651.0	141.4	23.03	38.81	191.2	7.53	.0129	1.01436
110.0	28.19	1.006	2045.0	2840.0	143.1	22.57	36.79	198.7	7.83	.0131	1.01333
115.0	26.41	.9427	2171.0	3020.0	144.7	22.22	35.37	205.6	8.14	.0134	1.01248
120.0	24.89	.8886	2294.0	3194.0	146.2	21.97	34.34	212.0	8.44	.0137	1.01176
125.0	23.58	.8416	2413.0	3364.0	147.6	21.77	33.55	218.1	8.75	.0140	1.01114
130.0	22.42	.8002	2530.0	3530.0	148.9	21.61	32.93	223.8	9.05	.0144	1.01059
135.0	21.39	.7635	2646.0	3693.0	150.1	21.49	32.44	229.3	9.36	.0148	1.01010
140.0	20.46	.7304	2759.0	3855.0	151.3	21.39	32.04	234.5	9.66	.0151	1.00966
145.0	19.62	.7005	2872.0	4014.0	152.4	21.31	31.71	239.6	9.96	.0155	1.00926
150.0	18.86	.6732	2983.0	4172.0	153.5	21.24	31.44	244.5	10.3	.0159	1.00890
155.0	18.16	.6482	3094.0	4328.0	154.5	21.18	31.20	249.2	10.6	.0163	1.00857
160.0	17.51	.6251	3204.0	4484.0	155.5	21.14	31.01	253.8	10.8	.0167	1.00826
165.0	16.92	.6038	3314.0	4638.0	156.4	21.10	30.84	258.3	11.1	.0171	1.00798
170.0	16.36	.5841	3423.0	4792.0	157.4	21.07	30.69	262.7	11.4	.0175	1.00772
175.0	15.85	.5656	3531.0	4945.0	158.2	21.04	30.56	267.0	11.7	.0179	1.00747
180.0	15.36	.5484	3639.0	5098.0	159.1	21.01	30.44	271.2	12.0	.0183	1.00724
185.0	14.91	.5323	3747.0	5250.0	159.9	20.99	30.34	275.3	12.3	.0187	1.00703
190.0	14.49	.5172	3854.0	5401.0	160.7	20.97	30.25	279.3	12.6	.0190	1.00683
195.0	14.09	.5029	3962.0	5552.0	161.5	20.96	30.17	283.2	12.8	.0194	1.00664
200.0	13.71	.4895	4069.0	5703.0	162.3	20.94	30.10	287.1	13.1	.0198	1.00646
210.0	13.02	.4647	4282.0	6003.0	163.8	20.92	29.98	294.7	13.6	.0206	1.00613
220.0	12.40	.4425	4495.0	6303.0	165.2	20.90	29.88	302.0	14.2	.0213	1.00584
230.0	11.83	.4223	4707.0	6601.0	166.5	20.89	29.79	309.1	14.7	.0221	1.00557
240.0	11.32	.4040	4919.0	6899.0	167.7	20.87	29.72	316.0	15.2	.0228	1.00533
250.0	10.85	.3873	5130.0	7196.0	169.0	20.87	29.66	322.8	15.7	.0235	1.00511
260.0	10.42	.3719	5341.0	7492.0	170.1	20.86	29.61	329.4	16.2	.0242	1.00490
270.0	10.02	.3577	5551.0	7788.0	171.2	20.85	29.57	335.8	16.7	.0249	1.00472
280.0	9.655	.3446	5762.0	8083.0	172.3	20.85	29.53	342.1	17.1	.0256	1.00454
290.0	9.314	.3325	5972.0	8378.0	173.3	20.85	29.50	348.3	17.6	.0263	1.00438
300.0	8.997	.3212	6182.0	8673.0	174.3	20.85	29.47	354.3	18.1	.0270	1.00423
310.0	8.702	.3106	6392.0	8968.0	175.3	20.85	29.45	360.3	18.5	.0276	1.00409
320.0	8.425	.3007	6602.0	9262.0	176.2	20.85	29.43	366.1	19.0	.0283	1.00396
330.0	8.166	.2915	6812.0	9556.0	177.2	20.85	29.41	371.8	19.4	.0289	1.00384
340.0	7.923	.2828	7022.0	9851.0	178.0	20.86	29.40	377.4	19.8	.0296	1.00373
350.0	7.694	.2746	7231.0	10140.0	178.9	20.87	29.40	382.9	20.3	.0302	1.00362
360.0	7.478	.2669	7441.0	10440.0	179.7	20.88	29.39	388.4	20.7	.0309	1.00352
370.0	7.274	.2596	7651.0	10730.0	180.5	20.89	29.39	393.7	21.1	.0315	1.00342
380.0	7.081	.2527	7861.0	11030.0	181.3	20.90	29.39	399.0	21.5	.0322	1.00333
390.0	6.898	.2462	8071.0	11320.0	182.1	20.92	29.40	404.2	21.9	.0328	1.00324
400.0	6.724	.2400	8281.0	11610.0	182.8	20.94	29.41	409.3	22.3	.0334	1.00316
420.0	6.402	.2285	8702.0	12200.0	184.2	20.98	29.44	419.2	23.1	.0347	1.00301
440.0	6.109	.2181	9123.0	12790.0	185.6	21.04	29.47	428.9	23.9	.0359	1.00287
460.0	5.843	.2085	9546.0	13380.0	186.9	21.10	29.52	438.4	24.7	.0371	1.00274
480.0	5.599	.1998	9969.0	13970.0	188.2	21.17	29.59	447.6	25.4	.0384	1.00263
500.0	5.374	.1918	10390.0	14570.0	189.4	21.25	29.66	456.5	26.1	.0396	1.00252
520.0	5.167	.1844	10820.0	15160.0	190.6	21.34	29.74	465.2	26.8	.0408	1.00242
540.0	4.975	.1776	11250.0	15750.0	191.7	21.44	29.83	473.8	27.5	.0420	1.00233
560.0	4.798	.1712	11680.0	16350.0	192.8	21.55	29.93	482.1	28.2	.0432	1.00225
580.0	4.632	.1653	12110.0	16950.0	193.8	21.66	30.03	490.2	28.9	.0444	1.00217
600.0	4.478	.1598	12550.0	17550.0	194.8	21.77	30.14	498.2	29.6	.0456	1.00210
620.0	4.333	.1547	12990.0	18160.0	195.8	21.89	30.26	506.0	30.3	.0468	1.00203
640.0	4.198	.1498	13420.0	18760.0	196.8	22.02	30.38	513.7	30.9	.0480	1.00197
660.0	4.071	.1453	13870.0	19370.0	197.7	22.14	30.50	521.2	31.6	.0492	1.00191
680.0	3.951	.1410	14310.0	19980.0	198.6	22.28	30.63	528.5	32.2	.0504	1.00185

Thermophysical properties of nitrogen—Continued

T K	Density kg/m^3	Density mol/dm^3	E J/mol	H J/mol	S J/(mol·K)	C_v J/(mol·K)	C_p	Sound m/s	Visc. μPa·s	Therm. W/(m·K)	Diel. const.
700.0	3.838	.1370	14760.0	20600.0	199.5	22.41	30.76	535.8	32.8	.0516	1.00180
720.0	3.732	.1332	15210.0	21210.0	200.4	22.54	30.89	542.9	33.5	.0528	1.00175
740.0	3.631	.1296	15660.0	21830.0	201.2	22.67	31.02	549.9	34.1	.0539	1.00170
760.0	3.536	.1262	16120.0	22460.0	202.1	22.81	31.15	556.8	34.7	.0551	1.00165
780.0	3.445	.1230	16570.0	23080.0	202.9	22.94	31.29	563.6	35.3	.0563	1.00161
800.0	3.359	.1199	17030.0	23710.0	203.7	23.08	31.42	570.3	35.9	.0575	1.00157
820.0	3.277	.1170	17500.0	24340.0	204.5	23.21	31.55	576.9	36.5	.0586	1.00153
840.0	3.199	.1142	17960.0	24970.0	205.2	23.34	31.68	583.4	37.0	.0598	1.00150
860.0	3.125	.1115	18430.0	25600.0	206.0	23.47	31.81	589.9	37.6	.0609	1.00146
880.0	3.054	.1090	18900.0	26240.0	206.7	23.60	31.93	596.2	38.2	.0621	1.00143
900.0	2.986	.1066	19380.0	26880.0	207.4	23.73	32.06	602.5	38.8	.0632	1.00140
920.0	2.922	.1043	19850.0	27520.0	208.1	23.85	32.18	608.7	39.3	.0644	1.00137
940.0	2.860	.1021	20330.0	28170.0	208.8	23.97	32.30	614.9	39.9	.0655	1.00134
1000.0	2.688	.09596	21780.0	30120.0	210.8	24.32	32.65	632.9	41.5	.0688	1.00126
1050.0	2.561	.09140	23000.0	31760.0	212.4	24.60	32.93	647.5	42.9	.0716	1.00119
1100.0	2.444	.08725	24240.0	33410.0	214.0	24.86	33.18	661.8	44.2	.0743	1.00114
1150.0	2.338	.08347	25490.0	35070.0	215.4	25.11	33.43	675.8	45.5	.0770	1.00109
1200.0	2.241	.08000	26750.0	36750.0	216.9	25.33	33.66	689.5	46.8	.0796	1.00104
1250.0	2.152	.07680	28020.0	38440.0	218.3	25.55	33.87	702.9	48.0	.0822	1.00100
1300.0	2.069	.07385	29310.0	40140.0	219.6	25.75	34.07	716.0	49.3	.0848	1.00096
1350.0	1.993	.07112	30600.0	41850.0	220.9	25.94	34.26	729.0	50.5	.0873	1.00093
1400.0	1.922	.06859	31900.0	43560.0	222.1	26.11	34.43	741.7	51.7	.0898	1.00089
1450.0	1.855	.06623	33210.0	45290.0	223.3	26.28	34.60	754.2	52.9	.0922	1.00086
1500.0	1.794	.06402	34530.0	47020.0	224.5	26.43	34.75	766.5	54.0	.0946	1.00083
1550.0	1.736	.06196	35850.0	48760.0	225.7	26.58	34.89	778.6	55.2	.0970	1.00081
1600.0	1.682	.06003	37190.0	50510.0	226.8	26.71	35.03	790.5	56.3	.0993	1.00078
1650.0	1.631	.05821	38520.0	52270.0	227.8	26.84	35.15	802.3	57.5	.102	1.00076
1700.0	1.583	.05650	39870.0	54030.0	228.9	26.95	35.27	813.9	58.6	.104	1.00073
1750.0	1.538	.05489	41220.0	55790.0	229.9	27.06	35.38	825.4	59.7	.106	1.00071
1800.0	1.495	.05337	42580.0	57570.0	230.9	27.17	35.48	836.6	60.8	.108	1.00069
1850.0	1.455	.05193	43940.0	59340.0	231.9	27.27	35.58	847.8	61.9	.111	1.00067
1900.0	1.417	.05056	45300.0	61120.0	232.8	27.36	35.67	858.8	62.9	.113	1.00066
					1.00 MPa isobar						
63.37[a]	868.2	30.99	−4209.0	−4176.0	68.06	27.31	54.62	1311.0	282.0	.152	1.46883
70.0	842.4	30.07	−3838.0	−3805.0	73.64	28.89	56.91	1092.0	206.0	.145	1.45355
80.0	798.6	28.50	−3267.0	−3232.0	81.29	27.76	57.66	898.4	140.0	.132	1.42760
90.0	748.6	26.72	−2685.0	−2648.0	88.16	26.72	59.45	752.0	100.0	.118	1.39824
100.0	690.2	24.63	−2073.0	−2032.0	94.64	26.23	64.36	616.7	73.1	.102	1.36417
103.748[b]	664.6	23.72	−1827.0	−1785.0	97.07	26.15	67.79	564.8	64.8	.0953	1.34942
103.748[b]	41.23	1.472	1787.0	2467.0	138.1	24.08	45.02	183.2	7.61	.0139	1.01954
105.0	40.24	1.436	1826.0	2523.0	138.6	23.87	43.77	185.5	7.68	.0139	1.01907
106.0	39.49	1.410	1856.0	2566.0	139.0	23.71	42.90	187.4	7.74	.0139	1.01871
108.0	38.13	1.361	1915.0	2650.0	139.8	23.42	41.41	190.9	7.85	.0139	1.01806
110.0	36.89	1.317	1972.0	2732.0	140.5	23.18	40.17	194.2	7.97	.0139	1.01747
115.0	34.26	1.223	2109.0	2926.0	142.3	22.69	37.86	201.9	8.26	.0141	1.01621
120.0	32.08	1.145	2238.0	3111.0	143.9	22.33	36.25	208.9	8.55	.0144	1.01518
125.0	30.24	1.079	2363.0	3290.0	145.3	22.06	35.08	215.4	8.85	.0146	1.01430
130.0	28.65	1.023	2485.0	3463.0	146.7	21.85	34.19	221.5	9.15	.0149	1.01354
135.0	27.25	.9726	2604.0	3632.0	147.9	21.69	33.50	227.3	9.45	.0153	1.01288
140.0	26.00	.9282	2721.0	3798.0	149.2	21.56	32.94	232.8	9.75	.0156	1.01229
145.0	24.89	.8883	2836.0	3961.0	150.3	21.45	32.49	238.1	10.0	.0160	1.01176
150.0	23.88	.8523	2950.0	4123.0	151.4	21.36	32.12	243.2	10.3	.0163	1.01128
155.0	22.96	.8194	3062.0	4283.0	152.4	21.29	31.81	248.1	10.6	.0167	1.01084
160.0	22.11	.7893	3174.0	4441.0	153.4	21.23	31.55	252.9	10.9	.0171	1.01044
165.0	21.34	.7616	3285.0	4598.0	154.4	21.18	31.32	257.5	11.2	.0175	1.01007
170.0	20.62	.7360	3396.0	4754.0	155.3	21.14	31.13	262.0	11.5	.0178	1.00973
175.0	19.95	.7122	3505.0	4910.0	156.2	21.10	30.96	266.4	11.8	.0182	1.00942
180.0	19.33	.6900	3615.0	5064.0	157.1	21.07	30.81	270.6	12.1	.0186	1.00912
185.0	18.75	.6693	3724.0	5218.0	158.0	21.05	30.68	274.8	12.3	.0190	1.00885
190.0	18.21	.6499	3832.0	5371.0	158.8	21.02	30.56	278.9	12.6	.0194	1.00859

Thermophysical properties of nitrogen—Continued

T K	Density kg/m^3	Density mol/dm^3	E J/mol	H J/mol	S J/(mol·K)	C_v J/(mol·K)	C_p J/(mol·K)	Sound m/s	Visc. μPa·s	Therm. W/(m·K)	Diel. const.
195.0	17.70	.6316	3940.0	5523.0	159.6	21.00	30.46	282.9	12.9	.0197	1.00835
200.0	17.21	.6145	4048.0	5675.0	160.3	20.98	30.37	286.9	13.2	.0201	1.00812
210.0	16.33	.5829	4263.0	5978.0	161.8	20.95	30.21	294.6	13.7	.0209	1.00770
220.0	15.54	.5547	4477.0	6280.0	163.2	20.93	30.08	302.0	14.2	.0216	1.00732
230.0	14.82	.5291	4690.0	6580.0	164.6	20.91	29.97	309.2	14.7	.0223	1.00699
240.0	14.17	.5060	4903.0	6879.0	165.8	20.90	29.88	316.2	15.2	.0230	1.00668
250.0	13.58	.4848	5115.0	7178.0	167.0	20.89	29.81	323.0	15.7	.0238	1.00640
260.0	13.04	.4654	5327.0	7475.0	168.2	20.88	29.74	329.6	16.2	.0245	1.00614
270.0	12.54	.4476	5538.0	7772.0	169.3	20.87	29.69	336.1	16.7	.0251	1.00590
280.0	12.08	.4311	5749.0	8069.0	170.4	20.86	29.64	342.4	17.2	.0258	1.00569
290.0	11.65	.4158	5960.0	8365.0	171.5	20.86	29.60	348.6	17.6	.0265	1.00548
300.0	11.25	.4016	6171.0	8661.0	172.5	20.86	29.56	354.7	18.1	.0272	1.00529
310.0	10.88	.3883	6381.0	8956.0	173.4	20.86	29.53	360.6	18.6	.0278	1.00512
320.0	10.53	.3759	6592.0	9252.0	174.4	20.86	29.51	366.5	19.0	.0285	1.00496
330.0	10.21	.3643	6802.0	9547.0	175.3	20.86	29.49	372.2	19.4	.0292	1.00480
340.0	9.902	.3534	7012.0	9841.0	176.1	20.87	29.47	377.8	19.9	.0298	1.00466
350.0	9.614	.3432	7222.0	10140.0	177.0	20.88	29.46	383.4	20.3	.0304	1.00452
360.0	9.344	.3335	7432.0	10430.0	177.8	20.89	29.45	388.8	20.7	.0311	1.00439
370.0	9.088	.3244	7642.0	10730.0	178.6	20.90	29.45	394.2	21.1	.0317	1.00427
380.0	8.846	.3158	7853.0	11020.0	179.4	20.91	29.45	399.5	21.5	.0324	1.00416
390.0	8.617	.3076	8063.0	11310.0	180.2	20.93	29.45	404.6	21.9	.0330	1.00405
400.0	8.400	.2998	8273.0	11610.0	180.9	20.95	29.45	409.8	22.4	.0336	1.00395
420.0	7.997	.2854	8694.0	12200.0	182.4	20.99	29.48	419.8	23.1	.0348	1.00376
440.0	7.631	.2724	9116.0	12790.0	183.7	21.04	29.51	429.5	23.9	.0361	1.00358
460.0	7.298	.2605	9539.0	13380.0	185.1	21.11	29.56	438.9	24.7	.0373	1.00343
480.0	6.993	.2496	9964.0	13970.0	186.3	21.18	29.62	448.1	25.4	.0385	1.00328
500.0	6.712	.2396	10390.0	14560.0	187.5	21.26	29.69	457.0	26.1	.0398	1.00315
520.0	6.453	.2303	10820.0	15160.0	188.7	21.35	29.76	465.8	26.9	.0410	1.00303
540.0	6.214	.2218	11250.0	15750.0	189.8	21.45	29.85	474.3	27.6	.0422	1.00292
560.0	5.992	.2139	11680.0	16350.0	190.9	21.55	29.95	482.6	28.3	.0434	1.00281
580.0	5.785	.2065	12110.0	16950.0	192.0	21.66	30.05	490.8	28.9	.0446	1.00271
600.0	5.592	.1996	12540.0	17550.0	193.0	21.78	30.16	498.7	29.6	.0458	1.00262
620.0	5.412	.1932	12980.0	18160.0	194.0	21.90	30.28	506.5	30.3	.0470	1.00254
640.0	5.243	.1871	13420.0	18760.0	194.9	22.02	30.39	514.2	30.9	.0482	1.00246
660.0	5.084	.1815	13860.0	19370.0	195.9	22.15	30.52	521.7	31.6	.0494	1.00238
680.0	4.935	.1761	14310.0	19990.0	196.8	22.28	30.64	529.1	32.2	.0506	1.00231
700.0	4.794	.1711	14760.0	20600.0	197.7	22.41	30.77	536.3	32.8	.0517	1.00225
720.0	4.661	.1664	15210.0	21220.0	198.5	22.54	30.90	543.4	33.5	.0529	1.00218
740.0	4.535	.1619	15660.0	21840.0	199.4	22.68	31.03	550.4	34.1	.0541	1.00212
760.0	4.416	.1576	16110.0	22460.0	200.2	22.81	31.16	557.3	34.7	.0553	1.00207
780.0	4.303	.1536	16570.0	23080.0	201.0	22.95	31.29	564.1	35.3	.0564	1.00201
800.0	4.196	.1498	17030.0	23710.0	201.8	23.08	31.43	570.8	35.9	.0576	1.00196
820.0	4.093	.1461	17500.0	24340.0	202.6	23.21	31.56	577.4	36.5	.0588	1.00191
840.0	3.996	.1426	17960.0	24970.0	203.4	23.34	31.69	583.9	37.1	.0599	1.00187
860.0	3.903	.1393	18430.0	25610.0	204.1	23.47	31.81	590.4	37.6	.0611	1.00182
880.0	3.815	.1362	18900.0	26240.0	204.8	23.60	31.94	596.7	38.2	.0622	1.00178
900.0	3.730	.1332	19370.0	26880.0	205.6	23.73	32.06	603.0	38.8	.0634	1.00174
920.0	3.649	.1303	19850.0	27530.0	206.3	23.85	32.19	609.2	39.3	.0645	1.00171
940.0	3.572	.1275	20330.0	28170.0	207.0	23.97	32.31	615.3	39.9	.0657	1.00167
1000.0	3.358	.1199	21780.0	30120.0	209.0	24.33	32.66	633.4	41.5	.0690	1.00157
1050.0	3.199	.1142	23000.0	31760.0	210.6	24.60	32.93	648.0	42.9	.0718	1.00149
1100.0	3.054	.1090	24240.0	33410.0	212.1	24.86	33.19	662.2	44.2	.0745	1.00142
1150.0	2.921	.1043	25490.0	35080.0	213.6	25.11	33.43	676.2	45.5	.0772	1.00136
1200.0	2.800	.09993	26750.0	36760.0	215.0	25.34	33.66	689.9	46.8	.0798	1.00130
1250.0	2.688	.09595	28020.0	38440.0	216.4	25.55	33.87	703.3	48.0	.0824	1.00125
1300.0	2.585	.09226	29310.0	40140.0	217.7	25.75	34.07	716.4	49.3	.0849	1.00120
1350.0	2.489	.08886	30600.0	41850.0	219.0	25.94	34.26	729.4	50.5	.0875	1.00116
1400.0	2.401	.08569	31900.0	43570.0	220.3	26.12	34.43	742.1	51.7	.0899	1.00112
1450.0	2.318	.08274	33210.0	45300.0	221.5	26.28	34.60	754.6	52.9	.0924	1.00108
1500.0	2.241	.07999	34530.0	47030.0	222.7	26.43	34.75	766.9	54.0	.0948	1.00104
1550.0	2.169	.07741	35850.0	48770.0	223.8	26.58	34.89	779.0	55.2	.0971	1.00101
1600.0	2.101	.07500	37180.0	50520.0	224.9	26.71	35.03	790.9	56.3	.0995	1.00098
1650.0	2.038	.07273	38520.0	52270.0	226.0	26.84	35.15	802.7	57.5	.102	1.00095

Thermophysical properties of nitrogen—Continued

T K	Density kg/m^3	Density mol/dm^3	E J/mol	H J/mol	S J/(mol·K)	C_v J/(mol·K)	C_p J/(mol·K)	Sound m/s	Visc. μPa·s	Therm. W/(m·K)	Diel. const.
1700.0	1.978	.07060	39870.0	54030.0	227.0	26.95	35.27	814.3	58.6	.104	1.00092
1750.0	1.921	.06859	41220.0	55800.0	228.1	27.06	35.38	825.7	59.7	.106	1.00089
1800.0	1.868	.06668	42570.0	57570.0	229.1	27.17	35.48	837.0	60.8	.109	1.00087
1850.0	1.818	.06489	43940.0	59350.0	230.0	27.27	35.58	848.1	61.9	.111	1.00084
1900.0	1.770	.06318	45300.0	61130.0	231.0	27.36	35.67	859.1	62.9	.113	1.00082
1.50 MPa isobar											
63.48[a]	868.4	31.00	−4207.0	−4159.0	68.09	27.61	54.58	1304.0	281.0	.152	1.46892
70.0	843.3	30.10	−3844.0	−3794.0	73.56	29.09	56.77	1092.0	207.0	.146	1.45402
80.0	799.8	28.55	−3275.0	−3222.0	81.19	27.91	57.46	900.9	141.0	.133	1.42834
90.0	750.6	26.79	−2697.0	−2641.0	88.03	26.84	59.10	756.9	101.0	.118	1.39936
100.0	693.3	24.75	−2091.0	−2031.0	94.46	26.31	63.58	624.8	74.2	.103	1.36598
105.0	659.8	23.55	−1767.0	−1703.0	97.66	26.20	67.91	557.5	63.3	.0942	1.34662
110.381[b]	617.2	22.03	−1385.0	−1317.0	101.2	26.22	76.59	478.7	52.5	.0847	1.32218
110.381[b]	64.02	2.285	1755.0	2412.0	135.0	25.19	56.22	181.2	8.45	.0168	1.03046
115.0	57.64	2.057	1921.0	2650.0	137.1	24.18	48.01	191.3	8.65	.0164	1.02739
116.0	56.52	2.017	1954.0	2698.0	137.5	24.01	46.82	193.2	8.70	.0163	1.02685
118.0	54.48	1.945	2018.0	2789.0	138.3	23.70	44.83	196.9	8.80	.0163	1.02588
120.0	52.66	1.879	2079.0	2877.0	139.1	23.44	43.21	200.4	8.90	.0163	1.02500
125.0	48.80	1.742	2224.0	3085.0	140.8	22.92	40.22	208.4	9.16	.0163	1.02315
130.0	45.67	1.630	2361.0	3281.0	142.3	22.53	38.19	215.7	9.43	.0164	1.02165
132.0	44.56	1.591	2414.0	3357.0	142.9	22.41	37.55	218.4	9.54	.0165	1.02113
134.0	43.53	1.554	2466.0	3431.0	143.4	22.29	36.98	221.0	9.66	.0166	1.02063
136.0	42.55	1.519	2517.0	3505.0	144.0	22.19	36.47	223.6	9.77	.0167	1.02017
138.0	41.63	1.486	2568.0	3577.0	144.5	22.10	36.02	226.1	9.88	.0168	1.01973
140.0	40.77	1.455	2618.0	3649.0	145.0	22.01	35.61	228.6	9.99	.0169	1.01931
145.0	38.79	1.384	2741.0	3825.0	146.3	21.83	34.74	234.5	10.3	.0171	1.01837
150.0	37.03	1.322	2862.0	3997.0	147.4	21.69	34.05	240.1	10.6	.0174	1.01753
155.0	35.46	1.266	2980.0	4165.0	148.5	21.57	33.48	245.4	10.8	.0178	1.01678
160.0	34.04	1.215	3097.0	4332.0	149.6	21.48	33.02	250.5	11.1	.0181	1.01610
165.0	32.75	1.169	3213.0	4496.0	150.6	21.40	32.63	255.5	11.4	.0184	1.01549
170.0	31.57	1.127	3327.0	4658.0	151.6	21.33	32.30	260.3	11.7	.0187	1.01493
175.0	30.48	1.088	3440.0	4819.0	152.5	21.27	32.01	264.9	11.9	.0191	1.01441
180.0	29.47	1.052	3552.0	4978.0	153.4	21.22	31.77	269.4	12.2	.0194	1.01393
185.0	28.54	1.019	3664.0	5136.0	154.3	21.18	31.55	273.8	12.5	.0198	1.01349
190.0	27.67	.9877	3775.0	5294.0	155.1	21.15	31.36	278.1	12.8	.0201	1.01307
195.0	26.86	.9588	3886.0	5450.0	155.9	21.11	31.20	282.3	13.0	.0205	1.01269
200.0	26.10	.9316	3996.0	5606.0	156.7	21.09	31.05	286.4	13.3	.0209	1.01233
205.0	25.38	.9061	4105.0	5761.0	157.5	21.06	30.92	290.4	13.6	.0212	1.01199
210.0	24.71	.8820	4214.0	5915.0	158.2	21.04	30.80	294.3	13.8	.0216	1.01167
220.0	23.47	.8379	4432.0	6222.0	159.6	21.01	30.60	301.9	14.3	.0223	1.01108
230.0	22.36	.7982	4648.0	6527.0	161.0	20.98	30.43	309.3	14.8	.0230	1.01055
240.0	21.36	.7624	4863.0	6831.0	162.3	20.96	30.29	316.5	15.3	.0237	1.01007
250.0	20.45	.7298	5078.0	7133.0	163.5	20.94	30.17	323.4	15.8	.0243	1.00964
260.0	19.61	.7001	5291.0	7434.0	164.7	20.92	30.07	330.2	16.3	.0250	1.00925
270.0	18.85	.6728	5505.0	7734.0	165.8	20.91	29.98	336.7	16.8	.0257	1.00888
280.0	18.14	.6476	5717.0	8034.0	166.9	20.90	29.91	343.2	17.3	.0264	1.00855
290.0	17.49	.6243	5930.0	8332.0	168.0	20.90	29.84	349.4	17.7	.0270	1.00824
300.0	16.89	.6027	6142.0	8631.0	169.0	20.89	29.79	355.6	18.2	.0277	1.00795
310.0	16.32	.5826	6354.0	8928.0	170.0	20.89	29.74	361.6	18.6	.0283	1.00769
320.0	15.80	.5639	6565.0	9225.0	170.9	20.89	29.70	367.5	19.1	.0290	1.00744
330.0	15.30	.5463	6777.0	9522.0	171.8	20.89	29.67	373.3	19.5	.0296	1.00721
340.0	14.84	.5298	6988.0	9819.0	172.7	20.90	29.64	378.9	19.9	.0303	1.00699
350.0	14.41	.5144	7199.0	10120.0	173.6	20.90	29.62	384.5	20.4	.0309	1.00678
360.0	14.00	.4998	7410.0	10410.0	174.4	20.91	29.60	390.0	20.8	.0315	1.00659
370.0	13.62	.4860	7621.0	10710.0	175.2	20.92	29.58	395.4	21.2	.0322	1.00641
380.0	13.25	.4730	7832.0	11000.0	176.0	20.93	29.58	400.7	21.6	.0328	1.00623
390.0	12.91	.4607	8043.0	11300.0	176.8	20.95	29.57	405.9	22.0	.0334	1.00607
400.0	12.58	.4491	8254.0	11590.0	177.5	20.97	29.57	411.0	22.4	.0340	1.00592
420.0	11.98	.4274	8676.0	12190.0	179.0	21.01	29.58	421.0	23.2	.0353	1.00563
440.0	11.43	.4078	9100.0	12780.0	180.3	21.06	29.60	430.8	24.0	.0365	1.00537

Thermophysical properties of nitrogen—Continued

T K	Density kg/m³	Density mol/dm³	E J/mol	H J/mol	S J/(mol·K)	C_v J/(mol·K)	C_p J/(mol·K)	Sound m/s	Visc. μPa·s	Therm. W/(m·K)	Diel. const.
460.0	10.93	.3900	9524.0	13370.0	181.7	21.12	29.64	440.2	24.7	.0377	1.00513
480.0	10.47	.3736	9949.0	13960.0	182.9	21.19	29.69	449.4	25.5	.0389	1.00492
500.0	10.05	.3586	10380.0	14560.0	184.1	21.27	29.75	458.4	26.2	.0402	1.00472
520.0	9.660	.3448	10800.0	15150.0	185.3	21.36	29.83	467.1	26.9	.0414	1.00453
540.0	9.301	.3320	11230.0	15750.0	186.4	21.46	29.91	475.6	27.6	.0426	1.00437
560.0	8.968	.3201	11660.0	16350.0	187.5	21.56	30.00	484.0	28.3	.0438	1.00421
580.0	8.659	.3091	12100.0	16950.0	188.6	21.67	30.10	492.1	29.0	.0450	1.00406
600.0	8.370	.2988	12530.0	17550.0	189.6	21.79	30.21	500.1	29.7	.0462	1.00393
620.0	8.100	.2891	12970.0	18160.0	190.6	21.91	30.32	507.9	30.3	.0474	1.00380
640.0	7.847	.2801	13410.0	18770.0	191.5	22.03	30.43	515.5	31.0	.0486	1.00368
660.0	7.610	.2716	13850.0	19380.0	192.5	22.16	30.55	523.0	31.6	.0497	1.00357
680.0	7.387	.2637	14300.0	19990.0	193.4	22.29	30.68	530.4	32.2	.0509	1.00346
700.0	7.176	.2561	14750.0	20600.0	194.3	22.42	30.80	537.6	32.9	.0521	1.00336
720.0	6.977	.2490	15200.0	21220.0	195.2	22.55	30.93	544.7	33.5	.0533	1.00327
740.0	6.789	.2423	15650.0	21840.0	196.0	22.69	31.06	551.7	34.1	.0545	1.00318
760.0	6.611	.2360	16110.0	22460.0	196.8	22.82	31.19	558.6	34.7	.0556	1.00309
780.0	6.442	.2299	16560.0	23090.0	197.7	22.95	31.32	565.4	35.3	.0568	1.00301
800.0	6.281	.2242	17030.0	23720.0	198.4	23.09	31.45	572.1	35.9	.0580	1.00294
820.0	6.128	.2187	17490.0	24350.0	199.2	23.22	31.58	578.7	36.5	.0591	1.00287
840.0	5.983	.2136	17960.0	24980.0	200.0	23.35	31.71	585.2	37.1	.0603	1.00280
860.0	5.844	.2086	18420.0	25610.0	200.7	23.48	31.83	591.6	37.7	.0615	1.00273
880.0	5.712	.2039	18900.0	26250.0	201.5	23.61	31.96	597.9	38.2	.0626	1.00267
900.0	5.585	.1994	19370.0	26890.0	202.2	23.73	32.08	604.2	38.8	.0637	1.00261
920.0	5.464	.1950	19850.0	27540.0	202.9	23.86	32.20	610.4	39.4	.0649	1.00255
940.0	5.348	.1909	20320.0	28180.0	203.6	23.98	32.32	616.5	39.9	.0660	1.00250
1000.0	5.029	.1795	21770.0	30130.0	205.6	24.33	32.67	634.5	41.6	.0694	1.00235
1050.0	4.790	.1710	23000.0	31770.0	207.2	24.61	32.94	649.1	42.9	.0721	1.00224
1100.0	4.573	.1632	24240.0	33420.0	208.7	24.87	33.20	663.3	44.2	.0749	1.00213
1150.0	4.375	.1562	25490.0	35090.0	210.2	25.11	33.44	677.3	45.5	.0775	1.00204
1200.0	4.193	.1497	26750.0	36770.0	211.6	25.34	33.67	690.9	46.8	.0802	1.00195
1250.0	4.026	.1437	28020.0	38460.0	213.0	25.55	33.88	704.3	48.0	.0828	1.00187
1300.0	3.872	.1382	29300.0	40160.0	214.4	25.76	34.08	717.5	49.3	.0853	1.00180
1350.0	3.729	.1331	30600.0	41870.0	215.6	25.94	34.27	730.4	50.5	.0878	1.00173
1400.0	3.596	.1284	31900.0	43580.0	216.9	26.12	34.44	743.1	51.7	.0903	1.00167
1450.0	3.473	.1240	33210.0	45310.0	218.1	26.28	34.60	755.5	52.9	.0927	1.00161
1500.0	3.357	.1198	34530.0	47040.0	219.3	26.44	34.76	767.8	54.1	.0951	1.00156
1550.0	3.249	.1160	35850.0	48780.0	220.4	26.58	34.90	779.9	55.2	.0975	1.00151
1600.0	3.148	.1124	37180.0	50530.0	221.5	26.71	35.03	791.8	56.4	.0998	1.00146
1650.0	3.053	.1090	38520.0	52290.0	222.6	26.84	35.16	803.6	57.5	.102	1.00142
1700.0	2.964	.1058	39870.0	54050.0	223.7	26.96	35.27	815.2	58.6	.104	1.00137
1750.0	2.879	.1028	41220.0	55810.0	224.7	27.07	35.38	826.6	59.7	.107	1.00133
1800.0	2.799	.09992	42570.0	57590.0	225.7	27.17	35.49	837.9	60.8	.109	1.00130
1850.0	2.724	.09723	43940.0	59360.0	226.7	27.27	35.58	849.0	61.9	.111	1.00126
1900.0	2.652	.09468	45300.0	61140.0	227.6	27.36	35.67	860.0	62.9	.113	1.00123
				2.00 MPa isobar							
63.59[a]	868.6	31.00	−4205.0	−4141.0	68.11	27.91	54.54	1297.0	281.0	.152	1.46901
70.0	844.1	30.13	−3849.0	−3783.0	73.48	29.28	56.64	1093.0	208.0	.146	1.45448
80.0	801.1	28.59	−3283.0	−3213.0	81.09	28.06	57.27	903.3	142.0	.133	1.42906
90.0	752.5	26.86	−2708.0	−2634.0	87.90	26.95	58.77	761.8	102.0	.119	1.40046
100.0	696.3	24.86	−2109.0	−2029.0	94.28	26.39	62.87	632.6	75.3	.103	1.36773
110.0	626.4	22.36	−1447.0	−1357.0	100.7	26.23	73.27	498.2	54.6	.0867	1.32741
115.0	579.4	20.68	−1057.0	−960.3	104.2	26.39	87.89	417.5	45.0	.0773	1.30067
115.571[b]	572.9	20.45	−1007.0	−909.5	104.6	26.43	90.89	406.8	43.8	.0761	1.29703
115.571[b]	90.49	3.230	1676.0	2295.0	132.4	26.34	74.92	178.6	9.31	.0202	1.04325
120.0	79.36	2.833	1875.0	2581.0	134.8	24.98	57.29	190.2	9.39	.0190	1.03785
122.0	75.76	2.704	1952.0	2691.0	135.7	24.54	53.16	194.6	9.46	.0187	1.03611
124.0	72.68	2.594	2023.0	2794.0	136.5	24.18	50.07	198.7	9.53	.0185	1.03462
126.0	69.98	2.498	2091.0	2892.0	137.3	23.87	47.65	202.5	9.62	.0184	1.03332
128.0	67.58	2.412	2156.0	2985.0	138.1	23.60	45.71	206.0	9.70	.0183	1.03217
130.0	65.43	2.335	2219.0	3075.0	138.7	23.36	44.11	209.4	9.79	.0182	1.03113
132.0	63.47	2.266	2279.0	3162.0	139.4	23.15	42.77	212.6	9.89	.0182	1.03019

Thermophysical properties of nitrogen—Continued

T K	Density kg/m^3	Density mol/dm^3	E J/mol	H J/mol	S J/(mol·K)	C_v J/(mol·K)	C_p J/(mol·K)	Sound m/s	Visc. μPa·s	Therm. W/(m·K)	Diel. const.
134.0	61.68	2.202	2338.0	3246.0	140.0	22.97	41.63	215.7	9.98	.0182	1.02933
136.0	60.03	2.143	2395.0	3329.0	140.7	22.81	40.65	218.7	10.1	.0182	1.02853
138.0	58.50	2.088	2451.0	3409.0	141.2	22.66	39.79	221.5	10.2	.0182	1.02780
140.0	57.08	2.037	2506.0	3488.0	141.8	22.53	39.04	224.3	10.3	.0183	1.02712
142.0	55.75	1.990	2560.0	3565.0	142.4	22.42	38.38	227.0	10.4	.0183	1.02648
144.0	54.50	1.945	2613.0	3641.0	142.9	22.31	37.79	229.6	10.5	.0184	1.02588
146.0	53.33	1.903	2666.0	3716.0	143.4	22.21	37.26	232.1	10.6	.0185	1.02531
148.0	52.22	1.864	2717.0	3790.0	143.9	22.13	36.79	234.6	10.7	.0185	1.02478
150.0	51.17	1.826	2768.0	3864.0	144.4	22.05	36.35	237.0	10.8	.0186	1.02428
155.0	48.76	1.741	2894.0	4043.0	145.6	21.87	35.44	242.8	11.1	.0189	1.02313
160.0	46.63	1.664	3017.0	4218.0	146.7	21.73	34.70	248.4	11.3	.0191	1.02211
165.0	44.71	1.596	3137.0	4390.0	147.7	21.62	34.09	253.6	11.6	.0194	1.02119
170.0	42.98	1.534	3256.0	4559.0	148.8	21.52	33.59	258.7	11.9	.0197	1.02036
175.0	41.40	1.478	3373.0	4726.0	149.7	21.44	33.16	263.6	12.1	.0200	1.01961
180.0	39.96	1.426	3489.0	4891.0	150.7	21.38	32.80	268.4	12.4	.0203	1.01892
185.0	38.62	1.379	3603.0	5054.0	151.5	21.32	32.49	272.9	12.7	.0206	1.01828
190.0	37.39	1.335	3717.0	5216.0	152.4	21.27	32.21	277.4	12.9	.0209	1.01769
195.0	36.24	1.294	3830.0	5376.0	153.2	21.23	31.97	281.8	13.2	.0213	1.01714
200.0	35.17	1.255	3943.0	5536.0	154.1	21.19	31.76	286.0	13.4	.0216	1.01663
205.0	34.17	1.220	4054.0	5694.0	154.8	21.16	31.58	290.1	13.7	.0219	1.01616
210.0	33.23	1.186	4165.0	5852.0	155.6	21.13	31.41	294.2	14.0	.0223	1.01571
215.0	32.35	1.155	4276.0	6008.0	156.3	21.10	31.26	298.1	14.2	.0226	1.01529
220.0	31.51	1.125	4386.0	6164.0	157.0	21.08	31.12	302.0	14.5	.0229	1.01489
230.0	29.98	1.070	4605.0	6474.0	158.4	21.04	30.89	309.6	15.0	.0236	1.01416
240.0	28.60	1.021	4823.0	6782.0	159.7	21.01	30.70	316.9	15.5	.0243	1.01351
250.0	27.36	.9765	5040.0	7088.0	161.0	20.99	30.54	323.9	15.9	.0249	1.01291
260.0	26.22	.9359	5256.0	7393.0	162.2	20.97	30.40	330.8	16.4	.0256	1.01237
270.0	25.18	.8988	5471.0	7696.0	163.3	20.96	30.28	337.5	16.9	.0262	1.01188
280.0	24.22	.8647	5686.0	7999.0	164.4	20.94	30.18	344.0	17.4	.0269	1.01143
290.0	23.34	.8332	5900.0	8300.0	165.5	20.93	30.09	350.3	17.8	.0275	1.01101
300.0	22.53	.8040	6113.0	8601.0	166.5	20.93	30.02	356.5	18.3	.0282	1.01062
310.0	21.77	.7769	6326.0	8900.0	167.5	20.92	29.95	362.6	18.7	.0288	1.01026
320.0	21.06	.7517	6539.0	9200.0	168.4	20.92	29.90	368.6	19.2	.0295	1.00992
330.0	20.40	.7281	6751.0	9498.0	169.4	20.92	29.85	374.4	19.6	.0301	1.00961
340.0	19.78	.7060	6964.0	9797.0	170.2	20.92	29.81	380.1	20.0	.0307	1.00932
350.0	19.20	.6852	7176.0	10090.0	171.1	20.93	29.77	385.7	20.4	.0314	1.00904
360.0	18.65	.6657	7388.0	10390.0	171.9	20.93	29.74	391.2	20.9	.0320	1.00878
370.0	18.13	.6473	7599.0	10690.0	172.8	20.94	29.72	396.6	21.3	.0326	1.00854
380.0	17.65	.6299	7811.0	10990.0	173.6	20.95	29.70	401.9	21.7	.0332	1.00831
390.0	17.19	.6134	8023.0	11280.0	174.3	20.97	29.69	407.2	22.1	.0338	1.00809
400.0	16.75	.5978	8235.0	11580.0	175.1	20.99	29.68	412.3	22.5	.0344	1.00788
420.0	15.94	.5689	8659.0	12170.0	176.5	21.03	29.68	422.3	23.3	.0357	1.00750
440.0	15.21	.5428	9083.0	12770.0	177.9	21.08	29.69	432.1	24.0	.0369	1.00715
460.0	14.54	.5190	9508.0	13360.0	179.2	21.14	29.72	441.6	24.8	.0381	1.00683
480.0	13.93	.4972	9934.0	13960.0	180.5	21.21	29.77	450.8	25.5	.0393	1.00655
500.0	13.37	.4772	10360.0	14550.0	181.7	21.29	29.82	459.7	26.2	.0405	1.00628
520.0	12.85	.4587	10790.0	15150.0	182.9	21.37	29.89	468.5	27.0	.0417	1.00604
540.0	12.37	.4417	11220.0	15750.0	184.0	21.47	29.97	477.0	27.7	.0429	1.00581
560.0	11.93	.4259	11650.0	16350.0	185.1	21.57	30.05	485.3	28.3	.0441	1.00560
580.0	11.52	.4112	12090.0	16950.0	186.2	21.68	30.15	493.5	29.0	.0453	1.00541
600.0	11.14	.3975	12520.0	17550.0	187.2	21.80	30.25	501.4	29.7	.0465	1.00522
620.0	10.78	.3847	12960.0	18160.0	188.2	21.92	30.36	509.2	30.4	.0477	1.00505
640.0	10.44	.3727	13400.0	18770.0	189.1	22.04	30.47	516.8	31.0	.0489	1.00490
660.0	10.13	.3614	13850.0	19380.0	190.1	22.17	30.59	524.3	31.6	.0501	1.00475
680.0	9.828	.3508	14290.0	19990.0	191.0	22.30	30.71	531.7	32.3	.0513	1.00461
700.0	9.548	.3408	14740.0	20610.0	191.9	22.43	30.83	538.9	32.9	.0525	1.00447
720.0	9.283	.3314	15190.0	21230.0	192.8	22.56	30.96	546.0	33.5	.0536	1.00435
740.0	9.033	.3224	15640.0	21850.0	193.6	22.69	31.09	553.0	34.1	.0548	1.00423
760.0	8.796	.3140	16100.0	22470.0	194.4	22.83	31.21	559.9	34.7	.0560	1.00412
780.0	8.572	.3060	16560.0	23090.0	195.3	22.96	31.34	566.6	35.3	.0571	1.00401
800.0	8.358	.2983	17020.0	23720.0	196.0	23.09	31.47	573.3	35.9	.0583	1.00391
820.0	8.155	.2911	17480.0	24350.0	196.8	23.22	31.60	579.9	36.5	.0595	1.00382
840.0	7.962	.2842	17950.0	24990.0	197.6	23.36	31.73	586.4	37.1	.0606	1.00372

Thermophysical properties of nitrogen—Continued

T K	Density kg/m³	Density mol/dm³	E J/mol	H J/mol	S J/(mol·K)	C_v J/(mol·K)	C_p J/(mol·K)	Sound m/s	Visc. μPa·s	Therm. W/(m·K)	Diel. const.
860.0	7.778	.2776	18420.0	25620.0	198.3	23.49	31.85	592.8	37.7	.0618	1.00364
880.0	7.602	.2713	18890.0	26260.0	199.1	23.61	31.98	599.1	38.3	.0629	1.00355
900.0	7.433	.2653	19360.0	26900.0	199.8	23.74	32.10	605.4	38.8	.0641	1.00347
920.0	7.273	.2596	19840.0	27540.0	200.5	23.86	32.22	611.6	39.4	.0652	1.00340
940.0	7.119	.2541	20320.0	28190.0	201.2	23.98	32.34	617.7	39.9	.0663	1.00333
1000.0	6.693	.2389	21770.0	30140.0	203.2	24.34	32.68	635.7	41.6	.0697	1.00313
1050.0	6.376	.2276	22990.0	31780.0	204.8	24.61	32.95	650.2	42.9	.0725	1.00298
1100.0	6.088	.2173	24230.0	33440.0	206.3	24.87	33.21	664.4	44.2	.0752	1.00284
1150.0	5.824	.2079	25480.0	35100.0	207.8	25.11	33.45	678.4	45.5	.0779	1.00272
1200.0	5.583	.1993	26740.0	36780.0	209.3	25.34	33.68	692.0	46.8	.0805	1.00260
1250.0	5.361	.1913	28020.0	38470.0	210.6	25.56	33.89	705.4	48.1	.0831	1.00250
1300.0	5.155	.1840	29300.0	40170.0	212.0	25.76	34.09	718.5	49.3	.0856	1.00240
1350.0	4.965	.1772	30590.0	41880.0	213.3	25.95	34.27	731.4	50.5	.0882	1.00231
1400.0	4.789	.1709	31900.0	43600.0	214.5	26.12	34.45	744.1	51.7	.0906	1.00223
1450.0	4.624	.1651	33210.0	45320.0	215.7	26.29	34.61	756.5	52.9	.0931	1.00215
1500.0	4.471	.1596	34520.0	47060.0	216.9	26.44	34.76	768.8	54.1	.0955	1.00208
1550.0	4.327	.1545	35850.0	48800.0	218.0	26.58	34.90	780.9	55.2	.0978	1.00201
1600.0	4.193	.1497	37180.0	50550.0	219.1	26.72	35.03	792.8	56.4	.100	1.00195
1650.0	4.066	.1451	38520.0	52300.0	220.2	26.84	35.16	804.5	57.5	.102	1.00189
1700.0	3.947	.1409	39870.0	54060.0	221.3	26.96	35.28	816.0	58.6	.105	1.00183
1750.0	3.835	.1369	41220.0	55830.0	222.3	27.07	35.39	827.5	59.7	.107	1.00178
1800.0	3.729	.1331	42570.0	57600.0	223.3	27.17	35.49	838.7	60.8	.109	1.00173
1850.0	3.628	.1295	43940.0	59380.0	224.3	27.27	35.59	849.8	61.9	.111	1.00168
1900.0	3.533	.1261	45300.0	61160.0	225.2	27.36	35.68	860.8	63.0	.114	1.00163

2.50 MPa isobar

T K	Density kg/m³	Density mol/dm³	E J/mol	H J/mol	S J/(mol·K)	C_v J/(mol·K)	C_p J/(mol·K)	Sound m/s	Visc. μPa·s	Therm. W/(m·K)	Diel. const.
63.70[a]	868.8	31.01	−4204.0	−4123.0	68.14	28.20	54.51	1290.0	281.0	.153	1.46910
70.0	844.9	30.16	−3854.0	−3772.0	73.40	29.47	56.50	1093.0	209.0	.146	1.45494
80.0	802.3	28.64	−3290.0	−3203.0	80.99	28.21	57.09	905.8	143.0	.134	1.42978
90.0	754.3	26.92	−2720.0	−2627.0	87.78	27.05	58.46	766.5	103.0	.120	1.40153
100.0	699.3	24.96	−2126.0	−2026.0	94.10	26.46	62.22	640.1	76.4	.104	1.36941
110.0	631.8	22.55	−1477.0	−1366.0	100.4	26.26	71.24	510.8	55.9	.0879	1.33049
115.0	588.2	21.00	−1105.0	−985.8	103.8	26.35	82.38	436.3	46.6	.0790	1.30567
116.0	577.9	20.63	−1023.0	−901.6	104.5	26.39	86.19	419.5	44.8	.0771	1.29984
118.0	554.5	19.79	−845.1	−718.8	106.1	26.55	97.71	382.2	40.8	.0730	1.28667
119.894[b]	526.5	18.79	−648.4	−515.4	107.8	26.85	120.5	339.3	36.7	.0688	1.27096
119.894[b]	123.5	4.407	1542.0	2110.0	129.7	27.62	114.1	175.7	10.3	.0248	1.05934
120.0	122.7	4.380	1551.0	2121.0	129.8	27.56	111.7	176.1	10.3	.0247	1.05897
121.0	116.6	4.161	1623.0	2224.0	130.6	27.00	94.70	179.9	10.3	.0236	1.05596
122.0	111.7	3.986	1686.0	2313.0	131.3	26.54	84.00	183.2	10.2	.0229	1.05356
123.0	107.6	3.840	1742.0	2393.0	132.0	26.15	76.52	186.2	10.2	.0224	1.05156
124.0	104.0	3.714	1793.0	2467.0	132.6	25.81	70.95	189.0	10.2	.0220	1.04984
126.0	98.19	3.505	1887.0	2600.0	133.7	25.25	63.09	194.0	10.2	.0213	1.04698
128.0	93.42	3.334	1971.0	2721.0	134.6	24.78	57.76	198.5	10.2	.0209	1.04466
129.0	91.32	3.260	2010.0	2777.0	135.0	24.58	55.67	200.6	10.2	.0207	1.04365
130.0	89.39	3.191	2048.0	2832.0	135.5	24.40	53.86	202.7	10.3	.0206	1.04271
132.0	85.91	3.066	2121.0	2937.0	136.3	24.07	50.88	206.6	10.3	.0203	1.04102
134.0	82.84	2.957	2190.0	3036.0	137.0	23.78	48.52	210.2	10.4	.0202	1.03953
136.0	80.10	2.859	2257.0	3131.0	137.7	23.54	46.60	213.6	10.5	.0201	1.03820
138.0	77.63	2.771	2320.0	3223.0	138.4	23.32	45.00	216.9	10.6	.0200	1.03701
140.0	75.38	2.691	2382.0	3311.0	139.0	23.13	43.65	220.0	10.6	.0199	1.03592
142.0	73.32	2.617	2442.0	3397.0	139.6	22.96	42.49	223.1	10.7	.0199	1.03492
144.0	71.41	2.549	2500.0	3481.0	140.2	22.81	41.49	226.0	10.8	.0199	1.03400
146.0	69.64	2.486	2558.0	3563.0	140.8	22.67	40.61	228.8	10.9	.0199	1.03315
148.0	67.99	2.427	2614.0	3644.0	141.3	22.55	39.84	231.5	11.0	.0199	1.03236
150.0	66.45	2.372	2669.0	3723.0	141.9	22.43	39.15	234.1	11.1	.0199	1.03161
152.0	65.00	2.320	2723.0	3800.0	142.4	22.33	38.54	236.7	11.2	.0200	1.03091
154.0	63.63	2.271	2776.0	3877.0	142.9	22.24	37.99	239.2	11.3	.0200	1.03026
156.0	62.34	2.225	2829.0	3952.0	143.4	22.15	37.49	241.6	11.4	.0201	1.02963
160.0	59.95	2.140	2932.0	4101.0	144.3	22.01	36.63	246.4	11.6	.0202	1.02849
165.0	57.28	2.044	3059.0	4281.0	145.4	21.85	35.74	252.0	11.8	.0204	1.02720

Thermophysical properties of nitrogen—Continued

T K	Density kg/m^3	Density mol/dm^3	E J/mol	H J/mol	S J/(mol·K)	C_v J/(mol·K)	C_p J/(mol·K)	Sound m/s	Visc. μPa·s	Therm. W/(m·K)	Diel. const.
170.0	54.89	1.959	3182.0	4458.0	146.5	21.73	35.02	257.4	12.1	.0207	1.02606
175.0	52.74	1.883	3304.0	4632.0	147.5	21.62	34.42	262.5	12.3	.0209	1.02503
180.0	50.79	1.813	3423.0	4803.0	148.4	21.53	33.91	267.5	12.6	.0212	1.02409
185.0	49.00	1.749	3542.0	4971.0	149.4	21.46	33.48	272.2	12.8	.0214	1.02323
190.0	47.35	1.690	3658.0	5137.0	150.3	21.39	33.11	276.9	13.1	.0217	1.02245
195.0	45.84	1.636	3774.0	5302.0	151.1	21.34	32.79	281.4	13.4	.0220	1.02172
200.0	44.43	1.586	3889.0	5465.0	151.9	21.29	32.51	285.8	13.6	.0223	1.02104
205.0	43.11	1.539	4003.0	5627.0	152.7	21.25	32.26	290.0	13.9	.0226	1.02042
210.0	41.89	1.495	4116.0	5788.0	153.5	21.22	32.04	294.2	14.1	.0229	1.01983
215.0	40.74	1.454	4229.0	5948.0	154.3	21.18	31.84	298.3	14.4	.0233	1.01928
220.0	39.66	1.416	4340.0	6107.0	155.0	21.16	31.67	302.2	14.6	.0236	1.01877
230.0	37.68	1.345	4563.0	6422.0	156.4	21.11	31.36	310.0	15.1	.0242	1.01782
240.0	35.91	1.282	4783.0	6734.0	157.7	21.07	31.11	317.4	15.6	.0248	1.01698
250.0	34.31	1.225	5003.0	7044.0	159.0	21.04	30.91	324.6	16.1	.0255	1.01622
260.0	32.86	1.173	5221.0	7352.0	160.2	21.02	30.73	331.5	16.5	.0261	1.01552
270.0	31.53	1.126	5438.0	7659.0	161.4	21.00	30.58	338.3	17.0	.0268	1.01489
280.0	30.32	1.082	5654.0	7964.0	162.5	20.98	30.45	344.9	17.5	.0274	1.01432
290.0	29.20	1.042	5869.0	8268.0	163.5	20.97	30.34	351.3	17.9	.0280	1.01378
300.0	28.17	1.005	6084.0	8571.0	164.6	20.96	30.25	357.6	18.4	.0287	1.01329
310.0	27.21	.9712	6299.0	8873.0	165.5	20.95	30.16	363.7	18.8	.0293	1.01284
320.0	26.32	.9394	6513.0	9174.0	166.5	20.95	30.09	369.7	19.2	.0299	1.01241
330.0	25.48	.9096	6726.0	9475.0	167.4	20.95	30.03	375.5	19.7	.0305	1.01202
340.0	24.70	.8818	6940.0	9775.0	168.3	20.95	29.97	381.3	20.1	.0312	1.01165
350.0	23.97	.8557	7153.0	10070.0	169.2	20.95	29.93	386.9	20.5	.0318	1.01130
360.0	23.29	.8312	7365.0	10370.0	170.0	20.96	29.89	392.4	20.9	.0324	1.01097
370.0	22.64	.8081	7578.0	10670.0	170.9	20.96	29.86	397.9	21.3	.0330	1.01067
380.0	22.03	.7862	7791.0	10970.0	171.6	20.98	29.83	403.2	21.7	.0336	1.01038
390.0	21.45	.7656	8003.0	11270.0	172.4	20.99	29.81	408.4	22.1	.0342	1.01010
400.0	20.90	.7461	8216.0	11570.0	173.2	21.00	29.80	413.6	22.5	.0348	1.00984
410.0	20.38	.7275	8428.0	11860.0	173.9	21.02	29.79	418.7	22.9	.0354	1.00959
420.0	19.89	.7099	8641.0	12160.0	174.6	21.04	29.78	423.7	23.3	.0361	1.00936
440.0	18.97	.6772	9066.0	12760.0	176.0	21.09	29.79	433.4	24.1	.0373	1.00893
460.0	18.14	.6474	9492.0	13350.0	177.3	21.15	29.81	442.9	24.8	.0385	1.00853
480.0	17.38	.6202	9919.0	13950.0	178.6	21.22	29.84	452.1	25.6	.0397	1.00817
500.0	16.68	.5952	10350.0	14550.0	179.8	21.30	29.89	461.1	26.3	.0409	1.00784
520.0	16.03	.5722	10780.0	15150.0	181.0	21.39	29.95	469.9	27.0	.0421	1.00753
540.0	15.44	.5509	11210.0	15750.0	182.1	21.48	30.02	478.4	27.7	.0433	1.00725
560.0	14.88	.5312	11640.0	16350.0	183.2	21.58	30.10	486.7	28.4	.0445	1.00699
580.0	14.37	.5129	12080.0	16950.0	184.3	21.69	30.20	494.8	29.1	.0456	1.00675
600.0	13.89	.4958	12510.0	17550.0	185.3	21.81	30.29	502.8	29.7	.0468	1.00652
620.0	13.44	.4798	12950.0	18160.0	186.3	21.93	30.40	510.6	30.4	.0480	1.00631
640.0	13.02	.4649	13390.0	18770.0	187.3	22.05	30.51	518.2	31.0	.0492	1.00611
660.0	12.63	.4508	13840.0	19380.0	188.2	22.17	30.62	525.7	31.7	.0504	1.00592
680.0	12.26	.4376	14280.0	20000.0	189.1	22.30	30.74	533.0	32.3	.0516	1.00575
700.0	11.91	.4251	14730.0	20610.0	190.0	22.43	30.86	540.2	32.9	.0528	1.00558
720.0	11.58	.4133	15180.0	21230.0	190.9	22.57	30.99	547.3	33.6	.0539	1.00543
740.0	11.27	.4022	15640.0	21850.0	191.7	22.70	31.11	554.3	34.2	.0551	1.00528
760.0	10.97	.3917	16090.0	22470.0	192.6	22.83	31.24	561.1	34.8	.0563	1.00514
780.0	10.69	.3817	16550.0	23100.0	193.4	22.97	31.37	567.9	35.4	.0575	1.00501
800.0	10.43	.3722	17010.0	23730.0	194.2	23.10	31.49	574.6	36.0	.0586	1.00488
820.0	10.17	.3632	17480.0	24360.0	195.0	23.23	31.62	581.1	36.6	.0598	1.00476
840.0	9.933	.3546	17940.0	24990.0	195.7	23.36	31.75	587.6	37.1	.0609	1.00465
860.0	9.704	.3464	18410.0	25630.0	196.5	23.49	31.87	594.0	37.7	.0621	1.00454
880.0	9.484	.3385	18880.0	26270.0	197.2	23.62	31.99	600.4	38.3	.0632	1.00444
900.0	9.275	.3310	19360.0	26910.0	197.9	23.74	32.12	606.6	38.9	.0644	1.00434
920.0	9.074	.3239	19840.0	27550.0	198.6	23.87	32.24	612.8	39.4	.0655	1.00424
950.0	8.789	.3137	20550.0	28520.0	199.7	24.05	32.41	621.9	40.2	.0672	1.00411
1000.0	8.352	.2981	21770.0	30150.0	201.3	24.34	32.70	636.8	41.6	.0700	1.00390
1050.0	7.957	.2840	22990.0	31790.0	202.9	24.62	32.96	651.4	42.9	.0728	1.00371
1100.0	7.598	.2712	24230.0	33450.0	204.5	24.87	33.22	665.6	44.3	.0755	1.00354
1150.0	7.269	.2595	25480.0	35110.0	206.0	25.12	33.46	679.5	45.6	.0782	1.00339
1200.0	6.968	.2487	26740.0	36790.0	207.4	25.35	33.68	693.1	46.8	.0808	1.00325
1250.0	6.691	.2388	28010.0	38480.0	208.8	25.56	33.90	706.4	48.1	.0834	1.00312

Thermophysical properties of nitrogen—Continued

T K	Density kg/m³	Density mol/dm³	E J/mol	H J/mol	S J/(mol·K)	C_v J/(mol·K)	C_p J/(mol·K)	Sound m/s	Visc. μPa·s	Therm. W/(m·K)	Diel. const.
1300.0	6.435	.2297	29300.0	40180.0	210.1	25.76	34.09	719.5	49.3	.0859	1.00300
1350.0	6.198	.2212	30590.0	41890.0	211.4	25.95	34.28	732.4	50.5	.0885	1.00288
1400.0	5.978	.2134	31890.0	43610.0	212.6	26.12	34.45	745.0	51.7	.0909	1.00278
1450.0	5.773	.2061	33200.0	45340.0	213.9	26.29	34.61	757.5	52.9	.0934	1.00268
1500.0	5.582	.1992	34520.0	47070.0	215.0	26.44	34.76	769.7	54.1	.0958	1.00259
1550.0	5.403	.1928	35850.0	48810.0	216.2	26.58	34.91	781.8	55.2	.0981	1.00251
1600.0	5.235	.1869	37180.0	50560.0	217.3	26.72	35.04	793.7	56.4	.100	1.00243
1650.0	5.077	.1812	38520.0	52320.0	218.4	26.84	35.16	805.4	57.5	.103	1.00235
1700.0	4.928	.1759	39870.0	54080.0	219.4	26.96	35.28	816.9	58.6	.105	1.00228
1750.0	4.788	.1709	41220.0	55840.0	220.4	27.07	35.39	828.3	59.7	.107	1.00222
1800.0	4.656	.1662	42570.0	57620.0	221.4	27.17	35.49	839.6	60.8	.110	1.00216
1850.0	4.531	.1617	43930.0	59390.0	222.4	27.27	35.59	850.7	61.9	.112	1.00210
1900.0	4.412	.1575	45300.0	61180.0	223.4	27.36	35.68	861.6	63.0	.114	1.00204
					3.00 MPa isobar						
63.81[a]	868.9	31.02	-4202.0	-4105.0	68.17	28.48	54.47	1284.0	280.0	.153	1.46920
70.0	845.7	30.19	-3860.0	-3760.0	73.32	29.66	56.37	1093.0	210.0	.147	1.45540
80.0	803.6	28.68	-3298.0	-3194.0	80.89	28.35	56.91	908.3	144.0	.134	1.43048
90.0	756.1	26.99	-2731.0	-2620.0	87.65	27.16	58.17	771.2	104.0	.120	1.40258
100.0	702.1	25.06	-2143.0	-2023.0	93.93	26.53	61.62	647.4	77.4	.105	1.37103
110.0	636.8	22.73	-1505.0	-1373.0	100.1	26.29	69.54	522.6	57.2	.0891	1.33335
115.0	595.9	21.27	-1147.0	-1006.0	103.4	26.32	78.38	453.0	48.2	.0805	1.31003
120.0	541.4	19.32	-722.7	-567.4	107.1	26.63	101.9	367.4	38.9	.0709	1.27926
122.0	509.4	18.18	-504.2	-339.2	109.0	26.99	131.2	321.3	34.5	.0664	1.26142
123.0	486.9	17.38	-364.5	-191.9	110.2	27.34	169.0	291.6	31.8	.0640	1.24901
123.5	471.8	16.84	-276.1	-98.0	111.0	27.62	211.6	273.0	30.1	.0628	1.24068
123.6	468.2	16.71	-255.7	-76.2	111.1	27.69	225.2	268.7	29.7	.0626	1.23869
123.620[b]	467.6	16.69	-252.1	-72.4	111.2	27.71	227.6	268.0	29.6	.0626	1.23835
123.620[b]	171.1	6.107	1314.0	1805.0	126.4	29.17	250.7	173.3	11.9	.0332	1.08291
125.0	150.5	5.373	1492.0	2051.0	128.3	27.94	137.3	179.9	11.4	.0284	1.07268
125.5	145.8	5.203	1539.0	2116.0	128.9	27.62	122.3	181.9	11.3	.0275	1.07032
126.0	141.7	5.057	1581.0	2174.0	129.3	27.33	111.4	183.7	11.2	.0268	1.06831
127.0	134.9	4.816	1654.0	2277.0	130.1	26.84	96.22	187.0	11.1	.0258	1.06498
127.5	132.1	4.714	1687.0	2324.0	130.5	26.62	90.72	188.5	11.1	.0254	1.06356
128.0	129.4	4.620	1719.0	2368.0	130.8	26.43	86.12	190.0	11.0	.0250	1.06227
129.0	124.8	4.455	1777.0	2450.0	131.5	26.07	78.82	192.8	11.0	.0245	1.05999
130.0	120.8	4.311	1830.0	2526.0	132.1	25.75	73.26	195.4	11.0	.0240	1.05802
131.0	117.2	4.184	1880.0	2597.0	132.6	25.47	68.85	197.9	11.0	.0236	1.05628
132.0	114.1	4.071	1927.0	2664.0	133.1	25.21	65.27	200.2	11.0	.0233	1.05472
133.0	111.2	3.968	1972.0	2728.0	133.6	24.98	62.28	202.4	11.0	.0230	1.05331
134.0	108.6	3.875	2015.0	2789.0	134.1	24.77	59.75	204.5	11.0	.0228	1.05203
135.0	106.1	3.788	2056.0	2847.0	134.5	24.58	57.58	206.6	11.0	.0225	1.05085
136.0	103.9	3.708	2095.0	2904.0	134.9	24.40	55.69	208.6	11.0	.0224	1.04976
138.0	99.87	3.565	2171.0	3012.0	135.7	24.09	52.56	212.4	11.0	.0221	1.04779
140.0	96.32	3.438	2242.0	3115.0	136.4	23.81	50.07	215.9	11.1	.0219	1.04606
142.0	93.15	3.325	2310.0	3213.0	137.1	23.57	48.03	219.3	11.1	.0217	1.04452
144.0	90.28	3.223	2376.0	3307.0	137.8	23.36	46.33	222.5	11.2	.0216	1.04313
146.0	87.68	3.130	2440.0	3398.0	138.4	23.17	44.89	225.6	11.3	.0215	1.04187
148.0	85.29	3.044	2501.0	3487.0	139.0	23.01	43.66	228.6	11.3	.0214	1.04071
150.0	83.08	2.965	2561.0	3573.0	139.6	22.86	42.59	231.5	11.4	.0214	1.03964
152.0	81.03	2.892	2620.0	3657.0	140.2	22.72	41.65	234.3	11.5	.0214	1.03865
154.0	79.12	2.824	2677.0	3740.0	140.7	22.60	40.83	236.9	11.6	.0214	1.03772
156.0	77.34	2.760	2734.0	3821.0	141.2	22.49	40.09	239.6	11.7	.0214	1.03686
158.0	75.66	2.701	2789.0	3900.0	141.7	22.39	39.43	242.1	11.8	.0214	1.03605
160.0	74.08	2.644	2844.0	3978.0	142.2	22.29	38.84	244.6	11.9	.0214	1.03528
162.0	72.58	2.591	2898.0	4055.0	142.7	22.21	38.31	247.0	12.0	.0214	1.03456
165.0	70.49	2.516	2977.0	4169.0	143.4	22.10	37.59	250.6	12.1	.0215	1.03355
170.0	67.33	2.403	3106.0	4355.0	144.5	21.94	36.60	256.2	12.3	.0217	1.03203
175.0	64.51	2.303	3233.0	4536.0	145.6	21.80	35.78	261.6	12.6	.0219	1.03067
180.0	61.97	2.212	3357.0	4713.0	146.6	21.69	35.11	266.8	12.8	.0221	1.02945
185.0	59.68	2.130	3478.0	4887.0	147.5	21.60	34.54	271.7	13.1	.0223	1.02835

Thermophysical properties of nitrogen—Continued

T K	Density kg/m^3	Density mol/dm^3	E J/mol	H J/mol	S J/(mol·K)	C_v J/(mol·K)	C_p J/(mol·K)	Sound m/s	Visc. μPa·s	Therm. W/(m·K)	Diel. const.
190.0	57.58	2.055	3599.0	5058.0	148.4	21.52	34.06	276.5	13.3	.0226	1.02734
195.0	55.65	1.986	3717.0	5228.0	149.3	21.45	33.64	281.2	13.5	.0228	1.02641
200.0	53.87	1.923	3835.0	5395.0	150.1	21.39	33.28	285.7	13.8	.0231	1.02556
205.0	52.22	1.864	3951.0	5560.0	151.0	21.34	32.97	290.1	14.0	.0234	1.02477
210.0	50.68	1.809	4066.0	5725.0	151.8	21.30	32.69	294.3	14.3	.0236	1.02403
215.0	49.24	1.758	4181.0	5887.0	152.5	21.26	32.44	298.5	14.5	.0239	1.02334
220.0	47.90	1.710	4294.0	6049.0	153.3	21.23	32.22	302.6	14.8	.0242	1.02270
225.0	46.64	1.665	4407.0	6210.0	154.0	21.20	32.02	306.5	15.0	.0245	1.02209
230.0	45.45	1.622	4520.0	6369.0	154.7	21.17	31.84	310.4	15.2	.0248	1.02152
240.0	43.26	1.544	4743.0	6686.0	156.0	21.13	31.54	318.0	15.7	.0254	1.02048
250.0	41.30	1.474	4965.0	7000.0	157.3	21.09	31.28	325.3	16.2	.0260	1.01954
260.0	39.52	1.411	5185.0	7312.0	158.5	21.06	31.06	332.4	16.7	.0267	1.01869
270.0	37.90	1.353	5404.0	7622.0	159.7	21.04	30.88	339.2	17.1	.0273	1.01792
280.0	36.43	1.300	5622.0	7930.0	160.8	21.02	30.72	345.9	17.6	.0279	1.01722
290.0	35.07	1.252	5839.0	8236.0	161.9	21.00	30.59	352.3	18.0	.0285	1.01657
300.0	33.81	1.207	6056.0	8541.0	162.9	20.99	30.47	358.6	18.5	.0291	1.01597
310.0	32.65	1.165	6271.0	8846.0	163.9	20.98	30.37	364.8	18.9	.0297	1.01542
320.0	31.57	1.127	6487.0	9149.0	164.9	20.98	30.28	370.8	19.3	.0304	1.01490
330.0	30.56	1.091	6701.0	9451.0	165.8	20.97	30.21	376.7	19.8	.0310	1.01442
340.0	29.62	1.057	6916.0	9753.0	166.7	20.97	30.14	382.5	20.2	.0316	1.01398
350.0	28.74	1.026	7130.0	10050.0	167.6	20.97	30.08	388.2	20.6	.0322	1.01356
360.0	27.91	.9962	7343.0	10350.0	168.5	20.98	30.03	393.7	21.0	.0328	1.01316
370.0	27.13	.9684	7557.0	10650.0	169.3	20.99	29.99	399.2	21.4	.0334	1.01279
380.0	26.39	.9421	7770.0	10950.0	170.1	21.00	29.96	404.5	21.8	.0340	1.01244
400.0	25.04	.8938	8197.0	11550.0	171.6	21.02	29.91	414.9	22.6	.0352	1.01180
420.0	23.82	.8503	8623.0	12150.0	173.1	21.06	29.88	425.0	23.4	.0364	1.01122
440.0	22.72	.8110	9050.0	12750.0	174.5	21.11	29.87	434.8	24.1	.0376	1.01070
460.0	21.72	.7753	9477.0	13350.0	175.8	21.17	29.89	444.3	24.9	.0388	1.01022
480.0	20.81	.7427	9905.0	13940.0	177.1	21.24	29.91	453.5	25.6	.0400	1.00979
500.0	19.97	.7127	10330.0	14540.0	178.3	21.31	29.96	462.5	26.3	.0412	1.00939
520.0	19.20	.6852	10760.0	15140.0	179.5	21.40	30.01	471.2	27.0	.0424	1.00902
540.0	18.48	.6597	11200.0	15740.0	180.6	21.49	30.08	479.7	27.7	.0436	1.00869
560.0	17.82	.6361	11630.0	16350.0	181.7	21.59	30.16	488.1	28.4	.0448	1.00837
580.0	17.21	.6141	12060.0	16950.0	182.7	21.70	30.24	496.2	29.1	.0460	1.00808
600.0	16.63	.5937	12500.0	17560.0	183.8	21.82	30.34	504.1	29.8	.0471	1.00781
620.0	16.10	.5745	12940.0	18160.0	184.8	21.93	30.44	511.9	30.4	.0483	1.00756
640.0	15.59	.5566	13380.0	18770.0	185.7	22.06	30.55	519.5	31.1	.0495	1.00732
660.0	15.12	.5398	13830.0	19390.0	186.7	22.18	30.66	527.0	31.7	.0507	1.00709
680.0	14.68	.5240	14270.0	20000.0	187.6	22.31	30.78	534.3	32.3	.0519	1.00688
700.0	14.26	.5091	14720.0	20620.0	188.5	22.44	30.89	541.5	33.0	.0531	1.00669
720.0	13.87	.4950	15170.0	21240.0	189.4	22.57	31.02	548.6	33.6	.0542	1.00650
740.0	13.49	.4817	15630.0	21860.0	190.2	22.71	31.14	555.6	34.2	.0554	1.00632
760.0	13.14	.4691	16090.0	22480.0	191.0	22.84	31.26	562.4	34.8	.0566	1.00616
780.0	12.81	.4571	16540.0	23110.0	191.9	22.97	31.39	569.2	35.4	.0577	1.00600
800.0	12.49	.4458	17010.0	23740.0	192.7	23.11	31.52	575.8	36.0	.0589	1.00585
820.0	12.19	.4349	17470.0	24370.0	193.4	23.24	31.64	582.4	36.6	.0601	1.00570
840.0	11.90	.4247	17940.0	25000.0	194.2	23.37	31.77	588.9	37.2	.0612	1.00557
860.0	11.62	.4148	18410.0	25640.0	195.0	23.50	31.89	595.3	37.7	.0624	1.00544
880.0	11.36	.4055	18880.0	26280.0	195.7	23.62	32.01	601.6	38.3	.0635	1.00531
900.0	11.11	.3965	19350.0	26920.0	196.4	23.75	32.13	607.8	38.9	.0647	1.00520
920.0	10.87	.3880	19830.0	27560.0	197.1	23.87	32.25	614.0	39.4	.0658	1.00508
950.0	10.53	.3758	20550.0	28530.0	198.2	24.05	32.43	623.1	40.3	.0675	1.00492
1000.0	10.01	.3571	21760.0	30160.0	199.8	24.34	32.71	638.0	41.6	.0703	1.00467
1050.0	9.533	.3403	22990.0	31800.0	201.4	24.62	32.98	652.5	43.0	.0730	1.00445
1100.0	9.103	.3249	24230.0	33460.0	203.0	24.88	33.23	666.7	44.3	.0758	1.00425
1150.0	8.710	.3109	25480.0	35130.0	204.4	25.12	33.47	680.5	45.6	.0784	1.00406
1200.0	8.349	.2980	26740.0	36810.0	205.9	25.35	33.69	694.1	46.8	.0811	1.00389
1250.0	8.018	.2862	28010.0	38500.0	207.3	25.57	33.90	707.5	48.1	.0837	1.00373
1300.0	7.712	.2753	29300.0	40200.0	208.6	25.77	34.10	720.5	49.3	.0862	1.00359
1350.0	7.428	.2651	30590.0	41910.0	209.9	25.95	34.28	733.4	50.5	.0887	1.00346
1400.0	7.164	.2557	31890.0	43620.0	211.1	26.13	34.46	746.0	51.7	.0912	1.00333
1450.0	6.919	.2470	33200.0	45350.0	212.3	26.29	34.62	758.5	52.9	.0937	1.00322
1500.0	6.690	.2388	34520.0	47090.0	213.5	26.44	34.77	770.7	54.1	.0961	1.00311

Thermophysical properties of nitrogen—Continued

T K	Density kg/m^3	Density mol/dm^3	E J/mol	H J/mol	S J/(mol·K)	C_v J/(mol·K)	C_p J/(mol·K)	Sound m/s	Visc. μPa·s	Therm. W/(m·K)	Diel. const.
1550.0	6.476	.2311	35850.0	48830.0	214.7	26.59	34.91	782.7	55.3	.0984	1.00301
1600.0	6.275	.2240	37180.0	50580.0	215.8	26.72	35.04	794.6	56.4	.101	1.00291
1650.0	6.086	.2172	38520.0	52330.0	216.9	26.85	35.17	806.3	57.5	.103	1.00282
1700.0	5.908	.2109	39870.0	54090.0	217.9	26.96	35.28	817.8	58.6	.105	1.00274
1750.0	5.740	.2049	41220.0	55860.0	218.9	27.07	35.39	829.2	59.7	.108	1.00266
1800.0	5.581	.1992	42570.0	57630.0	219.9	27.18	35.49	840.4	60.8	.110	1.00258
1850.0	5.432	.1939	43930.0	59410.0	220.9	27.27	35.59	851.5	61.9	.112	1.00251
1900.0	5.289	.1888	45300.0	61190.0	221.8	27.36	35.68	862.5	63.0	.114	1.00245
					3.40 MPa isobar						
63.89[a]	869.1	31.02	-4201.0	-4091.0	68.19	28.70	54.44	1279.0	280.0	.152	1.46927
80.0	804.5	28.72	-3304.0	-3186.0	80.81	28.46	56.77	910.3	144.0	.133	1.43104
90.0	757.6	27.04	-2739.0	-2614.0	87.55	27.24	57.94	774.9	105.0	.118	1.40341
100.0	704.3	25.14	-2155.0	-2020.0	93.80	26.59	61.17	653.1	78.2	.103	1.37229
110.0	640.6	22.86	-1526.0	-1378.0	99.92	26.31	68.35	531.5	58.1	.0861	1.33550
115.0	601.5	21.47	-1177.0	-1019.0	103.1	26.31	75.86	465.1	49.3	.0773	1.31317
120.0	551.6	19.69	-776.7	-604.1	106.6	26.52	93.03	387.0	40.5	.0683	1.28501
122.0	525.1	18.74	-585.0	-403.6	108.3	26.74	109.2	348.4	36.6	.0647	1.27019
124.0	489.0	17.45	-347.5	-152.7	110.3	27.20	149.2	300.2	32.1	.0612	1.25014
124.5	476.7	17.02	-272.8	-72.9	111.0	27.40	171.5	285.3	30.7	.0604	1.24336
125.0	461.7	16.48	-185.0	21.4	111.7	27.68	209.6	268.1	29.1	.0597	1.23509
125.5	441.2	15.75	-72.1	143.8	112.7	28.09	293.5	247.3	27.1	.0593	1.22393
125.7	429.9	15.35	-12.4	209.2	113.2	28.34	367.2	237.0	26.0	.0596	1.21779
125.8	423.0	15.10	23.5	248.7	113.5	28.50	428.1	231.2	25.4	.0599	1.21403
125.9	414.6	14.80	66.2	295.9	113.9	28.69	524.2	224.6	24.7	.0606	1.20950
126.0	403.8	14.41	120.3	356.2	114.4	28.95	702.8	216.9	23.8	.0621	1.20367
126.1	387.7	13.84	199.6	445.3	115.1	29.33	1176.0	206.9	22.6	.0645	1.19499
126.2	324.5	11.58	507.0	800.5	117.9	30.56	89250.0	183.6	18.4	.231	1.16148
126.3	247.1	8.820	922.0	1308.0	121.9	30.58	1344.0	175.2	14.7	.0600	1.12134
126.4	233.8	8.344	1003.0	1410.0	122.7	30.34	812.5	175.3	14.2	.0484	1.11452
126.5	225.1	8.036	1057.0	1480.0	123.3	30.13	607.0	175.7	13.9	.0445	1.11013
126.6	218.7	7.805	1099.0	1534.0	123.7	29.96	494.5	176.2	13.6	.0420	1.10684
126.8	209.0	7.459	1164.0	1620.0	124.4	29.66	372.0	177.1	13.3	.0388	1.10194
127.0	201.7	7.201	1215.0	1687.0	124.9	29.40	305.0	178.0	13.1	.0366	1.09829
127.2	195.9	6.993	1257.0	1743.0	125.4	29.17	262.1	178.9	12.9	.0351	1.09535
127.5	188.8	6.740	1311.0	1815.0	125.9	28.87	219.9	180.2	12.7	.0333	1.09179
128.0	179.7	6.414	1383.0	1913.0	126.7	28.45	177.9	182.2	12.4	.0312	1.08721
129.0	166.7	5.950	1496.0	2067.0	127.9	27.76	134.8	185.9	12.1	.0287	1.08072
130.0	157.4	5.617	1584.0	2190.0	128.9	27.22	112.3	189.2	11.9	.0270	1.07607
132.0	144.0	5.141	1726.0	2387.0	130.4	26.38	88.29	195.0	11.7	.0250	1.06946
134.0	134.4	4.799	1841.0	2550.0	131.6	25.74	75.41	200.1	11.6	.0237	1.06473
140.0	115.6	4.127	2115.0	2939.0	134.4	24.43	57.28	212.8	11.5	.0217	1.05548
150.0	97.53	3.481	2469.0	3446.0	137.9	23.22	45.93	229.6	11.7	.0205	1.04665
152.0	94.86	3.386	2532.0	3536.0	138.5	23.05	44.63	232.5	11.8	.0204	1.04535
154.0	92.41	3.298	2594.0	3625.0	139.1	22.90	43.50	235.4	11.9	.0203	1.04416
156.0	90.13	3.217	2654.0	3711.0	139.7	22.77	42.51	238.1	12.0	.0203	1.04305
158.0	88.00	3.141	2712.0	3795.0	140.2	22.65	41.63	240.8	12.0	.0202	1.04202
160.0	86.02	3.070	2770.0	3877.0	140.7	22.54	40.85	243.4	12.1	.0202	1.04106
162.0	84.15	3.004	2826.0	3958.0	141.2	22.43	40.16	245.9	12.2	.0202	1.04015
164.0	82.40	2.941	2882.0	4038.0	141.7	22.34	39.53	248.4	12.3	.0203	1.03930
166.0	80.74	2.882	2936.0	4116.0	142.2	22.26	38.96	250.8	12.4	.0203	1.03850
168.0	79.16	2.826	2990.0	4194.0	142.6	22.18	38.44	253.2	12.5	.0204	1.03774
170.0	77.67	2.772	3044.0	4270.0	143.1	22.11	37.97	255.5	12.5	.0204	1.03701
175.0	74.25	2.650	3174.0	4457.0	144.2	21.95	36.96	261.1	12.8	.0206	1.03536
180.0	71.19	2.541	3302.0	4640.0	145.2	21.82	36.13	266.4	13.0	.0208	1.03388
185.0	68.44	2.443	3427.0	4819.0	146.2	21.71	35.44	271.5	13.2	.0210	1.03256
190.0	65.93	2.353	3550.0	4995.0	147.1	21.62	34.86	276.4	13.5	.0213	1.03135
195.0	63.65	2.272	3671.0	5168.0	148.0	21.54	34.36	281.2	13.7	.0215	1.03025
200.0	61.55	2.197	3791.0	5338.0	148.9	21.48	33.93	285.8	13.9	.0218	1.02924
205.0	59.61	2.128	3909.0	5507.0	149.7	21.42	33.55	290.2	14.2	.0221	1.02831
210.0	57.81	2.063	4026.0	5674.0	150.5	21.37	33.22	294.6	14.4	.0224	1.02744

Thermophysical properties of nitrogen—Continued

T K	Density kg/m^3	Density mol/dm^3	E J/mol	H J/mol	S J/(mol·K)	C_v J/(mol·K)	C_p J/(mol·K)	Sound m/s	Visc. μPa·s	Therm. W/(m·K)	Diel. const.
215.0	56.13	2.003	4142.0	5839.0	151.3	21.32	32.93	298.8	14.6	.0227	1.02664
220.0	54.56	1.947	4257.0	6003.0	152.1	21.29	32.67	302.9	14.9	.0230	1.02588
225.0	53.09	1.895	4372.0	6166.0	152.8	21.25	32.44	307.0	15.1	.0233	1.02518
230.0	51.71	1.846	4486.0	6328.0	153.5	21.22	32.23	310.9	15.4	.0236	1.02452
235.0	50.41	1.799	4599.0	6488.0	154.2	21.19	32.05	314.8	15.6	.0239	1.02389
240.0	49.18	1.755	4711.0	6648.0	154.9	21.17	31.88	318.6	15.8	.0242	1.02331
250.0	46.91	1.675	4935.0	6965.0	156.2	21.13	31.58	325.9	16.3	.0248	1.02222
260.0	44.87	1.602	5157.0	7280.0	157.4	21.10	31.33	333.1	16.7	.0254	1.02124
270.0	43.01	1.535	5378.0	7592.0	158.6	21.07	31.12	340.0	17.2	.0260	1.02035
280.0	41.32	1.475	5597.0	7902.0	159.7	21.05	30.94	346.7	17.7	.0266	1.01954
290.0	39.76	1.419	5815.0	8211.0	160.8	21.03	30.79	353.2	18.1	.0272	1.01880
300.0	38.33	1.368	6033.0	8518.0	161.8	21.02	30.65	359.5	18.5	.0278	1.01812
310.0	37.00	1.321	6250.0	8824.0	162.8	21.01	30.54	365.7	19.0	.0284	1.01748
320.0	35.77	1.277	6466.0	9129.0	163.8	21.00	30.44	371.8	19.4	.0290	1.01690
330.0	34.62	1.236	6681.0	9433.0	164.7	20.99	30.35	377.7	19.8	.0296	1.01635
340.0	33.55	1.197	6897.0	9736.0	165.6	20.99	30.27	383.5	20.2	.0302	1.01584
350.0	32.54	1.162	7111.0	10040.0	166.5	20.99	30.20	389.2	20.7	.0308	1.01536
360.0	31.60	1.128	7326.0	10340.0	167.4	21.00	30.15	394.7	21.1	.0314	1.01491
370.0	30.71	1.096	7540.0	10640.0	168.2	21.00	30.10	400.2	21.5	.0320	1.01449
380.0	29.88	1.066	7754.0	10940.0	169.0	21.01	30.06	405.6	21.9	.0326	1.01409
390.0	29.09	1.038	7968.0	11240.0	169.8	21.02	30.02	410.8	22.3	.0332	1.01372
400.0	28.34	1.012	8182.0	11540.0	170.5	21.04	30.00	416.0	22.7	.0338	1.01336
420.0	26.96	.9623	8609.0	12140.0	172.0	21.07	29.96	426.1	23.4	.0350	1.01270
440.0	25.71	.9177	9036.0	12740.0	173.4	21.12	29.95	435.9	24.2	.0361	1.01211
460.0	24.58	.8773	9464.0	13340.0	174.7	21.18	29.95	445.4	24.9	.0373	1.01157
480.0	23.54	.8403	9893.0	13940.0	176.0	21.25	29.97	454.6	25.7	.0385	1.01108
500.0	22.59	.8064	10320.0	14540.0	177.2	21.32	30.01	463.6	26.4	.0397	1.01063
520.0	21.72	.7752	10750.0	15140.0	178.4	21.41	30.06	472.3	27.1	.0408	1.01021
540.0	20.91	.7464	11190.0	15740.0	179.5	21.50	30.12	480.9	27.8	.0420	1.00983
560.0	20.16	.7196	11620.0	16340.0	180.6	21.60	30.20	489.2	28.5	.0432	1.00948
580.0	19.47	.6948	12060.0	16950.0	181.7	21.71	30.28	497.3	29.1	.0443	1.00915
600.0	18.82	.6716	12490.0	17560.0	182.7	21.82	30.37	505.2	29.8	.0455	1.00884
620.0	18.21	.6500	12930.0	18160.0	183.7	21.94	30.47	513.0	30.5	.0467	1.00855
640.0	17.64	.6298	13380.0	18770.0	184.7	22.06	30.58	520.6	31.1	.0478	1.00828
660.0	17.11	.6107	13820.0	19390.0	185.6	22.19	30.69	528.1	31.7	.0490	1.00803
680.0	16.61	.5928	14270.0	20000.0	186.5	22.32	30.80	535.4	32.4	.0502	1.00779
700.0	16.14	.5760	14720.0	20620.0	187.4	22.45	30.92	542.6	33.0	.0513	1.00757
720.0	15.69	.5601	15170.0	21240.0	188.3	22.58	31.04	549.7	33.6	.0525	1.00736
740.0	15.27	.5450	15620.0	21860.0	189.2	22.71	31.16	556.6	34.2	.0536	1.00716
760.0	14.87	.5308	16080.0	22490.0	190.0	22.85	31.28	563.5	34.8	.0548	1.00697
780.0	14.49	.5172	16540.0	23110.0	190.8	22.98	31.41	570.2	35.4	.0559	1.00679
800.0	14.13	.5044	17000.0	23740.0	191.6	23.11	31.53	576.8	36.0	.0571	1.00662
820.0	13.79	.4922	17470.0	24370.0	192.4	23.24	31.66	583.4	36.6	.0582	1.00646
840.0	13.46	.4805	17930.0	25010.0	193.2	23.37	31.78	589.9	37.2	.0594	1.00630
860.0	13.15	.4694	18400.0	25650.0	193.9	23.50	31.90	596.2	37.8	.0605	1.00616
880.0	12.86	.4589	18870.0	26280.0	194.6	23.63	32.03	602.5	38.3	.0616	1.00601
900.0	12.57	.4487	19350.0	26930.0	195.4	23.75	32.15	608.8	38.9	.0628	1.00588
920.0	12.30	.4391	19830.0	27570.0	196.1	23.88	32.26	614.9	39.5	.0639	1.00575
950.0	11.92	.4253	20550.0	28540.0	197.1	24.06	32.44	624.0	40.3	.0655	1.00557
1000.0	11.32	.4042	21760.0	30170.0	198.8	24.35	32.72	638.9	41.6	.0683	1.00529
1050.0	10.79	.3851	22980.0	31810.0	200.4	24.62	32.99	653.4	43.0	.0710	1.00504
1100.0	10.30	.3678	24220.0	33470.0	201.9	24.88	33.24	667.6	44.3	.0737	1.00481
1150.0	9.859	.3519	25470.0	35140.0	203.4	25.13	33.48	681.4	45.6	.0764	1.00460
1200.0	9.451	.3374	26740.0	36820.0	204.8	25.35	33.70	695.0	46.9	.0790	1.00441
1250.0	9.076	.3240	28010.0	38510.0	206.2	25.57	33.91	708.3	48.1	.0815	1.00423
1300.0	8.730	.3116	29290.0	40210.0	207.5	25.77	34.10	721.4	49.3	.0840	1.00406
1350.0	8.409	.3002	30590.0	41920.0	208.8	25.96	34.29	734.2	50.6	.0865	1.00391
1400.0	8.111	.2895	31890.0	43630.0	210.1	26.13	34.46	746.8	51.8	.0890	1.00377
1450.0	7.834	.2796	33200.0	45360.0	211.3	26.29	34.62	759.2	52.9	.0914	1.00364
1500.0	7.575	.2704	34520.0	47100.0	212.5	26.45	34.77	771.5	54.1	.0938	1.00352
1550.0	7.332	.2617	35850.0	48840.0	213.6	26.59	34.91	783.5	55.3	.0961	1.00340
1600.0	7.105	.2536	37180.0	50590.0	214.7	26.72	35.05	795.3	56.4	.0984	1.00330
1650.0	6.891	.2460	38520.0	52340.0	215.8	26.85	35.17	807.0	57.5	.101	1.00320

Thermophysical properties of nitrogen—Continued

T K	Density kg/m³	Density mol/dm³	E J/mol	H J/mol	S J/(mol·K)	C_v J/(mol·K)	C_p J/(mol·K)	Sound m/s	Visc. μPa·s	Therm. W/(m·K)	Diel. const.
1700.0	6.690	.2388	39870.0	54100.0	216.9	26.97	35.28	818.5	58.6	.103	1.00310
1750.0	6.500	.2320	41220.0	55870.0	217.9	27.08	35.39	829.9	59.8	.105	1.00301
1800.0	6.320	.2256	42570.0	57640.0	218.9	27.18	35.50	841.1	60.8	.107	1.00293
1850.0	6.151	.2195	43930.0	59420.0	219.9	27.28	35.59	852.2	61.9	.110	1.00285
1900.0	5.990	.2138	45300.0	61200.0	220.8	27.37	35.68	863.2	63.0	.112	1.00277
				3.45 MPa isobar							
63.90[a]	869.1	31.02	-4200.0	-4089.0	68.19	28.73	54.43	1278.0	280.0	.152	1.46928
80.0	804.7	28.72	-3305.0	-3185.0	80.81	28.47	56.75	910.5	144.0	.133	1.43111
90.0	757.7	27.05	-2740.0	-2613.0	87.54	27.25	57.91	775.3	105.0	.118	1.40351
100.0	704.5	25.15	-2157.0	-2020.0	93.79	26.60	61.12	653.8	78.3	.103	1.37244
110.0	641.0	22.88	-1529.0	-1378.0	99.90	26.31	68.21	532.6	58.2	.0863	1.33577
115.0	602.1	21.49	-1181.0	-1021.0	103.1	26.31	75.58	466.5	49.4	.0775	1.31355
120.0	552.8	19.73	-782.9	-608.0	106.6	26.51	92.17	389.2	40.7	.0684	1.28566
122.0	526.8	18.80	-593.6	-410.1	108.2	26.72	107.4	351.4	36.9	.0649	1.27111
124.0	492.0	17.56	-362.2	-165.7	110.2	27.15	143.2	304.6	32.4	.0614	1.25176
125.0	466.6	16.65	-209.1	-1.9	111.5	27.57	192.0	274.4	29.6	.0598	1.23779
125.5	448.8	16.02	-108.8	106.6	112.4	27.91	249.2	255.6	27.8	.0592	1.22808
125.7	439.8	15.70	-59.5	160.3	112.8	28.10	290.8	246.8	27.0	.0592	1.22314
125.8	434.6	15.51	-31.7	190.7	113.1	28.21	320.1	242.0	26.5	.0592	1.22030
126.0	422.0	15.06	34.0	263.0	113.6	28.49	412.6	231.3	25.4	.0596	1.21349
126.1	414.1	14.78	74.6	308.0	114.0	28.67	492.9	225.2	24.7	.0601	1.20920
126.2	404.2	14.43	124.2	363.3	114.4	28.90	626.5	218.2	23.9	.0612	1.20388
126.3	390.9	13.95	190.3	437.6	115.0	29.21	895.6	209.8	22.8	.0643	1.19670
126.4	369.1	13.17	296.7	558.6	116.0	29.70	1706.0	198.8	21.2	.0737	1.18506
126.5	312.4	11.15	577.0	886.4	118.6	30.62	5021.0	182.3	17.8	.112	1.15513
126.6	264.7	9.447	832.3	1197.0	121.0	30.67	1761.0	176.9	15.5	.0668	1.13036
126.7	246.8	8.808	936.1	1328.0	122.1	30.46	997.8	176.4	14.7	.0542	1.12117
126.8	235.9	8.421	1002.0	1412.0	122.7	30.26	714.8	176.5	14.3	.0485	1.11562
126.9	228.1	8.141	1051.0	1475.0	123.2	30.08	566.9	176.8	14.0	.0443	1.11162
127.0	221.9	7.920	1091.0	1527.0	123.6	29.92	475.4	177.1	13.8	.0421	1.10848
127.2	212.4	7.581	1155.0	1610.0	124.3	29.64	367.1	178.0	13.4	.0391	1.10366
127.4	205.1	7.322	1205.0	1677.0	124.8	29.39	304.3	178.8	13.2	.0370	1.09999
128.0	189.9	6.779	1318.0	1827.0	126.0	28.79	211.2	181.3	12.8	.0332	1.09234
128.5	181.1	6.464	1389.0	1922.0	126.7	28.38	173.6	183.2	12.5	.0313	1.08791
129.0	174.1	6.215	1448.0	2003.0	127.3	28.03	149.9	185.1	12.3	.0299	1.08441
130.0	163.4	5.831	1545.0	2137.0	128.4	27.44	121.2	188.5	12.1	.0279	1.07905
132.0	148.5	5.302	1696.0	2347.0	130.0	26.55	92.67	194.4	11.8	.0256	1.07169
134.0	138.2	4.932	1817.0	2516.0	131.3	25.88	78.10	199.5	11.7	.0241	1.06656
140.0	118.2	4.220	2098.0	2916.0	134.2	24.52	58.36	212.5	11.6	.0219	1.05676
150.0	99.41	3.548	2457.0	3430.0	137.7	23.27	46.39	229.4	11.8	.0206	1.04756
152.0	96.66	3.450	2521.0	3521.0	138.3	23.10	45.03	232.3	11.8	.0205	1.04623
154.0	94.13	3.360	2583.0	3610.0	138.9	22.94	43.86	235.2	11.9	.0204	1.04499
156.0	91.78	3.276	2643.0	3696.0	139.5	22.80	42.83	238.0	12.0	.0204	1.04385
158.0	89.60	3.198	2702.0	3781.0	140.0	22.68	41.92	240.7	12.1	.0203	1.04279
160.0	87.55	3.125	2760.0	3864.0	140.5	22.57	41.12	243.3	12.1	.0203	1.04180
162.0	85.64	3.057	2817.0	3946.0	141.0	22.46	40.40	245.8	12.2	.0203	1.04087
164.0	83.83	2.992	2873.0	4026.0	141.5	22.37	39.75	248.3	12.3	.0204	1.04000
166.0	82.13	2.932	2928.0	4105.0	142.0	22.28	39.17	250.7	12.4	.0204	1.03917
168.0	80.52	2.874	2982.0	4183.0	142.5	22.20	38.64	253.1	12.5	.0204	1.03839
170.0	78.99	2.819	3036.0	4259.0	142.9	22.13	38.15	255.4	12.6	.0205	1.03765
175.0	75.48	2.694	3167.0	4447.0	144.0	21.97	37.11	261.0	12.8	.0207	1.03596
180.0	72.36	2.583	3295.0	4631.0	145.1	21.84	36.26	266.3	13.0	.0209	1.03445
185.0	69.54	2.482	3421.0	4810.0	146.0	21.73	35.55	271.5	13.2	.0211	1.03309
190.0	66.99	2.391	3544.0	4987.0	147.0	21.63	34.96	276.4	13.5	.0213	1.03186
195.0	64.66	2.308	3665.0	5160.0	147.9	21.56	34.45	281.2	13.7	.0216	1.03074
200.0	62.52	2.231	3785.0	5331.0	148.7	21.49	34.01	285.8	14.0	.0219	1.02970
205.0	60.54	2.161	3904.0	5500.0	149.6	21.43	33.63	290.2	14.2	.0222	1.02875
210.0	58.70	2.095	4021.0	5667.0	150.4	21.38	33.29	294.6	14.4	.0224	1.02787
215.0	56.99	2.034	4137.0	5833.0	151.2	21.33	32.99	298.8	14.7	.0227	1.02705
220.0	55.40	1.977	4253.0	5997.0	151.9	21.29	32.73	303.0	14.9	.0230	1.02628
225.0	53.90	1.924	4367.0	6161.0	152.7	21.26	32.49	307.0	15.1	.0233	1.02557

Thermophysical properties of nitrogen—Continued

T K	Density kg/m^3	Density mol/dm^3	E J/mol	H J/mol	S J/(mol·K)	C_v J/(mol·K)	C_p J/(mol·K)	Sound m/s	Visc. μPa·s	Therm. W/(m·K)	Diel. const.
230.0	52.50	1.874	4481.0	6322.0	153.4	21.23	32.28	311.0	15.4	.0236	1.02489
235.0	51.17	1.827	4595.0	6483.0	154.1	21.20	32.09	314.9	15.6	.0239	1.02426
240.0	49.92	1.782	4707.0	6643.0	154.7	21.18	31.92	318.6	15.8	.0242	1.02366
250.0	47.62	1.700	4931.0	6961.0	156.0	21.13	31.62	326.0	16.3	.0248	1.02256
260.0	45.54	1.625	5153.0	7276.0	157.3	21.10	31.36	333.2	16.8	.0254	1.02156
270.0	43.65	1.558	5374.0	7588.0	158.4	21.07	31.15	340.1	17.2	.0260	1.02066
280.0	41.93	1.497	5594.0	7899.0	159.6	21.05	30.97	346.8	17.7	.0266	1.01984
290.0	40.35	1.440	5812.0	8208.0	160.7	21.03	30.81	353.3	18.1	.0272	1.01908
300.0	38.89	1.388	6030.0	8515.0	161.7	21.02	30.67	359.7	18.5	.0278	1.01838
310.0	37.54	1.340	6247.0	8821.0	162.7	21.01	30.56	365.9	19.0	.0284	1.01774
320.0	36.29	1.295	6463.0	9127.0	163.7	21.00	30.45	371.9	19.4	.0290	1.01715
330.0	35.13	1.254	6679.0	9431.0	164.6	21.00	30.36	377.8	19.8	.0296	1.01659
340.0	34.04	1.215	6894.0	9734.0	165.5	20.99	30.29	383.6	20.2	.0302	1.01607
350.0	33.02	1.179	7109.0	10040.0	166.4	20.99	30.22	389.3	20.7	.0308	1.01559
360.0	32.06	1.144	7324.0	10340.0	167.2	21.00	30.16	394.9	21.1	.0314	1.01513
370.0	31.16	1.112	7538.0	10640.0	168.1	21.00	30.11	400.3	21.5	.0320	1.01470
380.0	30.31	1.082	7752.0	10940.0	168.9	21.01	30.07	405.7	21.9	.0326	1.01430
390.0	29.51	1.053	7966.0	11240.0	169.6	21.03	30.04	411.0	22.3	.0332	1.01392
400.0	28.75	1.026	8180.0	11540.0	170.4	21.04	30.01	416.2	22.7	.0338	1.01356
420.0	27.35	.9763	8607.0	12140.0	171.9	21.08	29.97	426.3	23.4	.0350	1.01289
440.0	26.08	.9311	9035.0	12740.0	173.3	21.12	29.95	436.1	24.2	.0362	1.01229
460.0	24.93	.8900	9463.0	13340.0	174.6	21.18	29.96	445.6	24.9	.0373	1.01174
480.0	23.88	.8525	9892.0	13940.0	175.9	21.25	29.98	454.8	25.7	.0385	1.01124
500.0	22.92	.8181	10320.0	14540.0	177.1	21.32	30.02	463.7	26.4	.0397	1.01078
520.0	22.03	.7864	10750.0	15140.0	178.3	21.41	30.07	472.5	27.1	.0409	1.01036
540.0	21.21	.7572	11180.0	15740.0	179.4	21.50	30.13	481.0	27.8	.0420	1.00997
560.0	20.45	.7301	11620.0	16340.0	180.5	21.60	30.20	489.3	28.5	.0432	1.00961
580.0	19.75	.7049	12050.0	16950.0	181.6	21.71	30.28	497.4	29.1	.0444	1.00928
600.0	19.09	.6814	12490.0	17560.0	182.6	21.83	30.38	505.4	29.8	.0455	1.00897
620.0	18.47	.6594	12930.0	18160.0	183.6	21.94	30.48	513.1	30.5	.0467	1.00867
640.0	17.90	.6389	13370.0	18780.0	184.6	22.07	30.58	520.7	31.1	.0479	1.00840
660.0	17.36	.6196	13820.0	19390.0	185.5	22.19	30.69	528.2	31.7	.0490	1.00815
680.0	16.85	.6014	14270.0	20000.0	186.4	22.32	30.80	535.5	32.4	.0502	1.00790
700.0	16.37	.5843	14720.0	20620.0	187.3	22.45	30.92	542.7	33.0	.0513	1.00768
720.0	15.92	.5682	15170.0	21240.0	188.2	22.58	31.04	549.8	33.6	.0525	1.00746
740.0	15.49	.5529	15620.0	21860.0	189.0	22.71	31.16	556.7	34.2	.0537	1.00726
760.0	15.09	.5385	16080.0	22490.0	189.9	22.85	31.29	563.6	34.8	.0548	1.00707
780.0	14.70	.5247	16540.0	23110.0	190.7	22.98	31.41	570.3	35.4	.0560	1.00689
800.0	14.34	.5117	17000.0	23740.0	191.5	23.11	31.54	577.0	36.0	.0571	1.00671
820.0	13.99	.4993	17470.0	24370.0	192.3	23.24	31.66	583.5	36.6	.0582	1.00655
840.0	13.66	.4875	17930.0	25010.0	193.0	23.37	31.78	590.0	37.2	.0594	1.00639
860.0	13.34	.4763	18400.0	25650.0	193.8	23.50	31.91	596.4	37.8	.0605	1.00624
880.0	13.04	.4655	18870.0	26290.0	194.5	23.63	32.03	602.7	38.3	.0616	1.00610
900.0	12.75	.4553	19350.0	26930.0	195.2	23.75	32.15	608.9	38.9	.0628	1.00597
920.0	12.48	.4454	19830.0	27570.0	195.9	23.88	32.27	615.1	39.5	.0639	1.00584
950.0	12.09	.4315	20550.0	28540.0	197.0	24.06	32.44	624.2	40.3	.0656	1.00565
1000.0	11.49	.4101	21760.0	30170.0	198.7	24.35	32.72	639.0	41.7	.0683	1.00537
1050.0	10.95	.3907	22980.0	31810.0	200.3	24.62	32.99	653.5	43.0	.0710	1.00511
1100.0	10.45	.3731	24220.0	33470.0	201.8	24.88	33.24	667.7	44.3	.0737	1.00488
1150.0	10.00	.3570	25470.0	35140.0	203.3	25.13	33.48	681.5	45.6	.0764	1.00466
1200.0	9.589	.3423	26740.0	36820.0	204.7	25.35	33.70	695.1	46.9	.0790	1.00447
1250.0	9.209	.3287	28010.0	38510.0	206.1	25.57	33.91	708.4	48.1	.0815	1.00429
1300.0	8.857	.3161	29290.0	40210.0	207.4	25.77	34.11	721.5	49.3	.0840	1.00412
1350.0	8.532	.3045	30590.0	41920.0	208.7	25.96	34.29	734.3	50.6	.0865	1.00397
1400.0	8.229	.2937	31890.0	43640.0	210.0	26.13	34.46	746.9	51.8	.0890	1.00383
1450.0	7.948	.2837	33200.0	45360.0	211.2	26.29	34.62	759.3	52.9	.0914	1.00369
1500.0	7.685	.2743	34520.0	47100.0	212.4	26.45	34.77	771.5	54.1	.0938	1.00357
1550.0	7.439	.2655	35850.0	48840.0	213.5	26.59	34.91	783.6	55.3	.0961	1.00345
1600.0	7.208	.2573	37180.0	50590.0	214.6	26.72	35.05	795.4	56.4	.0984	1.00334
1650.0	6.991	.2496	38520.0	52340.0	215.7	26.85	35.17	807.1	57.5	.101	1.00324
1700.0	6.787	.2423	39870.0	54110.0	216.7	26.97	35.29	818.6	58.6	.103	1.00315
1750.0	6.595	.2354	41220.0	55870.0	217.8	27.08	35.39	830.0	59.8	.105	1.00306
1800.0	6.413	.2289	42570.0	57650.0	218.8	27.18	35.50	841.2	60.8	.107	1.00297

Thermophysical properties of nitrogen—Continued

T K	Density kg/m^3	Density mol/dm^3	E J/mol	H J/mol	S J/(mol·K)	C_v J/(mol·K)	C_p J/(mol·K)	Sound m/s	Visc. μPa·s	Therm. W/(m·K)	Diel. const.
1850.0	6.241	.2228	43930.0	59420.0	219.7	27.28	35.59	852.3	61.9	.110	1.00289
1900.0	6.077	.2169	45300.0	61200.0	220.7	27.37	35.68	863.2	63.0	.112	1.00281
				3.50 MPa isobar							
63.91[a]	869.1	31.02	−4200.0	−4087.0	68.19	28.76	54.43	1278.0	280.0	.152	1.46929
80.0	804.8	28.73	−3306.0	−3184.0	80.80	28.49	56.74	910.8	145.0	.133	1.43118
90.0	757.9	27.05	−2741.0	−2612.0	87.53	27.26	57.88	775.8	105.0	.119	1.40361
100.0	704.8	25.16	−2159.0	−2019.0	93.77	26.61	61.06	654.5	78.4	.103	1.37260
110.0	641.5	22.90	−1531.0	−1379.0	99.87	26.32	68.08	533.6	58.4	.0864	1.33603
115.0	602.8	21.52	−1185.0	−1022.0	103.0	26.31	75.30	468.0	49.6	.0776	1.31392
120.0	553.9	19.77	−788.9	−611.9	106.5	26.50	91.33	391.4	40.9	.0686	1.28630
122.0	528.4	18.86	−601.9	−416.3	108.1	26.70	105.7	354.2	37.1	.0651	1.27200
124.0	494.7	17.66	−376.0	−177.8	110.1	27.10	138.0	308.8	32.8	.0616	1.25328
125.0	471.0	16.81	−230.6	−22.4	111.3	27.47	178.5	280.2	30.1	.0600	1.24019
125.5	455.1	16.24	−139.1	76.3	112.1	27.76	220.7	262.9	28.5	.0593	1.23150
126.0	433.3	15.47	−19.9	206.4	113.2	28.21	314.9	241.8	26.4	.0590	1.21960
126.2	421.1	15.03	44.0	276.9	113.7	28.48	397.5	231.6	25.3	.0593	1.21300
126.3	413.6	14.76	82.7	319.8	114.1	28.65	464.7	225.8	24.7	.0597	1.20893
126.4	404.5	14.44	128.5	371.0	114.5	28.85	567.4	219.4	23.9	.0604	1.20404
126.5	393.0	14.03	186.1	435.6	115.0	29.12	742.0	212.1	23.0	.0617	1.19785
126.6	377.0	13.46	264.8	524.9	115.7	29.47	1090.0	203.6	21.8	.0661	1.18929
126.7	351.6	12.55	389.3	668.1	116.8	29.99	1876.0	193.2	20.1	.0757	1.17579
126.8	312.2	11.14	586.9	901.0	118.6	30.54	2531.0	183.3	17.8	.0836	1.15505
126.9	278.7	9.949	764.8	1117.0	120.3	30.64	1709.0	179.0	16.1	.0679	1.13762
127.0	259.2	9.251	875.1	1253.0	121.4	30.51	1101.0	177.8	15.2	.0568	1.12753
127.1	246.6	8.803	949.2	1347.0	122.2	30.34	798.8	177.5	14.7	.0508	1.12108
127.2	237.5	8.478	1005.0	1418.0	122.7	30.18	630.0	177.6	14.4	.0471	1.11644
127.4	224.6	8.018	1087.0	1523.0	123.5	29.87	451.3	178.1	13.9	.0420	1.10986
127.6	215.4	7.689	1148.0	1604.0	124.2	29.61	358.3	178.9	13.6	.0393	1.10519
127.8	208.2	7.432	1198.0	1669.0	124.7	29.37	301.0	179.7	13.3	.0373	1.10155
128.0	202.3	7.221	1240.0	1725.0	125.1	29.16	262.0	180.4	13.2	.0358	1.09857
128.5	190.9	6.815	1326.0	1840.0	126.0	28.70	202.9	182.4	12.8	.0332	1.09283
129.0	182.4	6.510	1394.0	1932.0	126.7	28.31	169.3	184.3	12.6	.0314	1.08855
130.0	169.8	6.062	1503.0	2081.0	127.9	27.68	131.8	187.7	12.3	.0290	1.08228
131.0	160.6	5.732	1590.0	2201.0	128.8	27.16	111.0	190.9	12.1	.0274	1.07767
132.0	153.3	5.471	1665.0	2305.0	129.6	26.72	97.57	193.8	11.9	.0262	1.07404
134.0	142.0	5.069	1791.0	2482.0	130.9	26.01	81.03	199.0	11.8	.0246	1.06847
140.0	120.9	4.315	2081.0	2892.0	133.9	24.60	59.49	212.1	11.6	.0222	1.05807
150.0	101.3	3.616	2445.0	3413.0	137.5	23.31	46.86	229.2	11.8	.0208	1.04849
152.0	98.48	3.515	2510.0	3505.0	138.1	23.14	45.45	232.1	11.9	.0206	1.04711
154.0	95.86	3.422	2572.0	3595.0	138.7	22.98	44.23	235.0	12.0	.0205	1.04584
156.0	93.44	3.335	2633.0	3682.0	139.3	22.84	43.16	237.8	12.0	.0205	1.04466
158.0	91.20	3.255	2693.0	3768.0	139.8	22.71	42.22	240.5	12.1	.0205	1.04357
160.0	89.10	3.180	2751.0	3851.0	140.4	22.60	41.39	243.1	12.2	.0204	1.04255
162.0	87.13	3.110	2808.0	3933.0	140.9	22.49	40.65	245.7	12.3	.0204	1.04159
164.0	85.28	3.044	2864.0	4014.0	141.4	22.39	39.98	248.2	12.3	.0205	1.04070
166.0	83.53	2.982	2919.0	4093.0	141.8	22.31	39.38	250.6	12.4	.0205	1.03985
168.0	81.88	2.923	2974.0	4172.0	142.3	22.23	38.83	253.0	12.5	.0205	1.03905
170.0	80.32	2.867	3028.0	4249.0	142.8	22.15	38.34	255.3	12.6	.0206	1.03829
172.0	78.83	2.814	3081.0	4325.0	143.2	22.08	37.88	257.6	12.7	.0206	1.03757
175.0	76.73	2.739	3160.0	4438.0	143.9	21.99	37.27	260.9	12.8	.0207	1.03656
180.0	73.53	2.625	3288.0	4622.0	144.9	21.85	36.39	266.3	13.0	.0209	1.03501
185.0	70.66	2.522	3414.0	4802.0	145.9	21.74	35.67	271.4	13.3	.0212	1.03362
190.0	68.05	2.429	3538.0	4979.0	146.8	21.65	35.06	276.4	13.5	.0214	1.03237
195.0	65.67	2.344	3659.0	5153.0	147.7	21.57	34.54	281.2	13.7	.0217	1.03122
200.0	63.49	2.266	3780.0	5324.0	148.6	21.50	34.09	285.8	14.0	.0219	1.03017
205.0	61.47	2.194	3898.0	5494.0	149.4	21.44	33.70	290.3	14.2	.0222	1.02920
210.0	59.60	2.127	4016.0	5661.0	150.2	21.39	33.36	294.6	14.4	.0225	1.02830
215.0	57.86	2.065	4132.0	5827.0	151.0	21.34	33.06	298.9	14.7	.0228	1.02747
220.0	56.24	2.007	4248.0	5992.0	151.8	21.30	32.79	303.0	14.9	.0231	1.02669
225.0	54.71	1.953	4363.0	6155.0	152.5	21.27	32.55	307.1	15.2	.0234	1.02596
230.0	53.28	1.902	4477.0	6317.0	153.2	21.23	32.33	311.0	15.4	.0237	1.02527

Thermophysical properties of nitrogen—Continued

T K	Density kg/m^3	Density mol/dm^3	E J/mol	H J/mol	S J/(mol·K)	C_v J/(mol·K)	C_p J/(mol·K)	Sound m/s	Visc. μPa·s	Therm. W/(m·K)	Diel. const.
235.0	51.94	1.854	4590.0	6478.0	153.9	21.21	32.14	314.9	15.6	.0240	1.02462
240.0	50.67	1.808	4703.0	6639.0	154.6	21.18	31.96	318.7	15.9	.0242	1.02401
250.0	48.32	1.725	4927.0	6957.0	155.9	21.14	31.65	326.1	16.3	.0249	1.02289
260.0	46.21	1.649	5150.0	7272.0	157.1	21.11	31.40	333.3	16.8	.0255	1.02188
270.0	44.29	1.581	5371.0	7585.0	158.3	21.08	31.18	340.2	17.2	.0261	1.02096
280.0	42.54	1.518	5591.0	7896.0	159.4	21.05	31.00	346.9	17.7	.0267	1.02013
290.0	40.93	1.461	5809.0	8205.0	160.5	21.04	30.84	353.4	18.1	.0273	1.01936
300.0	39.45	1.408	6027.0	8512.0	161.6	21.02	30.70	359.8	18.6	.0279	1.01865
310.0	38.09	1.359	6244.0	8819.0	162.6	21.01	30.58	366.0	19.0	.0285	1.01800
320.0	36.82	1.314	6461.0	9124.0	163.5	21.00	30.47	372.0	19.4	.0291	1.01739
330.0	35.63	1.272	6676.0	9428.0	164.5	21.00	30.38	378.0	19.8	.0297	1.01683
340.0	34.53	1.232	6892.0	9732.0	165.4	21.00	30.30	383.8	20.3	.0303	1.01630
350.0	33.49	1.195	7107.0	10030.0	166.3	21.00	30.23	389.4	20.7	.0309	1.01581
360.0	32.52	1.161	7321.0	10340.0	167.1	21.00	30.18	395.0	21.1	.0315	1.01535
370.0	31.61	1.128	7536.0	10640.0	167.9	21.01	30.13	400.5	21.5	.0321	1.01491
380.0	30.75	1.097	7750.0	10940.0	168.7	21.02	30.08	405.8	21.9	.0326	1.01450
390.0	29.93	1.068	7964.0	11240.0	169.5	21.03	30.05	411.1	22.3	.0332	1.01412
400.0	29.16	1.041	8178.0	11540.0	170.3	21.04	30.02	416.3	22.7	.0338	1.01375
420.0	27.74	.9902	8605.0	12140.0	171.7	21.08	29.98	426.4	23.4	.0350	1.01307
440.0	26.46	.9444	9033.0	12740.0	173.1	21.12	29.96	436.2	24.2	.0362	1.01246
460.0	25.29	.9027	9461.0	13340.0	174.5	21.18	29.97	445.7	24.9	.0374	1.01191
480.0	24.22	.8646	9890.0	13940.0	175.8	21.25	29.99	454.9	25.7	.0385	1.01140
500.0	23.25	.8298	10320.0	14540.0	177.0	21.33	30.02	463.9	26.4	.0397	1.01094
520.0	22.35	.7976	10750.0	15140.0	178.2	21.41	30.07	472.6	27.1	.0409	1.01051
540.0	21.52	.7680	11180.0	15740.0	179.3	21.50	30.13	481.1	27.8	.0420	1.01012
560.0	20.75	.7405	11620.0	16340.0	180.4	21.61	30.21	489.4	28.5	.0432	1.00975
580.0	20.03	.7149	12050.0	16950.0	181.4	21.71	30.29	497.6	29.1	.0444	1.00941
600.0	19.36	.6911	12490.0	17560.0	182.5	21.83	30.38	505.5	29.8	.0455	1.00909
620.0	18.74	.6688	12930.0	18160.0	183.5	21.94	30.48	513.3	30.5	.0467	1.00880
640.0	18.15	.6480	13370.0	18780.0	184.4	22.07	30.58	520.9	31.1	.0479	1.00852
660.0	17.61	.6284	13820.0	19390.0	185.4	22.19	30.69	528.3	31.8	.0490	1.00826
680.0	17.09	.6100	14270.0	20000.0	186.3	22.32	30.81	535.7	32.4	.0502	1.00802
700.0	16.60	.5927	14710.0	20620.0	187.2	22.45	30.92	542.8	33.0	.0514	1.00779
720.0	16.15	.5763	15170.0	21240.0	188.1	22.58	31.04	549.9	33.6	.0525	1.00757
740.0	15.71	.5608	15620.0	21860.0	188.9	22.71	31.17	556.9	34.2	.0537	1.00737
760.0	15.30	.5461	16080.0	22490.0	189.8	22.85	31.29	563.7	34.8	.0548	1.00717
780.0	14.91	.5322	16540.0	23110.0	190.6	22.98	31.41	570.5	35.4	.0560	1.00699
800.0	14.54	.5190	17000.0	23740.0	191.4	23.11	31.54	577.1	36.0	.0571	1.00681
820.0	14.19	.5065	17460.0	24380.0	192.1	23.24	31.66	583.6	36.6	.0583	1.00664
840.0	13.85	.4945	17930.0	25010.0	192.9	23.37	31.79	590.1	37.2	.0594	1.00649
860.0	13.53	.4831	18400.0	25650.0	193.7	23.50	31.91	596.5	37.8	.0605	1.00633
880.0	13.23	.4722	18870.0	26290.0	194.4	23.63	32.03	602.8	38.3	.0617	1.00619
900.0	12.94	.4618	19350.0	26930.0	195.1	23.76	32.15	609.0	38.9	.0628	1.00605
920.0	12.66	.4518	19830.0	27570.0	195.8	23.88	32.27	615.2	39.5	.0639	1.00592
950.0	12.26	.4377	20550.0	28540.0	196.9	24.06	32.44	624.3	40.3	.0656	1.00573
1000.0	11.65	.4160	21760.0	30170.0	198.5	24.35	32.72	639.1	41.7	.0683	1.00544
1050.0	11.10	.3963	22980.0	31810.0	200.1	24.62	32.99	653.6	43.0	.0710	1.00518
1100.0	10.60	.3785	24220.0	33470.0	201.7	24.88	33.24	667.8	44.3	.0737	1.00495
1150.0	10.15	.3621	25470.0	35140.0	203.2	25.13	33.48	681.6	45.6	.0764	1.00473
1200.0	9.726	.3472	26740.0	36820.0	204.6	25.35	33.70	695.2	46.9	.0790	1.00453
1250.0	9.341	.3334	28010.0	38510.0	206.0	25.57	33.91	708.5	48.1	.0815	1.00435
1300.0	8.984	.3207	29290.0	40210.0	207.3	25.77	34.11	721.6	49.3	.0841	1.00418
1350.0	8.654	.3089	30590.0	41920.0	208.6	25.96	34.29	734.4	50.6	.0865	1.00403
1400.0	8.348	.2980	31890.0	43640.0	209.8	26.13	34.46	747.0	51.8	.0890	1.00388
1450.0	8.062	.2878	33200.0	45360.0	211.1	26.29	34.62	759.4	52.9	.0914	1.00375
1500.0	7.795	.2783	34520.0	47100.0	212.2	26.45	34.77	771.6	54.1	.0938	1.00362
1550.0	7.546	.2693	35850.0	48840.0	213.4	26.59	34.91	783.7	55.3	.0961	1.00350
1600.0	7.312	.2610	37180.0	50590.0	214.5	26.72	35.05	795.5	56.4	.0984	1.00339
1650.0	7.092	.2531	38520.0	52350.0	215.6	26.85	35.17	807.2	57.5	.101	1.00329
1700.0	6.885	.2457	39860.0	54110.0	216.6	26.97	35.29	818.7	58.7	.103	1.00319
1750.0	6.690	.2388	41220.0	55870.0	217.6	27.08	35.39	830.1	59.8	.105	1.00310
1800.0	6.505	.2322	42570.0	57650.0	218.6	27.18	35.50	841.3	60.8	.107	1.00301
1850.0	6.330	.2260	43930.0	59420.0	219.6	27.28	35.59	852.4	61.9	.110	1.00293

Thermophysical properties of nitrogen—Continued

T K	Density kg/m^3	Density mol/dm^3	E J/mol	H J/mol	S J/(mol·K)	C_v J/(mol·K)	C_p J/(mol·K)	Sound m/s	Visc. μPa·s	Therm. W/(m·K)	Diel. const.
1900.0	6.165	.2201	45300.0	61210.0	220.6	27.37	35.68	863.3	63.0	.112	1.00285
				3.55 MPa isobar							
63.93[a]	869.2	31.02	-4200.0	-4086.0	68.20	28.78	54.42	1277.0	280.0	.152	1.46930
80.0	804.9	28.73	-3306.0	-3183.0	80.79	28.50	56.72	911.0	145.0	.133	1.43125
90.0	758.1	27.06	-2742.0	-2611.0	87.52	27.27	57.86	776.2	105.0	.119	1.40371
100.0	705.1	25.17	-2160.0	-2019.0	93.75	26.61	61.01	655.2	78.5	.103	1.37275
110.0	641.9	22.91	-1534.0	-1379.0	99.85	26.32	67.94	534.7	58.5	.0865	1.33628
115.0	603.4	21.54	-1188.0	-1023.0	103.0	26.31	75.03	469.4	49.7	.0777	1.31429
120.0	555.0	19.81	-794.8	-615.6	106.5	26.49	90.54	393.6	41.1	.0688	1.28692
122.0	530.0	18.92	-609.9	-422.3	108.1	26.68	104.2	357.0	37.3	.0653	1.27287
124.0	497.3	17.75	-389.1	-189.1	110.0	27.05	133.5	312.9	33.1	.0618	1.25471
125.0	474.9	16.95	-250.0	-40.6	111.2	27.39	167.8	285.5	30.5	.0602	1.24235
125.5	460.4	16.44	-165.2	50.8	111.9	27.64	200.5	269.4	29.0	.0594	1.23443
126.0	441.7	15.77	-60.4	164.8	112.8	28.01	263.6	250.5	27.2	.0589	1.22419
126.2	432.1	15.42	-8.4	221.7	113.3	28.21	309.2	241.7	26.3	.0588	1.21895
126.4	420.3	15.00	53.6	290.3	113.8	28.46	382.8	231.9	25.2	.0590	1.21255
126.6	404.7	14.45	133.2	379.0	114.5	28.81	519.9	220.6	23.9	.0597	1.20416
126.7	394.5	14.08	184.5	436.6	114.9	29.04	642.1	214.1	23.1	.0605	1.19868
126.8	381.5	13.62	249.2	509.9	115.5	29.32	840.1	206.9	22.2	.0629	1.19170
126.9	364.0	12.99	335.8	609.0	116.3	29.69	1168.0	198.9	20.9	.0667	1.18236
127.0	340.1	12.14	454.5	746.9	117.4	30.12	1583.0	190.8	19.4	.0722	1.16968
127.1	312.1	11.14	596.6	915.2	118.7	30.45	1700.0	184.4	17.8	.0727	1.15500
127.2	287.6	10.26	726.9	1073.0	120.0	30.56	1414.0	180.9	16.6	.0653	1.14219
127.3	269.5	9.618	827.4	1197.0	120.9	30.50	1073.0	179.4	15.7	.0577	1.13282
127.4	256.4	9.151	902.9	1291.0	121.7	30.37	829.3	178.8	15.2	.0523	1.12609
127.5	246.5	8.800	961.8	1365.0	122.3	30.23	668.5	178.7	14.8	.0486	1.12104
127.6	238.7	8.522	1010.0	1426.0	122.7	30.08	559.7	178.7	14.5	.0458	1.11706
127.8	226.9	8.100	1085.0	1523.0	123.5	29.81	425.3	179.2	14.0	.0418	1.11103
128.0	218.1	7.784	1144.0	1600.0	124.1	29.57	346.9	179.8	13.7	.0394	1.10654
128.2	211.0	7.532	1193.0	1664.0	124.6	29.35	295.6	180.5	13.5	.0376	1.10297
128.5	202.6	7.230	1253.0	1744.0	125.2	29.05	245.1	181.7	13.2	.0355	1.09869
129.0	191.8	6.847	1335.0	1853.0	126.1	28.61	195.0	183.5	12.9	.0331	1.09329
130.0	176.9	6.314	1458.0	2020.0	127.4	27.92	144.7	187.0	12.5	.0301	1.08581
131.0	166.4	5.939	1553.0	2151.0	128.4	27.36	119.0	190.2	12.2	.0283	1.08056
132.0	158.3	5.649	1633.0	2261.0	129.2	26.90	103.1	193.1	12.1	.0269	1.07652
134.0	146.0	5.212	1765.0	2446.0	130.6	26.15	84.21	198.5	11.9	.0251	1.07045
136.0	136.9	4.887	1876.0	2603.0	131.7	25.57	73.20	203.3	11.8	.0239	1.06594
140.0	123.6	4.412	2063.0	2868.0	133.7	24.69	60.67	211.7	11.7	.0225	1.05940
145.0	111.9	3.996	2260.0	3148.0	135.6	23.91	52.31	220.9	11.8	.0215	1.05368
150.0	103.2	3.685	2433.0	3396.0	137.3	23.36	47.34	228.9	11.9	.0209	1.04943
152.0	100.3	3.580	2498.0	3490.0	137.9	23.18	45.87	232.0	11.9	.0208	1.04800
154.0	97.61	3.484	2561.0	3580.0	138.5	23.02	44.60	234.9	12.0	.0207	1.04669
156.0	95.12	3.395	2623.0	3668.0	139.1	22.88	43.50	237.7	12.1	.0206	1.04547
158.0	92.81	3.313	2683.0	3754.0	139.6	22.75	42.53	240.4	12.1	.0206	1.04435
160.0	90.65	3.236	2741.0	3838.0	140.2	22.63	41.67	243.0	12.2	.0205	1.04330
162.0	88.63	3.164	2799.0	3921.0	140.7	22.52	40.90	245.6	12.3	.0205	1.04232
164.0	86.73	3.096	2855.0	4002.0	141.2	22.42	40.21	248.1	12.4	.0205	1.04140
166.0	84.94	3.032	2911.0	4082.0	141.7	22.33	39.59	250.5	12.5	.0206	1.04053
168.0	83.25	2.971	2966.0	4160.0	142.1	22.25	39.03	252.9	12.5	.0206	1.03971
170.0	81.65	2.914	3020.0	4238.0	142.6	22.17	38.52	255.3	12.6	.0207	1.03893
172.0	80.12	2.860	3073.0	4315.0	143.0	22.10	38.05	257.5	12.7	.0207	1.03820
175.0	77.97	2.783	3152.0	4428.0	143.7	22.01	37.42	260.9	12.8	.0208	1.03716
180.0	74.71	2.667	3281.0	4613.0	144.7	21.87	36.53	266.3	13.1	.0210	1.03558
185.0	71.77	2.562	3407.0	4793.0	145.7	21.76	35.79	271.4	13.3	.0212	1.03416
190.0	69.11	2.467	3531.0	4971.0	146.7	21.66	35.16	276.4	13.5	.0215	1.03288
195.0	66.68	2.380	3654.0	5145.0	147.6	21.58	34.63	281.2	13.8	.0217	1.03171
200.0	64.46	2.301	3774.0	5317.0	148.5	21.51	34.17	285.8	14.0	.0220	1.03064
205.0	62.40	2.227	3893.0	5487.0	149.3	21.45	33.77	290.3	14.2	.0223	1.02965
210.0	60.50	2.159	4011.0	5655.0	150.1	21.39	33.42	294.7	14.5	.0225	1.02873
215.0	58.73	2.096	4128.0	5821.0	150.9	21.35	33.12	298.9	14.7	.0228	1.02788
220.0	57.07	2.037	4243.0	5986.0	151.6	21.31	32.84	303.1	14.9	.0231	1.02709

Thermophysical properties of nitrogen—Continued

T K	Density kg/m^3	Density mol/dm^3	E J/mol	H J/mol	S J/(mol·K)	C_v J/(mol·K)	C_p J/(mol·K)	Sound m/s	Visc. μPa·s	Therm. W/(m·K)	Diel. const.
225.0	55.53	1.982	4358.0	6150.0	152.4	21.27	32.60	307.1	15.2	.0234	1.02634
230.0	54.07	1.930	4473.0	6312.0	153.1	21.24	32.38	311.1	15.4	.0237	1.02565
235.0	52.70	1.881	4586.0	6473.0	153.8	21.21	32.18	315.0	15.6	.0240	1.02499
240.0	51.41	1.835	4699.0	6634.0	154.5	21.19	32.00	318.8	15.9	.0243	1.02437
250.0	49.03	1.750	4924.0	6952.0	155.8	21.14	31.69	326.2	16.3	.0249	1.02323
260.0	46.88	1.673	5146.0	7268.0	157.0	21.11	31.43	333.4	16.8	.0255	1.02220
270.0	44.93	1.604	5368.0	7581.0	158.2	21.08	31.21	340.3	17.2	.0261	1.02127
280.0	43.15	1.540	5587.0	7892.0	159.3	21.06	31.02	347.0	17.7	.0267	1.02042
290.0	41.52	1.482	5806.0	8202.0	160.4	21.04	30.86	353.5	18.1	.0273	1.01964
300.0	40.02	1.428	6024.0	8510.0	161.4	21.03	30.72	359.9	18.6	.0279	1.01892
310.0	38.63	1.379	6241.0	8816.0	162.4	21.01	30.60	366.1	19.0	.0285	1.01826
320.0	37.34	1.333	6458.0	9122.0	163.4	21.01	30.49	372.2	19.4	.0291	1.01764
330.0	36.14	1.290	6674.0	9426.0	164.4	21.00	30.40	378.1	19.8	.0297	1.01707
340.0	35.02	1.250	6889.0	9730.0	165.3	21.00	30.32	383.9	20.3	.0303	1.01654
350.0	33.97	1.212	7104.0	10030.0	166.1	21.00	30.25	389.6	20.7	.0309	1.01604
360.0	32.98	1.177	7319.0	10330.0	167.0	21.00	30.19	395.1	21.1	.0315	1.01557
370.0	32.06	1.144	7534.0	10640.0	167.8	21.01	30.14	400.6	21.5	.0321	1.01513
380.0	31.18	1.113	7748.0	10940.0	168.6	21.02	30.10	406.0	21.9	.0327	1.01471
390.0	30.36	1.084	7962.0	11240.0	169.4	21.03	30.06	411.3	22.3	.0333	1.01432
400.0	29.58	1.056	8176.0	11540.0	170.2	21.04	30.03	416.4	22.7	.0339	1.01395
420.0	28.13	1.004	8604.0	12140.0	171.6	21.08	29.99	426.5	23.4	.0350	1.01326
440.0	26.83	.9577	9031.0	12740.0	173.0	21.13	29.97	436.3	24.2	.0362	1.01264
460.0	25.65	.9154	9460.0	13340.0	174.4	21.18	29.97	445.8	24.9	.0374	1.01208
480.0	24.56	.8768	9889.0	13940.0	175.6	21.25	29.99	455.1	25.7	.0385	1.01156
500.0	23.57	.8414	10320.0	14540.0	176.9	21.33	30.03	464.0	26.4	.0397	1.01109
520.0	22.66	.8089	10750.0	15140.0	178.0	21.41	30.08	472.8	27.1	.0409	1.01066
540.0	21.82	.7788	11180.0	15740.0	179.2	21.51	30.14	481.3	27.8	.0421	1.01026
560.0	21.04	.7509	11620.0	16340.0	180.3	21.61	30.21	489.6	28.5	.0432	1.00989
580.0	20.31	.7250	12050.0	16950.0	181.3	21.71	30.29	497.7	29.1	.0444	1.00954
600.0	19.63	.7008	12490.0	17560.0	182.4	21.83	30.39	505.6	29.8	.0456	1.00922
620.0	19.00	.6782	12930.0	18160.0	183.4	21.95	30.48	513.4	30.5	.0467	1.00892
640.0	18.41	.6571	13370.0	18780.0	184.3	22.07	30.59	521.0	31.1	.0479	1.00864
660.0	17.85	.6373	13820.0	19390.0	185.3	22.19	30.70	528.5	31.8	.0490	1.00838
680.0	17.33	.6186	14260.0	20000.0	186.2	22.32	30.81	535.8	32.4	.0502	1.00813
700.0	16.84	.6010	14710.0	20620.0	187.1	22.45	30.93	543.0	33.0	.0514	1.00790
720.0	16.37	.5844	15170.0	21240.0	188.0	22.58	31.05	550.0	33.6	.0525	1.00768
740.0	15.93	.5687	15620.0	21860.0	188.8	22.72	31.17	557.0	34.2	.0537	1.00747
760.0	15.52	.5538	16080.0	22490.0	189.6	22.85	31.29	563.8	34.8	.0548	1.00727
780.0	15.12	.5397	16540.0	23110.0	190.5	22.98	31.42	570.6	35.4	.0560	1.00708
800.0	14.75	.5263	17000.0	23740.0	191.2	23.11	31.54	577.2	36.0	.0571	1.00691
820.0	14.39	.5136	17460.0	24380.0	192.0	23.24	31.66	583.8	36.6	.0583	1.00674
840.0	14.05	.5015	17930.0	25010.0	192.8	23.37	31.79	590.2	37.2	.0594	1.00658
860.0	13.72	.4899	18400.0	25650.0	193.5	23.50	31.91	596.6	37.8	.0605	1.00642
880.0	13.41	.4788	18870.0	26290.0	194.3	23.63	32.03	602.9	38.3	.0617	1.00628
900.0	13.12	.4683	19350.0	26930.0	195.0	23.76	32.15	609.1	38.9	.0628	1.00614
920.0	12.84	.4582	19820.0	27570.0	195.7	23.88	32.27	615.3	39.5	.0639	1.00600
950.0	12.43	.4438	20550.0	28540.0	196.7	24.06	32.44	624.4	40.3	.0656	1.00581
1000.0	11.82	.4218	21760.0	30170.0	198.4	24.35	32.72	639.2	41.7	.0683	1.00552
1050.0	11.26	.4019	22980.0	31820.0	200.0	24.62	32.99	653.7	43.0	.0711	1.00526
1100.0	10.75	.3838	24220.0	33470.0	201.6	24.88	33.24	667.9	44.3	.0737	1.00502
1150.0	10.29	.3673	25470.0	35140.0	203.0	25.13	33.48	681.7	45.6	.0764	1.00480
1200.0	9.864	.3521	26740.0	36820.0	204.5	25.36	33.70	695.3	46.9	.0790	1.00460
1250.0	9.473	.3381	28010.0	38510.0	205.9	25.57	33.91	708.6	48.1	.0815	1.00441
1300.0	9.111	.3252	29290.0	40210.0	207.2	25.77	34.11	721.7	49.4	.0841	1.00424
1350.0	8.777	.3133	30590.0	41920.0	208.5	25.96	34.29	734.5	50.6	.0865	1.00408
1400.0	8.466	.3022	31890.0	43640.0	209.7	26.13	34.46	747.1	51.8	.0890	1.00394
1450.0	8.176	.2918	33200.0	45370.0	210.9	26.29	34.62	759.5	52.9	.0914	1.00380
1500.0	7.906	.2822	34520.0	47100.0	212.1	26.45	34.77	771.7	54.1	.0938	1.00367
1550.0	7.653	.2732	35850.0	48840.0	213.3	26.59	34.91	783.8	55.3	.0961	1.00355
1600.0	7.415	.2647	37180.0	50590.0	214.4	26.72	35.05	795.6	56.4	.0984	1.00344
1650.0	7.192	.2567	38520.0	52350.0	215.5	26.85	35.17	807.3	57.5	.101	1.00334
1700.0	6.982	.2492	39860.0	54110.0	216.5	26.97	35.29	818.8	58.7	.103	1.00324

Thermophysical properties of nitrogen—Continued

T K	Density kg/m^3	Density mol/dm^3	E J/mol	H J/mol	S J/(mol·K)	C_v J/(mol·K)	C_p J/(mol·K)	Sound m/s	Visc. μPa·s	Therm. W/(m·K)	Diel. const.
1750.0	6.784	.2422	41220.0	55880.0	217.5	27.08	35.39	830.2	59.8	.105	1.00314
1800.0	6.597	.2355	42570.0	57650.0	218.5	27.18	35.50	841.4	60.8	.107	1.00305
1850.0	6.420	.2292	43930.0	59430.0	219.5	27.28	35.59	852.5	61.9	.110	1.00297
1900.0	6.252	.2232	45300.0	61210.0	220.4	27.37	35.68	863.4	63.0	.112	1.00289
3.60 MPa isobar											
63.94[a]	869.2	31.02	−4200.0	−4084.0	68.20	28.81	54.42	1277.0	280.0	.152	1.46931
80.0	805.0	28.73	−3307.0	−3182.0	80.78	28.51	56.70	911.3	145.0	.133	1.43132
90.0	758.3	27.07	−2743.0	−2610.0	87.51	27.28	57.83	776.7	105.0	.119	1.40381
100.0	705.3	25.18	−2162.0	−2019.0	93.74	26.62	60.96	655.9	78.6	.103	1.37291
110.0	642.4	22.93	−1536.0	−1379.0	99.82	26.32	67.81	535.8	58.6	.0866	1.33654
115.0	604.1	21.56	−1192.0	−1025.0	103.0	26.31	74.77	470.8	49.8	.0779	1.31466
120.0	556.1	19.85	−800.6	−619.2	106.4	26.48	89.78	395.7	41.2	.0690	1.28754
122.0	531.5	18.97	−617.8	−428.0	108.0	26.66	102.7	359.7	37.5	.0655	1.27371
124.0	499.8	17.84	−401.5	−199.7	109.9	27.01	129.6	316.7	33.4	.0621	1.25607
125.0	478.5	17.08	−267.8	−57.1	111.0	27.31	159.0	290.5	30.9	.0604	1.24433
125.5	465.2	16.60	−188.3	28.5	111.7	27.54	185.4	275.3	29.5	.0596	1.23700
126.0	448.5	16.01	−93.4	131.5	112.5	27.85	231.2	258.0	27.8	.0589	1.22791
126.5	425.5	15.19	31.4	268.4	113.6	28.32	332.2	237.1	25.7	.0586	1.21539
126.7	412.7	14.73	98.1	342.5	114.2	28.60	416.5	227.2	24.6	.0589	1.20847
126.8	404.9	14.45	138.1	387.2	114.5	28.77	480.9	221.7	24.0	.0592	1.20426
127.0	384.6	13.73	240.4	502.7	115.4	29.21	699.2	209.5	22.4	.0610	1.19335
127.2	353.5	12.62	395.0	680.2	116.8	29.83	1102.0	196.1	20.3	.0656	1.17681
127.3	333.2	11.89	497.3	800.0	117.8	30.15	1272.0	190.0	19.0	.0678	1.16605
127.4	312.1	11.14	606.1	929.2	118.8	30.37	1284.0	185.5	17.8	.0665	1.15496
127.6	277.2	9.894	794.6	1158.0	120.6	30.44	972.7	181.0	16.1	.0572	1.13681
127.8	254.6	9.089	924.8	1321.0	121.9	30.24	675.1	179.9	15.1	.0495	1.12520
128.0	239.7	8.556	1016.0	1437.0	122.8	29.99	501.9	179.9	14.5	.0448	1.11755
128.2	228.9	8.169	1086.0	1526.0	123.5	29.75	399.5	180.3	14.1	.0415	1.11202
128.4	220.4	7.868	1142.0	1599.0	124.0	29.52	333.8	180.8	13.8	.0394	1.10774
129.0	202.8	7.238	1267.0	1764.0	125.3	28.93	230.4	182.9	13.2	.0352	1.09881
129.5	192.6	6.876	1344.0	1867.0	126.1	28.52	187.6	184.6	12.9	.0330	1.09370
130.0	184.7	6.593	1408.0	1954.0	126.8	28.17	160.7	186.3	12.7	.0314	1.08971
131.0	172.6	6.162	1513.0	2097.0	127.9	27.57	128.3	189.6	12.4	.0292	1.08367
132.0	163.6	5.838	1598.0	2215.0	128.8	27.08	109.3	192.6	12.2	.0277	1.07915
134.0	150.2	5.361	1738.0	2410.0	130.2	26.30	87.68	198.0	12.0	.0257	1.07251
136.0	140.4	5.012	1854.0	2572.0	131.4	25.69	75.46	202.9	11.9	.0244	1.06768
140.0	126.4	4.511	2045.0	2843.0	133.4	24.77	61.90	211.4	11.8	.0228	1.06076
145.0	114.2	4.076	2245.0	3129.0	135.4	23.98	53.04	220.6	11.8	.0217	1.05478
150.0	105.2	3.754	2421.0	3380.0	137.1	23.41	47.83	228.7	11.9	.0211	1.05038
152.0	102.2	3.646	2487.0	3474.0	137.7	23.23	46.30	231.8	12.0	.0209	1.04890
154.0	99.38	3.547	2550.0	3565.0	138.3	23.06	44.98	234.7	12.0	.0208	1.04754
156.0	96.81	3.456	2612.0	3654.0	138.9	22.91	43.84	237.5	12.1	.0207	1.04630
158.0	94.43	3.371	2673.0	3741.0	139.5	22.78	42.83	240.2	12.2	.0207	1.04514
160.0	92.22	3.292	2732.0	3825.0	140.0	22.66	41.95	242.9	12.2	.0206	1.04406
162.0	90.14	3.217	2790.0	3908.0	140.5	22.55	41.16	245.5	12.3	.0206	1.04305
164.0	88.19	3.148	2846.0	3990.0	141.0	22.45	40.45	248.0	12.4	.0206	1.04211
166.0	86.36	3.082	2902.0	4070.0	141.5	22.36	39.81	250.4	12.5	.0207	1.04122
168.0	84.62	3.021	2957.0	4149.0	142.0	22.27	39.23	252.8	12.6	.0207	1.04038
170.0	82.98	2.962	3012.0	4227.0	142.4	22.19	38.70	255.2	12.7	.0207	1.03958
172.0	81.42	2.906	3065.0	4304.0	142.9	22.12	38.22	257.5	12.7	.0208	1.03883
175.0	79.23	2.828	3145.0	4418.0	143.5	22.03	37.58	260.8	12.9	.0209	1.03776
180.0	75.89	2.709	3274.0	4603.0	144.6	21.89	36.66	266.2	13.1	.0211	1.03615
185.0	72.89	2.602	3401.0	4785.0	145.6	21.77	35.90	271.4	13.3	.0213	1.03470
190.0	70.17	2.505	3525.0	4963.0	146.5	21.67	35.26	276.4	13.5	.0215	1.03339
195.0	67.70	2.416	3648.0	5137.0	147.4	21.59	34.72	281.2	13.8	.0218	1.03220
200.0	65.43	2.335	3768.0	5310.0	148.3	21.52	34.25	285.8	14.0	.0220	1.03111
205.0	63.34	2.261	3888.0	5480.0	149.2	21.46	33.85	290.3	14.2	.0223	1.03010
210.0	61.40	2.192	4006.0	5648.0	150.0	21.40	33.49	294.7	14.5	.0226	1.02917
215.0	59.60	2.127	4123.0	5815.0	150.7	21.36	33.18	299.0	14.7	.0229	1.02830
220.0	57.91	2.067	4239.0	5980.0	151.5	21.32	32.90	303.1	14.9	.0232	1.02749
225.0	56.34	2.011	4354.0	6144.0	152.2	21.28	32.65	307.2	15.2	.0235	1.02673

Thermophysical properties of nitrogen—Continued

T K	Density kg/m³	Density mol/dm³	E J/mol	H J/mol	S J/(mol·K)	C_v J/(mol·K)	C_p J/(mol·K)	Sound m/s	Visc. μPa·s	Therm. W/(m·K)	Diel. const.
230.0	54.86	1.958	4468.0	6307.0	153.0	21.25	32.43	311.2	15.4	.0237	1.02602
235.0	53.47	1.908	4582.0	6469.0	153.7	21.22	32.23	315.1	15.6	.0240	1.02536
240.0	52.15	1.861	4695.0	6629.0	154.3	21.19	32.05	318.9	15.9	.0243	1.02473
250.0	49.73	1.775	4920.0	6948.0	155.6	21.15	31.73	326.3	16.3	.0249	1.02357
260.0	47.55	1.697	5143.0	7264.0	156.9	21.11	31.47	333.4	16.8	.0255	1.02252
270.0	45.57	1.627	5364.0	7577.0	158.1	21.09	31.24	340.4	17.2	.0261	1.02157
280.0	43.76	1.562	5584.0	7889.0	159.2	21.06	31.05	347.1	17.7	.0267	1.02071
290.0	42.11	1.503	5803.0	8199.0	160.3	21.04	30.88	353.6	18.1	.0273	1.01992
300.0	40.58	1.449	6021.0	8507.0	161.3	21.03	30.74	360.0	18.6	.0279	1.01919
310.0	39.17	1.398	6239.0	8813.0	162.3	21.02	30.62	366.2	19.0	.0285	1.01852
320.0	37.86	1.351	6455.0	9119.0	163.3	21.01	30.51	372.3	19.4	.0291	1.01789
330.0	36.64	1.308	6671.0	9424.0	164.2	21.00	30.42	378.2	19.9	.0297	1.01731
340.0	35.51	1.267	6887.0	9728.0	165.1	21.00	30.34	384.0	20.3	.0303	1.01677
350.0	34.44	1.229	7102.0	10030.0	166.0	21.00	30.27	389.7	20.7	.0309	1.01626
360.0	33.44	1.194	7317.0	10330.0	166.9	21.00	30.20	395.3	21.1	.0315	1.01578
370.0	32.50	1.160	7532.0	10630.0	167.7	21.01	30.15	400.7	21.5	.0321	1.01534
380.0	31.62	1.129	7746.0	10940.0	168.5	21.02	30.11	406.1	21.9	.0327	1.01492
390.0	30.78	1.099	7960.0	11240.0	169.3	21.03	30.07	411.4	22.3	.0333	1.01452
400.0	29.99	1.070	8174.0	11540.0	170.0	21.04	30.04	416.6	22.7	.0339	1.01414
420.0	28.52	1.018	8602.0	12140.0	171.5	21.08	30.00	426.7	23.5	.0351	1.01344
440.0	27.20	.9710	9030.0	12740.0	172.9	21.13	29.98	436.5	24.2	.0362	1.01282
460.0	26.00	.9281	9458.0	13340.0	174.2	21.18	29.98	446.0	25.0	.0374	1.01224
480.0	24.91	.8890	9887.0	13940.0	175.5	21.25	30.00	455.2	25.7	.0386	1.01172
500.0	23.90	.8531	10320.0	14540.0	176.7	21.33	30.03	464.2	26.4	.0397	1.01125
520.0	22.98	.8201	10750.0	15140.0	177.9	21.41	30.08	472.9	27.1	.0409	1.01081
540.0	22.12	.7896	11180.0	15740.0	179.1	21.51	30.14	481.4	27.8	.0421	1.01040
560.0	21.33	.7613	11620.0	16340.0	180.1	21.61	30.22	489.7	28.5	.0432	1.01003
580.0	20.59	.7350	12050.0	16950.0	181.2	21.72	30.30	497.8	29.2	.0444	1.00968
600.0	19.91	.7105	12490.0	17560.0	182.2	21.83	30.39	505.8	29.8	.0456	1.00935
620.0	19.27	.6877	12930.0	18160.0	183.2	21.95	30.49	513.5	30.5	.0467	1.00905
640.0	18.66	.6662	13370.0	18780.0	184.2	22.07	30.59	521.1	31.1	.0479	1.00876
660.0	18.10	.6461	13820.0	19390.0	185.1	22.19	30.70	528.6	31.8	.0491	1.00850
680.0	17.57	.6272	14260.0	20000.0	186.1	22.32	30.81	535.9	32.4	.0502	1.00824
700.0	17.07	.6093	14710.0	20620.0	187.0	22.45	30.93	543.1	33.0	.0514	1.00801
720.0	16.60	.5925	15170.0	21240.0	187.8	22.58	31.05	550.2	33.6	.0525	1.00778
740.0	16.15	.5766	15620.0	21860.0	188.7	22.72	31.17	557.1	34.2	.0537	1.00757
760.0	15.73	.5615	16080.0	22490.0	189.5	22.85	31.29	564.0	34.8	.0548	1.00737
780.0	15.33	.5472	16540.0	23120.0	190.3	22.98	31.42	570.7	35.4	.0560	1.00718
800.0	14.95	.5336	17000.0	23740.0	191.1	23.11	31.54	577.3	36.0	.0571	1.00700
820.0	14.59	.5207	17460.0	24380.0	191.9	23.25	31.67	583.9	36.6	.0583	1.00683
840.0	14.24	.5084	17930.0	25010.0	192.7	23.38	31.79	590.4	37.2	.0594	1.00667
860.0	13.92	.4967	18400.0	25650.0	193.4	23.50	31.91	596.7	37.8	.0605	1.00651
880.0	13.60	.4855	18870.0	26290.0	194.2	23.63	32.03	603.0	38.3	.0617	1.00636
900.0	13.30	.4748	19350.0	26930.0	194.9	23.76	32.15	609.3	38.9	.0628	1.00622
920.0	13.01	.4646	19820.0	27570.0	195.6	23.88	32.27	615.4	39.5	.0639	1.00609
950.0	12.61	.4500	20540.0	28540.0	196.6	24.06	32.44	624.5	40.3	.0656	1.00589
1000.0	11.98	.4277	21760.0	30170.0	198.3	24.35	32.72	639.4	41.7	.0683	1.00560
1050.0	11.42	.4075	22980.0	31820.0	199.9	24.63	32.99	653.8	43.0	.0711	1.00533
1100.0	10.90	.3891	24220.0	33470.0	201.4	24.88	33.24	668.0	44.3	.0737	1.00509
1150.0	10.43	.3724	25470.0	35140.0	202.9	25.13	33.48	681.8	45.6	.0764	1.00487
1200.0	10.00	.3570	26740.0	36820.0	204.4	25.36	33.70	695.4	46.9	.0790	1.00466
1250.0	9.605	.3428	28010.0	38510.0	205.7	25.57	33.91	708.7	48.1	.0816	1.00447
1300.0	9.238	.3298	29290.0	40210.0	207.1	25.77	34.11	721.8	49.4	.0841	1.00430
1350.0	8.899	.3176	30590.0	41920.0	208.4	25.96	34.29	734.6	50.6	.0866	1.00414
1400.0	8.584	.3064	31890.0	43640.0	209.6	26.13	34.46	747.2	51.8	.0890	1.00399
1450.0	8.290	.2959	33200.0	45370.0	210.8	26.29	34.62	759.6	52.9	.0914	1.00385
1500.0	8.016	.2861	34520.0	47100.0	212.0	26.45	34.77	771.8	54.1	.0938	1.00372
1550.0	7.760	.2770	35850.0	48840.0	213.1	26.59	34.91	783.9	55.3	.0961	1.00360
1600.0	7.519	.2684	37180.0	50590.0	214.3	26.72	35.05	795.7	56.4	.0984	1.00349
1650.0	7.293	.2603	38520.0	52350.0	215.3	26.85	35.17	807.4	57.5	.101	1.00338
1700.0	7.080	.2527	39860.0	54110.0	216.4	26.97	35.29	818.9	58.7	.103	1.00328
1750.0	6.879	.2455	41220.0	55880.0	217.4	27.08	35.39	830.3	59.8	.105	1.00319
1800.0	6.689	.2388	42570.0	57650.0	218.4	27.18	35.50	841.5	60.8	.107	1.00310

Thermophysical properties of nitrogen—Continued

T K	Density kg/m^3	Density mol/dm^3	E J/mol	H J/mol	S J/(mol·K)	C_v J/(mol·K)	C_p J/(mol·K)	Sound m/s	Visc. μPa·s	Therm. W/(m·K)	Diel. const.
1850.0	6.510	.2324	43930.0	59430.0	219.4	27.28	35.59	852.6	61.9	.110	1.00301
1900.0	6.340	.2263	45300.0	61210.0	220.3	27.37	35.68	863.5	63.0	.112	1.00293
				3.65 MPa isobar							
63.95[a]	869.2	31.02	-4200.0	-4082.0	68.20	28.84	54.42	1276.0	280.0	.152	1.46932
80.0	805.1	28.74	-3308.0	-3181.0	80.77	28.53	56.69	911.5	145.0	.133	1.43139
90.0	758.5	27.07	-2745.0	-2610.0	87.49	27.29	57.80	777.1	105.0	.119	1.40391
100.0	705.6	25.19	-2163.0	-2018.0	93.72	26.63	60.90	656.6	78.7	.103	1.37306
110.0	642.8	22.95	-1539.0	-1380.0	99.80	26.33	67.68	536.8	58.7	.0867	1.33680
115.0	604.7	21.59	-1195.0	-1026.0	102.9	26.31	74.51	472.2	50.0	.0780	1.31502
120.0	557.2	19.89	-806.3	-622.8	106.4	26.47	89.05	397.8	41.4	.0692	1.28814
122.0	532.9	19.02	-625.4	-433.6	107.9	26.64	101.3	362.4	37.8	.0657	1.27453
124.0	502.1	17.92	-413.3	-209.6	109.8	26.96	126.0	320.4	33.7	.0623	1.25736
125.0	481.8	17.20	-284.3	-72.1	110.9	27.25	151.7	295.2	31.3	.0606	1.24616
126.0	454.3	16.22	-121.5	103.6	112.3	27.71	208.7	264.6	28.4	.0590	1.23107
126.5	434.7	15.52	-12.6	222.7	113.2	28.10	275.8	245.9	26.5	.0585	1.22037
126.7	424.6	15.16	41.4	282.3	113.7	28.31	323.2	237.3	25.6	.0584	1.21489
126.8	418.8	14.95	72.0	316.1	113.9	28.43	355.4	232.7	25.1	.0584	1.21176
127.0	405.0	14.46	143.2	395.7	114.6	28.73	448.1	222.8	24.0	.0587	1.20433
127.2	386.9	13.81	235.4	499.7	115.4	29.12	606.5	211.8	22.6	.0597	1.19457
127.4	361.6	12.91	362.0	644.8	116.5	29.62	856.4	200.1	20.8	.0622	1.18109
127.6	328.9	11.74	527.7	838.7	118.1	30.13	1045.0	190.0	18.8	.0640	1.16377
127.8	296.4	10.58	698.6	1044.0	119.7	30.37	963.3	184.2	17.0	.0597	1.14680
128.0	271.3	9.684	838.5	1215.0	121.0	30.31	752.3	181.7	15.9	.0528	1.13378
128.2	253.5	9.047	943.6	1347.0	122.0	30.12	573.8	181.0	15.1	.0476	1.12459
128.4	240.5	8.583	1024.0	1449.0	122.8	29.89	454.2	181.0	14.6	.0439	1.11794
128.6	230.5	8.228	1088.0	1531.0	123.5	29.67	374.9	181.3	14.2	.0412	1.11285
129.0	215.9	7.705	1187.0	1660.0	124.5	29.26	280.5	182.4	13.7	.0378	1.10542
129.5	203.0	7.246	1280.0	1783.0	125.4	28.82	217.5	184.1	13.3	.0349	1.09892
130.0	193.4	6.903	1353.0	1882.0	126.2	28.43	180.7	185.7	13.0	.0329	1.09408
131.0	179.4	6.403	1470.0	2040.0	127.4	27.79	139.3	189.0	12.6	.0303	1.08704
132.0	169.2	6.038	1562.0	2167.0	128.4	27.27	116.5	192.0	12.4	.0285	1.08194
134.0	154.5	5.516	1710.0	2372.0	129.9	26.44	91.48	197.5	12.1	.0262	1.07467
136.0	144.0	5.142	1830.0	2540.0	131.1	25.81	77.89	202.4	12.0	.0248	1.06947
140.0	129.2	4.612	2027.0	2819.0	133.2	24.86	63.19	211.1	11.9	.0230	1.06214
145.0	116.5	4.158	2231.0	3109.0	135.2	24.04	53.79	220.4	11.9	.0219	1.05590
150.0	107.1	3.824	2409.0	3363.0	136.9	23.46	48.34	228.5	12.0	.0212	1.05133
152.0	104.0	3.713	2475.0	3458.0	137.6	23.27	46.74	231.6	12.0	.0211	1.04981
154.0	101.2	3.611	2539.0	3550.0	138.2	23.10	45.37	234.5	12.1	.0209	1.04841
156.0	98.51	3.516	2602.0	3640.0	138.7	22.95	44.19	237.4	12.1	.0208	1.04712
158.0	96.07	3.429	2662.0	3727.0	139.3	22.81	43.15	240.1	12.2	.0208	1.04593
160.0	93.79	3.348	2722.0	3812.0	139.8	22.69	42.23	242.8	12.3	.0208	1.04482
162.0	91.66	3.272	2780.0	3896.0	140.3	22.58	41.41	245.4	12.4	.0207	1.04379
164.0	89.66	3.200	2837.0	3978.0	140.8	22.48	40.68	247.9	12.4	.0207	1.04282
166.0	87.78	3.133	2894.0	4059.0	141.3	22.38	40.03	250.4	12.5	.0208	1.04191
168.0	86.00	3.070	2949.0	4138.0	141.8	22.30	39.43	252.8	12.6	.0208	1.04105
170.0	84.32	3.010	3004.0	4216.0	142.3	22.22	38.89	255.1	12.7	.0208	1.04023
172.0	82.73	2.953	3058.0	4294.0	142.7	22.14	38.40	257.4	12.8	.0209	1.03946
175.0	80.48	2.873	3137.0	4408.0	143.4	22.04	37.74	260.8	12.9	.0210	1.03837
180.0	77.07	2.751	3267.0	4594.0	144.4	21.90	36.80	266.2	13.1	.0212	1.03672
185.0	74.01	2.642	3394.0	4776.0	145.4	21.78	36.02	271.4	13.3	.0214	1.03524
190.0	71.24	2.543	3519.0	4955.0	146.4	21.68	35.37	276.4	13.6	.0216	1.03390
195.0	68.72	2.453	3642.0	5130.0	147.3	21.60	34.81	281.2	13.8	.0218	1.03269
200.0	66.41	2.370	3763.0	5303.0	148.2	21.53	34.34	285.9	14.0	.0221	1.03157
205.0	64.27	2.294	3882.0	5473.0	149.0	21.47	33.92	290.4	14.3	.0224	1.03055
210.0	62.30	2.224	4001.0	5642.0	149.8	21.41	33.56	294.8	14.5	.0226	1.02960
215.0	60.47	2.158	4118.0	5809.0	150.6	21.36	33.24	299.0	14.7	.0229	1.02872
220.0	58.75	2.097	4234.0	5975.0	151.4	21.32	32.96	303.2	15.0	.0232	1.02789
225.0	57.15	2.040	4350.0	6139.0	152.1	21.29	32.71	307.3	15.2	.0235	1.02712
230.0	55.65	1.986	4464.0	6302.0	152.8	21.25	32.48	311.2	15.4	.0238	1.02640
235.0	54.23	1.936	4578.0	6464.0	153.5	21.22	32.27	315.1	15.7	.0241	1.02572

Thermophysical properties of nitrogen—Continued

T K	Density kg/m^3	Density mol/dm^3	E J/mol	H J/mol	S J/(mol·K)	C_v J/(mol·K)	C_p J/(mol·K)	Sound m/s	Visc. μPa·s	Therm. W/(m·K)	Diel. const.
240.0	52.90	1.888	4691.0	6624.0	154.2	21.20	32.09	319.0	15.9	.0244	1.02508
250.0	50.44	1.800	4916.0	6944.0	155.5	21.15	31.77	326.4	16.4	.0250	1.02390
260.0	48.22	1.721	5139.0	7260.0	156.7	21.12	31.50	333.5	16.8	.0256	1.02284
270.0	46.21	1.649	5361.0	7574.0	157.9	21.09	31.27	340.5	17.3	.0262	1.02188
280.0	44.38	1.584	5581.0	7886.0	159.1	21.07	31.08	347.2	17.7	.0268	1.02100
290.0	42.69	1.524	5800.0	8195.0	160.1	21.05	30.91	353.8	18.1	.0274	1.02020
300.0	41.15	1.469	6019.0	8504.0	161.2	21.03	30.76	360.1	18.6	.0280	1.01946
310.0	39.72	1.418	6236.0	8811.0	162.2	21.02	30.64	366.3	19.0	.0286	1.01878
320.0	38.39	1.370	6453.0	9117.0	163.2	21.01	30.53	372.4	19.4	.0292	1.01814
330.0	37.15	1.326	6669.0	9421.0	164.1	21.01	30.44	378.3	19.9	.0298	1.01755
340.0	36.00	1.285	6885.0	9725.0	165.0	21.00	30.35	384.1	20.3	.0304	1.01700
350.0	34.92	1.246	7100.0	10030.0	165.9	21.00	30.28	389.8	20.7	.0309	1.01649
360.0	33.90	1.210	7315.0	10330.0	166.7	21.01	30.22	395.4	21.1	.0315	1.01600
370.0	32.95	1.176	7529.0	10630.0	167.6	21.01	30.17	400.9	21.5	.0321	1.01555
380.0	32.05	1.144	7744.0	10930.0	168.4	21.02	30.12	406.3	21.9	.0327	1.01512
390.0	31.20	1.114	7958.0	11240.0	169.2	21.03	30.08	411.5	22.3	.0333	1.01472
400.0	30.40	1.085	8172.0	11540.0	169.9	21.05	30.05	416.7	22.7	.0339	1.01433
420.0	28.92	1.032	8600.0	12140.0	171.4	21.08	30.01	426.8	23.5	.0351	1.01363
440.0	27.57	.9843	9028.0	12740.0	172.8	21.13	29.99	436.6	24.2	.0362	1.01299
460.0	26.36	.9408	9457.0	13340.0	174.1	21.19	29.99	446.1	25.0	.0374	1.01241
480.0	25.25	.9011	9886.0	13940.0	175.4	21.25	30.01	455.3	25.7	.0386	1.01188
500.0	24.23	.8648	10320.0	14540.0	176.6	21.33	30.04	464.3	26.4	.0398	1.01140
520.0	23.29	.8313	10750.0	15140.0	177.8	21.41	30.09	473.0	27.1	.0409	1.01096
540.0	22.42	.8004	11180.0	15740.0	178.9	21.51	30.15	481.6	27.8	.0421	1.01054
560.0	21.62	.7717	11610.0	16340.0	180.0	21.61	30.22	489.9	28.5	.0433	1.01016
580.0	20.87	.7451	12050.0	16950.0	181.1	21.72	30.30	498.0	29.2	.0444	1.00981
600.0	20.18	.7202	12490.0	17560.0	182.1	21.83	30.39	505.9	29.8	.0456	1.00948
620.0	19.53	.6971	12930.0	18170.0	183.1	21.95	30.49	513.7	30.5	.0467	1.00917
640.0	18.92	.6753	13370.0	18780.0	184.1	22.07	30.59	521.3	31.1	.0479	1.00888
660.0	18.35	.6549	13820.0	19390.0	185.0	22.19	30.70	528.7	31.8	.0491	1.00861
680.0	17.81	.6358	14260.0	20000.0	186.0	22.32	30.82	536.1	32.4	.0502	1.00836
700.0	17.30	.6177	14710.0	20620.0	186.8	22.45	30.93	543.2	33.0	.0514	1.00812
720.0	16.83	.6006	15160.0	21240.0	187.7	22.58	31.05	550.3	33.6	.0525	1.00789
740.0	16.37	.5845	15620.0	21860.0	188.6	22.72	31.17	557.3	34.2	.0537	1.00768
760.0	15.95	.5692	16080.0	22490.0	189.4	22.85	31.30	564.1	34.8	.0549	1.00747
780.0	15.54	.5547	16540.0	23120.0	190.2	22.98	31.42	570.8	35.4	.0560	1.00728
800.0	15.16	.5409	17000.0	23750.0	191.0	23.11	31.54	577.5	36.0	.0571	1.00710
820.0	14.79	.5278	17460.0	24380.0	191.8	23.25	31.67	584.0	36.6	.0583	1.00693
840.0	14.44	.5154	17930.0	25010.0	192.6	23.38	31.79	590.5	37.2	.0594	1.00676
860.0	14.11	.5035	18400.0	25650.0	193.3	23.50	31.91	596.9	37.8	.0606	1.00660
880.0	13.79	.4921	18870.0	26290.0	194.0	23.63	32.03	603.2	38.3	.0617	1.00645
900.0	13.48	.4813	19350.0	26930.0	194.8	23.76	32.15	609.4	38.9	.0628	1.00631
920.0	13.19	.4709	19820.0	27570.0	195.5	23.88	32.27	615.5	39.5	.0639	1.00617
950.0	12.78	.4562	20540.0	28550.0	196.5	24.06	32.45	624.6	40.3	.0656	1.00598
1000.0	12.15	.4336	21760.0	30170.0	198.2	24.35	32.72	639.5	41.7	.0684	1.00568
1050.0	11.57	.4131	22980.0	31820.0	199.8	24.63	32.99	654.0	43.0	.0711	1.00540
1100.0	11.05	.3945	24220.0	33470.0	201.3	24.88	33.24	668.1	44.3	.0738	1.00516
1150.0	10.58	.3775	25470.0	35140.0	202.8	25.13	33.48	682.0	45.6	.0764	1.00493
1200.0	10.14	.3619	26740.0	36820.0	204.2	25.36	33.70	695.5	46.9	.0790	1.00473
1250.0	9.737	.3475	28010.0	38510.0	205.6	25.57	33.91	708.8	48.1	.0816	1.00454
1300.0	9.365	.3343	29290.0	40210.0	207.0	25.77	34.11	721.9	49.4	.0841	1.00436
1350.0	9.022	.3220	30590.0	41920.0	208.2	25.96	34.29	734.7	50.6	.0866	1.00420
1400.0	8.702	.3106	31890.0	43640.0	209.5	26.13	34.46	747.3	51.8	.0890	1.00405
1450.0	8.404	.3000	33200.0	45370.0	210.7	26.29	34.62	759.7	52.9	.0914	1.00391
1500.0	8.127	.2901	34520.0	47100.0	211.9	26.45	34.77	771.9	54.1	.0938	1.00378
1550.0	7.866	.2808	35850.0	48850.0	213.0	26.59	34.92	784.0	55.3	.0961	1.00365
1600.0	7.623	.2721	37180.0	50590.0	214.1	26.72	35.05	795.8	56.4	.0984	1.00354
1650.0	7.393	.2639	38520.0	52350.0	215.2	26.85	35.17	807.5	57.5	.101	1.00343
1700.0	7.178	.2562	39860.0	54110.0	216.3	26.97	35.29	819.0	58.7	.103	1.00333
1750.0	6.974	.2489	41220.0	55880.0	217.3	27.08	35.40	830.4	59.8	.105	1.00323
1800.0	6.782	.2421	42570.0	57650.0	218.3	27.18	35.50	841.6	60.8	.107	1.00314
1850.0	6.600	.2356	43930.0	59430.0	219.3	27.28	35.59	852.6	61.9	.110	1.00305
1900.0	6.427	.2294	45300.0	61210.0	220.2	27.37	35.68	863.6	63.0	.112	1.00297

Thermophysical properties of nitrogen—Continued

T K	Density kg/m³	Density mol/dm³	E J/mol	H J/mol	S J/(mol·K)	C_v J/(mol·K)	C_p J/(mol·K)	Sound m/s	Visc. μPa·s	Therm. W/(m·K)	Diel. const.
					3.70 MPa isobar						
63.96[a]	869.2	31.03	–4199.0	–4080.0	68.20	28.86	54.41	1275.0	280.0	.152	1.46933
80.0	805.3	28.74	–3309.0	–3180.0	80.76	28.54	56.67	911.8	145.0	.133	1.43145
90.0	758.6	27.08	–2746.0	–2609.0	87.48	27.30	57.78	777.6	105.0	.119	1.40402
100.0	705.9	25.20	–2165.0	–2018.0	93.71	26.63	60.85	657.3	78.8	.103	1.37321
110.0	643.3	22.96	–1542.0	–1380.0	99.78	26.33	67.55	537.9	58.8	.0868	1.33705
115.0	605.4	21.61	–1199.0	–1028.0	102.9	26.31	74.26	473.6	50.1	.0782	1.31538
120.0	558.3	19.93	–811.9	–626.2	106.3	26.46	88.36	399.9	41.6	.0693	1.28873
122.0	534.4	19.07	–632.9	–438.9	107.9	26.62	100.0	364.9	38.0	.0659	1.27533
124.0	504.3	18.00	–424.6	–219.1	109.7	26.93	122.8	324.0	34.0	.0625	1.25860
125.0	484.9	17.31	–299.7	–86.0	110.7	27.19	145.5	299.6	31.7	.0608	1.24786
126.0	459.4	16.40	–146.1	79.6	112.0	27.60	192.0	270.7	28.9	.0592	1.23383
126.5	442.0	15.78	–48.1	186.4	112.9	27.93	240.6	253.5	27.2	.0585	1.22437
127.0	418.2	14.93	80.7	328.6	114.0	28.41	342.6	233.2	25.1	.0582	1.21141
127.2	405.1	14.46	148.5	404.3	114.6	28.69	420.2	223.9	24.0	.0582	1.20438
127.4	388.7	13.87	232.8	499.5	115.4	29.04	540.0	213.8	22.7	.0588	1.19552
128.0	311.9	11.13	624.9	957.2	118.9	30.21	868.1	187.6	17.9	.0596	1.15489
128.5	259.5	9.264	918.6	1318.0	121.8	30.09	556.6	182.3	15.4	.0479	1.12771
128.7	246.5	8.799	998.0	1419.0	122.5	29.90	454.4	182.1	14.8	.0445	1.12102
128.8	241.1	8.606	1032.0	1462.0	122.9	29.80	414.6	182.1	14.6	.0432	1.11826
129.0	231.9	8.278	1091.0	1538.0	123.5	29.59	352.1	182.4	14.3	.0409	1.11357
129.5	215.1	7.677	1206.0	1688.0	124.6	29.12	257.5	183.7	13.7	.0372	1.10502
130.0	203.2	7.253	1293.0	1803.0	125.5	28.71	206.2	185.3	13.3	.0347	1.09902
131.0	186.7	6.664	1424.0	1979.0	126.9	28.02	152.5	188.4	12.9	.0315	1.09072
132.0	175.2	6.252	1524.0	2116.0	127.9	27.46	124.6	191.5	12.6	.0294	1.08493
134.0	159.1	5.678	1681.0	2333.0	129.5	26.60	95.64	197.0	12.2	.0268	1.07692
136.0	147.8	5.275	1806.0	2507.0	130.8	25.93	80.49	202.0	12.1	.0253	1.07132
140.0	132.1	4.715	2009.0	2793.0	132.9	24.95	64.54	210.7	11.9	.0233	1.06356
145.0	118.8	4.241	2216.0	3089.0	135.0	24.11	54.57	220.1	11.9	.0221	1.05704
150.0	109.1	3.895	2396.0	3346.0	136.7	23.51	48.85	228.4	12.0	.0214	1.05230
152.0	105.9	3.780	2463.0	3442.0	137.4	23.31	47.19	231.4	12.1	.0212	1.05072
154.0	102.9	3.675	2528.0	3535.0	138.0	23.14	45.77	234.4	12.1	.0211	1.04928
156.0	100.2	3.578	2591.0	3625.0	138.5	22.99	44.54	237.2	12.2	.0210	1.04796
158.0	97.71	3.488	2652.0	3713.0	139.1	22.85	43.47	240.0	12.2	.0209	1.04673
160.0	95.37	3.404	2712.0	3799.0	139.6	22.72	42.52	242.7	12.3	.0209	1.04559
162.0	93.19	3.326	2771.0	3883.0	140.2	22.61	41.67	245.3	12.4	.0208	1.04453
164.0	91.14	3.253	2829.0	3966.0	140.7	22.50	40.92	247.8	12.5	.0208	1.04354
166.0	89.21	3.184	2885.0	4047.0	141.2	22.41	40.25	250.3	12.5	.0208	1.04260
168.0	87.39	3.119	2941.0	4127.0	141.6	22.32	39.63	252.7	12.6	.0209	1.04172
170.0	85.67	3.058	2996.0	4206.0	142.1	22.24	39.08	255.1	12.7	.0209	1.04088
172.0	84.04	3.000	3050.0	4283.0	142.6	22.16	38.57	257.4	12.8	.0210	1.04009
175.0	81.74	2.918	3130.0	4398.0	143.2	22.06	37.89	260.8	12.9	.0210	1.03898
180.0	78.26	2.793	3260.0	4585.0	144.3	21.92	36.93	266.2	13.1	.0212	1.03729
185.0	75.13	2.682	3388.0	4768.0	145.3	21.80	36.14	271.4	13.4	.0214	1.03578
190.0	72.31	2.581	3513.0	4947.0	146.2	21.70	35.47	276.4	13.6	.0217	1.03442
195.0	69.74	2.489	3636.0	5122.0	147.2	21.61	34.91	281.2	13.8	.0219	1.03318
200.0	67.38	2.405	3757.0	5296.0	148.0	21.54	34.42	285.9	14.1	.0222	1.03204
205.0	65.21	2.328	3877.0	5467.0	148.9	21.47	34.00	290.4	14.3	.0224	1.03100
210.0	63.20	2.256	3996.0	5636.0	149.7	21.42	33.63	294.8	14.5	.0227	1.03003
215.0	61.34	2.189	4113.0	5803.0	150.5	21.37	33.30	299.1	14.7	.0230	1.02913
220.0	59.59	2.127	4230.0	5969.0	151.2	21.33	33.02	303.3	15.0	.0233	1.02830
225.0	57.97	2.069	4345.0	6133.0	152.0	21.29	32.76	307.3	15.2	.0235	1.02751
230.0	56.44	2.014	4460.0	6297.0	152.7	21.26	32.53	311.3	15.4	.0238	1.02678
235.0	55.00	1.963	4574.0	6459.0	153.4	21.23	32.32	315.2	15.7	.0241	1.02609
240.0	53.64	1.915	4687.0	6620.0	154.1	21.20	32.13	319.0	15.9	.0244	1.02544
250.0	51.14	1.825	4912.0	6939.0	155.4	21.16	31.81	326.5	16.4	.0250	1.02424
260.0	48.89	1.745	5136.0	7256.0	156.6	21.12	31.53	333.6	16.8	.0256	1.02316
270.0	46.85	1.672	5358.0	7570.0	157.8	21.09	31.30	340.6	17.3	.0262	1.02218
280.0	44.99	1.606	5578.0	7882.0	158.9	21.07	31.10	347.3	17.7	.0268	1.02129
290.0	43.28	1.545	5797.0	8192.0	160.0	21.05	30.93	353.9	18.2	.0274	1.02048
300.0	41.71	1.489	6016.0	8501.0	161.1	21.03	30.79	360.2	18.6	.0280	1.01973

Thermophysical properties of nitrogen—Continued

T K	Density kg/m^3	Density mol/dm^3	*E* J/mol	*H* J/mol	*S* J/(mol·K)	C_v J/(mol·K)	C_p J/(mol·K)	Sound m/s	Visc. μPa·s	Therm. W/(m·K)	Diel. const.
310.0	40.26	1.437	6233.0	8808.0	162.1	21.02	30.66	366.5	19.0	.0286	1.01903
320.0	38.91	1.389	6450.0	9114.0	163.1	21.01	30.55	372.5	19.5	.0292	1.01839
330.0	37.66	1.344	6666.0	9419.0	164.0	21.01	30.45	378.5	19.9	.0298	1.01779
340.0	36.49	1.302	6882.0	9723.0	164.9	21.01	30.37	384.3	20.3	.0304	1.01723
350.0	35.39	1.263	7098.0	10030.0	165.8	21.01	30.30	390.0	20.7	.0310	1.01671
360.0	34.36	1.227	7313.0	10330.0	166.6	21.01	30.23	395.5	21.1	.0316	1.01622
370.0	33.40	1.192	7527.0	10630.0	167.5	21.01	30.18	401.0	21.5	.0322	1.01576
380.0	32.48	1.159	7742.0	10930.0	168.3	21.02	30.13	406.4	21.9	.0327	1.01533
390.0	31.62	1.129	7956.0	11230.0	169.0	21.03	30.09	411.7	22.3	.0333	1.01492
400.0	30.81	1.100	8170.0	11530.0	169.8	21.05	30.06	416.8	22.7	.0339	1.01453
420.0	29.31	1.046	8598.0	12140.0	171.3	21.08	30.02	427.0	23.5	.0351	1.01381
440.0	27.95	.9975	9027.0	12740.0	172.7	21.13	30.00	436.8	24.2	.0363	1.01317
460.0	26.71	.9535	9455.0	13340.0	174.0	21.19	30.00	446.3	25.0	.0374	1.01258
480.0	25.59	.9133	9884.0	13940.0	175.3	21.25	30.01	455.5	25.7	.0386	1.01205
500.0	24.55	.8764	10310.0	14540.0	176.5	21.33	30.05	464.4	26.4	.0398	1.01156
520.0	23.60	.8425	10750.0	15140.0	177.7	21.42	30.10	473.2	27.1	.0409	1.01110
540.0	22.73	.8111	11180.0	15740.0	178.8	21.51	30.16	481.7	27.8	.0421	1.01069
560.0	21.91	.7821	11610.0	16340.0	179.9	21.61	30.23	490.0	28.5	.0433	1.01030
580.0	21.16	.7551	12050.0	16950.0	181.0	21.72	30.31	498.1	29.2	.0444	1.00994
600.0	20.45	.7300	12490.0	17560.0	182.0	21.83	30.40	506.0	29.8	.0456	1.00961
620.0	19.79	.7064	12930.0	18170.0	183.0	21.95	30.50	513.8	30.5	.0468	1.00930
640.0	19.17	.6844	13370.0	18780.0	184.0	22.07	30.60	521.4	31.1	.0479	1.00900
660.0	18.60	.6638	13810.0	19390.0	184.9	22.20	30.71	528.9	31.8	.0491	1.00873
680.0	18.05	.6443	14260.0	20000.0	185.8	22.32	30.82	536.2	32.4	.0502	1.00847
700.0	17.54	.6260	14710.0	20620.0	186.7	22.45	30.94	543.4	33.0	.0514	1.00823
720.0	17.05	.6087	15160.0	21240.0	187.6	22.59	31.06	550.4	33.6	.0526	1.00800
740.0	16.60	.5924	15620.0	21860.0	188.5	22.72	31.18	557.4	34.2	.0537	1.00778
760.0	16.16	.5769	16080.0	22490.0	189.3	22.85	31.30	564.2	34.9	.0549	1.00758
780.0	15.75	.5622	16540.0	23120.0	190.1	22.98	31.42	571.0	35.4	.0560	1.00738
800.0	15.36	.5482	17000.0	23750.0	190.9	23.12	31.55	577.6	36.0	.0572	1.00720
820.0	14.99	.5350	17460.0	24380.0	191.7	23.25	31.67	584.1	36.6	.0583	1.00702
840.0	14.63	.5223	17930.0	25010.0	192.4	23.38	31.79	590.6	37.2	.0594	1.00685
860.0	14.30	.5103	18400.0	25650.0	193.2	23.51	31.91	597.0	37.8	.0606	1.00669
880.0	13.97	.4988	18870.0	26290.0	193.9	23.63	32.04	603.3	38.3	.0617	1.00654
900.0	13.67	.4878	19350.0	26930.0	194.7	23.76	32.16	609.5	38.9	.0628	1.00639
920.0	13.37	.4773	19820.0	27580.0	195.4	23.88	32.27	615.7	39.5	.0639	1.00625
950.0	12.95	.4623	20540.0	28550.0	196.4	24.06	32.45	624.7	40.3	.0656	1.00606
1000.0	12.31	.4394	21760.0	30180.0	198.1	24.35	32.73	639.6	41.7	.0684	1.00575
1050.0	11.73	.4187	22980.0	31820.0	199.7	24.63	32.99	654.1	43.0	.0711	1.00548
1100.0	11.20	.3998	24220.0	33470.0	201.2	24.88	33.24	668.2	44.3	.0738	1.00523
1150.0	10.72	.3826	25470.0	35140.0	202.7	25.13	33.48	682.1	45.6	.0764	1.00500
1200.0	10.28	.3668	26740.0	36820.0	204.1	25.36	33.70	695.6	46.9	.0790	1.00479
1250.0	9.869	.3523	28010.0	38510.0	205.5	25.57	33.91	708.9	48.1	.0816	1.00460
1300.0	9.492	.3388	29290.0	40210.0	206.8	25.77	34.11	722.0	49.4	.0841	1.00442
1350.0	9.144	.3264	30590.0	41920.0	208.1	25.96	34.29	734.8	50.6	.0866	1.00426
1400.0	8.820	.3148	31890.0	43640.0	209.4	26.13	34.46	747.4	51.8	.0890	1.00410
1450.0	8.518	.3041	33200.0	45370.0	210.6	26.30	34.62	759.8	53.0	.0914	1.00396
1500.0	8.237	.2940	34520.0	47100.0	211.8	26.45	34.77	772.0	54.1	.0938	1.00383
1550.0	7.973	.2846	35850.0	48850.0	212.9	26.59	34.92	784.0	55.3	.0961	1.00370
1600.0	7.726	.2758	37180.0	50600.0	214.0	26.72	35.05	795.9	56.4	.0984	1.00359
1650.0	7.494	.2675	38520.0	52350.0	215.1	26.85	35.17	807.6	57.5	.101	1.00348
1700.0	7.275	.2597	39860.0	54110.0	216.2	26.97	35.29	819.1	58.7	.103	1.00337
1750.0	7.069	.2523	41220.0	55880.0	217.2	27.08	35.40	830.4	59.8	.105	1.00328
1800.0	6.874	.2454	42570.0	57650.0	218.2	27.18	35.50	841.7	60.8	.107	1.00318
1850.0	6.689	.2388	43930.0	59430.0	219.2	27.28	35.59	852.7	61.9	.110	1.00310
1900.0	6.515	.2325	45300.0	61210.0	220.1	27.37	35.68	863.7	63.0	.112	1.00301
					3.75 MPa isobar						
63.97[a]	869.2	31.03	–4199.0	–4078.0	68.21	28.89	54.41	1275.0	280.0	.152	1.46934
80.0	805.4	28.75	–3309.0	–3179.0	80.75	28.55	56.65	912.0	145.0	.133	1.43152
100.0	706.1	25.20	–2166.0	–2017.0	93.69	26.64	60.80	657.9	78.9	.103	1.37336

Thermophysical properties of nitrogen—Continued

T K	Density kg/m³	Density mol/dm³	E J/mol	H J/mol	S J/(mol·K)	C_v J/(mol·K)	C_p J/(mol·K)	Sound m/s	Visc. μPa·s	Therm. W/(m·K)	Diel. const.
110.0	643.7	22.98	−1544.0	−1381.0	99.75	26.33	67.42	538.9	58.9	.0869	1.33730
115.0	606.0	21.63	−1202.0	−1029.0	102.9	26.31	74.01	475.0	50.2	.0783	1.31573
120.0	559.3	19.96	−817.4	−629.6	106.3	26.45	87.68	401.9	41.8	.0695	1.28931
122.0	535.8	19.12	−640.2	−444.1	107.8	26.60	98.81	367.5	38.2	.0661	1.27611
124.0	506.4	18.08	−435.4	−228.0	109.6	26.89	119.9	327.4	34.3	.0627	1.25977
125.0	487.8	17.41	−314.2	−98.8	110.6	27.13	140.2	303.8	32.0	.0610	1.24945
126.0	463.9	16.56	−168.1	58.4	111.9	27.51	179.0	276.3	29.4	.0594	1.23629
126.5	448.2	16.00	−78.1	156.3	112.6	27.78	216.1	260.3	27.8	.0586	1.22775
127.0	427.9	15.27	34.0	279.6	113.6	28.18	284.7	241.9	26.0	.0580	1.21669
127.2	417.6	14.90	89.2	340.8	114.1	28.39	330.6	233.7	25.1	.0579	1.21109
127.4	405.2	14.46	153.8	413.1	114.7	28.65	396.0	225.0	24.0	.0579	1.20442
127.6	390.1	13.92	231.8	501.1	115.3	28.96	489.4	215.7	22.9	.0581	1.19628
128.0	348.4	12.44	443.6	745.1	117.3	29.73	723.9	197.6	20.0	.0596	1.17409
128.2	323.8	11.56	570.9	895.3	118.4	30.04	762.6	191.0	18.5	.0588	1.16112
128.4	300.6	10.73	695.1	1045.0	119.6	30.18	718.9	186.9	17.3	.0558	1.14897
129.0	251.9	8.993	976.1	1393.0	122.3	29.88	446.9	183.2	15.1	.0449	1.12381
129.5	229.4	8.188	1121.0	1579.0	123.7	29.42	310.8	183.7	14.2	.0398	1.11229
130.0	214.4	7.654	1224.0	1714.0	124.8	28.98	238.8	185.0	13.7	.0367	1.10469
130.5	203.4	7.260	1306.0	1822.0	125.6	28.60	196.1	186.5	13.4	.0344	1.09912
131.0	194.7	6.951	1373.0	1913.0	126.3	28.25	168.2	188.0	13.1	.0328	1.09475
132.0	181.6	6.481	1484.0	2062.0	127.4	27.66	134.0	191.0	12.8	.0304	1.08814
134.0	163.8	5.847	1651.0	2292.0	129.2	26.75	100.2	196.6	12.4	.0275	1.07927
136.0	151.7	5.413	1781.0	2474.0	130.5	26.05	83.28	201.6	12.2	.0257	1.07323
140.0	135.0	4.820	1990.0	2768.0	132.6	25.04	65.95	210.4	12.0	.0237	1.06501
145.0	121.2	4.325	2201.0	3068.0	134.8	24.17	55.36	219.9	12.0	.0223	1.05819
150.0	111.1	3.967	2384.0	3329.0	136.5	23.56	49.38	228.2	12.1	.0215	1.05328
152.0	107.8	3.848	2451.0	3426.0	137.2	23.36	47.65	231.2	12.1	.0214	1.05165
154.0	104.8	3.739	2517.0	3520.0	137.8	23.18	46.18	234.2	12.2	.0212	1.05016
156.0	102.0	3.639	2580.0	3611.0	138.4	23.02	44.90	237.1	12.2	.0211	1.04880
158.0	99.37	3.547	2642.0	3699.0	138.9	22.88	43.79	239.9	12.3	.0210	1.04754
160.0	96.97	3.461	2702.0	3786.0	139.5	22.75	42.81	242.5	12.4	.0210	1.04637
162.0	94.72	3.381	2762.0	3871.0	140.0	22.64	41.94	245.2	12.4	.0209	1.04528
164.0	92.62	3.306	2820.0	3954.0	140.5	22.53	41.16	247.7	12.5	.0209	1.04426
166.0	90.65	3.236	2876.0	4035.0	141.0	22.43	40.47	250.2	12.6	.0209	1.04330
168.0	88.79	3.169	2932.0	4116.0	141.5	22.34	39.84	252.6	12.7	.0210	1.04239
170.0	87.02	3.106	2988.0	4195.0	142.0	22.26	39.27	255.0	12.7	.0210	1.04154
172.0	85.36	3.047	3042.0	4273.0	142.4	22.18	38.75	257.3	12.8	.0210	1.04073
174.0	83.77	2.990	3096.0	4350.0	142.9	22.12	38.28	259.6	12.9	.0211	1.03996
180.0	79.45	2.836	3253.0	4576.0	144.1	21.94	37.07	266.2	13.2	.0213	1.03787
185.0	76.26	2.722	3381.0	4759.0	145.1	21.81	36.26	271.4	13.4	.0215	1.03633
190.0	73.38	2.619	3507.0	4939.0	146.1	21.71	35.58	276.4	13.6	.0217	1.03494
195.0	70.76	2.526	3630.0	5115.0	147.0	21.62	35.00	281.2	13.8	.0220	1.03367
200.0	68.36	2.440	3752.0	5289.0	147.9	21.55	34.50	285.9	14.1	.0222	1.03251
205.0	66.15	2.361	3872.0	5460.0	148.7	21.48	34.07	290.4	14.3	.0225	1.03145
210.0	64.11	2.288	3991.0	5629.0	149.6	21.43	33.70	294.8	14.5	.0227	1.03047
215.0	62.21	2.220	4108.0	5797.0	150.3	21.38	33.37	299.1	14.8	.0230	1.02955
220.0	60.44	2.157	4225.0	5963.0	151.1	21.34	33.07	303.3	15.0	.0233	1.02870
225.0	58.78	2.098	4341.0	6128.0	151.8	21.30	32.81	307.4	15.2	.0236	1.02790
230.0	57.23	2.043	4455.0	6291.0	152.6	21.27	32.58	311.4	15.5	.0239	1.02716
235.0	55.76	1.990	4570.0	6454.0	153.3	21.24	32.37	315.3	15.7	.0242	1.02646
240.0	54.38	1.941	4683.0	6615.0	153.9	21.21	32.18	319.1	15.9	.0245	1.02579
250.0	51.85	1.851	4909.0	6935.0	155.2	21.16	31.84	326.6	16.4	.0250	1.02458
260.0	49.56	1.769	5132.0	7252.0	156.5	21.13	31.57	333.7	16.8	.0256	1.02348
270.0	47.49	1.695	5354.0	7567.0	157.7	21.10	31.33	340.7	17.3	.0262	1.02249
280.0	45.60	1.628	5575.0	7879.0	158.8	21.07	31.13	347.4	17.7	.0268	1.02159
290.0	43.87	1.566	5794.0	8189.0	159.9	21.05	30.96	354.0	18.2	.0274	1.02076
300.0	42.27	1.509	6013.0	8498.0	161.0	21.04	30.81	360.4	18.6	.0280	1.02000
310.0	40.80	1.456	6231.0	8806.0	162.0	21.03	30.68	366.6	19.0	.0286	1.01929
320.0	39.43	1.408	6448.0	9112.0	162.9	21.02	30.57	372.7	19.5	.0292	1.01864
330.0	38.16	1.362	6664.0	9417.0	163.9	21.01	30.47	378.6	19.9	.0298	1.01803
340.0	36.98	1.320	6880.0	9721.0	164.8	21.01	30.38	384.4	20.3	.0304	1.01747
350.0	35.86	1.280	7095.0	10020.0	165.7	21.01	30.31	390.1	20.7	.0310	1.01694
360.0	34.82	1.243	7310.0	10330.0	166.5	21.01	30.25	395.7	21.1	.0316	1.01644

Thermophysical properties of nitrogen—Continued

T K	Density kg/m³	Density mol/dm³	E J/mol	H J/mol	S J/(mol·K)	C_v J/(mol·K)	C_p J/(mol·K)	Sound m/s	Visc. μPa·s	Therm. W/(m·K)	Diel. const.
370.0	33.84	1.208	7525.0	10630.0	167.3	21.02	30.19	401.1	21.5	.0322	1.01597
380.0	32.92	1.175	7740.0	10930.0	168.1	21.03	30.15	406.5	21.9	.0328	1.01553
390.0	32.05	1.144	7954.0	11230.0	168.9	21.04	30.11	411.8	22.3	.0334	1.01512
400.0	31.22	1.114	8168.0	11530.0	169.7	21.05	30.07	417.0	22.7	.0339	1.01472
420.0	29.70	1.060	8597.0	12130.0	171.2	21.09	30.03	427.1	23.5	.0351	1.01400
440.0	28.32	1.011	9025.0	12730.0	172.6	21.13	30.01	436.9	24.2	.0363	1.01334
460.0	27.07	.9662	9454.0	13330.0	173.9	21.19	30.01	446.4	25.0	.0375	1.01275
480.0	25.93	.9254	9883.0	13940.0	175.2	21.26	30.02	455.6	25.7	.0386	1.01221
500.0	24.88	.8881	10310.0	14540.0	176.4	21.33	30.05	464.6	26.4	.0398	1.01171
520.0	23.92	.8537	10740.0	15140.0	177.6	21.42	30.10	473.3	27.1	.0410	1.01125
540.0	23.03	.8219	11180.0	15740.0	178.7	21.51	30.16	481.8	27.8	.0421	1.01083
560.0	22.20	.7925	11610.0	16340.0	179.8	21.61	30.23	490.1	28.5	.0433	1.01044
580.0	21.44	.7651	12050.0	16950.0	180.9	21.72	30.31	498.3	29.2	.0445	1.01007
600.0	20.72	.7397	12490.0	17560.0	181.9	21.83	30.40	506.2	29.8	.0456	1.00974
620.0	20.05	.7158	12930.0	18170.0	182.9	21.95	30.50	513.9	30.5	.0468	1.00942
640.0	19.43	.6935	13370.0	18780.0	183.9	22.07	30.60	521.6	31.1	.0479	1.00912
660.0	18.84	.6726	13810.0	19390.0	184.8	22.20	30.71	529.0	31.8	.0491	1.00884
680.0	18.29	.6529	14260.0	20000.0	185.7	22.32	30.82	536.3	32.4	.0503	1.00858
700.0	17.77	.6343	14710.0	20620.0	186.6	22.45	30.94	543.5	33.0	.0514	1.00834
720.0	17.28	.6168	15160.0	21240.0	187.5	22.59	31.06	550.6	33.6	.0526	1.00810
740.0	16.82	.6003	15620.0	21860.0	188.3	22.72	31.18	557.5	34.3	.0537	1.00788
760.0	16.38	.5846	16070.0	22490.0	189.2	22.85	31.30	564.4	34.9	.0549	1.00768
780.0	15.96	.5697	16530.0	23120.0	190.0	22.98	31.43	571.1	35.5	.0560	1.00748
800.0	15.56	.5555	17000.0	23750.0	190.8	23.12	31.55	577.7	36.0	.0572	1.00729
820.0	15.19	.5421	17460.0	24380.0	191.6	23.25	31.67	584.3	36.6	.0583	1.00711
840.0	14.83	.5293	17930.0	25010.0	192.3	23.38	31.79	590.7	37.2	.0594	1.00694
860.0	14.49	.5171	18400.0	25650.0	193.1	23.51	31.92	597.1	37.8	.0606	1.00678
880.0	14.16	.5054	18870.0	26290.0	193.8	23.63	32.04	603.4	38.4	.0617	1.00663
900.0	13.85	.4943	19350.0	26930.0	194.5	23.76	32.16	609.6	38.9	.0628	1.00648
920.0	13.55	.4836	19820.0	27580.0	195.3	23.88	32.27	615.8	39.5	.0639	1.00634
950.0	13.13	.4685	20540.0	28550.0	196.3	24.06	32.45	624.9	40.3	.0656	1.00614
1000.0	12.48	.4453	21760.0	30180.0	198.0	24.35	32.73	639.7	41.7	.0684	1.00583
1050.0	11.89	.4243	22980.0	31820.0	199.6	24.63	32.99	654.2	43.0	.0711	1.00555
1100.0	11.35	.4052	24220.0	33480.0	201.1	24.89	33.24	668.3	44.3	.0738	1.00530
1150.0	10.86	.3877	25470.0	35140.0	202.6	25.13	33.48	682.2	45.6	.0764	1.00507
1200.0	10.41	.3717	26730.0	36820.0	204.0	25.36	33.70	695.7	46.9	.0790	1.00485
1250.0	10.00	.3570	28010.0	38510.0	205.4	25.57	33.91	709.0	48.1	.0816	1.00466
1300.0	9.619	.3434	29290.0	40210.0	206.7	25.77	34.11	722.1	49.4	.0841	1.00448
1350.0	9.266	.3307	30590.0	41930.0	208.0	25.96	34.29	734.9	50.6	.0866	1.00431
1400.0	8.938	.3190	31890.0	43640.0	209.3	26.13	34.46	747.5	51.8	.0890	1.00416
1450.0	8.633	.3081	33200.0	45370.0	210.5	26.30	34.62	759.9	53.0	.0914	1.00401
1500.0	8.347	.2979	34520.0	47110.0	211.7	26.45	34.78	772.1	54.1	.0938	1.00388
1550.0	8.080	.2884	35850.0	48850.0	212.8	26.59	34.92	784.1	55.3	.0961	1.00375
1600.0	7.830	.2795	37180.0	50600.0	213.9	26.72	35.05	796.0	56.4	.0985	1.00363
1650.0	7.594	.2711	38520.0	52350.0	215.0	26.85	35.17	807.7	57.5	.101	1.00352
1700.0	7.373	.2632	39860.0	54110.0	216.0	26.97	35.29	819.2	58.7	.103	1.00342
1750.0	7.164	.2557	41220.0	55880.0	217.1	27.08	35.40	830.5	59.8	.105	1.00332
1800.0	6.966	.2486	42570.0	57650.0	218.1	27.18	35.50	841.7	60.9	.107	1.00323
1850.0	6.779	.2420	43930.0	59430.0	219.0	27.28	35.59	852.8	61.9	.110	1.00314
1900.0	6.602	.2357	45300.0	61210.0	220.0	27.37	35.68	863.8	63.0	.112	1.00305
				3.80 MPa isobar							
63.98[a]	869.2	31.03	–4199.0	–4077.0	68.21	28.92	54.40	1274.0	280.0	.152	1.46935
80.0	805.5	28.75	–3310.0	–3178.0	80.74	28.57	56.63	912.2	145.0	.133	1.43159
100.0	706.4	25.21	–2168.0	–2017.0	93.67	26.65	60.75	658.6	79.0	.103	1.37351
110.0	644.2	22.99	–1546.0	–1381.0	99.73	26.34	67.29	540.0	59.0	.0871	1.33755
115.0	606.6	21.65	–1206.0	–1030.0	102.9	26.31	73.77	476.3	50.4	.0785	1.31608
120.0	560.3	20.00	–822.8	–632.8	106.2	26.44	87.04	403.9	41.9	.0697	1.28988
122.0	537.1	19.17	–647.3	–449.1	107.7	26.59	97.65	370.0	38.4	.0662	1.27686
124.0	508.5	18.15	–445.8	–236.4	109.5	26.86	117.3	330.7	34.5	.0629	1.26090
125.0	490.5	17.51	–327.8	–110.8	110.5	27.08	135.5	307.9	32.3	.0612	1.25095
126.0	467.9	16.70	–188.1	39.4	111.7	27.42	168.6	281.6	29.8	.0596	1.23851

Thermophysical properties of nitrogen—Continued

T K	Density kg/m^3	Density mol/dm^3	E J/mol	H J/mol	S J/(mol·K)	C_v J/(mol·K)	C_p J/(mol·K)	Sound m/s	Visc. μPa·s	Therm. W/(m·K)	Diel. const.
126.5	453.6	16.19	−104.2	130.5	112.4	27.67	198.1	266.5	28.4	.0588	1.23068
127.0	435.7	15.55	−3.6	240.8	113.3	28.00	247.9	249.6	26.7	.0581	1.22093
127.5	411.4	14.68	127.2	386.0	114.4	28.49	344.9	230.2	24.6	.0576	1.20775
128.0	374.4	13.36	318.6	602.9	116.1	29.23	541.1	208.7	21.7	.0579	1.18787
128.2	354.5	12.65	419.9	720.2	117.0	29.57	627.4	200.7	20.4	.0582	1.17732
128.4	333.0	11.89	531.3	851.0	118.0	29.87	671.2	194.3	19.1	.0577	1.16594
128.6	311.8	11.13	643.5	984.9	119.1	30.06	660.0	189.8	17.9	.0557	1.15483
129.0	276.3	9.862	841.8	1227.0	121.0	30.05	538.0	185.3	16.2	.0495	1.13634
129.5	246.6	8.801	1022.0	1453.0	122.7	29.69	377.1	184.3	14.9	.0428	1.12105
130.0	227.4	8.117	1147.0	1615.0	124.0	29.26	279.9	185.0	14.2	.0389	1.11128
130.5	213.9	7.635	1242.0	1740.0	124.9	28.85	223.1	186.2	13.7	.0362	1.10442
131.0	203.6	7.267	1319.0	1842.0	125.7	28.49	187.1	187.6	13.4	.0342	1.09921
132.0	188.5	6.727	1441.0	2005.0	127.0	27.87	144.8	190.6	13.0	.0314	1.09159
133.0	177.4	6.333	1537.0	2137.0	127.9	27.35	120.8	193.4	12.7	.0296	1.08606
134.0	168.8	6.024	1619.0	2250.0	128.8	26.91	105.3	196.2	12.5	.0282	1.08174
136.0	155.7	5.556	1756.0	2439.0	130.2	26.18	86.27	201.2	12.3	.0262	1.07522
138.0	145.9	5.206	1870.0	2600.0	131.4	25.60	74.98	205.8	12.2	.0249	1.07036
140.0	138.0	4.927	1971.0	2742.0	132.4	25.13	67.43	210.1	12.1	.0240	1.06649
145.0	123.5	4.410	2186.0	3048.0	134.5	24.23	56.19	219.6	12.0	.0225	1.05936
150.0	113.2	4.039	2371.0	3312.0	136.3	23.61	49.92	228.0	12.1	.0217	1.05427
152.0	109.7	3.916	2439.0	3410.0	137.0	23.40	48.12	231.1	12.2	.0215	1.05259
154.0	106.6	3.804	2505.0	3504.0	137.6	23.22	46.59	234.1	12.2	.0213	1.05105
156.0	103.7	3.701	2570.0	3596.0	138.2	23.06	45.27	236.9	12.3	.0212	1.04965
158.0	101.0	3.606	2632.0	3686.0	138.8	22.92	44.11	239.7	12.3	.0211	1.04835
160.0	98.57	3.518	2693.0	3773.0	139.3	22.78	43.10	242.4	12.4	.0211	1.04715
162.0	96.27	3.436	2752.0	3858.0	139.8	22.67	42.21	245.1	12.5	.0210	1.04603
164.0	94.12	3.359	2810.0	3942.0	140.3	22.56	41.41	247.6	12.5	.0210	1.04498
166.0	92.09	3.287	2868.0	4024.0	140.8	22.46	40.69	250.1	12.6	.0210	1.04400
168.0	90.19	3.219	2924.0	4104.0	141.3	22.37	40.05	252.6	12.7	.0211	1.04307
170.0	88.38	3.155	2979.0	4184.0	141.8	22.28	39.46	254.9	12.8	.0211	1.04220
172.0	86.68	3.094	3034.0	4262.0	142.3	22.21	38.93	257.3	12.9	.0211	1.04137
174.0	85.06	3.036	3088.0	4340.0	142.7	22.13	38.44	259.6	12.9	.0212	1.04058
180.0	80.64	2.878	3246.0	4566.0	144.0	21.95	37.21	266.1	13.2	.0214	1.03845
185.0	77.39	2.762	3375.0	4750.0	145.0	21.83	36.38	271.4	13.4	.0216	1.03687
190.0	74.46	2.658	3501.0	4930.0	146.0	21.72	35.68	276.4	13.6	.0218	1.03545
195.0	71.79	2.562	3624.0	5107.0	146.9	21.63	35.09	281.3	13.9	.0220	1.03416
200.0	69.34	2.475	3746.0	5282.0	147.8	21.56	34.59	285.9	14.1	.0223	1.03299
205.0	67.09	2.395	3867.0	5453.0	148.6	21.49	34.15	290.5	14.3	.0225	1.03190
210.0	65.01	2.321	3986.0	5623.0	149.4	21.44	33.77	294.9	14.6	.0228	1.03090
215.0	63.08	2.252	4103.0	5791.0	150.2	21.39	33.43	299.2	14.8	.0231	1.02997
220.0	61.28	2.187	4220.0	5957.0	151.0	21.34	33.13	303.4	15.0	.0234	1.02910
225.0	59.60	2.127	4336.0	6122.0	151.7	21.31	32.87	307.5	15.2	.0236	1.02829
230.0	58.02	2.071	4451.0	6286.0	152.4	21.27	32.63	311.5	15.5	.0239	1.02754
235.0	56.53	2.018	4566.0	6449.0	153.1	21.24	32.41	315.4	15.7	.0242	1.02682
240.0	55.13	1.968	4679.0	6610.0	153.8	21.21	32.22	319.2	15.9	.0245	1.02615
250.0	52.55	1.876	4905.0	6931.0	155.1	21.17	31.88	326.7	16.4	.0251	1.02491
260.0	50.23	1.793	5129.0	7248.0	156.4	21.13	31.60	333.8	16.8	.0257	1.02380
270.0	48.13	1.718	5351.0	7563.0	157.6	21.10	31.36	340.8	17.3	.0263	1.02279
280.0	46.21	1.649	5572.0	7875.0	158.7	21.08	31.16	347.5	17.7	.0269	1.02188
290.0	44.45	1.587	5791.0	8186.0	159.8	21.06	30.98	354.1	18.2	.0275	1.02104
300.0	42.84	1.529	6010.0	8495.0	160.8	21.04	30.83	360.5	18.6	.0281	1.02026
310.0	41.34	1.476	6228.0	8803.0	161.8	21.03	30.70	366.7	19.0	.0287	1.01955
320.0	39.96	1.426	6445.0	9109.0	162.8	21.02	30.59	372.8	19.5	.0292	1.01889
330.0	38.67	1.380	6662.0	9415.0	163.8	21.01	30.49	378.7	19.9	.0298	1.01827
340.0	37.47	1.337	6878.0	9719.0	164.7	21.01	30.40	384.5	20.3	.0304	1.01770
350.0	36.34	1.297	7093.0	10020.0	165.5	21.01	30.33	390.2	20.7	.0310	1.01716
360.0	35.28	1.259	7308.0	10330.0	166.4	21.01	30.26	395.8	21.1	.0316	1.01666
370.0	34.29	1.224	7523.0	10630.0	167.2	21.02	30.20	401.3	21.5	.0322	1.01618
380.0	33.35	1.190	7738.0	10930.0	168.0	21.03	30.16	406.7	21.9	.0328	1.01574
390.0	32.47	1.159	7952.0	11230.0	168.8	21.04	30.12	411.9	22.3	.0334	1.01532
400.0	31.63	1.129	8166.0	11530.0	169.6	21.05	30.08	417.1	22.7	.0340	1.01492
420.0	30.09	1.074	8595.0	12130.0	171.0	21.09	30.04	427.2	23.5	.0351	1.01418

Thermophysical properties of nitrogen—Continued

T K	Density kg/m^3	Density mol/dm^3	E J/mol	H J/mol	S J/(mol·K)	C_v J/(mol·K)	C_p J/(mol·K)	Sound m/s	Visc. μPa·s	Therm. W/(m·K)	Diel. const.
440.0	28.69	1.024	9023.0	12730.0	172.4	21.13	30.02	437.0	24.2	.0363	1.01352
460.0	27.42	.9789	9452.0	13330.0	173.8	21.19	30.01	446.5	25.0	.0375	1.01292
480.0	26.27	.9376	9881.0	13930.0	175.0	21.26	30.03	455.8	25.7	.0386	1.01237
500.0	25.21	.8997	10310.0	14540.0	176.3	21.33	30.06	464.7	26.4	.0398	1.01186
520.0	24.23	.8649	10740.0	15140.0	177.5	21.42	30.11	473.5	27.1	.0410	1.01140
540.0	23.33	.8327	11180.0	15740.0	178.6	21.51	30.17	482.0	27.8	.0421	1.01097
560.0	22.49	.8029	11610.0	16340.0	179.7	21.61	30.24	490.3	28.5	.0433	1.01058
580.0	21.72	.7752	12050.0	16950.0	180.8	21.72	30.32	498.4	29.2	.0445	1.01021
600.0	20.99	.7494	12490.0	17560.0	181.8	21.83	30.41	506.3	29.8	.0456	1.00986
620.0	20.32	.7252	12930.0	18170.0	182.8	21.95	30.50	514.1	30.5	.0468	1.00954
640.0	19.68	.7026	13370.0	18780.0	183.8	22.07	30.61	521.7	31.1	.0480	1.00924
660.0	19.09	.6814	13810.0	19390.0	184.7	22.20	30.71	529.1	31.8	.0491	1.00896
680.0	18.53	.6615	14260.0	20010.0	185.6	22.33	30.83	536.5	32.4	.0503	1.00870
700.0	18.00	.6427	14710.0	20620.0	186.5	22.46	30.94	543.6	33.0	.0514	1.00845
720.0	17.51	.6249	15160.0	21240.0	187.4	22.59	31.06	550.7	33.6	.0526	1.00821
740.0	17.04	.6081	15620.0	21870.0	188.2	22.72	31.18	557.6	34.3	.0537	1.00799
760.0	16.59	.5922	16070.0	22490.0	189.1	22.85	31.30	564.5	34.9	.0549	1.00778
780.0	16.17	.5772	16530.0	23120.0	189.9	22.98	31.43	571.2	35.5	.0560	1.00758
800.0	15.77	.5628	17000.0	23750.0	190.7	23.12	31.55	577.9	36.0	.0572	1.00739
820.0	15.39	.5492	17460.0	24380.0	191.5	23.25	31.67	584.4	36.6	.0583	1.00721
840.0	15.02	.5362	17930.0	25010.0	192.2	23.38	31.80	590.9	37.2	.0595	1.00703
860.0	14.68	.5239	18400.0	25650.0	193.0	23.51	31.92	597.2	37.8	.0606	1.00687
880.0	14.35	.5121	18870.0	26290.0	193.7	23.63	32.04	603.5	38.4	.0617	1.00671
900.0	14.03	.5008	19350.0	26930.0	194.4	23.76	32.16	609.7	38.9	.0628	1.00656
920.0	13.73	.4900	19820.0	27580.0	195.1	23.88	32.28	615.9	39.5	.0640	1.00642
950.0	13.30	.4747	20540.0	28550.0	196.2	24.06	32.45	625.0	40.3	.0656	1.00622
1000.0	12.64	.4511	21750.0	30180.0	197.8	24.35	32.73	639.8	41.7	.0684	1.00591
1050.0	12.04	.4299	22980.0	31820.0	199.5	24.63	32.99	654.3	43.0	.0711	1.00562
1100.0	11.50	.4105	24220.0	33480.0	201.0	24.89	33.25	668.4	44.3	.0738	1.00537
1150.0	11.01	.3928	25470.0	35150.0	202.5	25.13	33.48	682.3	45.6	.0764	1.00513
1200.0	10.55	.3766	26730.0	36830.0	203.9	25.36	33.70	695.8	46.9	.0790	1.00492
1250.0	10.13	.3617	28010.0	38520.0	205.3	25.57	33.91	709.1	48.1	.0816	1.00472
1300.0	9.746	.3479	29290.0	40220.0	206.6	25.77	34.11	722.2	49.4	.0841	1.00454
1350.0	9.388	.3351	30590.0	41930.0	207.9	25.96	34.29	735.0	50.6	.0866	1.00437
1400.0	9.056	.3232	31890.0	43650.0	209.2	26.13	34.46	747.6	51.8	.0890	1.00421
1450.0	8.747	.3122	33200.0	45370.0	210.4	26.30	34.63	760.0	53.0	.0914	1.00407
1500.0	8.457	.3019	34520.0	47110.0	211.6	26.45	34.78	772.2	54.1	.0938	1.00393
1550.0	8.187	.2922	35850.0	48850.0	212.7	26.59	34.92	784.2	55.3	.0962	1.00380
1600.0	7.933	.2832	37180.0	50600.0	213.8	26.72	35.05	796.1	56.4	.0985	1.00368
1650.0	7.695	.2747	38520.0	52350.0	214.9	26.85	35.17	807.8	57.5	.101	1.00357
1700.0	7.470	.2666	39860.0	54120.0	215.9	26.97	35.29	819.3	58.7	.103	1.00346
1750.0	7.258	.2591	41220.0	55880.0	217.0	27.08	35.40	830.6	59.8	.105	1.00336
1800.0	7.058	.2519	42570.0	57660.0	218.0	27.18	35.50	841.8	60.9	.107	1.00327
1850.0	6.869	.2452	43930.0	59430.0	218.9	27.28	35.59	852.9	61.9	.110	1.00318
1900.0	6.689	.2388	45300.0	61210.0	219.9	27.37	35.68	863.8	63.0	.112	1.00309
					3.85 MPa isobar						
63.99[a]	869.3	31.03	−4199.0	−4075.0	68.21	28.94	54.40	1274.0	280.0	.152	1.46936
80.0	805.6	28.76	−3311.0	−3177.0	80.73	28.58	56.62	912.5	145.0	.133	1.43166
100.0	706.7	25.22	−2169.0	−2017.0	93.66	26.66	60.70	659.3	79.1	.104	1.37367
110.0	644.6	23.01	−1549.0	−1382.0	99.71	26.34	67.17	541.0	59.2	.0872	1.33780
115.0	607.2	21.67	−1209.0	−1031.0	102.8	26.31	73.53	477.7	50.5	.0786	1.31643
120.0	561.3	20.04	−828.2	−636.0	106.2	26.43	86.42	405.9	42.1	.0699	1.29044
122.0	538.4	19.22	−654.2	−453.9	107.7	26.57	96.56	372.4	38.6	.0664	1.27760
124.0	510.4	18.22	−455.8	−244.5	109.4	26.83	114.9	333.9	34.8	.0631	1.26199
125.0	493.1	17.60	−340.7	−122.0	110.4	27.04	131.4	311.7	32.7	.0614	1.25237
126.0	471.6	16.83	−206.5	22.2	111.5	27.34	160.0	286.5	30.3	.0598	1.24055
127.0	442.3	15.79	−35.3	208.6	113.0	27.85	222.2	256.4	27.3	.0582	1.22450
127.5	421.5	15.05	78.9	334.8	114.0	28.25	289.3	238.7	25.4	.0575	1.21321
128.0	392.2	14.00	233.0	508.0	115.3	28.84	416.9	218.9	23.1	.0570	1.19744
128.5	349.8	12.49	451.6	759.9	117.3	29.59	580.5	200.2	20.1	.0571	1.17484
128.7	330.5	11.80	552.5	878.9	118.2	29.83	603.2	194.7	19.0	.0562	1.16464

Thermophysical properties of nitrogen—Continued

T K	Density kg/m^3	Density mol/dm^3	E J/mol	H J/mol	S J/(mol·K)	C_v J/(mol·K)	C_p J/(mol·K)	Sound m/s	Visc. μPa·s	Therm. W/(m·K)	Diel. const.
129.0	303.0	10.81	701.0	1057.0	119.6	30.01	573.9	189.4	17.5	.0531	1.15019
129.5	266.7	9.521	908.7	1313.0	121.6	29.88	444.1	185.8	15.8	.0459	1.13142
130.0	242.5	8.655	1060.0	1504.0	123.1	29.51	328.8	185.5	14.8	.0415	1.11896
130.5	225.8	8.059	1171.0	1649.0	124.2	29.10	255.7	186.3	14.2	.0382	1.11045
131.0	213.4	7.619	1259.0	1765.0	125.1	28.73	209.7	187.5	13.8	.0358	1.10419
132.0	195.9	6.992	1394.0	1945.0	126.4	28.07	157.3	190.2	13.2	.0326	1.09533
133.0	183.5	6.550	1499.0	2087.0	127.5	27.53	128.7	193.0	12.9	.0304	1.08911
134.0	174.0	6.210	1586.0	2206.0	128.4	27.07	110.8	195.8	12.7	.0289	1.08434
136.0	159.8	5.704	1729.0	2404.0	129.9	26.31	89.49	200.9	12.4	.0267	1.07728
138.0	149.4	5.331	1848.0	2570.0	131.1	25.71	77.12	205.5	12.2	.0253	1.07209
140.0	141.1	5.037	1951.0	2715.0	132.1	25.22	68.98	209.8	12.2	.0243	1.06801
145.0	126.0	4.496	2171.0	3027.0	134.3	24.30	57.04	219.4	12.1	.0227	1.06055
150.0	115.2	4.112	2358.0	3294.0	136.1	23.66	50.47	227.8	12.2	.0219	1.05527
152.0	111.6	3.985	2427.0	3393.0	136.8	23.45	48.60	230.9	12.2	.0217	1.05353
154.0	108.4	3.870	2494.0	3489.0	137.4	23.26	47.01	233.9	12.2	.0215	1.05195
156.0	105.5	3.764	2559.0	3582.0	138.0	23.10	45.64	236.8	12.3	.0214	1.05050
158.0	102.7	3.666	2622.0	3672.0	138.6	22.95	44.45	239.6	12.4	.0213	1.04917
160.0	100.2	3.576	2683.0	3759.0	139.1	22.82	43.40	242.3	12.4	.0212	1.04793
162.0	97.82	3.492	2743.0	3845.0	139.7	22.69	42.48	245.0	12.5	.0212	1.04678
164.0	95.62	3.413	2801.0	3929.0	140.2	22.58	41.66	247.6	12.6	.0211	1.04571
166.0	93.54	3.339	2859.0	4012.0	140.7	22.48	40.92	250.1	12.6	.0211	1.04470
168.0	91.59	3.269	2916.0	4093.0	141.2	22.39	40.26	252.5	12.7	.0211	1.04375
170.0	89.75	3.204	2971.0	4173.0	141.6	22.30	39.66	254.9	12.8	.0212	1.04286
172.0	88.00	3.141	3026.0	4252.0	142.1	22.23	39.11	257.2	12.9	.0212	1.04201
174.0	86.35	3.082	3080.0	4330.0	142.5	22.15	38.61	259.5	13.0	.0213	1.04121
180.0	81.84	2.921	3239.0	4557.0	143.8	21.97	37.35	266.1	13.2	.0214	1.03902
185.0	78.52	2.803	3368.0	4742.0	144.8	21.84	36.50	271.4	13.4	.0216	1.03742
190.0	75.53	2.696	3494.0	4922.0	145.8	21.74	35.79	276.4	13.7	.0218	1.03597
195.0	72.81	2.599	3618.0	5100.0	146.7	21.65	35.19	281.3	13.9	.0221	1.03466
200.0	70.32	2.510	3741.0	5274.0	147.6	21.57	34.67	286.0	14.1	.0223	1.03346
205.0	68.03	2.428	3861.0	5447.0	148.5	21.50	34.22	290.5	14.3	.0226	1.03236
210.0	65.92	2.353	3981.0	5617.0	149.3	21.44	33.83	294.9	14.6	.0229	1.03134
215.0	63.95	2.283	4099.0	5785.0	150.1	21.39	33.49	299.2	14.8	.0231	1.03039
220.0	62.12	2.217	4216.0	5952.0	150.8	21.35	33.19	303.4	15.0	.0234	1.02951
225.0	60.41	2.156	4332.0	6117.0	151.6	21.31	32.92	307.5	15.3	.0237	1.02869
230.0	58.81	2.099	4447.0	6281.0	152.3	21.28	32.68	311.5	15.5	.0240	1.02792
235.0	57.30	2.045	4561.0	6444.0	153.0	21.25	32.46	315.5	15.7	.0243	1.02719
240.0	55.87	1.994	4675.0	6606.0	153.7	21.22	32.26	319.3	16.0	.0245	1.02651
250.0	53.26	1.901	4901.0	6926.0	155.0	21.17	31.92	326.7	16.4	.0251	1.02525
260.0	50.90	1.817	5125.0	7244.0	156.2	21.14	31.63	333.9	16.9	.0257	1.02412
270.0	48.77	1.741	5348.0	7559.0	157.4	21.11	31.39	340.9	17.3	.0263	1.02310
280.0	46.82	1.671	5569.0	7872.0	158.6	21.08	31.19	347.6	17.8	.0269	1.02217
290.0	45.04	1.608	5788.0	8183.0	159.7	21.06	31.01	354.2	18.2	.0275	1.02132
300.0	43.40	1.549	6007.0	8492.0	160.7	21.04	30.85	360.6	18.6	.0281	1.02053
310.0	41.89	1.495	6225.0	8800.0	161.7	21.03	30.72	366.8	19.1	.0287	1.01981
320.0	40.48	1.445	6442.0	9107.0	162.7	21.02	30.61	372.9	19.5	.0293	1.01914
330.0	39.17	1.398	6659.0	9412.0	163.6	21.02	30.51	378.8	19.9	.0299	1.01851
340.0	37.95	1.355	6875.0	9717.0	164.5	21.01	30.42	384.7	20.3	.0305	1.01793
350.0	36.81	1.314	7091.0	10020.0	165.4	21.01	30.34	390.4	20.7	.0310	1.01739
360.0	35.74	1.276	7306.0	10320.0	166.3	21.02	30.27	395.9	21.1	.0316	1.01688
370.0	34.73	1.240	7521.0	10630.0	167.1	21.02	30.22	401.4	21.5	.0322	1.01640
380.0	33.79	1.206	7736.0	10930.0	167.9	21.03	30.17	406.8	21.9	.0328	1.01594
390.0	32.89	1.174	7950.0	11230.0	168.7	21.04	30.13	412.1	22.3	.0334	1.01552
400.0	32.04	1.144	8165.0	11530.0	169.5	21.05	30.10	417.3	22.7	.0340	1.01511
420.0	30.48	1.088	8593.0	12130.0	170.9	21.09	30.05	427.4	23.5	.0352	1.01437
440.0	29.06	1.037	9022.0	12730.0	172.3	21.14	30.02	437.2	24.2	.0363	1.01370
460.0	27.78	.9915	9451.0	13330.0	173.7	21.19	30.02	446.7	25.0	.0375	1.01309
480.0	26.61	.9497	9880.0	13930.0	174.9	21.26	30.04	455.9	25.7	.0387	1.01253
500.0	25.53	.9114	10310.0	14530.0	176.2	21.33	30.07	464.9	26.4	.0398	1.01202
520.0	24.54	.8761	10740.0	15140.0	177.3	21.42	30.11	473.6	27.1	.0410	1.01155
540.0	23.63	.8435	11180.0	15740.0	178.5	21.51	30.17	482.1	27.8	.0422	1.01111
560.0	22.78	.8133	11610.0	16340.0	179.6	21.61	30.24	490.4	28.5	.0433	1.01071
580.0	22.00	.7852	12050.0	16950.0	180.6	21.72	30.32	498.5	29.2	.0445	1.01034

Thermophysical properties of nitrogen—Continued

T K	Density kg/m³	Density mol/dm³	E J/mol	H J/mol	S J/(mol·K)	C_v J/(mol·K)	C_p J/(mol·K)	Sound m/s	Visc. μPa·s	Therm. W/(m·K)	Diel. const.
600.0	21.27	.7590	12480.0	17560.0	181.7	21.83	30.41	506.5	29.8	.0456	1.00999
620.0	20.58	.7346	12920.0	18170.0	182.7	21.95	30.51	514.2	30.5	.0468	1.00967
640.0	19.94	.7117	13370.0	18780.0	183.6	22.07	30.61	521.8	31.1	.0480	1.00936
660.0	19.34	.6902	13810.0	19390.0	184.6	22.20	30.72	529.3	31.8	.0491	1.00908
680.0	18.77	.6700	14260.0	20010.0	185.5	22.33	30.83	536.6	32.4	.0503	1.00881
700.0	18.24	.6510	14710.0	20620.0	186.4	22.46	30.95	543.8	33.0	.0514	1.00856
720.0	17.73	.6330	15160.0	21240.0	187.3	22.59	31.06	550.8	33.6	.0526	1.00832
740.0	17.26	.6160	15620.0	21870.0	188.1	22.72	31.19	557.8	34.3	.0538	1.00809
760.0	16.81	.5999	16070.0	22490.0	189.0	22.85	31.31	564.6	34.9	.0549	1.00788
780.0	16.38	.5846	16530.0	23120.0	189.8	22.98	31.43	571.3	35.5	.0560	1.00768
800.0	15.97	.5701	17000.0	23750.0	190.6	23.12	31.55	578.0	36.0	.0572	1.00748
820.0	15.59	.5563	17460.0	24380.0	191.4	23.25	31.68	584.5	36.6	.0583	1.00730
840.0	15.22	.5432	17930.0	25020.0	192.1	23.38	31.80	591.0	37.2	.0595	1.00713
860.0	14.87	.5307	18400.0	25650.0	192.9	23.51	31.92	597.4	37.8	.0606	1.00696
880.0	14.53	.5187	18870.0	26290.0	193.6	23.63	32.04	603.6	38.4	.0617	1.00680
900.0	14.21	.5073	19340.0	26930.0	194.3	23.76	32.16	609.9	38.9	.0629	1.00665
920.0	13.91	.4964	19820.0	27580.0	195.0	23.88	32.28	616.0	39.5	.0640	1.00650
950.0	13.47	.4808	20540.0	28550.0	196.1	24.06	32.45	625.1	40.3	.0656	1.00630
1000.0	12.80	.4570	21750.0	30180.0	197.7	24.35	32.73	639.9	41.7	.0684	1.00598
1050.0	12.20	.4354	22980.0	31820.0	199.3	24.63	33.00	654.4	43.0	.0711	1.00570
1100.0	11.65	.4158	24220.0	33480.0	200.9	24.89	33.25	668.6	44.3	.0738	1.00544
1150.0	11.15	.3979	25470.0	35150.0	202.4	25.13	33.48	682.4	45.6	.0764	1.00520
1200.0	10.69	.3815	26730.0	36830.0	203.8	25.36	33.71	695.9	46.9	.0790	1.00498
1250.0	10.26	.3664	28010.0	38520.0	205.2	25.57	33.91	709.2	48.1	.0816	1.00478
1300.0	9.873	.3524	29290.0	40220.0	206.5	25.77	34.11	722.3	49.4	.0841	1.00460
1350.0	9.511	.3395	30590.0	41930.0	207.8	25.96	34.29	735.1	50.6	.0866	1.00443
1400.0	9.174	.3275	31890.0	43650.0	209.1	26.13	34.47	747.7	51.8	.0890	1.00427
1450.0	8.860	.3163	33200.0	45370.0	210.3	26.30	34.63	760.1	53.0	.0914	1.00412
1500.0	8.568	.3058	34520.0	47110.0	211.4	26.45	34.78	772.3	54.1	.0938	1.00398
1550.0	8.294	.2960	35850.0	48850.0	212.6	26.59	34.92	784.3	55.3	.0962	1.00385
1600.0	8.037	.2869	37180.0	50600.0	213.7	26.72	35.05	796.2	56.4	.0985	1.00373
1650.0	7.795	.2782	38520.0	52360.0	214.8	26.85	35.17	807.8	57.5	.101	1.00362
1700.0	7.568	.2701	39860.0	54120.0	215.8	26.97	35.29	819.4	58.7	.103	1.00351
1750.0	7.353	.2625	41220.0	55880.0	216.9	27.08	35.40	830.7	59.8	.105	1.00341
1800.0	7.150	.2552	42570.0	57660.0	217.9	27.18	35.50	841.9	60.9	.107	1.00331
1850.0	6.959	.2484	43930.0	59430.0	218.8	27.28	35.59	853.0	61.9	.110	1.00322
1900.0	6.777	.2419	45300.0	61220.0	219.8	27.37	35.68	863.9	63.0	.112	1.00313
					3.90 MPa isobar						
64.00[a]	869.3	31.03	–4199.0	–4073.0	68.22	28.97	54.40	1273.0	280.0	.152	1.46937
80.0	805.7	28.76	–3312.0	–3176.0	80.72	28.59	56.60	912.7	145.0	.133	1.43173
100.0	706.9	25.23	–2171.0	–2016.0	93.64	26.66	60.65	660.0	79.2	.104	1.37382
110.0	645.0	23.02	–1551.0	–1382.0	99.68	26.34	67.05	542.0	59.3	.0873	1.33805
115.0	607.8	21.70	–1212.0	–1033.0	102.8	26.31	73.30	479.0	50.6	.0787	1.31678
120.0	562.3	20.07	–833.4	–639.1	106.1	26.42	85.82	407.8	42.3	.0700	1.29100
122.0	539.7	19.27	–661.0	–458.6	107.6	26.56	95.52	374.8	38.8	.0666	1.27833
124.0	512.3	18.29	–465.5	–252.2	109.3	26.80	112.7	337.0	35.0	.0633	1.26304
126.0	475.1	16.96	–223.5	6.5	111.4	27.28	152.8	291.1	30.6	.0600	1.24244
127.0	448.0	15.99	–62.8	181.1	112.8	27.72	203.2	262.6	27.9	.0584	1.22759
127.5	429.7	15.34	39.7	294.0	113.6	28.06	252.6	246.2	26.2	.0576	1.21763
128.0	405.4	14.47	170.4	439.9	114.8	28.54	339.6	228.0	24.1	.0569	1.20449
128.5	371.3	13.25	348.0	642.3	116.4	29.18	474.4	209.4	21.6	.0565	1.18621
129.0	328.5	11.73	571.4	904.0	118.4	29.78	547.6	195.3	18.9	.0549	1.16359
129.5	289.1	10.32	788.5	1166.0	120.4	29.93	486.1	188.7	16.8	.0486	1.14297
130.0	259.8	9.272	962.0	1383.0	122.1	29.70	379.0	186.7	15.5	.0440	1.12783
130.5	239.3	8.541	1093.0	1549.0	123.4	29.33	293.3	186.7	14.7	.0404	1.11732
131.0	224.4	8.010	1194.0	1681.0	124.4	28.96	236.2	187.5	14.2	.0376	1.10976
132.0	204.0	7.280	1345.0	1881.0	125.9	28.28	171.7	190.0	13.5	.0338	1.09939
133.0	190.0	6.781	1459.0	2034.0	127.1	27.71	137.7	192.7	13.1	.0314	1.09236
134.0	179.5	6.406	1552.0	2161.0	128.0	27.23	117.0	195.4	12.9	.0296	1.08708
136.0	164.1	5.857	1702.0	2368.0	129.5	26.44	92.96	200.5	12.5	.0273	1.07941

Thermophysical properties of nitrogen—Continued

T K	Density kg/m^3	Density mol/dm^3	E J/mol	H J/mol	S J/(mol·K)	C_v J/(mol·K)	C_p J/(mol·K)	Sound m/s	Visc. μPa·s	Therm. W/(m·K)	Diel. const.
138.0	153.0	5.460	1825.0	2539.0	130.8	25.82	79.39	205.2	12.3	.0258	1.07387
140.0	144.3	5.149	1931.0	2688.0	131.9	25.31	70.60	209.5	12.2	.0247	1.06956
145.0	128.4	4.584	2155.0	3006.0	134.1	24.37	57.91	219.2	12.2	.0230	1.06175
150.0	117.3	4.186	2345.0	3277.0	135.9	23.71	51.04	227.6	12.2	.0220	1.05628
152.0	113.6	4.055	2415.0	3377.0	136.6	23.49	49.09	230.8	12.2	.0218	1.05448
154.0	110.3	3.936	2483.0	3473.0	137.2	23.31	47.43	233.8	12.3	.0216	1.05286
156.0	107.2	3.827	2548.0	3567.0	137.8	23.14	46.01	236.7	12.3	.0215	1.05137
158.0	104.4	3.727	2611.0	3658.0	138.4	22.99	44.78	239.5	12.4	.0214	1.04999
160.0	101.8	3.634	2673.0	3746.0	139.0	22.85	43.71	242.2	12.5	.0213	1.04872
162.0	99.39	3.548	2733.0	3833.0	139.5	22.72	42.75	244.9	12.5	.0213	1.04755
164.0	97.13	3.467	2792.0	3917.0	140.0	22.61	41.91	247.5	12.6	.0212	1.04644
166.0	95.00	3.391	2850.0	4000.0	140.5	22.51	41.15	250.0	12.7	.0212	1.04541
168.0	93.01	3.320	2907.0	4082.0	141.0	22.41	40.47	252.4	12.8	.0212	1.04444
170.0	91.12	3.252	2963.0	4162.0	141.5	22.33	39.85	254.8	12.8	.0213	1.04352
172.0	89.34	3.189	3018.0	4241.0	141.9	22.25	39.29	257.2	12.9	.0213	1.04266
174.0	87.64	3.128	3073.0	4319.0	142.4	22.17	38.78	259.5	13.0	.0213	1.04184
180.0	83.04	2.964	3232.0	4548.0	143.7	21.98	37.49	266.1	13.2	.0215	1.03961
185.0	79.66	2.843	3362.0	4733.0	144.7	21.86	36.62	271.4	13.5	.0217	1.03797
190.0	76.61	2.735	3488.0	4914.0	145.7	21.75	35.89	276.4	13.7	.0219	1.03649
195.0	73.84	2.636	3613.0	5092.0	146.6	21.66	35.28	281.3	13.9	.0221	1.03515
200.0	71.31	2.545	3735.0	5267.0	147.5	21.58	34.75	286.0	14.1	.0224	1.03393
205.0	68.98	2.462	3856.0	5440.0	148.3	21.51	34.30	290.6	14.4	.0226	1.03281
210.0	66.83	2.385	3975.0	5610.0	149.2	21.45	33.90	295.0	14.6	.0229	1.03177
215.0	64.83	2.314	4094.0	5779.0	150.0	21.40	33.56	299.3	14.8	.0232	1.03081
220.0	62.97	2.248	4211.0	5946.0	150.7	21.36	33.25	303.5	15.0	.0234	1.02991
225.0	61.23	2.185	4327.0	6112.0	151.5	21.32	32.97	307.6	15.3	.0237	1.02908
230.0	59.60	2.127	4443.0	6276.0	152.2	21.28	32.73	311.6	15.5	.0240	1.02829
235.0	58.07	2.073	4557.0	6439.0	152.9	21.25	32.50	315.5	15.7	.0243	1.02756
240.0	56.62	2.021	4671.0	6601.0	153.6	21.23	32.30	319.4	16.0	.0246	1.02686
250.0	53.96	1.926	4897.0	6922.0	154.9	21.18	31.96	326.8	16.4	.0252	1.02559
260.0	51.57	1.841	5122.0	7240.0	156.1	21.14	31.67	334.0	16.9	.0258	1.02444
270.0	49.41	1.764	5344.0	7556.0	157.3	21.11	31.42	341.0	17.3	.0263	1.02341
280.0	47.44	1.693	5565.0	7869.0	158.5	21.08	31.21	347.8	17.8	.0269	1.02246
290.0	45.63	1.629	5785.0	8180.0	159.5	21.06	31.03	354.3	18.2	.0275	1.02160
300.0	43.97	1.569	6004.0	8490.0	160.6	21.05	30.88	360.7	18.6	.0281	1.02080
310.0	42.43	1.514	6222.0	8798.0	161.6	21.03	30.74	366.9	19.1	.0287	1.02007
320.0	41.01	1.464	6440.0	9104.0	162.6	21.02	30.62	373.0	19.5	.0293	1.01939
330.0	39.68	1.416	6657.0	9410.0	163.5	21.02	30.52	379.0	19.9	.0299	1.01875
340.0	38.44	1.372	6873.0	9715.0	164.4	21.02	30.43	384.8	20.3	.0305	1.01816
350.0	37.29	1.331	7089.0	10020.0	165.3	21.02	30.36	390.5	20.7	.0311	1.01761
360.0	36.20	1.292	7304.0	10320.0	166.2	21.02	30.29	396.1	21.1	.0317	1.01709
370.0	35.18	1.256	7519.0	10620.0	167.0	21.02	30.23	401.6	21.5	.0322	1.01661
380.0	34.22	1.221	7734.0	10930.0	167.8	21.03	30.18	406.9	21.9	.0328	1.01615
390.0	33.31	1.189	7948.0	11230.0	168.6	21.04	30.14	412.2	22.3	.0334	1.01572
400.0	32.45	1.158	8163.0	11530.0	169.3	21.06	30.11	417.4	22.7	.0340	1.01531
420.0	30.87	1.102	8591.0	12130.0	170.8	21.09	30.06	427.5	23.5	.0352	1.01455
440.0	29.43	1.051	9020.0	12730.0	172.2	21.14	30.03	437.3	24.2	.0363	1.01387
460.0	28.13	1.004	9449.0	13330.0	173.5	21.19	30.03	446.8	25.0	.0375	1.01325
480.0	26.95	.9618	9879.0	13930.0	174.8	21.26	30.04	456.0	25.7	.0387	1.01269
500.0	25.86	.9230	10310.0	14530.0	176.1	21.34	30.07	465.0	26.4	.0398	1.01217
520.0	24.86	.8873	10740.0	15140.0	177.2	21.42	30.12	473.7	27.1	.0410	1.01170
540.0	23.93	.8542	11170.0	15740.0	178.4	21.51	30.18	482.3	27.8	.0422	1.01126
560.0	23.08	.8237	11610.0	16340.0	179.5	21.61	30.25	490.6	28.5	.0433	1.01085
580.0	22.28	.7952	12050.0	16950.0	180.5	21.72	30.33	498.7	29.2	.0445	1.01047
600.0	21.54	.7687	12480.0	17560.0	181.6	21.83	30.42	506.6	29.8	.0457	1.01012
620.0	20.84	.7440	12920.0	18170.0	182.6	21.95	30.51	514.4	30.5	.0468	1.00979
640.0	20.19	.7208	13370.0	18780.0	183.5	22.07	30.61	522.0	31.1	.0480	1.00948
660.0	19.58	.6990	13810.0	19390.0	184.5	22.20	30.72	529.4	31.8	.0491	1.00919
680.0	19.01	.6786	14260.0	20010.0	185.4	22.33	30.83	536.7	32.4	.0503	1.00892
700.0	18.47	.6593	14710.0	20620.0	186.3	22.46	30.95	543.9	33.0	.0515	1.00867
720.0	17.96	.6411	15160.0	21240.0	187.2	22.59	31.07	551.0	33.7	.0526	1.00842
740.0	17.48	.6239	15620.0	21870.0	188.0	22.72	31.19	557.9	34.3	.0538	1.00820
760.0	17.02	.6076	16070.0	22490.0	188.9	22.85	31.31	564.7	34.9	.0549	1.00798

Thermophysical properties of nitrogen—Continued

T K	Density kg/m^3	Density mol/dm^3	E J/mol	H J/mol	S J/(mol·K)	C_v J/(mol·K)	C_p	Sound m/s	Visc. μPa·s	Therm. W/(m·K)	Diel. const.
780.0	16.59	.5921	16530.0	23120.0	189.7	22.99	31.43	571.5	35.5	.0561	1.00777
800.0	16.18	.5774	16990.0	23750.0	190.5	23.12	31.56	578.1	36.1	.0572	1.00758
820.0	15.79	.5635	17460.0	24380.0	191.2	23.25	31.68	584.7	36.6	.0583	1.00739
840.0	15.41	.5502	17930.0	25020.0	192.0	23.38	31.80	591.1	37.2	.0595	1.00722
860.0	15.06	.5375	18400.0	25650.0	192.8	23.51	31.92	597.5	37.8	.0606	1.00705
880.0	14.72	.5254	18870.0	26290.0	193.5	23.63	32.04	603.8	38.4	.0617	1.00689
900.0	14.39	.5138	19340.0	26930.0	194.2	23.76	32.16	610.0	38.9	.0629	1.00673
920.0	14.08	.5027	19820.0	27580.0	194.9	23.88	32.28	616.1	39.5	.0640	1.00659
950.0	13.64	.4870	20540.0	28550.0	196.0	24.06	32.45	625.2	40.3	.0656	1.00638
1000.0	12.97	.4629	21750.0	30180.0	197.6	24.35	32.73	640.1	41.7	.0684	1.00606
1050.0	12.36	.4410	22980.0	31820.0	199.2	24.63	33.00	654.5	43.0	.0711	1.00577
1100.0	11.80	.4212	24220.0	33480.0	200.8	24.89	33.25	668.7	44.3	.0738	1.00551
1150.0	11.29	.4030	25470.0	35150.0	202.3	25.13	33.48	682.5	45.6	.0764	1.00527
1200.0	10.83	.3864	26730.0	36830.0	203.7	25.36	33.71	696.1	46.9	.0790	1.00505
1250.0	10.40	.3711	28010.0	38520.0	205.1	25.57	33.92	709.3	48.1	.0816	1.00484
1300.0	10.000	.3569	29290.0	40220.0	206.4	25.77	34.11	722.4	49.4	.0841	1.00466
1350.0	9.633	.3438	30590.0	41930.0	207.7	25.96	34.29	735.2	50.6	.0866	1.00448
1400.0	9.292	.3317	31890.0	43650.0	208.9	26.13	34.47	747.8	51.8	.0890	1.00432
1450.0	8.974	.3203	33200.0	45380.0	210.2	26.30	34.63	760.2	53.0	.0915	1.00417
1500.0	8.678	.3097	34520.0	47110.0	211.3	26.45	34.78	772.4	54.1	.0938	1.00403
1550.0	8.400	.2998	35850.0	48850.0	212.5	26.59	34.92	784.4	55.3	.0962	1.00390
1600.0	8.140	.2905	37180.0	50600.0	213.6	26.73	35.05	796.3	56.4	.0985	1.00378
1650.0	7.895	.2818	38520.0	52360.0	214.7	26.85	35.17	807.9	57.5	.101	1.00366
1700.0	7.665	.2736	39860.0	54120.0	215.7	26.97	35.29	819.4	58.7	.103	1.00355
1750.0	7.448	.2658	41220.0	55890.0	216.7	27.08	35.40	830.8	59.8	.105	1.00345
1800.0	7.242	.2585	42570.0	57660.0	217.7	27.18	35.50	842.0	60.9	.107	1.00335
1850.0	7.048	.2516	43930.0	59440.0	218.7	27.28	35.59	853.1	61.9	.110	1.00326
1900.0	6.864	.2450	45300.0	61220.0	219.7	27.37	35.68	864.0	63.0	.112	1.00318
				4.00 MPa isobar							
64.02[a]	869.3	31.03	−4198.0	−4070.0	68.22	29.03	54.39	1272.0	279.0	.152	1.46939
80.0	806.0	28.77	−3313.0	−3174.0	80.70	28.62	56.57	913.2	146.0	.134	1.43187
100.0	707.4	25.25	−2174.0	−2015.0	93.61	26.68	60.55	661.3	79.4	.104	1.37412
110.0	645.9	23.05	−1556.0	−1383.0	99.64	26.35	66.80	544.1	59.5	.0875	1.33854
115.0	609.0	21.74	−1219.0	−1035.0	102.7	26.31	72.85	481.7	50.9	.0790	1.31746
120.0	564.2	20.14	−843.7	−645.1	106.0	26.41	84.69	411.6	42.6	.0704	1.29208
122.0	542.2	19.35	−674.1	−467.5	107.5	26.53	93.60	379.4	39.2	.0670	1.27973
124.0	515.9	18.41	−483.8	−266.6	109.1	26.75	108.8	343.0	35.5	.0637	1.26502
126.0	481.2	17.18	−254.3	−21.4	111.1	27.16	141.2	299.7	31.3	.0604	1.24584
127.0	457.5	16.33	−109.3	135.6	112.3	27.52	176.8	273.7	28.8	.0588	1.23279
127.5	442.4	15.79	−22.2	231.1	113.1	27.78	207.2	259.2	27.4	.0580	1.22458
128.0	423.9	15.13	81.3	345.6	114.0	28.12	254.5	243.4	25.7	.0571	1.21453
128.5	400.1	14.28	210.0	490.1	115.1	28.57	328.5	226.6	23.7	.0562	1.20164
129.0	368.9	13.17	373.9	677.6	116.6	29.12	420.5	210.4	21.4	.0553	1.18496
129.5	332.7	11.88	565.4	902.2	118.3	29.61	463.8	198.4	19.2	.0520	1.16579
130.0	298.6	10.66	753.4	1129.0	120.1	29.78	432.2	191.9	17.3	.0484	1.14793
131.0	250.2	8.932	1045.0	1493.0	122.9	29.35	296.3	188.7	15.2	.0414	1.12293
132.0	222.3	7.934	1235.0	1739.0	124.7	28.69	206.5	190.0	14.2	.0366	1.10867
133.0	204.3	7.292	1371.0	1920.0	126.1	28.08	158.9	192.3	13.6	.0335	1.09956
134.0	191.3	6.830	1478.0	2064.0	127.2	27.56	131.2	194.9	13.2	.0313	1.09304
136.0	173.2	6.181	1645.0	2292.0	128.9	26.71	100.7	199.9	12.8	.0285	1.08394
138.0	160.5	5.728	1777.0	2476.0	130.2	26.04	84.34	204.6	12.6	.0266	1.07761
140.0	150.8	5.381	1890.0	2633.0	131.3	25.50	74.08	209.0	12.4	.0254	1.07278
145.0	133.4	4.763	2124.0	2963.0	133.7	24.50	59.75	218.8	12.3	.0234	1.06422
150.0	121.5	4.336	2319.0	3242.0	135.5	23.81	52.20	227.3	12.3	.0224	1.05834
152.0	117.6	4.196	2391.0	3344.0	136.2	23.59	50.09	230.5	12.3	.0221	1.05642
154.0	114.0	4.070	2459.0	3442.0	136.9	23.39	48.31	233.5	12.4	.0219	1.05469
156.0	110.8	3.955	2526.0	3537.0	137.5	23.21	46.79	236.5	12.4	.0217	1.05311
158.0	107.8	3.849	2590.0	3630.0	138.1	23.06	45.47	239.3	12.5	.0216	1.05166
160.0	105.1	3.751	2653.0	3719.0	138.6	22.91	44.33	242.1	12.5	.0215	1.05032
162.0	102.5	3.660	2714.0	3807.0	139.2	22.78	43.31	244.7	12.6	.0215	1.04908
164.0	100.2	3.575	2774.0	3893.0	139.7	22.67	42.42	247.3	12.7	.0214	1.04792
166.0	97.94	3.496	2833.0	3977.0	140.2	22.56	41.62	249.9	12.7	.0214	1.04684

Thermophysical properties of nitrogen—Continued

T K	Density kg/m^3	Density mol/dm^3	E J/mol	H J/mol	S J/(mol·K)	C_v J/(mol·K)	C_p J/(mol·K)	Sound m/s	Visc. μPa·s	Therm. W/(m·K)	Diel. const.
168.0	95.85	3.421	2890.0	4059.0	140.7	22.46	40.90	252.3	12.8	.0214	1.04582
170.0	93.88	3.351	2947.0	4140.0	141.2	22.37	40.25	254.8	12.9	.0214	1.04486
172.0	92.02	3.284	3002.0	4220.0	141.6	22.29	39.66	257.1	13.0	.0215	1.04396
174.0	90.25	3.221	3057.0	4299.0	142.1	22.21	39.12	259.4	13.0	.0215	1.04310
176.0	88.58	3.162	3112.0	4377.0	142.5	22.14	38.63	261.7	13.1	.0215	1.04229
180.0	85.46	3.050	3218.0	4529.0	143.4	22.02	37.77	266.1	13.3	.0217	1.04077
185.0	81.94	2.925	3348.0	4716.0	144.4	21.88	36.86	271.4	13.5	.0218	1.03907
190.0	78.77	2.812	3476.0	4898.0	145.4	21.77	36.11	276.5	13.7	.0220	1.03754
195.0	75.90	2.709	3601.0	5077.0	146.3	21.68	35.47	281.4	13.9	.0223	1.03615
200.0	73.28	2.616	3724.0	5253.0	147.2	21.60	34.92	286.1	14.2	.0225	1.03488
205.0	70.87	2.530	3845.0	5427.0	148.1	21.53	34.45	290.7	14.4	.0228	1.03372
210.0	68.64	2.450	3965.0	5598.0	148.9	21.47	34.04	295.1	14.6	.0230	1.03265
215.0	66.58	2.377	4084.0	5767.0	149.7	21.42	33.68	299.4	14.9	.0233	1.03165
220.0	64.66	2.308	4202.0	5935.0	150.5	21.37	33.36	303.6	15.1	.0235	1.03073
225.0	62.86	2.244	4318.0	6101.0	151.2	21.33	33.08	307.8	15.3	.0238	1.02986
230.0	61.18	2.184	4434.0	6266.0	151.9	21.30	32.82	311.8	15.5	.0241	1.02905
235.0	59.60	2.127	4549.0	6429.0	152.6	21.26	32.60	315.7	15.8	.0244	1.02829
240.0	58.11	2.074	4663.0	6592.0	153.3	21.24	32.39	319.6	16.0	.0247	1.02758
245.0	56.71	2.024	4777.0	6753.0	154.0	21.21	32.20	323.3	16.2	.0250	1.02690
250.0	55.38	1.977	4890.0	6914.0	154.6	21.19	32.03	327.0	16.4	.0252	1.02627
260.0	52.92	1.889	5115.0	7232.0	155.9	21.15	31.73	334.2	16.9	.0258	1.02509
270.0	50.69	1.809	5338.0	7548.0	157.1	21.12	31.48	341.2	17.3	.0264	1.02402
280.0	48.66	1.737	5559.0	7862.0	158.2	21.09	31.27	348.0	17.8	.0270	1.02305
290.0	46.80	1.671	5779.0	8174.0	159.3	21.07	31.08	354.6	18.2	.0276	1.02216
300.0	45.09	1.610	5999.0	8484.0	160.4	21.05	30.92	361.0	18.7	.0282	1.02134
310.0	43.51	1.553	6217.0	8792.0	161.4	21.04	30.78	367.2	19.1	.0288	1.02058
320.0	42.05	1.501	6435.0	9100.0	162.4	21.03	30.66	373.3	19.5	.0294	1.01989
330.0	40.69	1.452	6652.0	9406.0	163.3	21.02	30.56	379.2	19.9	.0299	1.01924
340.0	39.42	1.407	6868.0	9711.0	164.2	21.02	30.47	385.1	20.3	.0305	1.01863
350.0	38.23	1.365	7084.0	10020.0	165.1	21.02	30.39	390.8	20.7	.0311	1.01806
360.0	37.12	1.325	7300.0	10320.0	165.9	21.02	30.32	396.3	21.2	.0317	1.01753
370.0	36.07	1.288	7515.0	10620.0	166.8	21.03	30.26	401.8	21.6	.0323	1.01703
380.0	35.09	1.252	7730.0	10920.0	167.6	21.04	30.21	407.2	22.0	.0329	1.01656
390.0	34.15	1.219	7944.0	11230.0	168.4	21.05	30.16	412.5	22.3	.0335	1.01612
400.0	33.27	1.188	8159.0	11530.0	169.1	21.06	30.13	417.7	22.7	.0341	1.01570
420.0	31.65	1.130	8588.0	12130.0	170.6	21.09	30.08	427.8	23.5	.0352	1.01492
440.0	30.18	1.077	9017.0	12730.0	172.0	21.14	30.05	437.6	24.3	.0364	1.01422
460.0	28.84	1.030	9446.0	13330.0	173.3	21.20	30.04	447.1	25.0	.0376	1.01359
480.0	27.63	.9861	9876.0	13930.0	174.6	21.26	30.06	456.3	25.7	.0387	1.01301
500.0	26.51	.9462	10310.0	14530.0	175.8	21.34	30.09	465.3	26.4	.0399	1.01248
520.0	25.48	.9096	10740.0	15140.0	177.0	21.42	30.13	474.0	27.1	.0410	1.01199
540.0	24.54	.8758	11170.0	15740.0	178.2	21.52	30.19	482.5	27.8	.0422	1.01154
560.0	23.66	.8444	11610.0	16340.0	179.3	21.62	30.26	490.8	28.5	.0434	1.01112
580.0	22.84	.8153	12040.0	16950.0	180.3	21.72	30.34	498.9	29.2	.0445	1.01074
600.0	22.08	.7881	12480.0	17560.0	181.3	21.84	30.42	506.9	29.9	.0457	1.01038
620.0	21.37	.7627	12920.0	18170.0	182.3	21.95	30.52	514.6	30.5	.0469	1.01004
640.0	20.70	.7390	13360.0	18780.0	183.3	22.08	30.62	522.2	31.2	.0480	1.00972
660.0	20.08	.7167	13810.0	19390.0	184.3	22.20	30.73	529.7	31.8	.0492	1.00943
680.0	19.49	.6957	14260.0	20010.0	185.2	22.33	30.84	537.0	32.4	.0503	1.00915
700.0	18.94	.6759	14710.0	20620.0	186.1	22.46	30.95	544.2	33.0	.0515	1.00888
720.0	18.41	.6573	15160.0	21250.0	187.0	22.59	31.07	551.2	33.7	.0526	1.00864
740.0	17.92	.6396	15610.0	21870.0	187.8	22.72	31.19	558.2	34.3	.0538	1.00840
760.0	17.45	.6229	16070.0	22490.0	188.6	22.85	31.31	565.0	34.9	.0549	1.00818
780.0	17.01	.6071	16530.0	23120.0	189.5	22.99	31.44	571.7	35.5	.0561	1.00797
800.0	16.59	.5920	16990.0	23750.0	190.3	23.12	31.56	578.4	36.1	.0572	1.00777
820.0	16.18	.5777	17460.0	24380.0	191.0	23.25	31.68	584.9	36.6	.0584	1.00758
840.0	15.80	.5640	17930.0	25020.0	191.8	23.38	31.80	591.4	37.2	.0595	1.00740
860.0	15.44	.5510	18400.0	25650.0	192.5	23.51	31.93	597.7	37.8	.0606	1.00723
880.0	15.09	.5386	18870.0	26290.0	193.3	23.64	32.05	604.0	38.4	.0618	1.00706
900.0	14.76	.5268	19340.0	26940.0	194.0	23.76	32.17	610.2	38.9	.0629	1.00691
920.0	14.44	.5154	19820.0	27580.0	194.7	23.88	32.28	616.4	39.5	.0640	1.00675
950.0	13.99	.4993	20540.0	28550.0	195.7	24.06	32.46	625.5	40.3	.0657	1.00654
1000.0	13.30	.4746	21750.0	30180.0	197.4	24.35	32.73	640.3	41.7	.0684	1.00621

Thermophysical properties of nitrogen—Continued

T K	Density kg/m³	Density mol/dm³	E J/mol	H J/mol	S J/(mol·K)	C_v J/(mol·K)	C_p J/(mol·K)	Sound m/s	Visc. μPa·s	Therm. W/(m·K)	Diel. const.
1050.0	12.67	.4522	22980.0	31830.0	199.0	24.63	33.00	654.8	43.0	.0711	1.00592
1100.0	12.10	.4318	24220.0	33480.0	200.6	24.89	33.25	668.9	44.3	.0738	1.00565
1150.0	11.58	.4132	25470.0	35150.0	202.0	25.13	33.49	682.7	45.6	.0765	1.00540
1200.0	11.10	.3962	26730.0	36830.0	203.5	25.36	33.71	696.3	46.9	.0791	1.00517
1250.0	10.66	.3805	28010.0	38520.0	204.9	25.57	33.92	709.6	48.1	.0816	1.00497
1300.0	10.25	.3660	29290.0	40220.0	206.2	25.77	34.11	722.6	49.4	.0841	1.00477
1350.0	9.877	.3526	30590.0	41930.0	207.5	25.96	34.30	735.4	50.6	.0866	1.00460
1400.0	9.528	.3401	31890.0	43650.0	208.7	26.13	34.47	748.0	51.8	.0891	1.00443
1450.0	9.202	.3285	33200.0	45380.0	209.9	26.30	34.63	760.4	53.0	.0915	1.00428
1500.0	8.898	.3176	34520.0	47110.0	211.1	26.45	34.78	772.6	54.1	.0938	1.00413
1550.0	8.614	.3075	35850.0	48860.0	212.3	26.59	34.92	784.6	55.3	.0962	1.00400
1600.0	8.347	.2979	37180.0	50600.0	213.4	26.73	35.05	796.4	56.4	.0985	1.00387
1650.0	8.096	.2890	38520.0	52360.0	214.5	26.85	35.17	808.1	57.5	.101	1.00376
1700.0	7.860	.2805	39860.0	54120.0	215.5	26.97	35.29	819.6	58.7	.103	1.00364
1750.0	7.637	.2726	41220.0	55890.0	216.5	27.08	35.40	831.0	59.8	.105	1.00354
1800.0	7.427	.2651	42570.0	57660.0	217.5	27.18	35.50	842.2	60.9	.107	1.00344
1850.0	7.227	.2580	43930.0	59440.0	218.5	27.28	35.59	853.2	61.9	.110	1.00335
1900.0	7.039	.2512	45300.0	61220.0	219.5	27.37	35.68	864.2	63.0	.112	1.00326

4.20 MPa isobar

T K	Density kg/m³	Density mol/dm³	E J/mol	H J/mol	S J/(mol·K)	C_v J/(mol·K)	C_p J/(mol·K)	Sound m/s	Visc. μPa·s	Therm. W/(m·K)	Diel. const.
64.07[a]	869.4	31.03	−4198.0	−4062.0	68.23	29.13	54.37	1270.0	279.0	.152	1.46943
80.0	806.4	28.79	−3316.0	−3170.0	80.66	28.67	56.50	914.2	146.0	.134	1.43214
100.0	708.5	25.29	−2180.0	−2014.0	93.55	26.70	60.35	664.0	79.8	.104	1.37471
110.0	647.6	23.11	−1566.0	−1384.0	99.55	26.36	66.34	548.1	59.9	.0880	1.33951
115.0	611.4	21.82	−1232.0	−1039.0	102.6	26.31	71.99	486.9	51.4	.0795	1.31879
120.0	567.9	20.27	−863.4	−656.2	105.9	26.38	82.64	418.9	43.2	.0710	1.29414
122.0	547.0	19.52	−699.0	−483.8	107.3	26.48	90.26	388.2	39.9	.0677	1.28236
124.0	522.4	18.65	−517.4	−292.2	108.9	26.66	102.5	354.1	36.4	.0645	1.26864
126.0	491.6	17.55	−306.5	−67.1	110.7	26.97	125.3	314.8	32.6	.0613	1.25156
127.0	472.1	16.85	−181.4	67.9	111.7	27.23	146.3	292.3	30.4	.0597	1.24079
128.0	447.4	15.97	−32.6	230.4	113.0	27.60	182.4	267.2	27.9	.0580	1.22729
129.0	413.7	14.77	158.7	443.1	114.6	28.18	250.0	239.5	24.9	.0560	1.20898
130.0	365.6	13.05	418.9	740.7	116.9	28.98	341.8	213.1	21.3	.0530	1.18320
130.5	337.9	12.06	569.0	917.2	118.3	29.31	359.0	203.9	19.6	.0506	1.16853
131.0	311.4	11.12	717.0	1095.0	119.7	29.46	347.4	198.0	18.1	.0479	1.15461
132.0	268.5	9.584	972.9	1411.0	122.1	29.26	280.3	193.4	16.0	.0426	1.13232
133.0	239.4	8.545	1165.0	1656.0	123.9	28.75	213.0	193.3	14.9	.0384	1.11738
134.0	219.5	7.834	1309.0	1845.0	125.3	28.20	168.0	194.8	14.2	.0352	1.10724
135.0	204.9	7.314	1423.0	1998.0	126.5	27.69	139.2	196.9	13.8	.0329	1.09987
136.0	193.6	6.912	1519.0	2126.0	127.4	27.25	119.9	199.2	13.4	.0311	1.09419
138.0	176.9	6.315	1675.0	2340.0	129.0	26.49	96.08	203.8	13.0	.0286	1.08580
140.0	164.7	5.879	1803.0	2517.0	130.3	25.89	82.09	208.1	12.8	.0270	1.07971
145.0	143.9	5.138	2058.0	2876.0	132.8	24.77	63.78	218.1	12.6	.0245	1.06941
150.0	130.1	4.645	2265.0	3170.0	134.8	24.01	54.70	226.8	12.5	.0231	1.06260
152.0	125.7	4.487	2340.0	3276.0	135.5	23.77	52.23	230.0	12.5	.0228	1.06042
154.0	121.7	4.345	2412.0	3379.0	136.1	23.56	50.16	233.1	12.6	.0225	1.05846
156.0	118.1	4.216	2481.0	3477.0	136.8	23.37	48.41	236.1	12.6	.0223	1.05669
158.0	114.8	4.098	2548.0	3573.0	137.4	23.20	46.92	239.0	12.6	.0221	1.05507
160.0	111.8	3.990	2612.0	3665.0	138.0	23.04	45.62	241.8	12.7	.0220	1.05359
162.0	109.0	3.889	2675.0	3755.0	138.5	22.90	44.48	244.5	12.8	.0219	1.05221
164.0	106.4	3.796	2737.0	3843.0	139.1	22.78	43.47	247.1	12.8	.0218	1.05094
166.0	103.9	3.709	2797.0	3929.0	139.6	22.66	42.58	249.7	12.9	.0218	1.04975
168.0	101.6	3.627	2856.0	4013.0	140.1	22.56	41.78	252.2	12.9	.0218	1.04863
170.0	99.47	3.551	2913.0	4096.0	140.6	22.46	41.06	254.6	13.0	.0218	1.04758
172.0	97.44	3.478	2970.0	4178.0	141.1	22.37	40.41	257.0	13.1	.0218	1.04659
174.0	95.53	3.410	3026.0	4258.0	141.5	22.29	39.82	259.4	13.2	.0218	1.04566
176.0	93.71	3.345	3081.0	4337.0	142.0	22.22	39.29	261.6	13.2	.0219	1.04478
180.0	90.33	3.224	3190.0	4492.0	142.8	22.08	38.34	266.1	13.4	.0219	1.04314
185.0	86.54	3.089	3322.0	4681.0	143.9	21.94	37.36	271.4	13.6	.0221	1.04129
190.0	83.13	2.967	3451.0	4866.0	144.9	21.82	36.54	276.6	13.8	.0223	1.03964
195.0	80.05	2.857	3577.0	5047.0	145.8	21.72	35.85	281.5	14.0	.0225	1.03815

Thermophysical properties of nitrogen—Continued

T K	Density kg/m^3	Density mol/dm^3	E J/mol	H J/mol	S J/(mol·K)	C_v J/(mol·K)	C_p J/(mol·K)	Sound m/s	Visc. μPa·s	Therm. W/(m·K)	Diel. const.
200.0	77.24	2.757	3702.0	5225.0	146.7	21.64	35.26	286.2	14.3	.0227	1.03679
205.0	74.67	2.665	3824.0	5400.0	147.6	21.57	34.76	290.9	14.5	.0230	1.03555
210.0	72.29	2.580	3945.0	5573.0	148.4	21.50	34.32	295.3	14.7	.0232	1.03440
215.0	70.10	2.502	4065.0	5743.0	149.2	21.45	33.94	299.7	14.9	.0235	1.03334
220.0	68.05	2.429	4183.0	5912.0	150.0	21.40	33.60	303.9	15.2	.0237	1.03236
225.0	66.14	2.361	4300.0	6079.0	150.7	21.36	33.29	308.1	15.4	.0240	1.03144
230.0	64.36	2.297	4417.0	6245.0	151.5	21.32	33.02	312.1	15.6	.0243	1.03058
235.0	62.68	2.237	4532.0	6409.0	152.2	21.29	32.78	316.0	15.8	.0246	1.02977
240.0	61.10	2.181	4647.0	6573.0	152.9	21.26	32.56	319.9	16.1	.0248	1.02901
245.0	59.61	2.128	4761.0	6735.0	153.5	21.23	32.36	323.7	16.3	.0251	1.02830
250.0	58.21	2.078	4875.0	6896.0	154.2	21.21	32.18	327.4	16.5	.0254	1.02762
260.0	55.60	1.985	5100.0	7217.0	155.4	21.17	31.87	334.7	16.9	.0260	1.02637
270.0	53.25	1.901	5324.0	7534.0	156.6	21.13	31.60	341.7	17.4	.0266	1.02524
280.0	51.11	1.824	5546.0	7849.0	157.8	21.11	31.37	348.4	17.8	.0271	1.02422
290.0	49.15	1.754	5767.0	8162.0	158.9	21.08	31.18	355.0	18.3	.0277	1.02328
300.0	47.35	1.690	5987.0	8473.0	159.9	21.07	31.01	361.5	18.7	.0283	1.02241
310.0	45.68	1.631	6206.0	8782.0	160.9	21.05	30.87	367.7	19.1	.0289	1.02162
320.0	44.14	1.576	6424.0	9090.0	161.9	21.04	30.74	373.8	19.5	.0295	1.02088
330.0	42.71	1.525	6642.0	9397.0	162.9	21.03	30.63	379.8	20.0	.0301	1.02020
340.0	41.38	1.477	6859.0	9702.0	163.8	21.03	30.53	385.6	20.4	.0306	1.01956
350.0	40.13	1.432	7075.0	10010.0	164.7	21.03	30.45	391.3	20.8	.0312	1.01896
360.0	38.95	1.390	7291.0	10310.0	165.5	21.03	30.37	396.9	21.2	.0318	1.01840
370.0	37.85	1.351	7506.0	10610.0	166.3	21.04	30.31	402.4	21.6	.0324	1.01788
380.0	36.82	1.314	7722.0	10920.0	167.2	21.04	30.26	407.8	22.0	.0330	1.01738
390.0	35.84	1.279	7937.0	11220.0	167.9	21.05	30.21	413.0	22.4	.0336	1.01692
400.0	34.91	1.246	8151.0	11520.0	168.7	21.07	30.17	418.2	22.8	.0341	1.01648
420.0	33.20	1.185	8581.0	12120.0	170.2	21.10	30.12	428.4	23.5	.0353	1.01566
440.0	31.66	1.130	9010.0	12730.0	171.6	21.15	30.09	438.2	24.3	.0365	1.01493
460.0	30.26	1.080	9440.0	13330.0	172.9	21.20	30.08	447.7	25.0	.0376	1.01426
480.0	28.98	1.034	9870.0	13930.0	174.2	21.27	30.09	456.9	25.7	.0388	1.01365
500.0	27.81	.9927	10300.0	14530.0	175.4	21.34	30.11	465.9	26.5	.0399	1.01310
520.0	26.73	.9542	10730.0	15130.0	176.6	21.43	30.15	474.6	27.2	.0411	1.01258
540.0	25.74	.9187	11170.0	15740.0	177.7	21.52	30.21	483.1	27.9	.0423	1.01211
560.0	24.82	.8858	11600.0	16340.0	178.8	21.62	30.28	491.4	28.5	.0434	1.01167
580.0	23.96	.8553	12040.0	16950.0	179.9	21.73	30.35	499.5	29.2	.0446	1.01127
600.0	23.16	.8268	12480.0	17560.0	180.9	21.84	30.44	507.4	29.9	.0458	1.01089
620.0	22.42	.8002	12920.0	18170.0	181.9	21.96	30.53	515.2	30.5	.0469	1.01053
640.0	21.72	.7752	13360.0	18780.0	182.9	22.08	30.64	522.8	31.2	.0481	1.01020
660.0	21.06	.7518	13810.0	19390.0	183.9	22.20	30.74	530.2	31.8	.0492	1.00989
680.0	20.45	.7298	14250.0	20010.0	184.8	22.33	30.85	537.5	32.4	.0504	1.00960
700.0	19.87	.7091	14700.0	20630.0	185.7	22.46	30.97	544.7	33.1	.0515	1.00932
720.0	19.32	.6895	15160.0	21250.0	186.5	22.59	31.08	551.8	33.7	.0527	1.00906
740.0	18.80	.6710	15610.0	21870.0	187.4	22.72	31.20	558.7	34.3	.0538	1.00882
760.0	18.31	.6535	16070.0	22500.0	188.2	22.86	31.32	565.5	34.9	.0550	1.00858
780.0	17.84	.6369	16530.0	23120.0	189.0	22.99	31.45	572.2	35.5	.0561	1.00836
800.0	17.40	.6211	16990.0	23750.0	189.8	23.12	31.57	578.9	36.1	.0573	1.00815
820.0	16.98	.6061	17460.0	24390.0	190.6	23.25	31.69	585.4	36.7	.0584	1.00795
840.0	16.58	.5918	17920.0	25020.0	191.4	23.38	31.81	591.9	37.2	.0595	1.00776
860.0	16.20	.5782	18390.0	25660.0	192.1	23.51	31.93	598.2	37.8	.0607	1.00758
880.0	15.83	.5651	18870.0	26300.0	192.9	23.64	32.05	604.5	38.4	.0618	1.00741
900.0	15.48	.5527	19340.0	26940.0	193.6	23.76	32.17	610.7	38.9	.0629	1.00725
920.0	15.15	.5408	19820.0	27580.0	194.3	23.89	32.29	616.9	39.5	.0640	1.00709
950.0	14.68	.5239	20540.0	28560.0	195.3	24.07	32.46	625.9	40.3	.0657	1.00686
1000.0	13.95	.4980	21750.0	30190.0	197.0	24.36	32.74	640.8	41.7	.0685	1.00652
1050.0	13.29	.4745	22980.0	31830.0	198.6	24.63	33.00	655.2	43.0	.0712	1.00621
1100.0	12.69	.4531	24220.0	33490.0	200.2	24.89	33.25	669.3	44.3	.0739	1.00593
1150.0	12.15	.4336	25470.0	35150.0	201.6	25.13	33.49	683.2	45.6	.0765	1.00567
1200.0	11.65	.4157	26730.0	36830.0	203.1	25.36	33.71	696.7	46.9	.0791	1.00543
1250.0	11.19	.3993	28010.0	38530.0	204.5	25.57	33.92	710.0	48.1	.0816	1.00521
1300.0	10.76	.3841	29290.0	40230.0	205.8	25.77	34.11	723.0	49.4	.0842	1.00501
1350.0	10.37	.3700	30590.0	41940.0	207.1	25.96	34.30	735.8	50.6	.0866	1.00482
1400.0	9.999	.3569	31890.0	43660.0	208.3	26.13	34.47	748.4	51.8	.0891	1.00465
1450.0	9.657	.3447	33200.0	45380.0	209.5	26.30	34.63	760.8	53.0	.0915	1.00449

Thermophysical properties of nitrogen—Continued

T K	Density kg/m^3	Density mol/dm^3	E J/mol	H J/mol	S J/(mol·K)	C_v J/(mol·K)	C_p J/(mol·K)	Sound m/s	Visc. μPa·s	Therm. W/(m·K)	Diel. const.
				4.60 MPa isobar							
64.15[a]	869.6	31.04	−4196.0	−4048.0	68.25	29.34	54.34	1265.0	279.0	.152	1.46951
80.0	807.4	28.82	−3322.0	−3162.0	80.59	28.78	56.37	916.2	147.0	.134	1.43268
100.0	710.5	25.36	−2192.0	−2010.0	93.43	26.76	59.97	669.3	80.6	.105	1.37587
110.0	650.9	23.23	−1584.0	−1386.0	99.37	26.39	65.48	555.9	60.8	.0888	1.34138
115.0	615.9	21.98	−1257.0	−1047.0	102.4	26.31	70.46	496.8	52.4	.0806	1.32133
120.0	574.7	20.51	−899.8	−675.6	105.5	26.34	79.24	432.6	44.4	.0723	1.29795
125.0	521.2	18.60	−484.6	−237.3	109.1	26.61	99.42	357.2	36.3	.0644	1.26796
127.0	492.5	17.58	−283.9	−22.3	110.8	26.87	117.4	321.3	32.7	.0613	1.25203
128.0	475.1	16.96	−169.4	101.9	111.8	27.07	131.8	301.5	30.8	.0598	1.24244
130.0	429.4	15.33	110.5	410.6	114.2	27.70	183.3	258.4	26.3	.0562	1.21749
132.0	361.7	12.91	495.4	851.7	117.6	28.64	250.7	219.4	21.2	.0506	1.18113
133.0	324.8	11.59	707.6	1104.0	119.5	28.90	249.9	208.7	19.0	.0469	1.16160
134.0	292.3	10.43	903.6	1344.0	121.3	28.85	227.8	203.4	17.3	.0434	1.14464
136.0	245.6	8.765	1215.0	1740.0	124.2	28.18	168.5	201.6	15.3	.0375	1.12053
138.0	216.6	7.731	1437.0	2032.0	126.3	27.38	127.1	204.1	14.4	.0334	1.10576
140.0	197.1	7.035	1606.0	2260.0	128.0	26.66	102.7	207.7	13.8	.0307	1.09592
142.0	182.8	6.526	1744.0	2449.0	129.3	26.06	87.51	211.6	13.5	.0288	1.08876
145.0	167.0	5.961	1917.0	2688.0	131.0	25.33	73.39	217.3	13.2	.0268	1.08085
150.0	148.7	5.306	2152.0	3019.0	133.2	24.43	60.38	226.1	13.0	.0248	1.07173
152.0	143.0	5.104	2235.0	3136.0	134.0	24.15	57.02	229.4	13.0	.0242	1.06893
154.0	138.0	4.924	2314.0	3248.0	134.7	23.90	54.27	232.5	13.0	.0238	1.06644
156.0	133.5	4.764	2388.0	3354.0	135.4	23.68	51.99	235.6	13.0	.0235	1.06423
158.0	129.4	4.618	2460.0	3456.0	136.1	23.48	50.06	238.5	13.0	.0232	1.06222
160.0	125.7	4.486	2529.0	3554.0	136.7	23.30	48.40	241.4	13.0	.0230	1.06040
162.0	122.3	4.364	2596.0	3650.0	137.3	23.15	46.97	244.2	13.1	.0228	1.05872
164.0	119.1	4.252	2660.0	3742.0	137.8	23.00	45.73	246.9	13.1	.0227	1.05718
166.0	116.2	4.148	2724.0	3833.0	138.4	22.87	44.63	249.5	13.2	.0226	1.05575
168.0	113.5	4.051	2785.0	3921.0	138.9	22.75	43.65	252.0	13.2	.0226	1.05442
170.0	110.9	3.960	2846.0	4007.0	139.4	22.64	42.78	254.5	13.3	.0225	1.05317
172.0	108.6	3.875	2905.0	4092.0	139.9	22.54	42.00	257.0	13.4	.0225	1.05201
174.0	106.3	3.794	2963.0	4175.0	140.4	22.45	41.29	259.4	13.4	.0225	1.05091
176.0	104.2	3.719	3020.0	4257.0	140.9	22.36	40.65	261.7	13.5	.0225	1.04987
178.0	102.2	3.647	3077.0	4338.0	141.3	22.29	40.06	264.0	13.6	.0225	1.04889
180.0	100.3	3.579	3132.0	4418.0	141.8	22.22	39.53	266.2	13.6	.0226	1.04796
185.0	95.88	3.422	3268.0	4612.0	142.8	22.06	38.38	271.7	13.8	.0227	1.04582
190.0	91.97	3.283	3400.0	4802.0	143.9	21.93	37.43	276.9	14.0	.0228	1.04392
195.0	88.45	3.157	3530.0	4987.0	144.8	21.81	36.63	281.9	14.2	.0230	1.04221
200.0	85.25	3.043	3656.0	5168.0	145.7	21.72	35.96	286.7	14.4	.0232	1.04066
205.0	82.33	2.939	3781.0	5346.0	146.6	21.64	35.38	291.4	14.6	.0234	1.03925
210.0	79.65	2.843	3904.0	5522.0	147.5	21.57	34.88	295.9	14.9	.0236	1.03795
215.0	77.17	2.755	4025.0	5695.0	148.3	21.51	34.45	300.3	15.1	.0239	1.03675
220.0	74.87	2.673	4145.0	5867.0	149.1	21.46	34.06	304.6	15.3	.0241	1.03564
225.0	72.74	2.596	4264.0	6036.0	149.8	21.41	33.73	308.7	15.5	.0244	1.03461
230.0	70.74	2.525	4382.0	6204.0	150.6	21.37	33.42	312.8	15.7	.0246	1.03365
235.0	68.86	2.458	4499.0	6370.0	151.3	21.33	33.15	316.8	16.0	.0249	1.03274
240.0	67.10	2.395	4615.0	6536.0	152.0	21.30	32.91	320.7	16.2	.0252	1.03189
245.0	65.44	2.336	4730.0	6700.0	152.6	21.27	32.69	324.5	16.4	.0254	1.03109
250.0	63.87	2.280	4845.0	6862.0	153.3	21.24	32.49	328.3	16.6	.0257	1.03034
260.0	60.98	2.177	5072.0	7185.0	154.6	21.20	32.13	335.5	17.1	.0263	1.02895
270.0	58.38	2.084	5298.0	7505.0	155.8	21.16	31.84	342.6	17.5	.0268	1.02770
280.0	56.01	1.999	5521.0	7822.0	156.9	21.13	31.59	349.4	17.9	.0274	1.02656
290.0	53.84	1.922	5744.0	8137.0	158.0	21.11	31.37	356.0	18.4	.0280	1.02552
300.0	51.85	1.851	5965.0	8450.0	159.1	21.09	31.19	362.5	18.8	.0286	1.02457
310.0	50.02	1.785	6185.0	8761.0	160.1	21.07	31.03	368.7	19.2	.0291	1.02369
320.0	48.32	1.725	6404.0	9071.0	161.1	21.06	30.89	374.9	19.6	.0297	1.02287
330.0	46.75	1.669	6622.0	9379.0	162.1	21.05	30.77	380.8	20.0	.0303	1.02212
340.0	45.28	1.616	6840.0	9686.0	163.0	21.05	30.66	386.7	20.4	.0309	1.02142
350.0	43.90	1.567	7057.0	9992.0	163.9	21.05	30.57	392.4	20.8	.0314	1.02076
360.0	42.62	1.521	7273.0	10300.0	164.7	21.05	30.48	398.0	21.3	.0320	1.02015
370.0	41.41	1.478	7490.0	10600.0	165.5	21.05	30.41	403.5	21.6	.0326	1.01957
380.0	40.27	1.437	7705.0	10910.0	166.4	21.06	30.35	408.9	22.0	.0332	1.01902

Thermophysical properties of nitrogen—Continued

T K	Density kg/m³	Density mol/dm³	E J/mol	H J/mol	S J/(mol·K)	C_v J/(mol·K)	C_p J/(mol·K)	Sound m/s	Visc. μPa·s	Therm. W/(m·K)	Diel. const.
390.0	39.20	1.399	7921.0	11210.0	167.1	21.07	30.30	414.2	22.4	.0337	1.01851
400.0	38.18	1.363	8136.0	11510.0	167.9	21.08	30.26	419.4	22.8	.0343	1.01803
410.0	37.22	1.329	8352.0	11810.0	168.7	21.10	30.22	424.5	23.2	.0349	1.01757
420.0	36.31	1.296	8567.0	12120.0	169.4	21.11	30.19	429.5	23.6	.0355	1.01713
440.0	34.62	1.236	8997.0	12720.0	170.8	21.16	30.15	439.3	24.3	.0366	1.01633
460.0	33.09	1.181	9428.0	13320.0	172.1	21.21	30.14	448.8	25.1	.0378	1.01560
480.0	31.69	1.131	9858.0	13930.0	173.4	21.28	30.14	458.0	25.8	.0389	1.01493
500.0	30.41	1.085	10290.0	14530.0	174.6	21.35	30.16	467.0	26.5	.0401	1.01432
520.0	29.23	1.043	10720.0	15130.0	175.8	21.44	30.20	475.7	27.2	.0412	1.01376
540.0	28.14	1.004	11160.0	15740.0	177.0	21.53	30.25	484.2	27.9	.0424	1.01325
560.0	27.13	.9685	11590.0	16340.0	178.1	21.63	30.32	492.5	28.6	.0436	1.01277
580.0	26.20	.9351	12030.0	16950.0	179.1	21.74	30.39	500.6	29.2	.0447	1.01232
600.0	25.32	.9039	12470.0	17560.0	180.2	21.85	30.47	508.5	29.9	.0459	1.01191
620.0	24.51	.8748	12910.0	18170.0	181.2	21.96	30.57	516.3	30.6	.0470	1.01152
640.0	23.75	.8476	13350.0	18780.0	182.1	22.09	30.66	523.9	31.2	.0482	1.01116
660.0	23.03	.8220	13800.0	19390.0	183.1	22.21	30.77	531.3	31.8	.0493	1.01082
680.0	22.36	.7980	14250.0	20010.0	184.0	22.34	30.88	538.6	32.5	.0505	1.01050
700.0	21.72	.7753	14700.0	20630.0	184.9	22.47	30.99	545.8	33.1	.0516	1.01020
720.0	21.12	.7540	15150.0	21250.0	185.8	22.60	31.11	552.8	33.7	.0528	1.00991
740.0	20.56	.7337	15610.0	21870.0	186.6	22.73	31.22	559.7	34.3	.0539	1.00964
760.0	20.02	.7146	16060.0	22500.0	187.5	22.86	31.34	566.6	34.9	.0551	1.00939
780.0	19.51	.6964	16520.0	23130.0	188.3	23.00	31.46	573.3	35.5	.0562	1.00915
800.0	19.03	.6792	16990.0	23760.0	189.1	23.13	31.59	579.9	36.1	.0574	1.00892
820.0	18.57	.6628	17450.0	24390.0	189.9	23.26	31.71	586.4	36.7	.0585	1.00870
840.0	18.13	.6472	17920.0	25030.0	190.6	23.39	31.83	592.9	37.3	.0596	1.00849
860.0	17.71	.6323	18390.0	25660.0	191.4	23.52	31.95	599.2	37.8	.0608	1.00830
880.0	17.31	.6180	18860.0	26300.0	192.1	23.64	32.07	605.5	38.4	.0619	1.00811
900.0	16.93	.6044	19340.0	26950.0	192.8	23.77	32.18	611.7	39.0	.0630	1.00793
920.0	16.57	.5914	19810.0	27590.0	193.5	23.89	32.30	617.8	39.5	.0641	1.00775
950.0	16.05	.5730	20540.0	28560.0	194.6	24.07	32.47	626.9	40.3	.0658	1.00751
1000.0	15.26	.5446	21750.0	30190.0	196.3	24.36	32.75	641.7	41.7	.0685	1.00713
1050.0	14.54	.5190	22980.0	31840.0	197.9	24.63	33.01	656.1	43.0	.0712	1.00679
1100.0	13.89	.4957	24210.0	33500.0	199.4	24.89	33.26	670.2	44.3	.0739	1.00648
1150.0	13.29	.4743	25470.0	35160.0	200.9	25.14	33.50	684.0	45.6	.0766	1.00620
1200.0	12.74	.4548	26730.0	36840.0	202.3	25.36	33.72	697.6	46.9	.0792	1.00594
1250.0	12.24	.4368	28000.0	38540.0	203.7	25.58	33.93	710.8	48.2	.0817	1.00570
1300.0	11.77	.4202	29290.0	40240.0	205.0	25.78	34.12	723.8	49.4	.0842	1.00548
1350.0	11.34	.4048	30580.0	41950.0	206.3	25.96	34.30	736.6	50.6	.0867	1.00528
1400.0	10.94	.3905	31890.0	43670.0	207.6	26.14	34.47	749.2	51.8	.0892	1.00509
1450.0	10.57	.3772	33200.0	45390.0	208.8	26.30	34.63	761.6	53.0	.0916	1.00491
1500.0	10.22	.3647	34520.0	47130.0	210.0	26.45	34.78	773.7	54.1	.0939	1.00475
1550.0	9.892	.3531	35840.0	48870.0	211.1	26.60	34.92	785.7	55.3	.0963	1.00459
1600.0	9.586	.3421	37180.0	50620.0	212.2	26.73	35.05	797.6	56.4	.0986	1.00445
1650.0	9.298	.3319	38520.0	52380.0	213.3	26.85	35.18	809.2	57.6	.101	1.00431
1700.0	9.027	.3222	39860.0	54140.0	214.3	26.97	35.29	820.7	58.7	.103	1.00419
1750.0	8.772	.3131	41210.0	55910.0	215.4	27.08	35.40	832.0	59.8	.105	1.00406
1800.0	8.530	.3045	42570.0	57680.0	216.4	27.18	35.50	843.2	60.9	.108	1.00395
1850.0	8.302	.2963	43930.0	59460.0	217.3	27.28	35.60	854.3	61.9	.110	1.00384
1900.0	8.085	.2886	45300.0	61240.0	218.3	27.37	35.69	865.2	63.0	.112	1.00374

4.80 MPa isobar

T K	Density kg/m³	Density mol/dm³	E J/mol	H J/mol	S J/(mol·K)	C_v J/(mol·K)	C_p J/(mol·K)	Sound m/s	Visc. μPa·s	Therm. W/(m·K)	Diel. const.
64.20[a]	869.6	31.04	−4196.0	−4041.0	68.26	29.45	54.32	1263.0	279.0	.152	1.46955
80.0	807.8	28.84	−3325.0	−3158.0	80.55	28.83	56.31	917.1	147.0	.134	1.43295
100.0	711.5	25.40	−2197.0	−2008.0	93.37	26.78	59.79	671.9	81.0	.105	1.37645
110.0	652.5	23.29	−1593.0	−1387.0	99.28	26.41	65.08	559.7	61.2	.0892	1.34229
115.0	618.0	22.06	−1269.0	−1051.0	102.3	26.32	69.77	501.6	52.9	.0811	1.32255
120.0	577.8	20.62	−916.8	−684.1	105.4	26.33	77.81	438.9	45.0	.0729	1.29972
125.0	526.8	18.80	−513.6	−258.3	108.9	26.54	95.09	366.8	37.1	.0651	1.27107
127.0	500.4	17.86	−324.0	−55.3	110.5	26.75	109.1	333.3	33.7	.0621	1.25639
128.0	484.9	17.31	−218.5	58.9	111.4	26.90	119.6	315.2	31.9	.0606	1.24782
130.0	446.4	15.94	27.2	328.4	113.5	27.37	153.5	276.1	28.0	.0573	1.22674
132.0	392.7	14.02	344.9	687.4	116.2	28.10	206.4	237.4	23.4	.0528	1.19764

Thermophysical properties of nitrogen—Continued

T K	Density kg/m³	Density mol/dm³	*E* J/mol	*H* J/mol	*S* J/(mol·K)	C_v J/(mol·K)	C_p	Sound m/s	Visc. μPa·s	Therm. W/(m·K)	Diel. const.
133.0	360.4	12.87	530.2	903.3	117.8	28.46	222.6	222.8	21.2	.0498	1.18044
134.0	328.6	11.73	717.1	1126.0	119.5	28.66	220.7	213.4	19.3	.0465	1.16360
136.0	275.8	9.846	1048.0	1536.0	122.5	28.41	184.8	206.0	16.7	.0406	1.13610
138.0	240.0	8.566	1302.0	1863.0	124.9	27.73	143.6	206.0	15.3	.0361	1.11767
140.0	215.8	7.703	1496.0	2119.0	126.8	27.02	114.7	208.6	14.5	.0328	1.10537
142.0	198.4	7.083	1651.0	2329.0	128.3	26.39	96.15	211.9	14.0	.0305	1.09660
145.0	179.7	6.413	1840.0	2589.0	130.1	25.61	78.99	217.3	13.6	.0281	1.08717
150.0	158.5	5.659	2093.0	2941.0	132.5	24.64	63.58	226.0	13.3	.0257	1.07663
152.0	152.1	5.430	2180.0	3064.0	133.3	24.34	59.68	229.3	13.2	.0250	1.07345
154.0	146.5	5.229	2262.0	3180.0	134.0	24.07	56.53	232.5	13.2	.0245	1.07066
156.0	141.5	5.051	2340.0	3290.0	134.7	23.84	53.93	235.5	13.2	.0241	1.06819
158.0	137.0	4.890	2414.0	3396.0	135.4	23.63	51.75	238.5	13.2	.0238	1.06596
160.0	132.9	4.744	2486.0	3498.0	136.1	23.44	49.90	241.4	13.2	.0235	1.06394
162.0	129.2	4.610	2555.0	3596.0	136.7	23.27	48.31	244.2	13.3	.0233	1.06210
164.0	125.7	4.487	2621.0	3691.0	137.2	23.11	46.92	246.9	13.3	.0232	1.06041
166.0	122.5	4.374	2686.0	3784.0	137.8	22.97	45.71	249.5	13.3	.0231	1.05885
168.0	119.6	4.268	2749.0	3874.0	138.4	22.85	44.63	252.1	13.4	.0230	1.05740
170.0	116.8	4.170	2811.0	3962.0	138.9	22.73	43.68	254.6	13.4	.0229	1.05605
172.0	114.2	4.078	2872.0	4049.0	139.4	22.63	42.82	257.1	13.5	.0229	1.05478
174.0	111.8	3.991	2931.0	4134.0	139.9	22.53	42.05	259.5	13.5	.0228	1.05359
176.0	109.5	3.909	2989.0	4217.0	140.3	22.44	41.36	261.8	13.6	.0228	1.05247
178.0	107.4	3.832	3047.0	4299.0	140.8	22.36	40.72	264.1	13.7	.0228	1.05142
180.0	105.3	3.759	3103.0	4380.0	141.3	22.28	40.14	266.4	13.7	.0229	1.05042
182.0	103.4	3.689	3159.0	4460.0	141.7	22.21	39.61	268.6	13.8	.0229	1.04947
185.0	100.6	3.591	3241.0	4577.0	142.3	22.11	38.90	271.8	13.9	.0230	1.04813
190.0	96.44	3.442	3375.0	4769.0	143.4	21.98	37.88	277.1	14.1	.0231	1.04609
195.0	92.69	3.308	3506.0	4957.0	144.3	21.86	37.03	282.1	14.3	.0233	1.04427
200.0	89.29	3.187	3634.0	5140.0	145.3	21.76	36.31	286.9	14.5	.0234	1.04262
205.0	86.19	3.077	3760.0	5320.0	146.2	21.68	35.70	291.6	14.7	.0236	1.04112
210.0	83.35	2.975	3884.0	5497.0	147.0	21.60	35.17	296.2	14.9	.0239	1.03974
215.0	80.73	2.882	4006.0	5672.0	147.8	21.54	34.71	300.6	15.1	.0241	1.03847
220.0	78.30	2.795	4127.0	5844.0	148.6	21.48	34.30	304.9	15.4	.0243	1.03730
225.0	76.05	2.714	4246.0	6015.0	149.4	21.44	33.94	309.1	15.6	.0246	1.03621
230.0	73.94	2.639	4365.0	6184.0	150.1	21.39	33.62	313.2	15.8	.0248	1.03519
235.0	71.96	2.569	4482.0	6351.0	150.9	21.35	33.34	317.2	16.0	.0251	1.03423
240.0	70.11	2.502	4599.0	6517.0	151.6	21.32	33.08	321.1	16.2	.0253	1.03334
245.0	68.36	2.440	4715.0	6682.0	152.2	21.29	32.85	324.9	16.5	.0256	1.03250
250.0	66.71	2.381	4830.0	6846.0	152.9	21.26	32.64	328.7	16.7	.0259	1.03170
260.0	63.68	2.273	5058.0	7170.0	154.2	21.22	32.27	336.0	17.1	.0264	1.03024
270.0	60.94	2.175	5284.0	7491.0	155.4	21.18	31.96	343.1	17.5	.0270	1.02892
280.0	58.45	2.086	5509.0	7809.0	156.5	21.15	31.70	349.9	18.0	.0275	1.02773
290.0	56.19	2.005	5732.0	8125.0	157.6	21.12	31.47	356.5	18.4	.0281	1.02664
300.0	54.10	1.931	5953.0	8439.0	158.7	21.10	31.28	363.0	18.8	.0287	1.02564
310.0	52.18	1.863	6174.0	8751.0	159.7	21.09	31.11	369.3	19.2	.0292	1.02472
320.0	50.41	1.799	6393.0	9061.0	160.7	21.07	30.96	375.4	19.7	.0298	1.02387
330.0	48.76	1.740	6612.0	9370.0	161.7	21.06	30.83	381.4	20.1	.0304	1.02308
340.0	47.22	1.686	6830.0	9678.0	162.6	21.06	30.72	387.2	20.5	.0310	1.02235
350.0	45.79	1.634	7048.0	9985.0	163.5	21.06	30.63	393.0	20.9	.0315	1.02166
360.0	44.44	1.586	7265.0	10290.0	164.3	21.06	30.54	398.6	21.3	.0321	1.02102
370.0	43.18	1.541	7481.0	10600.0	165.2	21.06	30.47	404.1	21.7	.0327	1.02041
380.0	41.99	1.499	7697.0	10900.0	166.0	21.07	30.40	409.5	22.1	.0333	1.01984
390.0	40.87	1.459	7913.0	11200.0	166.8	21.08	30.35	414.7	22.5	.0338	1.01931
400.0	39.81	1.421	8129.0	11510.0	167.5	21.09	30.30	419.9	22.8	.0344	1.01880
410.0	38.81	1.385	8345.0	11810.0	168.3	21.10	30.26	425.1	23.2	.0350	1.01832
420.0	37.86	1.351	8560.0	12110.0	169.0	21.12	30.23	430.1	23.6	.0356	1.01787
440.0	36.09	1.288	8991.0	12720.0	170.4	21.16	30.19	439.9	24.4	.0367	1.01703
460.0	34.50	1.231	9421.0	13320.0	171.8	21.22	30.17	449.4	25.1	.0379	1.01627
480.0	33.04	1.179	9853.0	13920.0	173.0	21.28	30.17	458.6	25.8	.0390	1.01557
500.0	31.70	1.132	10280.0	14530.0	174.3	21.36	30.19	467.6	26.5	.0402	1.01494
520.0	30.47	1.088	10720.0	15130.0	175.5	21.44	30.22	476.3	27.2	.0413	1.01435
540.0	29.34	1.047	11150.0	15740.0	176.6	21.53	30.27	484.8	27.9	.0425	1.01381
560.0	28.29	1.010	11590.0	16340.0	177.7	21.63	30.34	493.1	28.6	.0436	1.01331
580.0	27.31	.9749	12030.0	16950.0	178.8	21.74	30.41	501.2	29.3	.0448	1.01285

Thermophysical properties of nitrogen—Continued

T K	Density kg/m^3	Density mol/dm^3	E J/mol	H J/mol	S J/(mol·K)	C_v J/(mol·K)	C_p J/(mol·K)	Sound m/s	Visc. μPa·s	Therm. W/(m·K)	Diel. const.
600.0	26.40	.9424	12460.0	17560.0	179.8	21.85	30.49	509.1	29.9	.0459	1.01242
620.0	25.55	.9121	12910.0	18170.0	180.8	21.97	30.58	516.8	30.6	.0471	1.01201
640.0	24.76	.8837	13350.0	18780.0	181.8	22.09	30.68	524.4	31.2	.0482	1.01163
660.0	24.01	.8570	13800.0	19400.0	182.7	22.21	30.78	531.8	31.8	.0494	1.01128
680.0	23.31	.8320	14240.0	20010.0	183.6	22.34	30.89	539.1	32.5	.0505	1.01095
700.0	22.65	.8084	14690.0	20630.0	184.5	22.47	31.00	546.3	33.1	.0517	1.01063
720.0	22.02	.7861	15150.0	21250.0	185.4	22.60	31.12	553.3	33.7	.0528	1.01034
740.0	21.43	.7650	15600.0	21880.0	186.3	22.73	31.23	560.3	34.3	.0540	1.01006
760.0	20.87	.7451	16060.0	22500.0	187.1	22.87	31.35	567.1	34.9	.0551	1.00979
780.0	20.34	.7261	16520.0	23130.0	187.9	23.00	31.47	573.8	35.5	.0563	1.00954
800.0	19.84	.7082	16980.0	23760.0	188.7	23.13	31.59	580.4	36.1	.0574	1.00930
820.0	19.36	.6911	17450.0	24390.0	189.5	23.26	31.71	586.9	36.7	.0585	1.00907
840.0	18.90	.6748	17920.0	25030.0	190.3	23.39	31.84	593.4	37.3	.0597	1.00886
860.0	18.47	.6592	18390.0	25670.0	191.0	23.52	31.96	599.7	37.8	.0608	1.00865
880.0	18.05	.6444	18860.0	26310.0	191.8	23.64	32.07	606.0	38.4	.0619	1.00845
900.0	17.66	.6303	19340.0	26950.0	192.5	23.77	32.19	612.2	39.0	.0631	1.00827
920.0	17.28	.6167	19810.0	27600.0	193.2	23.89	32.31	618.3	39.5	.0642	1.00809
950.0	16.74	.5975	20530.0	28570.0	194.2	24.07	32.48	627.4	40.4	.0658	1.00783
1000.0	15.91	.5679	21750.0	30200.0	195.9	24.36	32.75	642.2	41.7	.0686	1.00744
1050.0	15.16	.5412	22970.0	31840.0	197.5	24.64	33.02	656.6	43.0	.0713	1.00708
1100.0	14.48	.5169	24210.0	33500.0	199.0	24.89	33.27	670.7	44.4	.0740	1.00676
1150.0	13.86	.4947	25470.0	35170.0	200.5	25.14	33.50	684.5	45.6	.0766	1.00647
1200.0	13.29	.4743	26730.0	36850.0	202.0	25.36	33.72	698.0	46.9	.0792	1.00620
1250.0	12.76	.4555	28000.0	38540.0	203.3	25.58	33.93	711.2	48.2	.0817	1.00595
1300.0	12.28	.4382	29290.0	40240.0	204.7	25.78	34.12	724.2	49.4	.0843	1.00572
1350.0	11.83	.4222	30580.0	41950.0	206.0	25.96	34.30	737.0	50.6	.0867	1.00551
1400.0	11.41	.4073	31890.0	43670.0	207.2	26.14	34.48	749.6	51.8	.0892	1.00531
1450.0	11.02	.3934	33200.0	45400.0	208.4	26.30	34.63	762.0	53.0	.0916	1.00512
1500.0	10.66	.3804	34520.0	47140.0	209.6	26.45	34.78	774.1	54.2	.0940	1.00495
1550.0	10.32	.3682	35840.0	48880.0	210.7	26.60	34.92	786.1	55.3	.0963	1.00479
1600.0	9.998	.3569	37180.0	50630.0	211.9	26.73	35.06	797.9	56.4	.0986	1.00464
1650.0	9.698	.3462	38520.0	52380.0	212.9	26.85	35.18	809.6	57.6	.101	1.00450
1700.0	9.415	.3361	39860.0	54150.0	214.0	26.97	35.29	821.1	58.7	.103	1.00437
1750.0	9.149	.3266	41210.0	55910.0	215.0	27.08	35.40	832.4	59.8	.105	1.00424
1800.0	8.897	.3176	42570.0	57690.0	216.0	27.18	35.50	843.6	60.9	.108	1.00412
1850.0	8.659	.3091	43930.0	59460.0	217.0	27.28	35.60	854.6	62.0	.110	1.00401
1900.0	8.433	.3010	45300.0	61250.0	217.9	27.37	35.69	865.5	63.0	.112	1.00390

5.00 MPa isobar

T K	Density kg/m^3	Density mol/dm^3	E J/mol	H J/mol	S J/(mol·K)	C_v J/(mol·K)	C_p J/(mol·K)	Sound m/s	Visc. μPa·s	Therm. W/(m·K)	Diel. const.
64.24[a]	869.7	31.04	−4195.0	−4034.0	68.28	29.55	54.30	1261.0	279.0	.152	1.46959
80.0	808.3	28.85	−3328.0	−3154.0	80.51	28.88	56.24	918.1	148.0	.134	1.43321
100.0	712.5	25.43	−2203.0	−2007.0	93.31	26.81	59.61	674.5	81.3	.105	1.37701
110.0	654.0	23.34	−1602.0	−1388.0	99.20	26.42	64.69	563.5	61.7	.0897	1.34318
115.0	620.1	22.13	−1280.0	−1054.0	102.2	26.32	69.12	506.3	53.3	.0816	1.32373
120.0	580.8	20.73	−933.1	−691.9	105.2	26.32	76.51	445.1	45.5	.0735	1.30141
125.0	531.9	18.99	−540.3	−276.9	108.6	26.48	91.53	375.7	37.8	.0657	1.27391
127.0	507.3	18.11	−359.5	−83.3	110.2	26.65	102.9	344.2	34.6	.0628	1.26022
128.0	493.2	17.60	−260.6	23.4	111.0	26.77	110.9	327.3	32.9	.0614	1.25240
130.0	459.5	16.40	−37.6	267.2	112.9	27.12	135.0	291.4	29.3	.0583	1.23390
132.0	415.1	14.82	235.9	573.4	115.2	27.69	173.1	254.7	25.2	.0544	1.20970
134.0	359.4	12.83	563.7	953.5	118.1	28.29	200.8	226.2	21.2	.0491	1.17989
136.0	305.9	10.92	889.7	1348.0	121.0	28.40	187.7	213.0	18.2	.0434	1.15173
138.0	265.0	9.459	1164.0	1692.0	123.5	27.96	155.9	209.5	16.3	.0386	1.13050
140.0	236.1	8.426	1380.0	1973.0	125.5	27.32	126.4	210.4	15.3	.0350	1.11566
142.0	215.3	7.684	1553.0	2203.0	127.2	26.69	105.2	212.9	14.6	.0323	1.10509
145.0	193.1	6.894	1761.0	2486.0	129.2	25.88	85.00	217.8	14.0	.0294	1.09392
150.0	168.9	6.027	2031.0	2861.0	131.7	24.85	66.99	226.1	13.6	.0266	1.08177
151.0	165.1	5.894	2078.0	2926.0	132.1	24.69	64.62	227.8	13.6	.0262	1.07990
152.0	161.6	5.770	2123.0	2990.0	132.5	24.53	62.51	229.4	13.5	.0259	1.07817
154.0	155.3	5.545	2209.0	3111.0	133.3	24.25	58.92	232.5	13.5	.0253	1.07504
156.0	149.8	5.346	2291.0	3226.0	134.1	24.00	55.98	235.6	13.4	.0248	1.07228

Thermophysical properties of nitrogen—Continued

T K	Density kg/m^3	Density mol/dm^3	E J/mol	H J/mol	S J/(mol·K)	C_v J/(mol·K)	C_p J/(mol·K)	Sound m/s	Visc. μPa·s	Therm. W/(m·K)	Diel. const.
158.0	144.8	5.168	2368.0	3335.0	134.8	23.77	53.54	238.6	13.4	.0244	1.06981
160.0	140.3	5.008	2442.0	3440.0	135.4	23.57	51.47	241.4	13.4	.0241	1.06759
162.0	136.2	4.862	2513.0	3542.0	136.1	23.39	49.70	244.2	13.4	.0238	1.06557
164.0	132.5	4.728	2582.0	3639.0	136.7	23.23	48.17	247.0	13.5	.0237	1.06372
166.0	129.0	4.604	2648.0	3734.0	137.2	23.08	46.83	249.6	13.5	.0235	1.06202
168.0	125.8	4.490	2713.0	3827.0	137.8	22.94	45.65	252.2	13.5	.0234	1.06044
170.0	122.8	4.383	2776.0	3917.0	138.3	22.82	44.61	254.7	13.6	.0233	1.05898
172.0	120.0	4.284	2838.0	4005.0	138.8	22.71	43.67	257.2	13.6	.0232	1.05761
174.0	117.4	4.190	2899.0	4092.0	139.3	22.61	42.84	259.6	13.7	.0232	1.05633
176.0	114.9	4.102	2958.0	4177.0	139.8	22.51	42.08	262.0	13.7	.0232	1.05512
178.0	112.6	4.020	3016.0	4260.0	140.3	22.43	41.39	264.3	13.8	.0232	1.05398
180.0	110.4	3.941	3074.0	4342.0	140.8	22.35	40.77	266.6	13.9	.0232	1.05291
182.0	108.3	3.867	3130.0	4423.0	141.2	22.27	40.20	268.8	13.9	.0232	1.05189
185.0	105.4	3.762	3214.0	4543.0	141.9	22.17	39.43	272.1	14.0	.0232	1.05046
190.0	100.9	3.603	3349.0	4737.0	142.9	22.03	38.34	277.3	14.2	.0234	1.04829
195.0	96.96	3.461	3482.0	4926.0	143.9	21.90	37.43	282.4	14.4	.0235	1.04634
200.0	93.36	3.332	3611.0	5112.0	144.8	21.80	36.67	287.3	14.6	.0237	1.04459
205.0	90.08	3.215	3738.0	5293.0	145.7	21.71	36.02	292.0	14.8	.0239	1.04300
210.0	87.07	3.108	3863.0	5472.0	146.6	21.64	35.45	296.5	15.0	.0241	1.04154
215.0	84.30	3.009	3986.0	5648.0	147.4	21.57	34.97	301.0	15.2	.0243	1.04020
220.0	81.74	2.918	4108.0	5822.0	148.2	21.51	34.54	305.3	15.4	.0245	1.03896
225.0	79.36	2.833	4228.0	5993.0	149.0	21.46	34.16	309.5	15.7	.0248	1.03781
230.0	77.15	2.754	4348.0	6163.0	149.7	21.42	33.82	313.6	15.9	.0250	1.03673
235.0	75.07	2.680	4466.0	6332.0	150.4	21.38	33.52	317.6	16.1	.0253	1.03573
240.0	73.12	2.610	4583.0	6499.0	151.2	21.34	33.25	321.5	16.3	.0255	1.03479
245.0	71.29	2.545	4699.0	6664.0	151.8	21.31	33.01	325.4	16.5	.0258	1.03390
250.0	69.56	2.483	4815.0	6829.0	152.5	21.28	32.79	329.2	16.7	.0260	1.03307
260.0	66.37	2.369	5044.0	7155.0	153.8	21.23	32.40	336.5	17.2	.0266	1.03154
270.0	63.50	2.267	5271.0	7477.0	155.0	21.19	32.08	343.6	17.6	.0271	1.03015
280.0	60.90	2.174	5496.0	7796.0	156.2	21.16	31.80	350.4	18.0	.0277	1.02890
290.0	58.53	2.089	5720.0	8113.0	157.3	21.13	31.57	357.0	18.4	.0282	1.02776
300.0	56.35	2.011	5942.0	8428.0	158.3	21.11	31.37	363.5	18.9	.0288	1.02672
310.0	54.35	1.940	6163.0	8741.0	159.4	21.10	31.19	369.8	19.3	.0294	1.02576
320.0	52.49	1.874	6383.0	9052.0	160.3	21.08	31.04	375.9	19.7	.0299	1.02487
330.0	50.77	1.812	6602.0	9361.0	161.3	21.07	30.90	381.9	20.1	.0305	1.02404
340.0	49.17	1.755	6821.0	9670.0	162.2	21.07	30.79	387.8	20.5	.0311	1.02327
350.0	47.67	1.702	7039.0	9977.0	163.1	21.06	30.68	393.5	20.9	.0316	1.02256
360.0	46.27	1.652	7256.0	10280.0	164.0	21.06	30.60	399.1	21.3	.0322	1.02189
370.0	44.95	1.605	7473.0	10590.0	164.8	21.07	30.52	404.6	21.7	.0328	1.02126
380.0	43.71	1.560	7689.0	10890.0	165.6	21.07	30.45	410.0	22.1	.0334	1.02066
390.0	42.54	1.519	7906.0	11200.0	166.4	21.08	30.39	415.3	22.5	.0339	1.02010
400.0	41.44	1.479	8122.0	11500.0	167.2	21.09	30.34	420.5	22.9	.0345	1.01958
410.0	40.40	1.442	8337.0	11810.0	167.9	21.11	30.30	425.6	23.3	.0351	1.01908
420.0	39.40	1.406	8553.0	12110.0	168.7	21.13	30.27	430.7	23.6	.0356	1.01860
440.0	37.57	1.341	8984.0	12710.0	170.1	21.17	30.22	440.5	24.4	.0368	1.01773
460.0	35.90	1.281	9415.0	13320.0	171.4	21.22	30.20	450.0	25.1	.0379	1.01693
480.0	34.38	1.227	9847.0	13920.0	172.7	21.29	30.20	459.2	25.8	.0391	1.01621
500.0	32.99	1.178	10280.0	14530.0	173.9	21.36	30.21	468.1	26.5	.0402	1.01555
520.0	31.71	1.132	10710.0	15130.0	175.1	21.45	30.25	476.9	27.2	.0414	1.01494
540.0	30.53	1.090	11150.0	15740.0	176.3	21.54	30.30	485.4	27.9	.0425	1.01438
560.0	29.44	1.051	11580.0	16340.0	177.4	21.64	30.36	493.6	28.6	.0437	1.01386
580.0	28.42	1.015	12020.0	16950.0	178.4	21.74	30.43	501.7	29.3	.0448	1.01337
600.0	27.48	.9808	12460.0	17560.0	179.5	21.86	30.51	509.6	29.9	.0460	1.01292
620.0	26.59	.9492	12900.0	18170.0	180.5	21.97	30.60	517.4	30.6	.0471	1.01250
640.0	25.77	.9197	13350.0	18780.0	181.4	22.09	30.69	525.0	31.2	.0483	1.01211
660.0	24.99	.8920	13790.0	19400.0	182.4	22.22	30.80	532.4	31.9	.0494	1.01174
680.0	24.26	.8659	14240.0	20010.0	183.3	22.34	30.90	539.7	32.5	.0506	1.01139
700.0	23.57	.8413	14690.0	20630.0	184.2	22.47	31.01	546.8	33.1	.0517	1.01107
720.0	22.92	.8182	15140.0	21260.0	185.1	22.60	31.13	553.9	33.7	.0529	1.01076
740.0	22.31	.7962	15600.0	21880.0	185.9	22.74	31.24	560.8	34.3	.0540	1.01047
760.0	21.73	.7755	16060.0	22500.0	186.8	22.87	31.36	567.6	34.9	.0552	1.01019
780.0	21.17	.7558	16520.0	23130.0	187.6	23.00	31.48	574.3	35.5	.0563	1.00993
800.0	20.65	.7371	16980.0	23760.0	188.4	23.13	31.60	580.9	36.1	.0575	1.00968

Thermophysical properties of nitrogen—Continued

T K	Density kg/m³	Density mol/dm³	E J/mol	H J/mol	S J/(mol·K)	C_v J/(mol·K)	C_p J/(mol·K)	Sound m/s	Visc. μPa·s	Therm. W/(m·K)	Diel. const.
820.0	20.15	.7193	17450.0	24400.0	189.2	23.26	31.72	587.4	36.7	.0586	1.00944
840.0	19.68	.7024	17910.0	25030.0	189.9	23.39	31.84	593.9	37.3	.0597	1.00922
860.0	19.22	.6862	18380.0	25670.0	190.7	23.52	31.96	600.2	37.8	.0609	1.00901
880.0	18.79	.6708	18860.0	26310.0	191.4	23.65	32.08	606.5	38.4	.0620	1.00880
900.0	18.38	.6560	19330.0	26950.0	192.1	23.77	32.20	612.7	39.0	.0631	1.00860
920.0	17.98	.6419	19810.0	27600.0	192.8	23.89	32.31	618.8	39.5	.0642	1.00842
950.0	17.42	.6219	20530.0	28570.0	193.9	24.07	32.48	627.8	40.4	.0659	1.00815
1000.0	16.56	.5912	21750.0	30200.0	195.6	24.36	32.76	642.6	41.7	.0686	1.00774
1050.0	15.78	.5634	22970.0	31850.0	197.2	24.64	33.02	657.0	43.1	.0713	1.00737
1100.0	15.07	.5381	24210.0	33500.0	198.7	24.90	33.27	671.1	44.4	.0740	1.00704
1150.0	14.43	.5150	25460.0	35170.0	200.2	25.14	33.50	684.9	45.7	.0766	1.00673
1200.0	13.83	.4938	26730.0	36850.0	201.6	25.37	33.72	698.4	46.9	.0792	1.00645
1250.0	13.29	.4742	28000.0	38550.0	203.0	25.58	33.93	711.7	48.2	.0818	1.00619
1300.0	12.78	.4562	29290.0	40250.0	204.3	25.78	34.13	724.7	49.4	.0843	1.00595
1350.0	12.31	.4395	30580.0	41960.0	205.6	25.97	34.31	737.4	50.6	.0868	1.00573
1400.0	11.88	.4240	31890.0	43680.0	206.9	26.14	34.48	750.0	51.8	.0892	1.00553
1450.0	11.47	.4095	33200.0	45410.0	208.1	26.30	34.64	762.3	53.0	.0916	1.00534
1500.0	11.10	.3960	34520.0	47140.0	209.3	26.45	34.79	774.5	54.2	.0940	1.00516
1550.0	10.74	.3834	35840.0	48880.0	210.4	26.60	34.93	786.5	55.3	.0963	1.00499
1600.0	10.41	.3716	37180.0	50630.0	211.5	26.73	35.06	798.3	56.4	.0986	1.00483
1650.0	10.10	.3604	38520.0	52390.0	212.6	26.86	35.18	809.9	57.6	.101	1.00468
1700.0	9.803	.3499	39860.0	54150.0	213.7	26.97	35.29	821.4	58.7	.103	1.00455
1750.0	9.526	.3400	41210.0	55920.0	214.7	27.08	35.40	832.7	59.8	.105	1.00441
1800.0	9.264	.3307	42570.0	57690.0	215.7	27.18	35.50	843.9	60.9	.108	1.00429
1850.0	9.016	.3218	43930.0	59470.0	216.7	27.28	35.60	855.0	62.0	.110	1.00417
1900.0	8.781	.3134	45300.0	61250.0	217.6	27.37	35.69	865.9	63.0	.112	1.00406

5.25 MPa isobar

T K	Density kg/m³	Density mol/dm³	E J/mol	H J/mol	S J/(mol·K)	C_v J/(mol·K)	C_p J/(mol·K)	Sound m/s	Visc. μPa·s	Therm. W/(m·K)	Diel. const.
64.29[a]	869.8	31.05	–4194.0	–4025.0	68.29	29.67	54.28	1258.0	279.0	.152	1.46964
80.0	808.9	28.87	–3331.0	–3149.0	80.47	28.95	56.17	919.3	148.0	.135	1.43354
100.0	713.7	25.47	–2210.0	–2004.0	93.23	26.84	59.40	677.7	81.8	.106	1.37771
110.0	655.9	23.41	–1613.0	–1389.0	99.09	26.44	64.23	568.0	62.2	.0902	1.34427
120.0	584.4	20.86	–952.5	–700.8	105.1	26.30	75.06	452.4	46.2	.0742	1.30343
125.0	537.7	19.19	–571.0	–297.4	108.4	26.42	87.87	386.1	38.7	.0665	1.27717
127.0	514.9	18.38	–398.9	–113.3	109.8	26.55	96.89	356.6	35.6	.0637	1.26445
128.0	502.1	17.92	–306.4	–13.4	110.6	26.64	103.0	341.0	34.0	.0623	1.25735
129.0	488.1	17.42	–208.2	93.2	111.4	26.76	110.5	324.8	32.4	.0608	1.24961
130.0	472.6	16.87	–103.0	208.2	112.3	26.90	120.0	308.2	30.7	.0593	1.24106
131.0	455.3	16.25	10.8	333.9	113.3	27.09	131.7	291.1	28.9	.0578	1.23155
132.0	435.7	15.55	134.8	472.4	114.3	27.31	145.7	274.1	27.1	.0560	1.22088
133.0	413.7	14.77	270.1	625.6	115.5	27.56	160.6	258.0	25.2	.0540	1.20897
134.0	389.7	13.91	415.4	792.9	116.8	27.82	173.1	243.9	23.4	.0516	1.19602
136.0	340.0	12.14	716.6	1149.0	119.4	28.16	178.4	224.7	20.1	.0464	1.16960
137.0	317.1	11.32	860.8	1325.0	120.7	28.15	171.8	219.5	18.8	.0439	1.15754
137.5	306.5	10.94	929.5	1409.0	121.3	28.10	167.0	217.8	18.3	.0427	1.15199
138.0	296.5	10.58	995.5	1492.0	121.9	28.03	161.6	216.5	17.8	.0416	1.14679
138.5	287.1	10.25	1059.0	1571.0	122.5	27.94	155.7	215.5	17.4	.0405	1.14194
139.0	278.4	9.938	1119.0	1647.0	123.0	27.83	149.6	214.9	17.0	.0395	1.13742
139.5	270.3	9.649	1176.0	1721.0	123.5	27.71	143.5	214.5	16.7	.0386	1.13324
140.0	262.8	9.380	1231.0	1791.0	124.0	27.58	137.5	214.4	16.4	.0377	1.12936
140.5	255.8	9.131	1283.0	1858.0	124.5	27.44	131.6	214.4	16.1	.0368	1.12577
141.0	249.3	8.900	1333.0	1922.0	125.0	27.30	126.0	214.6	15.9	.0360	1.12245
141.5	243.3	8.685	1380.0	1984.0	125.4	27.15	120.7	214.9	15.7	.0353	1.11937
142.0	237.7	8.486	1425.0	2043.0	125.8	27.01	115.8	215.3	15.5	.0346	1.11651
142.5	232.5	8.300	1467.0	2100.0	126.2	26.87	111.2	215.8	15.3	.0339	1.11385
143.0	227.7	8.126	1508.0	2154.0	126.6	26.73	106.9	216.3	15.1	.0333	1.11138
144.0	218.8	7.811	1585.0	2257.0	127.3	26.45	99.26	217.6	14.9	.0322	1.10689
145.0	211.0	7.533	1656.0	2353.0	128.0	26.19	92.76	219.0	14.7	.0312	1.10295
146.0	204.1	7.285	1723.0	2443.0	128.6	25.95	87.19	220.4	14.5	.0304	1.09944
147.0	197.9	7.062	1785.0	2528.0	129.2	25.72	82.40	222.0	14.3	.0296	1.09629
148.0	192.2	6.861	1843.0	2608.0	129.7	25.50	78.27	223.5	14.2	.0290	1.09346

Thermophysical properties of nitrogen—Continued

T K	Density kg/m^3	Density mol/dm^3	E J/mol	H J/mol	S J/(mol·K)	C_v J/(mol·K)	C_p	Sound m/s	Visc. μPa·s	Therm. W/(m·K)	Diel. const.
149.0	187.1	6.678	1898.0	2685.0	130.3	25.30	74.66	225.1	14.1	.0284	1.09088
150.0	182.4	6.510	1951.0	2758.0	130.7	25.11	71.51	226.7	14.0	.0279	1.08852
151.0	178.0	6.355	2002.0	2828.0	131.2	24.93	68.73	228.2	14.0	.0274	1.08635
152.0	174.0	6.212	2050.0	2895.0	131.7	24.77	66.26	229.8	13.9	.0270	1.08434
153.0	170.3	6.079	2097.0	2960.0	132.1	24.61	64.05	231.3	13.8	.0266	1.08248
154.0	166.8	5.954	2142.0	3023.0	132.5	24.46	62.08	232.9	13.8	.0262	1.08074
156.0	160.5	5.728	2228.0	3144.0	133.3	24.19	58.68	235.9	13.7	.0257	1.07759
158.0	154.9	5.527	2309.0	3259.0	134.0	23.95	55.87	238.8	13.7	.0252	1.07479
160.0	149.8	5.347	2386.0	3368.0	134.7	23.73	53.51	241.7	13.7	.0248	1.07229
162.0	145.2	5.184	2460.0	3473.0	135.3	23.54	51.51	244.5	13.7	.0245	1.07003
164.0	141.1	5.035	2531.0	3574.0	136.0	23.37	49.78	247.2	13.7	.0243	1.06796
166.0	137.2	4.898	2600.0	3672.0	136.6	23.21	48.28	249.9	13.7	.0241	1.06607
168.0	133.7	4.772	2667.0	3767.0	137.1	23.07	46.96	252.5	13.7	.0239	1.06433
170.0	130.4	4.655	2732.0	3860.0	137.7	22.93	45.80	255.0	13.8	.0238	1.06271
172.0	127.3	4.545	2796.0	3951.0	138.2	22.82	44.77	257.5	13.8	.0237	1.06120
174.0	124.5	4.443	2858.0	4039.0	138.7	22.71	43.84	259.9	13.9	.0237	1.05980
176.0	121.8	4.347	2918.0	4126.0	139.2	22.61	43.01	262.3	13.9	.0236	1.05848
178.0	119.3	4.257	2978.0	4211.0	139.7	22.51	42.25	264.6	14.0	.0236	1.05724
180.0	116.9	4.172	3037.0	4295.0	140.2	22.43	41.57	266.9	14.0	.0236	1.05607
182.0	114.6	4.091	3094.0	4378.0	140.6	22.35	40.95	269.1	14.1	.0236	1.05496
184.0	112.5	4.015	3151.0	4459.0	141.1	22.28	40.38	271.3	14.2	.0236	1.05391
190.0	106.6	3.806	3317.0	4697.0	142.3	22.09	38.92	277.7	14.4	.0237	1.05106
195.0	102.3	3.653	3451.0	4889.0	143.3	21.96	37.94	282.8	14.5	.0238	1.04896
200.0	98.47	3.515	3583.0	5076.0	144.3	21.85	37.12	287.7	14.7	.0240	1.04708
205.0	94.96	3.389	3711.0	5260.0	145.2	21.76	36.42	292.4	14.9	.0242	1.04537
210.0	91.74	3.275	3837.0	5441.0	146.1	21.68	35.81	297.0	15.1	.0244	1.04380
215.0	88.79	3.169	3962.0	5618.0	146.9	21.61	35.29	301.5	15.3	.0246	1.04237
220.0	86.06	3.072	4085.0	5794.0	147.7	21.55	34.84	305.8	15.5	.0248	1.04104
225.0	83.53	2.981	4206.0	5967.0	148.5	21.49	34.43	310.0	15.7	.0250	1.03982
230.0	81.17	2.897	4326.0	6138.0	149.2	21.45	34.07	314.1	16.0	.0252	1.03867
235.0	78.96	2.818	4445.0	6308.0	150.0	21.40	33.76	318.2	16.2	.0255	1.03761
240.0	76.89	2.745	4563.0	6476.0	150.7	21.37	33.47	322.1	16.4	.0257	1.03661
245.0	74.95	2.675	4680.0	6642.0	151.4	21.33	33.21	326.0	16.6	.0260	1.03567
250.0	73.11	2.610	4796.0	6808.0	152.0	21.30	32.98	329.7	16.8	.0262	1.03478
260.0	69.74	2.489	5027.0	7135.0	153.3	21.25	32.57	337.1	17.2	.0268	1.03316
270.0	66.71	2.381	5255.0	7459.0	154.5	21.21	32.23	344.2	17.7	.0273	1.03169
280.0	63.96	2.283	5481.0	7780.0	155.7	21.18	31.94	351.1	18.1	.0279	1.03037
290.0	61.46	2.194	5705.0	8098.0	156.8	21.15	31.69	357.7	18.5	.0284	1.02917
300.0	59.16	2.112	5928.0	8414.0	157.9	21.13	31.48	364.2	18.9	.0290	1.02806
310.0	57.05	2.036	6150.0	8728.0	158.9	21.11	31.29	370.5	19.3	.0295	1.02705
320.0	55.10	1.967	6370.0	9040.0	159.9	21.10	31.13	376.6	19.7	.0301	1.02611
330.0	53.28	1.902	6590.0	9351.0	160.9	21.09	30.99	382.6	20.2	.0306	1.02524
340.0	51.60	1.842	6809.0	9660.0	161.8	21.08	30.87	388.5	20.6	.0312	1.02443
350.0	50.02	1.785	7027.0	9968.0	162.7	21.08	30.76	394.2	21.0	.0318	1.02368
360.0	48.55	1.733	7245.0	10280.0	163.5	21.08	30.66	399.8	21.4	.0323	1.02297
370.0	47.16	1.683	7463.0	10580.0	164.4	21.08	30.58	405.3	21.8	.0329	1.02231
380.0	45.86	1.637	7679.0	10890.0	165.2	21.08	30.51	410.7	22.1	.0335	1.02169
390.0	44.63	1.593	7896.0	11190.0	166.0	21.09	30.45	416.0	22.5	.0340	1.02110
400.0	43.47	1.552	8112.0	11500.0	166.8	21.10	30.40	421.2	22.9	.0346	1.02054
410.0	42.37	1.512	8328.0	11800.0	167.5	21.12	30.35	426.4	23.3	.0352	1.02002
420.0	41.33	1.475	8544.0	12100.0	168.2	21.14	30.32	431.4	23.7	.0357	1.01952
440.0	39.41	1.407	8976.0	12710.0	169.6	21.18	30.27	441.2	24.4	.0369	1.01860
460.0	37.66	1.344	9408.0	13310.0	171.0	21.23	30.24	450.7	25.1	.0380	1.01777
480.0	36.06	1.287	9840.0	13920.0	172.3	21.30	30.23	459.9	25.9	.0392	1.01701
500.0	34.60	1.235	10270.0	14520.0	173.5	21.37	30.25	468.9	26.6	.0403	1.01631
520.0	33.26	1.187	10710.0	15130.0	174.7	21.45	30.28	477.6	27.3	.0415	1.01567
540.0	32.02	1.143	11140.0	15730.0	175.8	21.54	30.32	486.1	28.0	.0426	1.01508
560.0	30.88	1.102	11580.0	16340.0	176.9	21.64	30.38	494.4	28.6	.0438	1.01454
580.0	29.81	1.064	12020.0	16950.0	178.0	21.75	30.45	502.4	29.3	.0449	1.01403
600.0	28.82	1.029	12460.0	17560.0	179.0	21.86	30.53	510.3	30.0	.0461	1.01356
620.0	27.89	.9956	12900.0	18170.0	180.0	21.98	30.62	518.1	30.6	.0472	1.01312
640.0	27.03	.9646	13340.0	18780.0	181.0	22.10	30.71	525.6	31.2	.0484	1.01270
660.0	26.21	.9356	13790.0	19400.0	182.0	22.22	30.81	533.1	31.9	.0495	1.01232

Thermophysical properties of nitrogen—Continued

T K	Density kg/m³	Density mol/dm³	E J/mol	H J/mol	S J/(mol·K)	C_v J/(mol·K)	C_p J/(mol·K)	Sound m/s	Visc. μPa·s	Therm. W/(m·K)	Diel. const.
680.0	25.45	.9082	14240.0	20020.0	182.9	22.35	30.92	540.3	32.5	.0507	1.01195
700.0	24.72	.8825	14690.0	20640.0	183.8	22.48	31.03	547.5	33.1	.0518	1.01161
720.0	24.04	.8582	15140.0	21260.0	184.7	22.61	31.14	554.5	33.7	.0530	1.01129
740.0	23.40	.8352	15600.0	21880.0	185.5	22.74	31.26	561.4	34.3	.0541	1.01098
760.0	22.79	.8134	16050.0	22510.0	186.4	22.87	31.38	568.2	34.9	.0552	1.01069
780.0	22.21	.7928	16510.0	23140.0	187.2	23.00	31.49	574.9	35.5	.0564	1.01042
800.0	21.66	.7732	16980.0	23770.0	188.0	23.14	31.61	581.5	36.1	.0575	1.01016
820.0	21.14	.7545	17440.0	24400.0	188.8	23.27	31.73	588.1	36.7	.0587	1.00991
840.0	20.64	.7368	17910.0	25040.0	189.5	23.40	31.85	594.5	37.3	.0598	1.00967
860.0	20.17	.7198	18380.0	25680.0	190.3	23.52	31.97	600.8	37.9	.0609	1.00945
880.0	19.71	.7037	18850.0	26320.0	191.0	23.65	32.09	607.1	38.4	.0620	1.00923
900.0	19.28	.6882	19330.0	26960.0	191.7	23.77	32.21	613.3	39.0	.0631	1.00903
920.0	18.87	.6734	19810.0	27600.0	192.4	23.90	32.32	619.4	39.5	.0643	1.00883
950.0	18.28	.6524	20530.0	28580.0	193.5	24.08	32.49	628.4	40.4	.0659	1.00855
1000.0	17.38	.6202	21740.0	30210.0	195.2	24.37	32.77	643.2	41.7	.0687	1.00812
1050.0	16.56	.5911	22970.0	31850.0	196.8	24.64	33.03	657.6	43.1	.0714	1.00774
1100.0	15.82	.5645	24210.0	33510.0	198.3	24.90	33.27	671.7	44.4	.0740	1.00739
1150.0	15.14	.5403	25460.0	35180.0	199.8	25.14	33.51	685.5	45.7	.0767	1.00706
1200.0	14.51	.5181	26730.0	36860.0	201.2	25.37	33.73	698.9	46.9	.0793	1.00677
1250.0	13.94	.4976	28000.0	38550.0	202.6	25.58	33.93	712.2	48.2	.0818	1.00650
1300.0	13.41	.4787	29290.0	40250.0	203.9	25.78	34.13	725.2	49.4	.0843	1.00625
1350.0	12.92	.4612	30580.0	41970.0	205.2	25.97	34.31	737.9	50.6	.0868	1.00602
1400.0	12.46	.4449	31880.0	43680.0	206.5	26.14	34.48	750.5	51.8	.0893	1.00580
1450.0	12.04	.4298	33200.0	45410.0	207.7	26.30	34.64	762.8	53.0	.0917	1.00560
1500.0	11.64	.4156	34520.0	47150.0	208.9	26.46	34.79	775.0	54.2	.0940	1.00541
1550.0	11.27	.4023	35840.0	48890.0	210.0	26.60	34.93	787.0	55.3	.0964	1.00524
1600.0	10.92	.3899	37180.0	50640.0	211.1	26.73	35.06	798.8	56.5	.0987	1.00507
1650.0	10.60	.3782	38520.0	52400.0	212.2	26.86	35.18	810.4	57.6	.101	1.00492
1700.0	10.29	.3672	39860.0	54160.0	213.2	26.97	35.30	821.9	58.7	.103	1.00477
1750.0	9.997	.3568	41210.0	55930.0	214.3	27.08	35.40	833.2	59.8	.105	1.00463
1800.0	9.722	.3470	42570.0	57700.0	215.3	27.19	35.51	844.4	60.9	.108	1.00450
1850.0	9.462	.3377	43930.0	59480.0	216.2	27.28	35.60	855.4	62.0	.110	1.00438
1900.0	9.216	.3289	45300.0	61260.0	217.2	27.37	35.69	866.3	63.0	.112	1.00426

5.50 MPa isobar

T K	Density kg/m³	Density mol/dm³	E J/mol	H J/mol	S J/(mol·K)	C_v J/(mol·K)	C_p J/(mol·K)	Sound m/s	Visc. μPa·s	Therm. W/(m·K)	Diel. const.
64.35[a]	869.9	31.05	−4193.0	−4016.0	68.30	29.80	54.26	1256.0	278.0	.152	1.46969
80.0	809.5	28.89	−3335.0	−3144.0	80.42	29.01	56.09	920.6	149.0	.135	1.43387
100.0	714.9	25.52	−2217.0	−2002.0	93.16	26.88	59.19	680.8	82.3	.106	1.37840
110.0	657.8	23.48	−1623.0	−1389.0	98.99	26.46	63.80	572.5	62.7	.0907	1.34533
120.0	587.8	20.98	−971.0	−708.8	104.9	26.29	73.76	459.4	46.8	.0749	1.30535
125.0	543.1	19.38	−599.2	−315.5	108.1	26.37	84.84	395.8	39.5	.0673	1.28016
127.0	521.7	18.62	−434.2	−138.8	109.5	26.47	92.22	367.9	36.5	.0645	1.26822
128.0	509.9	18.20	−346.5	−44.3	110.3	26.54	97.00	353.3	35.0	.0631	1.26167
129.0	497.2	17.75	−254.4	55.5	111.0	26.63	102.8	338.3	33.5	.0617	1.25461
130.0	483.4	17.25	−157.2	161.6	111.9	26.74	109.7	322.9	31.9	.0603	1.24696
131.0	468.2	16.71	−53.7	275.5	112.7	26.87	118.2	307.1	30.3	.0589	1.23861
132.0	451.4	16.11	57.1	398.5	113.7	27.04	128.1	291.3	28.6	.0573	1.22943
133.0	432.9	15.45	176.1	532.1	114.7	27.23	139.2	275.9	26.9	.0555	1.21933
134.0	412.5	14.73	303.3	676.8	115.8	27.44	150.2	261.4	25.2	.0535	1.20833
135.0	390.8	13.95	437.4	831.6	116.9	27.65	158.8	248.8	23.5	.0513	1.19665
136.0	368.5	13.15	574.9	993.0	118.1	27.81	163.1	238.6	22.0	.0489	1.18472
137.0	346.6	12.37	711.7	1156.0	119.3	27.91	162.6	231.1	20.6	.0465	1.17308
138.0	325.9	11.63	844.0	1317.0	120.5	27.91	158.2	225.8	19.4	.0442	1.16215
139.0	306.8	10.95	969.5	1472.0	121.6	27.83	151.1	222.4	18.4	.0421	1.15218
139.5	298.0	10.64	1029.0	1546.0	122.1	27.76	146.9	221.3	18.0	.0411	1.14759
140.0	289.7	10.34	1087.0	1619.0	122.6	27.67	142.4	220.4	17.6	.0402	1.14327
140.5	281.9	10.06	1142.0	1689.0	123.1	27.57	137.7	219.8	17.3	.0393	1.13922
141.0	274.6	9.800	1195.0	1756.0	123.6	27.46	133.0	219.4	17.0	.0384	1.13542
141.5	267.7	9.554	1246.0	1822.0	124.1	27.35	128.3	219.2	16.7	.0376	1.13186
142.0	261.2	9.323	1295.0	1885.0	124.5	27.22	123.7	219.2	16.4	.0369	1.12853
142.5	255.1	9.107	1341.0	1945.0	124.9	27.10	119.3	219.3	16.2	.0361	1.12542
143.0	249.5	8.904	1386.0	2004.0	125.4	26.97	115.0	219.5	16.0	.0355	1.12250

Thermophysical properties of nitrogen—Continued

T K	Density kg/m^3	Density mol/dm^3	E J/mol	H J/mol	S J/(mol·K)	C_v J/(mol·K)	C_p J/(mol·K)	Sound m/s	Visc. μPa·s	Therm. W/(m·K)	Diel. const.
143.5	244.1	8.714	1429.0	2060.0	125.8	26.84	111.0	219.8	15.8	.0348	1.11978
144.0	239.1	8.535	1470.0	2115.0	126.1	26.71	107.1	220.1	15.7	.0342	1.11722
145.0	230.0	8.209	1548.0	2218.0	126.8	26.46	100.1	221.1	15.4	.0331	1.11255
146.0	221.8	7.918	1621.0	2315.0	127.5	26.22	93.98	222.2	15.1	.0321	1.10841
147.0	214.5	7.658	1688.0	2407.0	128.1	25.98	88.61	223.5	14.9	.0312	1.10471
148.0	208.0	7.423	1752.0	2493.0	128.7	25.76	83.91	224.8	14.8	.0305	1.10139
149.0	202.0	7.210	1812.0	2575.0	129.3	25.55	79.80	226.2	14.6	.0298	1.09838
150.0	196.6	7.016	1869.0	2653.0	129.8	25.35	76.19	227.6	14.5	.0292	1.09564
151.0	191.6	6.838	1923.0	2727.0	130.3	25.17	72.99	229.1	14.4	.0286	1.09313
152.0	187.0	6.674	1975.0	2799.0	130.8	24.99	70.16	230.6	14.3	.0281	1.09081
153.0	182.7	6.521	2024.0	2868.0	131.2	24.83	67.64	232.1	14.2	.0277	1.08868
154.0	178.7	6.380	2072.0	2934.0	131.6	24.67	65.37	233.5	14.2	.0273	1.08669
155.0	175.0	6.248	2118.0	2998.0	132.1	24.52	63.34	235.0	14.1	.0269	1.08484
156.0	171.6	6.124	2163.0	3061.0	132.5	24.38	61.50	236.5	14.1	.0266	1.08311
158.0	165.2	5.898	2248.0	3181.0	133.2	24.13	58.31	239.3	14.0	.0260	1.07996
160.0	159.6	5.697	2329.0	3294.0	133.9	23.90	55.64	242.2	14.0	.0256	1.07714
162.0	154.5	5.515	2406.0	3403.0	134.6	23.69	53.39	244.9	13.9	.0252	1.07462
164.0	149.9	5.350	2480.0	3508.0	135.3	23.51	51.45	247.6	13.9	.0249	1.07232
166.0	145.6	5.199	2551.0	3609.0	135.9	23.34	49.78	250.3	13.9	.0247	1.07023
168.0	141.8	5.060	2620.0	3707.0	136.5	23.19	48.32	252.9	14.0	.0245	1.06830
170.0	138.2	4.931	2687.0	3803.0	137.0	23.05	47.03	255.4	14.0	.0243	1.06653
172.0	134.8	4.812	2753.0	3896.0	137.6	22.92	45.89	257.9	14.0	.0242	1.06487
174.0	131.7	4.700	2816.0	3986.0	138.1	22.80	44.87	260.3	14.1	.0241	1.06334
176.0	128.8	4.596	2879.0	4075.0	138.6	22.70	43.96	262.7	14.1	.0241	1.06190
178.0	126.0	4.498	2939.0	4162.0	139.1	22.60	43.13	265.0	14.1	.0240	1.06055
180.0	123.4	4.406	2999.0	4248.0	139.6	22.51	42.39	267.3	14.2	.0240	1.05927
182.0	121.0	4.318	3058.0	4332.0	140.0	22.43	41.71	269.5	14.2	.0240	1.05807
184.0	118.7	4.235	3116.0	4415.0	140.5	22.35	41.09	271.7	14.3	.0240	1.05694
186.0	116.5	4.157	3173.0	4496.0	140.9	22.28	40.52	273.9	14.4	.0240	1.05586
190.0	112.4	4.011	3285.0	4656.0	141.8	22.15	39.51	278.1	14.5	.0241	1.05386
195.0	107.8	3.846	3421.0	4851.0	142.8	22.01	38.45	283.2	14.7	.0242	1.05160
200.0	103.6	3.699	3554.0	5041.0	143.8	21.90	37.57	288.2	14.8	.0243	1.04958
205.0	99.86	3.565	3684.0	5227.0	144.7	21.80	36.82	292.9	15.0	.0244	1.04775
210.0	96.44	3.442	3812.0	5409.0	145.6	21.72	36.18	297.5	15.2	.0246	1.04608
215.0	93.29	3.330	3937.0	5589.0	146.4	21.64	35.62	302.0	15.4	.0248	1.04455
220.0	90.39	3.226	4061.0	5766.0	147.2	21.58	35.13	306.3	15.6	.0250	1.04314
225.0	87.70	3.130	4183.0	5940.0	148.0	21.52	34.71	310.6	15.8	.0252	1.04184
230.0	85.20	3.041	4304.0	6113.0	148.8	21.47	34.33	314.7	16.0	.0255	1.04062
235.0	82.86	2.958	4424.0	6284.0	149.5	21.43	33.99	318.8	16.2	.0257	1.03949
240.0	80.67	2.879	4543.0	6453.0	150.2	21.39	33.68	322.7	16.5	.0259	1.03843
245.0	78.61	2.806	4660.0	6621.0	150.9	21.36	33.41	326.6	16.7	.0262	1.03743
250.0	76.68	2.737	4777.0	6787.0	151.6	21.33	33.16	330.4	16.9	.0264	1.03650
260.0	73.12	2.610	5009.0	7116.0	152.9	21.27	32.73	337.7	17.3	.0270	1.03478
270.0	69.92	2.496	5238.0	7442.0	154.1	21.23	32.37	344.8	17.7	.0275	1.03324
280.0	67.02	2.392	5465.0	7764.0	155.3	21.20	32.07	351.7	18.1	.0280	1.03184
290.0	64.38	2.298	5690.0	8083.0	156.4	21.17	31.81	358.4	18.6	.0286	1.03057
300.0	61.97	2.212	5914.0	8400.0	157.5	21.14	31.59	364.9	19.0	.0291	1.02941
310.0	59.75	2.133	6136.0	8715.0	158.5	21.12	31.39	371.2	19.4	.0297	1.02834
320.0	57.69	2.059	6358.0	9028.0	159.5	21.11	31.22	377.3	19.8	.0302	1.02735
330.0	55.79	1.991	6578.0	9340.0	160.4	21.10	31.07	383.3	20.2	.0308	1.02644
340.0	54.02	1.928	6797.0	9650.0	161.4	21.09	30.94	389.2	20.6	.0313	1.02559
350.0	52.37	1.869	7016.0	9959.0	162.3	21.09	30.83	394.9	21.0	.0319	1.02480
360.0	50.82	1.814	7234.0	10270.0	163.1	21.09	30.73	400.6	21.4	.0325	1.02406
370.0	49.37	1.762	7452.0	10570.0	164.0	21.09	30.65	406.1	21.8	.0330	1.02336
380.0	48.00	1.713	7669.0	10880.0	164.8	21.09	30.57	411.5	22.2	.0336	1.02271
390.0	46.71	1.667	7886.0	11180.0	165.6	21.10	30.51	416.8	22.6	.0342	1.02209
400.0	45.50	1.624	8103.0	11490.0	166.3	21.11	30.45	422.0	22.9	.0347	1.02151
410.0	44.35	1.583	8319.0	11790.0	167.1	21.13	30.40	427.1	23.3	.0353	1.02096
420.0	43.26	1.544	8536.0	12100.0	167.8	21.14	30.36	432.1	23.7	.0359	1.02044
440.0	41.24	1.472	8968.0	12700.0	169.2	21.19	30.31	441.9	24.4	.0370	1.01947
460.0	39.41	1.407	9400.0	13310.0	170.6	21.24	30.28	451.4	25.2	.0381	1.01860
480.0	37.74	1.347	9833.0	13920.0	171.9	21.30	30.27	460.6	25.9	.0393	1.01780
500.0	36.21	1.293	10270.0	14520.0	173.1	21.38	30.28	469.6	26.6	.0404	1.01707

Thermophysical properties of nitrogen—Continued

T K	Density kg/m³	Density mol/dm³	E J/mol	H J/mol	S J/(mol·K)	C_v J/(mol·K)	C_p J/(mol·K)	Sound m/s	Visc. μPa·s	Therm. W/(m·K)	Diel. const.
520.0	34.81	1.242	10700.0	15130.0	174.3	21.46	30.31	478.3	27.3	.0416	1.01640
540.0	33.51	1.196	11140.0	15730.0	175.4	21.55	30.35	486.8	28.0	.0427	1.01579
560.0	32.31	1.153	11570.0	16340.0	176.5	21.65	30.40	495.1	28.7	.0438	1.01522
580.0	31.20	1.114	12010.0	16950.0	177.6	21.75	30.47	503.1	29.3	.0450	1.01468
600.0	30.16	1.076	12450.0	17560.0	178.7	21.87	30.55	511.0	30.0	.0461	1.01419
620.0	29.19	1.042	12890.0	18170.0	179.7	21.98	30.64	518.8	30.6	.0473	1.01373
640.0	28.28	1.009	13340.0	18790.0	180.6	22.10	30.73	526.3	31.3	.0484	1.01330
660.0	27.43	.9791	13780.0	19400.0	181.6	22.23	30.83	533.7	31.9	.0496	1.01289
680.0	26.63	.9505	14230.0	20020.0	182.5	22.35	30.93	541.0	32.5	.0507	1.01251
700.0	25.87	.9235	14680.0	20640.0	183.4	22.48	31.04	548.2	33.1	.0519	1.01215
720.0	25.16	.8981	15140.0	21260.0	184.3	22.61	31.16	555.2	33.8	.0530	1.01181
740.0	24.49	.8741	15590.0	21880.0	185.1	22.74	31.27	562.1	34.4	.0542	1.01149
760.0	23.85	.8513	16050.0	22510.0	186.0	22.88	31.39	568.9	35.0	.0553	1.01119
780.0	23.24	.8297	16510.0	23140.0	186.8	23.01	31.51	575.6	35.6	.0564	1.01090
800.0	22.67	.8092	16970.0	23770.0	187.6	23.14	31.62	582.2	36.1	.0576	1.01063
820.0	22.12	.7897	17440.0	24410.0	188.4	23.27	31.74	588.7	36.7	.0587	1.01037
840.0	21.60	.7711	17910.0	25040.0	189.1	23.40	31.86	595.1	37.3	.0598	1.01013
860.0	21.11	.7534	18380.0	25680.0	189.9	23.53	31.98	601.4	37.9	.0610	1.00989
880.0	20.63	.7365	18850.0	26320.0	190.6	23.65	32.10	607.7	38.4	.0621	1.00967
900.0	20.18	.7203	19330.0	26960.0	191.3	23.78	32.21	613.9	39.0	.0632	1.00945
920.0	19.75	.7049	19810.0	27610.0	192.0	23.90	32.33	620.0	39.6	.0643	1.00924
950.0	19.13	.6829	20530.0	28580.0	193.1	24.08	32.50	629.0	40.4	.0660	1.00895
1000.0	18.19	.6492	21740.0	30210.0	194.8	24.37	32.77	643.8	41.7	.0687	1.00851
1050.0	17.33	.6187	22970.0	31860.0	196.4	24.64	33.03	658.2	43.1	.0714	1.00810
1100.0	16.56	.5909	24210.0	33520.0	197.9	24.90	33.28	672.2	44.4	.0741	1.00773
1150.0	15.85	.5656	25460.0	35190.0	199.4	25.14	33.51	686.0	45.7	.0767	1.00740
1200.0	15.19	.5423	26730.0	36870.0	200.8	25.37	33.73	699.5	46.9	.0793	1.00709
1250.0	14.59	.5209	28000.0	38560.0	202.2	25.58	33.94	712.7	48.2	.0819	1.00680
1300.0	14.04	.5011	29290.0	40260.0	203.5	25.78	34.13	725.7	49.4	.0844	1.00654
1350.0	13.53	.4828	30580.0	41970.0	204.8	25.97	34.31	738.4	50.6	.0869	1.00630
1400.0	13.05	.4658	31880.0	43690.0	206.1	26.14	34.48	751.0	51.8	.0893	1.00607
1450.0	12.61	.4499	33200.0	45420.0	207.3	26.31	34.64	763.3	53.0	.0917	1.00586
1500.0	12.19	.4351	34520.0	47160.0	208.5	26.46	34.79	775.5	54.2	.0941	1.00567
1550.0	11.80	.4213	35840.0	48900.0	209.6	26.60	34.93	787.4	55.3	.0964	1.00548
1600.0	11.44	.4082	37180.0	50650.0	210.7	26.73	35.06	799.2	56.5	.0987	1.00531
1650.0	11.09	.3960	38520.0	52400.0	211.8	26.86	35.18	810.8	57.6	.101	1.00515
1700.0	10.77	.3845	39860.0	54170.0	212.9	26.97	35.30	822.3	58.7	.103	1.00500
1750.0	10.47	.3736	41210.0	55930.0	213.9	27.08	35.41	833.6	59.8	.105	1.00485
1800.0	10.18	.3634	42570.0	57710.0	214.9	27.19	35.51	844.8	60.9	.108	1.00472
1850.0	9.908	.3537	43930.0	59480.0	215.9	27.28	35.60	855.8	62.0	.110	1.00459
1900.0	9.650	.3444	45300.0	61270.0	216.8	27.37	35.69	866.7	63.0	.112	1.00447
					5.75 MPa isobar						
64.40[a]	870.0	31.05	−4192.0	−4007.0	68.32	29.92	54.24	1253.0	278.0	.153	1.46974
80.0	810.0	28.91	−3338.0	−3139.0	80.38	29.07	56.01	921.8	149.0	.135	1.43420
100.0	716.1	25.56	−2224.0	−1999.0	93.09	26.91	58.99	683.9	82.8	.106	1.37908
110.0	659.6	23.54	−1634.0	−1389.0	98.89	26.47	63.39	576.9	63.2	.0912	1.34637
120.0	591.0	21.10	−988.7	−716.2	104.7	26.28	72.59	466.2	47.5	.0755	1.30718
125.0	548.0	19.56	−625.4	−331.5	107.9	26.33	82.29	404.9	40.3	.0681	1.28292
130.0	492.5	17.58	−203.7	123.4	111.5	26.61	102.3	336.1	33.0	.0612	1.25200
131.0	478.9	17.09	−107.6	228.8	112.3	26.71	108.7	321.4	31.5	.0598	1.24448
132.0	464.0	16.56	−6.1	341.0	113.1	26.83	116.0	306.6	29.9	.0584	1.23634
133.0	447.9	15.99	101.5	461.2	114.0	26.98	124.4	291.9	28.3	.0568	1.22751
134.0	430.4	15.36	215.6	589.9	115.0	27.14	133.1	277.8	26.8	.0550	1.21797
135.0	411.5	14.69	335.7	727.2	116.0	27.32	141.2	264.8	25.2	.0531	1.20777
136.0	391.7	13.98	460.5	871.7	117.1	27.48	147.4	253.5	23.7	.0510	1.19713
138.0	351.8	12.56	713.4	1171.0	119.3	27.68	150.0	237.1	21.0	.0466	1.17584
139.0	332.9	11.88	836.0	1320.0	120.3	27.69	146.8	231.9	19.9	.0445	1.16586
140.0	315.3	11.26	953.4	1464.0	121.4	27.62	141.6	228.3	19.0	.0425	1.15662
141.0	299.2	10.68	1064.0	1603.0	122.3	27.49	135.1	226.0	18.2	.0407	1.14822
142.0	284.7	10.16	1168.0	1734.0	123.3	27.32	127.8	224.6	17.5	.0391	1.14066

Thermophysical properties of nitrogen—Continued

T K	Density kg/m^3	Density mol/dm^3	E J/mol	H J/mol	S J/(mol·K)	C_v J/(mol·K)	C_p J/(mol·K)	Sound m/s	Visc. μPa·s	Therm. W/(m·K)	Diel. const.
143.0	271.7	9.696	1265.0	1858.0	124.1	27.12	120.3	224.0	17.0	.0376	1.13391
144.0	260.0	9.280	1355.0	1975.0	125.0	26.90	113.1	223.9	16.5	.0362	1.12791
145.0	249.6	8.909	1439.0	2084.0	125.7	26.67	106.2	224.2	16.2	.0350	1.12257
146.0	240.3	8.577	1517.0	2188.0	126.4	26.44	99.98	224.9	15.8	.0339	1.11781
147.0	231.9	8.278	1590.0	2285.0	127.1	26.21	94.34	225.8	15.6	.0329	1.11354
148.0	224.4	8.009	1659.0	2376.0	127.7	25.99	89.30	226.8	15.4	.0320	1.10971
149.0	217.6	7.766	1723.0	2463.0	128.3	25.78	84.81	227.9	15.2	.0312	1.10624
150.0	211.3	7.544	1784.0	2546.0	128.9	25.58	80.82	229.2	15.0	.0305	1.10309
151.0	205.6	7.340	1842.0	2625.0	129.4	25.38	77.27	230.5	14.9	.0299	1.10021
152.0	200.4	7.154	1897.0	2701.0	129.9	25.20	74.11	231.8	14.8	.0293	1.09757
153.0	195.6	6.981	1950.0	2774.0	130.4	25.03	71.28	233.2	14.7	.0288	1.09514
154.0	191.1	6.821	2001.0	2844.0	130.8	24.87	68.74	234.5	14.6	.0283	1.09289
155.0	186.9	6.672	2049.0	2911.0	131.2	24.71	66.46	235.9	14.5	.0279	1.09079
156.0	183.0	6.533	2096.0	2977.0	131.7	24.57	64.40	237.3	14.4	.0275	1.08884
158.0	175.9	6.280	2186.0	3102.0	132.5	24.30	60.82	240.1	14.3	.0269	1.08529
160.0	169.7	6.056	2271.0	3220.0	133.2	24.06	57.84	242.9	14.3	.0263	1.08215
162.0	164.0	5.854	2351.0	3333.0	133.9	23.84	55.32	245.6	14.2	.0259	1.07934
164.0	158.9	5.672	2428.0	3442.0	134.6	23.64	53.17	248.2	14.2	.0256	1.07680
166.0	154.3	5.506	2502.0	3546.0	135.2	23.47	51.32	250.8	14.2	.0253	1.07449
168.0	150.0	5.353	2573.0	3647.0	135.8	23.30	49.70	253.4	14.2	.0250	1.07237
170.0	146.0	5.213	2642.0	3745.0	136.4	23.16	48.29	255.9	14.2	.0249	1.07042
172.0	142.4	5.083	2709.0	3840.0	137.0	23.02	47.04	258.4	14.2	.0247	1.06862
174.0	139.0	4.962	2774.0	3933.0	137.5	22.90	45.92	260.8	14.2	.0246	1.06694
176.0	135.8	4.848	2838.0	4024.0	138.0	22.79	44.93	263.2	14.3	.0245	1.06537
178.0	132.9	4.742	2901.0	4113.0	138.5	22.69	44.03	265.5	14.3	.0245	1.06391
180.0	130.1	4.642	2962.0	4200.0	139.0	22.59	43.22	267.8	14.4	.0244	1.06253
182.0	127.4	4.548	3022.0	4286.0	139.5	22.50	42.48	270.0	14.4	.0244	1.06123
184.0	124.9	4.459	3081.0	4370.0	139.9	22.42	41.81	272.3	14.5	.0244	1.06000
186.0	122.5	4.374	3139.0	4453.0	140.4	22.35	41.19	274.4	14.5	.0244	1.05884
190.0	118.2	4.218	3253.0	4616.0	141.2	22.21	40.11	278.7	14.6	.0244	1.05669
195.0	113.2	4.041	3391.0	4814.0	142.3	22.07	38.97	283.8	14.8	.0245	1.05427
200.0	108.8	3.883	3525.0	5006.0	143.2	21.95	38.03	288.7	15.0	.0246	1.05211
205.0	104.8	3.741	3657.0	5194.0	144.2	21.85	37.23	293.5	15.2	.0247	1.05015
210.0	101.2	3.611	3786.0	5378.0	145.1	21.76	36.54	298.1	15.3	.0249	1.04837
215.0	97.81	3.491	3913.0	5560.0	145.9	21.68	35.95	302.6	15.5	.0251	1.04675
220.0	94.73	3.381	4038.0	5738.0	146.7	21.61	35.43	306.9	15.7	.0253	1.04525
225.0	91.88	3.280	4161.0	5914.0	147.5	21.55	34.98	311.2	15.9	.0255	1.04386
230.0	89.23	3.185	4283.0	6088.0	148.3	21.50	34.58	315.3	16.1	.0257	1.04258
235.0	86.76	3.097	4403.0	6260.0	149.0	21.46	34.22	319.4	16.3	.0259	1.04138
240.0	84.45	3.014	4523.0	6430.0	149.8	21.42	33.90	323.3	16.5	.0262	1.04026
245.0	82.28	2.937	4641.0	6599.0	150.4	21.38	33.61	327.2	16.7	.0264	1.03921
250.0	80.24	2.864	4759.0	6766.0	151.1	21.35	33.35	331.0	17.0	.0267	1.03822
255.0	78.31	2.795	4875.0	6933.0	151.8	21.32	33.12	334.7	17.2	.0269	1.03729
260.0	76.49	2.730	4991.0	7098.0	152.4	21.29	32.90	338.4	17.4	.0272	1.03640
270.0	73.12	2.610	5222.0	7425.0	153.7	21.25	32.52	345.5	17.8	.0277	1.03478
280.0	70.08	2.501	5449.0	7748.0	154.8	21.21	32.20	352.4	18.2	.0282	1.03331
290.0	67.31	2.402	5675.0	8069.0	156.0	21.18	31.93	359.1	18.6	.0287	1.03197
300.0	64.77	2.312	5900.0	8387.0	157.0	21.16	31.69	365.6	19.0	.0293	1.03075
310.0	62.44	2.229	6123.0	8703.0	158.1	21.14	31.49	371.9	19.4	.0298	1.02963
320.0	60.29	2.152	6345.0	9017.0	159.1	21.12	31.31	378.0	19.8	.0304	1.02860
330.0	58.29	2.081	6566.0	9329.0	160.0	21.11	31.16	384.1	20.3	.0309	1.02764
340.0	56.44	2.015	6786.0	9640.0	161.0	21.10	31.02	389.9	20.7	.0315	1.02675
350.0	54.71	1.953	7005.0	9950.0	161.9	21.10	30.90	395.7	21.0	.0320	1.02592
360.0	53.09	1.895	7224.0	10260.0	162.7	21.10	30.80	401.3	21.4	.0326	1.02514
370.0	51.57	1.841	7442.0	10570.0	163.6	21.10	30.71	406.8	21.8	.0331	1.02441
380.0	50.14	1.790	7660.0	10870.0	164.4	21.10	30.63	412.2	22.2	.0337	1.02373
390.0	48.79	1.742	7877.0	11180.0	165.2	21.11	30.56	417.5	22.6	.0343	1.02308
400.0	47.52	1.696	8094.0	11480.0	166.0	21.12	30.50	422.7	23.0	.0348	1.02247
410.0	46.32	1.653	8311.0	11790.0	166.7	21.13	30.45	427.8	23.4	.0354	1.02190
420.0	45.18	1.613	8527.0	12090.0	167.4	21.15	30.41	432.9	23.7	.0360	1.02135
440.0	43.07	1.537	8960.0	12700.0	168.9	21.19	30.35	442.7	24.5	.0371	1.02034
460.0	41.15	1.469	9393.0	13310.0	170.2	21.25	30.31	452.2	25.2	.0382	1.01943
480.0	39.41	1.407	9826.0	13910.0	171.5	21.31	30.30	461.4	25.9	.0394	1.01860

Thermophysical properties of nitrogen—Continued

T K	Density kg/m^3	Density mol/dm^3	E J/mol	H J/mol	S J/(mol·K)	C_v J/(mol·K)	C_p J/(mol·K)	Sound m/s	Visc. μPa·s	Therm. W/(m·K)	Diel. const.
500.0	37.82	1.350	10260.0	14520.0	172.7	21.38	30.31	470.3	26.6	.0405	1.01783
520.0	36.35	1.297	10690.0	15130.0	173.9	21.46	30.33	479.0	27.3	.0416	1.01713
540.0	35.00	1.249	11130.0	15730.0	175.1	21.56	30.37	487.5	28.0	.0428	1.01649
560.0	33.74	1.204	11570.0	16340.0	176.2	21.65	30.43	495.8	28.7	.0439	1.01589
580.0	32.58	1.163	12010.0	16950.0	177.2	21.76	30.49	503.9	29.3	.0451	1.01534
600.0	31.49	1.124	12450.0	17560.0	178.3	21.87	30.57	511.7	30.0	.0462	1.01482
620.0	30.48	1.088	12890.0	18170.0	179.3	21.99	30.65	519.5	30.6	.0474	1.01434
640.0	29.53	1.054	13330.0	18790.0	180.3	22.11	30.75	527.0	31.3	.0485	1.01389
660.0	28.65	1.022	13780.0	19400.0	181.2	22.23	30.85	534.4	31.9	.0496	1.01347
680.0	27.81	.9926	14230.0	20020.0	182.1	22.36	30.95	541.7	32.5	.0508	1.01307
700.0	27.02	.9645	14680.0	20640.0	183.0	22.49	31.06	548.8	33.2	.0519	1.01269
720.0	26.28	.9380	15130.0	21260.0	183.9	22.62	31.17	555.9	33.8	.0531	1.01234
740.0	25.57	.9129	15590.0	21890.0	184.8	22.75	31.28	562.8	34.4	.0542	1.01201
760.0	24.91	.8891	16050.0	22510.0	185.6	22.88	31.40	569.5	35.0	.0554	1.01169
780.0	24.28	.8666	16510.0	23140.0	186.4	23.01	31.52	576.2	35.6	.0565	1.01139
800.0	23.68	.8451	16970.0	23770.0	187.2	23.14	31.63	582.8	36.2	.0576	1.01111
820.0	23.11	.8248	17440.0	24410.0	188.0	23.27	31.75	589.3	36.7	.0588	1.01083
840.0	22.56	.8054	17910.0	25050.0	188.8	23.40	31.87	595.7	37.3	.0599	1.01058
860.0	22.05	.7869	18380.0	25680.0	189.5	23.53	31.99	602.1	37.9	.0610	1.01033
880.0	21.55	.7692	18850.0	26320.0	190.2	23.66	32.11	608.3	38.5	.0621	1.01010
900.0	21.08	.7524	19330.0	26970.0	191.0	23.78	32.22	614.5	39.0	.0633	1.00987
920.0	20.63	.7362	19800.0	27610.0	191.7	23.90	32.34	620.6	39.6	.0644	1.00966
950.0	19.98	.7133	20530.0	28590.0	192.7	24.08	32.51	629.6	40.4	.0660	1.00935
1000.0	19.00	.6781	21740.0	30220.0	194.4	24.37	32.78	644.4	41.8	.0688	1.00889
1050.0	18.11	.6463	22970.0	31860.0	196.0	24.64	33.04	658.7	43.1	.0715	1.00846
1100.0	17.29	.6173	24210.0	33520.0	197.5	24.90	33.28	672.8	44.4	.0741	1.00808
1150.0	16.55	.5908	25460.0	35190.0	199.0	25.14	33.52	686.5	45.7	.0768	1.00773
1200.0	15.87	.5666	26720.0	36870.0	200.5	25.37	33.74	700.0	47.0	.0794	1.00740
1250.0	15.25	.5442	28000.0	38570.0	201.8	25.58	33.94	713.2	48.2	.0819	1.00711
1300.0	14.67	.5236	29280.0	40270.0	203.2	25.78	34.13	726.2	49.4	.0844	1.00683
1350.0	14.13	.5044	30580.0	41980.0	204.5	25.97	34.32	738.9	50.6	.0869	1.00658
1400.0	13.63	.4867	31880.0	43700.0	205.7	26.14	34.49	751.5	51.8	.0893	1.00634
1450.0	13.17	.4701	33190.0	45430.0	206.9	26.31	34.64	763.8	53.0	.0917	1.00613
1500.0	12.74	.4546	34510.0	47160.0	208.1	26.46	34.79	775.9	54.2	.0941	1.00592
1550.0	12.33	.4401	35840.0	48910.0	209.2	26.60	34.93	787.9	55.3	.0964	1.00573
1600.0	11.95	.4266	37180.0	50660.0	210.4	26.73	35.06	799.7	56.5	.0987	1.00555
1650.0	11.59	.4138	38520.0	52410.0	211.4	26.86	35.18	811.3	57.6	.101	1.00538
1700.0	11.26	.4018	39860.0	54170.0	212.5	26.98	35.30	822.8	58.7	.103	1.00522
1750.0	10.94	.3904	41210.0	55940.0	213.5	27.09	35.41	834.1	59.8	.105	1.00507
1800.0	10.64	.3797	42570.0	57710.0	214.5	27.19	35.51	845.2	60.9	.108	1.00493
1850.0	10.35	.3695	43930.0	59490.0	215.5	27.28	35.60	856.2	62.0	.110	1.00479
1900.0	10.08	.3599	45300.0	61270.0	216.4	27.38	35.69	867.1	63.0	.112	1.00467
6.00 MPa isobar											
64.46[a]	870.1	31.06	−4191.0	−3998.0	68.33	30.05	54.21	1251.0	278.0	.153	1.46980
80.0	810.6	28.93	−3342.0	−3134.0	80.33	29.14	55.94	923.0	150.0	.135	1.43452
100.0	717.2	25.60	−2231.0	−1996.0	93.02	26.94	58.80	687.0	83.2	.107	1.37975
110.0	661.4	23.61	−1644.0	−1390.0	98.80	26.49	62.99	581.2	63.7	.0917	1.34739
120.0	594.2	21.21	−1006.0	−722.8	104.6	26.28	71.52	472.7	48.1	.0762	1.30894
125.0	552.6	19.73	−650.0	−345.8	107.7	26.30	80.11	413.5	41.0	.0688	1.28550
130.0	500.5	17.86	−244.7	91.1	111.1	26.50	96.60	348.1	34.0	.0621	1.25642
131.0	488.0	17.42	−154.3	190.2	111.9	26.58	101.6	334.3	32.5	.0607	1.24953
132.0	474.6	16.94	−59.6	294.5	112.6	26.68	107.3	320.4	31.1	.0594	1.24216
133.0	460.2	16.43	39.8	405.0	113.5	26.79	113.7	306.5	29.6	.0579	1.23426
134.0	444.7	15.87	144.2	522.1	114.4	26.91	120.6	292.9	28.1	.0563	1.22579
135.0	428.2	15.28	253.6	646.2	115.3	27.05	127.5	280.1	26.6	.0546	1.21677
136.0	410.6	14.66	367.5	776.9	116.2	27.19	133.6	268.3	25.2	.0528	1.20728
138.0	374.1	13.35	602.6	1052.0	118.3	27.42	140.0	249.5	22.5	.0487	1.18767
139.0	356.0	12.71	719.9	1192.0	119.3	27.48	139.7	242.7	21.4	.0467	1.17807
140.0	338.7	12.09	834.3	1331.0	120.3	27.48	137.3	237.7	20.3	.0447	1.16891
141.0	322.4	11.51	944.6	1466.0	121.2	27.42	133.3	234.0	19.4	.0429	1.16033
142.0	307.3	10.97	1050.0	1597.0	122.1	27.32	128.2	231.5	18.7	.0411	1.15242

Thermophysical properties of nitrogen—Continued

T K	Density kg/m^3	Density mol/dm^3	E J/mol	H J/mol	S J/(mol·K)	C_v J/(mol·K)	C_p J/(mol·K)	Sound m/s	Visc. μPa·s	Therm. W/(m·K)	Diel. const.
143.0	293.4	10.47	1149.0	1722.0	123.0	27.17	122.5	229.8	18.0	.0396	1.14520
144.0	280.8	10.02	1243.0	1842.0	123.9	27.00	116.5	228.9	17.5	.0381	1.13865
145.0	269.4	9.616	1331.0	1955.0	124.6	26.80	110.4	228.5	17.0	.0368	1.13275
146.0	259.1	9.248	1414.0	2063.0	125.4	26.60	104.6	228.5	16.6	.0356	1.12743
147.0	249.7	8.914	1491.0	2164.0	126.1	26.39	99.07	228.9	16.3	.0345	1.12264
148.0	241.3	8.612	1564.0	2261.0	126.7	26.18	93.99	229.5	16.0	.0336	1.11831
149.0	233.6	8.338	1633.0	2353.0	127.3	25.97	89.35	230.3	15.8	.0327	1.11439
150.0	226.6	8.088	1698.0	2440.0	127.9	25.77	85.15	231.3	15.6	.0319	1.11082
151.0	220.2	7.859	1760.0	2523.0	128.5	25.58	81.36	232.4	15.4	.0312	1.10757
152.0	214.3	7.649	1818.0	2603.0	129.0	25.40	77.94	233.5	15.2	.0305	1.10458
153.0	208.9	7.455	1874.0	2679.0	129.5	25.22	74.86	234.7	15.1	.0299	1.10184
154.0	203.8	7.276	1928.0	2752.0	130.0	25.05	72.09	236.0	15.0	.0294	1.09930
155.0	199.2	7.109	1979.0	2823.0	130.4	24.89	69.58	237.2	14.9	.0289	1.09695
156.0	194.8	6.954	2029.0	2892.0	130.9	24.74	67.30	238.5	14.8	.0285	1.09476
157.0	190.8	6.809	2077.0	2958.0	131.3	24.60	65.24	239.8	14.8	.0281	1.09271
158.0	186.9	6.673	2123.0	3022.0	131.7	24.46	63.36	241.1	14.7	.0278	1.09080
160.0	180.0	6.424	2211.0	3146.0	132.5	24.21	60.07	243.8	14.6	.0272	1.08730
162.0	173.7	6.201	2295.0	3263.0	133.2	23.98	57.29	246.4	14.5	.0267	1.08418
164.0	168.1	6.001	2375.0	3375.0	133.9	23.78	54.93	249.0	14.5	.0262	1.08138
166.0	163.0	5.819	2452.0	3483.0	134.6	23.59	52.89	251.6	14.4	.0259	1.07884
168.0	158.4	5.652	2525.0	3587.0	135.2	23.42	51.12	254.1	14.4	.0256	1.07652
170.0	154.1	5.499	2596.0	3687.0	135.8	23.27	49.57	256.6	14.4	.0254	1.07439
172.0	150.1	5.358	2665.0	3785.0	136.4	23.13	48.21	259.1	14.4	.0252	1.07243
174.0	146.4	5.226	2732.0	3880.0	136.9	23.00	46.99	261.5	14.5	.0251	1.07060
176.0	143.0	5.104	2798.0	3973.0	137.4	22.88	45.91	263.8	14.5	.0250	1.06890
178.0	139.8	4.989	2861.0	4064.0	137.9	22.77	44.94	266.1	14.5	.0249	1.06732
180.0	136.8	4.882	2924.0	4153.0	138.4	22.67	44.06	268.4	14.5	.0248	1.06583
182.0	133.9	4.780	2985.0	4240.0	138.9	22.58	43.27	270.6	14.6	.0248	1.06443
184.0	131.2	4.684	3045.0	4326.0	139.4	22.49	42.54	272.9	14.6	.0248	1.06311
186.0	128.7	4.594	3104.0	4411.0	139.9	22.41	41.88	275.0	14.7	.0248	1.06186
188.0	126.3	4.508	3163.0	4494.0	140.3	22.34	41.27	277.2	14.7	.0248	1.06067
190.0	124.0	4.426	3220.0	4576.0	140.7	22.27	40.71	279.3	14.8	.0248	1.05955
195.0	118.7	4.238	3360.0	4776.0	141.8	22.12	39.50	284.4	14.9	.0248	1.05696
200.0	114.0	4.070	3497.0	4971.0	142.8	22.00	38.49	289.3	15.1	.0249	1.05465
205.0	109.8	3.918	3630.0	5161.0	143.7	21.89	37.63	294.1	15.3	.0250	1.05257
210.0	105.9	3.780	3760.0	5348.0	144.6	21.80	36.91	298.7	15.5	.0252	1.05068
215.0	102.3	3.653	3888.0	5530.0	145.5	21.72	36.28	303.2	15.6	.0254	1.04895
220.0	99.09	3.537	4014.0	5710.0	146.3	21.65	35.73	307.6	15.8	.0255	1.04736
225.0	96.08	3.429	4138.0	5888.0	147.1	21.59	35.25	311.8	16.0	.0257	1.04590
230.0	93.28	3.330	4261.0	6063.0	147.9	21.53	34.83	316.0	16.2	.0259	1.04454
235.0	90.68	3.237	4382.0	6236.0	148.6	21.48	34.45	320.0	16.4	.0262	1.04327
240.0	88.24	3.150	4503.0	6408.0	149.3	21.44	34.12	324.0	16.6	.0264	1.04209
245.0	85.95	3.068	4622.0	6577.0	150.0	21.40	33.81	327.9	16.8	.0266	1.04098
250.0	83.80	2.991	4740.0	6746.0	150.7	21.37	33.54	331.7	17.0	.0269	1.03994
255.0	81.78	2.919	4857.0	6913.0	151.4	21.34	33.29	335.4	17.2	.0271	1.03896
260.0	79.86	2.851	4974.0	7079.0	152.0	21.31	33.06	339.1	17.4	.0273	1.03803
270.0	76.33	2.724	5205.0	7407.0	153.2	21.27	32.67	346.2	17.9	.0278	1.03632
280.0	73.13	2.610	5434.0	7732.0	154.4	21.23	32.33	353.1	18.3	.0284	1.03478
290.0	70.23	2.507	5661.0	8054.0	155.6	21.20	32.05	359.8	18.7	.0289	1.03338
300.0	67.57	2.412	5886.0	8373.0	156.6	21.17	31.80	366.3	19.1	.0294	1.03210
310.0	65.13	2.325	6110.0	8690.0	157.7	21.15	31.59	372.6	19.5	.0300	1.03092
320.0	62.88	2.244	6332.0	9005.0	158.7	21.13	31.40	378.8	19.9	.0305	1.02984
330.0	60.79	2.170	6554.0	9319.0	159.6	21.12	31.24	384.8	20.3	.0311	1.02883
340.0	58.85	2.101	6774.0	9630.0	160.6	21.11	31.10	390.7	20.7	.0316	1.02790
350.0	57.04	2.036	6994.0	9941.0	161.5	21.11	30.98	396.4	21.1	.0322	1.02703
360.0	55.35	1.976	7213.0	10250.0	162.3	21.11	30.87	402.0	21.5	.0327	1.02622
370.0	53.77	1.919	7432.0	10560.0	163.2	21.11	30.77	407.5	21.9	.0333	1.02546
380.0	52.27	1.866	7650.0	10870.0	164.0	21.11	30.69	413.0	22.3	.0338	1.02474
390.0	50.87	1.816	7867.0	11170.0	164.8	21.12	30.62	418.3	22.6	.0344	1.02407
400.0	49.54	1.768	8085.0	11480.0	165.6	21.13	30.56	423.5	23.0	.0349	1.02343
410.0	48.28	1.723	8302.0	11780.0	166.3	21.14	30.50	428.6	23.4	.0355	1.02283
420.0	47.10	1.681	8519.0	12090.0	167.1	21.16	30.46	433.6	23.8	.0361	1.02226
440.0	44.89	1.602	8952.0	12700.0	168.5	21.20	30.39	443.4	24.5	.0372	1.02121

Thermophysical properties of nitrogen—Continued

T K	Density kg/m^3	Density mol/dm^3	E J/mol	H J/mol	S J/(mol·K)	C_v J/(mol·K)	C_p J/(mol·K)	Sound m/s	Visc. μPa·s	Therm. W/(m·K)	Diel. const.
460.0	42.90	1.531	9385.0	13300.0	169.8	21.25	30.35	452.9	25.2	.0383	1.02026
480.0	41.08	1.466	9819.0	13910.0	171.1	21.32	30.34	462.1	25.9	.0394	1.01939
500.0	39.42	1.407	10250.0	14520.0	172.4	21.39	30.34	471.1	26.7	.0406	1.01859
520.0	37.89	1.352	10690.0	15120.0	173.6	21.47	30.36	479.8	27.3	.0417	1.01786
540.0	36.48	1.302	11120.0	15730.0	174.7	21.56	30.40	488.2	28.0	.0429	1.01719
560.0	35.17	1.255	11560.0	16340.0	175.8	21.66	30.45	496.5	28.7	.0440	1.01657
580.0	33.96	1.212	12000.0	16950.0	176.9	21.76	30.52	504.6	29.4	.0451	1.01599
600.0	32.83	1.172	12440.0	17560.0	177.9	21.87	30.59	512.4	30.0	.0463	1.01545
620.0	31.77	1.134	12880.0	18170.0	178.9	21.99	30.67	520.2	30.7	.0474	1.01495
640.0	30.79	1.099	13330.0	18790.0	179.9	22.11	30.76	527.7	31.3	.0486	1.01448
660.0	29.86	1.066	13770.0	19400.0	180.8	22.23	30.86	535.1	31.9	.0497	1.01404
680.0	28.99	1.035	14220.0	20020.0	181.8	22.36	30.96	542.4	32.6	.0509	1.01362
700.0	28.17	1.005	14680.0	20640.0	182.7	22.49	31.07	549.5	33.2	.0520	1.01323
720.0	27.39	.9777	15130.0	21270.0	183.5	22.62	31.18	556.5	33.8	.0531	1.01287
740.0	26.66	.9516	15580.0	21890.0	184.4	22.75	31.30	563.4	34.4	.0543	1.01252
760.0	25.97	.9268	16040.0	22520.0	185.2	22.88	31.41	570.2	35.0	.0554	1.01219
780.0	25.31	.9033	16500.0	23150.0	186.0	23.01	31.53	576.9	35.6	.0566	1.01188
800.0	24.68	.8810	16970.0	23780.0	186.8	23.14	31.65	583.5	36.2	.0577	1.01158
820.0	24.09	.8598	17430.0	24410.0	187.6	23.28	31.76	590.0	36.8	.0588	1.01130
840.0	23.52	.8396	17900.0	25050.0	188.4	23.40	31.88	596.4	37.3	.0599	1.01103
860.0	22.98	.8203	18370.0	25690.0	189.1	23.53	32.00	602.7	37.9	.0611	1.01077
880.0	22.47	.8019	18850.0	26330.0	189.9	23.66	32.11	608.9	38.5	.0622	1.01053
900.0	21.97	.7844	19320.0	26970.0	190.6	23.78	32.23	615.1	39.0	.0633	1.01029
920.0	21.50	.7676	19800.0	27620.0	191.3	23.91	32.34	621.2	39.6	.0644	1.01007
950.0	20.83	.7437	20520.0	28590.0	192.4	24.08	32.51	630.2	40.4	.0661	1.00975
1000.0	19.81	.7070	21740.0	30220.0	194.0	24.37	32.78	644.9	41.8	.0688	1.00927
1050.0	18.88	.6738	22960.0	31870.0	195.6	24.65	33.04	659.3	43.1	.0715	1.00882
1100.0	18.03	.6436	24210.0	33530.0	197.2	24.90	33.29	673.4	44.4	.0742	1.00842
1150.0	17.26	.6161	25460.0	35200.0	198.7	25.15	33.52	687.1	45.7	.0768	1.00806
1200.0	16.55	.5908	26720.0	36880.0	200.1	25.37	33.74	700.6	47.0	.0794	1.00772
1250.0	15.90	.5675	28000.0	38570.0	201.5	25.59	33.95	713.8	48.2	.0819	1.00741
1300.0	15.30	.5460	29280.0	40270.0	202.8	25.79	34.14	726.7	49.4	.0845	1.00713
1350.0	14.74	.5260	30580.0	41990.0	204.1	25.97	34.32	739.5	50.6	.0869	1.00686
1400.0	14.22	.5075	31880.0	43710.0	205.4	26.15	34.49	752.0	51.8	.0894	1.00662
1450.0	13.73	.4902	33190.0	45430.0	206.6	26.31	34.65	764.3	53.0	.0918	1.00639
1500.0	13.28	.4741	34510.0	47170.0	207.7	26.46	34.79	776.4	54.2	.0941	1.00617
1550.0	12.86	.4590	35840.0	48910.0	208.9	26.60	34.93	788.4	55.3	.0965	1.00597
1600.0	12.46	.4448	37180.0	50660.0	210.0	26.74	35.06	800.1	56.5	.0988	1.00579
1650.0	12.09	.4315	38520.0	52420.0	211.1	26.86	35.19	811.8	57.6	.101	1.00561
1700.0	11.74	.4190	39860.0	54180.0	212.1	26.98	35.30	823.2	58.7	.103	1.00544
1750.0	11.41	.4072	41210.0	55950.0	213.2	27.09	35.41	834.5	59.8	.106	1.00529
1800.0	11.09	.3960	42570.0	57720.0	214.2	27.19	35.51	845.7	60.9	.108	1.00514
1850.0	10.80	.3854	43930.0	59500.0	215.1	27.29	35.60	856.7	62.0	.110	1.00500
1900.0	10.52	.3754	45300.0	61280.0	216.1	27.38	35.69	867.5	63.0	.112	1.00487

6.25 MPa isobar

T K	Density kg/m^3	Density mol/dm^3	E J/mol	H J/mol	S J/(mol·K)	C_v J/(mol·K)	C_p J/(mol·K)	Sound m/s	Visc. μPa·s	Therm. W/(m·K)	Diel. const.
64.51[a]	870.2	31.06	–4190.0	–3989.0	68.34	30.17	54.19	1248.0	278.0	.153	1.46985
80.0	811.2	28.95	–3345.0	–3129.0	80.29	29.20	55.86	924.2	150.0	.136	1.43485
100.0	718.4	25.64	–2237.0	–1994.0	92.95	26.97	58.61	690.1	83.7	.107	1.38041
110.0	663.1	23.67	–1654.0	–1390.0	98.70	26.51	62.62	585.5	64.2	.0921	1.34838
120.0	597.1	21.31	–1022.0	–728.8	104.4	26.27	70.55	479.0	48.7	.0768	1.31063
125.0	556.9	19.88	–673.0	–358.6	107.5	26.27	78.21	421.6	41.7	.0695	1.28792
130.0	507.6	18.12	–281.5	63.4	110.8	26.42	92.09	359.2	34.9	.0629	1.26036
132.0	483.8	17.27	–106.1	255.8	112.2	26.55	100.7	332.9	32.1	.0603	1.24719
133.0	470.7	16.80	–13.0	359.0	113.0	26.64	105.8	319.8	30.7	.0589	1.23999
134.0	456.7	16.30	84.1	467.5	113.8	26.73	111.2	306.8	29.3	.0574	1.23233
135.0	441.9	15.77	185.3	581.6	114.7	26.84	116.9	294.2	27.9	.0559	1.22422
136.0	426.2	15.21	290.4	701.2	115.6	26.96	122.3	282.4	26.5	.0542	1.21569
138.0	393.0	14.03	509.0	954.6	117.4	27.17	130.1	262.3	23.9	.0505	1.19778
140.0	359.5	12.83	729.9	1217.0	119.3	27.29	131.1	248.1	21.7	.0467	1.17991
141.0	343.5	12.26	837.4	1347.0	120.2	27.28	129.2	243.2	20.7	.0448	1.17144
142.0	328.4	11.72	941.5	1475.0	121.1	27.23	126.0	239.5	19.9	.0431	1.16345

Thermophysical properties of nitrogen—Continued

T K	Density kg/m^3	Density mol/dm^3	E J/mol	H J/mol	S J/(mol·K)	C_v J/(mol·K)	C_p J/(mol·K)	Sound m/s	Visc. μPa·s	Therm. W/(m·K)	Diel. const.
143.0	314.2	11.21	1042.0	1599.0	122.0	27.14	122.0	236.8	19.1	.0415	1.15600
144.0	301.0	10.74	1137.0	1719.0	122.8	27.01	117.4	235.0	18.5	.0400	1.14913
145.0	288.9	10.31	1227.0	1834.0	123.6	26.86	112.4	233.8	18.0	.0386	1.14282
146.0	277.8	9.915	1313.0	1943.0	124.4	26.69	107.4	233.2	17.5	.0373	1.13707
147.0	267.6	9.553	1394.0	2048.0	125.1	26.51	102.4	233.0	17.1	.0362	1.13183
148.0	258.4	9.222	1471.0	2148.0	125.8	26.32	97.62	233.1	16.7	.0351	1.12706
149.0	249.9	8.920	1543.0	2244.0	126.4	26.12	93.10	233.5	16.4	.0342	1.12271
150.0	242.1	8.643	1612.0	2335.0	127.0	25.93	88.89	234.1	16.2	.0333	1.11875
151.0	235.0	8.390	1677.0	2422.0	127.6	25.75	85.01	234.9	16.0	.0325	1.11512
152.0	228.5	8.157	1739.0	2505.0	128.2	25.56	81.47	235.8	15.8	.0318	1.11179
153.0	222.5	7.941	1798.0	2585.0	128.7	25.39	78.23	236.7	15.6	.0311	1.10873
154.0	216.9	7.742	1854.0	2661.0	129.2	25.22	75.28	237.8	15.5	.0305	1.10590
155.0	211.7	7.558	1908.0	2735.0	129.7	25.06	72.59	238.9	15.4	.0300	1.10328
156.0	206.9	7.386	1960.0	2807.0	130.1	24.90	70.15	240.1	15.2	.0295	1.10085
157.0	202.4	7.225	2011.0	2876.0	130.6	24.76	67.91	241.3	15.2	.0291	1.09857
158.0	198.2	7.075	2059.0	2942.0	131.0	24.62	65.88	242.5	15.1	.0287	1.09645
160.0	190.5	6.800	2152.0	3071.0	131.8	24.36	62.30	245.0	14.9	.0280	1.09259
162.0	183.7	6.556	2239.0	3192.0	132.5	24.12	59.28	247.5	14.8	.0274	1.08915
164.0	177.5	6.337	2322.0	3308.0	133.3	23.91	56.70	250.0	14.8	.0269	1.08607
166.0	172.0	6.138	2401.0	3419.0	133.9	23.71	54.48	252.5	14.7	.0266	1.08329
168.0	166.9	5.957	2477.0	3526.0	134.6	23.53	52.56	255.0	14.7	.0262	1.08076
170.0	162.2	5.790	2550.0	3629.0	135.2	23.37	50.87	257.4	14.7	.0260	1.07844
172.0	157.9	5.637	2621.0	3730.0	135.8	23.23	49.39	259.8	14.7	.0258	1.07630
174.0	153.9	5.495	2690.0	3827.0	136.3	23.09	48.08	262.2	14.7	.0256	1.07433
176.0	150.2	5.363	2757.0	3922.0	136.9	22.97	46.91	264.6	14.7	.0255	1.07249
178.0	146.8	5.239	2822.0	4015.0	137.4	22.85	45.86	266.9	14.7	.0254	1.07078
180.0	143.5	5.124	2886.0	4106.0	137.9	22.75	44.91	269.1	14.7	.0253	1.06917
182.0	140.5	5.015	2948.0	4195.0	138.4	22.65	44.06	271.3	14.8	.0252	1.06767
184.0	137.6	4.912	3010.0	4282.0	138.9	22.56	43.28	273.5	14.8	.0252	1.06625
186.0	134.9	4.815	3070.0	4368.0	139.3	22.48	42.57	275.7	14.8	.0252	1.06491
188.0	132.3	4.723	3129.0	4452.0	139.8	22.41	41.92	277.8	14.9	.0251	1.06364
190.0	129.9	4.636	3187.0	4535.0	140.2	22.33	41.32	279.9	14.9	.0251	1.06243
195.0	124.3	4.435	3330.0	4739.0	141.3	22.18	40.03	285.0	15.1	.0252	1.05967
200.0	119.3	4.257	3468.0	4936.0	142.3	22.04	38.95	290.0	15.2	.0252	1.05721
205.0	114.7	4.095	3602.0	5129.0	143.2	21.93	38.04	294.7	15.4	.0253	1.05500
210.0	110.6	3.949	3734.0	5317.0	144.1	21.84	37.27	299.4	15.6	.0255	1.05300
215.0	106.9	3.816	3863.0	5501.0	145.0	21.75	36.61	303.9	15.7	.0256	1.05117
220.0	103.5	3.693	3991.0	5683.0	145.8	21.68	36.03	308.2	15.9	.0258	1.04949
225.0	100.3	3.579	4116.0	5862.0	146.7	21.62	35.53	312.5	16.1	.0260	1.04794
230.0	97.34	3.474	4239.0	6038.0	147.4	21.56	35.08	316.6	16.3	.0262	1.04651
235.0	94.59	3.376	4362.0	6213.0	148.2	21.51	34.69	320.7	16.5	.0264	1.04517
240.0	92.03	3.285	4483.0	6385.0	148.9	21.47	34.33	324.7	16.7	.0266	1.04393
245.0	89.63	3.199	4603.0	6556.0	149.6	21.43	34.01	328.6	16.9	.0268	1.04276
250.0	87.37	3.119	4721.0	6725.0	150.3	21.39	33.73	332.4	17.1	.0271	1.04166
255.0	85.24	3.043	4839.0	6893.0	151.0	21.36	33.47	336.1	17.3	.0273	1.04063
260.0	83.24	2.971	4957.0	7060.0	151.6	21.33	33.23	339.8	17.5	.0275	1.03966
270.0	79.53	2.839	5189.0	7390.0	152.9	21.28	32.81	346.9	17.9	.0280	1.03787
280.0	76.19	2.719	5418.0	7717.0	154.0	21.24	32.47	353.8	18.3	.0285	1.03625
290.0	73.15	2.611	5646.0	8040.0	155.2	21.21	32.17	360.5	18.7	.0291	1.03478
300.0	70.37	2.512	5872.0	8360.0	156.3	21.19	31.91	367.0	19.1	.0296	1.03344
310.0	67.82	2.421	6096.0	8678.0	157.3	21.16	31.69	373.3	19.5	.0301	1.03221
320.0	65.47	2.337	6319.0	8994.0	158.3	21.15	31.49	379.5	19.9	.0307	1.03108
330.0	63.29	2.259	6541.0	9308.0	159.3	21.13	31.33	385.5	20.3	.0312	1.03003
340.0	61.26	2.187	6763.0	9621.0	160.2	21.12	31.18	391.4	20.7	.0317	1.02906
350.0	59.38	2.119	6983.0	9932.0	161.1	21.12	31.05	397.2	21.1	.0323	1.02815
360.0	57.61	2.056	7202.0	10240.0	162.0	21.12	30.93	402.8	21.5	.0328	1.02730
370.0	55.96	1.997	7421.0	10550.0	162.8	21.12	30.83	408.3	21.9	.0334	1.02651
380.0	54.40	1.942	7640.0	10860.0	163.6	21.12	30.75	413.7	22.3	.0339	1.02576
390.0	52.94	1.890	7858.0	11170.0	164.4	21.13	30.67	419.0	22.7	.0345	1.02506
400.0	51.55	1.840	8075.0	11470.0	165.2	21.14	30.61	424.2	23.1	.0350	1.02440
410.0	50.25	1.793	8293.0	11780.0	166.0	21.15	30.55	429.3	23.4	.0356	1.02377
420.0	49.01	1.749	8510.0	12080.0	166.7	21.17	30.50	434.3	23.8	.0362	1.02317
440.0	46.71	1.667	8944.0	12690.0	168.1	21.21	30.43	444.1	24.5	.0373	1.02208

Thermophysical properties of nitrogen—Continued

T K	Density kg/m^3	Density mol/dm^3	E J/mol	H J/mol	S J/(mol·K)	C_v J/(mol·K)	C_p J/(mol·K)	Sound m/s	Visc. μPa·s	Therm. W/(m·K)	Diel. const.
460.0	44.64	1.593	9378.0	13300.0	169.5	21.26	30.39	453.6	25.3	.0384	1.02108
480.0	42.74	1.526	9812.0	13910.0	170.8	21.32	30.37	462.8	26.0	.0395	1.02018
500.0	41.01	1.464	10250.0	14520.0	172.0	21.39	30.37	471.8	26.7	.0407	1.01935
520.0	39.42	1.407	10680.0	15120.0	173.2	21.48	30.39	480.5	27.4	.0418	1.01859
540.0	37.95	1.355	11120.0	15730.0	174.4	21.57	30.43	489.0	28.1	.0429	1.01789
560.0	36.59	1.306	11560.0	16340.0	175.5	21.66	30.48	497.2	28.7	.0441	1.01724
580.0	35.33	1.261	11990.0	16950.0	176.5	21.77	30.54	505.3	29.4	.0452	1.01664
600.0	34.16	1.219	12440.0	17560.0	177.6	21.88	30.61	513.1	30.0	.0464	1.01608
620.0	33.06	1.180	12880.0	18170.0	178.6	21.99	30.69	520.9	30.7	.0475	1.01556
640.0	32.03	1.143	13320.0	18790.0	179.5	22.11	30.78	528.4	31.3	.0486	1.01507
660.0	31.07	1.109	13770.0	19410.0	180.5	22.24	30.88	535.8	32.0	.0498	1.01461
680.0	30.16	1.077	14220.0	20020.0	181.4	22.36	30.98	543.1	32.6	.0509	1.01418
700.0	29.31	1.046	14670.0	20650.0	182.3	22.49	31.09	550.2	33.2	.0521	1.01377
720.0	28.50	1.017	15120.0	21270.0	183.2	22.62	31.20	557.2	33.8	.0532	1.01339
740.0	27.74	.9902	15580.0	21890.0	184.1	22.75	31.31	564.1	34.4	.0543	1.01303
760.0	27.02	.9644	16040.0	22520.0	184.9	22.89	31.42	570.8	35.0	.0555	1.01269
780.0	26.34	.9400	16500.0	23150.0	185.7	23.02	31.54	577.5	35.6	.0566	1.01236
800.0	25.69	.9168	16970.0	23780.0	186.5	23.15	31.66	584.1	36.2	.0577	1.01205
820.0	25.07	.8948	17430.0	24420.0	187.3	23.28	31.77	590.6	36.8	.0589	1.01176
840.0	24.48	.8737	17900.0	25050.0	188.1	23.41	31.89	597.0	37.3	.0600	1.01148
860.0	23.92	.8537	18370.0	25690.0	188.8	23.54	32.01	603.3	37.9	.0611	1.01121
880.0	23.38	.8346	18840.0	26330.0	189.5	23.66	32.12	609.5	38.5	.0622	1.01096
900.0	22.87	.8163	19320.0	26980.0	190.3	23.79	32.24	615.7	39.0	.0634	1.01071
920.0	22.38	.7988	19800.0	27620.0	191.0	23.91	32.35	621.8	39.6	.0645	1.01048
950.0	21.68	.7740	20520.0	28600.0	192.0	24.09	32.52	630.8	40.4	.0661	1.01015
1000.0	20.62	.7358	21730.0	30230.0	193.7	24.38	32.79	645.5	41.8	.0689	1.00964
1050.0	19.65	.7013	22960.0	31870.0	195.3	24.65	33.05	659.9	43.1	.0716	1.00919
1100.0	18.77	.6699	24200.0	33530.0	196.8	24.91	33.29	673.9	44.4	.0742	1.00877
1150.0	17.96	.6412	25460.0	35200.0	198.3	25.15	33.53	687.6	45.7	.0768	1.00839
1200.0	17.23	.6149	26720.0	36890.0	199.8	25.37	33.74	701.1	47.0	.0794	1.00804
1250.0	16.55	.5907	28000.0	38580.0	201.1	25.59	33.95	714.3	48.2	.0820	1.00772
1300.0	15.92	.5683	29280.0	40280.0	202.5	25.79	34.14	727.2	49.4	.0845	1.00742
1350.0	15.34	.5476	30580.0	41990.0	203.8	25.97	34.32	740.0	50.7	.0870	1.00714
1400.0	14.80	.5283	31880.0	43710.0	205.0	26.15	34.49	752.5	51.9	.0894	1.00689
1450.0	14.30	.5103	33190.0	45440.0	206.2	26.31	34.65	764.8	53.0	.0918	1.00665
1500.0	13.83	.4935	34510.0	47180.0	207.4	26.46	34.80	776.9	54.2	.0942	1.00643
1550.0	13.39	.4778	35840.0	48920.0	208.6	26.60	34.94	788.8	55.3	.0965	1.00622
1600.0	12.97	.4631	37170.0	50670.0	209.7	26.74	35.07	800.6	56.5	.0988	1.00602
1650.0	12.59	.4493	38510.0	52430.0	210.7	26.86	35.19	812.2	57.6	.101	1.00584
1700.0	12.22	.4362	39860.0	54190.0	211.8	26.98	35.30	823.7	58.7	.103	1.00567
1750.0	11.88	.4239	41210.0	55960.0	212.8	27.09	35.41	834.9	59.8	.106	1.00551
1800.0	11.55	.4123	42570.0	57730.0	213.8	27.19	35.51	846.1	60.9	.108	1.00535
1850.0	11.24	.4013	43930.0	59510.0	214.8	27.29	35.60	857.1	62.0	.110	1.00521
1900.0	10.95	.3908	45300.0	61290.0	215.7	27.38	35.69	868.0	63.1	.112	1.00507

6.50 MPa isobar

T K	Density kg/m^3	Density mol/dm^3	E J/mol	H J/mol	S J/(mol·K)	C_v J/(mol·K)	C_p J/(mol·K)	Sound m/s	Visc. μPa·s	Therm. W/(m·K)	Diel. const.
64.57[a]	870.3	31.07	−4189.0	−3980.0	68.36	30.29	54.17	1246.0	278.0	.153	1.46990
80.0	811.7	28.97	−3348.0	−3124.0	80.24	29.26	55.79	925.4	150.0	.136	1.43517
100.0	719.5	25.68	−2244.0	−1991.0	92.88	27.00	58.42	693.1	84.1	.108	1.38106
110.0	664.8	23.73	−1663.0	−1389.0	98.61	26.53	62.26	589.6	64.7	.0926	1.34936
120.0	600.0	21.42	−1038.0	−734.3	104.3	26.27	69.66	485.0	49.3	.0775	1.31225
125.0	561.0	20.02	−694.8	−370.2	107.3	26.25	76.54	429.3	42.4	.0702	1.29020
130.0	514.1	18.35	−315.0	39.3	110.5	26.35	88.41	369.5	35.7	.0637	1.26393
132.0	491.9	17.56	−147.4	222.9	111.9	26.45	95.47	344.5	33.1	.0611	1.25163
134.0	467.0	16.67	32.3	422.2	113.4	26.59	104.0	319.5	30.4	.0584	1.23796
135.0	453.5	16.19	127.0	528.5	114.2	26.67	108.7	307.4	29.1	.0570	1.23057
136.0	439.3	15.68	225.1	639.6	115.0	26.77	113.4	295.8	27.7	.0555	1.22281
138.0	409.1	14.60	429.4	874.5	116.7	26.95	121.1	275.1	25.2	.0521	1.20642
140.0	377.8	13.48	638.8	1121.0	118.5	27.09	124.3	259.2	22.9	.0485	1.18964
142.0	347.6	12.41	844.2	1368.0	120.2	27.10	122.2	248.5	21.0	.0449	1.17358
143.0	333.4	11.90	942.9	1489.0	121.1	27.05	119.7	244.8	20.2	.0433	1.16611

Thermophysical properties of nitrogen—Continued

T K	Density kg/m^3	Density mol/dm^3	E J/mol	H J/mol	S J/(mol·K)	C_v J/(mol·K)	C_p J/(mol·K)	Sound m/s	Visc. μPa·s	Therm. W/(m·K)	Diel. const.
144.0	320.1	11.42	1038.0	1607.0	121.9	26.97	116.4	242.1	19.5	.0418	1.15909
145.0	307.6	10.98	1130.0	1722.0	122.7	26.86	112.6	240.1	18.9	.0403	1.15255
146.0	296.0	10.56	1217.0	1832.0	123.5	26.72	108.5	238.7	18.4	.0390	1.14650
147.0	285.2	10.18	1300.0	1939.0	124.2	26.57	104.3	237.9	17.9	.0378	1.14092
148.0	275.3	9.827	1379.0	2041.0	124.9	26.40	100.00	237.5	17.5	.0367	1.13579
149.0	266.2	9.501	1455.0	2139.0	125.5	26.23	95.83	237.4	17.2	.0356	1.13108
150.0	257.8	9.202	1526.0	2232.0	126.2	26.06	91.83	237.6	16.9	.0347	1.12676
151.0	250.1	8.925	1594.0	2322.0	126.8	25.88	88.04	238.0	16.6	.0338	1.12279
152.0	242.9	8.671	1659.0	2409.0	127.3	25.70	84.51	238.6	16.4	.0331	1.11913
153.0	236.3	8.435	1721.0	2491.0	127.9	25.53	81.22	239.3	16.2	.0323	1.11576
154.0	230.2	8.217	1780.0	2571.0	128.4	25.37	78.19	240.2	16.0	.0317	1.11265
155.0	224.5	8.014	1837.0	2648.0	128.9	25.21	75.39	241.1	15.8	.0311	1.10976
156.0	219.2	7.825	1891.0	2722.0	129.4	25.05	72.83	242.1	15.7	.0306	1.10708
157.0	214.3	7.649	1944.0	2794.0	129.8	24.90	70.47	243.1	15.6	.0301	1.10457
158.0	209.7	7.484	1995.0	2863.0	130.3	24.76	68.31	244.2	15.5	.0296	1.10224
160.0	201.3	7.184	2091.0	2996.0	131.1	24.49	64.48	246.5	15.3	.0288	1.09799
162.0	193.8	6.917	2182.0	3121.0	131.9	24.25	61.24	248.9	15.2	.0282	1.09422
164.0	187.1	6.678	2268.0	3241.0	132.6	24.03	58.46	251.2	15.1	.0277	1.09086
166.0	181.0	6.462	2350.0	3355.0	133.3	23.83	56.07	253.6	15.0	.0272	1.08782
168.0	175.5	6.265	2428.0	3465.0	134.0	23.64	53.99	256.0	15.0	.0269	1.08507
170.0	170.5	6.085	2503.0	3572.0	134.6	23.48	52.18	258.4	14.9	.0266	1.08255
172.0	165.9	5.920	2576.0	3674.0	135.2	23.32	50.58	260.8	14.9	.0263	1.08024
174.0	161.6	5.767	2647.0	3774.0	135.8	23.18	49.17	263.1	14.9	.0261	1.07811
176.0	157.6	5.625	2715.0	3871.0	136.3	23.05	47.91	265.4	14.9	.0260	1.07613
178.0	153.9	5.492	2782.0	3966.0	136.9	22.94	46.78	267.7	14.9	.0258	1.07428
180.0	150.4	5.368	2847.0	4058.0	137.4	22.83	45.77	269.9	14.9	.0257	1.07256
182.0	147.1	5.251	2911.0	4149.0	137.9	22.73	44.86	272.1	14.9	.0257	1.07094
184.0	144.1	5.142	2974.0	4238.0	138.4	22.63	44.02	274.3	15.0	.0256	1.06942
186.0	141.2	5.038	3035.0	4325.0	138.8	22.55	43.27	276.5	15.0	.0255	1.06799
188.0	138.4	4.940	3095.0	4411.0	139.3	22.47	42.57	278.6	15.1	.0255	1.06663
190.0	135.8	4.847	3154.0	4495.0	139.7	22.39	41.94	280.7	15.1	.0255	1.06535
195.0	129.8	4.634	3299.0	4702.0	140.8	22.23	40.56	285.8	15.2	.0255	1.06241
200.0	124.5	4.445	3439.0	4901.0	141.8	22.09	39.41	290.7	15.4	.0256	1.05979
205.0	119.7	4.274	3575.0	5096.0	142.8	21.97	38.45	295.4	15.5	.0256	1.05745
210.0	115.4	4.119	3708.0	5286.0	143.7	21.87	37.64	300.1	15.7	.0258	1.05533
215.0	111.5	3.978	3839.0	5473.0	144.6	21.79	36.94	304.6	15.9	.0259	1.05339
220.0	107.8	3.849	3967.0	5656.0	145.4	21.71	36.33	308.9	16.0	.0261	1.05162
225.0	104.5	3.730	4093.0	5836.0	146.2	21.65	35.80	313.2	16.2	.0262	1.04999
230.0	101.4	3.619	4218.0	6014.0	147.0	21.59	35.33	317.4	16.4	.0264	1.04848
235.0	98.51	3.516	4341.0	6189.0	147.8	21.54	34.92	321.4	16.6	.0266	1.04708
240.0	95.82	3.420	4463.0	6363.0	148.5	21.49	34.55	325.4	16.8	.0268	1.04577
245.0	93.31	3.330	4583.0	6535.0	149.2	21.45	34.21	329.3	17.0	.0270	1.04454
250.0	90.94	3.246	4703.0	6705.0	149.9	21.41	33.91	333.1	17.2	.0273	1.04339
255.0	88.71	3.166	4821.0	6874.0	150.6	21.38	33.64	336.8	17.4	.0275	1.04231
260.0	86.61	3.091	4939.0	7042.0	151.2	21.35	33.39	340.5	17.6	.0277	1.04129
270.0	82.73	2.953	5172.0	7373.0	152.5	21.30	32.96	347.6	18.0	.0282	1.03941
280.0	79.24	2.828	5403.0	7701.0	153.7	21.26	32.60	354.6	18.4	.0287	1.03772
290.0	76.06	2.715	5631.0	8025.0	154.8	21.23	32.28	361.3	18.8	.0292	1.03619
300.0	73.16	2.612	5858.0	8347.0	155.9	21.20	32.02	367.8	19.2	.0297	1.03479
310.0	70.50	2.517	6083.0	8666.0	156.9	21.18	31.79	374.1	19.6	.0303	1.03350
320.0	68.05	2.429	6307.0	8983.0	157.9	21.16	31.58	380.3	20.0	.0308	1.03232
330.0	65.78	2.348	6529.0	9298.0	158.9	21.15	31.41	386.3	20.4	.0313	1.03123
340.0	63.67	2.273	6751.0	9611.0	159.8	21.14	31.25	392.2	20.8	.0319	1.03021
350.0	61.71	2.202	6972.0	9923.0	160.7	21.13	31.12	397.9	21.2	.0324	1.02926
360.0	59.87	2.137	7192.0	10230.0	161.6	21.13	31.00	403.5	21.6	.0330	1.02838
370.0	58.14	2.075	7411.0	10540.0	162.5	21.13	30.90	409.1	22.0	.0335	1.02755
380.0	56.53	2.018	7630.0	10850.0	163.3	21.13	30.81	414.5	22.3	.0341	1.02678
390.0	55.00	1.963	7848.0	11160.0	164.1	21.14	30.73	419.8	22.7	.0346	1.02604
400.0	53.56	1.912	8066.0	11470.0	164.9	21.15	30.66	425.0	23.1	.0352	1.02535
410.0	52.20	1.863	8284.0	11770.0	165.6	21.16	30.60	430.1	23.5	.0357	1.02470
420.0	50.91	1.817	8501.0	12080.0	166.4	21.18	30.55	435.1	23.8	.0363	1.02408
440.0	48.53	1.732	8936.0	12690.0	167.8	21.22	30.47	444.9	24.6	.0374	1.02294
460.0	46.37	1.655	9370.0	13300.0	169.1	21.27	30.43	454.4	25.3	.0385	1.02191

Thermophysical properties of nitrogen—Continued

T K	Density kg/m^3	Density mol/dm^3	E J/mol	H J/mol	S J/(mol·K)	C_v J/(mol·K)	C_p J/(mol·K)	Sound m/s	Visc. μPa·s	Therm. W/(m·K)	Diel. const.
480.0	44.40	1.585	9804.0	13910.0	170.4	21.33	30.40	463.6	26.0	.0396	1.02097
500.0	42.61	1.521	10240.0	14510.0	171.7	21.40	30.40	472.5	26.7	.0408	1.02011
520.0	40.95	1.462	10670.0	15120.0	172.9	21.48	30.42	481.2	27.4	.0419	1.01932
540.0	39.43	1.407	11110.0	15730.0	174.0	21.57	30.45	489.7	28.1	.0430	1.01859
560.0	38.02	1.357	11550.0	16340.0	175.1	21.67	30.50	497.9	28.7	.0442	1.01792
580.0	36.71	1.310	11990.0	16950.0	176.2	21.77	30.56	506.0	29.4	.0453	1.01729
600.0	35.48	1.267	12430.0	17560.0	177.2	21.88	30.63	513.9	30.1	.0464	1.01671
620.0	34.34	1.226	12870.0	18180.0	178.2	22.00	30.71	521.6	30.7	.0476	1.01617
640.0	33.28	1.188	13320.0	18790.0	179.2	22.12	30.80	529.1	31.3	.0487	1.01566
660.0	32.28	1.152	13770.0	19410.0	180.2	22.24	30.89	536.5	32.0	.0499	1.01518
680.0	31.34	1.119	14220.0	20030.0	181.1	22.37	31.00	543.7	32.6	.0510	1.01473
700.0	30.45	1.087	14670.0	20650.0	182.0	22.50	31.10	550.9	33.2	.0521	1.01431
720.0	29.61	1.057	15120.0	21270.0	182.9	22.63	31.21	557.9	33.8	.0533	1.01391
740.0	28.82	1.029	15580.0	21900.0	183.7	22.76	31.32	564.7	34.4	.0544	1.01354
760.0	28.07	1.002	16040.0	22520.0	184.6	22.89	31.43	571.5	35.0	.0555	1.01318
780.0	27.36	.9766	16500.0	23150.0	185.4	23.02	31.55	578.2	35.6	.0567	1.01284
800.0	26.69	.9526	16960.0	23790.0	186.2	23.15	31.67	584.7	36.2	.0578	1.01252
820.0	26.04	.9296	17430.0	24420.0	187.0	23.28	31.78	591.2	36.8	.0589	1.01222
840.0	25.43	.9078	17900.0	25060.0	187.7	23.41	31.90	597.6	37.4	.0601	1.01193
860.0	24.85	.8870	18370.0	25700.0	188.5	23.54	32.02	603.9	37.9	.0612	1.01165
880.0	24.29	.8672	18840.0	26340.0	189.2	23.66	32.13	610.2	38.5	.0623	1.01139
900.0	23.76	.8482	19320.0	26980.0	189.9	23.79	32.25	616.3	39.1	.0634	1.01113
920.0	23.25	.8300	19800.0	27630.0	190.6	23.91	32.36	622.4	39.6	.0645	1.01089
950.0	22.53	.8042	20520.0	28600.0	191.7	24.09	32.53	631.4	40.4	.0662	1.01055
1000.0	21.42	.7646	21730.0	30230.0	193.4	24.38	32.80	646.1	41.8	.0689	1.01002
1050.0	20.42	.7288	22960.0	31880.0	195.0	24.65	33.05	660.5	43.1	.0716	1.00955
1100.0	19.50	.6962	24200.0	33540.0	196.5	24.91	33.30	674.5	44.4	.0743	1.00911
1150.0	18.67	.6664	25460.0	35210.0	198.0	25.15	33.53	688.2	45.7	.0769	1.00872
1200.0	17.90	.6390	26720.0	36890.0	199.4	25.38	33.75	701.6	47.0	.0795	1.00835
1250.0	17.20	.6139	28000.0	38580.0	200.8	25.59	33.95	714.8	48.2	.0820	1.00802
1300.0	16.55	.5906	29280.0	40290.0	202.1	25.79	34.14	727.8	49.5	.0845	1.00771
1350.0	15.94	.5691	30580.0	42000.0	203.4	25.97	34.32	740.5	50.7	.0870	1.00743
1400.0	15.38	.5491	31880.0	43720.0	204.7	26.15	34.49	753.0	51.9	.0894	1.00716
1450.0	14.86	.5304	33190.0	45450.0	205.9	26.31	34.65	765.3	53.0	.0918	1.00691
1500.0	14.37	.5130	34510.0	47180.0	207.1	26.46	34.80	777.4	54.2	.0942	1.00668
1550.0	13.91	.4967	35840.0	48930.0	208.2	26.60	34.94	789.3	55.4	.0965	1.00647
1600.0	13.49	.4814	37170.0	50680.0	209.3	26.74	35.07	801.1	56.5	.0988	1.00626
1650.0	13.08	.4670	38510.0	52430.0	210.4	26.86	35.19	812.7	57.6	.101	1.00607
1700.0	12.70	.4534	39860.0	54200.0	211.5	26.98	35.30	824.1	58.7	.103	1.00589
1750.0	12.34	.4406	41210.0	55960.0	212.5	27.09	35.41	835.4	59.8	.106	1.00572
1800.0	12.01	.4286	42570.0	57740.0	213.5	27.19	35.51	846.5	60.9	.108	1.00556
1850.0	11.69	.4171	43930.0	59520.0	214.5	27.29	35.61	857.5	62.0	.110	1.00541
1900.0	11.38	.4063	45300.0	61300.0	215.4	27.38	35.70	868.4	63.1	.112	1.00527
				7.00 MPa isobar							
64.67[a]	870.6	31.07	−4188.0	−3962.0	68.39	30.52	54.12	1241.0	277.0	.153	1.47001
80.0	812.8	29.01	−3355.0	−3114.0	80.16	29.38	55.65	927.8	151.0	.136	1.43580
100.0	721.7	25.76	−2257.0	−1985.0	92.74	27.06	58.07	699.0	85.0	.108	1.38234
110.0	668.1	23.85	−1682.0	−1388.0	98.42	26.56	61.59	597.7	65.7	.0935	1.35125
120.0	605.4	21.61	−1068.0	−743.8	104.0	26.27	68.07	496.6	50.4	.0787	1.31533
125.0	568.5	20.29	−735.3	−390.3	106.9	26.21	73.73	443.8	43.6	.0716	1.29442
130.0	525.4	18.75	−374.2	−1.0	110.0	26.24	82.75	388.2	37.3	.0651	1.27022
132.0	505.6	18.05	−218.5	169.4	111.3	26.29	87.78	365.3	34.8	.0627	1.25923
134.0	484.0	17.28	−54.4	350.8	112.6	26.37	93.74	342.3	32.3	.0602	1.24730
136.0	460.4	16.43	118.9	544.8	114.1	26.48	100.4	320.0	29.8	.0576	1.23432
138.0	434.8	15.52	301.0	752.0	115.6	26.61	106.7	299.4	27.5	.0547	1.22035
140.0	407.7	14.55	489.5	970.5	117.2	26.72	111.3	281.8	25.3	.0515	1.20571
142.0	380.4	13.58	679.6	1195.0	118.7	26.79	112.8	268.1	23.3	.0482	1.19102
144.0	354.0	12.64	865.5	1419.0	120.3	26.77	111.0	258.4	21.6	.0451	1.17698
145.0	341.5	12.19	955.4	1530.0	121.1	26.72	109.2	254.9	20.9	.0436	1.17038
146.0	329.6	11.77	1043.0	1638.0	121.8	26.65	106.8	252.1	20.2	.0422	1.16411

Thermophysical properties of nitrogen—Continued

T K	Density kg/m^3	Density mol/dm^3	E J/mol	H J/mol	S J/(mol·K)	C_v J/(mol·K)	C_p J/(mol·K)	Sound m/s	Visc. μPa·s	Therm. W/(m·K)	Diel. const.
147.0	318.4	11.36	1127.0	1743.0	122.5	26.55	104.2	250.0	19.7	.0409	1.15819
148.0	307.7	10.98	1209.0	1846.0	123.2	26.44	101.2	248.4	19.1	.0396	1.15263
149.0	297.8	10.63	1287.0	1946.0	123.9	26.32	98.12	247.3	18.7	.0385	1.14743
150.0	288.4	10.30	1362.0	2042.0	124.6	26.18	94.95	246.5	18.3	.0375	1.14258
151.0	279.7	9.985	1434.0	2135.0	125.2	26.04	91.77	246.1	17.9	.0365	1.13807
152.0	271.6	9.695	1504.0	2226.0	125.8	25.89	88.65	246.0	17.6	.0356	1.13387
153.0	264.0	9.424	1570.0	2313.0	126.3	25.74	85.62	246.0	17.3	.0348	1.12996
154.0	257.0	9.172	1634.0	2397.0	126.9	25.59	82.72	246.3	17.1	.0340	1.12633
155.0	250.4	8.937	1695.0	2478.0	127.4	25.44	79.96	246.7	16.9	.0333	1.12294
156.0	244.2	8.717	1754.0	2557.0	127.9	25.29	77.36	247.3	16.7	.0327	1.11979
157.0	238.4	8.511	1811.0	2633.0	128.4	25.14	74.92	248.0	16.5	.0321	1.11684
158.0	233.0	8.318	1865.0	2707.0	128.9	25.00	72.63	248.7	16.4	.0316	1.11408
159.0	228.0	8.136	1918.0	2778.0	129.3	24.87	70.50	249.6	16.2	.0311	1.11149
160.0	223.2	7.966	1969.0	2848.0	129.8	24.74	68.52	250.5	16.1	.0306	1.10907
162.0	214.4	7.654	2067.0	2981.0	130.6	24.49	64.95	252.4	15.9	.0298	1.10463
164.0	206.6	7.374	2159.0	3108.0	131.4	24.26	61.85	254.4	15.7	.0292	1.10067
166.0	199.5	7.122	2246.0	3229.0	132.1	24.05	59.16	256.5	15.6	.0286	1.09711
168.0	193.1	6.894	2330.0	3345.0	132.8	23.85	56.82	258.7	15.5	.0282	1.09389
170.0	187.3	6.686	2409.0	3456.0	133.5	23.67	54.76	260.9	15.5	.0278	1.09096
172.0	182.0	6.495	2486.0	3564.0	134.1	23.51	52.95	263.1	15.4	.0275	1.08828
174.0	177.0	6.319	2561.0	3668.0	134.7	23.36	51.35	265.3	15.4	.0272	1.08581
176.0	172.5	6.156	2632.0	3770.0	135.3	23.22	49.92	267.5	15.3	.0270	1.08353
178.0	168.2	6.005	2702.0	3868.0	135.8	23.09	48.64	269.7	15.3	.0268	1.08141
180.0	164.3	5.863	2770.0	3964.0	136.4	22.98	47.49	271.9	15.3	.0266	1.07944
182.0	160.6	5.731	2837.0	4058.0	136.9	22.87	46.45	274.0	15.3	.0265	1.07760
184.0	157.1	5.606	2902.0	4150.0	137.4	22.77	45.51	276.2	15.4	.0264	1.07586
186.0	153.8	5.489	2965.0	4240.0	137.9	22.68	44.66	278.3	15.4	.0264	1.07424
188.0	150.7	5.379	3027.0	4329.0	138.3	22.59	43.88	280.3	15.4	.0263	1.07270
190.0	147.8	5.274	3089.0	4416.0	138.8	22.51	43.16	282.4	15.4	.0263	1.07125
192.0	145.0	5.175	3149.0	4501.0	139.3	22.44	42.51	284.4	15.5	.0262	1.06987
195.0	141.1	5.035	3237.0	4628.0	139.9	22.33	41.62	287.4	15.5	.0262	1.06793
200.0	135.1	4.823	3381.0	4832.0	140.9	22.18	40.34	292.3	15.6	.0262	1.06500
205.0	129.8	4.633	3521.0	5031.0	141.9	22.06	39.28	297.0	15.8	.0263	1.06238
210.0	125.0	4.461	3656.0	5225.0	142.9	21.95	38.37	301.6	15.9	.0263	1.06002
215.0	120.6	4.305	3790.0	5415.0	143.8	21.86	37.60	306.1	16.1	.0264	1.05787
220.0	116.6	4.163	3920.0	5602.0	144.6	21.78	36.93	310.5	16.3	.0266	1.05591
225.0	112.9	4.031	4048.0	5785.0	145.4	21.70	36.35	314.7	16.4	.0267	1.05411
230.0	109.5	3.910	4175.0	5965.0	146.2	21.64	35.83	318.9	16.6	.0269	1.05244
235.0	106.4	3.797	4299.0	6143.0	147.0	21.59	35.38	322.9	16.8	.0271	1.05090
240.0	103.4	3.691	4423.0	6319.0	147.7	21.54	34.97	326.9	17.0	.0273	1.04946
245.0	100.7	3.593	4545.0	6493.0	148.4	21.50	34.61	330.8	17.2	.0275	1.04811
250.0	98.08	3.501	4666.0	6665.0	149.1	21.46	34.28	334.6	17.4	.0277	1.04685
255.0	95.64	3.414	4785.0	6836.0	149.8	21.42	33.99	338.4	17.5	.0279	1.04567
260.0	93.35	3.332	4904.0	7005.0	150.5	21.39	33.72	342.0	17.7	.0281	1.04455
270.0	89.13	3.181	5140.0	7340.0	151.7	21.34	33.25	349.2	18.1	.0286	1.04250
280.0	85.33	3.046	5372.0	7670.0	152.9	21.29	32.85	356.1	18.5	.0291	1.04066
290.0	81.88	2.923	5602.0	7997.0	154.1	21.26	32.52	362.8	18.9	.0295	1.03899
300.0	78.74	2.811	5830.0	8321.0	155.2	21.23	32.23	369.3	19.3	.0300	1.03747
310.0	75.86	2.708	6057.0	8642.0	156.2	21.20	31.98	375.6	19.7	.0306	1.03608
320.0	73.20	2.613	6282.0	8961.0	157.3	21.18	31.76	381.8	20.1	.0311	1.03480
330.0	70.75	2.525	6505.0	9277.0	158.2	21.17	31.57	387.8	20.5	.0316	1.03361
340.0	68.47	2.444	6728.0	9592.0	159.2	21.16	31.41	393.7	20.9	.0321	1.03251
350.0	66.35	2.368	6950.0	9905.0	160.1	21.15	31.26	399.4	21.3	.0327	1.03149
360.0	64.36	2.297	7170.0	10220.0	161.0	21.15	31.13	405.1	21.7	.0332	1.03053
370.0	62.51	2.231	7391.0	10530.0	161.8	21.15	31.02	410.6	22.0	.0337	1.02964
380.0	60.76	2.169	7610.0	10840.0	162.6	21.15	30.92	416.0	22.4	.0343	1.02880
390.0	59.12	2.110	7829.0	11150.0	163.4	21.15	30.84	421.3	22.8	.0348	1.02801
400.0	57.57	2.055	8048.0	11450.0	164.2	21.16	30.76	426.5	23.2	.0354	1.02727
410.0	56.10	2.003	8266.0	11760.0	165.0	21.18	30.70	431.6	23.5	.0359	1.02656
420.0	54.72	1.953	8484.0	12070.0	165.7	21.19	30.64	436.6	23.9	.0365	1.02590
440.0	52.15	1.861	8920.0	12680.0	167.1	21.23	30.56	446.4	24.6	.0376	1.02467
460.0	49.83	1.779	9355.0	13290.0	168.5	21.28	30.50	455.9	25.4	.0387	1.02356
480.0	47.71	1.703	9790.0	13900.0	169.8	21.34	30.47	465.1	26.1	.0398	1.02254

Thermophysical properties of nitrogen—Continued

T K	Density kg/m³	Density mol/dm³	E J/mol	H J/mol	S J/(mol·K)	C_v J/(mol·K)	C_p	Sound m/s	Visc. μPa·s	Therm. W/(m·K)	Diel. const.
500.0	45.78	1.634	10230.0	14510.0	171.0	21.41	30.46	474.0	26.8	.0409	1.02162
520.0	44.00	1.571	10660.0	15120.0	172.2	21.49	30.47	482.7	27.5	.0421	1.02077
540.0	42.36	1.512	11100.0	15730.0	173.4	21.58	30.50	491.1	28.1	.0432	1.01998
560.0	40.85	1.458	11540.0	16340.0	174.5	21.68	30.55	499.4	28.8	.0443	1.01926
580.0	39.44	1.408	11980.0	16950.0	175.6	21.78	30.60	507.4	29.5	.0454	1.01859
600.0	38.13	1.361	12420.0	17560.0	176.6	21.89	30.67	515.3	30.1	.0466	1.01796
620.0	36.90	1.317	12860.0	18180.0	177.6	22.01	30.75	523.0	30.8	.0477	1.01738
640.0	35.76	1.276	13310.0	18790.0	178.6	22.13	30.83	530.5	31.4	.0489	1.01683
660.0	34.68	1.238	13760.0	19410.0	179.5	22.25	30.93	537.9	32.0	.0500	1.01632
680.0	33.67	1.202	14210.0	20030.0	180.5	22.38	31.03	545.1	32.6	.0511	1.01584
700.0	32.72	1.168	14660.0	20650.0	181.4	22.50	31.13	552.2	33.3	.0523	1.01539
720.0	31.82	1.136	15110.0	21280.0	182.2	22.63	31.24	559.2	33.9	.0534	1.01496
740.0	30.97	1.106	15570.0	21900.0	183.1	22.76	31.35	566.1	34.5	.0545	1.01455
760.0	30.17	1.077	16030.0	22530.0	183.9	22.90	31.46	572.8	35.1	.0557	1.01417
780.0	29.41	1.050	16490.0	23160.0	184.8	23.03	31.57	579.5	35.7	.0568	1.01381
800.0	28.68	1.024	16960.0	23790.0	185.6	23.16	31.69	586.0	36.2	.0579	1.01346
820.0	27.99	.9992	17420.0	24430.0	186.3	23.29	31.80	592.5	36.8	.0590	1.01314
840.0	27.34	.9758	17890.0	25070.0	187.1	23.42	31.92	598.9	37.4	.0602	1.01282
860.0	26.71	.9534	18360.0	25700.0	187.9	23.54	32.03	605.2	38.0	.0613	1.01253
880.0	26.11	.9321	18840.0	26350.0	188.6	23.67	32.15	611.4	38.5	.0624	1.01224
900.0	25.54	.9117	19310.0	26990.0	189.3	23.79	32.26	617.5	39.1	.0635	1.01197
920.0	25.00	.8922	19790.0	27640.0	190.0	23.92	32.37	623.6	39.6	.0646	1.01171
950.0	24.22	.8645	20510.0	28610.0	191.1	24.09	32.54	632.6	40.5	.0663	1.01134
1000.0	23.03	.8220	21730.0	30240.0	192.7	24.38	32.81	647.3	41.8	.0690	1.01078
1050.0	21.95	.7836	22960.0	31890.0	194.4	24.65	33.06	661.6	43.1	.0717	1.01027
1100.0	20.97	.7485	24200.0	33550.0	195.9	24.91	33.31	675.6	44.5	.0744	1.00980
1150.0	20.07	.7165	25450.0	35220.0	197.4	25.15	33.54	689.3	45.7	.0770	1.00938
1200.0	19.25	.6872	26720.0	36900.0	198.8	25.38	33.76	702.7	47.0	.0796	1.00899
1250.0	18.49	.6602	27990.0	38600.0	200.2	25.59	33.96	715.9	48.2	.0821	1.00863
1300.0	17.80	.6352	29280.0	40300.0	201.5	25.79	34.15	728.8	49.5	.0846	1.00829
1350.0	17.15	.6120	30570.0	42010.0	202.8	25.98	34.33	741.5	50.7	.0871	1.00799
1400.0	16.54	.5905	31880.0	43730.0	204.1	26.15	34.50	754.0	51.9	.0895	1.00770
1450.0	15.98	.5705	33190.0	45460.0	205.3	26.31	34.66	766.3	53.1	.0919	1.00744
1500.0	15.46	.5518	34510.0	47200.0	206.5	26.47	34.80	778.3	54.2	.0943	1.00719
1550.0	14.97	.5342	35840.0	48940.0	207.6	26.61	34.94	790.3	55.4	.0966	1.00696
1600.0	14.51	.5178	37170.0	50690.0	208.7	26.74	35.07	802.0	56.5	.0989	1.00674
1650.0	14.07	.5023	38510.0	52450.0	209.8	26.86	35.19	813.6	57.6	.101	1.00653
1700.0	13.67	.4878	39860.0	54210.0	210.9	26.98	35.31	825.0	58.7	.103	1.00634
1750.0	13.28	.4740	41210.0	55980.0	211.9	27.09	35.41	836.3	59.8	.106	1.00616
1800.0	12.92	.4610	42570.0	57750.0	212.9	27.19	35.51	847.4	60.9	.108	1.00599
1850.0	12.57	.4488	43930.0	59530.0	213.9	27.29	35.61	858.4	62.0	.110	1.00582
1900.0	12.25	.4371	45300.0	61310.0	214.8	27.38	35.70	869.2	63.1	.112	1.00567

7.50 MPa isobar

T K	Density kg/m³	Density mol/dm³	E J/mol	H J/mol	S J/(mol·K)	C_v J/(mol·K)	C_p	Sound m/s	Visc. μPa·s	Therm. W/(m·K)	Diel. const.
64.78[a]	870.8	31.08	−4186.0	−3944.0	68.41	30.75	54.08	1237.0	277.0	.153	1.47011
80.0	813.9	29.05	−3362.0	−3104.0	80.07	29.50	55.50	930.2	152.0	.137	1.43643
100.0	723.9	25.84	−2269.0	−1979.0	92.61	27.12	57.74	704.7	85.9	.109	1.38359
110.0	671.3	23.96	−1700.0	−1387.0	98.25	26.60	60.97	605.5	66.6	.0944	1.35308
120.0	610.5	21.79	−1096.0	−751.6	103.8	26.28	66.71	507.5	51.4	.0798	1.31821
125.0	575.4	20.54	−772.2	−407.0	106.6	26.18	71.44	457.1	44.8	.0729	1.29826
130.0	535.2	19.10	−425.9	−33.2	109.5	26.16	78.56	404.9	38.7	.0665	1.27567
132.0	517.2	18.46	−278.7	127.6	110.7	26.18	82.34	383.6	36.3	.0641	1.26563
134.0	497.8	17.77	−125.5	296.6	112.0	26.22	86.72	362.3	33.9	.0617	1.25491
136.0	477.0	17.03	34.3	474.8	113.3	26.28	91.58	341.4	31.6	.0593	1.24342
138.0	454.7	16.23	200.8	663.0	114.7	26.36	96.53	321.6	29.4	.0567	1.23116
140.0	431.0	15.38	373.0	860.5	116.1	26.44	100.8	303.6	27.3	.0539	1.21829
145.0	370.4	13.22	809.5	1377.0	119.8	26.50	103.4	271.7	22.8	.0465	1.18570
147.0	347.7	12.41	977.1	1581.0	121.2	26.42	100.9	264.3	21.4	.0437	1.17364
148.0	337.0	12.03	1058.0	1681.0	121.8	26.36	99.16	261.6	20.8	.0425	1.16799
150.0	317.1	11.32	1213.0	1876.0	123.1	26.18	94.91	257.7	19.8	.0401	1.15750
151.0	307.9	10.99	1287.0	1969.0	123.8	26.07	92.55	256.4	19.4	.0391	1.15268

Thermophysical properties of nitrogen—Continued

T K	Density kg/m³	Density mol/dm³	E J/mol	H J/mol	S J/(mol·K)	C_v J/(mol·K)	C_p J/(mol·K)	Sound m/s	Visc. μPa·s	Therm. W/(m·K)	Diel. const.
152.0	299.1	10.68	1358.0	2061.0	124.4	25.96	90.11	255.4	19.0	.0381	1.14813
153.0	290.9	10.38	1427.0	2150.0	124.9	25.83	87.64	254.8	18.6	.0372	1.14385
154.0	283.2	10.11	1494.0	2236.0	125.5	25.71	85.17	254.4	18.3	.0363	1.13983
155.0	275.9	9.847	1558.0	2320.0	126.0	25.57	82.74	254.2	18.0	.0355	1.13605
156.0	269.0	9.601	1620.0	2401.0	126.6	25.44	80.36	254.2	17.8	.0348	1.13251
157.0	262.5	9.370	1680.0	2481.0	127.1	25.31	78.07	254.4	17.5	.0341	1.12917
158.0	256.4	9.153	1738.0	2558.0	127.6	25.18	75.88	254.7	17.3	.0335	1.12604
160.0	245.3	8.755	1849.0	2705.0	128.5	24.92	71.79	255.7	17.0	.0324	1.12032
162.0	235.3	8.399	1952.0	2845.0	129.4	24.68	68.13	257.0	16.7	.0315	1.11523
164.0	226.4	8.081	2050.0	2978.0	130.2	24.45	64.88	258.6	16.5	.0307	1.11069
166.0	218.3	7.794	2143.0	3105.0	130.9	24.23	62.00	260.3	16.3	.0301	1.10661
168.0	211.1	7.533	2231.0	3226.0	131.7	24.04	59.46	262.2	16.2	.0295	1.10292
170.0	204.4	7.297	2315.0	3343.0	132.4	23.85	57.22	264.1	16.0	.0290	1.09956
172.0	198.3	7.080	2396.0	3455.0	133.0	23.68	55.23	266.1	16.0	.0286	1.09650
174.0	192.7	6.880	2474.0	3564.0	133.7	23.52	53.46	268.2	15.9	.0283	1.09368
176.0	187.6	6.696	2549.0	3669.0	134.3	23.38	51.87	270.2	15.8	.0280	1.09109
178.0	182.8	6.524	2622.0	3772.0	134.8	23.24	50.46	272.3	15.8	.0278	1.08868
180.0	178.3	6.365	2693.0	3871.0	135.4	23.12	49.18	274.3	15.8	.0276	1.08645
182.0	174.2	6.216	2762.0	3968.0	135.9	23.00	48.03	276.4	15.8	.0274	1.08437
184.0	170.2	6.077	2829.0	4063.0	136.4	22.90	46.99	278.4	15.8	.0273	1.08241
186.0	166.6	5.946	2895.0	4156.0	136.9	22.80	46.04	280.5	15.8	.0272	1.08058
188.0	163.1	5.822	2959.0	4248.0	137.4	22.71	45.18	282.5	15.8	.0271	1.07886
190.0	159.8	5.705	3023.0	4337.0	137.9	22.62	44.38	284.5	15.8	.0270	1.07723
192.0	156.7	5.595	3085.0	4425.0	138.4	22.54	43.66	286.4	15.8	.0270	1.07569
194.0	153.8	5.490	3146.0	4512.0	138.8	22.47	42.99	288.4	15.8	.0269	1.07423
200.0	145.8	5.204	3323.0	4764.0	140.1	22.27	41.27	294.2	15.9	.0269	1.07027
205.0	139.9	4.994	3466.0	4968.0	141.1	22.14	40.09	298.9	16.1	.0269	1.06736
210.0	134.6	4.805	3605.0	5166.0	142.1	22.02	39.10	303.4	16.2	.0269	1.06475
215.0	129.8	4.634	3740.0	5359.0	143.0	21.92	38.26	307.8	16.3	.0270	1.06238
220.0	125.4	4.477	3873.0	5548.0	143.8	21.84	37.52	312.2	16.5	.0271	1.06022
225.0	121.4	4.333	4003.0	5734.0	144.7	21.76	36.89	316.4	16.6	.0272	1.05824
230.0	117.7	4.200	4132.0	5917.0	145.5	21.70	36.33	320.5	16.8	.0274	1.05642
235.0	114.2	4.077	4258.0	6098.0	146.3	21.64	35.83	324.6	17.0	.0275	1.05473
240.0	111.0	3.963	4383.0	6276.0	147.0	21.59	35.40	328.6	17.2	.0277	1.05316
245.0	108.0	3.856	4506.0	6452.0	147.7	21.54	35.00	332.4	17.3	.0279	1.05169
250.0	105.2	3.755	4629.0	6626.0	148.4	21.50	34.65	336.2	17.5	.0281	1.05032
255.0	102.6	3.661	4750.0	6798.0	149.1	21.46	34.33	340.0	17.7	.0283	1.04903
260.0	100.1	3.572	4870.0	6969.0	149.8	21.43	34.04	343.6	17.9	.0285	1.04782
270.0	95.51	3.409	5107.0	7307.0	151.1	21.37	33.53	350.8	18.3	.0289	1.04560
280.0	91.40	3.263	5341.0	7640.0	152.3	21.32	33.11	357.7	18.7	.0294	1.04360
290.0	87.68	3.130	5573.0	7969.0	153.4	21.29	32.75	364.4	19.1	.0299	1.04180
300.0	84.30	3.009	5803.0	8295.0	154.5	21.25	32.44	370.9	19.4	.0304	1.04015
310.0	81.19	2.898	6030.0	8618.0	155.6	21.23	32.17	377.2	19.8	.0309	1.03865
320.0	78.34	2.796	6257.0	8939.0	156.6	21.21	31.94	383.4	20.2	.0314	1.03727
330.0	75.70	2.702	6481.0	9257.0	157.6	21.19	31.74	389.4	20.6	.0319	1.03599
340.0	73.25	2.615	6705.0	9574.0	158.5	21.18	31.56	395.3	21.0	.0324	1.03481
350.0	70.97	2.533	6928.0	9888.0	159.4	21.17	31.40	401.0	21.4	.0329	1.03371
360.0	68.84	2.457	7149.0	10200.0	160.3	21.17	31.26	406.6	21.8	.0335	1.03268
370.0	66.85	2.386	7370.0	10510.0	161.2	21.16	31.14	412.2	22.1	.0340	1.03172
380.0	64.98	2.319	7591.0	10820.0	162.0	21.17	31.04	417.6	22.5	.0345	1.03082
390.0	63.21	2.256	7811.0	11130.0	162.8	21.17	30.94	422.8	22.9	.0351	1.02997
400.0	61.55	2.197	8030.0	11440.0	163.6	21.18	30.86	428.0	23.3	.0356	1.02917
410.0	59.98	2.141	8249.0	11750.0	164.4	21.19	30.79	433.1	23.6	.0361	1.02842
420.0	58.50	2.088	8467.0	12060.0	165.1	21.21	30.73	438.2	24.0	.0367	1.02771
440.0	55.75	1.990	8904.0	12670.0	166.5	21.24	30.64	447.9	24.7	.0378	1.02639
460.0	53.27	1.901	9340.0	13290.0	167.9	21.29	30.57	457.4	25.4	.0389	1.02520
480.0	51.01	1.821	9776.0	13900.0	169.2	21.35	30.54	466.6	26.1	.0400	1.02411
500.0	48.94	1.747	10210.0	14510.0	170.4	21.43	30.52	475.5	26.8	.0411	1.02312
520.0	47.04	1.679	10650.0	15120.0	171.6	21.51	30.53	484.2	27.5	.0422	1.02221
540.0	45.29	1.616	11090.0	15730.0	172.8	21.59	30.55	492.6	28.2	.0433	1.02137
560.0	43.67	1.559	11530.0	16340.0	173.9	21.69	30.59	500.8	28.8	.0445	1.02060
580.0	42.16	1.505	11970.0	16950.0	175.0	21.79	30.65	508.9	29.5	.0456	1.01988
600.0	40.76	1.455	12410.0	17570.0	176.0	21.90	30.71	516.7	30.2	.0467	1.01921

Thermophysical properties of nitrogen—Continued

T K	Density kg/m³	Density mol/dm³	E J/mol	H J/mol	S J/(mol·K)	C_v J/(mol·K)	C_p J/(mol·K)	Sound m/s	Visc. μPa·s	Therm. W/(m·K)	Diel. const.
620.0	39.45	1.408	12850.0	18180.0	177.0	22.02	30.79	524.4	30.8	.0479	1.01859
640.0	38.23	1.365	13300.0	18800.0	178.0	22.14	30.87	531.9	31.4	.0490	1.01800
660.0	37.08	1.324	13750.0	19420.0	178.9	22.26	30.96	539.2	32.1	.0501	1.01746
680.0	36.00	1.285	14200.0	20040.0	179.9	22.38	31.06	546.5	32.7	.0513	1.01694
700.0	34.99	1.249	14650.0	20660.0	180.8	22.51	31.16	553.6	33.3	.0524	1.01646
720.0	34.03	1.215	15110.0	21280.0	181.7	22.64	31.26	560.5	33.9	.0535	1.01600
740.0	33.12	1.182	15560.0	21910.0	182.5	22.77	31.37	567.4	34.5	.0547	1.01557
760.0	32.26	1.151	16020.0	22540.0	183.4	22.90	31.48	574.1	35.1	.0558	1.01516
780.0	31.44	1.122	16490.0	23170.0	184.2	23.03	31.59	580.8	35.7	.0569	1.01477
800.0	30.67	1.095	16950.0	23800.0	185.0	23.16	31.71	587.3	36.3	.0580	1.01440
820.0	29.93	1.068	17420.0	24440.0	185.8	23.29	31.82	593.8	36.9	.0592	1.01405
840.0	29.23	1.043	17890.0	25070.0	186.5	23.42	31.94	600.1	37.4	.0603	1.01372
860.0	28.57	1.020	18360.0	25710.0	187.3	23.55	32.05	606.4	38.0	.0614	1.01340
880.0	27.93	.9968	18830.0	26360.0	188.0	23.68	32.16	612.6	38.6	.0625	1.01310
900.0	27.32	.9751	19310.0	27000.0	188.7	23.80	32.28	618.8	39.1	.0636	1.01281
920.0	26.73	.9543	19790.0	27650.0	189.4	23.92	32.39	624.8	39.7	.0647	1.01253
950.0	25.91	.9247	20510.0	28620.0	190.5	24.10	32.55	633.8	40.5	.0664	1.01213
1000.0	24.63	.8793	21720.0	30250.0	192.2	24.39	32.82	648.5	41.8	.0691	1.01153
1050.0	23.48	.8381	22950.0	31900.0	193.8	24.66	33.08	662.7	43.2	.0718	1.01098
1100.0	22.43	.8007	24200.0	33560.0	195.3	24.92	33.32	676.7	44.5	.0744	1.01049
1150.0	21.48	.7665	25450.0	35230.0	196.8	25.16	33.55	690.4	45.8	.0771	1.01003
1200.0	20.60	.7352	26720.0	36920.0	198.2	25.38	33.76	703.8	47.0	.0796	1.00961
1250.0	19.79	.7063	27990.0	38610.0	199.6	25.60	33.97	716.9	48.3	.0822	1.00923
1300.0	19.04	.6796	29280.0	40310.0	201.0	25.79	34.16	729.8	49.5	.0847	1.00888
1350.0	18.35	.6549	30570.0	42030.0	202.2	25.98	34.34	742.5	50.7	.0872	1.00855
1400.0	17.70	.6319	31880.0	43750.0	203.5	26.15	34.50	755.0	51.9	.0896	1.00824
1450.0	17.10	.6105	33190.0	45480.0	204.7	26.32	34.66	767.2	53.1	.0920	1.00796
1500.0	16.54	.5905	34510.0	47210.0	205.9	26.47	34.81	779.3	54.2	.0943	1.00769
1550.0	16.02	.5717	35840.0	48960.0	207.0	26.61	34.95	791.2	55.4	.0967	1.00744
1600.0	15.52	.5541	37170.0	50710.0	208.1	26.74	35.07	802.9	56.5	.0990	1.00721
1650.0	15.06	.5376	38510.0	52460.0	209.2	26.87	35.20	814.5	57.6	.101	1.00699
1700.0	14.63	.5220	39860.0	54230.0	210.3	26.98	35.31	825.9	58.8	.103	1.00679
1750.0	14.21	.5074	41210.0	55990.0	211.3	27.09	35.42	837.2	59.9	.106	1.00659
1800.0	13.83	.4935	42570.0	57770.0	212.3	27.20	35.52	848.3	60.9	.108	1.00641
1850.0	13.46	.4803	43930.0	59550.0	213.3	27.29	35.61	859.2	62.0	.110	1.00623
1900.0	13.11	.4679	45300.0	61330.0	214.2	27.38	35.70	870.1	63.1	.112	1.00607

8.00 MPa isobar

T K	Density kg/m³	Density mol/dm³	E J/mol	H J/mol	S J/(mol·K)	C_v J/(mol·K)	C_p J/(mol·K)	Sound m/s	Visc. μPa·s	Therm. W/(m·K)	Diel. const.
64.89[a]	871.0	31.09	−4184.0	−3926.0	68.44	30.97	54.03	1233.0	277.0	.153	1.47022
80.0	815.0	29.09	−3368.0	−3093.0	79.98	29.61	55.37	932.6	153.0	.137	1.43705
100.0	726.0	25.91	−2282.0	−1973.0	92.47	27.18	57.42	710.4	86.8	.110	1.38481
110.0	674.4	24.07	−1717.0	−1385.0	98.07	26.64	60.40	613.1	67.5	.0953	1.35483
120.0	615.3	21.96	−1122.0	−757.9	103.5	26.28	65.51	517.9	52.5	.0810	1.32093
130.0	543.8	19.41	−471.8	−59.6	109.1	26.11	75.31	420.1	40.0	.0678	1.28049
132.0	527.2	18.82	−331.3	93.9	110.3	26.10	78.27	400.1	37.7	.0654	1.27119
134.0	509.5	18.19	−186.2	253.7	111.5	26.12	81.63	380.1	35.4	.0631	1.26135
136.0	490.7	17.51	−36.2	420.6	112.7	26.14	85.33	360.5	33.2	.0608	1.25094
138.0	470.7	16.80	119.0	595.1	114.0	26.18	89.17	341.5	31.1	.0585	1.23995
140.0	449.7	16.05	278.7	777.1	115.3	26.23	92.77	323.9	29.0	.0559	1.22843
145.0	394.6	14.09	687.8	1256.0	118.7	26.29	97.23	289.3	24.6	.0491	1.19863
147.0	373.1	13.32	848.9	1450.0	120.0	26.25	96.41	280.0	23.1	.0463	1.18710
148.0	362.7	12.95	927.6	1546.0	120.6	26.21	95.50	276.3	22.5	.0451	1.18156
150.0	342.9	12.24	1081.0	1734.0	121.9	26.09	92.88	270.6	21.3	.0427	1.17107
152.0	324.6	11.59	1226.0	1917.0	123.1	25.93	89.52	266.6	20.4	.0405	1.16144
154.0	307.9	10.99	1364.0	2092.0	124.3	25.74	85.72	264.1	19.6	.0386	1.15271
156.0	292.9	10.45	1494.0	2259.0	125.3	25.52	81.76	262.7	18.9	.0369	1.14485
158.0	279.3	9.968	1616.0	2419.0	126.4	25.29	77.82	262.2	18.4	.0355	1.13781
160.0	267.1	9.532	1732.0	2571.0	127.3	25.05	74.06	262.3	17.9	.0343	1.13150
162.0	256.1	9.140	1840.0	2715.0	128.2	24.82	70.55	262.9	17.6	.0332	1.12585
164.0	246.2	8.786	1943.0	2853.0	129.1	24.60	67.33	263.9	17.3	.0323	1.12077
166.0	237.2	8.467	2040.0	2985.0	129.9	24.39	64.42	265.1	17.0	.0315	1.11619

Thermophysical properties of nitrogen—Continued

T K	Density kg/m^3	Density mol/dm^3	E J/mol	H J/mol	S J/(mol·K)	C_v J/(mol·K)	C_p J/(mol·K)	Sound m/s	Visc. μPa·s	Therm. W/(m·K)	Diel. const.
168.0	229.1	8.176	2133.0	3111.0	130.6	24.19	61.79	266.5	16.8	.0309	1.11204
170.0	221.6	7.911	2221.0	3232.0	131.3	24.01	59.43	268.1	16.7	.0303	1.10827
172.0	214.8	7.669	2306.0	3349.0	132.0	23.83	57.32	269.8	16.5	.0299	1.10482
174.0	208.6	7.446	2387.0	3462.0	132.7	23.67	55.43	271.6	16.4	.0295	1.10166
176.0	202.8	7.240	2466.0	3571.0	133.3	23.52	53.72	273.5	16.4	.0291	1.09875
178.0	197.5	7.049	2542.0	3677.0	133.9	23.38	52.19	275.3	16.3	.0288	1.09605
180.0	192.5	6.871	2615.0	3780.0	134.5	23.25	50.81	277.3	16.2	.0286	1.09355
182.0	187.9	6.706	2687.0	3880.0	135.0	23.13	49.56	279.2	16.2	.0284	1.09122
184.0	183.5	6.551	2757.0	3978.0	135.5	23.02	48.42	281.1	16.2	.0282	1.08905
186.0	179.4	6.405	2825.0	4074.0	136.1	22.91	47.39	283.0	16.2	.0281	1.08700
188.0	175.6	6.268	2891.0	4168.0	136.6	22.82	46.44	285.0	16.2	.0279	1.08509
190.0	172.0	6.139	2956.0	4260.0	137.1	22.73	45.58	286.9	16.2	.0278	1.08328
192.0	168.6	6.017	3020.0	4350.0	137.5	22.64	44.79	288.8	16.2	.0278	1.08157
194.0	165.3	5.901	3083.0	4439.0	138.0	22.57	44.05	290.7	16.2	.0277	1.07995
196.0	162.2	5.791	3145.0	4526.0	138.4	22.49	43.38	292.6	16.2	.0276	1.07842
200.0	156.5	5.587	3265.0	4697.0	139.3	22.36	42.18	296.3	16.3	.0276	1.07557
205.0	150.1	5.357	3411.0	4905.0	140.3	22.22	40.90	300.9	16.3	.0275	1.07237
210.0	144.3	5.150	3553.0	5107.0	141.3	22.09	39.82	305.4	16.5	.0275	1.06951
215.0	139.0	4.962	3691.0	5303.0	142.2	21.99	38.90	309.8	16.6	.0276	1.06691
220.0	134.2	4.792	3826.0	5496.0	143.1	21.90	38.11	314.1	16.7	.0276	1.06455
225.0	129.9	4.635	3959.0	5685.0	144.0	21.82	37.42	318.2	16.9	.0277	1.06240
230.0	125.8	4.491	4089.0	5870.0	144.8	21.75	36.82	322.4	17.0	.0279	1.06041
235.0	122.1	4.358	4217.0	6053.0	145.6	21.69	36.29	326.4	17.2	.0280	1.05857
240.0	118.6	4.234	4343.0	6233.0	146.3	21.63	35.81	330.3	17.4	.0282	1.05687
245.0	115.4	4.118	4468.0	6411.0	147.1	21.58	35.39	334.2	17.5	.0283	1.05528
250.0	112.3	4.009	4592.0	6587.0	147.8	21.54	35.01	338.0	17.7	.0285	1.05379
255.0	109.5	3.908	4714.0	6761.0	148.5	21.50	34.67	341.7	17.9	.0287	1.05240
260.0	106.8	3.812	4835.0	6934.0	149.1	21.47	34.36	345.3	18.1	.0289	1.05109
265.0	104.3	3.722	4955.0	7105.0	149.8	21.43	34.07	348.9	18.2	.0291	1.04986
270.0	101.9	3.636	5075.0	7275.0	150.4	21.41	33.81	352.5	18.4	.0293	1.04869
280.0	97.46	3.479	5311.0	7610.0	151.6	21.36	33.36	359.3	18.8	.0297	1.04654
290.0	93.47	3.336	5544.0	7942.0	152.8	21.32	32.98	366.0	19.2	.0302	1.04460
300.0	89.83	3.207	5775.0	8270.0	153.9	21.28	32.64	372.5	19.6	.0307	1.04283
310.0	86.51	3.088	6004.0	8595.0	155.0	21.25	32.36	378.8	19.9	.0312	1.04122
320.0	83.45	2.979	6232.0	8917.0	156.0	21.23	32.11	385.0	20.3	.0317	1.03973
330.0	80.62	2.878	6458.0	9237.0	157.0	21.21	31.89	391.0	20.7	.0322	1.03837
340.0	78.01	2.784	6682.0	9555.0	157.9	21.20	31.70	396.9	21.1	.0327	1.03710
350.0	75.57	2.697	6906.0	9872.0	158.8	21.19	31.54	402.6	21.5	.0332	1.03592
360.0	73.30	2.616	7128.0	10190.0	159.7	21.19	31.39	408.2	21.8	.0337	1.03482
370.0	71.17	2.540	7350.0	10500.0	160.6	21.18	31.26	413.7	22.2	.0342	1.03380
380.0	69.17	2.469	7571.0	10810.0	161.4	21.18	31.15	419.1	22.6	.0348	1.03283
390.0	67.29	2.402	7792.0	11120.0	162.2	21.19	31.05	424.4	23.0	.0353	1.03193
400.0	65.52	2.339	8012.0	11430.0	163.0	21.20	30.96	429.6	23.3	.0358	1.03108
410.0	63.85	2.279	8231.0	11740.0	163.8	21.21	30.89	434.7	23.7	.0364	1.03027
420.0	62.26	2.222	8451.0	12050.0	164.5	21.22	30.82	439.7	24.1	.0369	1.02951
440.0	59.34	2.118	8888.0	12670.0	166.0	21.26	30.72	449.5	24.8	.0380	1.02810
460.0	56.69	2.023	9326.0	13280.0	167.3	21.31	30.65	458.9	25.5	.0391	1.02683
480.0	54.28	1.938	9763.0	13890.0	168.6	21.37	30.60	468.1	26.2	.0402	1.02567
500.0	52.08	1.859	10200.0	14500.0	169.9	21.44	30.58	477.0	26.9	.0413	1.02462
520.0	50.06	1.787	10640.0	15120.0	171.1	21.52	30.58	485.6	27.6	.0424	1.02365
540.0	48.19	1.720	11080.0	15730.0	172.2	21.60	30.60	494.1	28.2	.0435	1.02276
560.0	46.47	1.659	11520.0	16340.0	173.3	21.70	30.64	502.3	28.9	.0446	1.02193
580.0	44.87	1.602	11960.0	16950.0	174.4	21.80	30.69	510.3	29.6	.0457	1.02117
600.0	43.38	1.548	12400.0	17570.0	175.5	21.91	30.75	518.1	30.2	.0469	1.02045
620.0	41.99	1.499	12850.0	18180.0	176.5	22.03	30.82	525.8	30.8	.0480	1.01979
640.0	40.69	1.452	13290.0	18800.0	177.4	22.14	30.90	533.3	31.5	.0491	1.01917
660.0	39.47	1.409	13740.0	19420.0	178.4	22.27	30.99	540.6	32.1	.0503	1.01859
680.0	38.32	1.368	14190.0	20040.0	179.3	22.39	31.09	547.8	32.7	.0514	1.01804
700.0	37.24	1.329	14640.0	20660.0	180.2	22.52	31.18	554.9	33.3	.0525	1.01752
720.0	36.22	1.293	15100.0	21290.0	181.1	22.65	31.29	561.9	33.9	.0536	1.01704
740.0	35.25	1.258	15560.0	21910.0	182.0	22.78	31.40	568.7	34.5	.0548	1.01658
760.0	34.34	1.226	16020.0	22540.0	182.8	22.91	31.50	575.4	35.1	.0559	1.01614
780.0	33.47	1.195	16480.0	23170.0	183.6	23.04	31.62	582.1	35.7	.0570	1.01573

Thermophysical properties of nitrogen—Continued

T K	Density kg/m³	Density mol/dm³	E J/mol	H J/mol	S J/(mol·K)	C_v J/(mol·K)	C_p J/(mol·K)	Sound m/s	Visc. μPa·s	Therm. W/(m·K)	Diel. const.
800.0	32.65	1.165	16940.0	23810.0	184.4	23.17	31.73	588.6	36.3	.0581	1.01534
820.0	31.87	1.138	17410.0	24440.0	185.2	23.30	31.84	595.1	36.9	.0593	1.01496
840.0	31.12	1.111	17880.0	25080.0	186.0	23.43	31.95	601.4	37.5	.0604	1.01461
860.0	30.41	1.086	18350.0	25720.0	186.7	23.55	32.07	607.7	38.0	.0615	1.01427
880.0	29.73	1.061	18830.0	26360.0	187.5	23.68	32.18	613.9	38.6	.0626	1.01395
900.0	29.09	1.038	19300.0	27010.0	188.2	23.80	32.29	620.0	39.1	.0637	1.01364
920.0	28.47	1.016	19780.0	27660.0	188.9	23.93	32.40	626.1	39.7	.0648	1.01334
950.0	27.58	.9846	20500.0	28630.0	189.9	24.10	32.57	635.0	40.5	.0665	1.01292
1000.0	26.23	.9363	21720.0	30270.0	191.6	24.39	32.83	649.6	41.9	.0692	1.01228
1050.0	25.01	.8926	22950.0	31910.0	193.2	24.66	33.09	663.9	43.2	.0719	1.01170
1100.0	23.89	.8528	24190.0	33570.0	194.8	24.92	33.33	677.8	44.5	.0745	1.01117
1150.0	22.87	.8164	25450.0	35250.0	196.3	25.16	33.56	691.5	45.8	.0771	1.01069
1200.0	21.94	.7830	26710.0	36930.0	197.7	25.39	33.77	704.9	47.0	.0797	1.01024
1250.0	21.08	.7523	27990.0	38620.0	199.1	25.60	33.97	718.0	48.3	.0823	1.00983
1300.0	20.28	.7239	29280.0	40330.0	200.4	25.80	34.16	730.9	49.5	.0848	1.00946
1350.0	19.54	.6976	30570.0	42040.0	201.7	25.98	34.34	743.5	50.7	.0872	1.00911
1400.0	18.86	.6731	31880.0	43760.0	203.0	26.16	34.51	756.0	51.9	.0897	1.00878
1450.0	18.22	.6504	33190.0	45490.0	204.2	26.32	34.66	768.2	53.1	.0921	1.00848
1500.0	17.62	.6291	34510.0	47230.0	205.4	26.47	34.81	780.3	54.2	.0944	1.00820
1550.0	17.07	.6091	35840.0	48970.0	206.5	26.61	34.95	792.1	55.4	.0967	1.00793
1600.0	16.54	.5904	37170.0	50720.0	207.6	26.74	35.08	803.9	56.5	.0990	1.00768
1650.0	16.05	.5728	38510.0	52480.0	208.7	26.87	35.20	815.4	57.7	.101	1.00745
1700.0	15.58	.5562	39860.0	54240.0	209.7	26.99	35.31	826.8	58.8	.104	1.00723
1750.0	15.15	.5406	41210.0	56010.0	210.8	27.09	35.42	838.0	59.9	.106	1.00702
1800.0	14.73	.5258	42570.0	57780.0	211.8	27.20	35.52	849.1	61.0	.108	1.00683
1850.0	14.34	.5118	43930.0	59560.0	212.7	27.29	35.61	860.1	62.0	.110	1.00664
1900.0	13.97	.4986	45300.0	61340.0	213.7	27.38	35.70	870.9	63.1	.112	1.00647

8.50 MPa isobar

T K	Density kg/m³	Density mol/dm³	E J/mol	H J/mol	S J/(mol·K)	C_v J/(mol·K)	C_p J/(mol·K)	Sound m/s	Visc. μPa·s	Therm. W/(m·K)	Diel. const.
65.00[a]	871.2	31.10	−4182.0	−3909.0	68.47	31.19	53.98	1229.0	277.0	.153	1.47033
80.0	816.1	29.13	−3375.0	−3083.0	79.90	29.72	55.23	935.0	154.0	.137	1.43767
100.0	728.1	25.99	−2294.0	−1966.0	92.35	27.23	57.12	716.0	87.7	.110	1.38600
110.0	677.4	24.18	−1734.0	−1383.0	97.91	26.67	59.88	620.4	68.4	.0962	1.35653
120.0	619.8	22.12	−1147.0	−763.0	103.3	26.29	64.45	527.7	53.5	.0820	1.32349
130.0	551.6	19.69	−513.3	−81.6	108.7	26.07	72.70	434.1	41.2	.0690	1.28484
132.0	536.0	19.13	−378.1	66.2	109.9	26.05	75.08	415.2	38.9	.0667	1.27611
134.0	519.6	18.55	−239.4	218.9	111.0	26.04	77.74	396.3	36.8	.0645	1.26697
136.0	502.3	17.93	−96.8	377.3	112.2	26.04	80.65	377.7	34.6	.0622	1.25738
138.0	484.2	17.28	49.8	541.6	113.4	26.05	83.68	359.6	32.6	.0600	1.24734
140.0	465.1	16.60	200.0	712.0	114.6	26.07	86.64	342.5	30.6	.0576	1.23686
142.0	445.3	15.90	353.1	887.9	115.9	26.09	89.19	326.8	28.8	.0552	1.22604
145.0	414.9	14.81	585.5	1159.0	117.8	26.10	91.48	306.8	26.2	.0513	1.20954
147.0	394.8	14.09	739.5	1343.0	119.0	26.07	91.64	296.2	24.7	.0487	1.19868
148.0	384.9	13.74	815.5	1434.0	119.6	26.05	91.30	291.8	24.0	.0474	1.19338
150.0	365.7	13.05	964.4	1616.0	120.9	25.97	89.88	284.5	22.8	.0450	1.18317
152.0	347.6	12.41	1108.0	1793.0	122.0	25.85	87.67	279.0	21.8	.0428	1.17357
154.0	330.8	11.81	1246.0	1966.0	123.2	25.70	84.91	275.1	20.9	.0408	1.16469
156.0	315.3	11.25	1377.0	2133.0	124.2	25.53	81.82	272.5	20.1	.0390	1.15654
158.0	301.1	10.75	1502.0	2293.0	125.3	25.33	78.56	270.9	19.5	.0375	1.14911
160.0	288.1	10.28	1620.0	2447.0	126.2	25.13	75.30	270.1	18.9	.0361	1.14237
162.0	276.3	9.863	1732.0	2594.0	127.1	24.92	72.12	269.9	18.5	.0349	1.13627
164.0	265.6	9.480	1839.0	2735.0	128.0	24.71	69.10	270.1	18.1	.0339	1.13074
166.0	255.8	9.132	1940.0	2871.0	128.8	24.51	66.29	270.8	17.8	.0331	1.12572
168.0	246.9	8.814	2036.0	3001.0	129.6	24.32	63.69	271.7	17.6	.0323	1.12115
170.0	238.8	8.523	2128.0	3126.0	130.3	24.13	61.31	272.9	17.4	.0317	1.11699
172.0	231.3	8.257	2217.0	3246.0	131.0	23.96	59.15	274.3	17.2	.0311	1.11318
174.0	224.4	8.011	2301.0	3362.0	131.7	23.80	57.19	275.7	17.0	.0306	1.10968
176.0	218.1	7.784	2383.0	3475.0	132.4	23.64	55.41	277.3	16.9	.0302	1.10646
178.0	212.2	7.574	2462.0	3584.0	133.0	23.50	53.80	279.0	16.8	.0299	1.10348
180.0	206.7	7.379	2538.0	3690.0	133.6	23.37	52.33	280.7	16.7	.0296	1.10071
182.0	201.6	7.197	2613.0	3794.0	134.1	23.25	51.00	282.4	16.7	.0293	1.09813

Thermophysical properties of nitrogen—Continued

T K	Density kg/m^3	Density mol/dm^3	E J/mol	H J/mol	S J/(mol·K)	C_v J/(mol·K)	C_p J/(mol·K)	Sound m/s	Visc. μPa·s	Therm. W/(m·K)	Diel. const.
184.0	196.8	7.026	2685.0	3894.0	134.7	23.13	49.78	284.2	16.6	.0291	1.09573
186.0	192.4	6.866	2755.0	3993.0	135.2	23.02	48.68	286.0	16.6	.0289	1.09348
188.0	188.2	6.716	2824.0	4089.0	135.7	22.92	47.66	287.8	16.6	.0288	1.09136
190.0	184.2	6.574	2891.0	4184.0	136.2	22.83	46.73	289.6	16.6	.0287	1.08937
192.0	180.4	6.441	2956.0	4276.0	136.7	22.74	45.88	291.5	16.5	.0285	1.08749
194.0	176.9	6.314	3021.0	4367.0	137.2	22.66	45.09	293.3	16.5	.0285	1.08571
196.0	173.5	6.194	3084.0	4457.0	137.7	22.58	44.36	295.1	16.6	.0284	1.08403
200.0	167.3	5.970	3208.0	4631.0	138.5	22.44	43.07	298.7	16.6	.0283	1.08091
205.0	160.2	5.720	3357.0	4843.0	139.6	22.29	41.69	303.2	16.6	.0282	1.07741
210.0	153.9	5.495	3502.0	5049.0	140.6	22.16	40.53	307.6	16.7	.0281	1.07428
215.0	148.2	5.292	3642.0	5249.0	141.5	22.05	39.54	311.9	16.8	.0281	1.07146
220.0	143.1	5.107	3780.0	5444.0	142.4	21.96	38.69	316.1	17.0	.0282	1.06890
225.0	138.3	4.937	3914.0	5636.0	143.3	21.87	37.95	320.2	17.1	.0283	1.06656
230.0	134.0	4.782	4046.0	5824.0	144.1	21.80	37.30	324.3	17.2	.0284	1.06441
235.0	129.9	4.638	4176.0	6009.0	144.9	21.73	36.73	328.3	17.4	.0285	1.06242
240.0	126.2	4.504	4304.0	6191.0	145.7	21.68	36.22	332.2	17.5	.0286	1.06058
245.0	122.7	4.380	4430.0	6371.0	146.4	21.63	35.77	336.0	17.7	.0288	1.05887
250.0	119.4	4.263	4555.0	6549.0	147.1	21.58	35.37	339.8	17.9	.0289	1.05727
255.0	116.4	4.154	4679.0	6725.0	147.8	21.54	35.00	343.5	18.1	.0291	1.05577
260.0	113.5	4.051	4801.0	6899.0	148.5	21.50	34.67	347.1	18.2	.0293	1.05436
265.0	110.8	3.955	4922.0	7072.0	149.2	21.47	34.37	350.7	18.4	.0295	1.05303
270.0	108.2	3.863	5043.0	7243.0	149.8	21.44	34.09	354.2	18.6	.0297	1.05178
280.0	103.5	3.694	5280.0	7581.0	151.0	21.39	33.61	361.1	18.9	.0301	1.04947
290.0	99.23	3.542	5515.0	7915.0	152.2	21.34	33.20	367.7	19.3	.0305	1.04739
300.0	95.35	3.403	5748.0	8245.0	153.3	21.31	32.85	374.2	19.7	.0310	1.04550
310.0	91.80	3.277	5978.0	8572.0	154.4	21.28	32.55	380.5	20.1	.0315	1.04378
320.0	88.54	3.160	6207.0	8897.0	155.4	21.25	32.28	386.7	20.4	.0319	1.04219
330.0	85.53	3.053	6434.0	9218.0	156.4	21.24	32.05	392.7	20.8	.0324	1.04073
340.0	82.74	2.953	6660.0	9538.0	157.4	21.22	31.85	398.5	21.2	.0329	1.03938
350.0	80.15	2.861	6884.0	9855.0	158.3	21.21	31.67	404.3	21.6	.0334	1.03813
360.0	77.73	2.774	7108.0	10170.0	159.2	21.20	31.52	409.9	21.9	.0339	1.03696
370.0	75.47	2.694	7330.0	10490.0	160.0	21.20	31.38	415.4	22.3	.0345	1.03586
380.0	73.34	2.618	7552.0	10800.0	160.9	21.20	31.26	420.7	22.7	.0350	1.03484
390.0	71.35	2.547	7773.0	11110.0	161.7	21.21	31.15	426.0	23.0	.0355	1.03388
400.0	69.47	2.480	7994.0	11420.0	162.5	21.21	31.06	431.2	23.4	.0360	1.03297
410.0	67.69	2.416	8214.0	11730.0	163.2	21.22	30.98	436.3	23.8	.0366	1.03211
420.0	66.01	2.356	8434.0	12040.0	164.0	21.24	30.91	441.3	24.1	.0371	1.03130
440.0	62.90	2.245	8873.0	12660.0	165.4	21.27	30.79	451.0	24.8	.0382	1.02981
460.0	60.10	2.145	9311.0	13270.0	166.8	21.32	30.72	460.5	25.6	.0393	1.02846
480.0	57.54	2.054	9749.0	13890.0	168.1	21.38	30.67	469.6	26.2	.0404	1.02723
500.0	55.21	1.971	10190.0	14500.0	169.3	21.45	30.64	478.5	26.9	.0415	1.02611
520.0	53.06	1.894	10630.0	15110.0	170.5	21.53	30.64	487.1	27.6	.0426	1.02508
540.0	51.09	1.824	11060.0	15730.0	171.7	21.62	30.65	495.6	28.3	.0437	1.02413
560.0	49.26	1.758	11510.0	16340.0	172.8	21.71	30.69	503.8	28.9	.0448	1.02326
580.0	47.57	1.698	11950.0	16950.0	173.9	21.81	30.73	511.7	29.6	.0459	1.02245
600.0	45.99	1.642	12390.0	17570.0	174.9	21.92	30.79	519.6	30.2	.0470	1.02169
620.0	44.52	1.589	12840.0	18190.0	175.9	22.04	30.86	527.2	30.9	.0481	1.02099
640.0	43.14	1.540	13280.0	18800.0	176.9	22.15	30.94	534.7	31.5	.0493	1.02033
660.0	41.84	1.494	13730.0	19420.0	177.9	22.28	31.02	542.0	32.1	.0504	1.01971
680.0	40.63	1.450	14180.0	20040.0	178.8	22.40	31.12	549.2	32.8	.0515	1.01913
700.0	39.48	1.409	14640.0	20670.0	179.7	22.53	31.21	556.3	33.4	.0526	1.01858
720.0	38.40	1.371	15090.0	21290.0	180.6	22.66	31.31	563.2	34.0	.0538	1.01807
740.0	37.38	1.334	15550.0	21920.0	181.5	22.79	31.42	570.0	34.6	.0549	1.01758
760.0	36.41	1.300	16010.0	22550.0	182.3	22.92	31.53	576.8	35.2	.0560	1.01712
780.0	35.49	1.267	16470.0	23180.0	183.1	23.05	31.64	583.4	35.8	.0571	1.01668
800.0	34.62	1.236	16940.0	23810.0	183.9	23.18	31.75	589.9	36.3	.0583	1.01627
820.0	33.79	1.206	17400.0	24450.0	184.7	23.31	31.86	596.3	36.9	.0594	1.01587
840.0	33.01	1.178	17870.0	25090.0	185.5	23.43	31.97	602.7	37.5	.0605	1.01550
860.0	32.25	1.151	18350.0	25730.0	186.2	23.56	32.09	608.9	38.1	.0616	1.01514
880.0	31.53	1.126	18820.0	26370.0	187.0	23.69	32.20	615.1	38.6	.0627	1.01480
900.0	30.85	1.101	19300.0	27020.0	187.7	23.81	32.31	621.2	39.2	.0638	1.01447
920.0	30.19	1.078	19780.0	27670.0	188.4	23.93	32.42	627.3	39.7	.0649	1.01416
950.0	29.26	1.044	20500.0	28640.0	189.4	24.11	32.58	636.2	40.5	.0666	1.01371

Thermophysical properties of nitrogen—Continued

T K	Density kg/m³	Density mol/dm³	E J/mol	H J/mol	S J/(mol·K)	C_v J/(mol·K)	C_p J/(mol·K)	Sound m/s	Visc. μPa·s	Therm. W/(m·K)	Diel. const.
1000.0	27.82	.9931	21720.0	30280.0	191.1	24.40	32.84	650.8	41.9	.0693	1.01303
1050.0	26.52	.9468	22950.0	31920.0	192.7	24.67	33.10	665.0	43.2	.0720	1.01241
1100.0	25.34	.9046	24190.0	33590.0	194.3	24.92	33.34	679.0	44.5	.0746	1.01185
1150.0	24.26	.8661	25440.0	35260.0	195.8	25.16	33.56	692.6	45.8	.0772	1.01134
1200.0	23.27	.8308	26710.0	36940.0	197.2	25.39	33.78	705.9	47.1	.0798	1.01087
1250.0	22.36	.7982	27990.0	38640.0	198.6	25.60	33.98	719.0	48.3	.0823	1.01044
1300.0	21.52	.7681	29270.0	40340.0	199.9	25.80	34.17	731.9	49.5	.0848	1.01004
1350.0	20.74	.7402	30570.0	42050.0	201.2	25.99	34.35	744.5	50.7	.0873	1.00967
1400.0	20.01	.7143	31870.0	43770.0	202.5	26.16	34.51	757.0	51.9	.0897	1.00932
1450.0	19.33	.6901	33190.0	45500.0	203.7	26.32	34.67	769.2	53.1	.0921	1.00900
1500.0	18.70	.6676	34510.0	47240.0	204.8	26.47	34.82	781.2	54.3	.0945	1.00870
1550.0	18.11	.6464	35840.0	48980.0	206.0	26.61	34.95	793.1	55.4	.0968	1.00842
1600.0	17.55	.6266	37170.0	50740.0	207.1	26.75	35.08	804.8	56.5	.0991	1.00816
1650.0	17.03	.6079	38510.0	52490.0	208.2	26.87	35.20	816.3	57.7	.101	1.00791
1700.0	16.54	.5904	39860.0	54260.0	209.2	26.99	35.32	827.7	58.8	.104	1.00768
1750.0	16.08	.5738	41210.0	56020.0	210.3	27.10	35.42	838.9	59.9	.106	1.00746
1800.0	15.64	.5581	42570.0	57800.0	211.3	27.20	35.52	850.0	61.0	.108	1.00725
1850.0	15.22	.5433	43930.0	59580.0	212.2	27.30	35.62	861.0	62.0	.110	1.00705
1900.0	14.83	.5292	45300.0	61360.0	213.2	27.39	35.70	871.8	63.1	.112	1.00686
9.00 MPa isobar											
65.10[a]	871.4	31.10	−4180.0	−3891.0	68.50	31.41	53.93	1225.0	276.0	.154	1.47045
80.0	817.1	29.17	−3381.0	−3072.0	79.81	29.83	55.10	937.4	155.0	.138	1.43827
100.0	730.1	26.06	−2305.0	−1960.0	92.22	27.29	56.83	721.4	88.6	.111	1.38716
110.0	680.2	24.28	−1750.0	−1380.0	97.75	26.71	59.39	627.6	69.3	.0970	1.35818
120.0	624.1	22.28	−1171.0	−767.0	103.1	26.30	63.51	537.2	54.4	.0831	1.32593
130.0	558.6	19.94	−551.3	−100.0	108.4	26.04	70.56	447.1	42.3	.0702	1.28880
132.0	543.9	19.42	−420.5	43.0	109.5	26.00	72.51	429.1	40.1	.0679	1.28055
134.0	528.6	18.87	−286.8	190.2	110.6	25.98	74.67	411.1	38.0	.0657	1.27197
136.0	512.5	18.29	−150.1	341.8	111.7	25.96	77.00	393.4	35.9	.0635	1.26302
138.0	495.7	17.70	−10.4	498.3	112.9	25.95	79.45	376.2	34.0	.0614	1.25372
140.0	478.2	17.07	132.4	659.6	114.0	25.95	81.88	359.7	32.1	.0592	1.24406
142.0	460.1	16.42	277.6	825.6	115.2	25.95	84.09	344.2	30.3	.0569	1.23411
145.0	432.1	15.42	498.3	1082.0	117.0	25.94	86.47	323.8	27.7	.0532	1.21885
150.0	385.8	13.77	862.9	1517.0	119.9	25.84	86.54	298.9	24.3	.0472	1.19385
152.0	368.2	13.14	1004.0	1688.0	121.1	25.75	85.20	292.2	23.1	.0449	1.18447
154.0	351.6	12.55	1140.0	1857.0	122.2	25.64	83.28	287.1	22.1	.0429	1.17564
156.0	336.0	11.99	1271.0	2021.0	123.2	25.49	80.96	283.3	21.3	.0410	1.16739
158.0	321.5	11.47	1396.0	2181.0	124.3	25.33	78.37	280.6	20.6	.0394	1.15977
160.0	308.1	11.00	1516.0	2335.0	125.2	25.16	75.64	278.8	20.0	.0379	1.15275
162.0	295.7	10.56	1631.0	2483.0	126.2	24.97	72.89	277.8	19.5	.0367	1.14632
164.0	284.4	10.15	1740.0	2626.0	127.0	24.78	70.18	277.3	19.0	.0355	1.14044
166.0	274.0	9.780	1844.0	2764.0	127.9	24.60	67.57	277.4	18.6	.0346	1.13506
168.0	264.5	9.440	1943.0	2897.0	128.7	24.41	65.10	277.8	18.3	.0337	1.13014
170.0	255.7	9.126	2038.0	3024.0	129.4	24.24	62.80	278.5	18.1	.0330	1.12563
172.0	247.6	8.838	2130.0	3148.0	130.1	24.07	60.66	279.4	17.8	.0324	1.12149
174.0	240.2	8.572	2217.0	3267.0	130.8	23.91	58.69	280.5	17.7	.0318	1.11768
176.0	233.3	8.326	2302.0	3383.0	131.5	23.75	56.88	281.8	17.5	.0314	1.11416
178.0	226.9	8.097	2383.0	3495.0	132.1	23.61	55.22	283.2	17.4	.0310	1.11090
180.0	220.9	7.885	2462.0	3604.0	132.7	23.48	53.71	284.6	17.3	.0306	1.10788
182.0	215.3	7.686	2539.0	3710.0	133.3	23.35	52.32	286.2	17.2	.0303	1.10506
184.0	210.1	7.501	2613.0	3813.0	133.9	23.23	51.05	287.8	17.1	.0301	1.10243
186.0	205.3	7.327	2686.0	3914.0	134.4	23.12	49.88	289.4	17.0	.0299	1.09997
188.0	200.7	7.164	2756.0	4013.0	135.0	23.02	48.81	291.1	17.0	.0297	1.09766
190.0	196.4	7.010	2825.0	4109.0	135.5	22.92	47.83	292.8	17.0	.0295	1.09548
192.0	192.3	6.864	2893.0	4204.0	136.0	22.83	46.92	294.5	16.9	.0294	1.09343
194.0	188.4	6.726	2959.0	4297.0	136.4	22.75	46.08	296.2	16.9	.0292	1.09150
196.0	184.8	6.596	3024.0	4388.0	136.9	22.67	45.31	297.9	16.9	.0291	1.08966
198.0	181.3	6.472	3088.0	4478.0	137.4	22.59	44.59	299.7	16.9	.0290	1.08792
200.0	178.0	6.354	3150.0	4567.0	137.8	22.52	43.93	301.4	16.9	.0290	1.08627
205.0	170.4	6.083	3303.0	4783.0	138.9	22.36	42.46	305.7	17.0	.0288	1.08247
210.0	163.6	5.839	3451.0	4992.0	139.9	22.23	41.22	310.0	17.0	.0288	1.07908

Thermophysical properties of nitrogen—Continued

T K	Density kg/m^3	Density mol/dm^3	E J/mol	H J/mol	S J/(mol·K)	C_v J/(mol·K)	C_p J/(mol·K)	Sound m/s	Visc. μPa·s	Therm. W/(m·K)	Diel. const.
215.0	157.5	5.620	3594.0	5195.0	140.8	22.11	40.16	314.2	17.1	.0287	1.07602
220.0	151.9	5.421	3733.0	5394.0	141.8	22.01	39.25	318.3	17.2	.0287	1.07325
225.0	146.8	5.239	3870.0	5588.0	142.6	21.93	38.46	322.4	17.3	.0288	1.07073
230.0	142.1	5.072	4004.0	5778.0	143.5	21.85	37.77	326.4	17.5	.0289	1.06841
235.0	137.8	4.917	4135.0	5966.0	144.3	21.78	37.17	330.3	17.6	.0290	1.06627
240.0	133.7	4.774	4265.0	6150.0	145.0	21.72	36.63	334.2	17.8	.0291	1.06429
245.0	130.0	4.641	4393.0	6332.0	145.8	21.67	36.15	338.0	17.9	.0292	1.06245
250.0	126.5	4.516	4519.0	6512.0	146.5	21.62	35.72	341.7	18.1	.0294	1.06074
255.0	123.3	4.399	4644.0	6689.0	147.2	21.58	35.33	345.4	18.2	.0295	1.05913
260.0	120.2	4.290	4767.0	6865.0	147.9	21.54	34.97	349.0	18.4	.0297	1.05762
265.0	117.3	4.186	4889.0	7039.0	148.6	21.50	34.66	352.6	18.6	.0299	1.05620
270.0	114.6	4.089	5011.0	7212.0	149.2	21.47	34.36	356.0	18.7	.0300	1.05486
280.0	109.5	3.909	5250.0	7553.0	150.5	21.42	33.85	362.9	19.1	.0304	1.05240
290.0	105.0	3.747	5487.0	7889.0	151.6	21.37	33.42	369.5	19.5	.0309	1.05018
300.0	100.8	3.599	5721.0	8221.0	152.8	21.33	33.05	376.0	19.8	.0313	1.04817
310.0	97.07	3.465	5953.0	8550.0	153.8	21.30	32.73	382.3	20.2	.0318	1.04633
320.0	93.61	3.341	6182.0	8876.0	154.9	21.28	32.45	388.4	20.6	.0322	1.04465
330.0	90.41	3.227	6410.0	9199.0	155.9	21.26	32.21	394.4	20.9	.0327	1.04309
340.0	87.45	3.122	6637.0	9520.0	156.8	21.24	31.99	400.2	21.3	.0332	1.04166
350.0	84.71	3.023	6862.0	9839.0	157.8	21.23	31.81	405.9	21.7	.0337	1.04033
360.0	82.14	2.932	7087.0	10160.0	158.6	21.22	31.64	411.5	22.0	.0342	1.03908
370.0	79.75	2.846	7310.0	10470.0	159.5	21.22	31.50	417.0	22.4	.0347	1.03792
380.0	77.50	2.766	7533.0	10790.0	160.4	21.22	31.37	422.4	22.8	.0352	1.03684
390.0	75.39	2.691	7755.0	11100.0	161.2	21.22	31.26	427.7	23.1	.0357	1.03582
400.0	73.39	2.620	7976.0	11410.0	162.0	21.23	31.16	432.8	23.5	.0363	1.03485
410.0	71.51	2.553	8197.0	11720.0	162.7	21.24	31.07	437.9	23.8	.0368	1.03395
420.0	69.74	2.489	8417.0	12030.0	163.5	21.25	30.99	442.9	24.2	.0373	1.03309
440.0	66.45	2.372	8857.0	12650.0	164.9	21.29	30.87	452.6	24.9	.0384	1.03151
460.0	63.48	2.266	9296.0	13270.0	166.3	21.33	30.79	462.0	25.6	.0394	1.03008
480.0	60.78	2.170	9735.0	13880.0	167.6	21.39	30.73	471.2	26.3	.0405	1.02878
500.0	58.32	2.082	10170.0	14500.0	168.8	21.46	30.70	480.0	27.0	.0416	1.02759
520.0	56.06	2.001	10610.0	15110.0	170.0	21.54	30.69	488.7	27.7	.0427	1.02651
540.0	53.97	1.926	11050.0	15730.0	171.2	21.63	30.70	497.0	28.3	.0438	1.02551
560.0	52.04	1.858	11490.0	16340.0	172.3	21.72	30.73	505.2	29.0	.0449	1.02458
580.0	50.25	1.794	11940.0	16950.0	173.4	21.82	30.77	513.2	29.6	.0460	1.02372
600.0	48.58	1.734	12380.0	17570.0	174.4	21.93	30.83	521.0	30.3	.0472	1.02293
620.0	47.03	1.679	12830.0	18190.0	175.5	22.04	30.89	528.6	30.9	.0483	1.02218
640.0	45.57	1.627	13270.0	18810.0	176.4	22.16	30.97	536.1	31.6	.0494	1.02149
660.0	44.21	1.578	13720.0	19430.0	177.4	22.28	31.05	543.4	32.2	.0505	1.02083
680.0	42.93	1.532	14170.0	20050.0	178.3	22.41	31.14	550.6	32.8	.0516	1.02022
700.0	41.72	1.489	14630.0	20670.0	179.2	22.53	31.24	557.6	33.4	.0528	1.01964
720.0	40.58	1.448	15080.0	21300.0	180.1	22.66	31.34	564.6	34.0	.0539	1.01910
740.0	39.50	1.410	15540.0	21930.0	181.0	22.79	31.44	571.4	34.6	.0550	1.01858
760.0	38.48	1.373	16000.0	22560.0	181.8	22.92	31.55	578.1	35.2	.0561	1.01810
780.0	37.51	1.339	16470.0	23190.0	182.6	23.05	31.66	584.7	35.8	.0573	1.01763
800.0	36.59	1.306	16930.0	23820.0	183.4	23.18	31.77	591.2	36.4	.0584	1.01719
820.0	35.71	1.275	17400.0	24460.0	184.2	23.31	31.88	597.6	36.9	.0595	1.01678
840.0	34.88	1.245	17870.0	25100.0	185.0	23.44	31.99	603.9	37.5	.0606	1.01638
860.0	34.09	1.217	18340.0	25740.0	185.7	23.57	32.10	610.2	38.1	.0617	1.01600
880.0	33.33	1.190	18820.0	26380.0	186.5	23.69	32.21	616.4	38.6	.0628	1.01564
900.0	32.60	1.164	19290.0	27030.0	187.2	23.81	32.32	622.5	39.2	.0639	1.01530
920.0	31.91	1.139	19770.0	27670.0	187.9	23.94	32.43	628.5	39.7	.0650	1.01497
950.0	30.92	1.104	20500.0	28650.0	189.0	24.11	32.59	637.4	40.6	.0667	1.01450
1000.0	29.41	1.050	21710.0	30290.0	190.6	24.40	32.86	652.0	41.9	.0694	1.01378
1050.0	28.04	1.001	22940.0	31940.0	192.2	24.67	33.11	666.2	43.2	.0721	1.01312
1100.0	26.79	.9563	24190.0	33600.0	193.8	24.93	33.35	680.1	44.5	.0747	1.01253
1150.0	25.65	.9157	25440.0	35270.0	195.3	25.17	33.57	693.7	45.8	.0773	1.01199
1200.0	24.61	.8783	26710.0	36950.0	196.7	25.39	33.79	707.0	47.1	.0799	1.01149
1250.0	23.64	.8439	27980.0	38650.0	198.1	25.61	33.99	720.1	48.3	.0824	1.01104
1300.0	22.75	.8122	29270.0	40350.0	199.4	25.80	34.18	732.9	49.5	.0849	1.01061
1350.0	21.93	.7827	30570.0	42070.0	200.7	25.99	34.35	745.5	50.8	.0874	1.01022
1400.0	21.16	.7553	31870.0	43790.0	202.0	26.16	34.52	758.0	51.9	.0898	1.00986
1450.0	20.45	.7298	33190.0	45520.0	203.2	26.32	34.67	770.2	53.1	.0922	1.00952

Thermophysical properties of nitrogen—Continued

T K	Density kg/m^3	Density mol/dm^3	E J/mol	H J/mol	S J/(mol·K)	C_v J/(mol·K)	C_p J/(mol·K)	Sound m/s	Visc. μPa·s	Therm. W/(m·K)	Diel. const.
1500.0	19.78	.7060	34510.0	47250.0	204.4	26.48	34.82	782.2	54.3	.0945	1.00920
1550.0	19.15	.6836	35830.0	49000.0	205.5	26.62	34.96	794.0	55.4	.0969	1.00891
1600.0	18.57	.6627	37170.0	50750.0	206.6	26.75	35.08	805.7	56.6	.0992	1.00863
1650.0	18.01	.6430	38510.0	52510.0	207.7	26.87	35.21	817.2	57.7	.101	1.00837
1700.0	17.49	.6244	39860.0	54270.0	208.8	26.99	35.32	828.6	58.8	.104	1.00812
1750.0	17.00	.6069	41210.0	56040.0	209.8	27.10	35.42	839.8	59.9	.106	1.00789
1800.0	16.54	.5904	42570.0	57810.0	210.8	27.20	35.52	850.9	61.0	.108	1.00767
1850.0	16.10	.5747	43930.0	59590.0	211.8	27.30	35.62	861.8	62.1	.110	1.00746
1900.0	15.68	.5598	45300.0	61370.0	212.7	27.39	35.71	872.6	63.1	.112	1.00726
				10.00 MPa isobar							
65.32[a]	871.9	31.12	−4176.0	−3855.0	68.56	31.82	53.83	1217.0	276.0	.154	1.47067
80.0	819.2	29.24	−3393.0	−3051.0	79.65	30.05	54.84	942.1	157.0	.139	1.43947
100.0	734.0	26.20	−2328.0	−1946.0	91.98	27.39	56.30	732.0	90.2	.112	1.38942
120.0	632.1	22.56	−1215.0	−772.1	102.7	26.33	61.89	554.9	56.3	.0851	1.33047
130.0	571.2	20.39	−619.2	−128.7	107.8	26.00	67.20	470.8	44.4	.0725	1.29584
135.0	537.0	19.17	−305.5	216.2	110.4	25.88	70.87	429.5	39.3	.0670	1.27663
140.0	499.8	17.84	20.3	580.9	113.1	25.79	75.03	390.2	34.6	.0618	1.25592
145.0	459.8	16.41	356.5	965.7	115.8	25.71	78.67	355.2	30.4	.0565	1.23394
150.0	418.8	14.95	695.0	1364.0	118.5	25.61	80.10	327.8	26.9	.0510	1.21161
152.0	402.7	14.37	828.2	1524.0	119.5	25.55	79.80	319.3	25.7	.0488	1.20292
154.0	387.0	13.82	958.9	1683.0	120.6	25.47	79.02	312.3	24.6	.0468	1.19452
156.0	372.0	13.28	1087.0	1840.0	121.6	25.37	77.82	306.6	23.7	.0448	1.18647
158.0	357.6	12.77	1210.0	1994.0	122.6	25.25	76.30	302.1	22.8	.0431	1.17883
160.0	344.0	12.28	1330.0	2145.0	123.5	25.13	74.54	298.5	22.1	.0415	1.17164
162.0	331.3	11.82	1446.0	2292.0	124.4	24.98	72.62	295.9	21.4	.0400	1.16491
164.0	319.3	11.40	1558.0	2435.0	125.3	24.84	70.61	294.0	20.9	.0388	1.15862
166.0	308.1	11.00	1665.0	2574.0	126.1	24.68	68.56	292.7	20.4	.0376	1.15278
168.0	297.7	10.63	1768.0	2709.0	127.0	24.52	66.52	291.9	20.0	.0366	1.14735
170.0	288.1	10.28	1868.0	2840.0	127.7	24.37	64.53	291.5	19.6	.0358	1.14232
172.0	279.0	9.960	1964.0	2968.0	128.5	24.21	62.62	291.5	19.3	.0350	1.13765
174.0	270.7	9.661	2056.0	3091.0	129.2	24.06	60.79	291.8	19.0	.0343	1.13332
176.0	262.8	9.382	2145.0	3211.0	129.9	23.92	59.06	292.4	18.8	.0337	1.12929
178.0	255.6	9.122	2231.0	3327.0	130.5	23.78	57.43	293.1	18.6	.0332	1.12555
180.0	248.8	8.879	2314.0	3441.0	131.2	23.65	55.91	294.0	18.4	.0328	1.12206
182.0	242.4	8.652	2395.0	3551.0	131.8	23.52	54.50	295.0	18.2	.0324	1.11880
184.0	236.4	8.438	2474.0	3659.0	132.4	23.40	53.18	296.2	18.1	.0320	1.11575
186.0	230.8	8.238	2550.0	3764.0	132.9	23.29	51.96	297.4	18.0	.0317	1.11289
188.0	225.5	8.050	2624.0	3867.0	133.5	23.18	50.82	298.7	17.9	.0315	1.11021
190.0	220.6	7.872	2697.0	3967.0	134.0	23.08	49.77	300.1	17.8	.0312	1.10768
192.0	215.9	7.705	2768.0	4066.0	134.5	22.99	48.79	301.5	17.8	.0310	1.10530
194.0	211.4	7.546	2837.0	4162.0	135.0	22.90	47.89	303.0	17.7	.0308	1.10305
196.0	207.2	7.396	2905.0	4257.0	135.5	22.82	47.04	304.5	17.7	.0307	1.10092
198.0	203.2	7.253	2972.0	4351.0	136.0	22.74	46.26	306.0	17.7	.0305	1.09890
200.0	199.4	7.117	3037.0	4442.0	136.4	22.67	45.52	307.5	17.6	.0304	1.09698
205.0	190.6	6.805	3196.0	4666.0	137.5	22.50	43.90	311.4	17.6	.0302	1.09258
210.0	182.8	6.526	3349.0	4882.0	138.6	22.35	42.52	315.4	17.6	.0300	1.08866
215.0	175.8	6.275	3498.0	5091.0	139.6	22.23	41.35	319.3	17.7	.0299	1.08514
220.0	169.4	6.047	3642.0	5295.0	140.5	22.12	40.34	323.2	17.7	.0299	1.08195
225.0	163.6	5.839	3782.0	5495.0	141.4	22.03	39.46	327.1	17.8	.0299	1.07906
230.0	158.3	5.649	3920.0	5690.0	142.3	21.94	38.69	331.0	17.9	.0299	1.07640
235.0	153.3	5.473	4055.0	5882.0	143.1	21.87	38.01	334.8	18.0	.0299	1.07396
240.0	148.8	5.311	4187.0	6070.0	143.9	21.80	37.41	338.5	18.2	.0300	1.07170
245.0	144.6	5.160	4318.0	6256.0	144.7	21.75	36.87	342.2	18.3	.0301	1.06961
250.0	140.6	5.019	4447.0	6439.0	145.4	21.69	36.39	345.9	18.4	.0302	1.06766
255.0	136.9	4.887	4574.0	6620.0	146.1	21.65	35.96	349.5	18.6	.0303	1.06584
260.0	133.5	4.764	4700.0	6799.0	146.8	21.61	35.57	353.0	18.7	.0305	1.06413
265.0	130.2	4.647	4824.0	6976.0	147.5	21.57	35.22	356.5	18.9	.0306	1.06253
270.0	127.1	4.537	4947.0	7151.0	148.1	21.53	34.89	359.9	19.1	.0308	1.06101
280.0	121.5	4.335	5191.0	7497.0	149.4	21.47	34.33	366.7	19.4	.0311	1.05823
290.0	116.4	4.154	5430.0	7838.0	150.6	21.42	33.85	373.3	19.7	.0315	1.05574
300.0	111.7	3.989	5667.0	8174.0	151.7	21.38	33.44	379.6	20.1	.0319	1.05347

Thermophysical properties of nitrogen—Continued

T K	Density kg/m^3	Density mol/dm^3	E J/mol	H J/mol	S J/(mol·K)	C_v J/(mol·K)	C_p J/(mol·K)	Sound m/s	Visc. μPa·s	Therm. W/(m·K)	Diel. const.
310.0	107.5	3.838	5902.0	8507.0	152.8	21.35	33.08	385.9	20.4	.0324	1.05141
320.0	103.7	3.700	6134.0	8836.0	153.9	21.32	32.78	392.0	20.8	.0328	1.04953
330.0	100.1	3.573	6364.0	9163.0	154.9	21.30	32.51	397.9	21.1	.0333	1.04779
340.0	96.81	3.456	6593.0	9486.0	155.8	21.28	32.27	403.7	21.5	.0337	1.04619
350.0	93.75	3.346	6820.0	9808.0	156.8	21.27	32.07	409.4	21.9	.0342	1.04470
360.0	90.90	3.245	7046.0	10130.0	157.7	21.26	31.88	414.9	22.2	.0347	1.04331
370.0	88.24	3.149	7271.0	10450.0	158.5	21.26	31.72	420.4	22.6	.0352	1.04202
380.0	85.74	3.060	7495.0	10760.0	159.4	21.25	31.58	425.7	22.9	.0357	1.04081
390.0	83.40	2.977	7718.0	11080.0	160.2	21.26	31.46	431.0	23.3	.0362	1.03967
400.0	81.19	2.898	7941.0	11390.0	161.0	21.26	31.34	436.1	23.7	.0367	1.03860
410.0	79.10	2.823	8163.0	11700.0	161.8	21.27	31.25	441.2	24.0	.0372	1.03760
420.0	77.13	2.753	8384.0	12020.0	162.5	21.28	31.16	446.1	24.4	.0377	1.03664
440.0	73.49	2.623	8826.0	12640.0	164.0	21.31	31.02	455.8	25.1	.0388	1.03489
460.0	70.21	2.506	9267.0	13260.0	165.3	21.36	30.92	465.2	25.8	.0398	1.03330
480.0	67.22	2.399	9708.0	13880.0	166.7	21.42	30.85	474.3	26.4	.0409	1.03186
500.0	64.50	2.302	10150.0	14490.0	167.9	21.48	30.81	483.1	27.1	.0420	1.03055
520.0	61.99	2.213	10590.0	15110.0	169.1	21.56	30.80	491.7	27.8	.0431	1.02934
540.0	59.69	2.130	11030.0	15720.0	170.3	21.65	30.80	500.1	28.4	.0441	1.02824
560.0	57.56	2.054	11470.0	16340.0	171.4	21.74	30.82	508.2	29.1	.0452	1.02721
580.0	55.58	1.984	11920.0	16960.0	172.5	21.84	30.86	516.1	29.7	.0463	1.02626
600.0	53.74	1.918	12360.0	17570.0	173.5	21.95	30.90	523.9	30.4	.0474	1.02538
620.0	52.02	1.857	12810.0	18190.0	174.6	22.06	30.97	531.5	31.0	.0486	1.02456
640.0	50.41	1.799	13260.0	18810.0	175.5	22.18	31.04	538.9	31.6	.0497	1.02379
660.0	48.91	1.746	13710.0	19430.0	176.5	22.30	31.12	546.2	32.3	.0508	1.02306
680.0	47.49	1.695	14160.0	20060.0	177.4	22.42	31.20	553.4	32.9	.0519	1.02239
700.0	46.15	1.647	14610.0	20680.0	178.3	22.55	31.29	560.4	33.5	.0530	1.02175
720.0	44.90	1.603	15070.0	21310.0	179.2	22.68	31.39	567.3	34.1	.0541	1.02115
740.0	43.71	1.560	15530.0	21940.0	180.1	22.81	31.49	574.1	34.7	.0553	1.02058
760.0	42.58	1.520	15990.0	22570.0	180.9	22.94	31.60	580.7	35.3	.0564	1.02004
780.0	41.51	1.482	16450.0	23200.0	181.7	23.07	31.70	587.3	35.9	.0575	1.01953
800.0	40.49	1.445	16920.0	23840.0	182.5	23.19	31.81	593.8	36.4	.0586	1.01904
820.0	39.53	1.411	17390.0	24470.0	183.3	23.32	31.92	600.2	37.0	.0597	1.01858
840.0	38.61	1.378	17860.0	25110.0	184.1	23.45	32.03	606.5	37.6	.0608	1.01814
860.0	37.73	1.347	18330.0	25760.0	184.9	23.58	32.14	612.7	38.1	.0619	1.01772
880.0	36.89	1.317	18810.0	26400.0	185.6	23.70	32.25	618.9	38.7	.0630	1.01732
900.0	36.09	1.288	19280.0	27050.0	186.3	23.83	32.35	624.9	39.3	.0641	1.01694
920.0	35.33	1.261	19760.0	27690.0	187.0	23.95	32.46	630.9	39.8	.0652	1.01658
950.0	34.24	1.222	20490.0	28670.0	188.1	24.12	32.62	639.8	40.6	.0669	1.01606
1000.0	32.57	1.162	21700.0	30310.0	189.8	24.41	32.88	654.3	42.0	.0696	1.01526
1050.0	31.05	1.108	22940.0	31960.0	191.4	24.68	33.13	668.5	43.3	.0722	1.01454
1100.0	29.68	1.059	24180.0	33620.0	192.9	24.94	33.36	682.3	44.6	.0749	1.01389
1150.0	28.42	1.014	25440.0	35290.0	194.4	25.18	33.59	695.9	45.9	.0775	1.01329
1200.0	27.26	.9730	26700.0	36980.0	195.8	25.40	33.80	709.2	47.1	.0801	1.01274
1250.0	26.20	.9350	27980.0	38670.0	197.2	25.61	34.00	722.2	48.4	.0826	1.01223
1300.0	25.21	.8999	29270.0	40380.0	198.6	25.81	34.19	735.0	49.6	.0851	1.01176
1350.0	24.30	.8674	30560.0	42090.0	199.9	26.00	34.36	747.6	50.8	.0875	1.01133
1400.0	23.45	.8371	31870.0	43810.0	201.1	26.17	34.53	759.9	52.0	.0900	1.01093
1450.0	22.66	.8089	33180.0	45550.0	202.3	26.33	34.68	772.1	53.2	.0923	1.01055
1500.0	21.92	.7825	34500.0	47280.0	203.5	26.48	34.83	784.1	54.3	.0947	1.01020
1550.0	21.23	.7578	35830.0	49030.0	204.6	26.62	34.96	795.9	55.5	.0970	1.00987
1600.0	20.58	.7346	37170.0	50780.0	205.8	26.75	35.09	807.6	56.6	.0993	1.00957
1650.0	19.97	.7128	38510.0	52540.0	206.8	26.88	35.21	819.1	57.7	.102	1.00928
1700.0	19.40	.6923	39860.0	54300.0	207.9	26.99	35.32	830.4	58.8	.104	1.00900
1750.0	18.85	.6729	41210.0	56070.0	208.9	27.10	35.43	841.6	59.9	.106	1.00875
1800.0	18.34	.6546	42570.0	57840.0	209.9	27.21	35.53	852.6	61.0	.108	1.00850
1850.0	17.85	.6373	43930.0	59620.0	210.9	27.30	35.62	863.5	62.1	.110	1.00827
1900.0	17.39	.6208	45300.0	61410.0	211.8	27.39	35.71	874.3	63.1	.113	1.00805
				20.00 MPa isobar							
67.43[a]	876.7	31.29	−4134.0	−3495.0	69.15	34.90	52.68	1170.0	272.0	.157	1.47317
80.0	837.8	29.91	−3500.0	−2831.0	78.18	31.81	52.70	988.0	175.0	.146	1.45013
100.0	766.3	27.35	−2510.0	−1779.0	89.91	28.26	52.75	822.2	106.0	.124	1.40799

Thermophysical properties of nitrogen—Continued

T K	Density kg/m^3	Density mol/dm^3	E J/mol	H J/mol	S J/(mol·K)	C_v J/(mol·K)	C_p J/(mol·K)	Sound m/s	Visc. μPa·s	Therm. W/(m·K)	Diel. const.
120.0	688.0	24.56	–1527.0	–712.3	99.63	26.74	54.05	687.2	71.4	.101	1.36236
150.0	561.6	20.05	–55.3	942.4	111.9	25.21	56.05	527.4	44.0	.0720	1.29017
160.0	518.8	18.52	423.7	1504.0	115.6	24.78	56.11	487.6	38.4	.0646	1.26625
170.0	477.4	17.04	888.4	2062.0	118.9	24.37	55.44	456.1	34.0	.0580	1.24333
175.0	457.6	16.33	1113.0	2338.0	120.5	24.18	54.82	443.6	32.2	.0552	1.23247
180.0	438.6	15.66	1332.0	2610.0	122.1	23.99	54.01	433.2	30.6	.0527	1.22212
185.0	420.6	15.01	1545.0	2878.0	123.5	23.81	53.07	424.7	29.3	.0505	1.21233
190.0	403.5	14.40	1752.0	3141.0	124.9	23.64	52.04	417.8	28.2	.0486	1.20313
195.0	387.5	13.83	1952.0	3398.0	126.3	23.48	50.95	412.5	27.3	.0470	1.19452
200.0	372.5	13.29	2146.0	3650.0	127.6	23.32	49.83	408.4	26.4	.0456	1.18651
205.0	358.5	12.80	2333.0	3896.0	128.8	23.18	48.73	405.4	25.8	.0444	1.17906
210.0	345.4	12.33	2515.0	4137.0	129.9	23.04	47.65	403.4	25.2	.0434	1.17215
215.0	333.3	11.90	2692.0	4373.0	131.0	22.91	46.61	402.0	24.7	.0425	1.16574
220.0	321.9	11.49	2863.0	4603.0	132.1	22.80	45.62	401.3	24.3	.0418	1.15980
225.0	311.4	11.12	3030.0	4829.0	133.1	22.69	44.68	401.2	24.0	.0411	1.15428
230.0	301.6	10.76	3192.0	5050.0	134.1	22.59	43.81	401.5	23.7	.0406	1.14915
235.0	292.4	10.44	3351.0	5267.0	135.0	22.50	42.99	402.1	23.5	.0401	1.14438
240.0	283.8	10.13	3506.0	5480.0	135.9	22.41	42.23	403.1	23.3	.0397	1.13993
245.0	275.8	9.844	3658.0	5690.0	136.8	22.33	41.52	404.3	23.2	.0394	1.13577
250.0	268.3	9.575	3807.0	5896.0	137.6	22.26	40.86	405.7	23.1	.0391	1.13189
255.0	261.2	9.323	3953.0	6098.0	138.4	22.20	40.26	407.3	23.0	.0389	1.12824
260.0	254.5	9.085	4097.0	6298.0	139.2	22.14	39.69	409.1	23.0	.0387	1.12482
265.0	248.2	8.861	4238.0	6495.0	139.9	22.08	39.17	410.9	23.0	.0386	1.12160
270.0	242.3	8.649	4378.0	6690.0	140.7	22.03	38.69	412.9	22.9	.0385	1.11857
280.0	231.4	8.259	4651.0	7072.0	142.1	21.95	37.82	417.1	23.0	.0383	1.11299
290.0	221.5	7.907	4917.0	7447.0	143.4	21.87	37.07	421.6	23.1	.0383	1.10799
300.0	212.6	7.589	5179.0	7814.0	144.6	21.81	36.42	426.1	23.2	.0383	1.10347
310.0	204.5	7.299	5435.0	8175.0	145.8	21.75	35.85	430.8	23.3	.0384	1.09937
320.0	197.1	7.034	5688.0	8531.0	146.9	21.71	35.35	435.6	23.5	.0386	1.09562
330.0	190.2	6.790	5937.0	8883.0	148.0	21.67	34.91	440.3	23.7	.0388	1.09219
340.0	183.9	6.564	6183.0	9230.0	149.1	21.63	34.52	445.1	23.9	.0390	1.08902
350.0	178.1	6.356	6426.0	9573.0	150.1	21.61	34.17	449.8	24.2	.0393	1.08609
360.0	172.6	6.162	6667.0	9913.0	151.0	21.59	33.86	454.6	24.4	.0396	1.08338
370.0	167.6	5.981	6906.0	10250.0	151.9	21.57	33.58	459.3	24.7	.0399	1.08085
380.0	162.8	5.811	7143.0	10580.0	152.8	21.56	33.34	463.9	25.0	.0402	1.07849
390.0	158.4	5.653	7379.0	10920.0	153.7	21.55	33.12	468.5	25.2	.0406	1.07627
400.0	154.2	5.503	7613.0	11250.0	154.5	21.54	32.92	473.1	25.5	.0410	1.07420
410.0	150.2	5.363	7846.0	11580.0	155.3	21.54	32.74	477.6	25.8	.0413	1.07224
420.0	146.5	5.230	8078.0	11900.0	156.1	21.55	32.58	482.1	26.1	.0417	1.07040
440.0	139.7	4.985	8539.0	12550.0	157.6	21.57	32.31	490.9	26.7	.0426	1.06701
460.0	133.5	4.764	8997.0	13190.0	159.1	21.60	32.09	499.4	27.2	.0435	1.06396
480.0	127.9	4.564	9453.0	13840.0	160.4	21.64	31.93	507.8	27.8	.0444	1.06120
500.0	122.7	4.381	9907.0	14470.0	161.7	21.70	31.80	516.0	28.4	.0453	1.05869
520.0	118.1	4.214	10360.0	15110.0	163.0	21.76	31.70	523.9	29.0	.0462	1.05639
540.0	113.7	4.060	10810.0	15740.0	164.2	21.84	31.64	531.7	29.6	.0472	1.05428
560.0	109.8	3.918	11270.0	16370.0	165.3	21.93	31.60	539.3	30.2	.0482	1.05233
580.0	106.1	3.786	11720.0	17000.0	166.4	22.02	31.58	546.8	30.8	.0492	1.05052
600.0	102.6	3.663	12180.0	17640.0	167.5	22.12	31.58	554.1	31.4	.0502	1.04885
620.0	99.41	3.548	12630.0	18270.0	168.5	22.22	31.59	561.2	32.0	.0513	1.04728
640.0	96.41	3.441	13090.0	18900.0	169.5	22.33	31.62	568.2	32.6	.0523	1.04582
660.0	93.59	3.341	13550.0	19530.0	170.5	22.45	31.67	575.1	33.2	.0533	1.04445
680.0	90.95	3.246	14010.0	20170.0	171.5	22.57	31.72	581.9	33.7	.0544	1.04316
700.0	88.45	3.157	14470.0	20800.0	172.4	22.69	31.78	588.5	34.3	.0554	1.04195
720.0	86.10	3.073	14930.0	21440.0	173.3	22.81	31.85	595.0	34.9	.0565	1.04081
740.0	83.87	2.994	15390.0	22080.0	174.1	22.94	31.92	601.5	35.5	.0576	1.03973
760.0	81.76	2.918	15860.0	22710.0	175.0	23.06	32.00	607.8	36.0	.0586	1.03871
780.0	79.76	2.847	16330.0	23360.0	175.8	23.19	32.09	614.1	36.6	.0597	1.03774
800.0	77.86	2.779	16800.0	24000.0	176.6	23.31	32.17	620.2	37.1	.0608	1.03682
820.0	76.05	2.714	17270.0	24640.0	177.4	23.44	32.26	626.3	37.7	.0618	1.03595
840.0	74.32	2.653	17750.0	25290.0	178.2	23.56	32.35	632.3	38.2	.0629	1.03512
860.0	72.68	2.594	18230.0	25940.0	179.0	23.68	32.45	638.3	38.8	.0640	1.03432
880.0	71.11	2.538	18710.0	26590.0	179.7	23.81	32.54	644.1	39.3	.0650	1.03356
900.0	69.60	2.484	19190.0	27240.0	180.5	23.93	32.63	650.0	39.9	.0661	1.03284

Thermophysical properties of nitrogen—Continued

T K	Density kg/m^3	Density mol/dm^3	E J/mol	H J/mol	S J/(mol·K)	C_v J/(mol·K)	C_p J/(mol·K)	Sound m/s	Visc. μPa·s	Therm. W/(m·K)	Diel. const.
950.0	66.12	2.360	20400.0	28880.0	182.2	24.22	32.87	664.2	41.2	.0688	1.03116
1000.0	62.97	2.248	21630.0	30530.0	183.9	24.50	33.10	678.1	42.5	.0714	1.02965
1050.0	60.12	2.146	22870.0	32190.0	185.5	24.76	33.33	691.7	43.8	.0740	1.02828
1100.0	57.52	2.053	24120.0	33860.0	187.1	25.01	33.54	705.0	45.1	.0766	1.02703
1150.0	55.14	1.968	25380.0	35540.0	188.6	25.25	33.75	718.1	46.3	.0791	1.02588
1200.0	52.95	1.890	26650.0	37230.0	190.0	25.47	33.94	730.9	47.5	.0816	1.02484
1250.0	50.94	1.818	27930.0	38930.0	191.4	25.68	34.13	743.5	48.8	.0841	1.02387
1300.0	49.07	1.751	29230.0	40650.0	192.8	25.87	34.30	755.9	50.0	.0866	1.02297
1350.0	47.33	1.690	30530.0	42360.0	194.1	26.05	34.47	768.0	51.2	.0890	1.02214
1400.0	45.72	1.632	31840.0	44090.0	195.3	26.22	34.62	780.0	52.3	.0914	1.02137
1450.0	44.21	1.578	33150.0	45830.0	196.5	26.38	34.77	791.8	53.5	.0937	1.02065
1500.0	42.80	1.528	34480.0	47570.0	197.7	26.53	34.91	803.5	54.6	.0961	1.01998
1550.0	41.48	1.481	35810.0	49320.0	198.9	26.67	35.04	815.0	55.8	.0983	1.01935
1600.0	40.24	1.436	37150.0	51070.0	200.0	26.80	35.16	826.3	56.9	.101	1.01875
1650.0	39.07	1.395	38490.0	52830.0	201.1	26.92	35.27	837.5	58.0	.103	1.01820
1700.0	37.97	1.355	39840.0	54600.0	202.1	27.04	35.38	848.5	59.1	.105	1.01767
1750.0	36.93	1.318	41200.0	56370.0	203.1	27.14	35.48	859.4	60.2	.107	1.01717
1800.0	35.94	1.283	42560.0	58150.0	204.1	27.24	35.58	870.2	61.3	.109	1.01670
1850.0	35.01	1.250	43920.0	59930.0	205.1	27.34	35.67	880.8	62.3	.112	1.01626
1900.0	34.12	1.218	45290.0	61710.0	206.1	27.43	35.75	891.3	63.4	.114	1.01584

30.00 MPa isobar

T K	Density kg/m^3	Density mol/dm^3	E J/mol	H J/mol	S J/(mol·K)	C_v J/(mol·K)	C_p J/(mol·K)	Sound m/s	Visc. μPa·s	Therm. W/(m·K)	Diel. const.
69.49[a]	882.0	31.48	−4088.0	−3135.0	69.75	36.54	51.34	1154.0	270.0	.161	1.47594
80.0	853.5	30.47	−3582.0	−2597.0	76.96	33.03	50.98	1031.0	192.0	.153	1.45903
100.0	790.9	28.23	−2644.0	−1581.0	88.30	28.87	50.74	894.3	119.0	.133	1.42207
150.0	622.6	22.22	−384.9	965.1	108.9	25.52	50.80	644.9	55.3	.0856	1.32443
170.0	557.4	19.90	465.4	1973.0	115.3	24.66	49.90	579.3	44.4	.0724	1.28755
180.0	526.6	18.80	872.7	2469.0	118.1	24.28	49.17	553.9	40.5	.0670	1.27031
190.0	497.4	17.75	1266.0	2956.0	120.7	23.95	48.27	533.2	37.4	.0624	1.25411
200.0	470.1	16.78	1646.0	3434.0	123.2	23.66	47.24	516.9	34.9	.0587	1.23905
210.0	444.7	15.87	2011.0	3900.0	125.4	23.40	46.12	504.4	32.9	.0556	1.22521
220.0	421.4	15.04	2361.0	4356.0	127.6	23.17	44.97	495.2	31.4	.0531	1.21256
230.0	400.1	14.28	2699.0	4800.0	129.5	22.97	43.84	488.6	30.1	.0510	1.20107
240.0	380.7	13.59	3025.0	5233.0	131.4	22.80	42.75	484.2	29.2	.0494	1.19066
250.0	363.0	12.96	3340.0	5655.0	133.1	22.65	41.74	481.5	28.4	.0481	1.18123
260.0	346.8	12.38	3644.0	6068.0	134.7	22.52	40.80	480.1	27.9	.0470	1.17268
270.0	332.1	11.85	3941.0	6472.0	136.2	22.40	39.94	479.9	27.5	.0462	1.16492
280.0	318.6	11.37	4229.0	6867.0	137.7	22.30	39.16	480.5	27.2	.0456	1.15785
290.0	306.3	10.93	4511.0	7255.0	139.0	22.22	38.45	481.8	26.9	.0451	1.15140
300.0	294.9	10.53	4787.0	7636.0	140.3	22.14	37.81	483.6	26.8	.0448	1.14550
310.0	284.5	10.15	5057.0	8011.0	141.6	22.07	37.23	485.8	26.7	.0445	1.14007
320.0	274.8	9.809	5323.0	8381.0	142.7	22.02	36.71	488.4	26.7	.0444	1.13507
330.0	265.9	9.489	5584.0	8746.0	143.9	21.97	36.24	491.3	26.7	.0443	1.13045
340.0	257.5	9.192	5842.0	9106.0	144.9	21.92	35.81	494.3	26.8	.0443	1.12617
350.0	249.8	8.915	6097.0	9462.0	146.0	21.89	35.42	497.5	26.9	.0443	1.12219
360.0	242.5	8.656	6349.0	9814.0	147.0	21.86	35.07	500.9	27.0	.0444	1.11847
370.0	235.7	8.413	6598.0	10160.0	147.9	21.83	34.75	504.3	27.1	.0445	1.11501
380.0	229.3	8.185	6844.0	10510.0	148.8	21.81	34.46	507.8	27.3	.0447	1.11176
390.0	223.3	7.971	7089.0	10850.0	149.7	21.79	34.20	511.4	27.4	.0449	1.10870
400.0	217.7	7.769	7332.0	11190.0	150.6	21.78	33.96	515.0	27.6	.0451	1.10583
410.0	212.3	7.578	7573.0	11530.0	151.4	21.78	33.75	518.6	27.8	.0454	1.10313
420.0	207.2	7.397	7813.0	11870.0	152.2	21.77	33.55	522.3	28.0	.0457	1.10057
440.0	197.9	7.063	8289.0	12540.0	153.8	21.78	33.21	529.6	28.5	.0463	1.09585
460.0	189.4	6.761	8760.0	13200.0	155.3	21.80	32.94	536.9	28.9	.0470	1.09160
480.0	181.7	6.486	9229.0	13850.0	156.7	21.84	32.71	544.2	29.4	.0477	1.08774
500.0	174.7	6.235	9695.0	14510.0	158.0	21.88	32.53	551.3	29.9	.0485	1.08422
520.0	168.2	6.005	10160.0	15160.0	159.3	21.94	32.39	558.4	30.4	.0493	1.08100
540.0	162.3	5.792	10620.0	15800.0	160.5	22.01	32.28	565.4	30.9	.0502	1.07803
560.0	156.8	5.595	11080.0	16450.0	161.7	22.09	32.20	572.2	31.5	.0511	1.07529
580.0	151.6	5.412	11550.0	17090.0	162.8	22.18	32.14	579.0	32.0	.0520	1.07276
600.0	146.9	5.242	12010.0	17730.0	163.9	22.27	32.11	585.6	32.5	.0529	1.07039
620.0	142.4	5.083	12470.0	18370.0	164.9	22.37	32.10	592.2	33.1	.0538	1.06819

Thermophysical properties of nitrogen—Continued

T K	Density kg/m³	Density mol/dm³	E J/mol	H J/mol	S J/(mol·K)	C_v J/(mol·K)	C_p J/(mol·K)	Sound m/s	Visc. μPa·s	Therm. W/(m·K)	Diel. const.
640.0	138.2	4.934	12940.0	19020.0	166.0	22.48	32.10	598.7	33.6	.0548	1.06613
660.0	134.3	4.794	13400.0	19660.0	166.9	22.58	32.11	605.0	34.2	.0558	1.06420
680.0	130.6	4.662	13870.0	20300.0	167.9	22.70	32.14	611.3	34.7	.0568	1.06238
700.0	127.1	4.538	14330.0	20940.0	168.8	22.81	32.18	617.5	35.2	.0578	1.06067
720.0	123.9	4.421	14800.0	21590.0	169.7	22.93	32.23	623.6	35.8	.0588	1.05906
740.0	120.7	4.310	15270.0	22230.0	170.6	23.05	32.28	629.6	36.3	.0598	1.05753
760.0	117.8	4.205	15740.0	22880.0	171.5	23.18	32.34	635.5	36.8	.0608	1.05608
780.0	115.0	4.105	16220.0	23530.0	172.3	23.30	32.41	641.4	37.4	.0618	1.05471
800.0	112.3	4.009	16690.0	24180.0	173.1	23.42	32.48	647.2	37.9	.0628	1.05341
820.0	109.8	3.919	17170.0	24830.0	174.0	23.54	32.56	653.0	38.4	.0639	1.05217
840.0	107.4	3.833	17650.0	25480.0	174.7	23.66	32.63	658.6	39.0	.0649	1.05099
860.0	105.1	3.750	18130.0	26130.0	175.5	23.78	32.71	664.3	39.5	.0659	1.04986
880.0	102.9	3.671	18620.0	26790.0	176.3	23.90	32.80	669.8	40.0	.0670	1.04878
900.0	100.7	3.596	19100.0	27440.0	177.0	24.02	32.88	675.3	40.5	.0680	1.04775
950.0	95.83	3.420	20320.0	29090.0	178.8	24.30	33.09	688.9	41.8	.0706	1.04536
1000.0	91.39	3.262	21560.0	30750.0	180.5	24.58	33.30	702.2	43.1	.0731	1.04321
1050.0	87.36	3.118	22800.0	32420.0	182.1	24.84	33.50	715.2	44.3	.0757	1.04125
1100.0	83.68	2.987	24060.0	34100.0	183.7	25.08	33.70	727.9	45.6	.0782	1.03947
1150.0	80.30	2.866	25330.0	35790.0	185.2	25.32	33.89	740.5	46.8	.0807	1.03783
1200.0	77.19	2.755	26600.0	37490.0	186.6	25.53	34.07	752.8	48.0	.0832	1.03633
1250.0	74.32	2.653	27890.0	39200.0	188.0	25.74	34.25	764.9	49.2	.0856	1.03494
1300.0	71.66	2.558	29190.0	40920.0	189.4	25.93	34.41	776.9	50.4	.0880	1.03366
1350.0	69.19	2.470	30490.0	42640.0	190.7	26.11	34.57	788.6	51.6	.0904	1.03247
1400.0	66.88	2.387	31810.0	44370.0	191.9	26.28	34.71	800.2	52.7	.0928	1.03135
1450.0	64.72	2.310	33130.0	46110.0	193.1	26.43	34.85	811.7	53.9	.0951	1.03032
1500.0	62.71	2.238	34450.0	47860.0	194.3	26.58	34.98	822.9	55.0	.0974	1.02935
1550.0	60.81	2.171	35790.0	49610.0	195.5	26.72	35.11	834.1	56.1	.0996	1.02844
1600.0	59.03	2.107	37130.0	51370.0	196.6	26.84	35.22	845.1	57.2	.102	1.02758
1650.0	57.35	2.047	38480.0	53130.0	197.7	26.96	35.33	856.0	58.3	.104	1.02677
1700.0	55.76	1.990	39830.0	54900.0	198.7	27.08	35.43	866.7	59.4	.106	1.02601
1750.0	54.26	1.937	41190.0	56670.0	199.8	27.18	35.53	877.3	60.5	.108	1.02529
1800.0	52.84	1.886	42550.0	58450.0	200.8	27.28	35.62	887.8	61.6	.111	1.02461
1850.0	51.50	1.838	43920.0	60240.0	201.7	27.37	35.71	898.2	62.6	.113	1.02397
1900.0	50.22	1.792	45290.0	62020.0	202.7	27.46	35.79	908.4	63.7	.115	1.02335

40.00 MPa isobar

T K	Density kg/m³	Density mol/dm³	E J/mol	H J/mol	S J/(mol·K)	C_v J/(mol·K)	C_p J/(mol·K)	Sound m/s	Visc. μPa·s	Therm. W/(m·K)	Diel. const.
71.51[a]	887.6	31.68	−4039.0	−2776.0	70.35	37.18	49.87	1155.0	269.0	.164	1.47884
80.0	867.1	30.95	−3647.0	−2355.0	75.92	33.88	49.46	1073.0	208.0	.159	1.46668
100.0	811.0	28.95	−2749.0	−1367.0	86.94	29.33	49.36	955.7	132.0	.141	1.43354
150.0	663.2	23.67	−602.8	1087.0	106.9	25.86	48.44	733.5	64.6	.0967	1.34740
170.0	607.4	21.68	201.1	2046.0	112.9	25.02	47.40	671.8	53.1	.0837	1.31551
180.0	581.0	20.74	587.9	2517.0	115.5	24.65	46.76	646.8	48.8	.0783	1.30053
190.0	555.6	19.83	964.2	2981.0	118.1	24.32	46.06	625.4	45.2	.0735	1.28629
200.0	531.6	18.97	1330.0	3438.0	120.4	24.03	45.31	607.3	42.3	.0695	1.27282
210.0	508.8	18.16	1685.0	3887.0	122.6	23.77	44.53	592.2	39.9	.0659	1.26017
220.0	487.4	17.40	2029.0	4329.0	124.6	23.53	43.74	579.8	37.9	.0630	1.24833
230.0	467.3	16.68	2364.0	4762.0	126.6	23.33	42.95	569.8	36.2	.0604	1.23730
240.0	448.5	16.01	2689.0	5188.0	128.4	23.15	42.18	561.9	34.8	.0583	1.22705
250.0	431.0	15.38	3006.0	5606.0	130.1	22.99	41.43	555.8	33.7	.0565	1.21753
260.0	414.7	14.80	3314.0	6016.0	131.7	22.85	40.71	551.2	32.8	.0550	1.20872
270.0	399.5	14.26	3615.0	6420.0	133.2	22.73	40.04	547.8	32.1	.0537	1.20055
280.0	385.4	13.76	3909.0	6817.0	134.7	22.62	39.40	545.5	31.5	.0527	1.19297
290.0	372.2	13.29	4198.0	7208.0	136.0	22.52	38.81	544.1	31.0	.0518	1.18595
300.0	359.9	12.85	4480.0	7594.0	137.3	22.43	38.26	543.5	30.6	.0511	1.17942
310.0	348.5	12.44	4758.0	7974.0	138.6	22.36	37.75	543.5	30.3	.0506	1.17334
320.0	337.7	12.06	5031.0	8349.0	139.8	22.29	37.28	544.0	30.1	.0501	1.16768
330.0	327.7	11.70	5299.0	8719.0	140.9	22.23	36.84	544.9	29.9	.0498	1.16240
340.0	318.3	11.36	5564.0	9086.0	142.0	22.18	36.44	546.2	29.8	.0495	1.15745
350.0	309.4	11.04	5826.0	9448.0	143.1	22.14	36.07	547.9	29.7	.0494	1.15282
360.0	301.1	10.75	6085.0	9807.0	144.1	22.10	35.73	549.7	29.7	.0492	1.14847
370.0	293.2	10.47	6341.0	10160.0	145.1	22.07	35.41	551.8	29.7	.0492	1.14439

Thermophysical properties of nitrogen—Continued

T K	Density kg/m³	Density mol/dm³	E J/mol	H J/mol	S J/(mol·K)	C_v J/(mol·K)	C_p J/(mol·K)	Sound m/s	Visc. μPa·s	Therm. W/(m·K)	Diel. const.
380.0	285.8	10.20	6594.0	10520.0	146.0	22.04	35.12	554.1	29.8	.0492	1.14054
390.0	278.8	9.950	6845.0	10870.0	146.9	22.02	34.86	556.5	29.8	.0492	1.13690
400.0	272.1	9.712	7094.0	11210.0	147.8	22.00	34.61	559.1	29.9	.0493	1.13347
420.0	259.8	9.273	7587.0	11900.0	149.5	21.98	34.18	564.4	30.1	.0496	1.12714
440.0	248.7	8.876	8074.0	12580.0	151.0	21.97	33.82	570.1	30.4	.0500	1.12143
460.0	238.5	8.514	8556.0	13250.0	152.5	21.99	33.52	575.9	30.7	.0505	1.11625
480.0	229.3	8.183	9034.0	13920.0	154.0	22.01	33.27	581.9	31.1	.0510	1.11154
500.0	220.8	7.880	9509.0	14580.0	155.3	22.05	33.06	587.9	31.5	.0516	1.10722
520.0	212.9	7.600	9981.0	15240.0	156.6	22.10	32.89	593.9	31.9	.0523	1.10325
540.0	205.7	7.342	10450.0	15900.0	157.8	22.17	32.76	600.0	32.4	.0531	1.09959
560.0	199.0	7.102	10920.0	16550.0	159.0	22.24	32.66	606.0	32.8	.0538	1.09621
580.0	192.7	6.878	11390.0	17210.0	160.2	22.32	32.58	612.0	33.3	.0546	1.09306
600.0	186.9	6.670	11860.0	17860.0	161.3	22.41	32.53	617.9	33.8	.0555	1.09012
620.0	181.4	6.474	12330.0	18510.0	162.4	22.50	32.49	623.8	34.2	.0564	1.08738
640.0	176.2	6.291	12800.0	19160.0	163.4	22.60	32.48	629.6	34.7	.0572	1.08481
660.0	171.4	6.118	13270.0	19810.0	164.4	22.71	32.47	635.4	35.2	.0581	1.08240
680.0	166.9	5.956	13740.0	20460.0	165.4	22.82	32.49	641.1	35.7	.0591	1.08013
700.0	162.6	5.802	14210.0	21110.0	166.3	22.93	32.51	646.8	36.2	.0600	1.07799
720.0	158.5	5.657	14690.0	21760.0	167.2	23.05	32.54	652.4	36.7	.0609	1.07597
740.0	154.6	5.519	15160.0	22410.0	168.1	23.16	32.58	658.0	37.2	.0619	1.07406
760.0	151.0	5.389	15640.0	23060.0	169.0	23.28	32.63	663.5	37.7	.0629	1.07224
780.0	147.5	5.264	16120.0	23710.0	169.8	23.40	32.68	669.0	38.2	.0638	1.07052
800.0	144.2	5.146	16600.0	24370.0	170.6	23.52	32.74	674.4	38.7	.0648	1.06888
820.0	141.0	5.033	17080.0	25020.0	171.5	23.64	32.81	679.8	39.2	.0658	1.06732
840.0	138.0	4.925	17560.0	25680.0	172.3	23.76	32.87	685.1	39.7	.0668	1.06583
860.0	135.1	4.822	18040.0	26340.0	173.0	23.87	32.94	690.4	40.2	.0678	1.06441
900.0	129.7	4.629	19020.0	27660.0	174.5	24.10	33.09	700.9	41.2	.0698	1.06175
950.0	123.6	4.410	20250.0	29320.0	176.3	24.39	33.28	713.7	42.5	.0723	1.05873
1000.0	118.0	4.211	21490.0	30990.0	178.0	24.65	33.47	726.3	43.7	.0748	1.05600
1050.0	112.9	4.030	22740.0	32670.0	179.7	24.91	33.66	738.7	44.9	.0773	1.05351
1100.0	108.3	3.865	24000.0	34350.0	181.2	25.15	33.84	750.9	46.1	.0798	1.05125
1150.0	104.0	3.713	25280.0	36050.0	182.7	25.38	34.02	762.9	47.3	.0822	1.04917
1200.0	100.1	3.572	26560.0	37750.0	184.2	25.59	34.19	774.7	48.5	.0846	1.04726
1250.0	96.45	3.443	27850.0	39470.0	185.6	25.80	34.35	786.4	49.7	.0870	1.04549
1300.0	93.08	3.322	29150.0	41190.0	186.9	25.98	34.51	797.9	50.8	.0894	1.04385
1350.0	89.94	3.210	30460.0	42920.0	188.3	26.16	34.66	809.2	52.0	.0918	1.04233
1400.0	87.01	3.106	31780.0	44660.0	189.5	26.33	34.80	820.4	53.1	.0941	1.04090
1450.0	84.27	3.008	33100.0	46400.0	190.7	26.48	34.93	831.5	54.3	.0964	1.03958
1500.0	81.69	2.916	34430.0	48150.0	191.9	26.62	35.05	842.4	55.4	.0986	1.03833
1550.0	79.28	2.830	35770.0	49900.0	193.1	26.76	35.17	853.2	56.5	.101	1.03716
1600.0	77.00	2.748	37110.0	51670.0	194.2	26.89	35.28	863.9	57.6	.103	1.03606
1650.0	74.86	2.672	38460.0	53430.0	195.3	27.00	35.39	874.5	58.7	.105	1.03503
1700.0	72.83	2.599	39820.0	55200.0	196.3	27.12	35.48	884.9	59.7	.107	1.03405
1750.0	70.91	2.531	41180.0	56980.0	197.4	27.22	35.58	895.2	60.8	.110	1.03312
1800.0	69.09	2.466	42540.0	58760.0	198.4	27.32	35.66	905.5	61.9	.112	1.03224
1850.0	67.36	2.404	43910.0	60550.0	199.4	27.41	35.75	915.6	62.9	.114	1.03141
1900.0	65.71	2.346	45280.0	62340.0	200.3	27.50	35.82	925.6	63.9	.116	1.03062
				50.00 MPa isobar							
73.48[a]	893.2	31.88	-3988.0	-2419.0	70.93	37.18	48.37	1166.0	269.0	.168	1.48177
80.0	879.0	31.38	-3699.0	-2105.0	75.03	34.45	48.04	1113.0	223.0	.165	1.47338
200.0	575.9	20.56	1104.0	3536.0	118.4	24.40	44.12	683.0	48.9	.0788	1.29744
210.0	555.2	19.82	1451.0	3974.0	120.5	24.13	43.46	667.0	46.1	.0750	1.28576
220.0	535.4	19.11	1789.0	4405.0	122.5	23.89	42.81	653.2	43.8	.0718	1.27472
230.0	516.7	18.44	2119.0	4830.0	124.4	23.68	42.18	641.6	41.8	.0689	1.26431
240.0	499.0	17.81	2442.0	5249.0	126.2	23.49	41.55	631.8	40.2	.0664	1.25450
250.0	482.3	17.21	2757.0	5661.0	127.9	23.32	40.95	623.7	38.8	.0643	1.24527
260.0	466.5	16.65	3065.0	6068.0	129.5	23.17	40.37	617.0	37.6	.0624	1.23661
270.0	451.6	16.12	3367.0	6469.0	131.0	23.03	39.81	611.6	36.6	.0609	1.22847
280.0	437.5	15.62	3663.0	6864.0	132.4	22.91	39.29	607.3	35.7	.0595	1.22083
290.0	424.3	15.15	3953.0	7255.0	133.8	22.81	38.79	604.0	35.0	.0583	1.21366

Thermophysical properties of nitrogen—Continued

T K	Density kg/m^3	Density mol/dm^3	E J/mol	H J/mol	S J/(mol·K)	C_v J/(mol·K)	C_p J/(mol·K)	Sound m/s	Visc. μPa·s	Therm. W/(m·K)	Diel. const.
300.0	411.8	14.70	4239.0	7640.0	135.1	22.71	38.32	601.5	34.4	.0573	1.20693
310.0	400.1	14.28	4520.0	8021.0	136.3	22.63	37.87	599.7	33.9	.0565	1.20060
320.0	388.9	13.88	4796.0	8398.0	137.5	22.55	37.46	598.5	33.5	.0558	1.19464
330.0	378.4	13.51	5069.0	8770.0	138.7	22.48	37.07	597.9	33.2	.0552	1.18903
340.0	368.5	13.15	5338.0	9139.0	139.8	22.42	36.71	597.7	32.9	.0547	1.18374
350.0	359.1	12.82	5604.0	9505.0	140.8	22.37	36.37	597.9	32.7	.0544	1.17875
360.0	350.2	12.50	5867.0	9867.0	141.9	22.32	36.06	598.5	32.6	.0540	1.17403
380.0	333.7	11.91	6384.0	10580.0	143.8	22.25	35.49	600.5	32.4	.0536	1.16534
400.0	318.8	11.38	6893.0	11290.0	145.6	22.20	35.00	603.4	32.3	.0535	1.15753
420.0	305.3	10.90	7394.0	11980.0	147.3	22.17	34.58	606.9	32.3	.0535	1.15046
440.0	292.9	10.46	7889.0	12670.0	148.9	22.15	34.22	610.9	32.5	.0536	1.14405
460.0	281.6	10.05	8378.0	13350.0	150.4	22.16	33.92	615.3	32.7	.0539	1.13819
480.0	271.3	9.682	8864.0	14030.0	151.9	22.18	33.66	619.9	32.9	.0543	1.13283
500.0	261.7	9.341	9346.0	14700.0	153.2	22.21	33.45	624.8	33.2	.0548	1.12791
520.0	252.8	9.024	9825.0	15370.0	154.5	22.25	33.27	629.8	33.5	.0553	1.12336
540.0	244.6	8.730	10300.0	16030.0	155.8	22.31	33.12	634.8	33.9	.0559	1.11915
560.0	236.9	8.457	10780.0	16690.0	157.0	22.38	33.01	640.0	34.2	.0566	1.11524
580.0	229.8	8.201	11250.0	17350.0	158.1	22.45	32.92	645.1	34.6	.0573	1.11159
600.0	223.1	7.962	11730.0	18010.0	159.3	22.53	32.85	650.3	35.0	.0580	1.10819
620.0	216.8	7.737	12200.0	18660.0	160.3	22.63	32.81	655.5	35.5	.0588	1.10500
640.0	210.9	7.526	12680.0	19320.0	161.4	22.72	32.78	660.7	35.9	.0596	1.10201
660.0	205.3	7.327	13150.0	19980.0	162.4	22.82	32.77	665.9	36.3	.0605	1.09920
680.0	200.0	7.139	13630.0	20630.0	163.4	22.93	32.77	671.1	36.8	.0613	1.09654
700.0	195.0	6.961	14100.0	21290.0	164.3	23.04	32.78	676.2	37.2	.0622	1.09404
720.0	190.3	6.793	14580.0	21940.0	165.2	23.15	32.80	681.4	37.7	.0631	1.09167
740.0	185.8	6.633	15060.0	22600.0	166.1	23.26	32.83	686.5	38.2	.0640	1.08942
760.0	181.6	6.481	15540.0	23260.0	167.0	23.38	32.87	691.5	38.6	.0649	1.08728
780.0	177.5	6.336	16020.0	23910.0	167.9	23.49	32.92	696.6	39.1	.0658	1.08526
800.0	173.6	6.198	16500.0	24570.0	168.7	23.61	32.97	701.6	39.6	.0668	1.08332
820.0	169.9	6.066	16990.0	25230.0	169.5	23.73	33.02	706.6	40.1	.0677	1.08148
840.0	166.4	5.940	17480.0	25890.0	170.3	23.84	33.08	711.6	40.5	.0687	1.07973
900.0	156.7	5.593	18940.0	27880.0	172.6	24.19	33.27	726.3	42.0	.0716	1.07490
950.0	149.5	5.335	20180.0	29550.0	174.4	24.46	33.45	738.4	43.2	.0740	1.07132
1000.0	142.9	5.101	21430.0	31230.0	176.1	24.73	33.62	750.4	44.3	.0764	1.06807
1050.0	136.9	4.887	22680.0	32910.0	177.8	24.98	33.80	762.1	45.5	.0789	1.06512
1100.0	131.4	4.692	23950.0	34610.0	179.3	25.22	33.97	773.8	46.7	.0813	1.06242
1150.0	126.4	4.512	25230.0	36310.0	180.9	25.44	34.14	785.2	47.9	.0837	1.05994
1200.0	121.7	4.345	26520.0	38020.0	182.3	25.65	34.30	796.6	49.0	.0861	1.05766
1250.0	117.4	4.191	27810.0	39740.0	183.7	25.85	34.45	807.8	50.2	.0884	1.05554
1300.0	113.4	4.048	29120.0	41470.0	185.1	26.04	34.60	818.8	51.3	.0908	1.05358
1350.0	109.7	3.915	30430.0	43200.0	186.4	26.21	34.74	829.8	52.4	.0931	1.05175
1400.0	106.2	3.790	31750.0	44940.0	187.6	26.37	34.87	840.6	53.5	.0954	1.05005
1450.0	102.9	3.673	33070.0	46690.0	188.9	26.53	35.00	851.3	54.7	.0976	1.04845
1500.0	99.83	3.563	34410.0	48440.0	190.1	26.67	35.12	861.9	55.8	.0999	1.04695
1550.0	96.94	3.460	35750.0	50200.0	191.2	26.80	35.23	872.4	56.9	.102	1.04555
1600.0	94.21	3.363	37090.0	51960.0	192.3	26.93	35.34	882.7	57.9	.104	1.04422
1650.0	91.64	3.271	38450.0	53730.0	193.4	27.04	35.44	893.0	59.0	.106	1.04297
1700.0	89.20	3.184	39800.0	55510.0	194.5	27.15	35.53	903.1	60.1	.109	1.04179
1750.0	86.89	3.102	41170.0	57290.0	195.5	27.26	35.62	913.2	61.1	.111	1.04067
1800.0	84.70	3.023	42530.0	59070.0	196.5	27.35	35.71	923.1	62.2	.113	1.03961
1850.0	82.62	2.949	43900.0	60860.0	197.5	27.44	35.79	933.0	63.2	.115	1.03861
1900.0	80.64	2.878	45280.0	62650.0	198.5	27.53	35.86	942.7	64.2	.117	1.03765
				60.00 MPa isobar							
75.41[a]	898.7	32.08	−3936.0	−2065.0	71.49	36.77	46.87	1184.0	270.0	.172	1.48468
100.0	843.1	30.09	−2905.0	−911.4	84.73	29.93	47.33	1059.0	155.0	.155	1.45175
150.0	719.2	25.67	−895.4	1442.0	103.8	26.54	46.21	870.5	80.8	.115	1.37917
170.0	673.4	24.04	−140.5	2356.0	109.5	25.73	45.13	813.1	67.7	.102	1.35270
180.0	651.7	23.26	224.6	2804.0	112.1	25.37	44.55	788.9	62.8	.0965	1.34021
200.0	610.7	21.80	930.2	3683.0	116.7	24.75	43.34	748.6	55.0	.0871	1.31681
220.0	573.0	20.45	1605.0	4538.0	120.8	24.23	42.17	717.6	49.3	.0798	1.29553
240.0	538.8	19.23	2250.0	5370.0	124.4	23.81	41.06	694.2	45.2	.0740	1.27631

Thermophysical properties of nitrogen—Continued

T K	Density kg/m^3	Density mol/dm^3	E J/mol	H J/mol	S J/(mol·K)	C_v J/(mol·K)	C_p J/(mol·K)	Sound m/s	Visc. μPa·s	Therm. W/(m·K)	Diel. const.
260.0	507.7	18.12	2870.0	6181.0	127.7	23.47	40.03	676.9	42.1	.0694	1.25905
280.0	479.6	17.12	3467.0	6972.0	130.6	23.20	39.09	664.5	39.9	.0660	1.24357
300.0	454.3	16.21	4044.0	7745.0	133.3	22.98	38.24	656.0	38.2	.0633	1.22969
320.0	431.4	15.40	4605.0	8502.0	135.7	22.80	37.48	650.4	36.9	.0613	1.21725
340.0	410.6	14.66	5151.0	9245.0	138.0	22.65	36.81	647.2	36.0	.0599	1.20606
360.0	391.9	13.99	5685.0	9975.0	140.1	22.54	36.21	645.8	35.4	.0588	1.19597
380.0	374.8	13.38	6208.0	10690.0	142.0	22.45	35.69	645.7	35.0	.0581	1.18685
400.0	359.2	12.82	6722.0	11400.0	143.8	22.39	35.23	646.8	34.7	.0576	1.17856
420.0	344.9	12.31	7229.0	12100.0	145.5	22.35	34.83	648.7	34.6	.0574	1.17101
440.0	331.8	11.84	7729.0	12800.0	147.1	22.32	34.48	651.2	34.6	.0573	1.16411
460.0	319.7	11.41	8224.0	13480.0	148.7	22.32	34.19	654.2	34.6	.0574	1.15777
480.0	308.5	11.01	8715.0	14160.0	150.1	22.33	33.94	657.6	34.8	.0576	1.15194
500.0	298.2	10.64	9203.0	14840.0	151.5	22.35	33.72	661.4	34.9	.0579	1.14655
520.0	288.5	10.30	9687.0	15510.0	152.8	22.39	33.54	665.3	35.2	.0583	1.14156
540.0	279.6	9.979	10170.0	16180.0	154.1	22.44	33.40	669.4	35.4	.0588	1.13691
560.0	271.2	9.679	10650.0	16850.0	155.3	22.50	33.28	673.7	35.7	.0593	1.13259
580.0	263.3	9.399	11130.0	17510.0	156.5	22.57	33.18	678.1	36.0	.0599	1.12855
600.0	255.9	9.136	11610.0	18180.0	157.6	22.65	33.11	682.6	36.4	.0606	1.12476
620.0	249.0	8.888	12090.0	18840.0	158.7	22.74	33.06	687.1	36.7	.0613	1.12121
640.0	242.5	8.654	12570.0	19500.0	159.7	22.83	33.02	691.6	37.1	.0620	1.11786
660.0	236.3	8.434	13040.0	20160.0	160.7	22.93	33.00	696.2	37.5	.0627	1.11471
680.0	230.4	8.225	13520.0	20820.0	161.7	23.03	33.00	700.8	37.9	.0635	1.11174
700.0	224.9	8.027	14000.0	21480.0	162.7	23.14	33.00	705.5	38.3	.0643	1.10892
720.0	219.6	7.839	14490.0	22140.0	163.6	23.25	33.02	710.1	38.7	.0652	1.10625
740.0	214.6	7.661	14970.0	22800.0	164.5	23.36	33.04	714.7	39.2	.0660	1.10372
760.0	209.9	7.491	15450.0	23460.0	165.4	23.47	33.07	719.4	39.6	.0669	1.10131
780.0	205.3	7.328	15940.0	24120.0	166.3	23.58	33.11	724.0	40.0	.0678	1.09902
800.0	201.0	7.174	16420.0	24790.0	167.1	23.70	33.16	728.6	40.5	.0687	1.09684
820.0	196.8	7.026	16910.0	25450.0	167.9	23.81	33.21	733.3	40.9	.0696	1.09475
850.0	190.9	6.815	17640.0	26450.0	169.1	23.98	33.29	740.2	41.6	.0710	1.09179
900.0	181.9	6.493	18870.0	28110.0	171.0	24.26	33.44	751.6	42.7	.0733	1.08727
950.0	173.8	6.202	20120.0	29790.0	172.8	24.53	33.59	763.0	43.9	.0756	1.08320
1000.0	166.3	5.937	21370.0	31470.0	174.6	24.79	33.76	774.2	45.0	.0780	1.07950
1050.0	159.5	5.695	22630.0	33170.0	176.2	25.04	33.92	785.4	46.1	.0804	1.07613
1100.0	153.3	5.472	23900.0	34870.0	177.8	25.28	34.09	796.5	47.3	.0828	1.07304
1150.0	147.6	5.267	25180.0	36570.0	179.3	25.50	34.25	807.4	48.4	.0851	1.07019
1200.0	142.3	5.078	26480.0	38290.0	180.8	25.71	34.40	818.3	49.5	.0875	1.06757
1250.0	137.3	4.902	27770.0	40010.0	182.2	25.90	34.55	829.0	50.7	.0898	1.06514
1300.0	132.7	4.738	29080.0	41750.0	183.5	26.09	34.69	839.7	51.8	.0921	1.06288
1350.0	128.5	4.585	30400.0	43480.0	184.8	26.26	34.82	850.2	52.9	.0944	1.06078
1400.0	124.5	4.442	31720.0	45230.0	186.1	26.42	34.95	860.7	54.0	.0966	1.05881
1450.0	120.7	4.308	33050.0	46980.0	187.3	26.57	35.07	871.0	55.1	.0988	1.05697
1500.0	117.2	4.182	34390.0	48730.0	188.5	26.71	35.18	881.2	56.2	.101	1.05524
1550.0	113.8	4.063	35730.0	50500.0	189.7	26.84	35.29	891.4	57.2	.103	1.05361
1600.0	110.7	3.951	37080.0	52260.0	190.8	26.97	35.39	901.4	58.3	.105	1.05208
1650.0	107.7	3.846	38430.0	54040.0	191.9	27.08	35.49	911.4	59.4	.108	1.05063
1700.0	104.9	3.745	39790.0	55810.0	193.0	27.19	35.58	921.3	60.4	.110	1.04926
1750.0	102.3	3.650	41160.0	57590.0	194.0	27.29	35.66	931.0	61.5	.112	1.04796
1800.0	99.74	3.560	42520.0	59380.0	195.0	27.39	35.75	940.7	62.5	.114	1.04673
1850.0	97.33	3.474	43900.0	61170.0	196.0	27.47	35.82	950.3	63.5	.116	1.04556
1900.0	95.04	3.392	45270.0	62960.0	196.9	27.56	35.90	959.8	64.5	.118	1.04445
				80.00 MPa isobar							
79.15[a]	909.5	32.46	−3832.0	−1368.0	72.49	35.37	44.08	1230.0	271.0	.179	1.49025
100.0	868.5	31.00	−3016.0	−435.4	82.94	30.26	45.65	1145.0	177.0	.167	1.46601
150.0	759.0	27.09	−1093.0	1860.0	101.6	27.15	45.18	977.7	95.0	.129	1.40169
170.0	718.8	25.66	−364.6	2753.0	107.2	26.38	44.12	923.3	80.5	.117	1.37831
180.0	699.7	24.98	−11.3	3192.0	109.7	26.04	43.57	899.8	75.0	.111	1.36726
200.0	663.6	23.69	674.4	4052.0	114.2	25.41	42.45	859.5	66.2	.102	1.34643
220.0	630.1	22.49	1333.0	4890.0	118.2	24.87	41.39	827.4	59.6	.0940	1.32728
240.0	599.2	21.39	1968.0	5708.0	121.8	24.42	40.42	801.9	54.6	.0876	1.30973
260.0	570.7	20.37	2581.0	6508.0	125.0	24.04	39.53	781.8	50.7	.0823	1.29368

Thermophysical properties of nitrogen—Continued

T K	Density kg/m^3	Density mol/dm^3	*E* J/mol	*H* J/mol	*S* J/(mol·K)	C_v J/(mol·K)	C_p J/(mol·K)	Sound m/s	Visc. μPa·s	Therm. W/(m·K)	Diel. const.
280.0	544.5	19.44	3174.0	7290.0	127.9	23.73	38.73	766.3	47.8	.0781	1.27901
300.0	520.4	18.58	3751.0	8057.0	130.5	23.47	38.02	754.3	45.4	.0747	1.26560
320.0	498.3	17.79	4313.0	8811.0	132.9	23.26	37.38	745.3	43.6	.0720	1.25333
340.0	477.9	17.06	4863.0	9553.0	135.2	23.08	36.80	738.7	42.2	.0698	1.24210
360.0	459.0	16.38	5401.0	10280.0	137.3	22.94	36.29	734.1	41.1	.0681	1.23179
380.0	441.6	15.76	5930.0	11010.0	139.2	22.83	35.84	731.0	40.3	.0668	1.22232
400.0	425.5	15.19	6450.0	11720.0	141.0	22.74	35.44	729.2	39.7	.0658	1.21360
420.0	410.6	14.65	6964.0	12420.0	142.8	22.68	35.08	728.4	39.2	.0651	1.20555
440.0	396.7	14.16	7471.0	13120.0	144.4	22.64	34.77	728.4	38.9	.0646	1.19809
460.0	383.7	13.70	7974.0	13810.0	145.9	22.62	34.50	729.2	38.7	.0643	1.19118
480.0	371.7	13.27	8472.0	14500.0	147.4	22.61	34.27	730.5	38.6	.0642	1.18476
500.0	360.4	12.86	8966.0	15190.0	148.8	22.63	34.07	732.2	38.5	.0642	1.17877
520.0	349.8	12.49	9458.0	15860.0	150.1	22.65	33.90	734.4	38.6	.0643	1.17317
540.0	339.9	12.13	9947.0	16540.0	151.4	22.69	33.76	736.8	38.6	.0645	1.16794
560.0	330.5	11.80	10430.0	17220.0	152.6	22.74	33.65	739.5	38.8	.0648	1.16303
580.0	321.7	11.48	10920.0	17890.0	153.8	22.80	33.55	742.4	38.9	.0652	1.15842
600.0	313.4	11.19	11410.0	18560.0	154.9	22.87	33.48	745.5	39.1	.0656	1.15407
620.0	305.6	10.91	11890.0	19230.0	156.0	22.95	33.42	748.7	39.4	.0661	1.14998
640.0	298.1	10.64	12380.0	19890.0	157.1	23.04	33.38	752.1	39.6	.0667	1.14610
660.0	291.0	10.39	12860.0	20560.0	158.1	23.13	33.36	755.6	39.9	.0673	1.14244
680.0	284.3	10.15	13350.0	21230.0	159.1	23.22	33.35	759.1	40.2	.0679	1.13897
700.0	278.0	9.921	13830.0	21900.0	160.1	23.32	33.35	762.8	40.6	.0686	1.13567
720.0	271.9	9.704	14320.0	22560.0	161.0	23.43	33.36	766.5	40.9	.0693	1.13253
740.0	266.1	9.497	14810.0	23230.0	161.9	23.53	33.37	770.2	41.2	.0701	1.12955
760.0	260.5	9.299	15300.0	23900.0	162.8	23.64	33.40	774.0	41.6	.0708	1.12670
800.0	250.2	8.929	16280.0	25240.0	164.5	23.86	33.47	781.7	42.3	.0724	1.12139
850.0	238.4	8.509	17510.0	26910.0	166.6	24.13	33.58	791.5	43.3	.0745	1.11538
900.0	227.7	8.128	18750.0	28590.0	168.5	24.40	33.71	801.3	44.3	.0767	1.10996
950.0	218.0	7.783	20000.0	30280.0	170.3	24.67	33.85	811.3	45.4	.0789	1.10506
1000.0	209.2	7.467	21260.0	31980.0	172.1	24.92	33.99	821.3	46.4	.0811	1.10059
1050.0	201.1	7.177	22540.0	33680.0	173.7	25.16	34.14	831.3	47.5	.0834	1.09650
1100.0	193.6	6.910	23820.0	35390.0	175.3	25.39	34.29	841.2	48.5	.0856	1.09274
1150.0	186.7	6.663	25100.0	37110.0	176.8	25.61	34.44	851.2	49.6	.0879	1.08927
1200.0	180.3	6.434	26400.0	38840.0	178.3	25.81	34.58	861.1	50.6	.0901	1.08606
1250.0	174.3	6.221	27710.0	40570.0	179.7	26.00	34.71	871.0	51.7	.0924	1.08308
1300.0	168.7	6.022	29020.0	42310.0	181.1	26.18	34.84	880.8	52.8	.0946	1.08030
1350.0	163.5	5.836	30340.0	44050.0	182.4	26.35	34.96	890.6	53.8	.0968	1.07771
1400.0	158.6	5.661	31670.0	45800.0	183.7	26.50	35.08	900.3	54.9	.0990	1.07528
1450.0	154.0	5.497	33010.0	47560.0	184.9	26.65	35.19	910.0	56.0	.101	1.07301
1500.0	149.7	5.343	34350.0	49320.0	186.1	26.79	35.30	919.5	57.0	.103	1.07086
1550.0	145.6	5.197	35700.0	51090.0	187.3	26.92	35.40	929.1	58.0	.105	1.06884
1600.0	141.7	5.059	37050.0	52860.0	188.4	27.04	35.49	938.5	59.1	.108	1.06694
1650.0	138.1	4.928	38410.0	54640.0	189.5	27.15	35.58	947.9	60.1	.110	1.06513
1700.0	134.6	4.805	39770.0	56420.0	190.6	27.26	35.67	957.2	61.1	.112	1.06343
1750.0	131.3	4.687	41140.0	58210.0	191.6	27.36	35.75	966.5	62.2	.114	1.06181
1800.0	128.2	4.575	42510.0	60000.0	192.6	27.45	35.82	975.7	63.2	.116	1.06027
1850.0	125.2	4.468	43890.0	61790.0	193.6	27.54	35.89	984.8	64.2	.118	1.05880
1900.0	122.3	4.367	45260.0	63580.0	194.5	27.62	35.96	993.8	65.2	.120	1.05741

100.00 MPa isobar

T K	Density kg/m^3	Density mol/dm^3	*E* J/mol	*H* J/mol	*S* J/(mol·K)	C_v J/(mol·K)	C_p J/(mol·K)	Sound m/s	Visc. μPa·s	Therm. W/(m·K)	Diel. const.
82.76[a]	919.6	32.83	−3731.0	−684.7	73.36	33.78	41.68	1282.0	274.0	.186	1.49544
100.0	889.6	31.75	−3096.0	52.8	81.45	30.39	44.04	1221.0	197.0	.178	1.47770
150.0	790.3	28.21	−1238.0	2306.0	99.72	27.72	44.64	1067.0	108.0	.142	1.41930
170.0	753.8	26.91	−527.4	3189.0	105.2	26.99	43.61	1015.0	92.3	.130	1.39798
180.0	736.5	26.29	−181.4	3623.0	107.7	26.64	43.07	992.1	86.2	.125	1.38790
200.0	703.6	25.11	491.4	4473.0	112.2	26.01	42.00	952.4	76.4	.115	1.36885
220.0	673.0	24.02	1140.0	5303.0	116.2	25.45	40.99	919.7	69.0	.107	1.35123
240.0	644.6	23.01	1767.0	6113.0	119.7	24.97	40.07	893.2	63.3	.0998	1.33495
260.0	618.2	22.06	2374.0	6907.0	122.9	24.57	39.25	871.7	58.8	.0941	1.31991
280.0	593.6	21.19	2965.0	7684.0	125.7	24.22	38.51	854.4	55.2	.0893	1.30602
300.0	570.8	20.38	3539.0	8447.0	128.4	23.93	37.85	840.5	52.4	.0854	1.29318

Thermophysical properties of nitrogen—Continued

T K	Density kg/m³	Density mol/dm³	E J/mol	H J/mol	S J/(mol·K)	C_v J/(mol·K)	C_p J/(mol·K)	Sound m/s	Visc. μPa·s	Therm. W/(m·K)	Diel. const.
320.0	549.6	19.62	4101.0	9198.0	130.8	23.69	37.27	829.4	50.1	.0821	1.28130
340.0	529.9	18.91	4651.0	9938.0	133.0	23.48	36.75	820.7	48.2	.0794	1.27030
360.0	511.5	18.26	5191.0	10670.0	135.1	23.32	36.28	813.9	46.8	.0772	1.26009
380.0	494.3	17.64	5722.0	11390.0	137.1	23.18	35.87	808.7	45.6	.0754	1.25061
400.0	478.2	17.07	6245.0	12100.0	138.9	23.07	35.51	804.9	44.6	.0739	1.24179
420.0	463.2	16.53	6762.0	12810.0	140.6	22.99	35.19	802.1	43.8	.0728	1.23357
440.0	449.1	16.03	7273.0	13510.0	142.3	22.93	34.91	800.3	43.2	.0719	1.22590
460.0	435.9	15.56	7780.0	14210.0	143.8	22.89	34.66	799.3	42.8	.0712	1.21872
480.0	423.4	15.11	8282.0	14900.0	145.3	22.88	34.45	798.9	42.4	.0708	1.21200
500.0	411.7	14.70	8781.0	15590.0	146.7	22.87	34.27	799.1	42.2	.0704	1.20569
520.0	400.6	14.30	9277.0	16270.0	148.0	22.89	34.11	799.8	42.0	.0703	1.19976
540.0	390.2	13.93	9771.0	16950.0	149.3	22.92	33.98	800.8	42.0	.0702	1.19417
560.0	380.3	13.58	10260.0	17630.0	150.5	22.96	33.88	802.2	41.9	.0703	1.18890
580.0	371.0	13.24	10750.0	18310.0	151.7	23.01	33.79	803.8	42.0	.0705	1.18393
600.0	362.1	12.92	11240.0	18980.0	152.9	23.07	33.72	805.7	42.0	.0707	1.17922
620.0	353.7	12.62	11730.0	19650.0	154.0	23.14	33.67	807.9	42.1	.0711	1.17476
640.0	345.6	12.34	12220.0	20330.0	155.1	23.22	33.63	810.2	42.3	.0714	1.17053
660.0	338.0	12.06	12710.0	21000.0	156.1	23.31	33.61	812.6	42.5	.0719	1.16652
680.0	330.7	11.80	13200.0	21670.0	157.1	23.40	33.60	815.3	42.7	.0724	1.16269
700.0	323.8	11.56	13690.0	22340.0	158.1	23.49	33.59	818.0	42.9	.0729	1.15905
720.0	317.1	11.32	14180.0	23020.0	159.0	23.59	33.60	820.8	43.1	.0735	1.15558
750.0	307.7	10.98	14920.0	24020.0	160.4	23.74	33.63	825.3	43.6	.0745	1.15066
800.0	293.2	10.47	16150.0	25710.0	162.6	24.00	33.70	833.0	44.3	.0762	1.14316
850.0	280.2	10.00	17400.0	27390.0	164.6	24.27	33.80	841.2	45.1	.0780	1.13640
900.0	268.3	9.577	18650.0	29090.0	166.5	24.53	33.92	849.6	46.0	.0800	1.13029
950.0	257.5	9.190	19910.0	30790.0	168.4	24.79	34.05	858.2	46.9	.0821	1.12473
1000.0	247.5	8.835	21170.0	32490.0	170.1	25.03	34.19	867.0	47.9	.0841	1.11964
1050.0	238.4	8.508	22450.0	34210.0	171.8	25.27	34.33	875.9	48.8	.0863	1.11497
1100.0	229.9	8.206	23740.0	35930.0	173.4	25.49	34.47	884.9	49.8	.0884	1.11066
1150.0	222.0	7.925	25030.0	37650.0	174.9	25.70	34.60	893.9	50.8	.0906	1.10668
1200.0	214.7	7.664	26340.0	39390.0	176.4	25.90	34.73	902.9	51.8	.0928	1.10299
1250.0	207.9	7.421	27650.0	41130.0	177.8	26.09	34.86	912.0	52.8	.0949	1.09955
1300.0	201.5	7.193	28970.0	42870.0	179.2	26.26	34.98	921.1	53.8	.0971	1.09633
1350.0	195.5	6.980	30300.0	44620.0	180.5	26.43	35.09	930.1	54.8	.0992	1.09333
1400.0	189.9	6.779	31630.0	46380.0	181.8	26.58	35.20	939.1	55.9	.101	1.09051
1450.0	184.6	6.590	32970.0	48140.0	183.0	26.73	35.31	948.1	56.9	.103	1.08786
1500.0	179.6	6.411	34310.0	49910.0	184.2	26.86	35.41	957.1	57.9	.106	1.08536
1550.0	174.9	6.243	35670.0	51680.0	185.4	26.99	35.50	966.0	58.9	.108	1.08300
1600.0	170.4	6.083	37020.0	53460.0	186.5	27.11	35.59	974.9	59.9	.110	1.08078
1650.0	166.2	5.932	38380.0	55240.0	187.6	27.22	35.67	983.8	60.9	.112	1.07866
1700.0	162.1	5.788	39750.0	57030.0	188.7	27.32	35.75	992.6	61.9	.114	1.07666
1750.0	158.3	5.651	41120.0	58820.0	189.7	27.42	35.83	1001.0	62.9	.116	1.07476
1800.0	154.7	5.520	42500.0	60610.0	190.7	27.51	35.90	1010.0	63.9	.118	1.07295
1850.0	151.2	5.396	43880.0	62410.0	191.7	27.59	35.96	1019.0	64.9	.120	1.07123
1900.0	147.8	5.277	45260.0	64210.0	192.7	27.67	36.03	1027.0	65.8	.122	1.06958
				150.00 MPa isobar							
91.28[a]	942.2	33.63	−3492.0	968.6	75.08	30.64	37.54	1411.0	281.0	.203	1.50670
200.0	775.4	27.68	199.4	5619.0	108.5	27.34	41.66	1136.0	99.9	.143	1.40884
250.0	713.4	25.46	1752.0	7643.0	117.5	25.95	39.39	1065.0	80.3	.123	1.37288
300.0	660.4	23.57	3204.0	9567.0	124.6	24.94	37.69	1019.0	68.8	.110	1.34247
350.0	614.7	21.94	4583.0	11420.0	130.3	24.25	36.46	988.3	61.5	.100	1.31649
400.0	574.9	20.52	5910.0	13220.0	135.1	23.80	35.57	968.7	56.8	.0935	1.29408
450.0	540.1	19.28	7200.0	14980.0	139.2	23.54	34.94	956.2	53.6	.0890	1.27462
500.0	509.3	18.18	8465.0	16720.0	142.9	23.43	34.51	948.8	51.6	.0862	1.25757
550.0	481.9	17.20	9714.0	18430.0	146.2	23.43	34.22	944.8	50.3	.0845	1.24256
600.0	457.5	16.33	10960.0	20140.0	149.1	23.52	34.06	943.4	49.5	.0837	1.22924
650.0	435.6	15.55	12190.0	21840.0	151.8	23.67	33.99	943.9	49.1	.0835	1.21735
700.0	415.8	14.84	13430.0	23540.0	154.4	23.86	33.98	945.9	49.0	.0839	1.20669
750.0	397.8	14.20	14680.0	25240.0	156.7	24.08	34.02	949.0	49.2	.0846	1.19706
800.0	381.4	13.62	15930.0	26940.0	158.9	24.32	34.10	953.0	49.5	.0857	1.18834
900.0	352.7	12.59	18450.0	30360.0	162.9	24.81	34.31	963.1	50.4	.0884	1.17312

Thermophysical properties of nitrogen—Continued

T K	Density kg/m^3	Density mol/dm^3	E J/mol	H J/mol	S J/(mol·K)	C_v J/(mol·K)	C_p J/(mol·K)	Sound m/s	Visc. μPa·s	Therm. W/(m·K)	Diel. const.
1000.0	328.2	11.72	21000.0	33810.0	166.6	25.28	34.56	975.1	51.7	.0917	1.16027
1100.0	307.1	10.96	23590.0	37280.0	169.9	25.72	34.81	988.4	53.2	.0954	1.14928
1200.0	288.7	10.31	26210.0	40770.0	172.9	26.11	35.05	1002.0	54.9	.0992	1.13976
1300.0	272.6	9.729	28870.0	44280.0	175.7	26.45	35.27	1017.0	56.6	.103	1.13143
1400.0	258.2	9.216	31540.0	47820.0	178.3	26.76	35.47	1032.0	58.4	.107	1.12406
1500.0	245.3	8.757	34250.0	51380.0	180.8	27.02	35.64	1047.0	60.2	.111	1.11750
1600.0	233.8	8.344	36970.0	54950.0	183.1	27.26	35.80	1063.0	62.0	.115	1.11161
1700.0	223.3	7.969	39710.0	58540.0	185.3	27.46	35.94	1078.0	63.9	.119	1.10630
1800.0	213.7	7.628	42470.0	62140.0	187.3	27.64	36.07	1093.0	65.7	.123	1.10147
1900.0	205.0	7.316	45250.0	65750.0	189.3	27.80	36.18	1109.0	67.6	.127	1.09707

200.00 MPa isobar

T K	Density kg/m^3	Density mol/dm^3	E J/mol	H J/mol	S J/(mol·K)	C_v J/(mol·K)	C_p J/(mol·K)	Sound m/s	Visc. μPa·s	Therm. W/(m·K)	Diel. const.
99.21[a]	961.6	34.32	−3265.0	2562.0	76.37	29.03	35.37	1525.0	289.0	.219	1.51599
120.0	935.5	33.39	−2622.0	3367.0	83.72	30.01	41.49	1458.0	221.0	.211	1.50067
140.0	906.8	32.37	−1955.0	4224.0	90.32	30.16	43.67	1401.0	181.0	.200	1.48383
200.0	826.7	29.51	32.8	6811.0	105.7	28.49	41.79	1277.0	122.0	.167	1.43693
250.0	771.0	27.52	1572.0	8840.0	114.8	26.97	39.47	1209.0	98.4	.146	1.40450
300.0	722.9	25.80	3017.0	10770.0	121.8	25.82	37.74	1162.0	84.3	.131	1.37674
350.0	680.9	24.30	4393.0	12620.0	127.5	24.99	36.51	1129.0	75.1	.120	1.35260
400.0	643.6	22.97	5719.0	14420.0	132.3	24.44	35.64	1105.0	68.7	.112	1.33137
450.0	610.4	21.79	7010.0	16190.0	136.5	24.09	35.03	1089.0	64.2	.106	1.31256
500.0	580.5	20.72	8279.0	17930.0	140.2	23.91	34.63	1078.0	61.0	.102	1.29577
550.0	553.5	19.76	9533.0	19660.0	143.5	23.86	34.38	1070.0	58.8	.0988	1.28072
600.0	529.1	18.88	10780.0	21370.0	146.4	23.90	34.24	1065.0	57.2	.0968	1.26715
700.0	486.4	17.36	13270.0	24790.0	151.7	24.18	34.21	1060.0	55.4	.0951	1.24370
800.0	450.5	16.08	15780.0	28220.0	156.3	24.59	34.35	1062.0	54.9	.0955	1.22416
900.0	419.8	14.99	18320.0	31660.0	160.4	25.05	34.57	1067.0	55.1	.0971	1.20764
1000.0	393.3	14.04	20890.0	35130.0	164.0	25.50	34.82	1074.0	55.9	.0995	1.19349
1100.0	370.2	13.21	23490.0	38630.0	167.3	25.91	35.06	1084.0	56.9	.102	1.18122
1200.0	349.9	12.49	26130.0	42150.0	170.4	26.28	35.29	1095.0	58.2	.106	1.17049
1300.0	331.7	11.84	28790.0	45680.0	173.2	26.62	35.50	1107.0	59.6	.109	1.16100
1400.0	315.5	11.26	31490.0	49240.0	175.9	26.91	35.68	1119.0	61.1	.113	1.15256
1500.0	300.9	10.74	34200.0	52820.0	178.3	27.16	35.85	1132.0	62.7	.116	1.14498
1600.0	287.7	10.27	36940.0	56410.0	180.7	27.39	35.99	1145.0	64.3	.120	1.13815
1700.0	275.6	9.839	39690.0	60020.0	182.8	27.58	36.12	1159.0	66.0	.124	1.13194
1800.0	264.6	9.445	42460.0	63640.0	184.9	27.76	36.23	1172.0	67.7	.127	1.12629
1900.0	254.5	9.082	45240.0	67270.0	186.9	27.91	36.33	1186.0	69.4	.131	1.12110

400.00 MPa isobar

T K	Density kg/m^3	Density mol/dm^3	E J/mol	H J/mol	S J/(mol·K)	C_v J/(mol·K)	C_p J/(mol·K)	Sound m/s	Visc. μPa·s	Therm. W/(m·K)	Diel. const.
126.90[a]	1022.0	36.46	−2359.0	8610.0	79.97	29.08	33.62	1843.0	323.0	.275	1.54192
140.0	1012.0	36.11	−1996.0	9081.0	83.49	31.11	37.92	1794.0	292.0	.272	1.53622
160.0	994.0	35.48	−1393.0	9881.0	88.83	32.38	41.61	1735.0	254.0	.265	1.52601
200.0	956.2	34.13	−134.6	11580.0	98.33	31.90	42.70	1657.0	204.0	.246	1.50421
250.0	912.6	32.57	1393.0	13670.0	107.7	30.10	40.67	1598.0	167.0	.224	1.47902
300.0	874.4	31.21	2837.0	15650.0	114.9	28.49	38.65	1558.0	143.0	.206	1.45693
350.0	840.4	30.00	4212.0	17550.0	120.7	27.28	37.16	1527.0	127.0	.191	1.43731
400.0	809.7	28.90	5536.0	19380.0	125.6	26.41	36.12	1503.0	115.0	.179	1.41962
500.0	755.6	26.97	8093.0	22920.0	133.5	25.42	35.00	1467.0	99.4	.161	1.38860
600.0	709.0	25.31	10600.0	26400.0	139.9	25.11	34.63	1442.0	89.3	.149	1.36205
700.0	668.3	23.85	13090.0	29860.0	145.2	25.18	34.65	1425.0	82.8	.141	1.33898
800.0	632.2	22.57	15610.0	33340.0	149.9	25.45	34.86	1414.0	78.6	.136	1.31875
900.0	600.2	21.42	18170.0	36840.0	154.0	25.79	35.14	1407.0	75.8	.134	1.30085
1000.0	571.5	20.40	20750.0	40360.0	157.7	26.16	35.42	1405.0	74.1	.133	1.28493
1100.0	545.6	19.47	23380.0	43920.0	161.1	26.51	35.69	1405.0	73.2	.133	1.27067
1200.0	522.2	18.64	26040.0	47500.0	164.2	26.83	35.93	1407.0	72.9	.133	1.25783
1300.0	500.8	17.88	28730.0	51100.0	167.1	27.12	36.13	1412.0	72.9	.135	1.24621
1400.0	481.3	17.18	31440.0	54730.0	169.8	27.37	36.31	1417.0	73.3	.136	1.23564
1500.0	463.4	16.54	34180.0	58370.0	172.3	27.59	36.46	1424.0	73.9	.139	1.22598
1600.0	446.9	15.95	36940.0	62020.0	174.6	27.79	36.58	1431.0	74.7	.141	1.21712
1700.0	431.6	15.41	39720.0	65680.0	176.9	27.96	36.69	1440.0	75.6	.143	1.20896

Thermophysical properties of nitrogen—Continued

T K	Density kg/m³	Density mol/dm³	E J/mol	H J/mol	S J/(mol·K)	C_v J/(mol·K)	C_p J/(mol·K)	Sound m/s	Visc. μPa·s	Therm. W/(m·K)	Diel. const.
1800.0	417.4	14.90	42510.0	69360.0	179.0	28.12	36.78	1448.0	76.7	.146	1.20142
1900.0	404.3	14.43	45320.0	73040.0	180.9	28.25	36.85	1457.0	77.8	.149	1.19442
				600.00 MPa isobar							
150.40[a]	1067.0	38.10	−1385.0	14360.0	82.79	30.33	33.42	2062.0	362.0	.324	1.55849
160.0	1063.0	37.93	−1124.0	14700.0	84.94	31.79	36.07	2026.0	344.0	.323	1.55574
180.0	1050.0	37.49	−545.1	15460.0	89.42	33.31	39.74	1968.0	311.0	.319	1.54887
200.0	1037.0	37.00	57.1	16270.0	93.71	33.58	41.47	1925.0	284.0	.311	1.54109
300.0	967.4	34.53	3014.0	20390.0	110.4	30.42	39.68	1821.0	203.0	.271	1.50184
400.0	909.6	32.47	5721.0	24200.0	121.4	27.85	36.82	1775.0	163.0	.239	1.46893
500.0	860.7	30.72	8275.0	27800.0	129.5	26.54	35.42	1743.0	139.0	.216	1.44111
600.0	818.0	29.20	10770.0	31320.0	135.9	26.01	34.93	1719.0	123.0	.199	1.41683
700.0	780.0	27.84	13250.0	34810.0	141.3	25.93	34.92	1699.0	112.0	.187	1.39524
800.0	745.6	26.62	15760.0	38310.0	145.9	26.08	35.13	1684.0	104.0	.179	1.37585
1000.0	686.1	24.49	20890.0	45390.0	153.8	26.65	35.75	1665.0	94.6	.168	1.34241
1200.0	636.1	22.70	26170.0	52600.0	160.4	27.23	36.30	1659.0	89.5	.163	1.31458
1400.0	593.5	21.18	31580.0	59910.0	166.0	27.72	36.72	1660.0	87.2	.162	1.29111
1600.0	556.8	19.87	37090.0	67280.0	171.0	28.09	37.00	1667.0	86.5	.163	1.27103
1800.0	524.8	18.73	42670.0	74700.0	175.3	28.39	37.20	1678.0	86.9	.166	1.25368
				1000.00 MPa isobar							
190.40[a]	1140.0	40.68	702.4	25280.0	87.36	30.27	31.60	2394.0	448.0	.414	1.57789
250.0	1117.0	39.88	2358.0	27440.0	97.15	33.56	39.00	2253.0	376.0	.405	1.56565
300.0	1092.0	38.99	3771.0	29420.0	104.4	32.66	39.92	2203.0	329.0	.388	1.55224
400.0	1043.0	37.23	6473.0	33330.0	115.6	29.92	38.07	2161.0	266.0	.351	1.52555
500.0	999.6	35.68	9021.0	37050.0	123.9	28.18	36.41	2140.0	225.0	.320	1.50176
600.0	961.2	34.31	11500.0	40640.0	130.5	27.33	35.63	2122.0	198.0	.296	1.48061
800.0	895.0	31.95	16440.0	47740.0	140.7	27.03	35.58	2092.0	162.0	.262	1.44402
1000.0	838.8	29.94	21520.0	54910.0	148.7	27.39	36.16	2070.0	141.0	.241	1.41295
1200.0	790.2	28.20	26750.0	62210.0	155.4	27.84	36.77	2057.0	128.0	.227	1.38608
1400.0	747.4	26.68	32130.0	69610.0	161.1	28.23	37.24	2050.0	119.0	.218	1.36258
1600.0	709.6	25.33	37610.0	77090.0	166.1	28.55	37.57	2050.0	114.0	.213	1.34187
1800.0	675.9	24.13	43180.0	84630.0	170.5	28.79	37.80	2053.0	111.0	.210	1.32350

[a]At melting line.
[b]At liquid–vapor boundary.

Appendix J. Thermophysical Properties of Nitrogen Trifluoride

Thermophysical properties of coexisting gaseous and liquid nitrogen trifluoride

T K	Pres. MPa	Density kg/m^3	Density mol/dm^3	E J/mol	H J/mol	S J/(mol·K)	C_v J/(mol·K)	C_p J/(mol·K)	Sound m/s
66.360[a]	.00000	1869.0	26.32	−12350.0	−12350.0	92.59	47.45	75.36	1507.0
66.360[a]	.00000	.00002	.0000013	1633.0	5162.0	311.7	24.97	33.29	101.8
70.0	.00000	1855.0	26.12	−12090.0	−12090.0	96.51	45.76	71.35	1359.0
70.0	.00000	.00408	.0000585	1724.0	1754.0	270.3	25.01	33.33	104.5
75.0	.00000	1835.0	25.84	−11730.0	−11730.0	101.4	45.64	69.95	1235.0
75.0	.00001	.00574	.0000818	1849.0	1908.0	269.2	25.07	33.39	108.1
80.0	.00002	1816.0	25.56	−11380.0	−11380.0	105.9	45.99	70.16	1163.0
80.0	.00002	.00429	.0000614	1975.0	2247.0	273.2	25.15	33.46	111.6
85.0	.00005	1796.0	25.28	−11030.0	−11030.0	110.2	46.08	70.63	1116.0
85.0	.00005	.00274	.0000396	2101.0	3500.0	278.5	25.25	33.56	115.0
90.0	.00016	1775.0	25.00	−10680.0	−10680.0	114.2	45.81	70.98	1079.0
90.0	.00016	.00636	.0000905	2227.0	3982.0	273.0	25.38	33.70	118.3
95.0	.00041	1755.0	24.71	−10320.0	−10320.0	118.1	45.26	71.15	1046.0
95.0	.00041	.02354	.0003324	2354.0	3579.0	263.4	25.55	33.88	121.4
100.0	.00095	1734.0	24.42	−9966.0	−9966.0	121.7	44.56	71.16	1015.0
100.0	.00095	.06661	.0009390	2481.0	3489.0	256.1	25.75	34.10	124.4
105.0	.00201	1713.0	24.12	−9611.0	−9611.0	125.2	43.82	71.10	983.7
105.0	.00202	.1527	.002151	2608.0	3545.0	250.5	26.00	34.39	127.3
110.0	.00397	1692.0	23.82	−9256.0	−9256.0	128.5	43.08	71.00	953.2
110.0	.00397	.3045	.004289	2735.0	3661.0	245.9	26.30	34.74	130.0
115.0	.00732	1671.0	23.52	−8903.0	−8903.0	131.6	42.41	70.93	923.2
115.0	.00733	.5511	.007761	2862.0	3806.0	242.1	26.64	35.17	132.6
120.0	.01275	1649.0	23.22	−8550.0	−8549.0	134.6	41.83	70.90	893.7
120.0	.01275	.9284	.01307	2988.0	3964.0	238.9	27.03	35.67	135.0
125.0	.02113	1627.0	22.90	−8197.0	−8196.0	137.5	41.34	70.95	864.8
125.0	.02113	1.480	.02084	3114.0	4127.0	236.1	27.48	36.27	137.3
130.0	.03350	1604.0	22.59	−7843.0	−7841.0	140.3	40.94	71.10	836.5
130.0	.03350	2.258	.03180	3238.0	4291.0	233.6	27.97	36.95	139.3
135.0	.05111	1581.0	22.27	−7488.0	−7485.0	143.0	40.63	71.35	808.6
135.0	.05111	3.323	.04679	3360.0	4452.0	231.4	28.52	37.74	141.2
140.0	.07538	1558.0	21.94	−7131.0	−7127.0	145.6	40.41	71.74	781.1
140.0	.07538	4.743	.06679	3479.0	4608.0	229.4	29.12	38.65	142.9
145.0	.1079	1534.0	21.60	−6771.0	−6766.0	148.1	40.26	72.28	753.8
145.0	.1079	6.598	.09290	3596.0	4758.0	227.5	29.77	39.68	144.4
150.0	.1504	1510.0	21.26	−6408.0	−6401.0	150.5	40.18	72.97	726.4
150.0	.1504	8.974	.1264	3710.0	4900.0	225.9	30.48	40.85	145.6
155.0	.2048	1485.0	20.91	−6041.0	−6031.0	153.0	40.16	73.83	698.9
155.0	.2048	11.97	.1685	3820.0	5035.0	224.3	31.23	42.18	146.6
160.0	.2730	1459.0	20.55	−5670.0	−5656.0	155.3	40.19	74.89	670.9
160.0	.2730	15.68	.2208	3924.0	5161.0	222.9	32.05	43.68	147.3
165.0	.3570	1433.0	20.18	−5293.0	−5275.0	157.6	40.28	76.16	642.3
165.0	.3570	20.24	.2850	4024.0	5277.0	221.6	32.91	45.38	147.8
170.0	.4591	1406.0	19.80	−4911.0	−4887.0	159.9	40.42	77.66	613.2
170.0	.4591	25.76	.3627	4118.0	5384.0	220.3	33.83	47.31	148.0
175.0	.5813	1378.0	19.40	−4522.0	−4492.0	162.2	40.62	79.43	583.2
175.0	.5813	32.39	.4560	4205.0	5480.0	219.1	34.80	49.51	147.8
180.0	.7261	1349.0	18.99	−4125.0	−4087.0	164.4	40.87	81.50	552.4
180.0	.7261	40.28	.5672	4285.0	5565.0	218.0	35.82	52.04	147.5
185.0	.8956	1318.0	18.56	−3721.0	−3673.0	166.6	41.17	83.94	520.7
185.0	.8956	49.64	.6990	4356.0	5637.0	216.9	36.90	54.96	146.8

Thermophysical properties of coexisting gaseous and liquid nitrogen trifluoride—Continued

T K	Pres. MPa	Density kg/m³	Density mol/dm³	E J/mol	H J/mol	S J/(mol·K)	C_v J/(mol·K)	C_p J/(mol·K)	Sound m/s
190.0	1.092	1287.0	18.11	−3307.0	−3247.0	168.8	41.54	86.83	488.0
190.0	1.092	60.71	.8548	4418.0	5695.0	215.9	38.03	58.40	145.8
195.0	1.319	1253.0	17.64	−2882.0	−2807.0	171.1	41.99	90.30	454.3
195.0	1.319	73.80	1.039	4468.0	5737.0	214.9	39.21	62.54	144.4
200.0	1.578	1217.0	17.13	−2445.0	−2353.0	173.3	42.52	94.56	419.5
200.0	1.578	89.37	1.258	4504.0	5758.0	213.8	40.47	67.66	142.7
205.0	1.871	1178.0	16.59	−1992.0	−1879.0	175.5	43.17	99.97	383.5
205.0	1.871	108.1	1.522	4521.0	5751.0	212.8	41.82	74.29	140.6
210.0	2.203	1135.0	15.98	−1520.0	−1382.0	177.8	43.96	107.2	346.1
210.0	2.203	130.8	1.842	4514.0	5709.0	211.6	43.27	83.42	138.0
215.0	2.577	1087.0	15.31	−1021.0	−852.9	180.2	44.94	117.8	307.0
215.0	2.577	159.2	2.241	4472.0	5622.0	210.3	44.87	97.11	134.8
220.0	2.995	1032.0	14.53	−485.1	−278.9	182.7	46.19	135.2	265.5
220.0	2.995	195.5	2.752	4382.0	5470.0	208.9	46.68	120.4	130.9
225.0	3.464	962.5	13.55	113.5	369.1	185.5	47.89	171.7	220.9
225.0	3.464	245.3	3.454	4207.0	5210.0	207.0	48.83	171.2	126.3
230.0	3.991	863.2	12.15	851.2	1180.0	188.9	50.49	304.7	171.3
230.0	3.991	323.6	4.557	3855.0	4731.0	204.3	51.60	361.6	121.2
234.00[b]	4.461	562.5	7.920	2543.0	3106.0	197.0			

[a]Triple point.
[b]Critical point.

Thermophysical properties of nitrogen trifluoride on the melting line

T K	Pres. MPa	Density kg/m³	Density mol/dm³	E J/mol	H J/mol	S J/(mol·K)	C_v J/(mol·K)	C_p J/(mol·K)	Sound m/s
66.36[a]	.00000	1869.0	26.32	−12350.0	−12350.0	92.59	47.45	75.36	1507.0
66.6	1.333	1857.0	26.14	−12230.0	−12180.0	94.37	30.41	75.64	1990.0
66.8	2.403	1857.0	26.14	−12230.0	−12140.0	94.44	32.12	75.25	1911.0
67.0	3.475	1857.0	26.15	−12220.0	−12090.0	94.52	33.76	74.88	1840.0
67.2	4.550	1857.0	26.15	−12220.0	−12040.0	94.60	35.35	74.53	1775.0
67.4	5.628	1857.0	26.15	−12210.0	−12000.0	94.68	36.88	74.20	1715.0
67.6	6.708	1857.0	26.15	−12210.0	−11950.0	94.76	38.34	73.88	1661.0
67.8	7.791	1858.0	26.16	−12200.0	−11900.0	94.85	39.75	73.59	1610.0
68.0	8.877	1858.0	26.16	−12200.0	−11860.0	94.93	41.09	73.30	1564.0
68.2	9.966	1858.0	26.17	−12190.0	−11810.0	95.02	42.36	73.03	1521.0
68.4	11.06	1859.0	26.17	−12180.0	−11760.0	95.11	43.58	72.77	1480.0
68.6	12.15	1859.0	26.18	−12180.0	−11710.0	95.20	44.72	72.52	1443.0
68.8	13.25	1859.0	26.18	−12170.0	−11670.0	95.29	45.80	72.27	1407.0
69.0	14.35	1860.0	26.19	−12170.0	−11620.0	95.39	46.81	72.03	1374.0
69.2	15.45	1860.0	26.19	−12160.0	−11570.0	95.48	47.76	71.80	1343.0
69.4	16.55	1861.0	26.20	−12150.0	−11520.0	95.58	48.64	71.57	1313.0
69.6	17.66	1861.0	26.21	−12150.0	−11470.0	95.68	49.45	71.34	1286.0
69.8	18.77	1862.0	26.21	−12140.0	−11420.0	95.78	50.20	71.11	1259.0
70.0	19.89	1862.0	26.22	−12130.0	−11370.0	95.88	50.88	70.89	1234.0
70.2	21.00	1863.0	26.23	−12120.0	−11320.0	95.98	51.49	70.66	1210.0
70.4	22.12	1863.0	26.24	−12120.0	−11270.0	96.08	52.04	70.43	1187.0
70.6	23.24	1864.0	26.25	−12110.0	−11220.0	96.19	52.52	70.20	1165.0
70.8	24.36	1865.0	26.26	−12100.0	−11170.0	96.29	52.94	69.96	1143.0
71.0	25.49	1865.0	26.27	−12090.0	−11120.0	96.40	53.29	69.73	1123.0
71.2	26.62	1866.0	26.27	−12090.0	−11070.0	96.50	53.58	69.49	1103.0
71.4	27.75	1867.0	26.28	−12080.0	−11020.0	96.61	53.81	69.25	1084.0
71.6	28.89	1867.0	26.30	−12070.0	−10970.0	96.71	53.97	69.01	1066.0
71.8	30.02	1868.0	26.31	−12060.0	−10920.0	96.82	54.08	68.76	1048.0
72.0	31.16	1869.0	26.32	−12050.0	−10870.0	96.92	54.12	68.52	1030.0
72.2	32.31	1870.0	26.33	−12050.0	−10820.0	97.03	54.11	68.29	1013.0
72.4	33.45	1871.0	26.34	−12040.0	−10770.0	97.13	54.04	68.06	996.4
72.6	34.60	1871.0	26.35	−12030.0	−10720.0	97.23	53.92	67.84	980.0
72.8	35.75	1872.0	26.36	−12020.0	−10670.0	97.33	53.74	67.64	963.9
73.0	36.90	1873.0	26.38	−12020.0	−10620.0	97.43	53.51	67.46	948.2
73.2	38.06	1874.0	26.39	−12010.0	−10570.0	97.53	53.23	67.30	932.6
73.4	39.22	1875.0	26.40	−12000.0	−10510.0	97.63	52.90	67.18	917.4

[a]Triple point.

Thermophysical properties of nitrogen trifluoride

T K	Density kg/m³	Density mol/dm³	E J/mol	H J/mol	S J/(mol·K)	C_v J/(mol·K)	C_p J/(mol·K)	Sound m/s
				0.02 MPa isobar				
66.35[a]	1869.0	26.32	−12350.0	−12350.0	92.58	47.40	75.38	1509.0
68.0	1863.0	26.23	−12230.0	−12230.0	94.40	46.48	73.06	1430.0
70.0	1855.0	26.12	−12090.0	−12090.0	96.49	45.92	71.34	1355.0
80.0	1816.0	25.57	−11390.0	−11390.0	105.9	46.08	70.14	1162.0
90.0	1776.0	25.01	−10680.0	−10680.0	114.2	45.79	70.96	1082.0
100.0	1735.0	24.43	−9970.0	−9969.0	121.7	44.51	71.13	1018.0
110.0	1693.0	23.83	−9259.0	−9259.0	128.4	43.05	70.98	955.7
120.0	1649.0	23.22	−8550.0	−8549.0	134.6	41.82	70.90	894.1
124.449[b]	1629.0	22.94	−8236.0	−8235.0	137.2	41.39	70.94	868.0
124.449[b]	1.409	.01984	3100.0	4108.0	236.4	27.43	36.20	137.0
125.0	1.383	.01947	3116.0	4143.0	236.7	27.46	36.21	137.3
130.0	1.328	.01869	3255.0	4325.0	238.1	27.78	36.48	139.9
140.0	1.230	.01732	3538.0	4693.0	240.8	28.53	37.14	144.9
150.0	1.146	.01614	3829.0	5068.0	243.4	29.38	37.93	149.6
160.0	1.073	.01511	4128.0	5452.0	245.9	30.31	38.82	154.1
170.0	1.009	.01421	4437.0	5844.0	248.3	31.31	39.78	158.3
180.0	.9525	.01341	4756.0	6247.0	250.6	32.35	40.79	162.4
190.0	.9019	.01270	5085.0	6660.0	252.8	33.42	41.84	166.4
200.0	.8564	.01206	5426.0	7084.0	255.0	34.52	42.93	170.2
210.0	.8153	.01148	5777.0	7519.0	257.1	35.63	44.02	173.9
220.0	.7780	.01096	6139.0	7965.0	259.2	36.75	45.13	177.5
230.0	.7440	.01048	6512.0	8422.0	261.2	37.86	46.24	181.0
240.0	.7128	.01004	6897.0	8889.0	263.2	38.97	47.34	184.5
250.0	.6842	.009635	7292.0	9368.0	265.1	40.06	48.43	187.8
260.0	.6578	.009263	7699.0	9858.0	267.1	41.14	49.50	191.2
270.0	.6333	.008919	8116.0	10360.0	268.9	42.21	50.56	194.4
280.0	.6107	.008600	8543.0	10870.0	270.8	43.24	51.59	197.6
300.0	.5698	.008025	9428.0	11920.0	274.4	45.25	53.59	203.8
320.0	.5342	.007522	10350.0	13010.0	277.9	47.14	55.48	209.9
340.0	.5027	.007079	11310.0	14140.0	281.4	48.92	57.26	215.7
360.0	.4747	.006685	12310.0	15300.0	284.7	50.58	58.91	221.5
380.0	.4497	.006333	13340.0	16500.0	287.9	52.12	60.45	227.1
400.0	.4272	.006016	14390.0	17720.0	291.1	53.55	61.88	232.6
420.0	.4068	.005729	15480.0	18970.0	294.1	54.87	63.20	237.9
440.0	.3883	.005469	16590.0	20250.0	297.1	56.09	64.42	243.2
460.0	.3714	.005231	17720.0	21550.0	300.0	57.22	65.54	248.3
480.0	.3559	.005013	18880.0	22870.0	302.8	58.27	66.59	253.4
500.0	.3417	.004812	20050.0	24210.0	305.5	59.25	67.57	258.4
				.04 MPa isobar				
66.36[a]	1869.0	26.32	−12350.0	−12350.0	92.59	47.41	75.37	1508.0
68.0	1863.0	26.23	−12230.0	−12230.0	94.40	46.49	73.06	1430.0
70.0	1855.0	26.12	−12090.0	−12090.0	96.49	45.94	71.34	1355.0
80.0	1816.0	25.57	−11390.0	−11390.0	105.9	46.08	70.14	1162.0
90.0	1776.0	25.01	−10680.0	−10680.0	114.2	45.79	70.95	1082.0
100.0	1735.0	24.43	−9970.0	−9969.0	121.7	44.51	71.13	1018.0
110.0	1693.0	23.83	−9260.0	−9258.0	128.4	43.04	70.98	955.9
120.0	1649.0	23.22	−8551.0	−8549.0	134.6	41.82	70.89	894.3
130.0	1604.0	22.58	−7841.0	−7839.0	140.3	40.96	71.11	835.5
132.063[b]	1595.0	22.46	−7696.0	−7695.0	141.4	40.80	71.19	824.9
132.063[b]	2.659	.03744	3288.0	4357.0	232.6	28.19	37.27	140.1
135.0	2.577	.03629	3373.0	4475.0	233.5	28.38	37.39	141.7
140.0	2.480	.03492	3517.0	4663.0	234.9	28.74	37.66	144.2
150.0	2.307	.03249	3811.0	5042.0	237.5	29.54	38.33	149.0
160.0	2.158	.03039	4113.0	5430.0	240.0	30.44	39.13	153.6
170.0	2.027	.02855	4424.0	5825.0	242.4	31.40	40.03	157.9
180.0	1.912	.02692	4745.0	6230.0	244.7	32.42	41.00	162.1
190.0	1.809	.02548	5075.0	6645.0	247.0	33.48	42.02	166.1
200.0	1.717	.02418	5417.0	7071.0	249.2	34.57	43.07	169.9

Thermophysical properties of nitrogen trifluoride—Continued

T K	Density kg/m^3	Density mol/dm^3	E J/mol	H J/mol	S J/(mol·K)	C_v J/(mol·K)	C_p J/(mol·K)	Sound m/s
210.0	1.634	.02301	5769.0	7507.0	251.3	35.67	44.14	173.7
220.0	1.559	.02195	6132.0	7954.0	253.4	36.78	45.23	177.3
230.0	1.490	.02099	6506.0	8412.0	255.4	37.89	46.32	180.8
240.0	1.428	.02010	6891.0	8880.0	257.4	38.99	47.41	184.3
250.0	1.370	.01929	7286.0	9360.0	259.3	40.08	48.49	187.7
260.0	1.317	.01855	7693.0	9850.0	261.3	41.16	49.56	191.0
270.0	1.268	.01785	8110.0	10350.0	263.2	42.22	50.61	194.3
280.0	1.222	.01721	8538.0	10860.0	265.0	43.26	51.64	197.5
300.0	1.140	.01606	9424.0	11910.0	268.7	45.26	53.63	203.7
320.0	1.069	.01505	10350.0	13010.0	272.2	47.15	55.51	209.8
340.0	1.006	.01416	11310.0	14130.0	275.6	48.93	57.28	215.7
360.0	.9498	.01337	12310.0	15300.0	278.9	50.58	58.93	221.4
380.0	.8996	.01267	13330.0	16490.0	282.1	52.13	60.47	227.1
400.0	.8546	.01203	14390.0	17720.0	285.3	53.55	61.89	232.5
420.0	.8138	.01146	15480.0	18970.0	288.3	54.87	63.21	237.9
440.0	.7767	.01094	16590.0	20240.0	291.3	56.09	64.43	243.2
460.0	.7429	.01046	17720.0	21540.0	294.2	57.22	65.56	248.3
480.0	.7119	.01003	18870.0	22860.0	297.0	58.27	66.60	253.4
500.0	.6834	.009624	20050.0	24210.0	299.7	59.25	67.58	258.4
				.06 MPa isobar				
66.36[a]	1869.0	26.32	−12350.0	−12350.0	92.59	47.43	75.36	1508.0
68.0	1863.0	26.23	−12230.0	−12230.0	94.40	46.51	73.05	1430.0
70.0	1855.0	26.12	−12090.0	−12090.0	96.49	45.95	71.34	1354.0
80.0	1816.0	25.57	−11390.0	−11380.0	105.9	46.09	70.13	1162.0
90.0	1776.0	25.01	−10680.0	−10680.0	114.2	45.79	70.95	1082.0
100.0	1735.0	24.43	−9970.0	−9968.0	121.7	44.51	71.13	1019.0
110.0	1693.0	23.83	−9260.0	−9257.0	128.4	43.04	70.97	956.1
120.0	1649.0	23.22	−8551.0	−8548.0	134.6	41.82	70.89	894.5
130.0	1604.0	22.58	−7841.0	−7839.0	140.3	40.95	71.11	835.6
137.029[b]	1572.0	22.13	−7343.0	−7340.0	144.0	40.53	71.50	797.5
137.029[b]	3.852	.05424	3409.0	4515.0	230.5	28.76	38.10	141.9
140.0	3.752	.05283	3496.0	4632.0	231.4	28.95	38.21	143.5
150.0	3.484	.04906	3793.0	5016.0	234.0	29.70	38.75	148.4
160.0	3.254	.04582	4098.0	5407.0	236.5	30.56	39.46	153.1
170.0	3.054	.04301	4411.0	5806.0	239.0	31.50	40.29	157.5
180.0	2.878	.04053	4733.0	6213.0	241.3	32.50	41.21	161.7
190.0	2.722	.03833	5065.0	6630.0	243.5	33.55	42.19	165.7
200.0	2.583	.03636	5407.0	7057.0	245.7	34.62	43.21	169.6
210.0	2.457	.03459	5760.0	7495.0	247.9	35.71	44.27	173.4
220.0	2.343	.03299	6124.0	7943.0	250.0	36.81	45.34	177.1
230.0	2.239	.03153	6499.0	8401.0	252.0	37.92	46.41	180.6
240.0	2.145	.03020	6884.0	8871.0	254.0	39.02	47.49	184.1
250.0	2.058	.02898	7281.0	9351.0	256.0	40.10	48.56	187.5
260.0	1.978	.02785	7688.0	9842.0	257.9	41.18	49.62	190.9
270.0	1.904	.02681	8105.0	10340.0	259.8	42.23	50.66	194.1
280.0	1.835	.02584	8533.0	10860.0	261.6	43.27	51.69	197.4
300.0	1.712	.02411	9420.0	11910.0	265.3	45.27	53.67	203.6
320.0	1.604	.02259	10350.0	13000.0	268.8	47.16	55.54	209.7
340.0	1.509	.02125	11310.0	14130.0	272.2	48.93	57.31	215.6
360.0	1.425	.02007	12300.0	15290.0	275.5	50.59	58.96	221.4
380.0	1.350	.01901	13330.0	16490.0	278.8	52.13	60.49	227.0
400.0	1.282	.01805	14390.0	17710.0	281.9	53.55	61.91	232.5
420.0	1.221	.01719	15470.0	18960.0	285.0	54.87	63.22	237.9
440.0	1.165	.01641	16580.0	20240.0	287.9	56.09	64.44	243.2
460.0	1.114	.01569	17720.0	21540.0	290.8	57.22	65.57	248.3
480.0	1.068	.01504	18870.0	22860.0	293.6	58.27	66.61	253.4
500.0	1.025	.01444	20050.0	24200.0	296.4	59.25	67.59	258.4
				.10 MPa isobar				
66.37[a]	1869.0	26.32	−12350.0	−12350.0	92.60	47.46	75.35	1506.0

Thermophysical properties of nitrogen trifluoride—Continued

T K	Density kg/m³	Density mol/dm³	E J/mol	H J/mol	S J/(mol·K)	C_v J/(mol·K)	C_p J/(mol·K)	Sound m/s
68.0	1863.0	26.23	−12230.0	−12230.0	94.40	46.54	73.05	1429.0
70.0	1855.0	26.12	−12090.0	−12090.0	96.49	45.98	71.34	1354.0
80.0	1816.0	25.57	−11390.0	−11380.0	105.9	46.09	70.13	1162.0
90.0	1776.0	25.01	−10680.0	−10680.0	114.2	45.79	70.95	1082.0
100.0	1735.0	24.43	−9971.0	−9967.0	121.7	44.50	71.12	1019.0
110.0	1693.0	23.83	−9260.0	−9256.0	128.4	43.03	70.97	956.5
120.0	1649.0	23.22	−8551.0	−8547.0	134.6	41.81	70.89	894.8
130.0	1604.0	22.59	−7842.0	−7837.0	140.3	40.95	71.10	836.0
140.0	1558.0	21.93	−7128.0	−7124.0	145.6	40.42	71.77	780.0
143.920[b]	1539.0	21.67	−6849.0	−6844.0	147.5	40.28	72.15	759.7
143.920[b]	6.156	.08669	3571.0	4725.0	227.9	29.62	39.45	144.1
145.0	6.114	.08609	3604.0	4766.0	228.2	29.69	39.48	144.6
150.0	5.887	.08289	3757.0	4963.0	229.5	30.03	39.64	147.2
160.0	5.484	.07721	4067.0	5362.0	232.1	30.81	40.14	152.0
170.0	5.136	.07232	4384.0	5767.0	234.6	31.70	40.83	156.6
180.0	4.833	.06805	4710.0	6179.0	236.9	32.66	41.64	160.9
190.0	4.565	.06428	5044.0	6600.0	239.2	33.67	42.54	165.1
200.0	4.327	.06093	5389.0	7030.0	241.4	34.72	43.51	169.1
210.0	4.113	.05792	5744.0	7470.0	243.5	35.80	44.51	172.9
220.0	3.920	.05520	6109.0	7921.0	245.6	36.88	45.55	176.6
230.0	3.745	.05273	6485.0	8381.0	247.7	37.98	46.59	180.2
240.0	3.585	.05048	6872.0	8852.0	249.7	39.06	47.65	183.8
250.0	3.439	.04842	7269.0	9334.0	251.7	40.15	48.70	187.2
260.0	3.304	.04652	7677.0	9826.0	253.6	41.21	49.74	190.6
270.0	3.179	.04477	8095.0	10330.0	255.5	42.26	50.77	193.9
280.0	3.064	.04315	8524.0	10840.0	257.4	43.29	51.78	197.1
300.0	2.857	.04023	9411.0	11900.0	261.0	45.28	53.74	203.4
320.0	2.677	.03769	10340.0	12990.0	264.5	47.17	55.60	209.6
340.0	2.518	.03545	11300.0	14120.0	267.9	48.94	57.36	215.5
360.0	2.377	.03347	12300.0	15280.0	271.3	50.60	59.00	221.3
380.0	2.251	.03170	13320.0	16480.0	274.5	52.13	60.53	226.9
400.0	2.138	.03010	14380.0	17700.0	277.6	53.56	61.94	232.5
420.0	2.036	.02866	15470.0	18960.0	280.7	54.88	63.25	237.9
440.0	1.943	.02735	16580.0	20230.0	283.7	56.09	64.46	243.1
460.0	1.858	.02616	17710.0	21530.0	286.6	57.22	65.59	248.3
480.0	1.780	.02507	18870.0	22860.0	289.4	58.27	66.63	253.4
500.0	1.709	.02406	20040.0	24200.0	292.1	59.25	67.61	258.4

.101325 MPa isobar

T K	Density kg/m³	Density mol/dm³	E J/mol	H J/mol	S J/(mol·K)	C_v J/(mol·K)	C_p J/(mol·K)	Sound m/s
66.37[a]	1869.0	26.32	−12350.0	−12350.0	92.60	47.46	75.35	1506.0
68.0	1863.0	26.23	−12230.0	−12230.0	94.40	46.54	73.05	1429.0
70.0	1855.0	26.12	−12090.0	−12090.0	96.49	45.98	71.34	1354.0
80.0	1816.0	25.57	−11390.0	−11380.0	105.9	46.09	70.13	1162.0
90.0	1776.0	25.01	−10680.0	−10680.0	114.2	45.79	70.95	1082.0
100.0	1735.0	24.43	−9971.0	−9967.0	121.7	44.50	71.12	1019.0
110.0	1693.0	23.83	−9260.0	−9256.0	128.4	43.03	70.97	956.5
120.0	1649.0	23.22	−8551.0	−8547.0	134.6	41.81	70.89	894.9
130.0	1604.0	22.59	−7842.0	−7837.0	140.3	40.95	71.10	836.0
140.0	1558.0	21.93	−7128.0	−7124.0	145.6	40.42	71.77	780.0
144.108[b]	1538.0	21.66	−6835.0	−6831.0	147.6	40.28	72.17	758.7
144.108[b]	6.232	.08775	3576.0	4731.0	227.8	29.65	39.49	144.1
145.0	6.199	.08728	3603.0	4764.0	228.1	29.71	39.51	144.6
150.0	5.967	.08403	3756.0	4962.0	229.4	30.04	39.67	147.1
160.0	5.558	.07826	4066.0	5361.0	232.0	30.82	40.17	152.0
170.0	5.206	.07330	4383.0	5766.0	234.4	31.71	40.85	156.6
180.0	4.898	.06897	4709.0	6178.0	236.8	32.66	41.66	160.9
190.0	4.627	.06515	5044.0	6599.0	239.1	33.68	42.56	165.1
200.0	4.385	.06175	5388.0	7029.0	241.3	34.73	43.52	169.0
210.0	4.168	.05869	5743.0	7470.0	243.4	35.80	44.52	172.9
220.0	3.973	.05594	6108.0	7920.0	245.5	36.89	45.55	176.6
230.0	3.795	.05344	6484.0	8381.0	247.6	37.98	46.60	180.2
240.0	3.633	.05116	6871.0	8852.0	249.6	39.07	47.65	183.8

Thermophysical properties of nitrogen trifluoride—Continued

T K	Density kg/m^3	Density mol/dm^3	E J/mol	H J/mol	S J/(mol·K)	C_v J/(mol·K)	C_p J/(mol·K)	Sound m/s
250.0	3.484	.04907	7268.0	9334.0	251.5	40.15	48.70	187.2
260.0	3.348	.04714	7676.0	9826.0	253.5	41.21	49.74	190.6
270.0	3.222	.04537	8095.0	10330.0	255.4	42.26	50.77	193.9
280.0	3.105	.04372	8524.0	10840.0	257.2	43.29	51.78	197.1
300.0	2.895	.04077	9411.0	11900.0	260.9	45.28	53.74	203.4
320.0	2.712	.03819	10340.0	12990.0	264.4	47.17	55.61	209.6
340.0	2.551	.03593	11300.0	14120.0	267.8	48.94	57.36	215.5
360.0	2.408	.03391	12300.0	15280.0	271.2	50.60	59.00	221.3
380.0	2.281	.03212	13320.0	16480.0	274.4	52.13	60.53	226.9
400.0	2.166	.03050	14380.0	17700.0	277.5	53.56	61.94	232.5
420.0	2.063	.02904	15470.0	18960.0	280.6	54.88	63.25	237.9
440.0	1.968	.02772	16580.0	20230.0	283.6	56.09	64.47	243.1
460.0	1.882	.02651	17710.0	21530.0	286.4	57.22	65.59	248.3
480.0	1.804	.02540	18870.0	22860.0	289.3	58.27	66.63	253.4
500.0	1.731	.02438	20040.0	24200.0	292.0	59.25	67.61	258.4
				.20 MPa isobar				
66.39[a]	1869.0	26.32	-12350.0	-12350.0	92.61	47.53	75.31	1503.0
68.0	1863.0	26.23	-12230.0	-12230.0	94.39	46.62	73.05	1427.0
70.0	1855.0	26.13	-12090.0	-12080.0	96.48	46.04	71.33	1352.0
80.0	1816.0	25.57	-11390.0	-11380.0	105.9	46.11	70.13	1162.0
100.0	1735.0	24.43	-9972.0	-9964.0	121.7	44.49	71.12	1020.0
120.0	1649.0	23.22	-8553.0	-8544.0	134.6	41.80	70.88	895.8
140.0	1558.0	21.94	-7130.0	-7121.0	145.6	40.41	71.75	780.9
154.618[b]	1487.0	20.94	-6069.0	-6060.0	152.8	40.16	73.76	701.0
154.618[b]	11.71	.1649	3811.0	5024.0	224.4	31.17	42.07	146.5
155.0	11.73	.1652	3823.0	5034.0	224.5	31.20	42.08	146.7
165.0	10.89	.1533	4150.0	5455.0	227.1	31.85	42.15	151.9
170.0	10.52	.1481	4315.0	5666.0	228.4	32.23	42.31	154.3
180.0	9.853	.1387	4650.0	6091.0	230.8	33.07	42.81	159.0
190.0	9.277	.1306	4992.0	6523.0	233.1	34.00	43.49	163.4
200.0	8.771	.1235	5342.0	6961.0	235.4	34.99	44.28	167.6
210.0	8.321	.1172	5702.0	7409.0	237.6	36.01	45.16	171.6
220.0	7.918	.1115	6071.0	7865.0	239.7	37.06	46.09	175.5
230.0	7.554	.1064	6450.0	8330.0	241.8	38.12	47.06	179.2
240.0	7.224	.1017	6840.0	8806.0	243.8	39.19	48.05	182.9
250.0	6.922	.09747	7239.0	9291.0	245.8	40.25	49.04	186.4
260.0	6.645	.09357	7649.0	9787.0	247.7	41.30	50.04	189.9
270.0	6.391	.08999	8070.0	10290.0	249.6	42.34	51.03	193.3
280.0	6.156	.08668	8500.0	10810.0	251.5	43.35	52.02	196.6
300.0	5.735	.08075	9390.0	11870.0	255.2	45.33	53.93	203.0
320.0	5.368	.07559	10320.0	12960.0	258.7	47.20	55.76	209.2
340.0	5.047	.07107	11280.0	14100.0	262.1	48.96	57.49	215.2
360.0	4.762	.06706	12280.0	15260.0	265.5	50.61	59.11	221.1
380.0	4.508	.06348	13310.0	16460.0	268.7	52.15	60.62	226.8
400.0	4.281	.06027	14370.0	17690.0	271.8	53.57	62.02	232.3
420.0	4.075	.05738	15450.0	18940.0	274.9	54.88	63.32	237.8
440.0	3.888	.05475	16570.0	20220.0	277.9	56.10	64.52	243.1
460.0	3.718	.05235	17700.0	21520.0	280.8	57.22	65.64	248.3
480.0	3.562	.05015	18860.0	22840.0	283.6	58.27	66.68	253.4
500.0	3.418	.04813	20030.0	24190.0	286.3	59.25	67.65	258.4
				.30 MPa isobar				
66.41[a]	1869.0	26.32	-12350.0	-12340.0	92.62	47.61	75.27	1500.0
68.0	1863.0	26.23	-12230.0	-12220.0	94.38	46.70	73.04	1425.0
70.0	1855.0	26.13	-12090.0	-12080.0	96.47	46.11	71.33	1351.0
80.0	1816.0	25.58	-11390.0	-11380.0	105.8	46.13	70.12	1162.0
100.0	1735.0	24.43	-9973.0	-9961.0	121.6	44.48	71.11	1021.0
120.0	1649.0	23.23	-8554.0	-8541.0	134.6	41.78	70.86	896.7
140.0	1558.0	21.94	-7132.0	-7118.0	145.5	40.40	71.73	781.8

Thermophysical properties of nitrogen trifluoride—Continued

T K	Density kg/m^3	Density mol/dm^3	E J/mol	H J/mol	S J/(mol·K)	C_v J/(mol·K)	C_p J/(mol·K)	Sound m/s
160.0	1460.0	20.55	-5672.0	-5657.0	155.3	40.18	74.87	671.5
161.734[b]	1450.0	20.42	-5540.0	-5525.0	156.1	40.22	75.31	661.0
161.734[b]	17.16	.2417	3960.0	5201.0	222.4	32.34	44.24	147.5
165.0	16.79	.2365	4071.0	5339.0	223.3	32.51	44.12	149.3
170.0	16.17	.2277	4242.0	5559.0	224.6	32.80	44.01	152.0
180.0	15.08	.2124	4587.0	6000.0	227.1	33.51	44.12	157.0
190.0	14.15	.1992	4937.0	6443.0	229.5	34.35	44.52	161.7
200.0	13.34	.1879	5294.0	6891.0	231.8	35.26	45.11	166.1
210.0	12.63	.1778	5659.0	7345.0	234.0	36.23	45.84	170.3
220.0	12.00	.1689	6032.0	7808.0	236.2	37.24	46.66	174.4
230.0	11.43	.1609	6415.0	8279.0	238.2	38.27	47.54	178.2
240.0	10.92	.1537	6807.0	8759.0	240.3	39.31	48.46	182.0
250.0	10.45	.1472	7210.0	9248.0	242.3	40.35	49.40	185.6
260.0	10.03	.1412	7622.0	9747.0	244.2	41.38	50.35	189.2
270.0	9.635	.1357	8044.0	10260.0	246.2	42.41	51.31	192.6
280.0	9.275	.1306	8476.0	10770.0	248.0	43.42	52.26	196.0
290.0	8.942	.1259	8918.0	11300.0	249.9	44.41	53.20	199.3
300.0	8.632	.1215	9369.0	11840.0	251.7	45.37	54.12	202.5
320.0	8.075	.1137	10300.0	12940.0	255.3	47.23	55.91	208.8
340.0	7.587	.1068	11260.0	14070.0	258.7	48.99	57.61	214.9
360.0	7.156	.1008	12260.0	15240.0	262.0	50.63	59.21	220.8
380.0	6.772	.09536	13290.0	16440.0	265.3	52.16	60.71	226.6
400.0	6.428	.09051	14350.0	17670.0	268.4	53.57	62.10	232.2
420.0	6.117	.08614	15440.0	18920.0	271.5	54.89	63.39	237.7
440.0	5.836	.08217	16550.0	20200.0	274.5	56.10	64.59	243.0
460.0	5.579	.07856	17690.0	21510.0	277.4	57.22	65.69	248.3
480.0	5.344	.07525	18840.0	22830.0	280.2	58.27	66.73	253.4
500.0	5.129	.07222	20020.0	24180.0	282.9	59.25	67.70	258.4

.40 MPa isobar

T K	Density kg/m^3	Density mol/dm^3	E J/mol	H J/mol	S J/(mol·K)	C_v J/(mol·K)	C_p J/(mol·K)	Sound m/s
66.43[a]	1869.0	26.32	-12350.0	-12340.0	92.64	47.68	75.24	1497.0
68.0	1863.0	26.24	-12230.0	-12220.0	94.37	46.78	73.04	1423.0
70.0	1856.0	26.13	-12090.0	-12080.0	96.46	46.18	71.32	1349.0
80.0	1816.0	25.58	-11390.0	-11370.0	105.8	46.15	70.12	1162.0
100.0	1735.0	24.44	-9974.0	-9957.0	121.6	44.47	71.10	1021.0
120.0	1650.0	23.23	-8556.0	-8538.0	134.6	41.77	70.85	897.7
140.0	1558.0	21.94	-7134.0	-7116.0	145.5	40.38	71.71	782.7
160.0	1460.0	20.56	-5674.0	-5655.0	155.3	40.17	74.83	672.5
167.233[b]	1421.0	20.01	-5123.0	-5103.0	158.7	40.34	76.80	629.4
167.233[b]	22.58	.3179	4067.0	5325.0	221.0	33.32	46.21	147.9
170.0	22.15	.3119	4165.0	5447.0	221.7	33.44	46.00	149.5
175.0	21.31	.3000	4343.0	5676.0	223.0	33.69	45.72	152.3
180.0	20.54	.2893	4522.0	5904.0	224.3	33.99	45.59	154.9
190.0	19.20	.2703	4881.0	6360.0	226.8	34.71	45.65	159.9
200.0	18.05	.2541	5244.0	6818.0	229.1	35.55	46.01	164.6
210.0	17.04	.2400	5615.0	7281.0	231.4	36.46	46.57	169.0
220.0	16.16	.2276	5993.0	7750.0	233.6	37.43	47.26	173.2
230.0	15.37	.2165	6379.0	8227.0	235.7	38.42	48.04	177.2
240.0	14.67	.2065	6775.0	8711.0	237.8	39.43	48.89	181.1
250.0	14.03	.1975	7180.0	9205.0	239.8	40.45	49.77	184.8
260.0	13.45	.1893	7594.0	9707.0	241.7	41.47	50.67	188.5
270.0	12.91	.1818	8018.0	10220.0	243.7	42.48	51.59	192.0
280.0	12.42	.1749	8452.0	10740.0	245.6	43.48	52.50	195.4
290.0	11.97	.1685	8895.0	11270.0	247.4	44.46	53.41	198.8
300.0	11.55	.1626	9347.0	11810.0	249.3	45.42	54.32	202.1
320.0	10.80	.1520	10280.0	12910.0	252.8	47.27	56.07	208.5
340.0	10.14	.1428	11250.0	14050.0	256.3	49.01	57.74	214.6
360.0	9.559	.1346	12250.0	15220.0	259.6	50.65	59.32	220.6
380.0	9.043	.1273	13280.0	16420.0	262.9	52.17	60.80	226.4
400.0	8.580	.1208	14340.0	17650.0	266.0	53.58	62.18	232.1
420.0	8.164	.1150	15430.0	18910.0	269.1	54.89	63.46	237.6

Thermophysical properties of nitrogen trifluoride—Continued

T K	Density kg/m^3	Density mol/dm^3	E J/mol	H J/mol	S J/(mol·K)	C_v J/(mol·K)	C_p J/(mol·K)	Sound m/s
440.0	7.786	.1096	16540.0	20190.0	272.1	56.10	64.65	243.0
460.0	7.443	.1048	17680.0	21490.0	275.0	57.23	65.75	248.2
480.0	7.128	.1004	18830.0	22820.0	277.8	58.27	66.78	253.4
500.0	6.840	.09631	20010.0	24160.0	280.5	59.25	67.74	258.5
				.50 MPa isobar				
66.44[a]	1869.0	26.32	−12350.0	−12330.0	92.65	47.76	75.20	1494.0
68.0	1863.0	26.24	−12240.0	−12220.0	94.36	46.86	73.04	1421.0
70.0	1856.0	26.13	−12090.0	−12070.0	96.45	46.24	71.32	1347.0
80.0	1817.0	25.58	−11390.0	−11370.0	105.8	46.17	70.11	1162.0
100.0	1736.0	24.44	−9975.0	−9954.0	121.6	44.46	71.09	1022.0
120.0	1650.0	23.23	−8557.0	−8535.0	134.6	41.75	70.84	898.6
140.0	1559.0	21.95	−7136.0	−7113.0	145.5	40.37	71.69	783.6
160.0	1461.0	20.57	−5677.0	−5653.0	155.3	40.16	74.79	673.4
170.0	1407.0	19.81	−4917.0	−4892.0	159.9	40.40	77.55	615.2
171.783[b]	1396.0	19.66	−4773.0	−4747.0	160.7	40.49	78.26	602.6
171.783[b]	27.99	.3941	4150.0	5418.0	219.9	34.17	48.06	147.9
175.0	27.33	.3848	4268.0	5567.0	220.7	34.28	47.69	149.9
180.0	26.27	.3699	4453.0	5805.0	222.1	34.50	47.26	152.7
185.0	25.31	.3564	4637.0	6040.0	223.4	34.78	47.01	155.5
190.0	24.44	.3441	4822.0	6275.0	224.6	35.10	46.90	158.1
200.0	22.90	.3224	5193.0	6744.0	227.0	35.85	46.98	163.1
210.0	21.57	.3038	5570.0	7216.0	229.3	36.70	47.34	167.7
220.0	20.42	.2875	5952.0	7692.0	231.5	37.62	47.89	172.0
230.0	19.39	.2731	6343.0	8174.0	233.7	38.58	48.57	176.2
240.0	18.48	.2602	6742.0	8663.0	235.8	39.56	49.33	180.2
250.0	17.65	.2486	7149.0	9161.0	237.8	40.56	50.14	184.0
260.0	16.91	.2381	7566.0	9666.0	239.8	41.56	51.00	187.7
270.0	16.22	.2284	7992.0	10180.0	241.7	42.56	51.87	191.4
280.0	15.60	.2196	8427.0	10700.0	243.6	43.54	52.75	194.9
290.0	15.02	.2115	8872.0	11240.0	245.5	44.51	53.64	198.3
300.0	14.49	.2040	9326.0	11780.0	247.3	45.46	54.51	201.6
320.0	13.53	.1906	10260.0	12880.0	250.9	47.30	56.23	208.1
340.0	12.70	.1789	11230.0	14030.0	254.4	49.03	57.88	214.3
360.0	11.97	.1685	12230.0	15200.0	257.7	50.66	59.43	220.4
380.0	11.32	.1594	13260.0	16400.0	261.0	52.18	60.90	226.2
400.0	10.74	.1512	14330.0	17630.0	264.1	53.59	62.26	231.9
420.0	10.21	.1438	15410.0	18890.0	267.2	54.90	63.53	237.5
440.0	9.739	.1371	16530.0	20170.0	270.2	56.11	64.71	242.9
460.0	9.308	.1311	17660.0	21480.0	273.1	57.23	65.80	248.2
480.0	8.914	.1255	18820.0	22810.0	275.9	58.27	66.82	253.4
500.0	8.552	.1204	20000.0	24150.0	278.6	59.25	67.78	258.5
				.60 MPa isobar				
66.46[a]	1869.0	26.32	−12350.0	−12330.0	92.66	47.83	75.17	1491.0
68.0	1863.0	26.24	−12240.0	−12210.0	94.36	46.94	73.03	1419.0
70.0	1856.0	26.13	−12090.0	−12070.0	96.45	46.31	71.31	1346.0
80.0	1817.0	25.58	−11390.0	−11370.0	105.8	46.18	70.11	1162.0
100.0	1736.0	24.44	−9976.0	−9951.0	121.6	44.44	71.09	1023.0
120.0	1650.0	23.23	−8558.0	−8533.0	134.5	41.74	70.83	899.5
140.0	1559.0	21.95	−7138.0	−7110.0	145.5	40.36	71.67	784.5
160.0	1461.0	20.57	−5680.0	−5651.0	155.2	40.15	74.76	674.4
170.0	1408.0	19.82	−4921.0	−4890.0	159.9	40.38	77.50	616.2
175.702[b]	1374.0	19.35	−4466.0	−4435.0	162.5	40.65	79.70	578.9
175.702[b]	33.42	.4705	4217.0	5492.0	219.0	34.94	49.85	147.8
180.0	32.30	.4548	4381.0	5700.0	220.1	35.06	49.19	150.5
185.0	31.03	.4370	4571.0	5944.0	221.5	35.26	48.63	153.4
190.0	29.89	.4209	4761.0	6187.0	222.8	35.51	48.29	156.2
200.0	27.91	.3930	5141.0	6668.0	225.2	36.17	48.03	161.5
210.0	26.22	.3692	5524.0	7149.0	227.6	36.95	48.17	166.3

Thermophysical properties of nitrogen trifluoride—Continued

T K	Density kg/m³	Density mol/dm³	E J/mol	H J/mol	S J/(mol·K)	C_v J/(mol·K)	C_p J/(mol·K)	Sound m/s
220.0	24.77	.3487	5911.0	7632.0	229.8	37.81	48.56	170.9
230.0	23.49	.3307	6306.0	8120.0	232.0	38.74	49.12	175.2
240.0	22.35	.3147	6708.0	8615.0	234.1	39.69	49.79	179.3
250.0	21.33	.3003	7118.0	9116.0	236.2	40.67	50.53	183.2
260.0	20.41	.2874	7538.0	9626.0	238.2	41.65	51.33	187.0
270.0	19.57	.2756	7966.0	10140.0	240.1	42.63	52.16	190.7
280.0	18.80	.2648	8403.0	10670.0	242.0	43.60	53.01	194.3
290.0	18.10	.2549	8849.0	11200.0	243.9	44.56	53.86	197.8
300.0	17.45	.2457	9304.0	11750.0	245.7	45.51	54.71	201.2
320.0	16.29	.2293	10240.0	12860.0	249.3	47.33	56.39	207.7
340.0	15.28	.2151	11210.0	14000.0	252.8	49.06	58.01	214.1
360.0	14.39	.2026	12220.0	15180.0	256.1	50.68	59.54	220.2
380.0	13.60	.1915	13250.0	16380.0	259.4	52.19	60.99	226.1
400.0	12.90	.1816	14310.0	17620.0	262.6	53.60	62.34	231.8
420.0	12.27	.1727	15400.0	18880.0	265.6	54.90	63.60	237.4
440.0	11.70	.1647	16520.0	20160.0	268.6	56.11	64.77	242.9
460.0	11.18	.1574	17650.0	21470.0	271.5	57.23	65.86	248.2
480.0	10.70	.1507	18810.0	22790.0	274.4	58.27	66.87	253.4
500.0	10.26	.1445	19990.0	24140.0	277.1	59.25	67.82	258.5
			.80 MPa isobar					
66.50[a]	1869.0	26.32	-12350.0	-12320.0	92.69	47.98	75.10	1484.0
68.0	1864.0	26.24	-12240.0	-12210.0	94.34	47.09	73.02	1416.0
70.0	1856.0	26.13	-12090.0	-12060.0	96.43	46.44	71.31	1343.0
80.0	1817.0	25.58	-11390.0	-11360.0	105.8	46.22	70.10	1161.0
100.0	1736.0	24.44	-9978.0	-9945.0	121.6	44.42	71.07	1025.0
120.0	1650.0	23.24	-8561.0	-8527.0	134.5	41.71	70.81	901.4
140.0	1560.0	21.96	-7142.0	-7105.0	145.5	40.33	71.64	786.3
160.0	1462.0	20.58	-5685.0	-5646.0	155.2	40.12	74.68	676.2
170.0	1409.0	19.83	-4927.0	-4887.0	159.8	40.36	77.39	618.3
180.0	1351.0	19.02	-4137.0	-4095.0	164.3	40.82	81.26	555.7
182.282[b]	1335.0	18.80	-3942.0	-3899.0	165.4	41.00	82.57	538.0
182.282[b]	44.36	.6246	4318.0	5599.0	217.5	36.31	53.32	147.2
185.0	43.39	.6109	4428.0	5738.0	218.3	36.33	52.63	149.0
190.0	41.55	.5851	4631.0	5998.0	219.7	36.43	51.61	152.2
195.0	39.93	.5622	4832.0	6255.0	221.0	36.61	50.91	155.3
200.0	38.46	.5416	5031.0	6508.0	222.3	36.85	50.44	158.2
210.0	35.92	.5058	5428.0	7010.0	224.7	37.47	50.01	163.5
220.0	33.77	.4755	5827.0	7509.0	227.1	38.23	50.01	168.5
230.0	31.91	.4494	6230.0	8011.0	229.3	39.07	50.29	173.1
240.0	30.28	.4264	6640.0	8516.0	231.4	39.96	50.76	177.4
250.0	28.84	.4061	7056.0	9026.0	233.5	40.89	51.35	181.6
260.0	27.54	.3878	7480.0	9543.0	235.5	41.83	52.03	185.6
270.0	26.37	.3713	7912.0	10070.0	237.5	42.78	52.76	189.4
280.0	25.31	.3563	8353.0	10600.0	239.5	43.73	53.53	193.2
290.0	24.33	.3426	8803.0	11140.0	241.3	44.67	54.32	196.8
300.0	23.43	.3300	9260.0	11680.0	243.2	45.60	55.12	200.3
310.0	22.61	.3183	9727.0	12240.0	245.0	46.51	55.92	203.7
320.0	21.84	.3075	10200.0	12800.0	246.8	47.39	56.72	207.0
340.0	20.46	.2881	11180.0	13950.0	250.3	49.10	58.27	213.5
360.0	19.25	.2711	12180.0	15130.0	253.7	50.71	59.77	219.7
380.0	18.19	.2561	13220.0	16340.0	256.9	52.22	61.18	225.7
400.0	17.24	.2427	14280.0	17580.0	260.1	53.62	62.50	231.6
420.0	16.38	.2307	15380.0	18840.0	263.2	54.91	63.74	237.2
440.0	15.61	.2199	16490.0	20130.0	266.2	56.12	64.89	242.8
460.0	14.91	.2100	17630.0	21440.0	269.1	57.23	65.96	248.2
480.0	14.28	.2010	18790.0	22770.0	271.9	58.27	66.97	253.4
500.0	13.69	.1928	19970.0	24120.0	274.7	59.25	67.91	258.6
			1.00 MPa isobar					
66.54[a]	1869.0	26.32	-12350.0	-12310.0	92.72	48.13	75.02	1478.0

Thermophysical properties of nitrogen trifluoride—Continued

T K	Density kg/m^3	Density mol/dm^3	E J/mol	H J/mol	S J/(mol·K)	C_v J/(mol·K)	C_p J/(mol·K)	Sound m/s
68.0	1864.0	26.24	−12240.0	−12200.0	94.33	47.25	73.01	1412.0
70.0	1856.0	26.13	−12090.0	−12060.0	96.41	46.57	71.30	1340.0
80.0	1817.0	25.59	−11390.0	−11350.0	105.8	46.25	70.09	1161.0
100.0	1736.0	24.45	−9980.0	−9939.0	121.6	44.40	71.06	1026.0
120.0	1651.0	23.25	−8564.0	−8521.0	134.5	41.68	70.78	903.2
140.0	1560.0	21.97	−7145.0	−7100.0	145.5	40.30	71.60	788.0
160.0	1463.0	20.59	−5691.0	−5642.0	155.2	40.10	74.61	678.1
170.0	1410.0	19.85	−4934.0	−4884.0	159.8	40.33	77.28	620.3
180.0	1352.0	19.04	−4145.0	−4093.0	164.3	40.79	81.08	558.1
185.0	1321.0	18.60	−3735.0	−3681.0	166.6	41.12	83.60	524.6
187.747[b]	1301.0	18.32	−3495.0	−3440.0	167.8	41.37	85.46	502.8
187.747[b]	55.49	.7814	4391.0	5671.0	216.4	37.51	56.78	146.2
190.0	54.46	.7668	4487.0	5791.0	217.0	37.49	55.97	147.9
195.0	52.01	.7323	4702.0	6067.0	218.4	37.50	54.45	151.4
200.0	49.85	.7020	4912.0	6337.0	219.8	37.61	53.39	154.6
205.0	47.94	.6750	5120.0	6602.0	221.1	37.80	52.65	157.7
210.0	46.22	.6508	5327.0	6863.0	222.4	38.05	52.14	160.6
220.0	43.22	.6086	5739.0	7382.0	224.8	38.67	51.64	166.0
230.0	40.69	.5729	6152.0	7898.0	227.1	39.41	51.58	170.9
240.0	38.49	.5420	6569.0	8414.0	229.3	40.24	51.81	175.6
250.0	36.57	.5149	6992.0	8935.0	231.4	41.11	52.22	180.0
260.0	34.85	.4908	7422.0	9459.0	233.5	42.01	52.76	184.2
270.0	33.32	.4692	7859.0	9990.0	235.5	42.93	53.39	188.2
280.0	31.93	.4496	8303.0	10530.0	237.4	43.86	54.07	192.0
290.0	30.67	.4318	8756.0	11070.0	239.3	44.78	54.80	195.8
300.0	29.51	.4155	9216.0	11620.0	241.2	45.69	55.54	199.4
310.0	28.44	.4005	9685.0	12180.0	243.0	46.58	56.29	202.9
320.0	27.46	.3866	10160.0	12750.0	244.8	47.46	57.05	206.3
340.0	25.69	.3618	11140.0	13910.0	248.3	49.15	58.54	212.9
360.0	24.15	.3401	12150.0	15090.0	251.7	50.75	59.99	219.3
380.0	22.80	.3210	13190.0	16300.0	255.0	52.24	61.37	225.4
400.0	21.59	.3041	14260.0	17540.0	258.2	53.63	62.66	231.3
420.0	20.52	.2889	15350.0	18810.0	261.3	54.92	63.88	237.1
440.0	19.54	.2752	16470.0	20100.0	264.3	56.12	65.01	242.7
460.0	18.66	.2628	17600.0	21410.0	267.2	57.23	66.07	248.1
480.0	17.86	.2515	18770.0	22740.0	270.0	58.27	67.06	253.4
500.0	17.12	.2411	19950.0	24090.0	272.8	59.24	68.00	258.6
				1.20 MPa isobar				
66.58[a]	1869.0	26.32	−12340.0	−12300.0	92.74	48.27	74.95	1472.0
68.0	1864.0	26.24	−12240.0	−12190.0	94.31	47.40	73.01	1408.0
70.0	1856.0	26.14	−12100.0	−12050.0	96.40	46.70	71.29	1337.0
80.0	1817.0	25.59	−11390.0	−11350.0	105.8	46.29	70.08	1161.0
100.0	1737.0	24.45	−9982.0	−9933.0	121.6	44.37	71.04	1028.0
120.0	1651.0	23.25	−8567.0	−8515.0	134.5	41.65	70.76	905.1
140.0	1561.0	21.98	−7149.0	−7095.0	145.4	40.28	71.56	789.8
160.0	1463.0	20.61	−5696.0	−5638.0	155.1	40.07	74.54	679.9
170.0	1411.0	19.86	−4941.0	−4880.0	159.7	40.31	77.17	622.4
180.0	1353.0	19.05	−4154.0	−4091.0	164.2	40.76	80.91	560.5
185.0	1322.0	18.62	−3745.0	−3680.0	166.5	41.08	83.37	527.3
190.0	1289.0	18.15	−3322.0	−3256.0	168.8	41.48	86.41	492.1
192.465[b]	1270.0	17.88	−3099.0	−3032.0	169.9	41.75	88.46	471.5
192.465[b]	66.88	.9418	4444.0	5718.0	215.4	38.60	60.34	145.1
195.0	65.39	.9208	4558.0	5861.0	216.1	38.54	59.11	147.1
200.0	62.29	.8771	4783.0	6151.0	217.6	38.48	57.09	150.8
205.0	59.60	.8392	5003.0	6433.0	219.0	38.53	55.68	154.3
210.0	57.22	.8058	5219.0	6709.0	220.3	38.67	54.68	157.5
215.0	55.10	.7759	5434.0	6980.0	221.6	38.88	53.98	160.5
220.0	53.19	.7489	5647.0	7249.0	222.8	39.14	53.50	163.4
230.0	49.84	.7018	6071.0	7781.0	225.2	39.78	53.01	168.7
240.0	47.00	.6618	6497.0	8311.0	227.5	40.53	52.95	173.7
250.0	44.53	.6270	6927.0	8841.0	229.6	41.34	53.15	178.3

Thermophysical properties of nitrogen trifluoride—Continued

T K	Density kg/m³	Density mol/dm³	E J/mol	H J/mol	S J/(mol·K)	C_v J/(mol·K)	C_p J/(mol·K)	Sound m/s
260.0	42.36	.5964	7362.0	9374.0	231.7	42.20	53.54	182.7
270.0	40.42	.5692	7804.0	9912.0	233.7	43.09	54.05	186.9
280.0	38.68	.5447	8252.0	10460.0	235.7	43.99	54.64	190.9
290.0	37.11	.5225	8708.0	11010.0	237.6	44.88	55.29	194.8
300.0	35.67	.5022	9172.0	11560.0	239.5	45.78	55.97	198.5
310.0	34.35	.4837	9644.0	12120.0	241.4	46.66	56.68	202.1
320.0	33.14	.4666	10120.0	12690.0	243.2	47.52	57.39	205.6
340.0	30.97	.4361	11110.0	13860.0	246.7	49.20	58.82	212.4
360.0	29.09	.4096	12120.0	15050.0	250.1	50.78	60.22	218.9
380.0	27.44	.3863	13160.0	16270.0	253.4	52.26	61.56	225.1
400.0	25.97	.3657	14230.0	17510.0	256.6	53.65	62.83	231.1
420.0	24.66	.3473	15320.0	18780.0	259.7	54.93	64.02	236.9
440.0	23.48	.3307	16440.0	20070.0	262.7	56.13	65.14	242.6
460.0	22.42	.3156	17580.0	21380.0	265.6	57.24	66.18	248.1
480.0	21.44	.3020	18740.0	22720.0	268.5	58.27	67.16	253.4
500.0	20.56	.2895	19920.0	24070.0	271.2	59.24	68.08	258.7
				1.40 MPa isobar				
66.61[a]	1869.0	26.32	−12340.0	−12290.0	92.77	48.42	74.89	1466.0
68.0	1864.0	26.25	−12240.0	−12190.0	94.29	47.56	73.00	1404.0
70.0	1856.0	26.14	−12100.0	−12040.0	96.38	46.83	71.28	1334.0
80.0	1818.0	25.59	−11400.0	−11340.0	105.7	46.32	70.07	1161.0
100.0	1737.0	24.46	−9984.0	−9926.0	121.5	44.35	71.03	1030.0
120.0	1652.0	23.26	−8569.0	−8509.0	134.5	41.62	70.74	906.9
140.0	1561.0	21.98	−7153.0	−7089.0	145.4	40.25	71.53	791.5
160.0	1464.0	20.62	−5702.0	−5634.0	155.1	40.05	74.47	681.7
170.0	1411.0	19.87	−4947.0	−4877.0	159.7	40.28	77.07	624.4
180.0	1354.0	19.07	−4162.0	−4089.0	164.2	40.73	80.75	562.8
190.0	1291.0	18.17	−3333.0	−3256.0	168.7	41.44	86.11	495.1
195.0	1255.0	17.67	−2896.0	−2817.0	171.0	41.93	89.85	457.9
196.642[b]	1241.0	17.48	−2740.0	−2660.0	171.8	42.15	91.60	443.0
196.642[b]	78.61	1.107	4481.0	5746.0	214.5	39.62	64.09	143.9
200.0	76.07	1.071	4641.0	5948.0	215.6	39.47	61.96	146.7
205.0	72.31	1.018	4876.0	6251.0	217.0	39.36	59.48	150.6
210.0	69.08	.9727	5104.0	6544.0	218.5	39.37	57.75	154.2
215.0	66.25	.9328	5328.0	6829.0	219.8	39.47	56.52	157.6
220.0	63.74	.8975	5550.0	7109.0	221.1	39.65	55.65	160.7
225.0	61.48	.8656	5769.0	7386.0	222.3	39.89	55.03	163.7
230.0	59.43	.8368	5987.0	7660.0	223.5	40.17	54.61	166.5
240.0	55.83	.7861	6423.0	8204.0	225.9	40.83	54.19	171.7
250.0	52.75	.7427	6860.0	8745.0	228.1	41.58	54.15	176.6
260.0	50.06	.7049	7302.0	9288.0	230.2	42.40	54.36	181.2
270.0	47.69	.6715	7748.0	9833.0	232.3	43.25	54.74	185.6
280.0	45.57	.6416	8201.0	10380.0	234.3	44.12	55.23	189.8
290.0	43.66	.6147	8661.0	10940.0	236.2	44.99	55.80	193.8
300.0	41.92	.5903	9127.0	11500.0	238.1	45.87	56.42	197.6
310.0	40.34	.5680	9602.0	12070.0	240.0	46.73	57.07	201.3
320.0	38.88	.5475	10080.0	12640.0	241.8	47.59	57.74	204.9
330.0	37.54	.5286	10570.0	13220.0	243.6	48.42	58.42	208.4
340.0	36.29	.5111	11070.0	13810.0	245.3	49.24	59.10	211.8
360.0	34.06	.4796	12080.0	15000.0	248.7	50.81	60.45	218.4
380.0	32.10	.4520	13130.0	16230.0	252.0	52.29	61.75	224.8
400.0	30.37	.4276	14200.0	17470.0	255.2	53.66	62.99	230.9
420.0	28.82	.4058	15300.0	18750.0	258.3	54.94	64.16	236.8
440.0	27.43	.3863	16420.0	20040.0	261.4	56.13	65.26	242.5
460.0	26.18	.3686	17560.0	21360.0	264.3	57.24	66.29	248.1
480.0	25.04	.3525	18720.0	22690.0	267.1	58.27	67.26	253.5
500.0	23.99	.3378	19900.0	24050.0	269.9	59.24	68.17	258.8
				1.60 MPa isobar				
66.65[a]	1869.0	26.32	−12340.0	−12280.0	92.80	48.56	74.82	1460.0

Thermophysical properties of nitrogen trifluoride—Continued

T K	Density kg/m^3	Density mol/dm^3	E J/mol	H J/mol	S J/(mol·K)	C_v J/(mol·K)	C_p J/(mol·K)	Sound m/s
68.0	1864.0	26.25	-12240.0	-12180.0	94.28	47.72	72.99	1401.0
70.0	1857.0	26.14	-12100.0	-12040.0	96.37	46.96	71.27	1331.0
80.0	1818.0	25.60	-11400.0	-11340.0	105.7	46.36	70.06	1160.0
100.0	1737.0	24.46	-9985.0	-9920.0	121.5	44.33	71.01	1031.0
120.0	1652.0	23.26	-8572.0	-8503.0	134.4	41.59	70.72	908.7
140.0	1562.0	21.99	-7157.0	-7084.0	145.4	40.23	71.49	793.3
160.0	1465.0	20.63	-5707.0	-5629.0	155.1	40.03	74.40	683.5
170.0	1412.0	19.89	-4954.0	-4873.0	159.7	40.26	76.97	626.4
180.0	1356.0	19.09	-4170.0	-4087.0	164.2	40.70	80.58	565.2
190.0	1292.0	18.20	-3344.0	-3256.0	168.6	41.40	85.82	498.0
195.0	1257.0	17.70	-2909.0	-2818.0	170.9	41.88	89.45	461.2
200.0	1218.0	17.15	-2453.0	-2360.0	173.2	42.49	94.24	421.6
200.406[b]	1214.0	17.09	-2409.0	-2315.0	173.5	42.57	94.95	416.6
200.406[b]	90.76	1.278	4506.0	5758.0	213.8	40.58	68.13	142.5
205.0	86.37	1.216	4736.0	6051.0	215.2	40.30	64.45	146.6
210.0	81.98	1.154	4980.0	6366.0	216.7	40.14	61.57	150.7
215.0	78.23	1.102	5216.0	6669.0	218.1	40.12	59.58	154.4
220.0	74.96	1.056	5447.0	6963.0	219.5	40.20	58.16	157.9
225.0	72.08	1.015	5674.0	7251.0	220.8	40.36	57.14	161.1
230.0	69.49	.9784	5899.0	7535.0	222.0	40.58	56.41	164.2
240.0	65.01	.9154	6346.0	8094.0	224.4	41.14	55.56	169.8
250.0	61.23	.8622	6792.0	8648.0	226.7	41.83	55.23	175.0
260.0	57.97	.8163	7240.0	9200.0	228.8	42.60	55.24	179.8
270.0	55.12	.7762	7692.0	9753.0	230.9	43.41	55.47	184.3
280.0	52.59	.7405	8149.0	10310.0	233.0	44.25	55.84	188.6
290.0	50.32	.7086	8612.0	10870.0	234.9	45.10	56.32	192.8
300.0	48.27	.6797	9082.0	11440.0	236.8	45.96	56.87	196.7
310.0	46.40	.6534	9559.0	12010.0	238.7	46.81	57.47	200.5
320.0	44.69	.6293	10040.0	12590.0	240.5	47.65	58.09	204.2
330.0	43.12	.6072	10530.0	13170.0	242.3	48.48	58.73	207.8
340.0	41.67	.5867	11030.0	13760.0	244.1	49.29	59.38	211.3
360.0	39.06	.5500	12050.0	14960.0	247.5	50.84	60.68	218.0
380.0	36.79	.5180	13100.0	16190.0	250.9	52.31	61.95	224.5
400.0	34.78	.4897	14170.0	17440.0	254.1	53.68	63.16	230.7
420.0	32.99	.4646	15270.0	18710.0	257.2	54.95	64.30	236.6
440.0	31.39	.4420	16390.0	20010.0	260.2	56.14	65.38	242.4
460.0	29.94	.4216	17530.0	21330.0	263.1	57.24	66.40	248.0
480.0	28.63	.4031	18700.0	22670.0	266.0	58.27	67.36	253.5
500.0	27.43	.3863	19880.0	24020.0	268.7	59.24	68.26	258.8

1.80 MPa isobar

T K	Density kg/m^3	Density mol/dm^3	E J/mol	H J/mol	S J/(mol·K)	C_v J/(mol·K)	C_p J/(mol·K)	Sound m/s
66.69[a]	1869.0	26.32	-12340.0	-12270.0	92.82	48.70	74.75	1454.0
68.0	1864.0	26.25	-12240.0	-12170.0	94.26	47.87	72.98	1397.0
70.0	1857.0	26.14	-12100.0	-12030.0	96.35	47.09	71.26	1327.0
80.0	1818.0	25.60	-11400.0	-11330.0	105.7	46.39	70.06	1160.0
100.0	1738.0	24.47	-9987.0	-9914.0	121.5	44.31	71.00	1033.0
120.0	1653.0	23.27	-8575.0	-8497.0	134.4	41.57	70.69	910.5
140.0	1562.0	22.00	-7161.0	-7079.0	145.3	40.21	71.46	795.0
160.0	1466.0	20.64	-5712.0	-5625.0	155.0	40.01	74.33	685.3
170.0	1413.0	19.90	-4960.0	-4870.0	159.6	40.24	76.86	628.3
180.0	1357.0	19.11	-4179.0	-4084.0	164.1	40.67	80.42	567.5
190.0	1294.0	18.22	-3355.0	-3256.0	168.6	41.36	85.54	500.8
195.0	1259.0	17.73	-2922.0	-2820.0	170.9	41.83	89.06	464.5
200.0	1221.0	17.19	-2469.0	-2364.0	173.2	42.42	93.66	425.5
203.842[b]	1187.0	16.72	-2098.0	-1991.0	175.0	43.01	98.59	392.0
203.842[b]	103.4	1.456	4519.0	5755.0	213.0	41.50	72.58	141.1
205.0	102.3	1.440	4578.0	5828.0	213.4	41.40	71.34	142.2
210.0	96.20	1.355	4843.0	6172.0	215.0	41.02	66.52	146.9
215.0	91.22	1.284	5095.0	6496.0	216.6	40.84	63.35	151.1
220.0	87.00	1.225	5338.0	6807.0	218.0	40.80	61.16	154.9
225.0	83.33	1.173	5575.0	7109.0	219.3	40.87	59.59	158.5
230.0	80.09	1.128	5808.0	7404.0	220.6	41.02	58.46	161.8

Thermophysical properties of nitrogen trifluoride—Continued

T K	Density kg/m³	Density mol/dm³	*E* J/mol	*H* J/mol	*S* J/(mol·K)	C_v J/(mol·K)	C_p J/(mol·K)	Sound m/s
235.0	77.19	1.087	6038.0	7694.0	221.9	41.22	57.65	164.9
240.0	74.58	1.050	6266.0	7981.0	223.1	41.48	57.07	167.8
245.0	72.19	1.017	6494.0	8265.0	224.3	41.77	56.66	170.6
250.0	70.01	.9857	6721.0	8547.0	225.4	42.09	56.40	173.3
260.0	66.11	.9309	7176.0	9110.0	227.6	42.80	56.17	178.3
270.0	62.73	.8833	7634.0	9672.0	229.7	43.57	56.23	183.0
280.0	59.76	.8414	8096.0	10240.0	231.8	44.38	56.48	187.5
290.0	57.10	.8040	8563.0	10800.0	233.8	45.21	56.87	191.8
300.0	54.71	.7704	9037.0	11370.0	235.7	46.05	57.34	195.9
310.0	52.55	.7399	9516.0	11950.0	237.6	46.88	57.88	199.8
320.0	50.57	.7121	10000.0	12530.0	239.4	47.71	58.46	203.6
330.0	48.76	.6866	10500.0	13120.0	241.3	48.53	59.06	207.2
340.0	47.09	.6630	11000.0	13710.0	243.0	49.33	59.67	210.8
360.0	44.10	.6210	12020.0	14920.0	246.5	50.87	60.92	217.6
380.0	41.50	.5843	13070.0	16150.0	249.8	52.33	62.14	224.2
400.0	39.21	.5521	14140.0	17400.0	253.0	53.69	63.32	230.5
420.0	37.18	.5235	15240.0	18680.0	256.1	54.96	64.45	236.5
440.0	35.36	.4979	16370.0	19980.0	259.2	56.15	65.51	242.4
460.0	33.72	.4748	17510.0	21300.0	262.1	57.25	66.51	248.0
480.0	32.23	.4538	18670.0	22640.0	264.9	58.27	67.45	253.5
500.0	30.87	.4347	19860.0	24000.0	267.7	59.24	68.35	258.9
				2.00 MPa isobar				
66.72[a]	1869.0	26.32	−12340.0	−12260.0	92.85	48.85	74.68	1448.0
68.0	1864.0	26.25	−12240.0	−12170.0	94.25	48.02	72.97	1393.0
70.0	1857.0	26.15	−12100.0	−12020.0	96.34	47.22	71.25	1324.0
80.0	1818.0	25.60	−11400.0	−11320.0	105.7	46.42	70.05	1160.0
100.0	1738.0	24.47	−9989.0	−9908.0	121.5	44.28	70.98	1034.0
120.0	1653.0	23.27	−8577.0	−8492.0	134.4	41.54	70.67	912.4
140.0	1563.0	22.01	−7164.0	−7073.0	145.3	40.18	71.42	796.7
160.0	1467.0	20.65	−5718.0	−5621.0	155.0	39.99	74.27	687.1
170.0	1414.0	19.92	−4967.0	−4866.0	159.6	40.22	76.76	630.3
180.0	1358.0	19.12	−4187.0	−4082.0	164.1	40.65	80.26	569.7
190.0	1296.0	18.24	−3366.0	−3256.0	168.5	41.32	85.27	503.6
195.0	1261.0	17.75	−2934.0	−2821.0	170.8	41.78	88.69	467.7
200.0	1223.0	17.22	−2484.0	−2367.0	173.1	42.35	93.11	429.2
205.0	1180.0	16.62	−2008.0	−1888.0	175.5	43.09	99.23	387.4
207.010[b]	1161.0	16.35	−1805.0	−1682.0	176.5	43.47	102.6	368.7
207.010[b]	116.7	1.643	4521.0	5739.0	212.3	42.38	77.58	139.6
210.0	112.2	1.580	4690.0	5956.0	213.4	42.04	73.30	142.7
215.0	105.5	1.485	4962.0	6309.0	215.0	41.66	68.17	147.6
220.0	100.00	1.408	5220.0	6641.0	216.5	41.47	64.81	151.8
225.0	95.34	1.342	5469.0	6959.0	218.0	41.43	62.48	155.7
230.0	91.31	1.286	5711.0	7267.0	219.3	41.49	60.82	159.3
235.0	87.75	1.236	5949.0	7568.0	220.6	41.62	59.62	162.6
240.0	84.57	1.191	6184.0	7863.0	221.9	41.82	58.74	165.7
245.0	81.71	1.150	6417.0	8155.0	223.1	42.07	58.11	168.7
250.0	79.09	1.114	6649.0	8445.0	224.2	42.36	57.66	171.5
260.0	74.49	1.049	7112.0	9019.0	226.5	43.01	57.16	176.8
270.0	70.53	.9931	7575.0	9589.0	228.6	43.74	57.04	181.8
280.0	67.07	.9444	8042.0	10160.0	230.7	44.52	57.15	186.4
290.0	64.00	.9012	8514.0	10730.0	232.7	45.32	57.44	190.8
300.0	61.26	.8625	8990.0	11310.0	234.7	46.14	57.83	195.0
310.0	58.78	.8276	9473.0	11890.0	236.6	46.96	58.30	199.0
320.0	56.52	.7959	9962.0	12480.0	238.4	47.78	58.83	202.9
330.0	54.46	.7668	10460.0	13070.0	240.3	48.58	59.39	206.7
340.0	52.56	.7400	10960.0	13660.0	242.0	49.38	59.97	210.3
360.0	49.17	.6924	11990.0	14870.0	245.5	50.91	61.16	217.3
380.0	46.24	.6511	13040.0	16110.0	248.8	52.35	62.34	223.9
400.0	43.66	.6148	14110.0	17370.0	252.1	53.71	63.49	230.3
420.0	41.38	.5826	15220.0	18650.0	255.2	54.97	64.59	236.4
440.0	39.34	.5539	16340.0	19950.0	258.2	56.15	65.63	242.3

Thermophysical properties of nitrogen trifluoride—Continued

T K	Density kg/m^3	Density mol/dm^3	E J/mol	H J/mol	S J/(mol·K)	C_v J/(mol·K)	C_p J/(mol·K)	Sound m/s
460.0	37.50	.5280	17490.0	21270.0	261.2	57.25	66.62	248.0
480.0	35.83	.5046	18650.0	22620.0	264.0	58.27	67.55	253.6
500.0	34.32	.4832	19840.0	23980.0	266.8	59.23	68.43	259.0
			3.00 MPa isobar					
66.91[a]	1869.0	26.32	-12330.0	-12210.0	92.98	49.54	74.35	1419.0
68.0	1865.0	26.26	-12250.0	-12130.0	94.17	48.79	72.92	1375.0
70.0	1858.0	26.16	-12100.0	-11990.0	96.26	47.86	71.20	1309.0
80.0	1819.0	25.62	-11410.0	-11290.0	105.6	46.58	70.00	1159.0
100.0	1739.0	24.49	-9999.0	-9876.0	121.4	44.17	70.91	1042.0
120.0	1655.0	23.30	-8591.0	-8462.0	134.3	41.41	70.56	921.3
140.0	1566.0	22.05	-7183.0	-7047.0	145.2	40.07	71.25	805.2
160.0	1470.0	20.70	-5744.0	-5599.0	154.8	39.89	73.94	695.8
170.0	1419.0	19.98	-4998.0	-4848.0	159.4	40.12	76.29	639.8
180.0	1364.0	19.21	-4226.0	-4070.0	163.8	40.53	79.52	580.7
190.0	1303.0	18.35	-3417.0	-3254.0	168.3	41.15	84.03	517.0
195.0	1270.0	17.89	-2994.0	-2826.0	170.5	41.56	86.99	482.9
200.0	1234.0	17.38	-2555.0	-2382.0	172.7	42.06	90.70	446.9
205.0	1195.0	16.82	-2096.0	-1917.0	175.0	42.69	95.54	408.4
210.0	1150.0	16.20	-1609.0	-1424.0	177.4	43.49	102.3	366.6
212.0	1131.0	15.92	-1404.0	-1216.0	178.4	43.88	106.0	348.7
214.0	1109.0	15.62	-1192.0	-999.4	179.4	44.33	110.5	329.8
216.0	1086.0	15.29	-969.1	-773.0	180.5	44.85	116.3	309.8
218.0	1060.0	14.93	-734.0	-533.0	181.6	45.46	124.1	288.3
220.0	1031.0	14.51	-480.7	-274.0	182.7	46.22	135.8	264.7
220.054[b]	1031.0	14.52	-479.1	-272.4	182.7	46.21	135.5	265.1
220.054[b]	195.9	2.759	4381.0	5468.0	208.8	46.70	120.8	130.9
225.0	174.9	2.462	4787.0	6005.0	211.3	45.28	93.80	138.9
226.0	171.8	2.419	4857.0	6097.0	211.7	45.10	90.85	140.2
228.0	166.3	2.342	4993.0	6274.0	212.4	44.79	86.01	142.7
230.0	161.5	2.273	5122.0	6442.0	213.2	44.54	82.21	145.0
232.0	157.1	2.212	5247.0	6603.0	213.9	44.34	79.15	147.2
235.0	151.3	2.130	5427.0	6835.0	214.9	44.11	75.54	150.2
240.0	143.1	2.015	5712.0	7201.0	216.4	43.90	71.23	154.7
245.0	136.2	1.918	5985.0	7549.0	217.8	43.83	68.25	158.8
250.0	130.3	1.835	6250.0	7885.0	219.2	43.87	66.11	162.6
255.0	125.1	1.762	6509.0	8212.0	220.5	43.98	64.54	166.1
260.0	120.5	1.697	6764.0	8531.0	221.7	44.15	63.37	169.4
265.0	116.4	1.639	7016.0	8846.0	222.9	44.37	62.50	172.5
270.0	112.7	1.586	7265.0	9157.0	224.1	44.63	61.84	175.4
280.0	106.1	1.494	7762.0	9770.0	226.3	45.22	61.01	181.0
290.0	100.4	1.414	8257.0	10380.0	228.5	45.89	60.61	186.1
300.0	95.52	1.345	8753.0	10980.0	230.5	46.60	60.50	190.9
310.0	91.18	1.284	9252.0	11590.0	232.5	47.34	60.59	195.5
320.0	87.30	1.229	9755.0	12200.0	234.4	48.09	60.81	199.8
330.0	83.81	1.180	10260.0	12810.0	236.3	48.85	61.13	203.9
340.0	80.63	1.135	10780.0	13420.0	238.1	49.59	61.51	207.9
350.0	77.73	1.094	11290.0	14040.0	239.9	50.33	61.94	211.8
360.0	75.06	1.057	11820.0	14660.0	241.7	51.06	62.39	215.5
380.0	70.30	.9899	12880.0	15910.0	245.1	52.46	63.36	222.6
400.0	66.18	.9319	13970.0	17190.0	248.3	53.78	64.35	229.4
420.0	62.57	.8810	15080.0	18490.0	251.5	55.02	65.32	235.9
440.0	59.36	.8358	16220.0	19800.0	254.6	56.18	66.27	242.1
460.0	56.49	.7954	17370.0	21140.0	257.5	57.26	67.17	248.1
480.0	53.91	.7591	18540.0	22490.0	260.4	58.27	68.04	253.9
500.0	51.57	.7261	19730.0	23860.0	263.2	59.22	68.87	259.5
			4.00 MPa isobar					
67.10[a]	1869.0	26.32	-12320.0	-12170.0	93.12	50.21	74.02	1391.0

Thermophysical properties of nitrogen trifluoride—Continued

T K	Density kg/m³	Density mol/dm³	E J/mol	H J/mol	S J/(mol·K)	C_v J/(mol·K)	C_p J/(mol·K)	Sound m/s
68.0	1866.0	26.28	−12250.0	−12100.0	94.10	49.54	72.87	1356.0
70.0	1859.0	26.17	−12110.0	−11960.0	96.18	48.48	71.15	1294.0
80.0	1821.0	25.63	−11410.0	−11260.0	105.5	46.74	69.95	1157.0
100.0	1741.0	24.51	−10010.0	−9845.0	121.3	44.06	70.84	1050.0
120.0	1657.0	23.33	−8604.0	−8433.0	134.2	41.28	70.45	930.1
140.0	1568.0	22.09	−7201.0	−7020.0	145.0	39.96	71.09	813.4
160.0	1474.0	20.76	−5769.0	−5576.0	154.7	39.80	73.64	704.2
170.0	1424.0	20.05	−5029.0	−4830.0	159.2	40.03	75.84	648.9
180.0	1370.0	19.29	−4264.0	−4057.0	163.6	40.43	78.85	591.1
190.0	1311.0	18.46	−3466.0	−3249.0	168.0	41.01	82.93	529.5
195.0	1279.0	18.01	−3050.0	−2828.0	170.2	41.39	85.55	496.9
200.0	1245.0	17.53	−2621.0	−2393.0	172.4	41.84	88.73	462.8
205.0	1207.0	17.00	−2175.0	−1940.0	174.6	42.38	92.70	426.9
210.0	1166.0	16.42	−1707.0	−1464.0	176.9	43.05	97.91	388.8
215.0	1120.0	15.77	−1210.0	−956.8	179.3	43.91	105.3	347.6
217.0	1099.0	15.48	−1001.0	−742.3	180.3	44.32	109.3	330.1
218.0	1088.0	15.32	−892.9	−631.9	180.8	44.55	111.7	321.0
220.0	1065.0	15.00	−669.9	−403.1	181.8	45.05	117.3	302.0
222.0	1039.0	14.63	−434.8	−161.4	182.9	45.64	124.8	281.9
224.0	1010.0	14.22	−183.0	98.3	184.1	46.35	135.5	260.2
225.0	993.5	13.99	−48.6	237.4	184.7	46.76	142.9	248.6
226.0	975.5	13.74	93.7	384.9	185.4	47.24	152.6	236.2
227.0	955.2	13.45	246.4	543.8	186.1	47.78	166.0	222.9
228.0	931.8	13.12	414.2	719.0	186.8	48.44	186.0	208.3
228.5	918.3	12.93	506.2	815.5	187.3	48.83	200.6	200.4
229.0	903.2	12.72	606.0	920.5	187.7	49.27	220.6	191.9
229.5	885.7	12.47	717.0	1038.0	188.2	49.80	250.4	182.5
230.0	864.4	12.17	845.6	1174.0	188.8	50.45	300.6	172.0
230.093[b]	860.8	12.12	867.7	1198.0	188.9	50.55	310.6	170.3
230.093[b]	324.9	4.575	3850.0	4724.0	204.3	51.64	365.7	121.2
235.0	255.7	3.601	4611.0	5722.0	208.6	48.28	135.1	133.7
236.0	248.9	3.504	4711.0	5852.0	209.1	47.94	125.8	135.5
238.0	237.4	3.342	4893.0	6089.0	210.1	47.39	112.5	138.7
240.0	227.9	3.209	5058.0	6305.0	211.0	46.95	103.4	141.5
242.0	219.8	3.095	5212.0	6505.0	211.8	46.61	96.77	144.1
244.0	212.8	2.996	5358.0	6693.0	212.6	46.33	91.65	146.5
246.0	206.5	2.908	5497.0	6872.0	213.4	46.11	87.59	148.8
248.0	200.9	2.829	5630.0	7044.0	214.1	45.93	84.30	150.9
250.0	195.9	2.758	5759.0	7210.0	214.7	45.80	81.57	152.9
252.0	191.2	2.692	5885.0	7370.0	215.4	45.69	79.28	154.8
255.0	184.9	2.603	6068.0	7604.0	216.3	45.58	76.46	157.6
260.0	175.8	2.475	6361.0	7977.0	217.7	45.50	72.93	161.8
265.0	168.0	2.365	6644.0	8335.0	219.1	45.52	70.38	165.6
270.0	161.2	2.270	6920.0	8682.0	220.4	45.62	68.49	169.2
275.0	155.2	2.185	7190.0	9021.0	221.6	45.78	67.07	172.6
280.0	149.8	2.109	7457.0	9353.0	222.8	45.98	65.98	175.8
285.0	144.9	2.041	7720.0	9681.0	224.0	46.22	65.15	178.8
290.0	140.5	1.978	7982.0	10000.0	225.1	46.48	64.51	181.7
300.0	132.6	1.867	8503.0	10650.0	227.3	47.08	63.66	187.2
310.0	125.8	1.772	9022.0	11280.0	229.4	47.72	63.21	192.3
320.0	119.9	1.688	9541.0	11910.0	231.4	48.40	63.03	197.1
330.0	114.6	1.614	10060.0	12540.0	233.3	49.10	63.04	201.6
340.0	109.9	1.548	10590.0	13170.0	235.2	49.81	63.18	205.9
350.0	105.7	1.488	11120.0	13800.0	237.0	50.51	63.41	210.1
360.0	101.8	1.433	11650.0	14440.0	238.8	51.20	63.70	214.1
370.0	98.24	1.383	12190.0	15080.0	240.6	51.89	64.05	217.9
380.0	94.96	1.337	12730.0	15720.0	242.3	52.56	64.42	221.7
400.0	89.12	1.255	13830.0	17020.0	245.6	53.84	65.23	228.8
420.0	84.05	1.183	14950.0	18330.0	248.8	55.06	66.07	235.6
440.0	79.58	1.121	16090.0	19660.0	251.9	56.20	66.91	242.1
460.0	75.62	1.065	17250.0	21010.0	254.9	57.27	67.73	248.3
480.0	72.06	1.015	18430.0	22370.0	257.8	58.27	68.53	254.3
500.0	68.85	.9695	19620.0	23750.0	260.6	59.22	69.30	260.1

Thermophysical properties of nitrogen trifluoride—Continued

T K	Density kg/m^3	Density mol/dm^3	E J/mol	H J/mol	S J/(mol·K)	C_v J/(mol·K)	C_p J/(mol·K)	Sound m/s
			4.40 MPa isobar					
67.17[a]	1869.0	26.32	−12320.0	−12150.0	93.17	50.47	73.90	1381.0
68.0	1866.0	26.28	−12260.0	−12090.0	94.07	49.84	72.85	1349.0
70.0	1859.0	26.18	−12110.0	−11940.0	96.15	48.73	71.12	1288.0
80.0	1821.0	25.64	−11420.0	−11240.0	105.5	46.79	69.93	1157.0
100.0	1741.0	24.52	−10010.0	−9833.0	121.2	44.02	70.81	1053.0
120.0	1658.0	23.34	−8609.0	−8421.0	134.1	41.23	70.41	933.5
140.0	1570.0	22.10	−7208.0	−7009.0	145.0	39.92	71.02	816.7
160.0	1476.0	20.78	−5779.0	−5567.0	154.6	39.77	73.52	707.5
170.0	1425.0	20.07	−5041.0	−4822.0	159.1	40.00	75.67	652.4
180.0	1372.0	19.32	−4279.0	−4051.0	163.5	40.39	78.59	595.1
190.0	1314.0	18.50	−3485.0	−3247.0	167.9	40.97	82.53	534.3
195.0	1282.0	18.05	−3072.0	−2828.0	170.1	41.33	85.03	502.2
200.0	1249.0	17.58	−2646.0	−2396.0	172.3	41.76	88.03	468.8
205.0	1212.0	17.07	−2204.0	−1947.0	174.5	42.28	91.73	433.8
210.0	1172.0	16.51	−1743.0	−1477.0	176.7	42.91	96.49	396.8
215.0	1128.0	15.88	−1256.0	−978.8	179.1	43.70	103.0	357.3
217.0	1108.0	15.60	−1051.0	−769.5	180.0	44.07	106.4	340.6
218.0	1098.0	15.46	−946.8	−662.1	180.5	44.28	108.4	332.0
220.0	1076.0	15.15	−731.4	−440.9	181.6	44.72	113.0	314.3
222.0	1052.0	14.81	−506.4	−209.4	182.6	45.23	118.8	295.7
224.0	1026.0	14.44	−269.1	35.6	183.7	45.83	126.5	276.0
226.0	995.9	14.02	−14.9	298.9	184.9	46.54	137.5	254.8
227.0	979.2	13.79	121.0	440.1	185.5	46.95	145.1	243.4
228.0	960.9	13.53	264.7	589.9	186.2	47.42	154.9	231.5
229.0	940.3	13.24	418.8	751.1	186.9	47.96	168.4	218.7
230.0	916.6	12.91	587.8	928.7	187.6	48.60	188.1	204.9
230.5	903.1	12.72	680.1	1026.0	188.1	48.97	202.1	197.4
231.0	888.1	12.50	779.7	1132.0	188.5	49.39	220.8	189.5
231.5	870.9	12.26	889.3	1248.0	189.0	49.88	247.2	181.0
232.0	850.5	11.98	1014.0	1381.0	189.6	50.46	287.9	171.7
232.2	841.1	11.84	1069.0	1441.0	189.8	50.73	311.4	167.6
232.4	830.7	11.70	1130.0	1506.0	190.1	51.02	341.9	163.4
232.6	818.9	11.53	1197.0	1578.0	190.4	51.36	383.4	158.9
232.8	805.2	11.34	1273.0	1661.0	190.8	51.74	443.4	154.0
233.0	788.6	11.10	1362.0	1758.0	191.2	52.20	539.4	148.6
233.1	778.7	10.96	1414.0	1816.0	191.5	52.47	613.9	145.6
233.2	767.1	10.80	1475.0	1882.0	191.7	52.78	723.0	142.5
233.3	752.9	10.60	1547.0	1962.0	192.1	53.15	900.8	138.9
233.4	734.2	10.34	1642.0	2068.0	192.5	53.61	1254.0	134.8
235.0	349.1	4.916	3930.0	4825.0	204.3	51.65	315.3	124.4
235.5	336.9	4.744	4041.0	4969.0	204.9	51.17	266.1	125.8
236.0	326.9	4.603	4138.0	5094.0	205.5	50.77	233.9	127.1
236.5	318.5	4.484	4223.0	5204.0	205.9	50.41	210.9	128.3
237.0	311.1	4.381	4301.0	5305.0	206.4	50.10	193.5	129.5
238.0	298.8	4.207	4440.0	5486.0	207.1	49.57	168.7	131.6
239.0	288.6	4.064	4563.0	5645.0	207.8	49.13	151.9	133.5
240.0	279.9	3.941	4674.0	5791.0	208.4	48.75	139.5	135.3
241.0	272.3	3.835	4778.0	5925.0	209.0	48.42	130.0	136.9
242.0	265.6	3.740	4875.0	6051.0	209.5	48.14	122.4	138.5
244.0	254.1	3.578	5054.0	6284.0	210.4	47.66	111.1	141.4
246.0	244.4	3.441	5219.0	6498.0	211.3	47.29	103.0	144.1
248.0	236.1	3.324	5374.0	6698.0	212.1	46.98	96.92	146.6
250.0	228.7	3.221	5520.0	6887.0	212.9	46.74	92.14	148.9
252.0	222.2	3.129	5661.0	7067.0	213.6	46.55	88.30	151.1
254.0	216.3	3.046	5796.0	7240.0	214.3	46.39	85.14	153.1
256.0	210.9	2.970	5926.0	7408.0	214.9	46.27	82.50	155.1
260.0	201.5	2.837	6178.0	7729.0	216.2	46.11	78.36	158.7
265.0	191.5	2.696	6479.0	8111.0	217.6	46.03	74.62	163.0
270.0	182.9	2.575	6769.0	8477.0	219.0	46.06	71.92	166.8

Thermophysical properties of nitrogen trifluoride—Continued

T K	Density kg/m^3	Density mol/dm^3	E J/mol	H J/mol	S J/(mol·K)	C_v J/(mol·K)	C_p J/(mol·K)	Sound m/s
275.0	175.4	2.470	7050.0	8831.0	220.3	46.15	69.91	170.4
280.0	168.9	2.378	7326.0	9177.0	221.6	46.30	68.39	173.8
285.0	163.0	2.295	7598.0	9516.0	222.8	46.50	67.22	177.0
290.0	157.6	2.219	7867.0	9850.0	223.9	46.73	66.32	180.1
295.0	152.8	2.151	8134.0	10180.0	225.1	46.99	65.62	183.0
300.0	148.3	2.088	8399.0	10510.0	226.1	47.27	65.08	185.8
310.0	140.4	1.976	8927.0	11150.0	228.3	47.88	64.37	191.1
320.0	133.5	1.879	9453.0	11790.0	230.3	48.53	63.99	196.1
330.0	127.4	1.794	9981.0	12430.0	232.3	49.20	63.86	200.8
340.0	122.0	1.718	10510.0	13070.0	234.2	49.89	63.88	205.2
350.0	117.1	1.649	11040.0	13710.0	236.0	50.58	64.02	209.5
360.0	112.7	1.587	11580.0	14350.0	237.8	51.26	64.25	213.6
370.0	108.7	1.530	12120.0	15000.0	239.6	51.93	64.53	217.6
380.0	105.0	1.478	12670.0	15640.0	241.3	52.59	64.86	221.4
400.0	98.41	1.386	13770.0	16950.0	244.7	53.87	65.59	228.7
420.0	92.71	1.305	14900.0	18270.0	247.9	55.08	66.37	235.6
440.0	87.72	1.235	16040.0	19600.0	251.0	56.21	67.16	242.1
460.0	83.30	1.173	17200.0	20950.0	254.0	57.27	67.95	248.5
480.0	79.34	1.117	18380.0	22320.0	256.9	58.27	68.72	254.5
500.0	75.78	1.067	19580.0	23700.0	259.7	59.21	69.48	260.4
				4.45 MPa isobar				
67.18[a]	1869.0	26.32	−12320.0	−12150.0	93.17	50.50	73.88	1379.0
68.0	1866.0	26.28	−12260.0	−12090.0	94.06	49.88	72.85	1348.0
70.0	1859.0	26.18	−12110.0	−11940.0	96.15	48.76	71.12	1287.0
80.0	1821.0	25.64	−11420.0	−11240.0	105.5	46.80	69.93	1157.0
100.0	1741.0	24.52	−10010.0	−9831.0	121.2	44.01	70.81	1054.0
120.0	1658.0	23.34	−8610.0	−8419.0	134.1	41.23	70.41	934.0
140.0	1570.0	22.10	−7209.0	−7008.0	145.0	39.92	71.01	817.1
160.0	1476.0	20.78	−5780.0	−5566.0	154.6	39.77	73.50	707.9
170.0	1426.0	20.08	−5043.0	−4821.0	159.1	40.00	75.65	652.8
180.0	1372.0	19.32	−4281.0	−4051.0	163.5	40.39	78.56	595.6
190.0	1314.0	18.50	−3487.0	−3246.0	167.9	40.96	82.48	534.9
195.0	1283.0	18.06	−3074.0	−2828.0	170.1	41.32	84.97	502.9
200.0	1249.0	17.59	−2649.0	−2396.0	172.2	41.76	87.95	469.6
205.0	1213.0	17.08	−2208.0	−1947.0	174.5	42.27	91.62	434.7
210.0	1173.0	16.52	−1748.0	−1478.0	176.7	42.90	96.32	397.8
215.0	1129.0	15.89	−1261.0	−981.4	179.1	43.68	102.7	358.5
217.0	1109.0	15.62	−1058.0	−772.6	180.0	44.05	106.1	341.9
218.0	1099.0	15.47	−953.2	−665.6	180.5	44.25	108.0	333.4
220.0	1077.0	15.17	−738.6	−445.2	181.5	44.69	112.5	315.8
222.0	1053.0	14.83	−514.8	−214.8	182.6	45.19	118.1	297.3
224.0	1028.0	14.47	−279.0	28.6	183.7	45.77	125.6	277.8
226.0	998.2	14.06	−27.0	289.7	184.8	46.46	136.1	256.9
228.0	963.9	13.57	249.0	576.9	186.1	47.32	152.4	234.0
229.0	943.9	13.29	400.3	735.2	186.8	47.83	164.8	221.5
230.0	921.1	12.97	565.1	908.2	187.5	48.44	182.5	208.1
231.0	894.0	12.59	750.1	1104.0	188.4	49.19	210.6	193.3
231.5	878.0	12.36	854.0	1214.0	188.8	49.63	232.2	185.2
232.0	859.5	12.10	969.7	1337.0	189.4	50.15	263.7	176.4
232.5	837.2	11.79	1103.0	1481.0	190.0	50.78	314.7	166.8
232.7	826.7	11.64	1164.0	1547.0	190.3	51.08	345.5	162.6
232.8	820.9	11.56	1197.0	1582.0	190.4	51.24	364.7	160.4
233.0	808.1	11.38	1269.0	1660.0	190.8	51.60	414.1	155.8
233.2	793.1	11.17	1351.0	1749.0	191.2	52.01	488.0	150.8
233.4	774.5	10.91	1450.0	1858.0	191.6	52.52	612.4	145.2
233.5	763.0	10.74	1510.0	1924.0	191.9	52.82	714.2	142.2
233.6	749.2	10.55	1581.0	2003.0	192.2	53.17	872.6	138.8
233.7	731.5	10.30	1671.0	2103.0	192.7	53.59	1157.0	135.1
233.8	706.0	9.941	1799.0	2247.0	193.3	54.16	1840.0	130.6
233.9	649.3	9.143	2083.0	2570.0	194.7	55.15	7243.0	124.1
234.0	456.1	6.422	3167.0	3860.0	200.2	54.66	2236.0	119.1

Thermophysical properties of nitrogen trifluoride—Continued

T K	Density kg/m^3	Density mol/dm^3	E J/mol	H J/mol	S J/(mol·K)	C_v J/(mol·K)	C_p J/(mol·K)	Sound m/s
234.1	435.3	6.130	3303.0	4029.0	200.9	54.21	1341.0	119.6
234.2	421.9	5.941	3395.0	4144.0	201.4	53.88	1003.0	120.0
234.3	411.8	5.798	3467.0	4235.0	201.8	53.60	818.7	120.4
234.4	403.5	5.682	3527.0	4310.0	202.1	53.37	700.7	120.9
234.6	390.4	5.498	3625.0	4435.0	202.6	52.97	556.0	121.6
234.8	380.2	5.353	3705.0	4537.0	203.1	52.64	469.2	122.3
235.0	371.6	5.233	3774.0	4624.0	203.4	52.35	410.6	123.0
235.2	364.3	5.130	3834.0	4702.0	203.8	52.09	368.0	123.6
235.4	357.9	5.040	3889.0	4772.0	204.1	51.86	335.5	124.2
236.0	342.3	4.820	4029.0	4952.0	204.8	51.27	271.2	125.9
236.5	332.0	4.675	4127.0	5079.0	205.4	50.86	238.0	127.2
237.0	323.3	4.552	4214.0	5192.0	205.8	50.51	214.2	128.4
237.5	315.7	4.446	4293.0	5294.0	206.3	50.20	196.3	129.5
238.0	309.0	4.351	4366.0	5389.0	206.7	49.91	182.3	130.6
239.0	297.6	4.190	4498.0	5560.0	207.4	49.43	161.5	132.6
240.0	288.0	4.055	4616.0	5714.0	208.0	49.02	146.8	134.5
241.0	279.7	3.938	4725.0	5855.0	208.6	48.66	135.8	136.2
242.0	272.4	3.835	4826.0	5986.0	209.2	48.36	127.2	137.8
244.0	260.0	3.661	5011.0	6227.0	210.2	47.85	114.5	140.8
246.0	249.7	3.516	5181.0	6447.0	211.1	47.45	105.6	143.5
248.0	240.9	3.392	5339.0	6651.0	211.9	47.13	98.92	146.0
250.0	233.2	3.284	5488.0	6843.0	212.7	46.87	93.77	148.4
252.0	226.4	3.188	5631.0	7027.0	213.4	46.66	89.66	150.6
254.0	220.3	3.101	5767.0	7202.0	214.1	46.50	86.30	152.7
256.0	214.7	3.023	5900.0	7372.0	214.7	46.37	83.50	154.7
258.0	209.6	2.951	6029.0	7537.0	215.4	46.27	81.14	156.6
260.0	204.9	2.885	6154.0	7697.0	216.0	46.19	79.13	158.4
265.0	194.5	2.739	6457.0	8082.0	217.5	46.10	75.21	162.6
270.0	185.7	2.615	6749.0	8451.0	218.8	46.11	72.39	166.5
275.0	178.1	2.507	7032.0	8807.0	220.2	46.20	70.30	170.2
280.0	171.3	2.412	7310.0	9155.0	221.4	46.34	68.71	173.6
285.0	165.3	2.327	7583.0	9495.0	222.6	46.53	67.49	176.8
290.0	159.8	2.250	7852.0	9830.0	223.8	46.76	66.55	179.9
295.0	154.8	2.180	8120.0	10160.0	224.9	47.02	65.83	182.8
300.0	150.3	2.116	8386.0	10490.0	226.0	47.29	65.27	185.7
310.0	142.2	2.002	8915.0	11140.0	228.1	47.90	64.52	191.0
320.0	135.2	1.904	9442.0	11780.0	230.2	48.54	64.12	196.0
330.0	129.0	1.817	9971.0	12420.0	232.1	49.21	63.96	200.7
340.0	123.5	1.739	10500.0	13060.0	234.1	49.90	63.97	205.2
350.0	118.6	1.670	11030.0	13700.0	235.9	50.58	64.10	209.4
360.0	114.1	1.606	11570.0	14340.0	237.7	51.27	64.32	213.5
370.0	110.0	1.549	12110.0	14990.0	239.5	51.94	64.59	217.5
380.0	106.2	1.496	12660.0	15630.0	241.2	52.60	64.91	221.3
400.0	99.57	1.402	13770.0	16940.0	244.6	53.87	65.63	228.6
420.0	93.80	1.321	14890.0	18260.0	247.8	55.08	66.41	235.6
440.0	88.74	1.250	16030.0	19600.0	250.9	56.21	67.20	242.2
460.0	84.26	1.186	17200.0	20950.0	253.9	57.27	67.98	248.5
480.0	80.25	1.130	18380.0	22310.0	256.8	58.27	68.75	254.5
500.0	76.64	1.079	19570.0	23700.0	259.6	59.21	69.50	260.4
				4.50 MPa isobar				
67.19[a]	1869.0	26.32	−12320.0	−12140.0	93.18	50.53	73.87	1378.0
68.0	1866.0	26.28	−12260.0	−12080.0	94.06	49.91	72.84	1347.0
70.0	1859.0	26.18	−12110.0	−11940.0	96.14	48.79	71.12	1287.0
80.0	1821.0	25.64	−11420.0	−11240.0	105.5	46.81	69.93	1157.0
100.0	1742.0	24.52	−10010.0	−9830.0	121.2	44.01	70.80	1054.0
120.0	1658.0	23.35	−8611.0	−8418.0	134.1	41.22	70.40	934.4
140.0	1570.0	22.10	−7210.0	−7006.0	145.0	39.91	71.01	817.5
160.0	1476.0	20.78	−5782.0	−5565.0	154.6	39.77	73.49	708.3
170.0	1426.0	20.08	−5044.0	−4820.0	159.1	39.99	75.63	653.3
180.0	1373.0	19.33	−4283.0	−4050.0	163.5	40.39	78.53	596.0

Thermophysical properties of nitrogen trifluoride—Continued

T K	Density kg/m^3	Density mol/dm^3	E J/mol	H J/mol	S J/(mol·K)	C_v J/(mol·K)	C_p	Sound m/s
190.0	1314.0	18.51	−3489.0	−3246.0	167.9	40.95	82.43	535.5
195.0	1283.0	18.07	−3077.0	−2828.0	170.0	41.32	84.91	503.5
200.0	1249.0	17.59	−2652.0	−2396.0	172.2	41.75	87.87	470.3
205.0	1213.0	17.08	−2212.0	−1948.0	174.4	42.26	91.50	435.5
210.0	1174.0	16.53	−1752.0	−1480.0	176.7	42.88	96.16	398.8
215.0	1130.0	15.90	−1267.0	−983.9	179.0	43.65	102.5	359.7
217.0	1110.0	15.63	−1064.0	−775.7	180.0	44.02	105.8	343.2
218.0	1100.0	15.49	−959.6	−669.0	180.5	44.22	107.7	334.7
220.0	1078.0	15.18	−745.8	−449.5	181.5	44.65	112.0	317.2
222.0	1055.0	14.85	−523.1	−220.1	182.5	45.14	117.5	298.9
224.0	1029.0	14.49	−288.7	21.8	183.6	45.71	124.7	279.6
226.0	1000.0	14.09	−38.8	280.7	184.8	46.39	134.8	258.9
228.0	966.8	13.61	233.8	564.4	186.0	47.22	150.1	236.4
229.0	947.4	13.34	382.6	719.9	186.7	47.72	161.5	224.2
230.0	925.4	13.03	543.6	889.0	187.4	48.30	177.5	211.1
231.0	899.6	12.67	722.5	1078.0	188.2	49.00	202.0	196.9
231.5	884.6	12.46	821.8	1183.0	188.7	49.41	220.2	189.1
232.0	867.5	12.22	930.7	1299.0	189.2	49.88	245.5	180.8
232.5	847.5	11.93	1054.0	1431.0	189.8	50.44	283.6	171.8
232.7	838.3	11.80	1108.0	1489.0	190.0	50.70	305.0	168.0
232.8	833.4	11.73	1137.0	1521.0	190.2	50.84	317.7	166.0
233.0	822.8	11.58	1199.0	1587.0	190.4	51.14	348.6	161.8
233.2	810.8	11.42	1266.0	1661.0	190.8	51.47	390.2	157.4
233.4	796.9	11.22	1343.0	1744.0	191.1	51.85	449.2	152.8
233.6	780.3	10.99	1433.0	1842.0	191.5	52.29	540.3	147.7
233.7	770.5	10.85	1485.0	1900.0	191.8	52.55	607.7	144.9
233.8	759.2	10.69	1544.0	1965.0	192.1	52.84	701.0	142.0
233.9	745.9	10.50	1613.0	2041.0	192.4	53.17	839.3	138.9
240.0	296.5	4.175	4555.0	5633.0	207.6	49.30	155.3	133.6
234.1	707.2	9.958	1809.0	2261.0	193.3	54.05	1504.0	131.6
234.2	672.6	9.471	1984.0	2459.0	194.2	54.69	2679.0	127.0
234.3	590.8	8.319	2407.0	2948.0	196.3	55.56	9069.0	121.4
234.4	489.9	6.898	2982.0	3634.0	199.2	55.09	3383.0	119.5
234.5	459.1	6.464	3175.0	3871.0	200.2	54.58	1750.0	119.7
234.6	441.3	6.214	3292.0	4017.0	200.8	54.21	1226.0	120.1
234.7	428.6	6.035	3379.0	4125.0	201.3	53.91	964.1	120.5
234.8	418.6	5.894	3449.0	4213.0	201.6	53.66	805.1	120.8
234.9	410.3	5.778	3509.0	4288.0	202.0	53.44	697.6	121.2
235.0	403.3	5.678	3561.0	4353.0	202.2	53.23	619.6	121.6
235.2	391.5	5.513	3649.0	4466.0	202.7	52.88	513.2	122.3
235.4	382.0	5.379	3724.0	4561.0	203.1	52.58	443.4	122.9
235.6	373.9	5.265	3790.0	4644.0	203.5	52.31	393.7	123.5
235.8	366.9	5.166	3848.0	4719.0	203.8	52.06	356.4	124.1
236.0	360.7	5.078	3901.0	4787.0	204.1	51.84	327.1	124.7
236.5	347.6	4.894	4018.0	4937.0	204.7	51.36	275.5	126.1
237.0	337.0	4.745	4117.0	5066.0	205.3	50.95	241.6	127.4
237.5	328.0	4.619	4206.0	5180.0	205.8	50.60	217.3	128.5
238.0	320.2	4.509	4286.0	5284.0	206.2	50.28	199.0	129.7
239.0	307.2	4.326	4429.0	5469.0	207.0	49.74	173.0	131.7
240.0	296.5	4.175	4555.0	5633.0	207.6	49.30	155.3	133.6
241.0	287.4	4.047	4669.0	5781.0	208.3	48.91	142.3	135.4
242.0	279.5	3.935	4775.0	5918.0	208.8	48.58	132.4	137.1
243.0	272.5	3.836	4874.0	6047.0	209.4	48.29	124.5	138.6
244.0	266.2	3.748	4967.0	6168.0	209.9	48.04	118.1	140.1
246.0	255.2	3.593	5142.0	6394.0	210.8	47.61	108.3	142.9
248.0	245.9	3.462	5303.0	6603.0	211.6	47.27	101.0	145.5
250.0	237.8	3.349	5455.0	6799.0	212.4	47.00	95.48	147.9
252.0	230.7	3.248	5600.0	6985.0	213.2	46.78	91.07	150.1
254.0	224.3	3.158	5739.0	7164.0	213.9	46.60	87.50	152.2
256.0	218.5	3.076	5873.0	7336.0	214.5	46.46	84.53	154.2
258.0	213.2	3.002	6003.0	7502.0	215.2	46.35	82.04	156.2
260.0	208.3	2.933	6130.0	7664.0	215.8	46.27	79.93	158.0

Thermophysical properties of nitrogen trifluoride—Continued

T K	Density kg/m^3	Density mol/dm^3	E J/mol	H J/mol	S J/(mol·K)	C_v J/(mol·K)	C_p J/(mol·K)	Sound m/s
265.0	197.6	2.783	6436.0	8053.0	217.3	46.17	75.81	162.3
270.0	188.6	2.655	6729.0	8424.0	218.7	46.17	72.87	166.2
275.0	180.7	2.544	7014.0	8783.0	220.0	46.24	70.69	169.9
280.0	173.8	2.447	7293.0	9132.0	221.3	46.38	69.04	173.4
285.0	167.6	2.360	7567.0	9474.0	222.5	46.57	67.77	176.6
290.0	162.0	2.281	7838.0	9810.0	223.6	46.79	66.79	179.7
295.0	156.9	2.210	8106.0	10140.0	224.8	47.04	66.04	182.7
300.0	152.3	2.145	8373.0	10470.0	225.9	47.32	65.45	185.5
310.0	144.1	2.028	8903.0	11120.0	228.0	47.92	64.67	190.9
320.0	136.9	1.928	9431.0	11770.0	230.1	48.56	64.24	195.9
330.0	130.6	1.839	9960.0	12410.0	232.0	49.23	64.07	200.6
340.0	125.0	1.761	10490.0	13050.0	233.9	49.91	64.06	205.1
350.0	120.0	1.690	11030.0	13690.0	235.8	50.59	64.18	209.4
360.0	115.5	1.626	11560.0	14330.0	237.6	51.27	64.39	213.5
370.0	111.3	1.567	12110.0	14980.0	239.4	51.94	64.65	217.5
380.0	107.5	1.514	12650.0	15620.0	241.1	52.60	64.97	221.3
400.0	100.7	1.418	13760.0	16930.0	244.5	53.88	65.68	228.6
420.0	94.89	1.336	14880.0	18250.0	247.7	55.08	66.44	235.6
440.0	89.76	1.264	16030.0	19590.0	250.8	56.21	67.23	242.2
460.0	85.22	1.200	17190.0	20940.0	253.8	57.27	68.01	248.5
480.0	81.16	1.143	18370.0	22310.0	256.7	58.27	68.77	254.6
500.0	77.51	1.091	19570.0	23690.0	259.5	59.21	69.52	260.5
				4.55 MPa isobar				
67.20[a]	1869.0	26.32	−12310.0	−12140.0	93.19	50.56	73.85	1377.0
68.0	1866.0	26.28	−12260.0	−12080.0	94.06	49.95	72.84	1346.0
70.0	1859.0	26.18	−12110.0	−11940.0	96.14	48.82	71.11	1286.0
80.0	1821.0	25.64	−11420.0	−11240.0	105.5	46.81	69.93	1157.0
100.0	1742.0	24.52	−10010.0	−9828.0	121.2	44.00	70.80	1054.0
120.0	1658.0	23.35	−8611.0	−8416.0	134.1	41.21	70.40	934.8
140.0	1570.0	22.11	−7211.0	−7005.0	145.0	39.91	71.00	817.9
160.0	1476.0	20.79	−5783.0	−5564.0	154.6	39.76	73.48	708.7
170.0	1426.0	20.08	−5046.0	−4819.0	159.1	39.99	75.61	653.7
180.0	1373.0	19.33	−4285.0	−4049.0	163.5	40.38	78.50	596.5
190.0	1315.0	18.51	−3492.0	−3246.0	167.9	40.95	82.39	536.0
195.0	1283.0	18.07	−3080.0	−2828.0	170.0	41.31	84.84	504.2
200.0	1250.0	17.60	−2655.0	−2397.0	172.2	41.74	87.78	471.0
205.0	1214.0	17.09	−2215.0	−1949.0	174.4	42.25	91.39	436.3
210.0	1174.0	16.54	−1756.0	−1481.0	176.7	42.86	95.99	399.8
215.0	1130.0	15.92	−1272.0	−986.4	179.0	43.63	102.2	360.8
217.0	1111.0	15.64	−1070.0	−778.8	180.0	43.99	105.5	344.4
218.0	1101.0	15.50	−965.9	−672.4	180.5	44.19	107.3	336.0
220.0	1080.0	15.20	−753.0	−453.7	181.4	44.61	111.6	318.6
222.0	1056.0	14.87	−531.2	−225.4	182.5	45.10	116.9	300.5
224.0	1031.0	14.52	−298.3	15.1	183.6	45.66	123.9	281.4
226.0	1003.0	14.12	−50.4	272.0	184.7	46.32	133.5	261.0
228.0	969.6	13.65	219.1	552.4	185.9	47.12	148.0	238.8
229.0	950.7	13.39	365.5	705.4	186.6	47.60	158.5	226.9
230.0	929.4	13.09	523.1	870.8	187.3	48.16	173.1	214.1
231.0	904.7	12.74	696.7	1054.0	188.1	48.82	194.7	200.3
231.5	890.5	12.54	792.1	1155.0	188.6	49.21	210.2	192.8
232.0	874.7	12.32	895.5	1265.0	189.0	49.64	231.0	184.9
232.5	856.4	12.06	1010.0	1387.0	189.6	50.15	260.9	176.5
233.0	834.6	11.75	1141.0	1529.0	190.2	50.75	307.5	167.2
233.2	824.4	11.61	1201.0	1593.0	190.4	51.03	334.7	163.2
233.4	813.1	11.45	1266.0	1663.0	190.7	51.35	370.1	159.0
233.6	800.2	11.27	1338.0	1742.0	191.1	51.70	418.5	154.6
233.8	785.1	11.06	1420.0	1832.0	191.5	52.10	488.7	149.8
240.0	305.6	4.303	4490.0	5547.0	207.2	49.59	165.0	132.8
234.1	755.8	10.64	1576.0	2004.0	192.2	52.85	684.4	141.9
234.2	743.0	10.46	1643.0	2078.0	192.5	53.16	803.5	139.0
234.3	727.5	10.24	1722.0	2166.0	192.9	53.52	983.6	135.8

Thermophysical properties of nitrogen trifluoride—Continued

T K	Density kg/m³	Density mol/dm³	E J/mol	H J/mol	S J/(mol·K)	C_v J/(mol·K)	C_p J/(mol·K)	Sound m/s
234.4	708.0	9.969	1822.0	2278.0	193.4	53.94	1282.0	132.4
234.5	681.3	9.594	1957.0	2431.0	194.0	54.45	1843.0	128.8
234.6	641.1	9.027	2163.0	2667.0	195.0	55.04	3011.0	124.9
234.7	578.7	8.149	2494.0	3052.0	196.7	55.48	4526.0	121.6
234.8	517.6	7.288	2840.0	3465.0	198.4	55.27	3324.0	120.2
234.9	482.4	6.793	3054.0	3724.0	199.5	54.85	2018.0	120.1
235.0	460.9	6.491	3191.0	3892.0	200.2	54.49	1417.0	120.3
235.1	445.8	6.277	3291.0	4016.0	200.8	54.18	1099.0	120.6
235.2	434.0	6.111	3371.0	4116.0	201.2	53.92	905.8	121.0
235.3	424.5	5.977	3438.0	4199.0	201.6	53.68	775.9	121.3
235.4	416.4	5.863	3496.0	4272.0	201.9	53.47	682.6	121.6
235.6	403.1	5.676	3593.0	4395.0	202.4	53.10	557.2	122.3
235.8	392.5	5.527	3675.0	4498.0	202.8	52.79	476.3	122.9
236.0	383.6	5.402	3745.0	4587.0	203.2	52.51	419.5	123.5
236.2	376.0	5.294	3807.0	4667.0	203.5	52.26	377.3	124.1
236.4	369.2	5.199	3863.0	4739.0	203.8	52.03	344.6	124.7
237.0	352.7	4.967	4008.0	4924.0	204.6	51.44	279.0	126.3
237.5	341.8	4.813	4109.0	5054.0	205.2	51.03	244.6	127.5
238.0	332.7	4.684	4199.0	5170.0	205.7	50.68	219.9	128.7
238.5	324.7	4.572	4280.0	5275.0	206.1	50.36	201.3	129.8
239.0	317.6	4.473	4355.0	5372.0	206.5	50.08	186.7	130.8
240.0	305.6	4.303	4490.0	5547.0	207.2	49.59	165.0	132.8
241.0	295.6	4.162	4611.0	5704.0	207.9	49.17	149.7	134.6
242.0	286.9	4.040	4722.0	5848.0	208.5	48.82	138.3	136.3
243.0	279.3	3.933	4825.0	5982.0	209.0	48.51	129.3	138.0
244.0	272.5	3.837	4921.0	6107.0	209.6	48.24	122.1	139.5
246.0	260.9	3.673	5101.0	6340.0	210.5	47.78	111.2	142.3
248.0	251.0	3.535	5267.0	6554.0	211.4	47.42	103.3	144.9
250.0	242.5	3.415	5422.0	6754.0	212.2	47.13	97.28	147.4
252.0	235.0	3.309	5569.0	6944.0	212.9	46.89	92.56	149.6
254.0	228.4	3.215	5710.0	7125.0	213.6	46.71	88.75	151.8
256.0	222.3	3.131	5846.0	7299.0	214.3	46.56	85.61	153.8
258.0	216.8	3.053	5978.0	7468.0	215.0	46.44	82.98	155.8
260.0	211.8	2.982	6106.0	7631.0	215.6	46.35	80.74	157.6
265.0	200.8	2.827	6414.0	8024.0	217.1	46.23	76.43	162.0
270.0	191.4	2.695	6710.0	8398.0	218.5	46.22	73.35	166.0
275.0	183.4	2.582	6996.0	8759.0	219.8	46.29	71.08	169.7
280.0	176.3	2.482	7276.0	9110.0	221.1	46.42	69.37	173.1
285.0	169.9	2.393	7551.0	9453.0	222.3	46.60	68.05	176.4
290.0	164.2	2.313	7823.0	9791.0	223.5	46.82	67.04	179.5
295.0	159.1	2.240	8092.0	10120.0	224.6	47.07	66.25	182.5
300.0	154.3	2.173	8359.0	10450.0	225.7	47.34	65.64	185.3
310.0	145.9	2.054	8891.0	11110.0	227.9	47.93	64.82	190.7
320.0	138.6	1.952	9420.0	11750.0	229.9	48.57	64.37	195.7
330.0	132.2	1.862	9950.0	12390.0	231.9	49.24	64.17	200.5
340.0	126.6	1.782	10480.0	13030.0	233.8	49.92	64.15	205.0
350.0	121.5	1.710	11020.0	13680.0	235.7	50.60	64.26	209.3
360.0	116.8	1.645	11550.0	14320.0	237.5	51.28	64.45	213.4
370.0	112.6	1.586	12100.0	14970.0	239.3	51.95	64.71	217.4
380.0	108.8	1.532	12640.0	15610.0	241.0	52.61	65.02	221.3
400.0	101.9	1.435	13750.0	16920.0	244.3	53.88	65.72	228.6
420.0	95.97	1.351	14880.0	18240.0	247.6	55.08	66.48	235.6
440.0	90.78	1.278	16020.0	19580.0	250.7	56.21	67.26	242.2
460.0	86.18	1.213	17180.0	20930.0	253.7	57.27	68.04	248.5
480.0	82.07	1.156	18370.0	22300.0	256.6	58.27	68.80	254.6
500.0	78.37	1.104	19560.0	23690.0	259.4	59.21	69.54	260.5
				4.60 MPa isobar				
67.21[a]	1869.0	26.32	−12310.0	−12140.0	93.19	50.60	73.84	1375.0
68.0	1867.0	26.28	−12260.0	−12080.0	94.05	49.99	72.84	1345.0
70.0	1859.0	26.18	−12110.0	−11940.0	96.14	48.85	71.11	1285.0

Thermophysical properties of nitrogen trifluoride—Continued

T K	Density kg/m^3	Density mol/dm^3	E J/mol	H J/mol	S J/(mol·K)	C_v J/(mol·K)	C_p J/(mol·K)	Sound m/s
80.0	1821.0	25.64	-11420.0	-11240.0	105.5	46.82	69.92	1156.0
100.0	1742.0	24.52	-10010.0	-9826.0	121.2	44.00	70.80	1055.0
120.0	1658.0	23.35	-8612.0	-8415.0	134.1	41.21	70.39	935.3
140.0	1570.0	22.11	-7212.0	-7004.0	145.0	39.90	70.99	818.3
160.0	1476.0	20.79	-5784.0	-5563.0	154.6	39.76	73.46	709.1
170.0	1426.0	20.08	-5047.0	-4818.0	159.1	39.98	75.59	654.1
180.0	1373.0	19.33	-4287.0	-4049.0	163.5	40.38	78.47	597.0
190.0	1315.0	18.52	-3494.0	-3246.0	167.8	40.94	82.34	536.6
195.0	1284.0	18.08	-3082.0	-2828.0	170.0	41.30	84.78	504.8
200.0	1250.0	17.61	-2658.0	-2397.0	172.2	41.73	87.70	471.7
205.0	1214.0	17.10	-2219.0	-1950.0	174.4	42.24	91.28	437.1
210.0	1175.0	16.55	-1761.0	-1483.0	176.7	42.85	95.83	400.7
215.0	1131.0	15.93	-1278.0	-988.8	179.0	43.61	102.0	362.0
217.0	1112.0	15.66	-1076.0	-781.8	179.9	43.96	105.2	345.6
218.0	1102.0	15.52	-972.2	-675.7	180.4	44.16	107.0	337.3
220.0	1081.0	15.22	-760.0	-457.8	181.4	44.58	111.1	320.0
222.0	1058.0	14.89	-539.3	-230.5	182.4	45.06	116.3	302.1
224.0	1033.0	14.54	-307.7	8.6	183.5	45.61	123.1	283.1
226.0	1005.0	14.15	-61.7	263.5	184.6	46.25	132.3	263.0
228.0	972.3	13.69	204.8	540.8	185.9	47.03	146.0	241.2
229.0	953.8	13.43	348.9	691.4	186.5	47.50	155.8	229.4
230.0	933.2	13.14	503.4	853.5	187.2	48.03	169.1	217.0
231.0	909.5	12.81	672.4	1032.0	188.0	48.66	188.4	203.6
231.5	896.1	12.62	764.4	1129.0	188.4	49.02	201.8	196.4
232.0	881.2	12.41	863.3	1234.0	188.9	49.42	219.3	188.8
232.5	864.3	12.17	971.5	1349.0	189.4	49.89	243.5	180.8
233.0	844.6	11.89	1093.0	1479.0	189.9	50.43	278.9	172.1
233.5	820.7	11.56	1234.0	1632.0	190.6	51.08	336.8	162.6
233.7	809.3	11.40	1299.0	1703.0	190.9	51.39	371.9	158.5
233.8	803.1	11.31	1334.0	1741.0	191.1	51.56	393.6	156.3
234.0	789.2	11.11	1411.0	1825.0	191.4	51.92	449.6	151.9
234.2	772.9	10.88	1500.0	1923.0	191.8	52.35	532.2	147.1
234.4	752.7	10.60	1607.0	2041.0	192.3	52.85	665.3	141.9
234.5	740.4	10.43	1671.0	2112.0	192.6	53.14	767.1	139.2
234.6	726.0	10.22	1746.0	2196.0	193.0	53.47	910.6	136.3
234.7	708.5	9.976	1836.0	2297.0	193.4	53.84	1123.0	133.3
234.8	686.4	9.665	1948.0	2424.0	194.0	54.26	1453.0	130.1
234.9	657.6	9.259	2097.0	2593.0	194.7	54.73	1960.0	127.0
235.0	619.7	8.726	2294.0	2821.0	195.7	55.15	2599.0	124.1
235.1	575.1	8.097	2536.0	3104.0	196.9	55.37	2948.0	122.1
235.2	533.2	7.508	2773.0	3386.0	198.1	55.27	2590.0	121.1
235.3	501.6	7.062	2962.0	3613.0	199.0	54.98	1962.0	120.8
235.4	479.0	6.745	3102.0	3784.0	199.8	54.67	1490.0	120.8
235.5	462.3	6.509	3210.0	3917.0	200.3	54.39	1185.0	121.0
235.6	449.1	6.324	3297.0	4024.0	200.8	54.13	982.0	121.2
235.7	438.4	6.174	3370.0	4115.0	201.2	53.89	840.8	121.5
235.8	429.4	6.047	3433.0	4194.0	201.5	53.68	737.6	121.8
236.0	414.8	5.840	3538.0	4326.0	202.1	53.31	597.7	122.4
236.2	403.1	5.676	3626.0	4436.0	202.5	52.98	507.5	123.0
236.4	393.4	5.539	3700.0	4531.0	202.9	52.69	444.4	123.6
236.6	385.1	5.422	3767.0	4615.0	203.3	52.43	397.7	124.1
236.8	377.8	5.320	3826.0	4691.0	203.6	52.20	361.7	124.7
237.0	371.3	5.229	3880.0	4760.0	203.9	51.98	333.0	125.2
237.5	357.7	5.036	3999.0	4913.0	204.5	51.51	281.5	126.5
238.0	346.6	4.880	4102.0	5044.0	205.1	51.10	247.0	127.7
238.5	337.2	4.748	4192.0	5161.0	205.6	50.75	222.1	128.9
239.0	329.0	4.633	4275.0	5267.0	206.0	50.43	203.3	130.0
240.0	315.4	4.441	4421.0	5456.0	206.8	49.90	176.5	132.0
241.0	304.2	4.284	4549.0	5623.0	207.5	49.45	158.2	133.9
242.0	294.7	4.150	4666.0	5774.0	208.1	49.06	144.8	135.6
243.0	286.5	4.034	4773.0	5914.0	208.7	48.73	134.6	137.3
244.0	279.2	3.931	4874.0	6044.0	209.2	48.44	126.4	138.8
246.0	266.7	3.755	5059.0	6284.0	210.2	47.95	114.3	141.7

Thermophysical properties of nitrogen trifluoride—Continued

T K	Density kg/m^3	Density mol/dm^3	E J/mol	H J/mol	S J/(mol·K)	C_v J/(mol·K)	C_p J/(mol·K)	Sound m/s
248.0	256.3	3.609	5229.0	6504.0	211.1	47.57	105.7	144.4
250.0	247.3	3.483	5388.0	6708.0	211.9	47.26	99.18	146.9
252.0	239.5	3.372	5537.0	6902.0	212.7	47.01	94.12	149.2
254.0	232.5	3.274	5681.0	7086.0	213.4	46.82	90.05	151.3
256.0	226.2	3.186	5818.0	7262.0	214.1	46.66	86.72	153.4
258.0	220.5	3.105	5951.0	7433.0	214.8	46.53	83.94	155.4
260.0	215.3	3.032	6081.0	7598.0	215.4	46.44	81.59	157.3
265.0	203.9	2.871	6392.0	7994.0	216.9	46.30	77.07	161.6
270.0	194.3	2.736	6690.0	8371.0	218.3	46.28	73.85	165.7
275.0	186.0	2.619	6978.0	8734.0	219.7	46.34	71.49	169.4
280.0	178.8	2.517	7259.0	9087.0	220.9	46.47	69.70	172.9
285.0	172.3	2.426	7536.0	9432.0	222.2	46.64	68.34	176.2
290.0	166.5	2.344	7808.0	9771.0	223.3	46.85	67.28	179.3
295.0	161.2	2.269	8078.0	10110.0	224.5	47.10	66.46	182.3
300.0	156.3	2.201	8346.0	10440.0	225.6	47.37	65.83	185.2
310.0	147.8	2.081	8879.0	11090.0	227.7	47.95	64.97	190.6
320.0	140.4	1.976	9409.0	11740.0	229.8	48.59	64.49	195.6
330.0	133.9	1.885	9940.0	12380.0	231.8	49.25	64.28	200.4
340.0	128.1	1.804	10470.0	13020.0	233.7	49.93	64.24	204.9
350.0	122.9	1.731	11010.0	13670.0	235.6	50.61	64.34	209.2
360.0	118.2	1.665	11550.0	14310.0	237.4	51.29	64.52	213.4
370.0	113.9	1.604	12090.0	14960.0	239.1	51.96	64.78	217.4
380.0	110.0	1.549	12640.0	15610.0	240.9	52.61	65.08	221.2
400.0	103.1	1.451	13740.0	16910.0	244.2	53.88	65.77	228.6
420.0	97.06	1.367	14870.0	18240.0	247.5	55.08	66.52	235.6
440.0	91.80	1.293	16020.0	19570.0	250.6	56.21	67.29	242.2
460.0	87.14	1.227	17180.0	20930.0	253.6	57.27	68.06	248.5
480.0	82.98	1.168	18360.0	22300.0	256.5	58.27	68.82	254.6
500.0	79.24	1.116	19560.0	23680.0	259.3	59.21	69.56	260.5
				4.65 MPa isobar				
67.22[a]	1869.0	26.32	-12310.0	-12140.0	93.20	50.63	73.82	1374.0
68.0	1867.0	26.28	-12260.0	-12080.0	94.05	50.02	72.84	1345.0
70.0	1859.0	26.18	-12110.0	-11940.0	96.13	48.88	71.11	1284.0
80.0	1821.0	25.64	-11420.0	-11240.0	105.5	46.83	69.92	1156.0
100.0	1742.0	24.53	-10010.0	-9825.0	121.2	43.99	70.79	1055.0
120.0	1658.0	23.35	-8613.0	-8414.0	134.1	41.20	70.39	935.7
140.0	1570.0	22.11	-7213.0	-7002.0	145.0	39.90	70.98	818.7
160.0	1477.0	20.79	-5785.0	-5562.0	154.6	39.75	73.45	709.5
170.0	1427.0	20.09	-5049.0	-4817.0	159.1	39.98	75.57	654.6
180.0	1373.0	19.34	-4288.0	-4048.0	163.5	40.37	78.44	597.5
190.0	1315.0	18.52	-3496.0	-3245.0	167.8	40.94	82.29	537.2
195.0	1284.0	18.08	-3085.0	-2828.0	170.0	41.30	84.72	505.5
200.0	1251.0	17.61	-2661.0	-2397.0	172.2	41.72	87.62	472.5
205.0	1215.0	17.11	-2222.0	-1951.0	174.4	42.22	91.17	438.0
210.0	1176.0	16.56	-1765.0	-1484.0	176.6	42.83	95.67	401.7
215.0	1132.0	15.94	-1283.0	-991.3	178.9	43.58	101.7	363.1
217.0	1113.0	15.67	-1081.0	-784.7	179.9	43.94	104.9	346.9
218.0	1103.0	15.53	-978.4	-679.0	180.4	44.13	106.6	338.6
220.0	1082.0	15.23	-767.0	-461.8	181.4	44.54	110.7	321.4
222.0	1059.0	14.91	-547.3	-235.5	182.4	45.02	115.8	303.6
224.0	1034.0	14.56	-317.0	2.3	183.5	45.56	122.3	284.8
226.0	1007.0	14.17	-72.8	255.2	184.6	46.19	131.2	264.9
228.0	974.9	13.73	190.8	529.6	185.8	46.95	144.1	243.5
229.0	956.9	13.47	333.0	678.1	186.5	47.40	153.3	231.9
230.0	936.9	13.19	484.6	837.1	187.1	47.91	165.5	219.8
231.0	914.1	12.87	649.4	1011.0	187.9	48.50	182.8	206.7
232.0	887.1	12.49	833.6	1206.0	188.7	49.22	209.6	192.5
232.5	871.4	12.27	936.4	1315.0	189.2	49.65	229.6	184.8
233.0	853.4	12.02	1050.0	1437.0	189.7	50.14	257.7	176.6
233.5	832.2	11.72	1179.0	1575.0	190.3	50.72	300.2	167.7
233.7	822.4	11.58	1236.0	1638.0	190.6	50.98	324.2	163.9

Thermophysical properties of nitrogen trifluoride—Continued

T K	Density kg/m³	Density mol/dm³	E J/mol	H J/mol	S J/(mol·K)	C_v J/(mol·K)	C_p J/(mol·K)	Sound m/s
233.8	817.1	11.51	1267.0	1671.0	190.7	51.12	338.4	162.0
234.0	805.7	11.34	1332.0	1742.0	191.0	51.43	372.9	158.0
234.2	792.8	11.16	1404.0	1821.0	191.4	51.77	418.8	153.7
234.4	777.9	10.95	1486.0	1911.0	191.8	52.15	482.9	149.3
234.6	760.2	10.70	1582.0	2016.0	192.2	52.60	578.1	144.5
234.8	738.1	10.39	1698.0	2146.0	192.8	53.12	731.3	139.4
234.9	724.6	10.20	1768.0	2224.0	193.1	53.42	846.3	136.8
235.0	708.8	9.980	1850.0	2316.0	193.5	53.75	1003.0	134.0
235.1	689.8	9.714	1948.0	2427.0	194.0	54.11	1218.0	131.3
235.2	666.8	9.390	2067.0	2562.0	194.5	54.49	1502.0	128.6
235.3	639.0	8.998	2212.0	2729.0	195.2	54.86	1831.0	126.1
235.4	607.0	8.547	2383.0	2927.0	196.1	55.14	2104.0	124.1
235.5	573.3	8.073	2567.0	3143.0	197.0	55.25	2186.0	122.7
235.6	541.9	7.630	2746.0	3356.0	197.9	55.19	2029.0	121.9
235.7	515.3	7.255	2903.0	3544.0	198.7	55.00	1734.0	121.5
235.8	494.0	6.955	3034.0	3702.0	199.4	54.76	1437.0	121.4
235.9	477.0	6.717	3141.0	3834.0	199.9	54.52	1198.0	121.5
236.0	463.2	6.523	3231.0	3944.0	200.4	54.28	1017.0	121.6
236.1	451.8	6.361	3308.0	4038.0	200.8	54.06	880.5	121.8
236.2	442.0	6.224	3374.0	4121.0	201.2	53.85	776.6	122.1
236.4	426.2	6.001	3486.0	4261.0	201.7	53.48	630.8	122.6
236.6	413.6	5.823	3578.0	4377.0	202.2	53.15	534.7	123.1
236.8	403.1	5.676	3657.0	4477.0	202.7	52.86	466.8	123.7
237.0	394.2	5.550	3727.0	4565.0	203.0	52.60	416.5	124.2
237.2	386.4	5.441	3789.0	4644.0	203.4	52.36	377.7	124.7
237.5	376.3	5.299	3873.0	4750.0	203.8	52.04	333.7	125.5
238.0	362.5	5.104	3992.0	4904.0	204.5	51.57	283.1	126.8
238.5	351.2	4.945	4096.0	5036.0	205.0	51.17	248.8	127.9
239.0	341.6	4.810	4187.0	5154.0	205.5	50.81	223.9	129.1
239.5	333.3	4.693	4270.0	5261.0	206.0	50.50	205.0	130.1
240.0	326.0	4.590	4346.0	5360.0	206.4	50.22	190.1	131.2
241.0	313.4	4.413	4484.0	5538.0	207.1	49.73	167.9	133.1
242.0	303.0	4.266	4607.0	5698.0	207.8	49.31	152.2	134.9
243.0	294.0	4.139	4720.0	5844.0	208.4	48.96	140.4	136.6
244.0	286.1	4.028	4825.0	5979.0	208.9	48.65	131.2	138.2
246.0	272.7	3.840	5016.0	6227.0	209.9	48.13	117.7	141.1
248.0	261.7	3.685	5191.0	6453.0	210.9	47.72	108.2	143.8
250.0	252.2	3.552	5353.0	6662.0	211.7	47.40	101.2	146.4
252.0	244.0	3.436	5505.0	6858.0	212.5	47.14	95.75	148.7
254.0	236.8	3.334	5651.0	7045.0	213.2	46.92	91.41	150.9
256.0	230.2	3.242	5790.0	7225.0	213.9	46.76	87.87	153.0
258.0	224.3	3.158	5925.0	7397.0	214.6	46.62	84.93	155.0
260.0	218.9	3.082	6056.0	7565.0	215.2	46.52	82.46	156.9
265.0	207.1	2.916	6370.0	7964.0	216.8	46.37	77.71	161.3
270.0	197.2	2.777	6670.0	8344.0	218.2	46.34	74.36	165.4
275.0	188.7	2.657	6960.0	8709.0	219.5	46.39	71.90	169.1
280.0	181.3	2.552	7242.0	9064.0	220.8	46.51	70.04	172.7
285.0	174.7	2.459	7520.0	9411.0	222.0	46.68	68.63	176.0
290.0	168.7	2.375	7793.0	9751.0	223.2	46.88	67.53	179.1
295.0	163.3	2.299	8064.0	10090.0	224.4	47.12	66.68	182.1
300.0	158.4	2.230	8333.0	10420.0	225.5	47.39	66.02	185.0
310.0	149.6	2.107	8866.0	11070.0	227.6	47.97	65.12	190.5
320.0	142.1	2.001	9398.0	11720.0	229.7	48.60	64.62	195.5
330.0	135.5	1.908	9929.0	12370.0	231.7	49.26	64.38	200.3
340.0	129.6	1.825	10460.0	13010.0	233.6	49.94	64.33	204.9
350.0	124.4	1.751	11000.0	13650.0	235.4	50.62	64.42	209.2
360.0	119.6	1.684	11540.0	14300.0	237.3	51.29	64.59	213.3
370.0	115.3	1.623	12080.0	14950.0	239.0	51.96	64.84	217.3
380.0	111.3	1.567	12630.0	15600.0	240.8	52.62	65.13	221.2
400.0	104.2	1.468	13740.0	16900.0	244.1	53.88	65.81	228.6
420.0	98.15	1.382	14860.0	18230.0	247.4	55.09	66.56	235.6
440.0	92.82	1.307	16010.0	19570.0	250.5	56.21	67.32	242.2
460.0	88.11	1.241	17170.0	20920.0	253.5	57.27	68.09	248.6

Thermophysical properties of nitrogen trifluoride—Continued

T K	Density kg/m³	Density mol/dm³	*E* J/mol	*H* J/mol	*S* J/(mol·K)	C_v J/(mol·K)	C_p J/(mol·K)	Sound m/s
480.0	83.89	1.181	18350.0	22290.0	256.4	58.27	68.85	254.7
500.0	80.10	1.128	19550.0	23680.0	259.2	59.21	69.58	260.6
				4.70 MPa isobar				
67.23[a]	1869.0	26.32	−12310.0	−12130.0	93.21	50.66	73.80	1373.0
68.0	1867.0	26.28	−12260.0	−12080.0	94.04	50.06	72.83	1344.0
70.0	1859.0	26.18	−12110.0	−11930.0	96.13	48.91	71.11	1284.0
80.0	1821.0	25.65	−11420.0	−11230.0	105.5	46.83	69.92	1156.0
100.0	1742.0	24.53	−10010.0	−9823.0	121.2	43.99	70.79	1056.0
120.0	1658.0	23.35	−8613.0	−8412.0	134.1	41.20	70.38	936.1
140.0	1570.0	22.11	−7214.0	−7001.0	145.0	39.89	70.97	819.1
160.0	1477.0	20.79	−5787.0	−5561.0	154.6	39.75	73.43	709.9
170.0	1427.0	20.09	−5050.0	−4816.0	159.1	39.98	75.55	655.0
180.0	1374.0	19.34	−4290.0	−4047.0	163.5	40.37	78.41	598.0
190.0	1316.0	18.53	−3499.0	−3245.0	167.8	40.93	82.24	537.8
200.0	1251.0	17.62	−2664.0	−2397.0	172.2	41.71	87.54	473.2
205.0	1216.0	17.12	−2226.0	−1951.0	174.4	42.21	91.06	438.8
210.0	1177.0	16.57	−1769.0	−1485.0	176.6	42.82	95.52	402.6
215.0	1133.0	15.95	−1288.0	−993.6	178.9	43.56	101.5	364.2
217.0	1114.0	15.69	−1087.0	−787.7	179.9	43.91	104.6	348.1
218.0	1104.0	15.55	−984.6	−682.2	180.4	44.10	106.3	339.8
220.0	1083.0	15.25	−773.9	−465.8	181.3	44.51	110.3	322.8
222.0	1061.0	14.93	−555.1	−240.4	182.4	44.97	115.2	305.1
224.0	1036.0	14.59	−326.1	−3.9	183.4	45.50	121.5	286.5
226.0	1009.0	14.20	−83.7	247.2	184.5	46.12	130.1	266.8
228.0	977.5	13.76	177.3	518.8	185.7	46.86	142.3	245.7
229.0	959.9	13.52	317.5	665.2	186.4	47.30	150.9	234.4
230.0	940.4	13.24	466.6	821.5	187.1	47.79	162.2	222.5
231.0	918.4	12.93	627.6	991.1	187.8	48.36	177.9	209.7
232.0	892.7	12.57	805.9	1180.0	188.6	49.04	201.4	196.0
232.5	877.9	12.36	904.3	1285.0	189.1	49.43	218.3	188.6
233.0	861.2	12.13	1012.0	1399.0	189.6	49.88	241.2	180.8
233.5	842.0	11.86	1131.0	1527.0	190.1	50.40	274.1	172.4
234.0	818.8	11.53	1269.0	1676.0	190.7	51.02	325.5	163.4
234.2	808.0	11.38	1331.0	1744.0	191.0	51.31	355.2	159.5
234.4	795.9	11.21	1400.0	1819.0	191.4	51.63	393.7	155.5
234.6	782.2	11.01	1476.0	1903.0	191.7	51.98	445.2	151.3
234.8	766.3	10.79	1563.0	1998.0	192.1	52.38	517.2	146.8
235.0	747.2	10.52	1665.0	2112.0	192.6	52.83	623.5	142.2
235.2	723.4	10.19	1790.0	2252.0	193.2	53.36	789.7	137.3
235.3	709.0	9.983	1866.0	2336.0	193.6	53.66	908.1	134.8
235.4	692.3	9.748	1952.0	2434.0	194.0	53.98	1058.0	132.3
235.5	672.9	9.476	2053.0	2549.0	194.5	54.30	1239.0	129.9
235.6	650.6	9.161	2170.0	2683.0	195.0	54.62	1438.0	127.7
235.7	625.6	8.809	2302.0	2836.0	195.7	54.89	1616.0	125.8
235.8	599.0	8.434	2446.0	3004.0	196.4	55.07	1727.0	124.4
235.9	572.3	8.058	2595.0	3178.0	197.1	55.14	1740.0	123.3
236.0	547.2	7.705	2738.0	3348.0	197.9	55.09	1652.0	122.6
236.1	524.8	7.390	2870.0	3506.0	198.5	54.96	1497.0	122.2
236.2	505.7	7.120	2987.0	3647.0	199.1	54.78	1319.0	122.1
236.3	489.4	6.892	3088.0	3770.0	199.6	54.58	1151.0	122.1
236.4	475.7	6.698	3176.0	3878.0	200.1	54.37	1008.0	122.1
236.5	464.0	6.533	3253.0	3973.0	200.5	54.17	890.2	122.3
236.6	453.9	6.391	3321.0	4057.0	200.8	53.98	794.7	122.4
236.8	437.2	6.156	3437.0	4201.0	201.5	53.62	653.3	122.9
237.0	423.8	5.967	3533.0	4321.0	202.0	53.30	555.9	123.4
237.2	412.6	5.810	3616.0	4425.0	202.4	53.01	485.7	123.9
237.4	403.2	5.677	3689.0	4517.0	202.8	52.74	433.0	124.4
237.6	394.9	5.560	3754.0	4599.0	203.1	52.50	392.1	124.9
237.8	387.6	5.457	3813.0	4674.0	203.5	52.28	359.5	125.4
238.0	381.0	5.365	3867.0	4743.0	203.7	52.08	332.9	125.9
238.5	367.0	5.168	3987.0	4896.0	204.4	51.62	283.8	127.1

Thermophysical properties of nitrogen trifluoride—Continued

T K	Density kg/m^3	Density mol/dm^3	E J/mol	H J/mol	S J/(mol·K)	C_v J/(mol·K)	C_p J/(mol·K)	Sound m/s
239.0	355.6	5.007	4091.0	5029.0	204.9	51.22	250.0	128.2
239.5	345.9	4.870	4183.0	5148.0	205.4	50.87	225.3	129.3
240.0	337.5	4.752	4266.0	5256.0	205.9	50.56	206.4	130.4
241.0	323.3	4.552	4415.0	5448.0	206.7	50.03	179.3	132.3
242.0	311.7	4.389	4546.0	5617.0	207.4	49.58	160.6	134.2
243.0	301.9	4.250	4665.0	5770.0	208.0	49.19	146.9	135.9
244.0	293.3	4.130	4774.0	5912.0	208.6	48.86	136.5	137.5
245.0	285.7	4.024	4876.0	6044.0	209.1	48.57	128.1	139.1
246.0	279.0	3.928	4972.0	6169.0	209.6	48.31	121.4	140.5
248.0	267.2	3.763	5151.0	6400.0	210.6	47.88	110.9	143.3
250.0	257.3	3.623	5317.0	6614.0	211.4	47.54	103.3	145.9
252.0	248.7	3.502	5472.0	6815.0	212.2	47.26	97.46	148.2
254.0	241.1	3.394	5620.0	7005.0	213.0	47.04	92.82	150.5
256.0	234.3	3.299	5762.0	7187.0	213.7	46.86	89.07	152.6
258.0	228.1	3.212	5898.0	7362.0	214.4	46.71	85.96	154.6
260.0	222.5	3.133	6031.0	7531.0	215.0	46.60	83.35	156.5
265.0	210.4	2.962	6348.0	7934.0	216.6	46.44	78.38	161.0
270.0	200.2	2.819	6650.0	8317.0	218.0	46.39	74.88	165.1
275.0	191.4	2.696	6941.0	8685.0	219.4	46.44	72.31	168.9
280.0	183.8	2.588	7225.0	9041.0	220.6	46.55	70.39	172.4
285.0	177.0	2.493	7504.0	9389.0	221.9	46.71	68.92	175.8
290.0	170.9	2.407	7778.0	9731.0	223.1	46.92	67.78	179.0
295.0	165.4	2.329	8050.0	10070.0	224.2	47.15	66.90	182.0
300.0	160.4	2.259	8319.0	10400.0	225.3	47.41	66.21	184.9
310.0	151.5	2.133	8854.0	11060.0	227.5	47.99	65.28	190.3
320.0	143.8	2.025	9387.0	11710.0	229.6	48.62	64.75	195.4
330.0	137.1	1.931	9919.0	12350.0	231.5	49.28	64.49	200.2
340.0	131.2	1.847	10450.0	13000.0	233.5	49.95	64.42	204.8
350.0	125.8	1.772	10990.0	13640.0	235.3	50.63	64.50	209.1
360.0	121.0	1.704	11530.0	14290.0	237.2	51.30	64.66	213.3
370.0	116.6	1.642	12070.0	14940.0	238.9	51.97	64.90	217.3
380.0	112.6	1.585	12620.0	15590.0	240.7	52.62	65.19	221.2
400.0	105.4	1.484	13730.0	16900.0	244.0	53.89	65.86	228.6
420.0	99.24	1.397	14860.0	18220.0	247.3	55.09	66.59	235.6
440.0	93.84	1.321	16000.0	19560.0	250.4	56.21	67.36	242.2
460.0	89.07	1.254	17170.0	20910.0	253.4	57.27	68.12	248.6
480.0	84.80	1.194	18350.0	22280.0	256.3	58.27	68.87	254.7
500.0	80.97	1.140	19550.0	23670.0	259.1	59.21	69.60	260.6
			4.80 MPa isobar					
67.25[a]	1869.0	26.32	−12310.0	−12130.0	93.22	50.72	73.77	1370.0
68.0	1867.0	26.28	−12260.0	−12070.0	94.04	50.14	72.83	1342.0
70.0	1859.0	26.18	−12110.0	−11930.0	96.12	48.97	71.10	1282.0
80.0	1821.0	25.65	−11420.0	−11230.0	105.5	46.85	69.91	1156.0
100.0	1742.0	24.53	−10020.0	−9820.0	121.2	43.98	70.78	1056.0
120.0	1659.0	23.35	−8615.0	−8409.0	134.1	41.18	70.37	937.0
140.0	1571.0	22.12	−7215.0	−6998.0	144.9	39.89	70.96	819.9
160.0	1477.0	20.80	−5789.0	−5558.0	154.6	39.74	73.40	710.7
170.0	1427.0	20.10	−5053.0	−4814.0	159.1	39.97	75.51	655.9
180.0	1374.0	19.35	−4294.0	−4046.0	163.5	40.36	78.35	599.0
190.0	1317.0	18.54	−3503.0	−3244.0	167.8	40.92	82.15	538.9
200.0	1252.0	17.63	−2670.0	−2398.0	172.1	41.70	87.38	474.6
205.0	1217.0	17.13	−2233.0	−1953.0	174.3	42.19	90.84	440.4
210.0	1178.0	16.59	−1778.0	−1488.0	176.6	42.79	95.21	404.5
215.0	1135.0	15.98	−1299.0	−998.3	178.9	43.52	101.0	366.5
220.0	1086.0	15.28	−787.6	−473.5	181.3	44.45	109.5	325.5
222.0	1063.0	14.97	−570.6	−250.0	182.3	44.90	114.2	308.1
224.0	1039.0	14.63	−344.0	−16.0	183.3	45.41	120.1	289.9
226.0	1013.0	14.26	−104.9	231.7	184.4	46.00	128.0	270.6
228.0	982.4	13.83	151.2	498.2	185.6	46.70	139.1	250.0
229.0	965.5	13.59	287.9	640.9	186.2	47.11	146.7	239.1

Thermophysical properties of nitrogen trifluoride—Continued

T K	Density kg/m^3	Density mol/dm^3	E J/mol	H J/mol	S J/(mol·K)	C_v J/(mol·K)	C_p J/(mol·K)	Sound m/s
230.0	947.0	13.33	432.3	792.3	186.9	47.57	156.4	227.6
231.0	926.3	13.04	586.9	954.9	187.6	48.09	169.5	215.5
232.0	902.7	12.71	755.5	1133.0	188.4	48.70	188.1	202.5
233.0	874.7	12.32	944.8	1335.0	189.2	49.44	217.1	188.4
233.5	858.3	12.09	1051.0	1448.0	189.7	49.88	238.7	180.8
234.0	839.4	11.82	1169.0	1575.0	190.3	50.38	269.1	172.8
234.5	817.1	11.51	1303.0	1720.0	190.9	50.96	314.8	164.2
235.0	789.3	11.11	1462.0	1894.0	191.6	51.68	390.7	154.9
235.2	775.9	10.92	1538.0	1977.0	192.0	52.01	436.7	150.9
235.4	760.5	10.71	1622.0	2070.0	192.4	52.38	497.8	146.8
235.6	742.7	10.46	1718.0	2177.0	192.8	52.79	580.9	142.7
235.8	721.4	10.16	1832.0	2305.0	193.4	53.25	695.7	138.4
236.0	695.6	9.794	1968.0	2458.0	194.0	53.74	850.4	134.2
236.2	664.4	9.355	2134.0	2647.0	194.8	54.24	1034.0	130.4
236.4	628.2	8.845	2328.0	2870.0	195.8	54.66	1192.0	127.4
236.6	589.8	8.305	2539.0	3117.0	196.8	54.89	1254.0	125.3
236.8	553.4	7.792	2748.0	3364.0	197.9	54.89	1202.0	124.0
237.0	521.5	7.344	2939.0	3592.0	198.8	54.70	1069.0	123.5
237.2	495.3	6.975	3102.0	3790.0	199.7	54.42	913.5	123.3
237.4	474.2	6.677	3240.0	3959.0	200.4	54.11	773.8	123.4
237.6	457.0	6.435	3356.0	4102.0	201.0	53.80	662.0	123.7
237.8	442.8	6.235	3455.0	4225.0	201.5	53.51	575.4	124.0
238.0	430.8	6.066	3542.0	4333.0	201.9	53.23	508.4	124.4
238.2	420.4	5.920	3619.0	4429.0	202.4	52.98	455.8	124.8
238.4	411.4	5.793	3688.0	4516.0	202.7	52.74	413.8	125.3
238.6	403.4	5.680	3750.0	4596.0	203.0	52.52	379.7	125.7
239.0	389.6	5.486	3862.0	4736.0	203.6	52.12	327.8	126.6
239.5	375.6	5.289	3981.0	4888.0	204.3	51.69	282.7	127.7
240.0	364.0	5.125	4085.0	5021.0	204.8	51.30	250.7	128.8
240.5	354.1	4.986	4178.0	5140.0	205.3	50.97	226.7	129.8
241.0	345.4	4.864	4262.0	5249.0	205.8	50.66	208.2	130.9
242.0	330.9	4.659	4412.0	5443.0	206.6	50.14	181.2	132.8
243.0	318.9	4.491	4545.0	5614.0	207.3	49.69	162.5	134.6
244.0	308.8	4.348	4665.0	5769.0	207.9	49.31	148.7	136.3
245.0	300.0	4.224	4776.0	5912.0	208.5	48.98	138.0	137.9
246.0	292.2	4.114	4879.0	6046.0	209.0	48.69	129.6	139.4
248.0	278.9	3.927	5069.0	6292.0	210.0	48.21	116.9	142.2
250.0	267.8	3.770	5243.0	6516.0	210.9	47.82	107.9	144.9
252.0	258.3	3.637	5405.0	6725.0	211.8	47.51	101.1	147.3
254.0	250.0	3.520	5558.0	6922.0	212.6	47.26	95.85	149.6
256.0	242.6	3.416	5704.0	7109.0	213.3	47.06	91.60	151.8
258.0	235.9	3.322	5844.0	7289.0	214.0	46.90	88.13	153.8
260.0	229.9	3.238	5979.0	7462.0	214.7	46.77	85.23	155.8
262.0	224.4	3.160	6111.0	7630.0	215.3	46.68	82.79	157.7
265.0	216.9	3.055	6302.0	7874.0	216.2	46.58	79.76	160.4
270.0	206.2	2.903	6609.0	8262.0	217.7	46.51	75.94	164.5
275.0	197.0	2.773	6904.0	8635.0	219.0	46.53	73.17	168.4
280.0	188.9	2.660	7191.0	8995.0	220.3	46.63	71.10	172.0
285.0	181.8	2.560	7472.0	9347.0	221.6	46.78	69.51	175.4
290.0	175.5	2.471	7748.0	9691.0	222.8	46.98	68.29	178.6
295.0	169.7	2.390	8022.0	10030.0	223.9	47.21	67.34	181.6
300.0	164.5	2.316	8293.0	10360.0	225.1	47.46	66.61	184.6
310.0	155.3	2.186	8830.0	11030.0	227.2	48.03	65.59	190.1
320.0	147.3	2.075	9364.0	11680.0	229.3	48.65	65.00	195.2
330.0	140.4	1.977	9898.0	12330.0	231.3	49.30	64.70	200.0
340.0	134.2	1.890	10430.0	12970.0	233.2	49.97	64.61	204.6
350.0	128.7	1.813	10970.0	13620.0	235.1	50.64	64.65	209.0
360.0	123.8	1.743	11510.0	14270.0	236.9	51.31	64.80	213.2
370.0	119.2	1.679	12060.0	14920.0	238.7	51.98	65.02	217.2
380.0	115.1	1.621	12600.0	15570.0	240.4	52.63	65.30	221.1
400.0	107.7	1.517	13720.0	16880.0	243.8	53.89	65.95	228.5
420.0	101.4	1.428	14840.0	18210.0	247.0	55.09	66.67	235.6
440.0	95.89	1.350	15990.0	19550.0	250.2	56.22	67.42	242.2

Thermophysical properties of nitrogen trifluoride—Continued

T K	Density kg/m³	Density mol/dm³	E J/mol	H J/mol	S J/(mol·K)	C_v J/(mol·K)	C_p J/(mol·K)	Sound m/s
460.0	90.99	1.281	17160.0	20900.0	253.2	57.27	68.17	248.6
480.0	86.63	1.220	18340.0	22270.0	256.1	58.27	68.92	254.8
500.0	82.70	1.164	19540.0	23660.0	258.9	59.21	69.65	260.7
				4.90 MPa isobar				
67.27[a]	1869.0	26.32	-12310.0	-12130.0	93.23	50.79	73.74	1367.0
68.0	1867.0	26.29	-12260.0	-12070.0	94.03	50.21	72.82	1340.0
70.0	1859.0	26.18	-12120.0	-11930.0	96.11	49.03	71.09	1281.0
80.0	1822.0	25.65	-11420.0	-11230.0	105.5	46.86	69.91	1156.0
100.0	1742.0	24.53	-10020.0	-9817.0	121.2	43.97	70.78	1057.0
120.0	1659.0	23.36	-8616.0	-8406.0	134.1	41.17	70.36	937.8
140.0	1571.0	22.12	-7217.0	-6996.0	144.9	39.88	70.94	820.7
160.0	1478.0	20.80	-5792.0	-5556.0	154.5	39.74	73.38	711.5
170.0	1428.0	20.10	-5056.0	-4812.0	159.0	39.96	75.47	656.7
180.0	1375.0	19.36	-4297.0	-4044.0	163.4	40.35	78.29	599.9
190.0	1317.0	18.55	-3508.0	-3243.0	167.8	40.91	82.05	540.1
200.0	1253.0	17.65	-2676.0	-2399.0	172.1	41.68	87.23	476.0
205.0	1218.0	17.15	-2240.0	-1954.0	174.3	42.17	90.63	442.0
210.0	1179.0	16.60	-1786.0	-1491.0	176.5	42.76	94.91	406.4
215.0	1137.0	16.00	-1309.0	-1003.0	178.8	43.48	100.6	368.6
220.0	1088.0	15.32	-800.9	-481.0	181.2	44.38	108.7	328.2
222.0	1066.0	15.01	-585.7	-259.3	182.2	44.82	113.2	311.0
224.0	1042.0	14.68	-361.4	-27.5	183.3	45.32	118.8	293.1
226.0	1016.0	14.31	-125.4	217.0	184.4	45.89	126.1	274.2
228.0	987.0	13.90	126.2	478.8	185.5	46.56	136.3	254.2
230.0	953.1	13.42	400.3	765.4	186.8	47.37	151.5	232.5
231.0	933.6	13.15	549.5	922.2	187.4	47.85	162.6	220.9
232.0	911.6	12.84	710.3	1092.0	188.2	48.41	177.8	208.6
233.0	886.2	12.48	887.5	1280.0	189.0	49.06	200.1	195.4
233.5	871.7	12.27	984.7	1384.0	189.4	49.44	215.7	188.4
234.0	855.4	12.05	1090.0	1497.0	189.9	49.86	236.1	181.0
234.5	837.0	11.79	1206.0	1621.0	190.4	50.34	264.1	173.2
235.0	815.4	11.48	1336.0	1763.0	191.0	50.90	304.7	165.0
235.5	789.2	11.11	1489.0	1930.0	191.7	51.56	368.0	156.3
236.0	755.5	10.64	1677.0	2138.0	192.6	52.36	476.0	147.1
236.2	738.9	10.40	1769.0	2239.0	193.1	52.73	540.5	143.3
236.4	719.7	10.13	1872.0	2355.0	193.6	53.12	621.7	139.6
236.6	697.6	9.823	1990.0	2489.0	194.1	53.54	719.4	136.0
236.8	672.4	9.467	2126.0	2644.0	194.8	53.95	825.0	132.7
237.0	644.2	9.071	2278.0	2818.0	195.5	54.30	917.6	129.9
237.2	614.5	8.653	2442.0	3008.0	196.3	54.56	973.5	127.8
237.4	584.8	8.235	2610.0	3205.0	197.1	54.69	981.8	126.3
237.6	556.8	7.840	2773.0	3398.0	198.0	54.67	946.4	125.4
237.8	531.4	7.482	2926.0	3581.0	198.7	54.55	879.2	124.9
238.0	509.1	7.168	3065.0	3749.0	199.4	54.35	795.9	124.6
238.2	489.9	6.898	3189.0	3899.0	200.1	54.12	711.0	124.6
238.4	473.4	6.666	3299.0	4034.0	200.6	53.87	633.4	124.7
238.6	459.3	6.467	3396.0	4153.0	201.1	53.61	566.4	124.9
238.8	447.0	6.294	3482.0	4261.0	201.6	53.37	510.0	125.2
239.0	436.3	6.143	3560.0	4358.0	202.0	53.14	463.1	125.6
239.2	426.8	6.009	3631.0	4447.0	202.4	52.91	423.9	125.9
239.4	418.3	5.889	3696.0	4528.0	202.7	52.70	391.1	126.3
240.0	397.3	5.594	3863.0	4739.0	203.6	52.13	319.1	127.5
240.5	383.4	5.399	3981.0	4888.0	204.2	51.72	278.8	128.5
241.0	371.8	5.235	4084.0	5020.0	204.7	51.36	249.2	129.5
241.5	361.8	5.094	4177.0	5139.0	205.2	51.03	226.6	130.5
242.0	353.0	4.970	4261.0	5247.0	205.7	50.74	208.7	131.4
243.0	338.1	4.761	4413.0	5442.0	206.5	50.22	182.3	133.3
244.0	325.9	4.589	4547.0	5615.0	207.2	49.79	163.8	135.0
245.0	315.5	4.442	4668.0	5771.0	207.8	49.41	150.0	136.7
246.0	306.4	4.315	4780.0	5916.0	208.4	49.09	139.3	138.2
247.0	298.4	4.202	4884.0	6051.0	209.0	48.80	130.8	139.7

Thermophysical properties of nitrogen trifluoride—Continued

T K	Density kg/m^3	Density mol/dm^3	E J/mol	H J/mol	S J/(mol·K)	C_v J/(mol·K)	C_p J/(mol·K)	Sound m/s
248.0	291.3	4.101	4983.0	6178.0	209.5	48.54	123.8	141.2
250.0	278.8	3.926	5166.0	6414.0	210.4	48.11	113.1	143.9
252.0	268.3	3.778	5335.0	6632.0	211.3	47.77	105.2	146.4
254.0	259.2	3.650	5494.0	6836.0	212.1	47.49	99.14	148.8
256.0	251.2	3.537	5644.0	7029.0	212.9	47.27	94.34	151.0
258.0	244.0	3.436	5788.0	7214.0	213.6	47.09	90.45	153.1
260.0	237.5	3.345	5927.0	7392.0	214.3	46.95	87.23	155.1
262.0	231.6	3.262	6061.0	7563.0	214.9	46.83	84.53	157.0
265.0	223.7	3.150	6256.0	7812.0	215.9	46.72	81.21	159.8
270.0	212.2	2.989	6567.0	8207.0	217.3	46.62	77.06	164.0
275.0	202.5	2.852	6866.0	8584.0	218.7	46.63	74.06	167.9
280.0	194.1	2.733	7156.0	8949.0	220.0	46.71	71.82	171.6
285.0	186.7	2.629	7439.0	9303.0	221.3	46.86	70.13	175.0
290.0	180.1	2.535	7718.0	9651.0	222.5	47.04	68.82	178.2
295.0	174.1	2.451	7993.0	9992.0	223.7	47.26	67.80	181.3
300.0	168.6	2.375	8266.0	10330.0	224.8	47.51	67.00	184.3
310.0	159.1	2.240	8805.0	10990.0	227.0	48.07	65.90	189.8
320.0	150.9	2.124	9342.0	11650.0	229.1	48.68	65.26	195.0
330.0	143.7	2.023	9877.0	12300.0	231.1	49.33	64.92	199.9
340.0	137.3	1.934	10410.0	12950.0	233.0	49.99	64.79	204.5
350.0	131.7	1.854	10950.0	13600.0	234.9	50.66	64.81	208.9
360.0	126.5	1.782	11490.0	14240.0	236.7	51.33	64.94	213.1
370.0	121.9	1.716	12040.0	14890.0	238.5	51.99	65.15	217.2
380.0	117.6	1.656	12590.0	15550.0	240.2	52.64	65.41	221.1
400.0	110.1	1.550	13700.0	16860.0	243.6	53.90	66.04	228.5
420.0	103.6	1.459	14830.0	18190.0	246.8	55.09	66.75	235.6
440.0	97.93	1.379	15980.0	19530.0	250.0	56.22	67.49	242.3
460.0	92.92	1.308	17140.0	20890.0	253.0	57.28	68.23	248.7
480.0	88.45	1.245	18330.0	22260.0	255.9	58.27	68.97	254.8
500.0	84.43	1.189	19530.0	23650.0	258.7	59.21	69.69	260.8
			5.00 MPa isobar					
67.28[a]	1869.0	26.32	−12310.0	−12120.0	93.25	50.85	73.71	1365.0
68.0	1867.0	26.29	−12260.0	−12070.0	94.02	50.28	72.82	1338.0
70.0	1860.0	26.18	−12120.0	−11920.0	96.11	49.09	71.09	1279.0
80.0	1822.0	25.65	−11420.0	−11230.0	105.4	46.88	69.90	1156.0
100.0	1742.0	24.53	−10020.0	−9814.0	121.2	43.95	70.77	1058.0
120.0	1659.0	23.36	−8617.0	−8403.0	134.1	41.16	70.35	938.7
140.0	1571.0	22.12	−7219.0	−6993.0	144.9	39.87	70.93	821.5
160.0	1478.0	20.81	−5794.0	−5554.0	154.5	39.73	73.35	712.3
170.0	1428.0	20.11	−5059.0	−4810.0	159.0	39.96	75.43	657.6
180.0	1375.0	19.36	−4301.0	−4043.0	163.4	40.35	78.23	600.9
190.0	1318.0	18.56	−3512.0	−3243.0	167.7	40.90	81.96	541.2
200.0	1254.0	17.66	−2682.0	−2399.0	172.1	41.66	87.07	477.4
205.0	1219.0	17.16	−2247.0	−1956.0	174.3	42.15	90.42	443.6
210.0	1181.0	16.62	−1794.0	−1493.0	176.5	42.73	94.62	408.2
215.0	1138.0	16.03	−1319.0	−1007.0	178.8	43.43	100.1	370.8
220.0	1090.0	15.35	−814.0	−488.3	181.2	44.32	108.0	330.8
222.0	1069.0	15.05	−600.5	−268.2	182.2	44.75	112.2	313.9
224.0	1045.0	14.72	−378.3	−38.6	183.2	45.23	117.5	296.3
226.0	1020.0	14.36	−145.2	203.0	184.3	45.78	124.4	277.8
228.0	991.4	13.96	102.4	460.5	185.4	46.42	133.7	258.2
230.0	958.8	13.50	370.2	740.5	186.6	47.19	147.3	237.2
231.0	940.3	13.24	514.7	892.4	187.3	47.64	156.9	226.0
232.0	919.7	12.95	669.2	1055.0	188.0	48.15	169.6	214.3
233.0	896.2	12.62	837.0	1233.0	188.7	48.74	187.3	201.8
234.0	868.7	12.23	1024.0	1433.0	189.6	49.44	214.1	188.3
234.5	852.7	12.01	1128.0	1545.0	190.1	49.85	233.3	181.2
235.0	834.7	11.75	1242.0	1667.0	190.6	50.31	259.1	173.8
235.5	813.9	11.46	1369.0	1805.0	191.2	50.84	295.1	165.9
236.0	789.1	11.11	1515.0	1965.0	191.9	51.45	348.5	157.7
236.5	758.2	10.68	1690.0	2159.0	192.7	52.16	431.7	149.2

Thermophysical properties of nitrogen trifluoride—Continued

T K	Density kg/m^3	Density mol/dm^3	*E* J/mol	*H* J/mol	*S* J/(mol·K)	C_v J/(mol·K)	C_p J/(mol·K)	Sound m/s
236.7	743.6	10.47	1772.0	2249.0	193.1	52.49	477.5	145.8
236.8	735.6	10.36	1816.0	2299.0	193.3	52.65	503.6	144.1
237.0	718.3	10.11	1911.0	2405.0	193.7	53.00	562.4	140.8
237.2	699.0	9.842	2016.0	2524.0	194.2	53.35	628.0	137.6
237.4	677.6	9.542	2132.0	2656.0	194.8	53.69	695.1	134.7
238.0	605.7	8.528	2532.0	3118.0	196.7	54.41	813.7	128.6
238.2	581.7	8.190	2670.0	3281.0	197.4	54.48	808.4	127.5
238.4	558.9	7.870	2805.0	3440.0	198.1	54.46	783.1	126.7
238.6	537.9	7.574	2933.0	3593.0	198.7	54.37	742.9	126.2
238.8	518.8	7.305	3052.0	3737.0	199.3	54.23	693.5	126.0
239.0	501.7	7.064	3162.0	3870.0	199.9	54.05	640.3	125.9
239.2	486.5	6.850	3263.0	3993.0	200.4	53.85	588.0	125.9
239.4	473.0	6.660	3355.0	4106.0	200.9	53.64	539.2	126.0
239.6	461.0	6.491	3439.0	4209.0	201.3	53.43	495.3	126.2
239.8	450.3	6.340	3515.0	4304.0	201.7	53.22	456.6	126.4
240.0	440.6	6.204	3586.0	4392.0	202.1	53.01	422.7	126.7
240.5	420.3	5.918	3741.0	4586.0	202.9	52.54	356.0	127.5
241.0	404.0	5.688	3872.0	4751.0	203.5	52.11	308.2	128.4
241.5	390.4	5.497	3986.0	4896.0	204.1	51.73	272.9	129.3
242.0	378.9	5.335	4088.0	5025.0	204.7	51.39	246.0	130.3
242.5	368.9	5.194	4180.0	5143.0	205.2	51.08	225.0	131.2
243.0	360.0	5.070	4265.0	5251.0	205.6	50.79	208.1	132.1
244.0	345.0	4.858	4416.0	5446.0	206.4	50.30	182.7	133.9
245.0	332.6	4.683	4551.0	5619.0	207.1	49.87	164.6	135.5
246.0	321.9	4.533	4673.0	5776.0	207.8	49.50	150.9	137.2
247.0	312.7	4.403	4786.0	5922.0	208.4	49.18	140.3	138.7
248.0	304.5	4.288	4891.0	6058.0	208.9	48.89	131.8	140.2
250.0	290.5	4.091	5085.0	6307.0	209.9	48.42	118.9	142.9
252.0	278.8	3.926	5262.0	6536.0	210.8	48.04	109.7	145.5
254.0	268.8	3.785	5427.0	6748.0	211.7	47.73	102.7	147.9
256.0	260.1	3.662	5582.0	6948.0	212.4	47.48	97.31	150.2
258.0	252.3	3.553	5731.0	7138.0	213.2	47.28	92.95	152.4
260.0	245.4	3.455	5873.0	7320.0	213.9	47.12	89.37	154.4
262.0	239.1	3.366	6010.0	7496.0	214.6	46.99	86.38	156.4
264.0	233.3	3.285	6144.0	7666.0	215.2	46.90	83.86	158.2
270.0	218.5	3.076	6525.0	8151.0	217.0	46.74	78.22	163.5
275.0	208.2	2.932	6828.0	8533.0	218.4	46.73	74.98	167.4
280.0	199.4	2.807	7121.0	8902.0	219.8	46.80	72.58	171.1
285.0	191.6	2.698	7407.0	9260.0	221.0	46.93	70.75	174.6
290.0	184.7	2.601	7688.0	9610.0	222.2	47.10	69.35	177.9
295.0	178.5	2.513	7964.0	9954.0	223.4	47.32	68.26	181.0
300.0	172.8	2.433	8238.0	10290.0	224.6	47.56	67.41	184.0
310.0	162.9	2.294	8781.0	10960.0	226.7	48.11	66.22	189.6
320.0	154.4	2.174	9319.0	11620.0	228.8	48.71	65.52	194.8
330.0	147.0	2.070	9857.0	12270.0	230.8	49.35	65.14	199.7
340.0	140.4	1.978	10390.0	12920.0	232.8	50.01	64.98	204.3
350.0	134.6	1.895	10930.0	13570.0	234.7	50.68	64.98	208.8
360.0	129.3	1.821	11480.0	14220.0	236.5	51.34	65.08	213.0
370.0	124.6	1.754	12020.0	14870.0	238.3	52.00	65.27	217.1
380.0	120.2	1.692	12570.0	15530.0	240.0	52.65	65.52	221.0
390.0	116.2	1.636	13130.0	16180.0	241.7	53.29	65.81	224.8
400.0	112.4	1.583	13690.0	16840.0	243.4	53.91	66.13	228.5
420.0	105.8	1.490	14820.0	18170.0	246.6	55.10	66.82	235.6
440.0	99.98	1.408	15970.0	19520.0	249.8	56.22	67.55	242.3
460.0	94.85	1.335	17130.0	20880.0	252.8	57.28	68.29	248.7
480.0	90.27	1.271	18320.0	22250.0	255.7	58.27	69.02	254.9
500.0	86.16	1.213	19520.0	23640.0	258.5	59.21	69.73	260.8
				5.20 MPa isobar				
67.32[a]	1869.0	26.32	-12310.0	-12110.0	93.27	50.97	73.65	1359.0
68.0	1867.0	26.29	-12260.0	-12060.0	94.01	50.43	72.80	1335.0

Thermophysical properties of nitrogen trifluoride—Continued

T K	Density kg/m³	Density mol/dm³	E J/mol	H J/mol	S J/(mol·K)	C_v J/(mol·K)	C_p J/(mol·K)	Sound m/s
70.0	1860.0	26.19	-12120.0	-11920.0	96.09	49.21	71.08	1276.0
80.0	1822.0	25.65	-11420.0	-11220.0	105.4	46.90	69.89	1156.0
100.0	1743.0	24.54	-10020.0	-9808.0	121.2	43.93	70.76	1059.0
120.0	1659.0	23.36	-8620.0	-8397.0	134.0	41.14	70.33	940.4
140.0	1572.0	22.13	-7223.0	-6988.0	144.9	39.85	70.90	823.1
160.0	1479.0	20.82	-5799.0	-5549.0	154.5	39.72	73.29	713.9
170.0	1429.0	20.12	-5065.0	-4807.0	159.0	39.94	75.35	659.3
180.0	1376.0	19.38	-4308.0	-4040.0	163.4	40.33	78.11	602.8
190.0	1319.0	18.58	-3521.0	-3241.0	167.7	40.88	81.78	543.5
200.0	1256.0	17.69	-2694.0	-2400.0	172.0	41.63	86.77	480.2
205.0	1221.0	17.19	-2261.0	-1958.0	174.2	42.11	90.02	446.8
210.0	1183.0	16.66	-1811.0	-1498.0	176.4	42.67	94.05	411.8
215.0	1142.0	16.07	-1339.0	-1016.0	178.7	43.36	99.29	375.0
220.0	1094.0	15.41	-839.5	-502.1	181.0	44.21	106.6	335.9
222.0	1074.0	15.12	-629.1	-285.1	182.0	44.61	110.5	319.5
224.0	1051.0	14.80	-410.9	-59.5	183.0	45.06	115.3	302.4
226.0	1027.0	14.46	-182.9	176.8	184.1	45.57	121.3	284.6
228.0	999.7	14.08	57.5	426.9	185.2	46.16	129.2	265.8
230.0	969.3	13.65	314.6	695.6	186.4	46.85	140.3	246.0
231.0	952.3	13.41	451.7	839.5	187.0	47.25	147.7	235.5
232.0	933.8	13.15	596.2	991.6	187.6	47.69	157.1	224.6
233.0	913.3	12.86	750.2	1155.0	188.3	48.19	169.3	213.2
234.0	890.1	12.53	916.9	1332.0	189.1	48.76	186.1	201.2
235.0	863.1	12.15	1101.0	1529.0	189.9	49.43	210.6	188.5
236.0	830.5	11.69	1313.0	1758.0	190.9	50.23	249.1	174.9
236.5	811.1	11.42	1434.0	1889.0	191.5	50.70	277.8	167.8
237.0	788.6	11.10	1569.0	2037.0	192.1	51.23	316.5	160.5
237.5	762.2	10.73	1723.0	2208.0	192.8	51.83	369.8	153.1
238.0	730.4	10.28	1904.0	2410.0	193.7	52.48	441.0	145.9
238.5	692.4	9.750	2118.0	2651.0	194.7	53.14	524.0	139.4
239.0	649.0	9.138	2362.0	2931.0	195.8	53.71	590.9	134.4
239.5	603.9	8.503	2622.0	3234.0	197.1	54.03	611.9	131.1
240.0	561.5	7.906	2878.0	3535.0	198.4	54.06	587.0	129.3
240.5	524.4	7.384	3112.0	3816.0	199.5	53.86	533.1	128.5
241.0	493.5	6.948	3318.0	4067.0	200.6	53.52	469.2	128.4
241.5	468.0	6.590	3497.0	4286.0	201.5	53.13	408.8	128.6
242.0	447.1	6.296	3651.0	4477.0	202.3	52.74	357.8	129.1
242.5	429.7	6.051	3786.0	4645.0	203.0	52.36	316.6	129.8
243.0	415.0	5.843	3905.0	4795.0	203.6	52.01	283.8	130.5
243.5	402.3	5.665	4012.0	4930.0	204.1	51.68	257.5	131.2
244.0	391.3	5.509	4110.0	5054.0	204.6	51.38	236.2	132.0
244.5	381.5	5.372	4199.0	5167.0	205.1	51.10	218.7	132.8
245.0	372.7	5.248	4282.0	5273.0	205.5	50.85	204.2	133.6
246.0	357.7	5.036	4432.0	5465.0	206.3	50.39	181.4	135.2
247.0	345.0	4.858	4567.0	5638.0	207.0	49.99	164.6	136.8
248.0	334.1	4.704	4690.0	5795.0	207.7	49.64	151.6	138.3
249.0	324.6	4.570	4804.0	5942.0	208.3	49.33	141.3	139.8
250.0	316.1	4.451	4910.0	6079.0	208.8	49.06	133.0	141.2
252.0	301.5	4.246	5107.0	6331.0	209.8	48.59	120.3	143.8
254.0	289.4	4.075	5286.0	6562.0	210.7	48.22	111.0	146.3
256.0	279.0	3.928	5453.0	6777.0	211.6	47.92	104.0	148.7
258.0	269.9	3.800	5611.0	6979.0	212.4	47.68	98.50	150.9
260.0	261.8	3.686	5761.0	7172.0	213.1	47.48	94.07	153.1
262.0	254.5	3.584	5905.0	7356.0	213.8	47.32	90.42	155.1
264.0	247.9	3.491	6044.0	7534.0	214.5	47.20	87.38	157.0
266.0	241.9	3.406	6179.0	7706.0	215.1	47.10	84.80	158.9
270.0	231.2	3.256	6439.0	8036.0	216.4	46.98	80.69	162.4
275.0	219.9	3.096	6750.0	8430.0	217.8	46.93	76.92	166.5
280.0	210.1	2.959	7050.0	8807.0	219.2	46.97	74.15	170.3
285.0	201.6	2.839	7341.0	9172.0	220.5	47.07	72.06	173.8
290.0	194.1	2.733	7626.0	9528.0	221.7	47.23	70.46	177.2
295.0	187.4	2.638	7906.0	9878.0	222.9	47.43	69.21	180.4
300.0	181.3	2.552	8184.0	10220.0	224.0	47.66	68.24	183.4

Thermophysical properties of nitrogen trifluoride—Continued

T K	Density kg/m³	Density mol/dm³	E J/mol	H J/mol	S J/(mol·K)	C_v J/(mol·K)	C_p J/(mol·K)	Sound m/s
310.0	170.6	2.402	8731.0	10900.0	226.3	48.18	66.88	189.1
320.0	161.5	2.274	9274.0	11560.0	228.4	48.77	66.05	194.4
330.0	153.6	2.163	9815.0	12220.0	230.4	49.40	65.58	199.4
340.0	146.7	2.066	10360.0	12870.0	232.3	50.05	65.36	204.1
350.0	140.5	1.978	10900.0	13530.0	234.2	50.71	65.30	208.6
360.0	134.9	1.900	11440.0	14180.0	236.1	51.37	65.37	212.8
370.0	129.9	1.829	11990.0	14830.0	237.9	52.02	65.52	217.0
380.0	125.3	1.764	12540.0	15490.0	239.6	52.67	65.74	220.9
390.0	121.1	1.705	13100.0	16150.0	241.3	53.30	66.01	224.8
400.0	117.1	1.650	13660.0	16810.0	243.0	53.92	66.31	228.5
420.0	110.2	1.551	14790.0	18140.0	246.3	55.11	66.97	235.6
440.0	104.1	1.465	15940.0	19490.0	249.4	56.23	67.68	242.4
460.0	98.70	1.390	17110.0	20850.0	252.4	57.28	68.40	248.8
480.0	93.92	1.322	18290.0	22230.0	255.3	58.27	69.11	255.0
500.0	89.62	1.262	19490.0	23610.0	258.2	59.20	69.82	261.0
				5.40 MPa isobar				
67.36[a]	1869.0	26.32	−12310.0	−12100.0	93.30	51.10	73.59	1354.0
68.0	1867.0	26.29	−12260.0	−12060.0	93.99	50.58	72.79	1331.0
70.0	1860.0	26.19	−12120.0	−11910.0	96.08	49.33	71.06	1273.0
80.0	1822.0	25.66	−11420.0	−11210.0	105.4	46.93	69.88	1155.0
100.0	1743.0	24.54	−10020.0	−9801.0	121.1	43.91	70.74	1061.0
120.0	1660.0	23.37	−8622.0	−8391.0	134.0	41.11	70.31	942.1
140.0	1572.0	22.14	−7226.0	−6982.0	144.9	39.83	70.86	824.7
160.0	1479.0	20.83	−5804.0	−5545.0	154.5	39.70	73.23	715.5
170.0	1430.0	20.13	−5071.0	−4803.0	159.0	39.93	75.27	661.0
180.0	1377.0	19.39	−4315.0	−4037.0	163.3	40.32	77.99	604.7
190.0	1321.0	18.60	−3530.0	−3240.0	167.6	40.87	81.60	545.7
200.0	1258.0	17.71	−2706.0	−2401.0	171.9	41.61	86.48	482.9
205.0	1223.0	17.22	−2274.0	−1961.0	174.1	42.07	89.63	449.8
210.0	1186.0	16.70	−1827.0	−1503.0	176.3	42.62	93.51	415.3
215.0	1145.0	16.12	−1359.0	−1024.0	178.6	43.28	98.50	379.2
220.0	1099.0	15.47	−864.2	−515.1	180.9	44.10	105.3	340.9
222.0	1078.0	15.18	−656.6	−300.9	181.9	44.48	108.9	324.8
224.0	1057.0	14.88	−441.9	−78.9	182.9	44.91	113.2	308.2
226.0	1033.0	14.55	−218.5	152.7	183.9	45.39	118.6	291.0
228.0	1007.0	14.18	15.7	396.4	185.0	45.94	125.4	273.0
230.0	978.7	13.78	264.2	656.0	186.1	46.57	134.7	254.1
232.0	946.0	13.32	532.3	937.7	187.3	47.31	147.9	234.0
233.0	927.5	13.06	676.5	1090.0	188.0	47.74	157.0	223.5
234.0	907.1	12.77	829.9	1253.0	188.7	48.23	168.8	212.4
235.0	884.2	12.45	995.2	1429.0	189.4	48.78	184.5	200.9
236.0	857.9	12.08	1177.0	1624.0	190.3	49.41	206.7	188.8
237.0	826.7	11.64	1382.0	1846.0	191.2	50.15	239.6	176.2
238.0	788.1	11.10	1623.0	2109.0	192.3	51.03	291.4	163.1
238.5	764.8	10.77	1762.0	2264.0	193.0	51.53	327.9	156.6
239.0	738.1	10.39	1919.0	2438.0	193.7	52.07	372.3	150.3
239.5	707.7	9.965	2095.0	2637.0	194.5	52.61	421.1	144.5
240.0	673.9	9.489	2290.0	2859.0	195.5	53.10	464.7	139.7
240.5	638.2	8.987	2498.0	3099.0	196.5	53.47	490.9	136.1
241.0	602.7	8.487	2709.0	3346.0	197.5	53.67	494.3	133.6
241.5	569.2	8.015	2916.0	3590.0	198.5	53.70	478.4	132.1
242.0	538.9	7.589	3110.0	3822.0	199.4	53.58	449.3	131.2
242.5	512.2	7.213	3289.0	4038.0	200.3	53.35	413.5	130.9
243.0	489.0	6.886	3451.0	4235.0	201.2	53.07	376.3	130.9
243.5	469.0	6.604	3597.0	4414.0	201.9	52.76	341.2	131.1
244.0	451.7	6.361	3728.0	4577.0	202.6	52.45	309.9	131.5
244.5	436.7	6.149	3847.0	4725.0	203.2	52.14	282.9	132.1
245.0	423.5	5.964	3955.0	4861.0	203.7	51.85	260.0	132.6
245.5	411.9	5.800	4055.0	4986.0	204.2	51.57	240.5	133.3
246.0	401.5	5.653	4146.0	5102.0	204.7	51.31	224.0	134.0
247.0	383.7	5.402	4312.0	5312.0	205.6	50.84	197.7	135.4

Thermophysical properties of nitrogen trifluoride—Continued

T K	Density kg/m³	Density mol/dm³	E J/mol	H J/mol	S J/(mol·K)	C_v J/(mol·K)	C_p J/(mol·K)	Sound m/s
248.0	368.8	5.193	4460.0	5499.0	206.3	50.42	178.1	136.8
249.0	356.2	5.015	4593.0	5670.0	207.0	50.05	162.9	138.2
250.0	345.2	4.860	4715.0	5826.0	207.6	49.73	151.0	139.6
251.0	335.5	4.724	4829.0	5972.0	208.2	49.44	141.3	141.0
252.0	326.9	4.602	4936.0	6109.0	208.8	49.18	133.3	142.3
254.0	311.9	4.392	5134.0	6363.0	209.8	48.74	121.0	144.9
256.0	299.4	4.216	5315.0	6595.0	210.7	48.38	111.9	147.3
258.0	288.7	4.065	5483.0	6812.0	211.5	48.09	104.9	149.6
260.0	279.2	3.932	5643.0	7016.0	212.3	47.85	99.42	151.8
262.0	270.9	3.814	5795.0	7210.0	213.0	47.66	94.97	153.9
264.0	263.3	3.708	5940.0	7397.0	213.7	47.51	91.30	155.9
266.0	256.5	3.612	6081.0	7576.0	214.4	47.39	88.22	157.8
268.0	250.3	3.524	6217.0	7750.0	215.1	47.29	85.61	159.6
270.0	244.5	3.443	6350.0	7919.0	215.7	47.22	83.38	161.4
275.0	231.9	3.265	6670.0	8324.0	217.2	47.13	78.99	165.6
280.0	221.2	3.114	6977.0	8711.0	218.6	47.14	75.81	169.5
285.0	211.9	2.983	7273.0	9083.0	219.9	47.22	73.43	173.1
290.0	203.7	2.868	7563.0	9446.0	221.2	47.36	71.61	176.5
295.0	196.4	2.766	7848.0	9800.0	222.4	47.54	70.20	179.8
300.0	189.8	2.673	8128.0	10150.0	223.5	47.75	69.10	182.9
305.0	183.9	2.589	8406.0	10490.0	224.7	47.99	68.23	185.8
310.0	178.4	2.512	8682.0	10830.0	225.8	48.26	67.55	188.7
320.0	168.7	2.376	9228.0	11500.0	227.9	48.83	66.60	194.0
330.0	160.3	2.258	9773.0	12160.0	230.0	49.45	66.03	199.1
340.0	153.0	2.154	10320.0	12820.0	231.9	50.09	65.74	203.8
350.0	146.4	2.062	10860.0	13480.0	233.8	50.74	65.63	208.4
360.0	140.6	1.979	11410.0	14140.0	235.7	51.39	65.66	212.7
370.0	135.3	1.904	11960.0	14790.0	237.5	52.04	65.78	216.9
380.0	130.4	1.836	12510.0	15450.0	239.2	52.69	65.97	220.9
390.0	126.0	1.774	13070.0	16110.0	240.9	53.31	66.21	224.7
400.0	121.9	1.716	13630.0	16780.0	242.6	53.93	66.49	228.5
420.0	114.5	1.613	14760.0	18110.0	245.9	55.11	67.13	235.6
440.0	108.2	1.523	15920.0	19460.0	249.0	56.23	67.81	242.4
460.0	102.6	1.444	17090.0	20820.0	252.1	57.28	68.51	248.9
480.0	97.57	1.374	18270.0	22200.0	255.0	58.27	69.21	255.2
500.0	93.09	1.311	19470.0	23590.0	257.8	59.20	69.91	261.2
			5.60 MPa isobar					
67.39[a]	1870.0	26.32	−12310.0	−12090.0	93.32	51.22	73.53	1349.0
68.0	1867.0	26.29	−12260.0	−12050.0	93.98	50.72	72.78	1327.0
70.0	1860.0	26.19	−12120.0	−11900.0	96.06	49.45	71.05	1270.0
80.0	1822.0	25.66	−11420.0	−11210.0	105.4	46.96	69.87	1155.0
100.0	1743.0	24.54	−10020.0	−9795.0	121.1	43.89	70.73	1062.0
120.0	1660.0	23.38	−8625.0	−8385.0	134.0	41.09	70.29	943.8
140.0	1573.0	22.14	−7230.0	−6977.0	144.8	39.81	70.83	826.2
160.0	1480.0	20.84	−5809.0	−5540.0	154.4	39.69	73.18	717.1
170.0	1431.0	20.15	−5077.0	−4799.0	158.9	39.92	75.19	662.6
180.0	1378.0	19.41	−4322.0	−4034.0	163.3	40.31	77.88	606.6
190.0	1322.0	18.61	−3539.0	−3238.0	167.6	40.85	81.43	547.9
200.0	1260.0	17.74	−2717.0	−2401.0	171.9	41.58	86.20	485.6
205.0	1225.0	17.25	−2287.0	−1963.0	174.1	42.03	89.26	452.9
210.0	1188.0	16.73	−1842.0	−1508.0	176.2	42.57	93.00	418.8
215.0	1148.0	16.16	−1378.0	−1031.0	178.5	43.21	97.76	383.2
220.0	1103.0	15.53	−888.0	−527.3	180.8	44.00	104.2	345.6
222.0	1083.0	15.25	−683.0	−315.8	181.8	44.37	107.5	329.9
224.0	1062.0	14.95	−471.6	−97.0	182.7	44.77	111.4	313.8
226.0	1039.0	14.63	−252.3	130.5	183.8	45.22	116.2	297.1
228.0	1014.0	14.28	−23.4	368.7	184.8	45.73	122.2	279.8
230.0	987.3	13.90	217.8	620.6	185.9	46.31	130.1	261.7
232.0	956.8	13.47	475.3	890.9	187.1	46.98	140.9	242.7
234.0	921.4	12.97	755.9	1187.0	188.3	47.79	156.8	222.5

Thermophysical properties of nitrogen trifluoride—Continued

T K	Density kg/m^3	Density mol/dm^3	E J/mol	H J/mol	S J/(mol·K)	C_v J/(mol·K)	C_p J/(mol·K)	Sound m/s
235.0	901.3	12.69	908.3	1350.0	189.0	48.26	168.0	211.8
236.0	878.7	12.37	1072.0	1525.0	189.8	48.78	182.6	200.8
237.0	853.1	12.01	1250.0	1717.0	190.6	49.38	202.5	189.4
238.0	823.2	11.59	1449.0	1932.0	191.5	50.06	230.5	177.6
239.0	787.4	11.09	1677.0	2182.0	192.5	50.85	271.1	165.6
240.0	743.4	10.47	1945.0	2480.0	193.8	51.73	327.2	154.2
240.5	717.8	10.11	2097.0	2651.0	194.5	52.17	358.9	149.0
241.0	690.0	9.716	2262.0	2838.0	195.3	52.59	388.1	144.5
242.0	630.9	8.884	2615.0	3246.0	197.0	53.20	419.3	138.0
243.0	574.3	8.086	2969.0	3661.0	198.7	53.35	406.0	134.7
243.5	548.8	7.728	3135.0	3860.0	199.5	53.27	387.8	133.9
244.0	525.7	7.402	3292.0	4049.0	200.3	53.11	365.6	133.5
244.5	504.9	7.109	3438.0	4225.0	201.0	52.91	341.6	133.3
245.0	486.4	6.849	3572.0	4390.0	201.7	52.67	317.6	133.4
245.5	469.9	6.616	3697.0	4543.0	202.3	52.42	294.9	133.6
246.0	455.2	6.410	3812.0	4685.0	202.9	52.16	274.0	134.0
246.5	442.1	6.225	3918.0	4818.0	203.4	51.91	255.3	134.4
247.0	430.3	6.058	4017.0	4941.0	203.9	51.67	238.8	134.9
248.0	409.9	5.772	4195.0	5166.0	204.8	51.20	211.4	136.0
249.0	393.0	5.534	4354.0	5366.0	205.6	50.79	190.1	137.3
250.0	378.6	5.331	4497.0	5547.0	206.3	50.41	173.4	138.6
251.0	366.2	5.156	4628.0	5714.0	207.0	50.08	160.1	139.9
252.0	355.2	5.002	4749.0	5868.0	207.6	49.78	149.3	141.2
253.0	345.5	4.865	4862.0	6013.0	208.2	49.51	140.3	142.4
254.0	336.8	4.743	4969.0	6149.0	208.7	49.27	132.9	143.7
256.0	321.7	4.530	5166.0	6403.0	209.7	48.85	121.1	146.1
258.0	308.9	4.350	5348.0	6636.0	210.6	48.51	112.3	148.4
260.0	297.9	4.194	5518.0	6853.0	211.5	48.23	105.5	150.7
262.0	288.2	4.058	5679.0	7059.0	212.3	48.01	100.1	152.8
264.0	279.6	3.937	5832.0	7254.0	213.0	47.82	95.65	154.8
266.0	271.8	3.828	5979.0	7442.0	213.7	47.67	91.99	156.8
268.0	264.8	3.728	6121.0	7623.0	214.4	47.56	88.91	158.7
270.0	258.3	3.638	6258.0	7798.0	215.0	47.47	86.30	160.5
275.0	244.3	3.440	6588.0	8216.0	216.6	47.34	81.22	164.8
280.0	232.5	3.274	6902.0	8612.0	218.0	47.31	77.58	168.7
285.0	222.4	3.131	7205.0	8993.0	219.4	47.37	74.88	172.4
290.0	213.5	3.006	7500.0	9362.0	220.6	47.48	72.82	175.9
295.0	205.6	2.896	7788.0	9722.0	221.9	47.65	71.23	179.2
300.0	198.6	2.796	8072.0	10080.0	223.1	47.85	69.99	182.4
305.0	192.2	2.706	8353.0	10420.0	224.2	48.08	69.01	185.4
310.0	186.3	2.624	8631.0	10770.0	225.3	48.33	68.24	188.3
320.0	176.0	2.478	9183.0	11440.0	227.5	48.89	67.15	193.7
330.0	167.1	2.353	9730.0	12110.0	229.5	49.50	66.49	198.8
340.0	159.3	2.243	10280.0	12770.0	231.5	50.13	66.13	203.6
350.0	152.4	2.146	10820.0	13430.0	233.4	50.77	65.96	208.2
360.0	146.2	2.059	11370.0	14090.0	235.3	51.42	65.94	212.6
370.0	140.6	1.980	11920.0	14750.0	237.1	52.07	66.03	216.8
380.0	135.6	1.909	12480.0	15410.0	238.8	52.70	66.20	220.8
390.0	130.9	1.843	13040.0	16080.0	240.6	53.33	66.42	224.7
400.0	126.6	1.783	13600.0	16740.0	242.3	53.94	66.68	228.5
420.0	118.9	1.675	14740.0	18080.0	245.5	55.12	67.28	235.7
440.0	112.3	1.581	15890.0	19430.0	248.7	56.23	67.94	242.5
460.0	106.4	1.498	17060.0	20800.0	251.7	57.28	68.62	249.1
480.0	101.2	1.425	18250.0	22180.0	254.6	58.27	69.31	255.3
500.0	96.55	1.359	19450.0	23570.0	257.5	59.20	69.99	261.4
				5.80 MPa isobar				
67.43[a]	1870.0	26.32	−12300.0	−12080.0	93.35	51.34	73.47	1344.0
68.0	1867.0	26.30	−12260.0	−12040.0	93.96	50.87	72.77	1324.0
70.0	1860.0	26.19	−12120.0	−11900.0	96.05	49.57	71.04	1267.0
80.0	1823.0	25.66	−11430.0	−11200.0	105.4	46.98	69.86	1155.0
100.0	1743.0	24.55	−10030.0	−9789.0	121.1	43.87	70.72	1064.0

Thermophysical properties of nitrogen trifluoride—Continued

T K	Density kg/m^3	Density mol/dm^3	E J/mol	H J/mol	S J/(mol·K)	C_v J/(mol·K)	C_p J/(mol·K)	Sound m/s
120.0	1660.0	23.38	−8628.0	−8379.0	134.0	41.07	70.27	945.5
140.0	1573.0	22.15	−7233.0	−6971.0	144.8	39.80	70.80	827.8
160.0	1481.0	20.85	−5814.0	−5535.0	154.4	39.68	73.13	718.6
170.0	1432.0	20.16	−5082.0	−4795.0	158.9	39.91	75.11	664.3
180.0	1379.0	19.42	−4329.0	−4031.0	163.3	40.29	77.77	608.4
190.0	1323.0	18.63	−3548.0	−3237.0	167.5	40.83	81.26	550.0
200.0	1261.0	17.76	−2728.0	−2402.0	171.8	41.55	85.92	488.3
205.0	1227.0	17.28	−2301.0	−1965.0	174.0	42.00	88.89	455.8
210.0	1191.0	16.77	−1858.0	−1512.0	176.2	42.53	92.50	422.2
215.0	1151.0	16.21	−1396.0	−1038.0	178.4	43.15	97.05	387.1
220.0	1107.0	15.58	−911.1	−538.9	180.7	43.91	103.1	350.3
222.0	1087.0	15.31	−708.6	−329.7	181.6	44.26	106.2	334.9
224.0	1067.0	15.02	−500.0	−113.9	182.6	44.64	109.8	319.2
226.0	1045.0	14.71	−284.4	109.9	183.6	45.07	114.1	303.0
228.0	1021.0	14.38	−60.2	343.2	184.6	45.55	119.4	286.2
230.0	995.1	14.01	174.7	588.6	185.7	46.08	126.2	268.8
232.0	966.4	13.61	423.4	849.6	186.8	46.70	135.3	250.7
234.0	933.9	13.15	690.9	1132.0	188.0	47.42	147.9	231.6
235.0	915.6	12.89	834.1	1284.0	188.7	47.83	156.3	221.6
236.0	895.6	12.61	985.5	1445.0	189.4	48.28	166.9	211.4
237.0	873.5	12.30	1147.0	1619.0	190.1	48.78	180.4	200.9
238.0	848.6	11.95	1322.0	1808.0	190.9	49.34	198.2	190.1
239.0	820.1	11.55	1515.0	2017.0	191.8	49.97	222.1	179.1
240.0	786.7	11.08	1731.0	2254.0	192.8	50.67	254.2	168.1
241.0	747.3	10.52	1977.0	2528.0	193.9	51.43	295.0	157.7
242.0	701.3	9.875	2258.0	2845.0	195.2	52.17	337.3	148.8
243.0	651.0	9.166	2565.0	3198.0	196.7	52.75	363.6	142.4
244.0	601.1	8.464	2878.0	3563.0	198.2	53.01	363.2	138.5
245.0	555.9	7.827	3176.0	3917.0	199.6	52.97	342.0	136.5
245.5	535.6	7.542	3315.0	4084.0	200.3	52.85	326.9	136.1
246.0	517.0	7.280	3447.0	4244.0	201.0	52.70	310.5	135.8
246.5	500.1	7.042	3571.0	4395.0	201.6	52.51	293.6	135.8
247.0	484.7	6.825	3688.0	4537.0	202.1	52.31	277.0	135.9
247.5	470.7	6.627	3797.0	4672.0	202.7	52.10	261.2	136.1
248.0	457.9	6.448	3899.0	4799.0	203.2	51.89	246.4	136.4
249.0	435.7	6.135	4086.0	5032.0	204.1	51.47	220.5	137.2
250.0	417.0	5.871	4254.0	5241.0	205.0	51.07	199.3	138.2
251.0	401.0	5.646	4405.0	5432.0	205.7	50.71	182.1	139.3
252.0	387.2	5.452	4543.0	5607.0	206.4	50.37	168.0	140.4
253.0	375.1	5.281	4670.0	5769.0	207.1	50.07	156.4	141.6
254.0	364.3	5.130	4789.0	5920.0	207.7	49.80	146.8	142.8
255.0	354.7	4.994	4901.0	6063.0	208.2	49.55	138.7	144.0
256.0	346.0	4.871	5007.0	6198.0	208.8	49.33	131.8	145.2
258.0	330.8	4.657	5205.0	6450.0	209.7	48.94	120.8	147.5
260.0	317.8	4.475	5387.0	6683.0	210.6	48.62	112.4	149.7
262.0	306.6	4.317	5557.0	6901.0	211.5	48.36	105.8	151.8
264.0	296.7	4.178	5719.0	7107.0	212.3	48.14	100.5	153.9
266.0	287.9	4.054	5873.0	7303.0	213.0	47.97	96.13	155.9
268.0	280.0	3.942	6021.0	7492.0	213.7	47.83	92.52	157.8
270.0	272.8	3.841	6163.0	7674.0	214.4	47.71	89.46	159.7
272.0	266.1	3.747	6302.0	7850.0	215.0	47.63	86.86	161.4
275.0	257.2	3.621	6504.0	8105.0	216.0	47.54	83.60	164.0
280.0	244.2	3.438	6826.0	8512.0	217.4	47.49	79.45	168.0
285.0	233.1	3.283	7135.0	8902.0	218.8	47.52	76.39	171.8
290.0	223.5	3.147	7435.0	9278.0	220.1	47.61	74.08	175.4
295.0	215.0	3.028	7728.0	9643.0	221.4	47.76	72.30	178.7
300.0	207.4	2.921	8016.0	10000.0	222.6	47.95	70.91	181.9
305.0	200.6	2.824	8299.0	10350.0	223.7	48.17	69.82	185.0
310.0	194.3	2.737	8580.0	10700.0	224.9	48.41	68.95	187.9
320.0	183.4	2.582	9136.0	11380.0	227.0	48.95	67.72	193.4
330.0	173.9	2.449	9688.0	12060.0	229.1	49.55	66.96	198.6
340.0	165.7	2.333	10240.0	12720.0	231.1	50.17	66.52	203.4
350.0	158.4	2.231	10790.0	13390.0	233.0	50.80	66.30	208.1

Thermophysical properties of nitrogen trifluoride—Continued

T K	Density kg/m^3	Density mol/dm^3	*E* J/mol	*H* J/mol	*S* J/(mol·K)	C_v J/(mol·K)	C_p J/(mol·K)	Sound m/s
360.0	151.9	2.139	11340.0	14050.0	234.9	51.45	66.24	212.5
370.0	146.1	2.057	11890.0	14710.0	236.7	52.09	66.29	216.7
380.0	140.7	1.981	12450.0	15380.0	238.5	52.72	66.43	220.7
390.0	135.8	1.913	13010.0	16040.0	240.2	53.34	66.62	224.7
400.0	131.3	1.849	13570.0	16710.0	241.9	53.95	66.86	228.5
420.0	123.3	1.737	14710.0	18050.0	245.2	55.13	67.43	235.7
440.0	116.4	1.639	15870.0	19410.0	248.3	56.24	68.07	242.6
460.0	110.3	1.553	17040.0	20770.0	251.4	57.28	68.73	249.2
480.0	104.9	1.477	18230.0	22160.0	254.3	58.27	69.40	255.5
500.0	100.0	1.408	19430.0	23550.0	257.1	59.20	70.08	261.6
				6.00 MPa isobar				
67.47[a]	1870.0	26.32	−12300.0	−12070.0	93.38	51.46	73.41	1339.0
68.0	1868.0	26.30	−12260.0	−12040.0	93.95	51.01	72.76	1320.0
70.0	1860.0	26.20	−12120.0	−11890.0	96.03	49.69	71.03	1264.0
80.0	1823.0	25.67	−11430.0	−11190.0	105.4	47.01	69.85	1155.0
100.0	1744.0	24.55	−10030.0	−9783.0	121.1	43.85	70.70	1065.0
120.0	1661.0	23.39	−8630.0	−8374.0	133.9	41.05	70.25	947.1
140.0	1574.0	22.16	−7237.0	−6966.0	144.8	39.78	70.77	829.4
160.0	1481.0	20.86	−5818.0	−5531.0	154.4	39.66	73.07	720.2
170.0	1432.0	20.17	−5088.0	−4791.0	158.9	39.90	75.04	666.0
180.0	1381.0	19.44	−4336.0	−4028.0	163.2	40.28	77.66	610.3
190.0	1325.0	18.65	−3557.0	−3235.0	167.5	40.82	81.09	552.2
200.0	1263.0	17.78	−2740.0	−2402.0	171.8	41.53	85.65	490.9
205.0	1229.0	17.31	−2314.0	−1967.0	173.9	41.97	88.54	458.8
210.0	1193.0	16.80	−1873.0	−1516.0	176.1	42.48	92.03	425.5
215.0	1154.0	16.25	−1415.0	−1045.0	178.3	43.09	96.38	390.9
220.0	1110.0	15.63	−933.5	−549.8	180.6	43.82	102.1	354.7
225.0	1061.0	14.94	−422.1	−20.5	183.0	44.72	110.2	316.5
227.0	1039.0	14.63	−206.2	204.0	184.0	45.15	114.5	300.5
228.0	1027.0	14.46	−95.1	319.7	184.5	45.38	117.0	292.3
230.0	1002.0	14.12	134.4	559.4	185.5	45.88	123.0	275.6
232.0	975.3	13.73	375.8	812.7	186.6	46.44	130.6	258.2
234.0	944.8	13.30	632.8	1084.0	187.8	47.09	141.0	240.0
236.0	910.0	12.81	911.2	1380.0	189.0	47.86	155.7	221.0
238.0	868.6	12.23	1221.0	1712.0	190.4	48.77	178.0	201.1
239.0	844.5	11.89	1393.0	1897.0	191.2	49.29	193.8	190.9
240.0	817.2	11.51	1579.0	2101.0	192.1	49.87	214.2	180.6
241.0	786.0	11.07	1785.0	2327.0	193.0	50.50	240.0	170.5
242.0	750.1	10.56	2014.0	2582.0	194.1	51.17	270.6	161.0
243.0	709.6	9.992	2268.0	2869.0	195.2	51.81	301.3	152.8
244.0	665.8	9.375	2542.0	3182.0	196.5	52.34	322.3	146.5
245.0	621.7	8.754	2822.0	3508.0	197.9	52.65	326.7	142.3
246.0	580.2	8.170	3096.0	3830.0	199.2	52.72	316.0	139.9
247.0	543.2	7.648	3352.0	4137.0	200.4	52.59	295.9	138.6
248.0	511.0	7.195	3587.0	4420.0	201.6	52.32	271.7	138.2
249.0	483.5	6.808	3799.0	4680.0	202.6	51.99	247.2	138.3
250.0	460.1	6.479	3989.0	4916.0	203.5	51.63	224.7	138.8
251.0	440.1	6.197	4162.0	5130.0	204.4	51.27	205.0	139.5
252.0	422.9	5.954	4319.0	5326.0	205.2	50.93	188.3	140.4
253.0	407.9	5.743	4463.0	5507.0	205.9	50.61	174.2	141.3
254.0	394.7	5.557	4596.0	5676.0	206.6	50.31	162.3	142.4
255.0	383.0	5.392	4720.0	5833.0	207.2	50.04	152.3	143.5
256.0	372.5	5.245	4837.0	5981.0	207.8	49.80	143.8	144.6
258.0	354.4	4.990	5052.0	6254.0	208.8	49.36	130.3	146.8
260.0	339.2	4.776	5248.0	6504.0	209.8	49.01	120.1	148.9
262.0	326.2	4.593	5430.0	6736.0	210.7	48.71	112.1	151.1
264.0	314.8	4.433	5600.0	6954.0	211.5	48.46	105.8	153.1
266.0	304.8	4.292	5762.0	7160.0	212.3	48.26	100.7	155.1
268.0	295.9	4.166	5917.0	7357.0	213.0	48.10	96.43	157.0
270.0	287.8	4.052	6065.0	7546.0	213.7	47.96	92.88	158.9
272.0	280.4	3.948	6209.0	7729.0	214.4	47.86	89.87	160.7

Thermophysical properties of nitrogen trifluoride—Continued

T K	Density kg/m³	Density mol/dm³	E J/mol	H J/mol	S J/(mol·K)	C_v J/(mol·K)	C_p J/(mol·K)	Sound m/s
274.0	273.6	3.853	6349.0	7906.0	215.1	47.78	87.30	162.5
280.0	256.2	3.608	6748.0	8411.0	216.9	47.66	81.42	167.4
285.0	244.2	3.438	7064.0	8809.0	218.3	47.66	77.98	171.2
290.0	233.7	3.291	7369.0	9192.0	219.6	47.74	75.40	174.8
295.0	224.6	3.162	7667.0	9564.0	220.9	47.87	73.41	178.3
300.0	216.4	3.048	7958.0	9927.0	222.1	48.04	71.86	181.5
305.0	209.1	2.945	8245.0	10280.0	223.3	48.25	70.64	184.6
310.0	202.5	2.851	8529.0	10630.0	224.4	48.48	69.68	187.5
320.0	190.8	2.687	9090.0	11320.0	226.6	49.01	68.30	193.1
330.0	180.8	2.546	9645.0	12000.0	228.7	49.59	67.43	198.3
340.0	172.1	2.424	10200.0	12670.0	230.7	50.21	66.92	203.3
350.0	164.5	2.316	10750.0	13340.0	232.6	50.84	66.64	207.9
360.0	157.6	2.220	11300.0	14010.0	234.5	51.47	66.53	212.4
370.0	151.5	2.133	11860.0	14670.0	236.3	52.11	66.55	216.6
380.0	145.9	2.054	12420.0	15340.0	238.1	52.74	66.65	220.7
390.0	140.8	1.982	12980.0	16000.0	239.8	53.36	66.83	224.7
400.0	136.1	1.916	13540.0	16670.0	241.5	53.96	67.05	228.5
420.0	127.7	1.799	14680.0	18020.0	244.8	55.13	67.58	235.8
440.0	120.5	1.697	15840.0	19380.0	248.0	56.24	68.20	242.7
460.0	114.1	1.607	17020.0	20750.0	251.0	57.28	68.84	249.3
480.0	108.5	1.528	18210.0	22130.0	254.0	58.27	69.50	255.7
500.0	103.5	1.457	19410.0	23530.0	256.8	59.20	70.16	261.8
				6.25 MPa isobar				
67.52[a]	1870.0	26.33	−12300.0	−12060.0	93.41	51.61	73.33	1332.0
68.0	1868.0	26.30	−12260.0	−12030.0	93.93	51.19	72.74	1316.0
70.0	1861.0	26.20	−12120.0	−11880.0	96.01	49.84	71.01	1261.0
80.0	1823.0	25.67	−11430.0	−11180.0	105.3	47.04	69.84	1154.0
100.0	1744.0	24.56	−10030.0	−9775.0	121.1	43.82	70.69	1067.0
120.0	1661.0	23.39	−8633.0	−8366.0	133.9	41.02	70.22	949.2
140.0	1574.0	22.17	−7241.0	−6959.0	144.8	39.76	70.73	831.3
160.0	1482.0	20.87	−5824.0	−5525.0	154.3	39.65	73.00	722.1
170.0	1433.0	20.18	−5095.0	−4786.0	158.8	39.88	74.94	668.0
180.0	1382.0	19.46	−4345.0	−4024.0	163.2	40.27	77.52	612.5
190.0	1326.0	18.67	−3567.0	−3233.0	167.4	40.80	80.89	554.8
200.0	1265.0	17.81	−2753.0	−2403.0	171.7	41.50	85.33	494.1
205.0	1232.0	17.35	−2329.0	−1969.0	173.8	41.93	88.12	462.3
210.0	1196.0	16.84	−1892.0	−1520.0	176.0	42.43	91.47	429.6
215.0	1157.0	16.30	−1437.0	−1053.0	178.2	43.02	95.59	395.6
220.0	1115.0	15.70	−960.7	−562.6	180.5	43.72	100.9	360.1
225.0	1067.0	15.02	−456.8	−40.7	182.8	44.57	108.3	323.0
227.0	1046.0	14.72	−244.9	179.6	183.8	44.97	112.1	307.5
228.0	1034.0	14.57	−136.3	292.8	184.3	45.19	114.3	299.6
230.0	1011.0	14.23	87.3	526.4	185.3	45.65	119.5	283.5
232.0	985.3	13.87	320.9	771.4	186.4	46.16	125.9	266.9
234.0	957.1	13.48	567.4	1031.0	187.5	46.75	134.2	249.7
236.0	925.4	13.03	830.5	1310.0	188.7	47.42	145.4	231.9
238.0	888.9	12.52	1117.0	1616.0	189.9	48.19	161.3	213.4
240.0	845.7	11.91	1436.0	1960.0	191.4	49.10	184.7	194.5
241.0	820.6	11.55	1612.0	2153.0	192.2	49.62	200.6	185.0
242.0	792.5	11.16	1803.0	2363.0	193.1	50.16	219.7	175.7
243.0	761.2	10.72	2010.0	2593.0	194.0	50.73	241.6	167.0
244.0	726.5	10.23	2235.0	2846.0	195.0	51.29	263.7	159.2
246.0	650.8	9.163	2723.0	3405.0	197.3	52.15	289.8	148.0
247.0	613.3	8.635	2971.0	3695.0	198.5	52.34	288.1	144.8
248.0	578.3	8.143	3211.0	3979.0	199.6	52.36	278.2	142.8
249.0	546.6	7.696	3438.0	4250.0	200.7	52.24	263.2	141.7
250.0	518.5	7.300	3648.0	4505.0	201.8	52.03	245.9	141.2
251.0	493.8	6.952	3842.0	4741.0	202.7	51.77	228.1	141.2
252.0	472.2	6.648	4021.0	4961.0	203.6	51.47	211.2	141.5
253.0	453.2	6.382	4185.0	5164.0	204.4	51.18	195.8	142.0

Thermophysical properties of nitrogen trifluoride—Continued

T K	Density kg/m³	Density mol/dm³	E J/mol	H J/mol	S J/(mol·K)	C_v J/(mol·K)	C_p J/(mol·K)	Sound m/s
254.0	436.6	6.148	4336.0	5353.0	205.1	50.89	182.2	142.7
255.0	421.9	5.941	4477.0	5529.0	205.8	50.61	170.3	143.5
256.0	408.8	5.756	4608.0	5694.0	206.5	50.35	160.0	144.4
257.0	397.0	5.591	4732.0	5850.0	207.1	50.10	151.0	145.4
258.0	386.4	5.441	4848.0	5997.0	207.6	49.88	143.3	146.4
260.0	368.0	5.181	5064.0	6270.0	208.7	49.48	130.6	148.4
262.0	352.4	4.962	5261.0	6521.0	209.7	49.14	120.8	150.4
264.0	338.9	4.772	5445.0	6755.0	210.5	48.86	113.1	152.4
266.0	327.2	4.607	5618.0	6974.0	211.4	48.63	106.9	154.4
268.0	316.8	4.460	5782.0	7183.0	212.2	48.43	101.7	156.3
270.0	307.4	4.329	5938.0	7382.0	212.9	48.28	97.51	158.2
272.0	299.0	4.210	6089.0	7573.0	213.6	48.15	93.94	160.0
274.0	291.3	4.102	6234.0	7758.0	214.3	48.05	90.90	161.7
276.0	284.3	4.003	6376.0	7937.0	214.9	47.97	88.29	163.4
280.0	271.7	3.826	6648.0	8281.0	216.2	47.88	84.05	166.7
285.0	258.3	3.637	6973.0	8691.0	217.6	47.85	80.08	170.6
290.0	246.8	3.476	7286.0	9084.0	219.0	47.90	77.11	174.3
295.0	236.8	3.334	7589.0	9464.0	220.3	48.01	74.85	177.7
300.0	227.9	3.209	7886.0	9833.0	221.5	48.16	73.09	181.0
305.0	220.0	3.097	8177.0	10200.0	222.7	48.36	71.71	184.1
310.0	212.8	2.996	8465.0	10550.0	223.9	48.58	70.61	187.1
315.0	206.2	2.904	8749.0	10900.0	225.0	48.82	69.73	190.0
320.0	200.2	2.819	9031.0	11250.0	226.1	49.09	69.03	192.8
330.0	189.5	2.668	9591.0	11930.0	228.2	49.65	68.04	198.1
340.0	180.2	2.538	10150.0	12610.0	230.2	50.25	67.42	203.1
350.0	172.1	2.423	10700.0	13280.0	232.2	50.87	67.07	207.8
360.0	164.8	2.321	11260.0	13950.0	234.1	51.50	66.90	212.3
370.0	158.3	2.229	11820.0	14620.0	235.9	52.13	66.87	216.6
380.0	152.4	2.146	12380.0	15290.0	237.7	52.76	66.94	220.7
390.0	147.0	2.070	12940.0	15960.0	239.4	53.37	67.08	224.7
400.0	142.1	2.000	13510.0	16630.0	241.1	53.98	67.28	228.5
420.0	133.3	1.876	14650.0	17980.0	244.4	55.14	67.78	235.9
440.0	125.7	1.769	15810.0	19340.0	247.6	56.25	68.36	242.9
460.0	119.0	1.675	16990.0	20720.0	250.6	57.29	68.98	249.5
480.0	113.1	1.592	18180.0	22100.0	253.6	58.27	69.62	255.9
500.0	107.8	1.518	19380.0	23500.0	256.4	59.19	70.27	262.0

6.50 MPa isobar

T K	Density kg/m³	Density mol/dm³	E J/mol	H J/mol	S J/(mol·K)	C_v J/(mol·K)	C_p J/(mol·K)	Sound m/s
67.56[a]	1870.0	26.33	−12300.0	−12050.0	93.44	51.76	73.26	1326.0
68.0	1868.0	26.30	−12270.0	−12020.0	93.91	51.37	72.73	1311.0
70.0	1861.0	26.20	−12120.0	−11880.0	95.99	49.99	70.99	1257.0
80.0	1823.0	25.67	−11430.0	−11180.0	105.3	47.07	69.83	1154.0
100.0	1744.0	24.56	−10030.0	−9767.0	121.0	43.80	70.67	1069.0
120.0	1662.0	23.40	−8637.0	−8359.0	133.9	40.99	70.20	951.3
140.0	1575.0	22.18	−7245.0	−6952.0	144.7	39.74	70.70	833.2
160.0	1483.0	20.89	−5830.0	−5519.0	154.3	39.64	72.94	724.0
170.0	1435.0	20.20	−5102.0	−4781.0	158.8	39.87	74.85	670.0
180.0	1383.0	19.47	−4354.0	−4020.0	163.1	40.25	77.39	614.8
190.0	1328.0	18.70	−3578.0	−3230.0	167.4	40.78	80.69	557.4
200.0	1267.0	17.84	−2767.0	−2403.0	171.6	41.47	85.02	497.2
205.0	1234.0	17.38	−2345.0	−1971.0	173.8	41.89	87.72	465.9
210.0	1199.0	16.88	−1910.0	−1525.0	175.9	42.38	90.93	433.5
215.0	1161.0	16.35	−1458.0	−1061.0	178.1	42.95	94.85	400.1
220.0	1119.0	15.76	−987.0	−574.5	180.3	43.62	99.84	365.4
225.0	1073.0	15.10	−489.8	−59.4	182.6	44.44	106.6	329.1
227.0	1052.0	14.81	−281.7	157.1	183.6	44.81	110.0	314.1
228.0	1041.0	14.66	−175.2	268.1	184.1	45.01	112.0	306.5
230.0	1019.0	14.34	43.3	496.4	185.1	45.44	116.5	291.0
232.0	994.5	14.00	270.4	734.6	186.1	45.92	122.0	275.1
234.0	968.0	13.63	508.3	985.2	187.2	46.45	128.8	258.7
236.0	938.7	13.22	759.6	1251.0	188.3	47.04	137.8	241.8

Thermophysical properties of nitrogen trifluoride—Continued

T K	Density kg/m³	Density mol/dm³	E J/mol	H J/mol	S J/(mol·K)	C_v J/(mol·K)	C_p J/(mol·K)	Sound m/s
238.0	905.8	12.75	1029.0	1538.0	189.5	47.72	149.7	224.5
240.0	868.0	12.22	1322.0	1853.0	190.9	48.50	166.3	206.8
242.0	823.3	11.59	1647.0	2208.0	192.3	49.39	189.7	189.1
244.0	769.7	10.84	2018.0	2618.0	194.0	50.37	220.7	172.3
246.0	706.6	9.950	2438.0	3091.0	195.9	51.31	251.2	158.6
248.0	639.3	9.002	2887.0	3609.0	198.0	51.92	262.4	149.9
249.0	607.0	8.547	3109.0	3870.0	199.1	52.03	258.4	147.3
250.0	576.8	8.121	3324.0	4124.0	200.1	52.03	249.7	145.6
251.0	549.2	7.733	3528.0	4368.0	201.1	51.93	238.1	144.7
252.0	524.2	7.382	3719.0	4600.0	202.0	51.75	225.1	144.2
253.0	501.9	7.067	3899.0	4818.0	202.9	51.54	211.6	144.1
254.0	482.0	6.787	4066.0	5023.0	203.7	51.30	198.6	144.3
255.0	464.3	6.537	4221.0	5216.0	204.4	51.05	186.4	144.7
256.0	448.4	6.314	4367.0	5397.0	205.1	50.81	175.4	145.2
257.0	434.1	6.113	4504.0	5567.0	205.8	50.57	165.4	145.9
258.0	421.3	5.932	4632.0	5728.0	206.4	50.34	156.6	146.7
260.0	399.1	5.620	4869.0	6026.0	207.6	49.92	141.8	148.4
262.0	380.5	5.358	5084.0	6297.0	208.6	49.56	130.2	150.2
264.0	364.7	5.135	5282.0	6548.0	209.6	49.25	121.0	152.1
266.0	350.9	4.942	5467.0	6782.0	210.5	48.98	113.5	153.9
268.0	338.9	4.772	5641.0	7003.0	211.3	48.77	107.5	155.8
270.0	328.2	4.621	5806.0	7213.0	212.1	48.58	102.5	157.6
272.0	318.5	4.485	5964.0	7413.0	212.8	48.43	98.31	159.4
274.0	309.8	4.362	6116.0	7606.0	213.5	48.31	94.76	161.2
276.0	301.8	4.250	6263.0	7793.0	214.2	48.22	91.73	162.9
280.0	287.8	4.052	6545.0	8149.0	215.5	48.09	86.83	166.2
285.0	272.9	3.843	6880.0	8572.0	217.0	48.03	82.28	170.1
290.0	260.3	3.665	7200.0	8974.0	218.4	48.06	78.91	173.8
295.0	249.3	3.510	7510.0	9362.0	219.7	48.14	76.35	177.3
300.0	239.6	3.374	7812.0	9739.0	221.0	48.28	74.36	180.6
305.0	231.0	3.252	8108.0	10110.0	222.2	48.46	72.81	183.8
310.0	223.2	3.143	8399.0	10470.0	223.4	48.67	71.57	186.8
315.0	216.2	3.044	8687.0	10820.0	224.5	48.90	70.58	189.7
320.0	209.7	2.953	8972.0	11170.0	225.6	49.16	69.79	192.5
330.0	198.3	2.792	9537.0	11870.0	227.7	49.71	68.65	197.9
340.0	188.4	2.653	10100.0	12550.0	229.8	50.30	67.93	202.9
350.0	179.7	2.531	10660.0	13220.0	231.7	50.91	67.50	207.7
360.0	172.1	2.423	11220.0	13900.0	233.6	51.54	67.28	212.2
370.0	165.2	2.326	11780.0	14570.0	235.5	52.16	67.20	216.5
380.0	158.9	2.238	12340.0	15240.0	237.3	52.78	67.23	220.7
390.0	153.2	2.158	12900.0	15920.0	239.0	53.39	67.34	224.7
400.0	148.0	2.084	13470.0	16590.0	240.7	53.99	67.51	228.6
420.0	138.8	1.954	14620.0	17940.0	244.0	55.15	67.97	236.0
440.0	130.8	1.842	15780.0	19310.0	247.2	56.25	68.52	243.0
460.0	123.8	1.743	16960.0	20690.0	250.2	57.29	69.12	249.7
480.0	117.6	1.656	18150.0	22070.0	253.2	58.27	69.74	256.1
500.0	112.1	1.579	19360.0	23480.0	256.1	59.19	70.37	262.3
			6.75 MPa isobar					
67.61[a]	1870.0	26.33	−12300.0	−12040.0	93.47	51.90	73.19	1320.0
68.0	1868.0	26.31	−12270.0	−12010.0	93.89	51.55	72.71	1307.0
70.0	1861.0	26.21	−12120.0	−11870.0	95.97	50.13	70.98	1253.0
80.0	1824.0	25.68	−11430.0	−11170.0	105.3	47.10	69.82	1154.0
100.0	1745.0	24.57	−10030.0	−9759.0	121.0	43.77	70.66	1071.0
120.0	1662.0	23.41	−8640.0	−8351.0	133.9	40.97	70.18	953.4
140.0	1576.0	22.19	−7250.0	−6945.0	144.7	39.72	70.66	835.1
160.0	1484.0	20.90	−5836.0	−5513.0	154.3	39.62	72.87	725.9
170.0	1436.0	20.21	−5110.0	−4776.0	158.7	39.86	74.76	672.0
180.0	1384.0	19.49	−4362.0	−4016.0	163.1	40.24	77.26	617.0
190.0	1329.0	18.72	−3589.0	−3228.0	167.3	40.76	80.50	560.0
200.0	1269.0	17.87	−2781.0	−2403.0	171.6	41.45	84.71	500.3
205.0	1237.0	17.41	−2361.0	−1973.0	173.7	41.86	87.33	469.3

Thermophysical properties of nitrogen trifluoride—Continued

T K	Density kg/m³	Density mol/dm³	E J/mol	H J/mol	S J/(mol·K)	C_v J/(mol·K)	C_p J/(mol·K)	Sound m/s
210.0	1202.0	16.92	−1928.0	−1529.0	175.8	42.34	90.42	437.4
215.0	1164.0	16.40	−1479.0	−1068.0	178.0	42.89	94.15	404.5
220.0	1123.0	15.82	−1012.0	−585.7	180.2	43.54	98.84	370.4
225.0	1078.0	15.18	−521.5	−76.8	182.5	44.31	105.0	335.1
227.0	1058.0	14.90	−316.7	136.4	183.4	44.67	108.2	320.5
228.0	1048.0	14.75	−212.1	245.4	183.9	44.86	109.9	313.1
230.0	1026.0	14.45	1.9	469.1	184.9	45.26	113.9	298.1
232.0	1003.0	14.12	223.5	701.5	185.9	45.70	118.6	282.8
234.0	977.9	13.77	454.1	944.4	186.9	46.19	124.5	267.1
236.0	950.5	13.38	696.0	1200.0	188.0	46.73	131.8	251.0
238.0	920.3	12.96	952.0	1473.0	189.2	47.33	141.2	234.6
240.0	886.3	12.48	1226.0	1767.0	190.4	48.01	153.7	217.9
242.0	847.4	11.93	1525.0	2091.0	191.8	48.78	170.4	201.1
244.0	802.0	11.29	1855.0	2452.0	193.2	49.62	192.4	184.9
246.0	749.2	10.55	2222.0	2862.0	194.9	50.49	217.5	170.3
248.0	690.2	9.719	2624.0	3318.0	196.8	51.24	236.7	159.1
250.0	630.4	8.876	3037.0	3798.0	198.7	51.67	239.7	152.0
252.0	575.6	8.105	3434.0	4267.0	200.6	51.72	227.6	148.5
253.0	551.1	7.760	3620.0	4490.0	201.4	51.63	218.3	147.6
254.0	528.8	7.446	3797.0	4703.0	202.3	51.49	208.1	147.1
255.0	508.5	7.160	3963.0	4906.0	203.1	51.32	197.6	147.0
256.0	490.2	6.902	4121.0	5099.0	203.8	51.12	187.3	147.1
257.0	473.5	6.668	4269.0	5281.0	204.5	50.91	177.6	147.4
258.0	458.5	6.456	4408.0	5454.0	205.2	50.70	168.5	147.8
259.0	444.8	6.263	4540.0	5618.0	205.8	50.50	160.1	148.4
260.0	432.4	6.088	4666.0	5774.0	206.5	50.30	152.6	149.0
262.0	410.5	5.781	4899.0	6066.0	207.6	49.93	139.6	150.5
264.0	392.0	5.520	5112.0	6335.0	208.6	49.60	129.1	152.1
266.0	376.1	5.296	5309.0	6584.0	209.5	49.32	120.6	153.9
268.0	362.2	5.100	5494.0	6818.0	210.4	49.08	113.6	155.6
270.0	349.9	4.927	5669.0	7039.0	211.2	48.88	107.8	157.4
272.0	338.9	4.773	5835.0	7250.0	212.0	48.71	102.9	159.1
274.0	329.1	4.634	5994.0	7451.0	212.7	48.57	98.85	160.8
276.0	320.1	4.508	6148.0	7645.0	213.4	48.46	95.36	162.5
278.0	311.9	4.392	6296.0	7833.0	214.1	48.37	92.36	164.2
280.0	304.4	4.286	6440.0	8015.0	214.8	48.30	89.76	165.8
285.0	287.9	4.054	6785.0	8450.0	216.3	48.21	84.60	169.7
290.0	274.0	3.858	7114.0	8863.0	217.8	48.21	80.79	173.4
295.0	262.0	3.690	7430.0	9260.0	219.1	48.28	77.91	176.9
300.0	251.5	3.542	7738.0	9643.0	220.4	48.40	75.68	180.3
305.0	242.2	3.410	8038.0	10020.0	221.6	48.56	73.94	183.5
310.0	233.8	3.293	8333.0	10380.0	222.8	48.76	72.55	186.5
315.0	226.3	3.186	8625.0	10740.0	224.0	48.99	71.45	189.5
320.0	219.3	3.089	8913.0	11100.0	225.1	49.23	70.56	192.3
330.0	207.1	2.917	9483.0	11800.0	227.2	49.77	69.27	197.7
340.0	196.6	2.769	10050.0	12490.0	229.3	50.35	68.45	202.8
350.0	187.5	2.640	10610.0	13170.0	231.3	50.95	67.94	207.6
360.0	179.3	2.525	11170.0	13840.0	233.2	51.57	67.66	212.2
370.0	172.0	2.422	11730.0	14520.0	235.0	52.19	67.53	216.5
380.0	165.5	2.330	12300.0	15200.0	236.8	52.80	67.52	220.7
390.0	159.5	2.246	12870.0	15870.0	238.6	53.41	67.60	224.8
400.0	154.0	2.169	13440.0	16550.0	240.3	54.01	67.74	228.7
420.0	144.3	2.032	14590.0	17910.0	243.6	55.16	68.16	236.2
440.0	136.0	1.914	15750.0	19280.0	246.8	56.26	68.68	243.2
460.0	128.7	1.812	16930.0	20650.0	249.9	57.29	69.26	249.9
480.0	122.2	1.721	18120.0	22050.0	252.8	58.26	69.86	256.4
500.0	116.4	1.639	19330.0	23450.0	255.7	59.19	70.48	262.5
				7.00 MPa isobar				
67.65[a]	1870.0	26.33	−12290.0	−12030.0	93.50	52.04	73.11	1314.0
68.0	1868.0	26.31	−12270.0	−12000.0	93.88	51.73	72.70	1302.0
70.0	1861.0	26.21	−12130.0	−11860.0	95.96	50.28	70.96	1249.0

Thermophysical properties of nitrogen trifluoride—Continued

T K	Density kg/m^3	Density mol/dm^3	E J/mol	H J/mol	S J/(mol·K)	C_v J/(mol·K)	C_p J/(mol·K)	Sound m/s
80.0	1824.0	25.68	−11430.0	−11160.0	105.3	47.13	69.80	1153.0
100.0	1745.0	24.57	−10040.0	−9751.0	121.0	43.75	70.64	1073.0
120.0	1663.0	23.41	−8643.0	−8344.0	133.8	40.94	70.15	955.4
140.0	1576.0	22.20	−7254.0	−6939.0	144.7	39.70	70.62	837.0
160.0	1485.0	20.91	−5842.0	−5507.0	154.2	39.61	72.81	727.8
170.0	1437.0	20.23	−5117.0	−4771.0	158.7	39.85	74.67	674.0
180.0	1386.0	19.51	−4371.0	−4012.0	163.0	40.23	77.13	619.2
190.0	1331.0	18.74	−3599.0	−3225.0	167.3	40.75	80.31	562.5
200.0	1271.0	17.90	−2794.0	−2403.0	171.5	41.42	84.42	503.3
205.0	1239.0	17.45	−2376.0	−1974.0	173.6	41.83	86.95	472.7
210.0	1205.0	16.96	−1945.0	−1532.0	175.7	42.29	89.93	441.1
215.0	1168.0	16.44	−1500.0	−1074.0	177.9	42.83	93.49	408.7
220.0	1127.0	15.88	−1037.0	−596.1	180.1	43.46	97.91	375.3
225.0	1083.0	15.25	−551.9	−92.9	182.3	44.20	103.7	340.8
230.0	1033.0	14.54	−37.2	444.1	184.7	45.09	111.6	304.8
232.0	1011.0	14.23	179.5	671.4	185.7	45.50	115.8	290.0
234.0	986.9	13.90	404.1	907.8	186.7	45.95	120.8	274.9
236.0	961.2	13.53	638.2	1155.0	187.8	46.45	127.0	259.5
238.0	933.1	13.14	883.9	1417.0	188.9	47.00	134.6	243.9
240.0	902.0	12.70	1144.0	1695.0	190.0	47.60	144.4	228.0
242.0	867.1	12.21	1423.0	1996.0	191.3	48.27	157.0	212.1
244.0	827.4	11.65	1725.0	2326.0	192.6	49.01	173.2	196.4
246.0	782.1	11.01	2055.0	2691.0	194.1	49.79	192.5	181.8
248.0	731.1	10.29	2415.0	3095.0	195.8	50.54	211.3	169.3
250.0	676.7	9.528	2796.0	3531.0	197.5	51.13	222.3	160.1
252.0	623.2	8.775	3178.0	3976.0	199.3	51.43	220.9	154.3
254.0	574.6	8.091	3543.0	4408.0	201.0	51.45	209.8	151.3
256.0	532.5	7.498	3879.0	4812.0	202.6	51.25	194.1	150.0
258.0	497.0	6.998	4183.0	5184.0	204.0	50.94	177.4	149.9
260.0	467.2	6.578	4458.0	5522.0	205.3	50.59	161.8	150.4
262.0	442.1	6.225	4708.0	5832.0	206.5	50.24	148.4	151.5
264.0	420.8	5.925	4936.0	6117.0	207.6	49.92	137.0	152.7
266.0	402.5	5.668	5147.0	6382.0	208.6	49.63	127.6	154.2
268.0	386.6	5.444	5343.0	6629.0	209.5	49.38	119.7	155.8
270.0	372.7	5.247	5527.0	6861.0	210.4	49.16	113.2	157.4
272.0	360.3	5.073	5702.0	7082.0	211.2	48.98	107.7	159.1
274.0	349.1	4.916	5869.0	7293.0	212.0	48.82	103.1	160.7
276.0	339.1	4.775	6029.0	7495.0	212.7	48.69	99.15	162.3
278.0	330.0	4.646	6183.0	7690.0	213.4	48.59	95.76	164.0
280.0	321.6	4.528	6333.0	7878.0	214.1	48.51	92.82	165.6
285.0	303.4	4.272	6689.0	8327.0	215.7	48.39	87.00	169.4
290.0	288.1	4.057	7026.0	8751.0	217.2	48.36	82.74	173.1
295.0	275.1	3.873	7349.0	9156.0	218.5	48.41	79.52	176.7
300.0	263.7	3.713	7662.0	9547.0	219.9	48.52	77.04	180.0
305.0	253.6	3.571	7967.0	9928.0	221.1	48.67	75.10	183.2
310.0	244.6	3.444	8267.0	10300.0	222.3	48.85	73.57	186.3
315.0	236.5	3.330	8562.0	10660.0	223.5	49.07	72.34	189.3
320.0	229.1	3.226	8853.0	11020.0	224.6	49.30	71.35	192.2
330.0	216.1	3.043	9428.0	11730.0	226.8	49.82	69.91	197.6
340.0	204.9	2.886	9997.0	12420.0	228.9	50.39	68.97	202.7
350.0	195.2	2.749	10560.0	13110.0	230.8	50.99	68.38	207.6
360.0	186.6	2.628	11130.0	13790.0	232.8	51.60	68.04	212.2
370.0	179.0	2.520	11690.0	14470.0	234.6	52.21	67.86	216.6
380.0	172.0	2.422	12260.0	15150.0	236.4	52.82	67.82	220.8
390.0	165.8	2.334	12830.0	15830.0	238.2	53.43	67.86	224.9
400.0	160.0	2.253	13400.0	16510.0	239.9	54.02	67.98	228.8
420.0	149.9	2.110	14550.0	17870.0	243.2	55.17	68.35	236.3
440.0	141.1	1.987	15720.0	19240.0	246.4	56.26	68.84	243.4
460.0	133.5	1.880	16900.0	20620.0	249.5	57.29	69.40	250.2
480.0	126.8	1.785	18100.0	22020.0	252.5	58.26	69.98	256.6
500.0	120.8	1.700	19310.0	23420.0	255.3	59.19	70.59	262.8

Thermophysical properties of nitrogen trifluoride—Continued

T K	Density kg/m³	Density mol/dm³	E J/mol	H J/mol	S J/(mol·K)	C_v J/(mol·K)	C_p J/(mol·K)	Sound m/s
				7.50 MPa isobar				
67.75[a]	1870.0	26.33	−12290.0	−12000.0	93.57	52.32	72.97	1301.0
68.0	1869.0	26.32	−12270.0	−11990.0	93.84	52.09	72.67	1293.0
70.0	1862.0	26.22	−12130.0	−11840.0	95.92	50.57	70.93	1242.0
80.0	1824.0	25.69	−11440.0	−11140.0	105.2	47.18	69.78	1152.0
100.0	1746.0	24.58	−10040.0	−9736.0	121.0	43.69	70.61	1077.0
120.0	1664.0	23.43	−8649.0	−8329.0	133.8	40.89	70.10	959.5
140.0	1578.0	22.21	−7263.0	−6925.0	144.6	39.66	70.55	840.8
160.0	1487.0	20.93	−5854.0	−5496.0	154.1	39.58	72.68	731.5
170.0	1439.0	20.26	−5130.0	−4760.0	158.6	39.83	74.50	677.9
180.0	1388.0	19.55	−4388.0	−4004.0	162.9	40.21	76.89	623.5
190.0	1334.0	18.78	−3620.0	−3220.0	167.2	40.72	79.95	567.5
200.0	1275.0	17.96	−2820.0	−2402.0	171.4	41.38	83.86	509.3
210.0	1210.0	17.04	−1979.0	−1539.0	175.6	42.22	89.01	448.5
215.0	1174.0	16.53	−1540.0	−1086.0	177.7	42.73	92.27	417.0
220.0	1135.0	15.98	−1084.0	−615.1	179.9	43.32	96.23	384.7
225.0	1093.0	15.39	−609.4	−122.0	182.1	44.00	101.2	351.5
230.0	1046.0	14.72	−109.7	399.7	184.4	44.81	107.8	317.4
232.0	1025.0	14.43	98.9	618.6	185.3	45.17	111.1	303.5
234.0	1003.0	14.12	313.7	844.7	186.3	45.56	115.0	289.4
236.0	979.8	13.80	535.5	1079.0	187.3	45.99	119.6	275.1
238.0	954.9	13.45	765.7	1323.0	188.3	46.45	125.0	260.7
240.0	927.9	13.07	1006.0	1580.0	189.4	46.95	131.6	246.1
242.0	898.5	12.65	1258.0	1851.0	190.5	47.49	139.6	231.5
244.0	866.1	12.20	1524.0	2139.0	191.7	48.08	149.4	217.1
246.0	830.3	11.69	1808.0	2449.0	193.0	48.71	161.0	203.1
248.0	790.7	11.13	2111.0	2784.0	194.3	49.35	174.0	190.1
250.0	747.4	10.52	2432.0	3145.0	195.8	49.96	186.3	178.7
252.0	701.7	9.881	2768.0	3527.0	197.3	50.48	194.8	169.7
254.0	656.0	9.237	3108.0	3920.0	198.8	50.83	196.7	163.2
256.0	612.5	8.625	3440.0	4310.0	200.4	50.99	192.2	159.1
258.0	573.0	8.069	3756.0	4686.0	201.8	50.97	183.3	156.8
260.0	538.1	7.577	4051.0	5041.0	203.2	50.82	172.3	155.7
262.0	507.6	7.148	4325.0	5374.0	204.5	50.60	160.8	155.4
264.0	481.2	6.776	4578.0	5685.0	205.7	50.35	149.9	155.7
266.0	458.3	6.454	4812.0	5974.0	206.8	50.10	140.0	156.5
268.0	438.3	6.172	5030.0	6246.0	207.8	49.86	131.3	157.4
270.0	420.8	5.925	5235.0	6500.0	208.7	49.64	123.8	158.6
272.0	405.3	5.706	5427.0	6741.0	209.6	49.44	117.3	159.9
274.0	391.4	5.512	5609.0	6970.0	210.4	49.27	111.7	161.3
276.0	379.0	5.337	5783.0	7189.0	211.2	49.12	106.9	162.7
278.0	367.8	5.178	5950.0	7398.0	212.0	49.00	102.8	164.2
280.0	357.5	5.034	6110.0	7600.0	212.7	48.90	99.17	165.7
282.0	348.2	4.903	6265.0	7795.0	213.4	48.82	96.03	167.2
284.0	339.6	4.781	6416.0	7984.0	214.1	48.75	93.27	168.6
290.0	317.3	4.468	6845.0	8524.0	216.0	48.66	86.80	173.0
295.0	301.9	4.251	7183.0	8947.0	217.4	48.67	82.88	176.4
300.0	288.6	4.064	7508.0	9354.0	218.8	48.74	79.86	179.8
305.0	276.9	3.899	7824.0	9747.0	220.1	48.86	77.51	183.0
310.0	266.6	3.754	8132.0	10130.0	221.3	49.03	75.65	186.1
315.0	257.3	3.623	8434.0	10500.0	222.5	49.22	74.16	189.1
320.0	248.9	3.505	8732.0	10870.0	223.7	49.44	72.96	192.0
325.0	241.2	3.397	9026.0	11230.0	224.8	49.68	71.99	194.8
330.0	234.2	3.298	9318.0	11590.0	225.9	49.94	71.20	197.5
340.0	221.7	3.122	9895.0	12300.0	228.0	50.48	70.04	202.7
350.0	210.9	2.969	10470.0	12990.0	230.0	51.06	69.28	207.6
360.0	201.3	2.835	11040.0	13680.0	232.0	51.66	68.81	212.3
370.0	192.9	2.716	11610.0	14370.0	233.8	52.26	68.53	216.7
380.0	185.2	2.608	12180.0	15060.0	235.7	52.86	68.40	221.0
390.0	178.4	2.511	12750.0	15740.0	237.4	53.46	68.38	225.1
400.0	172.1	2.423	13330.0	16420.0	239.2	54.05	68.45	229.1
410.0	166.3	2.341	13910.0	17110.0	240.9	54.62	68.57	232.9

Thermophysical properties of nitrogen trifluoride—Continued

T K	Density kg/m^3	Density mol/dm^3	E J/mol	H J/mol	S J/(mol·K)	C_v J/(mol·K)	C_p J/(mol·K)	Sound m/s
420.0	161.0	2.267	14490.0	17800.0	242.5	55.19	68.74	236.7
440.0	151.5	2.133	15660.0	19170.0	245.7	56.27	69.16	243.8
460.0	143.2	2.016	16840.0	20560.0	248.8	57.29	69.67	250.6
480.0	135.9	1.913	18040.0	21960.0	251.8	58.26	70.22	257.1
500.0	129.4	1.822	19250.0	23370.0	254.7	59.18	70.79	263.4
				8.00 MPa isobar				
67.84[a]	1870.0	26.33	-12280.0	-11980.0	93.63	52.59	72.82	1289.0
68.0	1869.0	26.32	-12270.0	-11970.0	93.80	52.44	72.63	1284.0
70.0	1862.0	26.22	-12130.0	-11830.0	95.88	50.85	70.89	1235.0
80.0	1825.0	25.70	-11440.0	-11130.0	105.2	47.24	69.75	1152.0
100.0	1746.0	24.59	-10050.0	-9720.0	120.9	43.64	70.58	1080.0
120.0	1665.0	23.44	-8655.0	-8314.0	133.7	40.84	70.05	963.6
140.0	1579.0	22.23	-7271.0	-6911.0	144.5	39.62	70.48	844.6
160.0	1488.0	20.96	-5866.0	-5484.0	154.1	39.56	72.56	735.2
170.0	1441.0	20.29	-5144.0	-4750.0	158.5	39.81	74.33	681.7
180.0	1391.0	19.58	-4404.0	-3995.0	162.8	40.19	76.65	627.7
190.0	1337.0	18.83	-3640.0	-3215.0	167.0	40.70	79.60	572.3
200.0	1279.0	18.01	-2845.0	-2401.0	171.2	41.34	83.34	515.0
210.0	1215.0	17.11	-2012.0	-1544.0	175.4	42.15	88.17	455.4
215.0	1180.0	16.62	-1578.0	-1096.0	177.5	42.64	91.17	424.8
220.0	1143.0	16.09	-1129.0	-631.8	179.6	43.19	94.75	393.5
225.0	1102.0	15.51	-663.2	-147.5	181.8	43.83	99.14	361.6
230.0	1057.0	14.88	-176.0	361.6	184.1	44.57	104.7	329.0
235.0	1007.0	14.18	338.7	903.1	186.4	45.43	112.3	295.8
240.0	949.0	13.36	890.9	1490.0	188.9	46.45	123.0	262.0
242.0	923.1	13.00	1126.0	1741.0	189.9	46.91	128.6	248.4
244.0	895.2	12.61	1370.0	2005.0	191.0	47.40	135.2	235.0
246.0	865.0	12.18	1626.0	2283.0	192.1	47.92	142.9	221.8
248.0	832.3	11.72	1895.0	2577.0	193.3	48.45	151.6	209.2
250.0	796.9	11.22	2177.0	2890.0	194.6	49.00	160.8	197.5
252.0	759.1	10.69	2471.0	3220.0	195.9	49.52	169.3	187.1
255.0	699.6	9.852	2930.0	3742.0	197.9	50.16	177.4	174.9
257.0	660.2	9.297	3237.0	4098.0	199.3	50.44	177.7	169.2
258.0	641.2	9.028	3389.0	4275.0	200.0	50.52	176.4	167.1
260.0	604.9	8.518	3684.0	4623.0	201.4	50.60	171.5	164.0
262.0	571.8	8.052	3966.0	4959.0	202.6	50.58	164.3	162.1
264.0	542.1	7.633	4232.0	5280.0	203.9	50.47	156.2	161.2
266.0	515.6	7.259	4482.0	5584.0	205.0	50.32	147.8	160.8
268.0	492.0	6.928	4717.0	5871.0	206.1	50.14	139.8	161.0
270.0	471.1	6.634	4938.0	6143.0	207.1	49.96	132.3	161.5
272.0	452.5	6.372	5146.0	6401.0	208.1	49.78	125.6	162.3
274.0	435.9	6.138	5343.0	6646.0	209.0	49.62	119.6	163.2
276.0	421.0	5.928	5531.0	6880.0	209.8	49.47	114.2	164.3
278.0	407.5	5.739	5710.0	7104.0	210.6	49.34	109.6	165.5
280.0	395.3	5.566	5881.0	7319.0	211.4	49.23	105.4	166.7
282.0	384.2	5.409	6047.0	7526.0	212.1	49.14	101.8	168.0
284.0	373.9	5.265	6207.0	7726.0	212.8	49.06	98.58	169.4
286.0	364.5	5.133	6362.0	7920.0	213.5	49.00	95.74	170.7
290.0	347.7	4.896	6659.0	8293.0	214.8	48.93	90.97	173.4
295.0	329.7	4.643	7013.0	8736.0	216.3	48.91	86.34	176.7
300.0	314.3	4.426	7351.0	9159.0	217.7	48.95	82.78	180.0
305.0	300.9	4.237	7677.0	9565.0	219.1	49.05	80.00	183.1
310.0	289.1	4.071	7994.0	9960.0	220.4	49.19	77.81	186.2
315.0	278.6	3.923	8305.0	10340.0	221.6	49.37	76.05	189.2
320.0	269.1	3.789	8609.0	10720.0	222.8	49.57	74.63	192.1
325.0	260.5	3.668	8910.0	11090.0	223.9	49.80	73.47	194.9
330.0	252.6	3.557	9207.0	11460.0	225.0	50.04	72.53	197.6
340.0	238.7	3.361	9793.0	12170.0	227.2	50.57	71.13	202.8
350.0	226.7	3.192	10370.0	12880.0	229.2	51.13	70.20	207.8
360.0	216.2	3.044	10950.0	13580.0	231.2	51.72	69.59	212.5
370.0	206.9	2.913	11530.0	14270.0	233.1	52.31	69.21	217.0

Thermophysical properties of nitrogen trifluoride—Continued

T K	Density kg/m³	Density mol/dm³	E J/mol	H J/mol	S J/(mol·K)	C_v J/(mol·K)	C_p J/(mol·K)	Sound m/s
380.0	198.5	2.796	12100.0	14960.0	234.9	52.90	69.00	221.3
390.0	191.0	2.689	12680.0	15650.0	236.7	53.49	68.91	225.4
400.0	184.1	2.593	13260.0	16340.0	238.5	54.07	68.91	229.4
410.0	177.9	2.505	13840.0	17030.0	240.2	54.64	68.99	233.3
420.0	172.1	2.423	14420.0	17720.0	241.8	55.20	69.12	237.1
440.0	161.8	2.278	15600.0	19110.0	245.1	56.28	69.48	244.3
460.0	152.9	2.152	16780.0	20500.0	248.2	57.30	69.94	251.2
480.0	145.0	2.042	17990.0	21910.0	251.2	58.26	70.46	257.7
500.0	138.0	1.943	19200.0	23320.0	254.0	59.18	71.00	264.0
				9.00 MPa isobar				
68.02[a]	1870.0	26.33	−12280.0	−11930.0	93.76	53.11	72.54	1266.0
70.0	1863.0	26.23	−12140.0	−11790.0	95.81	51.41	70.82	1220.0
80.0	1826.0	25.71	−11450.0	−11100.0	105.1	47.34	69.70	1150.0
100.0	1748.0	24.61	−10050.0	−9689.0	120.8	43.54	70.51	1088.0
120.0	1666.0	23.46	−8668.0	−8284.0	133.6	40.74	69.96	971.5
140.0	1581.0	22.27	−7288.0	−6884.0	144.4	39.56	70.34	851.9
160.0	1492.0	21.01	−5888.0	−5460.0	153.9	39.52	72.32	742.3
180.0	1395.0	19.65	−4436.0	−3978.0	162.6	40.16	76.19	635.9
190.0	1343.0	18.91	−3679.0	−3203.0	166.8	40.67	78.95	581.6
200.0	1287.0	18.12	−2894.0	−2397.0	171.0	41.29	82.37	525.9
210.0	1225.0	17.25	−2075.0	−1553.0	175.1	42.05	86.67	468.6
215.0	1191.0	16.78	−1649.0	−1113.0	177.1	42.50	89.25	439.4
220.0	1156.0	16.28	−1212.0	−659.5	179.2	43.00	92.24	409.9
225.0	1118.0	15.74	−761.5	−189.8	181.3	43.56	95.75	380.0
230.0	1077.0	15.16	−294.3	299.3	183.5	44.20	100.00	349.8
235.0	1032.0	14.53	192.7	812.0	185.7	44.92	105.3	319.5
240.0	982.4	13.83	704.1	1355.0	188.0	45.73	112.1	289.1
245.0	926.6	13.05	1247.0	1937.0	190.4	46.65	121.1	259.2
250.0	862.8	12.15	1829.0	2570.0	192.9	47.66	132.7	230.7
255.0	790.1	11.13	2456.0	3265.0	195.7	48.69	145.2	205.8
260.0	711.7	10.02	3116.0	4014.0	198.6	49.52	152.8	187.4
265.0	635.7	8.952	3769.0	4775.0	201.5	49.97	149.8	176.9
267.0	608.0	8.561	4019.0	5071.0	202.6	50.03	146.1	174.5
268.0	594.9	8.376	4141.0	5216.0	203.2	50.03	144.0	173.7
270.0	570.0	8.027	4378.0	5499.0	204.2	50.02	139.3	172.4
272.0	547.2	7.705	4605.0	5773.0	205.2	49.97	134.3	171.6
274.0	526.3	7.411	4822.0	6036.0	206.2	49.90	129.3	171.3
276.0	507.2	7.142	5030.0	6290.0	207.1	49.82	124.4	171.3
278.0	489.7	6.896	5229.0	6534.0	208.0	49.74	119.8	171.5
280.0	473.7	6.671	5420.0	6769.0	208.8	49.66	115.5	172.0
282.0	459.1	6.464	5604.0	6997.0	209.6	49.58	111.6	172.6
284.0	445.6	6.275	5782.0	7216.0	210.4	49.51	107.9	173.4
286.0	433.2	6.100	5953.0	7429.0	211.2	49.45	104.6	174.3
288.0	421.7	5.939	6119.0	7635.0	211.9	49.40	101.6	175.2
290.0	411.1	5.789	6280.0	7835.0	212.6	49.36	98.89	176.2
295.0	387.6	5.458	6666.0	8315.0	214.2	49.31	93.12	178.9
300.0	367.7	5.178	7030.0	8768.0	215.7	49.32	88.60	181.7
305.0	350.6	4.937	7379.0	9202.0	217.2	49.39	85.02	184.6
310.0	335.6	4.726	7715.0	9620.0	218.5	49.50	82.17	187.5
315.0	322.4	4.539	8042.0	10020.0	219.8	49.64	79.87	190.3
320.0	310.5	4.373	8361.0	10420.0	221.1	49.82	78.00	193.1
325.0	299.9	4.223	8674.0	10810.0	222.3	50.02	76.48	195.8
330.0	290.2	4.086	8982.0	11180.0	223.4	50.24	75.23	198.5
340.0	273.2	3.847	9588.0	11930.0	225.6	50.73	73.34	203.7
350.0	258.8	3.644	10180.0	12650.0	227.7	51.27	72.04	208.6
360.0	246.2	3.467	10770.0	13370.0	229.8	51.83	71.16	213.3
370.0	235.1	3.311	11360.0	14080.0	231.7	52.40	70.56	217.8
380.0	225.3	3.172	11940.0	14780.0	233.6	52.98	70.18	222.2
390.0	216.4	3.047	12530.0	15480.0	235.4	53.55	69.96	226.4
400.0	208.4	2.935	13110.0	16180.0	237.2	54.12	69.85	230.4
410.0	201.1	2.832	13700.0	16880.0	238.9	54.68	69.83	234.4

Thermophysical properties of nitrogen trifluoride—Continued

T K	Density kg/m³	Density mol/dm³	E J/mol	H J/mol	S J/(mol·K)	C_v J/(mol·K)	C_p J/(mol·K)	Sound m/s
420.0	194.4	2.737	14290.0	17580.0	240.6	55.23	69.88	238.2
440.0	182.5	2.569	15470.0	18980.0	243.8	56.30	70.12	245.5
460.0	172.2	2.424	16670.0	20380.0	247.0	57.31	70.48	252.4
480.0	163.2	2.297	17880.0	21800.0	250.0	58.26	70.92	259.0
500.0	155.2	2.185	19100.0	23220.0	252.9	59.17	71.41	265.4
				10.00 MPa isobar				
68.21[a]	1870.0	26.34	-12270.0	-11890.0	93.88	53.60	72.26	1243.0
70.0	1864.0	26.25	-12140.0	-11760.0	95.74	51.95	70.73	1205.0
80.0	1827.0	25.73	-11450.0	-11060.0	105.0	47.43	69.65	1149.0
100.0	1749.0	24.63	-10060.0	-9657.0	120.7	43.45	70.45	1095.0
120.0	1668.0	23.49	-8680.0	-8254.0	133.5	40.65	69.87	979.4
140.0	1584.0	22.30	-7304.0	-6856.0	144.3	39.50	70.20	859.1
160.0	1495.0	21.05	-5911.0	-5436.0	153.8	39.49	72.10	749.2
180.0	1400.0	19.71	-4468.0	-3960.0	162.5	40.15	75.77	643.7
190.0	1349.0	18.99	-3717.0	-3190.0	166.6	40.65	78.35	590.4
200.0	1294.0	18.21	-2940.0	-2391.0	170.7	41.25	81.51	536.1
210.0	1234.0	17.37	-2133.0	-1558.0	174.8	41.98	85.37	480.8
220.0	1168.0	16.45	-1289.0	-680.9	178.9	42.86	90.19	424.7
225.0	1132.0	15.94	-850.0	-222.9	180.9	43.37	93.10	396.4
230.0	1094.0	15.41	-398.2	250.8	183.0	43.93	96.47	368.2
235.0	1053.0	14.83	68.6	742.9	185.1	44.56	100.5	340.0
240.0	1009.0	14.21	553.0	1257.0	187.3	45.25	105.3	312.0
245.0	960.4	13.52	1058.0	1798.0	189.5	46.00	111.2	284.7
250.0	907.0	12.77	1588.0	2371.0	191.8	46.82	118.4	258.4
255.0	848.3	11.94	2146.0	2983.0	194.2	47.67	126.4	234.2
260.0	784.5	11.05	2729.0	3634.0	196.8	48.47	133.7	213.7
270.0	655.4	9.229	3912.0	4995.0	201.9	49.52	134.9	188.8
275.0	598.8	8.431	4469.0	5655.0	204.3	49.69	128.4	183.6
277.0	578.4	8.144	4680.0	5908.0	205.3	49.71	125.2	182.4
278.0	568.7	8.008	4784.0	6033.0	205.7	49.71	123.5	182.0
280.0	550.3	7.749	4986.0	6276.0	206.6	49.70	120.2	181.4
282.0	533.2	7.507	5181.0	6514.0	207.4	49.69	116.9	181.1
284.0	517.2	7.282	5371.0	6744.0	208.2	49.67	113.7	181.1
286.0	502.3	7.073	5554.0	6968.0	209.0	49.64	110.6	181.2
288.0	488.5	6.878	5733.0	7186.0	209.8	49.62	107.6	181.5
290.0	475.6	6.697	5906.0	7399.0	210.5	49.60	104.9	181.9
295.0	447.0	6.293	6319.0	7908.0	212.3	49.58	98.80	183.5
300.0	422.6	5.950	6708.0	8389.0	213.9	49.59	93.78	185.5
305.0	401.6	5.655	7079.0	8847.0	215.4	49.65	89.67	187.8
310.0	383.3	5.397	7434.0	9287.0	216.8	49.74	86.31	190.2
315.0	367.2	5.171	7777.0	9711.0	218.2	49.87	83.56	192.7
320.0	352.9	4.969	8111.0	10120.0	219.5	50.03	81.31	195.2
325.0	340.1	4.789	8437.0	10520.0	220.7	50.21	79.44	197.8
330.0	328.5	4.626	8756.0	10920.0	221.9	50.42	77.89	200.3
335.0	318.0	4.477	9071.0	11300.0	223.1	50.64	76.61	202.8
340.0	308.3	4.341	9381.0	11680.0	224.2	50.88	75.53	205.2
350.0	291.2	4.101	9992.0	12430.0	226.4	51.39	73.88	210.1
360.0	276.5	3.893	10600.0	13160.0	228.4	51.93	72.72	214.7
370.0	263.6	3.712	11190.0	13890.0	230.4	52.48	71.91	219.2
380.0	252.2	3.551	11790.0	14600.0	232.3	53.05	71.36	223.5
390.0	242.0	3.407	12380.0	15310.0	234.2	53.61	71.00	227.7
400.0	232.7	3.277	12970.0	16020.0	236.0	54.17	70.78	231.8
410.0	224.4	3.159	13560.0	16730.0	237.7	54.72	70.66	235.7
420.0	216.7	3.051	14160.0	17440.0	239.4	55.27	70.63	239.5
440.0	203.1	2.860	15350.0	18850.0	242.7	56.32	70.74	246.8
460.0	191.5	2.696	16560.0	20270.0	245.9	57.32	71.01	253.8
480.0	181.3	2.552	17770.0	21690.0	248.9	58.26	71.38	260.5
500.0	172.3	2.426	19000.0	23120.0	251.8	59.17	71.81	266.9

Thermophysical properties of nitrogen trifluoride—Continued

T K	Density kg/m^3	Density mol/dm^3	E J/mol	H J/mol	S J/(mol·K)	C_v J/(mol·K)	C_p J/(mol·K)	Sound m/s
				20.00 MPa isobar				
70.02[a]	1874.0	26.39	-12180.0	-11430.0	95.08	56.52	69.49	1040.0
72.0	1868.0	26.30	-12050.0	-11290.0	97.00	53.87	68.70	1047.0
80.0	1838.0	25.89	-11510.0	-10740.0	104.2	47.83	69.13	1131.0
100.0	1762.0	24.81	-10150.0	-9343.0	119.8	42.54	69.95	1161.0
120.0	1685.0	23.73	-8795.0	-7953.0	132.5	39.96	69.08	1051.0
140.0	1606.0	22.61	-7458.0	-6573.0	143.1	39.17	69.06	923.6
160.0	1525.0	21.47	-6113.0	-5182.0	152.4	39.51	70.28	809.2
180.0	1440.0	20.28	-4740.0	-3754.0	160.8	40.40	72.66	707.7
200.0	1351.0	19.02	-3321.0	-2269.0	168.6	41.55	75.96	614.2
220.0	1254.0	17.66	-1844.0	-711.7	176.1	42.86	79.86	527.1
230.0	1204.0	16.95	-1083.0	97.0	179.7	43.57	81.90	486.4
240.0	1151.0	16.20	-308.2	926.3	183.2	44.32	83.95	448.0
260.0	1039.0	14.64	1279.0	2645.0	190.1	45.92	87.92	379.9
280.0	923.6	13.00	2901.0	4438.0	196.7	47.54	91.16	325.8
300.0	809.8	11.40	4523.0	6277.0	203.0	49.04	92.13	288.3
310.0	756.8	10.66	5318.0	7195.0	206.1	49.72	91.35	275.9
320.0	707.8	9.967	6095.0	8101.0	208.9	50.34	89.89	267.3
330.0	663.5	9.343	6851.0	8991.0	211.7	50.92	88.03	261.8
340.0	623.9	8.785	7585.0	9862.0	214.3	51.48	86.05	258.5
350.0	588.7	8.289	8300.0	10710.0	216.7	52.02	84.16	257.0
360.0	557.5	7.850	8998.0	11550.0	219.1	52.56	82.45	256.7
370.0	529.7	7.459	9681.0	12360.0	221.3	53.08	80.97	257.3
380.0	504.9	7.110	10350.0	13170.0	223.5	53.60	79.71	258.5
390.0	482.7	6.797	11010.0	13960.0	225.5	54.12	78.67	260.2
400.0	462.7	6.515	11670.0	14740.0	227.5	54.63	77.82	262.2
410.0	444.6	6.260	12320.0	15510.0	229.4	55.13	77.12	264.5
420.0	428.1	6.028	12960.0	16280.0	231.3	55.62	76.57	267.0
430.0	413.0	5.815	13610.0	17050.0	233.1	56.11	76.13	269.5
440.0	399.1	5.620	14250.0	17810.0	234.8	56.58	75.80	272.2
460.0	374.5	5.273	15520.0	19320.0	238.2	57.50	75.35	277.7
480.0	353.2	4.973	16800.0	20820.0	241.4	58.38	75.15	283.3
500.0	334.6	4.711	18080.0	22320.0	244.4	59.23	75.11	288.8
				30.00 MPa isobar				
71.80[a]	1882.0	26.50	-12110.0	-10970.0	96.10	55.24	66.72	838.0
72.0	1881.0	26.49	-12090.0	-10960.0	96.29	54.90	66.76	846.6
80.0	1850.0	26.05	-11570.0	-10420.0	103.4	47.22	68.79	1106.0
100.0	1774.0	24.98	-10230.0	-9026.0	119.0	41.74	69.62	1220.0
150.0	1588.0	22.36	-6940.0	-5598.0	146.8	39.50	68.48	914.8
170.0	1512.0	21.29	-5626.0	-4217.0	155.4	40.49	69.79	803.8
180.0	1473.0	20.74	-4960.0	-3514.0	159.4	41.12	70.75	754.7
200.0	1394.0	19.63	-3605.0	-2077.0	167.0	42.46	73.06	666.4
220.0	1312.0	18.47	-2214.0	-589.9	174.1	43.85	75.62	588.8
240.0	1227.0	17.28	-788.7	947.6	180.8	45.25	78.08	521.1
300.0	969.3	13.65	3602.0	5800.0	198.8	49.48	82.92	379.0
320.0	888.8	12.52	5067.0	7464.0	204.2	50.80	83.32	351.1
340.0	814.8	11.47	6513.0	9128.0	209.2	52.02	82.97	331.7
360.0	748.7	10.54	7933.0	10780.0	213.9	53.16	82.06	319.4
380.0	690.9	9.728	9325.0	12410.0	218.4	54.22	80.90	312.4
400.0	641.0	9.025	10690.0	14020.0	222.5	55.23	79.78	309.3
420.0	597.9	8.419	12040.0	15600.0	226.3	56.17	78.83	308.7
440.0	560.6	7.894	13370.0	17170.0	230.0	57.07	78.09	309.8
460.0	528.1	7.437	14690.0	18730.0	233.5	57.92	77.57	312.1
480.0	499.6	7.035	16010.0	20270.0	236.7	58.74	77.23	315.1
500.0	474.4	6.680	17330.0	21820.0	239.9	59.52	77.04	318.6
				40.00 MPa isobar				
80.0	1863.0	26.23	-11630.0	-10100.0	102.6	45.38	69.45	1068.0
90.0	1822.0	25.65	-10970.0	-9407.0	110.8	42.63	69.88	1255.0

Thermophysical properties of nitrogen trifluoride—Continued

T K	Density kg/m³	Density mol/dm³	E J/mol	H J/mol	S J/(mol·K)	C_v J/(mol·K)	C_p J/(mol·K)	Sound m/s
100.0	1785.0	25.13	−10300.0	−8710.0	118.1	41.01	69.46	1273.0
150.0	1607.0	22.63	−7073.0	−5305.0	145.8	40.00	67.71	957.4
200.0	1429.0	20.13	−3833.0	−1845.0	165.7	43.64	71.26	704.8
300.0	1065.0	15.00	3053.0	5719.0	196.2	50.66	78.92	440.3
320.0	997.3	14.04	4456.0	7305.0	201.3	51.90	79.56	412.7
340.0	933.4	13.14	5856.0	8900.0	206.2	53.05	79.88	391.3
360.0	874.1	12.31	7248.0	10500.0	210.7	54.13	79.92	375.3
380.0	819.8	11.54	8630.0	12090.0	215.1	55.14	79.72	364.0
400.0	770.7	10.85	10000.0	13690.0	219.1	56.07	79.38	356.3
420.0	726.5	10.23	11360.0	15270.0	223.0	56.95	78.99	351.6
440.0	686.9	9.672	12710.0	16850.0	226.7	57.77	78.62	349.2
460.0	651.4	9.172	14050.0	18410.0	230.2	58.55	78.30	348.4
480.0	619.5	8.723	15390.0	19980.0	233.5	59.29	78.07	348.8
500.0	590.7	8.318	16730.0	21540.0	236.7	60.01	77.92	350.1
				50.00 MPa isobar				
85.0	1853.0	26.09	−11360.0	−9446.0	105.9	41.89	70.98	1210.0
90.0	1832.0	25.80	−11030.0	−9093.0	110.0	41.37	70.36	1291.0
100.0	1795.0	25.27	−10370.0	−8394.0	117.3	40.34	69.49	1323.0
120.0	1725.0	24.29	−9080.0	−7022.0	129.8	39.42	67.76	1216.0
150.0	1625.0	22.88	−7191.0	−5005.0	144.8	40.66	67.13	994.5
200.0	1460.0	20.56	−4020.0	−1588.0	164.5	44.96	70.04	735.0
250.0	1295.0	18.24	−725.7	2016.0	180.6	48.85	74.02	579.7
300.0	1136.0	15.99	2665.0	5791.0	194.3	52.08	76.75	484.1
320.0	1076.0	15.15	4033.0	7333.0	199.3	53.24	77.43	457.7
340.0	1019.0	14.35	5403.0	8887.0	204.0	54.31	77.93	436.6
360.0	965.5	13.59	6772.0	10450.0	208.5	55.31	78.26	419.8
380.0	915.4	12.89	8138.0	12020.0	212.7	56.23	78.45	406.9
400.0	869.0	12.24	9500.0	13590.0	216.7	57.09	78.50	397.2
420.0	826.3	11.63	10860.0	15160.0	220.6	57.88	78.48	390.2
440.0	787.0	11.08	12210.0	16730.0	224.2	58.63	78.40	385.3
460.0	751.0	10.58	13560.0	18290.0	227.7	59.33	78.31	382.3
480.0	718.1	10.11	14910.0	19860.0	231.0	59.99	78.24	380.7
500.0	687.9	9.686	16260.0	21420.0	234.2	60.63	78.20	380.2

[a]At melting line.
[b]At liquid–vapor boundary.

Appendix K. Thermophysical Properties of Oxygen

Thermophysical properties of coexisting gaseous and liquid oxygen

T K	Pres. MPa	Density kg/m^3	Density mol/dm^3	E J/mol	H J/mol	S J/(mol·K)	C_v J/(mol·K)	C_p J/(mol·K)	Sound m/s	Visc. μPa·s	Therm. W/(m·K)	Diel. const.
54.359[a]	.00015	1306.0	40.82	−6183.0	−6183.0	67.10	36.29	53.24	1128.0	485.0	.192	1.56853
54.359[a]	.00015	.01063	.0003331	1120.0	1568.0	209.6	20.82	29.15	140.6	3.63	.00410	1.00000
56.0	.00024	1299.0	40.60	−6096.0	−6096.0	68.69	35.87	53.31	1127.0	468.0	.191	1.56493
56.0	.00025	.02192	.0006861	1154.0	1510.0	204.2	20.83	29.17	142.7	3.80	.00439	1.00001
58.0	.00043	1290.0	40.32	−5989.0	−5989.0	70.56	35.36	53.35	1121.0	446.0	.190	1.56052
58.0	.00043	.03650	.001142	1195.0	1572.0	200.7	20.84	29.19	145.2	3.99	.00469	1.00001
60.0	.00073	1281.0	40.04	−5882.0	−5882.0	72.37	34.87	53.37	1114.0	425.0	.188	1.55611
60.0	.00073	.05470	.001710	1236.0	1662.0	198.0	20.85	29.21	147.6	4.18	.00497	1.00002
62.0	.00119	1272.0	39.77	−5775.0	−5775.0	74.12	34.40	53.38	1105.0	404.0	.187	1.55168
62.0	.00119	.07996	.002500	1277.0	1753.0	195.5	20.86	29.24	150.0	4.36	.00522	1.00003
64.0	.00188	1264.0	39.49	−5669.0	−5668.0	75.82	33.94	53.39	1094.0	383.0	.185	1.54724
64.0	.00188	.1163	.003635	1318.0	1835.0	193.1	20.88	29.28	152.4	4.54	.00545	1.00004
66.0	.00288	1255.0	39.21	−5562.0	−5562.0	77.46	33.51	53.39	1082.0	364.0	.183	1.54279
66.0	.00288	.1682	.005259	1359.0	1906.0	190.6	20.90	29.32	154.7	4.71	.00567	1.00006
68.0	.00430	1246.0	38.93	−5455.0	−5455.0	79.05	33.10	53.40	1070.0	345.0	.181	1.53832
68.0	.00430	.2409	.007530	1399.0	1970.0	188.3	20.93	29.38	156.9	4.88	.00589	1.00009
70.0	.00625	1237.0	38.64	−5348.0	−5348.0	80.60	32.70	53.41	1057.0	327.0	.179	1.53383
70.0	.00625	.3400	.01063	1439.0	2028.0	186.0	20.97	29.46	159.1	5.05	.00610	1.00013
72.0	.00890	1227.0	38.36	−5241.0	−5241.0	82.11	32.32	53.43	1043.0	310.0	.176	1.52933
72.0	.00890	.4716	.01474	1479.0	2083.0	183.9	21.01	29.56	161.3	5.21	.00631	1.00018
74.0	.01240	1218.0	38.07	−5134.0	−5134.0	83.57	31.96	53.46	1029.0	294.0	.174	1.52480
74.0	.01240	.6424	.02008	1518.0	2136.0	181.8	21.06	29.67	163.3	5.38	.00653	1.00024
76.0	.01695	1209.0	37.78	−5028.0	−5027.0	84.99	31.61	53.50	1014.0	279.0	.171	1.52025
76.0	.01695	.8597	.02687	1557.0	2188.0	179.9	21.12	29.80	165.3	5.54	.00674	1.00032
78.0	.02277	1200.0	37.49	−4921.0	−4920.0	86.38	31.28	53.55	999.5	265.0	.169	1.51567
78.0	.02277	1.131	.03536	1595.0	2239.0	178.2	21.18	29.95	167.3	5.70	.00696	1.00042
80.0	.03009	1190.0	37.20	−4814.0	−4813.0	87.74	30.97	53.62	984.5	252.0	.166	1.51106
80.0	.03009	1.465	.04580	1632.0	2289.0	176.5	21.25	30.12	169.1	5.87	.00719	1.00055
82.0	.03919	1181.0	36.90	−4707.0	−4705.0	89.06	30.66	53.70	969.4	239.0	.163	1.50643
82.0	.03919	1.871	.05847	1669.0	2339.0	175.0	21.33	30.31	170.9	6.03	.00742	1.00070
84.0	.05035	1171.0	36.60	−4599.0	−4598.0	90.35	30.37	53.80	954.0	227.0	.161	1.50176
84.0	.05035	2.357	.07366	1705.0	2389.0	173.5	21.41	30.53	172.6	6.19	.00766	1.00088
86.0	.06387	1162.0	36.30	−4492.0	−4490.0	91.62	30.09	53.91	938.5	216.0	.158	1.49705
86.0	.06387	2.934	.09168	1740.0	2437.0	172.2	21.51	30.76	174.3	6.35	.00791	1.00109
88.0	.08008	1152.0	36.00	−4384.0	−4382.0	92.86	29.82	54.05	922.9	206.0	.155	1.49231
88.0	.08008	3.611	.1129	1775.0	2484.0	170.9	21.61	31.03	175.8	6.52	.00817	1.00134
90.0	.09931	1142.0	35.69	−4276.0	−4273.0	94.07	29.56	54.20	907.1	196.0	.152	1.48752
90.0	.09931	4.401	.1375	1808.0	2530.0	169.7	21.71	31.32	177.3	6.68	.00844	1.00164
92.0	.1219	1132.0	35.38	−4168.0	−4164.0	95.26	29.31	54.38	891.1	186.0	.149	1.48269
92.0	.1219	5.314	.1661	1841.0	2575.0	168.5	21.82	31.63	178.7	6.85	.00871	1.00197
94.0	.1483	1122.0	35.06	−4059.0	−4055.0	96.43	29.07	54.59	875.0	178.0	.146	1.47781
94.0	.1483	6.362	.1988	1872.0	2618.0	167.4	21.94	31.98	180.0	7.01	.00900	1.00236
96.0	.1789	1112.0	34.74	−3950.0	−3945.0	97.57	28.84	54.82	858.7	169.0	.143	1.47288
96.0	.1789	7.559	.2362	1903.0	2660.0	166.4	22.06	32.36	181.3	7.18	.00929	1.00281
98.0	.2140	1101.0	34.42	−3841.0	−3835.0	98.70	28.62	55.09	842.2	162.0	.141	1.46789
98.0	.2140	8.918	.2787	1932.0	2700.0	165.4	22.19	32.78	182.4	7.35	.00960	1.00331
100.0	.2540	1091.0	34.09	−3731.0	−3724.0	99.81	28.41	55.39	825.5	154.0	.138	1.46283
100.0	.2540	10.45	.3267	1960.0	2738.0	164.4	22.33	33.23	183.4	7.52	.00992	1.00389
102.0	.2994	1080.0	33.76	−3621.0	−3612.0	100.9	28.20	55.73	808.6	147.0	.135	1.45771
102.0	.2994	12.18	.3806	1987.0	2774.0	163.5	22.47	33.72	184.4	7.69	.0103	1.00453

Thermophysical properties of coexisting gaseous and liquid oxygen—Continued

T K	Pres. MPa	Density kg/m³	Density mol/dm³	E J/mol	H J/mol	S J/(mol·K)	C_v J/(mol·K)	C_p J/(mol·K)	Sound m/s	Visc. μPa·s	Therm. W/(m·K)	Diel. const.
104.0	.3506	1069.0	33.42	-3510.0	-3499.0	102.0	28.00	56.12	791.4	141.0	.132	1.45252
104.0	.3506	14.11	.4410	2012.0	2807.0	162.6	22.62	34.26	185.3	7.86	.0106	1.00525
106.0	.4081	1058.0	33.07	-3398.0	-3386.0	103.1	27.82	56.55	774.0	135.0	.129	1.44725
106.0	.4081	16.27	.5084	2036.0	2839.0	161.8	22.77	34.85	186.0	8.04	.0110	1.00605
108.0	.4722	1047.0	32.72	-3286.0	-3271.0	104.1	27.64	57.04	756.4	129.0	.126	1.44189
108.0	.4722	18.67	.5834	2059.0	2868.0	161.0	22.93	35.50	186.7	8.22	.0113	1.00695
110.0	.5434	1035.0	32.36	-3172.0	-3156.0	105.1	27.46	57.58	738.5	123.0	.124	1.43644
110.0	.5434	21.33	.6666	2079.0	2895.0	160.1	23.10	36.21	187.3	8.40	.0117	1.00794
112.0	.6222	1024.0	31.99	-3058.0	-3039.0	106.2	27.30	58.20	720.4	118.0	.121	1.43089
112.0	.6222	24.28	.7586	2098.0	2919.0	159.4	23.28	37.00	187.7	8.58	.0121	1.00904
114.0	.7090	1012.0	31.62	-2943.0	-2921.0	107.2	27.15	58.89	701.9	112.0	.118	1.42522
114.0	.7090	27.53	.8602	2115.0	2940.0	158.6	23.46	37.87	188.1	8.77	.0126	1.01025
116.0	.8043	999.4	31.23	-2827.0	-2801.0	108.2	27.00	59.66	683.1	108.0	.115	1.41944
116.0	.8043	31.11	.9722	2130.0	2958.0	157.9	23.65	38.83	188.4	8.96	.0130	1.01159
118.0	.9085	986.8	30.84	-2710.0	-2680.0	109.2	26.87	60.53	664.0	103.0	.113	1.41352
118.0	.9085	35.06	1.096	2143.0	2972.0	157.1	23.85	39.90	188.5	9.16	.0135	1.01307
120.0	1.022	973.9	30.43	-2591.0	-2558.0	110.2	26.74	61.51	644.6	98.4	.110	1.40746
120.0	1.022	39.39	1.231	2154.0	2984.0	156.4	24.05	41.10	188.6	9.37	.0140	1.01469
122.0	1.146	960.5	30.02	-2471.0	-2433.0	111.2	26.63	62.63	624.9	94.1	.107	1.40123
122.0	1.146	44.16	1.380	2162.0	2992.0	155.7	24.27	42.46	188.6	9.58	.0145	1.01648
124.0	1.280	946.8	29.59	-2350.0	-2306.0	112.2	26.52	63.89	604.7	90.0	.105	1.39483
124.0	1.280	49.40	1.544	2167.0	2996.0	155.0	24.50	43.99	188.4	9.80	.0150	1.01845
126.0	1.425	932.5	29.14	-2226.0	-2177.0	113.2	26.43	65.33	584.2	86.0	.102	1.38823
126.0	1.425	55.16	1.724	2169.0	2996.0	154.2	24.74	45.75	188.2	10.0	.0156	1.02061
128.0	1.581	917.8	28.68	-2101.0	-2046.0	114.2	26.36	66.99	563.3	82.1	.0994	1.38140
128.0	1.581	61.49	1.922	2168.0	2990.0	153.5	24.99	47.77	187.8	10.3	.0163	1.02300
130.0	1.749	902.4	28.20	-1973.0	-1911.0	115.2	26.30	68.91	542.0	78.4	.0968	1.37433
130.0	1.749	68.47	2.140	2163.0	2980.0	152.8	25.26	50.12	187.4	10.5	.0169	1.02563
132.0	1.929	886.4	27.70	-1843.0	-1773.0	116.2	26.26	71.16	520.1	74.8	.0942	1.36697
132.0	1.929	76.18	2.381	2154.0	2964.0	152.1	25.55	52.89	186.8	10.8	.0177	1.02854
134.0	2.123	869.6	27.18	-1710.0	-1631.0	117.2	26.23	73.81	497.8	71.2	.0916	1.35928
134.0	2.123	84.72	2.648	2140.0	2942.0	151.3	25.86	56.20	186.1	11.1	.0185	1.03177
136.0	2.330	851.9	26.62	-1573.0	-1485.0	118.2	26.23	77.01	475.0	67.7	.0889	1.35122
136.0	2.330	94.21	2.944	2122.0	2913.0	150.6	26.19	60.23	185.4	11.4	.0194	1.03537
138.0	2.551	833.2	26.04	-1432.0	-1334.0	119.3	26.26	80.92	451.5	64.3	.0863	1.34271
138.0	2.551	104.8	3.275	2097.0	2876.0	149.8	26.55	65.21	184.5	11.7	.0204	1.03941
140.0	2.788	813.2	25.41	-1287.0	-1177.0	120.3	26.32	85.84	427.3	60.9	.0837	1.33367
140.0	2.788	116.7	3.648	2065.0	2829.0	148.9	26.94	71.54	183.6	12.1	.0216	1.04396
142.0	3.039	791.6	24.74	-1135.0	-1012.0	121.4	26.42	92.25	402.2	57.5	.0811	1.32399
142.0	3.039	130.3	4.071	2024.0	2771.0	148.1	27.37	79.87	182.5	12.5	.0230	1.04914
144.0	3.307	768.2	24.01	-976.4	-838.6	122.6	26.58	101.0	376.2	54.1	.0786	1.31349
144.0	3.307	145.8	4.558	1973.0	2698.0	147.1	27.84	91.29	181.3	13.0	.0247	1.05512
146.0	3.592	742.2	23.19	-807.9	-653.0	123.8	26.79	113.6	348.8	50.6	.0762	1.30192
146.0	3.592	164.1	5.127	1907.0	2608.0	146.1	28.38	108.0	180.1	13.6	.0268	1.06214
148.0	3.896	712.6	22.27	-625.6	-450.7	125.0	27.11	133.7	319.9	47.0	.0739	1.28885
148.0	3.896	186.0	5.812	1823.0	2493.0	144.9	28.99	134.7	178.7	14.3	.0296	1.07064
150.0	4.219	677.5	21.17	-422.1	-222.8	126.5	27.56	171.1	288.6	43.2	.0718	1.27345
150.0	4.219	213.7	6.678	1709.0	2340.0	143.6	29.70	185.1	177.2	15.2	.0336	1.08144
152.0	4.563	631.8	19.74	-177.2	53.9	128.2	28.28	266.7	253.5	38.6	.0702	1.25357
152.0	4.563	252.4	7.888	1539.0	2117.0	141.8	30.59	316.6	175.6	16.5	.0403	1.09667
154.58[b]	5.043	436.1	13.63	697.7	1068.0	134.6						1.17088

[a]Triple point.
[b]Critical point.

Thermophysical properties of oxygen on the melting line

T K	Pres. MPa	Density kg/m^3	Density mol/dm^3	E J/mol	H J/mol	S J/(mol·K)	C_v J/(mol·K)	C_p J/(mol·K)	Sound m/s	Visc. μPa·s	Therm. W/(m·K)	Diel. const.
54.36[a]	.00015	1306.0	40.82	–6183.0	–6183.0	67.10	36.29	53.24	1128.0	485.0	.192	1.56853
56.0	14.44	1315.0	41.10	–6159.0	–5807.0	67.52	35.89	52.94	1164.0	529.0	.194	1.57310
56.1	15.33	1316.0	41.12	–6157.0	–5784.0	67.55	35.88	52.93	1166.0	531.0	.194	1.57337
56.2	16.22	1316.0	41.14	–6156.0	–5761.0	67.57	35.86	52.92	1168.0	534.0	.194	1.57365
56.3	17.11	1317.0	41.16	–6154.0	–5738.0	67.60	35.84	52.91	1170.0	537.0	.194	1.57392
56.4	18.00	1317.0	41.17	–6152.0	–5715.0	67.62	35.82	52.90	1173.0	539.0	.194	1.57419
56.5	18.90	1318.0	41.19	–6151.0	–5692.0	67.65	35.81	52.89	1175.0	542.0	.194	1.57446
56.6	19.80	1319.0	41.21	–6149.0	–5669.0	67.67	35.79	52.88	1177.0	544.0	.194	1.57473
56.7	20.69	1319.0	41.22	–6148.0	–5646.0	67.70	35.77	52.87	1180.0	547.0	.194	1.57500
56.8	21.59	1320.0	41.24	–6146.0	–5623.0	67.72	35.76	52.86	1182.0	549.0	.194	1.57527
56.9	22.49	1320.0	41.26	–6145.0	–5599.0	67.74	35.74	52.86	1185.0	552.0	.194	1.57554
57.0	23.39	1321.0	41.27	–6143.0	–5576.0	67.77	35.73	52.85	1187.0	554.0	.194	1.57581
57.1	24.30	1321.0	41.29	–6141.0	–5553.0	67.79	35.72	52.84	1189.0	557.0	.194	1.57607
57.2	25.20	1322.0	41.31	–6140.0	–5530.0	67.82	35.71	52.84	1192.0	559.0	.194	1.57634
57.3	26.11	1322.0	41.32	–6138.0	–5506.0	67.84	35.69	52.83	1194.0	562.0	.194	1.57660
57.4	27.01	1323.0	41.34	–6137.0	–5483.0	67.86	35.68	52.83	1196.0	564.0	.195	1.57687
57.5	27.92	1323.0	41.36	–6135.0	–5460.0	67.89	35.67	52.82	1199.0	567.0	.195	1.57713
57.6	28.83	1324.0	41.37	–6133.0	–5436.0	67.91	35.66	52.82	1201.0	569.0	.195	1.57740
57.7	29.74	1324.0	41.39	–6132.0	–5413.0	67.94	35.65	52.81	1204.0	571.0	.195	1.57766
57.8	30.65	1325.0	41.41	–6130.0	–5390.0	67.96	35.64	52.81	1206.0	574.0	.195	1.57792
57.9	31.56	1325.0	41.42	–6128.0	–5366.0	67.98	35.63	52.81	1209.0	576.0	.195	1.57818
58.0	32.47	1326.0	41.44	–6127.0	–5343.0	68.01	35.63	52.80	1211.0	579.0	.195	1.57844
58.1	33.39	1326.0	41.45	–6125.0	–5319.0	68.03	35.62	52.80	1213.0	581.0	.195	1.57870
58.2	34.30	1327.0	41.47	–6123.0	–5296.0	68.05	35.61	52.80	1216.0	583.0	.195	1.57896
58.3	35.22	1328.0	41.49	–6122.0	–5273.0	68.08	35.61	52.80	1218.0	586.0	.195	1.57922
58.4	36.14	1328.0	41.50	–6120.0	–5249.0	68.10	35.60	52.80	1221.0	588.0	.195	1.57948
58.5	37.06	1329.0	41.52	–6118.0	–5226.0	68.12	35.59	52.79	1223.0	590.0	.195	1.57974
58.6	37.98	1329.0	41.53	–6116.0	–5202.0	68.15	35.59	52.79	1225.0	593.0	.195	1.57999
58.7	38.90	1330.0	41.55	–6115.0	–5178.0	68.17	35.58	52.79	1228.0	595.0	.195	1.58025
58.8	39.83	1330.0	41.57	–6113.0	–5155.0	68.19	35.58	52.79	1230.0	597.0	.195	1.58050

[a]Triple point.

Thermophysical properties of oxygen

T K	Density kg/m³	Density mol/dm³	E J/mol	H J/mol	S J/(mol·K)	C_v J/(mol·K)	C_p J/(mol·K)	Sound m/s	Visc. μPa·s	Therm. W/(m·K)	Diel. const.
				.02 MPa isobar							
54.36[a]	1306.0	40.82	–6183.0	–6183.0	67.10	36.29	53.24	1129.0	485.0	.192	1.56854
60.0	1281.0	40.05	–5882.0	–5882.0	72.37	34.87	53.37	1114.0	425.0	.188	1.55614
77.109[b]	1204.0	37.62	–4968.0	–4968.0	85.77	31.43	53.53	1006.0	271.0	.170	1.51771
77.109[b]	1.003	.03135	1578.0	2216.0	179.0	21.15	29.88	166.4	5.63	.00687	1.00037
80.0	.9715	.03036	1639.0	2298.0	180.0	21.10	29.77	169.6	5.86	.00712	1.00036
85.0	.9128	.02853	1746.0	2447.0	181.8	21.02	29.63	175.0	6.26	.00758	1.00034
90.0	.8610	.02691	1851.0	2594.0	183.5	20.97	29.52	180.2	6.65	.00804	1.00032
95.0	.8148	.02546	1956.0	2742.0	185.1	20.93	29.44	185.3	7.04	.00852	1.00030
100.0	.7734	.02417	2061.0	2889.0	186.6	20.90	29.38	190.2	7.43	.00901	1.00029
105.0	.7361	.02301	2166.0	3036.0	188.0	20.88	29.33	195.0	7.82	.00950	1.00027
110.0	.7023	.02195	2271.0	3182.0	189.4	20.87	29.30	199.6	8.20	.01000	1.00026
115.0	.6715	.02099	2376.0	3329.0	190.7	20.85	29.27	204.1	8.58	.0105	1.00025
120.0	.6433	.02010	2480.0	3475.0	191.9	20.84	29.25	208.6	8.96	.0110	1.00024
130.0	.5934	.01855	2689.0	3767.0	194.3	20.83	29.21	217.2	9.72	.0120	1.00022
140.0	.5508	.01721	2897.0	4059.0	196.4	20.82	29.19	225.4	10.5	.0130	1.00021
150.0	.5139	.01606	3106.0	4351.0	198.5	20.81	29.18	233.4	11.2	.0139	1.00019
160.0	.4817	.01505	3314.0	4643.0	200.3	20.81	29.17	241.1	11.9	.0148	1.00018
170.0	.4532	.01417	3522.0	4934.0	202.1	20.81	29.16	248.5	12.6	.0158	1.00017
180.0	.4280	.01338	3731.0	5226.0	203.8	20.81	29.16	255.7	13.3	.0166	1.00016
190.0	.4054	.01267	3939.0	5518.0	205.3	20.81	29.15	262.8	14.0	.0175	1.00015
200.0	.3851	.01204	4147.0	5809.0	206.8	20.82	29.16	269.6	14.6	.0184	1.00014
210.0	.3667	.01146	4356.0	6101.0	208.3	20.83	29.16	276.3	15.3	.0192	1.00014
220.0	.3500	.01094	4564.0	6392.0	209.6	20.84	29.17	282.8	15.9	.0200	1.00013
230.0	.3348	.01046	4773.0	6684.0	210.9	20.85	29.18	289.1	16.5	.0208	1.00013
240.0	.3208	.01003	4981.0	6976.0	212.2	20.87	29.20	295.3	17.2	.0216	1.00012
250.0	.3080	.009626	5190.0	7268.0	213.4	20.89	29.22	301.3	17.8	.0224	1.00012
260.0	.2961	.009255	5399.0	7560.0	214.5	20.92	29.24	307.2	18.3	.0232	1.00011
280.0	.2750	.008594	5818.0	8146.0	216.7	20.98	29.31	318.7	19.5	.0247	1.00010
300.0	.2566	.008020	6239.0	8733.0	218.7	21.07	29.40	329.7	20.6	.0262	1.00010
320.0	.2406	.007519	6661.0	9322.0	220.6	21.18	29.50	340.3	21.7	.0278	1.00009
340.0	.2264	.007077	7086.0	9913.0	222.4	21.31	29.63	350.5	22.8	.0292	1.00008
360.0	.2138	.006683	7514.0	10510.0	224.1	21.46	29.78	360.3	23.8	.0307	1.00008
380.0	.2026	.006331	7945.0	11100.0	225.7	21.62	29.94	369.8	24.8	.0322	1.00008
400.0	.1924	.006015	8379.0	11700.0	227.2	21.79	30.11	378.9	25.8	.0337	1.00007
				.04 MPa isobar							
54.36[a]	1306.0	40.82	–6183.0	–6182.0	67.10	36.29	53.24	1129.0	485.0	.192	1.56855
60.0	1281.0	40.05	–5883.0	–5882.0	72.36	34.87	53.37	1114.0	425.0	.188	1.55615
80.0	1190.0	37.20	–4814.0	–4813.0	87.73	30.97	53.61	984.6	252.0	.166	1.51109
82.159[b]	1180.0	36.88	–4698.0	–4697.0	89.16	30.64	53.71	968.1	238.0	.163	1.50606
82.159[b]	1.906	.05958	1672.0	2343.0	174.9	21.34	30.33	171.1	6.04	.00744	1.00071
85.0	1.841	.05752	1733.0	2429.0	175.9	21.26	30.17	174.2	6.27	.00769	1.00068
90.0	1.734	.05418	1841.0	2579.0	177.6	21.15	29.95	179.5	6.66	.00814	1.00064
95.0	1.639	.05122	1947.0	2728.0	179.2	21.07	29.78	184.7	7.05	.00861	1.00061
100.0	1.554	.04858	2053.0	2877.0	180.8	21.01	29.66	189.6	7.44	.00909	1.00058
105.0	1.478	.04620	2159.0	3025.0	182.2	20.97	29.56	194.5	7.82	.00957	1.00055
110.0	1.410	.04406	2265.0	3172.0	183.6	20.94	29.49	199.2	8.21	.0101	1.00052
115.0	1.347	.04210	2370.0	3320.0	184.9	20.91	29.43	203.8	8.59	.0106	1.00050
120.0	1.290	.04032	2475.0	3467.0	186.1	20.89	29.38	208.3	8.97	.0110	1.00048
130.0	1.189	.03717	2684.0	3760.0	188.5	20.86	29.32	216.9	9.72	.0120	1.00044
140.0	1.104	.03449	2893.0	4053.0	190.7	20.84	29.27	225.2	10.5	.0130	1.00041
150.0	1.029	.03217	3102.0	4346.0	192.7	20.83	29.24	233.2	11.2	.0139	1.00038
160.0	.9645	.03014	3311.0	4638.0	194.6	20.82	29.22	240.9	11.9	.0149	1.00036
170.0	.9074	.02836	3520.0	4930.0	196.3	20.82	29.20	248.4	12.6	.0158	1.00034
180.0	.8567	.02677	3728.0	5222.0	198.0	20.82	29.19	255.6	13.3	.0167	1.00032
190.0	.8114	.02536	3937.0	5514.0	199.6	20.82	29.19	262.7	14.0	.0175	1.00030
200.0	.7707	.02409	4145.0	5806.0	201.1	20.82	29.18	269.5	14.6	.0184	1.00029

Thermophysical properties of oxygen—Continued

T K	Density kg/m³	Density mol/dm³	E J/mol	H J/mol	S J/(mol·K)	C_v J/(mol·K)	C_p J/(mol·K)	Sound m/s	Visc. μPa·s	Therm. W/(m·K)	Diel. const.
210.0	.7338	.02293	4353.0	6098.0	202.5	20.83	29.18	276.2	15.3	.0192	1.00027
220.0	.7004	.02189	4562.0	6390.0	203.9	20.84	29.19	282.7	15.9	.0200	1.00026
230.0	.6698	.02093	4771.0	6681.0	205.2	20.85	29.20	289.1	16.5	.0208	1.00025
240.0	.6419	.02006	4979.0	6974.0	206.4	20.87	29.21	295.3	17.2	.0216	1.00024
250.0	.6161	.01926	5188.0	7266.0	207.6	20.89	29.23	301.3	17.8	.0224	1.00023
260.0	.5924	.01851	5398.0	7558.0	208.7	20.92	29.26	307.2	18.4	.0232	1.00022
280.0	.5500	.01719	5817.0	8144.0	210.9	20.99	29.32	318.7	19.5	.0247	1.00020
300.0	.5133	.01604	6238.0	8731.0	212.9	21.07	29.41	329.7	20.6	.0263	1.00019
320.0	.4812	.01504	6660.0	9320.0	214.8	21.18	29.51	340.3	21.7	.0278	1.00018
340.0	.4528	.01415	7085.0	9912.0	216.6	21.31	29.64	350.5	22.8	.0293	1.00017
360.0	.4277	.01337	7513.0	10510.0	218.3	21.46	29.78	360.3	23.8	.0307	1.00016
380.0	.4051	.01266	7944.0	11100.0	219.9	21.62	29.94	369.8	24.8	.0322	1.00015
400.0	.3849	.01203	8378.0	11700.0	221.5	21.79	30.12	379.0	25.8	.0337	1.00014

.06 MPa isobar

T K	Density kg/m³	Density mol/dm³	E J/mol	H J/mol	S J/(mol·K)	C_v J/(mol·K)	C_p J/(mol·K)	Sound m/s	Visc. μPa·s	Therm. W/(m·K)	Diel. const.
54.37[a]	1306.0	40.82	−6183.0	−6182.0	67.10	36.29	53.24	1129.0	485.0	.192	1.56855
60.0	1281.0	40.05	−5883.0	−5881.0	72.36	34.87	53.37	1114.0	425.0	.188	1.55616
80.0	1190.0	37.20	−4814.0	−4812.0	87.73	30.97	53.61	984.7	252.0	.166	1.51111
85.465[b]	1164.0	36.38	−4521.0	−4519.0	91.28	30.16	53.88	942.7	219.0	.159	1.49832
85.465[b]	2.770	.08656	1731.0	2424.0	172.5	21.48	30.70	173.8	6.31	.00785	1.00103
90.0	2.619	.08183	1830.0	2563.0	174.1	21.34	30.39	178.8	6.67	.00824	1.00097
95.0	2.473	.07728	1938.0	2714.0	175.8	21.22	30.13	184.0	7.06	.00870	1.00092
100.0	2.343	.07323	2045.0	2865.0	177.3	21.13	29.94	189.1	7.44	.00916	1.00087
105.0	2.227	.06960	2152.0	3014.0	178.8	21.06	29.80	194.0	7.83	.00964	1.00083
110.0	2.122	.06633	2258.0	3163.0	180.1	21.01	29.68	198.8	8.21	.0101	1.00079
115.0	2.027	.06336	2364.0	3311.0	181.5	20.97	29.60	203.4	8.60	.0106	1.00075
120.0	1.941	.06065	2469.0	3459.0	182.7	20.94	29.52	207.9	8.98	.0111	1.00072
130.0	1.788	.05588	2680.0	3753.0	185.1	20.89	29.42	216.6	9.73	.0121	1.00066
140.0	1.658	.05183	2889.0	4047.0	187.3	20.87	29.35	225.0	10.5	.0130	1.00062
150.0	1.546	.04832	3099.0	4340.0	189.3	20.85	29.30	233.0	11.2	.0140	1.00057
160.0	1.448	.04527	3308.0	4633.0	191.2	20.84	29.27	240.8	11.9	.0149	1.00054
170.0	1.362	.04258	3517.0	4926.0	192.9	20.83	29.25	248.3	12.6	.0158	1.00051
180.0	1.286	.04019	3725.0	5218.0	194.6	20.83	29.23	255.6	13.3	.0167	1.00048
190.0	1.218	.03806	3934.0	5510.0	196.2	20.83	29.22	262.6	14.0	.0176	1.00045
200.0	1.157	.03615	4143.0	5803.0	197.7	20.83	29.21	269.5	14.6	.0184	1.00043
210.0	1.101	.03442	4351.0	6095.0	199.1	20.83	29.21	276.2	15.3	.0192	1.00041
220.0	1.051	.03285	4560.0	6387.0	200.5	20.84	29.21	282.7	15.9	.0200	1.00039
230.0	1.005	.03141	4769.0	6679.0	201.8	20.86	29.22	289.0	16.5	.0209	1.00037
240.0	.9631	.03010	4978.0	6971.0	203.0	20.87	29.23	295.2	17.2	.0216	1.00036
250.0	.9245	.02889	5187.0	7264.0	204.2	20.89	29.25	301.3	17.8	.0224	1.00034
260.0	.8888	.02778	5396.0	7556.0	205.4	20.92	29.27	307.2	18.4	.0232	1.00033
280.0	.8251	.02579	5815.0	8142.0	207.5	20.99	29.33	318.7	19.5	.0248	1.00031
300.0	.7700	.02407	6236.0	8730.0	209.6	21.07	29.42	329.7	20.6	.0263	1.00029
320.0	.7218	.02256	6659.0	9319.0	211.5	21.18	29.52	340.3	21.7	.0278	1.00027
340.0	.6793	.02123	7084.0	9910.0	213.2	21.31	29.65	350.5	22.8	.0293	1.00025
360.0	.6415	.02005	7512.0	10500.0	214.9	21.46	29.79	360.3	23.8	.0307	1.00024
380.0	.6077	.01899	7943.0	11100.0	216.6	21.62	29.95	369.8	24.8	.0322	1.00023
400.0	.5773	.01804	8377.0	11700.0	218.1	21.79	30.12	379.0	25.8	.0337	1.00021

.08 MPa isobar

T K	Density kg/m³	Density mol/dm³	E J/mol	H J/mol	S J/(mol·K)	C_v J/(mol·K)	C_p J/(mol·K)	Sound m/s	Visc. μPa·s	Therm. W/(m·K)	Diel. const.
54.37[a]	1306.0	40.82	−6183.0	−6181.0	67.10	36.29	53.24	1129.0	485.0	.192	1.56856
60.0	1282.0	40.05	−5883.0	−5881.0	72.36	34.88	53.37	1114.0	425.0	.188	1.55617
80.0	1190.0	37.20	−4814.0	−4812.0	87.73	30.97	53.61	984.8	252.0	.166	1.51112
87.992[b]	1152.0	36.00	−4385.0	−4382.0	92.85	29.82	54.04	922.9	206.0	.155	1.49233
87.992[b]	3.608	.1128	1775.0	2484.0	170.9	21.60	31.03	175.8	6.52	.00817	1.00134
90.0	3.516	.1099	1819.0	2547.0	171.6	21.52	30.85	178.1	6.67	.00834	1.00131
95.0	3.317	.1036	1929.0	2700.0	173.3	21.36	30.50	183.4	7.06	.00878	1.00123
100.0	3.140	.09813	2037.0	2852.0	174.8	21.24	30.24	188.5	7.45	.00924	1.00117
105.0	2.982	.09320	2145.0	3003.0	176.3	21.15	30.04	193.5	7.84	.00971	1.00111
110.0	2.840	.08877	2251.0	3153.0	177.7	21.08	29.88	198.3	8.22	.0102	1.00106

Thermophysical properties of oxygen—Continued

T K	Density kg/m^3	Density mol/dm^3	E J/mol	H J/mol	S J/(mol·K)	C_v J/(mol·K)	C_p J/(mol·K)	Sound m/s	Visc. μPa·s	Therm. W/(m·K)	Diel. const.
115.0	2.712	.08475	2358.0	3302.0	179.0	21.03	29.76	203.0	8.60	.0107	1.00101
120.0	2.595	.08110	2464.0	3450.0	180.3	20.99	29.67	207.6	8.98	.0111	1.00096
130.0	2.390	.07468	2675.0	3746.0	182.6	20.93	29.53	216.4	9.73	.0121	1.00089
140.0	2.215	.06922	2885.0	4041.0	184.8	20.89	29.43	224.8	10.5	.0131	1.00082
150.0	2.065	.06452	3095.0	4335.0	186.9	20.86	29.37	232.9	11.2	.0140	1.00077
160.0	1.934	.06043	3305.0	4629.0	188.8	20.85	29.32	240.6	11.9	.0149	1.00072
170.0	1.818	.05683	3514.0	4922.0	190.5	20.84	29.29	248.2	12.6	.0158	1.00068
180.0	1.716	.05364	3723.0	5214.0	192.2	20.83	29.27	255.5	13.3	.0167	1.00064
190.0	1.625	.05079	3932.0	5507.0	193.8	20.83	29.25	262.5	14.0	.0176	1.00060
200.0	1.543	.04823	4140.0	5799.0	195.3	20.83	29.24	269.4	14.6	.0184	1.00057
210.0	1.469	.04592	4349.0	6092.0	196.7	20.84	29.23	276.1	15.3	.0192	1.00055
220.0	1.402	.04381	4558.0	6384.0	198.1	20.85	29.23	282.6	15.9	.0201	1.00052
230.0	1.341	.04190	4767.0	6676.0	199.4	20.86	29.24	289.0	16.5	.0209	1.00050
240.0	1.285	.04014	4976.0	6969.0	200.6	20.88	29.25	295.2	17.2	.0217	1.00048
250.0	1.233	.03853	5185.0	7261.0	201.8	20.90	29.26	301.3	17.8	.0224	1.00046
260.0	1.185	.03704	5394.0	7554.0	203.0	20.92	29.28	307.2	18.4	.0232	1.00044
280.0	1.100	.03439	5814.0	8140.0	205.1	20.99	29.34	318.7	19.5	.0248	1.00041
300.0	1.027	.03209	6235.0	8728.0	207.2	21.08	29.43	329.7	20.6	.0263	1.00038
320.0	.9625	.03008	6658.0	9317.0	209.1	21.18	29.53	340.3	21.7	.0278	1.00036
340.0	.9058	.02831	7083.0	9909.0	210.9	21.31	29.65	350.5	22.8	.0293	1.00034
360.0	.8554	.02673	7511.0	10500.0	212.6	21.46	29.80	360.3	23.8	.0308	1.00032
380.0	.8103	.02532	7942.0	11100.0	214.2	21.62	29.95	369.8	24.8	.0322	1.00030
400.0	.7697	.02406	8376.0	11700.0	215.7	21.80	30.13	379.0	25.8	.0337	1.00029

.10 MPa isobar

T K	Density kg/m^3	Density mol/dm^3	E J/mol	H J/mol	S J/(mol·K)	C_v J/(mol·K)	C_p J/(mol·K)	Sound m/s	Visc. μPa·s	Therm. W/(m·K)	Diel. const.
54.37[a]	1306.0	40.82	–6183.0	–6181.0	67.11	36.29	53.23	1129.0	486.0	.192	1.56857
60.0	1282.0	40.05	–5883.0	–5880.0	72.36	34.88	53.37	1114.0	425.0	.188	1.55619
80.0	1190.0	37.20	–4814.0	–4812.0	87.73	30.97	53.61	984.8	252.0	.166	1.51114
90.066[b]	1142.0	35.68	–4273.0	–4270.0	94.11	29.55	54.21	906.5	195.0	.152	1.48737
90.066[b]	4.429	.1384	1809.0	2532.0	169.6	21.71	31.33	177.4	6.69	.00844	1.00165
95.0	4.171	.1303	1919.0	2686.0	171.3	21.51	30.88	182.8	7.07	.00887	1.00155
100.0	3.945	.1233	2029.0	2840.0	172.9	21.36	30.54	188.0	7.46	.00932	1.00147
105.0	3.744	.1170	2137.0	2992.0	174.4	21.24	30.29	193.0	7.84	.00978	1.00139
110.0	3.564	.1114	2245.0	3143.0	175.8	21.15	30.09	197.9	8.23	.0102	1.00132
115.0	3.401	.1063	2352.0	3293.0	177.1	21.09	29.94	202.6	8.61	.0107	1.00126
120.0	3.253	.1017	2458.0	3442.0	178.4	21.03	29.81	207.2	8.99	.0112	1.00121
130.0	2.994	.09356	2670.0	3739.0	180.8	20.96	29.64	216.1	9.74	.0122	1.00111
140.0	2.774	.08669	2881.0	4035.0	182.9	20.91	29.52	224.6	10.5	.0131	1.00103
150.0	2.585	.08077	3092.0	4330.0	185.0	20.88	29.43	232.7	11.2	.0140	1.00096
160.0	2.420	.07563	3301.0	4624.0	186.9	20.86	29.38	240.5	11.9	.0150	1.00090
170.0	2.275	.07111	3511.0	4917.0	188.7	20.85	29.33	248.0	12.6	.0159	1.00085
180.0	2.147	.06710	3720.0	5210.0	190.3	20.84	29.30	255.4	13.3	.0167	1.00080
190.0	2.033	.06353	3929.0	5503.0	191.9	20.84	29.28	262.4	14.0	.0176	1.00076
200.0	1.930	.06032	4138.0	5796.0	193.4	20.84	29.26	269.3	14.6	.0184	1.00072
210.0	1.837	.05742	4347.0	6089.0	194.8	20.84	29.26	276.0	15.3	.0193	1.00068
220.0	1.753	.05479	4556.0	6381.0	196.2	20.85	29.25	282.6	15.9	.0201	1.00065
230.0	1.676	.05239	4765.0	6674.0	197.5	20.86	29.26	289.0	16.6	.0209	1.00062
240.0	1.606	.05020	4974.0	6966.0	198.8	20.88	29.26	295.2	17.2	.0217	1.00060
250.0	1.542	.04818	5183.0	7259.0	199.9	20.90	29.28	301.3	17.8	.0225	1.00057
260.0	1.482	.04632	5393.0	7552.0	201.1	20.92	29.30	307.2	18.4	.0232	1.00055
280.0	1.376	.04299	5812.0	8138.0	203.3	20.99	29.36	318.7	19.5	.0248	1.00051
300.0	1.284	.04012	6234.0	8726.0	205.3	21.08	29.44	329.7	20.6	.0263	1.00048
320.0	1.203	.03760	6657.0	9316.0	207.2	21.19	29.54	340.3	21.7	.0278	1.00045
340.0	1.132	.03539	7082.0	9908.0	209.0	21.31	29.66	350.5	22.8	.0293	1.00042
360.0	1.069	.03342	7510.0	10500.0	210.7	21.46	29.80	360.4	23.8	.0308	1.00040
380.0	1.013	.03165	7941.0	11100.0	212.3	21.62	29.96	369.9	24.8	.0322	1.00038
400.0	.9622	.03007	8375.0	11700.0	213.9	21.80	30.13	379.1	25.8	.0337	1.00036

.101325 MPa isobar

T K	Density kg/m^3	Density mol/dm^3	E J/mol	H J/mol	S J/(mol·K)	C_v J/(mol·K)	C_p J/(mol·K)	Sound m/s	Visc. μPa·s	Therm. W/(m·K)	Diel. const.
54.37[a]	1306.0	40.82	–6183.0	–6180.0	67.11	36.29	53.23	1129.0	486.0	.192	1.56857

Thermophysical properties of oxygen—Continued

T K	Density kg/m³	Density mol/dm³	E J/mol	H J/mol	S J/(mol·K)	C_v J/(mol·K)	C_p J/(mol·K)	Sound m/s	Visc. μPa·s	Therm. W/(m·K)	Diel. const.
60.0	1282.0	40.05	−5883.0	−5880.0	72.36	34.88	53.37	1114.0	425.0	.188	1.55619
80.0	1190.0	37.20	−4814.0	−4812.0	87.73	30.97	53.61	984.8	252.0	.166	1.51114
90.191[b]	1141.0	35.66	−4266.0	−4263.0	94.18	29.54	54.22	905.5	195.0	.152	1.48706
90.191[b]	4.483	.1401	1811.0	2535.0	169.6	21.72	31.35	177.5	6.70	.00846	1.00167
95.0	4.228	.1321	1918.0	2685.0	171.2	21.52	30.90	182.7	7.07	.00887	1.00157
100.0	3.999	.1250	2028.0	2839.0	172.8	21.36	30.56	188.0	7.46	.00932	1.00149
105.0	3.795	.1186	2137.0	2991.0	174.3	21.25	30.30	193.0	7.85	.00978	1.00141
110.0	3.612	.1129	2244.0	3142.0	175.7	21.16	30.10	197.9	8.23	.0102	1.00134
115.0	3.447	.1077	2351.0	3292.0	177.0	21.09	29.95	202.6	8.61	.0107	1.00128
120.0	3.297	.1030	2458.0	3442.0	178.3	21.04	29.82	207.2	8.99	.0112	1.00122
130.0	3.034	.09481	2670.0	3739.0	180.6	20.96	29.64	216.1	9.74	.0122	1.00113
140.0	2.811	.08785	2881.0	4035.0	182.8	20.91	29.52	224.5	10.5	.0131	1.00104
150.0	2.619	.08185	3091.0	4329.0	184.9	20.88	29.44	232.7	11.2	.0140	1.00097
160.0	2.452	.07664	3301.0	4623.0	186.8	20.86	29.38	240.5	11.9	.0150	1.00091
170.0	2.306	.07206	3511.0	4917.0	188.5	20.85	29.34	248.0	12.6	.0159	1.00086
180.0	2.176	.06800	3720.0	5210.0	190.2	20.84	29.30	255.3	13.3	.0167	1.00081
190.0	2.060	.06438	3929.0	5503.0	191.8	20.84	29.28	262.4	14.0	.0176	1.00077
200.0	1.956	.06112	4138.0	5796.0	193.3	20.84	29.27	269.3	14.6	.0184	1.00073
210.0	1.862	.05819	4347.0	6088.0	194.7	20.84	29.26	276.0	15.3	.0193	1.00069
220.0	1.777	.05552	4556.0	6381.0	196.1	20.85	29.26	282.6	15.9	.0201	1.00066
230.0	1.699	.05309	4765.0	6674.0	197.4	20.86	29.26	289.0	16.6	.0209	1.00063
240.0	1.628	.05086	4974.0	6966.0	198.6	20.88	29.27	295.2	17.2	.0217	1.00060
250.0	1.562	.04882	5183.0	7259.0	199.8	20.90	29.28	301.3	17.8	.0225	1.00058
260.0	1.502	.04693	5393.0	7552.0	201.0	20.92	29.30	307.2	18.4	.0232	1.00056
280.0	1.394	.04356	5812.0	8138.0	203.2	20.99	29.36	318.7	19.5	.0248	1.00052
300.0	1.301	.04065	6233.0	8726.0	205.2	21.08	29.44	329.7	20.6	.0263	1.00048
320.0	1.219	.03810	6657.0	9316.0	207.1	21.19	29.54	340.3	21.7	.0278	1.00045
340.0	1.147	.03585	7082.0	9908.0	208.9	21.31	29.66	350.5	22.8	.0293	1.00043
360.0	1.083	.03386	7510.0	10500.0	210.6	21.46	29.80	360.4	23.8	.0308	1.00040
380.0	1.026	.03207	7941.0	11100.0	212.2	21.62	29.96	369.9	24.8	.0322	1.00038
400.0	.9749	.03047	8375.0	11700.0	213.7	21.80	30.13	379.1	25.8	.0337	1.00036
				.20 MPa isobar							
54.38[a]	1306.0	40.82	−6183.0	−6178.0	67.11	36.29	53.23	1129.0	486.0	.192	1.56860
60.0	1282.0	40.05	−5883.0	−5878.0	72.35	34.88	53.36	1115.0	426.0	.188	1.55625
80.0	1191.0	37.21	−4815.0	−4810.0	87.72	30.98	53.60	985.2	252.0	.166	1.51123
90.0	1142.0	35.69	−4277.0	−4271.0	94.06	29.57	54.19	907.4	196.0	.152	1.48761
97.237[b]	1105.0	34.54	−3883.0	−3877.0	98.27	28.70	54.98	848.5	165.0	.142	1.46980
97.237[b]	8.380	.2619	1921.0	2685.0	165.8	22.14	32.61	182.0	7.28	.00948	1.00311
100.0	8.104	.2533	1985.0	2775.0	166.7	21.97	32.21	185.1	7.50	.00970	1.00301
105.0	7.660	.2394	2099.0	2934.0	168.2	21.73	31.62	190.5	7.88	.0101	1.00285
110.0	7.268	.2271	2211.0	3091.0	169.7	21.54	31.18	195.7	8.26	.0106	1.00270
115.0	6.918	.2162	2321.0	3246.0	171.1	21.40	30.84	200.7	8.64	.0110	1.00257
120.0	6.603	.2063	2431.0	3400.0	172.4	21.29	30.58	205.5	9.02	.0115	1.00245
125.0	6.317	.1974	2539.0	3552.0	173.6	21.20	30.37	210.2	9.40	.0119	1.00235
130.0	6.057	.1893	2647.0	3704.0	174.8	21.13	30.20	214.7	9.77	.0124	1.00225
140.0	5.599	.1750	2861.0	4004.0	177.0	21.03	29.94	223.4	10.5	.0133	1.00208
150.0	5.208	.1628	3074.0	4303.0	179.1	20.97	29.77	231.8	11.2	.0142	1.00193
160.0	4.870	.1522	3285.0	4600.0	181.0	20.92	29.65	239.7	11.9	.0151	1.00181
170.0	4.574	.1429	3496.0	4896.0	182.8	20.90	29.56	247.4	12.6	.0160	1.00170
180.0	4.313	.1348	3707.0	5191.0	184.5	20.88	29.49	254.9	13.3	.0168	1.00160
190.0	4.080	.1275	3917.0	5486.0	186.1	20.87	29.44	262.0	14.0	.0177	1.00152
200.0	3.872	.1210	4127.0	5780.0	187.6	20.86	29.40	269.0	14.7	.0185	1.00144
210.0	3.684	.1151	4337.0	6074.0	189.0	20.86	29.38	275.8	15.3	.0193	1.00137
220.0	3.514	.1098	4546.0	6367.0	190.4	20.87	29.36	282.4	15.9	.0202	1.00130
230.0	3.359	.1050	4756.0	6661.0	191.7	20.87	29.35	288.8	16.6	.0210	1.00125
240.0	3.218	.1006	4965.0	6954.0	193.0	20.89	29.35	295.1	17.2	.0217	1.00119
250.0	3.087	.09649	5175.0	7248.0	194.2	20.91	29.36	301.2	17.8	.0225	1.00115
260.0	2.967	.09274	5385.0	7542.0	195.3	20.93	29.37	307.1	18.4	.0233	1.00110
280.0	2.754	.08606	5805.0	8129.0	197.5	21.00	29.41	318.7	19.5	.0248	1.00102
300.0	2.569	.08028	6227.0	8718.0	199.5	21.08	29.49	329.8	20.6	.0263	1.00095
320.0	2.407	.07524	6651.0	9309.0	201.4	21.19	29.58	340.4	21.7	.0278	1.00089

Thermophysical properties of oxygen—Continued

T K	Density kg/m^3	Density mol/dm^3	E J/mol	H J/mol	S J/(mol·K)	C_v J/(mol·K)	C_p J/(mol·K)	Sound m/s	Visc. μPa·s	Therm. W/(m·K)	Diel. const.
340.0	2.265	.07079	7076.0	9902.0	203.2	21.32	29.70	350.6	22.8	.0293	1.00084
360.0	2.139	.06684	7505.0	10500.0	204.9	21.46	29.84	360.5	23.8	.0308	1.00079
380.0	2.026	.06331	7936.0	11100.0	206.5	21.63	29.99	370.0	24.8	.0323	1.00075
400.0	1.924	.06014	8371.0	11700.0	208.1	21.80	30.16	379.2	25.8	.0337	1.00071
.40 MPa isobar											
54.41[a]	1306.0	40.83	–6182.0	–6173.0	67.11	36.28	53.23	1129.0	486.0	.192	1.56866
80.0	1191.0	37.22	–4817.0	–4806.0	87.70	30.99	53.58	985.9	252.0	.166	1.51140
90.0	1143.0	35.71	–4279.0	–4268.0	94.04	29.58	54.16	908.3	196.0	.152	1.48782
100.0	1091.0	34.10	–3733.0	–3721.0	99.80	28.42	55.36	826.1	154.0	.138	1.46300
105.733[b]	1060.0	33.12	–3413.0	–3401.0	102.9	27.84	56.49	776.4	135.0	.130	1.44796
105.733[b]	15.97	.4990	2033.0	2835.0	161.9	22.75	34.77	185.9	8.01	.0109	1.00594
110.0	15.17	.4741	2137.0	2981.0	163.2	22.40	33.82	191.0	8.34	.0112	1.00564
115.0	14.35	.4485	2256.0	3148.0	164.7	22.08	32.97	196.6	8.71	.0116	1.00534
120.0	13.63	.4259	2372.0	3311.0	166.1	21.83	32.33	201.9	9.09	.0120	1.00507
125.0	12.99	.4059	2486.0	3471.0	167.4	21.64	31.83	207.0	9.46	.0124	1.00483
130.0	12.41	.3878	2598.0	3629.0	168.7	21.49	31.44	211.9	9.83	.0128	1.00461
135.0	11.89	.3715	2709.0	3786.0	169.8	21.38	31.12	216.6	10.2	.0132	1.00442
140.0	11.41	.3567	2819.0	3941.0	171.0	21.28	30.87	221.2	10.6	.0137	1.00424
150.0	10.58	.3305	3037.0	4247.0	173.1	21.15	30.48	229.9	11.3	.0145	1.00393
160.0	9.862	.3082	3253.0	4551.0	175.0	21.06	30.22	238.3	12.0	.0154	1.00366
170.0	9.243	.2888	3467.0	4852.0	176.9	20.99	30.02	246.2	12.7	.0162	1.00343
180.0	8.700	.2719	3680.0	5151.0	178.6	20.95	29.88	253.8	13.4	.0171	1.00323
190.0	8.220	.2569	3892.0	5450.0	180.2	20.92	29.77	261.2	14.0	.0179	1.00305
200.0	7.792	.2435	4104.0	5747.0	181.7	20.91	29.69	268.3	14.7	.0187	1.00289
210.0	7.407	.2315	4315.0	6043.0	183.2	20.90	29.62	275.2	15.3	.0195	1.00275
220.0	7.060	.2206	4526.0	6339.0	184.5	20.90	29.58	281.9	16.0	.0203	1.00262
230.0	6.744	.2108	4737.0	6635.0	185.9	20.90	29.54	288.5	16.6	.0211	1.00250
240.0	6.456	.2018	4948.0	6930.0	187.1	20.91	29.52	294.8	17.2	.0219	1.00240
250.0	6.192	.1935	5158.0	7226.0	188.3	20.93	29.51	301.0	17.8	.0226	1.00230
260.0	5.949	.1859	5369.0	7521.0	189.5	20.95	29.51	307.0	18.4	.0234	1.00221
280.0	5.517	.1724	5791.0	8111.0	191.7	21.01	29.53	318.7	19.5	.0249	1.00205
300.0	5.144	.1608	6214.0	8702.0	193.7	21.10	29.59	329.8	20.7	.0264	1.00191
320.0	4.819	.1506	6638.0	9295.0	195.6	21.20	29.67	340.5	21.7	.0279	1.00179
340.0	4.533	.1416	7065.0	9889.0	197.4	21.33	29.77	350.8	22.8	.0294	1.00168
360.0	4.279	.1337	7494.0	10490.0	199.1	21.47	29.90	360.7	23.8	.0309	1.00159
380.0	4.052	.1266	7927.0	11090.0	200.7	21.63	30.05	370.3	24.8	.0323	1.00150
400.0	3.848	.1203	8362.0	11690.0	202.3	21.81	30.21	379.5	25.8	.0338	1.00143
.60 MPa isobar											
54.43[a]	1307.0	40.83	–6182.0	–6167.0	67.12	36.27	53.22	1130.0	487.0	.192	1.56873
80.0	1191.0	37.23	–4819.0	–4803.0	87.67	31.01	53.56	986.6	253.0	.166	1.51158
90.0	1143.0	35.72	–4281.0	–4265.0	94.01	29.60	54.12	909.1	197.0	.152	1.48804
100.0	1092.0	34.12	–3736.0	–3718.0	99.77	28.43	55.31	827.3	155.0	.138	1.46327
110.0	1036.0	32.37	–3173.0	–3155.0	105.1	27.47	57.56	738.9	123.0	.124	1.43653
111.457[b]	1027.0	32.09	–3089.0	–3071.0	105.9	27.34	58.02	725.3	119.0	.122	1.43241
111.457[b]	23.45	.7327	2093.0	2912.0	159.6	23.23	36.78	187.6	8.53	.0120	1.00873
115.0	22.43	.7009	2184.0	3040.0	160.7	22.86	35.66	192.1	8.80	.0122	1.00835
120.0	21.17	.6615	2308.0	3215.0	162.2	22.45	34.46	198.0	9.17	.0126	1.00788
125.0	20.07	.6272	2429.0	3385.0	163.6	22.13	33.57	203.6	9.53	.0129	1.00747
130.0	19.10	.5970	2547.0	3551.0	164.9	21.89	32.88	208.9	9.90	.0133	1.00711
135.0	18.24	.5701	2662.0	3714.0	166.1	21.70	32.33	214.0	10.3	.0137	1.00679
140.0	17.47	.5459	2776.0	3875.0	167.3	21.55	31.90	218.8	10.6	.0141	1.00650
150.0	16.12	.5038	3000.0	4191.0	169.5	21.33	31.26	228.1	11.3	.0149	1.00599
160.0	14.98	.4683	3220.0	4501.0	171.5	21.19	30.83	236.7	12.0	.0157	1.00557
170.0	14.01	.4379	3437.0	4808.0	173.3	21.09	30.51	245.0	12.7	.0165	1.00521
180.0	13.17	.4114	3653.0	5111.0	175.1	21.03	30.28	252.8	13.4	.0173	1.00489
190.0	12.42	.3882	3868.0	5413.0	176.7	20.98	30.11	260.4	14.1	.0181	1.00462
200.0	11.76	.3675	4081.0	5714.0	178.2	20.96	29.98	267.7	14.7	.0189	1.00437
210.0	11.17	.3490	4294.0	6013.0	179.7	20.94	29.88	274.7	15.4	.0197	1.00415

Thermophysical properties of oxygen—Continued

T K	Density kg/m³	Density mol/dm³	E J/mol	H J/mol	S J/(mol·K)	C_v J/(mol·K)	C_p J/(mol·K)	Sound m/s	Visc. μPa·s	Therm. W/(m·K)	Diel. const.
220.0	10.64	.3324	4506.0	6311.0	181.1	20.93	29.80	281.5	16.0	.0205	1.00395
230.0	10.15	.3173	4718.0	6609.0	182.4	20.93	29.74	288.1	16.6	.0212	1.00377
240.0	9.715	.3036	4930.0	6906.0	183.7	20.94	29.70	294.6	17.2	.0220	1.00361
250.0	9.313	.2910	5142.0	7203.0	184.9	20.95	29.67	300.8	17.8	.0228	1.00346
260.0	8.944	.2795	5353.0	7500.0	186.0	20.97	29.65	306.9	18.4	.0235	1.00332
280.0	8.289	.2590	5776.0	8093.0	188.2	21.03	29.65	318.7	19.6	.0250	1.00308
300.0	7.725	.2414	6201.0	8686.0	190.3	21.11	29.69	329.9	20.7	.0265	1.00287
320.0	7.234	.2261	6626.0	9280.0	192.2	21.21	29.76	340.7	21.8	.0280	1.00269
340.0	6.802	.2126	7054.0	9876.0	194.0	21.34	29.85	351.0	22.8	.0295	1.00252
360.0	6.420	.2006	7484.0	10470.0	195.7	21.48	29.97	360.9	23.8	.0310	1.00238
380.0	6.079	.1900	7917.0	11080.0	197.3	21.64	30.11	370.5	24.8	.0324	1.00226
400.0	5.772	.1804	8353.0	11680.0	198.9	21.82	30.27	379.8	25.8	.0339	1.00214

.80 MPa isobar

T K	Density kg/m³	Density mol/dm³	E J/mol	H J/mol	S J/(mol·K)	C_v J/(mol·K)	C_p J/(mol·K)	Sound m/s	Visc. μPa·s	Therm. W/(m·K)	Diel. const.
54.45[a]	1307.0	40.84	−6182.0	−6162.0	67.13	36.27	53.21	1130.0	488.0	.192	1.56879
80.0	1192.0	37.24	−4820.0	−4799.0	87.65	31.02	53.53	987.2	253.0	.166	1.51175
90.0	1144.0	35.74	−4284.0	−4261.0	93.99	29.61	54.09	910.0	197.0	.152	1.48825
100.0	1092.0	34.14	−3738.0	−3715.0	99.74	28.45	55.26	828.4	155.0	.138	1.46354
110.0	1036.0	32.39	−3177.0	−3153.0	105.1	27.48	57.47	740.5	123.0	.124	1.43689
115.914[b]	1000.0	31.25	−2832.0	−2807.0	108.2	27.01	59.62	683.9	108.0	.115	1.41969
115.914[b]	30.95	.9672	2130.0	2957.0	157.9	23.64	38.78	188.4	8.96	.0130	1.01153
120.0	29.34	.9169	2239.0	3112.0	159.2	23.15	37.14	193.8	9.25	.0132	1.01093
125.0	27.65	.8641	2368.0	3294.0	160.7	22.68	35.66	200.0	9.62	.0135	1.01030
130.0	26.20	.8186	2492.0	3469.0	162.1	22.33	34.56	205.8	9.98	.0138	1.00975
135.0	24.92	.7788	2612.0	3640.0	163.3	22.06	33.72	211.2	10.3	.0141	1.00928
140.0	23.79	.7434	2730.0	3807.0	164.6	21.84	33.06	216.4	10.7	.0145	1.00885
145.0	22.77	.7117	2846.0	3970.0	165.7	21.67	32.54	221.4	11.1	.0148	1.00847
150.0	21.85	.6829	2961.0	4132.0	166.8	21.53	32.12	226.1	11.4	.0152	1.00813
160.0	20.25	.6328	3186.0	4450.0	168.9	21.33	31.48	235.2	12.1	.0160	1.00753
170.0	18.89	.5902	3407.0	4762.0	170.8	21.20	31.03	243.7	12.8	.0167	1.00702
180.0	17.71	.5535	3626.0	5071.0	172.5	21.11	30.71	251.8	13.5	.0175	1.00659
190.0	16.68	.5214	3842.0	5377.0	174.2	21.05	30.46	259.6	14.1	.0183	1.00620
200.0	15.78	.4931	4058.0	5680.0	175.7	21.00	30.28	267.0	14.8	.0191	1.00587
210.0	14.97	.4678	4272.0	5982.0	177.2	20.98	30.14	274.2	15.4	.0198	1.00556
220.0	14.25	.4452	4486.0	6283.0	178.6	20.96	30.03	281.1	16.0	.0206	1.00529
230.0	13.59	.4247	4699.0	6583.0	179.9	20.96	29.94	287.8	16.7	.0214	1.00505
240.0	13.00	.4061	4912.0	6882.0	181.2	20.96	29.88	294.3	17.3	.0221	1.00483
250.0	12.45	.3891	5125.0	7181.0	182.4	20.97	29.83	300.7	17.9	.0229	1.00463
260.0	11.95	.3736	5337.0	7479.0	183.6	20.99	29.80	306.8	18.5	.0236	1.00444
280.0	11.07	.3460	5762.0	8074.0	185.8	21.04	29.77	318.7	19.6	.0251	1.00411
300.0	10.31	.3223	6188.0	8670.0	187.9	21.12	29.79	330.0	20.7	.0266	1.00383
320.0	9.653	.3017	6614.0	9266.0	189.8	21.22	29.84	340.8	21.8	.0281	1.00358
340.0	9.074	.2836	7043.0	9864.0	191.6	21.35	29.93	351.2	22.8	.0296	1.00337
360.0	8.562	.2676	7474.0	10460.0	193.3	21.49	30.04	361.2	23.8	.0310	1.00318
380.0	8.106	.2533	7907.0	11070.0	194.9	21.65	30.17	370.8	24.8	.0325	1.00301
400.0	7.696	.2405	8344.0	11670.0	196.5	21.82	30.32	380.1	25.8	.0340	1.00286

1.00 MPa isobar

T K	Density kg/m³	Density mol/dm³	E J/mol	H J/mol	S J/(mol·K)	C_v J/(mol·K)	C_p J/(mol·K)	Sound m/s	Visc. μPa·s	Therm. W/(m·K)	Diel. const.
54.47[a]	1307.0	40.84	−6181.0	−6157.0	67.13	36.26	53.21	1131.0	488.0	.192	1.56886
80.0	1192.0	37.25	−4822.0	−4795.0	87.63	31.03	53.51	987.9	254.0	.166	1.51192
90.0	1144.0	35.75	−4286.0	−4258.0	93.96	29.63	54.06	910.9	197.0	.152	1.48846
100.0	1093.0	34.15	−3741.0	−3712.0	99.71	28.46	55.20	829.6	156.0	.138	1.46381
110.0	1037.0	32.41	−3181.0	−3150.0	105.1	27.50	57.37	742.0	124.0	.124	1.43725
115.0	1007.0	31.46	−2891.0	−2859.0	107.6	27.09	59.10	694.9	110.0	.117	1.42288
119.623[b]	976.3	30.51	−2614.0	−2581.0	110.0	26.76	61.32	648.3	99.3	.111	1.40861
119.623[b]	38.54	1.204	2152.0	2982.0	156.5	24.01	40.86	188.6	9.33	.0139	1.01437
120.0	38.33	1.198	2163.0	2997.0	156.7	23.96	40.64	189.1	9.36	.0139	1.01430
125.0	35.84	1.120	2302.0	3194.0	158.3	23.30	38.26	196.1	9.71	.0141	1.01336
130.0	33.76	1.055	2433.0	3381.0	159.7	22.81	36.58	202.4	10.1	.0143	1.01258

Thermophysical properties of oxygen—Continued

T K	Density kg/m³	Density mol/dm³	E J/mol	H J/mol	S J/(mol·K)	C_v J/(mol·K)	C_p J/(mol·K)	Sound m/s	Visc. μPa·s	Therm. W/(m·K)	Diel. const.
135.0	31.97	.9991	2560.0	3561.0	161.1	22.44	35.33	208.3	10.4	.0146	1.01191
140.0	30.41	.9504	2683.0	3735.0	162.4	22.15	34.39	213.9	10.8	.0149	1.01133
145.0	29.03	.9072	2803.0	3905.0	163.5	21.93	33.65	219.1	11.1	.0152	1.01081
150.0	27.79	.8686	2920.0	4072.0	164.7	21.75	33.06	224.2	11.5	.0156	1.01035
155.0	26.68	.8337	3036.0	4236.0	165.8	21.60	32.58	229.0	11.8	.0159	1.00993
160.0	25.66	.8019	3151.0	4398.0	166.8	21.48	32.18	233.6	12.2	.0163	1.00955
170.0	23.87	.7460	3376.0	4716.0	168.7	21.31	31.58	242.5	12.8	.0170	1.00888
180.0	22.34	.6982	3598.0	5030.0	170.5	21.19	31.15	250.8	13.5	.0178	1.00831
190.0	21.01	.6567	3817.0	5340.0	172.2	21.11	30.83	258.8	14.2	.0185	1.00782
200.0	19.85	.6203	4035.0	5647.0	173.8	21.05	30.59	266.4	14.8	.0193	1.00738
210.0	18.81	.5879	4251.0	5952.0	175.2	21.02	30.40	273.7	15.5	.0200	1.00700
220.0	17.89	.5590	4466.0	6255.0	176.7	20.99	30.26	280.7	16.1	.0208	1.00665
230.0	17.05	.5329	4680.0	6557.0	178.0	20.98	30.14	287.5	16.7	.0215	1.00634
240.0	16.30	.5093	4894.0	6858.0	179.3	20.98	30.06	294.1	17.3	.0223	1.00606
250.0	15.61	.4878	5108.0	7158.0	180.5	20.99	29.99	300.5	17.9	.0230	1.00580
260.0	14.98	.4680	5321.0	7458.0	181.7	21.00	29.94	306.7	18.5	.0238	1.00557
270.0	14.40	.4499	5534.0	7757.0	182.8	21.03	29.91	312.8	19.1	.0245	1.00535
280.0	13.86	.4332	5748.0	8056.0	183.9	21.06	29.89	318.7	19.6	.0252	1.00515
300.0	12.91	.4033	6174.0	8654.0	186.0	21.13	29.89	330.1	20.7	.0267	1.00479
320.0	12.08	.3774	6602.0	9252.0	187.9	21.24	29.93	341.0	21.8	.0282	1.00448
340.0	11.35	.3547	7032.0	9851.0	189.7	21.36	30.00	351.4	22.8	.0296	1.00421
360.0	10.71	.3346	7463.0	10450.0	191.4	21.50	30.10	361.4	23.9	.0311	1.00397
380.0	10.13	.3167	7898.0	11060.0	193.1	21.66	30.23	371.0	24.9	.0326	1.00376
400.0	9.619	.3006	8335.0	11660.0	194.6	21.83	30.37	380.4	25.8	.0340	1.00357

1.20 MPa isobar

T K	Density kg/m³	Density mol/dm³	E J/mol	H J/mol	S J/(mol·K)	C_v J/(mol·K)	C_p J/(mol·K)	Sound m/s	Visc. μPa·s	Therm. W/(m·K)	Diel. const.
54.50[a]	1307.0	40.84	−6181.0	−6152.0	67.14	36.25	53.20	1131.0	489.0	.193	1.56892
80.0	1192.0	37.26	−4824.0	−4792.0	87.61	31.05	53.49	988.6	254.0	.166	1.51210
90.0	1144.0	35.76	−4288.0	−4254.0	93.94	29.64	54.03	911.8	198.0	.152	1.48868
100.0	1093.0	34.17	−3744.0	−3709.0	99.68	28.48	55.15	830.7	156.0	.138	1.46408
110.0	1038.0	32.44	−3185.0	−3148.0	105.0	27.52	57.28	743.6	124.0	.124	1.43761
115.0	1008.0	31.49	−2896.0	−2858.0	107.6	27.11	58.97	696.7	111.0	.117	1.42330
120.0	974.9	30.47	−2597.0	−2557.0	110.2	26.75	61.34	646.7	98.8	.110	1.40794
122.829[b]	954.9	29.84	−2421.0	−2381.0	111.6	26.58	63.13	616.6	92.4	.106	1.39860
122.829[b]	46.27	1.446	2164.0	2994.0	155.4	24.36	43.07	188.5	9.67	.0147	1.01727
125.0	44.80	1.400	2229.0	3086.0	156.1	24.01	41.59	191.8	9.82	.0148	1.01672
130.0	41.88	1.309	2370.0	3287.0	157.7	23.36	39.03	198.8	10.2	.0149	1.01563
135.0	39.46	1.233	2504.0	3477.0	159.1	22.87	37.23	205.3	10.5	.0151	1.01471
140.0	37.38	1.168	2633.0	3660.0	160.5	22.49	35.90	211.2	10.9	.0154	1.01394
145.0	35.57	1.111	2757.0	3837.0	161.7	22.20	34.89	216.8	11.2	.0157	1.01326
150.0	33.96	1.061	2879.0	4009.0	162.9	21.97	34.09	222.1	11.6	.0160	1.01266
155.0	32.53	1.016	2998.0	4178.0	164.0	21.79	33.46	227.2	11.9	.0163	1.01212
160.0	31.23	.9760	3115.0	4344.0	165.0	21.64	32.95	232.0	12.2	.0166	1.01163
165.0	30.05	.9392	3230.0	4508.0	166.0	21.52	32.52	236.7	12.6	.0169	1.01119
170.0	28.97	.9054	3344.0	4669.0	167.0	21.42	32.17	241.2	12.9	.0173	1.01079
180.0	27.06	.8456	3569.0	4988.0	168.8	21.27	31.62	249.8	13.6	.0180	1.01007
190.0	25.41	.7941	3791.0	5302.0	170.5	21.17	31.21	258.0	14.2	.0187	1.00946
200.0	23.97	.7491	4011.0	5613.0	172.1	21.10	30.91	265.7	14.9	.0194	1.00892
210.0	22.70	.7093	4229.0	5921.0	173.6	21.06	30.67	273.2	15.5	.0202	1.00844
220.0	21.56	.6738	4446.0	6226.0	175.0	21.03	30.49	280.3	16.1	.0209	1.00802
230.0	20.54	.6419	4661.0	6531.0	176.4	21.01	30.35	287.2	16.7	.0217	1.00764
240.0	19.62	.6131	4876.0	6834.0	177.7	21.00	30.24	293.9	17.3	.0224	1.00729
250.0	18.78	.5869	5091.0	7136.0	178.9	21.01	30.16	300.4	17.9	.0231	1.00698
260.0	18.01	.5630	5305.0	7437.0	180.1	21.02	30.09	306.7	18.5	.0239	1.00670
270.0	17.31	.5410	5519.0	7737.0	181.2	21.04	30.05	312.8	19.1	.0246	1.00643
280.0	16.66	.5207	5733.0	8038.0	182.3	21.07	30.02	318.7	19.7	.0253	1.00619
300.0	15.50	.4845	6161.0	8638.0	184.4	21.15	30.00	330.2	20.8	.0268	1.00576
320.0	14.50	.4532	6590.0	9238.0	186.3	21.25	30.02	341.1	21.8	.0283	1.00539
340.0	13.62	.4258	7021.0	9839.0	188.2	21.37	30.08	351.6	22.9	.0297	1.00506
360.0	12.85	.4016	7453.0	10440.0	189.9	21.51	30.17	361.6	23.9	.0312	1.00477
380.0	12.16	.3800	7888.0	11050.0	191.5	21.67	30.29	371.3	24.9	.0326	1.00451
400.0	11.54	.3607	8326.0	11650.0	193.1	21.84	30.42	380.6	25.8	.0341	1.00428

Thermophysical properties of oxygen—Continued

T K	Density kg/m³	Density mol/dm³	E J/mol	H J/mol	S J/(mol·K)	C_v J/(mol·K)	C_p J/(mol·K)	Sound m/s	Visc. μPa·s	Therm. W/(m·K)	Diel. const.
					1.40 MPa isobar						
54.52[a]	1307.0	40.85	–6181.0	–6147.0	67.14	36.25	53.20	1132.0	490.0	.193	1.56899
80.0	1193.0	37.28	–4825.0	–4788.0	87.59	31.06	53.47	989.3	255.0	.166	1.51227
90.0	1145.0	35.78	–4290.0	–4251.0	93.91	29.66	54.00	912.6	198.0	.152	1.48889
100.0	1094.0	34.19	–3747.0	–3706.0	99.65	28.49	55.10	831.8	156.0	.138	1.46435
110.0	1039.0	32.46	–3189.0	–3146.0	105.0	27.53	57.19	745.1	124.0	.124	1.43797
115.0	1009.0	31.52	–2900.0	–2856.0	107.6	27.12	58.84	698.5	111.0	.117	1.42372
120.0	976.0	30.50	–2602.0	–2556.0	110.1	26.77	61.15	648.9	99.1	.110	1.40845
125.0	940.1	29.38	–2290.0	–2243.0	112.7	26.48	64.50	595.3	88.1	.103	1.39174
125.672[b]	934.9	29.22	–2247.0	–2199.0	113.0	26.45	65.08	587.6	86.7	.102	1.38932
125.672[b]	54.17	1.693	2169.0	2996.0	154.4	24.70	45.44	188.2	9.99	.0155	1.02024
130.0	50.72	1.585	2301.0	3185.0	155.8	23.97	42.12	195.0	10.3	.0156	1.01894
135.0	47.46	1.483	2444.0	3388.0	157.4	23.33	39.51	202.0	10.6	.0157	1.01772
140.0	44.74	1.398	2580.0	3581.0	158.8	22.86	37.66	208.5	11.0	.0159	1.01670
145.0	42.41	1.325	2710.0	3766.0	160.1	22.49	36.30	214.5	11.3	.0161	1.01582
150.0	40.38	1.262	2835.0	3945.0	161.3	22.21	35.25	220.1	11.6	.0164	1.01506
155.0	38.58	1.206	2958.0	4119.0	162.4	21.98	34.43	225.4	12.0	.0167	1.01439
160.0	36.98	1.156	3078.0	4289.0	163.5	21.80	33.77	230.4	12.3	.0169	1.01378
165.0	35.52	1.110	3195.0	4457.0	164.5	21.65	33.23	235.3	12.6	.0173	1.01324
170.0	34.20	1.069	3312.0	4622.0	165.5	21.54	32.79	239.9	13.0	.0176	1.01274
180.0	31.87	.9960	3540.0	4946.0	167.4	21.36	32.11	248.8	13.6	.0182	1.01187
190.0	29.88	.9337	3765.0	5264.0	169.1	21.24	31.61	257.2	14.3	.0189	1.01113
200.0	28.15	.8796	3987.0	5579.0	170.7	21.15	31.24	265.1	14.9	.0196	1.01048
210.0	26.62	.8320	4207.0	5890.0	172.2	21.10	30.95	272.7	15.6	.0204	1.00991
220.0	25.27	.7897	4425.0	6198.0	173.7	21.06	30.73	279.9	16.2	.0211	1.00940
230.0	24.06	.7518	4642.0	6504.0	175.0	21.04	30.56	286.9	16.8	.0218	1.00895
240.0	22.96	.7176	4858.0	6809.0	176.3	21.03	30.43	293.7	17.4	.0225	1.00854
250.0	21.97	.6866	5074.0	7113.0	177.6	21.03	30.32	300.2	18.0	.0233	1.00817
260.0	21.07	.6583	5289.0	7416.0	178.8	21.04	30.24	306.6	18.6	.0240	1.00783
270.0	20.23	.6324	5504.0	7718.0	179.9	21.06	30.18	312.7	19.1	.0247	1.00752
280.0	19.47	.6085	5719.0	8019.0	181.0	21.08	30.14	318.7	19.7	.0255	1.00724
300.0	18.11	.5659	6148.0	8622.0	183.1	21.16	30.10	330.3	20.8	.0269	1.00673
320.0	16.93	.5292	6578.0	9224.0	185.0	21.26	30.11	341.3	21.9	.0284	1.00629
340.0	15.90	.4970	7009.0	9826.0	186.8	21.38	30.16	351.8	22.9	.0298	1.00591
360.0	15.00	.4686	7443.0	10430.0	188.6	21.52	30.24	361.9	23.9	.0313	1.00557
380.0	14.19	.4434	7879.0	11040.0	190.2	21.68	30.35	371.6	24.9	.0327	1.00527
400.0	13.46	.4208	8317.0	11640.0	191.8	21.85	30.48	380.9	25.9	.0342	1.00500
					1.60 MPa isobar						
54.54[a]	1307.0	40.85	–6180.0	–6141.0	67.15	36.24	53.19	1132.0	490.0	.193	1.56905
80.0	1193.0	37.29	–4827.0	–4784.0	87.57	31.07	53.45	990.0	255.0	.167	1.51244
90.0	1145.0	35.79	–4292.0	–4247.0	93.89	29.67	53.97	913.5	199.0	.153	1.48910
100.0	1095.0	34.21	–3750.0	–3703.0	99.63	28.51	55.05	832.9	157.0	.138	1.46462
110.0	1039.0	32.48	–3192.0	–3143.0	105.0	27.55	57.10	746.6	125.0	.124	1.43832
115.0	1009.0	31.55	–2905.0	–2854.0	107.5	27.14	58.72	700.3	111.0	.117	1.42414
120.0	977.1	30.53	–2607.0	–2555.0	110.1	26.78	60.97	651.1	99.5	.110	1.40896
125.0	941.5	29.42	–2297.0	–2243.0	112.6	26.49	64.22	598.0	88.5	.104	1.39237
128.237[b]	916.0	28.63	–2086.0	–2030.0	114.3	26.35	67.20	560.8	81.7	.0991	1.38058
128.237[b]	62.28	1.946	2167.0	2989.0	153.4	25.03	48.03	187.8	10.3	.0163	1.02330
130.0	60.46	1.889	2226.0	3072.0	154.1	24.67	46.16	190.8	10.4	.0163	1.02261
132.0	58.58	1.831	2289.0	3163.0	154.8	24.31	44.41	194.0	10.5	.0163	1.02190
135.0	56.09	1.753	2380.0	3293.0	155.8	23.86	42.31	198.6	10.7	.0163	1.02096
140.0	52.57	1.643	2524.0	3498.0	157.2	23.26	39.74	205.6	11.1	.0164	1.01964
145.0	49.62	1.551	2660.0	3691.0	158.6	22.81	37.91	212.0	11.4	.0166	1.01853
150.0	47.08	1.471	2790.0	3877.0	159.9	22.46	36.55	217.9	11.7	.0168	1.01757
155.0	44.87	1.402	2916.0	4057.0	161.0	22.19	35.50	223.5	12.1	.0170	1.01674
160.0	42.91	1.341	3040.0	4233.0	162.2	21.97	34.67	228.8	12.4	.0173	1.01601
165.0	41.15	1.286	3160.0	4404.0	163.2	21.80	34.00	233.8	12.7	.0176	1.01535
170.0	39.55	1.236	3279.0	4573.0	164.2	21.66	33.45	238.7	13.1	.0179	1.01475

Thermophysical properties of oxygen—Continued

T K	Density kg/m³	Density mol/dm³	E J/mol	H J/mol	S J/(mol·K)	C_v J/(mol·K)	C_p J/(mol·K)	Sound m/s	Visc. μPa·s	Therm. W/(m·K)	Diel. const.
175.0	38.10	1.191	3395.0	4739.0	165.2	21.54	33.00	243.3	13.4	.0182	1.01421
180.0	36.77	1.149	3511.0	4903.0	166.1	21.45	32.62	247.8	13.7	.0185	1.01371
190.0	34.42	1.076	3739.0	5226.0	167.9	21.30	32.02	256.4	14.3	.0192	1.01282
200.0	32.38	1.012	3963.0	5544.0	169.5	21.20	31.58	264.5	15.0	.0198	1.01206
210.0	30.59	.9560	4185.0	5858.0	171.0	21.14	31.24	272.2	15.6	.0205	1.01139
220.0	29.01	.9066	4404.0	6169.0	172.5	21.09	30.98	279.6	16.2	.0212	1.01080
230.0	27.60	.8625	4623.0	6478.0	173.8	21.06	30.78	286.6	16.8	.0220	1.01027
240.0	26.33	.8228	4840.0	6785.0	175.1	21.05	30.62	293.5	17.4	.0227	1.00980
250.0	25.18	.7869	5057.0	7090.0	176.4	21.05	30.49	300.1	18.0	.0234	1.00937
260.0	24.13	.7541	5273.0	7395.0	177.6	21.06	30.39	306.5	18.6	.0241	1.00898
270.0	23.17	.7241	5489.0	7698.0	178.7	21.07	30.32	312.7	19.2	.0248	1.00862
280.0	22.29	.6966	5704.0	8001.0	179.8	21.10	30.26	318.8	19.7	.0256	1.00829
300.0	20.72	.6475	6135.0	8606.0	181.9	21.17	30.20	330.4	20.8	.0270	1.00770
320.0	19.37	.6052	6566.0	9210.0	183.9	21.27	30.20	341.4	21.9	.0285	1.00720
340.0	18.18	.5683	6998.0	9814.0	185.7	21.39	30.23	352.0	22.9	.0299	1.00676
360.0	17.14	.5357	7432.0	10420.0	187.4	21.53	30.31	362.1	23.9	.0313	1.00637
380.0	16.22	.5068	7869.0	11030.0	189.1	21.69	30.41	371.9	24.9	.0328	1.00602
400.0	15.39	.4809	8308.0	11640.0	190.6	21.86	30.53	381.2	25.9	.0342	1.00571

1.80 MPa isobar

T K	Density kg/m³	Density mol/dm³	E J/mol	H J/mol	S J/(mol·K)	C_v J/(mol·K)	C_p J/(mol·K)	Sound m/s	Visc. μPa·s	Therm. W/(m·K)	Diel. const.
54.57[a]	1307.0	40.86	−6180.0	−6136.0	67.16	36.23	53.19	1133.0	491.0	.193	1.56912
80.0	1193.0	37.30	−4829.0	−4780.0	87.55	31.08	53.43	990.7	256.0	.167	1.51261
90.0	1146.0	35.80	−4294.0	−4244.0	93.87	29.68	53.94	914.3	199.0	.153	1.48931
100.0	1095.0	34.22	−3753.0	−3700.0	99.60	28.52	55.00	834.0	157.0	.138	1.46488
110.0	1040.0	32.51	−3196.0	−3141.0	104.9	27.56	57.02	748.1	125.0	.124	1.43867
115.0	1010.0	31.57	−2909.0	−2852.0	107.5	27.15	58.60	702.1	112.0	.117	1.42455
120.0	978.1	30.57	−2613.0	−2554.0	110.0	26.80	60.79	653.2	99.9	.111	1.40946
125.0	942.8	29.46	−2304.0	−2243.0	112.6	26.50	63.95	600.7	88.9	.104	1.39300
130.0	902.9	28.22	−1975.0	−1912.0	115.2	26.30	68.79	542.9	78.5	.0969	1.37455
130.581[b]	897.8	28.06	−1935.0	−1871.0	115.5	26.28	69.53	535.7	77.3	.0960	1.37222
130.581[b]	70.63	2.207	2161.0	2976.0	152.6	25.35	50.88	187.2	10.6	.0171	1.02645
135.0	65.50	2.047	2310.0	3189.0	154.2	24.45	45.86	194.9	10.9	.0170	1.02451
136.0	64.51	2.016	2342.0	3234.0	154.5	24.28	45.00	196.5	10.9	.0170	1.02413
140.0	60.94	1.904	2463.0	3409.0	155.8	23.71	42.23	202.5	11.2	.0170	1.02279
145.0	57.22	1.788	2607.0	3613.0	157.2	23.16	39.78	209.4	11.5	.0171	1.02139
150.0	54.10	1.691	2743.0	3808.0	158.6	22.74	38.01	215.7	11.8	.0173	1.02021
155.0	51.40	1.606	2874.0	3994.0	159.8	22.41	36.68	221.6	12.2	.0175	1.01920
160.0	49.04	1.532	3000.0	4175.0	160.9	22.15	35.65	227.1	12.5	.0177	1.01831
165.0	46.94	1.467	3124.0	4351.0	162.0	21.95	34.83	232.4	12.8	.0179	1.01752
170.0	45.05	1.408	3245.0	4523.0	163.0	21.78	34.16	237.4	13.1	.0182	1.01681
175.0	43.34	1.354	3364.0	4693.0	164.0	21.65	33.62	242.2	13.5	.0185	1.01617
180.0	41.78	1.306	3481.0	4860.0	165.0	21.54	33.16	246.8	13.8	.0188	1.01558
190.0	39.03	1.220	3712.0	5188.0	166.7	21.37	32.45	255.6	14.4	.0194	1.01455
200.0	36.66	1.146	3938.0	5509.0	168.4	21.25	31.93	263.8	15.0	.0201	1.01366
210.0	34.60	1.081	4162.0	5827.0	169.9	21.18	31.54	271.7	15.7	.0207	1.01289
220.0	32.79	1.025	4384.0	6140.0	171.4	21.12	31.23	279.2	16.3	.0214	1.01221
230.0	31.17	.9741	4604.0	6451.0	172.8	21.09	30.99	286.4	16.9	.0221	1.01161
240.0	29.72	.9287	4822.0	6760.0	174.1	21.07	30.81	293.3	17.5	.0228	1.01106
250.0	28.40	.8877	5040.0	7068.0	175.3	21.07	30.66	300.0	18.1	.0235	1.01057
260.0	27.21	.8503	5257.0	7374.0	176.5	21.08	30.54	306.4	18.6	.0242	1.01013
270.0	26.12	.8162	5473.0	7679.0	177.7	21.09	30.45	312.7	19.2	.0250	1.00972
280.0	25.12	.7849	5690.0	7983.0	178.8	21.11	30.39	318.8	19.8	.0257	1.00934
300.0	23.34	.7293	6122.0	8590.0	180.9	21.18	30.31	330.5	20.8	.0271	1.00868
320.0	21.80	.6814	6554.0	9195.0	182.8	21.28	30.28	341.6	21.9	.0285	1.00811
340.0	20.47	.6396	6987.0	9801.0	184.7	21.40	30.31	352.2	22.9	.0300	1.00761
360.0	19.29	.6028	7422.0	10410.0	186.4	21.54	30.37	362.4	24.0	.0314	1.00717
380.0	18.24	.5701	7859.0	11020.0	188.1	21.69	30.47	372.1	24.9	.0329	1.00678
400.0	17.31	.5409	8299.0	11630.0	189.6	21.86	30.58	381.5	25.9	.0343	1.00643

Thermophysical properties of oxygen—Continued

T K	Density kg/m^3	Density mol/dm^3	E J/mol	H J/mol	S J/(mol·K)	C_v J/(mol·K)	C_p J/(mol·K)	Sound m/s	Visc. μPa·s	Therm. W/(m·K)	Diel. const.
					2.00 MPa isobar						
54.59[a]	1307.0	40.86	-6180.0	-6131.0	67.16	36.23	53.18	1133.0	491.0	.193	1.56918
80.0	1194.0	37.31	-4830.0	-4777.0	87.53	31.10	53.41	991.4	256.0	.167	1.51278
90.0	1146.0	35.82	-4296.0	-4241.0	93.84	29.70	53.90	915.2	199.0	.153	1.48952
100.0	1096.0	34.24	-3755.0	-3697.0	99.57	28.54	54.95	835.2	157.0	.138	1.46515
110.0	1041.0	32.53	-3200.0	-3138.0	104.9	27.58	56.93	749.6	125.0	.124	1.43902
115.0	1011.0	31.60	-2913.0	-2850.0	107.5	27.17	58.48	703.9	112.0	.117	1.42496
120.0	979.2	30.60	-2618.0	-2553.0	110.0	26.81	60.62	655.4	100.0	.111	1.40995
125.0	944.2	29.51	-2310.0	-2242.0	112.5	26.52	63.69	603.3	89.2	.104	1.39361
130.0	904.7	28.27	-1984.0	-1913.0	115.1	26.31	68.34	546.2	78.9	.0971	1.37536
132.746[b]	880.2	27.51	-1793.0	-1721.0	116.6	26.24	72.09	511.9	73.4	.0932	1.36414
132.746[b]	79.26	2.477	2150.0	2957.0	151.8	25.66	54.06	186.6	10.9	.0180	1.02971
135.0	75.93	2.373	2232.0	3075.0	152.7	25.12	50.54	190.9	11.0	.0178	1.02845
136.0	74.61	2.332	2267.0	3125.0	153.0	24.91	49.26	192.7	11.1	.0178	1.02795
138.0	72.17	2.255	2334.0	3221.0	153.8	24.53	47.08	196.1	11.2	.0177	1.02703
140.0	69.98	2.187	2399.0	3313.0	154.4	24.20	45.30	199.3	11.3	.0177	1.02620
145.0	65.31	2.041	2551.0	3531.0	155.9	23.53	41.97	206.7	11.6	.0177	1.02443
150.0	61.47	1.921	2693.0	3735.0	157.3	23.03	39.67	213.5	11.9	.0178	1.02298
155.0	58.21	1.819	2829.0	3929.0	158.6	22.64	37.99	219.7	12.3	.0179	1.02176
160.0	55.39	1.731	2960.0	4115.0	159.8	22.34	36.72	225.4	12.6	.0181	1.02069
165.0	52.90	1.653	3086.0	4296.0	160.9	22.10	35.72	230.9	12.9	.0183	1.01976
170.0	50.69	1.584	3210.0	4473.0	162.0	21.91	34.92	236.1	13.2	.0185	1.01893
175.0	48.70	1.522	3331.0	4646.0	163.0	21.76	34.27	241.0	13.5	.0188	1.01818
180.0	46.89	1.465	3451.0	4816.0	163.9	21.63	33.73	245.8	13.8	.0191	1.01750
190.0	43.72	1.366	3685.0	5148.0	165.7	21.44	32.90	254.8	14.5	.0196	1.01631
200.0	41.01	1.282	3914.0	5474.0	167.4	21.31	32.29	263.2	15.1	.0203	1.01529
210.0	38.66	1.208	4140.0	5795.0	168.9	21.22	31.84	271.2	15.7	.0209	1.01441
220.0	36.60	1.144	4363.0	6111.0	170.4	21.16	31.49	278.8	16.3	.0216	1.01364
230.0	34.77	1.087	4584.0	6425.0	171.8	21.12	31.22	286.1	16.9	.0223	1.01295
240.0	33.13	1.035	4804.0	6736.0	173.1	21.10	31.00	293.1	17.5	.0230	1.01234
250.0	31.65	.9890	5023.0	7045.0	174.4	21.09	30.83	299.9	18.1	.0237	1.01178
260.0	30.30	.9470	5241.0	7353.0	175.6	21.09	30.70	306.4	18.7	.0244	1.01128
270.0	29.08	.9087	5458.0	7659.0	176.8	21.11	30.59	312.7	19.2	.0251	1.01082
280.0	27.95	.8736	5675.0	7965.0	177.9	21.13	30.51	318.8	19.8	.0258	1.01040
300.0	25.96	.8113	6108.0	8574.0	180.0	21.19	30.41	330.6	20.9	.0272	1.00966
320.0	24.24	.7577	6542.0	9181.0	181.9	21.29	30.37	341.8	21.9	.0286	1.00902
340.0	22.75	.7110	6976.0	9789.0	183.8	21.41	30.39	352.4	23.0	.0301	1.00846
360.0	21.44	.6700	7412.0	10400.0	185.5	21.55	30.44	362.6	24.0	.0315	1.00797
380.0	20.27	.6335	7850.0	11010.0	187.2	21.70	30.53	372.4	25.0	.0329	1.00753
400.0	19.23	.6010	8290.0	11620.0	188.7	21.87	30.64	381.8	25.9	.0344	1.00714
					2.50 MPa isobar						
54.65[a]	1308.0	40.87	-6179.0	-6118.0	67.18	36.21	53.17	1134.0	493.0	.193	1.56934
80.0	1195.0	37.34	-4835.0	-4768.0	87.48	31.13	53.36	993.1	257.0	.167	1.51321
90.0	1147.0	35.85	-4302.0	-4232.0	93.78	29.73	53.83	917.4	200.0	.153	1.49004
100.0	1097.0	34.28	-3762.0	-3689.0	99.50	28.58	54.83	837.9	158.0	.139	1.46581
110.0	1043.0	32.59	-3209.0	-3132.0	104.8	27.62	56.72	753.3	126.0	.125	1.43989
115.0	1013.0	31.67	-2924.0	-2845.0	107.4	27.20	58.19	708.2	113.0	.118	1.42598
120.0	981.8	30.68	-2631.0	-2550.0	109.9	26.84	60.21	660.6	101.0	.111	1.41117
125.0	947.4	29.61	-2326.0	-2242.0	112.4	26.55	63.06	609.7	90.2	.104	1.39512
130.0	908.9	28.40	-2005.0	-1917.0	114.9	26.33	67.29	554.3	80.0	.0975	1.37731
132.0	891.9	27.87	-1870.0	-1780.0	116.0	26.27	69.62	530.5	76.0	.0948	1.36951
134.0	873.8	27.31	-1729.0	-1638.0	117.0	26.23	72.52	505.4	72.1	.0920	1.36118
136.0	854.0	26.69	-1583.0	-1489.0	118.2	26.23	76.24	478.7	68.1	.0891	1.35219
137.547[b]	837.5	26.17	-1464.0	-1369.0	119.0	26.25	79.96	456.9	65.1	.0869	1.34468
137.547[b]	102.3	3.197	2103.0	2885.0	150.0	26.46	63.98	184.7	11.7	.0202	1.03845
140.0	96.78	3.024	2207.0	3033.0	151.0	25.74	57.53	190.1	11.7	.0198	1.03635
142.0	93.01	2.907	2285.0	3145.0	151.8	25.25	53.79	194.0	11.8	.0196	1.03492
144.0	89.72	2.804	2358.0	3249.0	152.5	24.83	50.89	197.7	11.9	.0195	1.03367
146.0	86.80	2.713	2427.0	3349.0	153.2	24.47	48.57	201.1	12.0	.0194	1.03256
148.0	84.18	2.631	2493.0	3444.0	153.9	24.15	46.67	204.4	12.1	.0193	1.03157
150.0	81.79	2.556	2557.0	3535.0	154.5	23.87	45.08	207.5	12.3	.0192	1.03066

Thermophysical properties of oxygen—Continued

T K	Density kg/m^3	Density mol/dm^3	E J/mol	H J/mol	S J/(mol·K)	C_v J/(mol·K)	C_p J/(mol·K)	Sound m/s	Visc. μPa·s	Therm. W/(m·K)	Diel. const.
155.0	76.64	2.395	2709.0	3753.0	155.9	23.30	42.05	214.6	12.5	.0192	1.02871
160.0	72.36	2.261	2852.0	3957.0	157.2	22.87	39.90	221.1	12.8	.0192	1.02709
165.0	68.70	2.147	2988.0	4153.0	158.4	22.53	38.30	227.2	13.1	.0193	1.02571
170.0	65.50	2.047	3120.0	4341.0	159.5	22.26	37.06	232.8	13.4	.0194	1.02450
175.0	62.68	1.959	3248.0	4524.0	160.6	22.05	36.08	238.2	13.7	.0196	1.02344
180.0	60.15	1.880	3372.0	4702.0	161.6	21.87	35.29	243.3	14.0	.0198	1.02249
185.0	57.87	1.809	3495.0	4877.0	162.6	21.73	34.64	248.2	14.4	.0200	1.02163
190.0	55.79	1.744	3615.0	5049.0	163.5	21.62	34.10	252.8	14.7	.0203	1.02084
200.0	52.14	1.629	3851.0	5385.0	165.2	21.44	33.25	261.7	15.3	.0208	1.01947
210.0	49.01	1.532	4082.0	5715.0	166.8	21.32	32.63	270.1	15.9	.0214	1.01829
220.0	46.28	1.446	4310.0	6038.0	168.3	21.24	32.15	278.0	16.5	.0220	1.01727
230.0	43.88	1.371	4535.0	6358.0	169.7	21.19	31.78	285.5	17.1	.0227	1.01637
240.0	41.75	1.305	4758.0	6674.0	171.1	21.16	31.50	292.7	17.6	.0233	1.01556
250.0	39.83	1.245	4980.0	6988.0	172.4	21.14	31.27	299.6	18.2	.0240	1.01485
260.0	38.10	1.191	5200.0	7300.0	173.6	21.14	31.08	306.3	18.8	.0247	1.01420
270.0	36.52	1.141	5420.0	7610.0	174.8	21.15	30.94	312.7	19.3	.0254	1.01361
280.0	35.08	1.096	5639.0	7919.0	175.9	21.16	30.83	319.0	19.9	.0261	1.01307
300.0	32.54	1.017	6075.0	8534.0	178.0	21.22	30.67	330.9	21.0	.0275	1.01211
320.0	30.36	.9489	6511.0	9146.0	180.0	21.32	30.60	342.2	22.0	.0289	1.01130
340.0	28.47	.8898	6948.0	9758.0	181.8	21.43	30.58	353.0	23.0	.0303	1.01059
360.0	26.81	.8379	7386.0	10370.0	183.6	21.57	30.61	363.2	24.0	.0317	1.00997
380.0	25.34	.7920	7826.0	10980.0	185.2	21.72	30.67	373.1	25.0	.0331	1.00942
400.0	24.03	.7511	8268.0	11600.0	186.8	21.89	30.77	382.6	26.0	.0346	1.00893

3.00 MPa isobar

T K	Density kg/m^3	Density mol/dm^3	E J/mol	H J/mol	S J/(mol·K)	C_v J/(mol·K)	C_p J/(mol·K)	Sound m/s	Visc. μPa·s	Therm. W/(m·K)	Diel. const.
54.70[a]	1308.0	40.88	−6178.0	−6105.0	67.19	36.20	53.16	1135.0	495.0	.193	1.56950
80.0	1196.0	37.36	−4839.0	−4758.0	87.42	31.16	53.31	994.8	258.0	.167	1.51364
90.0	1148.0	35.89	−4307.0	−4223.0	93.72	29.77	53.76	919.5	201.0	.153	1.49056
100.0	1098.0	34.33	−3769.0	−3682.0	99.43	28.61	54.72	840.7	159.0	.139	1.46646
110.0	1045.0	32.64	−3218.0	−3126.0	104.7	27.65	56.52	757.0	127.0	.125	1.44074
115.0	1015.0	31.73	−2935.0	−2840.0	107.3	27.24	57.92	712.5	114.0	.118	1.42697
120.0	984.3	30.76	−2644.0	−2546.0	109.8	26.88	59.82	665.7	102.0	.111	1.41235
125.0	950.6	29.71	−2342.0	−2241.0	112.3	26.57	62.48	615.9	91.1	.105	1.39658
130.0	913.0	28.53	−2025.0	−1919.0	114.8	26.34	66.35	562.1	81.0	.0980	1.37919
132.0	896.5	28.02	−1892.0	−1785.0	115.8	26.28	68.43	539.1	77.1	.0953	1.37162
134.0	879.0	27.47	−1755.0	−1645.0	116.9	26.24	70.99	515.0	73.2	.0926	1.36358
136.0	860.2	26.88	−1612.0	−1500.0	117.9	26.22	74.19	489.6	69.4	.0898	1.35497
138.0	839.6	26.24	−1462.0	−1348.0	119.0	26.24	78.38	462.5	65.5	.0869	1.34563
140.0	816.8	25.53	−1303.0	−1186.0	120.2	26.30	84.14	433.3	61.5	.0840	1.33531
141.697[b]	795.0	24.84	−1159.0	−1038.0	121.3	26.41	91.15	406.1	58.0	.0815	1.32550
141.697[b]	128.1	4.003	2031.0	2780.0	148.2	27.30	78.44	182.7	12.5	.0228	1.04831
145.0	117.2	3.663	2194.0	3013.0	149.8	26.13	64.13	190.8	12.5	.0218	1.04414
146.0	114.6	3.581	2238.0	3076.0	150.3	25.84	61.34	193.0	12.5	.0217	1.04314
148.0	109.9	3.435	2320.0	3194.0	151.1	25.34	56.92	197.0	12.6	.0213	1.04136
150.0	105.9	3.309	2397.0	3304.0	151.8	24.92	53.55	200.8	12.7	.0211	1.03982
152.0	102.3	3.198	2470.0	3408.0	152.5	24.55	50.88	204.3	12.8	.0209	1.03846
154.0	99.15	3.099	2540.0	3508.0	153.1	24.23	48.71	207.6	12.9	.0208	1.03725
160.0	91.24	2.851	2733.0	3785.0	154.9	23.47	44.09	216.6	13.2	.0205	1.03424
165.0	85.95	2.686	2882.0	3999.0	156.2	23.01	41.53	223.4	13.4	.0204	1.03224
170.0	81.47	2.546	3023.0	4202.0	157.4	22.65	39.65	229.6	13.7	.0204	1.03054
175.0	77.59	2.425	3159.0	4396.0	158.6	22.36	38.22	235.4	14.0	.0205	1.02907
180.0	74.18	2.318	3290.0	4584.0	159.6	22.14	37.09	240.8	14.3	.0206	1.02778
185.0	71.14	2.223	3418.0	4767.0	160.6	21.95	36.18	246.0	14.6	.0208	1.02663
190.0	68.41	2.138	3543.0	4946.0	161.6	21.81	35.43	251.0	14.9	.0209	1.02560
195.0	65.93	2.060	3666.0	5122.0	162.5	21.68	34.81	255.7	15.2	.0212	1.02466
200.0	63.66	1.989	3787.0	5295.0	163.4	21.58	34.29	260.3	15.4	.0214	1.02380
210.0	59.65	1.864	4024.0	5633.0	165.0	21.43	33.47	269.0	16.0	.0219	1.02229
220.0	56.19	1.756	4256.0	5965.0	166.6	21.33	32.85	277.2	16.6	.0225	1.02099
230.0	53.17	1.662	4485.0	6291.0	168.0	21.26	32.38	284.9	17.2	.0231	1.01985
240.0	50.51	1.578	4712.0	6613.0	169.4	21.21	32.01	292.3	17.8	.0237	1.01885
250.0	48.12	1.504	4936.0	6931.0	170.7	21.19	31.71	299.4	18.3	.0243	1.01796

Thermophysical properties of oxygen—Continued

T K	Density kg/m³	Density mol/dm³	E J/mol	H J/mol	S J/(mol·K)	C_v J/(mol·K)	C_p J/(mol·K)	Sound m/s	Visc. μPa·s	Therm. W/(m·K)	Diel. const.
260.0	45.98	1.437	5159.0	7247.0	171.9	21.18	31.48	306.2	18.9	.0250	1.01715
270.0	44.04	1.376	5381.0	7561.0	173.1	21.18	31.29	312.8	19.4	.0257	1.01642
280.0	42.27	1.321	5602.0	7873.0	174.2	21.20	31.15	319.1	20.0	.0263	1.01576
290.0	40.65	1.270	5822.0	8184.0	175.3	21.22	31.03	325.3	20.5	.0270	1.01515
300.0	39.16	1.224	6042.0	8494.0	176.4	21.25	30.94	331.2	21.0	.0277	1.01459
320.0	36.50	1.141	6481.0	9111.0	178.4	21.34	30.82	342.7	22.1	.0291	1.01359
340.0	34.20	1.069	6920.0	9727.0	180.2	21.46	30.77	353.5	23.1	.0305	1.01273
360.0	32.19	1.006	7360.0	10340.0	182.0	21.59	30.78	363.9	24.1	.0319	1.01198
380.0	30.42	.9505	7802.0	10960.0	183.7	21.74	30.82	373.8	25.1	.0333	1.01132
400.0	28.83	.9011	8246.0	11580.0	185.3	21.91	30.90	383.3	26.0	.0348	1.01072
					3.50 MPa isobar						
54.76[a]	1308.0	40.89	−6177.0	−6092.0	67.21	36.18	53.15	1137.0	496.0	.193	1.56966
80.0	1196.0	37.39	−4843.0	−4749.0	87.37	31.19	53.26	996.6	260.0	.167	1.51406
90.0	1149.0	35.92	−4312.0	−4215.0	93.67	29.80	53.68	921.6	202.0	.153	1.49108
100.0	1100.0	34.37	−3776.0	−3674.0	99.36	28.65	54.60	843.4	160.0	.139	1.46711
110.0	1046.0	32.70	−3227.0	−3120.0	104.6	27.69	56.32	760.6	128.0	.125	1.44158
115.0	1018.0	31.80	−2945.0	−2835.0	107.2	27.28	57.65	716.7	115.0	.118	1.42795
120.0	986.8	30.84	−2656.0	−2543.0	109.7	26.91	59.45	670.7	103.0	.112	1.41352
125.0	953.6	29.80	−2357.0	−2240.0	112.1	26.60	61.93	621.9	92.0	.105	1.39800
130.0	916.9	28.65	−2044.0	−1922.0	114.6	26.37	65.49	569.5	82.0	.0985	1.38099
132.0	900.9	28.15	−1913.0	−1789.0	115.6	26.29	67.37	547.2	78.1	.0958	1.37363
134.0	884.0	27.63	−1779.0	−1652.0	116.7	26.24	69.64	524.1	74.3	.0931	1.36585
136.0	865.9	27.06	−1639.0	−1510.0	117.7	26.21	72.43	499.8	70.5	.0904	1.35757
138.0	846.3	26.45	−1494.0	−1362.0	118.8	26.21	75.98	474.1	66.8	.0876	1.34867
140.0	824.9	25.78	−1341.0	−1205.0	119.9	26.25	80.67	446.7	62.9	.0848	1.33896
142.0	800.9	25.03	−1177.0	−1038.0	121.1	26.34	87.28	416.9	59.0	.0819	1.32815
144.0	773.1	24.16	−998.6	−853.7	122.4	26.52	97.56	383.5	54.8	.0790	1.31569
145.0	757.1	23.66	−900.4	−752.4	123.1	26.65	105.4	364.9	52.6	.0775	1.30854
145.2	753.6	23.55	−879.8	−731.2	123.3	26.69	107.3	361.0	52.1	.0772	1.30701
145.3	751.9	23.50	−869.3	−720.4	123.3	26.70	108.3	358.9	51.9	.0770	1.30623
145.365[b]	750.7	23.46	−862.6	−713.4	123.4	26.72	109.0	357.7	51.8	.0769	1.30573
145.365[b]	157.9	4.936	1930.0	2639.0	146.4	28.20	101.9	180.5	13.4	.0261	1.05978
150.0	136.3	4.261	2198.0	3019.0	149.0	26.29	69.18	193.1	13.3	.0238	1.05147
152.0	130.1	4.066	2290.0	3151.0	149.9	25.72	62.91	197.5	13.3	.0233	1.04907
154.0	124.8	3.901	2375.0	3272.0	150.7	25.24	58.35	201.5	13.4	.0229	1.04705
156.0	120.3	3.759	2454.0	3385.0	151.4	24.84	54.86	205.2	13.4	.0226	1.04531
158.0	116.2	3.633	2529.0	3492.0	152.1	24.48	52.10	208.7	13.5	.0223	1.04377
160.0	112.6	3.520	2600.0	3594.0	152.7	24.17	49.85	211.9	13.6	.0221	1.04239
162.0	109.4	3.418	2668.0	3692.0	153.3	23.90	47.97	215.1	13.6	.0220	1.04115
164.0	106.4	3.326	2734.0	3786.0	153.9	23.66	46.39	218.0	13.7	.0218	1.04001
170.0	98.81	3.088	2919.0	4053.0	155.5	23.08	42.83	226.3	14.0	.0216	1.03712
175.0	93.58	2.925	3065.0	4262.0	156.7	22.71	40.74	232.6	14.3	.0215	1.03513
180.0	89.07	2.784	3204.0	4461.0	157.8	22.42	39.16	238.4	14.5	.0215	1.03342
185.0	85.12	2.660	3338.0	4654.0	158.9	22.19	37.91	244.0	14.8	.0216	1.03192
190.0	81.61	2.550	3468.0	4841.0	159.9	22.01	36.92	249.2	15.1	.0217	1.03059
195.0	78.45	2.452	3596.0	5023.0	160.8	21.85	36.10	254.2	15.4	.0218	1.02939
200.0	75.60	2.362	3720.0	5202.0	161.8	21.73	35.42	259.0	15.7	.0220	1.02831
210.0	70.59	2.206	3964.0	5551.0	163.5	21.54	34.37	268.0	16.2	.0224	1.02642
220.0	66.34	2.073	4202.0	5890.0	165.0	21.42	33.59	276.4	16.8	.0229	1.02481
230.0	62.65	1.958	4435.0	6223.0	166.5	21.33	33.00	284.4	17.3	.0235	1.02342
240.0	59.40	1.856	4665.0	6550.0	167.9	21.27	32.54	292.0	17.9	.0241	1.02220
250.0	56.53	1.767	4893.0	6874.0	169.2	21.24	32.17	299.3	18.5	.0247	1.02111
260.0	53.95	1.686	5118.0	7194.0	170.5	21.22	31.89	306.2	19.0	.0253	1.02014
270.0	51.62	1.613	5342.0	7512.0	171.7	21.22	31.65	312.9	19.5	.0260	1.01927
280.0	49.51	1.547	5565.0	7827.0	172.8	21.23	31.47	319.3	20.1	.0266	1.01847
290.0	47.58	1.487	5788.0	8141.0	173.9	21.25	31.32	325.6	20.6	.0273	1.01775
300.0	45.81	1.432	6009.0	8454.0	175.0	21.28	31.20	331.6	21.1	.0280	1.01708
320.0	42.66	1.333	6451.0	9076.0	177.0	21.37	31.05	343.2	22.2	.0293	1.01590
340.0	39.95	1.248	6893.0	9696.0	178.9	21.48	30.96	354.1	23.2	.0307	1.01488
360.0	37.58	1.174	7335.0	10320.0	180.7	21.61	30.94	364.6	24.2	.0321	1.01399
380.0	35.49	1.109	7778.0	10930.0	182.3	21.76	30.97	374.5	25.1	.0335	1.01321

Thermophysical properties of oxygen—Continued

T K	Density kg/m^3	Density mol/dm^3	E J/mol	H J/mol	S J/(mol·K)	C_v J/(mol·K)	C_p J/(mol·K)	Sound m/s	Visc. μPa·s	Therm. W/(m·K)	Diel. const.
400.0	33.63	1.051	8224.0	11550.0	183.9	21.93	31.03	384.1	26.1	.0349	1.01251

4.00 MPa isobar

T K	Density kg/m^3	Density mol/dm^3	E J/mol	H J/mol	S J/(mol·K)	C_v J/(mol·K)	C_p J/(mol·K)	Sound m/s	Visc. μPa·s	Therm. W/(m·K)	Diel. const.
54.82[a]	1309.0	40.90	−6176.0	−6079.0	67.22	36.16	53.13	1138.0	498.0	.193	1.56982
80.0	1197.0	37.42	−4847.0	−4740.0	87.32	31.22	53.22	998.3	261.0	.167	1.51448
90.0	1150.0	35.95	−4317.0	−4206.0	93.61	29.84	53.61	923.7	203.0	.153	1.49159
100.0	1101.0	34.41	−3782.0	−3666.0	99.30	28.69	54.49	846.1	161.0	.139	1.46775
110.0	1048.0	32.75	−3236.0	−3114.0	104.6	27.73	56.14	764.2	129.0	.125	1.44241
115.0	1020.0	31.86	−2955.0	−2830.0	107.1	27.31	57.40	720.9	116.0	.118	1.42891
120.0	989.3	30.92	−2668.0	−2539.0	109.6	26.95	59.10	675.6	104.0	.112	1.41466
125.0	956.6	29.89	−2372.0	−2238.0	112.0	26.63	61.42	627.8	92.9	.105	1.39938
130.0	920.7	28.77	−2062.0	−1923.0	114.5	26.39	64.70	576.7	82.9	.0989	1.38272
132.0	905.1	28.29	−1933.0	−1792.0	115.5	26.31	66.41	555.1	79.1	.0963	1.37555
134.0	888.7	27.77	−1801.0	−1657.0	116.5	26.25	68.44	532.7	75.4	.0936	1.36801
136.0	871.2	27.23	−1665.0	−1518.0	117.5	26.21	70.91	509.4	71.7	.0910	1.36002
138.0	852.5	26.64	−1523.0	−1373.0	118.6	26.20	73.96	484.9	68.0	.0883	1.35150
140.0	832.2	26.01	−1375.0	−1222.0	119.7	26.22	77.88	459.1	64.3	.0855	1.34229
142.0	809.9	25.31	−1219.0	−1061.0	120.8	26.28	83.13	431.3	60.5	.0827	1.33219
144.0	784.6	24.52	−1051.0	−887.5	122.0	26.40	90.70	401.0	56.6	.0799	1.32085
145.0	770.5	24.08	−960.4	−794.3	122.7	26.49	95.97	384.6	54.6	.0785	1.31453
146.0	755.0	23.59	−864.5	−695.0	123.4	26.62	102.9	366.9	52.4	.0771	1.30762
147.0	737.6	23.05	−761.1	−587.5	124.1	26.79	112.6	347.6	50.1	.0756	1.29990
148.0	717.4	22.42	−646.4	−468.0	124.9	27.02	127.6	326.0	47.6	.0742	1.29098
148.5	705.8	22.06	−582.9	−401.6	125.3	27.18	138.8	314.0	46.3	.0735	1.28586
148.6	703.3	21.98	−569.6	−387.6	125.4	27.21	141.5	311.4	46.0	.0733	1.28476
148.659[b]	701.8	21.93	−561.5	−379.1	125.5	27.24	143.2	309.9	45.8	.0732	1.28409
148.659[b]	194.3	6.073	1789.0	2448.0	144.5	29.21	147.5	178.2	14.5	.0308	1.07389
150.0	181.0	5.656	1912.0	2619.0	145.7	28.30	112.6	183.3	14.3	.0288	1.06870
151.0	173.6	5.424	1987.0	2724.0	146.4	27.77	98.56	186.5	14.2	.0279	1.06582
152.0	167.4	5.231	2053.0	2818.0	147.0	27.31	88.89	189.4	14.1	.0271	1.06343
153.0	162.1	5.066	2113.0	2903.0	147.5	26.91	81.77	192.0	14.1	.0265	1.06139
154.0	157.5	4.922	2169.0	2982.0	148.0	26.56	76.25	194.5	14.1	.0260	1.05960
156.0	149.6	4.676	2270.0	3126.0	149.0	25.95	68.19	199.1	14.0	.0253	1.05657
158.0	143.1	4.473	2362.0	3256.0	149.8	25.44	62.53	203.2	14.1	.0247	1.05407
160.0	137.6	4.299	2446.0	3377.0	150.6	25.01	58.30	207.0	14.1	.0242	1.05193
162.0	132.7	4.148	2525.0	3490.0	151.3	24.64	55.01	210.6	14.1	.0238	1.05007
164.0	128.4	4.013	2600.0	3597.0	151.9	24.32	52.36	213.9	14.2	.0235	1.04842
166.0	124.5	3.892	2672.0	3700.0	152.6	24.03	50.19	217.1	14.3	.0233	1.04694
168.0	121.0	3.783	2741.0	3798.0	153.1	23.78	48.36	220.1	14.3	.0231	1.04559
170.0	117.8	3.682	2807.0	3893.0	153.7	23.56	46.81	223.1	14.4	.0229	1.04437
175.0	110.8	3.463	2964.0	4119.0	155.0	23.09	43.78	229.9	14.6	.0226	1.04169
180.0	104.9	3.280	3113.0	4332.0	156.2	22.73	41.57	236.1	14.9	.0225	1.03945
185.0	99.88	3.121	3254.0	4536.0	157.3	22.44	39.89	242.0	15.1	.0224	1.03752
190.0	95.44	2.983	3391.0	4732.0	158.4	22.22	38.57	247.5	15.4	.0224	1.03583
195.0	91.51	2.860	3523.0	4922.0	159.4	22.03	37.51	252.7	15.6	.0225	1.03434
200.0	87.98	2.749	3652.0	5107.0	160.3	21.88	36.65	257.7	15.9	.0226	1.03300
205.0	84.78	2.649	3779.0	5289.0	161.2	21.76	35.93	262.5	16.1	.0228	1.03178
210.0	81.86	2.558	3903.0	5467.0	162.1	21.66	35.32	267.1	16.4	.0230	1.03068
220.0	76.72	2.397	4146.0	5815.0	163.7	21.50	34.36	275.8	17.0	.0234	1.02873
230.0	72.30	2.259	4384.0	6155.0	165.2	21.40	33.64	284.0	17.5	.0239	1.02706
240.0	68.44	2.139	4618.0	6488.0	166.6	21.33	33.08	291.8	18.1	.0245	1.02560
250.0	65.04	2.033	4849.0	6817.0	167.9	21.29	32.64	299.2	18.6	.0251	1.02432
260.0	62.00	1.938	5077.0	7141.0	169.2	21.27	32.30	306.3	19.1	.0257	1.02317
270.0	59.27	1.852	5304.0	7463.0	170.4	21.26	32.02	313.0	19.7	.0263	1.02214
280.0	56.80	1.775	5529.0	7782.0	171.6	21.27	31.80	319.6	20.2	.0269	1.02121
290.0	54.55	1.705	5753.0	8099.0	172.7	21.28	31.62	325.9	20.7	.0276	1.02037
300.0	52.49	1.640	5976.0	8414.0	173.8	21.31	31.47	332.0	21.2	.0282	1.01959
320.0	48.84	1.526	6421.0	9042.0	175.8	21.39	31.27	343.7	22.3	.0296	1.01822
340.0	45.70	1.428	6865.0	9666.0	177.7	21.50	31.16	354.7	23.3	.0309	1.01704
360.0	42.96	1.343	7309.0	10290.0	179.5	21.63	31.11	365.3	24.2	.0323	1.01601
380.0	40.56	1.268	7755.0	10910.0	181.2	21.78	31.12	375.3	25.2	.0337	1.01511

Thermophysical properties of oxygen—Continued

T K	Density kg/m³	Density mol/dm³	E J/mol	H J/mol	S J/(mol·K)	C_v J/(mol·K)	C_p J/(mol·K)	Sound m/s	Visc. μPa·s	Therm. W/(m·K)	Diel. const.
400.0	38.42	1.201	8202.0	11530.0	182.8	21.95	31.16	384.9	26.2	.0351	1.01431
					5.00 MPa isobar						
54.93[a]	1309.0	40.92	-6175.0	-6053.0	67.25	36.14	53.11	1140.0	501.0	.193	1.57014
80.0	1199.0	37.47	-4855.0	-4721.0	87.22	31.28	53.12	1002.0	263.0	.167	1.51531
90.0	1153.0	36.02	-4327.0	-4189.0	93.49	29.91	53.48	928.0	205.0	.154	1.49261
100.0	1104.0	34.49	-3795.0	-3650.0	99.16	28.76	54.28	851.4	163.0	.139	1.46901
110.0	1051.0	32.86	-3253.0	-3101.0	104.4	27.80	55.78	771.1	130.0	.126	1.44403
120.0	994.0	31.06	-2692.0	-2531.0	109.4	27.01	58.45	685.0	105.0	.112	1.41686
125.0	962.3	30.07	-2400.0	-2234.0	111.8	26.69	60.50	639.0	94.6	.106	1.40204
130.0	927.8	28.99	-2097.0	-1925.0	114.2	26.43	63.31	590.3	84.7	.0997	1.38601
135.0	889.4	27.79	-1778.0	-1599.0	116.7	26.25	67.34	538.1	75.5	.0933	1.36833
136.0	881.1	27.54	-1712.0	-1531.0	117.2	26.22	68.37	527.2	73.8	.0920	1.36454
138.0	863.8	26.99	-1577.0	-1392.0	118.2	26.19	70.72	504.7	70.2	.0895	1.35663
140.0	845.3	26.42	-1437.0	-1248.0	119.2	26.17	73.59	481.2	66.7	.0868	1.34821
142.0	825.4	25.79	-1291.0	-1097.0	120.3	26.19	77.21	456.5	63.2	.0842	1.33917
144.0	803.5	25.11	-1137.0	-937.9	121.4	26.25	81.94	430.4	59.6	.0816	1.32933
146.0	779.2	24.35	-973.2	-767.9	122.6	26.36	88.47	402.3	55.9	.0789	1.31842
148.0	751.3	23.48	-794.9	-581.9	123.8	26.54	98.26	371.4	52.1	.0763	1.30597
149.0	735.4	22.98	-697.9	-480.3	124.5	26.68	105.4	354.5	50.0	.0750	1.29891
150.0	717.6	22.43	-593.3	-370.3	125.3	26.86	115.1	336.3	47.9	.0737	1.29105
151.0	697.1	21.79	-478.0	-248.5	126.1	27.10	129.7	316.2	45.5	.0723	1.28204
151.5	685.5	21.42	-414.6	-181.1	126.5	27.25	140.1	305.2	44.2	.0717	1.27693
152.0	672.4	21.01	-345.7	-107.8	127.0	27.44	154.1	293.3	42.8	.0710	1.27123
152.5	657.5	20.55	-269.4	-26.0	127.5	27.67	174.4	280.3	41.3	.0703	1.26472
153.0	639.7	19.99	-181.5	68.6	128.2	27.97	206.8	265.7	39.5	.0696	1.25699
153.2	631.4	19.73	-141.6	111.8	128.4	28.12	226.3	259.2	38.7	.0694	1.25339
153.4	622.0	19.44	-97.7	159.5	128.7	28.30	252.4	252.2	37.9	.0692	1.24936
153.6	611.3	19.10	-48.3	213.5	129.1	28.50	289.6	244.6	36.9	.0691	1.24473
153.8	598.5	18.70	9.4	276.7	129.5	28.76	348.1	236.0	35.8	.0691	1.23923
153.9	590.9	18.47	42.9	313.7	129.7	28.92	392.4	231.3	35.1	.0692	1.23598
154.0	582.2	18.19	81.0	355.9	130.0	29.10	455.9	226.0	34.4	.0694	1.23224
154.1	571.7	17.87	126.2	406.0	130.3	29.33	555.9	220.2	33.6	.0700	1.22777
154.2	558.2	17.45	183.2	469.8	130.8	29.62	741.7	213.3	32.5	.0711	1.22203
154.3	538.2	16.82	266.6	563.9	131.4	30.05	1246.0	204.4	31.0	.0716	1.21354
154.361[b]	509.1	15.91	386.0	700.3	132.3	30.67	3746.0	194.2	28.9	.0804	1.20126
154.361[b]	367.8	11.49	1003.0	1438.0	137.0	32.06	5610.0	175.7	21.1	.0767	1.14288
155.0	287.3	8.978	1435.0	1992.0	140.6	30.71	380.0	178.9	17.9	.0450	1.11051
155.2	278.8	8.714	1488.0	2061.0	141.1	30.45	318.0	180.0	17.6	.0431	1.10713
155.4	271.9	8.496	1532.0	2121.0	141.4	30.22	276.8	180.9	17.4	.0417	1.10436
155.6	265.9	8.310	1571.0	2173.0	141.8	30.01	247.1	181.8	17.2	.0405	1.10200
155.8	260.7	8.147	1606.0	2220.0	142.1	29.82	224.5	182.7	17.1	.0395	1.09994
156.0	256.1	8.002	1638.0	2263.0	142.4	29.63	206.7	183.5	16.9	.0387	1.09811
156.5	246.3	7.697	1708.0	2358.0	143.0	29.23	175.0	185.5	16.7	.0370	1.09424
157.0	238.3	7.447	1768.0	2440.0	143.5	28.87	153.8	187.3	16.5	.0357	1.09109
157.5	231.5	7.235	1822.0	2513.0	144.0	28.55	138.5	189.0	16.3	.0347	1.08842
158.0	225.6	7.050	1870.0	2579.0	144.4	28.27	126.8	190.6	16.2	.0338	1.08611
159.0	215.7	6.740	1955.0	2697.0	145.1	27.76	110.1	193.5	16.0	.0324	1.08222
160.0	207.5	6.485	2030.0	2801.0	145.8	27.32	98.64	196.2	15.8	.0314	1.07902
161.0	200.5	6.267	2097.0	2895.0	146.4	26.93	90.16	198.7	15.7	.0305	1.07630
162.0	194.5	6.078	2159.0	2982.0	146.9	26.58	83.61	201.1	15.6	.0298	1.07394
164.0	184.3	5.759	2271.0	3139.0	147.9	25.99	74.09	205.5	15.5	.0286	1.06997
166.0	175.9	5.497	2371.0	3280.0	148.7	25.49	67.45	209.6	15.4	.0277	1.06672
168.0	168.8	5.275	2462.0	3410.0	149.5	25.06	62.52	213.3	15.4	.0271	1.06397
170.0	162.6	5.082	2547.0	3531.0	150.2	24.70	58.70	216.8	15.4	.0265	1.06158
172.0	157.2	4.912	2627.0	3645.0	150.9	24.38	55.65	220.1	15.4	.0260	1.05947
174.0	152.3	4.759	2703.0	3754.0	151.5	24.09	53.14	223.3	15.5	.0257	1.05759
176.0	147.9	4.622	2776.0	3858.0	152.1	23.84	51.05	226.3	15.5	.0253	1.05589
178.0	143.9	4.496	2846.0	3958.0	152.7	23.62	49.28	229.2	15.6	.0251	1.05435
180.0	140.2	4.381	2914.0	4055.0	153.2	23.42	47.75	231.9	15.6	.0249	1.05293
185.0	132.1	4.128	3075.0	4286.0	154.5	23.00	44.73	238.5	15.8	.0245	1.04982
190.0	125.2	3.914	3226.0	4504.0	155.6	22.67	42.50	244.6	16.0	.0242	1.04720

Thermophysical properties of oxygen—Continued

T K	Density kg/m³	Density mol/dm³	E J/mol	H J/mol	S J/(mol·K)	C_v J/(mol·K)	C_p J/(mol·K)	Sound m/s	Visc. μPa·s	Therm. W/(m·K)	Diel. const.
195.0	119.3	3.729	3371.0	4712.0	156.7	22.41	40.78	250.3	16.2	.0241	1.04494
200.0	114.2	3.567	3511.0	4912.0	157.7	22.20	39.42	255.6	16.4	.0241	1.04296
205.0	109.5	3.423	3646.0	5107.0	158.7	22.03	38.32	260.8	16.6	.0241	1.04119
210.0	105.4	3.294	3778.0	5296.0	159.6	21.89	37.41	265.7	16.9	.0242	1.03961
215.0	101.6	3.176	3907.0	5481.0	160.5	21.78	36.65	270.4	17.1	.0243	1.03818
220.0	98.20	3.069	4033.0	5663.0	161.3	21.69	36.01	274.9	17.4	.0245	1.03688
230.0	92.14	2.880	4281.0	6017.0	162.9	21.54	34.99	283.5	17.9	.0249	1.03457
240.0	86.93	2.717	4523.0	6363.0	164.4	21.45	34.21	291.6	18.4	.0253	1.03259
250.0	82.38	2.575	4760.0	6702.0	165.7	21.39	33.62	299.2	18.9	.0258	1.03087
260.0	78.36	2.449	4994.0	7036.0	167.1	21.35	33.14	306.5	19.4	.0264	1.02935
270.0	74.77	2.337	5226.0	7365.0	168.3	21.34	32.76	313.5	19.9	.0269	1.02799
280.0	71.54	2.236	5455.0	7691.0	169.5	21.33	32.46	320.2	20.5	.0275	1.02677
290.0	68.61	2.144	5683.0	8015.0	170.6	21.35	32.21	326.7	21.0	.0281	1.02566
300.0	65.95	2.061	5910.0	8336.0	171.7	21.37	32.01	333.0	21.5	.0288	1.02465
310.0	63.50	1.984	6135.0	8655.0	172.8	21.40	31.85	339.0	22.0	.0294	1.02373
320.0	61.25	1.914	6361.0	8973.0	173.8	21.44	31.72	344.8	22.5	.0301	1.02288
340.0	57.23	1.788	6810.0	9605.0	175.7	21.55	31.54	356.1	23.4	.0314	1.02137
360.0	53.75	1.680	7258.0	10240.0	177.5	21.68	31.44	366.7	24.4	.0327	1.02006
380.0	50.70	1.584	7708.0	10860.0	179.2	21.82	31.41	376.8	25.4	.0341	1.01891
400.0	47.99	1.500	8158.0	11490.0	180.8	21.99	31.42	386.5	26.3	.0355	1.01789
				5.04 MPa isobar							
54.94[a]	1309.0	40.92	−6175.0	−6052.0	67.25	36.13	53.11	1140.0	501.0	.193	1.57016
80.0	1199.0	37.47	−4855.0	−4721.0	87.21	31.28	53.12	1002.0	263.0	.167	1.51534
90.0	1153.0	36.02	−4328.0	−4188.0	93.49	29.91	53.47	928.1	205.0	.154	1.49265
100.0	1104.0	34.49	−3796.0	−3650.0	99.16	28.76	54.27	851.6	163.0	.139	1.46906
110.0	1052.0	32.86	−3254.0	−3100.0	104.4	27.80	55.77	771.4	130.0	.126	1.44409
120.0	994.1	31.07	−2693.0	−2530.0	109.3	27.02	58.43	685.4	105.0	.112	1.41695
125.0	962.5	30.08	−2401.0	−2234.0	111.8	26.69	60.46	639.4	94.6	.106	1.40214
130.0	928.0	29.00	−2098.0	−1925.0	114.2	26.43	63.26	590.8	84.8	.0997	1.38614
135.0	889.7	27.81	−1780.0	−1599.0	116.7	26.25	67.26	538.8	75.6	.0934	1.36850
136.0	881.5	27.55	−1714.0	−1531.0	117.2	26.22	68.28	527.8	73.8	.0921	1.36471
138.0	864.2	27.01	−1579.0	−1392.0	118.2	26.19	70.60	505.4	70.3	.0895	1.35682
140.0	845.8	26.43	−1439.0	−1248.0	119.2	26.17	73.45	482.0	66.8	.0869	1.34843
142.0	825.9	25.81	−1293.0	−1098.0	120.3	26.19	77.02	457.5	63.3	.0843	1.33942
144.0	804.2	25.13	−1140.0	−939.6	121.4	26.24	81.68	431.4	59.7	.0816	1.32963
146.0	780.0	24.38	−977.0	−770.2	122.5	26.35	88.07	403.5	56.1	.0790	1.31879
148.0	752.4	23.51	−799.6	−585.3	123.8	26.53	97.60	372.9	52.2	.0764	1.30644
149.0	736.6	23.02	−703.3	−484.4	124.5	26.66	104.5	356.2	50.2	.0751	1.29946
150.0	719.1	22.47	−599.8	−375.5	125.2	26.83	113.8	338.2	48.0	.0738	1.29171
151.0	699.0	21.85	−486.1	−255.3	126.0	27.06	127.5	318.4	45.7	.0724	1.28287
151.5	687.6	21.49	−423.8	−189.3	126.4	27.21	137.2	307.6	44.4	.0718	1.27788
152.0	675.0	21.09	−356.5	−117.6	126.9	27.39	150.1	296.1	43.1	.0711	1.27235
152.5	660.6	20.65	−282.4	−38.3	127.4	27.61	168.3	283.4	41.6	.0704	1.26608
153.0	643.7	20.12	−198.2	52.3	128.0	27.89	196.3	269.4	39.9	.0697	1.25874
153.2	635.9	19.87	−160.4	93.1	128.3	28.02	212.5	263.3	39.2	.0695	1.25537
153.4	627.3	19.61	−119.4	137.6	128.6	28.18	233.4	256.7	38.4	.0692	1.25165
153.6	617.6	19.30	−74.1	187.0	128.9	28.36	261.9	249.6	37.5	.0690	1.24746
153.8	606.4	18.95	−22.7	243.2	129.3	28.58	303.1	241.8	36.5	.0689	1.24263
154.0	592.9	18.53	37.9	309.9	129.7	28.86	369.3	233.1	35.3	.0690	1.23682
154.1	584.8	18.28	73.5	349.2	130.0	29.02	420.7	228.2	34.6	.0691	1.23336
154.2	575.3	17.98	114.5	394.8	130.3	29.22	496.2	222.8	33.9	.0695	1.22931
154.3	563.7	17.62	164.0	450.1	130.6	29.47	619.9	216.7	32.9	.0702	1.22438
154.4	548.4	17.14	228.6	522.7	131.1	29.80	866.4	209.4	31.8	.0717	1.21786
154.5	523.6	16.36	331.3	639.3	131.8	30.33	1670.0	199.6	30.0	.0770	1.20738
154.566[b]	485.7	15.18	487.8	819.9	133.0	31.07	6894.0	188.8	27.4	.115	1.19147
154.566[b]	389.7	12.18	905.9	1320.0	136.2	32.04	9302.0	177.0	22.1	.123	1.15180
155.0	305.4	9.546	1338.0	1866.0	139.8	31.10	532.3	178.0	18.6	.0491	1.11775
155.1	298.9	9.342	1376.0	1915.0	140.1	30.94	459.2	178.5	18.3	.0474	1.11514
155.2	293.4	9.168	1409.0	1958.0	140.4	30.79	406.9	179.0	18.1	.0461	1.11292
155.4	284.2	8.880	1465.0	2032.0	140.8	30.53	336.1	180.0	17.8	.0440	1.10925
155.6	276.7	8.646	1512.0	2095.0	141.2	30.29	290.1	181.0	17.6	.0424	1.10628

Thermophysical properties of oxygen—Continued

T K	Density kg/m^3	Density mol/dm^3	E J/mol	H J/mol	S J/(mol·K)	C_v J/(mol·K)	C_p J/(mol·K)	Sound m/s	Visc. μPa·s	Therm. W/(m·K)	Diel. const.
155.8	270.3	8.449	1553.0	2149.0	141.6	30.08	257.4	181.9	17.4	.0412	1.10377
156.0	264.8	8.277	1589.0	2198.0	141.9	29.88	232.8	182.8	17.2	.0401	1.10158
156.5	253.6	7.924	1667.0	2303.0	142.6	29.44	191.4	184.8	16.9	.0381	1.09712
157.0	244.6	7.643	1733.0	2392.0	143.2	29.06	165.3	186.6	16.7	.0366	1.09357
157.5	237.1	7.410	1790.0	2470.0	143.6	28.72	147.0	188.4	16.5	.0355	1.09063
158.0	230.7	7.209	1841.0	2540.0	144.1	28.42	133.5	190.0	16.3	.0345	1.08810
159.0	220.0	6.875	1930.0	2663.0	144.9	27.89	114.6	193.0	16.1	.0330	1.08391
160.0	211.3	6.604	2008.0	2771.0	145.5	27.43	101.9	195.8	15.9	.0318	1.08051
161.0	204.0	6.375	2077.0	2868.0	146.2	27.03	92.65	198.3	15.8	.0309	1.07764
162.0	197.6	6.176	2141.0	2957.0	146.7	26.68	85.60	200.7	15.7	.0301	1.07517
163.0	192.0	6.001	2200.0	3040.0	147.2	26.36	80.01	203.0	15.6	.0295	1.07298
164.0	187.0	5.844	2255.0	3117.0	147.7	26.07	75.46	205.2	15.6	.0289	1.07103
166.0	178.3	5.573	2357.0	3261.0	148.6	25.56	68.47	209.3	15.5	.0280	1.06766
168.0	171.0	5.344	2449.0	3392.0	149.3	25.12	63.31	213.0	15.5	.0273	1.06482
170.0	164.7	5.146	2536.0	3515.0	150.1	24.75	59.34	216.6	15.5	.0267	1.06237
172.0	159.1	4.971	2616.0	3630.0	150.7	24.42	56.18	219.9	15.5	.0262	1.06021
174.0	154.1	4.815	2693.0	3740.0	151.4	24.14	53.59	223.1	15.5	.0258	1.05828
176.0	149.6	4.674	2767.0	3845.0	152.0	23.88	51.44	226.1	15.5	.0255	1.05654
178.0	145.5	4.546	2837.0	3946.0	152.5	23.65	49.61	229.0	15.6	.0252	1.05496
180.0	141.7	4.428	2905.0	4044.0	153.1	23.45	48.04	231.8	15.6	.0250	1.05351
185.0	133.5	4.171	3067.0	4276.0	154.4	23.03	44.95	238.4	15.8	.0245	1.05034
190.0	126.5	3.953	3220.0	4495.0	155.5	22.69	42.67	244.5	16.0	.0243	1.04768
195.0	120.5	3.766	3365.0	4703.0	156.6	22.43	40.92	250.2	16.2	.0242	1.04538
200.0	115.2	3.601	3505.0	4904.0	157.6	22.22	39.54	255.6	16.4	.0241	1.04337
205.0	110.6	3.455	3640.0	5099.0	158.6	22.04	38.42	260.7	16.7	.0241	1.04159
210.0	106.4	3.324	3773.0	5289.0	159.5	21.90	37.50	265.6	16.9	.0242	1.03998
215.0	102.6	3.205	3902.0	5474.0	160.4	21.79	36.73	270.3	17.1	.0243	1.03853
220.0	99.08	3.096	4029.0	5656.0	161.2	21.69	36.08	274.9	17.4	.0245	1.03721
230.0	92.95	2.905	4277.0	6012.0	162.8	21.55	35.04	283.5	17.9	.0249	1.03488
240.0	87.68	2.740	4519.0	6358.0	164.3	21.45	34.26	291.6	18.4	.0253	1.03288
250.0	83.08	2.596	4757.0	6698.0	165.7	21.39	33.65	299.2	18.9	.0258	1.03114
260.0	79.02	2.470	4991.0	7032.0	167.0	21.36	33.18	306.6	19.4	.0264	1.02960
270.0	75.40	2.356	5222.0	7361.0	168.2	21.34	32.79	313.6	20.0	.0270	1.02822
280.0	72.14	2.254	5452.0	7688.0	169.4	21.34	32.49	320.3	20.5	.0276	1.02699
290.0	69.18	2.162	5680.0	8011.0	170.5	21.35	32.24	326.7	21.0	.0282	1.02587
300.0	66.49	2.078	5907.0	8333.0	171.6	21.37	32.03	333.0	21.5	.0288	1.02486
310.0	64.02	2.001	6133.0	8652.0	172.7	21.40	31.87	339.0	22.0	.0294	1.02392
320.0	61.74	1.930	6358.0	8970.0	173.7	21.45	31.74	344.9	22.5	.0301	1.02307
340.0	57.69	1.803	6808.0	9603.0	175.6	21.55	31.56	356.1	23.5	.0314	1.02154
360.0	54.18	1.693	7256.0	10230.0	177.4	21.68	31.46	366.8	24.4	.0328	1.02022
380.0	51.10	1.597	7706.0	10860.0	179.1	21.83	31.42	376.9	25.4	.0341	1.01906
400.0	48.38	1.512	8156.0	11490.0	180.7	21.99	31.43	386.6	26.3	.0355	1.01804
					5.20 MPa isobar						
54.95[a]	1310.0	40.92	-6174.0	-6047.0	67.26	36.13	53.11	1141.0	501.0	.193	1.57021
80.0	1199.0	37.48	-4856.0	-4718.0	87.20	31.29	53.11	1002.0	263.0	.167	1.51548
90.0	1153.0	36.03	-4329.0	-4185.0	93.47	29.92	53.45	928.8	206.0	.154	1.49281
100.0	1104.0	34.51	-3798.0	-3647.0	99.14	28.77	54.24	852.5	163.0	.140	1.46926
110.0	1052.0	32.88	-3256.0	-3098.0	104.4	27.81	55.71	772.5	131.0	.126	1.44435
120.0	994.9	31.09	-2696.0	-2529.0	109.3	27.03	58.33	686.9	106.0	.113	1.41729
125.0	963.4	30.11	-2405.0	-2233.0	111.7	26.70	60.33	641.2	94.9	.106	1.40255
130.0	929.1	29.04	-2104.0	-1925.0	114.2	26.44	63.06	592.9	85.1	.0999	1.38664
135.0	891.2	27.85	-1787.0	-1600.0	116.6	26.25	66.94	541.3	75.9	.0935	1.36914
136.0	883.0	27.59	-1721.0	-1533.0	117.1	26.22	67.92	530.5	74.2	.0922	1.36539
138.0	865.9	27.06	-1587.0	-1395.0	118.1	26.19	70.17	508.3	70.6	.0897	1.35758
140.0	847.7	26.49	-1448.0	-1252.0	119.1	26.17	72.89	485.3	67.1	.0871	1.34930
142.0	828.1	25.88	-1304.0	-1103.0	120.2	26.18	76.28	461.1	63.7	.0845	1.34043
144.0	806.8	25.21	-1152.0	-946.1	121.3	26.23	80.65	435.6	60.1	.0819	1.33082
146.0	783.3	24.48	-991.6	-779.2	122.4	26.32	86.56	408.3	56.6	.0793	1.32023
148.0	756.5	23.64	-818.0	-598.1	123.7	26.48	95.15	378.7	52.8	.0767	1.30827
149.0	741.4	23.17	-724.5	-500.0	124.3	26.60	101.2	362.6	50.8	.0754	1.30158

Thermophysical properties of oxygen—Continued

T K	Density kg/m^3	Density mol/dm^3	E J/mol	H J/mol	S J/(mol·K)	C_v J/(mol·K)	C_p J/(mol·K)	Sound m/s	Visc. μPa·s	Therm. W/(m·K)	Diel. const.
150.0	724.8	22.65	-624.7	-395.1	125.0	26.75	109.1	345.5	48.8	.0741	1.29422
151.0	706.1	22.07	-516.4	-280.7	125.8	26.95	120.3	326.9	46.5	.0728	1.28597
152.0	684.3	21.38	-395.8	-152.6	126.6	27.22	137.3	306.3	44.1	.0715	1.27640
152.5	671.7	20.99	-328.7	-81.0	127.1	27.40	149.9	294.9	42.8	.0708	1.27089
153.0	657.4	20.54	-255.1	-2.0	127.6	27.61	167.3	282.7	41.3	.0701	1.26468
153.5	640.8	20.02	-171.9	87.7	128.2	27.88	193.5	269.1	39.7	.0694	1.25745
153.6	637.0	19.91	-153.8	107.4	128.3	27.95	200.4	266.2	39.3	.0693	1.25584
153.8	629.1	19.66	-115.4	149.1	128.6	28.08	216.8	260.2	38.6	.0690	1.25241
154.0	620.3	19.39	-73.8	194.5	128.9	28.24	237.9	253.7	37.8	.0688	1.24862
154.2	610.5	19.08	-27.8	244.7	129.2	28.43	266.2	246.9	36.9	.0686	1.24438
154.4	599.1	18.72	24.0	301.7	129.6	28.65	306.1	239.4	35.9	.0684	1.23951
154.6	585.7	18.30	84.4	368.5	130.0	28.91	367.0	231.1	34.8	.0684	1.23373
154.8	568.7	17.77	158.8	451.4	130.6	29.26	472.5	221.7	33.4	.0686	1.22648
154.9	558.0	17.44	204.6	502.8	130.9	29.48	560.5	216.4	32.6	.0690	1.22195
155.0	545.1	17.03	259.9	565.1	131.3	29.75	698.0	210.6	31.6	.0700	1.21645
155.1	528.3	16.51	330.7	645.7	131.8	30.09	936.9	203.9	30.4	.0719	1.20934
155.2	504.5	15.77	430.1	759.9	132.6	30.55	1403.0	196.3	28.7	.0759	1.19935
155.3	468.3	14.64	583.0	938.3	133.7	31.16	2187.0	188.0	26.4	.0835	1.18422
155.4	422.9	13.22	781.6	1175.0	135.2	31.66	2336.0	181.9	23.9	.0831	1.16544
155.5	387.0	12.09	948.0	1378.0	136.5	31.75	1696.0	179.5	22.1	.0717	1.15072
155.6	363.6	11.36	1062.0	1520.0	137.5	31.65	1188.0	178.9	21.0	.0634	1.14119
155.7	347.6	10.86	1144.0	1623.0	138.1	31.50	892.7	178.9	20.3	.0582	1.13470
155.8	335.7	10.49	1207.0	1702.0	138.6	31.34	714.3	179.1	19.8	.0548	1.12989
155.9	326.3	10.20	1258.0	1768.0	139.0	31.19	597.8	179.5	19.5	.0519	1.12611
156.0	318.5	9.954	1301.0	1823.0	139.4	31.04	516.3	179.9	19.2	.0501	1.12299
156.2	306.2	9.568	1371.0	1915.0	140.0	30.77	410.4	180.7	18.7	.0473	1.11804
156.4	296.5	9.266	1429.0	1990.0	140.5	30.53	344.6	181.6	18.4	.0453	1.11417
156.6	288.6	9.018	1477.0	2054.0	140.9	30.30	299.5	182.4	18.1	.0437	1.11100
156.8	281.8	8.806	1520.0	2110.0	141.2	30.09	266.7	183.2	17.9	.0424	1.10831
157.0	275.9	8.622	1558.0	2161.0	141.6	29.90	241.6	184.0	17.7	.0414	1.10597
157.5	263.8	8.244	1640.0	2270.0	142.3	29.47	198.6	185.9	17.3	.0393	1.10116
158.0	254.2	7.943	1708.0	2362.0	142.8	29.10	171.2	187.7	17.1	.0377	1.09736
158.5	246.2	7.693	1767.0	2443.0	143.3	28.76	152.0	189.4	16.8	.0364	1.09420
159.0	239.3	7.479	1820.0	2515.0	143.8	28.46	137.8	191.0	16.7	.0354	1.09150
160.0	228.0	7.124	1912.0	2642.0	144.6	27.93	117.9	193.9	16.4	.0338	1.08704
161.0	218.8	6.837	1992.0	2753.0	145.3	27.47	104.5	196.7	16.2	.0326	1.08343
162.0	211.0	6.595	2064.0	2852.0	145.9	27.07	94.80	199.2	16.1	.0316	1.08040
163.0	204.3	6.386	2129.0	2943.0	146.5	26.72	87.43	201.6	16.0	.0308	1.07778
164.0	198.5	6.202	2189.0	3028.0	147.0	26.39	81.60	203.9	15.9	.0301	1.07549
166.0	188.4	5.889	2299.0	3182.0	147.9	25.83	72.92	208.1	15.8	.0290	1.07159
168.0	180.1	5.629	2397.0	3321.0	148.7	25.36	66.72	212.0	15.7	.0281	1.06836
170.0	173.0	5.407	2488.0	3450.0	149.5	24.96	62.06	215.6	15.7	.0274	1.06560
172.0	166.8	5.214	2573.0	3570.0	150.2	24.61	58.41	219.0	15.7	.0268	1.06320
174.0	161.3	5.042	2652.0	3684.0	150.9	24.30	55.46	222.3	15.7	.0264	1.06108
176.0	156.4	4.888	2728.0	3792.0	151.5	24.03	53.04	225.4	15.7	.0260	1.05918
178.0	151.9	4.749	2801.0	3896.0	152.1	23.79	51.00	228.3	15.7	.0257	1.05746
180.0	147.9	4.621	2871.0	3996.0	152.6	23.57	49.26	231.2	15.8	.0254	1.05589
185.0	139.0	4.344	3037.0	4234.0	153.9	23.12	45.87	237.9	15.9	.0249	1.05248
190.0	131.6	4.112	3192.0	4457.0	155.1	22.77	43.40	244.1	16.1	.0246	1.04962
195.0	125.2	3.913	3340.0	4669.0	156.2	22.49	41.51	249.8	16.3	.0244	1.04718
200.0	119.6	3.738	3481.0	4872.0	157.3	22.27	40.03	255.3	16.5	.0244	1.04505
205.0	114.7	3.584	3619.0	5069.0	158.2	22.09	38.84	260.5	16.8	.0244	1.04316
210.0	110.3	3.446	3752.0	5261.0	159.2	21.94	37.86	265.4	17.0	.0244	1.04147
215.0	106.3	3.321	3882.0	5448.0	160.0	21.82	37.05	270.2	17.2	.0245	1.03994
220.0	102.6	3.207	4010.0	5632.0	160.9	21.72	36.36	274.7	17.5	.0247	1.03856
230.0	96.20	3.006	4260.0	5990.0	162.5	21.57	35.27	283.4	18.0	.0250	1.03611
240.0	90.70	2.834	4503.0	6338.0	164.0	21.47	34.45	291.6	18.5	.0255	1.03402
250.0	85.90	2.684	4742.0	6679.0	165.3	21.41	33.81	299.3	19.0	.0260	1.03220
260.0	81.67	2.552	4977.0	7015.0	166.7	21.37	33.31	306.6	19.5	.0265	1.03060
270.0	77.90	2.434	5210.0	7346.0	167.9	21.35	32.91	313.7	20.0	.0271	1.02917
280.0	74.51	2.329	5440.0	7673.0	169.1	21.35	32.59	320.4	20.5	.0276	1.02789
290.0	71.44	2.233	5669.0	7998.0	170.2	21.36	32.33	326.9	21.0	.0283	1.02673
300.0	68.65	2.145	5896.0	8320.0	171.3	21.38	32.12	333.2	21.5	.0289	1.02567

Thermophysical properties of oxygen—Continued

T K	Density kg/m³	Density mol/dm³	E J/mol	H J/mol	S J/(mol·K)	C_v J/(mol·K)	C_p J/(mol·K)	Sound m/s	Visc. μPa·s	Therm. W/(m·K)	Diel. const.
310.0	66.09	2.065	6123.0	8640.0	172.4	21.41	31.95	339.2	22.0	.0295	1.02471
320.0	63.73	1.992	6349.0	8959.0	173.4	21.45	31.81	345.1	22.5	.0302	1.02382
340.0	59.54	1.861	6799.0	9593.0	175.3	21.56	31.62	356.3	23.5	.0315	1.02224
360.0	55.91	1.747	7248.0	10220.0	177.1	21.68	31.51	367.0	24.4	.0328	1.02087
380.0	52.72	1.648	7698.0	10850.0	178.8	21.83	31.47	377.2	25.4	.0342	1.01967
400.0	49.91	1.560	8149.0	11480.0	180.4	22.00	31.47	386.9	26.3	.0356	1.01861
					5.30 MPa isobar						
54.97[a]	1310.0	40.93	-6174.0	-6045.0	67.26	36.13	53.10	1141.0	502.0	.193	1.57024
80.0	1199.0	37.49	-4857.0	-4716.0	87.19	31.30	53.10	1003.0	264.0	.167	1.51556
90.0	1153.0	36.04	-4330.0	-4183.0	93.46	29.93	53.44	929.2	206.0	.154	1.49291
100.0	1104.0	34.52	-3799.0	-3646.0	99.12	28.78	54.22	853.0	163.0	.140	1.46938
110.0	1052.0	32.89	-3258.0	-3097.0	104.4	27.82	55.68	773.2	131.0	.126	1.44451
120.0	995.3	31.11	-2699.0	-2528.0	109.3	27.03	58.27	687.8	106.0	.113	1.41751
125.0	963.9	30.12	-2408.0	-2232.0	111.7	26.71	60.24	642.2	95.1	.106	1.40281
130.0	929.8	29.06	-2107.0	-1925.0	114.1	26.45	62.93	594.2	85.3	.0999	1.38696
135.0	892.0	27.88	-1791.0	-1601.0	116.6	26.25	66.75	542.9	76.1	.0936	1.36954
136.0	883.9	27.62	-1726.0	-1534.0	117.1	26.23	67.71	532.2	74.4	.0923	1.36581
138.0	866.9	27.09	-1592.0	-1396.0	118.1	26.19	69.90	510.2	70.8	.0898	1.35805
140.0	848.9	26.53	-1454.0	-1254.0	119.1	26.17	72.55	487.3	67.4	.0872	1.34983
142.0	829.5	25.92	-1310.0	-1106.0	120.1	26.17	75.84	463.4	63.9	.0846	1.34104
144.0	808.4	25.26	-1160.0	-949.9	121.2	26.22	80.05	438.1	60.4	.0820	1.33154
146.0	785.2	24.54	-1000.0	-784.5	122.4	26.30	85.69	411.3	56.9	.0794	1.32110
148.0	758.9	23.72	-829.1	-605.6	123.6	26.45	93.76	382.2	53.2	.0768	1.30937
149.0	744.3	23.26	-737.0	-509.2	124.2	26.56	99.33	366.5	51.2	.0756	1.30284
150.0	728.1	22.76	-639.3	-406.4	124.9	26.70	106.6	349.8	49.2	.0743	1.29570
151.0	710.1	22.19	-533.9	-295.1	125.7	26.89	116.5	331.8	47.0	.0730	1.28775
152.0	689.4	21.55	-417.8	-171.8	126.5	27.13	131.1	312.1	44.7	.0717	1.27866
152.5	677.6	21.18	-354.0	-103.8	126.9	27.29	141.5	301.4	43.4	.0710	1.27350
153.0	664.5	20.77	-285.0	-29.8	127.4	27.47	155.2	289.9	42.1	.0703	1.26778
153.5	649.6	20.30	-208.7	52.4	127.9	27.70	174.6	277.5	40.6	.0697	1.26129
154.0	632.0	19.75	-121.8	146.6	128.6	28.00	204.3	263.7	38.9	.0690	1.25366
154.2	623.9	19.50	-82.8	189.0	128.8	28.14	221.2	257.7	38.1	.0687	1.25016
154.4	615.0	19.22	-40.5	235.3	129.1	28.30	242.8	251.3	37.3	.0685	1.24630
154.6	604.9	18.90	6.2	286.6	129.5	28.49	271.4	244.6	36.4	.0682	1.24199
154.8	593.4	18.54	58.8	344.6	129.8	28.71	311.2	237.2	35.4	.0681	1.23704
155.0	579.8	18.12	119.9	412.4	130.3	28.98	370.5	229.2	34.3	.0680	1.23121
155.2	562.9	17.59	194.0	495.3	130.8	29.31	467.4	220.3	33.0	.0681	1.22400
155.3	552.5	17.27	238.6	545.6	131.1	29.52	542.1	215.4	32.2	.0684	1.21961
155.4	540.3	16.89	290.9	604.7	131.5	29.77	647.5	210.2	31.3	.0691	1.21444
155.5	525.5	16.42	353.9	676.7	132.0	30.06	800.7	204.5	30.2	.0702	1.20817
155.6	506.9	15.84	432.6	767.1	132.6	30.41	1020.0	198.6	28.9	.0720	1.20035
155.7	483.6	15.11	531.8	882.5	133.3	30.81	1289.0	192.7	27.4	.0745	1.19057
155.8	456.0	14.25	650.5	1022.0	134.2	31.20	1481.0	187.6	25.7	.0761	1.17912
155.9	427.9	13.37	775.5	1172.0	135.2	31.47	1474.0	184.0	24.2	.0742	1.16748
156.0	402.8	12.59	891.0	1312.0	136.1	31.58	1310.0	182.0	22.9	.0696	1.15718
156.1	382.4	11.95	988.8	1432.0	136.8	31.57	1097.0	181.0	21.9	.0645	1.14882
156.2	366.2	11.44	1069.0	1532.0	137.5	31.49	906.2	180.6	21.2	.0603	1.14223
156.3	353.3	11.04	1135.0	1615.0	138.0	31.37	758.0	180.5	20.6	.0570	1.13699
156.4	342.7	10.71	1190.0	1685.0	138.5	31.25	646.8	180.6	20.2	.0545	1.13273
156.5	333.9	10.44	1237.0	1745.0	138.8	31.12	563.0	180.9	19.8	.0524	1.12919
156.6	326.4	10.20	1279.0	1798.0	139.2	30.99	498.7	181.2	19.5	.0505	1.12616
156.8	314.1	9.816	1348.0	1888.0	139.7	30.74	407.6	181.9	19.1	.0480	1.12122
157.0	304.3	9.508	1406.0	1963.0	140.2	30.51	346.8	182.6	18.7	.0460	1.11728
157.2	296.1	9.253	1455.0	2028.0	140.6	30.30	303.6	183.4	18.4	.0445	1.11400
157.4	289.1	9.034	1499.0	2085.0	141.0	30.10	271.3	184.1	18.2	.0432	1.11121
157.6	282.9	8.842	1538.0	2137.0	141.3	29.91	246.3	184.9	18.0	.0421	1.10877
158.0	272.6	8.519	1606.0	2228.0	141.9	29.57	210.0	186.3	17.7	.0403	1.10466
158.5	262.2	8.193	1678.0	2325.0	142.5	29.19	179.6	188.1	17.4	.0386	1.10052
159.0	253.5	7.923	1740.0	2409.0	143.1	28.85	158.6	189.7	17.1	.0373	1.09711
159.5	246.2	7.694	1795.0	2484.0	143.5	28.55	143.1	191.3	16.9	.0362	1.09421

Thermophysical properties of oxygen—Continued

T K	Density kg/m³	Density mol/dm³	E J/mol	H J/mol	S J/(mol·K)	C_v J/(mol·K)	C_p J/(mol·K)	Sound m/s	Visc. μPa·s	Therm. W/(m·K)	Diel. const.
160.0	239.8	7.494	1845.0	2553.0	144.0	28.27	131.2	192.8	16.8	.0353	1.09169
161.0	229.0	7.158	1934.0	2675.0	144.7	27.77	113.8	195.6	16.5	.0338	1.08745
162.0	220.2	6.881	2012.0	2782.0	145.4	27.33	101.8	198.2	16.3	.0326	1.08398
163.0	212.6	6.645	2082.0	2879.0	146.0	26.95	92.93	200.7	16.2	.0317	1.08103
164.0	206.1	6.441	2146.0	2969.0	146.5	26.61	86.05	203.0	16.1	.0309	1.07847
165.0	200.3	6.259	2205.0	3052.0	147.0	26.30	80.56	205.3	16.0	.0302	1.07620
166.0	195.1	6.097	2261.0	3130.0	147.5	26.02	76.05	207.4	16.0	.0296	1.07418
168.0	186.1	5.815	2363.0	3275.0	148.4	25.52	69.08	211.3	15.9	.0286	1.07067
170.0	178.4	5.577	2457.0	3408.0	149.2	25.09	63.91	215.0	15.8	.0279	1.06771
172.0	171.8	5.370	2544.0	3531.0	149.9	24.73	59.91	218.5	15.8	.0273	1.06514
174.0	166.0	5.188	2626.0	3648.0	150.5	24.41	56.71	221.8	15.8	.0268	1.06289
176.0	160.8	5.025	2704.0	3759.0	151.2	24.12	54.10	225.0	15.8	.0263	1.06087
178.0	156.1	4.878	2778.0	3864.0	151.8	23.87	51.91	228.0	15.8	.0260	1.05906
180.0	151.8	4.744	2849.0	3966.0	152.3	23.65	50.06	230.8	15.9	.0257	1.05740
185.0	142.5	4.454	3017.0	4207.0	153.7	23.18	46.47	237.6	16.0	.0251	1.05383
190.0	134.8	4.212	3175.0	4433.0	154.9	22.82	43.86	243.8	16.2	.0248	1.05086
195.0	128.2	4.005	3324.0	4647.0	156.0	22.53	41.89	249.6	16.4	.0246	1.04832
200.0	122.4	3.825	3467.0	4852.0	157.0	22.30	40.34	255.1	16.6	.0245	1.04610
205.0	117.3	3.665	3605.0	5051.0	158.0	22.12	39.10	260.4	16.8	.0245	1.04415
210.0	112.7	3.523	3739.0	5244.0	158.9	21.97	38.09	265.3	17.0	.0246	1.04241
215.0	108.6	3.394	3870.0	5432.0	159.8	21.84	37.25	270.1	17.3	.0247	1.04083
220.0	104.8	3.276	3999.0	5616.0	160.7	21.74	36.54	274.7	17.5	.0248	1.03940
230.0	98.24	3.070	4250.0	5976.0	162.3	21.59	35.41	283.4	18.0	.0251	1.03689
240.0	92.58	2.893	4494.0	6326.0	163.8	21.48	34.57	291.6	18.5	.0256	1.03474
250.0	87.67	2.740	4733.0	6668.0	165.2	21.42	33.91	299.3	19.0	.0260	1.03287
260.0	83.33	2.604	4969.0	7004.0	166.5	21.38	33.40	306.7	19.5	.0266	1.03123
270.0	79.47	2.483	5202.0	7336.0	167.7	21.36	32.99	313.7	20.0	.0271	1.02976
280.0	76.00	2.375	5433.0	7664.0	168.9	21.35	32.66	320.5	20.5	.0277	1.02845
290.0	72.86	2.277	5662.0	7990.0	170.1	21.36	32.39	327.0	21.0	.0283	1.02726
300.0	70.00	2.188	5890.0	8312.0	171.2	21.39	32.17	333.3	21.5	.0289	1.02618
310.0	67.39	2.106	6116.0	8633.0	172.2	21.42	32.00	339.3	22.0	.0296	1.02520
320.0	64.98	2.031	6343.0	8952.0	173.2	21.46	31.86	345.2	22.5	.0302	1.02429
340.0	60.69	1.897	6793.0	9587.0	175.1	21.56	31.66	356.5	23.5	.0315	1.02267
360.0	56.99	1.781	7243.0	10220.0	177.0	21.69	31.54	367.1	24.5	.0329	1.02127
380.0	53.74	1.679	7694.0	10850.0	178.7	21.84	31.49	377.3	25.4	.0342	1.02005
400.0	50.86	1.590	8145.0	11480.0	180.3	22.00	31.50	387.0	26.3	.0356	1.01897
				5.40 MPa isobar							
54.98[a]	1310.0	40.93	−6174.0	−6042.0	67.26	36.12	53.10	1141.0	502.0	.193	1.57027
80.0	1200.0	37.49	−4858.0	−4714.0	87.18	31.31	53.09	1003.0	264.0	.168	1.51564
90.0	1153.0	36.04	−4331.0	−4182.0	93.45	29.93	53.42	929.6	206.0	.154	1.49301
100.0	1105.0	34.52	−3800.0	−3644.0	99.11	28.79	54.19	853.5	163.0	.140	1.46951
110.0	1053.0	32.90	−3260.0	−3095.0	104.3	27.83	55.65	773.9	131.0	.126	1.44467
120.0	995.8	31.12	−2701.0	−2527.0	109.3	27.04	58.21	688.7	106.0	.113	1.41772
125.0	964.5	30.14	−2411.0	−2232.0	111.7	26.72	60.16	643.3	95.2	.106	1.40306
130.0	930.5	29.08	−2110.0	−1925.0	114.1	26.45	62.81	595.5	85.5	.100	1.38727
135.0	892.9	27.90	−1795.0	−1602.0	116.5	26.26	66.56	544.5	76.3	.0937	1.36993
136.0	884.8	27.65	−1730.0	−1535.0	117.0	26.23	67.50	533.8	74.6	.0924	1.36622
138.0	867.9	27.12	−1597.0	−1398.0	118.0	26.19	69.64	512.0	71.1	.0899	1.35852
140.0	850.0	26.56	−1459.0	−1256.0	119.1	26.16	72.22	489.3	67.6	.0873	1.35036
142.0	830.8	25.96	−1316.0	−1108.0	120.1	26.17	75.41	465.6	64.1	.0847	1.34165
144.0	810.0	25.31	−1167.0	−953.7	121.2	26.21	79.47	440.6	60.7	.0822	1.33225
146.0	787.1	24.60	−1009.0	−789.7	122.3	26.29	84.86	414.2	57.1	.0796	1.32195
148.0	761.3	23.79	−839.8	−612.8	123.5	26.43	92.47	385.6	53.5	.0770	1.31044
149.0	747.0	23.34	−749.2	−517.9	124.2	26.53	97.65	370.2	51.6	.0757	1.30405
150.0	731.3	22.85	−653.3	−417.0	124.8	26.66	104.3	354.0	49.6	.0745	1.29711
151.0	713.9	22.31	−550.6	−308.5	125.5	26.83	113.2	336.6	47.5	.0732	1.28944
152.0	694.2	21.69	−438.3	−189.4	126.3	27.05	125.9	317.6	45.3	.0719	1.28075
152.5	683.1	21.35	−377.3	−124.3	126.8	27.19	134.6	307.4	44.0	.0713	1.27589
153.0	670.9	20.97	−311.9	−54.3	127.2	27.35	145.9	296.6	42.7	.0706	1.27056
153.5	657.3	20.54	−240.7	22.2	127.7	27.55	161.0	285.0	41.4	.0699	1.26462
154.0	641.6	20.05	−161.6	107.7	128.3	27.80	182.5	272.4	39.8	.0692	1.25783

Thermophysical properties of oxygen—Continued

T K	Density kg/m^3	Density mol/dm^3	E J/mol	H J/mol	S J/(mol·K)	C_v J/(mol·K)	C_p J/(mol·K)	Sound m/s	Visc. μPa·s	Therm. W/(m·K)	Diel. const.
154.5	623.0	19.47	-70.7	206.6	128.9	28.11	215.9	258.4	38.1	.0685	1.24978
154.6	618.8	19.34	-50.6	228.6	129.1	28.19	224.9	255.4	37.7	.0684	1.24797
155.0	599.5	18.74	39.5	327.7	129.7	28.54	275.1	242.5	36.0	.0679	1.23968
155.2	587.9	18.37	92.5	386.4	130.1	28.76	314.0	235.4	35.0	.0677	1.23470
155.4	574.3	17.95	153.6	454.5	130.5	29.02	369.9	227.8	33.9	.0676	1.22887
155.6	557.7	17.43	226.5	536.3	131.0	29.35	455.3	219.4	32.6	.0676	1.22181
155.8	536.6	16.77	318.0	640.1	131.7	29.76	593.9	210.3	31.0	.0682	1.21284
155.9	523.5	16.36	374.2	704.3	132.1	30.01	693.4	205.5	30.1	.0688	1.20732
156.0	508.2	15.88	439.5	779.5	132.6	30.29	814.0	200.6	29.1	.0696	1.20089
156.1	490.6	15.33	515.1	867.3	133.2	30.59	942.1	195.9	27.9	.0706	1.19350
156.2	470.9	14.72	600.1	967.1	133.8	30.88	1046.0	191.7	26.7	.0713	1.18529
156.3	450.3	14.07	690.8	1075.0	134.5	31.14	1093.0	188.3	25.5	.0710	1.17672
156.4	430.0	13.44	781.8	1184.0	135.2	31.31	1078.0	185.8	24.3	.0693	1.16836
156.5	411.3	12.85	868.4	1289.0	135.9	31.41	1014.0	184.2	23.4	.0667	1.16064
156.6	394.6	12.33	947.7	1386.0	136.5	31.42	923.0	183.1	22.5	.0637	1.15381
156.7	380.1	11.88	1018.0	1473.0	137.0	31.39	823.1	182.5	21.9	.0607	1.14790
156.8	367.7	11.49	1080.0	1550.0	137.5	31.32	728.4	182.2	21.3	.0581	1.14285
156.9	357.1	11.16	1135.0	1619.0	138.0	31.23	645.4	182.1	20.8	.0558	1.13853
157.0	347.9	10.87	1183.0	1680.0	138.4	31.12	575.5	182.2	20.4	.0539	1.13482
157.2	332.8	10.40	1265.0	1784.0	139.0	30.90	469.2	182.6	19.8	.0508	1.12875
157.4	320.9	10.03	1331.0	1870.0	139.6	30.68	395.3	183.1	19.4	.0483	1.12396
157.6	311.2	9.725	1388.0	1943.0	140.0	30.47	342.3	183.8	19.0	.0465	1.12004
157.8	302.9	9.467	1437.0	2008.0	140.4	30.27	302.8	184.4	18.7	.0451	1.11674
158.0	295.8	9.244	1481.0	2065.0	140.8	30.08	272.4	185.1	18.5	.0438	1.11389
158.2	289.5	9.048	1520.0	2117.0	141.1	29.90	248.4	185.8	18.3	.0427	1.11140
158.5	281.4	8.793	1573.0	2187.0	141.6	29.65	220.4	186.9	18.0	.0414	1.10814
159.0	270.1	8.441	1649.0	2289.0	142.2	29.27	187.6	188.5	17.7	.0396	1.10367
159.5	260.9	8.153	1714.0	2376.0	142.8	28.93	164.9	190.2	17.4	.0381	1.10001
160.0	253.1	7.908	1772.0	2454.0	143.3	28.62	148.2	191.7	17.2	.0370	1.09692
160.5	246.3	7.696	1823.0	2525.0	143.7	28.34	135.4	193.2	17.0	.0360	1.09424
161.0	240.3	7.509	1871.0	2590.0	144.1	28.08	125.3	194.6	16.9	.0351	1.09187
162.0	230.0	7.189	1956.0	2707.0	144.8	27.61	110.1	197.3	16.6	.0338	1.08784
163.0	221.5	6.922	2032.0	2812.0	145.5	27.20	99.27	199.8	16.5	.0326	1.08449
164.0	214.2	6.693	2100.0	2907.0	146.1	26.83	91.10	202.2	16.3	.0317	1.08162
165.0	207.7	6.492	2163.0	2995.0	146.6	26.50	84.69	204.5	16.2	.0310	1.07911
166.0	202.0	6.314	2221.0	3077.0	147.1	26.20	79.51	206.7	16.1	.0303	1.07689
168.0	192.3	6.008	2329.0	3227.0	148.0	25.68	71.63	210.7	16.0	.0292	1.07307
170.0	184.0	5.751	2426.0	3365.0	148.8	25.23	65.89	214.5	16.0	.0284	1.06988
172.0	177.0	5.531	2515.0	3492.0	149.5	24.85	61.50	218.0	15.9	.0277	1.06713
174.0	170.8	5.337	2599.0	3611.0	150.2	24.51	58.03	221.4	15.9	.0272	1.06473
176.0	165.3	5.165	2679.0	3724.0	150.9	24.22	55.21	224.5	15.9	.0267	1.06260
178.0	160.3	5.010	2755.0	3832.0	151.5	23.96	52.87	227.6	16.0	.0263	1.06069
180.0	155.8	4.869	2827.0	3936.0	152.1	23.73	50.89	230.5	16.0	.0260	1.05895
182.0	151.7	4.741	2897.0	4036.0	152.6	23.52	49.20	233.3	16.0	.0257	1.05736
185.0	146.1	4.566	2998.0	4180.0	153.4	23.24	47.08	237.3	16.1	.0254	1.05521
190.0	138.0	4.314	3157.0	4409.0	154.6	22.87	44.34	243.6	16.3	.0250	1.05210
195.0	131.2	4.099	3308.0	4625.0	155.7	22.57	42.27	249.5	16.4	.0248	1.04946
200.0	125.2	3.912	3452.0	4832.0	156.8	22.34	40.66	255.0	16.7	.0247	1.04717
205.0	119.9	3.747	3591.0	5032.0	157.8	22.15	39.37	260.2	16.9	.0246	1.04515
210.0	115.2	3.600	3726.0	5226.0	158.7	21.99	38.32	265.2	17.1	.0247	1.04335
215.0	110.9	3.467	3858.0	5416.0	159.6	21.86	37.45	270.0	17.3	.0248	1.04173
220.0	107.1	3.346	3987.0	5601.0	160.5	21.76	36.72	274.6	17.6	.0249	1.04025
230.0	100.3	3.134	4239.0	5962.0	162.1	21.60	35.56	283.4	18.0	.0252	1.03767
240.0	94.48	2.953	4484.0	6313.0	163.6	21.50	34.68	291.6	18.5	.0257	1.03546
250.0	89.44	2.795	4724.0	6656.0	165.0	21.43	34.01	299.3	19.0	.0261	1.03354
260.0	84.99	2.656	4961.0	6994.0	166.3	21.39	33.49	306.7	19.6	.0266	1.03186
270.0	81.04	2.533	5194.0	7326.0	167.5	21.36	33.07	313.8	20.1	.0272	1.03036
280.0	77.49	2.422	5425.0	7655.0	168.7	21.36	32.73	320.6	20.6	.0278	1.02901
290.0	74.28	2.321	5655.0	7981.0	169.9	21.37	32.45	327.1	21.1	.0284	1.02780
300.0	71.36	2.230	5883.0	8305.0	171.0	21.39	32.23	333.4	21.6	.0290	1.02669
310.0	68.69	2.146	6110.0	8626.0	172.0	21.42	32.05	339.5	22.1	.0296	1.02568
320.0	66.23	2.070	6337.0	8946.0	173.0	21.46	31.90	345.3	22.5	.0303	1.02476
340.0	61.85	1.933	6788.0	9581.0	175.0	21.56	31.70	356.6	23.5	.0316	1.02311

Thermophysical properties of oxygen—Continued

T K	Density kg/m^3	Density mol/dm^3	E J/mol	H J/mol	S J/(mol·K)	C_v J/(mol·K)	C_p J/(mol·K)	Sound m/s	Visc. μPa·s	Therm. W/(m·K)	Diel. const.
360.0	58.06	1.815	7238.0	10210.0	176.8	21.69	31.58	367.3	24.5	.0329	1.02168
380.0	54.75	1.711	7689.0	10840.0	178.5	21.84	31.52	377.5	25.4	.0343	1.02043
400.0	51.82	1.619	8141.0	11480.0	180.1	22.00	31.52	387.2	26.3	.0357	1.01933

5.50 MPa isobar

T K	Density kg/m^3	Density mol/dm^3	E J/mol	H J/mol	S J/(mol·K)	C_v J/(mol·K)	C_p J/(mol·K)	Sound m/s	Visc. μPa·s	Therm. W/(m·K)	Diel. const.
54.99[a]	1310.0	40.93	–6174.0	–6040.0	67.27	36.12	53.10	1141.0	502.0	.193	1.57030
80.0	1200.0	37.50	–4859.0	–4712.0	87.17	31.31	53.08	1003.0	264.0	.168	1.51573
90.0	1154.0	36.05	–4332.0	–4180.0	93.44	29.94	53.41	930.1	206.0	.154	1.49311
100.0	1105.0	34.53	–3802.0	–3642.0	99.10	28.79	54.17	854.1	163.0	.140	1.46963
110.0	1053.0	32.91	–3261.0	–3094.0	104.3	27.83	55.61	774.5	131.0	.126	1.44483
120.0	996.2	31.13	–2703.0	–2527.0	109.3	27.05	58.15	689.6	106.0	.113	1.41793
125.0	965.0	30.16	–2414.0	–2231.0	111.7	26.72	60.08	644.4	95.4	.106	1.40331
130.0	931.2	29.10	–2114.0	–1925.0	114.1	26.45	62.69	596.7	85.6	.100	1.38758
135.0	893.7	27.93	–1799.0	–1603.0	116.5	26.26	66.37	546.0	76.5	.0938	1.37032
136.0	885.7	27.68	–1734.0	–1536.0	117.0	26.23	67.29	535.4	74.7	.0925	1.36664
138.0	869.0	27.16	–1602.0	–1399.0	118.0	26.19	69.39	513.7	71.3	.0900	1.35898
140.0	851.2	26.60	–1465.0	–1258.0	119.0	26.16	71.90	491.2	67.8	.0874	1.35088
142.0	832.2	26.01	–1323.0	–1111.0	120.1	26.16	74.99	467.8	64.4	.0849	1.34225
144.0	811.6	25.36	–1174.0	–957.3	121.1	26.20	78.91	443.1	60.9	.0823	1.33295
146.0	789.0	24.66	–1018.0	–794.6	122.2	26.27	84.06	417.0	57.4	.0797	1.32279
148.0	763.7	23.87	–850.2	–619.7	123.4	26.40	91.26	388.9	53.8	.0772	1.31147
149.0	749.6	23.43	–760.9	–526.2	124.1	26.50	96.09	373.8	51.9	.0759	1.30522
150.0	734.4	22.95	–666.8	–427.1	124.7	26.62	102.2	358.0	50.0	.0747	1.29846
151.0	717.6	22.43	–566.4	–321.1	125.4	26.77	110.3	341.1	48.0	.0734	1.29104
152.0	698.7	21.83	–457.5	–205.6	126.2	26.98	121.5	322.9	45.8	.0721	1.28271
153.0	676.7	21.15	–336.4	–76.3	127.0	27.25	138.3	302.8	43.4	.0708	1.27309
153.5	664.1	20.75	–269.3	–4.3	127.5	27.43	150.5	291.9	42.1	.0702	1.26758
154.0	649.9	20.31	–196.0	74.9	128.0	27.64	167.0	280.2	40.7	.0695	1.26140
154.5	633.5	19.80	–113.9	163.9	128.6	27.90	190.7	267.4	39.1	.0688	1.25430
155.0	613.8	19.18	–18.9	267.8	129.3	28.23	227.9	253.3	37.3	.0681	1.24583
155.5	588.8	18.40	97.6	396.6	130.1	28.69	294.5	237.4	35.1	.0674	1.23507
155.6	582.8	18.21	125.0	427.0	130.3	28.80	314.4	233.9	34.6	.0673	1.23249
155.8	569.3	17.79	185.5	494.7	130.7	29.06	365.5	226.6	33.5	.0671	1.22674
156.0	553.2	17.29	256.5	574.6	131.3	29.37	438.3	218.9	32.3	.0670	1.21990
156.2	533.5	16.67	342.3	672.1	131.9	29.74	543.2	210.7	30.9	.0673	1.21156
156.4	508.9	15.91	448.5	794.3	132.7	30.18	682.4	202.4	29.1	.0680	1.20121
156.6	479.2	14.98	577.5	944.8	133.6	30.65	814.9	195.0	27.2	.0685	1.18876
156.8	446.9	13.97	720.7	1115.0	134.7	31.04	864.9	189.5	25.3	.0675	1.17534
157.0	416.3	13.01	861.6	1284.0	135.8	31.24	820.1	186.2	23.7	.0644	1.16270
157.2	390.0	12.19	987.6	1439.0	136.8	31.25	720.1	184.5	22.4	.0603	1.15194
157.3	378.8	11.84	1043.0	1508.0	137.2	31.21	663.6	184.1	21.9	.0583	1.14735
157.4	368.7	11.52	1094.0	1572.0	137.6	31.15	608.5	183.9	21.4	.0564	1.14326
157.6	351.7	10.99	1183.0	1683.0	138.3	30.98	511.6	183.8	20.7	.0532	1.13637
157.8	338.0	10.56	1257.0	1778.0	138.9	30.80	435.5	184.1	20.1	.0507	1.13084
158.0	326.7	10.21	1320.0	1859.0	139.4	30.60	377.4	184.5	19.7	.0485	1.12629
158.2	317.2	9.914	1375.0	1930.0	139.9	30.41	332.8	185.0	19.3	.0469	1.12248
158.4	309.1	9.659	1423.0	1993.0	140.3	30.22	298.0	185.6	19.0	.0455	1.11921
158.6	302.0	9.436	1466.0	2049.0	140.6	30.05	270.3	186.3	18.8	.0443	1.11635
158.8	295.6	9.239	1506.0	2101.0	141.0	29.87	247.8	186.9	18.5	.0433	1.11383
159.0	290.0	9.062	1542.0	2149.0	141.3	29.71	229.3	187.5	18.4	.0423	1.11157
159.5	278.0	8.687	1621.0	2254.0	141.9	29.33	194.6	189.1	18.0	.0404	1.10679
160.0	268.1	8.380	1689.0	2345.0	142.5	28.99	170.6	190.7	17.7	.0389	1.10289
160.5	259.9	8.121	1748.0	2426.0	143.0	28.68	153.0	192.2	17.5	.0377	1.09960
161.0	252.7	7.896	1802.0	2499.0	143.5	28.40	139.4	193.6	17.3	.0367	1.09677
162.0	240.7	7.522	1897.0	2628.0	144.3	27.89	120.0	196.4	17.0	.0350	1.09205
163.0	230.9	7.217	1979.0	2741.0	144.9	27.45	106.6	199.0	16.7	.0337	1.08821
164.0	222.7	6.960	2052.0	2842.0	145.6	27.06	96.84	201.4	16.6	.0327	1.08497
165.0	215.6	6.737	2119.0	2935.0	146.1	26.71	89.32	203.8	16.4	.0318	1.08218
166.0	209.3	6.541	2181.0	3021.0	146.7	26.39	83.34	206.0	16.3	.0310	1.07973
168.0	198.7	6.208	2293.0	3179.0	147.6	25.84	74.41	210.1	16.2	.0298	1.07556
170.0	189.8	5.932	2393.0	3321.0	148.4	25.37	68.01	213.9	16.1	.0289	1.07212
172.0	182.3	5.696	2486.0	3452.0	149.2	24.97	63.19	217.5	16.1	.0282	1.06918

Thermophysical properties of oxygen—Continued

T K	Density kg/m^3	Density mol/dm^3	E J/mol	H J/mol	S J/(mol·K)	C_v J/(mol·K)	C_p J/(mol·K)	Sound m/s	Visc. μPa·s	Therm. W/(m·K)	Diel. const.
174.0	175.7	5.490	2572.0	3574.0	149.9	24.62	59.41	220.9	16.0	.0276	1.06663
176.0	169.8	5.308	2654.0	3690.0	150.6	24.32	56.37	224.1	16.0	.0271	1.06437
178.0	164.6	5.145	2731.0	3800.0	151.2	24.05	53.86	227.2	16.1	.0266	1.06235
180.0	159.9	4.997	2805.0	3905.0	151.8	23.80	51.75	230.1	16.1	.0263	1.06052
182.0	155.6	4.862	2876.0	4007.0	152.3	23.59	49.95	233.0	16.1	.0260	1.05885
185.0	149.7	4.679	2978.0	4153.0	153.1	23.31	47.71	237.0	16.2	.0256	1.05660
190.0	141.3	4.417	3139.0	4384.0	154.4	22.92	44.83	243.4	16.3	.0252	1.05337
195.0	134.2	4.193	3291.0	4603.0	155.5	22.62	42.66	249.3	16.5	.0250	1.05062
200.0	128.0	4.000	3437.0	4812.0	156.6	22.37	40.98	254.8	16.7	.0248	1.04825
205.0	122.5	3.829	3577.0	5013.0	157.6	22.18	39.65	260.1	16.9	.0248	1.04616
210.0	117.7	3.677	3713.0	5209.0	158.5	22.02	38.56	265.1	17.1	.0248	1.04430
215.0	113.3	3.540	3846.0	5399.0	159.4	21.89	37.65	270.0	17.4	.0249	1.04262
220.0	109.3	3.416	3976.0	5586.0	160.3	21.78	36.90	274.6	17.6	.0250	1.04111
230.0	102.3	3.198	4228.0	5948.0	161.9	21.62	35.70	283.4	18.1	.0253	1.03845
240.0	96.38	3.012	4474.0	6300.0	163.4	21.51	34.80	291.6	18.6	.0257	1.03618
250.0	91.21	2.850	4715.0	6645.0	164.8	21.44	34.11	299.4	19.1	.0262	1.03421
260.0	86.66	2.708	4952.0	6983.0	166.1	21.39	33.57	306.8	19.6	.0267	1.03249
270.0	82.61	2.582	5186.0	7317.0	167.4	21.37	33.14	313.9	20.1	.0273	1.03095
280.0	78.98	2.468	5418.0	7646.0	168.6	21.37	32.79	320.7	20.6	.0278	1.02958
290.0	75.70	2.366	5648.0	7973.0	169.7	21.38	32.51	327.2	21.1	.0284	1.02834
300.0	72.72	2.272	5876.0	8297.0	170.8	21.40	32.28	333.5	21.6	.0290	1.02721
310.0	69.98	2.187	6104.0	8619.0	171.9	21.43	32.10	339.6	22.1	.0297	1.02617
320.0	67.47	2.109	6330.0	8939.0	172.9	21.47	31.95	345.5	22.6	.0303	1.02523
340.0	63.01	1.969	6782.0	9575.0	174.8	21.57	31.73	356.8	23.5	.0316	1.02354
360.0	59.14	1.848	7233.0	10210.0	176.6	21.70	31.61	367.4	24.5	.0330	1.02208
380.0	55.76	1.743	7684.0	10840.0	178.3	21.84	31.55	377.6	25.4	.0343	1.02081
400.0	52.77	1.649	8136.0	11470.0	179.9	22.01	31.55	387.4	26.4	.0357	1.01969

5.60 MPa isobar

T K	Density kg/m^3	Density mol/dm^3	E J/mol	H J/mol	S J/(mol·K)	C_v J/(mol·K)	C_p J/(mol·K)	Sound m/s	Visc. μPa·s	Therm. W/(m·K)	Diel. const.
55.00[a]	1310.0	40.93	−6174.0	−6037.0	67.27	36.12	53.10	1142.0	502.0	.193	1.57033
80.0	1200.0	37.50	−4859.0	−4710.0	87.16	31.32	53.07	1004.0	264.0	.168	1.51581
90.0	1154.0	36.06	−4333.0	−4178.0	93.42	29.94	53.40	930.5	206.0	.154	1.49321
100.0	1105.0	34.54	−3803.0	−3641.0	99.08	28.80	54.15	854.6	164.0	.140	1.46975
110.0	1053.0	32.92	−3263.0	−3093.0	104.3	27.84	55.58	775.2	131.0	.126	1.44498
120.0	996.7	31.15	−2705.0	−2526.0	109.2	27.05	58.09	690.5	106.0	.113	1.41814
125.0	965.6	30.17	−2416.0	−2231.0	111.6	26.73	59.99	645.5	95.6	.106	1.40356
130.0	931.8	29.12	−2117.0	−1925.0	114.0	26.46	62.57	598.0	85.8	.100	1.38788
135.0	894.6	27.96	−1804.0	−1603.0	116.5	26.26	66.19	547.6	76.7	.0939	1.37071
136.0	886.6	27.71	−1739.0	−1537.0	117.0	26.23	67.09	537.1	74.9	.0926	1.36704
138.0	870.0	27.19	−1606.0	−1400.0	118.0	26.19	69.14	515.5	71.5	.0901	1.35944
140.0	852.3	26.64	−1470.0	−1260.0	119.0	26.16	71.59	493.2	68.0	.0876	1.35140
142.0	833.5	26.05	−1329.0	−1114.0	120.0	26.16	74.59	469.9	64.6	.0850	1.34284
144.0	813.1	25.41	−1181.0	−960.9	121.1	26.19	78.37	445.5	61.2	.0824	1.33363
146.0	790.8	24.71	−1026.0	−799.4	122.2	26.26	83.31	419.7	57.7	.0799	1.32360
148.0	765.9	23.94	−860.3	−626.4	123.4	26.38	90.12	392.1	54.1	.0774	1.31247
150.0	737.3	23.04	−679.8	−436.7	124.6	26.58	100.3	361.9	50.4	.0749	1.29977
151.0	721.0	22.53	−581.5	−332.9	125.3	26.72	107.6	345.5	48.4	.0736	1.29256
152.0	702.9	21.96	−475.6	−220.6	126.1	26.91	117.6	327.8	46.3	.0724	1.28455
153.0	682.0	21.31	−359.0	−96.3	126.9	27.16	132.1	308.6	44.0	.0711	1.27541
153.5	670.2	20.94	−295.2	−27.8	127.3	27.31	142.1	298.3	42.7	.0704	1.27025
154.0	657.1	20.54	−226.3	46.4	127.8	27.50	155.3	287.3	41.4	.0697	1.26455
154.5	642.3	20.07	−150.7	128.2	128.3	27.72	173.2	275.5	40.0	.0690	1.25814
155.0	625.2	19.54	−65.7	220.9	128.9	28.00	199.2	262.6	38.3	.0683	1.25073
155.5	604.5	18.89	33.4	329.9	129.6	28.35	240.1	248.5	36.5	.0676	1.24181
156.0	578.0	18.06	156.1	466.1	130.5	28.84	312.4	232.7	34.3	.0669	1.23044
156.2	564.7	17.65	215.6	532.9	130.9	29.08	358.0	225.9	33.2	.0666	1.22479
156.4	549.3	17.17	284.1	610.3	131.4	29.37	418.7	218.7	32.0	.0664	1.21824
156.6	531.0	16.60	364.2	701.7	132.0	29.71	497.8	211.3	30.7	.0665	1.21051
156.8	509.4	15.92	458.6	810.4	132.7	30.08	590.4	204.2	29.2	.0667	1.20140
157.0	484.6	15.14	567.4	937.2	133.5	30.47	673.4	197.7	27.6	.0667	1.19099
157.2	458.0	14.31	685.7	1077.0	134.4	30.80	715.1	192.7	26.0	.0659	1.17992

Thermophysical properties of oxygen—Continued

T K	Density kg/m³	Density mol/dm³	E J/mol	H J/mol	S J/(mol·K)	C_v J/(mol·K)	C_p J/(mol·K)	Sound m/s	Visc. μPa·s	Therm. W/(m·K)	Diel. const.
157.4	431.9	13.50	804.9	1220.0	135.3	31.02	706.7	189.2	24.6	.0639	1.16911
157.6	407.9	12.75	917.9	1357.0	136.2	31.11	660.8	187.1	23.3	.0611	1.15928
157.8	387.1	12.10	1020.0	1483.0	137.0	31.08	594.3	186.0	22.3	.0580	1.15075
158.0	369.4	11.55	1109.0	1595.0	137.7	30.98	523.0	185.5	21.5	.0551	1.14356
158.2	354.7	11.08	1187.0	1692.0	138.3	30.84	457.6	185.4	20.9	.0525	1.13756
158.4	342.2	10.69	1255.0	1778.0	138.9	30.68	402.2	185.6	20.3	.0504	1.13253
158.6	331.6	10.36	1314.0	1854.0	139.3	30.50	356.9	185.9	19.9	.0485	1.12826
158.8	322.5	10.08	1366.0	1922.0	139.8	30.33	320.2	186.4	19.6	.0470	1.12459
159.0	314.6	9.830	1413.0	1983.0	140.2	30.16	290.2	186.9	19.3	.0458	1.12140
159.2	307.5	9.610	1455.0	2038.0	140.5	29.99	265.6	187.5	19.0	.0447	1.11858
159.5	298.3	9.323	1512.0	2113.0	141.0	29.75	236.1	188.3	18.7	.0432	1.11490
160.0	285.6	8.927	1594.0	2222.0	141.7	29.38	200.6	189.8	18.3	.0412	1.10984
160.5	275.3	8.603	1664.0	2315.0	142.2	29.04	175.6	191.3	18.0	.0397	1.10572
161.0	266.6	8.330	1726.0	2398.0	142.8	28.73	157.3	192.8	17.7	.0384	1.10226
161.5	259.0	8.095	1782.0	2473.0	143.2	28.45	143.1	194.2	17.5	.0373	1.09927
162.0	252.4	7.887	1832.0	2542.0	143.6	28.19	131.9	195.6	17.3	.0364	1.09665
163.0	241.1	7.536	1922.0	2665.0	144.4	27.71	115.3	198.2	17.1	.0349	1.09221
164.0	231.8	7.245	2001.0	2774.0	145.1	27.29	103.4	200.7	16.9	.0337	1.08855
165.0	223.9	6.996	2072.0	2873.0	145.7	26.92	94.52	203.1	16.7	.0327	1.08543
166.0	216.9	6.780	2138.0	2964.0	146.2	26.59	87.59	205.3	16.6	.0318	1.08271
167.0	210.8	6.588	2198.0	3049.0	146.7	26.28	82.01	207.4	16.5	.0311	1.08031
168.0	205.3	6.416	2255.0	3128.0	147.2	26.01	77.42	209.5	16.4	.0305	1.07816
170.0	195.8	6.118	2360.0	3276.0	148.1	25.52	70.29	213.4	16.3	.0295	1.07443
172.0	187.7	5.865	2456.0	3411.0	148.9	25.10	64.99	217.0	16.2	.0286	1.07129
174.0	180.7	5.646	2544.0	3536.0	149.6	24.73	60.87	220.5	16.2	.0280	1.06857
176.0	174.5	5.454	2628.0	3655.0	150.3	24.42	57.58	223.7	16.2	.0274	1.06618
178.0	169.0	5.281	2707.0	3767.0	150.9	24.13	54.89	226.8	16.2	.0270	1.06404
180.0	164.0	5.126	2782.0	3874.0	151.5	23.88	52.64	229.8	16.2	.0266	1.06212
182.0	159.5	4.985	2854.0	3978.0	152.1	23.66	50.73	232.7	16.2	.0263	1.06037
185.0	153.4	4.794	2958.0	4126.0	152.9	23.37	48.36	236.7	16.3	.0259	1.05801
190.0	144.6	4.520	3121.0	4360.0	154.1	22.97	45.32	243.1	16.4	.0254	1.05464
195.0	137.2	4.289	3275.0	4581.0	155.3	22.66	43.06	249.1	16.6	.0251	1.05179
200.0	130.8	4.088	3422.0	4792.0	156.3	22.41	41.31	254.7	16.8	.0250	1.04933
205.0	125.2	3.912	3563.0	4994.0	157.3	22.21	39.92	260.0	17.0	.0249	1.04717
210.0	120.2	3.755	3700.0	5191.0	158.3	22.04	38.79	265.1	17.2	.0249	1.04525
215.0	115.7	3.614	3833.0	5383.0	159.2	21.91	37.86	269.9	17.4	.0250	1.04353
220.0	111.6	3.487	3964.0	5570.0	160.1	21.80	37.08	274.5	17.7	.0251	1.04197
230.0	104.4	3.263	4218.0	5934.0	161.7	21.63	35.84	283.4	18.1	.0254	1.03923
240.0	98.29	3.072	4465.0	6288.0	163.2	21.52	34.92	291.6	18.6	.0258	1.03690
250.0	92.99	2.906	4706.0	6633.0	164.6	21.45	34.22	299.4	19.1	.0263	1.03489
260.0	88.33	2.760	4944.0	6973.0	165.9	21.40	33.66	306.8	19.6	.0268	1.03312
270.0	84.19	2.631	5179.0	7307.0	167.2	21.38	33.22	313.9	20.1	.0273	1.03155
280.0	80.48	2.515	5411.0	7637.0	168.4	21.37	32.86	320.7	20.6	.0279	1.03014
290.0	77.12	2.410	5641.0	7965.0	169.5	21.38	32.57	327.3	21.1	.0285	1.02887
300.0	74.07	2.315	5870.0	8289.0	170.6	21.40	32.34	333.6	21.6	.0291	1.02772
310.0	71.28	2.228	6098.0	8611.0	171.7	21.43	32.15	339.7	22.1	.0297	1.02666
320.0	68.72	2.148	6324.0	8932.0	172.7	21.47	31.99	345.6	22.6	.0304	1.02570
340.0	64.16	2.005	6777.0	9570.0	174.6	21.57	31.77	356.9	23.6	.0317	1.02398
360.0	60.22	1.882	7228.0	10200.0	176.5	21.70	31.64	367.6	24.5	.0330	1.02249
380.0	56.78	1.774	7680.0	10840.0	178.2	21.85	31.58	377.8	25.4	.0344	1.02119
400.0	53.73	1.679	8132.0	11470.0	179.8	22.01	31.57	387.5	26.4	.0357	1.02004

5.70 MPa isobar

T K	Density kg/m³	Density mol/dm³	E J/mol	H J/mol	S J/(mol·K)	C_v J/(mol·K)	C_p J/(mol·K)	Sound m/s	Visc. μPa·s	Therm. W/(m·K)	Diel. const.
55.01[a]	1310.0	40.93	−6174.0	−6034.0	67.27	36.11	53.09	1142.0	503.0	.193	1.57037
80.0	1200.0	37.51	−4860.0	−4708.0	87.15	31.32	53.06	1004.0	265.0	.168	1.51589
90.0	1154.0	36.06	−4334.0	−4176.0	93.41	29.95	53.38	930.9	207.0	.154	1.49331
100.0	1105.0	34.55	−3804.0	−3639.0	99.07	28.81	54.13	855.1	164.0	.140	1.46988
110.0	1054.0	32.93	−3265.0	−3091.0	104.3	27.85	55.55	775.9	132.0	.126	1.44514
120.0	997.1	31.16	−2708.0	−2525.0	109.2	27.06	58.03	691.4	106.0	.113	1.41836
125.0	966.1	30.19	−2419.0	−2230.0	111.6	26.73	59.91	646.5	95.7	.107	1.40382
130.0	932.5	29.14	−2120.0	−1925.0	114.0	26.46	62.46	599.3	86.0	.100	1.38819
135.0	895.4	27.98	−1808.0	−1604.0	116.4	26.26	66.01	549.1	76.9	.0940	1.37109

Thermophysical properties of oxygen—Continued

T K	Density kg/m^3	Density mol/dm^3	*E* J/mol	*H* J/mol	*S* J/(mol·K)	C_v J/(mol·K)	C_p J/(mol·K)	Sound m/s	Visc. μPa·s	Therm. W/(m·K)	Diel. const.
136.0	887.5	27.73	−1743.0	−1537.0	116.9	26.23	66.89	538.7	75.1	.0927	1.36745
138.0	870.9	27.22	−1611.0	−1402.0	117.9	26.19	68.90	517.2	71.7	.0902	1.35989
140.0	853.4	26.67	−1475.0	−1262.0	118.9	26.16	71.29	495.1	68.2	.0877	1.35190
142.0	834.7	26.09	−1335.0	−1116.0	120.0	26.16	74.20	472.0	64.8	.0851	1.34342
144.0	814.6	25.46	−1188.0	−964.3	121.0	26.18	77.85	447.9	61.4	.0826	1.33431
146.0	792.6	24.77	−1034.0	−804.1	122.1	26.24	82.59	422.5	58.0	.0800	1.32440
148.0	768.1	24.00	−870.2	−632.8	123.3	26.36	89.05	395.3	54.5	.0775	1.31345
150.0	740.1	23.13	−692.3	−445.9	124.5	26.55	98.54	365.7	50.8	.0750	1.30102
151.0	724.3	22.64	−595.9	−344.1	125.2	26.68	105.2	349.7	48.8	.0738	1.29402
152.0	706.8	22.09	−492.7	−234.6	125.9	26.85	114.2	332.6	46.7	.0726	1.28629
153.0	686.9	21.47	−380.1	−114.5	126.7	27.07	126.8	314.1	44.5	.0713	1.27756
154.0	663.6	20.74	−253.7	21.1	127.6	27.38	146.1	293.8	42.1	.0700	1.26738
154.5	650.1	20.32	−183.0	97.5	128.1	27.57	160.2	282.8	40.7	.0693	1.26149
155.0	634.7	19.84	−105.1	182.2	128.7	27.80	179.6	270.9	39.3	.0686	1.25484
155.5	616.8	19.28	−17.1	278.6	129.3	28.10	207.8	258.1	37.6	.0678	1.24711
156.0	595.1	18.60	86.1	392.6	130.0	28.47	251.9	244.1	35.7	.0671	1.23778
156.5	567.3	17.73	214.1	535.6	130.9	28.97	327.3	228.6	33.5	.0663	1.22587
156.6	560.6	17.52	244.0	569.4	131.1	29.10	348.3	225.4	32.9	.0661	1.22304
156.8	545.9	17.06	309.7	643.8	131.6	29.36	398.2	218.8	31.8	.0658	1.21679
157.0	529.0	16.53	384.5	729.3	132.2	29.66	457.9	212.2	30.6	.0658	1.20964
157.2	509.7	15.93	469.5	827.3	132.8	29.99	522.2	205.8	29.3	.0657	1.20152
157.4	488.3	15.26	564.1	937.6	133.5	30.31	577.7	200.2	27.9	.0654	1.19255
157.6	465.7	14.55	665.1	1057.0	134.2	30.60	609.6	195.6	26.5	.0646	1.18313
157.8	443.2	13.85	767.9	1179.0	135.0	30.81	612.9	192.2	25.2	.0631	1.17378
158.0	421.8	13.18	867.9	1300.0	135.8	30.93	592.2	189.8	24.1	.0611	1.16498
158.2	402.4	12.57	961.9	1415.0	136.5	30.96	554.8	188.3	23.1	.0587	1.15698
158.4	385.1	12.03	1048.0	1522.0	137.2	30.91	508.1	187.5	22.3	.0563	1.14992
158.6	370.0	11.56	1125.0	1618.0	137.8	30.82	459.1	187.1	21.6	.0532	1.14378
158.8	356.9	11.15	1194.0	1705.0	138.3	30.69	412.7	187.0	21.0	.0515	1.13848
159.0	345.6	10.80	1256.0	1784.0	138.8	30.55	371.5	187.1	20.5	.0498	1.13390
159.2	335.8	10.49	1311.0	1854.0	139.3	30.39	336.0	187.4	20.1	.0484	1.12992
159.4	327.1	10.22	1361.0	1919.0	139.7	30.23	305.9	187.8	19.8	.0471	1.12643
159.6	319.4	9.982	1406.0	1977.0	140.1	30.07	280.5	188.2	19.5	.0459	1.12334
159.8	312.5	9.767	1447.0	2031.0	140.4	29.92	259.0	188.7	19.3	.0449	1.12058
160.0	306.3	9.573	1485.0	2081.0	140.7	29.77	240.6	189.3	19.1	.0440	1.11810
160.5	293.1	9.160	1570.0	2192.0	141.4	29.41	205.2	190.6	18.6	.0420	1.11281
161.0	282.3	8.821	1642.0	2288.0	142.0	29.08	179.8	192.0	18.3	.0404	1.10850
161.5	273.1	8.536	1705.0	2373.0	142.5	28.77	161.0	193.4	18.0	.0391	1.10487
162.0	265.3	8.290	1762.0	2449.0	143.0	28.49	146.4	194.8	17.8	.0380	1.10174
163.0	252.2	7.881	1861.0	2585.0	143.8	27.98	125.4	197.5	17.4	.0362	1.09656
164.0	241.6	7.549	1947.0	2702.0	144.5	27.54	111.0	200.0	17.2	.0348	1.09238
165.0	232.6	7.270	2024.0	2808.0	145.2	27.14	100.4	202.4	17.0	.0336	1.08887
166.0	225.0	7.030	2093.0	2904.0	145.8	26.79	92.32	204.7	16.8	.0327	1.08585
167.0	218.2	6.820	2157.0	2993.0	146.3	26.47	85.92	206.8	16.7	.0319	1.08321
168.0	212.2	6.632	2217.0	3076.0	146.8	26.18	80.71	208.9	16.6	.0312	1.08086
170.0	201.9	6.310	2326.0	3229.0	147.7	25.66	72.74	212.9	16.5	.0300	1.07683
172.0	193.2	6.039	2425.0	3369.0	148.5	25.23	66.89	216.6	16.4	.0291	1.07346
174.0	185.8	5.806	2516.0	3498.0	149.3	24.85	62.41	220.1	16.3	.0284	1.07056
176.0	179.3	5.602	2601.0	3619.0	150.0	24.52	58.86	223.3	16.3	.0278	1.06802
178.0	173.5	5.421	2682.0	3734.0	150.6	24.22	55.97	226.5	16.3	.0273	1.06577
180.0	168.2	5.258	2759.0	3843.0	151.2	23.97	53.57	229.5	16.3	.0269	1.06375
182.0	163.5	5.109	2833.0	3948.0	151.8	23.73	51.54	232.4	16.3	.0266	1.06191
184.0	159.2	4.974	2903.0	4049.0	152.4	23.53	49.80	235.1	16.4	.0263	1.06023
190.0	148.0	4.625	3103.0	4336.0	153.9	23.02	45.83	242.9	16.5	.0256	1.05593
195.0	140.3	4.385	3259.0	4559.0	155.0	22.70	43.47	248.9	16.7	.0253	1.05298
200.0	133.7	4.178	3407.0	4771.0	156.1	22.44	41.64	254.6	16.8	.0252	1.05043
205.0	127.9	3.996	3549.0	4976.0	157.1	22.23	40.20	259.9	17.0	.0251	1.04819
210.0	122.7	3.834	3687.0	5174.0	158.1	22.07	39.03	265.0	17.3	.0251	1.04621
215.0	118.0	3.689	3821.0	5366.0	159.0	21.93	38.07	269.8	17.5	.0251	1.04444
220.0	113.8	3.557	3952.0	5555.0	159.9	21.82	37.26	274.5	17.7	.0252	1.04283
230.0	106.5	3.327	4207.0	5920.0	161.5	21.65	35.99	283.3	18.2	.0255	1.04002
240.0	100.2	3.131	4455.0	6275.0	163.0	21.53	35.04	291.6	18.7	.0259	1.03763

Thermophysical properties of oxygen—Continued

T K	Density kg/m³	Density mol/dm³	E J/mol	H J/mol	S J/(mol·K)	C_v J/(mol·K)	C_p J/(mol·K)	Sound m/s	Visc. μPa·s	Therm. W/(m·K)	Diel. const.
250.0	94.77	2.962	4697.0	6622.0	164.4	21.46	34.32	299.5	19.2	.0264	1.03557
260.0	90.00	2.813	4936.0	6962.0	165.7	21.41	33.75	306.9	19.6	.0269	1.03375
270.0	85.77	2.680	5171.0	7297.0	167.0	21.39	33.29	314.0	20.1	.0274	1.03215
280.0	81.97	2.562	5403.0	7628.0	168.2	21.38	32.93	320.8	20.6	.0280	1.03071
290.0	78.54	2.455	5634.0	7956.0	169.4	21.39	32.63	327.4	21.1	.0285	1.02941
300.0	75.43	2.357	5863.0	8281.0	170.5	21.41	32.39	333.7	21.6	.0292	1.02823
310.0	72.58	2.268	6091.0	8604.0	171.5	21.44	32.20	339.8	22.1	.0298	1.02716
320.0	69.97	2.187	6318.0	8925.0	172.5	21.48	32.04	345.7	22.6	.0304	1.02617
340.0	65.32	2.041	6771.0	9564.0	174.5	21.58	31.81	357.0	23.6	.0317	1.02441
360.0	61.30	1.916	7223.0	10200.0	176.3	21.70	31.68	367.8	24.5	.0330	1.02290
380.0	57.79	1.806	7675.0	10830.0	178.0	21.85	31.61	378.0	25.5	.0344	1.02157
400.0	54.68	1.709	8128.0	11460.0	179.6	22.01	31.60	387.7	26.4	.0358	1.02040
				5.80 MPa isobar							
55.02[a]	1310.0	40.94	−6173.0	−6032.0	67.28	36.11	53.09	1142.0	503.0	.193	1.57040
80.0	1200.0	37.51	−4861.0	−4706.0	87.14	31.33	53.05	1004.0	265.0	.168	1.51597
90.0	1154.0	36.07	−4335.0	−4175.0	93.40	29.96	53.37	931.3	207.0	.154	1.49341
100.0	1106.0	34.56	−3805.0	−3638.0	99.06	28.81	54.11	855.6	164.0	.140	1.47000
110.0	1054.0	32.94	−3266.0	−3090.0	104.3	27.85	55.51	776.6	132.0	.126	1.44530
120.0	997.6	31.18	−2710.0	−2524.0	109.2	27.06	57.98	692.3	107.0	.113	1.41857
125.0	966.6	30.21	−2422.0	−2230.0	111.6	26.74	59.83	647.6	95.9	.107	1.40406
130.0	933.1	29.16	−2123.0	−1924.0	114.0	26.47	62.34	600.5	86.2	.100	1.38849
135.0	896.2	28.01	−1812.0	−1605.0	116.4	26.27	65.83	550.6	77.1	.0941	1.37147
140.0	854.5	26.71	−1481.0	−1263.0	118.9	26.16	70.99	497.0	68.5	.0878	1.35241
142.0	836.0	26.13	−1341.0	−1119.0	119.9	26.15	73.83	474.1	65.1	.0853	1.34399
144.0	816.1	25.50	−1195.0	−967.6	121.0	26.17	77.36	450.3	61.7	.0827	1.33497
146.0	794.3	24.82	−1042.0	−808.5	122.1	26.23	81.90	425.1	58.3	.0802	1.32519
148.0	770.2	24.07	−879.9	−638.9	123.2	26.34	88.04	398.3	54.8	.0777	1.31440
150.0	742.9	23.22	−704.4	−454.6	124.5	26.51	96.92	369.3	51.1	.0752	1.30223
151.0	727.5	22.74	−609.8	−354.7	125.1	26.64	103.1	353.7	49.2	.0740	1.29541
152.0	710.5	22.21	−509.0	−247.8	125.8	26.79	111.2	337.1	47.2	.0728	1.28794
153.0	691.5	21.61	−399.7	−131.4	126.6	27.00	122.3	319.4	45.0	.0715	1.27957
154.0	669.5	20.92	−278.7	−1.5	127.4	27.27	138.6	300.0	42.7	.0702	1.26995
154.5	657.0	20.53	−211.9	70.6	127.9	27.44	150.1	289.5	41.4	.0695	1.26448
155.0	642.9	20.09	−139.4	149.2	128.4	27.64	165.3	278.4	40.1	.0688	1.25840
155.5	627.0	19.59	−59.2	236.8	129.0	27.89	186.0	266.6	38.6	.0681	1.25150
156.0	608.4	19.01	31.8	336.8	129.6	28.20	216.1	253.8	36.9	.0673	1.24347
156.5	585.7	18.31	138.7	455.6	130.4	28.59	262.5	240.0	35.0	.0665	1.23376
157.0	556.9	17.40	270.7	604.0	131.3	29.10	337.2	225.1	32.7	.0656	1.22147
157.2	542.9	16.97	333.5	675.4	131.8	29.34	377.8	219.1	31.7	.0654	1.21553
157.4	527.2	16.48	403.4	755.5	132.3	29.61	423.4	213.1	30.5	.0651	1.20891
157.6	509.9	15.93	480.8	844.8	132.9	29.89	469.5	207.4	29.3	.0648	1.20159
157.8	491.0	15.35	564.8	942.8	133.5	30.17	508.3	202.4	28.1	.0644	1.19369
158.0	471.4	14.73	653.4	1047.0	134.1	30.42	532.5	198.2	26.9	.0636	1.18548
158.5	423.5	13.23	876.0	1314.0	135.8	30.78	520.8	191.5	24.2	.0581	1.16566
158.6	414.7	12.96	918.3	1366.0	136.2	30.81	509.0	190.8	23.8	.0573	1.16205
158.8	398.3	12.45	998.8	1465.0	136.8	30.80	479.4	189.7	22.9	.0556	1.15534
159.0	383.6	11.99	1073.0	1557.0	137.4	30.75	445.2	189.0	22.3	.0540	1.14931
159.2	370.4	11.58	1142.0	1643.0	137.9	30.66	409.6	188.7	21.7	.0524	1.14396
159.4	358.8	11.21	1204.0	1721.0	138.4	30.54	375.3	188.6	21.2	.0509	1.13923
159.6	348.4	10.89	1260.0	1793.0	138.8	30.41	343.9	188.7	20.7	.0495	1.13504
159.8	339.3	10.60	1312.0	1859.0	139.3	30.27	315.8	188.9	20.3	.0483	1.13133
160.0	331.1	10.35	1359.0	1920.0	139.6	30.13	291.2	189.2	20.0	.0471	1.12803
160.2	323.7	10.12	1402.0	1976.0	140.0	29.98	269.8	189.6	19.7	.0460	1.12506
160.5	313.9	9.810	1461.0	2052.0	140.5	29.77	242.8	190.3	19.4	.0446	1.12114
161.0	300.3	9.384	1547.0	2165.0	141.2	29.42	208.3	191.5	18.9	.0427	1.11567
161.5	289.0	9.033	1620.0	2262.0	141.8	29.10	183.1	192.8	18.5	.0410	1.11120
162.0	279.6	8.737	1685.0	2349.0	142.3	28.80	164.1	194.2	18.3	.0397	1.10742
162.5	271.4	8.481	1743.0	2427.0	142.8	28.52	149.2	195.5	18.0	.0386	1.10417
163.0	264.2	8.256	1796.0	2499.0	143.2	28.26	137.4	196.8	17.8	.0376	1.10132
164.0	252.0	7.875	1890.0	2627.0	144.0	27.79	119.7	199.3	17.5	.0359	1.09650
165.0	242.0	7.561	1973.0	2740.0	144.7	27.37	107.1	201.7	17.2	.0346	1.09253

Thermophysical properties of oxygen—Continued

T K	Density kg/m³	Density mol/dm³	E J/mol	H J/mol	S J/(mol·K)	C_v J/(mol·K)	C_p J/(mol·K)	Sound m/s	Visc. μPa·s	Therm. W/(m·K)	Diel. const.
166.0	233.4	7.294	2047.0	2842.0	145.3	26.99	97.60	204.1	17.1	.0336	1.08917
167.0	226.0	7.063	2114.0	2936.0	145.9	26.66	90.23	206.3	16.9	.0327	1.08626
168.0	219.4	6.858	2177.0	3023.0	146.4	26.35	84.31	208.4	16.8	.0319	1.08369
170.0	208.3	6.509	2291.0	3182.0	147.3	25.81	75.37	212.4	16.6	.0306	1.07932
172.0	199.0	6.219	2393.0	3326.0	148.2	25.36	68.92	216.1	16.5	.0297	1.07569
174.0	191.1	5.971	2487.0	3459.0	148.9	24.96	64.04	219.6	16.5	.0289	1.07260
176.0	184.1	5.754	2575.0	3583.0	149.7	24.62	60.19	223.0	16.4	.0282	1.06991
178.0	178.0	5.563	2657.0	3700.0	150.3	24.32	57.09	226.1	16.4	.0277	1.06753
180.0	172.5	5.391	2736.0	3811.0	150.9	24.05	54.52	229.2	16.4	.0272	1.06540
182.0	167.5	5.236	2810.0	3918.0	151.5	23.81	52.37	232.1	16.4	.0269	1.06348
184.0	163.0	5.094	2882.0	4021.0	152.1	23.60	50.54	234.9	16.5	.0265	1.06172
190.0	151.4	4.731	3085.0	4311.0	153.6	23.07	46.36	242.7	16.6	.0259	1.05724
195.0	143.4	4.482	3242.0	4536.0	154.8	22.74	43.88	248.8	16.7	.0255	1.05417
200.0	136.6	4.267	3392.0	4751.0	155.9	22.48	41.98	254.4	16.9	.0253	1.05153
205.0	130.5	4.080	3535.0	4957.0	156.9	22.26	40.48	259.8	17.1	.0252	1.04922
210.0	125.2	3.913	3674.0	5156.0	157.9	22.09	39.27	264.9	17.3	.0252	1.04718
215.0	120.4	3.763	3809.0	5350.0	158.8	21.95	38.28	269.8	17.5	.0253	1.04535
220.0	116.1	3.628	3940.0	5539.0	159.7	21.83	37.44	274.5	17.8	.0253	1.04370
230.0	108.5	3.392	4197.0	5907.0	161.3	21.66	36.14	283.3	18.2	.0256	1.04081
240.0	102.1	3.191	4445.0	6263.0	162.8	21.54	35.16	291.7	18.7	.0260	1.03836
250.0	96.55	3.017	4688.0	6611.0	164.2	21.47	34.42	299.5	19.2	.0264	1.03624
260.0	91.67	2.865	4927.0	6952.0	165.6	21.42	33.83	307.0	19.7	.0269	1.03439
270.0	87.35	2.730	5163.0	7288.0	166.8	21.39	33.37	314.1	20.2	.0275	1.03275
280.0	83.47	2.608	5396.0	7619.0	168.0	21.39	33.00	320.9	20.7	.0280	1.03128
290.0	79.97	2.499	5627.0	7948.0	169.2	21.39	32.69	327.5	21.2	.0286	1.02995
300.0	76.79	2.400	5857.0	8274.0	170.3	21.41	32.45	333.8	21.7	.0292	1.02875
310.0	73.88	2.309	6085.0	8597.0	171.4	21.44	32.24	339.9	22.1	.0298	1.02765
320.0	71.22	2.226	6312.0	8919.0	172.4	21.48	32.08	345.9	22.6	.0305	1.02664
340.0	66.48	2.077	6766.0	9558.0	174.3	21.58	31.85	357.2	23.6	.0318	1.02485
360.0	62.38	1.949	7218.0	10190.0	176.1	21.71	31.71	367.9	24.5	.0331	1.02330
380.0	58.80	1.838	7670.0	10830.0	177.8	21.85	31.64	378.1	25.5	.0344	1.02195
400.0	55.64	1.739	8123.0	11460.0	179.5	22.02	31.62	387.9	26.4	.0358	1.02076
					5.90 MPa isobar						
55.03[a]	1310.0	40.94	−6173.0	−6029.0	67.28	36.11	53.09	1142.0	503.0	.193	1.57043
80.0	1200.0	37.52	−4862.0	−4705.0	87.13	31.34	53.04	1005.0	265.0	.168	1.51605
90.0	1154.0	36.08	−4336.0	−4173.0	93.39	29.96	53.36	931.7	207.0	.154	1.49351
100.0	1106.0	34.56	−3807.0	−3636.0	99.05	28.82	54.09	856.2	164.0	.140	1.47012
110.0	1054.0	32.95	−3268.0	−3089.0	104.3	27.86	55.48	777.2	132.0	.126	1.44546
120.0	998.0	31.19	−2712.0	−2523.0	109.2	27.07	57.92	693.2	107.0	.113	1.41878
125.0	967.2	30.22	−2424.0	−2229.0	111.6	26.75	59.75	648.6	96.1	.107	1.40431
130.0	933.8	29.18	−2127.0	−1924.0	114.0	26.47	62.23	601.8	86.3	.100	1.38879
135.0	897.1	28.03	−1816.0	−1605.0	116.4	26.27	65.66	552.1	77.3	.0942	1.37185
140.0	855.6	26.74	−1486.0	−1265.0	118.9	26.16	70.71	498.9	68.7	.0879	1.35290
142.0	837.3	26.17	−1346.0	−1121.0	119.9	26.15	73.46	476.2	65.3	.0854	1.34455
144.0	817.5	25.55	−1202.0	−970.8	120.9	26.17	76.88	452.6	61.9	.0829	1.33562
146.0	796.0	24.88	−1050.0	−812.9	122.0	26.22	81.25	427.7	58.5	.0803	1.32595
148.0	772.3	24.14	−889.3	−644.8	123.2	26.32	87.09	401.4	55.1	.0779	1.31533
150.0	745.5	23.30	−716.1	−462.9	124.4	26.48	95.42	372.9	51.5	.0754	1.30340
151.0	730.5	22.83	−623.2	−364.7	125.0	26.60	101.1	357.7	49.6	.0742	1.29675
152.0	714.1	22.32	−524.5	−260.1	125.7	26.74	108.5	341.5	47.6	.0730	1.28951
153.0	695.8	21.75	−418.3	−147.0	126.5	26.93	118.3	324.3	45.5	.0717	1.28146
154.0	674.9	21.09	−301.7	−22.0	127.3	27.18	132.4	305.7	43.3	.0705	1.27232
154.5	663.2	20.73	−238.2	46.5	127.7	27.33	142.0	295.8	42.1	.0698	1.26719
155.0	650.2	20.32	−169.9	120.4	128.2	27.51	154.2	285.3	40.8	.0691	1.26156
156.0	619.3	19.35	−13.1	291.7	129.3	27.98	192.4	262.4	37.9	.0676	1.24816
156.5	599.9	18.75	80.6	395.3	130.0	28.29	223.9	249.8	36.2	.0668	1.23984
157.0	576.5	18.02	190.8	518.3	130.7	28.69	271.1	236.3	34.3	.0660	1.22981
157.5	547.1	17.10	325.3	670.3	131.7	29.20	341.4	222.3	32.0	.0649	1.21731
157.6	540.3	16.89	355.9	705.3	131.9	29.32	358.3	219.5	31.5	.0648	1.21443
157.8	525.8	16.43	421.4	780.5	132.4	29.56	393.4	214.1	30.5	.0645	1.20828
158.0	509.9	15.94	492.4	862.6	132.9	29.80	427.4	209.0	29.4	.0641	1.20162

Thermophysical properties of oxygen—Continued

T K	Density kg/m^3	Density mol/dm^3	E J/mol	H J/mol	S J/(mol·K)	C_v J/(mol·K)	C_p J/(mol·K)	Sound m/s	Visc. μPa·s	Therm. W/(m·K)	Diel. const.
158.5	466.9	14.59	687.5	1092.0	134.4	30.36	479.0	198.9	26.7	.0607	1.18362
159.0	424.7	13.27	885.9	1330.0	135.9	30.65	465.5	193.2	24.3	.0573	1.16617
159.2	409.4	12.80	960.6	1422.0	136.5	30.67	446.6	191.9	23.5	.0559	1.15988
159.4	395.3	12.35	1031.0	1509.0	137.0	30.65	423.2	191.1	22.9	.0545	1.15411
159.6	382.5	11.95	1097.0	1591.0	137.5	30.59	397.1	190.5	22.3	.0531	1.14885
159.8	370.8	11.59	1158.0	1668.0	138.0	30.50	370.2	190.3	21.7	.0517	1.14411
160.0	360.3	11.26	1215.0	1739.0	138.4	30.39	344.1	190.2	21.3	.0504	1.13984
160.2	350.8	10.96	1267.0	1805.0	138.9	30.27	319.5	190.3	20.9	.0492	1.13600
160.4	342.2	10.70	1315.0	1867.0	139.2	30.15	297.0	190.4	20.5	.0481	1.13254
161.0	321.0	10.03	1440.0	2028.0	140.2	29.75	242.7	191.4	19.7	.0452	1.12399
161.5	307.1	9.597	1526.0	2141.0	140.9	29.42	209.9	192.5	19.2	.0432	1.11841
162.0	295.6	9.237	1600.0	2239.0	141.6	29.11	185.3	193.7	18.8	.0416	1.11380
162.5	285.8	8.931	1666.0	2327.0	142.1	28.81	166.4	195.0	18.5	.0403	1.10990
163.0	277.3	8.667	1725.0	2406.0	142.6	28.54	151.6	196.3	18.3	.0391	1.10654
164.0	263.3	8.228	1829.0	2546.0	143.4	28.04	129.8	198.8	17.9	.0372	1.10096
165.0	251.9	7.872	1918.0	2668.0	144.2	27.60	114.6	201.2	17.6	.0357	1.09645
166.0	242.3	7.574	1998.0	2777.0	144.8	27.20	103.5	203.5	17.3	.0345	1.09269
167.0	234.1	7.317	2070.0	2876.0	145.4	26.85	94.99	205.7	17.2	.0335	1.08946
168.0	227.0	7.093	2135.0	2967.0	146.0	26.53	88.24	207.9	17.0	.0327	1.08664
170.0	214.9	6.715	2254.0	3133.0	147.0	25.96	78.21	211.9	16.8	.0313	1.08189
172.0	204.9	6.403	2361.0	3282.0	147.8	25.49	71.09	215.7	16.7	.0302	1.07799
174.0	196.4	6.139	2458.0	3419.0	148.6	25.08	65.75	219.3	16.6	.0293	1.07470
176.0	189.1	5.910	2547.0	3546.0	149.3	24.72	61.59	222.6	16.6	.0286	1.07184
178.0	182.6	5.708	2632.0	3666.0	150.0	24.41	58.26	225.8	16.5	.0281	1.06933
180.0	176.9	5.527	2712.0	3779.0	150.7	24.13	55.52	228.9	16.5	.0276	1.06709
182.0	171.7	5.365	2788.0	3888.0	151.3	23.88	53.23	231.8	16.5	.0272	1.06507
184.0	166.9	5.216	2861.0	3992.0	151.8	23.66	51.29	234.6	16.5	.0268	1.06323
190.0	154.8	4.838	3067.0	4286.0	153.4	23.13	46.89	242.5	16.7	.0261	1.05856
195.0	146.5	4.580	3225.0	4514.0	154.6	22.78	44.30	248.6	16.8	.0257	1.05537
200.0	139.4	4.358	3376.0	4730.0	155.7	22.51	42.33	254.3	17.0	.0255	1.05264
205.0	133.2	4.164	3521.0	4938.0	156.7	22.29	40.77	259.7	17.2	.0254	1.05026
210.0	127.7	3.992	3660.0	5138.0	157.7	22.12	39.52	264.8	17.4	.0253	1.04815
215.0	122.8	3.839	3796.0	5333.0	158.6	21.97	38.49	269.7	17.6	.0254	1.04627
220.0	118.4	3.700	3929.0	5524.0	159.5	21.85	37.63	274.4	17.8	.0255	1.04457
230.0	110.6	3.457	4186.0	5893.0	161.1	21.68	36.28	283.3	18.3	.0257	1.04160
240.0	104.0	3.251	4436.0	6250.0	162.6	21.56	35.29	291.7	18.7	.0261	1.03909
250.0	98.34	3.073	4679.0	6599.0	164.1	21.48	34.52	299.6	19.2	.0265	1.03692
260.0	93.35	2.917	4919.0	6941.0	165.4	21.43	33.92	307.0	19.7	.0270	1.03503
270.0	88.93	2.779	5155.0	7278.0	166.7	21.40	33.45	314.2	20.2	.0275	1.03335
280.0	84.97	2.655	5389.0	7611.0	167.9	21.39	33.06	321.0	20.7	.0281	1.03184
290.0	81.39	2.544	5620.0	7940.0	169.0	21.40	32.75	327.6	21.2	.0287	1.03049
300.0	78.15	2.442	5850.0	8266.0	170.1	21.42	32.50	333.9	21.7	.0293	1.02926
310.0	75.19	2.350	6079.0	8590.0	171.2	21.45	32.29	340.1	22.2	.0299	1.02814
320.0	72.47	2.265	6306.0	8912.0	172.2	21.49	32.13	346.0	22.7	.0305	1.02711
340.0	67.63	2.114	6760.0	9552.0	174.2	21.59	31.89	357.3	23.6	.0318	1.02528
360.0	63.46	1.983	7213.0	10190.0	176.0	21.71	31.74	368.1	24.6	.0331	1.02371
380.0	59.81	1.869	7665.0	10820.0	177.7	21.86	31.67	378.3	25.5	.0345	1.02233
400.0	56.59	1.769	8119.0	11450.0	179.3	22.02	31.65	388.0	26.4	.0358	1.02112

6.00 MPa isobar

T K	Density kg/m^3	Density mol/dm^3	E J/mol	H J/mol	S J/(mol·K)	C_v J/(mol·K)	C_p J/(mol·K)	Sound m/s	Visc. μPa·s	Therm. W/(m·K)	Diel. const.
55.05[a]	1310.0	40.94	−6173.0	−6027.0	67.28	36.11	53.09	1142.0	504.0	.193	1.57046
80.0	1201.0	37.52	−4863.0	−4703.0	87.12	31.34	53.03	1005.0	265.0	.168	1.51614
90.0	1155.0	36.08	−4337.0	−4171.0	93.38	29.97	53.34	932.1	207.0	.154	1.49361
100.0	1106.0	34.57	−3808.0	−3634.0	99.03	28.83	54.07	856.7	164.0	.140	1.47025
110.0	1055.0	32.96	−3269.0	−3087.0	104.2	27.87	55.45	777.9	132.0	.126	1.44561
120.0	998.5	31.20	−2714.0	−2522.0	109.2	27.08	57.86	694.1	107.0	.113	1.41898
125.0	967.7	30.24	−2427.0	−2228.0	111.6	26.75	59.68	649.7	96.2	.107	1.40456
130.0	934.4	29.20	−2130.0	−1924.0	113.9	26.48	62.11	603.0	86.5	.100	1.38909
135.0	897.9	28.06	−1820.0	−1606.0	116.3	26.27	65.49	553.6	77.5	.0943	1.37223
140.0	856.7	26.77	−1491.0	−1267.0	118.8	26.16	70.42	500.7	68.9	.0880	1.35340
142.0	838.5	26.20	−1352.0	−1123.0	119.8	26.14	73.10	478.3	65.5	.0855	1.34511

Thermophysical properties of oxygen—Continued

T K	Density kg/m³	Density mol/dm³	E J/mol	H J/mol	S J/(mol·K)	C_v J/(mol·K)	C_p J/(mol·K)	Sound m/s	Visc. μPa·s	Therm. W/(m·K)	Diel. const.
144.0	818.9	25.59	-1208.0	-973.9	120.9	26.16	76.41	454.9	62.2	.0830	1.33626
146.0	797.7	24.93	-1058.0	-817.1	122.0	26.21	80.62	430.3	58.8	.0805	1.32671
148.0	774.3	24.20	-898.5	-650.6	123.1	26.30	86.19	404.3	55.4	.0780	1.31624
150.0	748.1	23.38	-727.5	-470.8	124.3	26.45	94.02	376.4	51.8	.0756	1.30453
151.0	733.4	22.92	-636.0	-374.3	124.9	26.56	99.31	361.5	50.0	.0744	1.29805
152.0	717.5	22.42	-539.3	-271.7	125.6	26.70	106.0	345.8	48.0	.0732	1.29100
153.0	699.9	21.87	-435.8	-161.5	126.3	26.87	114.9	329.1	46.0	.0719	1.28325
154.0	680.0	21.25	-323.2	-40.8	127.1	27.09	127.2	311.2	43.8	.0707	1.27452
155.0	656.8	20.53	-197.5	94.8	128.0	27.39	145.4	291.7	41.5	.0694	1.26441
156.0	628.5	19.64	-51.6	253.8	129.0	27.79	175.3	270.2	38.8	.0679	1.25216
156.5	611.5	19.11	33.0	347.0	129.6	28.06	198.4	258.5	37.2	.0672	1.24481
157.0	591.5	18.49	129.2	453.8	130.3	28.38	230.8	246.1	35.5	.0663	1.23624
157.5	567.5	17.74	241.8	580.1	131.1	28.79	277.3	233.2	33.6	.0653	1.22599
158.0	538.0	16.81	377.0	733.9	132.1	29.28	340.0	220.1	31.4	.0643	1.21346
158.5	502.4	15.70	538.4	920.5	133.3	29.82	404.2	208.4	28.9	.0623	1.19847
159.0	463.3	14.48	717.9	1132.0	134.6	30.28	434.9	199.9	26.5	.0598	1.18214
159.5	425.7	13.30	896.9	1348.0	135.9	30.51	421.3	194.9	24.4	.0567	1.16656
159.6	418.7	13.08	931.1	1390.0	136.2	30.52	414.7	194.2	24.1	.0560	1.16368
159.8	405.4	12.67	997.4	1471.0	136.7	30.52	398.5	193.2	23.4	.0548	1.15821
160.0	393.0	12.28	1060.0	1549.0	137.2	30.49	379.6	192.5	22.8	.0535	1.15314
160.2	381.6	11.92	1120.0	1623.0	137.7	30.43	359.1	192.1	22.3	.0523	1.14849
160.4	371.1	11.60	1175.0	1693.0	138.1	30.34	338.2	191.8	21.8	.0511	1.14423
160.6	361.5	11.30	1227.0	1758.0	138.5	30.25	317.6	191.8	21.4	.0500	1.14034
161.0	344.8	10.78	1321.0	1877.0	139.3	30.02	279.8	192.0	20.7	.0479	1.13357
161.5	327.6	10.24	1421.0	2007.0	140.1	29.71	240.6	192.7	20.0	.0456	1.12665
162.0	313.6	9.800	1507.0	2120.0	140.8	29.40	210.2	193.6	19.5	.0437	1.12100
162.5	301.8	9.432	1582.0	2218.0	141.4	29.10	186.6	194.7	19.1	.0421	1.11630
163.0	291.8	9.120	1649.0	2307.0	141.9	28.82	168.1	195.9	18.8	.0408	1.11230
163.5	283.1	8.848	1709.0	2387.0	142.4	28.55	153.3	197.1	18.5	.0396	1.10884
164.0	275.5	8.609	1764.0	2461.0	142.8	28.29	141.4	198.3	18.3	.0386	1.10580
165.0	262.5	8.204	1861.0	2592.0	143.6	27.83	123.2	200.7	17.9	.0369	1.10065
166.0	251.8	7.869	1946.0	2709.0	144.3	27.42	110.1	203.0	17.6	.0355	1.09642
167.0	242.7	7.585	2023.0	2814.0	145.0	27.04	100.2	205.2	17.4	.0344	1.09284
168.0	234.8	7.339	2092.0	2910.0	145.6	26.71	92.54	207.4	17.3	.0335	1.08973
169.0	227.9	7.122	2157.0	2999.0	146.1	26.40	86.35	209.5	17.1	.0326	1.08700
170.0	221.7	6.928	2217.0	3083.0	146.6	26.12	81.26	211.5	17.0	.0319	1.08457
172.0	211.0	6.593	2327.0	3237.0	147.5	25.62	73.39	215.3	16.9	.0308	1.08037
174.0	202.0	6.311	2427.0	3378.0	148.3	25.20	67.56	218.9	16.8	.0298	1.07685
176.0	194.2	6.069	2520.0	3508.0	149.0	24.83	63.06	222.3	16.7	.0291	1.07382
178.0	187.4	5.856	2606.0	3631.0	149.7	24.50	59.48	225.5	16.7	.0284	1.07117
180.0	181.3	5.666	2688.0	3747.0	150.4	24.21	56.55	228.6	16.6	.0279	1.06881
182.0	175.8	5.495	2766.0	3857.0	151.0	23.96	54.12	231.6	16.6	.0275	1.06669
184.0	170.9	5.340	2840.0	3964.0	151.6	23.73	52.06	234.4	16.7	.0271	1.06477
186.0	166.3	5.198	2912.0	4066.0	152.1	23.53	50.30	237.1	16.7	.0268	1.06301
190.0	158.3	4.947	3048.0	4261.0	153.2	23.18	47.44	242.4	16.8	.0263	1.05990
195.0	149.7	4.679	3209.0	4491.0	154.4	22.83	44.73	248.5	16.9	.0259	1.05659
200.0	142.4	4.449	3361.0	4709.0	155.5	22.55	42.68	254.2	17.1	.0257	1.05376
205.0	136.0	4.249	3507.0	4919.0	156.5	22.32	41.06	259.6	17.2	.0255	1.05130
210.0	130.3	4.072	3647.0	5121.0	157.5	22.14	39.77	264.8	17.4	.0255	1.04913
215.0	125.2	3.914	3784.0	5317.0	158.4	21.99	38.70	269.7	17.6	.0255	1.04719
220.0	120.7	3.771	3917.0	5508.0	159.3	21.87	37.82	274.4	17.9	.0256	1.04544
230.0	112.7	3.522	4175.0	5879.0	160.9	21.69	36.43	283.4	18.3	.0258	1.04240
240.0	106.0	3.311	4426.0	6238.0	162.4	21.57	35.41	291.7	18.8	.0262	1.03982
250.0	100.1	3.129	4670.0	6588.0	163.9	21.49	34.62	299.6	19.3	.0266	1.03761
260.0	95.03	2.970	4911.0	6931.0	165.2	21.44	34.01	307.1	19.7	.0271	1.03566
270.0	90.51	2.829	5147.0	7268.0	166.5	21.41	33.52	314.3	20.2	.0276	1.03395
280.0	86.47	2.702	5381.0	7602.0	167.7	21.40	33.13	321.1	20.7	.0281	1.03241
290.0	82.82	2.588	5613.0	7931.0	168.9	21.41	32.81	327.7	21.2	.0287	1.03103
300.0	79.51	2.485	5843.0	8258.0	170.0	21.42	32.55	334.1	21.7	.0293	1.02977
310.0	76.49	2.390	6072.0	8582.0	171.0	21.45	32.34	340.2	22.2	.0299	1.02863
320.0	73.71	2.304	6300.0	8905.0	172.1	21.49	32.17	346.1	22.7	.0306	1.02758
340.0	68.79	2.150	6755.0	9546.0	174.0	21.59	31.92	357.5	23.6	.0318	1.02572
360.0	64.54	2.017	7208.0	10180.0	175.8	21.72	31.77	368.2	24.6	.0332	1.02412

Thermophysical properties of oxygen—Continued

T K	Density kg/m³	Density mol/dm³	E J/mol	H J/mol	S J/(mol·K)	C_v J/(mol·K)	C_p J/(mol·K)	Sound m/s	Visc. μPa·s	Therm. W/(m·K)	Diel. const.
380.0	60.83	1.901	7661.0	10820.0	177.5	21.86	31.70	378.4	25.5	.0345	1.02271
400.0	57.55	1.798	8115.0	11450.0	179.2	22.02	31.67	388.2	26.4	.0359	1.02148
				6.50 MPa isobar							
55.10[a]	1310.0	40.95	−6172.0	−6014.0	67.30	36.09	53.08	1144.0	505.0	.193	1.57062
80.0	1202.0	37.55	−4866.0	−4693.0	87.07	31.37	52.99	1007.0	266.0	.168	1.51655
90.0	1156.0	36.11	−4342.0	−4162.0	93.32	30.00	53.28	934.2	208.0	.154	1.49410
100.0	1108.0	34.61	−3814.0	−3626.0	98.97	28.86	53.98	859.3	165.0	.140	1.47086
110.0	1056.0	33.01	−3278.0	−3081.0	104.2	27.90	55.29	781.3	133.0	.126	1.44639
120.0	1001.0	31.27	−2725.0	−2517.0	109.1	27.11	57.59	698.5	108.0	.113	1.42002
125.0	970.3	30.32	−2440.0	−2225.0	111.5	26.78	59.30	654.8	97.0	.107	1.40578
130.0	937.6	29.30	−2145.0	−1923.0	113.8	26.50	61.58	609.1	87.4	.101	1.39056
135.0	901.9	28.18	−1839.0	−1608.0	116.2	26.29	64.69	561.0	78.4	.0947	1.37406
140.0	861.9	26.94	−1516.0	−1274.0	118.6	26.15	69.12	509.8	69.9	.0886	1.35577
142.0	844.4	26.39	−1380.0	−1134.0	119.6	26.13	71.47	488.2	66.6	.0861	1.34780
144.0	825.7	25.80	−1240.0	−988.2	120.6	26.13	74.31	465.9	63.4	.0836	1.33933
146.0	805.6	25.18	−1094.0	−836.2	121.7	26.16	77.82	442.6	60.1	.0812	1.33027
148.0	783.8	24.50	−941.6	−676.3	122.8	26.22	82.30	418.3	56.8	.0788	1.32048
150.0	759.7	23.74	−779.8	−506.0	123.9	26.33	88.26	392.5	53.4	.0764	1.30972
152.0	732.6	22.89	−605.6	−321.6	125.1	26.50	96.65	364.9	49.9	.0741	1.29765
153.0	717.4	22.42	−512.2	−222.3	125.8	26.62	102.3	350.3	48.1	.0729	1.29096
154.0	700.9	21.90	−413.3	−116.5	126.5	26.77	109.4	334.9	46.2	.0717	1.28368
155.0	682.6	21.33	−307.4	−2.7	127.2	26.95	118.8	318.7	44.2	.0705	1.27566
156.0	661.9	20.69	−192.1	122.1	128.0	27.18	131.5	301.5	42.1	.0693	1.26665
157.0	638.1	19.94	−63.9	262.1	128.9	27.49	149.6	283.3	39.8	.0679	1.25628
158.0	609.5	19.05	83.0	424.2	129.9	27.88	176.5	263.8	37.2	.0664	1.24397
158.5	592.9	18.53	166.1	516.9	130.5	28.12	194.8	253.7	35.8	.0655	1.23681
159.0	574.2	17.94	257.5	619.7	131.2	28.40	217.0	243.6	34.3	.0645	1.22881
159.5	553.1	17.29	358.4	734.4	131.9	28.71	242.3	233.7	32.7	.0634	1.21986
160.0	529.7	16.56	469.4	862.1	132.7	29.04	267.9	224.5	31.0	.0620	1.20996
161.0	478.3	14.95	714.0	1149.0	134.5	29.63	299.2	210.3	27.6	.0584	1.18837
162.0	428.3	13.38	960.8	1446.0	136.3	29.88	289.9	202.7	24.8	.0543	1.16764
162.5	406.1	12.69	1076.0	1588.0	137.2	29.85	274.5	200.8	23.7	.0523	1.15852
163.0	386.3	12.07	1182.0	1720.0	138.0	29.74	255.4	199.9	22.8	.0503	1.15041
163.5	368.8	11.53	1279.0	1843.0	138.8	29.57	235.0	199.5	22.0	.0485	1.14330
164.0	353.5	11.05	1367.0	1955.0	139.4	29.37	215.1	199.6	21.4	.0468	1.13710
164.5	340.2	10.63	1447.0	2058.0	140.1	29.15	196.9	200.0	20.8	.0453	1.13171
165.0	328.5	10.27	1519.0	2153.0	140.6	28.92	180.7	200.6	20.4	.0439	1.12699
165.5	318.2	9.943	1586.0	2239.0	141.2	28.68	166.7	201.3	20.0	.0427	1.12284
166.0	309.0	9.656	1646.0	2320.0	141.7	28.46	154.6	202.2	19.7	.0416	1.11916
167.0	293.3	9.167	1755.0	2464.0	142.5	28.02	135.1	204.0	19.2	.0397	1.11290
168.0	280.4	8.763	1850.0	2591.0	143.3	27.62	120.5	206.0	18.8	.0381	1.10775
169.0	269.5	8.421	1934.0	2706.0	144.0	27.25	109.3	207.9	18.5	.0368	1.10341
170.0	260.1	8.127	2011.0	2811.0	144.6	26.91	100.4	209.9	18.2	.0357	1.09968
171.0	251.8	7.869	2081.0	2907.0	145.2	26.60	93.21	211.9	18.1	.0347	1.09641
172.0	244.5	7.640	2147.0	2998.0	145.7	26.31	87.32	213.8	17.9	.0339	1.09352
174.0	231.9	7.247	2266.0	3163.0	146.6	25.80	78.22	217.4	17.7	.0325	1.08858
176.0	221.4	6.920	2373.0	3312.0	147.5	25.36	71.53	220.9	17.5	.0314	1.08447
178.0	212.5	6.640	2471.0	3450.0	148.3	24.98	66.40	224.3	17.4	.0305	1.08096
180.0	204.7	6.396	2562.0	3578.0	149.0	24.64	62.34	227.4	17.3	.0298	1.07791
182.0	197.8	6.180	2648.0	3700.0	149.7	24.35	59.05	230.5	17.2	.0291	1.07521
184.0	191.6	5.987	2729.0	3815.0	150.3	24.08	56.32	233.4	17.2	.0286	1.07280
186.0	186.0	5.812	2807.0	3925.0	150.9	23.85	54.02	236.3	17.2	.0282	1.07062
188.0	180.9	5.653	2881.0	4031.0	151.4	23.64	52.06	239.0	17.2	.0278	1.06864
190.0	176.2	5.506	2953.0	4134.0	152.0	23.44	50.37	241.6	17.2	.0275	1.06682
195.0	165.9	5.186	3123.0	4377.0	153.2	23.04	47.01	247.9	17.3	.0269	1.06285
200.0	157.3	4.915	3283.0	4605.0	154.4	22.73	44.51	253.8	17.4	.0265	1.05951
205.0	149.8	4.682	3434.0	4823.0	155.5	22.47	42.58	259.3	17.6	.0263	1.05663
210.0	143.3	4.478	3580.0	5032.0	156.5	22.27	41.05	264.6	17.7	.0262	1.05412
215.0	137.5	4.297	3721.0	5234.0	157.4	22.10	39.80	269.6	17.9	.0261	1.05188
220.0	132.3	4.134	3858.0	5430.0	158.3	21.97	38.78	274.4	18.1	.0262	1.04988
225.0	127.6	3.986	3991.0	5622.0	159.2	21.85	37.92	279.0	18.3	.0262	1.04808

Thermophysical properties of oxygen—Continued

T K	Density kg/m³	Density mol/dm³	*E* J/mol	*H* J/mol	*S* J/(mol·K)	C_v J/(mol·K)	C_p J/(mol·K)	Sound m/s	Visc. μPa·s	Therm. W/(m·K)	Diel. const.
230.0	123.3	3.852	4122.0	5809.0	160.0	21.76	37.19	283.5	18.5	.0263	1.04643
240.0	115.7	3.615	4377.0	6175.0	161.6	21.63	36.02	291.9	19.0	.0266	1.04353
250.0	109.2	3.411	4625.0	6531.0	163.0	21.53	35.14	299.9	19.5	.0270	1.04104
260.0	103.5	3.234	4869.0	6879.0	164.4	21.48	34.45	307.5	19.9	.0275	1.03888
270.0	98.47	3.077	5108.0	7220.0	165.7	21.44	33.91	314.7	20.4	.0279	1.03697
280.0	94.00	2.937	5344.0	7557.0	166.9	21.43	33.47	321.6	20.9	.0285	1.03527
290.0	89.97	2.812	5578.0	7890.0	168.1	21.43	33.11	328.3	21.4	.0290	1.03374
300.0	86.33	2.698	5810.0	8220.0	169.2	21.45	32.82	334.7	21.8	.0296	1.03236
310.0	83.01	2.594	6041.0	8547.0	170.3	21.48	32.59	340.8	22.3	.0302	1.03110
320.0	79.97	2.499	6271.0	8871.0	171.3	21.52	32.40	346.8	22.8	.0308	1.02994
340.0	74.58	2.331	6727.0	9516.0	173.3	21.61	32.11	358.2	23.7	.0321	1.02790
360.0	69.93	2.186	7183.0	10160.0	175.1	21.74	31.94	369.0	24.7	.0334	1.02615
380.0	65.89	2.059	7638.0	10790.0	176.8	21.88	31.84	379.3	25.6	.0347	1.02462
400.0	62.32	1.947	8093.0	11430.0	178.4	22.04	31.80	389.1	26.5	.0361	1.02327
				7.00 MPa isobar							
55.16[a]	1311.0	40.96	-6171.0	-6000.0	67.31	36.08	53.07	1145.0	507.0	.193	1.57078
80.0	1202.0	37.57	-4870.0	-4684.0	87.02	31.40	52.95	1009.0	268.0	.168	1.51695
90.0	1157.0	36.15	-4347.0	-4153.0	93.27	30.03	53.22	936.3	209.0	.154	1.49460
100.0	1109.0	34.65	-3820.0	-3618.0	98.90	28.89	53.88	861.8	166.0	.140	1.47146
110.0	1058.0	33.06	-3286.0	-3074.0	104.1	27.93	55.14	784.5	134.0	.127	1.44716
120.0	1003.0	31.34	-2736.0	-2513.0	109.0	27.14	57.32	702.9	108.0	.114	1.42103
125.0	972.9	30.40	-2452.0	-2222.0	111.3	26.81	58.94	659.9	97.8	.107	1.40697
130.0	940.7	29.40	-2160.0	-1922.0	113.7	26.53	61.07	615.0	88.2	.101	1.39200
135.0	905.7	28.30	-1857.0	-1610.0	116.1	26.30	63.95	568.0	79.3	.0951	1.37582
140.0	866.9	27.09	-1539.0	-1281.0	118.4	26.15	67.96	518.4	70.9	.0891	1.35803
142.0	850.0	26.56	-1407.0	-1143.0	119.4	26.12	70.04	497.6	67.7	.0867	1.35033
144.0	832.1	26.00	-1270.0	-1001.0	120.4	26.11	72.51	476.2	64.5	.0842	1.34220
146.0	812.9	25.41	-1128.0	-852.7	121.4	26.12	75.50	454.0	61.3	.0818	1.33356
148.0	792.4	24.76	-980.8	-698.1	122.5	26.16	79.20	431.0	58.1	.0795	1.32430
150.0	770.0	24.06	-826.1	-535.2	123.6	26.23	83.93	407.0	54.9	.0772	1.31428
152.0	745.2	23.29	-662.0	-361.4	124.7	26.36	90.21	381.6	51.6	.0749	1.30326
154.0	717.3	22.42	-485.0	-172.7	126.0	26.54	98.99	354.7	48.2	.0727	1.29089
155.0	701.7	21.93	-390.1	-70.9	126.6	26.67	104.8	340.4	46.4	.0715	1.28403
156.0	684.7	21.40	-289.7	37.5	127.3	26.82	112.2	325.6	44.6	.0704	1.27659
157.0	666.0	20.81	-182.2	154.1	128.1	27.01	121.5	310.2	42.6	.0692	1.26841
158.0	645.0	20.16	-65.9	281.4	128.9	27.24	133.7	294.2	40.6	.0679	1.25930
159.0	621.2	19.41	62.2	422.8	129.8	27.53	149.9	277.5	38.4	.0665	1.24897
160.0	593.5	18.55	205.4	582.8	130.8	27.89	171.1	260.6	36.0	.0648	1.23709
161.0	561.3	17.54	367.5	766.6	131.9	28.31	196.8	244.2	33.5	.0628	1.22332
162.0	524.5	16.39	549.0	976.0	133.2	28.75	221.0	229.9	30.8	.0604	1.20775
163.0	485.3	15.17	743.1	1205.0	134.6	29.12	233.6	219.1	28.3	.0574	1.19128
164.0	447.1	13.97	937.0	1438.0	136.1	29.30	230.7	212.3	26.1	.0542	1.17540
165.0	412.7	12.90	1120.0	1662.0	137.4	29.26	216.5	208.7	24.3	.0512	1.16121
165.5	397.3	12.42	1204.0	1768.0	138.1	29.18	206.8	207.7	23.5	.0497	1.15491
166.0	383.2	11.97	1284.0	1869.0	138.7	29.06	196.4	207.2	22.9	.0484	1.14914
166.5	370.3	11.57	1360.0	1964.0	139.2	28.91	185.7	207.0	22.3	.0471	1.14389
167.0	358.6	11.21	1430.0	2055.0	139.8	28.75	175.1	207.0	21.8	.0459	1.13913
168.0	338.2	10.57	1557.0	2220.0	140.8	28.39	155.7	207.7	21.0	.0437	1.13089
169.0	321.2	10.04	1670.0	2367.0	141.6	28.02	139.2	208.9	20.4	.0418	1.12405
170.0	306.9	9.590	1769.0	2499.0	142.4	27.66	125.6	210.3	19.9	.0402	1.11831
171.0	294.6	9.207	1859.0	2619.0	143.1	27.32	114.6	211.9	19.6	.0388	1.11341
172.0	284.0	8.875	1940.0	2729.0	143.8	27.00	105.6	213.5	19.3	.0376	1.10918
173.0	274.7	8.583	2015.0	2831.0	144.4	26.70	98.10	215.2	19.0	.0365	1.10547
174.0	266.4	8.324	2085.0	2926.0	144.9	26.42	91.86	216.9	18.8	.0356	1.10217
176.0	252.2	7.880	2211.0	3099.0	145.9	25.92	82.10	220.3	18.5	.0341	1.09655
178.0	240.4	7.511	2324.0	3256.0	146.8	25.47	74.84	223.6	18.2	.0328	1.09190
180.0	230.3	7.197	2427.0	3400.0	147.6	25.09	69.25	226.8	18.1	.0318	1.08794
182.0	221.6	6.924	2522.0	3533.0	148.3	24.75	64.83	229.9	17.9	.0310	1.08451
184.0	213.8	6.683	2612.0	3659.0	149.0	24.45	61.24	232.9	17.9	.0303	1.08149
186.0	207.0	6.468	2697.0	3779.0	149.7	24.18	58.27	235.8	17.8	.0297	1.07880

Thermophysical properties of oxygen—Continued

T K	Density kg/m^3	Density mol/dm^3	E J/mol	H J/mol	S J/(mol·K)	C_v J/(mol·K)	C_p J/(mol·K)	Sound m/s	Visc. μPa·s	Therm. W/(m·K)	Diel. const.
188.0	200.8	6.274	2777.0	3893.0	150.3	23.93	55.77	238.5	17.8	.0292	1.07637
190.0	195.1	6.097	2854.0	4002.0	150.8	23.72	53.65	241.2	17.7	.0288	1.07417
195.0	182.9	5.716	3035.0	4259.0	152.2	23.27	49.50	247.6	17.8	.0280	1.06943
200.0	172.8	5.399	3203.0	4499.0	153.4	22.91	46.48	253.6	17.8	.0274	1.06550
205.0	164.1	5.130	3361.0	4725.0	154.5	22.63	44.19	259.2	17.9	.0271	1.06215
210.0	156.6	4.895	3512.0	4942.0	155.6	22.40	42.39	264.5	18.1	.0269	1.05925
215.0	150.0	4.688	3657.0	5150.0	156.5	22.21	40.95	269.6	18.2	.0268	1.05670
220.0	144.1	4.504	3797.0	5352.0	157.5	22.06	39.77	274.5	18.4	.0268	1.05443
225.0	138.8	4.338	3934.0	5548.0	158.3	21.94	38.79	279.2	18.6	.0268	1.05239
230.0	134.0	4.187	4068.0	5740.0	159.2	21.84	37.96	283.7	18.8	.0269	1.05053
240.0	125.5	3.922	4328.0	6113.0	160.8	21.68	36.65	292.3	19.2	.0271	1.04729
250.0	118.3	3.696	4580.0	6474.0	162.2	21.58	35.66	300.3	19.7	.0274	1.04452
260.0	112.0	3.500	4827.0	6827.0	163.6	21.52	34.90	308.0	20.1	.0278	1.04212
270.0	106.5	3.327	5069.0	7172.0	164.9	21.48	34.29	315.3	20.6	.0283	1.04001
280.0	101.6	3.174	5307.0	7513.0	166.2	21.46	33.81	322.2	21.0	.0288	1.03814
290.0	97.16	3.036	5543.0	7849.0	167.4	21.46	33.42	328.9	21.5	.0293	1.03646
300.0	93.17	2.912	5777.0	8181.0	168.5	21.48	33.10	335.3	22.0	.0299	1.03495
310.0	89.55	2.798	6010.0	8511.0	169.6	21.50	32.83	341.5	22.4	.0305	1.03357
320.0	86.23	2.695	6241.0	8838.0	170.6	21.54	32.62	347.6	22.9	.0311	1.03232
340.0	80.37	2.512	6700.0	9487.0	172.6	21.63	32.30	359.0	23.8	.0323	1.03009
360.0	75.33	2.354	7158.0	10130.0	174.4	21.76	32.10	369.8	24.8	.0336	1.02818
380.0	70.94	2.217	7614.0	10770.0	176.1	21.90	31.98	380.1	25.7	.0349	1.02653
400.0	67.08	2.096	8071.0	11410.0	177.8	22.06	31.93	390.0	26.6	.0363	1.02507

7.50 MPa isobar

T K	Density kg/m^3	Density mol/dm^3	E J/mol	H J/mol	S J/(mol·K)	C_v J/(mol·K)	C_p J/(mol·K)	Sound m/s	Visc. μPa·s	Therm. W/(m·K)	Diel. const.
55.22[a]	1311.0	40.97	−6171.0	−5987.0	67.33	36.06	53.06	1146.0	508.0	.193	1.57094
80.0	1203.0	37.60	−4874.0	−4675.0	86.97	31.43	52.91	1010.0	269.0	.168	1.51736
90.0	1158.0	36.18	−4352.0	−4145.0	93.21	30.07	53.15	938.4	210.0	.154	1.49509
100.0	1110.0	34.69	−3827.0	−3610.0	98.84	28.93	53.79	864.4	167.0	.140	1.47207
110.0	1060.0	33.11	−3294.0	−3067.0	104.0	27.97	54.99	787.8	134.0	.127	1.44792
120.0	1005.0	31.41	−2746.0	−2508.0	108.9	27.17	57.07	707.1	109.0	.114	1.42203
125.0	975.3	30.48	−2465.0	−2219.0	111.2	26.84	58.60	664.8	98.6	.108	1.40814
130.0	943.7	29.49	−2175.0	−1921.0	113.6	26.55	60.60	620.8	89.0	.102	1.39339
135.0	909.4	28.42	−1875.0	−1612.0	115.9	26.32	63.26	574.8	80.2	.0955	1.37753
140.0	871.6	27.24	−1562.0	−1287.0	118.3	26.16	66.91	526.6	71.9	.0896	1.36019
145.0	829.0	25.91	−1229.0	−939.8	120.7	26.09	72.18	475.4	64.0	.0836	1.34081
146.0	819.7	25.62	−1160.0	−867.0	121.2	26.09	73.53	464.7	62.4	.0825	1.33661
148.0	800.2	25.01	−1017.0	−716.9	122.2	26.11	76.65	442.8	59.3	.0801	1.32781
150.0	779.1	24.35	−867.9	−559.9	123.3	26.16	80.52	420.2	56.2	.0779	1.31838
152.0	756.2	23.63	−711.5	−394.1	124.4	26.25	85.44	396.6	53.1	.0757	1.30816
154.0	730.9	22.84	−545.4	−217.1	125.5	26.38	91.93	371.8	49.9	.0735	1.29692
156.0	702.4	21.95	−366.4	−24.7	126.8	26.58	100.9	345.7	46.6	.0713	1.28433
158.0	669.3	20.92	−169.4	189.2	128.1	26.86	113.9	318.1	43.1	.0691	1.26984
160.0	629.6	19.68	54.0	435.2	129.7	27.27	133.5	289.0	39.3	.0665	1.25261
161.0	606.4	18.95	179.3	575.1	130.6	27.52	146.7	274.3	37.2	.0651	1.24261
162.0	580.4	18.14	315.7	729.2	131.5	27.82	161.8	260.0	35.1	.0633	1.23149
163.0	551.7	17.24	463.8	898.8	132.6	28.14	177.2	246.8	32.9	.0614	1.21925
164.0	520.8	16.28	621.7	1082.0	133.7	28.44	189.3	235.6	30.8	.0591	1.20619
166.0	458.3	14.32	945.9	1470.0	136.0	28.79	193.0	221.0	26.9	.0540	1.18006
167.0	429.7	13.43	1101.0	1659.0	137.2	28.78	185.6	217.3	25.4	.0515	1.16823
168.0	404.1	12.63	1246.0	1840.0	138.2	28.65	174.6	215.2	24.1	.0491	1.15767
169.0	381.5	11.92	1379.0	2008.0	139.2	28.44	161.7	214.3	23.1	.0470	1.14844
170.0	361.8	11.31	1500.0	2163.0	140.2	28.17	148.7	214.3	22.3	.0450	1.14045
171.0	344.8	10.78	1609.0	2305.0	141.0	27.87	136.4	214.8	21.6	.0433	1.13357
172.0	330.1	10.31	1709.0	2436.0	141.8	27.57	125.4	215.7	21.1	.0417	1.12762
173.0	317.2	9.912	1800.0	2557.0	142.5	27.27	115.8	216.8	20.6	.0403	1.12244
174.0	305.8	9.558	1884.0	2668.0	143.1	26.98	107.6	218.1	20.2	.0391	1.11790
175.0	295.8	9.244	1961.0	2772.0	143.7	26.70	100.5	219.5	19.9	.0380	1.11388
176.0	286.8	8.962	2033.0	2870.0	144.2	26.44	94.48	220.9	19.7	.0371	1.11029
178.0	271.3	8.479	2164.0	3049.0	145.3	25.96	84.72	223.9	19.3	.0355	1.10413
180.0	258.4	8.075	2281.0	3210.0	146.2	25.53	77.28	227.0	19.0	.0341	1.09902
182.0	247.4	7.731	2389.0	3359.0	147.0	25.15	71.47	229.9	18.8	.0330	1.09466

Thermophysical properties of oxygen—Continued

T K	Density kg/m³	Density mol/dm³	E J/mol	H J/mol	S J/(mol·K)	C_v J/(mol·K)	C_p J/(mol·K)	Sound m/s	Visc. μPa·s	Therm. W/(m·K)	Diel. const.
184.0	237.8	7.432	2488.0	3497.0	147.7	24.81	66.83	232.9	18.6	.0321	1.09089
186.0	229.4	7.168	2580.0	3627.0	148.4	24.51	63.05	235.7	18.5	.0314	1.08758
188.0	221.9	6.933	2668.0	3750.0	149.1	24.24	59.92	238.5	18.4	.0307	1.08462
190.0	215.1	6.722	2751.0	3867.0	149.7	24.00	57.27	241.2	18.3	.0302	1.08197
192.0	208.9	6.529	2830.0	3979.0	150.3	23.78	55.02	243.8	18.3	.0297	1.07956
194.0	203.3	6.353	2906.0	4087.0	150.9	23.58	53.08	246.4	18.3	.0293	1.07736
200.0	188.8	5.902	3120.0	4391.0	152.4	23.09	48.59	253.6	18.3	.0284	1.07173
205.0	178.9	5.591	3286.0	4627.0	153.6	22.78	45.90	259.3	18.3	.0280	1.06787
210.0	170.3	5.323	3442.0	4851.0	154.7	22.53	43.81	264.7	18.4	.0277	1.06455
215.0	162.8	5.089	3592.0	5066.0	155.7	22.32	42.15	269.9	18.6	.0275	1.06165
220.0	156.2	4.881	3737.0	5273.0	156.6	22.16	40.80	274.8	18.7	.0274	1.05908
225.0	150.2	4.695	3877.0	5474.0	157.5	22.02	39.69	279.5	18.9	.0274	1.05679
230.0	144.9	4.527	4014.0	5670.0	158.4	21.91	38.76	284.0	19.1	.0274	1.05471
240.0	135.5	4.233	4278.0	6050.0	160.0	21.74	37.29	292.7	19.4	.0276	1.05110
250.0	127.5	3.984	4535.0	6417.0	161.5	21.63	36.19	300.8	19.9	.0278	1.04804
260.0	120.6	3.768	4784.0	6775.0	162.9	21.56	35.34	308.5	20.3	.0282	1.04540
270.0	114.5	3.579	5029.0	7125.0	164.2	21.52	34.68	315.9	20.7	.0286	1.04308
280.0	109.2	3.412	5270.0	7469.0	165.5	21.49	34.15	322.9	21.2	.0291	1.04104
290.0	104.4	3.262	5508.0	7808.0	166.7	21.49	33.72	329.6	21.7	.0296	1.03921
300.0	100.0	3.126	5744.0	8143.0	167.8	21.50	33.37	336.1	22.1	.0302	1.03756
310.0	96.10	3.003	5978.0	8476.0	168.9	21.53	33.08	342.3	22.6	.0307	1.03606
320.0	92.51	2.891	6211.0	8805.0	169.9	21.56	32.84	348.3	23.0	.0313	1.03469
340.0	86.16	2.693	6673.0	9458.0	171.9	21.66	32.49	359.8	24.0	.0325	1.03228
360.0	80.72	2.523	7132.0	10110.0	173.8	21.78	32.26	370.7	24.9	.0338	1.03022
380.0	75.99	2.375	7591.0	10750.0	175.5	21.92	32.12	381.0	25.8	.0351	1.02843
400.0	71.83	2.245	8050.0	11390.0	177.2	22.08	32.05	390.8	26.7	.0365	1.02686

8.00 MPa isobar

T K	Density kg/m³	Density mol/dm³	E J/mol	H J/mol	S J/(mol·K)	C_v J/(mol·K)	C_p J/(mol·K)	Sound m/s	Visc. μPa·s	Therm. W/(m·K)	Diel. const.
55.27[a]	1311.0	40.98	−6170.0	−5974.0	67.34	36.05	53.05	1147.0	510.0	.193	1.57110
80.0	1204.0	37.63	−4878.0	−4665.0	86.92	31.45	52.86	1012.0	270.0	.168	1.51776
100.0	1111.0	34.73	−3833.0	−3602.0	98.78	28.96	53.70	866.9	168.0	.141	1.47266
110.0	1061.0	33.16	−3301.0	−3060.0	103.9	28.00	54.84	791.0	135.0	.127	1.44866
120.0	1007.0	31.47	−2757.0	−2503.0	108.8	27.20	56.83	711.3	110.0	.114	1.42300
125.0	977.8	30.56	−2477.0	−2215.0	111.1	26.87	58.28	669.6	99.4	.108	1.40928
130.0	946.6	29.58	−2190.0	−1919.0	113.5	26.57	60.16	626.4	89.8	.102	1.39474
135.0	913.0	28.53	−1893.0	−1612.0	115.8	26.34	62.63	581.4	81.0	.0959	1.37917
140.0	876.1	27.38	−1584.0	−1291.0	118.1	26.16	65.96	534.5	72.8	.0900	1.36225
145.0	834.9	26.09	−1257.0	−950.6	120.5	26.07	70.65	485.0	65.0	.0842	1.34349
146.0	826.0	25.81	−1189.0	−879.4	121.0	26.07	71.83	474.8	63.5	.0831	1.33946
148.0	807.4	25.23	−1050.0	−733.1	122.0	26.07	74.51	453.9	60.5	.0808	1.33105
150.0	787.5	24.61	−906.0	−581.0	123.0	26.10	77.74	432.4	57.5	.0785	1.32211
152.0	766.0	23.94	−755.8	−421.6	124.1	26.16	81.73	410.1	54.5	.0764	1.31252
154.0	742.7	23.21	−598.0	−253.3	125.2	26.26	86.78	387.1	51.4	.0743	1.30214
156.0	716.9	22.40	−430.6	−73.5	126.3	26.40	93.36	363.1	48.3	.0722	1.29074
158.0	688.0	21.50	−250.5	121.6	127.6	26.60	102.2	338.1	45.1	.0701	1.27802
160.0	654.8	20.46	−53.4	337.6	128.9	26.87	114.4	312.2	41.8	.0678	1.26353
162.0	615.9	19.25	166.7	582.4	130.4	27.24	131.2	285.9	38.2	.0652	1.24668
164.0	569.9	17.81	415.5	864.7	132.2	27.70	151.3	261.2	34.4	.0620	1.22698
166.0	518.0	16.19	690.1	1184.0	134.1	28.14	166.5	241.4	30.8	.0580	1.20500
168.0	465.7	14.55	971.2	1521.0	136.1	28.35	167.4	229.1	27.5	.0536	1.18311
169.0	441.4	13.79	1106.0	1686.0	137.1	28.34	162.8	225.5	26.2	.0515	1.17303
170.0	419.0	13.09	1235.0	1846.0	138.1	28.24	156.0	223.2	25.1	.0494	1.16379
171.0	398.6	12.46	1356.0	1998.0	138.9	28.08	147.7	221.9	24.1	.0476	1.15545
172.0	380.4	11.89	1468.0	2141.0	139.8	27.88	138.9	221.3	23.3	.0458	1.14801
173.0	364.1	11.38	1572.0	2275.0	140.6	27.64	130.1	221.3	22.6	.0442	1.14139
174.0	349.6	10.93	1669.0	2401.0	141.3	27.39	121.7	221.7	22.1	.0428	1.13551
175.0	336.7	10.52	1759.0	2519.0	142.0	27.13	113.9	222.4	21.6	.0415	1.13029
176.0	325.1	10.16	1842.0	2629.0	142.6	26.88	107.0	223.4	21.2	.0403	1.12564
177.0	314.7	9.836	1920.0	2733.0	143.2	26.63	100.7	224.4	20.8	.0392	1.12146
178.0	305.4	9.543	1993.0	2831.0	143.7	26.39	95.22	225.6	20.6	.0383	1.11770
180.0	289.0	9.033	2126.0	3012.0	144.7	25.94	86.02	228.2	20.1	.0366	1.11118

Thermophysical properties of oxygen—Continued

T K	Density kg/m³	Density mol/dm³	E J/mol	H J/mol	S J/(mol·K)	C_v J/(mol·K)	C_p J/(mol·K)	Sound m/s	Visc. μPa·s	Therm. W/(m·K)	Diel. const.
182.0	275.3	8.604	2247.0	3177.0	145.6	25.53	78.78	230.8	19.7	.0353	1.10572
184.0	263.5	8.236	2357.0	3328.0	146.5	25.16	72.99	233.6	19.5	.0341	1.10105
186.0	253.3	7.915	2458.0	3469.0	147.2	24.83	68.31	236.3	19.3	.0332	1.09699
188.0	244.2	7.632	2554.0	3602.0	147.9	24.54	64.45	239.0	19.1	.0323	1.09342
190.0	236.1	7.380	2643.0	3727.0	148.6	24.27	61.22	241.6	19.0	.0317	1.09023
192.0	228.9	7.152	2729.0	3847.0	149.2	24.03	58.50	244.2	18.9	.0311	1.08737
194.0	222.3	6.946	2810.0	3962.0	149.8	23.82	56.16	246.7	18.8	.0306	1.08478
196.0	216.2	6.757	2888.0	4072.0	150.4	23.62	54.15	249.2	18.8	.0301	1.08241
200.0	205.5	6.422	3036.0	4282.0	151.5	23.28	50.84	254.0	18.8	.0294	1.07821
205.0	194.1	6.066	3209.0	4528.0	152.7	22.93	47.69	259.7	18.8	.0288	1.07377
210.0	184.4	5.762	3371.0	4760.0	153.8	22.66	45.29	265.1	18.8	.0284	1.06999
215.0	175.9	5.498	3526.0	4981.0	154.8	22.43	43.39	270.3	18.9	.0282	1.06672
220.0	168.5	5.266	3675.0	5194.0	155.8	22.25	41.87	275.2	19.0	.0280	1.06384
225.0	161.9	5.059	3819.0	5400.0	156.7	22.10	40.61	280.0	19.2	.0279	1.06127
230.0	155.9	4.872	3959.0	5601.0	157.6	21.98	39.57	284.5	19.3	.0279	1.05896
235.0	150.5	4.703	4095.0	5796.0	158.5	21.88	38.69	289.0	19.5	.0280	1.05687
240.0	145.5	4.548	4229.0	5988.0	159.3	21.80	37.94	293.2	19.7	.0280	1.05496
250.0	136.8	4.274	4489.0	6361.0	160.8	21.68	36.72	301.4	20.1	.0283	1.05160
260.0	129.2	4.038	4742.0	6723.0	162.2	21.60	35.79	309.2	20.5	.0286	1.04870
270.0	122.6	3.832	4990.0	7077.0	163.5	21.55	35.07	316.5	20.9	.0290	1.04618
280.0	116.8	3.650	5233.0	7425.0	164.8	21.53	34.49	323.6	21.4	.0294	1.04395
290.0	111.6	3.488	5474.0	7768.0	166.0	21.52	34.02	330.3	21.8	.0299	1.04196
300.0	106.9	3.341	5711.0	8106.0	167.2	21.53	33.64	336.8	22.3	.0305	1.04017
310.0	102.7	3.208	5947.0	8441.0	168.3	21.55	33.32	343.1	22.7	.0310	1.03855
320.0	98.79	3.087	6181.0	8772.0	169.3	21.58	33.06	349.1	23.2	.0316	1.03708
340.0	91.96	2.874	6646.0	9430.0	171.3	21.68	32.68	360.7	24.1	.0328	1.03448
360.0	86.11	2.691	7108.0	10080.0	173.2	21.80	32.42	371.6	25.0	.0340	1.03226
380.0	81.04	2.532	7568.0	10730.0	174.9	21.94	32.26	381.9	25.9	.0353	1.03034
400.0	76.58	2.393	8029.0	11370.0	176.6	22.10	32.17	391.8	26.8	.0366	1.02865

8.50 MPa isobar

T K	Density kg/m³	Density mol/dm³	E J/mol	H J/mol	S J/(mol·K)	C_v J/(mol·K)	C_p J/(mol·K)	Sound m/s	Visc. μPa·s	Therm. W/(m·K)	Diel. const.
55.33[a]	1312.0	40.99	−6169.0	−5961.0	67.35	36.04	53.04	1149.0	511.0	.193	1.57125
80.0	1205.0	37.65	−4882.0	−4656.0	86.87	31.48	52.82	1014.0	271.0	.168	1.51816
100.0	1113.0	34.77	−3839.0	−3594.0	98.71	28.99	53.61	869.5	168.0	.141	1.47325
110.0	1063.0	33.21	−3309.0	−3053.0	103.9	28.03	54.71	794.2	136.0	.127	1.44941
120.0	1009.0	31.53	−2767.0	−2497.0	108.7	27.23	56.59	715.4	111.0	.114	1.42397
125.0	980.2	30.63	−2489.0	−2211.0	111.0	26.89	57.97	674.3	100.0	.108	1.41040
130.0	949.5	29.67	−2204.0	−1917.0	113.3	26.60	59.74	631.9	90.6	.102	1.39607
135.0	916.4	28.64	−1910.0	−1613.0	115.6	26.35	62.04	587.8	81.8	.0963	1.38077
140.0	880.5	27.52	−1604.0	−1295.0	118.0	26.17	65.10	542.0	73.7	.0905	1.36423
145.0	840.5	26.27	−1284.0	−960.0	120.3	26.06	69.31	494.2	66.1	.0848	1.34603
146.0	831.9	26.00	−1217.0	−890.2	120.8	26.05	70.34	484.3	64.6	.0836	1.34214
148.0	814.1	25.44	−1081.0	−747.2	121.8	26.04	72.67	464.2	61.6	.0814	1.33407
150.0	795.1	24.85	−941.3	−599.2	122.8	26.06	75.43	443.7	58.7	.0792	1.32554
152.0	774.9	24.22	−796.2	−445.1	123.8	26.09	78.74	422.6	55.8	.0770	1.31647
154.0	753.1	23.54	−644.9	−283.7	124.8	26.16	82.81	400.9	52.8	.0750	1.30676
156.0	729.4	22.80	−486.1	−113.2	125.9	26.26	87.89	378.6	49.9	.0729	1.29626
158.0	703.4	21.98	−317.9	68.7	127.1	26.40	94.39	355.6	46.9	.0709	1.28477
160.0	674.3	21.07	−137.8	265.6	128.3	26.60	102.8	331.9	43.8	.0688	1.27204
162.0	641.4	20.04	57.8	481.8	129.7	26.86	113.8	308.0	40.6	.0666	1.25772
164.0	603.8	18.87	272.2	722.6	131.1	27.18	127.3	284.6	37.3	.0639	1.24150
166.0	561.3	17.54	506.7	991.3	132.8	27.54	141.0	263.6	34.0	.0607	1.22335
168.0	515.7	16.12	755.7	1283.0	134.5	27.84	149.5	247.2	30.8	.0570	1.20404
170.0	470.7	14.71	1005.0	1583.0	136.3	27.96	149.0	236.6	28.0	.0532	1.18522
172.0	430.0	13.44	1241.0	1874.0	138.0	27.85	141.0	230.9	25.9	.0495	1.16831
173.0	411.7	12.87	1352.0	2012.0	138.8	27.72	135.3	229.3	25.0	.0479	1.16081
174.0	395.0	12.34	1456.0	2144.0	139.6	27.56	129.1	228.5	24.2	.0463	1.15395
175.0	379.7	11.87	1554.0	2270.0	140.3	27.36	122.7	228.1	23.5	.0449	1.14773
176.0	365.9	11.43	1646.0	2390.0	141.0	27.16	116.3	228.2	23.0	.0435	1.14209
177.0	353.3	11.04	1733.0	2503.0	141.6	26.94	110.2	228.5	22.5	.0423	1.13699
178.0	341.8	10.68	1815.0	2610.0	142.2	26.72	104.5	229.1	22.1	.0412	1.13236
180.0	321.9	10.06	1964.0	2809.0	143.3	26.28	94.46	230.8	21.4	.0392	1.12434

Thermophysical properties of oxygen—Continued

T K	Density kg/m^3	Density mol/dm^3	E J/mol	H J/mol	S J/(mol·K)	C_v J/(mol·K)	C_p J/(mol·K)	Sound m/s	Visc. μPa·s	Therm. W/(m·K)	Diel. const.
182.0	305.2	9.537	2098.0	2989.0	144.3	25.87	86.16	232.9	20.9	.0376	1.11763
184.0	290.9	9.092	2220.0	3155.0	145.2	25.49	79.38	235.2	20.5	.0362	1.11193
186.0	278.6	8.707	2331.0	3308.0	146.0	25.14	73.83	237.6	20.2	.0350	1.10703
188.0	267.8	8.370	2435.0	3451.0	146.8	24.82	69.24	240.1	19.9	.0341	1.10275
190.0	258.3	8.071	2532.0	3585.0	147.5	24.54	65.41	242.6	19.7	.0332	1.09896
192.0	249.7	7.804	2624.0	3713.0	148.2	24.28	62.18	245.1	19.6	.0325	1.09559
194.0	242.0	7.563	2710.0	3834.0	148.8	24.04	59.43	247.5	19.5	.0319	1.09255
196.0	235.0	7.344	2793.0	3951.0	149.4	23.83	57.06	250.0	19.4	.0314	1.08979
198.0	228.6	7.144	2873.0	4063.0	150.0	23.64	55.01	252.3	19.3	.0309	1.08726
200.0	222.7	6.959	2949.0	4171.0	150.5	23.46	53.21	254.7	19.3	.0305	1.08495
205.0	209.7	6.554	3130.0	4427.0	151.8	23.09	49.57	260.3	19.2	.0298	1.07987
210.0	198.8	6.212	3300.0	4668.0	153.0	22.79	46.82	265.7	19.2	.0293	1.07559
215.0	189.3	5.916	3460.0	4897.0	154.0	22.55	44.67	270.9	19.3	.0289	1.07191
220.0	181.0	5.657	3613.0	5115.0	155.0	22.35	42.96	275.8	19.4	.0287	1.06869
225.0	173.7	5.428	3761.0	5327.0	156.0	22.19	41.56	280.6	19.5	.0286	1.06584
230.0	167.1	5.222	3904.0	5531.0	156.9	22.06	40.40	285.2	19.6	.0285	1.06328
235.0	161.1	5.035	4043.0	5731.0	157.7	21.95	39.42	289.6	19.8	.0285	1.06098
240.0	155.7	4.866	4179.0	5926.0	158.6	21.86	38.59	293.9	20.0	.0285	1.05888
250.0	146.1	4.566	4443.0	6305.0	160.1	21.72	37.26	302.1	20.3	.0287	1.05519
260.0	137.9	4.310	4700.0	6672.0	161.6	21.64	36.25	309.9	20.7	.0290	1.05204
270.0	130.8	4.087	4951.0	7031.0	162.9	21.58	35.46	317.3	21.1	.0294	1.04929
280.0	124.5	3.890	5197.0	7382.0	164.2	21.56	34.83	324.3	21.5	.0298	1.04688
290.0	118.9	3.714	5439.0	7727.0	165.4	21.55	34.32	331.1	22.0	.0302	1.04473
300.0	113.8	3.557	5679.0	8069.0	166.6	21.55	33.90	337.6	22.4	.0307	1.04280
310.0	109.2	3.414	5916.0	8406.0	167.7	21.58	33.56	343.9	22.8	.0313	1.04106
320.0	105.1	3.284	6151.0	8740.0	168.7	21.61	33.28	350.0	23.3	.0318	1.03947
340.0	97.75	3.055	6619.0	9401.0	170.7	21.70	32.86	361.5	24.2	.0330	1.03668
360.0	91.50	2.859	7083.0	10060.0	172.6	21.81	32.58	372.4	25.1	.0342	1.03430
380.0	86.07	2.690	7545.0	10710.0	174.4	21.96	32.40	382.8	26.0	.0355	1.03225
400.0	81.32	2.541	8007.0	11350.0	176.0	22.11	32.30	392.7	26.8	.0368	1.03044

9.00 MPa isobar

T K	Density kg/m^3	Density mol/dm^3	E J/mol	H J/mol	S J/(mol·K)	C_v J/(mol·K)	C_p J/(mol·K)	Sound m/s	Visc. μPa·s	Therm. W/(m·K)	Diel. const.
55.39[a]	1312.0	41.00	−6168.0	−5948.0	67.37	36.02	53.03	1150.0	513.0	.193	1.57141
80.0	1206.0	37.68	−4886.0	−4647.0	86.82	31.51	52.78	1015.0	272.0	.168	1.51856
100.0	1114.0	34.81	−3845.0	−3586.0	98.65	29.02	53.52	872.0	169.0	.141	1.47384
110.0	1064.0	33.26	−3317.0	−3046.0	103.8	28.06	54.57	797.3	137.0	.127	1.45014
120.0	1011.0	31.60	−2777.0	−2492.0	108.6	27.26	56.37	719.5	112.0	.115	1.42492
125.0	982.5	30.71	−2500.0	−2207.0	110.9	26.92	57.67	679.0	101.0	.108	1.41150
130.0	952.2	29.76	−2217.0	−1915.0	113.2	26.62	59.34	637.2	91.4	.103	1.39736
135.0	919.8	28.74	−1926.0	−1613.0	115.5	26.37	61.49	594.1	82.7	.0967	1.38232
140.0	884.6	27.64	−1624.0	−1299.0	117.8	26.18	64.31	549.4	74.6	.0909	1.36613
145.0	845.8	26.43	−1309.0	−968.2	120.1	26.05	68.11	502.9	67.0	.0853	1.34844
146.0	837.5	26.17	−1243.0	−899.6	120.6	26.04	69.03	493.4	65.6	.0842	1.34468
148.0	820.3	25.64	−1111.0	−759.6	121.5	26.02	71.08	474.0	62.7	.0820	1.33689
150.0	802.2	25.07	−974.1	−615.1	122.5	26.02	73.46	454.3	59.8	.0798	1.32872
152.0	783.0	24.47	−833.3	−465.5	123.5	26.04	76.27	434.2	57.0	.0777	1.32009
154.0	762.5	23.83	−687.4	−309.7	124.5	26.08	79.63	413.7	54.1	.0756	1.31093
156.0	740.4	23.14	−535.4	−146.5	125.6	26.15	83.70	392.6	51.3	.0736	1.30113
158.0	716.5	22.39	−376.2	25.8	126.7	26.26	88.72	371.1	48.4	.0717	1.29056
160.0	690.4	21.57	−207.9	209.2	127.8	26.40	94.97	349.2	45.6	.0697	1.27906
162.0	661.5	20.67	−28.7	406.7	129.1	26.58	102.8	327.1	42.6	.0676	1.26643
164.0	629.2	19.66	163.7	621.3	130.4	26.81	112.2	305.2	39.6	.0653	1.25245
166.0	593.4	18.54	370.7	856.1	131.8	27.08	122.6	284.6	36.6	.0627	1.23703
168.0	554.4	17.32	591.3	1111.0	133.3	27.35	131.7	266.7	33.6	.0596	1.22037
170.0	513.8	16.06	819.2	1380.0	134.9	27.55	136.3	252.9	30.8	.0562	1.20325
172.0	474.4	14.83	1045.0	1652.0	136.5	27.60	135.0	243.6	28.5	.0528	1.18674
174.0	438.3	13.70	1259.0	1917.0	138.0	27.49	129.0	238.2	26.5	.0495	1.17175
176.0	406.5	12.70	1458.0	2166.0	139.5	27.24	120.2	235.5	25.0	.0466	1.15868
178.0	379.3	11.85	1638.0	2397.0	140.8	26.90	110.5	234.7	23.8	.0440	1.14754
180.0	356.1	11.13	1800.0	2608.0	141.9	26.52	101.1	235.1	22.9	.0418	1.13814
182.0	336.5	10.52	1946.0	2802.0	143.0	26.13	92.67	236.3	22.2	.0400	1.13020
184.0	319.7	9.990	2079.0	2980.0	144.0	25.76	85.39	238.0	21.6	.0384	1.12344

Thermophysical properties of oxygen—Continued

T K	Density kg/m^3	Density mol/dm^3	E J/mol	H J/mol	S J/(mol·K)	C_v J/(mol·K)	C_p J/(mol·K)	Sound m/s	Visc. μPa·s	Therm. W/(m·K)	Diel. const.
186.0	305.2	9.537	2201.0	3144.0	144.9	25.41	79.23	239.9	21.2	.0370	1.11762
188.0	292.5	9.142	2313.0	3297.0	145.7	25.08	74.04	242.1	20.8	.0359	1.11257
190.0	281.4	8.794	2418.0	3441.0	146.4	24.79	69.67	244.3	20.6	.0349	1.10813
192.0	271.5	8.484	2516.0	3577.0	147.2	24.51	65.96	246.6	20.4	.0340	1.10419
194.0	262.6	8.205	2608.0	3705.0	147.8	24.26	62.80	248.9	20.2	.0333	1.10066
196.0	254.5	7.953	2696.0	3828.0	148.5	24.04	60.08	251.2	20.0	.0327	1.09747
198.0	247.2	7.724	2781.0	3946.0	149.1	23.83	57.72	253.5	19.9	.0321	1.09457
200.0	240.4	7.514	2861.0	4059.0	149.6	23.64	55.66	255.7	19.9	.0316	1.09192
205.0	225.8	7.056	3051.0	4327.0	150.9	23.24	51.51	261.3	19.7	.0307	1.08615
210.0	213.5	6.671	3227.0	4576.0	152.1	22.92	48.40	266.6	19.7	.0301	1.08133
215.0	202.9	6.342	3393.0	4812.0	153.3	22.66	45.99	271.7	19.7	.0297	1.07722
220.0	193.8	6.055	3550.0	5037.0	154.3	22.44	44.07	276.6	19.7	.0294	1.07364
225.0	185.7	5.802	3702.0	5253.0	155.3	22.27	42.52	281.4	19.8	.0292	1.07048
230.0	178.4	5.576	3848.0	5462.0	156.2	22.13	41.24	286.0	19.9	.0291	1.06767
235.0	171.9	5.372	3990.0	5666.0	157.1	22.01	40.16	290.4	20.1	.0290	1.06514
240.0	166.0	5.186	4129.0	5864.0	157.9	21.91	39.25	294.7	20.2	.0290	1.06284
250.0	155.5	4.861	4398.0	6249.0	159.5	21.77	37.80	302.9	20.6	.0292	1.05882
260.0	146.7	4.583	4658.0	6621.0	160.9	21.68	36.70	310.7	20.9	.0294	1.05540
270.0	138.9	4.342	4911.0	6984.0	162.3	21.62	35.84	318.1	21.3	.0297	1.05243
280.0	132.2	4.130	5160.0	7339.0	163.6	21.59	35.16	325.2	21.7	.0301	1.04982
290.0	126.1	3.942	5404.0	7688.0	164.8	21.57	34.62	331.9	22.1	.0306	1.04751
300.0	120.7	3.772	5646.0	8032.0	166.0	21.58	34.17	338.5	22.6	.0310	1.04544
310.0	115.8	3.619	5885.0	8371.0	167.1	21.60	33.80	344.8	23.0	.0316	1.04357
320.0	111.4	3.480	6122.0	8708.0	168.2	21.63	33.50	350.8	23.4	.0321	1.04186
340.0	103.5	3.236	6592.0	9373.0	170.2	21.72	33.05	362.4	24.3	.0332	1.03888
360.0	96.88	3.027	7058.0	10030.0	172.1	21.83	32.74	373.4	25.2	.0345	1.03634
380.0	91.11	2.847	7522.0	10680.0	173.8	21.97	32.54	383.7	26.1	.0357	1.03415
400.0	86.05	2.689	7986.0	11330.0	175.5	22.13	32.42	393.6	26.9	.0370	1.03223

9.50 MPa isobar

T K	Density kg/m^3	Density mol/dm^3	E J/mol	H J/mol	S J/(mol·K)	C_v J/(mol·K)	C_p J/(mol·K)	Sound m/s	Visc. μPa·s	Therm. W/(m·K)	Diel. const.
55.44[a]	1312.0	41.01	−6167.0	−5935.0	67.38	36.01	53.02	1151.0	514.0	.193	1.57157
80.0	1206.0	37.70	−4889.0	−4637.0	86.77	31.54	52.74	1017.0	273.0	.169	1.51896
100.0	1115.0	34.84	−3850.0	−3578.0	98.59	29.05	53.43	874.5	170.0	.141	1.47442
110.0	1066.0	33.31	−3324.0	−3039.0	103.7	28.10	54.44	800.5	137.0	.128	1.45086
120.0	1013.0	31.66	−2787.0	−2487.0	108.5	27.29	56.16	723.5	112.0	.115	1.42585
125.0	984.8	30.78	−2512.0	−2203.0	110.8	26.95	57.39	683.5	102.0	.109	1.41257
130.0	955.0	29.84	−2231.0	−1912.0	113.1	26.65	58.96	642.4	92.2	.103	1.39862
135.0	923.1	28.85	−1942.0	−1613.0	115.4	26.39	60.98	600.1	83.5	.0971	1.38382
140.0	888.6	27.77	−1644.0	−1301.0	117.7	26.19	63.58	556.4	75.5	.0914	1.36796
145.0	850.9	26.59	−1333.0	−975.3	119.9	26.05	67.03	511.2	68.0	.0858	1.35073
150.0	808.8	25.28	−1005.0	−629.0	122.3	25.99	71.76	464.4	60.9	.0804	1.33170
152.0	790.5	24.70	−867.7	−483.1	123.3	26.00	74.18	445.1	58.1	.0783	1.32344
154.0	771.0	24.09	−726.3	−332.0	124.2	26.02	77.01	425.5	55.3	.0763	1.31473
156.0	750.3	23.45	−579.9	−174.7	125.3	26.07	80.36	405.5	52.6	.0743	1.30550
158.0	728.1	22.75	−427.7	−10.1	126.3	26.14	84.38	385.3	49.9	.0724	1.29565
160.0	704.1	22.00	−268.4	163.3	127.4	26.24	89.22	364.7	47.1	.0705	1.28508
162.0	678.0	21.19	−100.9	347.4	128.5	26.38	95.06	344.1	44.4	.0686	1.27365
164.0	649.5	20.30	76.3	544.3	129.7	26.54	102.0	323.6	41.6	.0665	1.26123
166.0	618.3	19.32	264.3	756.0	131.0	26.74	109.8	303.8	38.8	.0641	1.24772
168.0	584.4	18.26	463.2	983.3	132.4	26.96	117.5	285.7	36.0	.0615	1.23318
170.0	548.6	17.14	670.5	1225.0	133.8	27.15	123.4	270.3	33.4	.0585	1.21791
172.0	512.2	16.01	880.9	1474.0	135.3	27.27	125.7	258.5	30.9	.0554	1.20257
174.0	477.2	14.91	1088.0	1725.0	136.7	27.27	124.1	250.4	28.8	.0523	1.18789
176.0	444.8	13.90	1285.0	1969.0	138.1	27.16	119.3	245.3	27.0	.0494	1.17444
178.0	415.8	12.99	1470.0	2201.0	139.4	26.93	112.5	242.5	25.6	.0467	1.16249
180.0	390.4	12.20	1639.0	2418.0	140.6	26.64	104.9	241.3	24.5	.0444	1.15208
182.0	368.3	11.51	1795.0	2620.0	141.8	26.31	97.37	241.3	23.6	.0423	1.14309
184.0	349.2	10.91	1937.0	2808.0	142.8	25.97	90.31	242.0	22.9	.0406	1.13534
186.0	332.6	10.39	2068.0	2982.0	143.7	25.63	84.01	243.3	22.3	.0390	1.12864
188.0	318.1	9.941	2189.0	3145.0	144.6	25.31	78.51	244.9	21.9	.0377	1.12279
190.0	305.3	9.541	2301.0	3297.0	145.4	25.01	73.76	246.8	21.5	.0366	1.11767

Thermophysical properties of oxygen—Continued

T K	Density kg/m^3	Density mol/dm^3	E J/mol	H J/mol	S J/(mol·K)	C_v J/(mol·K)	C_p J/(mol·K)	Sound m/s	Visc. μPa·s	Therm. W/(m·K)	Diel. const.
192.0	293.9	9.186	2406.0	3440.0	146.2	24.73	69.67	248.7	21.2	.0356	1.11313
194.0	283.8	8.868	2504.0	3576.0	146.9	24.47	66.15	250.8	21.0	.0347	1.10907
196.0	274.6	8.581	2598.0	3705.0	147.5	24.23	63.10	252.9	20.8	.0340	1.10543
198.0	266.3	8.321	2687.0	3828.0	148.1	24.01	60.46	255.1	20.6	.0334	1.10212
200.0	258.7	8.084	2772.0	3947.0	148.7	23.82	58.14	257.2	20.5	.0328	1.09911
205.0	242.2	7.569	2970.0	4225.0	150.1	23.39	53.49	262.5	20.3	.0317	1.09261
210.0	228.5	7.140	3153.0	4484.0	151.4	23.04	50.01	267.7	20.2	.0310	1.08721
215.0	216.8	6.775	3325.0	4727.0	152.5	22.76	47.33	272.8	20.1	.0304	1.08263
220.0	206.7	6.459	3487.0	4958.0	153.6	22.54	45.21	277.6	20.1	.0301	1.07868
225.0	197.8	6.181	3643.0	5180.0	154.6	22.35	43.49	282.3	20.2	.0298	1.07520
230.0	189.9	5.934	3793.0	5394.0	155.5	22.20	42.09	286.9	20.3	.0297	1.07212
235.0	182.8	5.711	3938.0	5601.0	156.4	22.07	40.91	291.3	20.4	.0296	1.06935
240.0	176.3	5.510	4079.0	5803.0	157.2	21.97	39.92	295.6	20.5	.0295	1.06685
250.0	165.0	5.158	4352.0	6194.0	158.8	21.82	38.34	303.8	20.8	.0296	1.06249
260.0	155.5	4.858	4616.0	6571.0	160.3	21.72	37.15	311.6	21.1	.0298	1.05878
270.0	147.2	4.599	4872.0	6938.0	161.7	21.65	36.23	319.0	21.5	.0301	1.05558
280.0	139.9	4.371	5123.0	7296.0	163.0	21.62	35.50	326.1	21.9	.0305	1.05278
290.0	133.4	4.169	5370.0	7648.0	164.2	21.60	34.91	332.9	22.3	.0309	1.05030
300.0	127.6	3.988	5613.0	7995.0	165.4	21.60	34.43	339.4	22.7	.0313	1.04808
310.0	122.4	3.825	5854.0	8337.0	166.5	21.62	34.04	345.7	23.1	.0318	1.04608
320.0	117.7	3.677	6092.0	8676.0	167.6	21.65	33.72	351.8	23.6	.0324	1.04427
330.0	113.3	3.541	6329.0	9012.0	168.7	21.69	33.45	357.7	24.0	.0329	1.04261
340.0	109.3	3.417	6565.0	9345.0	169.6	21.74	33.23	363.4	24.4	.0335	1.04109
360.0	102.3	3.195	7033.0	10010.0	171.5	21.85	32.90	374.3	25.3	.0347	1.03839
380.0	96.13	3.004	7500.0	10660.0	173.3	21.99	32.67	384.7	26.2	.0359	1.03606
400.0	90.78	2.837	7965.0	11310.0	175.0	22.15	32.54	394.6	27.0	.0372	1.03402

10.00 MPa isobar

T K	Density kg/m^3	Density mol/dm^3	E J/mol	H J/mol	S J/(mol·K)	C_v J/(mol·K)	C_p J/(mol·K)	Sound m/s	Visc. μPa·s	Therm. W/(m·K)	Diel. const.
55.50[a]	1313.0	41.02	−6166.0	−5923.0	67.40	36.00	53.01	1152.0	516.0	.193	1.57172
80.0	1207.0	37.73	−4893.0	−4628.0	86.72	31.56	52.70	1019.0	274.0	.169	1.51936
100.0	1116.0	34.88	−3856.0	−3570.0	98.53	29.09	53.35	876.9	171.0	.141	1.47500
110.0	1067.0	33.36	−3332.0	−3032.0	103.7	28.13	54.31	803.5	138.0	.128	1.45158
120.0	1015.0	31.72	−2796.0	−2481.0	108.4	27.32	55.95	727.4	113.0	.115	1.42677
130.0	957.6	29.93	−2244.0	−1909.0	113.0	26.67	58.61	647.5	92.9	.103	1.39985
135.0	926.2	28.95	−1957.0	−1612.0	115.3	26.41	60.50	606.0	84.3	.0974	1.38528
140.0	892.4	27.89	−1662.0	−1304.0	117.5	26.20	62.91	563.2	76.3	.0918	1.36972
145.0	855.7	26.74	−1355.0	−981.5	119.8	26.05	66.06	519.2	68.9	.0863	1.35291
150.0	815.0	25.47	−1034.0	−641.2	122.1	25.97	70.27	473.9	61.9	.0809	1.33449
152.0	797.4	24.92	−899.9	−498.6	123.0	25.96	72.38	455.3	59.2	.0789	1.32656
154.0	778.9	24.34	−762.3	−351.5	124.0	25.97	74.81	436.5	56.5	.0769	1.31824
156.0	759.2	23.73	−620.6	−199.1	125.0	26.00	77.63	417.5	53.8	.0749	1.30948
160.0	716.1	22.38	−321.9	125.0	127.0	26.12	84.81	378.9	48.6	.0712	1.29037
162.0	692.2	21.63	−163.2	299.1	128.1	26.22	89.38	359.4	45.9	.0694	1.27985
164.0	666.4	20.82	2.8	483.0	129.2	26.34	94.69	340.0	43.3	.0674	1.26856
166.0	638.5	19.95	177.1	678.2	130.4	26.49	100.6	321.2	40.7	.0653	1.25645
168.0	608.5	19.02	359.9	885.7	131.6	26.65	106.8	303.5	38.1	.0630	1.24350
170.0	576.6	18.02	550.2	1105.0	132.9	26.81	112.4	287.6	35.6	.0604	1.22985
175.0	494.8	15.46	1038.0	1684.0	136.3	27.00	116.5	260.1	30.0	.0533	1.19526
176.0	479.3	14.98	1133.0	1800.0	137.0	26.97	115.3	256.8	29.1	.0519	1.18876
178.0	450.0	14.06	1316.0	2027.0	138.2	26.85	111.3	252.0	27.5	.0492	1.17658
180.0	423.4	13.23	1489.0	2245.0	139.5	26.65	105.9	249.2	26.2	.0468	1.16561
182.0	399.7	12.49	1650.0	2450.0	140.6	26.39	99.78	247.8	25.1	.0446	1.15588
184.0	378.8	11.84	1799.0	2644.0	141.7	26.09	93.59	247.4	24.2	.0427	1.14733
186.0	360.3	11.26	1937.0	2825.0	142.6	25.79	87.68	247.8	23.5	.0410	1.13984
188.0	344.1	10.75	2065.0	2995.0	143.5	25.48	82.25	248.8	23.0	.0396	1.13326
190.0	329.7	10.30	2184.0	3154.0	144.4	25.19	77.39	250.1	22.5	.0383	1.12747
192.0	316.9	9.904	2295.0	3305.0	145.2	24.91	73.09	251.7	22.1	.0372	1.12232
194.0	305.5	9.546	2399.0	3447.0	145.9	24.65	69.33	253.4	21.8	.0362	1.11774
196.0	295.2	9.224	2498.0	3582.0	146.6	24.41	66.03	255.2	21.5	.0354	1.11362
198.0	285.8	8.932	2592.0	3711.0	147.3	24.18	63.14	257.2	21.3	.0346	1.10989
200.0	277.3	8.666	2681.0	3835.0	147.9	23.98	60.60	259.1	21.2	.0340	1.10650
202.0	269.5	8.422	2767.0	3954.0	148.5	23.79	58.37	261.1	21.0	.0335	1.10341

Thermophysical properties of oxygen—Continued

T K	Density kg/m³	Density mol/dm³	E J/mol	H J/mol	S J/(mol·K)	C_v J/(mol·K)	C_p J/(mol·K)	Sound m/s	Visc. μPa·s	Therm. W/(m·K)	Diel. const.
205.0	258.9	8.092	2889.0	4125.0	149.3	23.53	55.48	264.1	20.8	.0328	1.09922
210.0	243.7	7.617	3079.0	4392.0	150.6	23.17	51.64	269.2	20.7	.0319	1.09322
215.0	230.9	7.215	3256.0	4642.0	151.8	22.87	48.68	274.1	20.6	.0312	1.08815
220.0	219.8	6.868	3424.0	4880.0	152.9	22.63	46.35	278.8	20.5	.0308	1.08380
225.0	210.1	6.565	3583.0	5107.0	153.9	22.43	44.48	283.5	20.6	.0305	1.07999
230.0	201.4	6.295	3737.0	5325.0	154.9	22.27	42.94	288.0	20.6	.0303	1.07663
235.0	193.7	6.054	3885.0	5537.0	155.8	22.14	41.67	292.4	20.7	.0301	1.07362
240.0	186.8	5.836	4029.0	5742.0	156.6	22.03	40.59	296.6	20.8	.0301	1.07090
250.0	174.6	5.456	4306.0	6139.0	158.2	21.86	38.89	304.8	21.1	.0301	1.06618
260.0	164.3	5.134	4573.0	6521.0	159.7	21.75	37.61	312.6	21.4	.0302	1.06219
270.0	155.4	4.856	4833.0	6892.0	161.1	21.68	36.62	319.9	21.7	.0305	1.05875
280.0	147.6	4.613	5086.0	7254.0	162.5	21.64	35.84	327.0	22.1	.0308	1.05575
290.0	140.7	4.397	5335.0	7609.0	163.7	21.63	35.21	333.8	22.5	.0312	1.05310
300.0	134.5	4.205	5581.0	7959.0	164.9	21.63	34.70	340.3	22.9	.0316	1.05073
310.0	129.0	4.031	5823.0	8304.0	166.0	21.64	34.28	346.6	23.3	.0321	1.04860
320.0	123.9	3.874	6063.0	8645.0	167.1	21.67	33.93	352.7	23.7	.0326	1.04667
330.0	119.3	3.730	6301.0	8983.0	168.1	21.71	33.65	358.6	24.1	.0332	1.04491
340.0	115.1	3.598	6538.0	9318.0	169.1	21.76	33.41	364.3	24.6	.0337	1.04329
360.0	107.6	3.363	7009.0	9982.0	171.0	21.87	33.05	375.3	25.4	.0349	1.04043
380.0	101.1	3.161	7477.0	10640.0	172.8	22.01	32.81	385.7	26.3	.0361	1.03796
400.0	95.49	2.984	7944.0	11300.0	174.5	22.16	32.66	395.5	27.1	.0374	1.03581

15.00 MPa isobar

T K	Density kg/m³	Density mol/dm³	E J/mol	H J/mol	S J/(mol·K)	C_v J/(mol·K)	C_p J/(mol·K)	Sound m/s	Visc. μPa·s	Therm. W/(m·K)	Diel. const.
56.06[a]	1316.0	41.12	-6158.0	-5793.0	67.54	35.88	52.93	1165.0	530.0	.194	1.57327
80.0	1215.0	37.97	-4929.0	-4534.0	86.25	31.81	52.34	1036.0	286.0	.170	1.52321
100.0	1128.0	35.24	-3911.0	-3486.0	97.94	29.38	52.61	900.8	179.0	.143	1.48053
110.0	1082.0	33.80	-3401.0	-2957.0	103.0	28.42	53.20	832.7	146.0	.130	1.45835
120.0	1033.0	32.28	-2885.0	-2420.0	107.6	27.61	54.23	763.6	120.0	.117	1.43528
130.0	981.3	30.67	-2360.0	-1871.0	112.0	26.92	55.83	693.3	100.0	.106	1.41091
140.0	925.1	28.91	-1820.0	-1301.0	116.3	26.36	58.17	621.8	83.8	.0953	1.38476
145.0	894.9	27.97	-1543.0	-1007.0	118.3	26.13	59.67	585.8	76.8	.0903	1.37084
150.0	863.0	26.97	-1260.0	-704.1	120.4	25.94	61.44	550.0	70.4	.0855	1.35623
155.0	829.1	25.91	-970.7	-391.8	122.4	25.78	63.51	514.5	64.4	.0809	1.34086
160.0	793.1	24.79	-673.6	-68.4	124.5	25.66	65.91	479.6	58.9	.0766	1.32463
165.0	754.8	23.59	-368.1	267.9	126.6	25.57	68.66	445.7	53.8	.0726	1.30748
180.0	625.4	19.55	597.6	1365.0	132.9	25.43	77.19	357.8	40.6	.0611	1.25078
190.0	535.7	16.74	1248.0	2144.0	137.1	25.18	77.40	321.7	34.1	.0527	1.21244
195.0	494.8	15.46	1555.0	2525.0	139.1	24.94	74.81	311.6	31.7	.0490	1.19522
200.0	458.2	14.32	1843.0	2891.0	141.0	24.63	71.12	305.7	29.8	.0459	1.17995
205.0	426.1	13.32	2110.0	3236.0	142.7	24.29	67.02	302.7	28.3	.0434	1.16669
210.0	398.3	12.45	2356.0	3561.0	144.2	23.95	62.96	301.9	27.2	.0413	1.15529
215.0	374.4	11.70	2584.0	3866.0	145.7	23.63	59.24	302.4	26.4	.0397	1.14551
220.0	353.6	11.05	2796.0	4154.0	147.0	23.35	55.95	304.0	25.7	.0385	1.13707
225.0	335.4	10.48	2995.0	4426.0	148.2	23.09	53.10	306.1	25.2	.0375	1.12975
230.0	319.5	9.985	3183.0	4686.0	149.4	22.87	50.67	308.7	24.9	.0367	1.12333
235.0	305.4	9.544	3362.0	4934.0	150.4	22.69	48.59	311.5	24.6	.0361	1.11768
240.0	292.8	9.150	3533.0	5172.0	151.4	22.53	46.82	314.6	24.4	.0356	1.11264
245.0	281.5	8.796	3697.0	5402.0	152.4	22.39	45.30	317.7	24.3	.0352	1.10812
250.0	271.2	8.475	3855.0	5625.0	153.3	22.28	43.99	320.9	24.2	.0349	1.10405
260.0	253.3	7.915	4159.0	6054.0	155.0	22.11	41.88	327.4	24.2	.0345	1.09695
270.0	238.1	7.441	4449.0	6465.0	156.5	21.99	40.25	333.9	24.2	.0344	1.09097
280.0	225.1	7.033	4728.0	6860.0	157.9	21.92	38.99	340.3	24.4	.0344	1.08584
290.0	213.6	6.677	4999.0	7245.0	159.3	21.87	37.98	346.6	24.6	.0345	1.08136
300.0	203.6	6.362	5263.0	7621.0	160.6	21.86	37.16	352.7	24.8	.0347	1.07742
310.0	194.6	6.081	5522.0	7989.0	161.8	21.86	36.49	358.7	25.1	.0350	1.07391
320.0	186.5	5.828	5777.0	8351.0	162.9	21.87	35.94	364.5	25.3	.0353	1.07077
330.0	179.2	5.599	6029.0	8708.0	164.0	21.90	35.48	370.2	25.7	.0357	1.06792
340.0	172.5	5.390	6278.0	9061.0	165.1	21.94	35.09	375.7	26.0	.0362	1.06533
360.0	160.7	5.023	6770.0	9756.0	167.1	22.05	34.49	386.3	26.7	.0371	1.06079
380.0	150.7	4.709	7256.0	10440.0	168.9	22.18	34.07	396.5	27.4	.0382	1.05691
400.0	142.0	4.437	7739.0	11120.0	170.7	22.33	33.77	406.2	28.1	.0394	1.05356

Thermophysical properties of oxygen—Continued

T K	Density kg/m³	Density mol/dm³	E J/mol	H J/mol	S J/(mol·K)	C_v J/(mol·K)	C_p J/(mol·K)	Sound m/s	Visc. μPa·s	Therm. W/(m·K)	Diel. const.
					20.00 MPa isobar						
56.62[a]	1319.0	41.21	-6149.0	-5664.0	67.68	35.79	52.88	1178.0	545.0	.194	1.57479
80.0	1223.0	38.21	-4962.0	-4438.0	85.80	32.04	52.03	1053.0	297.0	.171	1.52690
100.0	1138.0	35.57	-3962.0	-3400.0	97.39	29.64	52.00	923.2	187.0	.144	1.48567
120.0	1049.0	32.78	-2962.0	-2352.0	106.9	27.87	52.97	795.7	127.0	.119	1.44281
130.0	1001.0	31.29	-2457.0	-1818.0	111.2	27.16	53.96	732.0	106.0	.108	1.42027
140.0	950.8	29.71	-1944.0	-1271.0	115.3	26.56	55.35	668.7	90.3	.0981	1.39668
150.0	897.0	28.03	-1423.0	-709.1	119.1	26.06	57.15	606.5	77.2	.0889	1.37178
160.0	839.2	26.23	-889.6	-127.0	122.9	25.65	59.32	547.0	66.2	.0805	1.34540
170.0	777.3	24.29	-345.2	478.1	126.6	25.32	61.71	491.8	57.1	.0732	1.31753
180.0	711.9	22.25	207.8	1107.0	130.2	25.04	63.95	443.2	49.4	.0668	1.28849
200.0	579.6	18.11	1300.0	2405.0	137.0	24.42	64.52	375.4	38.1	.0543	1.23108
205.0	549.0	17.16	1559.0	2724.0	138.6	24.23	63.34	365.5	36.1	.0517	1.21807
210.0	520.4	16.26	1807.0	3037.0	140.1	24.02	61.75	358.1	34.5	.0493	1.20596
215.0	493.9	15.43	2046.0	3341.0	141.5	23.81	59.88	352.8	33.0	.0473	1.19483
220.0	469.6	14.67	2273.0	3636.0	142.9	23.60	57.88	349.2	31.9	.0456	1.18467
225.0	447.4	13.98	2490.0	3920.0	144.1	23.39	55.86	346.9	30.9	.0442	1.17546
230.0	427.2	13.35	2696.0	4194.0	145.4	23.20	53.90	345.8	30.1	.0430	1.16712
235.0	408.9	12.78	2894.0	4459.0	146.5	23.02	52.06	345.5	29.4	.0420	1.15958
240.0	392.2	12.26	3083.0	4715.0	147.6	22.86	50.35	345.8	28.9	.0411	1.15275
245.0	377.0	11.78	3265.0	4963.0	148.6	22.72	48.80	346.7	28.4	.0404	1.14654
250.0	363.1	11.35	3441.0	5204.0	149.6	22.60	47.39	348.0	28.1	.0398	1.14089
260.0	338.6	10.58	3775.0	5665.0	151.4	22.40	44.98	351.4	27.6	.0390	1.13100
270.0	317.8	9.931	4091.0	6105.0	153.0	22.26	43.04	355.5	27.3	.0384	1.12262
280.0	299.8	9.370	4392.0	6527.0	154.6	22.16	41.48	360.1	27.1	.0380	1.11542
290.0	284.2	8.880	4683.0	6935.0	156.0	22.10	40.21	365.0	27.0	.0379	1.10917
300.0	270.4	8.449	4965.0	7332.0	157.3	22.06	39.18	370.0	27.1	.0378	1.10368
310.0	258.1	8.065	5239.0	7719.0	158.6	22.05	38.32	375.0	27.1	.0379	1.09881
320.0	247.0	7.720	5508.0	8099.0	159.8	22.06	37.61	380.1	27.3	.0381	1.09446
330.0	237.1	7.409	5773.0	8472.0	161.0	22.08	37.01	385.1	27.5	.0383	1.09053
340.0	228.0	7.126	6033.0	8840.0	162.1	22.11	36.51	390.1	27.7	.0386	1.08697
350.0	219.8	6.868	6290.0	9202.0	163.1	22.15	36.09	395.0	27.9	.0390	1.08373
360.0	212.2	6.630	6545.0	9561.0	164.1	22.21	35.72	399.8	28.2	.0394	1.08075
380.0	198.7	6.208	7048.0	10270.0	166.0	22.33	35.15	409.3	28.7	.0403	1.07548
400.0	187.0	5.844	7546.0	10970.0	167.8	22.47	34.73	418.5	29.3	.0413	1.07094
					30.00 MPa isobar						
57.73[a]	1325.0	41.39	-6131.0	-5406.0	67.94	35.65	52.81	1204.0	572.0	.195	1.57773
100.0	1158.0	36.18	-4051.0	-3222.0	96.38	30.11	51.06	965.0	203.0	.147	1.49505
120.0	1076.0	33.63	-3092.0	-2200.0	105.7	28.34	51.23	851.0	139.0	.123	1.45581
140.0	991.0	30.97	-2137.0	-1168.0	113.6	26.97	52.08	743.0	102.0	.103	1.41541
160.0	900.5	28.14	-1179.0	-113.3	120.7	25.91	53.42	643.4	77.6	.0863	1.37336
200.0	709.4	22.17	713.1	2066.0	132.8	24.39	54.89	493.2	50.1	.0635	1.28735
210.0	663.2	20.73	1165.0	2613.0	135.5	24.07	54.35	468.9	46.0	.0593	1.26711
220.0	619.5	19.36	1602.0	3152.0	138.0	23.77	53.31	450.0	42.6	.0556	1.24815
230.0	578.9	18.09	2020.0	3678.0	140.4	23.49	51.88	436.1	39.9	.0526	1.23076
240.0	542.0	16.94	2417.0	4188.0	142.5	23.23	50.21	426.3	37.8	.0502	1.21506
250.0	508.9	15.90	2795.0	4682.0	144.5	23.01	48.47	419.9	36.1	.0483	1.20106
260.0	479.2	14.98	3155.0	5158.0	146.4	22.83	46.78	416.1	34.9	.0468	1.18865
270.0	452.8	14.15	3498.0	5618.0	148.1	22.68	45.22	414.1	33.9	.0457	1.17766
280.0	429.2	13.41	3826.0	6063.0	149.8	22.57	43.82	413.7	33.1	.0448	1.16791
290.0	408.2	12.76	4143.0	6495.0	151.3	22.49	42.58	414.3	32.6	.0442	1.15924
300.0	389.2	12.16	4449.0	6915.0	152.7	22.44	41.50	415.8	32.2	.0438	1.15150
310.0	372.2	11.63	4746.0	7325.0	154.0	22.41	40.57	417.8	31.9	.0435	1.14455
320.0	356.7	11.15	5036.0	7727.0	155.3	22.40	39.76	420.3	31.7	.0434	1.13827
330.0	342.7	10.71	5319.0	8121.0	156.5	22.41	39.06	423.2	31.6	.0434	1.13258
340.0	329.8	10.31	5598.0	8508.0	157.7	22.43	38.46	426.2	31.5	.0434	1.12740
360.0	307.2	9.599	6142.0	9267.0	159.9	22.51	37.48	432.9	31.6	.0438	1.11830
380.0	287.8	8.994	6674.0	10010.0	161.9	22.62	36.74	440.0	31.8	.0444	1.11056
400.0	271.0	8.470	7196.0	10740.0	163.7	22.75	36.18	447.3	32.1	.0452	1.10389

Thermophysical properties of oxygen—Continued

T K	Density kg/m^3	Density mol/dm^3	E J/mol	H J/mol	S J/(mol·K)	C_v J/(mol·K)	C_p J/(mol·K)	Sound m/s	Visc. μPa·s	Therm. W/(m·K)	Diel. const.
					40.00 MPa isobar						
58.82[a]	1330.0	41.57	−6113.0	−5150.0	68.20	35.58	52.79	1231.0	598.0	.195	1.58055
80.0	1250.0	39.05	−5077.0	−4052.0	84.16	32.79	51.12	1118.0	342.0	.175	1.54023
100.0	1175.0	36.72	−4128.0	−3038.0	95.47	30.52	50.37	1003.0	218.0	.149	1.50346
120.0	1099.0	34.36	−3199.0	−2035.0	104.6	28.76	50.07	898.4	151.0	.126	1.46690
140.0	1022.0	31.95	−2284.0	−1032.0	112.3	27.37	50.20	802.2	111.0	.106	1.43021
160.0	943.6	29.49	−1382.0	−25.0	119.1	26.27	50.56	715.0	86.7	.0906	1.39325
200.0	784.0	24.50	372.0	2005.0	130.4	24.66	50.60	579.6	58.8	.0691	1.32044
220.0	708.4	22.14	1202.0	3009.0	135.2	24.05	49.65	534.6	50.8	.0618	1.28687
240.0	639.6	19.99	1985.0	3986.0	139.4	23.56	48.01	503.7	45.3	.0565	1.25682
250.0	608.4	19.01	2357.0	4461.0	141.4	23.36	47.01	492.7	43.2	.0545	1.24333
260.0	579.4	18.11	2717.0	4926.0	143.2	23.19	45.97	484.3	41.5	.0528	1.23090
270.0	552.6	17.27	3064.0	5381.0	144.9	23.05	44.93	478.0	40.1	.0515	1.21949
280.0	528.0	16.50	3401.0	5825.0	146.5	22.94	43.93	473.4	38.9	.0504	1.20906
290.0	505.3	15.79	3726.0	6259.0	148.1	22.85	42.98	470.3	38.0	.0495	1.19952
300.0	484.5	15.14	4043.0	6685.0	149.5	22.79	42.11	468.3	37.3	.0489	1.19080
310.0	465.3	14.54	4351.0	7102.0	150.9	22.75	41.32	467.3	36.7	.0484	1.18282
320.0	447.7	13.99	4653.0	7512.0	152.2	22.73	40.61	467.1	36.2	.0481	1.17550
340.0	416.4	13.01	5237.0	8311.0	154.6	22.74	39.40	468.4	35.5	.0478	1.16258
360.0	389.5	12.17	5803.0	9089.0	156.8	22.80	38.44	471.3	35.2	.0478	1.15155
380.0	366.1	11.44	6354.0	9850.0	158.9	22.89	37.68	475.4	35.0	.0482	1.14204
400.0	345.7	10.80	6895.0	10600.0	160.8	23.02	37.09	480.2	35.0	.0488	1.13375
					50.00 MPa isobar						
59.89[a]	1336.0	41.74	−6094.0	−4896.0	68.45	35.57	52.80	1256.0	622.0	.196	1.58325
80.0	1262.0	39.43	−5125.0	−3857.0	83.41	33.10	50.82	1150.0	364.0	.177	1.54620
200.0	836.7	26.15	132.2	2044.0	128.6	24.99	48.24	648.0	65.8	.0732	1.34420
220.0	770.2	24.07	924.4	3002.0	133.2	24.39	47.42	603.6	57.5	.0663	1.31427
240.0	708.5	22.14	1681.0	3939.0	137.3	23.91	46.25	570.4	51.5	.0611	1.28688
260.0	652.6	20.40	2399.0	4850.0	140.9	23.55	44.85	546.7	47.2	.0574	1.26244
280.0	603.1	18.85	3079.0	5732.0	144.2	23.30	43.38	530.7	44.1	.0548	1.24102
300.0	559.6	17.49	3727.0	6586.0	147.1	23.14	41.99	520.6	42.0	.0531	1.22244
320.0	521.7	16.30	4346.0	7413.0	149.8	23.06	40.76	514.8	40.4	.0520	1.20637
340.0	488.5	15.27	4943.0	8218.0	152.2	23.05	39.73	512.1	39.4	.0515	1.19244
360.0	459.4	14.36	5521.0	9003.0	154.5	23.09	38.87	511.6	38.7	.0514	1.18031
380.0	433.8	13.56	6085.0	9773.0	156.6	23.17	38.17	512.8	38.3	.0516	1.16969
400.0	411.0	12.84	6638.0	10530.0	158.5	23.28	37.61	515.1	38.0	.0520	1.16032
					60.00 MPa isobar						
60.95[a]	1341.0	41.90	−6074.0	−4642.0	68.69	35.61	52.84	1281.0	644.0	.197	1.58585
70.0	1308.0	40.87	−5639.0	−4170.0	75.90	34.62	51.52	1236.0	501.0	.189	1.56923
80.0	1273.0	39.78	−5169.0	−3660.0	82.72	33.40	50.60	1180.0	386.0	.178	1.55179
100.0	1205.0	37.65	−4254.0	−2661.0	93.87	31.23	49.42	1072.0	247.0	.154	1.51812
150.0	1038.0	32.45	−2082.0	−233.6	113.6	27.53	47.89	858.0	114.0	.104	1.43777
160.0	1006.0	31.43	−1665.0	244.3	116.7	27.00	47.70	822.5	102.0	.0969	1.42227
200.0	877.9	27.44	−52.1	2135.0	127.2	25.35	46.74	705.3	71.9	.0766	1.36293
220.0	817.7	25.55	714.3	3062.0	131.6	24.74	45.97	661.9	63.2	.0698	1.33555
240.0	761.3	23.79	1450.0	3972.0	135.6	24.27	45.00	627.9	56.8	.0648	1.31023
260.0	709.3	22.17	2155.0	4862.0	139.1	23.90	43.90	602.2	52.2	.0611	1.28722
280.0	662.3	20.70	2829.0	5728.0	142.4	23.64	42.76	583.3	48.7	.0584	1.26661
300.0	620.1	19.38	3476.0	6572.0	145.3	23.47	41.65	569.9	46.2	.0566	1.24830
320.0	582.4	18.20	4098.0	7395.0	147.9	23.38	40.65	560.8	44.4	.0554	1.23210
340.0	548.8	17.15	4700.0	8199.0	150.4	23.35	39.77	555.0	43.0	.0548	1.21779
360.0	518.8	16.21	5286.0	8986.0	152.6	23.37	39.02	551.7	42.0	.0545	1.20512
380.0	492.0	15.38	5858.0	9760.0	154.7	23.44	38.39	550.4	41.3	.0546	1.19385
400.0	467.9	14.62	6420.0	10520.0	156.7	23.53	37.88	550.4	40.9	.0549	1.18380

Thermophysical properties of oxygen—Continued

T K	Density kg/m³	Density mol/dm³	E J/mol	H J/mol	S J/(mol·K)	C_v J/(mol·K)	C_p J/(mol·K)	Sound m/s	Visc. μPa·s	Therm. W/(m·K)	Diel. const.
					80.00 MPa isobar						
63.03[a]	1350.0	42.20	-6033.0	-4137.0	69.15	35.81	52.93	1328.0	684.0	.198	1.59075
70.0	1326.0	41.45	-5704.0	-3774.0	74.63	35.10	51.55	1291.0	560.0	.192	1.57852
80.0	1293.0	40.42	-5244.0	-3265.0	81.42	33.96	50.32	1237.0	431.0	.182	1.56202
100.0	1230.0	38.45	-4355.0	-2275.0	92.48	31.86	48.83	1134.0	275.0	.157	1.53070
120.0	1169.0	36.54	-3498.0	-1309.0	101.3	30.16	47.80	1046.0	192.0	.135	1.50070
140.0	1110.0	34.68	-2668.0	-361.4	108.6	28.79	46.97	970.5	145.0	.116	1.47181
160.0	1052.0	32.87	-1863.0	571.0	114.8	27.68	46.27	904.8	115.0	.102	1.44400
180.0	995.3	31.10	-1082.0	1490.0	120.2	26.77	45.62	847.7	95.8	.0904	1.41734
200.0	940.9	29.40	-325.1	2396.0	125.0	26.03	44.95	798.7	82.4	.0819	1.39193
220.0	889.0	27.78	407.7	3287.0	129.2	25.42	44.22	757.5	72.8	.0753	1.36795
240.0	839.9	26.25	1116.0	4164.0	133.1	24.94	43.42	723.6	65.8	.0704	1.34555
260.0	794.0	24.81	1800.0	5024.0	136.5	24.57	42.59	696.3	60.6	.0667	1.32484
280.0	751.5	23.49	2461.0	5868.0	139.6	24.30	41.76	674.6	56.6	.0641	1.30584
300.0	712.4	22.26	3102.0	6695.0	142.5	24.10	40.96	657.7	53.6	.0622	1.28852
320.0	676.6	21.15	3723.0	7506.0	145.1	23.98	40.22	644.6	51.2	.0609	1.27279
340.0	643.9	20.12	4329.0	8304.0	147.5	23.93	39.56	634.8	49.4	.0601	1.25853
360.0	614.0	19.19	4920.0	9090.0	149.8	23.92	38.98	627.6	48.1	.0598	1.24559
380.0	586.7	18.33	5501.0	9864.0	151.9	23.95	38.49	622.4	47.0	.0597	1.23385
400.0	561.7	17.55	6072.0	10630.0	153.8	24.02	38.07	619.0	46.2	.0599	1.22317
					100.00 MPa isobar						
65.06[a]	1359.0	42.49	-5990.0	-3636.0	69.61	36.11	53.03	1369.0	718.0	.199	1.59534
70.0	1343.0	41.97	-5760.0	-3377.0	73.44	35.61	51.80	1343.0	621.0	.195	1.58701
80.0	1312.0	40.99	-5307.0	-2868.0	80.24	34.50	50.22	1290.0	475.0	.184	1.57123
100.0	1253.0	39.15	-4437.0	-1883.0	91.24	32.44	48.44	1190.0	302.0	.161	1.54178
120.0	1196.0	37.39	-3601.0	-926.6	99.96	30.77	47.25	1105.0	211.0	.138	1.51386
140.0	1142.0	35.68	-2794.0	8.4	107.2	29.41	46.28	1034.0	159.0	.120	1.48718
160.0	1089.0	34.03	-2013.0	925.5	113.3	28.31	45.44	972.9	127.0	.106	1.46170
180.0	1038.0	32.44	-1256.0	1826.0	118.6	27.40	44.67	920.0	106.0	.0945	1.43742
200.0	989.1	30.91	-522.5	2713.0	123.3	26.66	43.94	874.2	91.4	.0860	1.41438
220.0	942.5	29.45	189.2	3584.0	127.4	26.05	43.22	835.0	81.0	.0796	1.39262
240.0	898.3	28.07	879.5	4441.0	131.2	25.57	42.50	801.9	73.4	.0747	1.37220
260.0	856.8	26.77	1549.0	5284.0	134.5	25.18	41.79	774.3	67.6	.0711	1.35316
280.0	817.8	25.56	2200.0	6113.0	137.6	24.89	41.10	751.5	63.2	.0684	1.33549
300.0	781.5	24.42	2834.0	6928.0	140.4	24.68	40.44	732.9	59.7	.0665	1.31916
320.0	747.8	23.37	3452.0	7731.0	143.0	24.54	39.85	717.8	57.1	.0652	1.30411
340.0	716.5	22.39	4057.0	8523.0	145.4	24.46	39.31	705.7	54.9	.0644	1.29026
360.0	687.6	21.49	4650.0	9304.0	147.6	24.43	38.83	696.0	53.3	.0640	1.27751
380.0	660.8	20.65	5234.0	10080.0	149.7	24.44	38.43	688.5	52.0	.0639	1.26578
400.0	635.9	19.87	5809.0	10840.0	151.7	24.48	38.08	682.6	50.9	.0641	1.25498
					120.00 MPa isobar						
67.04[a]	1368.0	42.75	-5945.0	-3138.0	70.04	36.48	53.13	1407.0	746.0	.199	1.59966
80.0	1328.0	41.52	-5361.0	-2470.0	79.15	35.04	50.27	1338.0	519.0	.187	1.57964
100.0	1273.0	39.78	-4505.0	-1488.0	90.12	33.00	48.19	1241.0	329.0	.163	1.55170
120.0	1220.0	38.12	-3686.0	-538.1	98.79	31.33	46.88	1159.0	230.0	.141	1.52547
140.0	1169.0	36.54	-2896.0	388.7	105.9	29.99	45.82	1090.0	173.0	.123	1.50052
160.0	1120.0	35.01	-2132.0	1296.0	112.0	28.89	44.91	1032.0	138.0	.109	1.47679
180.0	1073.0	33.54	-1392.0	2186.0	117.2	27.98	44.08	981.7	115.0	.0979	1.45425
200.0	1028.0	32.14	-674.0	3059.0	121.8	27.24	43.32	938.3	99.6	.0895	1.43291
220.0	985.7	30.81	23.2	3919.0	125.9	26.63	42.60	900.8	88.3	.0831	1.41275
240.0	945.1	29.54	700.9	4764.0	129.6	26.13	41.91	868.7	80.0	.0783	1.39379
260.0	906.7	28.34	1361.0	5595.0	132.9	25.74	41.26	841.3	73.8	.0746	1.37601
280.0	870.6	27.21	2004.0	6414.0	136.0	25.43	40.65	818.1	68.9	.0720	1.35940
300.0	836.6	26.14	2632.0	7222.0	138.8	25.20	40.08	798.6	65.1	.0701	1.34392
320.0	804.8	25.15	3247.0	8018.0	141.3	25.04	39.56	782.3	62.2	.0687	1.32953
340.0	775.0	24.22	3850.0	8805.0	143.7	24.94	39.10	768.8	59.8	.0679	1.31615
360.0	747.1	23.35	4443.0	9582.0	145.9	24.89	38.69	757.7	57.9	.0675	1.30373
380.0	721.1	22.54	5028.0	10350.0	148.0	24.88	38.34	748.5	56.4	.0674	1.29219
400.0	696.8	21.77	5605.0	11120.0	150.0	24.91	38.04	741.1	55.1	.0675	1.28146

[a]At melting line.

[b]At liquid–vapor boundary.

Appendix L. Coefficients for Equations for Computing Thermophysical Properties

Coefficients for Argon

Argon Coefficients for MBWR eq

$G(1)= -.65697312940\times 10^{-4}$
$G(2)= .18229578010\times 10^{-1}$
$G(3)= -.36494701410$
$G(4)= .12320121070\times 10^{2}$
$G(5)= -.86135782740\times 10^{3}$
$G(6)= .79785796910\times 10^{-5}$
$G(7)= -.29114891100\times 10^{-2}$
$G(8)= .75818217580$
$G(9)= .87804881690\times 10^{3}$
$G(10)= .14231459890\times 10^{-7}$
$G(11)= .16741461310\times 10^{-3}$
$G(12)= -.32004479090\times 10^{-1}$
$G(13)= .25617663720\times 10^{-5}$
$G(14)= -.54759349410\times 10^{-4}$
$G(15)= -.45050320580\times 10^{-1}$
$G(16)= .20132546530\times 10^{-5}$
$G(17)= -.16789412730\times 10^{-7}$
$G(18)= .42073292710\times 10^{-4}$
$G(19)= -.54442129960\times 10^{-6}$
$G(20)= -.80048550110\times 10^{3}$
$G(21)= -.13193042010\times 10^{5}$
$G(22)= -.49549239300\times 10^{1}$
$G(23)= .80921321770\times 10^{4}$
$G(24)= -.98701040610\times 10^{-2}$
$G(25)= .20204415620$
$G(26)= -.16374172050\times 10^{-4}$
$G(27)= -.70389441360\times 10^{-1}$
$G(28)= -.11543245390\times 10^{-7}$
$G(29)= .15559901170\times 10^{-5}$
$G(30)= -.14921785360\times 10^{-10}$
$G(31)= -.10013560710\times 10^{-8}$
$G(32)= .29339632160\times 10^{-7}$
$\gamma = -.0055542372$

Argon Coefficients for Vapor Pressure

$V_p(1)= 3.4151115519$ $V_p(2)= 1.1910812519$
$V_p(3)= -.3407632334$ $V_p(5)= .89555855251$
$V_p(6)= 1.5$ $V_p(9)= .06890606625$
$V_p(7)= 83.80$ $V_p(8)= 150.86$
$V_p(4)= 0.0$

Argon Coefficients for Saturated Liquid and Vapor Densities

$A(1)= -.270262923777\times 10^{2}$
$A(2)= .131040241866$
$A(3)= -.267486438128\times 10^{1}$
$A(4)= .300176804406\times 10^{2}$
$A(5)= -.875899149326\times 10^{2}$
$A(6)= -.408267436456\times 10^{2}$
$A(7)= .104268066451\times 10^{3}$
$A(8)= -.671278555379\times 10^{2}$
$A(9)= .151002935701\times 10^{2}$
$A(10)= -.331243536637\times 10^{2}$
$A(11)= .633146212581\times 10^{2}$
$A(12)= -.427149706899\times 10^{2}$
$A(13)= .100599900030\times 10^{2}$
$A(14)= .137682084900\times 10^{2}$
$A(15)= -.664630363191\times 10^{-1}$
$A(16)= .133368782730\times 10^{1}$
$A(17)= -.144371463244\times 10^{2}$
$A(18)= .601938472000\times 10^{2}$
$A(19)= -.230888463887\times 10^{2}$
$A(20)= .465318358887\times 10^{1}$

Argon Coefficients for Ideal Gas C_p

$G_i(1)=G_i(2)=G_i(3)=0$
$G_i(4)=G_i(9)=2.5$
$G_i(5)=G_i(6)=G_i(7)=0$
$G_i(8)=0$

Argon Coefficients for First Term of Viscosity eq
Viscosity in μPa·s, Density in g/ml

$G_v(1)= .61145472787\times 10^{4}$ $G_v(2)= -.10394390312\times 10^{5}$
$G_v(3)= .67594614619\times 10^{4}$ $G_v(4)= -.22536509380\times 10^{4}$
$G_v(5)= .42593950138\times 10^{3}$ $G_v(6)= -.47252671093\times 10^{2}$
$G_v(7)= .31795275425\times 10^{1}$ $G_v(8)= -.11629083780$
$G_v(9)= .18043010592\times 10^{-2}$

Argon Coefficients for First Term of Thermal Conductivity eq
Thermal Conductivity in W/(m·K), Density in g/ml

$G_t(1)= .62777703742\times 10^{1}$ $G_t(2)= -.96096376637\times 10^{1}$
$G_t(3)= .58887549191\times 10^{1}$ $G_t(4)= -.18920926320\times 10^{1}$
$G_t(5)= .34886571437$ $G_t(6)= -.38016786193\times 10^{-1}$
$G_t(7)= .25207283167\times 10^{-2}$ $G_t(8)= -.91098744478\times 10^{-4}$
$G_t(9)= .13990842942\times 10^{-5}$

Argon Coefficients for Second Term of Viscosity eq

$F_v(1)= .14653652433$ $F_v(2)= -.77487424965\times 10^{-1}$
$F_v(3)= .14000000000\times 10^{1}$ $F_v(4)= .15280000000\times 10^{3}$

Argon Coefficients for Second Term of Thermal Conductivity eq

$F_t(1)= .24142103270\times 10^{-1}$ $F_t(2)= .75696234255\times 10^{-2}$
$F_t(3)= .10000000000\times 10^{1}$ $F_t(4)= .15280000000\times 10^{3}$

Argon Coefficients for Third Term of Viscosity eq

$E_v(1)= -.12313579086\times 10^{2}$ $E_v(2)= .20694685712$
$E_v(3)= .16029145122\times 10^{2}$ $E_v(4)= .11717461351\times 10^{4}$
$E_v(5)= -.56995898780\times 10^{3}$ $E_v(6)= .40136071933\times 10^{2}$
$E_v(7)= .39870122403\times 10^{2}$ $E_v(8)= .53700000000$

Argon Coefficients for Third Term of Thermal Conductivity eq

$E_t(1)= -.33327027332\times 10^{2}$ $E_t(2)= 0.0$
$E_t(3)= .30694859971\times 10^{2}$ $E_t(4)= 0.0$
$E_t(5)= .22956551674\times 10^{4}$ $E_t(6)= -.35559415848\times 10^{3}$
$E_t(7)= 0.0$ $E_t(8)= 1.0$

Argon Coefficients for Melting eq

$A= -.210562165\times 10^{3}$
$B= .177760527$
$C= .159817868\times 10^{1}$

Coefficients for Ethylene

Ethylene Coefficients for MBWR eq

$G(1)= -.2146684366683\times10^{-2}$
$G(2)= .1791433722534$
$G(3)= -.3675315603930\times10^{1}$
$G(4)= .3707178934669\times10^{3}$
$G(5)= -.3198282566709\times10^{5}$
$G(6)= .5809379774732\times10^{-4}$
$G(7)= -.7895570824899\times10^{-1}$
$G(8)= .1148620375835\times10^{2}$
$G(9)= .2713774629193\times10^{5}$
$G(10)= -.8647124319107\times10^{-5}$
$G(11)= .1617727266385\times10^{-1}$
$G(12)= -.2731527496271\times10^{1}$
$G(13)= -.2672283641459\times10^{-3}$
$G(14)= -.4752381331990\times10^{-2}$
$G(15)= -.6255637346217\times10^{1}$
$G(16)= .4576234964434\times10^{-3}$
$G(17)= -.7534839269320\times10^{-5}$
$G(18)= .1638171982209\times10^{-1}$
$G(19)= -.3563090740740\times10^{-3}$
$G(20)= -.1833000783170\times10^{5}$
$G(21)= -.1805074209985\times10^{7}$
$G(22)= -.4794587918874\times10^{3}$
$G(23)= .3531948274957\times10^{7}$
$G(24)= -.2562571039155\times10^{1}$
$G(25)= .1044308253292\times10^{3}$
$G(26)= -.1695303363659\times10^{-1}$
$G(27)= -.1710334224958\times10^{3}$
$G(28)= -.2054114462372\times10^{-4}$
$G(29)= .6727558766661\times10^{-2}$
$G(30)= -.1557168403328\times10^{-6}$
$G(31)= -.1229814736077\times10^{-4}$
$G(32)= .4234325938573\times10^{-4}$
$\gamma = -.0172$

Ethylene Coefficients for Vapor Pressure

$V_p(1)= 8.2095798$ $V_p(2)= 4.315424145$
$V_p(3)= -1.692585975$ $V_p(5)= 3.446501098$
$V_p(6)= 1.5$ $V_p(9)= .00012129514$
$V_p(7)= 103.986$ $V_p(8)= 282.3428$
$V_p(4)= -.1976495575$

Ethylene Coefficients for Saturated Liquid and Vapor Densities

$A(1)= -.609621515594\times10^{2}$
$A(2)= .203185312702\times10^{-1}$
$A(3)= -.925441265813$
$A(4)= .243630795888\times10^{2}$
$A(5)= -.854745622888\times10^{3}$
$A(6)= .123927868183\times10^{4}$
$A(7)= -.142710711789\times10^{4}$
$A(8)= .837358670405\times10^{3}$
$A(9)= .432203696552\times10^{3}$
$A(10)= -.137917541161\times10^{4}$
$A(11)= .126858600124\times10^{4}$
$A(12)= -.571552321713\times10^{3}$
$A(13)= .106012234360\times10^{3}$
$A(14)= -.479047060183\times10^{1}$
$A(15)= .151381345283\times10^{-1}$
$A(16)= -.403456079445$
$A(17)= .508683920225\times10^{1}$
$A(18)= -.246711997987\times10^{2}$
$A(19)= .980030915247\times10^{1}$
$A(20)= -.216846516122\times10^{1}$

Ethylene Coefficients for Ideal Gas Heat C_p

$G_i(1)= .5603615762\times10^{6}$
$G_i(2)= -.2141069802\times10^{5}$
$G_i(3)= .2532008897\times10^{3}$
$G_i(4)= .3554495281\times10^{1}$
$G_i(5)= -.9951927478\times10^{-2}$
$G_i(6)= .5108931070\times10^{-4}$
$G_i(7)= -.1928667482\times10^{-7}$
$G_i(8)= -.2061703241\times10^{2}$
$G_i(9)= .3\times10^{4}$

Ethylene Coefficients for Melting eq

$A= -.357923875\times10^{3}$
$B= .245332143\times10^{-1}$
$C= .206450000\times10^{1}$

Coefficients for Parahydrogen

Parahydrogen Coefficients for MBWR eq

$G(1)= .4675528393416\times10^{-4}$
$G(2)= .4289274251454\times10^{-2}$
$G(3)= -.5164085596504\times10^{-1}$
$G(4)= .2961790279801$
$G(5)= -.3027194968412\times10^{1}$
$G(6)= .1908100320379\times10^{-5}$
$G(7)= -.1339776859288\times10^{-3}$
$G(8)= .3056473115421\times10^{-1}$
$G(9)= .5161197159532\times10^{1}$
$G(10)= .1999981550224\times10^{-7}$
$G(11)= .2896367059356\times10^{-4}$
$G(12)= -.2257803939041\times10^{-2}$
$G(13)= -.2287392761826\times10^{-6}$
$G(14)= .2446261478645\times10^{-5}$
$G(15)= -.1718181601119\times10^{-3}$
$G(16)= -.5465142603459\times10^{-7}$
$G(17)= .4051941401315\times10^{-9}$
$G(18)= .1157595123961\times10^{-6}$
$G(19)= -.1269162728389\times10^{-8}$
$G(20)= -.4983023605519\times10^{1}$
$G(21)= -.1606676092098\times10^{2}$
$G(22)= -.1926799185310\times10^{-1}$
$G(23)= .9319894638928$
$G(24)= -.3222596554434\times10^{-4}$
$G(25)= .1206839307669\times10^{-3}$
$G(26)= -.3841588197470\times10^{-7}$
$G(27)= -.4036157453608\times10^{-5}$
$G(28)= -.1250868123513\times10^{-10}$
$G(29)= .1976107321888\times10^{-9}$
$G(30)= -.2411883474011\times10^{-13}$
$G(31)= -.4127551498251\times10^{-13}$
$G(32)= .8917972883610\times10^{-12}$
$\gamma = -.0041$

Parahydrogen Coefficients for Vapor Pressure

$V_p(1)=$ 3.05300134164 $V_p(2)=$ 2.80810925813
$V_p(3)=$ −0.655461216567 $V_p(5)=$ 1.59514439374
$V_p(6)=$ 1.5814454428 $V_p(9)=$ 0.0070420875
$V_p(7)=$ 13.8 $V_p(8)=$ 32.938
$V_p(4)=$ 0.0

Parahydrogen Coefficients Saturated Liquid and Vapor Densities

$A(1)= .916617720187\times10^{2}$
$A(2)= -.179492524446$
$A(3)= .454671158395\times10^{1}$
$A(4)= -.658499589788\times10^{2}$
$A(5)= .734466804535\times10^{3}$
$A(6)= -.682501045175\times10^{3}$
$A(7)= .631783674710\times10^{3}$
$A(8)= -.539408873282\times10^{3}$
$A(9)= .430923811783\times10^{3}$
$A(10)= -.300295738811\times10^{3}$
$A(11)= .156567165346\times10^{3}$
$A(12)= -.504103608225\times10^{2}$
$A(13)= .720706926514\times10^{1}$
$A(14)= -.123944440318\times10^{3}$
$A(15)= .140334800142\times10^{1}$
$A(16)= -.211023804313\times10^{2}$
$A(17)= .173254622817\times10^{3}$
$A(18)= -.444294580871\times10^{3}$
$A(19)= .138699365355\times10^{3}$
$A(20)= -.235774161015\times10^{2}$

Parahydrogen Coefficients for Ideal Gas C_p for Temperatures > 140 K

$G_i(1)= .5262185164597\times10^{8}$
$G_i(2)= -.1487906248823\times10^{7}$
$G_i(3)= .1601391392264\times10^{5}$
$G_i(4)= -.8031235938946\times10^{2}$
$G_i(5)= .2307407941873$
$G_i(6)= -.3176386248370\times10^{-3}$
$G_i(7)= .1643857271214\times10^{-6}$
$G_i(8)= .9230816464058\times10^{1}$
$G_i(9)= .3\times10^{4}$

Parahydrogen Coefficients for Ideal Gas C_p for Temperatures < 22 K

$G_i(1)=$ 0.0
$G_i(2)=$ 0.0
$G_i(3)=$ 0.0
$G_i(4)=$ 2.5000315
$G_i(5)=$ 0.0
$G_i(6)=$ 0.0
$G_i(7)=$ 0.0
$G_i(8)=$ 0.0
$G_i(9)=$ 0.0

Parahydrogen Coefficients for Ideal Gas C_p for Temperatures $40 < T < 140$ K

$G_i(1)= .2905965792270\times10^{6}$
$G_i(2)= -.2831103639248\times10^{5}$
$G_i(3)= .1050424877391\times10^{4}$
$G_i(4)= -.1535751501769\times10^{2}$
$G_i(5)= .1218941696566$
$G_i(6)= -.2599406479908\times10^{-4}$
$G_i(7)= -.1288757333406\times10^{-5}$
$G_i(8)= .1717441975231\times10^{6}$
$G_i(9)= .3\times10^{4}$

Parahydrogen Coefficients for Melting Pressure for Temperatures < 22 K

A= −21.281484395
B= .125746643
C= 1.955

Parahydrogen Coefficients for Melting Pressure for Temperatures > 22 K

A= −26.5289115
B= .248578596
C= 1.764739

Parahydrogen Coefficients for Dielectric Constant

$A= .20245443\times10^{-2}$
$B= .37171832\times10^{-6}$
$C= -.92085013\times10^{-8}$
$D= -.34065328\times10^{-11}$
E= .0
F= .0

Coefficients for Methane

Methane Coefficients for MBWR eq

$G(1) = -.189506118654\times10^{-2}$
$G(2) = .104756987190$
$G(3) = -.157446511658\times10^{1}$
$G(4) = .782544605655\times10^{2}$
$G(5) = -.382099919632\times10^{4}$
$G(6) = .858039193151\times10^{-4}$
$G(7) = -.502993394999\times10^{-1}$
$G(8) = .881435651333\times10^{1}$
$G(9) = -.327098978594\times10^{4}$
$G(10) = -.401087946417\times10^{-5}$
$G(11) = .270307051249\times10^{-2}$
$G(12) = -.308038191105$
$G(13) = .194123002261\times10^{-4}$
$G(14) = -.198179473576\times10^{-4}$
$G(15) = .615529077453$
$G(16) = -.536626852203\times10^{-4}$
$G(17) = .154281787794\times10^{-5}$
$G(18) = -.111409049265\times10^{-2}$
$G(19) = .193931540966\times10^{-4}$
$G(20) = .391590731296\times10^{4}$
$G(21) = -.160023162832\times10^{6}$
$G(22) = .197857473814\times10^{2}$
$G(23) = .168195529711\times10^{6}$
$G(24) = .611041574405\times10^{-1}$
$G(25) = .381473591215\times10^{1}$
$G(26) = .127257796890\times10^{-3}$
$G(27) = -.348122335444\times10^{1}$
$G(28) = -.548112616636\times10^{-6}$
$G(29) = .188081779935\times10^{-3}$
$G(30) = .780999906720\times10^{-9}$
$G(31) = -.290669323872\times10^{-6}$
$G(32) = .377310956389\times10^{-5}$
$\gamma = -.0096$

Methane Coefficients for Vapor Pressure

$V_p(1) = 4.77748580$	$V_p(2) = 1.76065363$
$V_p(3) = -.56788894$	$V_p(5) = 1.32786231$
$V_p(6) = 1.5$	$V_p(9) = .01174350$
$V_p(7) = 90.68$	$V_p(8) = 190.555$
$V_p(4) = 0.0$	

Methane Coefficients for Saturated Liquid and Vapor Densities

$A(1) = .183603246136\times10^{2}$
$A(2) = -.182553840603\times10^{-1}$
$A(3) = .586623807178$
$A(4) = -.106005894683\times10^{2}$
$A(5) = .167157622432\times10^{3}$
$A(6) = -.155569173217\times10^{3}$
$A(7) = .106609415022\times10^{3}$
$A(8) = -.341087933790\times10^{2}$
$A(9) = .973203073452\times10^{1}$
$A(10) = -.388410018388\times10^{2}$
$A(11) = .529573454771\times10^{2}$
$A(12) = -.291075304738\times10^{2}$
$A(13) = .585307647478\times10^{1}$
$A(14) = -.637276532186$
$A(15) = .994444109622\times10^{-3}$
$A(16) = -.392314821657\times10^{-1}$
$A(17) = .728820880748$
$A(18) = -.625821815315\times10^{1}$
$A(19) = .295561390641\times10^{1}$
$A(20) = -.763972649504$

Methane Coefficients for Ideal Gas C_p

$G_i(1) = -1.8044750507\times10^{6}$
$G_i(2) = 7.7426666393\times10^{4}$
$G_i(3) = -1.3241658754\times10^{3}$
$G_i(4) = 1.5438149595\times10^{1}$
$G_i(5) = -5.1479005257\times10^{-2}$
$G_i(6) = 1.0809172196\times10^{-4}$
$G_i(7) = -6.5501783437\times10^{-8}$
$G_i(8) = -6.7490056171$
$G_i(9) = .3\times10^{4}$

Methane Coefficients for First Term of Viscosity eq
Viscosity in μPa·s, Density in g/ml

$G_v(1) = -.20909747942\times10^{5}$	$G_v(2) = .26472692181\times10^{5}$
$G_v(3) = -.14728175613\times10^{5}$	$G_v(4) = .47167401921\times10^{4}$
$G_v(5) = -.94918721789\times10^{3}$	$G_v(6) = .12199792872\times10^{3}$
$G_v(7) = -.96279935575\times10^{1}$	$G_v(8) = .42741516570$
$G_v(9) = -.81415307247\times10^{-2}$	

Methane Coefficients for First Term of Thermal Conductivity eq
Thermal Conductivity in W/(m·K), Density in g/ml

$G_t(1) = -.21476213125\times10^{3}$	$G_t(2) = .21904610575\times10^{3}$
$G_t(3) = -.86180973719\times10^{2}$	$G_t(4) = .14960986936\times10^{2}$
$G_t(5) = -.47306603177$	$G_t(6) = -.23311779643$
$G_t(7) = .37784390759\times10^{-1}$	$G_t(8) = -.23204806092\times10^{-2}$
$G_t(9) = .53117637687\times10^{-4}$	

Methane Coefficients for Second Term of Viscosity eq

$F_v(1) = .16969859271$	$F_v(2) = -.13337234608\times10^{-1}$
$F_v(3) = .14000000000\times10^{1}$	$F_v(4) = .16800000000\times10^{3}$

Methane Coefficients for Second Term of Thermal Conductivity eq

$F_t(1) = -.25276292100\times10^{-3}$	$F_t(2) = .33432859310\times10^{-3}$
$F_t(3) = .11200000000\times10^{1}$	$F_t(4) = .16800000000\times10^{3}$

Methane Coefficients for Third Term of Viscosity eq

$E_v(1) = -.12653191679\times10^{2}$	$E_v(2) = .42903609488\times10^{-1}$
$E_v(3) = .17571599671\times10^{2}$	$E_v(4) = .61276818706\times10^{4}$
$E_v(5) = -.30193918656\times10^{4}$	$E_v(6) = .18873011594\times10^{3}$
$E_v(7) = .14529023444\times10^{3}$	$E_v(8) = .16200000000$

Methane Coefficients for Third Term of Thermal Conductivity eq

$E_t(1) = -.13948119270\times10^{2}$	$E_t(2) = .74421462902$
$E_t(3) = .12319512908\times10^{2}$	$E_t(4) = .22209758501\times10^{4}$
$E_t(5) = -.88525979933\times10^{3}$	$E_t(6) = .72835897919\times10^{2}$
$E_t(7) = -.29706914540\times10^{1}$	$E_t(8) = .16280000000$

Methane Coefficients for Melting Pressure eq

$A = -.190926942\times10^{3}$
$B = .456559760\times10^{-1}$
$C = .185000000\times10^{1}$

Methane Coefficients for Dielectric Constant eq

$A = .6535957240472\times10^{-2}$	$D = .0$
$B = .6695777944274\times10^{-5}$	$E = -.3049517489092\times10^{-6}$
$C = -.2343581387584\times10^{-6}$	$F = .0$

Coefficients for Nitrogen

Nitrogen Coefficients for MBWR eq

$G(1) = .138029747465691 \times 10^{-3}$
$G(2) = .108450650134880 \times 10^{-1}$
$G(3) = -.247132406436209$
$G(4) = .345525798080709 \times 10^{1}$
$G(5) = -.427970769066595 \times 10^{3}$
$G(6) = .106491156699760 \times 10^{-4}$
$G(7) = -.114086707973499 \times 10^{-2}$
$G(8) = .144490249728747 \times 10^{-4}$
$G(9) = .187145756755327 \times 10^{4}$
$G(10) = .821887688683079 \times 10^{-8}$
$G(11) = .236099049334759 \times 10^{-3}$
$G(12) = -.514480308120135 \times 10^{-1}$
$G(13) = .491454501366803 \times 10^{-5}$
$G(14) = -.115162716239893 \times 10^{-3}$
$G(15) = -.716803724664983 \times 10^{-1}$
$G(16) = .761666761949981 \times 10^{-5}$
$G(17) = -.113093006621295 \times 10^{-6}$
$G(18) = .373683116683089 \times 10^{-4}$
$G(19) = -.203985150758086 \times 10^{-6}$
$G(20) = -.171966200898966 \times 10^{4}$
$G(21) = -.121305519974777 \times 10^{5}$
$G(22) = -.988139914142789 \times 10^{1}$
$G(23) = .561988689351085 \times 10^{4}$
$G(24) = -.182304396411845 \times 10^{-1}$
$G(25) = -.259982649847705$
$G(26) = -.419189342315742 \times 10^{-4}$
$G(27) = -.259640667053023 \times 10^{-1}$
$G(28) = -.125868320192119 \times 10^{-7}$
$G(29) = .104928659940046 \times 10^{-5}$
$G(30) = -.545836930515201 \times 10^{-10}$
$G(31) = -.767451167059717 \times 10^{-9}$
$G(32) = .593123287099439 \times 10^{-8}$
$\gamma = -.0056$

Nitrogen Coefficients for Vapor Pressure eq

$V_p(1) = 5.1113192094$ $V_p(2) = .6482667539$
$V_p(3) = -.15108730916$ $V_p(5) = .74028493342$
$V_p(6) = 1.5$ $V_p(9) = .012462975$
$V_p(7) = 63.15$ $V_p(8) = 126.26$
$V_p(4) = 0.0$

Nitrogen Coefficients for Saturated Liquid and Vapor Densities

$A(1) = -.158453465507 \times 10^{2}$
$A(2) = .419136911423 \times 10^{-1}$
$A(3) = -.101965371660 \times 10^{1}$
$A(4) = .134763743799 \times 10^{2}$
$A(5) = -.109930399087 \times 10^{3}$
$A(6) = .925518835497 \times 10^{2}$
$A(7) = -.956233831320 \times 10^{2}$
$A(8) = .100104366710 \times 10^{3}$
$A(9) = -.701857937398 \times 10^{2}$
$A(10) = .900076998647 \times 10^{1}$
$A(11) = .286981120347 \times 10^{2}$
$A(12) = -.216767601780 \times 10^{2}$
$A(13) = .496558226471 \times 10^{1}$
$A(14) = .218307928477 \times 10^{2}$
$A(15) = -.126493309807$
$A(16) = .241544188633 \times 10^{1}$
$A(17) = -.245256871794 \times 10^{2}$
$A(18) = .935925207124 \times 10^{2}$
$A(19) = -.360938251632 \times 10^{2}$
$A(20) = .757453271989 \times 10^{1}$

Nitrogen Coefficients for Ideal Gas C_p

$G_i(1) = -0.735210401157252 \times 10^{3}$
$G_i(2) = 0.342239980411978 \times 10^{2}$
$G_i(3) = -0.557648284567620$
$G_i(4) = 0.350404228308756 \times 10^{1}$
$G_i(5) = -0.173390185081005 \times 10^{-4}$
$G_i(6) = 0.174650849766463 \times 10^{-7}$
$G_i(7) = -0.356892033544348 \times 10^{-11}$
$G_i(8) = 0.100538722808834 \times 10^{1}$
$G_i(9) = 0.335340610000000 \times 10^{4}$

Nitrogen Coefficients for First Term of Viscosity eq
Viscosity in μPa·s, Density in g/ml

$G_v(1) = -.18224240000 \times 10^{5}$ $G_v(2) = .19915327374 \times 10^{5}$
$G_v(3) = -.91542324494 \times 10^{4}$ $G_v(4) = .23255484059 \times 10^{4}$
$G_v(5) = -.36307214228 \times 10^{3}$ $G_v(6) = .36457506811 \times 10^{2}$
$G_v(7) = -.22261880817 \times 10^{1}$ $G_v(8) = .78053904895 \times 10^{-1}$
$G_v(9) = -.11894029104 \times 10^{-2}$

Nitrogen Coefficients for First Term of Thermal Conductivity eq
Thermal Conductivity in W/(m·K), Density in g/ml

$G_t(1) = -.20029573972 \times 10^{2}$ $G_t(2) = .49765746684 \times 10^{1}$
$G_t(3) = .80188959378 \times 10^{1}$ $G_t(4) = -.55022716888 \times 10^{1}$
$G_t(5) = .15363738965 \times 10^{1}$ $G_t(6) = -.22974737257$
$G_t(7) = .19360547346 \times 10^{-1}$ $G_t(8) = -.85677385768 \times 10^{-3}$
$G_t(9) = .15564670935 \times 10^{-4}$

Nitrogen Coefficients for Second Term of Viscosity eq

$F_v(1) = -.11217739623$ $F_v(2) = .32912317244 \times 10^{-1}$
$F_v(3) = .14000000000 \times 10^{1}$ $F_v(4) = .11800000000 \times 10^{3}$

Nitrogen Coefficients for Second Term of Thermal Conductivity eq

$F_t(1) = .53875666637 \times 10^{-1}$ $F_t(2) = .61027911104 \times 10^{-2}$
$F_t(3) = .12000000000 \times 10^{1}$ $F_t(4) = .11800000000 \times 10^{3}$

Nitrogen Coefficients for Third Term of Viscosity eq

$E_v(1) = -.12128154129 \times 10^{2}$ $E_v(2) = .57156092139$
$E_v(3) = .16094611148 \times 10^{2}$ $E_v(4) = .36954086158 \times 10^{4}$
$E_v(5) = -.80889801180 \times 10^{3}$ $E_v(6) = .68464435640 \times 10^{2}$
$E_v(7) = -.21241135912 \times 10^{1}$ $E_v(8) = .31500000000$

Nitrogen Coefficients for Third Term of Thermal Conductivity eq

$E_t(1) = -.38613291627 \times 10^{2}$ $E_t(2) = .0$
$E_t(3) = .37201743333 \times 10^{2}$ $E_t(4) = .0$
$E_t(5) = -.39013509079 \times 10^{2}$ $E_t(6) = -.31826109485 \times 10^{2}$
$E_t(7) = .0$ $E_t(8) = 1.0$

Nitrogen Coefficients for Melting eq

A = $-.160000281 \times 10^{3}$
B = $.938575502 \times 10^{-1}$
C = $.179500000 \times 10^{1}$

Nitrogen Coefficients for Dielectric Constant

A = $.43993836 \times 10^{-2}$ D = $-.31450178 \times 10^{-8}$
B = $.18932096 \times 10^{-5}$ E = $-.28592703 \times 10^{-6}$
C = 0.0   F = $-.44666034 \times 10^{-7}$

Coefficients for Nitrogen Trifluoride

Nitrogen Trifluoride Coefficients for MBWR eq

$G(1) = .1774353868\times10^{-2}$
$G(2) = -.5409379418\times10^{-1}$
$G(3) = .3976634466$
$G(4) = -.5209476694\times10^{2}$
$G(5) = -.3286322888\times10^{4}$
$G(6) = -.5990517411\times10^{-4}$
$G(7) = .9217525601\times10^{-1}$
$G(8) = -.4848977075\times10^{2}$
$G(9) = -.4235892691\times10^{6}$
$G(10) = -.9824248063\times10^{-6}$
$G(11) = .5432235989\times10^{-2}$
$G(12) = -.1462388500\times10^{1}$
$G(13) = -.3366180440\times10^{-3}$
$G(14) = .2801374599\times10^{-1}$
$G(15) = .8435288597$
$G(16) = -.1324421452\times10^{-2}$
$G(17) = .1875604377\times10^{-4}$
$G(18) = .2959643991\times10^{-1}$
$G(19) = -.7009976870\times10^{-3}$
$G(20) = .4365820912\times10^{6}$
$G(21) = -.1111397536\times10^{7}$
$G(22) = .2411866612\times10^{4}$
$G(23) = .3179136276\times10^{6}$
$G(24) = .6166849090\times10^{1}$
$G(25) = .4260854720\times10^{1}$
$G(26) = .1090598789\times10^{-1}$
$G(27) = -.3340951059\times10^{1}$
$G(28) = .8597429644\times10^{-5}$
$G(29) = .1240544214\times10^{-3}$
$G(30) = .1286224248\times10^{-7}$
$G(31) = -.8941104276\times10^{-7}$
$G(32) = .3353054595\times10^{-5}$
$\gamma = -.0056$

Nitrogen Trifluoride Coefficients for Vapor Pressure

$V_p(1) = 11.593879492$ $V_p(2) = 9.6548502312$
$V_p(3) = -2.8727732815$ $V_p(5) = 7.3112441673$
$V_p(6) = 1.5$ $V_p(9) = .18537827\times10^{-6}$
$V_p(7) = 66.36$ $V_p(8) = 234.0$
$V_p(4) = -1.37977007$

Nitrogen Trifluoride Coefficients for Saturated Liquid and Vapor Densities

$A(1) = .131285181636\times10^{3}$
$A(2) = -.226998555536\times10^{1}$
$A(3) = .384708389498\times10^{2}$
$A(4) = -.309827268239\times10^{3}$
$A(5) = -.635348526635\times10^{3}$
$A(6) = .926979028357\times10^{3}$
$A(7) = .371498011259\times10^{3}$
$A(8) = -.182291654470\times10^{4}$
$A(9) = .105592403853\times10^{4}$
$A(10) = .121673895344\times10^{4}$
$A(11) = -.207208928323\times10^{4}$
$A(12) = .113463394710\times10^{4}$
$A(13) = -.226642137140\times10^{3}$
$A(14) = .105087295173\times10^{1}$
$A(15) = -.131686474246\times10^{-2}$
$A(16) = .390141331900\times10^{-1}$
$A(17) = -.587796597975$
$A(18) = .400900521017\times10^{1}$
$A(19) = -.203385386977\times10^{1}$
$A(20) = .400434364424$

Nitrogen Trifluoride Coefficients for Ideal Gas C_p

$G_i(1) = .7427518245951\times10^{6}$
$G_i(2) = -.4389825372134\times10^{5}$
$G_i(3) = .1012629224351\times10^{4}$
$G_i(4) = -.7140693612211\times10^{1}$
$G_i(5) = .5481339146452\times10^{-1}$
$G_i(6) = -.7677196006769\times10^{-4}$
$G_i(7) = .4203630864340\times10^{-7}$
$G_i(8) = -.6328752997967$
$G_i(9) = .3\times10^{4}$

Nitrogen Trifluoride Coefficients for Melting

A $= -.190939971\times10^{3}$
B $= .813750194\times10^{-1}$
C $= .185000000\times10^{1}$

Coefficients for Oxygen

Oxygen Coefficients for MBWR eq

$G(1) = -.4365859650\times10^{-4}$
$G(2) = .2005820677\times10^{-1}$
$G(3) = -.4197909916$
$G(4) = .1878215317\times10^{2}$
$G(5) = -.1287473398\times10^{4}$
$G(6) = .1556745888\times10^{-5}$
$G(7) = .1343639359\times10^{-3}$
$G(8) = -.2228415518$
$G(9) = .4767792275\times10^{3}$
$G(10) = .4790846641\times10^{-7}$
$G(11) = .2462611107\times10^{-3}$
$G(12) = -.1921891680\times10^{-1}$
$G(13) = -.6978320847\times10^{-6}$
$G(14) = -.6214145909\times10^{-4}$
$G(15) = -.1860852567\times10^{-1}$
$G(16) = .2609791417\times10^{-5}$
$G(17) = -.2447611408\times10^{-7}$
$G(18) = .1457743352\times10^{-4}$
$G(19) = -.1726492873\times10^{-6}$
$G(20) = -.2384892520\times10^{3}$
$G(21) = -.2301807796\times10^{5}$
$G(22) = -.2790303526\times10^{1}$
$G(23) = .9400577575\times10^{4}$
$G(24) = -.4169449637\times10^{-2}$
$G(25) = .2008497853$
$G(26) = -.1256076520\times10^{-4}$
$G(27) = -.6406362964\times10^{-1}$
$G(28) = -.2475580168\times10^{-8}$
$G(29) = .1346309703\times10^{-5}$
$G(30) = -.1161502470\times10^{-10}$
$G(31) = -.1034699798\times10^{-8}$
$G(32) = .2365936964\times10^{-7}$
$\gamma = -.0056$

Oxygen Coefficients for Vapor Pressure

$V_p(1) = 7.568956$	$V_p(2) = 5.004836$
$V_p(3) = -2.13746$	$V_p(5) = 3.454481$
$V_p(6) = 1.514$	$V_p(9) = .0001479953$
$V_p(7) = 54.359$	$V_p(8) = 154.581$
$V_p(4) = 0.0$	

Oxygen Coefficients for Saturated Liquid and Vapor Densities

$A(1) = .581394753076\times10^{2}$
$A(2) = -.490241196133\times10^{-1}$
$A(3) = .168328893252\times10^{1}$
$A(4) = -.325161223398\times10^{2}$
$A(5) = .550300989872\times10^{3}$
$A(6) = -.510968506115\times10^{3}$
$A(7) = .315091559049\times10^{3}$
$A(8) = -.232566659258\times10^{2}$
$A(9) = -.488425479359\times10^{2}$
$A(10) = -.150624217523\times10^{3}$
$A(11) = .280441603851\times10^{3}$
$A(12) = -.176693896861\times10^{3}$
$A(13) = .403247747449\times10^{2}$
$A(14) = .252198688365\times10^{1}$
$A(15) = -.136098316472\times10^{-1}$
$A(16) = .282316159403$
$A(17) = -.286645905341\times10^{1}$
$A(18) = .617024212284\times10^{1}$
$A(19) = -.810220795462$
$A(20) = -.279601068969$

Oxygen Coefficients for Ideal Gas C_p

$G_i(1) = -0.498199853711943\times10^{4}$
$G_i(2) = 0.230247779995218\times10^{3}$
$G_i(3) = -0.345565323510732\times10^{1}$
$G_i(4) = 0.352187677367116\times10^{1}$
$G_i(5) = -0.435420216024420\times10^{-4}$
$G_i(6) = 0.134635345013162\times10^{-7}$
$G_i(7) = 0.162059825959105\times10^{-10}$
$G_i(8) = 0.103146851572565\times10^{1}$
$G_i(9) = 0.223918105000000\times10^{4}$

Oxygen Coefficients for First Term of Viscosity eq
Viscosity in μPa·s, Density in g/ml

$G_v(1) = -.97076378593\times10^{4}$	$G_v(2) = .82801254201\times10^{4}$
$G_v(3) = -.24668758803\times10^{4}$	$G_v(4) = .21324360243\times10^{3}$
$G_v(5) = .37851049522\times10^{2}$	$G_v(6) = -.10487216090\times10^{2}$
$G_v(7) = .11134441304\times10^{1}$	$G_v(8) = -.53676093757\times10^{-1}$
$G_v(9) = .10279379641\times10^{-2}$	

Oxygen Coefficients for First Term of Thermal Conductivity eq
Thermal Conductivity in W/(m·K), Density in g/ml

$G_t(1) = -.20395052193\times10^{3}$	$G_t(2) = .24088141709\times10^{3}$
$G_t(3) = -.12014175183\times10^{3}$	$G_t(4) = .32954949190\times10^{2}$
$G_t(5) = -.54244239598\times10^{1}$	$G_t(6) = .54734865540$
$G_t(7) = -.32854821539\times10^{-1}$	$G_t(8) = .10753572103\times10^{-2}$
$G_t(9) = -.14610986820\times10^{-4}$	

Oxygen Coefficients for Second Term of Viscosity eq

$F_v(1) = .43526515153$	$F_v(2) = -.20361263878$
$F_v(3) = .14000000000\times10^{1}$	$F_v(4) = .10000000000\times10^{3}$

Oxygen Coefficients for Second Term of Thermal Conductivity eq

$F_t(1) = .30600000000\times10^{-1}$	$F_t(2) = .27850000000\times10^{-1}$
$F_t(3) = .11200000000\times10^{1}$	$F_t(4) = .10000000000\times10^{3}$

Oxygen Coefficients for Third Term of Viscosity eq

$E_v(1) = -.14454972110\times10^{2}$	$E_v(2) = -.31421728994$
$E_v(3) = .18201161468\times10^{2}$	$E_v(4) = .27390429525\times10^{3}$
$E_v(5) = -.27498956948\times10^{4}$	$E_v(6) = .24340689667\times10^{3}$
$E_v(7) = .11911504104\times10^{3}$	$E_v(8) = .43500000000$

Oxygen Coefficients for Third Term of Thermal Conductivity eq

$E_t(1) = -.21520741137\times10^{2}$	$E_t(2) = .0$
$E_t(3) = .16799504261\times10^{2}$	$E_t(4) = .0$
$E_t(5) = -.29944878721\times10^{4}$	$E_t(6) = .47350508788\times10^{3}$
$E_t(7) = .0$	$E_t(8) = 1.0$

Oxygen Coefficients for Melting

$A = -.267226854\times10^{3}$
$B = .227606348$
$C = .176900000\times10^{1}$

Oxygen Coefficients for Dielectric Constant

$A = .39608100\times10^{-2}$	$D = 0.0$
$B = .29700000\times10^{-6}$	$E = 0.0$
$C = -.41300000\times10^{-7}$	$F = -.21400000\times10^{-7}$

Journal of Physical and Chemical Reference Data Cumulative Listing of Reprints and Supplements

Reprints from Volume 1

1. Gaseous Diffusion Coefficients, *T.R. Marrero and E.A. Mason,* Vol. 1, No. 1, pp. 1–118 (1972) $7.00
2. Selected Values of Critical Supersaturation for Nucleation of Liquids from the Vapor, *G.M. Pound,* Vol. 1, No. 1, pp. 119–134 (1972) $3.00
3. Selected Values of Evaporation and Condensation Coefficients for Simple Substances, *G.M. Pound,* Vol. 1, No. 1, pp. 135–146 (1972) $3.00
4. Atlas of the Observed Absorption Spectrum of Carbon Monoxide between 1060 and 1900 Å, *S.G. Tilford and J.D. Simmons,* Vol. 1, No. 1, pp. 147–188 (1972) $4.50
5. Tables of Molecular Vibrational Frequencies, Part 5, *T. Shimanouchi,* Vol. 1, No. 1, pp. 189–216 (1972) (superseded by No.103) $4.00
6. Selected Values of Heats of Combustion and Heats of Formation of Organic Compounds Containing the Elements C, H, N, O, P, and S, *Eugene S. Domalski,* Vol. 1, No. 2, pp. 221–278 (1972) $5.00
7. Thermal Conductivity of the Elements, *C.Y. Ho, R.W. Powell, and P.E. Liley,* Vol. 1, No. 2, pp. 279–422 (1972) $7.50
8. The Spectrum of Molecular Oxygen, *Paul H. Krupenie,* Vol. 1, No. 2, pp. 423–534 (1972) $6.50
9. A Critical Review of the Gas-Phase Reaction Kinetics of the Hydroxyl Radical, *Wm. E. Wilson, Jr.,* Vol. 1, No. 2, pp. 535–574 (1972) $4.50
10. Molten Salts: Volume 3, Nitrates, Nitrites, and Mixtures, Electrical Conductance, Density, Viscosity, and Surface Tension Data, *G.J. Janz, Ursula Krebs, H.F. Siegenthaler, and R.P.T. Tomkins,* Vol. 1, No. 3, pp. 581–746 (1972) $8.50
11. High Temperature Properties and Decomposition of Inorganic Salts—Part 3. Nitrates and Nitrites, *Kurt H. Stern,* Vol. 1, No. 3, pp. 747–772 (1972) $4.00
12. High-Pressure Calibration: A Critical Review, *D.L. Decker, W.A. Bassett, L. Merrill, H.T. Hall, and J.D. Barnett,* Vol. 1, No. 3, pp. 773–836 (1972) $5.00
13. The Surface Tension of Pure Liquid Compounds, *Joseph J. Jasper,* Vol. 1, No. 4, pp. 841–1009 (1972) $8.50
14. Microwave Spectra of Molecules of Astrophysical Interest, I. Formaldehyde, Formamide, and Thioformaldehyde, *Donald R. Johnson, Frank J. Lovas, and William H. Kirchhoff,* Vol. 1, No. 4, pp. 1011–1046 (1972) $4.50
15. Osmotic Coefficients and Mean Activity Coefficients of Uni-univalent Electrolytes in Water at 25° C, *Walter J. Hamer and Yung-Chi Wu,* Vol. 1, No. 4, pp. 1047-1099 (1972) $5.00
16. The Viscosity and Thermal Conductivity Coefficients of Gaseous and Liquid Fluorine, *H.J.M. Hanley and R. Prydz,* Vol. 1, No. 4, pp. 1101-1113 (1972) $3.00

Reprints from Volume 2

17. Microwave Spectra of Molecules of Astrophysical Interest, II. Methylenimine, *William H. Kirchhoff, Donald R. Johnson, and Frank J. Lovas,* Vol. 2, No. 1, pp. 1-10 (1973) $3.00
18. Analysis of Specific Heat Data in the Critical Region of Magnetic Solids, *F.J. Cook,* Vol. 2, No. 1, pp. 11-24 (1973) $3.00
19. Evaluated Chemical Kinetic Rate Constants for Various Gas Phase Reactions, *Keith Schofield,* Vol. 2, No. 1, pp. 25–84 (1973) $5.00
20. Atomic Transition Probabilities for Forbidden Lines of the Iron Group Elements. (A Critical Data Compilation for Selected Lines), *M.W. Smith and W.L. Wiese,* Vol. 2, No. 1, pp. 85–120 (1973) $4.50
21. Tables of Molecular Vibrational Frequencies, Part 6, *T. Shimanouchi,* Vol. 2, No. 1, pp. 121–162 (1973) (superseded by No. 103) $4.50
22. Compilation of Energy Band Gaps in Elemental and Binary Compound Semiconductors and Insulators, *W.H. Strehlow and E.L. Cook,* Vol. 2, No. 1, pp. 163–200 (1973) $4.50
23. Microwave Spectra of Molecules of Astrophysical Interest, III. Methanol, *R.M. Lees, F.J. Lovas, W.H. Kirchhoff, and D.R. Johnson,* Vol. 2, No. 2, pp. 205–214 (1973) $3.00
24. Microwave Spectra of Molecules of Astrophysical Interest, IV. Hydrogen Sulfide, *Paul Helminger, Frank C. De Lucia, and William H. Kirchhoff,* Vol. 2, No. 2, pp. 215–224 (1973) $3.00
25. Tables of Molecular Vibrational Frequencies, Part 7, *T. Shimanouchi,* Vol. 2, No. 2, pp. 225–256 (1973) (superseded by No. 103) $4.00

Journal of Physical and Chemical Reference Data
Reprint and Supplement Orders

To: American Chemical Society
Business Operations
Books and Journals Division
1155 Sixteenth Street, N.W.
Washington, D.C. 20036

Name: ____________________

Title: ____________________

Organization: ____________________

Address: ____________________

City: ____________ State: ____________

Country: ____________ Zip: ____________

I am a member of ____________________
(ACS or AIP Member or Affiliated Society)

ORDERS FOR REPRINTS AND SUPPLEMENTS MUST BE PREPAID. Please add $2.00 extra for foreign postage and handling. California residents add 6% state use tax. Make checks payable to the American Chemical Society.

BULK RATES: Subtract 20% from the listed price for orders of 50 or more of any one item.

Please ship the following reprints and supplements:

Reprint No./Package ____, ____ copies, $ ______

Reprint No./Package ____, ____ copies, $ ______

Reprint No./Package ____, ____ copies, $ ______

Reprint No./Package ____, ____ copies, $ ______

Reprint No./Package ____, ____ copies, $ ______

Reprint No./Package ____, ____ copies, $ ______

Suppl. 1 Vol. 2 ☐ Hardcover ☐ Softcover ____ copies $ ______

Suppl. 1 Vol. 3 ☐ Hardcover ☐ Softcover ____ copies $ ______

Suppl. 1 Vol. 6 ☐ Hardcover ☐ Softcover ____ copies $ ______

Suppl. 1 Vol. 10 ☐ Hardcover only ____ copies $ ______

Total Enclosed $ ______

(Continuation of Cumulative Listing of Reprints)

26.Energy Levels of Neutral Helium (^{4}He I), *W.C. Martin,* Vol. 2, No. 2, pp. 257–266 (1973) $3.00

27.Survey of Photochemical and Rate Data for Twenty-eight Reactions of Interest in Atmospheric Chemistry, *R.F. Hampson, Editor, W. Braun, R.L. Brown, D. Garvin, J.T. Herron, R.E. Huie, M.J. Kurylo, A.H. Laufer, J.D. McKinley, H. Okabe, M.D. Scheer, W. Tsang, and D.H. Stedman,* Vol. 2, No. 2, pp. 267–312 (1973) $4.50

28.Compilation of the Static Dielectric Constant of Inorganic Solids, *K.F. Young and H.P.R. Frederikse,* Vol. 2, No. 2, pp. 313–410 (1973) $6.50

29.Soft X-Ray Emission Spectra of Metallic Solids: Critical Review of Selected Systems, *A.J. McAlister, R.C. Dobbyn, J.R. Cuthill, and M.L. Williams,* Vol. 2, No. 2, pp. 411–426 (1973) $3.00

30.Ideal Gas Thermodynamic Properties of Ethane and Propane, *J. Chao, R.C. Wilhoit, and B.J. Zwolinski,* Vol. 2, No. 2, pp. 427–438 (1973) $3.00

31.An Analysis of Coexistence Curve Data for Several Binary Liquid Mixtures Near Their Critical Points, *A. Stein and G.F. Allen,* Vol. 2, No. 3, pp. 443–466 (1973) $4.00

32.Rate Constants for the Reactions of Atomic Oxygen (O ^{3}P) with Organic Compounds in the Gas Phase, *John T. Herron and Robert E. Huie,* Vol. 2, No. 3, pp. 467–518 (1973) $5.00

33.First Spectra of Neon, Argon, and Xenon 136 in the 1.2-4.0 μm Region, *Curtis J. Humphreys,* Vol. 2, No. 3, pp. 519–530 (1973) $3.00

34.Elastic Properties of Metals and Alloys, I. Iron, Nickel, and Iron-Nickel Alloys, *H.M. Ledbetter and R.P. Reed,* Vol. 2, No. 3, pp. 531–618 (1973) $6.00

35.The Viscosity and Thermal Conductivity Coefficients of Dilute Argon, Krypton, and Xenon, *H.J.M. Hanley,* Vol. 2, No. 3, pp. 619–642 (1973) $4.00

36.Diffusion in Copper and Copper Alloys, Part I. Volume and Surface Self-Diffusion in Copper, *Daniel B. Butrymowicz, John R. Manning, and Michael E. Read,* Vol. 2, No. 3, pp. 643–656 (1973) $3.00

37.The 1973 Least-Squares Adjustment of the Fundamental Constants, *E. Richard Cohen and B.N. Taylor,* Vol. 2, No. 4, pp. 663–734 (1973) $5.50

38.The Viscosity and Thermal Conductivity Coefficients of Dilute Nitrogen and Oxygen, *H.J.M. Hanley and James F. Ely,* Vol. 2, No. 4, pp. 735–756 (1973) $4.00

39.Thermodynamic Properties of Nitrogen Including Liquid and Vapor Phases from 63 K to 2000 K with Pressures to 10,000 Bar, *Richard T. Jacobsen and Richard B. Stewart,* Vol. 2, No. 4, pp. 757–922 (1973) $8.50

40.Thermodynamic Properties of Helium 4 from 2 to 1500 K at Pressures to 10^8 Pa, *Robert T. McCarty,* Vol. 2, No. 4, pp. 923–1042 (1973) $7.00

Reprints from Volume 3

41.Molten Salts: Volume 4, Part 1, Fluorides and Mixtures, Electrical Conductance, Density, Viscosity, and Surface Tension Data, *G.J. Janz, G.L. Gardner, Ursula Krebs, and R.P.T. Tomkins,* Vol. 3, No. 1, pp. 1–115 (1974) $7.00

42.Ideal Gas Thermodynamic Properties of Eight Chloro- and Fluoromethanes, *A.S. Rodgers, J. Chao, R. C. Wilhoit, and B.J. Zwolinski,* Vol. 3, No. 1, pp. 117–140 (1974) $4.00

43.Ideal Gas Thermodynamic Properties of Six Chloroethanes, *J. Chao, A.S. Rodgers, R.C. Wilhoit, and B.J. Zwolinski,* Vol. 3, No. 1, pp. 141–162 (1974) $4.00

44.Critical Analysis of Heat-Capacity Data and Evaluation of Thermodynamic Properties of Ruthenium, Rhodium, Palladium, Iridium, and Platinum from 0 to 300 K. A Survey of the Literature Data on Osmium, *George T. Furukawa, Martin L. Reilly, and John S. Gallagher,* Vol. 3, No. 1, pp. 163–209 (1974) $4.50

45.Microwave Spectra of Molecules of Astrophysical Interest, V. Water Vapor, *Frank C. De Lucia, Paul Helminger, and William H. Kirchhoff,* Vol. 3, No. 1, pp. 211–219 (1974) $3.00

46.Microwave Spectra of Molecules of Astrophysical Interest, VI. Carbonyl Sulfide and Hydrogen Cyanide, *Arthur G. Maki,* Vol. 3, No. 1, pp. 221–244 (1974) $4.00

47.Microwave Spectra of Molecules of Astrophysical Interest, VII. Carbon Monoxide, Carbon Monosulfide, and Silicon Monoxide, *Frank J. Lovas and Paul H. Krupenie,* Vol. 3, No. 1, pp. 245–257 (1974) $3.00

48.Microwave Spectra of Molecules of Astrophysical Interest, VIII. Sulfur Monoxide, *Eberhard Tiemann,* Vol. 3, No. 1, pp. 259–268 (1974) $3.00

49.Tables of Molecular Vibrational Frequencies, Part 8, *T. Shimanouchi,* Vol. 3, No. 1, pp. 269–308 (1974) (superseded by No. 103) $4.50

50.JANAF Thermochemical Tables, 1974 Supplement, *M.W. Chase, J.L. Curnutt, A.T. Hu, H. Prophet, A.N. Syverud, and L.C. Walker,* Vol. 3, No. 2, pp. 311–480 (1974) $8.50

51.High Temperature Properties and Decomposition of Inorganic Salts, Part 4. Oxy-Salts of the Halogens, *Kurt H. Stern,* Vol. 3, No. 2, pp. 481–526 (1974) $4.50

52.Diffusion in Copper and Copper Alloys, Part II. Copper-Silver and Copper-Gold Systems, *Daniel B. Butrymowicz, John R. Manning, and Michael E. Read,* Vol. 3, No. 2, pp. 527–602 (1974) $5.50

53.Microwave Spectral Tables I. Diatomic Molecules, *Frank J. Lovas and Eberhard Tiemann,* Vol. 3, No. 3, pp. 609–770 (1974) $8.50

54.Ground Levels and Ionization Potentials for Lanthanide and Actinide Atoms and Ions, *W.C. Martin, Lucy Hagan, Joseph Reader, and Jack Sugar,* Vol. 3, No. 3, pp. 771–780 (1974) $3.00

55.Behavior of the Elements at High Pressures, *John Francis Cannon,* Vol. 3, No. 3, pp. 781–824 (1974) $4.50

56.Reference Wavelengths from Atomic Spectra in the Range 15 Å to 25000 Å, *Victor Kaufman and Bengt Edlén,* Vol. 3, No. 4, pp. 825–895 (1974) $5.50

57.Elastic Properties of Metals and Alloys. II. Copper, *H.M. Ledbetter and E.R. Naimon,* Vol. 3, No. 4, pp.897–935 (1974) $4.50

58.A Critical Review of H-Atom Transfer in the Liquid Phase: Chlorine Atom, Alkyl, Trichloromethyl, Alkoxy, and Alkylperoxy Radicals, *D.G. Hendry, T. Mill, L. Piszkiewicz, J.A. Howard, and H.K. Eigenmann,* Vol. 3, No. 4, pp. 937–978 (1974) $4.50

59.The Viscosity and Thermal Conductivity Coefficients for Dense Gaseous and Liquid Argon, Krypton, Xenon, Nitrogen, and Oxygen, *H.J.M. Hanley, R.D. McCarty, and W.M. Haynes,* Vol. 3, No. 4, pp. 979–1017 (1974) $4.50

Reprints from Volume 4

60.JANAF Thermochemical Tables, 1975 Supplement, *M.W. Chase, J.L. Curnutt, H. Prophet, R.A. McDonald, and A.N. Syverud,* Vol. 4, No. 1, pp. 1–175 (1975) $8.50

61.Diffusion in Copper and Copper Alloys, Part III. Diffusion in Systems Involving Elements of the Groups IA, IIA, IIIB, IVB, VB, VIB, and VIIB, *Daniel B. Butrymowicz, John R. Manning, and Michael E. Read,* Vol. 4, No. 1, pp. 177–249 (1975) $6.00

62.Ideal Gas Thermodynamic Properties of Ethylene and Propylene, *Jing Chao and Bruno J. Zwolinski,* Vol. 4, No. 1, pp. 251–261 (1975) $3.00

63.Atomic Transition Probabilities for Scandium and Titanium (A Critical Data Compilation of Allowed Lines), *W.L. Wiese and J.R. Fuhr,* Vol. 4, No. 2, pp. 263–352 (1975) $6.00

(Continuation of Cumulative Listing of Reprints)

64. Energy Levels of Iron, Fe I through Fe XXVI, *Joseph Reader and Jack Sugar,* Vol. 4, No. 2, pp. 353–440 (1975) $6.00

65. Ideal Gas Thermodynamic Properties of Six Fluoroethanes, *S.S. Chen, A.S. Rodgers, J. Chao, R.C. Wilhoit, and B.J. Zwolinski,* Vol. 4, No. 2, pp. 441–456 (1975) $3.00

66. Ideal Gas Thermodynamic Properties of the Eight Bromo- and Iodomethanes, *S.A. Kudchadker and A.P. Kudchadker,* Vol. 4, No. 2, pp. 457–470 (1975) $3.00

67. Atomic Form Factors, Incoherent Scattering Functions, and Photon Scattering Cross Sections, *J.H. Hubbell, Wm.J. Veigele, E.A. Briggs, R.T. Brown, D.T. Cromer, and R.J. Howerton,* Vol. 4, No. 3, pp. 471–538 (1975) $5.50

68. Binding Energies in Atomic Negative Ions, *H. Hotop and W.C. Lineberger,* Vol. 4, No. 3, pp. 539–576 (1975) $4.50

69. A Survey of Electron Swarm Data, *J. Dutton,* Vol. 4, No. 3, pp. 577–856 (1975) $12.00

70. Ideal Gas Thermodynamic Properties and Isomerization of *n*-Butane and Isobutane, *S.S. Chen, R.C. Wilhoit, and B.J. Zwolinski,* Vol. 4, No. 4, pp. 859–869 (1975) $3.00

71. Molten Salts: Volume 4, Part 2, Chlorides and Mixtures, Electrical Conductance, Density, Viscosity, and Surface Tension Data, *G.J. Janz, R.P.T. Tomkins, C.B. Allen, J.R. Downey, Jr., G.L. Gardner, U. Krebs, and S.K. Singer,* Vol. 4, No. 4, pp. 871–1178 (1975) $13.00

72. Property Index to Volumes 1–4 (1972–1975), Vol. 4, No. 4, pp. 1179–1192 (1975) $3.00

Reprints from Volume 5

73. Scaled Equation of State Parameters for Gases in the Critical Region, *J.M.H. Levelt Sengers, W.L. Greer, and J.V. Sengers,* Vol. 5, No. 1, pp. 1–51 (1976) $5.00

74. Microwave Spectra of Molecules of Astrophysical Interest, IX. Acetaldehyde, *A. Bauder, F.J. Lovas, and D.R. Johnson,* Vol. 5, No. 1, pp. 53–77 (1976) $4.00

75. Microwave Spectra of Molecules of Astrophysical Interest, X. Isocyanic Acid, *G. Winnewisser, W.H. Hocking, and M.C.L. Gerry,* Vol. 5, No. 1, pp. 79–101 (1976) $4.00

76. Diffusion in Copper and Copper Alloys, Part IV. Diffusion in Systems Involving Elements of Group VIII, *Daniel B. Butrymowicz, John R. Manning, and Michael E. Read,* Vol. 5, No. 1, pp. 103–200 (1976) $6.50

77. A Critical Review of the Stark Widths and Shifts of Spectral Lines from Non-Hydrogenic Atoms, *N. Konjevic and J.R. Roberts,* Vol. 5, No. 2, pp. 209–257 (1976) $5.00

78. Experimental Stark Widths and Shifts for Non-Hydrogenic Spectral Lines of Ionized Atoms (A Critical Review and Tabulation of Selected Data), *N. Konjevic and W.L. Wiese,* Vol. 5, No. 2, pp. 259–308 (1976) $5.00

79. Atlas of the Absorption Spectrum of Nitric Oxide (NO) between 1420 and 1250 Å, *E. Miescher and F. Alberti,* Vol. 5, No. 2, pp. 309–317 (1976) $3.00

80. Ideal Gas Thermodynamic Properties of Propanone and 2-Butanone, *Jing Chao and Bruno J. Zwolinski,* Vol. 5, No. 2, pp. 319–328 (1976) $3.00

81. Refractive Index of Alkali Halides and Its Wavelength and Temperature Derivatives, *H.H. Li,* Vol. 5, No. 2, pp. 329–528 (1976) $9.50

82. Tables of Critically Evaluated Oscillator Strengths for the Lithium Isoelectronic Sequence, *G.A. Martin and W.L. Wiese,* Vol. 5, No. 3, pp. 537–570 (1976) $4.50

83. Ideal Gas Thermodynamic Properties of Six Chlorofluoromethanes, *S.S. Chen, R. C. Wilhoit, and B.J. Zwolinski,* Vol. 5, No. 3, pp. 571–580 (1976) $3.00

84. Survey of Superconductive Materials and Critical Evaluation of Selected Properties, *B.W. Roberts,* Vol. 5, No. 3, pp. 581–821 (1976) $12.50

85. Nuclear Spins and Moments, *Gladys H. Fuller,* Vol. 5, No. 4, pp. 835–1092 (1976) $11.50

86. Nuclear Moments and Moment Ratios as Determined by Mössbauer Spectroscopy, *J.G. Stevens and B.D. Dunlap,* Vol. 5, No. 4, pp. 1093–1121 (1976) $4.00

87. Rate Coefficients for Ion-Molecule Reactions, I. Ions Containing C and H, *L. Wayne Sieck and Sharon G. Lias,* Vol. 5, No. 4, pp. 1123–1146 (1976) $4.00

88. Microwave Spectra of Molecules of Astrophysical Interest, XI. Silicon Sulfide, *Eberhard Tiemann,* Vol. 5, No. 4, pp. 1147–1156 (1976) $3.00

89. Property Index and Author Index to Volumes 1–5 (1972–1976), Vol. 5, No. 4, pp. 1161–1183 $4.00

Reprints from Volume 6

90. Diffusion in Copper and Copper Alloys, Part V. Diffusion in Systems Involving Elements of Group VA, *Daniel B. Butrymowicz, John R. Manning, and Michael E. Read,* Vol. 6, No. 1, pp. 1–50 (1977) $5.00

91. The Calculated Thermodynamic Properties of Superfluid Helium-4, *James S. Brooks and Russell J. Donnelly,* Vol. 6, No. 1, pp. 51–104 (1977) $5.00

92. Thermodynamic Properties of Normal and Deuterated Methanols, *S.S. Chen, R.C. Wilhoit, and B.J. Zwolinski,* Vol. 6, No. 1, pp. 105–112 (1977) $3.00

93. The Spectrum of Molecular Nitrogen, *Alf Lofthus and Paul H. Krupenie,* Vol. 6, No. 1, pp. 113–307 (1977) $9.50

94. Energy Levels of Chromium, Cr I through Cr XXIV, *Jack Sugar and Charles Corliss,* Vol. 6, No. 2, pp. 317–383 (1977) $5.50

95. The Activity and Osmotic Coefficients of Aqueous Calcium Chloride at 298.15 K, *Bert R. Staples and Ralph L. Nuttall,* Vol. 6, No. 2, pp. 385–407 (1977) $4.00

96. Molten Salts: Volume 4, Part 3, Bromides and Mixtures; Iodides and Mixtures–Electrical Conductance, Density, Viscosity, and Surface Tension Data, *G.J. Janz, R.P.T. Tomkins, C.B. Allen, J.R. Downey, Jr., and S.K. Singer,* Vol. 6, No. 2, pp. 409–596 (1977) $9.00

97. The Viscosity and Thermal Conductivity Coefficients for Dense Gaseous and Liquid Methane, *H.J.M. Hanley, W.M. Haynes, and R.D. McCarty,* Vol. 6, No. 2, pp. 597–609 (1977) $3.00

98. Phase Diagrams and Thermodynamic Properties of Ternary Copper-Silver Systems, *Y. Austin Chang, Daniel Goldberg, and Joachim P. Neumann,* Vol. 6, No. 3, pp. 621–673 (1977) $5.00

99. Crystal Data Space-Group Tables, *Alan D. Mighell, Helen M. Ondik, and Bettijoyce Breen Molino,* Vol. 6, No. 3, pp. 675–829 (1977) $8.00

100. Energy Levels of One-Electron Atoms, *Glen W. Erickson,* Vol. 6, No. 3, pp. 831–869 (1977) $4.50

101. Rate Constants for Reactions of ClO_x of Atmospheric Interest, *R.T. Watson,* Vol. 6, No. 3, pp. 871–917 (1977) $4.50

102. NMR Spectral Data: A Compilation of Aromatic Proton Chemical Shifts in Mono- and Di-Substituted Benzenes, *B.L. Shapiro and L.E. Mohrmann,* Vol. 6, No. 3, pp. 919–991 (1977) $5.50

103. Tables of Molecular Vibrational Frequencies. Consolidated Volume II. *T. Shimanouchi,* Vol. 6, No. 3, pp. 993–1102 (1977) (supersedes Nos. 5, 21, 25, 49) $6.50

104. Effects of Isotopic Composition, Temperature, Pressure, and Dissolved Gases on the Density of Liquid Water, *George S. Kell,* Vol. 6, No. 4, pp. 1109–1131 (1977) $4.00

105. Viscosity of Water Substance–New International Formulation and Its Background, *A. Nagashima,* Vol. 6, No. 4, pp. 1133–1166 (1977) $4.50

106.A Correlation of the Existing Viscosity and Thermal Conductivity Data of Gaseous and Liquid Ethane, *H.J.M. Hanley, K.E. Gubbins, and S. Murad,* Vol. 6, No. 4, pp. 1167–1180 (1977) $3.00

107.Elastic Properties of Zinc: A Compilation and a Review, *H.M. Ledbetter,* Vol. 6, No. 4, pp. 1181–1203 (1977) $4.00

108.Behavior of the AB-Type Compounds at High Pressures and High Temperatures, *Leo Merrill,* Vol. 6, No. 4, pp. 1205–1252 (1977) $4.50

109.Energy Levels of Manganese, Mn I through Mn XXV, *Charles Corliss and Jack Sugar,* Vol. 6, No. 4, pp. 1253–1329 (1977) $5.50

Reprints from Volume 7

110.Tables of Atomic Spectral Lines for the 10 000 Å to 40 000 Å Region, *Michael Outred,* Vol. 7, No. 1, pp. 1-262 (1978) $11.50

111.Evaluated Activity and Osmotic Coefficients for Aqueous Solutions: The Alkaline Earth Metal Halides, *R.N. Goldberg and R.L. Nuttall,* Vol. 7, No. 1, pp. 263–310 (1978) $4.50

112.Microwave Spectra of Molecules of Astrophysical Interest XII. Hydroxyl Radical, *Robert A. Beaudet and Robert L. Poynter,* Vol. 7, No. 1, pp. 311-362 (1978) $5.00

113.Ideal Gas Thermodynamic Properties of Methanoic and Ethanoic Acids, *Jing Chao and Bruno J. Zwolinski,* Vol. 7, No. 1, pp. 363–377 (1978) $3.00

114.Critical Review of Hydrolysis of Organic Compounds in Water Under Environmental Conditions, *W. Mabey and T. Mill,* Vol. 7, No. 2, pp. 383–415 (1978) $4.50

115.Ideal Gas Thermodynamic Properties of Phenol and Creosols, *S.A. Kudchadker, A.P. Kudchadker, R.C. Wilhoit, and B.J. Zwolinski,* Vol. 7, No. 2, pp. 417–423 (1978) $3.00

116.Densities of Liquid $CH_{4-a}X_a$ (X = Br,I) and $CH_{4-(a+b+c+d)}F_aCl_bBr_cI_d$ Halomethanes, *A.P. Kudchadker, S.A.Kudchadker, P.R. Patnaik, and P.P. Mishra,* Vol. 7, No. 2, pp. 425–439 (1978) $3.00

117.Microwave Spectra of Molecules of Astrophysical Interest XIII. Cyanoacetylene, *W.J. Lafferty and F.J. Lovas,* Vol. 7, No. 2, pp. 441–493 (1978) $5.00

118.Atomic Transition Probabilities for Vanadium, Chromium, and Manganese (A Critical Data Compilation of Allowed Lines), *S.M. Younger, J.R. Fuhr, G.A. Martin, and W.L. Wiese,* Vol. 7, No. 2, pp. 495–629 (1978) $7.50

119.Thermodynamic Properties of Ammonia, *Lester Haar and John S. Gallagher,* Vol. 7, No. 3, pp. 635–792 (1978) $8.00

120.JANAF Thermochemical Tables, 1978 Supplement, *M.W. Chase, Jr., J.L. Curnutt, R.A. McDonald, and A.N. Syverud,* Vol. 7, No. 3, pp. 793–940 (1978) $8.00

121.Viscosity of Liquid Water in the Range −8 °C to 150 °C, *Joseph Kestin, Mordechai Sokolov, and William A. Wakeham,* Vol. 7, No. 3, pp. 941–948 (1978) $3.00

122.The Molar Volume (Density) of Solid Oxygen in Equilibrium with Vapor, *H.M. Roder,* Vol. 7, No. 3, pp. 949–957 (1978) $3.00

123.Thermal Conductivity of Ten Selected Binary Alloy Systems, *C.Y. Ho, M.W. Ackerman, K.Y. Wu, S.G. Oh, and T.N. Havill,* Vol. 7, No. 3, pp. 959–1177 (1978) $10.00

124.Semi-Empirical Extrapolation and Estimation of Rate Constants for Abstraction of H from Methane by H, O, HO, and O_2, *Robert Shaw,* Vol. 7, No. 3, pp. 1179–1190 (1978) $3.00

125.Energy Levels of Vanadium, V I through V XXIII, *Jack Sugar and Charles Corliss,* Vol. 7, No. 3, pp. 1191–1262 (1978) $5.50

126.Recommended Atomic Electron Binding Energies, $1s$ to $6p_{3/2}$, for the Heavy Elements, $Z = 84$ to 103, *F.T. Porter and M.S. Freedman,* Vol. 7, No. 4, pp. 1267–1284 (1978) $4.00

127.Ideal Gas Thermodynamic Properties of $CH_{4-(a+b+c+d)}F_aCl_bBr_cI_d$ Halomethanes, *Shanti A. Kudchadker and Arvind P. Kudchadker,* Vol. 7, No. 4, pp. 1285–1307 (1978) $4.00

128.Critical Review of Vibrational Data and Force Field Constants for Polyethylene, *John Barnes and Bruno Fanconi,* Vol. 7, No. 4, pp. 1309–1321 (1978) $3.00

129.Tables of Molecular Vibrational Frequencies, Part 9, *Takehiko Shimanouchi, Hiroatsu Matsuura, Yoshiki Ogawa, and Issei Harada,* Vol. 7, No. 4, pp. 1323–1443 (1978) $7.00

130.Microwave Spectral Tables. II. Triatomic Molecules, *Frank J. Lovas,* Vol. 7, No. 4, pp. 1445–1750 (1978) $13.00

Reprints from Volume 8

131.Energy Levels of Titanium, Ti I through Ti XXII, *Charles Corliss and Jack Sugar,* Vol. 8, No. 1, pp. 1–62 (1979) $5.00

132.The Spectrum and Energy Levels of the Neutral Atom of Boron (B I), *G.A. Odintzova and A.R. Striganov,* Vol. 8, No. 1, pp. 63–67 (1979) $3.00

133.Relativistic Atomic Form Factors and Photon Coherent Scattering Cross Sections, *J.H. Hubbell and I. Øverbø,* Vol. 8, No. 1, pp. 69–105 (1979) $4.50

134.Microwave Spectra of Molecules of Astrophysical Interest. XIV. Vinyl Cyanide (Acrylonitrile), *M.C.L. Gerry, K. Yamada, and G. Winnewisser,* Vol. 8, No. 1, pp. 107–123 (1979) $4.00

135.Molten Salts: Volume 4, Part 4, Mixed Halide Melts. Electrical Conductance, Density, Viscosity, and Surface Tension Data, *G.J. Janz, R.P.T. Tomkins, and C.B. Allen,* Vol. 8, No. 1, pp. 125–302 (1979) $9.00

136.Atomic Radiative and Radiationless Yields for K and L Shells, *M.O. Krause,* Vol. 8, No. 2, pp. 307–327 (1979) $4.00

137.Natural Widths of Atomic K and L Levels, $K\alpha$ X-ray Lines and Several KLL Auger Lines, *M.O. Krause and J.H. Oliver,* Vol. 8, No. 2, pp. 329–338 (1979) $3.00

138.Electrical Resistivity of Alkali Elements, *T.C. Chi,* Vol. 8, No. 2, pp. 339–438 (1979) $6.50

139.Electrical Resistivity of Alkaline Earth Elements, *T.C. Chi,* Vol. 8, No. 2, pp. 439–497 (1979) $5.00

140.Vapor Pressures and Boiling Points of Selected Halomethanes, *A.P. Kudchadker, S.A. Kudchadker, R.P. Shukla, and P.R. Patnaik,* Vol. 8, No. 2, pp. 499–517 (1979) $4.00

141.Ideal Gas Thermodynamic Properties of Selected Bromoethanes and Iodoethane, *S.A. Kudchadker and A.P. Kudchadker,* Vol. 8, No. 2, pp. 519–526 (1979) $3.00

142.Thermodynamic Properties of Normal and Deuterated Naphthalenes, *S.S. Chen, S.A. Kudchadker, and R.C. Wilhoit,* Vol. 8, No. 2, pp. 527–535 (1979) $3.00

143.Microwave Spectra of Molecules of Astrophysical Interest. XV. Propyne, *A. Bauer, D. Boucher, J. Burie, J. Demaison, and A. Dubrulle,* Vol. 8, No. 2, pp. 537–558 (1979) $4.00

144.A Correlation of the Viscosity and Thermal Conductivity Data of Gaseous and Liquid Propane, *P.M. Holland, H.J.M. Hanley, K.E. Gubbins, and J.M. Haile,* Vol. 8, No. 2, pp. 559–575 (1979) $4.00

145.Microwave Spectra of Molecules of Astrophysical Interest. XVI. Methyl Formate, *A. Bauder,* Vol. 8, No. 3, pp. 583–618 (1979) $4.50

(Continuation of Cumulative Listing of Reprints)

146. Molecular Structures of Gas-Phase Polyatomic Molecules Determined by Spectroscopic Methods, *Marlin D. Harmony, Victor W. Laurie, Robert L. Kuczkowski, R.H. Schwendeman, D.A. Ramsay, Frank J. Lovas, Walter J. Lafferty, and Arthur G. Maki,* Vol. 8, No. 3, pp. 619–721 (1979) $6.50

147. Critically Evaluated Rate Constants for Gaseous Reactions of Several Electronically Excited Species, *Keith Schofield,* Vol. 8, No. 3, pp. 723–798 (1979) $5.50

148. A Review, Evaluation, and Correlation of the Phase Equilibria, Heat of Mixing, and Change in Volume on Mixing for Liquid Mixtures of Methane + Ethane, *M.J. Hiza, R.C. Miller, and A.J. Kidnay,* Vol. 8, No. 3, pp. 799–816 (1979) $4.00

149. Energy Levels of Aluminum, Al I through Al XIII, *W.C. Martin and Romuald Zalubas,* Vol. 8, No. 3, pp. 817–864 (1979) $4.50

150. Energy Levels of Calcium, Ca I through Ca XX, *Jack Sugar and Charles Corliss,* Vol. 8, No. 3, pp. 865–916 (1979) $5.00

151. Evaluated Activity and Osmotic Coefficients for Aqueous Solutions: Iron Chloride and the Bi-univalent Compounds of Nickel and Cobalt, *R.N. Goldberg, R.L. Nutall, and B.R. Staples,* Vol. 8, No. 4, pp. 923–1003 (1979) $6.00

152. Evaluated Activity and Osmotic Coefficients for Aqueous Solutions: Bi-univalent Compounds of Lead, Copper, Manganese, and Uranium, *R.N. Goldberg,* Vol. 8, No. 4, pp. 1005–1050 (1979) $4.50

153. Microwave Spectra of Molecules of Astrophysical Interest. XVII. Dimethyl Ether, *F.J. Lovas, H. Lutz, and H. Dreizler,* Vol. 8, No. 4, pp. 1051–1107 (1979) $5.00

154. Energy Levels of Potassium, K I through K XIX, *Charles Corliss and Jack Sugar,* Vol. 8, No. 4, pp. 1109–1145 (1979) $4.50

155. Electrical Resistivity of Copper, Gold, Palladium, and Silver, *R.A. Matula,* Vol. 8, No. 4, pp. 1147–1298 (1979) $8.00

Reprints from Volume 9

156. Energy Levels of Magnesium, Mg I through Mg XII, *W.C. Martin and Romuald Zalubas,* Vol 9, No. 1, pp. 1–58(1980) $6.00

157. Microwave Spectra of Molecules of Astrophysical Interest. XVIII. Formic Acid, *Edmond Willemot, Didier Dangoisse, Nicole Monnanteuil, and Jean Bellet,* Vol. 9, No. 1, pp. 59–160 (1980) $7.50

158. Refractive Index of Alkaline Earth Halides and Its Wavelength and Temperature Derivatives, *H.H. Li,* Vol. 9, No. 1, pp. 161–289 (1980). $8.50

159. Evaluated Kinetic and Photochemical Data for Atmospheric Chemistry, *D.L. Baulch, R.A. Cox, R.F. Hampson, Jr., J.A. Kerr, J. Troe, and R.L. Watson,* Vol. 9, No. 2, pp. 295–471 (1980) $10.00

160. Energy Levels of Scandium, Sc I through Sc XXII, *Jack Sugar and Charles Corliss,* Vol. 9, No. 2, pp. 473–511 (1980) $5.50

161. A Compilation of Kinetic Parameters for the Thermal Degradation of *n*-Alkane Molecules, *D.L. Allara and Robert Shaw,* Vol. 9, No. 3, pp. 523–559 (1980) $5.50

162. Refractive Index of Silicon and Germanium and Its Wavelength and Temperature Derivatives, *H.H. Li,* Vol. 9, No. 3, pp. 561–658 (1980) $7.50

163. Microwave Spectra of Molecules of Astrophysical Interest XIX. *Methyl Cyanide, D. Boucher, J. Burie, A. Bauer, A. Dubrulle, and J. Demaison,* Vol. 9, No. 3, pp. 659–719 (1980). $6.00

164. A Review, Evaluation, and Correlation of the Phase Equilibria, Heat of Mixing, and Change in Volume on Mixing for Liquid Mixtures of Methane + Propane, *R.C. Miller, A.J. Kidnay, and M.J. Hiza,* Vol. 9, No. 3, pp. 721–734 (1980) $4.00

165. Saturation States of Heavy Water, *P.G. Hill and R.D. Chris MacMillan,* Vol. 9, No. 3, pp. 735–749(1980) $4.00

166. The Solubility of Some Sparingly Soluble Lead Salts: An Evaluation of the Solubility in Water and Aqueous Electrolyte Solution, *H. Lawrence Clever and Francis J. Johnston,* Vol. 9, No. 3, pp. 751-784 (1980) $5.50

167. Molten Salts Data as Reference Standards for Density, Surface Tension, Viscosity, and Electrical Conductance: KNO_3 and NaCl, *George J. Janz,* Vol. 9, No. 4, pp. 791–829 (1980) $5.50

168. Molten Salts: Volume 5, Part 1, Additional Single and Multi-Component Salt Systems. Electrical Conductance, Density, Viscosity, and Surface Tension Data, *G.J. Janz and R.P. Tomkins,* Vol. 9, No. 4, pp. 831–1021 (1980) $10.50

169. Pair, Triplet, and Total Atomic Cross Sections (and Mass Attenuation Coefficients) for 1 MeV–100 GeV Photons in Elements $Z=1$ to 100, *J.H. Hubbell, H.A. Gimm, and I. Øverbø,* Vol. 9, No. 4, pp. 1023–1147 (1980) $8.00

170. Tables of Molecular Vibrational Frequencies, Part 10, *Takehiko Shimanouchi, Hiroatsu Matsuura, Yoshiki Ogawa, and Issei Harada,* Vol. 9, No. 4, pp. 1149–1254 (1980) $7.50

171. An Improved Representative Equation for the Dynamic Viscosity of Water Substance, *J.T.R. Watson, R.S. Basu, and J.V. Sengers,* Vol. 9, No. 4, pp. 1255–1290 (1980) $5.50

172. Static Dielectric Constant of Water and Steam, *M. Uematsu and E. U. Franck,* Vol. 9, No. 4, pp. 1291–1306 (1980) $4.00

173. Compilation and Evaluation of Solubility Data in the Mercury (I) Chloride-Water System, *Y. Marcus,* Vol. 9, No. 4, pp. 1307–1329 (1980) $5.00

Reprints from Volume 10

174. Evaluated Activity and Osmotic Coefficients for Aqueous Solutions: Bi-Univalent Compounds of Zinc, Cadmium, and Ethylene Bis(Trimethylammonium) Chloride and Iodide, *R. N. Goldberg,* Vol. 10, No. 1, pp. 1–55 (1981) $6.00

175. Tables of the Dynamic and Kinematic Viscosity of Aqueous KCl Solutions in the Temperature Range 25–150 °C and the Pressure Range 0.1–35 MPa, *Joseph Kestin, H. Ezzat Khalifa, and Robert J. Correia,* Vol. 10, No. 1, pp. 57–70 (1981) $4.00

176. Tables of the Dynamic and Kinematic Viscosity of Aqueous NaCl Solutions in the Temperature Range 20–150 °C and the Pressure Range 0.1–35 MPa, *Joseph Kestin, H. Ezzat Khalifa, and Robert J. Correia,* Vol. 10, No. 1, pp. 71–87 (1981) $5.00

177. Heat Capacity and Other Thermodynamic Properties of Linear Macromolecules. I. Selenium, *Umesh Gaur, Hua-Cheng Shu, Aspy Mehta, and Bernhard Wunderlich,* Vol. 10, No. 1, pp. 89–117 (1981) $5.00

178. Heat Capacity and Other Thermodynamic Properties of Linear Macromolecules. II. Polyethylene, *Umesh Gaur and Bernhard Wunderlich,* Vol. 10, No. 1, pp. 119–152 (1981) $5.50

179. Energy Levels of Sodium, Na I through Na XI, *W. C. Martin and Romuald Zalubas,* Vol. 10, No. 1, pp. 153–195 (1981) $5.50

180. Energy Levels of Nickel, Ni I through Ni XXVIII, *Charles Corliss and Jack Sugar,* Vol. 10, No. 1, pp. 197–289 (1981) $7.00

181. Ion Product of Water Substance, 0–1000 °C, 1–10,000 bars New International Formulation and Its Background, *William L. Marshall and E. U. Franck*, Vol. 10, No. 2, pp. 295–304 (1981) $4.00

182. Atomic Transition Probabilities for Iron, Cobalt, and Nickel (A Critical Data Compilation of Allowed Lines), *J. R. Fuhr, G. A. Martin, W. L. Wiese, and S. M. Younger*, Vol. 10, No. 2, pp. 305–565 (1981) $12.50

183. Thermodynamic Tabulations for Selected Phases in the System $CaO\text{-}Al_2O_3SiO_2\text{-}H_2O$ at 101.325 kPa (1 atm) between 273.15 and 1800 K, *John L. Haas, Jr., Gilpin R. Robinson, Jr., and Bruce S. Hemingway*, Vol. 10, No. 3, pp. 575–669 (1981) $7.00

184. Evaluated Activity and Osmotic Coefficients for Aqueous Solutions: Thirty-Six Uni-Bivalent Electrolytes, R. N. Goldberg, Vol. 10, No. 3, pp. 671–764 (1981) $7.00

185. Activity and Osmotic Coefficients of Aqueous Alkali Metal Nitrites, *Bert R. Staples*, Vol. 10, No. 3, pp. 765–778 (1981) $4.00

186. Activity and Osmotic Coefficients of Aqueous Sulfuric Acid at 298.15 K, *Bert R. Staples*, Vol. 10, No. 3, pp. 779–798 (1981) $5.00

187. Rate Constants for the Decay and Reactions of the Lowest Electronically Excited Singlet State of Molecular Oxygen in Solution, *Francis Wilkinson and James G. Brummer*, Vol. 10, No. 4, pp. 809–999 (1981) $10.00

188. Heat Capacity and Other Thermodynamic Properties of Linear Macromolecules. III. Polyoxides, *Umesh Gaur and Bernhard Wunderlich*, Vol. 10, No. 4, pp. 1001–1049 (1981) $5.50

189. Heat Capacity and Other Thermodynamic Properties of Linear Macromolecules. IV. Polypropylene, *Umesh Gaur and Bernhard Wunderlich*, Vol. 10, No. 4, pp. 1051–1064 (1981) $4.00

190. Tables of N_2O Absorption Lines for the Calibration of Tunable Infrared Lasers from 522 cm^{-1} to 657 cm^{-1} and from 1115 cm^{-1} to 1340 cm^{-1}, *W. B. Olson, A. G. Maki, and W. J. Lafferty*, Vol. 10, No. 4, pp. 1065–1084 (1981) $5.00

191. Microwave Spectra of Molecules of Astrophysical Interest. XX. Methane, *I. Ozier, M. C. L. Gerry, and A. G. Robiette*, Vol. 10, No. 4, pp. 1085–1095 (1981) $4.00

192. Energy Levels of Cobalt, Co I through Co XXVII,, *Jack Sugar and Charles Corliss*, Vol. 10, No. 4, pp. 1097–1174 (1981) $6.50

193. A Critical Review of Henry's Law Constants for Chemicals of Environmental Interest, *Donald Mackay and Wan Ying Shiu*, Vol. 10, No. 4, pp. 1175–1199 (1981) $5.00

194. Property, Materials, and Author Indexes to the Journal of Physical and Chemical Reference Data, Vol. 1–10, pp. 1205–1225 (1972–1981) $5.00

Reprints from Volume 11

195. A Fundamental Equation of State for Heavy Water, *P. G. Hill, R. D. Chris MacMillan, and V. Lee*, Vol. 11, No. 1, pp. 1–14 (1982) $5.00

196. Volumetric Properties of Aqueous Sodium Chloride Solutions, *P. S. Z. Rogers and Kenneth S. Pitzer*, Vol. 11, No. 1, pp. 15–81 (1982) $9.00

197. Ideal Gas Thermodynamic Properties of CH_3, CD_3, CD_4, C_2D_2, C_2D_4, C_2D_6, C_2H_6, $CH_3N_2CH_3$, and $CD_3N_2CD_3$, *Krishna M. Pamidimukkala, David Rogers, and Gordon B. Skinner*, Vol. 11, No. 1, pp. 83–99 (1982) $6.00

198. Peak Absorption Coefficients of Microwave Absorption Lines of Carbonyl Sulphide, *Z. Kisiel and D. J. Millen*, Vol. 11, No. 1, pp. 99–116 (1982) $6.00

199. Vibrational Contributions to Molecular Dipole Polarizabilities, *David M. Bishop and Lap M. Cheung*, Vol. 11, No. 1, pp. 119–133 (1982) $5.00

200. Energy Levels of Iron, Fe I through Fe XXVI, *Charles Corliss and Jack Sugar*, Vol. 11, No. 1, pp. 135–241 (1982) $11.00

201. Microwave Spectra of Molecules of Astrophysical Interest. XXI. Ethanol(C_2H_5OH) and Propionitrile (C_2H_5CN), *Frank J. Lovas*, Vol. 11, No. 2, pp. 251–312 (1982) $8.00

202. Heat Capacity and Other Thermodynamic Properties of Linear Macromolecules, V. Polystyrene, *Umesh Gaur and Bernhard Wunderlich*, Vol. 11, No. 2, pp. 313–325 (1982) $5.00

203. Evaluated Kinetic and Photochemical Data for Atmospheric Chemistry: Supplement 1, CODATA Task Group on Chemical Kinetics, *D. L. Baulch, R. A. Cox, P. J. Crutzen, R. F. Hampson, Jr., J. A. Kerr (Chairman), J. Troe, and R. T. Watson*, Vol. 11, No. 2, pp. 327–496 (1982) $15.00

204. Molten Salts Data: Diffusion Coefficients in Single and Multi-Component Salt Systems, *G. J. Janz and N. P. Bansal*, Vol. 11, No. 3, pp. 505–693 (1982) $16.00

205. JANAF Thermochemical Tables, 1982 Supplement, *M. W. Chase, Jr., J. L. Curnutt, J. R. Downey, Jr., R. A. McDonald, A. N. Syverud, and E. A. Valenzuela*, Vol. 11, No. 3, pp. 695–940 (1982) $20.00

206. Critical Evaluation of Vapor-Liquid Equilibrium, Heat of Mixing, and Volume Change of Mixing Data. General Procedures, *Buford D. Smith, Ol Muthu, Ashok Dewan, and Matthew Gierlach*, Vol. 11, No. 3, pp. 941–951 (1982) $5.00

207. Rate Coefficients for Vibrational Energy Transfer Involving the Hydrogen Halides, *Stephen R. Leone*, Vol. 11, No. 3, pp. 953–996 (1982) $7.00

208. Behavior of the AB_2-Type Compounds at High Pressures and High Temperatures, *Leo Merrill*, Vol. 11, No. 4, pp. 1005–1064 (1982) $8.00

209. Heat Capacity and Other Thermodynamic Properties of Linear Macromolecules. VI. Acrylic Polymers, *Umesh Gaur, Suk-fai Lau, Brent B. Wunderlich, and Bernhard Wunderlich*, Vol. 11, No. 4, pp. 1065–1089 (1982) $6.00

210. Molecular Form Factors and Photon Coherent Scattering Cross Sections of Water, *L. R. M. Morin*, Vol. 11, No. 4, pp. 1091–1098 (1982) $5.00

211. Evaluation of Binary *PTxy* Vapor–Liquid Equilibrium Data for C_6 Hydrocarbons. Benzene + Cyclohexane, *Buford D. Smith, Ol Muthu, Ashok Dewan, and Matthew Gierlach*, Vol. 11, No. 4, pp. 1099–1126 (1982) $6.00

212. Evaluation of Binary Excess Enthalpy Data for C_6 Hydrocarbons. Benzene + Cyclohexane, *Buford D. Smith, Ol Muthu, Ashok Dewan, and Matthew Gierlach*, Vol. 11, No. 4, pp. 1127–1149 (1982) $6.00

213. Evaluation of Binary Excess Volume Data for C_6 Hydrocarbons. Benzene + Cyclohexane, *Buford D. Smith, Ol Muthu, Ashok Dewan, and Matthew Gierlach*, Vol. 11, No. 4, pp. 1151–1169 (1982) $6.00

Special Reprints Packages

These special reprint packages offer selected articles in specific subject areas from JOURNAL OF PHYSICAL AND CHEMICAL REFERENCE DATA. And, they are offered at a better rate than when purchased individually.

You will have available, at a fraction of the cost, a complete library of literature for your specific requirements without purchasing back issues of the journal.

Look over the reprints packages available — they are listed by subject area. Inside this brochure you will find the titles corresponding to the reprint numbers. You are sure to find building your information bank in this manner to be thorough and economical.

Package A1 (4 Parts) MOLECULAR VIBRATIONAL FREQUENCIES. Consisting of Reprint Nos. 103, 129, 170, NSRDS 39
If purchased individually $26.10
Special package price **$22.00**

Package A2 (4 Parts) JANAF THERMOCHEMICAL TABLES. Consisting of Reprint Nos. 50, 60, 120, NSRDS 37
If purchased individually $40.60
Special package price **$32.00**

Package A3 (19 Parts) ATOMIC ENERGY LEVELS. Consisting of Reprint Nos. 26, 54, 64, 68, 94, 100, 109, 125, 126, 131, 132, 149, 150, 154, 156, 160, 179, 180, 192
If purchased individually $94.00
Special package price **$66.00**

Package A4 (6 Parts) ATOMIC SPECTRA. Consisting of Reprint Nos. 33, 56, 77, 78, 110, 132
If purchased individually $33.00
Special package price **$27.00**

Package A5 (5 Parts) ATOMIC TRANSITION PROBABILITIES. Consisting of Reprint Nos. 20, 63, 82, 118, 182
If purchased individually $35.00
Special package price **$28.00**

Package A6 (20 Parts) MOLECULES IN SPACE. Consisting of Reprint Nos. 14, 17, 23, 24, 45, 46, 47, 48, 74, 75, 88, 112, 117, 134, 143, 145, 153, 157, 163, 191
If purchased individually $82.50
Special package price **$58.00**

Package A7 (9 Parts) CHEMICAL DATA RELEVANT TO AIR POLLUTION. Consisting of Reprint Nos. 1, 9, 19, 27, 32, 87, 101, 147, 159
If purchased individually $50.00
Special package price **$40.00**

Package A8 (7 Parts) MOLECULAR SPECTRA. Consisting of Reprint Nos. 4, 8, 53, 79, 93, 130, 146
If purchased individually $51.50
Special package price **$41.00**

Package A9 (9 Parts) THERMODYNAMIC PROPERTIES OF ELECTROLYTE SOLUTIONS. Consisting of Reprint Nos. 15, 95, 111, 151, 152, 174, 184, 185, 186
If purchased individually $46.00
Special package price **$37.00**

Package B1 (12 Parts) IDEAL GAS THERMODYNAMIC PROPERTIES. Consisting of Reprint Nos. 30, 42, 43, 62, 65, 66, 70, 80, 83, 113, 115, 141
If purchased individually $38.00
Special package price **$31.00**

Package B2 (3 Parts) ELASTIC PROPERTIES OF METALS AND ALLOYS. Consisting of Reprint Nos. 34, 57, 107
If purchased individually $14.50
Special package price **$12.00**

Package B3 (3 Parts) RESISTIVITY. Consisting of Reprint Nos. 138, 139, 155
If purchased individually $19.50
Special package price **$16.00**

Package B4 (7 Parts) MOLTEN SALTS. Consisting of Reprint Nos. 10, 41, 71, 96, 135, 167, 168
If purchased individually $62.50
Special package price **$44.00**

Package B5 (3 Parts) REFRACTIVE INDEX. Consisting of Reprint Nos. 81, 158, 162
If purchased individually $25.50
Special package price **$21.00**

Package B6 (11 Parts) VISCOSITY AND THERMAL CONDUCTIVITY OF FLUIDS. Consisting of Reprint Nos. 16, 35, 38, 59, 97, 106, 121, 144, 171, 175, 176
If purchased individually $43.00
Special package price **$35.00**

Supplements to JPCRD

When the topic demands it, and the quality of the data justifies it, the JOURNAL OF PHYSICAL AND CHEMICAL REFERENCE DATA issues a special Supplement. Each Supplement is a monograph — collected tables of highly significant physical or chemical property data in one complete volume.

Listed below are the special Supplements to JPCRD that have been published. Each is a valuable resource for the physical chemist and chemical physicist. You can use the order form on the last page of this brochure to order the Supplements important to you.

PHYSICAL AND THERMODYNAMIC PROPERTIES OF ALIPHATIC ALCOHOLS by R.C. Wilhoit and B.J. Zwolinski. (Supplement No. 1 to Volume 2) 1973, 420 pages.
Hardcover $33.00
Softcover $30.00

THERMAL CONDUCTIVITY OF THE ELEMENTS: A COMPREHENSIVE REVIEW by C.Y. Ho, R.W. Powell, and P.E. Liley. (Supplement No. 1 to Volume 3) 1974, 796 pages.
Hardcover $60.00
Softcover $55.00

ENERGETICS OF GASEOUS IONS by H.M. Rosenstock, K. Draxl, B.W. Steiner, and J.T. Herron. (Supplement No. 1 to Volume 6) 1977, 783 pages.
Hardcover $70.00
Softcover $65.00

EVALUATED KINETIC DATA FOR HIGH TEMPERATURE REACTIONS: VOLUME 4, HOMOGENEOUS GAS PHASE REACTIONS OF HALOGEN- AND CYANIDE-CONTAINING SPECIES by D.L. Baulch, J. Duxbury, S.J. Grant and D.C. Montague. (Supplement No. 1 to Volume 10) 1981, 721 pages.
Hardcover only $80.00